Physical constants

Acceleration due to gravity (standard),
$g = 32.174$ ft/s^2 (9.80665 m/s^2)(value varies with latitude)
Avogadro's number, $N = 6.02283 \times 10^{23}$ molecules/g-mole
Boltzmann's constant, $k = 1.3805 \times 10^{-23}$ J/°K
British thermal unit (Btu at 60°F) = 778 ft-lb = 1054.54 J
Faraday's constant, $F = 96,487$ C(abs)/g-equiv.
Gas law constants
$\quad\quad\quad R = 1543$ ft · lb/(lb-mol) · °R (universal) $\quad = .082054\ \text{latm}/K\,\text{mol}$
$\quad\quad\quad R = 53.3$ ft · lb/(lb-air) · °R (engineering gas constant for air)
$\quad\quad\quad R = 0.729$ ft^3 · atm/(lb-mol) · °R
Latent heat of vaporization of water (100°C and 1 atm) = 971 Btu/lb = 2258 J/g
Molecular mass of dry air = 28.97 lb/lb-mol (g/g-mol)
One acre = 43,560 ft^2 (2.4698 hectare)
One angstrom, A = 10^{-10} m
One bar = 14.504 lb$_f$/in^2 (10^5 N/m^2)
One horsepower = 550 ft · lb/s = 0.7457 kW
One pound per square inch = 2.31 ft of water
One torr = 1/760 standard atmosphere = 1 mm Hg
Specific weight of water (68°F) = 62.31 lb/ft^3 (see Table C-1)
Standard atmosphere = 14.7 lb/in^2 [101.325 kPa (kN/m^2)]
$\quad\quad\quad\quad\quad = 33.899$ ft (10.333 m) of water
$\quad\quad\quad\quad\quad = 29.92$ in Hg = 760 mm Hg
Standard conditions
$\quad\quad\quad$ Specific gravity of solids and liquids is referred to water at 39.2°F (4°C)
$\quad\quad\quad$ Gases are referred to air free of carbon dioxide and hydrogen at 32°F (0°C)
$\quad\quad\quad$ General scientific = 32°F and 14.7 lb/in^2
$\quad\quad\quad$ Compressors and blowers = 68°F, 14.7 lb/in^2, and 36% relative humidity
$\quad\quad\quad$ Natural gas industry = 60°F and 14.7 lb/in^2
Temperature (absolute)
$\quad\quad\quad$ Rankine, °R = 459.6 + °F
$\quad\quad\quad$ Kelvin, °K = 273.0 + °C
Velocity of light, $c = (2.99776 \times 10^8$ m/s)
Volume occupied by an ideal gas [32°F (0°C) and 1 atm] = 359 ft^3/lb-mol $\quad -°385 \frac{ft^3}{lb\,mol}$
$\quad\quad\quad\quad\quad\quad\quad\quad\quad\quad\quad\quad = 22.4146$ L/g-mol

1

WASTEWATER ENGINEERING
Treatment, Disposal, and Reuse

McGraw-Hill Series in Water Resources and Environmental Engineering

Rolf Eliasen, Paul H. King, and Ray K. Linsley — *Consulting Editors*

Bailey and Ollis: *Biochemical Engineering Fundamentals*
Bishop: *Marine Pollution and Its Control*
Biswas: *Models for Water Quality Management*
Bouwer: *Groundwater Hydrology*
Canter: *Environmental Impact Assessment*
Chanlett: *Environmental Protection*
Chow, Maidment and Mays: *Applied Hydrology*
Eckenfelder: *Industrial Water Pollution Control*
Linsley and Franzini: *Water Resources Engineering*
Metcalf & Eddy, Inc.: *Wastewater Engineering: Collection and Pumping of Wastewater*
Metcalf & Eddy, Inc.: *Wastewater Engineering: Treatment, Disposal, Reuse*
Peavy, Rowe, and Tchobanoglous: *Environmental Engineering*
Rich: *Low-Maintenance, Mechanically-Simple Wastewater Treatment Systems*
Sawyer and McCarty: *Chemistry for Environmental Engineering*
Steel and McGhee: *Water Supply and Sewerage*
Tchobanoglous, Theisen, and Eliassen: *Solid Wastes: Engineering Principles and Management Issues*

WASTEWATER ENGINEERING
Treatment, Disposal, and Reuse

Third Edition

METCALF & EDDY, INC.

Revised by

George Tchobanoglous
Professor of Civil Engineering
University of California, Davis

Franklin L. Burton
Vice President, Retired
Metcalf & Eddy, Inc.

McGraw-Hill, Inc.

New York   St. Louis   San Francisco   Auckland   Bogotá   Caracas
Hamburg   Lisbon   London   Madrid   Mexico   Milan   Montreal
New Delhi   Paris   San Juan   São Paulo
Singapore   Sydney   Tokyo   Toronto

This book was set in Times Roman by Publication Services.
The editors were B. J. Clark and John M. Morriss;
the production supervisor was Louise Karam.
The cover was designed by Metcalf & Eddy, Inc.
New drawings were done by Publication Services.
Project supervision was done by Publication Services.
Arcata Graphics/Halliday was printer and binder.

WASTEWATER ENGINEERING
Treatment, Disposal, and Reuse

1 2 3 4 5 6 7 8 9 0   HAL   HAL   9 5 4 3 2 1 0

ISBN 0-07-041690-7

Library of Congress Cataloging-in-Publication Data

Tchobanoglous, George.
    Wastewater engineering: treatment, disposal, and reuse / Metcalf & Eddy, Inc.—3rd
ed. / revised by George Tchobanoglous, Frank Burton.
       p.   cm.—(McGraw-Hill series in water resources and environmental engineering)
    Rev. ed of: Wastewater engineering / Metcalf & Eddy, Inc. 2nd ed. c 1979.
    Includes bibliographical references.
    ISBN 0-07-041690-7
    1. Sewerage.   2. Sewage disposal.   3. Water reuse.   I. Burton, Frank,
(date).   II. Metcalf & Eddy.   Wastewater engineering.
III. Title   IV. Series.
TD645.T34   1991
628'.2—dc20   89-77500

ABOUT THE PRINCIPAL AUTHORS

George Tchobanoglous is a professor of civil engineering at the University of California at Davis. He received a B.S. degree in civil engineering from the University of the Pacific, an M.S. degree in sanitary engineering from the University of California at Berkeley, and a Ph.D. in environmental engineering from Stanford University. His principal research interests are in the areas of wastewater treatment, wastewater filtration, aquatic wastewater management systems, individual onsite treatment systems, and solid waste management. He has authored or coauthored over 200 technical publications and 6 textbooks. He is the principal writer and editor for the first two editions of this textbook. Professor Tchobanoglous serves nationally and internationally as consultant to both governmental agencies and private concerns. An active member of numerous professional societies, he is past president of the Association of Environmental Engineering Professors. He is a registered civil engineer in California.

Franklin L. Burton spent 30 years with Metcalf & Eddy serving as vice president and chief engineer in their Western Regional Office in Palo Alto, California. He received a B.S. degree in mechanical engineering from Lehigh University and a M.S. degree in civil engineering from the University of Michigan. He has been in charge of the planning, design, and technical review of over 40 wastewater treatment plants and was the principal-in-charge of Metcalf & Eddy's textbook *Wastewater Engineering: Collection and Pumping of Wastewater*. He also served on Metcalf & Eddy's Technical Practice Board which was responsible for the technical oversight of the firm's activities. He is a Fellow of the American Society of Civil Engineers and is a member of several technical societies. He is a registered civil and mechanical engineer in California. He retired from Metcalf & Eddy in 1986 and is in private practice in Los Altos, California.

CONTENTS

Chapter 9 Design of Facilities for Physical and Chemical Treatment of Wastewater 445

Chapter 10 Design of Facilities for the Biological Treatment of Wastewater 529

Chapter 11 Advanced Wastewater Treatment 663

Chapter 12 Design of Facilities for the Treatment and Disposal of Sludge 765

Chapter 13   Natural Treatment Systems   927

Chapter 14   Small Wastewater Treatment Systems   1017

Chapter 15   Management of Wastewater
from Combined Sewers   1103

PREFACE

In 1914, shortly after the turn of the century, Metcalf & Eddy published its world-famous three volume treatise on wastewater engineering, *American Sewerage Practice*. The third volume, *Disposal of Sewage,* was revised in 1916 and again in 1935. Subsequently, the three volumes were combined into a single text, *Sewerage and Sewage Disposal,* in 1922 and a second edition was published in 1930. In 1972, a new version of the textbook was published, *Wastewater Engineering: Collection, Treatment, Disposal,* followed in 1979 by the second edition, *Wastewater Engineering: Treatment, Disposal, Reuse.* As with the earlier textbooks, both editions have enjoyed widespread use in colleges and universities and by practicing engineers in both the public and the private sectors. A companion textbook, *Wastewater Engineering: Collection and Pumping of Wastewater,* was also published in 1981.

Since the publication of the second edition in 1979, many significant changes have occurred in the field of wastewater engineering, resulting from the increased knowledge of the environmental effects of wastewater discharges, improved methods of treatment, changing regulations, and the increasing importance of reclaimed wastewater as a water resource. The third edition shares many of the objectives of the previous editions: (1) to keep pace with the technical developments in the field of environmental engineering that have occurred in the past 10 years, (2) to reflect the impact of changing federal legislation on water quality control and sludge management, (3) to provide information on other types of wastewater systems, such as small systems and overflows from combined sewers, that are receiving increased attention, and (4) to continue to provide useful information for students, teachers, practicing engineers, and other users. With the publication of the third edition, the collective publications of Metcalf & Eddy now span almost an entire century during which Metcalf & Eddy has shared its wastewater knowledge with the profession.

Because of the slowing trend in the United States in adopting metric units, this edition features U.S. customary units as the primary method of expression. Many of the users of the first edition have continued to use that edition as their reference of choice because it was written using U.S. customary units. The decision to change from the International System of Units (or SI, for short) was a difficult one, but alternative equations expressed in SI units have been provided in many cases. Conversion factors have also been provided for the data tables and are also included in the appendix. Some problems using SI units have been included to maintain familiarity with metric computations.

All of the chapters have been revised extensively and expanded to provide up-to-date information. Two completely new chapters, dealing with individual and small wastewater treatment systems and the management of wastewater from combined sewers, have been added. Wastewater disposal and reuse, previously covered in one chapter, are now separate chapters, reflecting the increased importance of these two subjects. To enhance the transfer of information and data, 410 figures, 265 data tables, most of which are new, and 90 worked example problems are included. To enhance the utility of this text for teaching, a total of 284 discussion topics and problems have been developed. Finally, the appendix has also been expanded to provide useful data tables and information for problem solving.

Changes and pending changes in federal legislation continue to have major impacts on wastewater engineering. With increased knowledge of toxic substances and their impacts on the environment, new regulations reflect concerns on their control. Regulations for the control of air emissions including volatile organic compounds (VOCs) and for the control of toxic substances in sludge are of equal concern as the regulations for water quality. Wastewater engineering has to consider the total environmental effects of the proposed solutions to wastewater problems.

Sludge treatment and disposal, one of the most difficult problems in environmental engineering, is further complicated by the shrinking disposal options. Landfilling, the most popular method of sludge disposal, is becoming limited due to diminishing capacities at existing landfills and the difficulty in obtaining new sites. Ocean disposal of sludge is no longer permitted. Because of potential air emission problems, incineration is losing its popularity for large installations. Sludge reuse offers many interesting possibilities if the sludge is "clean." As a result, composting and other technologies are receiving considerable attention by many communities, including large cities. Discussions dealing with composting and land application of sludge have been expanded to reflect the increased use of both methods of sludge treatment.

The authors would like to dedicate their efforts to two individuals who have have made significant contributions to the field of environmental engineering and to Metcalf & Eddy: Harrison P. Eddy, Jr., and Dr. Rolf Eliassen. Mr. Eddy, son of the founder of Metcalf & Eddy, passed away in 1989 after serving as Senior Partner and President of Metcalf & Eddy for many years. Mr. Eddy took a personal interest in the first two editions of this text, and it was through his efforts that the books were published. Dr. Eliassen, Chairman Emeritus of Metcalf & Eddy, Inc., has been a mentor to the authors for many years and his encouragement throughout our careers and in preparation of these texts is an inspiration to achieve excellence in engineering.

John G. Chalas
Senior Vice President
Metcalf & Eddy, Inc.
Director of Technology

George Tchobanoglous
Professor of Civil Engineering
University of California, Davis
Principal Author

Franklin L. Burton
Vice President, Retired
Metcalf & Eddy, Inc.
Principal Author

ACKNOWLEDGMENTS

Metcalf & Eddy, Inc. is fortunate in having the services of Dr. George Tchobanoglous of the University of California, Davis, and Franklin L. Burton, a retired Vice President of Metcalf & Eddy, as principal authors of the third edition. It is hoped that this edition will continue the firm's contribution to the profession and the practice of wastewater engineering, which spans nearly the entire century.

The principal authors were responsible for writing, editing, coordination, and responding to reviewers' comments for the third edition. Both authors have been involved in previous editions of the Metcalf & Eddy textbooks. Dr. Tchobanoglous was the principal author of the first and second editions of this text and the companion volume, *Wastewater Engineering: Collection and Pumping of Wastewater*. Mr. Burton was a technical reviewer for the second edition and was Metcalf & Eddy's principal reviewer and coordinator for the companion text.

Other authors for the text were Dr. Robert G. Smith of the University of California, Davis, for Chap. 13, "Natural Treatment Systems"; David R. Bingham of Metcalf & Eddy for Chap. 15, "Management of Wastewater from Combined Sewers"; Dr. Takashi Asano of the California Water Resources Control Board for Chap. 16, "Wastewater Reclamation and Reuse"; and Dr. Dominique N. Brocard of Metcalf & Eddy for Chap. 17, "Effluent Disposal." Ms. Terry Poxon of the University of California, Davis also assisted in the preparation of Chap. 8, "Biological Unit Processes."

No undertaking of this magnitude can be accomplished alone. It is with grateful appreciation that we acknowledge the contribution of the following Metcalf & Eddy staff:

Bradley W. Behrman   John A. Lager
Stephen L. Bishop   Richard H. Marshall, Jr.
Steven Biuso   Jon R. Pearson
David P. Bova   Winfield Peterson III
Paul T. Bowen   Charles E. Pound

Edward J. Boyajian	Robert J. Reimold
Patricia M. Caton	George Ross
William I. Douglass	James A. Ryan, Jr.
Alan C. Ford	George K. Tozer
Allen F. Goulart	Cesar M. Vincenty
Eugene S. Grafton	Mark Voorhees
Frank M. Gunby, Jr.	Stanley Wagher
Terry L. Krause	Thomas K. Walsh

This effort constitutes a unique combination of talent from the academic community, regulatory agencies, and practicing engineers and scientists. The group represents hundreds of years of experience in wastewater engineering.

Special appreciation is extended to David P. Bova, Director of Wastewater Technology, and Stephen L. Bishop, Director of Water Technology, of Metcalf & Eddy for their assistance and advice during preparation of this text.

Thanks also to Janus Alleman, Purdue University and Paul King, Northeastern University for their reviews of the manuscript.

Finally, we acknowledge our gratitude to Eckardt C. Beck, Chairman of the Board of Metcalf & Eddy Companies, Inc.; Richard T. Dewling, Chairman of the Board of Metcalf & Eddy, Inc.; and George A. Bicher, President of Metcalf & Eddy, Inc., for their professional encouragement and support that made this undertaking possible.

John G. Chalas
Senior Vice President
Metcalf & Eddy, Inc.
Director of Technology

WASTEWATER ENGINEERING
Treatment, Disposal, and Reuse

CHAPTER

1

WASTEWATER ENGINEERING: AN OVERVIEW

Every community produces both liquid and solid wastes. The liquid portion—wastewater—is essentially the water supply of the community after it has been fouled by a variety of uses. From the standpoint of sources of generation, wastewater may be defined as a combination of the liquid- or water-carried wastes removed from residences, institutions, and commercial and industrial establishments, together with such groundwater, surface water, and stormwater as may be present.

If untreated wastewater is allowed to accumulate, the decomposition of the organic materials it contains can lead to the production of large quantities of malodorous gases. In addition, untreated wastewater usually contains numerous pathogenic, or disease-causing, microorganisms that dwell in the human intestinal tract or that may be present in certain industrial waste. Wastewater also contains nutrients, which can stimulate the growth of aquatic plants, and it may contain toxic compounds. For these reasons, the immediate and nuisance-free removal of wastewater from its sources of generation, followed by treatment and disposal, is not only desirable but also necessary in an industrialized society. In the United States, it is now mandated by numerous federal and state laws.

Wastewater engineering is that branch of environmental engineering in which the basic principles of science and engineering are applied to the problems of water pollution control. The ultimate goal—wastewater management—is the protection of the environment in a manner commensurate with public health, economic, social, and political concerns.

1

To provide an initial perspective of the treatment, disposal, and reuse of wastewater, a brief review of the historical background, current status, and expected new directions in these areas of wastewater engineering is presented in this chapter. Although the subjects of source control, collection, transmission, and pumping will not be covered (see Preface), the role of the engineer in the overall field of wastewater engineering is discussed at the end of this chapter.

1-1 WASTEWATER TREATMENT

Wastewater collected from municipalities and communities must ultimately be returned to receiving waters or to the land. The complex question of which contaminants in wastewater must be removed to protect the environment—and to what extent—must be answered specifically for each case. The answer to this question requires analyses of local conditions and needs, together with the application of scientific knowledge, engineering judgment based on past experience, and consideration of federal and state requirements and regulations.

Background

Although the collection of stormwater and drainage dates from ancient times, the collection of wastewater can be traced only to the early 1800s. The systematic treatment of wastewater followed in the late 1800s and early 1900s. Development of the germ theory by Koch and Pasteur in the latter half of the nineteenth century marked the beginning of a new era in sanitation. Before that time, the relationship of pollution to disease had been only faintly understood, and the science of bacteriology, then in its infancy, had not been applied to the subject of wastewater treatment.

In the United States, the treatment and disposal of wastewater did not receive much attention in the late 1800s because the extent of the nuisance caused by the discharge of untreated wastewater into the relatively large bodies of water (compared to those in Europe) was not severe, and because large areas of land suitable for disposal were available. By the early 1900s, however, nuisance and health conditions brought about an increasing demand for more effective means of wastewater management. The impracticability of procuring sufficient areas for the disposal of untreated wastewater on land, particularly for larger cities, led to the adoption of more intensive methods of treatment.

Current Status

Methods of treatment in which the application of physical forces predominates are known as *unit operations.* Methods of treatment in which the removal of contaminants is brought about by chemical or biological reactions are known as *unit processes.* At the present time, unit operations and processes are grouped together to provide what is known as *primary, secondary,* and *advanced* (or *tertiary*) treatment. In primary treatment, physical operations such as screening and sedimentation are used to remove the floating and settleable solids found in wastewater. In secondary treatment, biological and chemical processes are used to remove most of the organic matter.

In advanced treatment, additional combinations of unit operations and processes are used to remove other constituents, such as nitrogen and phosphorus, that are not reduced significantly by secondary treatment. Land treatment processes, now more commonly termed "natural systems," combine physical, chemical, and biological treatment mechanisms and produce water with quality similar to or better than that from advanced wastewater treatment.

Over the last 40 years, the number of treatment plants serving municipalities and communities has nearly tripled. Implementation of the federal Clean Water Act (discussed in Chap. 4) brought about substantial changes in water pollution control to achieve "fishable and swimmable" waters. Over 15,000 facilities are in operation, according to the U.S. Environmental Protection Agency's recent needs survey [7]. An analysis of the data on the sizes of treatment plants reported in Table 1-1 shows that approximately 81 percent of all publicly owned treatment works with treatment needs are smaller than 1 Mgal/d (43.8 L/s); 16 percent are in the range between 1 and 10 Mgal/d (43.8 and 438.1 L/s); and about 3 percent are larger than 10 Mgal/d (438 L/s). Correspondingly, approximately 9 percent of the total facility design capacity at publicly owned treatment works is in plants with a design flow of less than 1 Mgal/d (43.8 L/s); 25 percent is in plants with a design flow between 1 and 10 Mgal/d (43.8 and 438.1 L/s); and 66 percent is in plants larger than 10 Mgal/d (438 L/s). These findings are essentially the same as those reported in 1974 and are not expected to change appreciably over the next 20 years [7,8].

Data on the number of treatment facilities categorized by the level of treatment are reported in Table 1-2. In 1988, approximately 11 percent of the treatment facilities had less than secondary treatment, 76 percent had secondary treatment or greater, and 12 percent had no discharge. The number of primary treatment plants had been reduced significantly from 1974, when over 2800 were reported in operation [8]. During the next 20 years, the number of treatment plants in the United States is expected to increase by about 10 percent [6]. However, the number of plants being upgraded to higher levels of treatment is significantly greater. As shown in Table 1-2, nearly

TABLE 1-1
Number of treatment facilities by flow range[a]

Flow ranges, Mgal/d[b]	Number of facilities		Total capacity, Mgal/d[b]	
	1988	When needs met	1988	When needs met
0.01–0.10	5,983	5,497	259	267
0.11–1.00	6,589	7,681	2,307	2,683
1.01–10.0	2,427	3,376	7,178	10,535
>10	446	739	18,992	30,805
Other[c]	146	81	0	0
Total	15,591[d]	17,374	28,736	44,290

[a] Adapted from Ref. 7.

[b] Mgal/d × 43.8126 = L/s; Mgal/d × 0.043813 = m³/s.

[c] Flow data unavailable.

[d] Not including 117 untreated discharges.

TABLE 1-2
Number of treatment facilities by level of treatment[a]

Level of treatment	Number of facilities		
	1988	When needs met	Increase
Less than secondary	1,789	48[b]	(1,741)
Secondary	8,536	9,659	1,123
Greater than secondary	3,412	5,293	1,881
No discharge	1,854	2,363	509
Other[c]	117	11	(106)
Total	15,708	17,374	1,666

[a] Adapted from Ref. 7.
[b] Waiver of secondary treatment applied for and tentatively approved.
[c] Level of treatment information unavailable.

all of the facilities providing less than secondary treatment in 1988 will be replaced or upgraded to higher levels of treatment. The number of facilities providing treatment greater than secondary will increase by 55 percent when the needs are met for the documented facilities. Therefore, the emphasis in the future will be on upgrading wastewater treatment plants to provide secondary and advanced wastewater treatment processes.

New Directions and Concerns

With the passage of the Federal Water Pollution Control Act Amendments of 1972 (Public Law 92-500), Congress established a far-reaching program for the control of pollution in U.S. waterways. The implications of this important legislation, subsequent amendments and laws, and the corresponding regulations and guidelines are discussed in Chap. 4. New directions and concerns are also evident in various specific areas of wastewater treatment, including (1) the changing nature of the wastewater to be treated; (2) the problem of industrial wastes; (3) the impact of stormwater and nonpoint sources of pollution; (4) combined sewer overflows; (5) treatment operations, processes, and concepts; (6) health and environmental concerns; (7) treatment process effectiveness; and (8) small systems including individual onsite systems.

Changing Wastewater Characteristics. The number of organic compounds that have been synthesized since the turn of the century now exceeds half a million, and some 10,000 new compounds are added each year. As a result, many of these compounds are now found in the wastewater from most municipalities and communities. Currently, the release of volatile organic compounds (VOC) and volatile toxic organic compounds (VTOC) found in wastewater is of great concern in the operation of both collection systems and treatment plants. The total emission of VTOCs from municipal wastewater treatment plants in California has been estimated to be as high as 800 tons/yr (725 Mt/yr) [1].

The control of odors and in particular the control of hydrogen sulfide generation is of concern in collection systems and at treatment plants. Some of the increase in sulfide generation observed in collection systems has been attributed to the decrease of metals in industrial waste discharges. With the implementation of effective industrial pretreatment programs (for the control and treatment of industrial wastes prior to discharge into municipal collection systems), the quantity of metals present in municipal wastewater has decreased significantly. Concomitant with this decrease in metals, an increase in the release of hydrogen sulfide to the atmosphere above sewers and at treatment plant headworks has been observed in a number of locations. The sulfide produced in sewers, which is now released as hydrogen sulfide, had reacted previously with the metals present in the wastewater to form metallic sulfides (e.g., ferrous sulfide). The release of excess hydrogen sulfide has led to the accelerated corrosion of concrete sewers and headwork structures and to the release of odors. The control of odors is of increasing concern as residential and commercial development approaches existing treatment plant locations and in the siting of new facilities.

The Problem of Industrial Wastes. The number of industries that now discharge wastes to domestic sewers has increased significantly during the past 20 to 30 years. In view of the toxic effects often caused by the presence of these wastes, even at very low concentrations, the general practice of combining pretreated or partially pretreated industrial and domestic wastes is being reevaluated by a number of communities. In the future, many municipalities may either provide separate treatment facilities for these wastes or require that they be treated to a higher degree at the point of origin to render them harmless before allowing their discharge to domestic sewers.

Impact of Stormwater and Nonpoint Sources of Pollution. As the number of treatment plants providing secondary or greater than secondary treatment continues to increase, the importance of stormwater and nonpoint sources of pollution (e.g., runoff from agricultural areas) in limiting the quality of the nation's streams and rivers is increasing. In many river basins, additional treatment beyond secondary treatment will have essentially no impact on stream quality until the stormwater discharges and nonpoint sources of pollution are controlled.

Combined Sewer Overflows. Overflows from combined sewers have been recognized as a difficult problem requiring solution, especially for many of the older cities in the United States. Combined sewers carry a mixture of wastewater and stormwater runoff and, when the capacity of the interceptors is reached, overflows occur to the receiving waters. Large overflows can significantly impact the water quality and can prevent attainment of the mandated standards. Methods of control may involve significant modifications to the collection system, construction of storage facilities for containing all or a portion of the peak flows, or provision of additional and special treatment facilities. Many of these methods of control are very costly to implement, and little governmental financial assistance is available to local municipalities. Combined sewer overflows and control technologies are addressed in Chap. 15.

Treatment Operations, Processes, and Concepts. At the present time, most of the unit operations and processes used for wastewater treatment are undergoing continual and intensive investigation from the standpoint of implementation and application. As a result, many modifications and new operations and processes have been developed and implemented; more need to be made to meet the increasingly stringent requirements for environmental enhancement of water courses [6]. In addition to the developments taking place with conventional treatment methods, alternative treatment systems and technologies, such as those involving the use of aquatic plants, are also under development. If significant improvements are to be made in the analysis and application of both existing and new processes, improved methods of wastewater characterization must be developed [3].

Although most of the organic compounds found in wastewater can be treated readily, the number of such compounds that are not amenable to treatment or that are only slightly amenable to treatment with the conventional processes presently used is increasing. Moreover, in many cases, little or no information is available on the long-term environmental effects caused by the presence of these compounds. As these effects become more clearly understood, it is anticipated that more emphasis will be placed on advanced treatment for the removal of specific contaminants. The unregulated release of VOCs and VTOCs from wastewater treatment plants may necessitate the covering of treatment plant headworks and primary treatment facilities and the installing of special treatment facilities to process the compounds that are released. In some cases, improved source control may be necessary prior to discharge to the collection system to eliminate these compounds.

Because of the changing characteristics of the wastewater, studies of wastewater treatability are increasing, especially with reference to specific compounds. Such studies are especially important where new treatment methods are being proposed. Therefore, the engineer must understand the general approach and methodology involved in (1) the assessment of the treatability of a wastewater (domestic or industrial); (2) the conduct of laboratory and pilot plant studies; and (3) the translation of experimental data into design parameters.

The relationship between the design of collection systems and wastewater treatment is receiving more attention. As wastewater is transported in collection systems, it undergoes both biological and chemical transformations [5,6]. To a large extent, the nature of these transformations depends on the types of wastes being discharged and the design of the collection system. In the future, as the importance of these transformations becomes understood more clearly with respect to wastewater treatment, it is anticipated that the design of wastewater collection systems and treatment facilities will be coordinated to a much greater extent than in the past.

Health and Environmental Concerns. In meeting the requirements of the Clean Water Act and its amendments, public health and environmental concerns have come to play an increasingly important part in the selection and design of both collection and treatment facilities. Discharge of contaminants to the environment is receiving close scrutiny. For example, the release of VOCs and VTOCs from collection and treatment facilities, as noted earlier, is becoming of greater concern to regulatory agencies. Odors are one of the most serious environmental concerns to the public. New techniques for odor measurement are now being used to quantify the development and movement of odors that may emanate from wastewater facilities, and special efforts are being made to design facilities that minimize the development of odors, contain them effectively, and provide proper treatment for their destruction.

Treatment Process Effectiveness. As the federal grant programs are being phased out, many municipalities are having to make difficult decisions with respect to the financing of improvements to wastewater management facilities. Therefore, the effectiveness of any proposed improvements and facilities is being examined in detail, especially with respect to treatment plant performance, energy and resource use, operation and maintenance costs, and capital costs.

Over the past 15 years, a significant amount of money has been spent to construct wastewater treatment plants. Unfortunately, the performance of many of these facilities has not fulfilled the requirements of the discharge permits. In many cases, newly constructed plants have had to be retrofitted or modified at considerable expense to meet the discharge requirements and to provide more reliable performance. Currently, performance certification is required in a number of states before final payment is made on projects financed with government grants. Treatment plants of improved design that are easier to operate and maintain will be required to meet existing and more stringent discharge requirements.

The need to conserve energy and resources is well-documented. Detailed energy analyses are now becoming an important part of any project analysis. More attention is being given to the selection of processes that conserve energy and resources. There is an increasing trend to minimize power use in the design of wastewater treatment plants by paying more careful attention to plant siting and by designing facilities to recover energy for in-plant heating and power generation.

Operation and maintenance costs are extremely important to operating agencies and, especially, to small communities with limited budgets because these costs are funded totally with local moneys. Thus, the operability of treatment plants is receiving renewed attention. Value engineering, where an outside party or firm not associated with the project is asked to review the proposed improvements, is gaining popularity. Value engineering reviews have also led to considerable savings in capital costs.

Small and Individual Onsite Systems. During the past 15 years, interest in small treatment systems has often been overshadowed by concern over the design, construction, and operation of large regional systems. Small systems were often designed and constructed as small-scale models of large plants. As a consequence, many are operationally energy and resource intensive. Because of economic, environmental, and energy concerns, the design, construction, and operation of small systems are coming under careful review. New and innovative designs continue to be developed, and alternative treatment processes are being used. Small systems are given special attention in Chap. 14.

Because the ratio of the number of people discharging to wastewater collection systems to the number discharging to individual onsite systems has not changed significantly over the past 20 years, greater attention is now being focused on the design, operation, and maintenance of individual onsite systems. The health and pollutional hazards, including groundwater contamination, caused by their use and the limits of their application must be defined and quantified. The organization of local operation and maintenance districts for onsite systems is another recent development.

1-2   SLUDGE DISPOSAL AND REUSE

The ultimate disposal of the solid and semisolid residuals (sludge) and concentrated contaminants removed by treatment has been and continues to be one of the most difficult and expensive problems in the field of wastewater engineering. Recent legislation banning the ocean discharge of sludge has eliminated one disposal option used by some large coastal cities. Because of the concerns about air and groundwater pollution, the disposal of sludge by incineration, and by application on land or in landfills is receiving special attention. New regulations that restrict the discharge of contaminants to the environment are being promulgated. The number and capacity of landfills have been reduced and new landfill locations that meet environmental, social, and economic requirements are increasingly difficult to find. As a result, the treatment and disposal of sludge has become one of the most significant challenges for the environmental engineer.

Background

In the early 1900s, wastewater from most communities was discharged directly to streams and rivers in storm sewers. Accumulations of sludge and the development of offensive odors and unsightly conditions resulted. To overcome these problems, separate sewers were built and wastewater treatment was instituted. The disposal of sludge became a problem with the application of the more intensive methods of treatment, which resulted in the production of large volumes of sludge.

Current Status

When the U.S. Environmental Protection Agency established secondary treatment as the minimum acceptable level of treatment prior to surface water discharge, the quantity of sludge requiring disposal increased significantly. Data on sludge disposal methods now in use are reported in Table 1-3. As noted in Table 1-3, some form of landfilling or land application is the most commonly used method for the disposal of sludge. Land application of sludge is used extensively as a means of disposal, as a means of reclaiming marginal land for productive uses, and as a means of utilizing the nutrient content in sludge. However, as stated previously, landfilling or land application of sludge is becoming more strictly regulated, and landfill sites for the disposal of sludge are more difficult to locate. Because of potential landfill limitations, composting is becoming a popular means of stabilizing and distributing sludge for reuse as a soil amendment. Significant advances in composting technology have occurred in recent years. Incineration of sludge by large municipalities is used extensively, but incineration operation and emission control are subject to greater regulatory restrictions.

TABLE 1-3
Summary data on sludge disposal methods[a]

| | Percent using indicated method | | |
| | Plant size, Mgal/d[b] | | |
Sludge disposal method	<1.0	1.01 to 10.0	>10.0
Land application	39	39	21
Landfill burial	31	35	12
Incineration	1	1	32
Distribution and marketing	11	13	19
Ocean discharge	1	0	4
Other	17	12	12
Total	100	100	100

[a] Adapted from Ref. 9.
[b] Mgal/d $\times$ 43.8126 = L/s; Mgal/d $\times$ 3.7854 $\times$ 10^3 = m^3/d.

New Directions and Concerns

The increase in sludge production resulting from the construction of additional secondary and advanced wastewater treatment plants will clearly tax the capacity of existing sludge processing and disposal methods. Improved treatment methods will be needed to provide higher levels of treatment not only for routine wastewater constituents but also for the removal of specific compounds (e.g., metals, VTOCs, etc.). The removal of these constituents, in turn, will lead to the production of larger volumes of sludge that will require disposal. The continuing search for better methods for the processing, disposal, and reuse of sludge such as thermal processing and composting will remain high, if not highest, on the list of priorities in the future.

At the time of the writing of this text (1989), new regulations for the use and disposal of wastewater sludges have been proposed by the EPA. The regulations establish pollutant numerical limits and management practices for (1) land application, (2) distribution and marketing, (3) monofilling, (4) incineration, and (5) surface application [2]. Although the impact of the regulations on future sludge disposal practices cannot be fully assessed at this time, the proposed restrictive limits on certain constituents, such as copper, may potentially decrease land application and reuse opportunities. In order to limit pollutant concentrations in sludges that are applied to land or marketed and reused, more intensive monitoring and industrial pretreatment will be required. Additional discussion regarding the effects of sludge regulations on wastewater treatment and sludge disposal is provided in Chaps. 4 and 12.

1-3 WASTEWATER RECLAMATION AND REUSE

Although secondary treatment is a sufficient level of treatment for a majority of applications, advanced treatment will be required in a number of locations (see Table

1-2). Where advanced treatment is required, the opportunities for reuse are improved and are being evaluated in many of the facilities' plans.

Background

In the past, the disposal of wastewater in most municipalities and communities was carried out by the easiest method possible, without much regard to unpleasant conditions produced at the place of disposal. Irrigation was probably the first method of wastewater disposal, although dilution was the earliest method adopted by most municipalities. With increased industrial and urban development, effluent disposal and its effects on the environment now require special consideration.

Current Status

Surface water discharge remains the most common method of wastewater disposal. To protect the aquatic environment, however, the individual states, in conjunction with the federal government, have developed receiving water standards for the streams, rivers, and estuarial and coastal waters of the United States. Many states have adopted more stringent requirements than those prescribed by the federal government. In a number of places, treatment plants have been designed and located so that a portion of the treated effluent can be disposed of by land application in conjunction with a variety of reuse applications such as golf course irrigation, use as industrial cooling water, and groundwater recharge. This trend is expected to continue to increase in the future, especially in the arid and semiarid areas and in locations where fresh water is in short supply.

New Directions and Concerns

In many locations where the available supply of fresh water has become inadequate to meet water needs, it is clear that the once-used water collected from communities and municipalities must be viewed not as a waste to be disposed of but as a resource. This concept is expected to become more widely adopted as other parts of the country experience water shortages. The use of dual water systems, such as those now used in St. Petersburg, Florida and Rancho Viejo, California, is expected to increase in the future. In both locations, treated effluent is used for landscape watering and other nonpotable uses. Because water reuse is expected to become of even greater importance in the future, the impact of water reuse on wastewater management planning and several potential reuse applications are considered in Chap. 16.

1-4   EFFLUENT DISPOSAL

After treatment, wastewater must either be reused, as discussed above, or disposed of to the environment. The most common means of treated wastewater disposal is by discharge and dilution into streams, rivers, lakes, estuaries, or the ocean. If adverse environmental impacts are to be avoided, the quality of the treated and dispersed effluent must be consistent with local water quality objectives.

Background

For many years, effluent disposal to receiving waters was accomplished by an open pipe. Mixing was accomplished variably, depending upon the natural characteristics of the receiving water. An important aspect of effluent disposal was that of the assimilative capacity of the receiving waters, often representing the amount of organic matter that could be discharged without excessively taxing the dissolved oxygen resources. Greater attention is now being paid to the environmental effects of other constituents, such as suspended solids, nutrients, and toxic compounds, and how they can be safely assimilated into the aquatic environment.

New Directions and Concerns

Effluent disposal focuses on the transport of contaminants in the environment and the transformation processes that occur. To ensure that effluent disposal is accomplished in conformance with the environmental requirements, a rigorous analysis must be performed in many cases. Mathematical modeling techniques are used and involve the application of material balances for transport analysis and kinetic expressions to describe the response of the physical system. By modeling the river and estuarine systems, it is possible to assess the assimilative capacity of these systems and thus to predict the impacts of the proposed discharge. Some of the important transformations that occur include oxidation, bacterial conversions, natural decay, and photosynthesis and respiration. Techniques used in analyzing effluent disposal and its potential environmental impacts are discussed in Chap. 17.

1-5 THE ROLE OF THE ENGINEER

Practicing wastewater engineers are involved in the conception, planning, evaluation, design, construction, and operation and maintenance of the systems that are needed to meet wastewater management objectives. The major elements of wastewater systems and the associated engineering tasks are identified in Table 1-4.

Knowledge of the methods used for the determination of wastewater flowrates and characteristics (Chaps. 2 and 3) is essential to an understanding of all aspects of wastewater engineering. The subjects of source control, collection, and transmission and pumping, covered in a companion text [5], must also be studied by the engineer if truly integrated wastewater systems are to be designed.

The primary focus of this book (Chaps. 4 to 17) is on two elements listed in Table 1-4: (1) treatment, and (2) disposal and reuse. These areas of wastewater engineering, like the others, have been and continue to be in a dynamic period of development. Old ideas are being reevaluated, and new concepts are being formulated. To play an active role in the development of this field, the engineer must know the fundamentals

TABLE 1-4
Major elements of wastewater management systems and associated engineering tasks

Element	Engineering task	See Chapter
Wastewater generation	Estimation of the quantities of wastewater, evaluation of techniques for the reduction of wastewater, and determination of wastewater characteristics	2, 3
Source control (pretreatment)	Design of systems to provide partial treatment of wastewater before it is discharged to collection systems (principally involves industrial dischargers)	[a]
Collection system	Design of sewers used to remove wastewater from the various sources of wastewater generation	[b]
Transmission and pumping	Design of large sewers (often called trunk and interceptor sewers), pumping stations, and force mains for transporting wastewater to treatment facilities or to other locations for processing	[b]
Treatment (wastewater and sludge)	Selection, analysis, and design of treatment operation and processes to meet specified treatment objectives related to the removal of wastewater contaminants of concern	4–13, 15
Disposal and reuse (wastewater and sludge)	Design of facilities used for the disposal and reuse of treated effluent in the aquatic and land environment, and the disposal and reuse of sludge	12, 13, 16, 17
Small systems[c]	Design of facilities for the collection, treatment, and disposal and reuse of wastewater from individual residences and small communities	14

[a] Although the design of industrial pretreatment facilities is not covered specifically in this text, the material presented in Chaps. 4–13 is applicable.
[b] Not covered in the text; see companion text, Ref. 5.
[c] Small systems include all of the elements listed in this table.

on which it is based. The delineation of these fundamentals is the main purpose of this book.

REFERENCES

1. Corsi, R. L., D. P. Y. Chang, E. D. Schroeder, and Q. Qiu: "Emissions of Volatile and Potentially Toxic Organic Compounds From Municipal Wastewater Treatment Plants," presented at the 80th Annual Meeting of APCA, New York, June 21–26, 1987.
2. *Federal Register*: "Standards for the Disposal of Sewage Sludge," 40 CFR Parts 257 and 503, February 6, 1989.
3. Levine, A. D., G. Tchobanoglous, and T. Asano: "Characterization of the Size Distribution of Contaminants in Wastewater: Treatment and Reuse Implications," *Journal WPCF,* vol. 57, no. 7, July 1985.
4. Metcalf & Eddy, Inc.: *Wastewater Engineering: Treatment, Disposal, Reuse,* 2nd ed., McGraw-Hill, New York, 1979.
5. Metcalf & Eddy, Inc.: *Wastewater Engineering: Collection and Pumping of Wastewater,* McGraw-Hill, New York, 1981.

6. Parker, D. S.: Wastewater Technology Innovation for the Year 2000, *Journal of Environmental Engineering, ASCE,* vol. 114, No. 3, pp. 487–506, June 1988.

7. U.S. Environmental Protection Agency: *Assessment of Needed Publicly Owned Wastewater Treatment Facilities in the United States, 1988 Needs Survey Report to Congress,* EPA 430/09-89-001, February 1989.

8. U.S. Environmental Protection Agency: *Cost Estimates for Construction of Publicly Owned Wastewater Treatment Facilities, 1974 Needs Survey, Final Report to Congress,* Washington, DC, 1975.

9. U.S. Environmental Protection Agency: *Environmental Regulations and Technology, Use and Disposal of Municipal Wastewater Sludge,* EPA 625/10-84-003, September 1984.

10. U.S. Environmental Protection Agency: *National Water Quality Inventory, 1986 Report to Congress,* EPA 440/4-87-008, November 1987.

11. U.S. Environmental Protection Agency: *Sludge Composting, Distribution and Marketing Requirements in the United States,* 1986.

CHAPTER
2

WASTEWATER
FLOWRATES

Determining the rates of wastewater flow is a fundamental step in the design of wastewater collection, treatment, and disposal facilities. Reliable data on existing and projected flows must be available if these facilities are to be designed properly and if the associated costs are to be minimized and also shared equitably when the facilities serve more than one community or district. In situations where wastewater flowrate data are limited or unavailable, wastewater-flowrate estimates have to be developed from water consumption records and other information.

The purpose of this chapter is to develop a basis for properly assessing wastewater flowrates for a community. The subjects considered include (1) definition of the various components that make up the wastewater from a community, (2) water supply data and its relationship to wastewater flowrates, (3) wastewater sources and flowrates, (4) analysis of flowrate data, and (5) methods of reducing wastewater flowrates. For information on determining flowrates for sewer design and for the measurement of wastewater flows, the companion text, Ref. 6, should be consulted. Methods of metering flowrates at a wastewater treatment plant are discussed in Chap. 6.

2-1 COMPONENTS
OF WASTEWATER FLOWS

The components that make up the wastewater flow from a community depend on the type of collection system used and may include the following:

1. *Domestic* (also called *sanitary*) *wastewater*. Wastewater discharged from residences and from commercial, institutional, and similar facilities.

2. *Industrial wastewater.* Wastewater in which industrial wastes predominate.
3. *Infiltration/inflow (I/I).* Water that enters the sewer system through indirect and direct means. Infiltration is extraneous water that enters the sewer system through leaking joints, cracks and breaks, or porous walls. Inflow is stormwater that enters the sewer system from storm drain connections (catch basins), roof leaders, foundation and basement drains, or through manhole covers.
4. *Storm water.* Runoff resulting from rainfall and snowmelt.

Three types of sewer systems are used for the removal of wastewater and stormwater: sanitary sewer systems, storm sewer systems, and combined sewer systems. Where separate sewers are used for the collection of wastewater (sanitary sewers) and stormwater (storm sewers), wastewater flows in sanitary sewers consist of three major components: (1) domestic wastewater, (2) industrial wastewater, and (3) infiltration/inflow. Where only one sewer system (combined sewer) is used, wastewater flows consist of these three components plus stormwater. In both cases, the percentage of the wastewater components varies with local conditions and the time of the year.

For areas now served with sewers, wastewater flowrates are commonly determined from existing records or by direct field measurements. For new developments, wastewater flowrates are derived from an analysis of population data and corresponding projected unit rates of water consumption or from estimates of per capita wastewater flowrates from similar communities. These subjects are considered further in this chapter.

2-2 ESTIMATING WASTEWATER FLOWRATES FROM WATER SUPPLY DATA

If field measurements of wastewater flowrates are not possible and actual wastewater flowrate data are not available, water supply records can often be used as an aid to estimate wastewater flowrates. The types of water-use data available and how the data can be analyzed and applied for estimating wastewater flowrates are discussed in this section. Where water records are not available, useful data for various types of establishments and water-using devices are provided for making estimates of wastewater flowrates.

Municipal Water Use

Municipal water use is generally divided into four categories: (1) domestic (water used for sanitary and general purposes), (2) industrial (nondomestic purposes), (3) public service (water used for fire fighting, system maintenance, and municipal landscape irrigation), and (4) unaccounted for system losses and leakage. Typical per capita values for these uses are reported in Table 2-1. The importance of categorizing water use for the purposes of estimating wastewater flows is discussed in this section.

TABLE 2-1
Typical municipal water use in the United States[a]

| Use | Flow, gal/capita · d | | |
	Range	Average	Percent based on average flow
Domestic	40–130	60	36.4
Industrial (nondomestic)	10–100	70	42.4
Public service	5–20	10	6.0
Unaccounted system losses and leakage	10–40	25	15.2
	65–290	165	100.0

[a] Ref. 8.
Note: gal × 3.7854 = L

Domestic Water Use. Domestic water use encompasses the water supplied to residential areas, commercial districts, institutional facilities, and recreational facilities, as measured by individual water meters. The uses to which this water is put include drinking, washing, bathing, culinary, waste removal, and yard watering. Using the average flow values reported in Table 2-1, over one-third of the water used in a municipal water supply system is for domestic purposes.

Residential areas. Water used by residential households consists of water for interior use such as showers and toilets and water for exterior use such as lawn watering and car washing. Typical data for interior water use are presented in Table 2-2. Water use for exterior applications varies widely depending upon the geographic location, climate, and time of year and mainly consists of landscape irrigation.

TABLE 2-2
Typical distribution of residential interior water use[a,b]

Use	% of total
Baths	8.9
Dishwashers	3.1
Faucets	11.7
Showers	21.2
Toilets	28.4
Toilet leakage	5.5
Washing machines	21.2
	100.0

[a] Adapted from Ref. 9.
[b] Without water-conserving fixtures.

Commercial facilities. The water used by commercial facilities for sanitary purposes will vary widely depending on the type of activity (e.g., an office as compared to a restaurant). Typical water-use values for various types of commercial facilities are reported in Table 2-3. For large commercial water-using facilities such as laundries and car washes, careful estimates of actual water use should be made.

Institutional facilities. Water used by facilities such as hospitals, schools, and rest homes is usually based on some measure of the size of the facility and the type of housing function provided (e.g., per student or per bed). Water use for schools will vary significantly depending on whether the students are housed on campus or are day students. Representative water-use values for institutional facilities are reported in Table 2-4.

TABLE 2-3
Typical rates of water use for commercial facilities[a]

User	Unit	Flow, gal/unit · d Range	Typical
Airport	Passenger	4–5	3
Apartment house	Person	100–200	100
Automobile service station	Employee	8–15	13
	Vehicle served	8–15	10
Boarding house	Person	25–50	40
Department store	Toilet room	400–600	550
	Employee	8–13	10
Hotel	Guest	40–60	50
	Employee	8–13	10
Lodging house and tourist home	Guest	30–50	40
Motel	Guest	25–40	35
Motel with kitchen	Guest	25–60	40
Laundry (self-service)	Machine	400–650	550
	Wash	45–55	50
Office	Employee	8–20	15
Public lavatory	User	3–6	5
Restaurant (including toilet)			
Conventional	Customer	8–10	9
Short-order	Customer	3–8	6
Bar and cocktail lounge	Customer	2–4	3
	Seat	15–25	20
Shopping center	Parking space	1–3	2
	Employee	8–13	10
Theater			
Indoor	Seat	2–4	3
Outdoor	Car	3–5	4

[a] Adapted in part from Refs. 7 and 8.
Note: gal × 3.7854 = L

TABLE 2-4
Typical water-use values for institutional facilities[a]

		Flow, gal/unit · d	
User	Unit	Range	Typical
Assembly hall	Seat	2–4	3
Hospital, medical	Bed	130–260	150
	Employee	5–15	10
Hospital, mental	Bed	80–150	120
	Employee	5–15	10
Prison	Inmate	80–150	120
	Employee	5–15	90
Rest home	Resident	5–120	90
	Employee	5–15	10
School, day			
With cafeteria, gym, and showers	Student	15–30	25
With cafeteria only	Student	10–20	15
Without cafeteria and gym	Student	5–15	10
School, boarding	Student	50–100	75

[a] Adapted in part from Refs. 7 and 8.
Note: gal × 3.7854 = L

Recreational facilities. Recreational facilities such as swimming pools, bowling alleys, camps, resorts, and country clubs perform a wide range of functions involving water use. Typical water-use values are reported in Table 2-5.

Industrial (Nondomestic) Water Use. The amount of water supplied by municipal agencies to industries for process (nondomestic) purposes is highly variable. Large water-using industries such as canneries, chemical plants, and refineries usually have their own supply and are not dependent on public agencies. Other industries such as those involved in "high technology," which have more modest process water requirements, may depend wholly on municipal supplies. Typical data on the magnitude of water use to be expected from various industrial operations are presented in Table 2-6. Because industrial water use varies widely, it is therefore desirable in practical design work to inspect the plant concerned and to make careful estimates of the quantities of both water used from all sources and the wastes produced.

Public Service and System Maintenance. Public service water represents the smallest component of municipal water use. Public service water uses include water used for public buildings, fire fighting, irrigating public parks and greenbelts, and system maintenance. System maintenance water uses include water for disinfecting new water lines and storage reservoirs, line and hydrant flushing, and hydraulic flushing of sewers. Only small amounts of water used for these purposes reach the sanitary sewer system, except that from public buildings.

TABLE 2-5
Typical water-use values for recreational facilities[a,b]

User	Unit	Flow, gal/unit · d	
		Range	Typical
Apartment, resort	Person	50–70	60
Bowling alley	Alley	150–250	200
Camp			
Pioneer type	Person	15–30	25
Children's central toilet			
and bath	Person	35–50	45
Day, with meals	Person	10–20	15
Day, without meals	Person	8–18	13
Luxury, private bath	Person	75–100	350
Trailer	Trailer	75–150	125
Campground, developed	Person	20–40	30
Country club	Member		
	present	60–125	100
	Employee	10–15	50
Dormitory (bunk house)	Person	20–45	35
Fairground	Visitor	1–2	3
Picnic park, with flush toilets	Visitor	5–10	8
Swimming pool and beach	Customer	5–15	10
	Employee	8–15	10
Visitor center	Visitor	4–8	6

[a] Adapted in part from Refs. 7 and 8.

[b] It is assumed that water under pressure, flush toilets, and washbasins are provided unless otherwise indicated.

Note: gal × 3.7854 = L

Unaccounted System Losses and Leakage. Unaccounted system losses include unauthorized use, incorrect meter calibration or readings, improper meter sizing, and inadequate system controls. Leakage is due to system age, materials of construction, and lack of system maintenance. Unaccounted system losses and leakage may range from 10 to 12 percent of production for newer distribution systems (less than 25 years old) and from 15 to 30 percent for older systems. In small water systems, unaccounted losses and leakage may account for as much as 50 percent of production. As much as 40 to 60 percent of the unaccounted water may be attributed to meter error [1]. Therefore, while water records may be useful in forecasting wastewater flowrates, the accuracy of the records must be checked carefully.

Estimating Water Consumption From Water Supply Records. Water records of various types are kept by water supply agencies. These records usually include information on the amount of water produced or withdrawn and discharged to the water supply system and the amount of water actually used (consumed). The distinction is

TABLE 2-6
Typical rates of water use for various industries

Industry	Range of flow, gal/ton product
Cannery	
Green beans	12,000–17,000
Peaches and pears	3,600–4,800
Other fruits and vegetables	960–8,400
Chemical	
Ammonia	24,000–72,000
Carbon dioxide	14,400–21,600
Lactose	144,000–192,000
Sulfur	1,920–2,400
Food and beverage	
Beer	2,400–3,840
Bread	480–960
Meat packing	3,600–4,800[a]
Milk products	2,400–4,800
Whisky	14,400–19,200
Pulp and paper	
Pulp	60,000–190,000
Paper	29,000–38,000
Textile	
Bleaching	48,000–72,000[b]
Dyeing	7,200–14,400[b]

[a] Live weight.
[b] Cotton.
Note: gal/U.S. ton (short) × 0.00417 = $m^3/10^3$ kg

important because more water is produced than is actually used by the consumer. The difference between these two values is the amount of water lost or unaccounted for in the distribution system plus the amount used for various public services that may be unmetered. Therefore, in using water supply records to estimate wastewater flowrates, it is necessary to determine the amount of water actually used by the customers. Unaccounted water and losses do not reach the wastewater system and have to be excluded in making flow estimates. Data from municipal water-use records are analyzed in Example 2-1 to determine consumption and unaccounted system losses.

Example 2-1 Estimating water consumption from water supply data. A small community water supply agency furnishes water to 147 customers from a well supply. Water records are kept showing the amount of water pumped to the system. The agency recently installed meters for all customers and total water sales records are also kept. The following data were obtained:

Month	Production, gal/mo	Sales, gal/mo
May	1,414,100	1,033,600
June	1,421,000	1,104,300
July	1,407,600	1,086,300
Total	4,242,700	3,224,200
Average, gal/d	46,116	35,046

From the water supply data, determine the amount of water consumed (gal/capita · d) and the amount of water that is unaccounted system loss (as a percent of production). The average household size as determined by the local planning agency is 2.43 persons per service.

Solution

1. Determine the average daily per capita water consumption for the period of record. Use the sales records because that is the actual water measured as used by the customers.

$$\text{Daily consumption} = \frac{35,046 \text{ gal/d}}{(147 \text{ services})(2.43 \text{ persons/service})}$$
$$= 98 \text{ gal/capita} \cdot \text{d}$$

2. Determine unaccounted system losses. The difference between the production rate and sales represents unaccounted system losses and leakage.

$$\text{Unaccounted system losses} = \frac{(46,116 - 35,046)}{46,116} \times 100\%$$
$$= 24\%$$

Comment. Metering errors often account for a large percentage of system losses, and records of meter calibration need to be checked. Differences in production and consumption as large as those in the example are significant and require investigation. If water production records are used without investigating unaccounted losses, the computed consumption rates may be in error.

Water Use by Various Devices and Appliances. Typical rates of water use for various devices and appliances are presented in Table 2-7. Although these rates vary widely, they are useful in estimating total water use when no other data are available.

Variations in Water Use

A direct comparison of water supply records from different municipalities or communities is likely to be misleading. In some municipalities, large quantities of water used for industrial purposes may be obtained from privately owned supplies; in other municipalities, the industries may mainly use municipal supplies. The unaccounted for system losses may also vary widely as discussed above. Other factors affect the variations in water use and they are discussed in this section.

TABLE 2-7
Typical rates of water use for various devices and appliances[a]

Device/appliance	Unit	Range of flow	Typical
Automatic home-type washing machine	gal/load	20–50	30
Automatic home-type dishwasher	gal/load	4–10	6
Bathtub	gal/use	20–30	24
Continuous-flowing drinking fountain	gal/min	1–2	1
Dishwashing machine, commercial:			
Conveyor type, at 15 lb_f/in^2	gal/min	4–6	5
Stationary rack type, at 15 lb_f/in^2	gal/min	6–9	8
Fire hose, $1\frac{1}{2}$ in, $\frac{1}{2}$ in nozzle, 65 ft head	gal/min	35–40	38
Fire hose, home, 125 ft head, $\frac{3}{4}$ in	gal/min	8–12	10
Garbage grinder, home-type	gal/person · d	0.5–1.0	0.75
Garden hose, $\frac{5}{8}$ in, 25 ft head	gal/min	2.5–4	3.5
Garden hose, $\frac{3}{4}$ in, 25 ft head	gal/min	4–6	5
Sprinkler	gal/min	1–3	2
Lawn sprinkler, 3,000 ft^2 lawn, 1 in/wk	gal/wk	1500–1900	1800
Shower head, $\frac{5}{8}$ in, 25 ft head	gal/use	10–25	18
Washbasin	gal/use	0.75–2	1.25
Water closet, flush valve, 25 lb_f/in^2	gal/min	20–30	25
Water closet, tank	gal/use	4–6	5

[a] Adapted in part from Ref. 7.

Note: gal × 3.7854 = L
     in × 25.4 = mm
    ft × 0.3048 = m
   ft^2 × 0.00929 = m^2
 lb_f/in^2 × 6.8948 = kN/m^2

Factors Affecting Municipal Water Use. Factors that affect water use in a community water system include climate, size of the community, density of development, economics, dependability and quality of the supply, water conservation, and the extent of metered services.

Climate. Climatic effects such as temperature and precipitation can significantly impact consumption. Water use is at its peak when it is hot and dry, due largely to increased need for exterior use such as landscape irrigation. The ecological seasons may also vary in different parts of the world and may also affect consumption patterns.

Community size. Community size affects not only the average per capita water use but also the peak rate of use. The rate of use fluctuates over a wider range in small communities with higher peak flows (as compared to average use) and lower minimum flows.

Density of development. The density of development (i.e., single-family housing, condominiums, and apartments) affects both interior and exterior water use. Single-family homes may have more water-using appliances such as washing machines and dishwashers than apartments. Exterior water use for condominiums and apartments is generally much less than single-family homes because of reduced needs for landscape watering.

Economics. The affluence or economic capabilities of a community affects water use (and resulting wastewater flows). As the assessed value of property increases, so does water use and wastewater flowrates [2]. Part of the increase in consumption may be due to the greater use of water-using appliances such as dishwashers, garbage grinders, and washing machines.

Dependability and quality of supply. A water supply that is dependable and of good quality will encourage use by its customers. Supplies that are not dependable in terms of poor pressure and limited quantities during peak or dry periods or that have objectionable taste or mineral content may have lower water use.

Water conservation. Water conservation may take different forms: (1) the cutback of water use during emergencies, such as droughts, to achieve a short-term reduction, or (2) the institution of a long-range program including the installation of water-conserving fixtures to effect a permanent reduction in water use. In emergencies, voluntary or mandatory conservation may be required for supplies impacted by drought or dry period occurrences. For example, in the Oakland area during the California drought of 1977 and 1978, total water use was reduced from 25 to 35 percent by water conservation measures [3]. A major share of the reduction was due to the decrease in exterior water use. The use of low-flow toilets is now specified in many local building codes. In the future, the use of water-conserving devices and appliances is expected to increase significantly.

For estimating wastewater flowrates from water use, the effect of conservation on interior water use is of particular interest. The effect of the installation of water-conserving fixtures on interior water use and resulting wastewater flowrates is discussed in Section 2-5. The extent of the water savings actually achieved depends on the overall scope of the water conservation measures. For information on the effectiveness of specific water conservation programs, Ref. 9 may be consulted.

Metered services. Water agencies with metered services usually charge their customers based on the water used. Systems with unmetered services charge customers some form of a flat rate for unlimited water use. Metering the individual consumer's supply and billing at established meter rates indirectly prevents waste of water by users and tends to reduce actual water use. The waste and unaccounted-for water in metered systems ranges from 10 to 20 percent of the total water entering the distribution system. The corresponding range in unmetered systems is much higher (typically 30 percent).

Fluctuations in Water Use. Although it is important to know the average rates of water use, it is equally important to have data on the fluctuations in rates of use. Representative data on the typical fluctuations in water use are reported in Table 2-8.

TABLE 2-8
Typical fluctuations in water use in community systems[a]

	Percentage of average for year	
	Range	Typical
Daily average in maximum month	110–140	120
Daily average in maximum week	120–170	140
Maximum day	160–220	180
Maximum hour	225–320	270[b]

[a] Ref. 8.
[b] 1.5 × maximum day value.

The maximum use usually occurs during two seasons: (1) in summer months, when water is in demand for garden and lawn irrigation, and (2) in winter months, when large quantities are wasted to prevent freezing pipes and fixtures.

Hourly variations in water consumption also affect the rate of wastewater flow. In general, the wastewater-discharge curve closely parallels the water-consumption curve, but with a lag of several hours. In some cities, large quantities of water used by industries and obtained from sources other than the municipal supply are discharged into sewers during the working hours of the day. When such discharge occurs, the peak flow is higher than the amount resulting from the normal variation in the draft on the municipal supply.

Proportion of Municipal Water Supply Reaching the Collection System as Wastewater

Because wastewater consists primarily of used water, the portion of the water supplied that reaches the collection system must be estimated. A considerable portion of the water produced does not reach the sanitary sewer system and includes (1) product water used by manufacturing establishments; (2) water used for landscape irrigation, system maintenance, and extinguishing fires; (3) water used by consumers whose facilities are not connected to sewers; and (4) leakage from water mains and service pipes (unaccounted for losses).

About 60 to 85 percent of the per capita consumption of water becomes wastewater (the lower percentages are applicable to the semiarid region of the southwestern United States). Application of appropriate percentages to records from metered water use generally can be used to obtain a reasonable estimate of wastewater flowrates, excluding infiltration/inflow. In some cases, however, excessive infiltration, roof water, and water used by industries that is obtained from privately owned water supplies make the quantity of wastewater larger than the water consumption from the public supply. If a community has well-constructed sewers and if stormwater and snowmelt drainage are excluded and there is no substantial change in the industrial uses of water, the variation from year to year in the ratio of wastewater to water supply will not be great.

2-3   WASTEWATER SOURCES AND FLOWRATES

Data that can be used to estimate average wastewater flowrates from various domestic and industrial sources and the infiltration/inflow contribution are presented in this section. Variations in the flowrates that must be established before collection systems and treatment facilities are designed are also discussed.

Domestic Wastewater Sources and Flowrates

The principal sources of domestic wastewater in a community are the residential areas and commercial districts. Other important sources include institutional and recreational facilities. For existing developments, flowrate data should be obtained by direct measurement. For areas being developed, methods of estimating flowrates are considered in the following discussion. Wastewater flowrates for small systems (systems with 1000 people or less) may differ significantly from larger systems and are discussed in Chap. 14.

Residential Areas. For many residential areas, wastewater flowrates are commonly determined on the basis of population density and the average per capita contribution of wastewater. Data on ranges and typical flowrate values are given in Table 2-9. For large residential areas, it is often advisable to develop flowrates on the basis of land-use areas and anticipated population densities. Where possible, these rates should be based on actual flow data from selected similar residential areas, preferably in the same locale.

In the past, the preparation of population projections for use in estimating wastewater flowrates was often the responsibility of the engineer, but today such data are usually available from local, regional, and state planning agencies. If the data are not available and population projections have to be prepared, Ref. 5 may be consulted for population forecasting methodology.

Commercial Districts. Commercial wastewater flowrates are generally expressed in gal/acre · d (m³/ha · d) and are based on existing or anticipated future development or comparative data. Average unit-flowrate allowances for commercial developments normally range from 800 to 1500 gal/acre · d (7.5 to 14 m³/ha · d). Because unit flowrates can vary widely for commercial facilities, every effort should be made to obtain records from similar facilities. Estimates for certain commercial sources may also be made from the data in Table 2-10.

Institutional Facilities. Some typical flowrates from institutional facilities, essentially domestic in nature, are shown in Table 2-11. Again, it is stressed that flowrates

TABLE 2-9
Typical wastewater flowrates from residential sources[a]

| Source | Unit | Flow, gal/unit · d | |
		Range	Typical
Apartment:			
High-rise	Person	35–75	50
Low-rise	Person	50–80	65
Hotel	Guest	30–55	45
Individual residence:			
Typical home	Person	45–90	70
Better home	Person	60–100	80
Luxury home	Person	75–150	95
Older home	Person	30–60	45
Summer cottage	Person	25–50	40
Motel:			
With kitchen	Unit	90–180	100
Without kitchen	Unit	75–150	95
Trailer park	Person	30–50	40

[a] Adapted in part from Ref. 7.
Note: gal × 3.7854 = L

vary with the region, climate, and type of facility. The actual records of institutions are the best sources of flow data for design purposes.

Recreational Facilities. Wastewater flowrates from many recreational facilities are highly seasonal. Typical data on wastewater flowrates from recreational facilities are presented in Table 2-12.

Sources and Rates of Industrial (Nondomestic) Wastewater Flows

Nondomestic wastewater flowrates from industrial sources vary with the type and size of the facility, the degree of water reuse, and the onsite wastewater treatment methods, if any. Extremely high peak flowrates may be reduced by the use of detention tanks and equalization basins. Typical design values for estimating the flows from industrial areas that have no or little wet-process type industries are 1000 to 1500 gal/acre · d (9 to 14 m^3/ha · d) for light industrial developments and 1500 to 3000 gal/acre · d (14 to 28 m^3/ha · d) for medium industrial developments. Alternatively, for estimating industrial flowrates where the nature of the industry is known, data such as those reported in Table 2-6 can be used. For industries without internal recycling or reuse programs, it can be assumed that about 85 to 95 percent of the water used in the various operations and processes will become wastewater. For large industries with internal water-reuse programs, separate estimates must be made. Average domestic

TABLE 2-10
Typical wastewater flowrates from commercial sources[a]

Source	Unit	Flow, gal/unit · d	
		Range	Typical
Airport	Passenger	2–4	3
Automobile service station	Vehicle served	7–13	10
	Employee	9–15	12
Bar	Customer	1–5	3
	Employee	10–16	13
Department store	Toilet room	400–600	500
	Employee	8–12	10
Hotel	Guest	40–56	48
	Employee	7–13	10
Industrial building (sanitary waste only)	Employee	7–16	13
Laundry (self-service)	Machine	450–650	550
	Wash	45–55	50
Office	Employee	7–16	13
Restaurant	Meal	2–4	3
Shopping center	Employee	7–13	10
	Parking space	1–2	2

[a] Adapted in part from Ref. 2.
Note: gal × 3.7854 = L

TABLE 2-11
Typical wastewater flowrates from institutional sources[a]

Source	Unit	Flow, gal/unit · d	
		Range	Typical
Hospital, medical	Bed	125–240	165
	Employee	5–15	10
Hospital, mental	Bed	75–140	100
	Employee	5–15	10
Prison	Inmate	75–150	115
	Employee	5–15	10
Rest home	Resident	50–120	85
School, day			
With cafeteria, gym, and showers	Student	15–30	25
With cafeteria only	Student	10–20	15
Without cafeteria and gym	Student	5–17	11
School, boarding	Student	50–100	75

[a] Adapted in part from Ref. 2.
Note: gal × 3.7854 = L

TABLE 2-12
Typical wastewater flowrates from recreational facilities[a]

Facility	Unit	Flow, gal/unit · d	
		Range	Typical
Apartment, resort	Person	50–70	60
Cabin, resort	Person	8–50	40
Cafeteria	Customer	1–3	2
	Employee	8–12	10
Campground (developed)	Person	20–40	30
Cocktail lounge	Seat	12–25	20
Coffee shop	Customer	4–8	6
	Employee	8–12	10
Country club	Member present	60–130	100
	Employee	10–15	13
Day camp (no meals)	Person	10–15	13
Dining hall	Meal served	4–10	7
Dormitory, bunkhouse	Person	20–50	40
Hotel, resort	Person	40–60	50
Store, resort	Customer	1–4	3
	Employee	8–12	10
Swimming pool	Customer	5–12	10
	Employee	8–12	10
Theatre	Seat	2–4	3
Visitor Center	Visitor	4–8	5

[a] Adapted in part from Ref. 7.
Note: gal × 3.7854 = L

(sanitary) wastewater contributed from industrial facilities may vary from 8 to 25 gal/capita · d (30 to 95 L/capita · d).

Infiltration/Inflow

Extraneous flows in sewers, described as infiltration and inflow, are illustrated in Fig. 2-1 and are defined as follows:

Infiltration. Water entering a sewer system, including sewer service connections, from the ground through such means as defective pipes, pipe joints, connections, or manhole walls.

Steady inflow. Water discharged from cellar and foundation drains, cooling-water discharges, and drains from springs and swampy areas. This type of inflow is steady and is identified and measured along with infiltration.

Direct inflow. Those types of inflow that have a direct stormwater runoff connection to the sanitary sewer and cause an almost immediate increase in wastewater flows. Possible sources are roof leaders, yard and areaway drains, manhole covers, cross connections from storm drains and catch basins, and combined sewers.

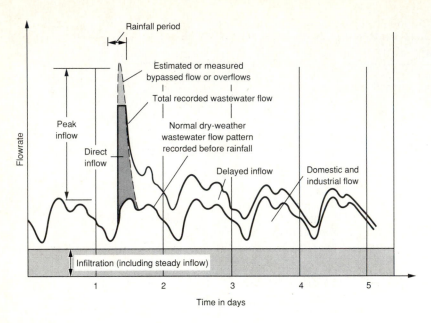

FIGURE 2-1
Graphic identification of infiltration/inflow.

Total inflow. The sum of the direct inflow at any point in the system plus any flow discharged from the system upstream through overflows, pumping station bypasses, and the like.

Delayed inflow. Stormwater that may require several days or more to drain through the sewer system. This category can include the discharge of sump pumps from cellar drainage as well as the slowed entry of surface water through manholes in ponded areas.

The initial impetus for defining and identifying infiltration/inflow was the Federal Water Pollution Control Act Amendments of 1972. As a condition of receiving a federal grant for the design and construction of wastewater treatment facilities, grant applicants have to demonstrate that their wastewater collection systems are not subject to excessive infiltration/inflow. A cost-effectiveness analysis has to be made to determine if it is more economical to make repairs to the collection system to correct infiltration/inflow or to design the treatment facilities for larger flows. By correcting infiltration/inflow problems and "tightening" the collection system, the community benefits with (1) no overloaded or surcharged sewers and the associated problems of wastewater backups and overflows, (2) more efficient operation of wastewater treatment facilities, and (3) the use of the collection system hydraulic capacity for wastewater requiring treatment instead of for infiltration/inflow.

Detailed procedures for making an analysis of infiltration/inflow, including an example cost-effectiveness analysis, are provided in the companion text, Ref. 6. Because an understanding of the effects of infiltration/inflow is important in determining treatment plant flowrates, a discussion of excessive infiltration/inflow is included in this section. For additional information on infiltration/inflow requirements, Ref. 12 may be consulted.

Infiltration into Sewers. One portion of the rainfall in a given area runs quickly into the storm sewers or other drainage channels; another portion evaporates or is absorbed by vegetation; and the remainder percolates into the ground, becoming groundwater. The proportion of the rainfall that percolates into the ground depends on the character of the surface and soil formation and on the rate and distribution of the precipitation. Any reduction in permeability, such as that due to buildings, pavements, or frost, decreases the opportunity for precipitation to become groundwater and increases the surface runoff correspondingly.

The amount of groundwater flowing from a given area may vary from a negligible amount for a highly impervious district or a district with a dense subsoil to 25 or 30 percent of the rainfall for a semipervious district with a sandy subsoil permitting rapid passage of water. The percolation of water through the ground from rivers or other bodies of water sometimes has considerable effect on the groundwater table, which rises and falls continually.

The presence of high groundwater results in leakage into the sewers and in an increase in the quantity of wastewater and the expense of disposing of it. The amount of flow that can enter a sewer from groundwater, or infiltration, may range from 100 to 10,000 gal/d · in-mi (0.0094 to 0.94 m^3/d · mm-km) or more. The number of inch-miles (millimeter-kilometers) in a wastewater collection system is the sum of the products of sewer diameters, in inches (millimeters), times the lengths, in miles (kilometers), of sewers of corresponding diameters. Expressed another way, infiltration may range from 20 to 3000 gal/acre · d (0.2 to 28 m^3/ha · d). During heavy rains, when there may be leakage through manhole covers, or inflow, as well as infiltration, the rate may exceed 50,000 gal/acre · d (470 m^3/ha · d). Infiltration/inflow is a variable part of the wastewater, depending on the quality of the material and workmanship in constructing the sewers and building connections, the character of the maintenance, and the elevation of the groundwater compared with that of the sewers.

The sewers first built in a community usually follow the watercourses in the bottoms of valleys, close to (and occasionally below) the beds of streams. As a result, these old sewers may receive comparatively large quantities of groundwater, whereas sewers built later at high elevations will receive relatively small quantities of groundwater. With an increase in the percentage of area in a community that is paved or built over comes (1) an increase in the percentage of stormwater conducted rapidly to the storm sewers and watercourses and (2) a decrease in the percentage of the stormwater that can percolate into the earth and tend to infiltrate the sanitary sewers.

The rate and quantity of infiltration depend on the length of sewers, the area served, the soil and topographic conditions, and, to a certain extent, the population density (which affects the number and total length of house connections). Although the elevation of the water table varies with the quantity of rain and melting snow percolating into the ground, the leakage through defective joints, porous concrete, and cracks has been large enough, in some cases, to lower the groundwater table to the level of the sewer.

Most of the pipe sewers built during the first half of this century were laid with cement mortar joints or hot-poured bituminous compound joints. Manholes were almost always constructed of brick masonry. Deterioration of pipe joints, pipe-to-manhole joints, and the waterproofing of brickwork has resulted in a high potential for infiltration into these old sewers. The use of high-quality pipe with dense walls, precast manhole sections, and joints sealed with rubber or synthetic gaskets is standard practice in modern sewer design. The use of these improved materials has greatly reduced infiltration into newly constructed sewers, and it is expected that the increase of infiltration rates with time will be much slower than has been the case with the older sewers.

Inflow into Sewers. As described previously, the type of inflow that causes a "steady flow" cannot be identified separately, so it is included in the measured infiltration. The direct inflow can cause an almost immediate increase in flowrates in sanitary sewers. The effects of inflow on peak flowrates that must be handled by a wastewater treatment plant are shown in Example 2-2.

Example 2-2   Computing infiltration/inflow from wastewater flow records. A large city has measured high flowrates during the wet season of the year. The flowrate during the dry period of the year, when rainfall is rare and groundwater infiltration is negligible, averages 33.8 Mgal/d (128,000 m³/d). During the wet period when groundwater levels are elevated, the flowrate averaged 63.4 Mgal/d (240,000 m³/d) excluding those days during and following any significant rainfall events. During a recent storm, hourly flowrates were recorded during the peak flow period, as well as several days following the storm. The flowrate plots are shown in the accompanying figure. Compute the infiltration and inflow and determine if the infiltration is excessive. Excessive infiltration is defined by the local regulatory agency as rates over 8000 gal/d · in-mi (0.752 m³/d · mm-km) of sewer. The composite diameter-length of the sewer system is 6600 in-mi (270,000 mm-km).

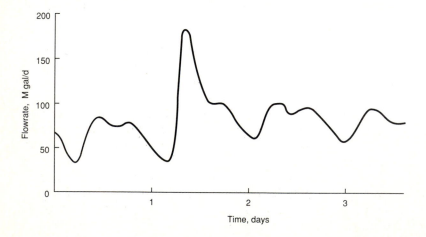

Time, days

Solution

1. Determine the infiltration and inflow components during the wet season.

   (a) Because infiltration is low during dry periods, high groundwater infiltration is computed as peak flowrate minus base (dry-weather) flowrate:

$$\text{Infiltration} = (63.4 - 33.8) \text{ Mgal/d}$$
$$= 29.6 \text{ Mgal/d } (112,000 \text{ m}^3/\text{d})$$

   (b) The maximum hourly inflow is graphically determined from the accompanying figure as the difference between the maximum hourly wet-weather flowrate during the storm and the comparable flowrate on the preceding day. In this case, the maximum inflow is

$$\text{Inflow} = (180 - 80) \text{ Mgal/d}$$
$$= 100 \text{ Mgal/d } (378,500 \text{ m}^3/\text{d})$$

2. Determine if the infiltration is excessive.

   (a) Calculate the infiltration by dividing the calculated flowrate in gal/d by the composite diameter-length of the sewer system.

$$\text{Infiltration} = \frac{29,600,000 \text{ gal/d}}{6,600 \text{ in-mi}}$$
$$= 4485 \text{ gal/d} \cdot \text{in-mi } (0.415 \text{ m}^3/\text{d} \cdot \text{mm-km})$$

   (b) Using the regulatory agency criterion of 8000 gal/d · in-mi (0.752 m^3/d · mm-km), the infiltration is not excessive.

   Comment. In this example, the peak flowrate during the storm period was 5.3 times the average dry-weather flowrate. As discussed in Chap. 5, the peak flowrate factor is high for a system of this size. Because inflow represents over 50 percent of the peak flow and requires oversizing of the hydraulic capacity of the treatment plant, methods of inflow reduction should be investigated to decrease the hydraulic load on the sewer system and treatment facilities.

Variations in Wastewater Flowrates

Short-term, seasonal, and industrial variations in wastewater flowrates are briefly discussed here. The analysis of flowrate data and the definition of flowrate variations are discussed in Sec. 2-4.

Short-Term Variations. The variations in wastewater flowrates observed at treatment plants tend to follow a somewhat diurnal pattern, as shown in Fig. 2-2. Minimum flows occur during the early morning hours when water consumption is lowest and when the base flow consists of infiltration and small quantities of sanitary wastewater. The first peak flow generally occurs in the late morning when wastewater from the peak morning water use reaches the treatment plant. A second peak flow generally occurs in the early evening between 7 and 9 p.m., but this varies with the size of the community and the length of the sewers.

   When extraneous flows are minimal, wastewater-discharge curves closely parallel water-consumption curves, but with a lag of several hours. In the absence of a day when home laundering is done, the variation in weekday flowrates is negligible. A plot of typical weekly flowrates for both wet and dry periods is shown in Fig. 2-3.

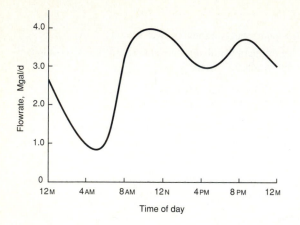

FIGURE 2-2
Typical hourly variation in domestic wastewater flowrates.

Seasonal Variations. Seasonal variations in domestic wastewater flows are commonly observed at resort areas, in small communities with college campuses, and in communities with seasonal commercial and industrial activities. The magnitude of the variations to be expected depends on both the size of the community and the seasonal activity. An example of a seasonal recreational variation as well as the influence of infiltration/inflow is Lake Arrowhead, California, as illustrated in Fig. 2-4. Flowrates increase in the summer because of higher recreational occupancy rates. The high flow periods occur in the winter and early spring when groundwater levels rise and infiltration increases.

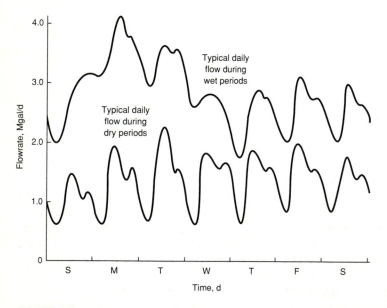

FIGURE 2-3
Typical daily and weekly variations in domestic wastewater flowrates.

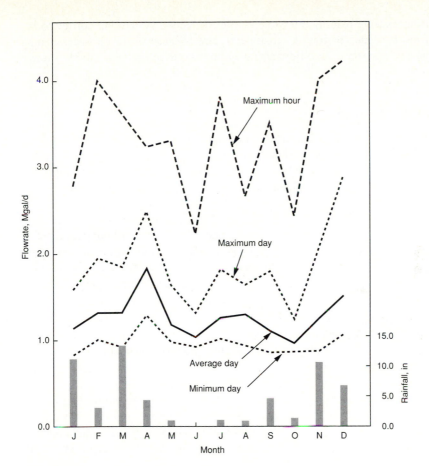

FIGURE 2-4
Monthly flowrates and rainfall at Lake Arrowhead, California.

Industrial Variations. Industrial wastewater discharges are difficult to predict. Many manufacturing facilities generate relatively constant flowrates during production, but the flowrates change markedly during cleanup and shutdown. Although internal process changes may lead to reduced discharge rates, plant expansion and increased production may lead to increased wastewater generation. Where joint treatment facilities are to be constructed, special attention should be given to industrial flowrate projections, whether they are prepared by the industry or jointly with the city's staff or engineering consultant. Industrial discharges are most troublesome in smaller wastewater treatment plants where there is limited capacity to absorb shock loadings.

2-4 ANALYSIS OF WASTEWATER FLOWRATE DATA

Because the hydraulic design of both collection and treatment facilities is affected by variations in wastewater flowrates, the flowrate characteristics have to be analyzed carefully from existing records. Flowrates in the collection system may differ some-what from the flowrate entering the treatment plant because of the flow-dampening effect of the sewer system. Peak flowrates may be attenuated by the available storage capacity in the sewer system.

Flowrates for Design

Where flow records are kept for treatment plants and pumping stations, at least two years of the most recent data should be analyzed. Longer-term records may be analyzed to determine changes or trends in wastewater generation rates. Important information that needs to be obtained through the analysis of wastewater flowrate data includes the following:

Average daily flow. The average flowrate occurring over a 24-hour period based on total annual flowrate data. Average flowrate is used in evaluating treatment plant capacity and in developing flowrate ratios used in design. The average flowrate may also be used to estimate such items as pumping and chemical costs, sludge solids, and organic-loading rates.

Maximum daily flow. The maximum flowrate that occurs over a 24-hour period based on annual operating data. The maximum daily flowrate is important particularly in the design of facilities involving retention time such as equalization basins and chlorine-contact tanks.

Peak hourly flow. The peak sustained hourly flowrate occurring during a 24-hour period based on annual operating data. Data on peak hourly flows are needed for the design of collection and interceptor sewers, wastewater-pumping stations, wastewater flowmeters, grit chambers, sedimentation tanks, chlorine-contact tanks, and conduits or channels in the treatment plant. The use of peaking factors to determine peak flowrates is discussed in Chap. 5.

Minimum daily flow. The minimum flowrate that occurs over a 24-hour period based on annual operating data. Minimum flowrates are important in the sizing of conduits where solids deposition might occur at low flowrates.

Minimum hourly flow. The minimum sustained hourly flowrate occurring over a 24-hour period based on annual operating data. Data on the minimum hourly flowrate are needed to determine possible process effects and for sizing of wastewater flowmeters, particularly those that pace chemical-feed systems. At some treatment facilities, such as those using trickling filters, recirculation of effluent is required to sustain the process during low-flow periods. For wastewater pumping, minimum flowrates are important to ensure that the pumping systems have adequate turndown to match the low flowrates.

Sustained flow. The flowrate value sustained or exceeded for a specified number of consecutive days based on annual operating data. Data on sustained flowrates may be used in sizing equalization basins and other plant hydraulic components. An ex-ample of a plot of sustained peak and low flowrates is shown in Fig. 2-5. When devel-oping plots similar to Fig. 2-5, the longest available period of record should be used.

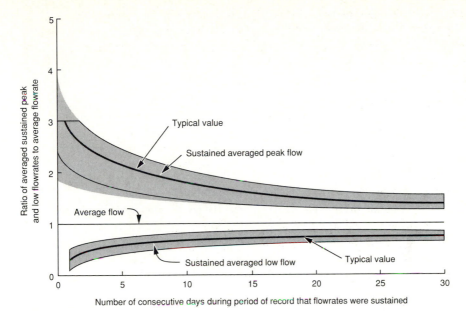

Number of consecutive days during period of record that flowrates were sustained

FIGURE 2-5
Typical ratios of averaged sustained peak and low daily flowrates to average annual daily flowrates for time periods up to 30 days.

Statistical Analysis of Wastewater Flowrates

In developing wastewater management systems, it is often necessary to determine the statistical characteristics of wastewater flowrates. The first step in assessing the statistical characteristics of a series of observations is to determine whether the observations are distributed normally or are skewed. For most practical purposes, the type of the distribution can be determined by plotting the data on both arithmetic- and log-probability paper and noting whether or not the data can be fitted with a straight line. If the distribution is normal, statistical measures used to describe the distribution include the mean, median, mode, standard deviation, coefficient of variation, coefficient of skewness, and coefficient of kurtosis [4,13,14]. If the distribution is skewed, the geometric mean and standard deviation are noted. The determination of statistical measures for wastewater flowrate data is illustrated in Example 2-3.

Example 2-3 Statistical analysis of wastewater flowrate data. Determine the statistical characteristics of the following weekly flowrate data obtained from an industrial discharger for a calendar quarter of operation. Using these data, predict also the maximum weekly flowrate that will occur during a full year's operation.

Week No.	Flowrate, Mgal/wk	Week No.	Flowrate, Mgal/wk
1	0.768	8	0.971
2	0.803	9	1.007
3	0.985	10	0.912
4	0.888	11	0.863
5	0.996	12	0.840
6	1.078	13	0.828
7	1.061		

Solution

1. Determine graphically, using probability paper, whether the flowrate data are distributed normally or are skewed (log-normal).

   (*a*) Set up a data analysis table with three columns, as described below.

       i. In column 1, enter the rank serial number starting with number 1.

       ii. In column 2, arrange the flowrate data in ascending order.

       iii. In column 3, enter the probability plotting position.

Rank serial no., m	Flowrate, Mgal/wk	Plotting position,[a] %
1	0.768	7.1
2	0.803	14.3
3	0.828	21.4
4	0.840	28.6
5	0.863	35.7
6	0.888	42.9
7	0.912	50.0
8	0.936	57.1
9	0.971	64.3
10	0.996	71.4
11	1.007	78.6
12	1.061	85.7
13	1.078	92.9

[a] Plotting position $= (m/n + 1)100$

The term $(n + 1)$ is used to obtain the plotting positions as opposed to just n because there may be an observation that is either larger or smaller than the largest or smallest in the data set.

(*b*) On both arithmetic- and log-probability paper, plot the weekly flowrates expressed in Mgal/wk versus the plotting position determined above. The resulting plots are presented below. Because the data fall on a straight line in both plots, the flowrate data can be described adequately by either type of distribution. This fact can be taken as an indication that the distribution is not skewed significantly and that normal statistics can be applied.

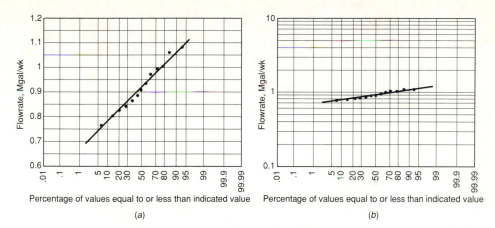

Percentage of values equal to or less than indicated value

(a)

Percentage of values equal to or less than indicated value

(b)

Probability plots: (a) arithmetic and (b) logarithmic, Example 2-3, step 1b.

2. Determine the statistical characteristics of the flowrate data.

(a) Set up a data analysis table to obtain the quantities needed to determine the statistical characteristics.

Flowrate Mgal/wk	$(x - \bar{x})$	$(x - \bar{x})^2$	$(x - \bar{x})^3$ 10^{-3}	$(x - \bar{x})^4$ 10^{-4}
0.768	−0.151	0.023	−3.443	5.199
0.803	−0.116	0.013	−1.561	1.811
0.828	−0.091	0.008	−0.754	0.686
0.840	−0.079	0.006	−0.493	0.390
0.863	−0.056	0.003	−0.174	0.098
0.888	−0.031	0.001	−0.300	0.009
0.912	−0.007	0.000	0.000	0.000
0.936	0.017	0.000	0.005	0.001
0.971	0.052	0.003	0.141	0.073
0.996	0.073	0.006	0.457	0.352
1.007	0.088	0.008	0.682	0.600
1.061	0.142	0.020	2.863	4.066
1.078	0.159	0.025	4.020	6.391
11.951		0.116	1.713	19.676

(b) Determine the statistical characteristics.

i. Mean

$$\bar{x} = \frac{\sum x}{n}$$

$$\bar{x} = \frac{11.951}{13} = 0.919 \text{ Mgal/wk}$$

ii. Median (the middle-most value)

$$\text{Median} = 0.912 \text{ Mgal/wk (see data table above)}$$

iii. Mode

$$\text{Mode} = 3(\text{Med}) - 2(\bar{x}) = 3(0.912) - 2(0.919)$$
$$= 0.898 \text{ Mgal/wk}$$

iv. Standard deviation

$$s = \sqrt{\frac{\sum (x - \bar{x})^2}{n - 1}}$$

$$s = \sqrt{\frac{0.116}{12}} = 0.098 \text{ Mgal/wk}$$

v. Coefficient of variation

$$\text{CV} = \frac{100s}{\bar{x}}$$

$$\text{CV} = \frac{100(0.098)}{0.919} = 10.7\%$$

vi. Coefficient of skewness

$$\alpha_3 = \frac{\sum (x - \bar{x})^3 / n - 1}{s^3}$$

$$\alpha_3 = \frac{1.713 \times 10^{-3} / 12}{(0.098)^3} = 0.152$$

vii. Coefficient of kurtosis

$$\alpha_4 = \frac{\sum (x - \bar{x})^4 / n - 1}{s^4}$$

$$\alpha_4 = \frac{19.67 \times 10^{-4} / 12}{(0.098)^4} = 1.78$$

Reviewing the statistical characteristics, it can be seen that the distribution is somewhat skewed ($\alpha_3 = 0.152$ versus 0 for a normal distribution) and is considerably flatter than a normal distribution would be ($\alpha_4 = 1.78$ versus 3.0 for a normal distribution).

3. Determine the probable annual maximum weekly flowrate.

   (a) Determine the probability factor.

$$\text{Peak week} = \frac{m}{n + 1} = \frac{52}{52 + 1} = 0.981$$

   (b) Determine the flowrate from preceding figure (a) at the 98.1 percentile.

$$\text{Peak weekly flowrate} = 1.002 \text{ Mgal/wk}$$

Comment. The statistical analysis of data is important in establishing the design conditions for wastewater treatment plants. The application of statistical analysis to other design parameters is examined in Chap. 5.

2-5 REDUCTION OF WASTEWATER FLOWRATES

Because of the importance of conserving both resources and energy, various means for reducing wastewater flowrates and pollutant loadings from domestic sources are gaining increasing attention. The reduction of wastewater flowrates from domestic sources results directly from the reduction in interior water use. Therefore, the terms "interior water use" and "domestic wastewater flowrates" are used interchangeably.

A comparison of residential interior water use (and resulting per capita wastewater flowrates) for homes without and with water-conserving fixtures is given in Table 2-13. Two levels of water-conserving fixtures are included in Table 2-13: Level 1, which includes retrofit devices such as flow restrictors and toilet dams, and Level 2, which uses water-conserving devices and appliances such as low-flush toilets and low water-use washing machines.

The principal devices and appliances that are used to reduce domestic water use and wastewater flowrates are described in Table 2-14. The actual flowrate reduction that is possible using these devices and appliances as compared with the flowrates from the conventional devices is reported in Table 2-15. Another method of achieving flowrate reductions is to restrict the use of appliances that tend to increase water consumption, such as automatic dishwashers and garbage disposal units.

TABLE 2-13
Comparisons of interior water use without and with conservation devices[a]

	Flow, gal/capita · d		
		With conservation devices	
Use	Without conservation devices	Level 1[b]	Level 2[b]
Baths	7	7	7
Dishwashers	2	1	1
Faucets	9	9	8
Showers	16	12	8
Toilets	22	19	14
Toilet leakage	4	4	8
Washing machines	16	14	13
Total	76	66	59

[a] Adapted from Ref. 9.

[b] Level 1 uses retrofit devices such as flow restrictors and toilet dams. Level 2 uses water-conserving devices and appliances such as low-flush toilets and low water-use washing machines.

Note: gal × 3.7854 = L

TABLE 2-14
Flow reduction devices and appliances

Device/appliance	Description and/or application
Faucet aerators	Increases the rinsing power of water by adding air and concentrating flow, thus reducing the amount of wash water used
Limiting-flow shower heads	Restricts and concentrates water passage by means of orifices that limit and divert shower flow for optimum use by the bather
Low-flush toilets	Reduces the discharge of lower amounts of water per flush
Pressure-reducing valve	Maintains home water pressure at a lower level than that of the water distribution system; decreases the probability of leaks and dripping faucets
Retrofit kits for bathroom fixtures	Kits may consist of shower-flow restrictors, toilet dams, or displacement bags, and toilet leak detector tablets
Toilet dam	A partition in the water closet that reduces the amount of water per flush
Toilet leak detectors	Tablets that dissolve in the water closet and release dye to indicate leakage of the flush valve
Water-efficient dishwasher	Reduces the water used
Water-efficient clothes washer	Reduces the water used

TABLE 2-15
Reductions achieved by flow reduction devices and appliances[a,b]

Device/appliance	Flow reduction, gal/capita · d or unit
Faucet aerator	0.5
Limiting-flow shower heads	
3 gal/min	7
0.5 gal/min	14
Low-flush toilets	
3.4 gal/flush	8
0.5 gal/flush	20
Pressure-reducing valve, %	3–6
Retrofit kits for bathroom fixtures	4–7
Toilet dam	4
Toilet leak detectors, gal/d/toilet	24
Water-efficient dishwasher	1
Water-efficient clothes washer	1.5

[a] Adapted from Refs. 9, 10, and 11.

[b] As compared to non-conserving devices or appliances.

Note: gal × 3.7854 = L

In many communities, the use of one or more of the flow reduction devices is now specified for all new residential dwellings; in others, the use of garbage grinders has been limited in new housing developments. Further, many individuals who are concerned about conservation have installed such devices as a means of reducing water consumption.

DISCUSSION TOPICS AND PROBLEMS

2-1. The flowrate variation for city A is shown in the following figure. If the detention time (volume/flow rate) of a storage basin under average flow conditions is 6 h, what would be the average detention time for the period of 8 a.m. to 2 a.m.?

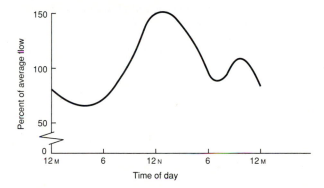

2-2. Sewers are to be installed in a camping area that contains a developed campground for 200 persons, lodges and cabins for 100 persons, and resort apartments for 50 persons. Assume that persons staying in lodges use the dining hall for 3 meals per day and that a 5-seat cafeteria with 4 employees and an estimated 100 customers per day has been constructed. Daily attendance at visitor centers is expected to be 50 percent capacity. Other facilities include a 10-machine laundromat, a 20-seat cocktail lounge, and three gas stations — 265 gal/d per station. Determine the average wastewater flowrate in gal/d using the unit flows.

2-3. Obtain data from your local water agency for three years of operation on the amount of water withdrawn, the amount of water sold (consumed), and the number of customers. Determine the amount of water lost or unaccounted for and compare it to the amount sold. Obtain data on the average household size from your local planning agency, and compute the average per capita water consumption.

2-4. Using the data from Prob. 2-3, can you discern a trend in water use? Identify the pertinent factors that may cause a change in the water-use trend. If a drought or dry period occurred during the period for which flow records were available, what was the reduction in water use during the dry period?

2-5. Obtain an annual report or one year of flow data from your local wastewater treatment facility. From these records, determine the average, maximum, and minimum daily flowrates and the maximum and minimum hourly flowrates. Compute the maximum- and minimum-to-average flowrate ratios.

2-6. Using the data obtained in Prob. 2-3 and 2-5, determine the ratio of the water withdrawn for the water supply to the measured wastewater flow for the same period. How does this value compare with the values reported in the text?

2-7. From the data obtained for Prob. 2-5, compute the peak-to-average wastewater flowrate ratio and compare your value to Fig. 2-5.

2-8. From the flowrate data obtained for Prob. 2-5, prepare a statistical analysis of the data and plot on probability paper. Determine the mean, the standard deviation, and the maximum one day per year occurrence.

2-9. Estimate the ratio of the peak hourly flowrate during the wet period to the average daily flowrate for the dry period for the curve given in Fig. 2-3.

2-10. If the average water use in Example 2-1 is representative of the average annual use, compute the expected water demand for the maximum day and maximum hour.

2-11. If the community in Example 2-1 instituted a water conservation program and provided one retrofit kit per household, consisting of a faucet aerator, toilet dam, and a 2.9 gal/min limiting-flow shower head, (a) compute the water savings if all of the households installed the retrofit kit, and (b) compute the maximum day and maximum hour water demand for the reduced consumption and compare results with Prob. 2-10.

2-12. The wastewater treatment plant has been experiencing high wastewater flowrates during the wet-weather months. The average monthly flows, in Mgal/d, are reported below. The rapid increase in flows during the winter months is due mainly to increased infiltration/inflow. Infiltration is estimated to be 67 percent of the excess flow. It has been determined that 206 mi of the sewers need to be repaired at an average cost of $100,000/mi and the repair will be effective in reducing the infiltration by 30 percent. How many years from now will it take to pay back the cost of the sewer repair program based on the annual savings in treatment cost, assuming that the future annual flowrates are equal to those in the table below? The current cost of treatment is $0.95/1000 gal and the future cost of treatment is estimated to escalate at 6 percent per year. Assume that the sewer repair will be complete in three years.

Month	Average monthly flow, Mgal/d
January	77.40
February	86.65
March	73.70
April	56.00
May	38.57
June	28.53
July	25.10
August	23.51
September	24.57
October	29.32
November	34.87
December	40.68

2-13. Land use in an area is given in the first table at the top of the next page. The school has 1500 students. The average flowrate is 75 L/student · d and the peaking factor (ratio of peak flow to average flow) is 4.0. Average flowrate allowances and peaking factors for the other developments are shown in the second table at the top of the next page.

Type of development	Area, ha
Residential	125
Commercial	11
School	4
Industrial	8

Type of development	Average flowrate, $m^3/ha \cdot d$	Peaking factor
Residential	40	3.0
Commercial	20	2.0
Industrial	30	2.5

Determine the peak wastewater flowrate from the area.

2-14. Estimate the wastewater flows from a large industrial development covering an area of 200 ha. From water-meter readings, it has been determined that the annual use of water within the area is 4.24×10^6 m^3. Twenty percent of the gross area of the development has been landscaped. The average water demand for irrigation of landscape areas is estimated to be 1.3 m/yr.

Assuming that 85 percent of the nonirrigation water consumption ultimately reaches the sewer, estimate the annual wastewater production within the area. Assuming that all industries within the area operate concurrently for 12 h/d, 5 d/wk throughout the year and that the wastewater production during the hours of operation is essentially constant, estimate the maximum wastewater flowrate. Also, compute the average annual wastewater production in cubic meters per day, and determine the value of the peaking factor that relates peak flow to average annual flow. Ignore infiltration and inflow.

REFERENCES

1. Brainard, F. S., Jr.: "Importance of Large Meters in Unaccounted for Water Analysis," Proceedings AWWA Distribution System Symposium, September 1984.
2. Geyer, J. C., and Lentz, J. J.: "Evaluation of Sanitary Sewer System Designs," The Johns Hopkins University School of Engineering, 1962.
3. Harnett, J. S.: "Effects of the California Drought on the East Bay Municipal District," *Journal AWWA*, vol. 70, p. 69, 1978.
4. McCuen, R. H.: *Statistical Methods for Engineers*, Prentice-Hall, Englewood Cliffs, NJ, 1985.
5. McJunkin, F. E.: "Population Forecasting by Sanitary Engineers," *J. Sanitary Engineering Division, ASCE*, vol. 90, no. SA4, 1964.
6. Metcalf & Eddy, Inc.: *Wastewater Engineering: Collection and Pumping of Wastewater*, McGraw-Hill, New York, 1981.
7. Salvato, J. A.: *Environmental Engineering and Sanitation*, 3rd ed., Wiley, New York, 1982.
8. Tchobanoglous, G., and Schroeder, E. D.: *Water Quality*, Addison-Wesley, Reading, MA, 1985.
9. U.S. Department of Housing and Urban Development: *Residential Water Conservation Projects, Summary Report*, June 1984.

10. U.S. Department of Housing and Urban Development: *Water Saved by Low-Flush Toilets and Low-Flow Shower Heads,* March 1984.
11. U.S. Department of Housing and Urban Development: *Survey of Water Fixture Use,* March 1984.
12. U.S. Environmental Protection Agency: *Construction Grants 1985* (CG-85), 430/9-84-004, July 1984.
13. Velz, C. J., *Graphical Approach to Statistics: Water & Sewage Works, Reference and Data Issue,* 1952.
14. Waugh, A. E.: *Elements of Statistical Analysis,* 2nd ed., McGraw-Hill, New York, 1943.

CHAPTER
3

WASTEWATER CHARACTERISTICS

An understanding of the nature of wastewaters is essential in the design and operation of collection, treatment, and disposal facilities and in the engineering management of environmental quality. To promote this understanding, the information in this chapter is presented in six sections dealing with (1) an introduction to the physical, chemical, and biological characteristics of wastewater; (2) the definition and application of physical characteristics; (3) the definition and application of chemical characteristics; (4) the definition and application of biological characteristics;. (5) data on wastewater composition; and (6) wastewater characterization studies.

3-1 PHYSICAL, CHEMICAL, AND BIOLOGICAL CHARACTERISTICS OF WASTEWATER

The following discussion will briefly introduce the physical, chemical, and biological constituents of wastewater; the contaminants of concern in wastewater treatment; the methods of analysis; and the units of expression used to characterize the contaminants in wastewater.

Constituents Found in Wastewater

Wastewater is characterized in terms of its physical, chemical, and biological composition. The principal physical properties and the chemical and biological constituents of wastewater and their sources are reported in Table 3-1. It should be noted that many of the parameters listed in Table 3-1 are interrelated. For example, temperature, a physical property, affects both the biological activity in the wastewater and the amounts of gases dissolved in the wastewater.

47

TABLE 3-1
Physical, chemical, and biological characteristics of wastewater and their sources

Characteristic	Sources
Physical properties:	
Color	Domestic and industrial wastes, natural decay of organic materials
Odor	Decomposing wastewater, industrial wastes
Solids	Domestic water supply, domestic and industrial wastes, soil erosion, inflow/infiltration
Temperature	Domestic and industrial wastes
Chemical constituents:	
Organic:	
Carbohydrates	Domestic, commercial, and industrial wastes
Fats, oils, and grease	Domestic, commercial, and industrial wastes
Pesticides	Agricultural wastes
Phenols	Industrial wastes
Proteins	Domestic, commercial, and industrial wastes
Priority pollutants	Domestic, commercial, and industrial wastes
Surfactants	Domestic, commercial, and industrial wastes
Volatile organic compounds	Domestic, commercial, and industrial wastes
Other	Natural decay of organic materials
Inorganic:	
Alkalinity	Domestic wastes, domestic water supply, groundwater infiltration
Chlorides	Domestic wastes, domestic water supply, groundwater infiltration
Heavy metals	Industrial wastes
Nitrogen	Domestic and agricultural wastes
pH	Domestic, commercial, and industrial wastes
Phosphorus	Domestic, commercial, and industrial wastes; natural runoff
Priority pollutants	Domestic, commercial, and industrial wastes
Sulfur	Domestic water supply; domestic, commercial, and industrial wastes
Gases:	
Hydrogen sulfide	Decomposition of domestic wastes
Methane	Decomposition of domestic wastes
Oxygen	Domestic water supply, surface-water infiltration
Biological constituents:	
Animals	Open watercourses and treatment plants
Plants	Open watercourses and treatment plants
Protists:	
Eubacteria	Domestic wastes, surface-water infiltration, treatment plants
Archaebacteria	Domestic wastes, surface-water infiltration, treatment plants
Viruses	Domestic wastes

Contaminants of Concern
in Wastewater Treatment

The important contaminants of concern in wastewater treatment are listed in Table 3-2. Secondary treatment standards for wastewater are concerned with the removal of biodegradable organics, suspended solids, and pathogens. Many of the more stringent standards that have been developed recently deal with the removal of nutrients and priority pollutants. When wastewater is to be reused, standards normally include requirements for the removal of refractory organics, heavy metals, and in some cases dissolved inorganic solids.

Analytical Methods

The analyses used to characterize wastewater vary from precise quantitative chemical determinations to the more qualitative biological and physical determinations.

TABLE 3-2
Important contaminants of concern in wastewater treatment

Contaminants	Reason for importance
Suspended solids	Suspended solids can lead to the development of sludge deposits and anaerobic conditions when untreated wastewater is discharged in the aquatic environment.
Biodegradable organics	Composed principally of proteins, carbohydrates, and fats, biodegradable organics are measured most commonly in terms of BOD (biochemical oxygen demand) and COD (chemical oxygen demand.) If discharged untreated to the environment, their biological stabilization can lead to the depletion of natural oxygen resources and to the development of septic conditions.
Pathogens	Communicable diseases can be transmitted by the pathogenic organisms in wastewater.
Nutrients	Both nitrogen and phosphorus, along with carbon, are essential nutrients for growth. When discharged to the aquatic environment, these nutrients can lead to the growth of undesirable aquatic life. When discharged in excessive amounts on land, they can also lead to the pollution of groundwater.
Priority pollutants	Organic and inorganic compounds selected on the basis of their known or suspected carcinogenicity, mutagenicity, teratogenicity, or high acute toxicity. Many of these compounds are found in wastewater.
Refractory organics	These organics tend to resist conventional methods of wastewater treatment. Typical examples include surfactants, phenols, and agricultural pesticides.
Heavy metals	Heavy metals are usually added to wastewater from commercial and industrial activities and may have to be removed if the wastewater is to be reused.
Dissolved inorganics	Inorganic constituents such as calcium, sodium, and sulfate are added to the original domestic water supply as a result of water use and may have to be removed if the wastewater is to be reused.

The quantitative methods of analysis are either gravimetric, volumetric, or physicochemical. In the physicochemical methods, properties other than mass or volume are measured. Instrumental methods of analysis such as turbidity, colorimetry, potentiometry, polarography, adsorption spectrometry, fluorometry, spectroscopy, and nuclear radiation are representative of the physicochemical analyses. Details concerning the various analyses may be found in *Standard Methods* [18], the accepted reference that details the conduct of water and wastewater analyses.

Units of Measurement for Physical and Chemical Parameters

The results of the analysis of wastewater samples are expressed in terms of physical and chemical units of measurement. The most common units are reported in Table 3-3. Measurements of chemical parameters are usually expressed in the physical unit of milligrams per liter (mg/L) or grams per cubic meter (g/m^3). The concentration of trace constituents is usually expressed as micrograms per liter (μg/L). For the dilute systems in which one liter weighs approximately one kilogram, such as those encountered in natural waters and wastewater, the units of mg/L or g/m^3 are interchangeable with parts per million (ppm), which is a mass-to-mass ratio.

Dissolved gases, considered to be chemical constituents, are measured in units of mg/L or g/m^3. Gases evolved as a by-product of wastewater treatment, such as carbon dioxide and methane (anaerobic decomposition), are measured in terms of ft^3 (m^3 or L). Results of tests and parameters such as temperature, odor, hydrogen ion, and biological organisms are expressed in other units as explained in Secs. 3-2, 3-3, and 3-4.

3-2 PHYSICAL CHARACTERISTICS: DEFINITION AND APPLICATION

The most important physical characteristic of wastewater is its total solids content, which is composed of floating matter, settleable matter, colloidal matter, and matter in solution. Other important physical characteristics include odor, temperature, density, color, and turbidity.

Total Solids

Analytically, the total solids content of a wastewater is defined as all the matter that remains as residue upon evaporation at 103 to 105°C (see Fig. 3-1). Matter that has a significant vapor pressure at this temperature is lost during evaporation and is not defined as a solid. Settleable solids are those solids that will settle to the bottom of a cone-shaped container (called an Imhoff cone) in a 60-minute period (see Fig. 3-2). Settleable solids, expressed as mL/L, are an approximate measure of the quantity of sludge that will be removed by primary sedimentation. Total solids, or residue upon evaporation, can be further classified as nonfilterable (suspended) or filterable by passing a known volume of liquid through a filter (see Fig. 3-3). A glass-fiber

TABLE 3-3
Units commonly used to express analytical results

Basis	Application	Unit
Physical analyses:		
Density	$\dfrac{\text{Mass of solution}}{\text{Unit volume}}$	kg/m^3
Percent by volume	$\dfrac{\text{Volume of solute} \times 100}{\text{Total volume of solution}}$	% (by vol)
Percent by mass	$\dfrac{\text{Mass of solute} \times 100}{\text{Combined mass of solute} + \text{solvent}}$	% (by mass)
Volume ratio	$\dfrac{\text{Milliliters}}{\text{Liter}}$	mL/L
Mass per unit volume	$\dfrac{\text{Micrograms}}{\text{Liter of solution}}$	$\mu g/L$
	$\dfrac{\text{Milligrams}}{\text{Liter of solution}}$	mg/L
	$\dfrac{\text{Grams}}{\text{Cubic meter of solution}}$	g/m^3
Mass ratio	$\dfrac{\text{Milligrams}}{10^6 \text{ milligrams}}$	ppm
Chemical analyses:		
Molality	$\dfrac{\text{Moles of solute}}{1000 \text{ grams of solvent}}$	mol/kg
Molarity	$\dfrac{\text{Moles of solute}}{\text{Liter of solution}}$	mol/L
Normality	$\dfrac{\text{Equivalents of solute}}{\text{Liter of solution}}$	equiv/L
	$\dfrac{\text{Milliequivalents of solute}}{\text{Liter of solution}}$	meq/L

Note: mg/L = g/m^3.

filter (Whatman GF/C) with a nominal pore size of about 1.2 micrometers (μm) is used most commonly for this separation step. Polycarbonate membrane filters are also used. It should be noted that the results obtained with glass fiber and polycarbonate filters with the same nominal pore size will be somewhat different because of the structure of the filter (see Fig. 3-4).

The filterable-solids fraction consists of colloidal and dissolved solids. The colloidal fraction consists of the particulate matter with an approximate size range of from 0.001 to 1 μm. The dissolved solids consist of both organic and inorganic molecules and ions that are present in true solution in water. The colloidal fraction cannot be

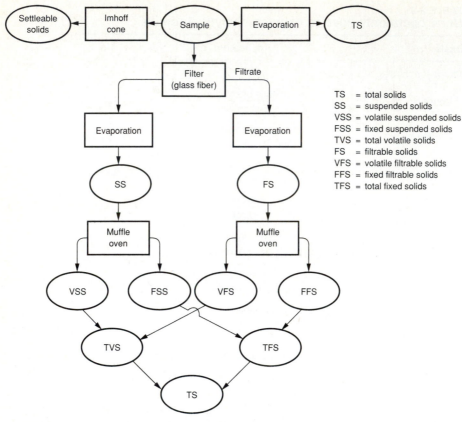

TS = total solids
SS = suspended solids
VSS = volatile suspended solids
FSS = fixed suspended solids
TVS = total volatile solids
FS = filtrable solids
VFS = volatile filtrable solids
FFS = fixed filtrable solids
TFS = total fixed solids

FIGURE 3-1
Interrelationships of solids found in water and wastewater. In much of the water-quality literature, the solids passing through the filter are called dissolved solids [23].

removed by settling. Generally, biological oxidation or coagulation, followed by sedimentation, is required to remove these particles from suspension. The principal types of materials that comprise the nonfilterable and filterable solids in wastewater and their approximate size range are reported in Fig. 3-5.

Each of the categories of solids may be further classified on the basis of their volatility at $550 \pm 50°C$. The organic fraction will oxidize and will be driven off as gas at this temperature, and the inorganic fraction remains behind as ash. Thus the terms "volatile suspended solids" and "fixed suspended solids" refer, respectively, to the organic and inorganic (or mineral) content of the suspended solids. At $550 \pm 50°C$, the decomposition of inorganic salts is restricted to magnesium carbonate, which decomposes into magnesium oxide and carbon dioxide at $350°C$. Calcium carbonate, the major component of the inorganic salts, is stable up to a temperature of $825°C$. The volatile-solids analysis is applied most commonly to wastewater sludges to measure their biological stability. The solids content of a medium-strength wastewater may be classified approximately as shown in Fig. 3-6.

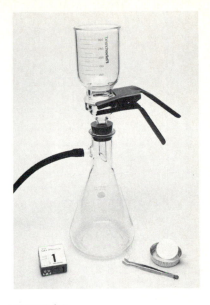

FIGURE 3-2
Imhoff cone used to determine settleable solids in wastewater. Solids that accumulate in the bottom of the cone are reported as mL/L.

FIGURE 3-3
Apparatus used for the determination of suspended solids. After the waste-water sample has been filtered, the previously tared glass-fiber filter is placed in an aluminum dish, to be dried before weighing.

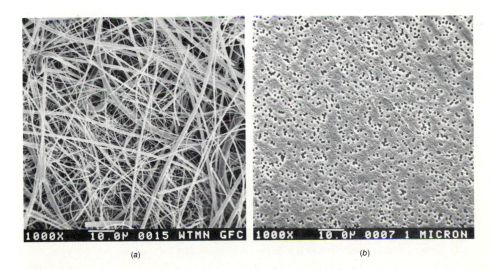

(a)   (b)

FIGURE 3-4
Micrographs of two laboratory filters used for the measurement of suspended solids in wastewater: (a) glass fiber filter with a nominal pore size of 1.2 μm and (b) polycarbonate membrane filter with a nominal pore size of 1.0 μm [6].

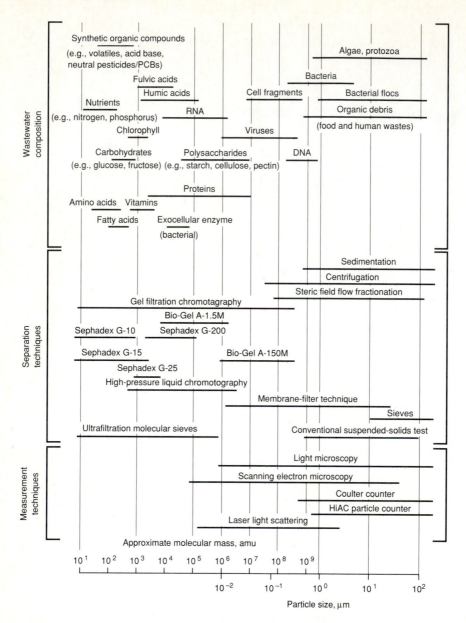

FIGURE 3-5
Size ranges of organic contaminants in wastewater and size separation and measurement techniques used for their quantification (adapted from Ref. 6).

Example 3-1   Analysis of solids data. The following test results were obtained for a wastewater sample taken at the headworks to a wastewater treatment plant. All of the tests were performed using a sample size of 50 mL. Determine the concentration of total solids, total volatile solids, suspended solids, and volatile suspended solids. The samples used in the solids analyses were all either evaporated, dried, or ignited to constant weight.

Tare mass of evaporating dish = 53.5433 g

Mass of evaporating dish plus residue after evaporation at 105°C = 53.5793 g

Mass of evaporating dish plus residue after ignition at 550°C = 53.5772 g

Tare mass of Whatman GF/C filter = 1.5433 g

Residue on Whatman GF/C filter after drying at 105°C = 1.5553 g

Residue on Whatman GF/C filter after ignition at 550°C = 1.5531 g

Solution

1. Determine total solids

$$TS = \frac{\left(\begin{array}{c}\text{mass of evaporating} \\ \text{dish plus residue, g}\end{array} - \begin{array}{c}\text{mass of evaporating} \\ \text{dish, g}\end{array}\right) \times 1000 \text{ mg/g}}{\text{sample size, L}}$$

$$TS = \frac{(53.5793 - 53.5433) \times 1000 \text{ mg/g}}{0.050 \text{ L}} = 720 \text{ mg/L}$$

2. Determine volatile solids

$$VS = \frac{(53.5793 - 53.5772) \times 1000 \text{ mg/g}}{0.050 \text{ L}} = 42 \text{ mg/L}$$

3. Determine the suspended solids

$$SS = \frac{(1.5553 - 1.5433) \times 1000 \text{ mg/g}}{0.050 \text{ L}} = 240 \text{ mg/L}$$

4. Determine the volatile suspended solids

$$VSS = \frac{(1.5553 - 1.5531) \times 1000 \text{ mg/g}}{0.050 \text{ L}} = 44 \text{ mg/L}$$

Typical data on the distribution of the filterable solids in wastewater are reported in Table 3-4. As noted in Table 3-4, the total filterable solids were determined using polycarbonate membrane filters. The smallest pore size used was 0.1 μm. It is interesting to note that there is a significant amount of material in the size range between 0.1 and 1.0 μm that is now not measured. Based on the results of a recent study, it was suggested that a filter with a pore size of 0.1 μm would be a better delimiter of the filterable solids in wastewater [6]. In the future, it is anticipated that information on the size distribution of the solids in wastewater will play a greater role in the design of both collection systems and treatment facilities.

Odors

Odors in domestic wastewater usually are caused by gases produced by the decomposition of organic matter or by substances added to the wastewater. Fresh wastewater has a distinctive, somewhat disagreeable odor, which is less objectionable than the odor of wastewater that has undergone anaerobic (devoid of oxygen) decomposition. The most characteristic odor of stale or septic wastewater is that of hydrogen sul-

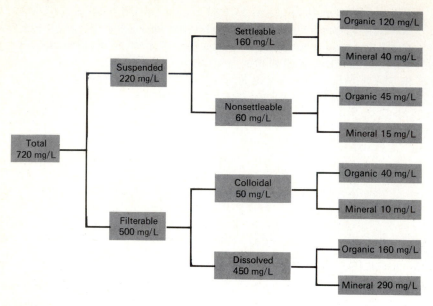

FIGURE 3-6
Classification of solids found in medium-strength wastewater.

fide, which is produced by anaerobic microorganisms that reduce sulfate to sulfide. Industrial wastewater may contain either odorous compounds or compounds that produce odors during the process of wastewater treatment.

Odors have been rated as the first concern of the public relative to the implementation of wastewater treatment facilities [13]. Within the past few years, the control of odors has become a major consideration in the design and operation of wastewater collection, treatment, and disposal facilities, especially with respect to

TABLE 3-4
Typical data on the distribution of filterable solids in various untreated wastewater samples

Sample[a] (date, time)	Conc.,[b] mg/L	Percent of mass retained in indicated μm size range					
		>0.1 <1.0	>1.0 <3.0	>3.0 <5.0	>5.0 <8.0	>8.0 <12.0	>12.0
UCD (7/14/82, 11 A.M.)	62.2	12.5	12.9	5.8	3.8	6.1	58.8
UCD (7/14/82, 11 A.M.)	129.9	16.1	25.1	0.0	0.0	0.0	58.8
LV (8/3/83, 2 P.M.)	284.0	1.8	32.6	11.5	11.1	1.8	41.2
LV (8/3/83, 8:30 P.M.)	146.1	14.2	32.4	6.9	0.0	6.5	40.0
LB (8/8/83, 2 P.M.)	268.0	20.5	18.7	6.7	3.0	10.1	41.0

[a] Samples collected at University of California, Davis, CA, Las Vegas, NV, and Los Banos, CA.
[b] Mass of solids retained on a polycarbonate membrane filter with a pore size of 0.1 μm.

the public acceptance of these facilities. In many areas, projects have been rejected because of the fear of potential odors. In view of the importance of odors in the field of wastewater management, it is appropriate to consider the effects they produce, how they are detected, and their characterization and measurement.

Effects of Odors. The importance of odors at low concentrations in human terms is related primarily to the psychological stress they produce rather than to the harm they do to the body. Offensive odors can cause poor appetite for food, lowered water consumption, impaired respiration, nausea and vomiting, and mental perturbation. In extreme situations, offensive odors can lead to the deterioration of personal and community pride, interfere with human relations, discourage capital investment, lower socio-economic status, and deter growth. These problems can result in a decline in market and rental property values, tax revenues, payrolls, and sales.

Detection of Odors. The malodorous compounds responsible for producing psychological stress in humans are detected by the olfactory system, but the precise mechanism involved is at present not well understood. Since 1870, more than 30 theories have been proposed to explain olfaction. One of the difficulties in developing a universal theory has been the inadequate explanation of why compounds with similar structures may have different odors and why compounds with very different structures may have similar odors. At present, there appears to be some general agreement that the odor of a molecule must be related to the molecule as a whole.

Over the years, a number of attempts have been made to classify odors in a systematic fashion. The major categories of offensive odors and the compounds involved are listed in Table 3-5. All these compounds may be found or may develop in domestic wastewater, depending on local conditions. The odor detection and recognition thresholds for specific malodorous compounds associated with untreated wastewater are listed in Table 3-6.

TABLE 3-5
Odorous compounds associated with untreated wastewater

Odorous compound	Chemical formula	Odor, quality
Amines	CH_3NH_2, $(CH_3)_3H$	Fishy
Ammonia	NH_3	Ammoniacal
Diamines	$NH_2(CH_2)_4NH_2$, $NH_2(CH_2)_5NH_2$	Decayed flesh
Hydrogen sulfide	H_2S	Rotten eggs
Mercaptans (e.g., methyl and ethyl)	CH_3SH, $CH_3(CH_2)SH$	Decayed cabbage
Mercaptans (e.g., T=butyl and crotyl)	$(CH_3)_3CSH$, $CH_3(CH_2)_3SH$	Skunk
Organic sulfides	$(CH_3)_2S$, $(C_6H_5)_2S$	Rotten cabbage
Skatole	C_9H_9N	Fecal matter

TABLE 3-6
Odor thresholds of odorous compounds associated with untreated wastewater[a]

Odorous compound	Chemical formula	Odor threshold, ppmV[b]	
		Detection	Recognition
Ammonia	NH_3	17	37
Chlorine	Cl_2	0.080	0.314
Dimethyl sulfide	$(CH_3)_2S$	0.001	0.001
Diphenyl sulfide	$(C_6H_5)_2S$	0.0001	0.0021
Ethyl mercaptan	CH_3CH_2SH	0.0003	0.001
Hydrogen sulfide	H_2S	<0.00021	0.00047
Indole	C_8H_7N	0.0001	—
Methyl amine	CH_3NH_2	4.7	—
Methyl mercaptan	CH_3SH	0.0005	0.001
Skatole	C_9H_9N	0.001	0.019

[a] Adapted in part from Refs. 13, 33.
[b] Parts per million by volume.

Odor Characterization and Measurement. It has been suggested that four independent factors are required for the complete characterization of an odor: intensity, character, hedonics, and detectability (see Table 3-7). To date, detectability is the only factor that has been used in the development of statutory regulations for nuisance odors.

TABLE 3-7
Factors that must be considered for the complete characterization of an odor

Factor	Description
Character	Relates to the mental associations made by the subject in sensing the odor; determination can be quite subjective
Detectability	The number of dilutions required to reduce an odor to its minimum detectable threshold odor concentration (MDTOC)
Hedonics	The relative pleasantness or unpleasantness of the odor sensed by the subject
Intensity	The perceived strength of the odor; usually measured by the butanol olfactometer or calculated from the dilutions to threshold (D/T) when the relationship is established

Odor can be measured by sensory methods, and specific odorant concentrations can be measured by instrumental methods. It has been shown that, under carefully controlled conditions, the sensory (organoleptic) measurement of odors by the human olfactory system can provide meaningful and reliable information. Therefore, the sensory method is often used to measure the odors emanating from wastewater treatment facilities. The availability of a direct-reading meter for hydrogen sulfide (see Fig. 3-7) that can be used to detect concentrations as low as 1 ppb is a significant development.

In the sensory method, human subjects (often a panel of subjects) are exposed to odors that have been diluted with odor-free air, and the number of dilutions required to reduce an odor to its minimum detectable threshold odor concentration (MDTOC) are noted. The detectable odor concentration is reported as the dilutions to the MDTOC, commonly called D/T (dilutions to threshold). Thus, if four volumes of diluted air must be added to 1 unit volume of sampled air to reduce the odorant to its MDTOC, the odor concentration would be reported as four dilutions to MDTOC. Other terminology commonly used to measure odor strength is ED_{50}. The ED_{50} value represents the number of times an odorous air sample must be diluted before the average person (50 percentile) can barely detect an odor in the diluted sample. Details of the test procedure are provided in Ref. 2. However, the sensory determination of this minimum threshold concentration can be subject to a number of errors. Adaptation and cross adaptation, synergism, subjectivity, and sample modification (see Table 3-8) are the principal errors. To avoid errors in sample modification during storage in sample collection containers, direct-reading olfactometers have been developed to measure odors at their source without using sampling containers.

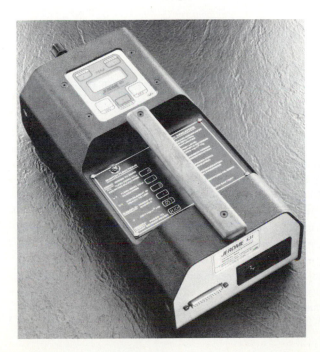

FIGURE 3-7
Portable H_2S meter used for field odor studies (*from Arizona Instrument Corporation, Jerome Instrument Division*).

TABLE 3-8
Types of errors in the sensory detection of odors[a]

Type of error	Description
Adaptation and cross adaptation	When exposed continually to a background concentration of an odor, the subject is unable to detect the presence of that odor at low concentrations. When removed from the background odor concentration, the subject's olfactory system will recover quickly. Ultimately, a subject with an adapted olfactory system will be unable to detect the presence of an odor to which his system has adapted.
Sample modification	Both the concentration and composition of odorous gases and vapors can be modified in sample-collection containers and in odor-detection devices. To minimize problems associated with sample modification, the period of odor containment should be minimized or eliminated, and minimum contact should be allowed with any reactive surfaces.
Subjectivity	When the subject has knowledge of the presence of an odor, random error can be introduced in sensory measurements. Often, knowledge of the odor may be inferred from other sensory signals such as sound, sight, or touch.
Synergism	When more than one odorant is present in a sample, it has been observed that it is possible for a subject to exhibit increased sensitivity to a given odor because of the presence of another odor.

[a] Adapted from Ref. 10.

The threshold odor of a water or wastewater sample is determined by diluting the sample with odor-free water. The "threshold odor number" (TON) corresponds to the greatest dilution of the sample with odor-free water at which an odor is just perceptible. The recommended sample size is 200 mL. The numerical value of the TON is determined as follows:

$$\text{TON} = \frac{A + B}{A} \qquad (3\text{-}1)$$

where A = mL of sample and B = mL of odor-free water. The odor emanating from the liquid sample is determined with human subjects (often a panel of subjects) as discussed above. Details for this procedure, which was approved by the Standard Methods Committee in 1985, may be found in Ref. 18.

With regard to the instrumental measurement of odors, air-dilution olfactometry provides a reproducible method for measuring threshold odor concentrations. Equipment used to analyze odors include (1) the dynamic forced-choice triangle olfactometer, (2) the butanol wheel, and (3) the scentometer. The triangle olfactometer enables the operator to introduce the sample at different concentrations at six different cups (see Fig. 3-8). At each cup, two ports contain purified air and one port contains a diluted sample. Six dilution ratios are commonly used, varying from 4500 to 15x.

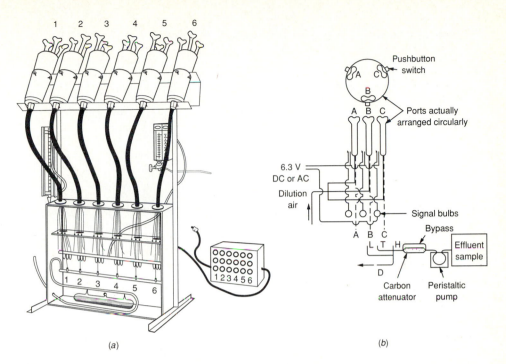

FIGURE 3-8
Dynamic forced-choice triangle olfactometer: (a) schematic and (b) flow diagram.

Higher dilution ratios can be achieved using the carbon attenuator. All sample dilutions and blanks are delivered continuously to sniffing cups at a rate of about 500 mL/min. Each odor panel member (usually six) then sniffs each of the three ports and selects a port that he or she believes to contain the sample. The butanol wheel is a device used to measure the intensity of the odor against a scale containing various concentrations of butanol. A scentometer (see Fig. 3-9) is a hand-held device in which malodorous air passes through graduated orifices and is mixed with air that has been purified by passing through activated carbon beds. The dilution ratios are determined by the ratio of the size of the malodorous to purified inlets. The scentometer is very useful in the field for making odor determinations over a large area surrounding a treatment plant. Often a mobile odor laboratory, which contains several types of olfactory and analytical equipment in a single van type vehicle, is used for field sites.

It is often desirable to know the specific compounds responsible for odor. Although gas chromatography has been used successfully for this purpose, it has not been used as successfully in the detection and quantification of odors derived from wastewater collection, treatment, and disposal facilities. Equipment developed and found useful in the chemical analysis of odors is the triple-stage quadrupole mass spectrometer. The spectrometer can be used as a conventional mass spectrometer to produce simple mass spectra or as a triple-stage quadrupole to produce collesionally

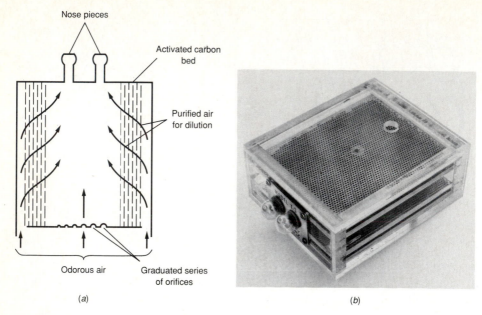

Nose pieces

Activated carbon
bed

Purified air
for dilution

Odorous air   Graduated series
of orifices

(a)   (b)

FIGURE 3-9
Scentometer used for field studies of odors: (a) schematic and (b) front view looking at nose pieces
(5 in × 6 in × 2.5 in, from Barnebey & Sutcliffe Corp.). *Note:* Odorous air that passes through graduated orifices is mixed with air from the same source, which is purified by passing through activated
carbon beds.

activated disassociation spectra. The former operating mode provides the masses
of molecular or parent ions present in samples, while the latter provides positive
identification of compounds. Types of compounds that can be identified include
ammonia, amino acids, and volatile organic compounds.

Temperature

The temperature of wastewater is commonly higher than that of the water supply,
because of the addition of warm water from households and industrial activities. As
the specific heat of water is much greater than that of air, the observed wastewater
temperatures are higher than the local air temperatures during most of the year and are
lower only during the hottest summer months. Depending on the geographic location,
the mean annual temperature of wastewater varies from about 10 to 21.1°C (50 to
70°F); 15.6°C (60°F) is a representative value. The variation that can be expected in
influent wastewater temperatures is illustrated in Fig. 3-10. Depending on the location
and time of year, the effluent temperatures can either be higher or lower than the
corresponding influent values.

    The temperature of water is a very important parameter because of its effect
on chemical reactions and reaction rates, aquatic life, and the suitability of the water

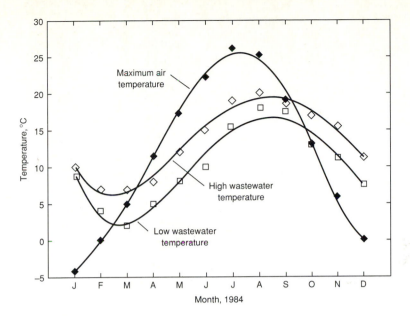

FIGURE 3-10
Typical variations in monthly wastewater temperatures.

for beneficial uses. Increased temperature, for example, can cause a change in the species of fish that can exist in the receiving water body. Industrial establishments that use surface water for cooling-water purposes are particularly concerned with the temperature of the intake water.

In addition, oxygen is less soluble in warm water than in cold water. The increase in the rate of biochemical reactions that accompanies an increase in temperature, combined with the decrease in the quantity of oxygen present in surface waters, can often cause serious depletions in dissolved oxygen concentrations in the summer months. When significantly large quantities of heated water are discharged to natural receiving waters, these effects are magnified. It should also be realized that a sudden change in temperature can result in a high rate of mortality of aquatic life. Moreover, abnormally high temperatures can foster the growth of undesirable water plants and wastewater fungus.

Optimum temperatures for bacterial activity are in the range from about 25 to 35°C. Aerobic digestion and nitrification stop when the temperature rises to 50°C. When the temperature drops to about 15°C, methane-producing bacteria become quite inactive, and at about 5°C, the autotrophic-nitrifying bacteria practically cease functioning. At 2°C, even the chemoheterotrophic bacteria acting on carbonaceous material become essentially dormant. The effects of temperature on the performance of biological treatment processes are considered in greater detail in Chaps. 8 and 10.

Density

The density of wastewater ρ_w is defined as its mass per unit volume expressed as slug/ft^3 (kg/m^3). Density is an important physical characteristic of wastewater because of the potential for the formation of density currents in sedimentation tanks and in other treatment units. The density of domestic wastewater that does not contain significant amounts of industrial waste is essentially the same as that of water at the same temperature. In some cases, the specific gravity of the wastewater s_w, defined as $s_w = \rho_w/\rho_o$ where ρ_o is the density of water, is used in place of the density. Both the density and specific gravity of wastewater are temperature dependent and will vary with the concentration of total solids in the wastewater. Some typical values of specific gravity for various types of wastewater sludge are reported in Chap. 9 in Table 9-9.

Color

Historically, the term "condition" was used along with composition and concentration to describe wastewater. Condition refers to the age of the wastewater, which is determined qualitatively by its color and odor. Fresh wastewater is usually a light brownish-gray color. However, as the travel time in the collection system increases, and more anaerobic conditions develop, the color of the wastewater changes sequentially from gray to dark gray and ultimately to black. When the color of the wastewater is black the wastewater is often described as septic. Some industrial wastewaters may also add color to domestic wastewater. In most cases, the gray, dark gray, and black color of the wastewater is due to the formation of metallic sulfides, which form as the sulfide produced under anaerobic conditions reacts with the metals in the wastewater.

Turbidity

Turbidity, a measure of the light-transmitting properties of water, is another test used to indicate the quality of waste discharges and natural waters with respect to colloidal and residual suspended matter. The measurement of turbidity is based on comparison of the intensity of light scattered by a sample as compared to the light scattered by a reference suspension under the same conditions [18]. Colloidal matter will scatter or absorb light and thus prevent its transmission. In general, there is no relationship between turbidity and the concentration of suspended solids in untreated wastewater. There is, however, a reasonable relationship between turbidity and suspended solids for the settled secondary effluent from the activated sludge process (see Eq. 6-39 in Chap. 6).

3-3   CHEMICAL CHARACTERISTICS: DEFINITION AND APPLICATION

This discussion of chemical characteristics of wastewater is presented in four parts: (1) organic matter; (2) the measurement of organic content; (3) inorganic matter; and

(4) gases. The measurement of organic content is discussed separately because of its importance in both the design and operation of wastewater treatment plants and the management of water quality.

Organic Matter

In a wastewater of medium strength, about 75 percent of the suspended solids and 40 percent of the filterable solids are organic in nature, as shown in Fig. 3-6. These solids are derived from both the animal and plant kingdoms and the activities of man as related to the synthesis of organic compounds. Organic compounds are normally composed of a combination of carbon, hydrogen, and oxygen, together with nitrogen in some cases. Other important elements, such as sulfur, phosphorus, and iron, may also be present. The principal groups of organic substances found in wastewater are proteins (40 to 60 percent), carbohydrates (25 to 50 percent), and fats and oils (10 percent). Urea, the chief constituent of urine, is another important organic compound contributing to wastewater. Because it decomposes so rapidly, undecomposed urea is seldom found in other than very fresh wastewater.

Along with the proteins, carbohydrates, fats and oils, and urea, wastewater contains small quantities of a large number of different synthetic organic molecules ranging from simple to extremely complex in structure. Typical examples, discussed in this section, include surfactants, organic priority pollutants, volatile organic compounds, and agricultural pesticides. Further, the number of such compounds is growing yearly as more and more organic molecules are being synthesized. The presence of these substances has, in recent years, complicated wastewater treatment because many of them either cannot be or are very slowly decomposed biologically.

Proteins. Proteins are the principal constituents of the animal organism. They occur to a lesser extent in plants. All raw animal and plant foodstuffs contain proteins. The amount present varies from small percentages in watery fruits such as tomatoes and in the fatty tissues of meat to quite high percentages in beans or lean meats. Proteins are complex in chemical structure and unstable, being subject to many forms of decomposition. Some are soluble in water; others are insoluble. The chemistry of the formation of proteins involves the combination or linking together of a large number of amino acids. The molecular weights of proteins are very high, ranging from about 20,000 to 20 million.

All proteins contain carbon, which is common to all organic substances, as well as hydrogen and oxygen. In addition they contain, as their distinguishing characteristic, a fairly high and constant proportion of nitrogen, about 16 percent. In many cases sulfur, phosphorus, and iron are also constituents. Urea and proteins are the chief sources of nitrogen in wastewater. When proteins are present in large quantities, extremely foul odors are apt to be produced by their decomposition.

Carbohydrates. Widely distributed in nature, carbohydrates include sugars, starches, cellulose, and wood fiber. All are found in wastewater. Carbohydrates contain carbon, hydrogen, and oxygen. The common carbohydrates contain six or a multiple of six carbon atoms in a molecule, and hydrogen and oxygen in the proportions in which these elements are found in water. Some carbohydrates, notably the sugars, are soluble in water; others, such as the starches, are insoluble. The sugars tend to decompose; the enzymes of certain bacteria and yeasts set up fermentation with the production of alcohol and carbon dioxide. The starches, on the other hand, are more stable but are converted into sugars by microbial activity as well as by dilute mineral acids. From the standpoint of bulk and resistance to decomposition, cellulose is the most important carbohydrate found in wastewater. The destruction of cellulose in the soil goes on readily, largely as a result of the activity of various fungi, particularly when acid conditions prevail.

Fats, Oils, and Grease. Fats and oils are the third major component of foodstuffs. The term "grease," as commonly used, includes the fats, oils, waxes, and other related constituents found in wastewater. Grease content is determined by extraction of the waste sample with trichlorotrifluoroethane (grease is soluble in trichlorotrifluoroethane). Other extractable substances include mineral oils, such as kerosene and lubricating and road oils.

Fats and oils are compounds (esters) of alcohol or glycerol (glycerin) with fatty acids. The glycerides of fatty acids that are liquid at ordinary temperatures are called oils, and those that are solids are called fats. They are quite similar, chemically, being composed of carbon, hydrogen, and oxygen in varying proportions.

Fats and oils are contributed to domestic wastewater in butter, lard, margarine, and vegetable fats and oils. Fats are also commonly found in meats, in the germinal area of cereals, in seeds, in nuts, and in certain fruits.

Fats are among the more stable of organic compounds and are not easily decomposed by bacteria. Mineral acids attack them, however, resulting in the formation of glycerin and fatty acid. In the presence of alkalies, such as sodium hydroxide, glycerin is liberated, and alkali salts of the fatty acids are formed. These alkali salts are known as soaps, and like the fats, they are stable. Common soaps are made by saponification of fats with sodium hydroxide. They are soluble in water, but in the presence of hardness constituents, the sodium salts are changed to calcium and magnesium salts of the fatty acids, or so-called mineral soaps. These are insoluble and are precipitated.

Kerosene, lubricating and road oils are derived from petroleum and coal tar and contain essentially carbon and hydrogen. These oils sometimes reach the sewers in considerable volume from shops, garages, and streets. For the most part, they float on the wastewater, although a portion is carried into the sludge on settling solids. To an even greater extent than fats, oils, and soaps, the mineral oils tend to coat surfaces. The particles interfere with biological action and cause maintenance problems.

As indicated in the foregoing discussion, the grease content of wastewater can cause many problems in both sewers and waste treatment plants. If grease is not removed before discharge of the waste, it can interfere with the biological life in the surface waters and create unsightly floating matter and films.

Surfactants. Surfactants, or surface-active agents, are large organic molecules that are slightly soluble in water and cause foaming in wastewater treatment plants and in the surface waters into which the waste effluent is discharged. Surfactants tend to collect at the air-water interface. During aeration of wastewater, these compounds collect on the surface of the air bubbles and thus create a very stable foam. The determination of surfactants is accomplished by measuring the color change in a standard solution of methylene blue dye. Another name for surfactant is methylene blue active substance (MBAS).

Before 1965, the type of surfactant present in synthetic detergents, called alkyl-benzene-sulfonate (ABS), was especially troublesome because it resisted breakdown by biological means. As a result of legislation in 1965, ABS has been replaced in detergents by linear-alkyl-sulfonate (LAS), which is biodegradable. Because surfactants come primarily from synthetic detergents, the foaming problem has been greatly reduced.

Priority Pollutants. The Environmental Protection Agency has identified approximately 129 priority pollutants in 65 classes to be regulated by categorical discharge standards [4]. Priority pollutants (both inorganic and organic) were selected on the basis of their known or suspected carcinogenicity, mutagenicity, teratogenicity, or high acute toxicity. Many of the organic priority pollutants are also classified as volatile organic compounds (VOCs). Representative examples of the priority pollutants are shown in Table 3-9.

Within a wastewater collection and treatment system, organic priority pollutants may be removed, transformed, generated, or simply transported through the system unchanged. Five primary mechanisms are involved: (1) volatilization (also gas stripping); (2) degradation; (3) sorption to particles and sludge; (4) pass-through (i.e., passage through the entire system); and (5) generation as result of chlorination or as byproducts of the degradation of precursor compounds. It is also important to note that these mechanisms are not mutually exclusive, as competition and simultaneous action can be significant [1].

Two types of standards are used to control pollutant discharges to publicly owned treatment works (POTWs). The first, "prohibited discharge standards," applies to all commercial and industrial establishments that discharge to POTWs. Prohibited standards restrict the discharge of pollutants that may create a fire or explosion hazard in sewers or treatment works, are corrosive (pH < 5.0), obstruct flow, upset treatment processes, or increase the temperature of the wastewater entering the plant to above 40°C. "Categorical Standards" apply to industrial and commercial discharges in 25 industrial categories ("categorical industries") and are intended to restrict the discharge of the 129 priority pollutants. It is anticipated that this list will continue to expand in the future.

Volatile Organic Compounds (VOCs). Organic compounds that have a boiling point ≤ 100°C and/or a vapor pressure >1mm Hg at 25°C are generally considered to be volatile organic compounds (VOCs). For example, vinyl chloride, which has a boiling point of −13.9°C and a vapor pressure of 2548 mm Hg at 20°C, is an example of an extremely volatile organic compound. Volatile organic compounds are of great concern because (1) once such compounds are in the vapor state they are much more mobile and, therefore, more likely to be released to the environment; (2) the presence

TABLE 3-9
Typical waste compounds produced by commercial industrial and agricultural activities that have been classified as priority pollutants

Name (Formula)	Use	Concern
Nonmetals		
Arsenic (As)	Alloying additive for metals, especially lead and copper as shot, battery grids, cable sheaths, boiler tubes. High purity (semiconductor) grade.	Carcinogen and mutagen. *Long term*—sometimes can cause fatigue and loss of energy; dermatitis.
Selenium (Se)	Electronics, xerographic plates, TV cameras, photocells, magnetic computer cores, solar batteries, rectifiers, relays, ceramics (colorant for glass) steel and copper, rubber accelerator, catalyst, trace element in animal feeds.	*Long term*—red staining of fingers, teeth, and hair; general weakness; depression; irritation of nose and mouth.
Metals		
Barium (Ba)	Getter alloys in vacuum tubes, deoxidizer for copper, Frary's metal, lubricant for anode rotors in X-ray tubes, spark-plug alloys.	Flammable at room temperature in powder form. *Long term*—increased blood pressure and nerve block.
Cadmium (Cd)	Electrodeposited and dipped coatings on metals, bearing and low-melting alloys, brazing alloys, fire protection system, nickel-cadmium storage batteries, power transmission wire, TV phosphors, basis of pigments used in ceramic glazes, machinery enamels, fungicide photography and lithography, selenium rectifiers, electrodes for cadmium-vapor lamps, and photoelectric cells.	Flammable in powder form. Toxic by inhalation of dust or fume. A carcinogen. Soluble compounds of cadmium are highly toxic. *Long term*—concentrates in the liver, kidneys, pancreas, and thyroid; hypertension suspected effect.
Chromium (Cr)	Alloying and plating element on metal and plastic substrates for corrosion resistance, chromium-containing and stainless steels, protective coating for automotive and equipment accessories, nuclear and high-temperature research, constituent of inorganic pigments.	Hexavalent chromium compounds are carcinogenic and corrosive on tissue. *Long term*—skin sensitization and kidney damage.
Lead (Pb)	Storage batteries, gasoline additive, cable covering, ammunition, piping, tank linings, solder and fusible alloys, vibration damping in heavy construction, foil, babbit and other bearing alloys.	Toxic by ingestion or inhalation of dust or fumes. *Long term*—brain and kidney damage; birth defects.
Mercury (Hg)	Amalgams, catalyst electrical apparatus, cathodes for production of chlorine and caustic soda, instruments, mercury vapor lamps, mirror coating, arc lamps, boilers.	Highly toxic by skin absorption and inhalation of fume or vapor. *Long term*—toxic to central nervous system; may cause birth defects.
Silver (Ag)	Manufacture of silver nitrate, silver bromide, photo chemicals; lining vats and other equipment for chemical reaction vessels, water distillation, etc.; mirrors, electric conductors, silver plating electronic equipment; sterilant; water purification; surgical cements; hydration and oxidation catalyst special batteries, solar cells, reflectors for solar towers; low-temperature brazing alloys; table cutlery; jewelry; dental, medical, and scientific equipment; electrical contacts; bearing metal; magnet windings; dental amalgams. Colloidal silver is used as a nucleating agent in photography and medicine, often combined with protein.	Toxic metal. *Long term*—permanent grey discoloration of skin, eyes, and mucus membranes.

Organic compounds

Compound	Uses	Hazards
Benzene (C_6H_6)	Manufacturing of ethylbenzene (for styrene monomer); dodecylbenzene (for detergents); cyclohexane (for nylon); phenol; nitrobenzene (for aniline); maleic anhydride; chlorobenzene hexachloride; benzene sulfonic acid; as a solvent.	A carcinogen. Highly toxic. Flammable, dangerous fire risk.
Ethylbenzene ($C_6H_5C_2H_5$)	Intermediate in production of styrene; solvent.	Toxic by ingestion, inhalation, and skin absorption; irritant to skin and eyes. Flammable, dangerous fire risk.
Toluene ($C_6HC_5H_3$)	Aviation gasoline and high-octane blending stock; benzene, phenol, and caprolactam; solvent for paints and coatings, gums, resins, most oils, rubber, vinyl organosols; diluent and thinner in nitrocellulose lacquers; adhesive solvent in plastic toys and model airplanes; chemicals (benzoic acid, benzyl and bezoyl derivatives, saccharine, medicines, dyes, perfumes); source of toluenediisocyanates (polyurethane resins); explosives (TNT); toluene sulfonates (detergents); scintillation counters.	Flammable, dangerous fire risk. Toxic by ingestion, inhalation, and skin absorption.

Halogenated compounds

Compound	Uses	Hazards
Chlorobenzene (C_6H_5Cl)	Phenol, chloronitrobenzene, aniline, solvent carrier for methylene diisocyanate, solvent, pesticide intermediate, heat transfer.	Moderate fire risk. Avoid inhalation and skin contact.
Chloroethene (CH_2CHCl)	Polyvinyl chloride and copolymers, organic synthesis, adhesives for plastics.	An extremely toxic and hazardous material by all avenues of exposure. A carcinogen.
Dichloromethane (CH_2Cl_2)	Paint removers, solvent degreasing, plastics processing, blowing agent in foams, solvent extraction, solvent for cellulose acetate, aerosol propellant.	Toxic. A carcinogen, narcotic.
Tetrachloroethene (CCl_2CCl_2)	Dry cleaning solvent, vapor-degreasing solvent, drying agent for metals and certain other solids, vermifuge, heat transfer medium, manufacture of fluorocarbons.	Irritant to eyes and skin.

Pesticides, herbicides, insecticides [a]

Compound	Uses	Hazards
Endrin ($C_{12}H_8OCl_6$)	Insecticide and fumigant.	Toxic by inhalation and skin absorption, carcinogen.
Lindane ($C_6H_6Cl_6$)	Pesticide.	Toxic by inhalation, ingestion, skin absorption.
Methoxychlor ($Cl_3CCH(C_6H_4OCH_3)_2$)	Insecticide.	Toxic material.
Toxaphene ($C_{10}H_{10}Cl_8$)	Insecticide and fumigant.	Toxic by ingestion, inhalation, skin absorption.
Silvex ($Cl_3C_6H_2OCH(CH_3)COOH$)	Herbicide, plant growth regulator.	Toxic material; use has been restricted.

[a] Pesticides, herbicides, and insecticides are listed by trade name. The compounds listed are also halogenated organic compounds.

of some of these compounds in the atmosphere may pose a significant public health risk; and (3) they contribute to a general increase in reactive hydrocarbons in the atmosphere, which can lead to the formation of photochemical oxidants. The release of these compounds in sewers and at treatment plants, especially at the headworks, is of particular concern with respect to the health of collection system and treatment plant workers. The release and control of VOCs is considered further in Chaps. 6 and 9. The physical phenomena involved in the release of VOCs are considered in detail in Ref. 24.

Pesticides and Agricultural Chemicals. Trace organic compounds, such as pesticides, herbicides, and other agricultural chemicals, are toxic to most life forms and therefore can be significant contaminants of surface waters. These chemicals are not common constituents of domestic wastewater but result primarily from surface runoff from agricultural, vacant, and park lands. Concentrations of these chemicals can result in fish kills, in contamination of the flesh of fish that decreases their value as a source of food, and in impairment of water supplies. Many of these chemicals are also classified as priority pollutants.

Measurement of Organic Content

Over the years, a number of different tests have been developed to determine the organic content of wastewaters. In general, the tests may be divided into those used to measure gross concentrations of organic matter greater than about 1 mg/L and those used to measure trace concentrations in the range of 10^{-12} to 10^{-3} mg/L. Laboratory methods commonly used today to measure gross amounts of organic matter (greater than 1 mg/L) in wastewater include: (1) biochemical oxygen demand (BOD); (2) chemical oxygen demand (COD); and (3) total organic carbon (TOC). Complementing these laboratory tests is the theoretical oxygen demand (ThOD), which is determined from the chemical formula of the organic matter.

Other methods used in the past included (1) total, albuminoid, organic, and ammonia nitrogen, and (2) oxygen consumed. These determinations, with the exception of albuminoid nitrogen and oxygen consumed, are still included in complete wastewater analyses. Their significance, however, has changed. Whereas formerly they were used almost exclusively to indicate organic matter, they are now used to determine the availability of nitrogen to sustain biological activity in industrial waste treatment processes and to foster undesirable algal growths in receiving water.

Trace organics in the range of 10^{-12} to 10^{-3} mg/L are determined using instrumental methods including gas chromotography and mass spectroscopy. Within the past 10 years, the sensitivity of the methods used for the detection of trace organic compounds has improved significantly and detection of concentrations in the range of 10^{-9} mg/L is now almost a routine matter.

The concentration of pesticides is typically measured by the carbon-chloroform extract method, which consists of separating the contaminants from the water by passing a water sample through an activated-carbon column and then extracting the contaminant from the carbon using chloroform. The chloroform can then be evaporated and the contaminants can be weighed. Pesticides and herbicides in concentrations of 1 part per billion (ppb) and less can be determined accurately by several methods, including gas chromatography and electron capture or coulometric detectors [18].

Biochemical Oxygen Demand. The most widely used parameter of organic pollution applied to both wastewater and surface water is the 5-day BOD (BOD_5). This determination involves the measurement of the dissolved oxygen used by microorganisms in the biochemical oxidation of organic matter. Despite the widespread use of the BOD test, it has a number of limitations (which are discussed later in this section). It is hoped that, through the continued efforts of workers in the field, one of the other measures of organic content, or perhaps a new measure, will ultimately be used in its place. Why, then, if the test suffers from serious limitations, is further space devoted to it in this text? The reason is that BOD test results are now used (1) to determine the approximate quantity of oxygen that will be required to biologically stabilize the organic matter present, (2) to determine the size of waste treatment facilities, and (3) to measure the efficiency of some treatment processes, and (4) to determine compliance with wastewater discharge permits. Because it is likely that the BOD test will continue to be used for some time, it is important to know as much as possible about the test and its limitations.

To ensure that meaningful results are obtained, the sample must be suitably diluted with a specially prepared dilution water so that adequate nutrients and oxygen will be available during the incubation period. Normally, several dilutions are prepared to cover the complete range of possible values. The ranges of BOD that can be measured with various dilutions based on percentage mixtures and direct pipetting are reported in Table 3-10. The general procedure for preparing the BOD bottles for incubation is illustrated in Fig. 3-11.

When the sample contains a large population of microorganisms (untreated wastewater, for example), seeding is not necessary. If required, the dilution water is "seeded" with a bacterial culture that has been acclimated to the organic matter or other materials that may be present in the wastewater. The seed culture that is used to prepare the dilution water for the BOD test is a mixed culture. Such cultures contain large numbers of saprophytic bacteria and other organisms that oxidize the organic matter. In addition, they contain certain autotrophic bacteria that oxidize noncarbonaceous matter. A variety of commercial seed preparations are also available.

The incubation period is usually five days at 20°C, but other lengths of time and temperatures can be used. Longer time periods (typically seven days), which correspond to work schedules, are often used, especially in small plants where the laboratory staff is not available on the weekends. The temperature, however, should be constant throughout the test. The dissolved oxygen of the samples is measured (see Fig. 3-12) before and after incubation, and the BOD is calculated using Eq. 3-2 or Eq. 3-3.

When dilution water is not seeded,

$$\text{BOD, mg/L} = \frac{D_1 - D_2}{P} \tag{3-2}$$

TABLE 3-10
BOD measurable with various dilutions of samples[a]

By using percent mixtures		By direct pipetting into 300 mL bottles	
% mixture	Range of BOD	mL	Range of BOD
0.01	20,000–70,000	0.02	30,000–105,000
0.02	10,000–35,000	0.05	12,000–42,000
0.05	4,000–14,000	0.10	6,000–21,000
0.1	2,000–7,000	0.20	3,000–10,500
0.2	1,000–3,500	0.50	1,200–4,200
0.5	400–1,400	1.0	600–2,100
1.0	200–700	2.0	300–1,050
2.0	100–350	5.0	120–420
5.0	40–140	10.0	60–210
10.0	20–70	20.0	30–105
20.0	10–35	50.0	12–42
50.0	4–14	100.0	6–21
100.0	0–7	300.0	0–7

[a] Ref. 32.

When dilution water is seeded,

$$\text{BOD, mg/L} = \frac{(D_1 - D_2) - (B_1 - B_2)f}{P} \qquad (3\text{-}3)$$

where D_1 = dissolved oxygen of diluted sample immediately after preparation, mg/L
D_2 = dissolved oxygen of diluted sample after 5 d incubation at 20°C, mg/L
P = decimal volumetric fraction of sample used
B_1 = dissolved oxygen of seed control before incubation, mg/L
B_2 = dissolved oxygen of seed control after incubation, mg/L
f = ratio of seed in sample to seed in control
= (% seed in D_1)/(% seed in B_1)

Biochemical oxidation is a slow process and theoretically takes an infinite time to go to completion. Within a 20-day period, the oxidation of the carbonaceous organic matter is about 95 to 99 percent complete, and in the 5-day period used for the BOD test, oxidation is from 60 to 70 percent complete. The 20°C temperature used is an average value for slow-moving streams in temperate climates and is easily duplicated in an incubator. Different results would be obtained at different temperatures because biochemical reaction rates are temperature-dependent.

The kinetics of the BOD reaction are, for practical purposes, formulated in accordance with first-order reaction kinetics and may be expressed as

$$\frac{dL_t}{dt} = -kL_t \qquad (3\text{-}4)$$

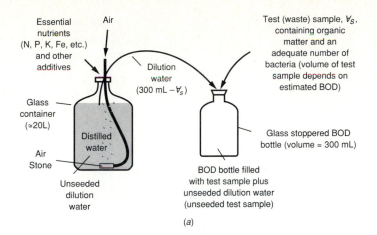

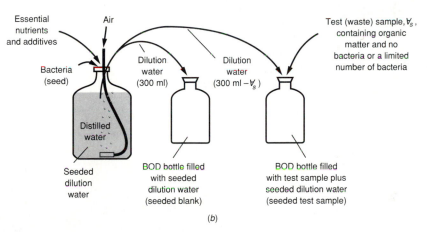

FIGURE 3-11
Procedure for setting up BOD test bottles: (a) with unseeded dilution water and (b) with seeded dilution water [23].

where L_t is the amount of the first-stage BOD remaining in the water at time t and k is the reaction rate constant. This equation can be integrated as

$$\ln L_t\big|_0^t = -kt \qquad (3\text{-}5)$$

$$\frac{L_t}{L} = e^{-kt} = 10^{-Kt} \qquad (3\text{-}6)$$

where L or BOD_L is the BOD remaining at time $t = 0$ (i.e., the total or ultimate first-stage BOD initially present). The relation between k (base e) and K (base 10) is as follows:

$$K(\text{base } 10) = \frac{k(\text{base } e)}{2.303} \qquad (3\text{-}7)$$

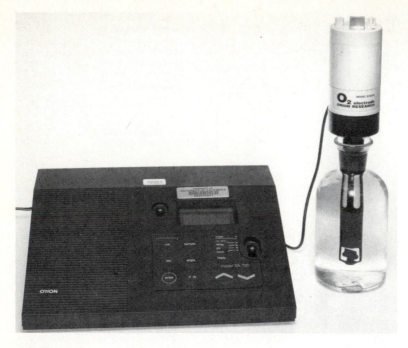

FIGURE 3-12
Measurement of oxygen in BOD bottle with a DO probe equipped with a stirring mechanism.

The amount of BOD remaining at time t equals

$$L_t = L(e^{-kt}) \tag{3-8}$$

and y, the amount of BOD that has been exerted at any time t, equals

$$y_t = L - L_t = L(1 - e^{-kt}) \tag{3-9}$$

Note that the 5-day BOD equals

$$y_5 = L - L_5 = L(1 - e^{-5k}) \tag{3-10}$$

This relationship is shown in Fig. 3-13. The use of the BOD equations is illustrated in Example 3-2.

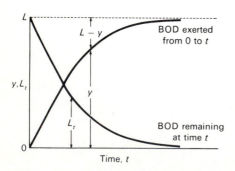

FIGURE 3-13
Formation of the first-stage BOD curve.

Example 3-2   Calculation of BOD. Determine the 1-day BOD and ultimate first-stage BOD for a wastewater whose 5-day, 20°C BOD is 200 mg/L. The reaction constant k(base e) = 0.23 d^{-1}.

Solution

1. Determine ultimate BOD.

$$L_t = Le^{-kt}$$

$$y_5 = L - L_5 = L(1 - e^{-kt})$$

$$200 = L(1 - e^{-5(0.23)}) = L(1 - 0.316)$$

$$L = 293 \text{ mg/L}$$

2. Determine 1-day BOD.

$$L_t = Le^{-kt}$$

$$y_1 = L - L_1 = 293(e^{-0.23(1)}) = 293(0.795) = 233 \text{ mg/L}$$

$$y_1 = L - L_1 = 293 - 233 = 60 \text{ mg/L}$$

For polluted water and wastewater, a typical value of k (base e, 20°C) is 0.23 d^{-1} (= 0.10 $d^{-1}k$ base 10). The value of reaction-rate constant varies significantly, however, with the type of waste. The range may be from 0.05 to 0.3 d^{-1} (base e) or more. For the same ultimate BOD, the oxygen uptake will vary with time and with different reaction-rate constant values (see Fig. 3-14).

As mentioned, the temperature at which the BOD of a wastewater sample is determined is usually 20°C. It is possible, however, to determine the reaction constant k at a temperature other than 20°C. The following approximate equation, which is derived from the van't Hoff-Arrhenius relationship, may be used:

$$k_T = k_{20}\theta^{(T-20)} \tag{3-11}$$

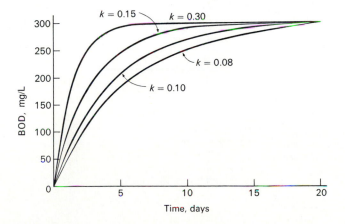

FIGURE 3-14
Effect of the rate constant k on BOD (for a given L value).

The value of θ has been found to vary from 1.056 in the temperature range between 20 and 30°C to 1.135 in the temperature range between 4 and 20°C [15]. A value of θ often quoted in the literature is 1.047 [14], but it has been observed that this value does not apply at cold temperatures (e.g., below 20°C) [15].

Nitrification in the BOD test. Noncarbonaceous matter, such as ammonia, is produced during the hydrolysis of proteins. Two groups of autotrophic bacteria are capable of oxidizing ammonia to nitrite and subsequently to nitrate. The generalized reactions are as follows:

(a) $\qquad NH_3 + \frac{3}{2}O_2 \xrightarrow{\text{nitrite-forming bacteria}} HNO_2 + H_2O \qquad$ (3-12)

(b) $\qquad HNO_2 + \frac{1}{2}O_2 \xrightarrow{\text{nitrate-forming bacteria}} HNO_3 \qquad$ (3-13)

$$NH_3 + 2O_2 \xrightarrow{\hspace{3cm}} HNO_3 + H_2O \qquad (3\text{-}14)$$

The oxygen demand associated with the oxidation of ammonia to nitrate is called the nitrogenous biochemical oxygen demand (NBOD). The normal exertion of the oxygen demand in a BOD test for a domestic wastewater is shown in Fig. 3-15. Because the reproductive rate of the nitrifying bacteria is slow, it normally takes from 6 to 10 days for them to reach significant numbers and to exert a measurable oxygen demand. However, if a sufficient number of nitrifying bacteria are present initially, the interference caused by nitrification can be significant.

When nitrification occurs in the BOD test, erroneous interpretations of treatment operating data are possible. For example, assume that the effluent BOD from a biological treatment process is 20 mg/L without nitrification and 40 mg/L with nitrification. If the influent BOD to the treatment process is 200 mg/L, then the cor-

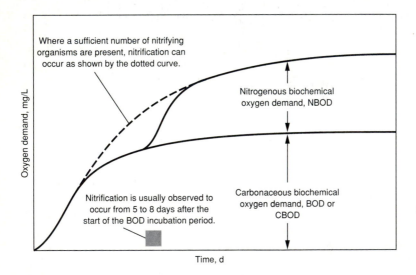

FIGURE 3-15
Definition sketch for the exertion of the carbonaceous and nitrogenous biochemical oxygen demand in a waste sample.

responding BOD removal efficiency would be reported as 90 and 80 percent without and with nitrification, respectively. Thus, if nitrification is occurring but is not suspected, it might be concluded that the treatment process is not performing well when in actuality it is performing quite well.

Carbonaceous biochemical oxygen demand (CBOD). The interference caused by the presence of nitrifying bacteria can be eliminated by pretreatment of the sample or by the use of inhibitory agents. Pretreatment procedures include pasteurization, chlorination, and acid treatment. Inhibitory agents are usually chemical in nature and include compounds such as methylene blue, thiourea and allylthiourea, 2-chlor-6 (trichloromethyl) pyridine, and other proprietary products [37]. Suppression of the nitrification reaction in the BOD test is listed as a standard procedure in the latest edition of *Standard Methods* [18]. The results of the suppressed BOD test should be reported as CBOD (carbonaceous biochemical oxygen demand). The CBOD test is now being used as substitue for the BOD test in discharge permits, especially where nitrification is known to occur.

Analysis of BOD data. The value of k is needed if the BOD_5 is to be used to obtain L, the ultimate or 20-day BOD. The usual procedure followed when these values are unknown is to determine k and L from a series of BOD measurements. There are several ways to do this, including the method of least-squares, the method of moments [11], the daily-difference method [27], the rapid-ratio method [16], the Thomas method [26], and the Fujimoto method [5]. The least-squares method and the Fujimoto method are illustrated in the following discussion.

The least-squares method involves fitting a curve through a set of data points, so that the sum of the squares of the residuals (the difference between the observed value and the value of the fitted curve) must be a minimum. Using this method, a variety of different types of curves can be fitted through a set of data points. For example, for a time series of BOD measurements on the same sample, the following equation may be written for each of the various n data points:

$$\frac{dy}{dt}\bigg|_{t=n} = k(L - y_n) \tag{3-15}$$

In this equation both k and L are unknown. If it is assumed that dy/dt represents the value of the slope of the curve to be fitted through all the data points for a given k and L value, then because of experimental error, the two sides of Eq. 3-15 will not be equal but will differ by an amount R. Rewriting Eq. 3-15 in terms of R for the general case yields

$$R = k(L - y) - \frac{dy}{dt} \tag{3-16}$$

Simplifying and using the notation y' for dy/dt gives

$$R = kL - ky - y' \tag{3-17}$$

Substituting a for kL and $-b$ for k gives

$$R = a + by - y' \tag{3-18}$$

Now, if the sum of the squares of the residuals R is to be a minimum, the following equations must hold:

$$\frac{\partial}{\partial a} \sum R^2 = \sum 2R \frac{\partial R}{\partial a} = 0 \tag{3-19}$$

$$\frac{\partial}{\partial b} \sum R^2 = \sum 2R \frac{\partial R}{\partial b} = 0 \tag{3-20}$$

If the indicated operations in Eqs. 3-19 and 3-20 are carried out using the value of the residual R defined by Eq. 3-18, the following set of equations result:

$$na + b\sum y - \sum y' = 0 \tag{3-21}$$

$$a\sum y + b\sum y^2 - \sum yy' = 0 \tag{3-22}$$

where n = number of data points
$\quad a = -bL$
$\quad b = -k(\text{base } e)$
$\quad L = -a/b$
$\quad y = y_t, \text{ mg/L}$
$\quad y' = \dfrac{y_{n+1} - y_{n-1}}{2\Delta t}$

Application of the least-squares method in the analysis of BOD data is illustrated in Example 3-3, which follows the discussion of the Fujimoto method.

In the Fujimoto method [5], an arithmetic plot is prepared of BOD_{t+1} versus BOD_t. The value at the intersection of the plot with a line of slope 1 corresponds to the the ultimate BOD. After the BOD_L has been determined, the rate constant is determined using Eq. 3-9 and one of the BOD values. The application of the Fujimoto method is illustrated in Example 3-3.

Example 3-3 Calculation of BOD constants using the least squares and the Fujimoto methods. Compute L and k using the least-squares and Fujimoto methods for the following BOD data reported for a stream receiving some treated effluent:

t, d	2	4	6	8	10
y, mg/L	11	18	22	24	26

Solution

1. Set up a computation table and perform the indicated steps.

Time	y	y^2	y'	yy'
2	11	121	4.50	49.5
4	18	324	2.75	49.5
6	22	484	1.50	33.0
8	24	576	1.00	24.0
	75	1,505	9.75	156.0

The slope y' is computed as follows:

$$\frac{dy}{dt} = y' = \frac{y_{n+1} - y_{n-1}}{2\Delta t}$$

2. Substitute the values computed in step 1 in Eqs. 3-19 and 3-20, and solve for a and b.

$$4a + 75b - 9.75 = 0$$

$$75a + 1505b - 156.0 = 0$$

$$a = 7.5 \text{ and } b = -0.271$$

3. Determine the values of k and L.

$$k = -b = 0.271 \text{ (base } e)$$

$$L = \frac{a}{b} = \frac{7.5}{0.271} = 27.7 \text{ mg/L}$$

4. Prepare an arithmetic plot of BOD_{t+1} versus BOD_t, and on the same plot draw a line with a slope of 1. The value at the intersection of the two lines (BOD = 27.8 mg/L) corresponds to the ultimate BOD.

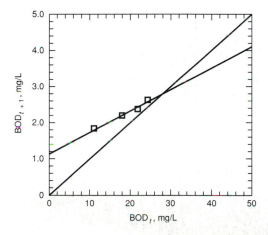

5. Determine the k value using Eq. 3-9.

$$y_6 = L - L_6 = L(1 - e^{-6K'})$$

$$22 = L - L_6 = 27.8(1 - e^{-6K'})$$

$$k = 0.293 \text{ d}^{-1}$$

Respirometric determination of BOD. Determination of the BOD value and the corresponding rate constant k can be accomplished more efficiently in the laboratory using an instrumented large-volume (1.0 L) electrolysis cell or a laboratory respirometer. An electrolysis cell (see Fig. 3-16a) may also be used to obtain a continuous BOD [38,39]. Within the cell, oxygen pressure over the sample is maintained constant by continually replacing the oxygen used by the micoorganisms. Oxygen

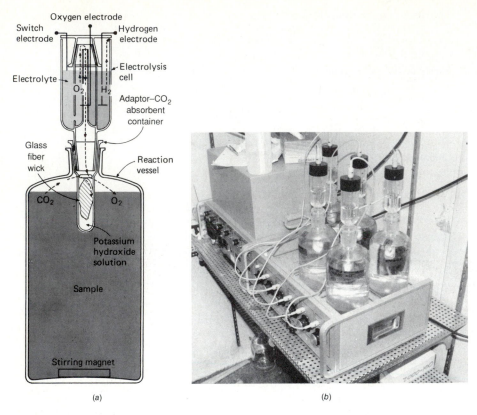

FIGURE 3-16
Electrolytic respirometer for BOD determination: (*a*) schematic [37,38] and (*b*) commercial respirometer with multiple electrolysis cells.

replacement is accomplished by means of an electrolysis reaction in which oxygen is produced in response to changes in the pressure. The BOD readings are determined by noting the length of time that the oxygen was generated and by correlating it to the amount of oxygen produced by the electrolysis reaction. Advantages of the electrolysis cell over a conventional laboratory respirometer are that (1) the use of a large (1 L) sample minimizes the errors of grab sampling and pipetting in dilutions, and (2) the value of the BOD is available directly. A typical example of a commercially available electrolytic respirometer with multiple electrolysis cells is also shown in Fig. 3-16*b*.

Limitations in the BOD test. The limitations of the BOD test are as follows: (1) a high concentration of active, acclimated seed bacteria is required; (2) pretreatment is needed when dealing with toxic wastes, and the effects of nitrifying organisms must be reduced; (3) only the biodegradable organics are measured; (4) the test does not have stoichiometric validity after the soluble organic matter present in solution has been used (see Fig. 3-17); and (5) an arbitrary, long period of time

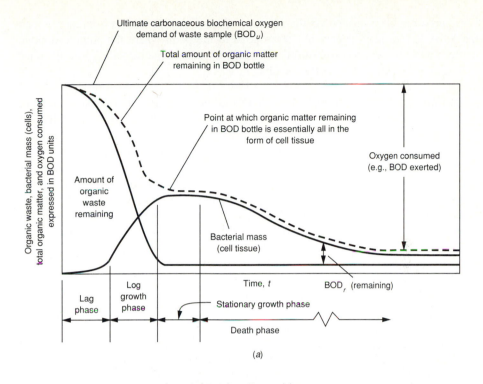

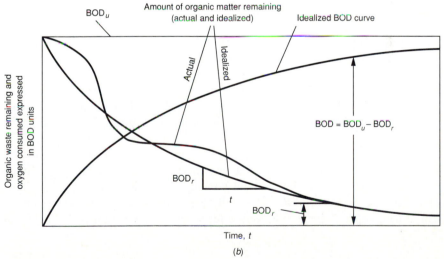

FIGURE 3-17
Functional analysis of the BOD test: (a) interrelationship of organic waste, bacterial mass (cell tissue), total organic waste, and oxygen consumed in BOD test and (b) idealized representation of the BOD test [23].

is required to obtain results. Of the above, perhaps the most serious limitation is that the 5-day period may or may not correspond to the point where the soluble organic matter that is present has been used. The lack of stoichiometric validity at all times reduces the usefulness of the test results.

Chemical Oxygen Demand. The COD test is used to measure the content of organic matter of both wastewater and natural waters. The oxygen equivalent of the organic matter that can be oxidized is measured by using a strong chemical oxidizing agent in an acidic medium. Potassium dichromate has been found to be excellent for this purpose. The test must be performed at an elevated temperature. A catalyst (silver sulfate) is required to aid the oxidation of certain classes of organic compounds. Since some inorganic compounds interfere with the test, care must be taken to eliminate them. The principal reaction using dichromate as the oxidizing agent may be represented in a general way by the following unbalanced equation:

$$\text{Organic matter } (C_aH_bO_c) + Cr_2O_7^{-2} + H^+ \xrightarrow[\text{heat}]{\text{catalyst}} Cr^{+3} + CO_2 + H_2O \quad (3\text{-}23)$$

The COD test is also used to measure the organic matter in industrial and municipal wastes that contain compounds that are toxic to biological life. The COD of a waste is, in general, higher than the BOD because more compounds can be chemically oxidized than can be biologically oxidized. For many types of wastes, it is possible to correlate COD with BOD. This can be very useful because the COD can be determined in three hours, compared with five days for the BOD. Once the correlation has been established, COD measurements can be used to good advantage for treatment-plant control and operation.

Total Organic Carbon. Another means for measuring the organic matter present in water is the TOC test, which is especially applicable to small concentrations of organic matter. The test is performed by injecting a known quantity of sample into a high-temperature furnace or chemically-oxidizing environment. The organic carbon is oxidized to carbon dioxide in the presence of a catalyst. The carbon dioxide that is produced is quantitatively measured by means of an infrared analyzer. Acidification and aeration of the sample prior to analysis eliminates errors due to the presence of inorganic carbon. If VOCs are known to be present, the aeration step is omitted to eliminate their removal by stripping. The test can be performed very rapidly and is becoming more popular. Certain resistant organic compounds may not be oxidized, however, and the measured TOC value will be slightly less than the actual amount present in the sample. Typical TOC values for wastewater are reported in Table 3-16 in Sec. 3-5.

Theoretical Oxygen Demand. Organic matter of animal or vegetable origin in wastewater is generally a combination of carbon, hydrogen, oxygen, and nitrogen. The principal groups of these elements present in wastewater are, as previously noted, carbohydrates, proteins, fats, and products of their decomposition. The biological decomposition of the substances is discussed in Chap. 8. If the chemical formula of the organic matter is known, the ThOD may be computed, as illustrated in Example 3-4.

Example 3-4 Calculation of ThOD. Determine the ThOD for glycine ($CH_2(NH_2)$ COOH) using the following assumptions:

1. In the first step, the organic carbon and nitrogen are converted to carbon dioxide (CO_2) and ammonia (NH_3), respectively.
2. In the second and third steps, the ammonia is oxidized sequentially to nitrite and nitrate.
3. The ThOD is the sum of the oxygen required for all three steps.

Solution

1. Write balanced reaction for the carbonaceous oxygen demand.

$$CH_2(NH_2)COOH + \tfrac{3}{2}O_2 \rightarrow NH_3 + 2CO_2 + H_2O$$

2. Write balanced reactions for the nitrogenous oxygen demand.

(*a*) $$NH_3 + \tfrac{3}{2}O_2 \rightarrow HNO_2 + H_2O$$

(*b*) $$HNO_2 + \tfrac{1}{2}O_2 \rightarrow HNO_3$$

$$\overline{\qquad\qquad\qquad\qquad\qquad\qquad}$$

$$NH_3 + 2O_2 \rightarrow HNO_3 + H_2O$$

3. Determine the ThOD.

$$\text{ThOD} = (\tfrac{3}{2} + \tfrac{4}{2}) \quad \text{mol } O_2/\text{mol glycine}$$
$$= 3.5 \text{ mol } O_2/\text{mol glycine} \times 32 \text{ g/mol } O_2$$
$$= 112 \text{ g } O_2/\text{mol glycine}$$

Correlation Among Gross Measures of Organic Content. Establishment of constant relationships among the various measures of organic content depends primarily on the nature of the wastewater and its source. Of all the measures, the most difficult to correlate to the others is the BOD_5 test, because of the problems cited previously (see BOD discussion). For typical untreated domestic wastes, however, the BOD_5/COD ratio varies from 0.4 to 0.8, and the BOD_5/TOC ratio varies from 1.0 to 1.6. It should also be noted that these ratios vary considerably with the degree of treatment the wastewater has undergone. Because of the rapidity with which the COD, TOC, and related tests can be conducted, it is anticipated that more use will be made of these tests in the future.

Inorganic Matter

Several inorganic components of wastewaters and natural waters are important in establishing and controlling water quality. The concentrations of inorganic substances in water are increased both by the geologic formation with which the water comes in contact and by the wastewaters, treated or untreated, that are discharged to it [17,20]. The natural waters dissolve some of the rocks and minerals with which they come in contact. Wastewaters, with the exception of some industrial wastes, are seldom treated for removal of the inorganic constituents that are added in the use cycle. Concentrations of inorganic constituents also are increased by the natural evaporation process, which removes some of the surface water and leaves the inorganic substance in the water. Because concentrations of various inorganic constituents can greatly affect the beneficial uses made of the waters, it is well to examine the nature of some of the constituents, particularly those added to surface water via the use cycle.

pH. The hydrogen-ion concentration is an important quality parameter of both natural waters and wastewaters. The concentration range suitable for the existence of most biological life is quite narrow and critical. Wastewater with an adverse concentration of hydrogen-ion is difficult to treat by biological means, and if the concentration is not altered before discharge, the wastewater effluent may alter the concentration in the natural waters.

The hydrogen-ion concentration in water is closely connected with the extent to which water molecules dissociate. Water will dissociate into hydrogen and hydroxyl ions as follows:

$$H_2O \longleftrightarrow H^+ + OH^- \tag{3-24}$$

Applying the law of mass action to this equation yields

$$\frac{[H^+][OH^-]}{[H_2O]} = K \tag{3-25}$$

where the brackets indicate the concentration of the constituents in moles per liter. Because the concentration of water in a dilute aqueous system is essentially constant, this concentration can be incorporated into the equilibrium constant K to give

$$[H^+][OH^-] = K_w \tag{3-26}$$

K_w is known as the ionization constant, or ion product, of water and is approximately equal to 1×10^{-14} at a temperature of 25°C. Equation 3-26 can be used to calculate the hydroxyl-ion concentration when the hydrogen-ion concentration is known, and vice versa.

The usual means of expressing the hydrogen-ion concentration is as pH, which is defined as the negative logarithm of the hydrogen-ion concentration.

$$pH = -\log_{10}[H^+] \tag{3-27}$$

With pOH, which is defined as the negative logarithm of the hydroxyl-ion concentration, it can be seen from Eq. 3-26 that, for water at 25°C,

$$pH + pOH = 14 \tag{3-28}$$

The pH of aqueous systems can be conveniently measured with a pH meter. Various pH papers and indicator solutions that change color at definite pH values are also used. The pH is determined by comparing the color of the paper or solution to a series of color of standard.

Chlorides. Another quality parameter of significance is the chloride concentration. Chlorides in natural water result from the leaching of chloride-containing rocks and soils with which the water comes in contact, and in coastal areas, from saltwater intrusion. In addition, agricultural, industrial, and domestic wastewaters discharged to surface waters are a source of chlorides.

Human excreta, for example, contain about 6 g of chlorides per person per day. In areas where the hardness of water is high, home regeneration type water softeners will also add large quantities of chlorides. Because conventional methods of waste treatment do not remove chloride to any significant extent, higher than usual chloride concentrations can be taken as an indication that the body of water is being used for waste disposal. Infiltration of groundwater into sewers adjacent to saltwater is also a potential source of high chlorides as well as sulfates.

Alkalinity. Alkalinity in wastewater results from the presence of the hydroxides, carbonates, and bicarbonates of elements such as calcium, magnesium, sodium, potassium, or ammonia. Of these, calcium and magnesium bicarbonates are most common. Borates, silicates, phosphates, and similar compounds can also contribute to the alkalinity. The alkalinity in wastewater helps to resist changes in pH caused by the addition of acids. Wastewater is normally alkaline, receiving its alkalinity from the water supply, the groundwater, and the materials added during domestic use. Alkalinity is determined by titrating against a standard acid; the results are expressed in terms of calcium carbonate, $CaCO_3$. The concentration of alkalinity in wastewater is important where chemical treatment is to be used (see Chaps. 9 and 11), in biological nutrient removal (see Chap. 11), and where ammonia is to be removed by air stripping (see Chap. 11).

Nitrogen. The elements nitrogen and phosphorus are essential to the growth of protista and plants and as such are known as nutrients or biostimulants. Trace quantities of other elements, such as iron, are also needed for biological growth, but nitrogen and phosphorus are, in most cases, the major nutrients of importance. Because nitrogen is an essential building block in the synthesis of protein, nitrogen data will be required to evaluate the treatability of wastewater by biological processes. Insufficient nitrogen can necessitate the addition of nitrogen to make the waste treatable. Nutrient requirements for biological waste treatment are discussed in Chaps. 8 and 10. Where control of algal growths in the receiving water is necessary to protect beneficial uses, removal or reduction of nitrogen in wastewaters prior to discharge may be desirable (see Chap. 11).

Forms of nitrogen. Total nitrogen is comprised of organic nitrogen, ammonia, nitrite, and nitrate. Organic nitrogen is determined by the Kjeldahl method. The aqueous sample is first boiled to drive off the ammonia, and then it is digested. During the digestion, the organic nitrogen is converted to ammonia. Total Kjeldahl nitrogen is determined in the same manner as organic nitrogen, except that the ammonia is not driven off before the digestion step. Kjeldahl nitrogen is, therefore, the total of the organic and ammonia nitrogen.

Ammonia nitrogen exists in aqueous solution as either the ammonium ion or ammonia, depending on the pH of the solution, in accordance with the following equilibrium reaction:

$$NH_3 + H_2O \longleftrightarrow NH_4^+ + OH^- \tag{3-29}$$

At pH levels above 7, the equilibrium is displaced to the left, at levels below pH 7, the ammonium ion is predominant. Ammonia is determined by raising the pH, distilling off the ammonia with the steam produced when the sample is boiled, and condensing the steam that absorbs the gaseous ammonia. The measurement is made colorimetrically, titrimetrically, or with specific-ion electrodes.

Nitrite nitrogen, determined colorimetrically, is relatively unstable and is easily oxidized to the nitrate form. It is an indicator of past pollution in the process of stabilization and seldom exceeds 1 mg/L in wastewater or 0.1 mg/L in surface waters or groundwaters. Although present in low concentrations, nitrite can be very important in wastewater or water-pollution studies because it is extremely toxic to most fish and other aquatic species. Nitrites present in wastewater effluents are oxidized by chlorine and thus increase the chlorine dosage requirements and the cost of disinfection.

Nitrate nitrogen is the most highly oxidized form of nitrogen found in wastewaters. Where secondary effluent is to be reclaimed for groundwater recharge, the nitrate concentration is important. The U.S. EPA interim drinking-water standards [28] limit it to 45 mg/L as NO_3^- because of its serious and occasionally fatal effects on infants. Nitrates may vary in concentration from 0 to 20 mg/L as N in wastewater effluents. A typical range is from 15 to 20 mg/L as N. The nitrate concentration is also usually determined by colorimetric methods.

Nitrogen pathways in nature. The various forms of nitrogen that are present in nature and the pathways by which the forms are changed are depicted in Fig. 3-18. The nitrogen present in fresh wastewater is primarily combined in proteinaceous matter and urea. Decomposition by bacteria readily changes the form to ammonia. The age of wastewater is indicated by the relative amount of ammonia that is present. In an aerobic environment, bacteria can oxidize the ammonia nitrogen to nitrites and nitrates (see Example 3-4). The predominance of nitrate nitrogen in wastewater indicates that the waste has been stabilized with respect to oxygen demand. Nitrates, however, can be used by animals to form animal protein. Death and decomposition of the plant and animal protein by bacteria again yields ammonia. Thus, if nitrogen in the form of nitrates can be reused to make protein by algae and other plants, it may be necessary to remove or to reduce the nitrogen that is present to prevent these growths.

Phosphorus. Phosphorus is also essential to the growth of algae and other biological organisms. Because of noxious algal blooms that occur in surface waters, there is presently much interest in controlling the amount of phosphorus compounds that enter surface waters in domestic and industrial waste discharges and natural runoff. Municipal wastewaters, for example, may contain from 4 to 15 mg/L of phosphorus as P.

The usual forms of phosphorus found in aqueous solutions include the orthophosphate, polyphosphate, and organic phosphate. The orthophosphates, for example, PO_4^{-3}, HPO_4^{-2}, $H_2PO_4^-$, H_3PO_4, are available for biological metabolism without further breakdown. The polyphosphates include those molecules with two or more phosphorus atoms, oxygen atoms, and in some cases, hydrogen atoms combined in a complex molecule. Polyphosphates undergo hydrolysis in aqueous solutions and

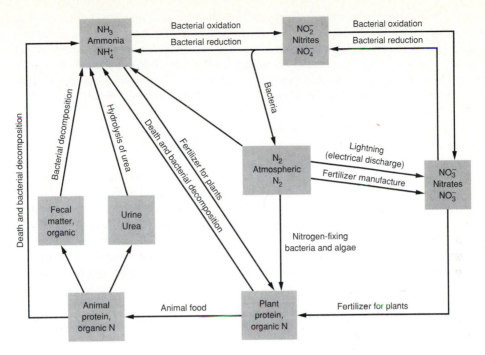

FIGURE 3-18
Generalized nitrogen cycle.

revert to the orthophosphate forms; however, this hydrolysis is usually quite slow. The organically bound phosphorus is usually of minor importance in most domestic wastes, but it can be an important constituent of industrial wastes and wastewater sludges.

Orthophosphate can be determined by directly adding a substance, such as ammonium molybdate, that will form a colored complex with the phosphate. The polyphosphates and organic phosphates must be converted to orthophosphates, using an acid digestion step, before they can be determined in a similar manner.

Sulfur. The sulfate ion occurs naturally in most water supplies and is present in wastewater as well. Sulfur is required in the synthesis of proteins and is released in their degradation. Sulfate is reduced biologically under anaerobic conditions to sulfide, which in turn can combine with hydrogen to form hydrogen sulfide (H_2S). The following generalized reactions are typical.

$$\text{Organic matter} + SO_4^{-2} \xrightarrow{\text{bacteria}} S^{-2} + H_2O + CO_2 \qquad (3\text{-}30)$$

$$S^{-2} + 2H^+ \longrightarrow H_2S \qquad (3\text{-}31)$$

Hydrogen sulfide released to the atmosphere above the wastewater in sewers that are not flowing full tends to accumulate at the crown of the pipe. The accumulated H_2S can then be oxidized biologically to sulfuric acid, which is corrosive to sewer pipes [9].

Sulfates are reduced to sulfides in sludge digesters and may upset the biological process if the sulfide concentration exceeds 200 mg/L. Fortunately, such concentrations are rare. The H_2S gas, which is evolved and mixed with the wastewater gas ($CH_4 + CO_2$), is corrosive to the gas piping, and, if burned in gas engines, the products of combustion can damage the engine and severely corrode exhaust-gas heat-recovery equipment, especially if allowed to cool below the dew point.

Toxic Inorganic Compounds. Because of their toxicity, certain cations are of great importance in the treatment and disposal of wastewaters. Many of these compounds are classified as priority pollutants (see Table 3-9). Copper, lead, silver, chromium, arsenic, and boron are toxic in varying degrees to microorganisms and therefore must be taken into consideration in the design of a biological treatment plant. Many plants have been upset by the introduction of these ions to the extent that the microorganisms were killed and treatment ceased. For instance, in sludge digesters, copper is toxic at a concentration of 100 mg/L, chromium and nickel are toxic at concentrations of 500 mg/L, and sodium is also toxic at high concentrations [8]. Other toxic cations include potassium and ammonium at 4000 mg/L. The alkalinity present in the digesting sludge will combine with and precipitate the calcium ions before the calcium concentration approaches the toxic level.

Some toxic anions, including cyanides and chromates, are also present in industrial wastes. These are found particularly in metal-plating wastes and should be removed by pretreatment at the site of the industry rather than be mixed with the municipal wastewater. Fluoride, another toxic anion, is found commonly in wastewater from electronics manufacturing facilities. Organic compounds present in some industrial wastes are also toxic.

Heavy Metals. Trace quantities of many metals, such as nickel (Ni), manganese (Mn), lead (Pb), chromium (Cr), cadmium (Cd), zinc (Zn), copper (Cu), iron (Fe), and mercury (Hg), are important constituents of most waters. Many of these metals are also classified as priority pollutants (see Table 3-9). Some of these metals are necessary for growth of biological life, and absence of sufficient quantities of them could limit growth of algae, for example. The presence of any of these metals in excessive quantities will interfere with many beneficial uses of the water because of their toxicity; therefore, it is frequently desirable to measure and control the concentrations of these substances. Methods for determining the concentrations of these substances vary in complexity according to the interfering substances that may be present [18]. In addition, quantities of many of these metals can be determined at very low concentrations by such instrumental methods as polarography and atomic absorption spectroscopy.

Gases

Gases commonly found in untreated wastewater include nitrogen (N_2), oxygen (O_2), carbon dioxide (CO_2), hydrogen sulfide (H_2S), ammonia (NH_3), and methane (CH_4).

The first three are common gases of the atmosphere and will be found in all waters exposed to air. The latter three are derived from the decomposition of the organic matter present in wastewater. Although not found in untreated wastewater, other gases with which the environmental engineer must be familiar include chlorine (Cl_2) and ozone (O_3) (disinfection and odor control), and the oxides of sulfur and nitrogen (combustion processes). The following discussion is limited to those gases that are of interest in untreated wastewater. Under most circumstances, the ammonia in untreated wastewater will be present as the ammonium ion (see "Nitrogen").

Dissolved Oxygen. Dissolved oxygen is required for the respiration of aerobic microorganisms as well as all other aerobic life forms. However, oxygen is only slightly soluble in water. The actual quantity of oxygen (other gases too) that can be present in solution is governed by (1) the solubility of the gas, (2) the partial pressure of the gas in the atmosphere, (3) the temperature, and (4) the purity (salinity, suspended solids, etc.) of the water. The interrelationship of these variables is delineated in Chap. 6 and is illustrated in Appendix D, where the effect of temperature and salinity on dissolved-oxygen concentration is presented.

Because the rate of biochemical reactions that use oxygen increases with increasing temperature, dissolved-oxygen levels tend to be more critical in the summer months. The problem is compounded in summer months because stream flows are usually lower, and thus the total quantity of oxygen available is also lower. The presence of dissolved oxygen in wastewater is desirable because it prevents the formation of noxious odors. The role of oxygen in wastewater treatment is discussed in Chaps. 8 and 10; its importance in water-quality management is discussed in Chap. 17.

Hydrogen Sulfide. Hydrogen sulfide is formed, as mentioned previously, from the anaerobic decomposition of organic matter containing sulfur or from the reduction of mineral sulfites and sulfates. It is not formed in the presence of an abundant supply of oxygen. This gas is a colorless, inflammable compound with the characteristic odor of rotten eggs. The blackening of wastewater and sludge usually results from the formation of hydrogen sulfide that has combined with the iron present to form ferrous sulfide (FeS). Various other metallic sulfides are also formed. Although hydrogen sulfide is the most important gas formed from the standpoint of odors, other volatile compounds such as indol, skatol, and mercaptans, which may also be formed during anaerobic decomposition, may cause odors far more offensive than that of hydrogen sulfide.

Methane. The principal by-product of the anaerobic decomposition of the organic matter in wastewater is methane gas (see Chaps. 8 and 12). Methane is a colorless, odorless, combustible hydrocarbon of high fuel value. Normally, large quantities are not encountered in untreated wastewater because even small amounts of oxygen tend to be toxic to the organisms responsible for the production of methane (see Chap. 8). Occasionally, however, as a result of anaerobic decay in accumulated bottom deposits, methane is produced. Because methane is highly combustible and the explosion hazard is high, manholes and sewer junctions or junction chambers where there is an opportunity for gas to collect should be ventilated with a portable blower during and before the time required for employees to work in them. In treatment plants, methane is produced from the anaerobic treatment process used to stabilize wastewater sludges (see Chap. 12). In treatment plants where methane is produced, notices should be posted about the plant warning of explosion hazards, and plant employees should be instructed in safety measures to be maintained while working in and about the structures where gas may be present.

3-4   BIOLOGICAL CHARACTERISTICS: DEFINITION AND APPLICATION

The environmental engineer must have considerable knowledge of the biological characteristics of wastewater. The engineer must know (1) the principal groups of microorganisms found in surface water and wastewaters as well as those responsible for biological treatment, (2) the pathogenic organisms found in wastewater, (3) the organisms used as indicators of pollution and their significance, (4) the methods used to enumerate the indicator organisms, and (5) the methods used to evaluate the toxicity of treated wastewaters. These matters are discussed in this section.

Microorganisms

The principal groups of organisms found in surface water and wastewater are classified as eucaryotes, eubacteria, and archaebacteria (see Table 3-11). As reported in Table 3-11, most bacteria are classified as eubacteria. The category protista, contained within

TABLE 3-11
Classification of microorganisms[a]

Group	Cell structure	Characterization	Representative members
Eucaryotes	Eucaryotic[b]	Multicellular with extensive differentiation of cells and tissue	Plants (seed plants, ferns, mosses) Animals (vertebrates, invertebrates)
		Unicellular or coenocytic or, mycelial; little or no tissue differentiation	Protists (algae, fungi, protozoa)
Eubacteria	Procaryotic[c]	Cell chemistry similar to eucaryotes	Most bacteria
Archaebacteria	Procaryotic[c]	Distinctive cell chemistry	Methanogens, halophiles, thermacidophiles

[a] Adapted from Ref. 19.
[b] Contain true nucleus.
[c] Contain no nuclear membrane.

the eucaryote classification, includes algae, fungi, and protozoa. Plants including seed plants, ferns, and mosses are classified as multicellular eucaryotes. Invertebrates and vertebrates are classified as multicellular eucaryotic animals [19]. Viruses, which are also found in wastewater, are classified separately according to the host infected. Because the organisms in the various groups are discussed in detail in the subsequent chapters of this book, the following discussion is meant to serve only as a general introduction to the various groups and their importance in the field of wastewater treatment and water quality management.

Bacteria. Bacteria are single-celled procaryotic eubacteria. Most bacteria can be grouped by form into four general categories: spheroid, rod, curved rod or spiral, and filamentous. Spherical bacteria, known as cocci (singular, coccus), are about 1 to 3 μm in diameter. The rod-shaped bacteria, known as bacilli (singular, bacillus) are quite variable in size, ranging from 0.3 to 1.5 μm in width (or diameter) and from 1.0 to 10.0 μm in length. *Escherichia coli,* a common organism found in human feces, is described as being 0.5 μm in width by 2 μm in length. Curved rod-shaped bacteria, known as vibrios, typically vary in size from 0.6 to 1.0 μm in width (or diameter) and from 2 to 6 μm in length. Spiral bacteria, known as spirilla (singular, spirillum), may be found in lengths up to 50 μm. Filamentous forms, known under a variety of names, can occur in lengths of 100 μm and longer.

Because of the extensive and fundamental role played by bacteria in the decomposition and stabilization of organic matter, both in nature and in treatment plants, their characteristics, functions, metabolism, and synthesis must be understood. These subjects are discussed extensively in Chap. 8. Coliform bacteria are also used as an indicator of pollution by human wastes. Their significance and some of the tests used to determine their presence are discussed in a subsequent section.

Fungi. Fungi are aerobic, multicellular, nonphotosynthetic, chemoheterotrophic, eucaryotic protists. Most fungi are saprophytes, obtaining their food from dead organic matter. Along with bacteria, fungi are the principal organisms responsible for the decomposition of carbon in the biosphere. Ecologically, fungi have two advantages over bacteria: They can grow in low-moisture areas and they can grow in low pH environments. Without the presence of fungi to break down organic material, the carbon cycle would soon cease to exist and organic matter would start to accumulate.

Algae. Algae can be a great nuisance in surface waters because, when conditions are right, they will rapidly reproduce and cover streams, lakes, and reservoirs in large floating colonies called blooms. Algal blooms are usually characteristic of what is called a eutrophic lake, or a lake with a high content of the compounds needed for biological growth. Because effluent from wastewater treatment plants is usually high in biological nutrients, discharge of the effluent to lakes causes enrichment and increases the rate of eutrophication. The same effects can also occur in streams.

The presence of algae affects the value of water for water supply because they often cause taste and odor problems. Algae can also alter the value of surface waters for the growth of certain kinds of fish and other aquatic life, for recreation, and for other beneficial uses. Determination of the concentration of algae in surface waters involves collecting the sample by one of several possible methods and microscopically counting them. Detailed procedures for algae counts are outlined in *Standard Methods* [18].

One of the most important problems facing the environmental engineering profession in terms of water quality management is how to treat wastes of various origins so that the effluents do not encourage the growth of algae and other aquatic plants. The solution may involve the removal of carbon, the removal of various forms of nitrogen and phosphorus, and possibly the removal of some of the trace elements, such as iron and cobalt.

Protozoa. Protozoa are single-celled eucaryotic microorganisms without cell walls. The majority of protozoa are aerobic or facultatively anaerobic chemoheterotrops, although some anaerobic types are known. Protozoa of importance to wastewater engineers include amoebas, flagellates, and free-swimming and stalked ciliates. Protozoa feed on bacteria and other microscopic microorganisms and are essential in the operation of biological treatment processes and in the purification of streams because they maintain a natural balance among the different groups of microorganisms. A number of protoza are also pathogenic. *Giardia lamblia,* the cause of giardiasis (often called hikers disease) and *cryptosporidium,* because of its importance as a causative agent in life-threatening infections in patients with acquired immune deficiency syndrome (AIDS), are of great concern in drinking water supplies.

Plants and Animals. Plants and animals of importance range in size from microscopic rotifers and worms to macroscopic crustaceans. A knowledge of these organisms is helpful in evaluating the condition of streams and lakes, in determining the toxicity of wastewaters discharged to the environment, and in observing the effectiveness of biological life in the secondary treatment processes used to destroy organic wastes.

From the standpoint of human health, a number of worms are of great concern. Two important worm phyla are the Platyhelminthes and the Aschelminthes. Members of the phylum Platyhelminthes are often referred to as flatworms. Flatworms of the class Tubellaria are present in ponds and streams all over the world. The class Trematoda, commonly known as flukes, and the class Cestoda, commonly known as tapeworms, are parasitic forms of great public health significance. The nematodes, comprising more than 10,000 species, are the most important members of the phylum Aschelminthes. The most serious parasitic forms are Trichinella, which causes trichinosis; Necator, which causes hookworm; Ascaris, which causes roundworm infestation; and Filaria, which causes filariasis [3].

Viruses. Viruses are obligate parasitic particles consisting of a strand of genetic material—deoxyribonucleic acid (DNA) or ribonucleic acid (RNA) with a protein coat. Viruses do not have the ability to synthesize new compounds. Instead they invade the living (host) cell where the viral genetic material redirects cell activities to the production of new viral particles at the expense of the host cell. When an infected cell dies, large numbers of viruses are released to infect other cells.

Viruses that are excreted by human beings may become a major hazard to public health. For example, from experimental studies, it has been found that from 10,000 to 100,000 infectious doses of hepatitis virus are emitted from each gram of feces of a patient ill with this disease [10]. It is known that some viruses will live as long as 41 days in water or wastewater at 20°C and for 6 days in a normal river. A number of outbreaks of infectious hepatitis have been attributed to transmission of the virus through water supplies. Much more study is required on the part of biologists and engineers to determine the mechanics of travel and removal of virus in soils, surface waters, and wastewater treatment plants.

Pathogenic Organisms

Pathogenic organisms found in wastewater may be discharged by human beings who are infected with disease or who are carriers of a particular disease. The principal categories of pathogenic organisms found in wastewater are, as reported in Table 3-12, bacteria, viruses, protozoa, and helminths. The usual bacterial pathogenic organisms that may be excreted by man cause diseases of the gastrointestinal tract such as typhoid and paratyphoid fever, dysentery, diarrhea, and cholera. Because these organisms are highly infectious, they are responsible for many thousands of deaths each year in areas with poor sanitation, especially in the tropics [3,7].

Use of Indicator Organisms

Because the numbers of pathogenic organisms present in wastes and polluted waters are few and difficult to isolate and identify, the coliform organism, which is more numerous and more easily tested for, is commonly used as an indicator organism. The intestinal tract of man contains countless rod-shaped bacteria known as coliform organisms. Each person discharges from 100 to 400 billion coliform organisms per day, in addition to other kinds of bacteria. Thus, the presence of coliform organisms is taken as an indication that pathogenic organisms may also be present, and the absence of coliform organisms is taken as an indication that the water is free from disease-producing organisms.

The coliform bacteria include the genera *Escherichia* and *Aerobacter*. The use of coliforms as indicator organisms is complicated by the fact that *Aerobacter* and certain *Escherichia* can grow in soil. Thus, the presence of coliforms does not always mean contamination with human wastes. Apparently, *Escherichia coli (E. coli)* are entirely of fecal origin. There is difficulty in determining *E. coli* to the exclusion of the soil coliforms; as a result, the entire coliform group is used as an indicator of fecal pollution.

Other organisms that have been proposed for use as indicators of pollution are summarized in Table 3-13. In recent years, tests have been developed that distinguish among total coliforms, fecal coliforms, and fecal streptococci; and all three are being reported in the literature. The use of the ratio of fecal coliforms to fecal streptococci is discussed later in this chapter. Indicator organisms that have been used to establish performance criteria for various water uses are reported in Table 3-14.

TABLE 3-12
Infectious agents potentially present in raw domestic wastewater[a]

Organism	Disease	Remarks
Bacteria		
Escherichia coli (enteropathogenic)	Gastroenteritis	Diarrhea
Legionella pneumophila	Legionellosis	Acute respiratory illness
Leptospira (150 spp.)	Leptospirosis	Jaundice, fever (Weil's disease)
Salmonella typhi	Typhoid fever	High fever, diarrhea, ulceration of small intestine
Salmonella (~1700 spp.)	Salmonellosis	Food poisoning
Shigella (4 spp.)	Shigellosis	Bacillary dysentery
Vibrio cholerae	Cholera	Extremely heavy diarrhea, dehydration
Yersinia enterolitica	Yersinosis	Diarrhea
Viruses		
Adenovirus (31 types)	Respiratory disease	
Enteroviruses (67 types, e.g., polio, echo, and Coxsackie viruses)	Gastroenteritis, heart anomalies, meningitis	
Hepatitis A	Infectious hepatitis	Jaundice, fever
Norwalk agent	Gastroenteritis	Vomiting
Reovirus	Gastroenteritis	
Rotavirus	Gastroenteritis	
Protozoa		
Balantidium coli	Balantidiasis	Diarrhea, dysentery
Cryptosporidium	Cryptosporidiosis	Diarrhea
Entamoeba histolytica	Amebiasis (amoebic dysentery)	Prolonged diarrhea with bleeding, abscesses of the liver and small intestine
Giardia lamblia	Giardiasis	Mild to severe diarrhea, nausea, indigestion
Helminths[b]		
Ascaris lumbricoides	Ascariasis	Roundworm infestation
Enterobius vericularis	Enterobiasis	Pinworm
Fasciola hepatica	Fascioliasis	Sheep liver fluke
Hymenolepis nana	Hymenolepiasis	Dwarf tapeworm
Taenia saginata	Taeniasis	Beef tapeworm
T. solium	Taeniasis	Pork tapeworm
Trichuris trichiura	Trichuriasis	Whipworm

[a] Adapted in part from Refs. 3 and 19.

[b] The helminths listed are those with a worldwide distribution.

TABLE 3-13
Specific organisms that have been used or proposed for use as indicators of human pollution[a]

Indicator organism	Characteristics
Coliform bacteria	Species of gram-negative rods that may ferment lactose with gas production (or produce a distinctive colony within 24 ± 2 h to 48 ± 3 h incubation on a suitable medium) at 35 ± 0.5°C. There are strains that do not conform to the definition. The total coliform group includes four genera in the Enterobacteriaceae family. These are *Escherichia, Klebisella, Citrobactor,* and *Enterobacter.* Of the group, the *Escherichia* genus (*E. coli* species) appears to be most representative of fecal contamination.
Fecal coliform bacteria	A fecal coliform bacteria group was established based on the ability to produce gas (or colonies) at an elevated incubation temperature (44.5 ± 0.2°C for 24 ± 2 h).
Klebisella	The total coliform population includes the genus *Klebisella.* The thermotolerant *Klebisella* are also included in the fecal coliform group. This group is cultured at 35 ± 0.5°C for 24 ± 2 h.
E. coli	The *E. coli* is one of the coliform bacteria population and is more representative of fecal sources than other coliform genera.
Fecal streptococci	This group had been used in conjunction with fecal coliforms to determine the source of recent fecal contamination (man or farm animals). Several apparently ubiquitous strains cannot be distinguished from the true fecal streptococci under usual analytical procedures, which detracts from their use as an indicator organism.
Enterococci	Two strains of fecal streptococci—*S. faecalis* and *S. faecium*—are the most human-specific members of the fecal streptococcus group. By eliminating the other strains through the analytical procedures, the two strains known as enterococci can be isolated and enumerated.
	The enterococci are generally found in lower numbers than other indicator organisms; however, they exhibit better survival in seawater.
Clostridium perfringens	This is a spore-forming anaerobic persistent bacteria, and its characteristics make it a desirable indicator where disinfection is employed, where pollution may have occurred in the past, or where the interval before analysis is protracted.
P. aeruginosa and *A. hydrophila*	These organisms may be present in sewage in large numbers. Both can be considered aquatic organisms and can be recovered in water in the absence of immediate sources of fecal pollution.

[a] Adapted in part from Refs. 3 and 7.

Enumeration of Coliform Organisms

The standard test for the coliform group may be carried out using either the multiple-tube fermentation technique or by the membrane filter technique. The complete multiple-tube fermentation procedure for total coliform involves three test phases, identified as the presumptive, confirmed, and completed test [18]. A similar procedure

TABLE 3-14
Indicator organisms used in establishing performance criteria for various water uses

Water use	Indicator organism
Drinking water	Total coliform
Freshwater recreation	Fecal coliform
	E. coli
	Enterococci
Saltwater recreation	Fecal coliform
	Total coliform
	Enterococci
Shellfish growing areas	Total coliform
	Fecal coliform
Agricultural irrigation	Total coliform
	(for reclaimed water)
Wastewater effluent disinfection	Total coliform
	Fecal coliform

is available for the fecal coliform group as well as for other bacterial groups [18]. The presumptive test is based on the ability of the coliform group to ferment lactose broth, producing gas. The confirmed test consists of growing cultures of coliform bacteria from the presumptive test on a medium that suppresses the growth of other organisms. The completed test is based on the ability for the cultures grown in the confirmed test to again ferment the lactose broth. For most routine wastewater analyses, only the presumptive test is performed.

Multiple-Tube Fermentation Technique. The multiple-tube fermentation technique is based on the principle of dilution to extinction. The test may be described as follows. First, a series of serial dilutions is made, as illustrated in Fig. 3-19. The next step is to transfer a 1-mL sample from each of the serial dilutions to each of five fermentation tubes containing a suitable lactose culture medium and an inverted gas collection tube. For total coliform the inoculated tubes are incubated in a water bath at $35 \pm 0.5°C$ for 24 ± 2 hr. The accumulation of gas in the inverted gas-collection tubes after 24 hr is considered to be a positive reaction. The results for each dilution are reported as a fraction, with the number of positive tubes over the total number of tubes. For example, the fraction $\frac{3}{5}$ denotes three positive tubes in a five-tube sample. The test for fecal coliform is similar with the exception that a different culture medium is used and the inoculated tubes are first incubated at $35 \pm 0.5°C$ for 3 hr and then incubated in a water bath at $44.4 \pm 0.2°C$ for 21 ± 2 hr. Estimation of bacterial numbers using the test results from the multiple-tube technique is considered subsequently.

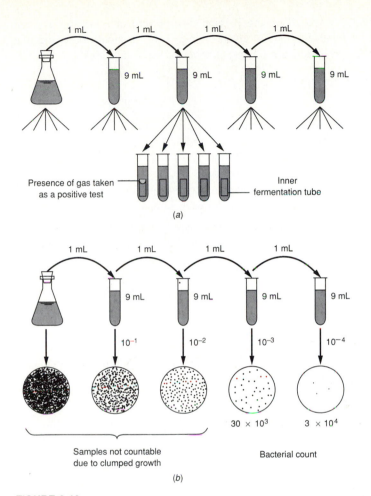

FIGURE 3-19
Illustration of methods used to obtain bacterial counts: (a) use of a liquid medium and (b) use of a solid medium [23].

Estimation of Coliform Densities. Concentrations of total coliform bacteria are most often reported as the "most probable number" per 100 mL (MPN/100mL). The MPN is based on the application of the Poisson distribution for extreme values to the analysis of the number of positive and negative results obtained when testing multiple portions of equal volume and in portions constituting a geometric series. It is emphasized that the MPN is not the absolute concentration of organisms that are present but only a statistical estimate of that concentration. The MPN can be determined using the Poisson distribution directly, MPN tables (see Appendix F) derived from the Poisson distribution, or the Thomas equation [25]. The application of these methods is illustrated in Example 3-5.

The joint probability (based on the Poisson distribution) of obtaining a given result from a series of three dilutions is given by Eq. 3-32. It should be noted that Eq. 3-32 can be expanded to account for any number of serial dilutions.

$$y = \frac{1}{a}[(1 - e^{-n_1\lambda})^{p_1}(e^{-n_1\lambda})^{q_1}][(1 - e^{-n_2\lambda})^{p_2}(e^{-n_2\lambda})^{q_2}]$$
$$[(1 - e^{-n_3\lambda})^{p_3}(e^{-n_3\lambda})^{q_3}] \quad (3\text{-}32)$$

where
y = probability of occurrence of a given result
a = constant for a given set of conditions
n_1, n_2, n_3 = sample size in each dilution
λ = coliform density, number/mL
p_1, p_2, p_3 = number of positive tubes in each sample dilution
q_1, q_2, q_3 = number of negative tubes in each sample dilution

When the Poisson equation or MPN tables are not available, the equation proposed by Thomas [25] can be used to estimate the MPN. The Thomas equation is

$$\text{MPN/100 mL} = \frac{\text{Number of positive tubes} \times 100}{\sqrt{\left(\begin{array}{c}\text{mL of sample in} \\ \text{negative tubes}\end{array}\right) \times \left(\begin{array}{c}\text{mL of sample in} \\ \text{all tubes}\end{array}\right)}} \quad (3\text{-}33)$$

In applying the Thomas equation in situations in which some of the dilutions have all five tubes positive, the count of positive tubes should begin with the highest dilution in which at least one negative result has occurred. The application of the Thomas equation is illustrated in Example 3-5.

Example 3-5 Calculation of MPN. A bacterial analysis for a surface water yielded the following results for the standard confirmed test for total coliform. Determine the coliform density (MPN) using the Poisson equation, the MPN tables in Appendix F, and the Thomas equation.

Size of portion, mL	Number positive	Number negative
10.0	4	1
1.0	4	1
0.1	2	3
0.01	0	5

Solution

1. Determine the MPN using the Poisson equation (Eq. 3-32). Substitute the appropriate values for n, p, and q and solve the Poisson equation by successive trials.

$$n = 10, \qquad p_1 = 4, \qquad q_1 = 1$$
$$n = 1.0, \qquad p_2 = 4, \qquad q_2 = 1$$
$$n = 0.1, \qquad p_3 = 2, \qquad q_3 = 3$$
$$n = 0.01, \qquad p_4 = 0, \qquad q_4 = 5$$

(*a*) Substitute the coefficient values in Eq. 3-32 and determine *ya* values for selected values of λ.

$$y = \frac{1}{a}[(1 - e^{-10\lambda})^4(e^{-10\lambda})^1][(1 - e^{-\lambda})^4(e^{-\lambda})^1]$$
$$[(1 - e^{-0.1\lambda})^2(e^{-0.1\lambda})^3][(1 - e^{-0.01\lambda})^0(e^{-0.01\lambda})^5]$$

λ	*ya*
0.44	1.919×10^{-7}
0.45	1.932×10^{-7}
0.46	1.938×10^{-7}
0.47	1.938×10^{-7}
0.48	1.932×10^{-7}
0.50	1.904×10^{-7}

(*b*) The maximum value of *ya* occurs for a λ value of about 0.46 or 0.47 organisms per milliliter. Thus, the MPN/100 mL is

$$\text{MPN/100 mL} = 100\,(0.46) = 46$$

2. From Appendix F, eliminating the portion with no positive tubes, as outlined, the MPN/100 mL is 47.
3. Determine the MPN using the Thomas equation (Eq. 3-33).

Number of positive tubes $(4 + 4 + 2) = 10$

$$\text{mL of sample in negative tubes} = [(1 \times 10) + (1 \times 1.0) + (3 \times 0.1) + (5 \times 0.01)]$$
$$= 11.35$$

$$\text{mL of sample in negative tubes} = [(5 \times 10) + (5 \times 1.0) + (5 \times 0.1) + (5 \times 0.01)]$$
$$= 55.55$$

$$\text{MPN/100 mL} = \frac{10 \times 100}{\sqrt{(11.35) \times (55.55)}} = 40/100 \text{ mL}$$

Comment. With the availability of powerful hand calculators, the use of the complete Poisson equation is no longer a major undertaking. However, because of established practice, many analytical laboratories are continuing to use the MPN tables in *Standard Methods* as the basis for reporting MPN values.

Membrane-Filter Technique. The membrane-filter technique can also be used to determine the number of coliform organisms that are present in water. The determination is accomplished by passing a known volume of water sample through a membrane filter that has a very small pore size (see Fig. 3-20). The bacteria are retained on the filter because they are larger than the pores. The bacteria are then contacted with an agar that contains nutrients necessary for the growth of the bacteria. After incubation, the coliform colonies can be counted and the concentration in the

FIGURE 3-20
Membrane-filter apparatus used to test for bacteria in relatively clean wastewaters. After the membrane filter is centered on the filter support, the funnel top is attached and the wastewater sample to be tested is poured into the funnel. To aid in the filtration process, a vacuum line is attached to the base of the filter apparatus. After the sample has been filtered, the membrane filter is placed in a petri dish containing a culture medium and then incubated. After incubation, the bacterial colonies are counted (see Fig. 3-19).

original water sample determined. The membrane-filter technique has the advantage of being faster than the MPN procedure and of giving a direct count of the number of coliforms. Both methods are subject to limitations, however. Detailed procedures are given for both methods in *Standard Methods* [18].

Ratio of Fecal Coliforms to Fecal Streptococci

It has been observed that the quantities of fecal coliforms and fecal streptococci that are discharged by human beings are significantly different from the quantities discharged by animals. Therefore, it has been suggested that the ratio of the fecal coliform (FC) count to the fecal streptococci (FS) count in a sample can be used to show whether the suspected contamination derives from human or from animal wastes. Typical data on the ratio of FC to FS counts for human beings and various animals are reported in Table 3-15. The FC/FS ratio for domestic animals is less than 1.0, whereas the ratio for human beings is more than 4.0.

If ratios are obtained in the range of 1 to 2, interpretation is uncertain. If the sample is collected near the suspected source of pollution, the most likely interpretation is that the pollution derives equally from human and animal sources. The foregoing interpretations are subject to the following constraints [7]:

TABLE 3-15
Estimated per capita contribution of indicator microorganisms from human beings and some animals[a]

| | Average indicator density/g of feces | | Average contribution/capita · 24h | | |
Animal	Fecal coliform, 10^6	Fecal streptococci, 10^6	Fecal coliform, 10^6	Fecal streptococci, 10^6	Ratio FC/FS
Chicken	1.3	3.4	240	620	0.4
Cow	0.23	1.3	5,400	31,000	0.2
Duck	33.0	54.0	11,000	18,000	0.6
Human	13.0	3.0	2,000	450	4.4
Pig	3.3	84.0	8,900	230,000	0.04
Sheep	16.0	38.0	18,000	43,000	0.4
Turkey	0.29	2.8	130	1,300	0.1

[a] Ref. 7.

Note: g × 0.0022 = lb.

1. The sample pH should be between 4 and 9 to exclude any adverse effects of pH on either group of microorganisms.
2. At least two counts should be made on each sample.
3. To minimize errors due to differential death rates, samples should not be taken farther downstream than 24 hours of flow time from the suspected source of pollution.
4. Only the fecal coliform count obtained at 44°C is to be used to compute the ratio.

Use of the FC/FS ratio can be very helpful in establishing the source of pollution in rainfall-runoff studies and in pollution studies conducted in rural areas, especially where septic tanks are used. In many situations where human pollution is suspected on the basis of coliform test results, the actual pollution may, in fact, be caused by animal discharges. Establishing the source of pollution can be very important, especially where it is proposed or implied that the implementation of conventional wastewater management facilities will eliminate the measured coliform values.

Toxicity Tests

Toxicity tests have been used to (1) assess the suitability of environmental conditions for aquatic life; (2) establish acceptable receiving water concentrations for conventional parameters (such as DO, pH, temperature, salinity, or turbidity); (3) study the effects of water quality parameters on wastewater toxicity; (4) assess the toxicity of wastewater to a variety of fresh, estuarine, and marine test species; (5) establish relative sensitivity of a group of standard aquatic organisms to effluent as well as standard toxicants; (6) assess the degree of wastewater treatment needed to meet water pollution control requirements; (7) determine the effectiveness of wastewater treatment methods; (8) establish permissible effluent discharge rates; and (9) determine compliance with federal and state water quality standards and water quality criteria associated with NPDES permits. Such tests provide results that are useful in protecting human health and aquatic life from impacts caused by the release of contaminants into surface waters.

During the past several decades, pollution control measures were focused primarily on conventional pollutants (oxygen-demanding materials, suspended solids, etc.) that were identified as causing water quality degradation. Recently, additional attention has been focused on the control of toxic substances, especially those contained in wastewater treatment plant discharges. The early requirements for monitoring and regulating toxic discharges were on a *chemical-specific* basis. The chemical-specific approach has many shortcomings, including the inability to identify synergistic effects or the bioavailability of the toxin. The more contemporary whole-effluent, or toxicity-based, approach to toxicity control involves the use of toxicity tests to measure the toxicity of treated wastewater discharges. The whole-effluent test procedure is used to determine the aggregate toxicity of unaltered effluent discharged into receiving waters—toxicity is the only parameter measured.

The national policy prohibiting the discharge of toxic pollutants in toxic amounts is documented in Section 101(a) (3) of the federal Clean Water Act. Because it is not economically feasible to determine the specific toxicity of each of the thousands of potentially toxic substances in complex effluents, whole-effluent toxicity testing using aquatic organisms is a direct, cost effective means of determining effluent toxicity. Whole-effluent toxicity testing involves the introduction of appropriate bioassay organisms into test aquariums (see Fig. 3-21) containing various concentrations of the effluent in question and observing their responses. General test procedures, the evaluation of test results and the application of the test results are described in the following discussion.

Toxicity Testing. Toxicity tests are classified according to (1) duration: short-term, intermediate, and/or long-term; (2) method of adding test solutions: static, recirculation, renewal, or flow-through; and (3) purpose: NPDES permit requirements, mixing zone determinations, etc. Detailed contemporary testing protocols are summarized in Refs. 29-32 and 34-35.

Toxicity testing has been widely validated in recent years. Even though organisms vary in sensitivity to effluent toxicity, the EPA has documented that (1) toxicity of effluents correlate well with toxicity measurements in the receiving waters when effluent dilution was measured; and (2) predictions of impacts from both effluent and receiving water toxicity tests compare favorably with ecological community responses in the receiving waters. The U.S. EPA has conducted nationwide tests with freshwater, estuarine, and marine ecosystems. Methods include both acute as well as chronic exposures.

Recent methods for the rapid conduct of bioassays [34] take four to seven days, as opposed to older tests requiring three or more weeks, and involve several different

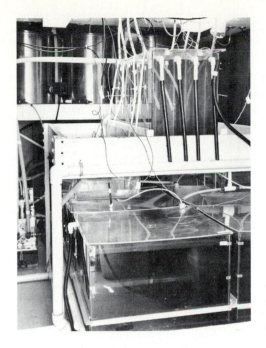

FIGURE 3-21
Laboratory setup for the conduct of flow-through whole-effluent toxicity tests. The device located above the test aquariums is used to achieve the desired wastewater dilutions.

phylogenic groups. The tests are based on species of nearly national distribution for which a large body of life history and toxicity sensitivity data are available. Proper testing protocol involves assessment of a range of sensitivities of test species to a particular effluent. Typically, two or three species are considered to eliminate uncertainty for this factor.

Common marine species include *Champia parvula*, the red alga; *Mysidopsis bahia*, the mysid shrimp; *Menidia beryllina*, the inland silversides; and *Cyrinidon variegatus*, the sheepshead minnow. Common freshwater species include *Pimephales promelas*, fathead minnow; and *Ceriodaphnia dubia*, the daphnid shrimp.

In the red algae test, male and female plants are exposed to a range of effluent concentrations for two days and then incubated in clean seawater. Cystocarps (the product of reproduction) become evident in five to seven days. The number are counted and compared with controls. The acute endpoint relates to no cystocarp production, whereas the chronic endpoint is determined as an impairment of the number of cystocarps formed compared to controls.

The shrimp tests are based on growth, reproduction, and survival. During the 7-day exposure to a range of effluent concentrations, juvenile mysids mature and mate. The acute test endpoint is the death of the shrimp. The chronic endpoint is the presence or absence of eggs in oviduct, and growth (measured as dry weight) of surviving animals at the end of the test.

The fish tests are based on larval growth and survival. Newly hatched fish are exposed to a range of effluents for seven days with daily renewals of test solutions.

At the termination of the test, survival is determined and growth is measured (as an increase in dry weight) compared with control. The acute endpoint is the death of the fish.

Evaluation of Toxicity Test Results. A number of terms are utilized in expressing toxicity test results. *Acute toxicity* is toxicity which is severe enough to produce a response rapidly (typically a response observed in 48 or 96 hours.) Acute means short, and does not necessarily imply mortality. The LC_{50} is the concentration of effluent in dilution water that causes mortality to 50 percent of the test population. The EC_{50} is the effluent concentration that causes a measurable negative effect on 50 percent of the test population. The NOAEL (no observed acute effect level) is defined as the highest tested effluent concentration that causes 10 percent or less mortality. *Chronic toxicity* is the toxicity impact that lingers or continues for a relatively long period of time, often 1/10 of the life span or more. Chronic effects could include mortality, reduced growth, or reduced reproduction. The NOEC (no observable effect concentration) is the highest measured continuous concentration of an effluent or toxicant that causes no observable effect based on the results of chronic testing. The LOEC (lowest observed effect concentration) is defined as the lowest observed concentration having any effect. The LOEC is determined by an analysis of variance techniques.

Toxicity data are analyzed using the procedures developed by Stephan [20]. The LC_{50} values are determined analytically using the Spearman Karber, moving average, binomial, and probit methods. Graphical methods, as illustrated in Example 3-6, can also be used to obtain estimated LC_{50} values. Typically, LC_{50} values are computed based on survival at both 48- and 96-hour exposures. Analysis of variance and Duncan's Multiple Comparison of Means typically are utilized to compare chronic test results.

Toxic Units. The toxic units (TU) approach has become widely accepted for utilizing the toxicity test results. Both federal and state standards and/or criteria have been or are being formulated on the toxic unit basis. In the toxic units approach [30], a TU concentration is established for the protection of aquatic life.

Toxic Unit Acute (TU_a) is the reciprocal of the effluent dilution that caused the acute effect by the end of the acute exposure period.

$$TU_a = 100/LC_{50} \qquad (3\text{-}34)$$

Toxic Unit Chronic (TU_c) is the reciprocal of the effluent dilution that caused no unacceptable effect on the test organisms by the end of the chronic exposure period:

$$TU_c = 100/NOEC \qquad (3\text{-}35)$$

where NOEC is the no observable effect concentration.

Formerly, acute to chronic ratios (ACR) were determined by the equation ACR = LC_{50}/NOEC. The chronic data were determined using extrapolation from acute data. Acute to chronic ratios have been found to vary tremendously between species and between different toxicants. Use of the whole effluent approach prevents implementation of overly stringent as well as overly lenient requirements.

Example 3-6 Analysis of toxicity data. Using the following hypothetical data determine the 48- and 96-hour LC_{50} values in percent by volume.

Concentration of waste, % by volume	No. of test animals	No. of test animals dead after[a]	
		48 h	96 h
40	20	17 (85)	20 (100)
20	20	12 (60)	20 (100)
10	20	6 (30)	14 (70)
5	20	0 (0)	7 (35)
3	20	0 (0)	4 (20)

[a] Percentage values are given in parentheses.

Solution

1. Plot the concentration of wastewater in percent by volume (log scale) against test animals surviving in percent (probability scale), as shown below.

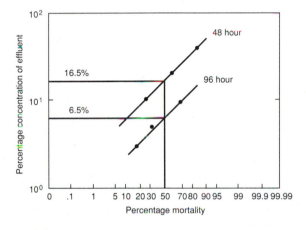

2. Fit a line to the data points by eye, giving most consideration to the points lying between 16 and 84 percent mortality.
3. Find the wastewater concentration causing 50 percent mortality.

   The estimated LC_{50} values, as shown in the above figure, are 16.5 percent for 48 h and 6.5 percent for 96 h.

   Comment. The estimated values of the LC_{50} concentrations obtained graphically are usually quite close to the values obtained with formal prohibit analysis [20]. It should be noted that confidence limits cannot be obtained in a graphical analysis [18].

Application of Toxicity Test Results.

Water quality criteria ensure protection of designated uses by including magnitude (quantity of toxicant allowable), duration (period of time over which instream concentration is averaged), and frequency (how often criteria can be exceeded without unacceptable receiving water ecological community impacts). Contemporary water quality criteria are designed to protect against short-term (acute) effects through use of the Criterion Maximum Concentration (CMC), and against long-term (chronic) effects through use of the Criterion Continous Concentration (CCC). These criteria generally apply after mixing. Typical water quality criteria contain a concentration limit, an averaging period, and a return frequency. The CMC is typically the four-day average concentration not to be exceeded more than once every three years, on the average, and the CCC is typically the one-hour average concentration not to be exceeded more than once every three years, on the average.

Protection against acute toxicity. For protection against acute toxicity, the Criterion Maximum Concentration must not exceed 0.3 acute toxic units (TU_a), as measured by the most sensitive result of the tests conducted.

$$CMC = TU_a/CID \leq 0.3\ TU_a \tag{3-36}$$

In Eq. 3-36, CID is the Critical Initial Dilution. In ocean discharges, the CID is defined as the dilution achieved given "worst case" ambient conditions within the region close to the discharge, where mixing and dilution of the effluent plume are determined by the initial momentum and bouyancy of the discharge. In river discharges, the dilution achieved at the boundary of a mixing zone is usually taken as the CID. Based on the results of numerous 96-hour effluent toxicity tests, it was found that a factor of 0.3 accounted for 91 percent of the observed LC_{50} to LC_1 ratios (LC_1 is equal to the concentration of effluent in the dilution water that causes mortality to 1 percent of the test population). Consequently, for acute protection, the CMC should not exceed 0.3 TU_a based on the most sensitive species tested. The acute criterion thus equals the CMC that approximates the LC_1.

Protection against chronic toxity. The Criterion Continous Concentration (CCC) prevents chronic effects from occuring outside the initial mixing zone, or zone of influence, resulting from the discharge. For chronic protection, the CCC must not exceed 1.0 chronic toxic units (TU_c), based on the results obtained with the most sensitive of at least three species tested.

$$CCC = TU_c/CID \leq 1.0\ Tu_c \tag{3-37}$$

Compliance is based on comparison of the toxicity criteria (both TU_a and TU_c), expressed in Toxicity Units, in the effluent with critical initial dilution to determine if the EPA's recommended criteria will be met. The application of these criteria is illustrated in Example 3-7.

Example 3-7 Application of toxicity test results. A critical initial dilution of 225:1 is achieved for a treated effluent discharged to marine receiving waters. Toxicity tests were conducted with the wastewater treatment plant effluent using three marine species. Based on the acute and chronic toxicity test results given below, it was found that *Champia parvula* exhibited the most sensitive species acute endpoint (2.59% effluent) as measured by the EC_{50}, and also the most sensitive species chronic endpoint (1.0%) as measured by the No Observed

Effect Concentration (NOEC). Using the given toxicity data, determine the compliance with the CMC and CCC criteria.

Results of acute toxicity tests

Species	Exposure, h	Control survival, %	Percent effluent	
			LC$_{50}$ or EC$_{50}$[a]	NOAEL
Mysidopis bahia	96	100	18.66	10.0
Cyprinodon variegatus	96	100	>100	50.0
Champia parvula	48/168	100	2.59	12.25

[a] EC$_{50}$ results based on reduction of cystocarp production.

Results of chronic toxicity tests

Species	Exposure, d	Control survival, %	Percent effluent	
			NOEC	LOEC
Mysidopis bahia	7	82	6.0	10.0
Cyprinodon variegatus	7	98.8	15.0	>15.0
Champia parvula	7	100	1.0	2.25

Solution

1. Check compliance with the CMC criterion.

   (a) Based on data for the most sensitive species tested, the number of acute toxic units (TU$_a$), from Eq. 3-34, is

   $$TU_a = 100/LC_{50} = 100/2.59 = 38.6 \text{ units}$$

   (b) For acute protection, the Criterion Maximum Concentration (CMC) must not exceed 0.3 acute toxic units (Eq. 3-36). Following an initial dilution of 225, the CMC is

   $$CMC = TU_a/CID \leq 0.3\ TU_a$$

   $$38.6/225 \leq 0.3(38.6)$$

   $$0.17 \leq 11.58$$

   The CMC (0.17) is considerably less than the value of 0.3 TU$_a$ (11.58) required for compliance with the CMC criterion.

2. Check compliance with the CCC criterion.

   (a) Based on data for the most sensitive species tested, the number of chronic toxic units (TU$_c$), from Eq. 3-35, is

   $$TU_c = 100/NOEC = 100/1.0 = 100 \text{ units}$$

   (b) For chronic protection, the Criterion Continuous Concentration (CCC) must not exceed 1.0 chronic toxic units (Eq. 3-37). Following an initial dilution of 225, the CCC is

$$CCC = TU_c/CID \leq 1.0\ TU_c$$

$$100/225 \leq 1.0(100)$$

$$0.44 \leq 100$$

The CCC (0.44) is considerably less than the value of 1.0 TU_c (100) required for compliance with the CCC criterion.

In summary, there are a number of advantages to the use of whole effluent toxicity testing. In this approach, the bioavailability of the toxics is measured and the effects of any synergistic interactions are also considered. Because the aggregate toxicity of all components of the wastewater effluent is determined, the toxic effect can be limited by limiting only one parameter, the effluent toxicity. Because contemporary receiving water management strategies are based on site-specific water quality criteria, toxicity testing facilitates comparison of effluent toxicity with site-specific water quality criteria designed to protect representative, sensitive species and allow for establishment of discharge limitations that will protect aquatic environments.

3-5 WASTEWATER COMPOSITION

Composition refers to the actual amounts of physical, chemical, and biological constituents present in wastewater. In this section, data on the constituents found in wastewater and septage are presented. Discussions are also included on the need to characterize wastewater more fully and on the mineral pickup resulting from water use. Variations in the composition of wastewater with time are discussed in Chap. 5.

Constituents in Wastewater and Septage

Typical data on the individual constituents found in domestic wastewater are reported in Table 3-16. Depending on the concentrations of these constituents, wastewater is classified as strong, medium, or weak. Both the constituents and the concentrations vary with the hour of the day, the day of the week, the month of the year, and other local conditions (see Chap. 5). Therefore, the data in Table 3-16 are intended to serve only as a guide and not as a basis for design. Septage is the sludge produced in individual onsite wastewater-disposal systems, principally septic tanks and cesspools. The actual quantities and constituents of septage vary widely. The greatest variations are found in communities that do not regulate the collection and disposal of septage. Some data on the constituents found in septage are given in Table 3-17.

Microorganisms in Wastewater

Representative data on the type and number of microorganisms commonly found in wastewater are reported in Table 3-18. The relatively wide variation in the reported range of values is characteristic of wastewater analyses. It has been estimated that up to 3 or 4 percent of the total coliform group are the pathogenic *E. coli* [18].

TABLE 3-16
Typical composition of untreated domestic wastewater

Contaminants	Unit	Concentration Weak	Medium	Strong
Solids, total (TS)	mg/L	350	720	1200
Dissolved, total (TDS)	mg/L	250	500	850
Fixed	mg/L	145	300	525
Volatile	mg/L	105	200	325
Suspended solids (SS)	mg/L	100	220	350
Fixed	mg/L	20	55	75
Volatile	mg/L	80	165	275
Settleable solids	mL/L	5	10	20
Biochemical oxygen demand, mg/L: 5-day, 20°C (BOD_5, 20°C)	mg/L	110	220	400
Total organic carbon (TOC)	mg/L	80	160	290
Chemical oxygen demand (COD)	mg/L	250	500	1000
Nitrogen (total as N)	mg/L	20	40	85
Organic	mg/L	8	15	35
Free ammonia	mg/L	12	25	50
Nitrites	mg/L	0	0	0
Nitrates	mg/L	0	0	0
Phosphorus (total as P)	mg/L	4	8	15
Organic	mg/L	1	3	5
Inorganic	mg/L	3	5	10
Chlorides[a]	mg/L	30	50	100
Sulfate[a]	mg/L	20	30	50
Alkalinity (as $CaCO_3$)	mg/L	50	100	200
Grease	mg/L	50	100	150
Total coliform[b]	no/100 mL	10^6–10^7	10^7–10^8	10^7–10^9
Volatile organic compounds (VOCs)	μg/L	<100	100–400	>400

[a] Values should be increased by amount present in domestic water supply.
[b] See Table 3-18 for typical values for other microorganisms.
Note: 1.8(°C) + 32 = °F.

Some organisms (shigella, helminth eggs, protozoa cysts) are almost never looked for in a routine analysis. Great care should be exercised when reviewing reported virus values. In recent years, refinements in virus detection and enumeration methods have rendered most early results suspect. Thus, the date of the study is almost as important as the reported concentration values.

Need for Specialized Analyses

In general, the constituents reported in Table 3-16 are those that are analyzed more or less routinely. In the past, it was believed that these constituents were sufficient

TABLE 3-17
Typical characteristics of septage

Constituent	Concentration, mg/L	
	Range	Typical
Total solids (TS)	5,000–100,000	40,000
Suspended solids (SS)	4,000–100,000	15,000
Volatile suspended solids (VSS)	1,200–14,000	7,000
5-day, 20°C BOD_5	2,000–30,000	6,000
Chemical oxygen demand	5,000–80,000	30,000
Total Kjedhal nitrogen (TKN as N)	100–1,600	700
Ammonia, NH_3, as N	100–800	400
Total phosphorus as P	50–800	250
Heavy metals[a]	100–1,000	300

Note: lb × 0.4536 = kg.
[a] Primarily iron (Fe), zinc (Zn), and aluminum (Al)

to characterize a wastewater for biological treatment, but as our understanding of the chemistry and microbiology of wastewater treatment and environmental quality has continued to expand, the importance of analyzing additional constituents is becoming more appreciated [12].

These additional constituents that are now analyzed include many of the metals necessary for the growth of microorganisms such as calcium, cobalt, copper, iron,

TABLE 3-18
Types and numbers of microorganisms typically found in untreated domestic wastewater[a]

Organism	Concentration, number/mL
Total coliform	$10^5–10^6$
Fecal coliform	$10^4–10^5$
Fecal streptococci	$10^3–10^4$
Enterococci	$10^2–10^3$
Shigella	Present[b]
Salmonella	$10^0–10^2$
Pseudomonas aeroginosa	$10^1–10^2$
Clostridium perfringens	$10^1–10^3$
Mycobacterium tuberculosis	Present[b]
Protozoan cysts	$10^1–10^3$
Giardia cysts	$10^{-1}–10^2$
Cryptosporidium cysts	$10^{-1}–10^1$
Helminth ova	$10^{-2}–10^1$
Enteric virus	$10^1–10^2$

[a] Adapted in part from Refs. 3,7.
[b] Results for these tests are usually reported as positive or negative rather than being quantified.

magnesium, manganese, and zinc. The presence or absence of hydrogen sulfide should be determined to assess whether corrosive conditions may develop and whether any trace metals necessary for the growth of microorganisms are being precipitated [36]. The concentration of sulfate should be determined to assess the suitability of anaerobic waste treatment. The presence of filamentous organisms in the wastewater should also be determined, especially if biological treatment is being considered. Priority pollutants must be analyzed to determine if special treatment and control methods will be required to minimize the release of these compounds to the environment.

Mineral Increase Resulting from Water Use

Data on the increase in the mineral content of wastewater resulting from water use and the variation of the increase within a sewage system are especially important in evaluating the reuse potential of wastewater. Increases in the mineral content of wastewater result from domestic use, from the addition of highly mineralized water from private wells and groundwater, and from industrial use. Domestic and industrial water softeners also contribute significantly to the increase in mineral content and, in some areas, may represent the major source. Occasionally, water added from private wells and groundwater infiltration will (because of its high quality) serve to dilute the mineral concentration in the wastewater. Typical data on the incremental increase in mineral content that can be expected in municipal wastewater resulting from domestic use are reported in Table 3-19.

3-6 WASTEWATER CHARACTERIZATION STUDIES

Wastewater characterization studies are conducted to determine (1) the physical, biological, and chemical characteristics, and the concentrations of constituents in the wastewater; and (2) the best means of reducing the pollutant concentrations. Procedures for wastewater sampling, methods for sample analysis, and expressions used to present the results are described in this section; methods for flow measurement are described in Chaps. 2 and 6.

Sampling

The sampling techniques used in a wastewater survey must ensure that representative samples are obtained, because the data from the analysis of the samples will ultimately serve as a basis for designing treatment facilities. There are no universal procedures for sampling; sampling programs must be individually tailored to fit each situation. Special procedures are necessary to handle problems when sampling wastes that vary considerably in composition. Thus suitable sampling locations must be selected, and the frequency and type of sample to be collected must be determined.

Sampling Locations. Examination of drawings that show sewers and manholes will help to determine sampling locations where flow conditions encourage a homogeneous

TABLE 3-19
Typical mineral increase from domestic water use[a]

Constituent	Increment range,[a] mg/L
Anions	
Bicarbonate (HCO_3)	50–100
Carbonate (CO_3)	0–10
Chloride (Cl)	20–50[b]
Nitrate (NO_3)	20–40
Phosphate (PO_4)	5–15
Sulfate (SO_4)	15–30
Cations	
Calcium (Ca)	6–16
Magnesium (Mg)	4–10
Potassium (K)	7–15
Sodium (Na)	40–70
Other constituents	
Aluminum (Al)	0.1–0.2
Boron (B)	0.1–0.4
Fluoride (F)	0.2–0.4
Manganese (Mn)	0.2–0.4
Silica (SiO_2)	2–10
Total alkalinity (as $CaCO_3$)	60–120
Total dissolved solids (TDS)	150–380

[a] Reported values do not include commercial and industrial additions.
[b] Excluding the addition from domestic water softeners.

mixture. In sewers and in deep, narrow channels, samples should be taken from a point one-third the water depth from the bottom. The collection point in wide channels should be rotated across the channel. The velocity of flow at the sample point should, at all times, be sufficient to prevent deposition of solids. When collecting samples, care should be taken to avoid creating excessive turbulence that may liberate dissolved gases and yield an unrepresentative sample.

Sampling Intervals. The degree of flowrate variation dictates the time interval for sampling, which must be short enough to provide a true representation of the flow. Even when flowrates vary only slightly, the concentration of waste products may vary widely. Frequent sampling (10- or 15-minute uniform intervals) allows estimation of the average concentration during the sampling period.

Sampling Equipment. Careful selection of sampling equipment is important if continuous or automatic sampling is appropriate. A typical automatic sampling device is shown in Fig. 3-22. The scope of this chapter does not permit a complete description of the many automatic devices suitable for sampling both domestic and industrial wastewaters. More detailed information may be found in Refs. 2, 11, and 22. A dis-

FIGURE 3-22
Typical automatic sampling device used at wastewater treatment plants.

cussion of precautions to be observed in taking samples and using sampling equipment is presented in Ref. 11.

Sample Preservation

A carefully performed sampling program will be worthless if the physical, chemical, and biological integrity of the samples is not maintained during interim periods between sample collection and sample analysis. Considerable research on the problem of sample preservation has failed to perfect a universal treatment or method or to formulate a set of fixed rules applicable to samples of all types. Prompt analysis is undoubtedly the most positive assurance against error due to sample deterioration. When analytical and testing conditions dictate a lag between collection and analysis, such as when a 24-hour composite sample is collected, provisions must be made for preserving samples. Current methods of sample preservation for the analysis of properties subject to deterioration must be used. Probable errors due to deterioration of the sample should be noted in reporting analytical data.

DISCUSSION TOPICS AND PROBLEMS

3-1. The following test results were obtained for a wastewater sample. The size of the sample was 85 mL. Determine the concentration of total and volatile solids expressed as mg/L.

$$
\begin{aligned}
\text{Tare mass of evaporating dish} &= 22.6435 \text{ g} \\
\text{Mass of evaporating dish plus residue after evaporation at 105°C} &= 22.6783 \text{ g} \\
\text{Mass of evaporating dish plus residue after ignition at 550°C} &= 22.6768 \text{ g}
\end{aligned}
$$

3-2. The suspended solids for a wastewater sample was found to be 175 mg/L. If the following test results were obtained, what size sample was used in the analysis?

$$
\begin{aligned}
\text{Tare mass of glass fiber filter} &= 1.5413 \text{ g} \\
\text{Residue on glass fiber filter after drying at 105°C} &= 1.5538 \text{ g}
\end{aligned}
$$

3-3. The following test results were obtained for a wastewater sample taken at an industrial facility. All of the tests were performed using a sample size of 100 mL. Determine the concentration of total solids, total volatile solids, suspended solids, and dissolved solids.

$$
\begin{aligned}
\text{Tare mass of evaporating dish} &= 52.1533 \text{ g} \\
\text{Mass of evaporating dish plus residue after evaporation at 105°C} &= 52.1890 \text{ g} \\
\text{Mass of evaporating dish plus residue after ignition at 550°C} &= 52.1863 \text{ g} \\
\text{Tare mass of Whatman GF/C filter} &= 1.5413 \text{ g} \\
\text{Residue on Whatman GF/C filter after drying at 105°C} &= 1.5541 \text{ g} \\
\text{Residue on Whatman GF/C filter after ignition at 550°C} &= 1.5519 \text{ g}
\end{aligned}
$$

3-4. The following test results were obtained for a wastewater sample taken at the headworks to a treatment plant. All of the tests were performed using a sample size of 50 mL. Determine the concentration of total solids, total volatile solids, suspended solids, and volatile suspended solids.

$$
\begin{aligned}
\text{Tare mass of evaporating dish} &= 62.003 \text{ g} \\
\text{Mass of evaporating dish plus residue after evaporation at 105°C} &= 62.039 \text{ g} \\
\text{Mass of evaporating dish plus residue after ignition at 550°C} &= 62.036 \text{ g} \\
\text{Tare mass of Whatman GF/C filter} &= 1.540 \text{ g} \\
\text{Residue on Whatman GF/C filter after drying at 105°C} &= 1.552 \text{ g} \\
\text{Residue on Whatman GF/C filter after ignition at 550°C} &= 1.549 \text{ g}
\end{aligned}
$$

3-5. The local Air Pollution Control District has threatened to fine and penalize the local wastewater management agency, your client, because of frequently recurring odor complaints from residents who live downwind of the plant. The plant manager, a full-time employee at the treatment plant, claims that no problem exists. He proves his point by consistently finding less than five dilutions to MDTOC at the plant boundary using a hand-held sniff dilution olfactometer as employed by the local Air Pollution Control District. You, however, live downwind of the plant and have frequently detected odors from it. Why do these differences exist? How would you resolve them objectively?

3-6. You have been asked to review an odor-control system that has apparently failed to adequately control odors from a sludge-dewatering building. The wastewater-management agency, your client, claims the system has failed to perform according to specifications. The engineering contractor who installed the system claims that the specifications were not adequate.

In your investigation you find that the agency employed a reputable odor consultant to develop the odor-control-system specifications. The consultant used the ASTM Panel Method for odor measurement, using evacuated glass cylinders for sample collection. Several measurements were made, and the maximum observed value was doubled to develop the control-system specifications. In this way a 90 percent odor-removal requirement was established to meet the desired final odor-emission limit of 2.8×10^4 odor units per minute (the product of airflow in m³/min and number of dilutions to MDTOC).

Using a direct-reading olfactometer, you find that the control system removes 99 percent of the odor, and that at a rate of 10^6 odor units per minute the final odor emission is 10^6 odor units per minute. What reasons might explain your findings? How would you resolve the problem?

3-7. You have been asked by a wastewater management agency to review the adequacy of their odor-control program. What would be your major considerations in making such a review?

3-8. In a BOD determination, 6 mL of wastewater are mixed with 294 mL of diluting water containing 8.6 mg/L of dissolved oxygen. After a 5-day incubation at 20°C, the dissolved oxygen content of the mixture is 5.4 mg/L. Calculate the BOD of the wastewater. Assume that the initial dissolved oxygen of wastewater is zero.

3-9. The BOD_5 of a waste sample was found to be 40.0 mg/L. The initial oxygen concentration of the BOD dilution water was equal to 9 mg/L, the DO concentration measured after incubation was equal to 2.74 mg/L, and the size of sample used was equal to 40 mL. If the volume of the BOD bottle used was equal to 300 mL, estimate the initial DO concentration in the waste sample.

3-10. What size of sample expressed as a percent is required if the 5-day BOD is 400 mg/L and the total oxygen consumed in the BOD bottle is limited to 2 mg/L?

3-11. A wastewater sample is diluted by a factor of 10 using seeded dilution water. If the following results are obtained, determine the 5-day BOD.

	Dissolved oxygen, mg/L	
Time	Diluted sample	Seeded sample
0	8.55	8.75
1	4.35	8.70
2	4.02	8.66
3	3.35	8.61
4	2.75	8.57
5	2.40	8.53
6	2.10	8.49

3-12. Using the data from Problem 3-11, determine the 4- and 6-day BOD.

3-13. The 5-day 20°C BOD of a wastewater is 210 mg/L. What will be the ultimate BOD? What will be the 10-day demand? If the bottle had been incubated at 30°C, what would the 5-day BOD have been? $k = 0.23$ d^{-1}.

3-14. The 5-day BOD at 20°C is equal to 250 mg/L for three different samples, but the 20°C k values are equal to 0.25 d^{-1}, 0.35 d^{-1}, and 0.46 d^{-1}. Determine the ultimate BOD of each sample.

3-15. The BOD value of a wastewater was measured at 2- and 8-day and found to be 125 and 225 mg/L, respectively. Determine the 5-day value using the first-order rate model.

3-16. The following BOD results were obtained on a sample of untreated wastewater at 20°C:

t, d	0	1	2	3	4	5
y, mg/L	0	65	109	138	158	172

Compute the reaction rate constant k and the ultimate first-stage BOD using both the least-squares and the Fujimoto methods.

3-17. The following BOD results were obtained for a stream sample at 26°C:

t, d	0	1	2	3	4	5	6	7	8	9
y, mg/L	0	3	5.4	7	8.3	9	9.6	9.8	10	10.1

Compute the reaction rate constant k at 20°C and the ultimate first-stage BOD using both the least-squares and the Fujimoto methods.

3-18. Given the following results determined for a wastewater sample at 20°C, determine the ultimate carbonaceous oxygen demand, the ultimate nitrogenous oxygen demand (NOD), the carbonaceous BOD reaction-rate constant (k), and the nitrogenous NOD reaction-rate constant (k_n). Determine $k(\theta = 1.05)$ and $k_n(\theta = 1.08)$ at 25°C.

Time, d	BOD, mg/L	Time, d	BOD, mg/L
0	0	11	63
1	10	12	69
2	18	13	74
3	23	14	77
4	26	16	82
5	29	18	85
6	31	20	87
7	32	25	89
8	33	30	90
9	46	40	90
10	56		

3-19. Compute the carbonaceous and nitrogenous oxygen demand of a waste represented by the formula $C_9N_2H_6O_2$. (N is converted to NH_3 in the first step.)

3-20. Determine the carbonaceous and nitrogenous oxygen demand in mg/L for a 1 L solution containing 300 mg of acetic acid ($CH_3(NH_2)COOH$).

3-21. The following data have been obtained from a waste characterization:

$$BOD_5 = 400 \text{ mg/L}$$
$$k = 0.29 \text{ d}^{-1}$$
$$NH_3 = 80 \text{ mg/L}$$

Estimate the total quantity of oxygen in mg/L that must be furnished to completely stabilize this wastewater. What is the COD and the ThOD for this waste?

3-22. An industrial wastewater is known to contain only stearic acid ($C_{18}H_{36}O_2$), glycine ($C_2H_5O_2N$), and glucose ($C_6H_{12}O_6$). The results of a laboratory analysis are as follows: organic nitrogen = 11 mg/L, organic carbon = 130 mg/L, and COD = 425 mg/L. Determine the concentration of each of the three constituents in mg/L.

3-23. The following analysis was obtained for an industrial wastewater sample. Using this information, estimate the ultimate oxygen demand.

Organic carbon = 300 mg/L
Organic nitrogen = 25 mg/L
Ammonia nitrogen = 15 mg/L
Nitrite nitrogen = 15 mg/L

3-24. How many mg/L of $Cr_2O_7^{-2}$ are consumed if the COD of a wastewater sample is found to 450 mg/L?

3-25. The dissolved oxygen of a tidal estuary must be maintained at 4.5 mg/L or more. The average temperature of the water during summer months is 24°C and the chloride concentration is 5000 mg/L. What percent saturation does this represent?

3-26. Bacteria have equivalent diameters of 2×10^{-6} m and densities of approximately 1 kg/L. Under optimal conditions, bacteria can divide every 30 min. Determine the mass of bacteria that would accumulate in 72 hr under continuing optimal growth conditions. Can this occur? Explain.

3-27. If the bacteria found in feces have an average volume of 2.0 μm^3, determine the concentration of suspended solids that would be represented by a bacterial density equal to 10^8 organisms/mL. Assume the density of the bacteria is 1.05 kg/L.

3-28. Derive, from fundamental considerations, an expression that can be used to compute the MPN based on a single sample comprised of five fermentation tubes.

3-29. A single coliform test was conducted using five-10mL portions. If two of the five tubes are positive, what is the MPN per 100 mL?

3-30. A single coliform test was conducted using ten-10mL portions. If seven of the ten tubes are positive, what is the MPN per 100mL?

3-31. Six weekly effluent samples have been analyzed for bacterial content using the standard confirmed test. Determine the coliform density, expressed as MPN, for the first three weekly samples using the Poisson equation (Eq. 3-32). Check the answers obtained using the standard MPN tables and the Thomas equation (Eq. 3-33).

Size of portion, mL	Sample number					
	1	2	3	4	5	6
100.0			5/5	5/5	5/5	5/5
10.0		4/5	4/5	5/5	5/5	5/5
1.0	4/5	5/5	5/5	5/5	5/5	5/5
0.1	3/5	3/5	3/5	2/5	1/5	5/5
0.01	1/5	2/5	2/5	3/5	2/5	5/5
0.001					0/5	1/5

3-32. Using the data given in Prob. 3-31, determine the coliform density, expressed as MPN, for the fourth, fifth, and sixth weekly samples using the Poisson equation (Eq. 3-32). Check the answers obtained using the standard MPN tables and with the Thomas equation (Eq. 3-33).

3-33. Discuss the advantages and disadvantages of using the fecal coliform and fecal streptococci tests to indicate bacteriological pollution.

3-34. The following bioassay data were obtained for a treatment effluent using fathead minnows. Determine the 60- and 96-h LC_{50} values.

Concentration of waste, % by volume	No. of test animals	No. of test animals surviving		
		After 24 h	After 60 h	After 96 h
12	20	8	2	0
10	20	10	5	0
8	20	13	8	0
6	20	16	11	0
4	20	20	16	5
2	20	20	20	14

3-35. Using the bioassay data given in Prob. 3-34, estimate the 48-h LC_{50} value.

3-36. Obtain wastewater composition data for your community. How does the data from your community compare with the data reported in Table 3-16? How would you classify the strength of your community's wastewater?

REFERENCES

1. Corsi, R. L., D. P. Y. Chang, E. D. Schroeder, and Q. Qiu: "Emissions of Volatile and Potentially Toxic Organic Compounds From Municipal Wastewater Treatment Plants," presented at the 80th Annual Meeting of APCA, New York, June 21–26, 1987.
2. *Determination Of Odor And Taste Thresholds By A Forced Choice Ascending Concentration Series Method Of Limits,* ASTM, E679-79.
3. Feachem, R. G., D. J. Bradley, H. Garelick, and D. D. Mara: *Sanitation And Disease: Health Aspects of Excreta and Wastewater Management,* Published for the World Bank by John Wiley & Sons, New York, 1983.
4. *Federal Register,* 46 CRF, 2264, January 8, 1981.
5. Fujimoto, Y.: "Graphical Use of First-Stage BOD Equation," *Journal WPCF,* vol. 36, no. 1, 1961.
6. Levine, A. D., G. Tchobanoglous, and T. Asano: "Characterization of the Size Distribution of Contaminants in Wastewater: Treatment and Reuse Implications," *Journal WPCF,* vol. 57, no. 7, 1985.
7. Mara, D. D.: *Bacteriology for Sanitary Engineers,* Churchill Livingston, Edinburgh, 1974.
8. McCarty, P. L.: "Anaerobic Waste Treatment Fundamentals," *Public Works,* vol. 95, no. 11, 1964.
9. Metcalf & Eddy, Inc.: *Collection and Pumping of Wastewater,* McGraw-Hill, New York, 1981.
10. Metcalf & Eddy, Inc.: *Wastewater Engineering: Treatment, Disposal, Reuse,* 2nd ed., McGraw-Hill, New York, 1979.
11. Moore, E. W., H. A. Thomas, and W. B. Snow: "Simplified Method for Analysis of BOD Data," *Sewage Ind. Wastes,* vol. 22, no. 10, 1950.
12. Parker, D. S.: "Wastewater Technology Innovation For The Year 2000," *Journal of Environmental Engineering, ASCE,* vol. 114, no. 3, pp. 487–506, June, 1988.
13. Patterson, R. G., R. C. Jain, and S. Robinson: "Odor Controls For Sewage Treatment Facilities," presented at the 77th Annual Meeting of the Air Pollution Control Association, San Francisco, June, 1984.
14. Phelps, E. B.: *Stream Sanitation,* Wiley, New York, 1944.
15. Schroepfer, G. J., M. L. Robins, and R. H. Susag: The Research Program on the Mississippi River in the Vicinity of Minneapolis and St. Paul, *Advances in Water Pollution Research,* vol.1, Pergamon, London, 1964.
16. Sheehy, J. P.: "Rapid Methods for Solving Monomolecular Equations," *Journal WPCF,* vol. 32, no. 6, 1960.
17. Snoeyink, V. L., and D. Jenkins, *Water Chemistry,* 2nd ed., John Wiley & Sons, New York, 1988.
18. *Standard Methods for the Examination of Water and Waste Water,* 17th ed., American Public Health Association, 1989.
19. Stanier, R. Y., J. L. Ingraham, M. L. Wheelis, and P. R. Painter: *The Microbial World,* 5th ed., Prentice-Hall, Englewood Cliffs, NJ, 1986.
20. Stephen, C. E.: "Methods for Calculating an LC_{50}," F. L. Mayer and J. L. Hamelink, eds: *Aquatic Toxicology and Hazard Evaluation,* ASTM STP 634, American Society for Testing and Materials, Philadelphia, pp. 65–84, 1982.
21. Stumm, W., and J. J. Morgan: *Aquatic Chemistry,* 2nd ed., Wiley-Interscience, New York, 1970.
22. Tchobanoglous, G., and R. Eliassen: "The Indirect Cycle of Water Reuse," *Water Wastes Eng.,* vol. 6, no. 2, 1969.
23. Tchobanoglous, G., and E. D. Schroeder: *Water Quality: Characteristics, Modeling, Modification,* Addison-Wesley, Reading, MA, 1985.

24. Thibodeaux, L. J.: *Chemodynamics: Environmental Movement Of Chemicals In Air, Water, and Soil,* John Wiley & Sons, New York, 1979.

25. Thomas, H. A., Jr.: "Bacterial Densities From Fermentation Tube Tests," *Journal AWWA,* vol. 34, no. 4, 1942.

26. Thomas, H. A., Jr.: "Graphical Determination of BOD Curve Constants," *Water & Sewage Works,* vol. 97, p.123, 1950.

27. Tsivoglou, E. C.: *Oxygen Relationships in Streams,* Robert A. Taft Sanitary Engineering Center, Technical Report W-58-2, 1958.

28. U.S. Environmental Protection Agency, *National Interim Primary Drinking Water Regulations,* EPA-570/9-76-003, Washington, DC, 1976.

29. U.S. Environmental Protection Agency, *Methods for Measuring the Acute Toxicity of Effluents to Freshwater and Marine Organisms,* U.S. EPA Environmental Monitoring and Support Laboratory, EPA-600/4-85/013, Cincinnati, OH 1985.

30. U.S. Environmental Protection Agency, *Technical Support Document for Water Quality-Based Toxics Control,* U.S. EPA Office of Water, EPA-440/4-85/032, Washington, DC, 1985.

31. U.S. Environmental Protection Agency: *Short Term Methods for Estimating Chronic Toxicity of Effluents and Receiving Waters to Freshwater Organisms,* EPA-660/4-85/014, 1985.

32. U.S. Environmental Protection Agency: *User's Guide to the Conduct and Interpretation of Complex Effluent Toxicity Tests at Estuarine/Marine Sites,* EPA-600/X-86/224, Washington, DC, 1985.

33. U.S. Environmental Protection Agency: Design Manual-Odor and Corrosion Control in Sanitary Sewerage Systems and Treatment Plants, EPA 1600/1-85/018, 1985.

34. U.S. Environmental Protection Agency: *Short Term Methods for Estimating the Chronic Toxicity of Effluents and Receiving Waters to Marine and Estuarine Organisms,* EPA-600/4-88/028, 1988.

35. U.S. Environmental Protection Agency: *Short Term Methods for Estimating Chronic Toxicity of Effluents and Receiving Waters to Freshwater Organisms,* EPA-660/2nd. ed., 1989.

36. Wood D. K., and G. Tchobanoglous: "Trace Elements in Biological Waste Treatment," *Journal WPCF,* vol. 47, no. 7, 1975.

37. Young, J. C.: "Chemical Methods for Nitrification Control," *Journal WPCF,* vol. 45, no. 4, 1973.

38. Young, J. C., and E. R. Baumann: "The Electrolytic Respirometer—I: Factors Affecting Oxygen Uptake Measurements," *Water Res.,* vol. 10, no. 11, 1976.

39. Young, J. C., and E. R. Baumann: "The Electrolytic Respirometer—II: Use in Water Pollution Control Plant Laboratories," *Water Res.,* vol. 10, no. 12, 1976.

CHAPTER
4

WASTEWATER TREATMENT OBJECTIVES, METHODS, AND IMPLEMENTATION CONSIDERATIONS

Since the early 1900s, when the field of environmental engineering was in its infancy in the United States, there has been a steady evolution and development in the methods used for wastewater treatment. Descriptions of the many methods and variations that have been tried to date would fill several large volumes. The approach followed in this text is to identify and discuss basic principles and their application to wastewater treatment.

This introductory chapter is intended to provide perspective and to illustrate how the subject matter to be presented in the following chapters fits into the overall scheme of the design, construction, operation and maintenance, implementation, and financing of wastewater treatment facilities. The following topics are covered: (1) wastewater treatment objectives and regulations, (2) classification of wastewater treatment methods, (3) application of methods for the treatment of wastewater and sludge, (4) selection of the treatment process, (5) implementation of wastewater management programs, and (6) financing.

4-1 WASTEWATER TREATMENT OBJECTIVES AND REGULATIONS

As noted in Chap. 1, methods of wastewater treatment were first developed in response to the concern for public health and the adverse conditions caused by the discharge of wastewater to the environment. Also important, as cities became larger in the United States, was the limited availability of land required for wastewater treatment and disposal, principally by irrigation and intermittent filtration, methods commonly used in the early 1900s [15]. The purpose of developing other methods of treatment was to accelerate the forces of nature under controlled conditions in treatment facilities of comparatively smaller size.

In general, from about 1900 to the early 1970s, treatment objectives were concerned with (1) the removal of suspended and floatable material, (2) the treatment of biodegradable organics, and (3) the elimination of pathogenic organisms. Unfortunately, these objectives were not uniformly met throughout the United States, as is evidenced by the many plants that were discharging partially treated wastewater well into the 1960s.

From the early 1970s to about 1980, wastewater treatment objectives were based primarily on aesthetic and environmental concerns. The earlier objectives of BOD, suspended solids, and pathogenic organisms reduction continued but at higher levels. Removal of nutrients such as nitrogen and phosphorus also began to be addressed, particularly in some of the inland streams and lakes. A major effort was undertaken by both state and federal agencies to achieve more effective and widespread treatment of wastewater to improve the quality of the surface waters. This effort resulted in part from (1) an increased understanding of the environmental effects caused by wastewater discharges; (2) a developing knowledge of the adverse long-term effects caused by the discharge of some of the specific constituents found in wastewater; (3) the development of national concern for environmental protection. The result of these efforts was a significant improvement in the quality of the surface waters.

Since 1980, because of increased scientific knowledge and an expanded information base, wastewater treatment has begun to focus on the health concerns related to toxic and potentially toxic chemicals released to the environment. The water-quality improvement objectives of the 1970s have continued, but the emphasis has shifted to the definition and removal of toxic and trace compounds that may cause long-term health effects. As a consequence, while the early treatment objectives remain valid today, the required degree of treatment has increased significantly, and additional treatment objectives and goals have been added. The removal of toxic compounds, such as refractory organics and heavy metals identified in Table 3-2, are examples of additional treatment objectives that are being considered. Therefore, treatment objectives must go hand-in-hand with the water quality objectives or standards established by the federal, state, and regional regulatory authorities.

Current Regulatory Environment

A significant event in the field of wastewater management was the passage of the Federal Water Pollution Control Act Amendments of 1972 (Public Law 92-500) often referred to as the Clean Water Act (CWA). Before that date, there were no specific national water pollution control goals or objectives. The CWA not only established national goals and objectives ("to restore and maintain the chemical, physical, and biological integrity of the Nation's waters") but also marked a change in water pollution control philosophy. No longer was the classification of the receiving stream of ultimate importance, as it had been before. It was decreed in the CWA that the quality of the nation's waters was to be improved by the imposition of specific effluent limitations. A National Pollution Discharge Elimination System (NPDES) program was established based on uniform technological minimums with which each point source discharger had to comply. To date, over 60,000 permits have been issued under the NPDES program [12].

Pursuant to Section 304(d) of Public Law 92-500, the U.S. Environmental Protection Agency published its definition of secondary treatment. This definition, originally issued in 1973, was amended in 1985 to allow for additional flexibility in applying the percent removal requirements of pollutants to treatment facilities serving separate sewer systems. The current definition of secondary treatment is reported in Table 4-1 [6,7]. The definition of secondary treatment includes three major effluent parameters: 5-day BOD, suspended solids, and pH. The substitution of 5-day carbonaceous BOD ($CBOD_5$) for BOD_5 may be made at the option of the NPDES permitting authority. Special interpretations of the definition of secondary treatment are permitted for publicly owned treatment works (1) served by combined sewer systems, (2) using waste stabilization ponds and trickling filters, (3) receiving industrial flows, or (4) receiving less concentrated influent wastewater from separate sewers. The secondary treatment regulations were amended further in 1989 to clarify the percent removal requirements during dry periods for treatment facilities served by combined sewers [7].

TABLE 4-1
Minimum national standards for secondary treatment [a,b]

Characteristic of discharge	Unit of measurement	Average 30-day concentration	Average 7-day concentration
BOD_5	mg/L	30 [c,d]	45 [c]
Suspended solids	mg/L	30 [c,d]	45 [c]
Hydrogen-ion concentration	pH units	Within the range of 6.0 to 9.0 at all times [e]	
$CBOD_5$ [f]	mg/L	25 [c,d]	40 [c]

[a] Refs. 6 and 7.

[b] Present standards allow stabilization ponds and trickling filters to have higher 30-day average concentrations (45 mg/L) and 7-day average concentrations (65 mg/L) BOD/suspended solids performance levels as long as the water quality of the receiving water is not adversely affected. Exceptions are also permitted for combined sewers, certain industrial categories, and less-concentrated wastewater from separate sewers. For precise requirements of exceptions, Ref. 6 should be consulted.

[c] Not to be exceeded.

[d] Average removal shall not be less than 85 percent.

[e] Only enforced if caused by industrial wastewater or by in-plant inorganic chemical addition.

[f] May be substituted for BOD_5 at the option of the NPDES permitting authority.

In 1987, Congress enacted the Water Quality Act of 1987 (WQA), the first major revision of the Clean Water Act. Important provisions of the WQA are (1) the strengthening of federal water quality regulations by providing changes in permitting and adding substantial penalties for permit violations, (2) significantly amending the CWA's formal sludge control program by emphasizing the identification and regulation of toxic pollutants in sludge, (3) providing funding for state and EPA studies for defining non-point and toxic sources of pollution, (4) establishing new deadlines for compliance including priorities and permit requirements for stormwater, and (5) a phase-out of the construction grants program as a method of financing publicly owned treatment works (POTW). Further discussion on financing wastewater treatment facilities is included in Sec. 4-6. For a summary of the provisions of the WQA, Ref. 13 may be consulted. Other useful references are Ref. 18, for wastes in the marine environment, and Ref. 22, for legal considerations, legislative history, and implementation.

In response to the provisions of the Water Quality Act, new regulations have been promulgated or proposed for controlling the disposal of sludge from wastewater treatment plants. The Ocean Dumping Ban Act of 1988 prohibits any dumping of wastewater sludge into ocean waters. In 1989, the EPA proposed new standards for the disposal of sludge from wastewater treatment plants [3,8]. The proposed regulations establish pollutant numerical limits and management practices for (1) application of sludge to agricultural and non-agricultural land, (2) distribution and marketing, (3) monofilling or surface disposal, and (4) incineration. The proposed regulations are under review at the time of writing of this text (1989). Additional regulations may be promulgated or current regulations amended as additional information becomes available on pollutants in wastewater sludge. The current regulatory environment for sludge disposal is in a state of flux; therefore, the design engineer must be aware of the current regulations and proposed changes when planning and designing sludge disposal facilities.

Trends in Regulations

Regulations are always subject to change as more information becomes available regarding the characteristics of wastewater, effectiveness of treatment processes, and environmental effects. It is anticipated that the focus of future regulations will be on the implementation of the Water Quality Act of 1987. Receiving the most attention will be control of the pollutional effects of stormwater and nonpoint sources, toxics in wastewater (priority pollutants), and as noted above the overall management of sludge, including the control of toxic substances. Nutrient removal, the control of pathogenic organisms, and the removal of organic and inorganic substances such as VOCs and total dissolved solids will also continue to receive attention in specific applications.

Other Regulatory Considerations

In addition to the requirements established under the 1987 Water Quality Act and enforced by the U.S. Environmental Protection Agency, other federal, state, and local regulations have to be considered in the planning, design, construction, and operation of wastewater treatment plants. Significant federal regulations include those prescribed by the Occupational Safety and Health Act (OSHA), which deals with safety provisions to be included in the facilities' design. State, regional, and local regulations may include water quality standards for the protection of public health and the beneficial uses of the receiving waters, air quality standards for the regulation of air emissions (including odor) from treatment facilities, and regulations for the disposal and reuse of sludge. Because all of these guidelines and regulations affect the design of wastewater treatment and disposal facilities, the practicing engineer must be thoroughly familiar with them and their interpretation and be aware of contemplated changes. Contemplated changes and current interpretations of the regulatory aspects of water pollution control are summarized in various weekly publications [4,5].

4-2 CLASSIFICATION OF WASTEWATER TREATMENT METHODS

After treatment objectives have been established for a specific project and the applicable state and federal regulations have been reviewed, the degree of treatment can be determined by comparing the influent wastewater characteristics to the required effluent wastewater characteristics. A number of different treatment and disposal or reuse alternatives are then developed and evaluated, and the best alternative is selected. It will therefore be helpful at this point to review the classification of the methods used for wastewater treatment (mentioned briefly in Chap. 1) and to consider the application of these methods in achieving treatment objectives.

The contaminants in wastewater are removed by physical, chemical, and biological means. The individual methods usually are classified as physical unit operations, chemical unit processes, and biological unit processes. Although these operations and processes occur in a variety of combinations in treatment systems, it has been found advantageous to study their scientific basis separately because the principles involved do not change.

Physical Unit Operations

Treatment methods in which the application of physical forces predominate are known as physical unit operations. Because most of these methods evolved directly from man's first observations of nature, they were the first to be used for wastewater treatment. Screening, mixing, flocculation, sedimentation, flotation, filtration, and gas transfer are typical unit operations. These methods are considered in detail in Chap. 6, and their application is discussed in Chap. 9.

Chemical Unit Processes

Treatment methods in which the removal or conversion of contaminants is brought about by the addition of chemicals or by other chemical reactions are known as chemical unit processes. Precipitation, adsorption, and disinfection are the most common examples used in wastewater treatment. In chemical precipitation, treatment is accomplished by producing a chemical precipitate that will settle. In most cases, the settled precipitate will contain both the constituents that may have reacted with the added chemicals and the constituents that were swept out of the wastewater as the precipitate settled. Adsorption involves the removal of specific compounds from the wastewater on solid surfaces using the forces of attraction between bodies. Chemical unit processes are considered in detail from a theoretical standpoint in Chap. 7, and their application is also discussed in Chap. 9.

Biological Unit Processes

Treatment methods in which the removal of contaminants is brought about by biological activity are known as biological unit processes. Biological treatment is used primarily to remove the biodegradable organic substances (colloidal or dissolved) in wastewater. Basically, these substances are converted into gases that can escape to the atmosphere and into biological cell tissue that can be removed by settling. Biological treatment is also used to remove nutrients (nitrogen and phosphorus) in wastewater. With proper environmental control, wastewater can be treated biologically in most cases. Therefore, the engineer should provide the proper environment so that the process can operate effectively. The fundamental principles of biological treatment are discussed in Chap. 8, and their application is discussed in Chap. 10. Biological nutrient removal is discussed in Chap. 11.

4-3   APPLICATION OF TREATMENT METHODS

The principal methods now used for the treatment of wastewater and sludge are identified in this section. Detailed descriptions of each method are not presented because the purpose here is only to introduce the many different ways in which treatment can be accomplished. The detailed descriptions are presented throughout the remainder of this book.

Wastewater Processing

It is noted in Chap. 1 that unit operations and processes are grouped together to provide various levels of treatment. Historically, the term "preliminary" and/or "primary" referred to physical unit operations; "secondary" referred to chemical and biological unit processes; and "advanced" or "tertiary" referred to combinations of all three. These terms are arbitrary, however, and in most cases of little value. A more rational approach is first to establish the level of contaminant removal (treatment) required before the wastewater can be reused or discharged to the environment. The required unit operations and processes necessary to achieve that required level of treatment can then be grouped together on the basis of fundamental considerations.

The contaminants of major interest in wastewater and the unit operations, processes, or methods applicable to the removal of these contaminants are shown in Table 4-2. Application of these operations, processes, and methods to perform specific functions is described in the following paragraphs.

TABLE 4-2

Unit operations and processes and wastewater treatment systems used to remove major contaminants found in wastewater

Contaminant	Unit operation, unit process, or treatment system	See Chapter
Suspended solids	Screening and comminution	6, 9
	Grit removal	6, 9
	Sedimentation	6, 9
	Filtration	6, 9
	Flotation	6, 9
	Chemical polymer addition	7, 9
	Coagulation/sedimentation	7, 9
	Natural systems (land treatment)	13
Biodegradable organics	Activated-sludge variations	8, 10
	Fixed-film reactor: trickling filters	8, 10
	Fixed-film reactor: rotating biological contactors	8, 10
	Lagoon variations	8, 10, 14
	Intermittent sand filtration	6, 14
	Physical-chemical systems	7, 9
	Natural systems	13
Volatile organics	Air stripping	6
	Off gas treatment	9
	Carbon adsorption	7, 11
Pathogens	Chlorination	7, 9
	Hypochlorination	7
	Bromine chloride	7
	Ozonation	7
	UV radiation	7
	Natural systems	13
Nutrients:		
Nitrogen	Suspended-growth nitrification and denitrification variations	11
	Fixed-film nitrification and denitrification variations	11
	Ammonia stripping	11
	Ion exchange	11
	Breakpoint chlorination	7, 11
	Natural systems	13
Phosphorus	Metal-salt addition	7, 11
	Lime coagulation/sedimentation	7, 11
	Biological phosphorus removal	8, 11
	Biological-chemical phosphorus removal	7, 11
	Natural systems	13
Nitrogen and Phosphorus	Biological nutrient removal	8, 11
Refractory organics	Carbon adsorption	6, 9
	Tertiary ozonation	11
	Natural systems	13
Heavy metals	Chemical precipitation	7, 9
	Ion exchange	11
	Natural systems	13
Dissolved organic solids	Ion exchange	11
	Reverse osmosis	11
	Electrodialysis	11

Preliminary Wastewater Treatment. Preliminary wastewater treatment is defined as the removal of wastewater constituents that may cause maintenance or operational problems with the treatment operations, processes, and ancillary systems. Examples of preliminary operations are screening and comminution for the removal of debris and rags, grit removal for the elimination of coarse suspended matter that may cause wear or clogging of equipment, and flotation for the removal of large quantities of oil and grease. Preliminary treatment in this text is distinguished from industrial pretreatment, where constituents are treated at their source before discharge to the sewer system.

Primary Wastewater Treatment. In primary treatment, a portion of the suspended solids and organic matter is removed from the wastewater. This removal is usually accomplished with physical operations such as screening and sedimentation. The effluent from primary treatment will ordinarily contain considerable organic matter and will have a relatively high BOD. Treatment plants using only primary treatment will be phased out in the future as implementation of the EPA secondary treatment requirements is completed. Only in rare instances (for those communities having a secondary treatment waiver) will primary treatment be used as the sole method of treatment. The principal function of primary treatment will continue to be as a precursor to secondary treatment.

Conventional Secondary Wastewater Treatment. Secondary treatment is directed principally toward the removal of biodegradable organics and suspended solids. Disinfection is included frequently in the definition of conventional secondary treatment. Conventional secondary treatment is defined as the combination of processes customarily used for the removal of these constituents and includes biological treatment by activated sludge, fixed-film reactors, or lagoon systems and sedimentation.

Nutrient Removal or Control. The removal or control of nutrients in wastewater treatment is important for several reasons. Nutrient removal or control is generally required for (1) discharges to confined bodies of water where eutrophication may be caused or accelerated, (2) discharges to flowing streams where nitrification can tax oxygen resources or where rooted aquatic plants can flourish, and (3) recharge of groundwaters that may be used indirectly for public water supplies. The nutrients of principal concern are nitrogen and phosphorus and may be removed by biological, chemical, or a combination of processes. In many cases, the nutrient removal processes are coupled with secondary treatment; for example, metal salts may be added to the aeration tank mixed liquor for the precipitation of phosphorus in the final sedimentation tanks, or biological denitrification may follow an activated sludge process that produces a nitrified effluent.

Advanced Wastewater Treatment/Wastewater Reclamation. Advanced wastewater treatment is a term that has many definitions. In the context of this book, advanced wastewater treatment is defined as the level of treatment required beyond conventional secondary treatment to remove constituents of concern including nutrients, toxic compounds, and increased amounts of organic material and suspended solids. In addition to the nutrient removal processes, unit operations or processes frequently employed in advanced wastewater treatment are chemical coagulation, flocculation, and sedimentation followed by filtration and activated carbon. Less used processes include ion exchange and reverse osmosis for specific ion removal or for the reduction in dissolved solids. Advanced wastewater treatment is also used in a variety of reuse applications where a high quality of water is required such as for industrial cooling water and groundwater recharge (see Chap. 16). Some form of natural treatment (formerly termed land treatment) may also be equivalent to advanced wastewater treatment in terms of effluent quality (see Chap. 13).

Toxic Waste Treatment/Specific Contaminant Removal. The removal of toxic substances and specific contaminants is a complex subject and is covered only generally in this book. For industrial waste discharges to municipal collection and treatment systems, the concentrations of toxic pollutants are usually controlled by pretreatment prior to discharge to the municipal system. In some cases, removal of toxic substances is done at the municipal treatment facilities. Many toxic substances such as heavy metals are reduced by some form of chemical-physical treatment such as chemical coagulation, flocculation, sedimentation, and filtration. Some degree of removal is also accomplished by conventional secondary treatment. Wastewaters containing volatile organic constituents may be treated by air stripping or by carbon adsorption. Small concentrations of specific contaminants may be removed by ion exchange.

Treatment of Combined Sewer Overflows. Combined sewer overflows consist of large, intermittent discharges of wastewater resulting from the mixture of stormwater with wastewater. Except for the initial runoff, or "first flush," the concentrations of the constituents of concern are relatively dilute when compared to wastewater from domestic and industrial sources. The treatment systems required for combined sewer overflows usually focus on the removal of suspended solids and pathogens. Suspended solids removal may be accomplished by grit removal and/or sedimentation. Disinfection customarily is done by chlorination. Combined sewer overflows are covered in Chap. 15.

Sludge Processing

For the most part, the methods and systems reported in Table 4-2 are used to treat the liquid portion of the wastewater. Of equal if not of more importance in the overall design of treatment facilities are the corresponding unit operations and processes or systems used to process the sludge removed from the liquid portion of the wastewater. The principal methods now in use are reported in Table 4-3. Because the processing and treatment of sludge has become so specialized, Chap. 12 is devoted entirely to this subject.

TABLE 4-3
Sludge-processing and disposal methods

Processing or disposal function	Unit operation, unit process, or treatment method	See Chapter
Preliminary operations	Sludge pumping	12
	Sludge grinding	12
	Sludge blending and storage	12
	Sludge degritting	12
Thickening	Gravity thickening	6, 12
	Flotation thickening	6, 12
	Centrifugation	12
	Gravity belt thickening	12
	Rotary drum thickening	12
Stabilization	Lime stabilization	12
	Heat treatment	12
	Anaerobic digestion	8, 12
	Aerobic digestion	8, 12
	Composting	12
Conditioning	Chemical conditioning	12
	Heat treatment	12
Disinfection	Pasteurization	12
	Long-term storage	12
Dewatering	Vacuum filter	12
	Centrifuge	12
	Belt press filter	12
	Filter press	12
	Sludge drying beds	12
	Lagoons	12
Heat drying	Dryer variations	12
	Multiple effect evaporator	12
Thermal reduction	Multiple hearth incineration	12
	Fluidized bed incineration	12
	Co-incineration with solid wastes	12
	Wet air oxidation	12
	Vertical deep well reactor	12
Ultimate disposal	Land application	12
	Distribution and marketing	12
	Landfill	12
	Lagooning	12
	Chemical fixation	12

4-4 SELECTION OF TREATMENT-PROCESS FLOW DIAGRAMS

Treatment plant design is one of the most challenging aspects of environmental engineering. Both theoretical knowledge and practical experience are necessary in the selection and analysis of the treatment-process flow diagrams. Process flow diagrams are graphical representations of combinations of unit operations and processes used to achieve specific treatment objectives. Examples of process flow diagrams are shown in Fig. 4-1. Practical experience is especially important in the design and layout of the physical facilities and their appurtenances and in the preparation of plans and specifications. The purpose of this section is to describe the major elements involved in the selection of the process flow diagrams. These elements are (1) needs of the owner of the facilities, (2) past experience, (3) regulatory agency requirements, (4) process analysis and selection, (5) compatibility with existing facilities, (6) cost considerations, (7) environmental considerations, and (8) other important considerations such as equipment, personnel, and energy.

Owner Needs

A factor often overlooked in the selection of a treatment process is the needs of the owner of the facilities. Owner needs may take the form of limitations of cost and the ability to pay for the project, operating capabilities where existing staff will be utilized, process preferences based upon personal experience, concerns about using proven processes or equipment and not experimenting, and considerations about possible environmental impacts. As will be discussed in Chap. 14, owner needs are especially important in small communities where there is no past history of construction and operation of treatment systems. For projects both large and small, it is important for the design engineer and the owner to reach an understanding about their mutual goals and objectives so that the needs of the owner are satisfied and the selected treatment process meets the basic purposes for its selection (i.e., meeting waste discharge regulations in the most cost-effective manner and mitigating adverse environmental impacts).

Past Experience

Increased emphasis is being placed on treatment plant performance and reliability in order to meet consistently more stringent wastewater discharge standards. Past experience in the design and operation of wastewater treatment systems is important in process selection so that the capabilities and limitations of various processes and their support systems can be assessed realistically. Dealing with known performance eliminates many of the uncertainties of design and prevents major miscalculations in terms of inadequate design. Information about performance, maintenance problems, ease or difficulty of control, and adaptability to changing conditions can be obtained from operating systems. Because new processes and systems do not have a history of performance, they have to be examined carefully through a series of progressive evaluations. With an increasingly stringent regulatory environment, the risks associated with a process have to be assessed carefully before the final selection is made.

Regulatory Agency Requirements

Many of the state and regional regulatory agencies not only establish the permit requirements for wastewater discharges but also issue design guidelines for specific

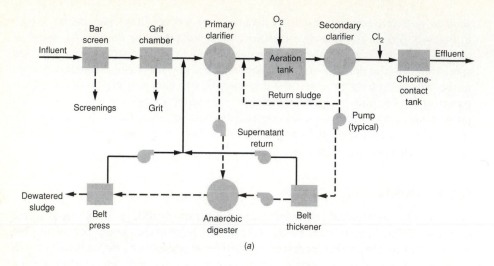

(a)

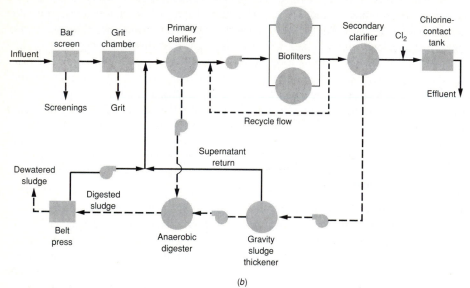

(b)

Legend:
——— wastewater
– – – sludge or solids

FIGURE 4-1
Examples of typical alternative treatment process flow diagrams: (a) activated sludge and (b) trickling filter.

processes. Well-known design standards include the so-called "Ten States Standards," published by the Great Lakes-Upper Mississippi River Board of State Sanitary Engineers [11], and the "Guides for the Design of Wastewater Treatment Works," published by the New England Interstate Water Pollution Control Commission [17]. For each state, the requirements of the regulatory agencies, including those having jurisdiction for public health, air quality, and solid waste management, have to be carefully investigated. The design engineer should review these requirements carefully and involve the state planning or regulatory agency in the early stages of a project to ensure conformance with the requirements.

Process Selection

Process analysis and selection is one of the most challenging aspects of treatment plant design. Both theoretical knowledge and practical experience are necessary in the consideration and evaluation of process alternatives. The principal elements of process analysis, discussed in detail in Chap. 5, include (1) development of the process flow diagram, (2) the establishing of process design criteria and sizing treatment units, (3) preparation of solids balances, (4) evaluation of the hydraulic requirements (hydraulic profile), and (5) site layout considerations.

Compatibility with Existing Facilities

An important consideration not to be overlooked in the expansion and upgrading of existing wastewater treatment facilities is the compatibility with existing process units. The introduction of a new operation or process into an existing facility represents new operating requirements and additional training of personnel for the proper operation and maintenance of the new unit. Often, equipment furnished by the same manufacturer as the existing installation may permit fewer spare parts to be kept on hand, provided the equipment has a good record of service.

Cost Considerations

Of major significance in the selection and design of alternative wastewater treatment facilities, especially to the client, is the question of cost—not only initial construction costs but also annual operation and maintenance costs. Although cost estimating is not covered in this text, a few comments about the preparation of cost estimates are in order.

Ordinarily, cost estimates are divided into three levels of detail: (1) order of magnitude estimates that are used for conceptual planning and are derived from cost curves and selected publications; (2) budget estimates (prepared during the preliminary design stage) derived from published or historical bid information, manufacturers' quotations, or limited quantity takeoffs; and (3) definitive estimates derived from detailed quantity takeoffs of completed plans and specifications. The accuracy of the estimates vary according to the level of detail; therefore contingencies of varying percentages are added to the estimates to account for undefined items and for unforeseen conditions.

Construction Cost Estimates. When preparing estimates of construction cost, the same basis of comparison should be used to evaluate all the alternatives and to project future costs. Methods commonly used for projecting costs are (1) escalation based on an assumed rate of inflation, or (2) a published cost index. The *Engineering News-Record* Construction Cost Index (ENRCCI), published in the magazine *ENR* (a McGraw-Hill publication), and the Sewer Construction and Sewage Treatment Plant Construction indexes of the U.S. Environmental Protection Agency are the indexes used most commonly in the field of wastewater engineering.

For purposes of comparison, data in engineering reports and in the literature can be adjusted to a common basis by using the following relationship:

$$\text{Current cost} = \frac{\text{Current value of index}}{\text{Value of index at time of estimate}} \times \text{Estimated cost} \qquad (4\text{-}1)$$

When possible, index values should also be adjusted to reflect current local costs. Both the ENRCCI and EPA indexes include costs for various geographical locations. The *ENR* publishes cost indexes for 20 cities and EPA publishes indexes for 25 cities. When using the ENRCCI, if the month of the year in which the facilities were built is not given, it is common practice to use the June end-of-the-month index value.

To project costs into the future, the following relationship can be used. The future value of the index is often projected to the one-third or mid-point of the construction period.

$$\text{Future cost} = \frac{\text{Projected future value of index}}{\text{Current value of index}} \times \text{Current cost} \qquad (4\text{-}2)$$

It should be noted, however, that updating or projecting costs for periods of more than three to five years can result in gross inaccuracies, especially if the index has increased or decreased significantly.

Operations and Maintenance Cost Estimates. The annual costs for operations and maintenance (O&M) are important factors in the evaluation of alternative treatment processes. The principal elements of O&M costs are labor, energy, chemicals, and materials and supplies. Where possible, each of these elements should be estimated separately because costs of each may escalate at different rates. For estimating staffing requirements for treatment plants to develop labor costs, Ref. 21 is useful. Energy costs should be estimated based on the estimated energy consumption by the process equipment and the appropriate energy rate obtained from the utility furnishing the energy. Chemical costs should be computed similarly based on the estimated amounts consumed and the appropriate unit price. Materials and supplies are estimated on predicted usage and should be included.

Cost Comparisons. In the evaluation of alternative treatment systems, costs may be compared using present worth, total annual costs, or life cycle costs. In a present worth analysis, all future expenditures are converted to a present worth cost at the beginning of the planning period. A discount rate is used in the analysis and represents the time value of money (the ability of money to earn interest). In a total annual cost comparison, the capital costs are amortized based on probable interest rates for bonds and the duration of the bond issue. The annual fixed (amortized) cost is added to the annual operating and maintenance costs to determine the total annual cost. Life cycle costs are used to determine the total cost of a facility over its total useful life and include the capital cost and the operating and maintenance costs. Life cycle costs are particularly useful in comparing the costs of a rehabilitated existing facility to those of a new facility. For additional information on economic evaluations, Refs. 9 and 14 may be consulted.

Environmental Considerations

The environmental impacts of a proposed wastewater treatment facility are as important, if not more so, as cost considerations. While detailed environmental review procedures are not covered in this text, a few comments regarding applicable environmental considerations that must also be addressed are appropriate.

The protocol for evaluation of environmental impacts is set forth in the National Environmental Policy Act (NEPA) of 1969 (42 USC 4321-4347 as amended). Environmental evaluations should focus on social, technical, ecological, economic, political, legal, and institutional (STEEPLI) criteria. Application of the NEPA regulations requires that an Environmental Impact Statement (EIS) be prepared for any proposed federal action that is determined to have a significant impact on the quality of the human environment. The development of an EIS is controlled by the Council on Environmental Quality (CEQ) Regulations for Implementing the Procedural Provisions of the National Environmental Policy Act (40 CFR 1500-1508).

The NEPA regulations ensure that the probable environmental effects are identified, that a reasonable number of alternative actions and their environmental impacts are considered, that the environmental information is available for public understanding and scrutiny, and that the public and governmental agencies participate as a part of the decision process. All pertinent regulations and the inherent protection afforded must be disclosed in the EIS. NEPA neither prohibits nor permits any action but requires full disclosure of environmental information and public participation in the decision making process.

Subpart E of EPA's regulations sets forth the procedures and requirements for implementing the NEPA regulations for Municipal Wastewater Treatment Construction Grants Program under the Clean Water Act. The basic elements of the process include the Environmental Information Document (EID), which is generated by the grantee (owner) as an integral part of a facilities plan, consistent with Section 201 of the Clean Water Act. The EID is the basis for agency review of the environmental impacts of the facilities plan and preparation of an Environmental Assessment (EA). The EA must be of sufficient detail so as to be an adequate basis for EPA's independent review and decision to issue a Finding of No Significant Impact (FNSI) or to issue a notice of intent for an EIS and subsequent Record of Decision. If an EIS is required, then following development of a draft EIS and input based on public hearings, a final EIS is prepared. In the resultant Record of Decision, the findings and the recommended actions selected are summarized.

To address these federal environmental considerations adequately, engineers should consult the most current version of these regulations that require an integrated federal, state, and grantee consultation process. In addition, appropriate state environmental regulatory agencies should be consulted regarding applicable state requirements. Refs. 4 and 5 are good sources of information for environmental considerations.

Other Important Considerations

Although the following important design considerations are beyond the scope of this book, they are introduced briefly in this section: (1) equipment availability, (2) personnel requirements, (3) energy and resource requirements.

Equipment Availability. The availability of equipment plays an important part in process selection because of (1) the need to provide redundant systems when there are long delivery times for spare parts and replacement units and (2) when equipment delivery is critical to the construction schedule. Most of the equipment used in wastewater treatment is custom manufactured, except for items such as small pumps, motors, and valves. Some items of equipment may be manufactured from alloy materials, such as stainless steel, that require special manufacturing techniques or are proprietary and only available from limited sources, perhaps even from overseas suppliers. Therefore, the design engineer should consider carefully the equipment components that make up the process or system to determine their potential effects upon the design, construction, and operation and maintenance of the facilities.

Personnel Requirements. The selection of a treatment process should consider not only the amount of operating and maintenance personnel needed but also the skills required. The simpler and less complex the process, the fewer highly skilled people are needed. For example, an aerated lagoon treatment system will require less highly skilled personnel than an activated-sludge plant. Where facilities are being added to an existing treatment plant, capabilities of the existing personnel should be evaluated so that the new facilities can be added without causing major staffing problems and the need for extensive retraining.

Some of the more complex processes require high levels of automatic controls utilizing electronic instruments and devices. Proper instrumentation and controls can save labor and even allow some of the small plants to operate unattended. However, complex instrumentation and control systems may require the on-staff services of highly skilled instrumentation technicians. Instrumentation specialists may be difficult to recruit and maintain on staff because of the high demand for well-qualified technicians. The extent and complexity of the control systems and the staffing levels required have to be evaluated carefully.

Energy and Resource Requirements. Concern over the rate of consumption of natural resources and energy has increased in recent years as shortages have occurred and worldwide demands have increased. Because the operation of wastewater management facilities depends on energy resources to a large extent, it is important to appraise the requirements realistically.

The operation of facilities is the main consumer of energy at treatment plants. Because energy consumption of different unit processes and operations varies greatly and because there are innumerable combinations possible, data must be available for each prospective treatment operation or process considered.

The main energy sources are (1) electric power, (2) either natural gas or propane, and (3) diesel fuel or gasoline. Electric power is used mainly for running the electric motors for the process equipment and for providing lighting and power for various ancillary support systems. Natural gas or propane is used for building and digester heating and is used as a fuel source for standby engine-generators. Diesel fuel or gasoline is used similarly for standby engine-generators and for vehicle fuel. Particular attention needs to be paid to the electrical energy costs because of the complex pricing structure used by utilities.

Electrical energy charges are commonly assessed based upon energy use, power factor charges, and demand charges. Power factor charges are concerns for plants having large electric-motor driven equipment. The demand charges are assessed by the utility companies when they commit sufficient power-generating capacity to meet the entire demands of the treatment system. Peak power use for as little as 15 minutes may establish a demand charge for up to 12 months. Demand charges can be reduced in some instances by providing power-generating capability at the treatment plant. The recovery and use of digester gas for meeting energy needs and reducing demand is one example of how both user charges and demand charges can be reduced with resulting cost savings to the treatment plant (see Fig. 4-2). Digester gas use is discussed in more detail in Chap. 12. As part of an energy cost evaluation, a sensitivity analysis should be considered to assess the impacts of future changes in energy costs on the overall cost of operation for the treatment facilities.

4-5 IMPLEMENTATION OF WASTEWATER MANAGEMENT PROGRAMS

A program for the implementation of a wastewater treatment project has several major steps, usually consisting of (1) facilities planning, (2) design, (3) value engineering, (4) construction, and (5) startup and operation. Most major projects having a construction cost over $10 million follow all steps. Smaller projects (less than $10 million) may not include the value-engineering step, although some simplified form of value engineering is highly desirable.

Facilities Planning

A facilities plan is a document established to analyze systematically the technical, economic, environmental, and financial factors necessary to select a cost-effective wastewater management plan. The facilities plan itself may include an environmental impact assessment; on major projects, the environmental assessment is usually a separate document. The scope of the facilities plan includes (1) defining the problem; (2) identifying design year needs (usually at least 20 years); (3) defining, developing and analyzing alternative treatment and disposal systems; (4) selecting a plan; and (5) outlining an implementation plan including financial arrangements and a schedule for

(a)   (b)

FIGURE 4-2
Large dual-fuel engines: (*a*) used to convert digester gas into electrical energy and heat and (*b*) used to power large pumps.

design and construction. The ultimate objective of a facilities plan is a well-defined, cost-effective, and environmentally sound project capable of being implemented and being acceptable to taxpayers and regulatory authorities. For more details about preparation of a facilities plan, Ref. 20 may be consulted.

Design

Following facilities planning, the approach generally used for designing a facility consists of conceptual design, preliminary design, special studies, and final design. The conceptual design is used to finalize the preliminary design criteria used in the facilities plan, to establish preliminary facilities layouts, and to define the necessary field investigations required such as surveys and geotechnical studies. The preliminary design is an expansion of the conceptual design and defines fully the facilities to be included in the project so that final design can proceed. Special studies may include field studies or testing necessary for the development of design criteria. The final design involves the production of the detailed contract plans and specifications used to bid for and build the project. Mitigation measures may also be included in the design to reduce or lessen unavoidable environmental impacts. Because the design approach varies with the type and size of the project, only a general outline of the design process is provided in this text.

Conceptual Design. The successful accomplishment of a project depends largely on the quality of thought and actions taken during the early stages of the project. During the conceptual and preliminary stages of a project, the principal engineering decisions are made, the equipment is selected, and the preliminary layout of the facilities is prepared. Tasks that are accomplished during the conceptual design include developing and finalizing the basic design data, preparing the process flow diagram, analyzing the plant hydraulics and preparing a hydraulic profile, defining operation and control strategies, and developing a facilities site layout. During this stage, topographic surveys and soil borings should also be made. Soils investigations are particularly important in defining the foundation conditions and in establishing structural design criteria.

Preliminary Design. The preliminary design stage represents about the 20 to 30 percent stage of the project. During the preliminary design, the site plan is finalized, the equipment requirements are defined, alternative mechanical equipment and piping arrangements are made, space requirements and architectural concepts are developed, and support systems and utilities requirements are determined. At this stage, the design should be well-enough developed that a preliminary cost estimate can be made to establish the project construction cost budget. On large projects, value engineering usually occurs at the completion of this stage.

Special Studies. Special studies are sometimes conducted before or during the preliminary design stage. These studies could include pilot plant testing of new equipment or processes (see Fig. 4-3), odor surveys to document background or existing conditions, or receiving water investigations to determine dispersion characteristics for outfall siting. It is important that these investigations be completed before the final design starts in order to eliminate uncertainties and costly redesign.

Final Design. The final step in the design of treatment plant facilities is the preparation of construction plans and specifications. This task is usually carried out by

FIGURE 4-3
Pilot plant test facilities for wastewater treatment.

consulting engineering firms. Some large cities and regional agencies have their own design staffs. The coordinated effort of specialists from many disciplines is involved. These will include engineers specializing in various fields (civil, environmental, chemical, mechanical, electrical, structural, soils, etc.), architects, designers, drafters, and other technical and support personnel. The plans and specifications become the official documents on which contractors base their bids for the construction of the facilities, and which construction managers or administrators use to hold the contractor responsible for the completion of the project as specified.

Value Engineering

Value engineering (VE) is an intensive review of a project in which a specialized cost control technique is used to identify unnecessary high costs in a project. The purpose of the VE analysis is to obtain the best project at the least cost without sacrificing quality or reliability. For projects receiving federal funding, the U.S. Environmental Protection Agency has mandated that all projects with a total construction cost over $10 million be subjected to a VE analysis. Depending on the size and complexity of the project, the VE effort may vary from one team and one review session to multiple teams and multiple reviews. For large projects, two review sessions are usually held, each lasting about one week: one at approximately the 20–30 percent stage of design completion, and a second at the 65–75 percent stage. The VE team members are senior professionals who are not involved with the design of the project. For detailed information about the VE process, Ref. 20 may be consulted.

Construction

The quality of the design plans and specifications are often measured by (1) ease of integration of new facilities into existing sites, (2) clarity of presentation that allows contractors to submit bids with small allowances for undefined or unforeseen conditions, (3) specification of high quality materials of construction to ensure a long useful life of the facilities, (4) timely completion of the work, and (5) a minimum of changes required during construction. Some of the construction considerations and management techniques for construction are discussed below.

Construction Considerations. In the preparation of the final plans and specifications, the design engineer must consider many of the details of construction. Some of the principal considerations are (1) how the plant will be built, (2) how it will interface with existing facilities, and (3) what the materials of construction will be. The "buildability" of a set of plans will be reflected in the bid price and the number of changes that must be made during construction. Numerous changes can result in costly change orders. Integrating a new facility with an existing one may present problems in (1) maintaining operations during construction, (2) continuing treatment at a level that will not violate discharge permit requirements, and (3) creating safety hazards to personnel. The construction contract must define clearly how these issues are addressed.

In selecting materials of construction, three principles are fundamental to the engineering design of process oriented facilities: (1) durability—the life of the equipment is expected to last at least 20 years and structures, 30 to 40 years; (2) good quality materials and equipment to minimize maintenance and replacement; and (3) environmental suitability, realizing that wastewater and its attendant chemicals are corrosive. For these reasons, most process structures are constructed of reinforced concrete and other materials of construction are selected based upon their corrosion-resistant properties. For information about materials of construction for wastewater treatment plants, Ref. 23 may be consulted.

Construction and Program Management. Management techniques used to ensure timely construction of the project in accordance with the plans and specifications include construction management and program management. Construction management usually provides for review of the contract plans and specifications and a management overseeing of the construction contractor's operations. The purposes of construction management are to (1) verify the technical adequacy, operability, and constructability of the plans and specifications before construction begins; (2) establish construction schedules consistent with the program objectives and to optimize cash resources; (3) review the contractor's operation to ensure conformance with the plans and specifications; and (4) control change orders and possible construction claims. Program management differs from construction management in that it provides a single source of responsibility and authority (accountable to the owner) for the management, planning, engineering, permitting, financing, construction, and startup operations of the total wastewater management program. Program management is often used in very large projects or projects that are privatized (see Sec. 4-6).

Startup and Operations

Some of the principal concerns in wastewater engineering relate to the startup, operation, and maintenance of treatment plants. The challenges facing the design engineer and the treatment plant operator include the following: (1) providing, operating, and maintaining a treatment plant that consistently meets its performance requirements; (2) managing operation and maintenance costs within the required performance levels; (3) maintaining equipment to ensure proper operation and service; and (4) training operating personnel. Therefore, the design has to be done with the operations in mind, and the plant has to be operated in accordance with the design concept. One of the principal tools used for plant startup, operation, and maintenance is the operations and maintenance (O&M) manual. The purpose of an O&M manual is to provide treatment system personnel with the proper understanding of recommended operating techniques and procedures, and the references necessary to efficiently operate and maintain their facilities. The design engineer usually has the lead responsibility in preparing the manual. Additional information about O&M manual preparation may be found in Ref. 19.

4-6   FINANCING

As discussed earlier in this chapter, the traditional funding sources for wastewater treatment plants have changed. The U.S. Government has provided grants for construction of treatment facilities for over 30 years. The 1987 Water Quality Act provides a 9-year transition program that phases out the construction grants program and phases in a state revolving loan fund program. The new revolving loan program pays only a portion of the costs; the wastewater agencies have to provide the balance. Therefore, cities, towns, and small communities have to investigate their funding options carefully to determine what is the most economical financing method for them. Alternative financing methods that are used commonly include (1) long-term municipal debt financing (with or without federal or state grants or loans), (2) non-debt financing, (3) leasing, and (4) private financing (privatization). Because financing is becoming more integrally involved with wastewater treatment design, construction, and operation, a brief discussion of the financing methods is provided in this section. For more information on financing alternatives, Refs. 1, 2, 10, and 22 may be consulted.

Long-Term Municipal Debt Financing

For projects with major capital expenditures, public agencies often use long-term debt to spread the cost of the project over a number of years. Long-term financing mechanisms include general obligation bonds, limited or special obligation bonds, revenue bonds, special assessment bonds, industrial development bonds, locally issued bonds, and small denomination bonds called "mini-bonds." Of these options, general obligation and revenue bonds are used most frequently. General obligation bonds are debt instruments backed by the full faith and credit of the issuing agency. The bonds are secured by an unconditional pledge of the issuing agency to levy unlimited taxes to meet the bond obligations. Revenue bonds are used to finance projects that generate revenue and are expected to be self-sustaining. Principal and interest charges are paid from the revenues; no taxes are levied. Tax exempt bonds result in lower interest rates as the earnings are not subject to federal or local taxes. The 1984 and 1986 tax acts substantially limit the ability of agencies to issue debt that is fully tax exempt by restricting the use of bond proceeds.

To increase the marketability of bonds and revenue, several features or variations may be added to the bond structure. Also, to reduce risk during periods of uncertain economic conditions, municipal bond insurance and letters of credit may be used to enhance the credit worthiness of the bonds.

Non-Debt Financing

Non-debt financing is a method of generating revenues from system charges and is sometimes called "pay-as-you-go" financing. The funds generated annually by rates or charges that are not used for operations and maintenance or for debt payments can be used to finance new construction. Techniques used in non-debt financing may be connection charges, special assessments, system development charges, and increasing rates in advance of construction. This method of financing may be limited to smaller projects, depending on the amount of funds that can be generated by these techniques.

Leasing

Leasing is an alternative form of facility financing that has limited application for wastewater treatment facilities. Leasing is complex, involving tax benefits to the lessor and tax implications to the lessee. The tax acts of 1984 and 1986 substantially reduced the benefits of tax-oriented leasing. Therefore, the legal and tax consequences have to be investigated carefully before undertaking a lease. In some cases, leases may be attractive for municipal agencies as a means of acquiring needed facilities and equipment where debt limitations restrict direct purchase and ownership. Many leases include an option to buy at the end of the contract period as an ultimate ownership feature.

Privatization

Privatization refers to private sector ownership and operation of facilities and services used by government entities in performing their public function [10]. The term *privatization* came into vogue after the federal income tax amendments of 1981. The tax act focused attention on private sector tax benefits that could be shared with the public sector, thereby lowering the cost of facilities for the public sector and reducing user fees. In addition to cost savings, privatization may offer advantages in construction and operating efficiencies and in meeting effluent standards. Construction efficiencies may be realized by reduced construction time, greater flexibility in flow-matching the sizing to meet current needs, and the increased use of modular designs. Examples of privatization of wastewater facilities include the construction of two wastewater treatment plants for Auburn, Alabama (see Fig. 4-4), and the construction of a sludge composting plant for Baltimore, Maryland (see Fig. 4-5).

FIGURE 4-4
Aerial view of a wastewater treatment facility financed by privatization (Auburn, AL, design average flowrate = 5.4 Mgal/d).

FIGURE 4-5
Sludge-composting facility financed by privatization (Baltimore, MD, design average capacity = 210 wet tons/d at 23 percent dewatered sludge).

The overall result of privatization is a reduction in life cycle cost, as much as 20 to 30 percent as compared to conventionally financed, constructed, and operated projects. Operating efficiencies may result under private operation by centralized administration, bulk ordering of chemicals and supplies, and sharing of key personnel among multiple facilities. Assurances in meeting effluent standards may be provided by the resources available from the private operator such as required management skills and trained operating personnel.

DISCUSSION TOPICS AND PROBLEMS

4-1. Prepare a brief summary of the history of wastewater treatment in your community. Identify major events that helped to bring about changes or improvements. Identify if any of the events were related to changes in the Clean Water Act or crisis situations (treatment process failure, overflow, lack of capacity, etc.).

4-2. Obtain a copy of the NPDES permit for your local wastewater treatment plant and a copy of the latest annual report of treatment plant performance. Compare the treatment plant performance to the permit requirements, and note any violations and their causes. At what percentage of design flowrate is the treatment plant operating?

4-3. Prepare a summary of the method of sludge disposal and the regulations covering sludge disposal for your local wastewater treatment plant. List the requirements covering toxic pollutants for sludge disposal, and compare the treatment plant performance data to the requirements. Comment on any current or potential future problems for sludge disposal in your community.

4-4. Obtain a copy of the OSHA requirements and list at least 10 of the requirements that apply to the design of wastewater treatment plants.

4-5. Visit your local wastewater treatment plant and prepare a summary of the unit processes and operations used. Use unit operations and processes in Tables 4-2 and 4-3 as a guide.

4-6. What consideration was given to energy conservation in the design of your local wastewater treatment facilities? Prepare a list of possible improvements that could be used to increase energy conservation at wastewater treatment plants.

4-7. How is stormwater collected and treated in your community? If it is not treated, are there any potential pollutional problems caused by the untreated discharges? If there are potential pollutional problems, how might they be mitigated?

4-8. Prepare a brief summary of the history of the regulation of toxic substances in your community. Include a copy of the ordinance or regulations covering the discharge or disposal of toxic substances. Identify any documented problems in the disposal of toxic materials, and list potential solutions to these problems.

4-9. Identify the industries in your community that are required to pretreat their industrial wastewater before discharge to the municipal sewer system. Select one of the industries and identify the types of unit operations or processes it is required to use for industrial pretreatment.

4-10. The construction for a small wastewater treatment plant was estimated to be $8 million in 1987. If the construction of the plant is to be delayed until 1992, estimate the cost in 1992 for this same plant. Use end-of-year *ENRCCI* values in making your projection.

4-11. Determine the year when your local wastewater treatment plant was constructed or expanded and its construction costs. What would the cost be to construct or expand the plant today? What has been the average rate of inflation from the time your plant was constructed to the present?

4-12. If a facilities plan was prepared for your local wastewater treatment plant, what alternative treatment processes were considered? What were the reasons given for selecting the unit operations and processes actually used?

4-13. Review the unit operations and processes used at your local wastewater treatment plant with the treatment plant operator, and identify operating and maintenance problems. How might these problems have been mitigated by the design?

4-14. Identify current and potential future reuse programs for treated wastewater effluent and sludge. What percentage of the existing wastewater and sludge production might be reused? Justify your answer. What unit operations or processes might be required to make the treated effluent and sludge acceptable for reuse?

4-15. If an EIR was prepared for the construction of your local wastewater treatment plant, obtain a copy and review it, specifically with respect to impacts of the selected alternative. What measures were recommended to mitigate the impacts and how were they implemented?

4-16. If an EIR was not prepared for the construction of your local wastewater treatment plant, would the preparation of one, in your judgment, have caused any significant differences in the implementation of the treatment facilities?

4-17. From the annual operating report for your local wastewater treatment plant, determine the total operating and maintenance costs, and compute the percentage of the total for the categories of labor, energy, chemicals, and materials and supplies.

4-18. Obtain an annual operating report from another community for a wastewater treatment plant of similar size using a different treatment process. Determine the operating and maintenance cost percentages similar to Prob. 4-17 and compare the results to Prob. 4-17. Analyze the reasons for the differences.

4-19. Prepare a summary of the financing used for your wastewater treatment plant. How much of the construction cost was financed by federal and state grants or loans? How was the local share of the construction cost financed?

4-20. For your local wastewater treatment plant project, obtain a copy of the revenue plan and summarize the rate structure used to pay the annual costs for debt service and for the operations and maintenance costs.

REFERENCES

1. American Public Works Association: "IWR White Paper on Privatization and Contracting for Wastewater Services," *APWA Reporter,* May 1988.
2. American Water Works Association: *Water Utility Capital Financing, AWWA Manual* M29, 1988.
3. *Biocycle*: "EPA Sludge Disposal Regulations Proposed," vol. 30, no. 44, March 1989.
4. Bureau of National Affairs, *The Environmental Reporter,* 1231 25th St., N.W., Washington, DC.
5. Environmental Law Institute, *Clean Water Deskbook,* 1616 P. Street, N.W., Washington, DC.
6. *Federal Register*: "Secondary Treatment Regulation," 40 CFR Part 133, July 1, 1988.
7. *Federal Register*: "Amendment to the Secondary Treatment Regulations: Percent Removal Requirements During Dry Weather Periods for Treatment Works Served by Combined Sewers," 40 CFR Part 133, January 27, 1989.
8. *Federal Register*: "Standards for the Disposal of Sewage Sludge," 40 CFR Parts 257 and 503, February 6, 1989.
9. Grant, E.L., and W.G. Ireson: *Principles of Engineering Economy,* 5th ed., Ronald Press, 1970.
10. Goldman, H., and S. Mokuvos: *The Privatization Book,* Arthur Young, New York, 1984.
11. Great Lakes-Upper Mississippi River Board of State Sanitary Engineers: *Recommended Standards for Sewage Works,* 1978 Edition.
12. Hegewald, M.: "Setting the Water Quality Agenda: 1988 and Beyond," *Journal WPCF,* vol. 60, no. 5, 1988.
13. Henrichs, R.: "Law, Literature Review," *Journal WPCF,* vol. 60, no. 6, 1988.
14. James, L.D., and R.R. Lee: *Economics of Water Resources,* McGraw-Hill, New York, 1971.
15. Metcalf & Eddy: *Sewage and Sewage Disposal, A Textbook,* 2nd ed., McGraw-Hill, New York, 1930.
16. Novick, S. M.: *Law of Environmental Protection 1988,* Environmental Law Institute, Clark Boardman, New York, 1988.
17. Technical Advisory Board of the New England Interstate Water Pollution Control Commission: *Guides for the Design of Wastewater Treatment Works,* 1980 Edition.
18. U.S. Congress, Office of Technology Assessment: *Wastes in the Marine Environment,* 07A-0334, U.S. GPO, Washington, DC, April, 1987.
19. U.S. Environmental Protection Agency: *Considerations for Preparation of Operation and Maintenance Manuals,* 1974.
20. U.S. Environmental Protection Agency: *Construction Grants,* 1985 (CG-85).
21. U.S. Environmental Protection Agency: "Estimating Staffing for Municipal Wastewater Treatment Facilities," Contract No. 68-01-0328, March 1973.
22. U.S. Environmental Protection Agency: *Study of the Future Federal Role in Municipal Wastewater Treatment,* December 1984.
23. Water Pollution Control Federation: *Wastewater Treatment Design,* Manual of Practice no. 8, 1977.

CHAPTER
5

INTRODUCTION TO WASTEWATER TREATMENT PLANT DESIGN

The nature of the wastewater to be treated, the general objectives and methodology of treatment, and the steps in an implementation program have been considered in the previous chapters. Many of the important factors that need to be considered in developing the actual design of a wastewater treatment plant are discussed in this chapter. The initial stages of a project, starting with the facilities plan and continuing through the conceptual and preliminary design phases, are considered critical to the overall design process. It is during these initial stages that the design flowrates and mass loadings are developed, process selection is made, the design criteria are developed, refined, and established, and the facilities layouts are prepared. At the completion of preliminary design, the project is fully defined so that preparation of the detailed plans and specifications can proceed expeditiously.

Important treatment plant design considerations that are typical to most projects covered in this chapter include (1) impact of flowrate and mass-loading factors on design, (2) evaluation and selection of design flowrates, (3) evaluation and selection of design mass loadings, (4) process selection, and (5) elements of conceptual process design. The specific principles and design features of the various unit operations and processes that comprise the treatment system are covered in the following chapters. Information on wastewater-peaking factors for collection systems may be found in the companion volume to this text [7].

5-1 IMPACT OF FLOWRATE
AND MASS-LOADING FACTORS ON DESIGN

The rated capacity of wastewater treatment plants is normally based on the average annual daily flowrate at the design year. As a practical matter, however, wastewater treatment plants have to be designed to meet a number of conditions that are influenced by flowrates, wastewater characteristics, and a combination of both (mass loading). Peaking conditions also have to be considered, including peak hydraulic flowrates and peak process mass-loading rates. Peak hydraulic flowrates are important so that the unit operations and processes and their interconnecting conduits can be sized appropriately to handle the applied flowrates. Peak process loading rates are important in sizing the process units and their support systems so that treatment plant performance objectives can be achieved consistently and reliably.

Additionally, periods of initial operation and low flows and loads must be taken into consideration in design. Typical flowrate and mass-loading factors that are important in the design and operation of wastewater treatment facilities are described in Table 5-1. The overall objective of wastewater treatment is to provide a wastewater treatment system that is capable of coping with a wide range of probable wastewater conditions while complying with the overall performance requirements. In fulfilling this objective, the influence of the flowrate and mass-loading factors must be fully understood.

5-2 EVALUATION AND SELECTION
OF DESIGN FLOWRATES

The procedure for evaluating and selecting design flowrates usually involves the development of average flowrates based on population projections, industrial flow contributions, and allowances for infiltration/inflow. The average flowrates are then multiplied by appropriate peaking factors to obtain the peak flowrates. In developing average flowrates and peaking factors, the following items must be considered: (1) the development and forecasting of average daily flowrates, (2) the rationale used in selecting flowrate factors, (3) application of peaking and minimum flowrate factors, and (4) upstream control of peak flowrates that may affect treatment plant design. The length of the design period, also important in flowrate projections, is discussed later in this chapter.

Forecasting Average Flowrates

The development and forecasting of average daily flowrates is necessary to determine the design capacity as well as the hydraulic requirements of the treatment system. Average flowrates need to be developed both for the initial period of operation and for the future (design) period. In determining the design flowrate, elements to be considered are (1) the current base flows; (2) estimated future flows for residential, commercial, institutional, and industrial sources, and (3) nonexcessive infiltration/inflow. (Infiltration/inflow is discussed in Chap. 2.) Existing base flows equal actual metered

TABLE 5-1
Typical flowrate and mass-loading factors used for the design and operation of wastewater treatment plant facilities

Factor	Application
Based on flowrate	
Peak hour	Sizing of pumping facilities and conduits; bar-rack sizing
	Sizing of physical unit operations: grit chambers, sedimentation tanks, and filters; sizing chlorine-contact tanks
Maximum day	Sizing of sludge-pumping system
Greater-than-one-day maximum	Screenings and grit storage
Maximum week	Recordkeeping and reporting
Maximum month	Recordkeeping and reporting; sizing of chemical storage facilities
Minimum hour	Sizing turndown of pumping facilities and low range of plant flowmeter
Minimum day	Sizing of influent channels to control solids deposition; sizing effluent recycle requirements for trickling filters
Minimum month	Selection of minimum number of operating units required during low-flow periods
Based on mass loading	
Maximum day	Sizing of selected biological processing units
Greater-than-one-day maximum	Sizing of sludge-thickening and -dewatering systems
Sustained peaks	Sizing of selected sludge processing units
Maximum month	Sizing of sludge storage facilities; sizing of composting requirements
Minimum month	Process turndown requirements
Minimum day	Sizing of trickling-filter recycle

flowrates minus excessive infiltration/inflow (defined as infiltration/inflow that can be controlled by cost-effective improvements to the collection system).

Many state agencies have also established design flowrates to be used where no actual flow measurements are available. One interstate agency has established a minimum average design flow of 70 gal/capita·d (270 L/capita·d) to be used where no flow data are available [11]. To this flow, an allowance for infiltration should be added. A total dry-weather base flow of 120 gal/capita · d (460 L/capita · d) has been established by EPA as a historical average where infiltration is not excessive. The base flow includes 70 gal/capita · d for domestic flows, 10 gal/capita · d (40 L/capita · d) for commercial and small industrial flows, and 40 gal/capita · d (150 L/capita · d) for infiltration [3].

Rationale for the Selection of Flowrate Factors

The rationale for selecting flowrate factors is based on hydraulic and process considerations. The process units and hydraulic conduits have to be sized to accommodate the anticipated peak flowrates that will pass through the treatment plant. Provisions have to be made to ensure bypassing of wastewater does not occur either in the collection system or at the treatment plant. Many of the process units are designed based on detention time or overflow rate (flowrate per unit of surface area) to achieve the desired removal rates of BOD and suspended solids (SS). Because the performance of these units can be affected significantly by varying flowrate conditions and mass loadings, minimum and peak flowrates must be considered in design.

Minimum and Peak Flowrate Factors

As noted in Table 5-1, low flowrates are also of concern in treatment plant design, particularly during the initial years of operation when the plant is operating well below the design capacity and in designing pumping stations. In cases where very low nighttime flow is expected, provisions for recycling treated effluent may have to be included to sustain the process. In absence of flow-measuring data, minimum daily flowrates may be assumed to range from 30 to 50 percent of average flowrates for small and medium-size communities, respectively [14].

The flowrate peaking factors (the ratio of peak flowrate to average flowrate) most frequently used in design are those for peak hour and maximum day (see Table 5-1). Peak hourly flowrates are used to size the hydraulic conveyance system and other facilities such as sedimentation tanks and chlorine contact tanks where little volume is available for flow dampening. Other peaking factors, such as maximum week or maximum month, may be used for treatment facilities, such as pond systems, that have long detention times or for sizing-sludge processing facilities that also have long detention times or storage. Peaking factors may be developed from flowrate records or based on published curves or data from similar communities.

Peaking Factors Developed from Flowrate Data. The most common method of determining peaking factors is from the analysis of flowrate data. Where flowrate records are available, at least two years of data should be analyzed to develop the peak-to-average flowrate factors. These factors may then be applied to estimated future average flowrates, adjusted for any anticipated future special conditions. Where commercial, institutional, or industrial wastewaters are expected to make up a significant portion of the average flowrates (say 25 percent or more of all flows, exclusive of infiltration), peaking factors for the various categories of flow should be estimated separately. Peak flows from each category most probably will not occur simultaneously; therefore, some adjustment may have to be made to the total peak flow to prevent overestimating the peaking conditions. If possible, peaking factors for industrial wastewater should be estimated on the basis of average water use, number of shifts worked, and pertinent details of plant operations.

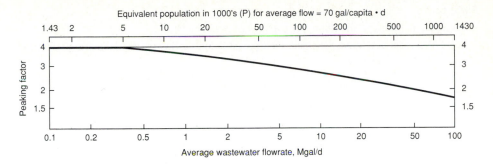

FIGURE 5-1
Hourly peaking factor for domestic wastewater flowrates. Peaking factor is the ratio of maximum hourly flowrate to average flowrate. *Note:* Mgal/d $\times$ 0.043813 = m^3/s.

Peaking Factor Estimates Based on Published Data. If flow measurement records are inadequate to establish peaking factors, the curve given in Fig. 5-1 may be used for estimating peak hourly flowrates from domestic sources. This curve was developed from analyses of the records of numerous communities throughout the United States. It is based on average residential flowrates, exclusive of extreme high flow occurrences (i.e., values greater than the 99 percentile value) and infiltration/inflow, and includes small amounts of commercial flows and industrial wastes.

Factors for peak hourly flowrates must consider the characteristics of the collection system serving the wastewater treatment plant. Improvements to or rehabilitation of the collection system may also increase or decrease the peaking factors. In cases where the wastewater is pumped to the treatment plant and recorded data are not available, the peak flowrate can be equated with the maximum pumping capacity. Where flow to the treatment plant is by gravity, the peak flowrate can be estimated based on the hydraulic capacity of the influent sewers.

Forecasting design flowrates, including the use of peaking factors, is illustrated in Example 5-1.

Example 5-1   Forecasting design flowrates. A residential community with a population of 15,000 is planning to expand its wastewater treatment plant. In 20 years, the population is estimated to increase to 25,000 residents, and 1,000 visitors per day are expected to attend a proposed junior college. A new industry will also move in and contribute an average flowrate of 0.22 Mgal/d and a peak flowrate of 0.33 Mgal/d for 24 h/d of operation. The present average daily wastewater flowrate is 1.60 Mgal/d and the infiltration/inflow has been determined to be nonexcessive. Infiltration is estimated to be 25 gal/capita · d at average flow and 37.5 gal/capita · d at peak flow. Residential water use in the new homes is expected to be 10 percent less than in the current residences because of the installation of water-saving appliances and fixtures. Compute the future average, peak, and minimum design flowrates. For peak residential flowrates, use Fig. 5-1. Assume that the peak industrial flowrate occurs during the day shift. For calculating minimum design flowrates, assume that the ratio of minimum to average flowrate is 0.35 for residential minimum flowrates, and assume that the industrial plant is shut down one day a week.

Solution

1. Compute the present and future per capita wastewater flowrates.

   (a) For present conditions, compute the average domestic flowrate excluding infiltration.

   i. Compute infiltration.

   $$\text{Infiltration} = 15,000 \times 25 \text{ gal/capita} \cdot \text{d} = 375,000 \text{ gal/d}$$

   ii. Compute average domestic flowrate.

   $$\begin{aligned}
   \text{Domestic flowrate, gal/d} &= \text{Total average flow} - \text{Infiltration} \\
   &= 1,600,000 - 375,000 \\
   &= 1,225,000 \text{ gal/d}
   \end{aligned}$$

   (b) Compute present per capita flowrate by dividing the existing domestic flowrate by the present population.

   $$\text{Per capita flowrate} = \frac{1,225,000 \text{ gal/d}}{15,000 \text{ persons}} = 81.7 \text{ gal/capita} \cdot \text{d}$$

   (c) For future conditions, reduce existing per capita flowrate by 10 percent.

   $$\text{Future per capita flowrate} = 81.7 \times 0.9 = 73.5 \text{ gal/capita} \cdot \text{d}$$

2. Compute future average flowrate.

   (a) Existing residents   = 1,225,000 gal/d

   (b) Future residents = $10,000 \times 73.5$ gal/capita $\cdot$ d   = 735,000 gal/d

   (c) Day students (assume 15 gal/capita $\cdot$ d from Table 2-4)
       $= 1,000 \times 15$ gal/capita $\cdot$ d   = 15,000 gal/d

     Subtotal   1,975,000 gal/d

   (d) Total domestic flowrate (converted to Mgal/d)   = 1.975

   (e) Industrial flowrate, Mgal/d   = 0.220

   (f) Infiltration, Mgal/d $= 25,000 \times 25$ gal/capita $\cdot$ d $\times 10^{-6}$ = 0.625

   (g) Total future average flowrate, Mgal/d   = 2.82

3. Compute future peak flowrate.

   (a) Residential peak flowrate: From Fig. 5-1,
       the peaking factor for 1.975 Mgal/d is 3.0.
       The peak hourly flowrate is 1.975 Mgal/d $\times$ 3.0   = 5.93 Mgal/d

   (b) Industrial peak flowrate   = 0.33 Mgal/d

   (c) Infiltration, Mgal/d $= 25,000 \times 37.5$ gal/capita $\cdot$ d $\times 10^{-6}$ = 0.94 Mgal/d

   (d) Total future peak flowrate   = 7.20 Mgal/d

4. Compute the minimum flowrate.

   (a) Residential minimum flowrate: As indicated in Fig. 2-2,
       the low flowrate usually occurs in the early morning hours.
       The minimum flowrate based on current flow
       $= 0.35 \times (1.60)$ = 0.56 Mgal/d

   (b) Industrial minimum flowrate   = 0.00 Mgal/d

   (c) Total minimum flowrate   = 0.56 Mgal/d

Comment. If wastewater flowrate records are not adequate, future average daily flowrates may be calculated based on the future population and unit wastewater flowrates, similar to those given in Table 2-7. Appropriate adjustments should be made in the calculations to account for any special conditions such as flow reduction, infiltration/inflow allowances, and industrial flows. When peak flowrates for more than one flow component are calculated, some adjustment in the total peak flowrate should be made if the peaks from the components do not occur simultaneously. The range in flowrates, as illustrated in this example, is reasonably representative of what can occur at a treatment plant and allowances for similar ranges have to be made in the process design.

Upstream Control of Peak Flowrates

Planning wastewater facilities to handle peak flowrates may involve several considerations including (1) improvements to the collection system to reduce peak flow related to infiltration/inflow (I/I), (2) installation of flow-equalization facilities to provide storage either in the collection system or at the treatment plant. Other alternatives for peak flowrate control at the treatment plant, namely, provision for flow-splitting and bypass facilities, are discussed under process selection.

Improvement to the collection system may involve a lengthy and costly process and may not have an immediate effect on significantly reducing peak flowrates. In some cases, the amount of flow reduction resulting from collection system rehabilitation has been less than anticipated, particularly if infiltration is a significant component of I/I. In some unusual circumstances, the flowrates have actually increased after completion of the collection system improvement program. Therefore, some safety factors should be considered when estimating possible peak flowrate reduction resulting from collection system improvements.

Flow equalization can be an effective measure in reducing peak flowrates. Benefits derived by upstream flow equalization include (1) reduced hydraulic loading on already overtaxed collection facilities, (2) reduced potential of overflows and possible resulting health hazards or pollution problems, and (3) reduced peak loading of the treatment plant. Equalization depends on available volume and may be of limited value in extreme peak flow conditions. Siting of equalization facilities in the collection system is often difficult because of limited available space at locations that are compatible with the system hydraulics. Operation and maintenance may also be difficult to manage, particularly in remote areas. Ease of operation, maintenance, control, and environmental factors are major reasons why many equalization facilities are located at treatment plants. The required analysis for sizing flow equalization facilities is presented in Chap. 6.

5-3 EVALUATION AND SELECTION OF DESIGN MASS LOADINGS

The evaluation and selection of design mass loadings involves determining (1) the variation in concentrations of wastewater constituents, (2) the analysis of mass loadings, including average and sustained peak mass loadings, and (3) the impacts of toxic and other inhibitory pollutants.

Variations In Concentrations of Wastewater Constituents

From the standpoint of treatment processes, one of the most serious deficiencies results when the design of a treatment plant is based on average flowrates and average BOD and SS loadings, with little or no recognition of peak conditions. In many communities, peak influent flowrates and BOD and SS loadings can reach two or more times average values. In general, peak flowrates and BOD and SS mass-loading rates do not occur at the same time. A design based on the concurrence of peak flowrates and constituent concentrations may result in excessive capacity. Analysis of current records is the best method of arriving at appropriate peak and sustained mass loadings. The statistical analysis of the data is the same as discussed in Chap. 2 in connection with the analysis of flow data.

The principal factors responsible for loading variations are (1) the established habits of community residents, which cause short-term (hourly, daily, and weekly) variations; (2) seasonal conditions, which usually cause longer term variations; and (3) industrial activities, which cause both long- and short-term variations. These same factors were also discussed in Chap. 2 in connection with wastewater-flowrate variations.

Short-Term Variations. Typical data for the hourly variations in domestic wastewater strength are shown in Fig. 5-2. The BOD variation follows the flow variation (same as Fig. 2-2). The peak BOD (organic matter) concentration often occurs in the evening around 9 p.m. Wastewater from combined sewer systems usually contains more inorganic matter than wastewater from sanitary sewer systems because of the larger quantities of storm drainage that enter the combined sewer system.

Seasonal Variations. For domestic flow only, and neglecting the effects of infiltration, the unit (per capita) loadings and the strength of the wastewater from most

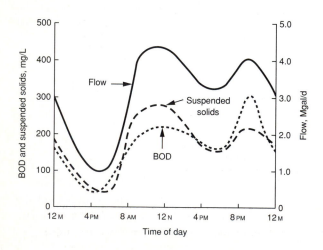

FIGURE 5-2
Typical hourly variation in flow and strength of domestic wastewater.

seasonal sources, such as resorts, will remain about the same on a daily basis throughout the year even though the total flowrate varies. The total mass of BOD and SS of the wastewater, however, will increase directly with the population served.

In combined sewers, seasonal variations in BOD and SS are primarily a function of the amount of stormwater that enters the system. In the presence of stormwater, average concentrations of these constituents generally will be lower than the corresponding concentrations in domestic wastewater alone. The seasonal BOD variation for the influent to the Calumet Sewage Treatment Works in Chicago [8] is illustrated in Fig. 5-3. The measured BOD values are below average during the spring and summer months, the period corresponding to the time of the spring thaw and the high summer rainfall.

Although the presence of stormwater usually means that the measured concentrations of most constituents will be lower, significantly higher BOD and SS loadings may occur during the early stages of a storm. This temporary increase is the result of the so-called "first-flush effect," which is most pronounced after a long dry period, when material deposited during the dry period is washed away when scouring velocities are attained in the collection system. The high initial concentrations are seldom sustained for more than two hours. After that, the dilution effect will be observed.

Infiltration/inflow, as explained in Chap. 2, is another source of water flow into the collection system. In most cases, the presence of this extraneous water tends to decrease the concentrations of BOD and SS, but this depends on the characteristics of the water entering the sewer. In some cases, concentrations of some inorganic

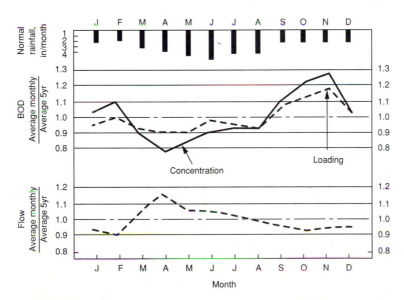

FIGURE 5-3
Variation in BOD concentration and flow for the Calumet Sewage Treatment Works, Chicago.

constituents may actually increase where the groundwater contains high levels of dissolved constituents.

Industrial Variations. The concentrations of both BOD and SS in industrial wastewater can vary significantly throughout the day. For example, the BOD and suspended solids concentrations contributed from vegetable-processing facilities during the noon wash-up period may far exceed those contributed during working hours. Problems with high short-term loadings most commonly occur in small treatment plants that have limited reserve capacity to handle these so-called "shock loadings." The seasonal impact of industrial wastes is clearly shown in Fig. 5-4, in which both the flow and BOD loading data are presented for a three-year period for the City of Modesto, California [1]. The variations result from the waste contributions of canneries and other industries related to agriculture.

As noted in Chap. 2, when industrial wastes are to be accommodated in municipal collection and treatment facilities, special attention must be given to developing adequate wastewater characterizations and flowrate projections. Further, any proposed future process changes should also be assessed to determine what effects they might have on the wastes to be discharged.

Analysis of Mass Loadings

The analysis of wastewater data involves the determination of the flowrate and mass-loading variations. The analysis may involve determining the simple average or flow-weighted average concentrations of specific constituents, mass loadings (flowrate times concentration), or sustained mass loadings. In almost all cases, a flow-weighted

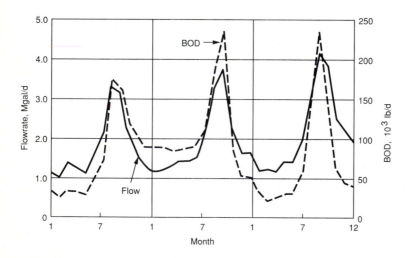

FIGURE 5-4
Seasonal variation in the flow and BOD mass loading at the Modesto wastewater treatment plant, Modesto, CA. *Note:* Mgal/d × 0.043813 = m³/s; lb × 0.4536 = kg.

average should be used because it is a more accurate method of analysis. A simple average may result in erroneous conclusions, as exemplified later in Example 5-2.

The computational methods used and the importance of mass loadings on the design of treatment processes are considered below. The statistical analysis of the data is the same as discussed in Chap. 2 for the analysis of flowrate data.

Simple Average. The simple or arithmetic average of a number of individual measurements is given by

$$x = \frac{1}{n} \sum_{i=1}^{n} x_i \qquad (5\text{-}1)$$

where x = arithmetic average concentration of the constituent
  n = number of observations
  x_i = average concentration of the constituent during the ith time period

Flow-Weighted Average. To obtain a more representative assessment of constituent concentrations in domestic wastewater, the flow-weighted average is computed by using Eq. 5-2:

$$x_w = \frac{\displaystyle\sum_{i=1}^{n} x_i q_i}{\displaystyle\sum_{i=1}^{n} q_i} \qquad (5\text{-}2)$$

where x_w = flow-weighted average concentration of the constituent
  n = number of observations
  x_i = average concentration of the constituent during ith time period
  q_i = average flowrate during ith time period

The application of Eqs. 5-1 and 5-2 is illustrated in Example 5-2. .

Example 5-2 Analyzing wastewater data using simple and flow-weighted averages. Develop simple and flow-weighted averages for BOD and SS data given in Fig. 5-2.

Solution

1. Compute simple average.
   (*a*) To analyze the BOD and SS data given in Fig. 5-2, divide the day's record into 24 one-hour increments, and record the hourly values for BOD and SS, as shown in columns (1), (2), and (3) in the table below.
   (*b*) Sum the 24 individual average hourly values, and divide by 24. For BOD, the average is 3918/24 = 163.3 mg/L. For SS, the average is 4032/24 = 168 mg/L.
2. Compute flow-weighted average.
   (*a*) To analyze the data given in Fig. 5-2, divide the day's record into 24 one-hour increments, similar to the simple-average computation, but include the hourly flowrate data.

(b) Multiply the corresponding hourly averages of the flowrate and concentration. Sum the 24 values and the values of the 24 individual flowrates, as shown in the last three columns of the table below.

(c) Divide the 24 values by the summed values of the flowrates. The weighted average for BOD is 30,615.50/164.90 = 186 mg/L, and, for SS, the weighted average is 31,867.60/164.90 = 193 mg/L.

Hour (1)	BOD, mg/L (2)	SS, mg/L (3)	Flowrate, q, Mgal/d (4)	BOD × q, columns (2) × (4)	SS × q, columns (3) × (4)
12 M	161	172	6.80	1,094.80	1,169.60
1 A.M.	132	143	5.30	699.60	757.90
2	93	105	3.90	362.70	409.50
3	64	77	3.20	204.80	246.40
4	41	47	2.50	102.50	117.50
5	45	40	2.30	103.50	92.00
6	59	42	3.30	135.70	96.60
7	108	85	3.40	367.20	289.00
8	139	196	6.40	889.60	1,254.40
9	180	251	8.90	1,602.00	2,233.90
10	202	263	9.60	1,939.20	2,524.80
11	211	274	9.80	2,067.80	2,685.20
12 N	213	261	9.60	2,044.80	2,505.60
1 P.M.	208	249	9.40	1,955.20	2,340.60
2	200	225	8.70	1,740.00	1,957.50
3	195	195	8.00	1,560.00	1,560.00
4	182	161	7.50	1,365.00	1,207.50
5	156	147	7.30	1,138.80	1,073.10
6	150	145	7.50	1,125.00	1,087.50
7	179	169	8.00	1,432.00	1,352.00
8	230	198	8.90	2,047.00	1,762.20
9	305	206	9.10	2,775.50	1,874.60
10	262	201	8.70	2,279.40	1,748.70
11	203	180	7.80	1,583.40	1,404.00
Totals	3,918	4,032	164.90	30,615.50	31,867.60
Weighted average				186	193

Comment. When comparing the computation of a simple average to a flow-weighted average in this example, the differences are significant. Varying flow conditions can significantly affect the calculations of the average concentrations if flow-weighting is not used. In these examples, if simple averages were used, the average BOD values could have been understated by about 22 mg/L and the SS by 24 mg/L, as compared to the flow-weighted average. If simple averages were used in establishing process-loading values in this case, the treatment facilities could be underdesigned by over 10 percent. Although simple arithmetic averages are still used, they may be of little value because the magnitude of the flow at the time of the measurement is not taken into account. If the flowrate remains constant, the use of a simple average is acceptable.

Average Mass Loadings. Constituent mass loadings are usually expressed in pounds per day (kilograms per day) and may be computed using Eq. 5-3, when the flowrate is expressed in million gallons per day, or Eq. 5-4, when the flowrate is expressed in cubic meters per day. Note that in the SI system of units, the concentration expressed in milligrams per liter is equivalent to grams per cubic meter.

$$\text{Mass loading, lb/d} = (\text{concentration, mg/L}) \, (\text{flowrate, Mgal/d}) \, [8.34 \, \text{lb/Mgal} \cdot (\text{mg/L})] \quad (5\text{-}3)$$

$$\text{Mass loading, kg/d} = \frac{(\text{concentration, g/m}^3) \, (\text{flowrate, m}^3/\text{d})}{10^3 \, (\text{g/kg})} \quad (5\text{-}4)$$

An example of a diurnal mass-loading curve is illustrated in Fig. 5-5. The wide variation in loading rates and the compounding effects, particularly during the high flow and concentration periods, is clearly illustrated in Fig. 5-5. The impact of these load variations is seen most dramatically in the effects on the biological system operating conditions. The maximum hourly BOD loading may vary as much as 3 to 4 times the minimum hourly BOD load in a 24 h period. Variations of this type have to be accounted for in the design of the biological treatment system.

Sustained Peak Mass Loadings. To design treatment processes that function properly under various loading conditions, data must be available for the sustained peak mass loadings of constituents that are to be expected. In the past, such informa-

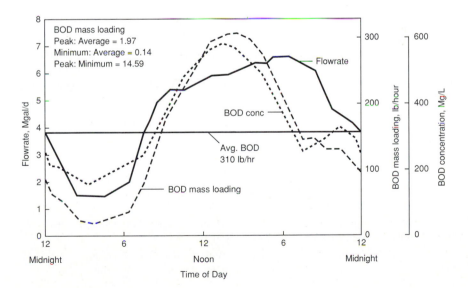

FIGURE 5-5
Illustration of diurnal wastewater flow, BOD, and mass-loading variability. *Note:* Mgal/d × 0.043813 = m³/s.

tion has seldom been available. When the data are not available, curves similar to those shown in Fig. 5-6 can be used. The curves for BOD, SS, TKN (Total Kjeldahl Nitrogen), NH3 (ammonia), and phosphorus were derived from an analysis of the records of over 50 treatment plants throughout the country. It should be noted that significant variations will be observed from plant to plant, depending on the size of the system, the percentage of combined wastewater, the size and slope of the interceptors, and the types of wastewater contributors.

The procedure used to develop the mass-loading curves shown in Fig. 5-6 is as follows. First, the average mass loading is determined for the period of record. Second, the records are reviewed for the highest and lowest sustained one-day mass loading. These values are divided by the average mass loading and the numbers are plotted. Third, the same procedures are followed for two-consecutive-days, three-consecutive-days, etc. sustained loadings until ratio values are found for the period of interest (usually 10 to 30 days).

The daily mass-loading rates for the various plants were developed using hourly data and the following expression:

Daily mass loading, lb/d

$$= \sum_{i=1}^{24} (\text{concentration, mg/L}) (\text{flowrate, Mgal/h}) [8.34 \text{ lb/Mgal} \cdot (\text{mg/L})] \quad (5\text{-}5)$$

The development of a sustained peak mass-loading curve is illustrated in Example 5-3. The application of this curve will be discussed in Chaps. 8 and 10.

Example 5-3 Development of sustained peak mass-loading curve for BOD.

Develop a sustained BOD peak mass-loading curve for a treatment plant with a design flowrate of 22.8 Mgal/d (1 m^3/s). Assume that the long-term daily average BOD concentration is 200 mg/L.

Solution

1. Compute the daily mass-loading value for BOD.

   Daily BOD mass loading, lb/d $= (200 \text{ mg/L})(22.8 \text{ Mgal/d})[8.34 \text{ lb/Mgal} \cdot (\text{mg/L})]$
   $= 38,030 \text{ lb/d } (17,266 \text{ kg/d})$

2. Set up a computation table for the development of the necessary information for the peak sustained BOD mass-loading curve (see following table).

3. Obtain peaking factors for the sustained peak BOD loading rate from Fig. 5-6a, and determine the sustained mass-loading rates for various time periods [see table, columns (1), (2), and (3)].

4. Develop data for the sustained mass-loading curve [see table, column (4)], and prepare a plot of the resulting data (see following figure).

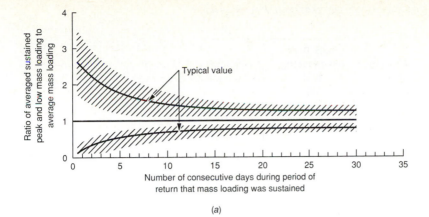

(a)

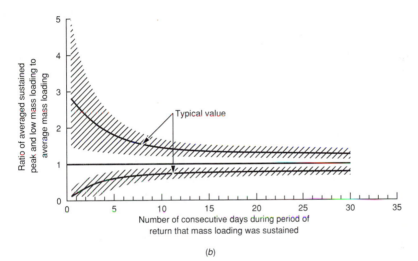

(b)

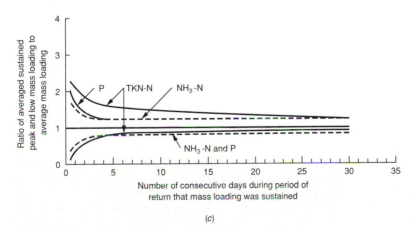

(c)

FIGURE 5-6

Typical information on the ratio of averaged sustained peak and low-constituent mass loadings to average mass loadings for: (a) BOD, (b) SS, and (c) nitrogen and phosphorus.

Length of sustained peak, d (1)	Peaking factor[a] (2)	Peak BOD mass loading, lb/d (3)	Total mass loading, lb[b] (4)
1	2.4	91,272	91,272
2	2.1	79,863	159,726
3	1.9	72,314	216,942
4	1.8	68,508	274,032
5	1.7	64,702	323,510
10	1.4	53,284	532,840
15	1.3	49,478	742,170
20	1.25	47,575	951,500
30	1.15	43,769	1,313,070
365	1.0	38,030	

[a] From Fig. 5-6a.

[b] Column 1 × column 3 = column 4

Note: lb × 0.4536 = kg

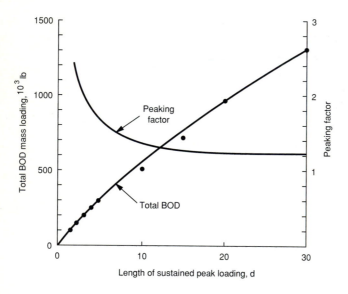

Comment. The interpretation of the curve plotted for this example is as follows. If the sustained peak loading period were to last for 10 days, the total amount of BOD that would be received at a treatment facility during the 10-day period would be 532,840 lbs. The corresponding amounts for sustained peak periods of 1 and 2 days would be 91,272 and 159,726 lb, respectively. Computations for an example of this type can be facilitated by using a personal computer spreadsheet program.

Impacts of Toxic and Other Inhibitory Pollutants

Heavy metals and nonmetallic wastes, including organic compounds, may enter the wastewater system and have an inhibitory or toxic effect upon the treatment system, particularly biological treatment processes. A list of pollutants, for example, that have an inhibitory effect on the activated-sludge process is presented in Table 5-2 [14]. Heavy metals, such as copper, zinc, nickel, lead, cadmium, and chromium, can react with the microbial enzymes to retard or completely inhibit metabolism. Heavy metals present as precipitates may be solubilized by a change in pH, causing a decrease in biological process efficiency. Biological degradation of certain organic compounds such as cyanide and humic acids, may cause the release of toxic metals from soluble complexes, also affecting biological activity. Nonmetallic wastes above specific concentrations can also cause a temporary loss in biological process efficiency if sufficient acclimatization is not provided. In those cases, process selection should

TABLE 5-2
Threshold concentrations of pollutants inhibitory to the activated-sludge process[a]

	Concentration, mg/L	
Pollutant	Carbonaceous removal	Nitrification
Aluminum	15–26	
Ammonia	480	
Arsenic	0.1	
Borate (Boron)	0.05–100	
Cadmium	10–100	
Calcium	2,500	
Chromium (hexavalent)	1–10	0.25
Chromium (trivalent)	50	
Copper	1.0	0.005–0.5
Cyanide	0.1–5	0.34
Iron	1,000	
Manganese	10	
Magnesium		50
Mercury	0.1–5.0	
Nickel	1.0–2.5	0.25
Silver	5	
Sulfate		500
Zinc	0.8–10	0.08–0.5
Phenols:		
Phenol	200	4–10
Cresol		4–16
2-4 Dinitrophenol		150

[a] Ref. 14.

consider how proper treatment can be accommodated so that permit requirements can be met consistently. Therefore, if the wastewater contains significant levels of heavy metals or other toxic materials that may inhibit or otherwise impact the treatment process, these materials will have to be removed at their source or at the treatment plant by appropriate pretreatment.

The presence of toxic materials in the wastewater influent may also result in their presence in the effluent or in the waste sludge. Significant concentrations in the effluent may result in NPDES-permit toxicity violations. High levels of heavy metals may inhibit anaerobic sludge digestion and may also make the sludge (and ash, if incinerated) unsuitable for disposal in landfills or for application on land. For additional information on the removal of toxic pollutants (including volatile organics, semivolatile organics, and heavy metals) by various wastewater treatment processes and resulting concentrations in sludges, Ref. 4 may be consulted. Because the identification of toxic materials and their effects on treatment and environmental systems continues to evolve, recent literature reviews, such as those prepared annually by the Water Pollution Control Federation, also should be consulted.

TABLE 5-3
Estimate of the components of total (dissolved and suspended) solids in wastewater

Component	Dry weight, lb/capita · d	
	Range	Typical
Water supply	0.02–0.04	0.03
Domestic wastes:		
Feces (solids, 23%)	0.07–0.15	0.09
Ground food wastes	0.07–0.18	0.10
Sinks, baths, laundries, and other sources of domestic wash waters	0.13–0.22	0.18
Toilet (including paper)	0.03–0.06	0.04
Urine (solids, 3.7%)	0.09–0.15	0.11
Water softeners	[a]	[a]
Total for domestic wastewater, excluding water softeners	0.41–0.80	0.55
Industrial wastes	0.33–0.88	0.44[b]
Total domestic and industrial wastes	0.74–1.68	0.99
Nonpoint sources	0.02–0.09	0.04[c]
Storm water	0.04–0.09	0.06[c]
Total for domestic, industrial, nonpoint, and storm water	0.80–1.86	1.09

[a] Variable.
[b] Varies with the type and size of facility.
[c] Varies with the season.
Note: lb × 453.59 = g

Unit Loading Factors

When it is impossible to conduct a wastewater characterization study and other data are unavailable, unit per capita loading factors are used to estimate the total waste loadings to be treated.

The total solids in wastewater are derived from the potable water supply; domestic, commercial, and industrial water use; various nonpoint sources; and groundwater infiltration. Domestic wastewater solids include those derived from toilets, sinks, baths, laundries, garbage grinders, and water softeners. Typical data on the daily per capita quantities of dry solids derived from these and the aforementioned sources are reported in Table 5-3. Assuming that the typical per capita wastewater flow is 100 gal/d (380 L/d) and using the total solids value reported in Table 3-16 for medium-strength wastewater (720 mg/L), the total solids contribution would be about 0.6 lb/capita · d (274 g/capita · d). Excluding industrial wastes, this value compares well with the data reported in Table 5-3.

From an analysis of data on the composition of wastewater from a number of municipalities, it has been possible to develop unit loading factors for the principal contaminants of concern in wastewater, as reported in Table 5-4. These values must be used with great care because wastewater constituents vary widely. Household garbage grinders, for example, can have a significant effect on wastewater characteristics, as illustrated in Example 5-4.

Example 5-4 Determining the effects of garbage grinders on wastewater characteristics. Determine the per capita characteristics for BOD and SS, if garbage grinders are installed in a community. Assume that the average per capita flow is 100 gal/d (380 L/d). How would the wastewater be classified, using the classifications listed in Table 3-16?

Solution

1. Determine the per capita BOD and SS contributions.

   (a) From Table 5-4, the typical average per capita contributions for domestic wastewater with ground kitchen wastes are

   $$BOD = 0.22 \text{ lbs/capita} \cdot d$$

   $$SS = 0.26 \text{ lbs/capita} \cdot d$$

   (b) Compute per capita BOD contribution.

   $$BOD, \text{mg/L} = \frac{0.22 \text{ lb/capita} \cdot d \times 10^6 \text{ gal/Mgal}}{[8.34 \text{ lb/Mgal} \cdot (\text{mg/L})] \times 100 \text{ gal/capita} \cdot d}$$
   $$= 264 \text{ mgl/L}$$

   (c) Compute per capita SS contribution.

   $$SS, \text{mg/L} = \frac{0.26 \text{ lb/capita} \cdot d \times 10^6 \text{ gal/Mgal}}{[8.34 \text{ lb/Mgal} \cdot (\text{mg/L})] \times 100 \text{ gal/capita} \cdot d}$$
   $$= 312 \text{ mg/L}$$

2. Determine the waste classification.

(a) Using the data given in Table 3-16, the wastewater would be classified as being about midway between medium and strong.

Comment. For the typical values reported in Table 5-4, garbage grinders increase the per capita BOD contribution from 0.18 to 0.22 lb/capita · d, or an increase of 25 percent. Similarly, the suspended solids will increase from 0.20 to 0.26 lb/capita · d, or 33 percent. Therefore, garbage grinders can have an appreciable effect on the wastewater strength if their use is widespread in a community. The resulting loads will increase the sizes of the solids-handling facilities and the biological treatment units. Existing treatment plants often become overloaded as new developments are served. In some communities, garbage grinders are not permitted because of their potential impact on the existing treatment plants.

5-4 PROCESS SELECTION

As discussed in Chap. 4, one of the most challenging aspects of treatment plant design is the analysis and selection of the treatment process capable of meeting the permit requirements. The methodology of process analysis resulting in process selection includes several evaluation steps that will vary depending upon the complexity of the project and the experience of the design engineer. Process analysis will need to

TABLE 5-4
Unit waste-loading factors

Constituent	Value, lb/capita · d	
	Range	Typical
Normal domestic wastewater without contribution from ground kitchen wastes		
BOD_5	0.13–0.24	0.18
SS	0.13–0.25	0.20
Nutrients[a]		
Ammonia nitrogen	0.004–0.008	0.007
Organic nitrogen	0.013–0.026	0.020
Total Kjeldahl nitrogen	0.020–0.031	0.027
Organic phosphorus	0.002–0.004	0.003
Inorganic phosphorus	0.004–0.007	0.006
Total phosphorus	0.007–0.011	0.008
Normal domestic wastewater with contribution from ground kitchen wastes[b]		
BOD_5	0.18–0.26	0.22
SS	0.20–0.33	0.26

[a] Values adapted from Ref. 13.

[b] Values for nutrients are approximately the same as those shown for wastewater without contribution from ground kitchen wastes.

Note: lb × 453.59 = g

consider (1) important factors in process selection, (2) process selection based on kinetic analysis, (3) process selection based on empirical relationships, (4) the impact of variations of wastewater flowrates and constituent loadings on process design, and (5) process reliability in meeting performance requirements.

Important Factors in Process Selection

In the earlier chapters of this text and in the preceding sections of this chapter, a variety of influent conditions have been described that must be considered in process selection. The importance of understanding the variability of the influent conditions cannot be overstated because the unit operations and processes must have the capability of handling these variations successfully. This capability has been termed "equilibrium" and has been defined as the inherent tolerance that wastewater processes have for the pollutant loads applied to the plant [17]. Therefore, selection of facilities that are compatible with the range of influent wastewater flows and loads and that produce a consistent effluent is one of the most important design considerations.

The various combinations of unit operations and processes in a treatment plant work as a system; therefore, the designer must use a "systems" approach in the facilities design. The major part of the selection process is the evaluation of various combinations of unit operations and processes and their interactions. Part of this selection process may include consideration of flow equalization in reducing loadings on the treatment units. The evaluation process is not limited to the wastewater treatment units alone; the interaction of the liquid with the sludge-processing alternatives must be done as an integral part of the evaluation. The mass-balance analysis then becomes a critical element of the evaluation.

The most important factors that must be considered in evaluating and selecting unit operations and processes are identified in Table 5-5. Each factor is important in its own right, but some factors require additional attention and explanation. The first factor, process applicability, stands out above all others and reflects directly upon the skill and experience of the design engineer. Many resources are available to the designer to determine applicability, including past experience in similar type projects. Available resources include performance data from operating installations, published information in technical journals, manuals of practice published by the Water Pollution Control Federation, process design manuals published by EPA, and pilot plant studies. Examples of published data for the performance of various unit operations and processes used in primary and secondary treatment are presented in Table 5-6. Where the applicability of a process to a given situation is unknown or uncertain, pilot plant studies must be conducted to determine performance capabilities and to obtain design data on which a full-scale design can be based.

Treatment plant performance is the measure of the success of the design, either in terms of effluent quality or of the percent removal obtained for the constituents of concern. For biological systems commonly used for the secondary treatment of wastewater, many factors can affect process performance. Examples of the factors that can affect the performance of activated-sludge, trickling-filters, and rotating biological

TABLE 5-5
Important factors that must be considered when evaluating and selecting unit operations and processes

Factor	Comment
1. Process applicability	The applicability of a process is evaluated on the basis of past experience, published data, data from full-scale plants, and from pilot plant studies. If new or unusual conditions are encountered, pilot plant studies are essential.
2. Applicable flow range	The process should be matched to the expected range of flowrates. For example, stabilization ponds are not suitable for extremely large flowrates.
3. Applicable flow variation	Most unit operations and processes have to be designed to operate over a wide range of flowrates. Most processes work best at a relatively constant flowrate. If the flow variation is too great, flow equalization may be necessary.
4. Influent-wastewater characteristics	The characteristics of the influent wastewater affect the types of processes to be used (e.g., chemical or biological) and the requirements for their proper operation.
5. Inhibiting and unaffected constituents	What constituents are present and may be inhibitory to the treatment processes? What constituents are not affected during treatment?
6. Climatic constraints	Temperature affects the rate of reaction of most chemical and biological processes. Temperature may also affect the physical operation of the facilities. Warm temperatures may accelerate odor generation and also limit atmospheric dispersion.
7. Reaction kinetics and reactor selection	Reactor sizing is based on the governing reaction kinetics. Data for kinetic expressions usually are derived from experience, published literature, and the results of pilot plant studies. The effect of reaction kinetics on reactor selection is addressed in Appendix G.
8. Performance	Performance is usually measured in terms of effluent quality, which must be consistent with the effluent-discharge requirements.
9. Treatment residuals	The types and amounts of solid, liquid, and gaseous residuals produced must be known or estimated. Often, pilot plant studies are used to identify and quantify residuals.
10. Sludge-processing	Are there any constraints that would make sludge processing and disposal infeasible or expensive? How might recycle loads from sludge processing affect the liquid unit operations or processes? The selection of the sludge-processing system should go hand-in-hand with the selection of the liquid treatment system.

TABLE 5-5
(continued)

Factor	Comment
11. Environmental constraints	Environmental factors, such as prevailing winds, wind directions and proximity to residential areas, may restrict or affect the use of certain processes, especially where odors may be produced. Noise and traffic may affect selection of a plant site. Receiving waters may have special limitations, requiring the removal of specific constituents such as nutrients.
12. Chemical requirements	What resources and what amounts must be committed for a long period of time for the successful operation of the unit operation or process? What effects might the addition of chemicals have on the characteristics of the treatment residuals and the cost of treatment?
13. Energy requirements	The energy requirements, as well as probable future energy cost, must be known if cost-effective treatment systems are to be designed.
14. Other resource requirements	What, if any, additional resources must be committed to the successful implementation of the proposed treatment system using the unit operation or the process under consideration?
15. Personnel requirements	How many people and what levels of skills are needed to operate the unit operation or process? Are these skills readily available? How much training will be required?
16. Operating and maintenance requirements	What special operating or maintenance requirements will need to be provided? What spare parts will be required and what will be their availability and cost?
17. Ancillary processes	What support processes are required? How do they affect the effluent quality, especially when they become inoperative?
18. Reliability	What is the long-term reliability of the unit operation or process under consideration? Is the operation or process easily upset? Can it stand periodic shock loadings? If so, how do such occurrences affect the quality of the effluent?
19. Complexity	How complex is the process to operate under routine or emergency conditions? What levels of training must the operators have to operate the process?
20. Compatibility	Can the unit operation or process be used successfully with existing facilities? Can plant expansion be accomplished easily?
21. Land availability	Is there sufficient space to accommodate not only the facilities currently under consideration but possible future expansion? How much of a buffer zone is available to provide landscaping to minimize visual and other impacts?

TABLE 5-6
Degree of treatment achieved by various unit operations and processes used in primary and secondary treatment[a]

Treatment units	Constituent removal efficiency, percent					
	BOD	COD	SS	P[b]	Org-N[c]	NH₃-N
Bar racks	nil	nil	nil	nil	nil	nil
Grit chambers	0–5[d]	0–5[d]	0–10[d]	nil	nil	nil
Primary sedimentation	30–40	30–40	50–65	10–20	10–20	0
Activated sludge (conventional)	80–95	80–85	80–90	10–25	15–50	8–15
Trickling filters						
High rate, rock media	65–80	60–80	60–85	8–12	15–50	8–15
Super rate, plastic media	65–85	65–85	65–85	8–12	15–50	8–15
Rotating biological contactors (RBCs)	80–85	80–85	80–85	10–25	15–50	8–15
Chlorination	nil	nil	nil	nil	nil	nil

[a] Adapted in part from Ref. 10 and 14.
[b] P = Total phosphorus.
[c] Org-N = Organic nitrogen.
[d] The higher numbers apply if grit washers are not used.

contactors (RBCs) processes are presented in Table 5-7. Therefore, in determining the process applicability and for making a selection, careful review of performance factors must be made.

Design provisions for flowrate variations, in addition to flow equalization, may include flow splitting and unit process bypassing under certain peak flowrate conditions. Minimum treatment requirements, if permitted by regulatory authorities, may include primary treatment and disinfection of the entire flow and secondary treatment of a portion of the flow. Advantages of a unit process flow-splitting and bypassing strategy are that (1) the biomass in the secondary treatment process can be preserved during peak storm conditions and not lost due to washout, (2) the quality of the treatment plant effluent can be restored shortly after the storm event, and (3) the entire treatment facilities need not be oversized to handle unusual events. A disadvantage is that the effluent quality may violate the discharge permit for short periods of time.

Process Selection Based on Kinetic Analysis

Wastewater treatment is carried out in tanks or basins of various types and shapes under controlled conditions. The biological or chemical transformations occur in reactors and the resulting products of the reactions are separated typically in settling basins. Each treatment plant will require the selection of at least one type of reactor for chemical or biological treatment, and, in most cases, require one or more settling basins. Particular emphasis is placed on reaction kinetics and reactor selection, which

TABLE 5-7
Factors affecting the performance of typical secondary treatment processes[a]

Process	Factors affecting performance
Activated-sludge	Reactor type Hydraulic detention time Hydraulic loading Organic loading Aeration capacity Mean cell-residence time (MCRT) Food/microorganisms ratio (F/M) Return sludge recirculation rate Nutrients Environmental factors (pH, temperature)
Trickling-filter	Media type and depth Hydraulic loading Organic loading Ventilation Filter staging Recirculation rate Flow distribution
RBCs	Number of stages Organic loading Hydraulic loading Drive mechanisms Media density Shaft selection Recirculation rate Submergence Rotational speed

[a] Adapted in part from Refs. 15 and 16.

are key ingredients in process selection. A brief introduction to reactor selection is provided in this section. Selection of settling basins used with reactors is included in Chaps. 9 and 10.

Consideration of Reactor Types. Containers, vessels, or tanks in which chemical and biological reactions are carried out are commonly called reactors. The principal types of reactors used for the treatment of wastewater are (1) the batch reactor; (2) the plug-flow reactor, also known as a tubular-flow reactor; (3) the complete-mix reactor, also known as a continuous-flow stirred-tank reactor; (4) complete-mix reactors in series; (5) the arbitrary-flow reactor; (6) the packed-bed reactor; and (7) the fluidized-bed reactor. Descriptions of these reactors are presented in Table 5-8. The classification of the first five reactors is based on their hydraulic characteristics. Homogeneous reactions are usually carried out in such reactors. Heterogeneous reactions are usually carried out in the latter two types of reactors.

Operational factors that must be considered in the type of reactor or reactors to be used in the treatment process include (1) the nature of the wastewater to be treated,

TABLE 5-8

Principal types of reactors used for the treatment of wastewater

Type of reactor	Identification sketch	Description and/or application
Batch		Flow is neither entering nor leaving the reactor. The liquid contents are mixed completely. For example, the BOD test discussed in Chap. 3 is carried out in a bottle batch reactor.
Plug-flow, also known as tubular-flow		Fluid particles pass through the tank and are discharged in the same sequence in which they enter. The particles retain their identity and remain in the tank for a time equal to the theoretical detention time. This type of flow is approximated in long tanks with a high length-to-width ratio in which longitudinal dispersion is minimal or absent.
Complete-mix, also known as continuous-flow stirred-tank		Complete mixing occurs when the particles entering the tank are dispersed immediately throughout the tank. The particles leave the tank in proportion to their statistical population. Complete mixing can be accomplished in round or square tanks if the contents of the tank are uniformly and continuously redistributed.
Arbitrary-flow		Arbitrary flow is any degree of partial mixing between plug-flow and complete mixing.
Complete-mix reactors in series		The series of complete-mix reactors is used to model the flow regime that exists between the hydraulic flow patterns corresponding to the complete-mix and plug-flow reactors. If the series is composed of one reactor, the complete-mix regime prevails. If the series consists of an infinite number of reactors in series, the plug-flow regime prevails.
Packed-bed	Packing medium	Packed-bed reactors are filled with some type of packing medium, such as rock, slag, ceramic, or plastic. With respect to flow, they can be completely filled (anaerobic filter) or intermittently dosed (trickling filter).
Fluidized-bed	Expanded packing medium	The fluidized-bed reactor is similar to the packed-bed reactor in many respects, but the packing medium is expanded by the upward movement of fluid (air or water) through the bed. The porosity of the packing can be varied by controlling the flowrate of the fluid.

(2) the reaction kinetics governing the treatment process, (3) process requirements, and (4) local environmental conditions. In practice, the construction costs and operation and maintenance costs also affect reactor selection. Because the relative importance of these factors varies with each application, each factor should be considered separately when the type of reactor is to be selected.

Reactor Flow Regimes and Reactor Combinations. Some of the more common alternative flow regimes and reactor combinations are shown schematically in Fig. 5-7. The flow regime shown in Fig. 5-7a is used to achieve intermediate levels of treatment by blending various amounts of treated or untreated wastewater. The flow regime used in Fig. 5-7b is often adopted to achieve greater process control and will be considered specifically in Chaps. 8 and 10. The flow regimes shown in Figs. 5-7c and 5-7d are used to reduce the loading applied to the head end of a plug-flow reactor. Each of these hydraulic regimes is considered further in the following chapters.

Among the numerous types of reactor combinations that are possible and that have been used, two combinations using a plug-flow reactor and a complete-mix reactor are shown in Fig. 5-8. In the arrangement shown in Fig. 5-8a, more complete

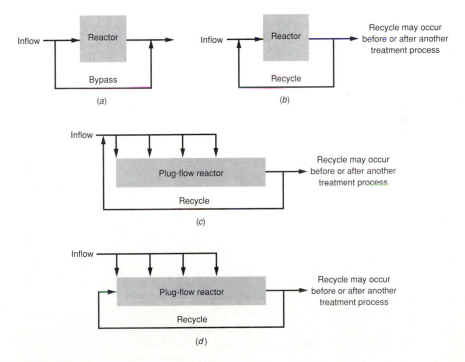

FIGURE 5-7
Flow regimes commonly used in the treatment of wastewater: (a) direct input with bypass flow (plug-flow or complete-mix reactor), (b) direct input with recycle flow (plug-flow or complete-mix reactor), (c) step input with or without recycle (plug-flow reactor, recycle type 1), and (d) step input with recycle (plug-flow reactor, recycle type 2).

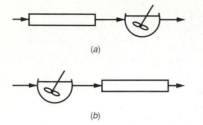

(a)

(b)

FIGURE 5-8
Hybrid reactor systems: (a) plug-flow reactor followed by complete-mix reactor, and (b) complete-mix reactor followed by plug-flow reactor.

mixing takes place later; in the arrangement shown in Fig. 5-8b, it occurs first. For example, if no reaction takes place and the reactors are used only to equalize temperature, the result will be identical. If a reaction is occurring, however, the product yields of the two reactor systems can be different. The use of such hybrid reactor systems will depend on the specific product requirements. Additional details on the analysis of such processes may be found in Refs. 2, 5, and 6.

Selection of Reaction-Rate Expressions. In treatment plant design, the unit processes may be designed on the basis of the rate at which the reaction proceeds rather than the equilibrium position of the reaction, because the reaction usually takes too long to go to completion. In this case, quantities of chemicals in excess of stoichemetric or exact reacting amounts may be used to accomplish the treatment step in a reasonable period of time. Therefore, selection of reaction-rate expressions for the process that is to be designed is based on (1) information obtained from the literature, (2) experience with the design and operation of similar systems, or (3) data derived from pilot plant studies. In cases where significantly different wastewater characteristics occur or new applications of existing technology or new processes are being considered, pilot plant testing is recommended. The various rate expressions that have been developed for biological waste treatment are considered in Chap. 8, and their application is illustrated in Chap. 10.

Application of Mass-Balance Analysis. A mass balance affords a convenient way of defining what occurs within treatment facilities as a function of time. To illustrate the basic concepts involved, a mass-balance analysis will be performed on the contents of the container shown schematically in Fig. 5-9. First, the system boundary must be established so that all the flows of mass into and out of the system can be identified. In Fig. 5-9, the boundary is shown by a dashed line. The proper selection of the system boundary is extremely important because, in many situations, it will be possible to simplify the mass-balance computations.

To apply a mass-balance analysis to the liquid contents of the complete-mix reactor shown in Fig. 5-9, it will be assumed that (1) the volumetric flowrate into and out of the container is constant; (2) the liquid within the reactor is not subject to evaporation (isothermal conditions); (3) the liquid within the container is mixed completely; (4) a chemical reaction involving the reactant C is occurring within the reactor; and (5) the rate of change in the concentration of the reactant C occurring within the reactor is governed by a first-order reaction ($r_c = -kC$). For the stated assumptions, the mass balance can be formulated as follows:

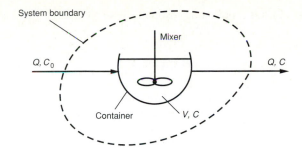

System boundary

Mixer

Q, C_0

Q, C

Container

V, C

FIGURE 5-9
Definition sketch for a mass-balance
analysis for a complete-mix reactor.

1. General word statement:

| Rate of accumulation of reactant within the system boundary | = | Rate of flow of reactant into the system boundary | − | Rate of flow of reactant out of the system boundary | + | Rate of generation (utilization) of reactant within the system boundary | (5-6) |

2. Simplified word statement:

$$\text{Accumulation} = \text{Inflow} - \text{Outflow} + \text{Generation} \qquad (5\text{-}7)$$

3. Symbolic representation (refer to Fig. 5-9):

$$V\frac{dC}{dt} = QC_o - QC + V(\text{rate of reaction}, r_c)$$

$$V\frac{dC}{dt} = QC_o - QC + V(-kC) \qquad (5\text{-}8)$$

where V = volume of reactor, L^3
$\frac{dC}{dt}$ = rate of change of reactant concentration within the reactor, $ML^{-3}T^{-1}$
Q = volumetric rate of flow into and out of the container, L^3T^{-1}
C_o = concentration of reactant in the influent, ML^{-3}
C = concentration of reactant in reactor and effluent, ML^{-3}
k = first-order reaction-rate constant, T^{-1}

In Eq. 5-6, a positive sign is used for the rate-of-utilization term because the necessary negative sign is part of the rate expression. Before substituting numerical values in any mass-balance expression, a unit check should always be made to assure that units of the individual quantities are consistent. The analytical procedures used for the solution of mass-balance equations usually are governed by the mathematical form of the final expression. The general non-steady state solution for Eq. 5-8 is presented in Appendix G.

Mass Balance for Batch Reactor. Before proceeding further, it will be instructive to explore the difference between the rate-of-change term that appears as part of the accumulation term and the rate-of-generation or decay term. In general, these terms are not equal, except in the special case when there is no inflow or outflow from the container or vessel in which the reaction is occurring. Such a container is known as a batch reactor (see Table 5-8). In this situation, Q is equal to zero and Eq. 5-8 becomes

$$\frac{dC}{dt} = (\text{rate of utilization}, r_u, \text{ or generation}, r_g) \tag{5-9}$$

The key point to remember is that when flow is not occurring, the concentration per unit volume is changing according to the applicable rate expression. On the other hand, when flow is occurring, the concentration in the reactor is also being modified by the inflow or outflow from the reactor.

Performance Comparisons. Performance comparisons are of great interest in reactor selection and design, where the engineer can influence results. For example, wastewater treatment plants are usually required to remove 85 percent or more of the entering BOD. In most systems, the removal is accomplished in two steps: a primary stage, in which about 30 percent of the BOD is removed, and a secondary stage, which removes over 55 percent. To achieve an overall performance of 85 percent BOD removal, either a complete-mix or plug-flow reactor can be used. It should be noted, however, that the total volume required for the two reactor types will be quite different depending on the removal kinetics. The total volume required for various removal efficiencies for first-order kinetics, using 1, 2, 4, 6, 8, or 10 complete-mix reactors in series as compared to a plug-flow reactor, is reported in Table 5-9. As indicated in Table 5-9, greater reactor volume is required for complete-mix reactors to achieve the same removal efficiencies as plug-flow reactors. Also, as more complete-mix reactors are added in series, the total required volume begins to approach the volume of the plug-flow reactor.

TABLE 5-9
Required reactor volumes expressed in terms of Q/k for complete-mix reactors in series and a plug-flow reactor for various removal efficiencies for first-order kinetics[a]

No. of reactors in series	Reactor volume $V = K(Q/k)$			
	85% removal efficiency	90% removal efficiency	95% removal efficiency	98% removal efficiency
1	5.67	9.00	19.00	49.00
2	3.18	4.32	6.96	12.14
4	2.48	3.10	4.48	6.64
6	2.22	2.82	3.90	5.50
8	2.16	2.64	3.60	5.04
10	2.10	2.60	3.50	4.80
Plug-flow	1.90	2.30	3.00	3.91

[a] Volume of individual reactors equals value in table divided by the number of reactors in series.

Process Selection Based on Empirical Relationships

If appropriate reaction-rate expressions cannot be developed, generalized loading criteria are frequently used. Early design loading criteria for activated-sludge systems were based on lb $BOD/10^3$ ft^3 of aeration tank capacity (kg BOD/m^3). For example, if a process that is loaded at 30 lb/10^3 ft^3 produces an acceptable effluent and one loaded at 60 lb/10^3 ft^3 does not, the successful experience tends to be repeated. Unfortunately, records often are not well-maintained, and the limits of such loading criteria are seldom defined. Examples of loading criteria are presented in the design chapters for unit operations and processes.

Impact of Variations of Wastewater Flowrates and Constituent Loadings on Process Selection

Almost all kinetic and empirical factors are based on constant wastewater flowrate and loading conditions. In practice, the flowrates and loadings vary, sometimes over a very wide range. It is necessary, therefore, to identify the anticipated range of flow and loading conditions and how they might affect the various alternative unit operations and processes being considered. Table 5-10 identifies critical design and sizing factors for secondary treatment plant facilities and describes potential performance impacts of flowrate and constituent mass-loading variations. For example, the solids removal capability of a primary sedimentation basin is related to the overflow rate, which is a function of flowrate and surface area of the basin. When high flowrates occur, the removal efficiency of the sedimentation basin decreases, and more solids and organic loads are passed on to the following secondary treatment process. The increased flowrate, therefore, not only impacts the performance of the sedimentation basin but the succeeding processes as well. All unit operations and processes have to be evaluated similarly to ensure that performance impacts are properly identified.

Reliability Considerations in Process Selection and Design

Important factors in process selection and design are treatment plant performance and reliability in meeting permit requirements. Most permits specify effluent-constituent requirements based on 7-day and 30-day average concentrations. The national standards for secondary treatment discussed in Chap. 4 exemplify these requirements. Because wastewater treatment effluent quality is variable for a number of reasons (varying loads, changing environmental conditions, etc.), it is necessary to ensure that the treatment system is designed to produce effluent concentrations equal to or less than the permit limits. Two approaches in process selection and design are (1) the use of arbitrary safety factors, and (2) statistical analysis of treatment plant performance to determine a functional relationship between effluent quality and the probable frequency of occurrence. The latter approach, termed the "reliability concept," is pre-

TABLE 5-10
Effect of flowrates and constituent mass loadings on the selection and sizing of secondary treatment plant facilities

Unit operation or process	Critical design factor(s)	Sizing criteria	Effects of design criteria on plant performance
Wastewater pumping and piping	Maximum hour flowrate	Flowrate	Wetwell may flood, collection system may surcharge, or treatment units may overflow if peak rate is exceeded.
Screening	Maximum hour flowrate	Flowrate	Headlosses through bar rack and screens increase at high flowrates.
	Minimum flowrate	Channel approach velocity	Solids may deposit in approach channel at low flowrates.
Grit removal	Maximum hour flowrate[a]	Overflow rate	At high flowrates, grit removal efficiency decreases in flow-through type grit chambers causing grit problems in other processes.
Primary sedimentation	Maximum hour flowrate[a]	Overflow rate	Solids removal efficiency decreases at high overflow rates; increases loading on secondary treatment system.
	Minimum hour flowrate	Detention time	At low flowrates, long detention times may cause the wastewater to be septic.
Activated sludge	Maximum hour flowrate[a]	Hydraulic residence time	Solids washout at high flowrates; may need effluent recycle at low flowrates.
	Maximum daily organic load	Food/microorganism ratio	High oxygen demand may exceed aeration capacity and cause poor treatment performance.
Trickling filters	Maximum hour flowrate[a]	Hydraulic loading	Solids washout at high flowrates may cause loss of process efficiency.
	Minimum hour flowrate	Hydraulic and organic loading	Increased recycle at low flowrates may be required to sustain process.
	Maximum daily organic load	Mass loading/media volume	Inadequate oxygen during peak load may result in loss of process efficiency and cause odors.
Secondary sedimentation	Maximum hour flowrate[a]	Overflow rate or detention time	Reduced solids removal efficiency at high overflow rates or short-detention times.
	Minimum hour flowrate	Detention time	Possible rising sludge at long-detention time.
	Maximum daily organic load	Solids loading rate	Solids loading to sedimentation tanks may be limiting.
Chlorine-contact tank	Maximum hour flowrate	Detention time	Reduced bacteria kill at reduced-detention time.

[a] Typically, the 99 percentile value is used.

ferred because it provides a consistent basis for analysis of uncertainty and a rational basis for the analysis of performance and reliability. The application of the reliability concept to process selection and design is discussed in this section and is based on material presented in Ref. 9.

Reliability Concept. Reliability of a system may be defined as the probability of adequate performance for at least a specified period of time under specified conditions, or, in terms of treatment plant performance, the percent of the time that effluent concentrations meet the permit requirements. For example, a treatment process with a reliability of 99 percent is expected to meet the performance requirements 99 percent of the time. For one percent of the time, or three to four times per year, the permit limits are expected to be exceeded. For each specific case where the reliability concept is to be employed, the levels of reliability must be evaluated, including the cost of the facilities required to achieve specified levels of reliability, associated operating and maintenance costs, and the cost of adverse environmental effects of a discharge violation.

The reliability concept has been applied to the analysis of the performance of 37 activated-sludge plants to form the basis of statistical correlation. The data analysis has resulted in the conclusion that the log-normal distribution for effluent BOD and SS may be used to predict the effluent quality performance and the reliability of wastewater treatment plants [9].

Application. Because of the variations in effluent quality performance, a treatment plant should be designed to produce an average effluent concentration below the permit requirements. The following question arises: what mean value guarantees that an effluent concentration is consistently less than a specified limit with a certain reliability? The approach involves the use of a Coefficient of Reliability (COR) that relates mean constituent values (design values) to the standard that must be achieved on a probability basis. The mean value, m_x, may be obtained by the relationship

$$m_x = X_s(\text{COR}) \tag{5-10}$$

where m_x = mean constituent value
 X_s = a fixed standard
 COR = coefficient of reliability

The coefficient of reliability is determined by

$$\text{COR} = (V_x^2 + 1)^{1/2} \times e\{-Z_{1-\alpha}ln(V_x^2 + 1)^{1/2}\} \tag{5-11}$$

where V_x = ratio of the standard deviation of existing distribution (σ_x) to the mean value of the existing distribution (m_x). V_x is termed the coefficient of variation.
 $Z_{1-\alpha}$ = number of standard deviations away from mean of a normal distribution
 $1 - \alpha$ = cumulative probability of occurrence (reliability level)

TABLE 5-11
Values of standardized normal distribution[a]

Cumulative probability $1 - \alpha$	Percentile $Z_{1-\alpha}$
99.9	3.090
99	2.326
98	2.054
95	1.645
92	1.405
90	1.282
80	0.842
70	0.525
60	0.253
50	0

[a] Ref. 9.

Values of $Z_{1-\alpha}$ for various cumulative probability levels, $1 - \alpha$, are given in Table 5-11. Values of COR for determining effluent concentrations for different coefficients of variation at different levels of reliability are reported in Table 5-12. Selection of an appropriate design value of V_x must be based on experience from operating facilities (actual or published data). The use of the reliability concept is illustrated in Example 5-5.

Example 5-5   Determining design effluent concentration based on the Coefficient of Reliability. An existing activated-sludge plant is required to be expanded and upgraded to meet new permit requirements. The new effluent requirements are

	30-day mean	7-day mean
BOD$_5$, mg/L	25	45
Suspended solids (SS), mg/L	30	45

Determine the mean design effluent BOD and SS concentrations required to meet 95 percent reliability level for the 30-day standard and 99 percent reliability for the 7-day standard. The coefficient of variation is estimated to be 0.70.

Solution

1. Determine the design effluent concentrations for 95 percent reliability for the 30-day standard.
   (*a*) From Table 5-12, the COR for $V_x = 0.70$ and 95 percent reliability is 0.43
   (*b*) Mean design BOD$_5$ = COR $\times X_s$ = 0.43 $\times$ 25 = 10.8 mg/L
   (*c*) Mean design SS = 0.43 $\times$ 30 = 12.9 mg/L
2. Determine the design effluent concentrations for 99 percent reliability for the 7-day standard.
   (*a*) From Table 5-12, the COR for $V_x = 0.70$ and 99 percent reliability is 0.28
   (*b*) Mean design BOD$_5$ = 0.28 $\times$ 45 = 12.6 mg/L
   (*c*) Mean design SS = 0.28 $\times$ 45 = 12.6 mg/L

3. Select the design effluent concentrations (lowest values).

$$BOD_5 = 10.8 \text{ mg/L}$$

$$SS = 12.6 \text{ mg/L}$$

Comment. When the concept of reliability is used, the mean effluent values selected for design may be significantly lower than permit requirements. In cases where the coefficient of variability is high and the reliability requirements are also high, additional unit operations or processes, such as filtration, may have to be used to meet permit requirements consistently.

Another method of determining design conditions to meet effluent standards is the graphical probability method, similar to the method used in Example 2-3 in Chap. 2. Plant performance data can be plotted on log-probability or arithmetic-probability paper to determine the distribution characteristics. For example, the peak day may be determined at the 99+ percentile, based on occurring once every 365 days. Values equal to or less than the indicated value can be determined at the appropriate percentiles. These values can be compared to the values obtained from using the Coefficient of Reliability approach for selecting the appropriate mean effluent concentrations for design.

5-5 ELEMENTS OF CONCEPTUAL PROCESS DESIGN

The purpose of this section is to identify and discuss the principal elements of conceptual process design: (1) establishing the design period for facilities, (2) development of the process flow diagram, (3) establishing process design criteria, (4) preliminary sizing of treatment units, (5) preparation of solids balances, (6) site layout considerations, and (7) evaluation of plant hydraulics (hydraulic profile).

TABLE 5-12
Coefficient of reliability as a function of V_x and reliability[a]

	Reliability, %							
V_x	50	80	90	92	95	98	99	99.9
0.3	1.04	0.81	0.71	0.69	0.64	0.57	0.53	0.42
0.4	1.08	0.78	0.66	0.63	0.57	0.49	0.44	0.33
0.5	1.12	0.75	0.61	0.58	0.51	0.42	0.37	0.26
0.6	1.17	0.73	0.57	0.54	0.47	0.37	0.32	0.21
0.7	1.22	0.72	0.54	0.50	0.43	0.33	0.28	0.17
0.8	1.28	0.71	0.52	0.48	0.40	0.30	0.25	0.15
0.9	1.35	0.70	0.50	0.46	0.38	0.28	0.22	0.12
1.0	1.41	0.70	0.49	0.44	0.36	0.26	0.20	0.11
1.2	1.56	0.70	0.46	0.41	0.33	0.22	0.17	0.08
1.5	1.80	0.70	0.45	0.39	0.30	0.19	0.14	0.06

[a] Ref. 9.

Design Period

The design period establishes the target date when the design capacity of the facilities will be reached. Design periods may vary for individual components, depending upon the ease or difficulty of expansion. Typical design periods for various types of facilities are given in Table 5-13. Longer periods are preferred for structures and hydraulic conduit systems, that cannot be easily expanded. The selection of the design period depends upon growth characteristics, environmental considerations, and the availabilty and source of construction funds.

Treatment Process Flow Diagrams

Treatment process flow diagrams are graphical representations of particular combinations of unit operations and processes. Depending on the constituents that must be removed, an almost limitless number of different flow diagrams can be developed by combining the unit operations and processes reported in Tables 4-2 and 4-3. Apart from the analysis of the suitability of the types of individual treatment units, the exact configuration of process units selected will also depend on factors such as (1) the designer's past experience, (2) design and regulatory agency policies on the application of specific treatment methods, (3) the availability of suppliers of equipment for specific treatment methods, (4) the maximum use that can be made of existing facilities, (5) initial construction costs, and (6) future operation and maintenance costs. A typical process flow diagram for the treatment of wastewater to meet secondary treatment standards, as defined by the U.S. Environmental Protection Agency (see Table 4-1), is shown in Fig. 5-10.

Process Design Criteria

After one or more preliminary process flow diagrams have been developed, the next step is to establish the process design criteria so that the size of the physical facilities

TABLE 5-13
Typical design periods for wastewater facilities

Facility	Planning period range, yrs
Collection systems	20–40
Pumping stations	
Structures	20–40
Pumping equipment	10–25
Treatment plants	
Process structures	20–40
Process equipment	10–20
Hydraulic conduits	20–40

can be determined. For example, if the hydraulic detention time in the aerated grit chamber shown in Fig. 5-10 is to be 3.5 min at a peak flowrate, the corresponding grit chamber volume required would be calculated. The hydraulic detention time would be an example of the process design criteria for the grit chamber. Similar procedures are followed for each unit operation and process.

When the computations have been completed, all the key design criteria should be listed in a summary table. Some of the design parameters commonly used in developing a summary table are listed in Appendix A-3. Because most treatment plants are designed to be effective for some time in the future (up to 20 years), design criteria are given for the time when the facilities will first be put into operation, and for the end of the design period. The latter will be influenced by projections of the population to be served and the economic studies of cost effectiveness for various design periods.

Preliminary Sizing

After the design criteria have been established, the next step is to determine the number and size of the physical facilities needed. In considering sizing, physical site

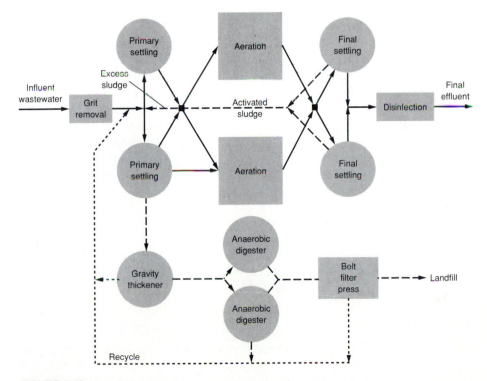

FIGURE 5-10
Process flow diagram for treatment plant designed to meet secondary treatment standards of the U.S. EPA.

constraints need to be considered: for example, will the site accommodate the use of round tanks or will rectangular tanks have to be used? Operational considerations, such as flow splitting and load balancing, will have to be evaluated, particularly in process trains that combine different numbers of unit operations or processes (e.g., two primary clarifiers and three aeration tanks). Maintenance factors have to be considered in selecting the number of units so that provisions are included for taking a unit out of service for maintenance and repair. In small plants where a single unit is being considered, maintenance of that unit may be a particular problem, unless special provisions, such as temporary storage, are included.

Solids Balance

After the design criteria are established and the preliminary sizing is completed, solids balances should be prepared for each process flow diagram. They should be prepared for the average load with appropriate peaking factors applied for maximum loads. Such information must be available to size (1) sludge-thickening and storage facilities, (2) sludge digestors, (3) sludge-dewatering facilities, (4) thermal reduction systems, (5) composting facilities, and (6) sludge-piping and -pumping equipment and other appurtenant facilities. The preparation of a solids balance is illustrated in Chap. 12.

Plant Layout

Plant layout refers to the spatial arrangement of the physical facilities required to achieve a given treatment objective. The overall plant layout includes the location of the control and administrative buildings and any other necessary structures. Several different layouts, using scaled cardboard cutouts of the various treatment facilities or computer-generated overlays, are normally evaluated before a final selection is made. Among the factors that must be considered when laying out a treatment plant are the following: (1) geometry of the available treatment plant sites, (2) topography, (3) soil and foundation conditions, (4) location of the influent sewer, (5) location of the point of discharge, (6) plant hydraulics, preferably with straight-flow paths between units to minimize head loss and provide symmetry for flow splits, (7) types of processes involved, (8) process performance and efficiency, (9) transportation access, (10) accessibility to operating personnel, (11) reliability and economy of operation, (12) aesthetics, (13) environmental control, and (14) provisions for future plant expansion, including additional area. The physical layouts of a variety of plants, both small and large, are shown in Figs. 5-11 through 5-13.

Plant Hydraulics

After the process flow diagram has been selected and the size of the corresponding physical facilities is determined, hydraulic computations and profiles are prepared for both average and peak flowrates. Hydraulic computations are made to size the interconnecting conduits and channels and to compute the headlosses through the plant. Typical ranges of headlosses through treatment units are given in Table

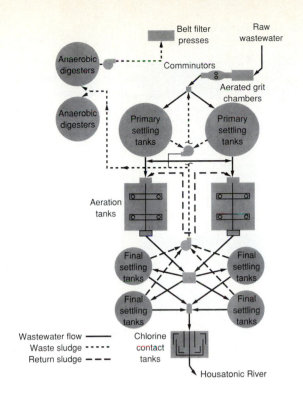

FIGURE 5-11
Layout and aerial view of Housatonic Wastewater Plant, Milford, CT. (Design average flowrate = 8 Mgal/d)

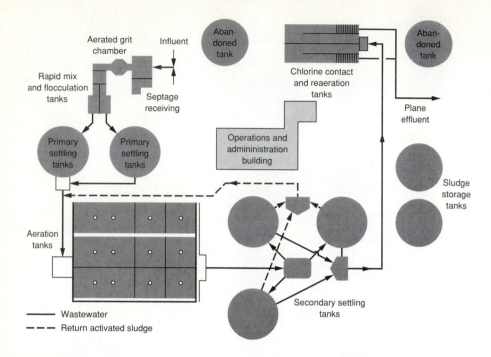

FIGURE 5-12
Layout and aerial view of Leominster Wastewater Treatment Plant, Leominster, MA. (Design average flowrate = 9.3 Mgal/d)

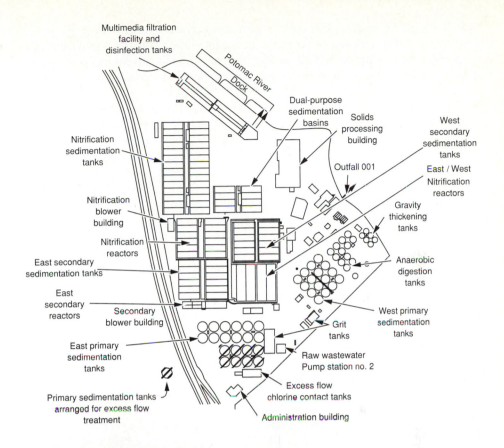

Multimedia filtration facility and disinfection tanks

Potomac River

Dock

Dual-purpose sedimentation basins

Solids processing building

West secondary sedimentation tanks

Nitrification sedimentation tanks

Outfall 001

East / West Nitrification reactors

Nitrification blower building

Gravity thickening tanks

Nitrification reactors

East secondary sedimentation tanks

Anaerobic digestion tanks

East secondary reactors

Secondary blower building

West primary sedimentation tanks

East primary sedimentation tanks

Grit tanks

Raw wastewater Pump station no. 2

Primary sedimentation tanks arranged for excess flow treatment

Excess flow chlorine contact tanks

Administration building

FIGURE 5-13
Layout and aerial view of Blue Plains Wastewater Treatment Plant, Washington, DC. (Design average flowrate = 309 Mgal/d)

5-14. In designing the plant hydraulic system, consideration needs to be given to (1) equalizing the flow splitting between the treatment units, (2) making provisions for bypassing secondary treatment units at extreme peak flows to prevent loss of biomass, and (3) minimizing the number of changes in direction of wastewater flow in conduits and channels.

   Hydraulic profiles are prepared for three reasons: (1) to ensure that the hydraulic gradient is adequate for the wastewater to flow through the treatment facilities by gravity, (2) to establish the head requirement for the pumps where pumping will be needed, and (3) to ensure that the plant facilities will not be flooded or backed up during periods of peak flow. Profiles for the flow diagram given in Fig. 5-10 are shown in Fig. 5-14. In preparing a hydraulic profile, distorted vertical and horizontal scales are commonly used to depict the physical facilities.

   Hydraulic profile computations involve the determination of the headloss as the wastewater flows through each of the physical facilities in the process flow diagram. Specific computational procedures may vary depending on local conditions. For example, if a downstream discharge condition is the control point, some designers prepare the hydraulic profile by working backward from the control point. Other designers prefer to work from the head end of the plant. Still others work from the center in each direction, adjusting the elevations at the end of the computations. The use of

TABLE 5-14
Typical headlosses across various treatment units[a]

Treatment unit	Headloss range, ft
Bar screen	0.5–1.0
Grit chambers	
Aerated	1.5–4.0
Velocity controlled	1.5–3.0
Primary sedimentation	1.5–3.0
Aeration tank	0.7–2.0
Trickling filter	
Low-rate	10.0–20.0
High-rate, rock media	6.0–16.0
High-rate, plastic media	16.0–40.0
Secondary sedimentation	1.5–3.0
Filtration	10.0–16.0
Carbon adsorption	10.0–20.0
Chlorine-contact tank	0.7–6.0

[a] Adapted in part from Refs. 10 and 17.

Note: ft × 0.3048 = m

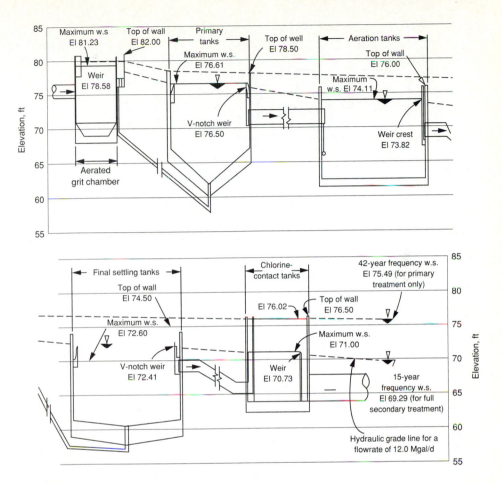

FIGURE 5-14
Hydraulic profile for treatment plant shown in Fig. 6-8. *Note:* ft × 0.3048 = m; Mgal/d × 0.043813 = m³/s; w.s. = water surface.

mathematical models and digital computers allows many possible hydraulic conditions to be analyzed. For information on head loss calculations through a treatment plant, Ref. 10 may be consulted.

DISCUSSION TOPICS AND PROBLEMS

5-1. With the data given in Fig. 5-2, compute the hourly BOD mass loadings and plot a mass-loading curve. At what time is the mass loading at its maximum? At its minimum? What are the ratios of the maximum and minimum mass loadings to the average?

5-2. The variations of influent flowrate and BOD with time are given in the following figure. Compute both the average and the flow-weighted BOD.

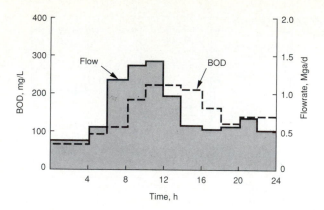

5-3. What explanation can you offer for the tracer curves shown below, obtained for the same plug-flow chlorine-contact basin?

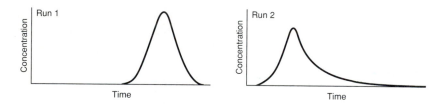

5-4. An industry currently discharges 44,000 lb of BOD_5 and 33,000 lb of suspended solids per day. If the unit waste-loading factors listed in Table 5-4 for domestic wastewater with ground kitchen wastes are used, determine the population equivalent for each parameter in the industrial waste discharge. (Note: Many cities establish their charges for the treatment of industrial wastes on the basis of population equivalence.)

5-5. Using the data shown in Fig. 5-4, develop a curve of the ratio of the sustained peak-to-average BOD loadings similar to that shown in Fig. 5-6. Develop the curve for a 24-month period using 1-month time increments.

5-6. Visit your local wastewater treatment plant and identify the types of reactors or reactor combinations used. Describe the type of flow regime used in the secondary treatment reactor. What other types of flow regime could be used effectively? Justify your answer.

5-7. For first-order removal kinetics, demonstrate that the maximum treatment efficiency in a series of complete-mix reactors occurs when all the reactors are of the same size.

5-8. Determine the number of completely mixed chlorine contact chambers each having a detention time of 30 min that would be required in a series arrangement to reduce the bacterial count of a polluted water sample from 10^6 organisms/mL to 14.5 organisms/mL if the first-order removal-rate constant is equal to 6.1 h^{-1}. If a plug-flow chlorine contact chamber were used with the same detention time as the series completely mixed chambers, what would the bacterial count be after treatment?

5-9. Obtain, or if necessary develop, the process flow diagram for your local wastewater treatment plant. How does it compare to Fig. 5-10? How does it compare in complexity to those shown in Chap. 11?

5-10. At the same time that you are obtaining the process flow diagram for your local wastewater treatment plant, obtain a copy of the hydraulic profile, if it is available. If so, how does it compare with the profile shown in Fig. 5-14?

5-11. If available, obtain a copy of the basic design data for your local wastewater treatment plant. Identify the flowrate and mass-loading factors that are critical for each unit operation or process.

5-12. Develop the hydraulic profile for average and peak flowrate conditions for the portion of the wastewater treatment plant shown in the following figure. Assume that the recycle sludge is returned directly to the aeration tank, that 90° v-notch weirs are used around periphery of the primary and secondary clarifiers and that the overflow weir in the aeration tank is a Francis type with two end contractions. Other pertinent data and information are as follows:

$$Q_{avg} = 1.0 \text{ Mgal/d plus } 100\% \text{ sludge recycle}$$

$$Q_{peak} = 2.0 \text{ Mgal/d plus } 50\% \text{ sludge recycle}$$

$$\text{Spacing of v-notch weirs} = 2 \text{ ft}$$

$$\text{Width of aeration tank effluent weir} = 4.5 \text{ ft}$$

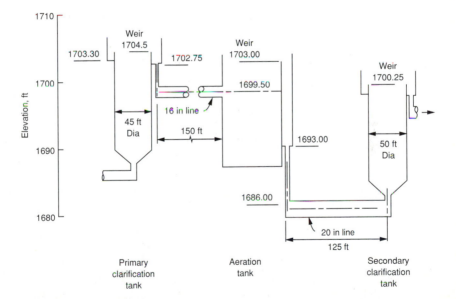

5-13. Develop a summary table from the literature, similar to Table 5-1, listing application factors used in the design of treatment plant process units. List information from at least three references, and cite the references. If different approaches for designing common unit operations or processes are cited in the literature, describe which one you would use and why.

5-14. If the community in Example 5-1 plans to add the following new facilities in addition to the new industry, what will be the future average, peak, and minimum flowrates? Use the typical wastewater flowrates in Table 2-10.

New facilities   Hotel—300 guests/d
                                40 employees
                         Restaurant—600 meals/d
                         Office building—200 employees
                         Self-service laundry—20 machines

5-15. Develop a sustained suspended-solids mass-loading curve for a treatment plant with a design flowrate of 1 m^3/s. Assume that the long-term average suspended-solids concentration is 220 mg/L. Use the suspended-solids peaking factors in Fig. 5-6b. Calculate the mass loadings in metric units and plot a mass-loading curve.

REFERENCES

1. City of Modesto, CA: *Annual Operating Data for Sewage Treatment Plant*, 1970–1972.
2. Denbigh, K. G. and J. C. R. Turner: *Chemical Reactor Theory*, 2nd ed., Cambridge, New York, 1965.
3. Federal Register: "Amendment to the Secondary Treatment Regulations: Percent Removal Requirements During Dry Weather Periods for Treatment Works Served by Combined Sewers," 40 CFR Part 133, January 27, 1989.
4. Hannah, S. A., B. M. Austern, A. E. Eralp, and R. H. Wise: "Comparative Removal of Toxic Pollutants by Six Wastewater Treatment Processes," *Journal WPCF*, vol. 58, no. 1, 1986.
5. Kafarou, V.: *Cybernetic Methods in Chemistry and Chemical Engineering*, MIR Publishers, Moscow, 1976.
6. Levenspiel, O.: *Chemical Reaction Engineering*, 2nd ed., Wiley, New York, 1972.
7. Metcalf & Eddy, Inc.: *Wastewater Engineering: Collection and Pumping of Wastewater*, McGraw-Hill, New York, 1981.
8. Metropolitan Sanitary District of Greater Chicago: *Annual Operating Data for Calumet Sewage Treatment Works*, Chicago, 1969–1973.
9. Niku, S., E. D. Schroeder, and F. J. Samaniego: "Performance of Activated Sludge Processes and Reliability-Based Design," *Journal WPCF*, vol. 51, no. 12, 1979.
10. Qasim, S. R.: *Wastewater Treatment Plants: Planning, Design, and Operation*, Holt, Rinehart and Winston, 1985.
11. Technical Advisory Board of the New England Interstate Water Pollution Control Commission: *Guides for the Design of Wastewater Treatment Works*, 1980.
12. U.S. Environmental Protection Agency: *Process Design Manual for Upgrading Wastewater Treatment Plants*, 1974.
13. Water Pollution Control Federation: *Nutrient Control*, Manual of Practice FD-7, 1983.
14. Water Pollution Control Federation, *Wastewater Treatment Plant Design*, Manual of Practice no. 8, 1977.
15. Water Pollution Control Federation: *Activated Sludge*, Manual of Practice OM-9, 1987.
16. Water Pollution Control Federation: *O & M of Trickling Filters, RBCs, and Related Processes*, Manual of Practice OM-10, 1988.
17. Water Pollution Control Federation: *Wastewater Treatment Plant Design*, Manual of Practice no. 8, Preliminary Draft, 1988.

CHAPTER
6

PHYSICAL
UNIT
OPERATIONS

Those operations used for the treatment of wastewater in which change is brought about by means of or through the application of physical forces are known as unit operations. Because they were derived originally from observations of the physical world, they were the first treatment methods to be used. Today, physical unit operations form the basis of most process flow diagrams. Those used in a typical flow diagrams for wastewater treatment are identified in Fig. 6-1.

The unit operations most commonly used in wastewater treatment include (1) flow metering, (2) screening, (3) comminution, (4) flow equalization, (5) mixing, (6) sedimentation, (7) accelerated gravity settling, (8) flotation, (9) filtration, (10) gas transfer, and (11) volatilization and gas stripping. The principal applications of these operations are summarized in Table 6-1. With the exception of comminution, which is discussed in Chap. 9, a separate section is devoted to each of these operations in this chapter. A discussion of comminution is not included here because comminutors are complete in themselves as supplied by the manufacturer, and no detailed theoretical analysis is possible. Microscreening, another unit operation that on occasion has been used to remove residual suspended solids, is discussed briefly in Chap. 11. Unit operations associated with the processing of sludge are discussed separately in Chap. 12.

In this chapter, each unit operation to be considered will be described, and the fundamentals involved in the engineering analysis of each one will be discussed. The

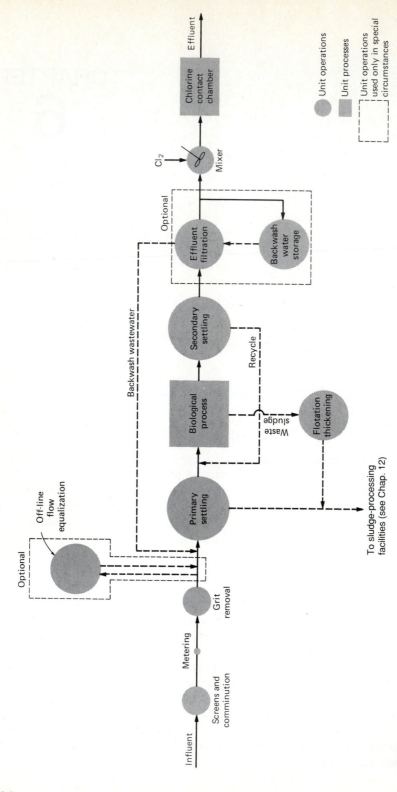

FIGURE 6-1
Location of physical unit operations in a wastewater treatment plant flow diagram.

194

TABLE 6-1
Applications of physical unit operations in wastewater treatment

Operation	Application	See Section
Flow metering	Process control, process monitoring, and discharge reports	6-1
Screening	Removal of coarse and settleable solids by interception (surface straining)	6-2
Comminution	Grinding of coarse solids to a more or less uniform size	9-2
Flow equalization	Equalization of flow and mass loadings of BOD and suspended solids	6-3
Mixing	Mixing chemicals and gases with wastewater, and maintaining solids in suspension	6-4
Flocculation	Promotes the aggregation of small particles into larger particles to enhance their removal by gravity sedimentation	6-5
Sedimentation	Removal of settleable solids and thickening of sludges	6-6
Flotation	Removal of finely divided suspended solids and particles with densities close to that of water. Also thickens biological sludges	6-7
Filtration	Removal of fine residual suspended solids remaining after biological or chemical treatment	6-8
Microscreening	Same as filtration. Also removal of algae from stabilization-pond effluent	11-4
Gas transfer	Addition and removal of gases	6-9
Volatilization and gas stripping	Emission of volatile and semi-volatile organic compounds from wastewaters	6-10

practical application of these operations in the design of facilities is detailed in Chaps. 9 and 10. This same approach will be used in Chaps. 7 and 8, which deal with the chemical and biological unit processes.

6-1 FLOW MEASUREMENT

The correct application, selection, and maintenance of flow-metering devices is critical to the efficient operation of a modern wastewater treatment facility. A complete flow measurement system consists of two elements: (1) a sensor or detector and (2) a converter device. The sensor or detector is exposed to or affected by the flow; the converter is the device used to translate the signal or reading from the sensor into a flow reading. Because of the importance of flow metering, the purpose of this section is to review the types of sensors or detectors that are available for measuring flowrates and their application, important criteria that should be considered in selecting the type of meter to be used and maintenance considerations. Because of the rapid advances made in the metering device electronics, current information on converters should be obtained from meter manufacturers.

Types and Application
of Flow-Metering Devices

A number of devices are available that can be used to measure flowrates in open channels and closed conduits. In the discussion that follows, only the device used to sense the flow is considered. The principal types of sensors or detectors used for the measurement of different flow streams in wastewater treatment facilities are identified in Table 6-2.

For Open Channels. In open channels, or partially filled conduits, the head generated by an obstruction, such as a flume or weir plate or the cross-sectional wetted area, and corresponding velocity are used to determine the flowrate. Perhaps the most widely used device for measuring the flowrate of untreated wastewater is the Parshall flume.

For Closed Conduits. Three techniques are commonly used for measuring flowrates in closed conduits: (1) insertion of an obstruction to create a predictable headloss or pressure difference, (2) measurement of the effect of the moving fluid (e.g., momentum change, sonic wave transmittance, magnetic field shift), and (3) measurement of incremental units of fluid volume. Flow tubes, orifices, pitot tubes, rotameters, and venturi tubes are all used to produce pressure differentials that can be converted into flowrate readings. Magnetic, target, ultrasonic, and vortex measuring devices are included in the second category. Included in the third category are turbine and propeller meters in which the speed of rotation of a sensing element can be correlated to velocity and the flowrate.

Selection Criteria for Metering Devices

Important criteria that must be considered in the selection of flow metering devices include type of application, proper sizing, fluid composition, accuracy, headloss, installation requirements, operating environment, and ease of maintenance. Additional details on these selection criteria are reported in Table 6-3.

Although all of the criteria listed in Table 6-3 are important, accuracy and repeatability are critical, especially where the readings from the metering device are to be used for process control. Instrumentation accuracy is usually expressed as a plus or minus ($\pm$) percentage of the maximum or actual flowrate. Thus, the accuracy of an element in the system must be evaluated in the context of the overall system accuracy. The overall system can be no more accurate (and is usually less accurate) than the least accurate element. Further, because the accuracy of some meters is affected by ambient temperature, power source voltage, electronic interference, and humidity, these factors should be considered carefully in selecting metering devices. The estimated range and accuracy of the metering devices used in wastewater treatment is presented in Table 6-4.

In many treatment plant applications, repeatability (the same value is measured each time) is often more important than accuracy. For example, when flow splitting

TABLE 6-2
Application of flow-metering devices in wastewater treatment facilities[a]

Metering device	Application							
	Raw wastewater	Primary effluent	Secondary effluent	Primary sludge	Return sludge	Thickened sludge	Mixed liquor	Process water
For open channels								
Head/area								
Flume	✓	✓	✓					✓
Weir		✓	✓					✓
Other								
Magnetic (insert type)			✓					✓
Velocity-head								
For closed conduits								
Head/pressure								
Flow tube	✓[b]	✓[b]	✓	✓[b]	✓[b]	✓[b,c]	✓	✓
Orifice								✓
Pitot tube								✓
Rotameter								✓
Venturi	✓[b]	✓[b]	✓	✓[b]	✓[b]	✓[b]	✓	✓
Moving fluid effects								
Magnetic (tube type)	✓	✓	✓	✓	✓	✓		✓
Magnetic (insert type)	✓	✓						
Target								
Ultrasonic (doppler)	✓	✓	✓	✓	✓	✓[d]	✓	✓
Ultrasonic (transmission)		✓	✓					✓
Vortex shedding		✓	✓					
Positive displacement								
Propeller								✓
Turbine			✓					✓

[a] Based on industry practice and engineering judgement.
[b] Flushing or diaphragm sealed connections recommended.
[c] Use with in-line reciprocating pumps not recommended.
[d] Solids content less than 4 percent.

197

TABLE 6-3
Typical criteria used in the selection of flow-metering devices

Criteria	Consideration
Application	Is the metering device suitable for open or closed conduit flow?
Sizing	Is the device appropriately sized for the range of flow that needs to be monitored?
	Are proper operating velocities maintained?
Fluid composition	Is the device compatible with the fluid being monitored?
	Is the solids content of the fluid compatible with the measuring device?
	Does the device have the recommended minimum clear opening for the fluid being monitored?
	Are the wetted components constructed of materials nonreactive with the fluid?
Accuracy and repeatability	Is the accuracy and repeatability of the device consistent with the application?
	Is the stated accuracy of the component consistent with overall system accuracy?
	Has the effect of environmental factors on the stated accuracy been considered?
Headloss	Is the headloss caused by the device within constraints of the hydraulic profile?
Installation requirements	Is sufficient straight length of pipe or channel provided ahead of the meter?
	Is the device located properly with respect to valves and pumps?
	Are the flow-meter devices accessible for service?
	Are quick disconnect couplings and bypass piping provided?
Operating environment	Is the equipment associated with the flow-metering device appropriately rated for its intended application to prevent explosion hazard?
	Where necessary, is the equipment resistent to moisture and corrosive gases?
	Have provisions been made to ensure operation of the device within an acceptable temperature range?
Provisions for maintenance	Are provisions made for flushing or rodding the meter and tap lines?

among process units, actual flow measurement is not as important as is repeatability. The repeatability of the flow-metering devices considered previously is reported in Table 6-4.

Maintenance of Flow-Metering Devices

To ensure proper performance from metering devices, proper cleaning, maintenance, calibration, and recordkeeping is essential. Provisions should be made for cleaning

TABLE 6-4
Characteristics of flow-metering devices used in wastewater treatment facilities[a]

Metering device	Range[b]	Accuracy,[b] percent of actual rate	Repeatability[b] percent of full scale	Straight upstream run in pipe diameters
For open channels				
Head/area				
Flume	10:1–75:1[c]	±5–10[d]	±0.5	
Weir	500:1	±5	±0.5[g]	
Other				
Magnetic (insert type)	10:1	±1–2[e]	±0.5	
Velocity-head				
For closed conduits				
Head/pressure				
Flow tube	4:1	±.3	±0.5	4–10[f]
Orifice	4:1	±1	±1	±5[g]
Pitot tube	3:1	±3	±1[g]	10[g]
Rotameter	10:1	0.5–10	1[g]	5[g]
Venturi meter	4:1	±1	±0.5	4–10[f]
Moving fluid effects				
Magnetic (tube type)	10:1	±1–2[e]	±0.5	5
Magnetic (insert type)	10:1	±1–2[e]	±0.5	5
Target	10:1	±5	1[f]	20
Ultrasonic (Doppler)	10:1	±3	±1	7–10
Ultrasonic (transmission)	10:1	±2	±1	7–10
Vortex shedding	15:1	±1	±0.5	10
Positive displacement				
Propeller	10:1	±2	±0.5	5
Turbine	10:1	±0.25	±0.05	10[h]

[a] Based on industry practice and engineering judgement.

[b] Based on both the primary element and primary conversion device.

[c] Depends on the type of flume.

[d] Parshall flumes ±5%, Palmer-Bowlus flume ±10%.

[e] Of full scale.

[f] Depends on the type of flow-disturbing obstruction.

[g] Estimated.

[h] Assuming that flow straightening is used (25 to 30 pipe diameters, otherwise).

the meter and tap lines by flushing or rodding. In sludge-metering applications where intermittent operation is expected, the capacity to flush the meter and associated piping and to fill them with clean water should be provided. Self-cleaning electrodes are available for use with magnetic flow meters using either high frequency ultrasonic waves or heat.

Flow meters should be calibrated in the field to ensure that specifications are met and to establish base line data that may be used for future monitoring and periodic maintenance calibration. Flow meters should be calibrated periodically by a factory representative to ensure that the meter is functioning properly.

In conjunction with meter maintenance and calibration, recordkeeping is an essential part of proper meter maintenance. In addition to the original calibration data, current operating and maintenance data should be kept on each meter. With adequate records, metering problems can be corrected before they become a problem.

6-2 SCREENING

The first unit operation encountered in wastewater treatment plants is screening. A screen is a device with openings, generally of uniform size, that is used to retain the coarse solids found in wastewater.

Description

The screening element may consist of parallel bars, rods or wires, grating, wire mesh, or perforated plate, and the openings may be of any shape but generally are circular or rectangular slots. A screen composed of parallel bars or rods is called a *bar rack* (sometimes called a *bar screen*). The term "screen" is used for screening devices consisting of perforated plates, wedge wire elements, and wire cloth. The materials removed by these devices are known as *screenings*. According to the method used to clean them, bar racks and screens are designated as hand-cleaned or mechanically cleaned. Typically, bar racks have clear openings (spaces between bars) of 5/8 in (15 mm) or more. Screens have openings of less than 5/8 in (15 mm). The principal types of screening devices now in use are described in Table 6-5 and illustrated in Figs. 6-2 and 6-3.

Bar Racks. In wastewater treatment, bar racks are used to protect pumps, valves, pipelines, and other appurtenances from damage or clogging by rags and large objects. Industrial waste plants may or may not need them, depending on the character of the wastes. A typical bar rack used for wastewater treatment is shown in Fig. 6-2.

Screens. Early screens were of the inclined disk or drum type, whose screening media consisted of bronze or copper plates with milled slots, and were installed in place of sedimentation tanks for primary treatment. Since the early 1970s, there has been a resurgence of interest in the field of wastewater treatment in the use of screens of all types. The applications range from primary treatment to the removal of the residual suspended solids from biological treatment processes. To a large extent, this renewed interest developed because better screening materials and better screening devices are now available, and research is continuing in this area. Typical screening devices used for wastewater treatment are shown in Fig. 6-3. The rotary disk screen shown in Fig. 6-3c has also been used as substitute for a primary sedimentation tank [20]. The use of screening devices is considered further in Chap. 9.

Analysis

The analysis associated with the use of screening devices involves the determination of the headloss through them. The approach used for bar racks differs from that used for screens; so they are discussed separately.

TABLE 6-5
Description of screening devices used in wastewater treatment

Type of screening device	Size classification	Screening surface		Application	See Figure
		Size range in[a]	Screen material		
Bar rack	Coarse	0.6–1.5	Steel, Stainless-steel	Pretreatment	6-2
Screens:					
Inclined (Fixed)	Medium	0.01–0.1	Stainless-steel wedge-wire screen	Primary treatment	6-3a
Inclined (Rotary)	Coarse	0.03 × 0.09 × 2	Milled bronze or copper plates	Pretreatment	
Drum (rotary)	Coarse	0.1–0.2	Stainless-steel wedge-wire screen	Pretreatment	
	Medium	0.01–0.1	Stainless-steel wedge-wire screen	Primary treatment	6-3b
	Fine	6–35 μm	Stainless-steel and polyester screen cloths	Removal of residual secondary suspended solids	
Rotary disk	Medium	0.01–0.4	Stainless-steel	Primary treatment	
	Fine	0.001–0.02	Stainless-steel	Primary treatment	6-3c
Centrifugal	Fine	0.002–0.02	Stainless-steel, polyester, and various other fabric screen cloths	Primary treatment, secondary treatment with settling tank, and the removal of residual secondary suspended solids	6-3d

[a] Unless otherwise noted

Note: mm × 0.03937 = in.

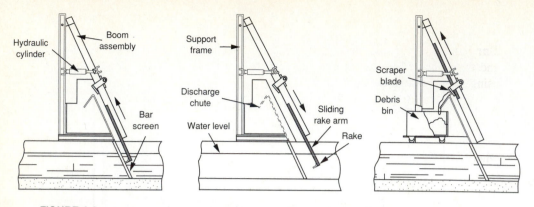

FIGURE 6-2
Typical mechanically cleaned bar rack used for wastewater treatment (from of Franklin Miller).

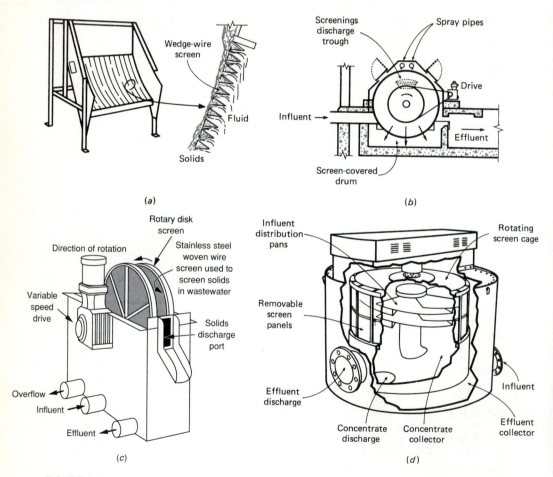

FIGURE 6-3
Typical screening devices used for wastewater treatment: (*a*) inclined fixed screen (shown with cover removed), and (*b*) rotary drum screen, (*c*) rotary disk screen, and (*d*) centrifugal screen.

Bar Racks. Hydraulic losses through bar racks are a function of approach velocity and the velocity through the bars. The headloss through bar racks can be estimated using the following equation.

$$h_L = \frac{1}{0.7}\left(\frac{V^2 - v^2}{2g}\right) \tag{6-1}$$

where h_L = headloss, ft (m)
0.7 = an empirical discharge coefficient to account for turbulence and eddy losses
V = velocity of flow through the openings of the bar rack, ft/s (m/s)
v = approach velocity in upstream channel, ft/s (m/s)
g = acceleration due to gravity, ft/s^2 (m/s^2)

The headloss calculated using Eq. 6-1 applies only when the bars are clean. Headloss increases with the degree of clogging.

Fine Screens. The clear-water headloss through screens may be obtained from manufacturers' rating tables, or it may be calculated by means of the common orifice formula:

$$h_L = \frac{1}{C(2g)}\left(\frac{Q}{A}\right)^2 \tag{6-2}$$

where h_L = headloss, ft (m)
C = coefficient of discharge for the screen
g = acceleration due to gravity, ft/s^2 (m/s^2)
Q = discharge through screen, ft^3/s (m^3/s)
A = effective open area of submerged screen, ft^2 (m^2)

Values of C and A depend on screen design factors, such as the size and milling of slots, the wire diameter and weave, and particularly the percent of open area, and must be determined experimentally. A typical value of C for a clean screen is 0.60. The headloss through a clean screen is relatively insignificant. The important determination is the headloss during operation, and this depends on the size and amount of solids in the wastewater, the size of the apertures, and the method and frequency of cleaning.

6-3 FLOW EQUALIZATION

The variations that are observed in the influent-wastewater flowrate and strength at almost all wastewater treatment facilities were discussed in Chaps. 2 and 3, respectively. Flow equalization is used to overcome the operational problems caused by flowrate variations, to improve the performance of the downstream processes, and to reduce the size and cost of downstream treatment facilities.

Description

Flow equalization simply is the damping of flowrate variations so that a constant or nearly constant flowrate is achieved. This technique can be applied in a number of different situations, depending on the characteristics of the collection system. The principal applications are for the equalization of [23]

1. Dry-weather flows
2. Wet-weather flows from separate sanitary sewers
3. Combined stormwater and sanitary wastewater flows

The application of flow equalization in wastewater treatment is illustrated in the two flow diagrams given in Fig. 6-4. In the in-line arrangement (Fig. 6-4a), all of the flow passes through the equalization basin. This arrangement can be used to achieve a considerable amount of constituent concentration and flowrate damping. In the off-line arrangement (Fig. 6-4b), only the flowrate above some predetermined flowrate is diverted into the equalization basin. Although pumping requirements are minimized

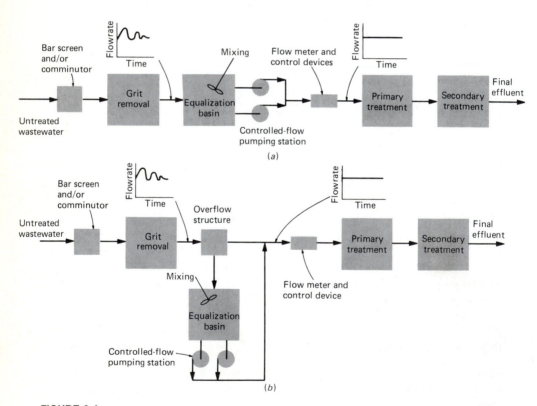

FIGURE 6-4
Typical wastewater treatment plant flowsheets incorporating flowrate equalization (adapted from Ref. 17): (a) in-line equalization and (b) off-line equalization.

in this arrangement, the amount of constituent-concentration damping is considerably reduced.

The principal benefits that are cited as deriving from application of flow equalization are as follows: (1) biological treatment is enhanced, because shock loadings are eliminated or can be minimized, inhibiting substances can be diluted, and pH can be stabilized; (2) the effluent quality and thickening performance of secondary sedimentation tanks following biological treatment is improved through constant solids loading; (3) effluent-filtration surface-area requirements are reduced, filter performance is improved, and more uniform filter-backwash cycles are possible; and (4) in chemical treatment, damping of mass loading improves chemical feed control and process reliability [16]. Apart from improving the performance of most treatment operations and processes, flow equalization is an attractive option for upgrading the performance of overloaded treatment plants.

Analysis

The theoretical analysis of flow equalization is concerned with the following questions:

1. Where in the treatment-process flowsheet should the equalization facilities be located?
2. What type of equalization flowsheet should be used—in-line or off-line?
3. What is the required basin volume?

The practical aspects of design (e.g., type of construction, degree of compartmentalization, type of mixing equipment, pumping and control methods, and sludge and scum removal) are discussed in Chap. 9.

Location of Equalization Facilities. The best location for equalization facilities must be determined for each system. Because the optimum location will vary with the type of treatment and the characteristics of the collection system and the wastewater, detailed studies should be performed for several locations throughout the system. Probably the most common location will continue to be at current and proposed treatment-plant sites. There is also a need to consider location of equalization facilities in the treatment-process flow diagram. In some cases, equalization after primary treatment and before biological treatment may be appropriate. Equalization after primary treatment causes fewer problems with sludge and scum. If flow-equalization systems are to be located ahead of primary settling and biological systems, the design must provide for sufficient mixing to prevent solids deposition and concentration variations, and sufficient aeration to prevent odor problems.

In-Line or Off-Line Equalization. As described earlier and shown in Fig. 6-4, it is possible to achieve considerable damping of constituent mass loadings to the downstream processes with in-line equalization, but only slight damping is achieved with off-line equalization. The analysis of the effect of in-line equalization on the constituent mass loading is illustrated in Example 6-1.

Volume Requirements for Equalization Basin. The volume required for flowrate equalization is determined by using an inflow mass diagram in which the cumulative inflow volume is plotted versus the time of day. The average daily flowrate, also plotted on the same diagram, is the straight line drawn from the origin to the endpoint of the diagram. Diagrams for two typical flowrate patterns are shown in Fig. 6-5.

To determine the required volume, a line parallel to the coordinate axis, defined by the average daily flowrate, is drawn tangent to the mass inflow curve. The required volume is then equal to the vertical distance from the point of tangency to the straight line representing the average flowrate. If the inflow mass curve goes above the line representing the average flowrate (flowrate pattern B), the inflow mass diagram must be bounded with two lines that are parallel to the average flowrate line and tangent to extremities of the inflow mass diagram. The required volume is then equal to the vertical distance between the two lines. The determination of the required volume for equalization is also illustrated in Example 6-1. This procedure is exactly the same as if the average hourly volume were subtracted from the volume flow occurring each hour, and the resulting cumulative volumes were plotted. In this case, the low and high points of the curve would be determined using a horizontal line.

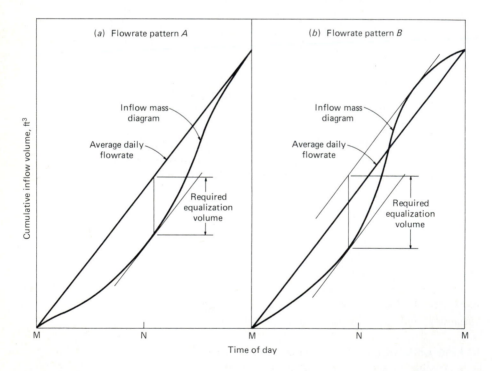

FIGURE 6-5
Schematic mass diagrams for the determination of the equalization volume required for two typical flowrate patterns.

The physical interpretation of the diagrams shown in Fig. 6-5 is as follows. At the low point of tangency (flowrate pattern A) the storage basin is empty. Beyond this point, the basin begins to fill because the slope of the inflow mass diagram is greater than that of the average daily flowrate. The basin continues to fill until it becomes full at midnight. For flowrate pattern B, the basin is filled at the upper point of tangency.

In practice, the volume of the equalization basin will be larger than that theoretically determined to account for the following factors [23]:

1. Continuous operation of aeration and mixing equipment will not allow complete drawdown, although special structures can be built.
2. Volume must be provided to accommodate the concentrated plant recycle streams that are expected, if such flows are returned to the equalization basin (a practice that is not recommended because of the potential to create odors).
3. Some contingency should be provided for unforeseen changes in diurnal flow.

Although no fixed value can be given, the additional volume will vary from 10 to 20 percent of the theoretical value.

Example 6-1 Determination of flowrate-equalization volume requirements and effects on BOD mass loading. For the flowrate and BOD concentration data given in the table on the following page, determine (1) the in-line storage volume required to equalize the flowrate, and (2) the effect of flow equalization on the BOD mass-loading rate.

Solution

1. Determine the volume of the basin required for the flow equalization.

   (a) The first step is to develop a cumulative mass curve of the wastewater flowrate expressed in cubic feet. This is accomplished by converting the average flowrate during each hourly period to cubic feet, using the following expression, and then cumulatively by summing the hourly values.

   $$\text{Volume, ft}^3 = (q_i, \text{ft}^3/\text{s})(3600 \text{ s/h})$$

   For example, for the first three time periods shown in the data table, the corresponding hourly volumes are as follows:
   For the time period M–1:

   $$V_{m-1} = (9.7 \text{ft}^3/\text{s})(3600 \text{ s/h})$$
   $$= 34,900 \text{ ft}^3$$

   For the time period 1–2:

   $$V_{1-2} = (7.8 \text{ft}^3/\text{s})(3600)$$
   $$= 28,100 \text{ ft}^3$$

   The cumulative flow, expressed in ft³, at the end of each time period is determined as follows:

	Given data		Derived data	
Time period	**Average flowrate during time period, ft^3/s**	**Average BOD concentration during time period, mg/L**	**Cumulative volume flow at end time period, 10^{3}ft^3**	**BOD mass loading during time period, lb/hr**
M–1	9.7	150	34.9	329
1–2	7.8	115	63.0	202
2–3	5.8	75	83.9	98
3–4	4.6	50	100.5	52
4–5	3.7	45	113.8	37
5–6	3.5	60	126.4	47
6–7	4.2	90	141.5	85
7–8	7.2	130	167.4	211
8–9	12.5	175	212.4	492
9–10	14.5	200	264.6	652
10–11	15.0	215	318.6	725
11–N	15.2	220	373.3	752
N–1	15.0	220	427.3	742
1–2	14.3	210	478.8	675
2–3	13.6	200	527.8	612
3–4	12.4	190	572.4	530
4–5	11.5	180	613.8	466
5–6	11.5	170	655.2	440
6–7	11.6	175	697.0	457
7–8	12.9	210	743.4	609
8–9	14.1	280	794.2	888
9–10	14.1	305	844.9	967
10–11	13.4	245	893.2	738
11–M	12.2	180	937.1	494
Average	10.8			471

Note: ft^3/s × 0.0283 = m^3/s
      ft^3 × 0.0283 = m^3
      lb × 0.4536 = kg

At the end of the first time period M–1:

$$V_1 = 34,900 \text{ ft}^3$$

At the end of the second time period 1–2:

$$V_2 = 34,900 + 28,100 = 63,000 \text{ ft}^3$$

The cumulative flows for all the hourly time periods are computed in a similar manner (see data table).

(b) The second step is to prepare a plot of the cumulative flow volume, as shown in the following mass diagram. As will be noted, the slope of the line drawn from the origin to the endpoint of the inflow mass diagram represents the average flowrate for the day, which in this case is equal to 10.84 ft^3/s.

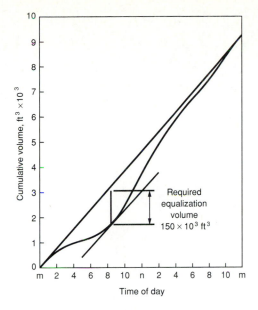

(c) The third step is to determine the required volume. This is done by drawing a line parallel to the average flowrate tangent to the low point of the inflow mass diagram. The required volume is represented by the vertical distance from the point of tangency to the straight line representing the average flowrate. In this case, the required volume is equal to

$$\text{Volume of equalization basin } V = 150{,}000 \text{ ft}^3 \ (4245 \text{ m}^3)$$

2. Determine the effect of the equalization basin on the BOD mass-loading rate. There are a number of ways to do this, but perhaps the simplest way is to perform the necessary computations, starting with the time period when the equalization basin is empty. Because the equalization basin is empty at about 8:30 a.m. (see following figure), the necessary computations will be performed starting with the 8–9 time period.

(a) The first step is to compute the liquid volume in the equalization basin at the end of each time period. This is done by subtracting the equalized hourly flowrate expressed as a volume from the inflow flowrate also expressed as a volume. The volume corresponding to the equalized flowrate for a period of 1 h is 10.84 ft^3/s × 3600 s/h = 39,000 ft^3. Using this value, the volume in storage is computed using the following expression:

$$V_{sc} = V_{sp} + V_{ic} - V_{oc}$$

where V_{sc} = volume in the equalization basin at the end of current time period
V_{sp} = volume in the equalization basin at the end of previous time period
V_{ic} = volume of inflow during the current time period
V_{oc} = volume of outflow during the current time period

Thus, using the values in the original data table, the volume in the equalization basin for the time period 8–9 is as follows:

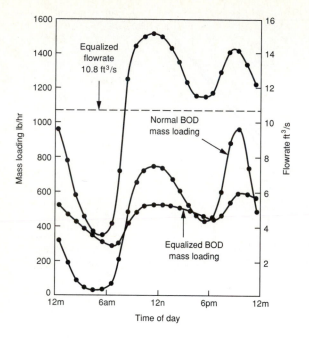

$$V_{sc} = 0 + 45 \times 10^3 - 39 \times 10^3 = 6.0 \times 10^3 \text{ ft}^3$$

For time period 9–10:

$$V_{sc} = 6.0 + 52.2 - 39.0 = 19.2 \times 10^3 \text{ ft}^3$$

The volume in storage at the end of each time period has been computed in a similar way (see following computation table).

(b) The second step is to compute the average concentration leaving the storage basin. This is done by using the following expression, which is based on the assumption that the contents of the equalization basin are mixed completely:

$$X_{oc} = \frac{(V_{ic})(X_{ic}) + (V_{sp})(X_{sp})}{V_{ic} + V_{sp}}$$

where X_{oc} = average concentration of BOD in the outflow from the storage basin during the current time period, mg/L

V_{ic} = volume of wastewater inflow during the current period, 10^3 ft^3

X_{ic} = average concentration of BOD in the inflow wastewater volume, mg/L

V_{sp} = volume of wastewater in storage basin at the end of the previous time period, 10^3 ft^3

X_{sp} = concentration of BOD in wastewater in storage basin at the end of the previous time period

Time period	Volume of flow during time period, $10^3 ft^3$	Volume in storage at end of time period, $10^3 ft^3$	Average BOD concentration during time period, mg/L	Equalized BOD concentration during time period, mg/L	Equalized BOD mass loading during time period, lb/hr
8–9	45.0	6.0	175	175	426
9–10	52.2	19.1	200	197	480
10–11	54.0	34.1	215	210	511
11–N	54.7	49.7	220	216	526
N–1	54.0	64.7	220	218	531
1–2	51.5	77.1	210	214	521
2–3	49.0	87.1	200	209	509
3–4	44.6	92.6	190	203	494
4–5	41.4	95.0	180	196	477
5–6	41.4	97.3	170	188	458
6–7	41.8	100.0	175	184	448
7–8	46.4	107.4	210	192	467
8–9	50.8	119.1	280	220	536
9–10	50.8	130.8	305	245	596
10–11	48.2	140.0	245	245	596
11–M	43.9	144.9	180	230	560
M–1	34.9	140.8	150	214	521
1–2	28.1	129.8	115	196	477
2–3	20.9	111.6	75	179	436
3–4	16.6	89.1	50	162	394
4–5	13.3	63.4	45	147	358
5–6	12.6	36.9	60	132	321
6–7	15.1	13.0	90	119	290
7–8	25.9	0	130	126	307
Average					468

Note: $ft^3 \times 0.0283 = m^3$
$lb \times 0.4536 = kg$

Using the data given in column 2 of the above computation table, the effluent concentration is computed as follows:
For the time period 8–9:

$$X_{oc} = \frac{(45.0)(175) + (0)(0)}{45}$$

$$= 175 \text{ mg/L}$$

For the time period 9–10:

$$X_{oc} = \frac{(52.2)(200) + (6.0)(175)}{(52.2 + 6.0)}$$

$$= 197 \text{ mg/L}$$

All the concentration values computed in a similar manner are reported in the above computation table.

(c) The third step is to compute the hourly mass-loading rate using the following expression:

$$\text{Mass-loading rate, lb/h} = \frac{(\text{mg/L})(\bar{q}_i, \text{ft}^3/\text{s})(3600 \text{ s/h})(7.48 \text{ gal/ft}^3)[8.34 \text{ lb/Mgal} \cdot (\text{mg/L})]}{10^6 \text{gal/Mgal}}$$

For example, for the time period 8–9, the mass-loading rate is

$$= \frac{(175 \text{ mg/L})(10.8 \text{ ft}^3/\text{s})(3600 \text{ s/h})(7.48)(8.34)}{10^6}$$

$$= 426 \text{ lb/h}$$

All the hourly values are summarized in the above computation table. The corresponding values without flow equalization are reported in the original data table.

(d) The effect of flow equalization can best be shown graphically by plotting the hourly unequalized and equalized BOD mass loading on the plot prepared in step 2. The following flowrate ratios, derived from the data presented in the table given in the problem statement and the computation table prepared in step 2a, are also helpful in assessing the benefits derived from flow equalization:

	BOD mass loading	
Ratio	Unequalized	Equalized
$\dfrac{\text{Peak}}{\text{Average}}$	$\dfrac{967}{471} = 2.05$	$\dfrac{596}{468} = 1.27$
$\dfrac{\text{Minimum}}{\text{Average}}$	$\dfrac{37}{471} = 0.08$	$\dfrac{290}{468} = 0.62$
$\dfrac{\text{Peak}}{\text{Minimum}}$	$\dfrac{967}{37} = 26.14$	$\dfrac{596}{290} = 2.06$

Comment. Where on-line equalization basins are used, additional damping of the BOD mass-loading rate can be obtained by increasing the volume of the basins. Although the flow to a treatment plant was equalized in this example, flow equalization would be used more realistically in locations with high infiltration/inflow or stormwater peaks.

6-4 MIXING

Mixing is an important unit operation in many phases of wastewater treatment, including (1) the mixing of one substance completely with another, (2) the mixing of liquid suspensions, (3) the blending of miscible liquids, (4) flocculation, and (5) heat transfer. An example is the mixing of chemicals with wastewater, as shown in Fig. 6-1, where chlorine or hypochlorite is mixed with the effluent from secondary settling tanks. In the activated-sludge process, the contents of the aeration tank must be mixed and air or pure oxygen must be supplied to provide the microorganisms with oxygen. Diffused air is often used to fulfill both the mixing and oxygen requirements. Alternatively, mechanical turbine-aerator mixers may be used. Chemicals are also mixed with sludge to improve its dewatering characteristics. In anaerobic digestion, mixing is used to accelerate the biological conversion process and to heat the contents of the digester uniformly.

Description/Application

Most mixing operations in wastewater can be classified as continuous rapid (30 s or less) or continuous. Continuous-rapid mixing is used most often where one substance is to be mixed with another. Continuous mixing is used where the contents of a reactor or holding tank or basin must be kept in suspension. Each of these types of mixing is considered in the following discussion.

Continuous-Rapid Mixing of Chemicals. In continuous-rapid mixing, the principal objective is to mix completely one substance in another. Rapid mixing ranges from a fraction of a second to about 30 seconds. The rapid mixing of chemicals in a liquid can be carried out in a number of different ways, including (1) in hydraulic jumps in open channels, (2) in Venturi flumes, (3) in pipelines, (4) by pumping, (5) with static mixers, and (6) with mechanical mixers. In the first four of these ways, mixing is accomplished as a result of turbulence that exists in the flow regime. In static mixers, turbulence is induced through the dissipation of energy. In mechanical mixing, turbulence is induced through the input of energy by means of rotating impellers such as propellers, turbines, and paddles. Typical devices used for mixing in wastewater treatment plants are shown in Fig. 6-6.

Continuous Mixing in Reactors and Holding Tanks. In continuous mixing, the principal objective is to maintain the contents of a reactor or holding tank in a completely mixed state. Continuous mixing can be accomplished in a number of different ways, including (1) with mechanical mixers, (2) pneumatically, (3) with static mixers, and (4) by pumping. In mechanical mixing, as noted above, turbulence is induced through the input of energy by means of rotating impellers such as propellers, turbines, and paddles. Pneumatic mixing, an important factor in the design of aeration channels in biological wastewater treatment (see Chap. 10) involves the injection of gases. A baffled over-and-under-flow channel is a form of a static mixer used for flocculation.

Energy Dissipation in Mixing

The power input per unit volume of liquid can be used as a rough measure of mixing effectiveness, based on the reasoning that more input power creates greater turbulence, and greater turbulence leads to better mixing. Camp and Stein [2] studied the establishment and effect of velocity gradients in coagulation tanks of various types and developed the following equations that can be used for the design and operation of mixing systems.

$$G = \sqrt{\frac{P}{\mu V}} \qquad (6\text{-}3)$$

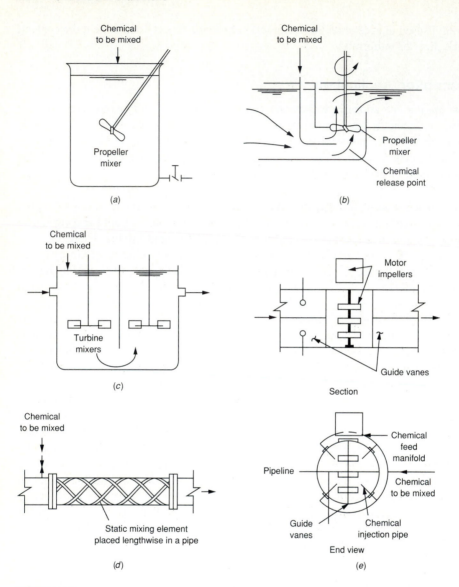

FIGURE 6-6
Typical mixers used in wastewater treatment plants: (*a, b*) propeller mixer, (*c*) turbine mixer, (*d*) static in-line mixer, and (*e*) in-line turbine mixer (section and view).

where G = mean velocity gradient, 1/s
$\quad\quad P$ = power requirement, ft · lb/s (W)
$\quad\quad \mu$ = dynamic viscosity, lb · s/ft^2 (N · s/m^2)
$\quad\quad V$ = flocculator volume, ft^3 (m^3)

In Eq. 6-3, G is a measure of the mean velocity gradient in the fluid. As shown, the value of G depends on the power input, the viscosity of the fluid, and the volume of the basin. Multiplying both sides of Eq. 6-3 by the theoretical detention time $t_d = V/Q$ yields

$$Gt_d = \frac{V}{Q}\sqrt{\frac{P}{\mu V}} = \frac{1}{Q}\sqrt{\frac{PV}{\mu}} \qquad (6\text{-}4)$$

where t_d = detention time, s
$\quad\quad Q$ = flowrate, ft^3/s (m^3/s)

Typical values for G for various mixing operations are reported in Table 6-6. The power required for various types of mixers is considered in the following discussion.

Power Requirements for Mixing

The power requirements for mixing using propeller and turbine mixers, paddle mixers, static mixers, and pneumatic mixing are delineated in the following discussion.

Using Propeller and Turbine Mixers. Mixing in wastewater processes usually occurs in the regime of turbulent flow in which inertial forces predominate. As a general rule, the higher the velocity and the greater the turbulence, the more efficient the mixing. On the basis of inertial and viscous forces, Rushton [16] has developed the following mathematical relationships for power requirements for laminar and turbulent conditions.

TABLE 6-6
Typical velocity gradient (G) and detention time values for wastewater treatment processes

Process	Range of values	
	Detention time	G value, s^{-1}
Mixing		
Typical rapid mixing operations in wastewater treatment	5–20 s	250–1,500
Rapid mixing in contact filtration processes	< 1–5 s	1,500–7,500
Flocculation		
Typical flocculation processes used in wastewater treatment	10–30 min	20–80
Flocculation in direct filtration processes	2–10 min	20–100
Flocculation in contact filtration processes[a]	2–5 min	30–150

[a] Flocculation occurs within granular-medium filter bed.

$$\text{Laminar: } P = k\mu n^2 D^3 \qquad (6\text{-}5)$$

$$\text{Turbulent: } P = k\rho n^3 D^5 \qquad (6\text{-}6)$$

where P = power requirement, ft-lb/s (W)
k = constant (see Table 6-6)
μ = dynamic viscosity of fluid, lb · s/ft^2 (N · s/m^2)
ρ = mass density of fluid, slug/ft^3 (kg/m^3)
D = diameter of impeller, ft (m)
n = revolutions per second, rev/s

Values of k, as developed by Rushton, are presented in Table 6-7. For the turbulent range, it is assumed that vortex conditions have been eliminated by four baffles at the tank wall, each 10 percent of the tank diameter, as shown in Fig. 6-7.

Equation 6-5 applies if the Reynolds number is less than 10, and Eq. 6-6 applies if the Reynolds number is greater than 10,000. For intermediate values of the Reynolds number, Ref. 14 should be consulted. The Reynolds number is given by

$$N_R = \frac{D^2 n \rho}{\mu} \qquad (6\text{-}7)$$

where D = diameter of impeller, ft (m)
n = rev/s
ρ = mass density of liquid, slug/ft^3 (kg/m^3)
μ = dynamic viscosity, lb · s/ft^2 (N · s/m^2)

Mixers are selected on the basis of laboratory or pilot plant tests or similar data provided by manufacturers. No satisfactory method exists for scaling up from an agitator of one design to a unit of a different design. Geometrical similarity should be preserved, and the power input per unit volume should be kept the same. Mixers with small impellers operating at high speeds are best for dispersing gases or small amounts of chemicals in wastewater. Mixers with slow-moving impellers are best for blending two fluid streams or for flocculation.

TABLE 6-7
Values of k for mixing power requirements [16]

Impeller	Laminar range, Eq. 6-5	Turbulent range, Eq. 6-6
Propeller, square pitch, 3 blades	41.0	0.32
Propeller, pitch of two, 3 blades	43.5	1.00
Turbine, 6 flat blades	71.0	6.30
Turbine, 6 curved blades	70.0	4.80
Fan turbine, 6 blades	70.0	1.65
Turbine, 6 arrowhead blades	71.0	4.00
Flat paddle, 6 blades	36.5	1.70
Shrouded turbine, 2 curved blades	97.5	1.08
Shrouded turbine with stator (no baffles)	172.5	1.12

Standard tank configuration used in the analysis of mixer performance.

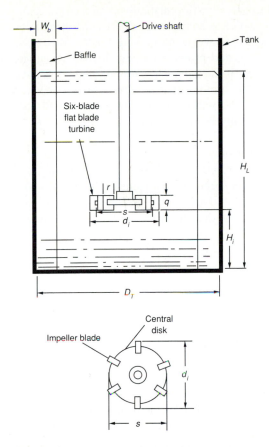

Notes: 1. The agitator is a six-blade flat turbine impeller
2. Impeller diameter, d_i = 1/3 tank diameter
3. Impeller height from bottom, H_i = 1.0 impeller diameter
4. Impeller blade width, q = 1/5 impeller diameter
5. Impeller blade length, r = 1/4 impeller diameter
6. Length of impeller blade mounted on the central disk = $r/2$ = 1/8 impeller diameter
7. Liquid height, H_L = 1.0 tank diameter
8. Number of baffles = 4 mounted vertically at tank wall and extending from the tank bottom to above the liquid surface
9. Baffle width, W_b = 1/10 tank diameter
10. Central disk diameter, s = 1/4 tank diameter

Source: Adapted from Ref. 16

FIGURE 6-7
Definition sketch for turbine mixer in baffled tank.

Vortexing, or mass swirling of the liquid, must be restricted with all types of impellers. Vortexing causes a reduction in the difference between the fluid velocity and the impeller velocity and thereby decreases the effectiveness of mixing. If the mixing vessel is fairly small, vortexing can be prevented by mounting the impellers off-center or at an angle with the vertical, or by having them enter the side of the basin at an angle. The usual method in both circular and rectangular tanks is to install four or more vertical baffles extending approximately one-tenth the diameter out from the wall. These effectively break up the mass rotary motion and promote vertical mixing. Concrete mixing tanks may be made square and the baffles may be omitted.

Using Paddle Mixers. Paddle mixers generally rotate slowly, as they apply a large surface to the liquid. Paddles are used as flocculation devices when coagulants, such as aluminum or ferric sulfate, and coagulant aids, such as polyelectrolytes and lime, are added to wastewater or sludges. Mechanically, flocculation is promoted by gentle stirring with slow-moving paddles. The action is sometimes aided by the installation of stationary slats or stator blades, located between the moving blades, that serve to break up the mass rotation of the liquid and promote mixing. Increased particle contact will promote floc growth; however, if the agitation is too vigorous, the shear forces that are set up will break up the floc into smaller particles. Agitation should be carefully controlled so that the floc particles will be of suitable size and will settle readily. The production of a good floc usually requires a detention time of 10 to 30 minutes.

Numerous experiments have been performed by equipment manufacturers and plant operators to determine the optimum configuration of paddle size, spacing, and velocity. It has been found that a paddle-tip speed of approximately 2 to 3 ft/s (0.6 to 0.9 m/s) achieves sufficient turbulence without breaking up the floc. Power in a mechanical paddle system can be related to the drag force on the paddles as follows.

$$F_D = \frac{C_D A \rho v_p^2}{2} \tag{6-8}$$

$$P = F_D v_p = \frac{C_D A \rho v_p^3}{2} \tag{6-9}$$

where F_D = drag force, lb_f (N)
C_D = coefficient of drag of paddle moving perpendicular to fluid
A = cross-sectional area of paddles, ft^2 (m^2)
ρ = mass fluid density, $slug/ft^3$ (kg/m^3)
v_p = relative velocity of paddles with respect to the fluid, ft/s (m/s), usually assumed to be 0.6 to 0.75 times the paddle-tip speed
P = power requirement, ft · lb/s (W)

The application of Eq. 6-9 is illustrated in Example 6-2.

Example 6-2 Power requirements and paddle area for a wastewater floccu-lator. Determine the theoretical power requirement and the paddle area required to achieve a G value of 50/s in a tank with a volume of 10^5 ft³ (2832 m³). Assume that the water temperature is 15°C (60°F), the coefficient of drag C_D for rectangular paddles is 1.8, the paddle-tip velocity v_p is 2 ft/s (0.6 m/s), and the relative velocity of the paddles v is 0.75 v_p.

Solution

1. Determine the theoretical power requirement using Eq. 6-3.

$$P = G^2 \mu V$$

μ at 15°C = 2.359 × 10^{-5} lb · s/ft² (see Appendix C)

$$= (50/s)^2 2.359 \times 10^{-5} \frac{lb \cdot s}{ft^2} (10^5 ft^3)$$

$$= 5898 \text{ ft} \cdot lb_f/s(8.1 \text{ kW})$$

2. Determine the required paddle area using Eq. 6-9.

$$A = \frac{2P}{C_D \rho v^3}$$

$$\rho \text{ at } 15°C = 1.938 \text{ slug/ft}^3 \text{ (see Appendix C)}$$

$$= \frac{2 \times 5898 \text{ ft} \cdot \text{lb/s}}{1.8(1.938 \text{ slug/ft}^3)(0.75 \times 2.0 \text{ ft/s})^3}$$

$$= 1002 \text{ ft}^2 \text{ (98.9 m}^2)$$

For Static Mixers. Static mixers are principally identified by their lack of moving parts. Typical examples include in-line static mixers, which contain elements that bring about sudden changes in the velocity patterns as well as momentum reversals (see Fig. 6-6e) and channels with closely-spaced over and under baffles. In-line static mixers are commonly used for the mixing of chemicals, whereas over and under baffled channels are used for flocculation.

The power consumed by static-mixing devices can be computed using the following equation.

$$P = \gamma Q h \tag{6-10}$$

where P = power dissipated, ft · lb/s (kW)
γ = specific weight of water, lb/ft^3 (kN/m^3)
Q = flowrate, ft^3/s (m^3/s)
h = headloss dissipated as liquid passes through device, ft (m)

For Pneumatic Mixing. In mixing tanks, flocculation tanks, and aerated channels, flocculation is achieved by introducing air bubbles in the bottom of the tank. When air is injected in mixing or flocculation tanks or channels, the power dissipated by the rising air bubbles can be estimated with the following equation [7].

$$P = p_a V_a \ln \frac{p_c}{p_a} \tag{6-11}$$

where P = power dissipated, ft · lb/s (kW)
p_a = atmospheric pressure, lb/ft^2 (kN/m^2)
V_a = volume of air at atmospheric pressure, ft^3/s (m^3/s)
p_c = air pressure at the point of discharge, lb/ft^2 (kN/m^2)

Equation 6-11 is derived from a consideration of the work done when the volume of air released under compressed conditions expands isothermally. If the flow of air at atmospheric pressure is expressed in terms of ft^3/min (m^3/min) and the pressure is expressed in terms of feet (meters) of water, Equation 6-11 can be written as follows.

$$P = KQ_a \ln\left(\frac{h + 34}{34}\right) \qquad \text{U.S. Customary units} \tag{6-12a}$$

$$P = KQ_a\ln\left(\frac{h + 10.33}{10.33}\right) \qquad \text{S.I. Units} \qquad (6\text{-}12b)$$

where K = constant = 81.5 (1.689 in S.I. Units)

Q_a = air flow rate at atmospheric pressure, ft³/min (m³/min)

h = air pressure at the point of discharge expressed in feet of water, ft (m)

The velocity gradient G, achieved in pneumatic mixing, is obtained by substituting P from Eq. 6-12 into Eq. 6-3.

6-5 SEDIMENTATION

Sedimentation is the separation from water, by gravitational settling, of suspended particles that are heavier than water. It is one of the most widely used unit operations in wastewater treatment. The terms *sedimentation* and *settling* are used interchangeably. A sedimentation basin may also be referred to as a sedimentation tank, settling basin, or settling tank (see Fig. 6-8).

Sedimentation is used for grit removal, particulate-matter removal in the primary settling basin, biological-floc removal in the activated-sludge settling basin, and chemical-floc removal when the chemical coagulation process is used. It is also used for solids concentration in sludge thickeners. In most cases, the primary purpose is to produce a clarified effluent, but it is also necessary to produce sludge with a solids concentration that can be easily handled and treated. In designing sedimentation basins (see Chap. 9), consideration must be given to production of both a clarified effluent and a concentrated sludge.

Description

On the basis of the concentration and the tendency of particles to interact, four types of settling can occur: discrete particle, flocculant, hindered (also called zone), and compression. These types of settling phenomena are described in Table 6-8. During a

FIGURE 6-8
Typical sedimentation tanks used for the removal of particulate matter in untreated wastewater (tanks in background) and biological floc removal in the activated-sludge process (tanks in foreground). (Westfield, MA, design average flowrate = 4 Mgal/d)

TABLE 6-8
Types of settling phenomena involved in wastewater treatment

Type of settling phenomenon	Description	Application/occurrence
Discrete particle (type 1)	Refers to the sedimentation of particles in a suspension of low solids concentration. Particles settle as individual entities, and there is no significant interaction with neighboring particles	Removes grit and sand particles from wastewater
Flocculant (type 2)	Refers to a rather dilute suspension of particles that coalesce, or flocculate, during the sedimentation operation. By coalescing, the particles increase in mass and settle at a faster rate	Removes a portion of the suspended solids in untreated wastewater in primary settling facilities, and in upper portions of secondary settling facilities. Also removes chemical floc in settling tanks
Hindered, also called zone (type 3)	Refers to suspensions of intermediate concentration, in which interparticle forces are sufficient to hinder the settling of neighboring particles. The particles tend to remain in fixed positions with respect to each other, and the mass of particles settles as a unit. A solids-liquid interface develops at the top of the settling mass	Occurs in secondary settling facilities used in conjunction with biological treatment facilities
Compression (type 4)	Refers to settling in which the particles are of such concentration that a structure is formed, and further settling can occur only by compression of the structure. Compression takes place from the weight of the particles, which are constantly being added to the structure by sedimentation from the supernatant liquid	Usually occurs in the lower layers of a deep sludge mass, such as in the bottom of deep secondary settling facilities and in sludge-thickening facilities

sedimentation operation, it is common to have more than one type of settling occurring at a given time, and it is possible to have all four occurring simultaneously.

Because of the fundamental importance of sedimentation in the treatment of wastewater, the analysis of each type of settling will be discussed separately. In addition, after the discussion of flocculant settling, a brief analysis of tube settlers (inclined small-diameter tubes used to improve efficiency of the sedimentation operation) will be presented. Both discrete and flocculant settling can occur in situations where tube settlers are used.

Analysis of Discrete Particle Settling (Type 1)

The settling of discrete, nonflocculating particles can be analyzed by means of the classic laws of sedimentation formed by Newton and Stokes. Newton's law yields the terminal particle velocity by equating the gravitational force of the particle with the frictional resistance, or drag. The gravitational force is given by

$$\text{Gravitational force} = (\rho_s - \rho)gV \tag{6-13}$$

where ρ_s = density of particle
ρ = density of fluid
g = acceleration due to gravity
V = volume of particle

The frictional drag force depends on the particle velocity, fluid density, fluid viscosity, and particle diameter and the drag coefficient C_D (dimensionless) and is defined by Eq. 6-14.

$$\text{Frictional drag force} = \frac{C_D A \rho v^2}{2} \tag{6-14}$$

where C_D = drag coefficient
A = cross-sectional or projected area of particles at right angles to v
v = particle velocity

Equating the gravitational force with the frictional drag force for spherical particles yields Newton's law:

$$V_c = \left[\frac{4}{3} \frac{g(\rho_s - \rho)d}{C_D \rho} \right]^{1/2} \tag{6-15}$$

where V_c = terminal velocity of particle
d = diameter of particle

The drag coefficient takes on different values depending on whether the flow regime surrounding the particle is laminar or turbulent. The drag coefficient is shown in Fig. 6-9 as a function of the Reynolds number. Although particle shape affects the value of the drag coefficient, for spherical particles the curve in Fig. 6-9 is approximated by the following equation (upper limit of $N_R = 10^4$) [7]:

$$C_D = \frac{24}{N_R} + \frac{3}{\sqrt{N_R}} + 0.34 \tag{6-16}$$

For Reynolds numbers less than 0.3, the first term in Eq. 6-16 predominates, and substitution of this drag term into Eq. 6-15 yields Stokes' law:

$$V_c = \frac{g(\rho_s - \rho)d^2}{18\mu} \tag{6-17}$$

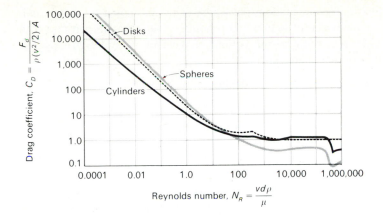

FIGURE 6-9
Drag coefficients of spheres, disks, and cylinders.

For laminar-flow conditions, Stokes found the drag force to be

$$F_d = 3\pi\mu vd \qquad (6\text{-}18)$$

Equating this force to the effective particle weight also yields Eq. 6-17.

In the design of sedimentation basins, the usual procedure is to select a particle with a terminal velocity V_c and to design the basin so that all particles that have a terminal velocity equal to or greater than V_c will be removed. The rate at which clarified water is produced is then

$$Q = AV_c \qquad (6\text{-}19)$$

where A is the surface of the sedimentation basin. Equation 6-19 yields

$$V_c = \frac{Q}{A} = \text{overflow rate, gal/ft}^2 \cdot d \ (m^3/m^2 \cdot d)$$

which shows that the overflow rate or surface-loading rate, a common basis of design, is equivalent to the settling velocity. Equation 6-19 also indicates that, for type 1 settling, the flow capacity is independent of the depth.

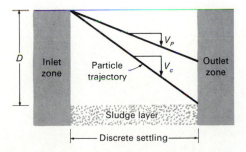

FIGURE 6-10
Type 1 settling in an ideal settling basin.

For continuous-flow sedimentation, the length of the basin and the time a unit volume of water is in the basin (detention time) should be such that all particles with the design velocity V_c will settle to the bottom of the tank. The design velocity, detention time, and basin depth are related as follows:

$$V_c = \frac{\text{depth}}{\text{detention time}} \tag{6-20}$$

In practice, design factors must be adjusted to allow for the effects of inlet and outlet turbulence, short circuiting, sludge storage, and velocity gradients due to the operation of sludge-removal equipment. These factors are discussed in Chap. 9. The discussion in this chapter refers to ideal settling in which the factors are omitted.

Type 1 settling in an ideal settling basin is shown in Fig. 6-10. A full-scale settling basin used in practice is shown in Fig. 9-17. Particles that have a velocity of fall less than V_c will not all be removed during the time provided for settling. Assuming that the particles of various sizes are uniformly distributed over the entire depth of the basin at the inlet, it can be seen from an analysis of the particle trajectory in Fig. 6-10 that particles with a settling velocity less than V_c will be removed in the ratio

$$X_r = \frac{V_p}{V_c} \tag{6-21}$$

where X_r is the fraction of the particles with settling velocity V_p that are removed.

In a typical suspension of particulate matter, a large gradation of particle sizes occurs. To determine the efficiency of removal for a given settling time, it is necessary to consider the entire range of settling velocities present in the system. This can be accomplished in two ways: (1) by use of sieve analysis and hydrometer tests combined with Eq. 6-17 or (2) by use of a settling column. With either method, a settling-velocity analysis curve can be constructed from the data. Such a curve is shown in Fig. 6-11.

For a given clarification rate Q where

$$Q = V_c A \tag{6-22}$$

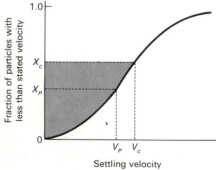

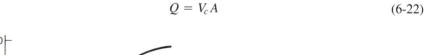

FIGURE 6-11
Definition sketch for the analysis of discrete particle settling.

only those particles with a velocity greater than V_c will b̶ remaining particles will be removed in the ratio V_p/V_c. Th̶tely removed. The removed is given by Eq. 6-23.

$$\text{Fraction removed} = (1 - X_c) + \int_0^{X_c} \frac{V_p}{V_c} dx \qquad (6\text{-}23)$$

where $1 - X_c$ = fraction particles with velocity V_p greater than V̶

$\int_0^{X_c} \dfrac{V_p}{V_c} dx$ = fraction of particles removed with V_p less than V_c

The use of Eq. 6-23 is illustrated in Example 6-3.

Example 6-3   Removal of discrete particles (type 1 settling). A particle-size distribution has been obtained from a sieve analysis of sand particles. For each weight fraction, an average settling velocity has been calculated. The data are as follows:

Settling velocity, ft/min	10.0	5.0	2.0	1.0	0.75	0.5
Settling velocity, m/min	3.0	1.5	0.60	0.03	0.23	0.15
Weight fraction remaining	0.55	0.46	0.35	0.21	0.11	0.03

What is the overall removal for an overflow rate of 10^5 gal/ft$^2 \cdot$ d (4075 m^3/m$^2 \cdot$ d)?

Solution

1. Draw a curve of the fraction remaining versus the settling-velocity curve as shown below.

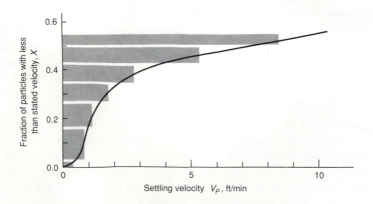

2. Compute the critical settling velocity V_c of the particles that will be removed completely when the rate of clarification is 10^5 gal/ft$^2 \cdot$ d.

$$\frac{10^5 \text{ gal/ft}^2 \cdot \text{d}}{7.48 \text{ gal/ft}^3 (24 \text{ hr/d})(60 \text{ min/hr})} = 9.3 \text{ ft/min} \ (2.8 \text{ m/min})$$

3. Determin ction of the particles removed using Eq. 6-23. From the curve plotted in
step 1, nd that 0.55 of the particles have a settling velocity less than 9.3 ft/min.
The gre integration of the second term on the right side of Eq. 6-23 is shown on the
settlir as a series of rectangles (shaded) and in the following tabulation. (Note that
beca   is constant, it is taken outside the integral or summation sign.)

	V_p	$V_p\,dx$
.05	0.4	0.020
0.13	0.8	0.104
0.10	1.1	0.110
0.09	1.7	0.153
0.07	2.8	0.196
0.07	5.2	0.364
0.04	8.3	0.332
0.55		1.297

$$\text{Fraction removed} = (1 - X_c) + \frac{1}{V_c}\Sigma V_p dx$$

$$= (1 - 0.55) + \frac{1.297}{9.3} = 0.588$$

Analysis of Flocculant Settling (Type 2)

Particles in relatively dilute solutions will not act as discrete particles but will coalesce
during sedimentation. As coalescence or flocculation occurs, the mass of the particle
increases and it settles faster. The extent to which flocculation occurs is dependent
on the opportunity for contact, which varies with the overflow rate, the depth of
the basin, the velocity gradients in the system, the concentration of particles, and
the range of particle sizes. The effects of these variables can be determined only by
sedimentation tests.

To determine the settling characteristics of a suspension of flocculant particles,
a settling column may be used. Such a column can be of any diameter but should be
equal in height to the depth of the proposed tank. Satisfactory results can be obtained
with a 6 in (150 mm) diameter plastic tube about 10 ft (3 m) high. Sampling ports
should be inserted at 2 ft (0.6 m) intervals. The solution containing the suspended
matter should be introduced into the column in such a way that a uniform distribution
of particle sizes occurs from top to bottom.

Care should be taken to ensure that a uniform temperature is maintained throughout the test to eliminate convection currents. Settling should take place under quiescent conditions. At various time intervals, samples are withdrawn from the ports and

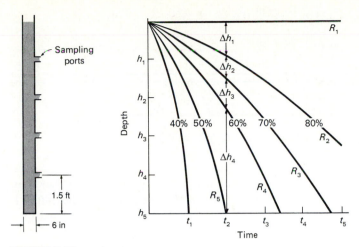

FIGURE 6-12
Settling column and settling curves for flocculant particles.

analyzed for suspended solids. The percent removal is computed for each sample analyzed and is plotted as a number against time and depth, as elevations are plotted on a survey grid. Between the plotted points, curves of equal percent removal are drawn. A settling column and the results of a sedimentation test are shown in Fig. 6-12. The resulting curves are shown, but the plotted numbers representing the individual samples have been omitted from the figure. Determination of the amount of material removed, using the curve given in Fig. 6-12, is illustrated in Example 6-4.

Example 6-4   Removal of flocculant suspended solids (type 2 settling). Using the results of the settling test shown in Fig. 6-12, determine the overall removal of solids if the detention time is t_2 and the depth is h_5.

Solution

1. Determine the percent removal.

$$\text{Percent removal} = \frac{\Delta h_1}{h_5} \times \frac{R_1 + R_2}{2} + \frac{\Delta h_2}{h_5} \times \frac{R_2 + R_3}{2}$$

$$+ \frac{\Delta h_3}{h_5} \times \frac{R_3 + R_4}{2} + \frac{\Delta h_4}{h_5} \times \frac{R_4 + R_5}{2}$$

2. For the curves shown in Fig. 6-12, the computations would be

$$\frac{\Delta h_n}{h_5} \times \frac{R_n + R_{n+1}}{2} = \text{percent removal}$$

$$0.20 \times \frac{100 + 80}{2} = \qquad 18.00$$

$$0.11 \times \frac{80 + 70}{2} = \qquad 8.25$$

$$0.15 \times \frac{70 + 60}{2} = \qquad 9.75$$

$$\underline{0.54} \times \frac{60 + 50}{2} = \qquad \underline{29.70}$$

$$1.00 \qquad\qquad\qquad 65.70$$

yielding a total removal for quiescent settling of 65.7 percent.

To account for the less than optimum conditions encountered in the field, the design settling velocity or overflow rate obtained from column studies often is multiplied by a factor of 0.65 to 0.85, and the detention times are multiplied by a factor of 1.25 to 1.5.

Analysis of Plate and Tube Settlers

In the analysis for the settling of discrete particles (type 1) presented earlier in this section, it was shown that the removal efficiency is related directly to the settling velocity and not to the depth of the basin. From this finding, it can be concluded that sedimentation basins should be constructed as shallow as possible to optimize the removal efficiency. Although this approach is correct theoretically, there are numerous practical considerations that limit the use of extremely shallow basins (see Chap. 9). Plate and tube settlers have been developed as an alternative to shallow basins and are used in conjunction with both existing and specially designed sedimentation basins. Plate and tube settlers are shallow settling devices consisting of stacked off-set trays or bundles of small plastic tubes of various geometries (see Fig. 6-13a). They are

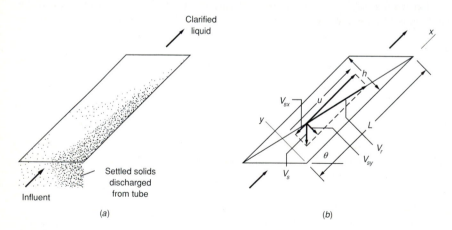

FIGURE 6-13
Typical inclined tube settler: (a) operation and (b) definition sketch.

used to enhance the settling characteristics of sedimentation basins. Although plate and tube settlers have been used in primary, secondary, and tertiary sedimentation applications, a number of problems have developed with their use. The principal problems are clogging and odors due to biological growths and the buildup of oil and grease.

The shape, hydraulic radii, angle of inclination, and length of the plate and tube settlers will vary according to the particular installation. Normal practice is to insert the plate or tube settlers in sedimentation basins (either rectangular or circular) of sufficient depth. The flow within the basin passes upward through the plate or tube modules and exits from the basin above the modules. The solids that settle out in within the plates or tubes move by means of gravity counter currently downward and out of the tube modules to the basin bottom (see Fig. 6-13a). To be self-cleaning, plate or tube settlers are usually set at an angle between 45 and 60° above the horizontal. When the angle is increased above 60°, the efficiency decreases. If the plates and tubes are inclined at angles less than 45°, sludge will tend to accumulate within the plates or tubes. To control biological growths and the production of odors, the accumulated solids must be flushed out periodically (usually with a high pressure hose). The need for flushing poses a problem with the use of plate and tube settlers where the characteristics of the solids to be removed vary from day to day.

Referring to the definition sketch presented in Fig. 6-13b, the analysis of the plate and tube settlers is as follows. For the inclined-coordinate system, the velocity components for the particle are

$$V_{sx} = U - V_s \sin \theta \tag{6-24}$$

$$V_{sy} = -V_s \cos \theta \tag{6-25}$$

where V_{sx} = velocity component in x direction
U = fluid velocity in x direction
V_s = normal settling velocity of particle
q = inclination angle for tube with horizontal axis
V_{sy} = settling velocity in y direction

For this system of coordinates, it can be seen that V_{sy} is the critical velocity component, and the analysis for the removal is the same as the analysis presented previously for discrete particles.

Analysis of Hindered Settling (Type 3)

In systems that contain high concentrations of suspended solids, both hindered or zone settling (type 3) and compression settling (type 4) usually occur in addition to discrete (free) and flocculant settling. The settling phenomenon that occurs when a concentrated suspension, initially of uniform concentration throughout, is placed in a graduated cylinder is shown in Fig. 6-14.

Because of the high concentration of particles, the liquid tends to move up through the interstices of the contacting particles. As a result, the contacting particles tend to settle as a zone, or "blanket," maintaining the same relative position with

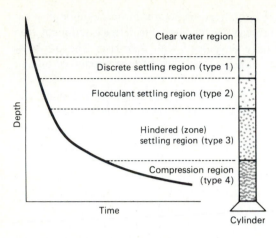

FIGURE 6-14
Schematic of settling regions for activated sludge.

respect to each other. The phenomenon is known as *hindered settling*. As the particles in this region settle, a relatively clear layer of water is produced above the particles in the settling region.

The scattered, relatively light particles remaining in this region usually settle as discrete or flocculant particles, as discussed previously in this chapter. In most cases, an identifiable interface develops between the more or less clear upper region and the hindered-settling region in Fig. 6-14. The rate of settling in the hindered-settling region is a function of the concentration of solids and their characteristics.

As settling continues, a compressed layer of particles begins to form on the bottom of the cylinder in the compression-settling region. The particles in this region apparently form a structure in which there is close physical contact between the particles. As the compression layer forms, regions containing successively lower concentrations of solids than those in the compression region extend upward in the cylinder. Thus, in actuality the hindered-settling region contains a gradation in solids concentration from that found at the interface of the settling region to that found in the compression-settling region. According to Dick and Ewing [6], the forces of physical interaction between the particles that are especially strong in the compression-settling region lessen progressively with height. They may exist to some extent in the hindered-settling region.

Because of the variability encountered, settling tests are usually required to determine the settling characteristics of suspensions where hindered and compression settling are important considerations. On the basis of data derived from column settling tests, two different design approaches can be used to obtain the required area for the settling/thickening facilities. In the first approach, the data derived from a single (batch) settling test are used. In the second approach, known as *solids-flux method*, data from a series of settling tests conducted at different solids concentrations are used. Both methods are described in the following discussion.

Area Requirement Based on Single-Batch Test Results. For purposes of design, the final overflow rate selected should be based on a consideration of the following factors: (1) The area needed for clarification, (2) the area needed for thickening, and (3) the rate of sludge withdrawal. Column settling tests, as previously described, can be used to determine the area needed for the free-settling region directly. However, because the area required for thickening is usually greater than the area required for the settling, the rate of free settling is rarely the controlling factor. In the case of the activated-sludge process where stray, light, fluffy floc particles may be present, it is conceivable that the free flocculant-settling velocity of these particles could control the design.

The area requirement for thickening is determined according to a method developed by Talmadge and Fitch [17]. A column of height H_0 is filled with a suspension of solids of uniform concentration C_0. The position of the interface as time elapses and the suspension settles is given in Fig. 6-15. The rate at which the interface subsides is then equal to the slope of the curve at that point in time. According to the procedure, the area required for thickening is given by Eq. 6-26:

$$A = \frac{Qt_u}{H_o} \qquad (6\text{-}26)$$

where A = area required for sludge thickening, ft^2 (m^2)
Q = flowrate into tank, ft^3/s (m^3/s)
H_o = initial height of interface in column, ft (m)
t_u = time to reach desired underflow concentration, s

(Note: Any consistent set of units may be used in Eq. 6-26.)

The critical concentration controlling the sludge-handling capability of the tank occurs at a height H_2 where the concentration is C_2. This point is determined by extending the tangents to the hindered-settling and compression regions of the subsidence curve to the point of intersection and by bisecting the angle thus formed, as shown in Fig. 6-15. The time t_u can be determined as follows:

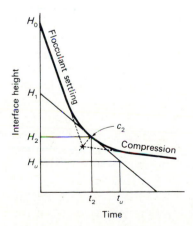

FIGURE 6-15
Graphical analysis of interface settling curve [21].

1. Construct a horizontal line at the depth H_u that corresponds to the depth at which the solids are at the desired underflow concentration C_u. The value of H_u is determined using the following expression:

$$H_u = \frac{C_o H_o}{C_u} \tag{6-27}$$

2. Construct a tangent to the settling curve at the point indicated by C_2.
3. Construct a vertical line from the point of intersection of the two lines drawn in steps 1 and 2 to the time axis to determine the value of t_u.

With this value of t_u, the area required for the thickening is computed using Eq. 6-26. The area required for clarification is then determined. The larger of the two areas is the controlling value. Application of this procedure is illustrated in Example 6-5.

Example 6-5 Calculations for sizing an activated-sludge settling tank. The settling curve shown in the following diagram was obtained for an activated sludge with an initial solids concentration C_o of 4000 mg/L. The initial height of the interface in the settling column was at 2.0 ft. Determine the area required to yield a thickened sludge concentration C_u of 12,000 mg/L with a total inflow of 0.1 Mgal/d (400 m^3/d). In addition, determine the solids loading in lb/ft$^2 \cdot$ d and the overflow rate in gal/ft$^2 \cdot$ d.

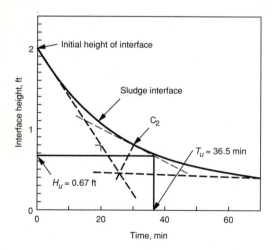

Solution

1. Determine the area required for thickening using Eq. 6-27.
   (a) Determine the value of H_u

$$H_u = \frac{C_o H_o}{C_u}$$

$$= \frac{4000 \text{ mg/L} \times 2.0 \text{ ft}}{12,000 \text{ mg/L}} = 0.67 \text{ ft}$$

On the settling curve, a horizontal line is constructed at $H_u = 0.67$ ft. A tangent is constructed to the settling curve at C_2, the midpoint of the region between hindered and compression settling. Bisecting the angle formed where the two tangents meet determines point C_2. The intersection of the tangent at C_2 and the line $H_u = 0.67$ ft determines t_u. Thus $t_u = 36.5$ min, and the required area is

$$A = \frac{Qt_u}{H_o} = \frac{(0.1 \text{ Mgal/d})[1.55\text{ft}^3/\text{s} \cdot (\text{Mgal/d})](36.5 \text{ min})(60 \text{ s/min})}{2.0 \text{ ft}}$$

$$= 170 \text{ ft}^2$$

2. Determine the area required for clarification.

   (a) Determine the interface subsidence velocity v. The subsidence velocity is determined by computing the slope of the tangent drawn from the initial portion of the interface settling curve. The computed velocity represents the unhindered settling rate of the sludge.

   $$v = \frac{2.0 \text{ ft} - 0.8 \text{ ft}}{20 \text{ min}}\left(60 \frac{\text{min}}{\text{h}}\right)$$

   $$= 3.6 \text{ ft/h}$$

   (b) Determine the clarification rate. Because the clarification rate is proportional to the liquid volume above the critical sludge zone, it may be computed as follows:

   $$Q_c = 0.1\text{Mgal/d} \times [1.55 \text{ ft}^3/\text{s} \cdot (\text{Mgal/d})]\frac{(2.0 \text{ ft} - 0.67 \text{ ft})}{2.0 \text{ ft}}$$

   $$= 0.103 \text{ ft}^3/\text{s}$$

   (c) Determine the area required for clarification. The required area is obtained by dividing the clarification rate by the settling velocity.

   $$A = \frac{Q_c}{V} = \frac{(0.103 \text{ ft}^3/\text{s}) \times 60 \text{ s/min} \times 60 \text{ min/h}}{3.6 \text{ ft/h}}$$

   $$= 103 \text{ ft}^2$$

3. The controlling area is the thickening area (170 ft^2) because it exceeds the area required for clarification (103 ft^2).

4. Determine the solids loading. The solids loading is computed as follows:

   $$\text{Solids, lb/d} = (0.1 \text{ Mgal/d})[8.34 \text{ Mgal/d} \cdot (\text{mg/L})](2000 \text{ mg/L})$$

   $$= 3340 \text{ lb/d}$$

   $$\text{Solids loading} = \frac{3340 \text{ lb/d}}{170 \text{ ft}^2} = 19.6 \text{ lb/ft}^2 \cdot \text{d}$$

5. Determine the hydraulic-loading rate.

   $$\text{Hydraulic-loading rate} = \frac{100,000 \text{ gal/d}}{170 \text{ ft}^2}$$

   $$= 588 \text{ gal/ft}^2 \cdot \text{d}$$

Area Requirements Based on Solids-Flux Analysis. An alternative method of arriving at the area required for hindered settling is based on an analysis of the solids (mass) flux [3,6,11,28]. Data derived from settling tests must be available when applying this method, which is based on an analysis of the mass flux (movement across a boundary) of the solids in the settling basin.

In a settling basin that is operating at steady state, a constant flux of solids is moving downward, as shown in Fig. 6-16. Within the tank, the downward flux of solids is brought about by gravity (hindered) settling and by bulk transport due to the underflow being pumped out and recycled. At any point in the tank, the mass flux of solids due to gravity (hindered) settling is

$$SF_g = kC_iV_i \tag{6-28}$$

where SF_g = solids flux due to gravity, $lb/ft^2 \cdot h$
$\qquad k = (1/16,030)$
$\qquad C_i$ = concentration of solids at the point in question, mg/L
$\qquad V_i$ = settling velocity of the solids at concentration C_i, ft/h

$$\text{Note: } \left(C_i\frac{mg}{L}\right)\left(V_i\frac{ft}{h}\right)\left(\frac{7.48 \text{ gal}}{ft^3}\right)\left[\frac{8.34 \text{ lb}}{Mgal \cdot (mg/L)}\right]\left(\frac{1.0 \text{ Mgal}}{10^6 \text{ gal}}\right) = \left(\frac{1}{16,030}C_iV_i\right) lb/ft^2 \cdot h$$

The mass flux of solids due to the bulk movement of the suspension is

$$SF_u = kC_iU_b \tag{6-29}$$

where SF_u = solids flux due to underflow, $lb/ft^2 \cdot h$
$\qquad k = (1/16,030)$
$\qquad U_b$ = bulk downward velocity, ft/h

The total-mass flux SF_t of solids is the sum of previous components and is given by

$$SF_t = SF_g + SF_u \tag{6-30}$$

$$SF_t = k(C_iV_i - C_iU_b) \tag{6-31}$$

In this equation, the flux of solids due to gravity (hindered) settling depends on the concentration of solids and the settling characteristics of the solids at that concentration. The procedure used to develop a solids-flux curve from column-

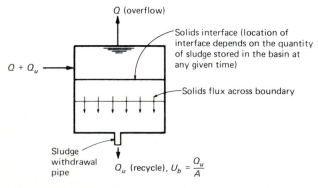

FIGURE 6-16
Definition sketch for a settling basin operating at steady state.

settling illustrated in Fig. 6-17. At low concentrations (below about 1000 mg/L), the movement of solids due to gravity is small because the settling velocity of the solids is more or less independent of concentration. If the velocity remains essentially the same as the solids concentration increases, the total flux due to gravity starts to increase as the solids concentration starts to increase. At very high solids concentrations, the hindered-settling velocity approaches zero, and the total solids flux due to gravity again becomes extremely low. Thus it can be concluded that the solids flux due to gravity must pass through a maximum value as the concentration is increased. This is shown schematically in Figs. 6-17c and 6-18.

The solids flux due to bulk transport is a linear function of the concentration with slope equal to U_b, the underflow velocity (see Fig. 6-18). The total flux, which is the sum of the gravity and the underflow flux, is also shown in Fig. 6-18. Increasing or decreasing the flowrate of the underflow causes the total-flux curve to shift upward or downward. Because the underflow velocity can be controlled, it is used for process control.

The required cross-sectional area of the thickener is determined as follows: As shown in Fig. 6-18, if a horizontal line is drawn tangent to the low point on the total-flux curve, its intersection with the vertical axis represents the limiting solids flux SF_L that can be processed in the settling basin. The corresponding underflow concentration is obtained by dropping a vertical line to the x axis from the intersection of the horizontal line and the underflow flux line. This can be done because the gravity flux is negligible at the bottom of the settling basin and the solids are removed by bulk flow. The fact that the gravity flux is negligible at the bottom of the tank can be verified by performing a materials balance around the portion of the settling tank that lies below the depth where the limiting solids flux occurs and comparing the gravity settling velocity of the sludge to the velocity in the sludge withdrawal pipe. If the quantity of solids fed to the settling basin is greater than the limiting solids-flux value defined in Fig. 6-18, the solids will build up in the settling basin and, if adequate storage capacity is not provided, ultimately overflow at the top. Using the limiting solids-flux value, the required area derived from a materials balance is given by

$$A = \frac{(Q + Q_u)C_o}{SF_L} \times 8.34 \tag{6-32}$$

$$= \frac{(1 + \alpha)QC_o}{SF_L} \times 8.34 \tag{6-33}$$

where
$$A = \text{cross-sectional area, ft}^2$$
$$(Q + Q_u) = \text{total volumetric flowrate to settling basin, Mgal/d}$$
$$C_o = \text{influent solids concentration, mg/L}$$
$$SF_L = \text{limiting solids flux, lb/ft}^2 \cdot \text{d}$$
$$\alpha = Q_u/Q$$

Referring to Fig. 6-18, if a thicker underflow concentration is required, the slope of the underflow flux line must be reduced. This, in turn, will lower the value of the limiting flux and increase the required settling area. In an actual design, the use of several different flowrates for the underflow should be evaluated. Typical values

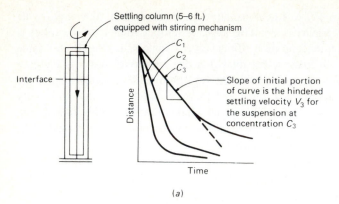

Settling column (5–6 ft.) equipped with stirring mechanism

Interface

Distance

Time

C_1
C_2
C_3

Slope of initial portion of curve is the hindered settling velocity V_3 for the suspension at concentration C_3

(a)

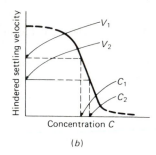

Hindered settling velocity

V_1
V_2
C_1
C_2

Concentration C

(b)

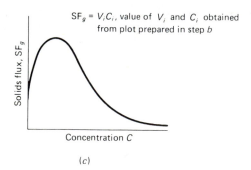

$SF_g = V_i C_i$, value of V_i and C_i obtained from plot prepared in step b

Solids flux, SF_g

Concentration C

(c)

FIGURE 6-17
Procedure for preparing plot of solid flux due to gravity as a function of solids concentration: (a) hindered settling velocities derived from column-settling tests for suspension at different concentrations, (b) plot of hindered settling velocities obtained in step a versus corresponding concentration, and (c) plot of computed value of solids flux versus corresponding concentration.

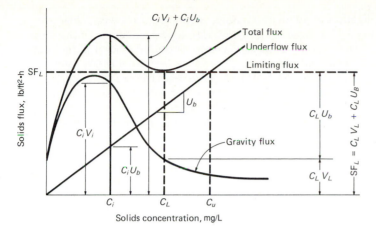

FIGURE 6-18
Definition sketch for the analysis of settling data using the solids-flux method of analysis.

for biological sludges are about 150 to 300 gal/ft$^2 \cdot$ d (7.1×10^{-5} to 1.4×10^{-4} m/s) [24]. The application of this method of analysis is illustrated in Example 6-6.

An alternative graphical method of analysis to that presented in Fig. 6-18 for determining the limiting solids flux is shown in Fig. 6-19. As shown, for a given underflow concentration, the value of the limiting flux on the ordinate is obtained by drawing a line tangent to the flux curve passing through the desired underflow and intersecting the ordinate. The geometric relationship of this method to that given in Fig. 6-18 is shown by the lightly dashed lines in Fig. 6-19. The method detailed in Fig. 6-19 is especially useful where the effect of the use of various underflow concentrations on the size of the treatment facilities (aerator and sedimentation basin) is to be evaluated. Application of the solids-flux method of analysis is illustrated in Examples 6-6 and 10-2.

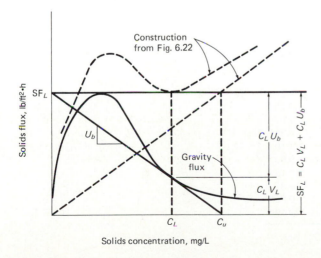

FIGURE 6-19
Alternative definition sketch for the analysis of settling data using the solids-flux method of analysis.

Example 6-6   Application of solids-flux analysis. Given the following settling data for a biological sludge, derived from a pure-oxygen activated-sludge pilot plant, estimate the maximum concentration of the aerator mixed-liquor biological suspended solids that can be maintained if the sedimentation-tank application rate $Q + Q_r$ has been fixed at 600 gal/ft² · d (24 m³/m² · d) and the sludge recycle rate Q_r is equal to 40 percent. The definition sketch for this problem is shown in the figure given below. As shown, settled and thickened biological solids from the sedimentation tank are returned to the aeration tank to maintain the desired level of biological solids in the aerator. Assume that the solids wasting rate Q_w is negligible in this example.

MLSS, mg/L	Initial settling velocity	
	ft/h	gal/ft² · d
2,000	14.0	2,520
3,000	11.5	2,070
4,000	9.1	1,638
5,000	7.0	1,260
6,000	4.2	756
7,000	3.0	540
8,000	2.2	396
9,000	1.6	288
10,000	1.2	216
15,000	0.5	90
20,000	0.2	36
30,000	0.1	18

Note: ft/h = ft³/ft² · d
ft/h × 180 = gal/ft² · d

Solution

1. Develop the gravity solids-flux curve from the given data and plot the curve.

   (*a*) Set up a computation table to determine the solids-flux values corresponding to the given solids concentrations.

MLSS, mg/L	Initial settling velocity, ft/h	Solids flux, lb/ft² · d
2,000	14.0	1.75[a]
3,000	11.5	2.15
4,000	9.1	2.27
5,000	7.0	2.18
6,000	4.2	1.57
7,000	3.0	1.31
8,000	2.2	1.10
9,000	1.6	0.90
10,000	1.2	0.75
15,000	0.5	0.47
20,000	0.2	0.25
30,000	0.1	0.19

[a] 1.75 = (2,000 × 14.0)/16,030 (see Eq. 6-28).

(b) Plot the gravity solids-flux curve (see following figure).

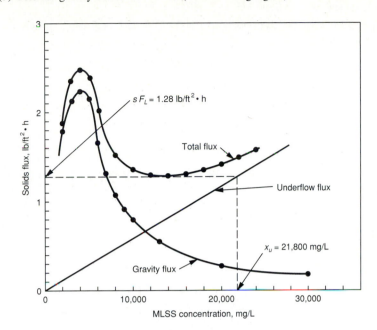

2. Determine the underflow bulk velocity. Referring to the definition sketch, the applied loading on the sedimentation facilities equals $(Q + Q_r)$, which per unit area is equal to 600 gal/ft$^2 \cdot$ d (3.33 ft/h). The underflow velocity is therefore equal to

$$U_b = [0.4Q/(Q + 0.4Q)](3.33 \text{ ft/h})$$
$$= 0.95 \text{ ft/h}$$

Note: $(600 \text{ gal/ft}^2 \cdot \text{d}) \left[\dfrac{1}{(7.48 \text{ gal/ft}^3)(24 \text{ h/d})} \right] = 3.33 \text{ ft/h}$

3. Develop the total-flux curve for the system, and determine the value of the limiting flux and maximum underflow concentration.

   (a) Plot the underflow curve on the solids-flux curve using the following relationship:

   $$SF_u = kX_i U_b$$

   where X_i = MLSS concentration, mg/L
   U_b = bulk underflow velocity, ft/h

   (b) At X_i = 10,000 mg/L, SF_u = 10,000 × 0.95/16,030 = 0.59 lb/ft$^2 \cdot$ d.

   (c) Plot the total-solids-flux curve by summing the values of the gravity and underflow solids-flux (see solids-flux curve).

   (d) From the solids-flux curve, the limiting solids flux is found to be equal to

   $$SF_L = 1.28 \text{ lb/ft}^2 \cdot \text{h}$$

   (e) From the solids-flux curve, the maximum underflow solids concentration is equal to 21,800 mg/L.

4. Estimate the maximum solids concentration that can be maintained in the reactor shown in the definition sketch.

(a) Write a mass balance for the system within the boundary neglecting the rate of cell growth within the reactor.

$$QX_o + Q_rX_t = (Q + Q_r)X$$

(b) Assuming the $X_o = 0(X_o \ll X_r)$ and that $Q_r/Q = 0.4$, solve for the concentration of MLSS in the aerator.

$$0.4Q\,(21,800 \text{ mg/L}) = (1 + 0.4)\,QX$$

$$X = 6229 \text{ mg/L}$$

Comment. As shown in this analysis, the concentration of the return solids will affect the maximum concentration of solids that can be maintained in the aerator. For this reason, the sedimentation tank should be considered an integral part of the design of an activated-sludge treatment process. This subject is considered in detail in Chap. 10, which deals with the design of biological treatment processes.

Analysis of Compression Settling (Type 4)

The volume required for the sludge in the compression region can also be determined by settling tests. The rate of consolidation in this region has been found to be proportional to the difference in the depth at time t and the depth to which the sludge will settle after a long period of time. This phenomenon can be represented as Eq. 6-34.

$$H_t - H_\infty = (H_2 - H_\infty)e^{-i(t-t_2)} \tag{6-34}$$

where H_t = sludge height time t
 H_∞ = sludge depth after long period, say 24 h
 H_2 = sludge height at time t_2
 i = constant for a given suspension

It has been observed that stirring serves to compact sludge in the compression region by breaking up the floc and permitting water to escape. Rakes are often used on sedimentation equipment to manipulate the sludge and thus produce better compaction. Dick and Ewing [6] found that stirring would produce better settling in the hindered-settling region also. With these facts in mind, it is apparent that, when appropriate, stirring should be investigated as an essential part of the settling tests if the proper areas and volumes are to be determined from the tests.

6-6 ACCELERATED GRAVITY SEPARATION

Sedimentation, as described in the previous section, occurs under the force of gravity in a constant acceleration field. The removal of settleable particles can also be accomplished by taking advantage of a changing acceleration field. The purpose of the following discussion is to introduce the subject of accelerated gravity solids separation.

Description

A number of devices that take advantage of both gravitational and centrifugal forces and induced velocities have been developed for the removal of grit from wastewater. The principles involved in one such device, known as the *Teacup separator*, are considered in the following discussion [27]. Another type of separator in which a velocity is induced to enhance the separation process, is considered in Sec. 9-3 in Chap. 9. In appearance, the Teacup separator looks like a squat tin can (see Fig. 6-20a). Wastewater from which grit is to be separated is introduced tangentially near the bottom and exits tangentially through the opening in the top of the unit. Grit is removed through the opening in the bottom of the unit.

Analysis

Because the top of the Teacup is enclosed, the rotating flow creates a free vortex within the Teacup (see Fig. 6-20b). The most important characteristic of a free vortex is that the product of the tangential velocity times the radius is a constant.

$$V_r = \text{constant} \tag{6-35}$$

where V = tangential velocity, ft/s (m/s)
r = radius, ft (m)

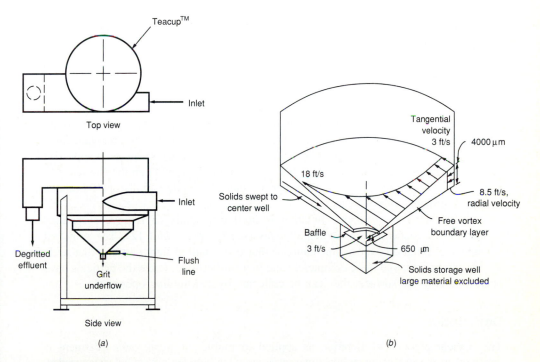

FIGURE 6-20
Teacup separator: (a) outline sketch of typical unit and (b) definition sketch (from Eutek Systems).

The significance of Eq. 6-35 can be illustrated by the following example. Assume that the tangential velocity in a Teacup with a 5 ft (1.5 m) radius is 3 ft/s (0.9 m/s). The product of the velocity times the radius at the outer edge of the Teacup is equal to 15 ft^2/s (1.35 m^2/s). If the discharge port has a radius of 1 ft (0.9 m), the tangential velocity at the entrance to the discharge port is 15 ft/s (4.5 m/s). The centrifugal force experienced by a particle within this flow field is equal to the square of the velocity divided by the radius. Because the centrifugal force is also proportional to the inverse of the radius, a fivefold decrease in the radius results in a 125-fold inrcease in the centrifugal force.

Because of the high centrifugal forces near the discharge port, some of the particles, depending on their size, density, and drag, are retained within the body of the free vortex near the center of the Teacup, while other particles are swept out of the unit. Grit and sand particles will be retained while organic particles are discharged from the unit. Organic particles with the same settling velocity as sand will typically be from four to eight times as large. The corresponding drag forces for these organic particles will be from 16 to 64 times as great. As a result, the organic particles tend to move with the fluid and are transported out of the Teacup. The particles held in the free vortex ultimately settle to the bottom of the unit under the force of gravity. Organic particles that sometimes settle usually consist of oil and grease attached to grit or sand particles. Referring to Fig. 6-20b, it can be seen that a free-vortex flow also creates a boundary layer within the Teacup. Particles that settle within this boundary layer are transported to the center of the Teacup by the radial velocity. The application of the Teacup separator is considered in Chap. 9.

6-7 FLOTATION

Flotation is a unit operation used to separate solid or liquid particles from a liquid phase. Separation is brought about by introducing fine gas (usually air) bubbles into the liquid phase. The bubbles attach to the particulate matter, and the buoyant force of the combined particle and gas bubbles is great enough to cause the particle to rise to the surface. Particles that have a higher density than the liquid can thus be made to rise. The rising of particles with lower density than the liquid can also be facilitated (e.g., oil suspension in water).

In wastewater treatment, flotation is used principally to remove suspended matter and to concentrate biological sludges (see Chaps. 9 and 12). The principal advantage of flotation over sedimentation is that very small or light particles that settle slowly can be removed more completely and in a shorter time. Once the particles have been floated to the surface, they can be collected by a skimming operation.

Description

The present practice of flotation as applied to municipal wastewater treatment is confined to the use of air as the flotation agent. Air bubbles are added or caused to form in one of the following methods:

1. Injection of air while the liquid is under pressure, followed by release of the pressure (dissolved-air flotation)
2. Aeration at atmospheric pressure (air flotation)
3. Saturation with air at atmospheric pressure, followed by application of a vacuum to the liquid (vacuum flotation)

Further, in all these systems the degree of removal can be enhanced through the use of various chemical additives.

Dissolved-Air Flotation. In dissolved-air flotation (DAF) systems, air is dissolved in the wastewater under a pressure of several atmospheres, followed by release of the pressure to the atmospheric level (see Fig. 6-21). In small pressure systems, the entire flow may be pressurized by means of a pump to 40 to 50 lb/in^2 gage (275 to 350 kPa) with compressed air added at the pump suction (see Fig. 6-21a). The entire flow is held in a retention tank under pressure for several minutes to allow time for the air to dissolve. The pressurized flow is then admitted through a pressure-reducing valve to the flotation tank where the air comes out of solution in minute bubbles throughout the entire volume of liquid.

In the larger units, a portion of the DAF effluent (15 to 120 percent) is recycled, pressurized, and semisaturated with air (Fig. 6-21b). The recycled flow is mixed with the unpressurized main stream just before admission to the flotation tank, with the result that the air comes out of solution in contact with particulate matter at the entrance to the tank. Pressure types of units have been used mainly for the treatment of industrial wastes and for the concentration of sludges.

Air Flotation. In air flotation systems, air bubbles are formed by introducing the gas phase directly into the liquid phase through a revolving impeller or through diffusers. Aeration alone for a short period is not particularly effective in bringing about flotation of solids. The provisions of aeration tanks for flotation of grease and other solids from normal wastewater is usually not warranted, although some success with these units has been experienced on certain scum-forming wastes.

Vacuum Flotation. Vacuum flotation consists of saturating the wastewater with air either (1) directly in an aeration tank or (2) by permitting air to enter on the suction side of a wastewater pump. A partial vacuum is applied, which causes the dissolved air to come out of solution as minute bubbles. The bubbles and the attached solid particles rise to the surface to form a scum blanket, which is removed by a skimming mechanism. Grit and other solids that settle to the bottom are raked to a central sludge sump for removal. If this unit is used for grit removal and if the sludge is to be digested, the grit must be separated from the sludge in a grit classifier before the sludge is pumped to the digesters.

The unit consists of a covered cylindrical tank in which a partial vacuum is maintained. The tank is equipped with scum- and sludge-removal mechanisms. The floating material is continuously swept to the tank periphery, automatically discharged into a scum trough, and removed from the unit to a pump also under partial vacuum.

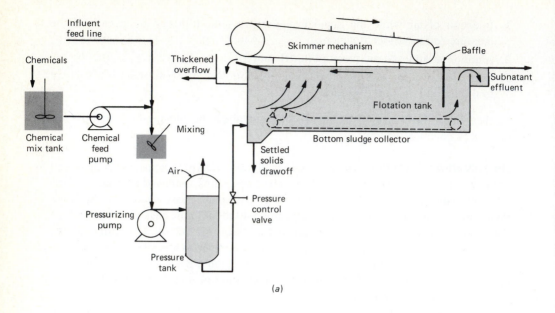

(a)

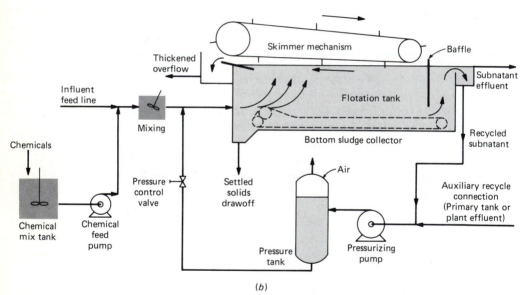

(b)

FIGURE 6-21
Schematic of dissolved-air flotation systems: (a) without recycle and (b) with recycle.

Auxiliary equipment includes an aeration tank for saturating the wastewater with air, a short-period detention tank for removal of large air bubbles, vacuum pumps, and sludge and scum pumps.

Chemical Additives. Chemicals are commonly used to aid the flotation process. These chemicals, for the most part, function to create a surface or a structure that can easily absorb or entrap air bubbles. Inorganic chemicals, such as the aluminum and ferric salts and activated silica, can be used to bind the particulate matter together and, in so doing, create a structure that can easily entrap air bubbles. Various organic polymers can be used to change the nature of either the air-liquid interface of the solid-liquid interface or both. These compounds usually collect on the interface to bring about the desired changes.

Analysis

Because flotation is very dependent on the type of surface of the particulate matter, laboratory and pilot plant tests must usually be performed to yield the necessary design criteria. Factors that must be considered in the design of flotation units include the concentration of particulate matter, quantity of air used, the particle-rise velocity, and the solids-loading rate. In the following analysis, dissolved-air flotation is discussed because it is the method most commonly used. The design of dissolved-air flotation systems is discussed in Chap. 9.

The performance of a dissolved-air flotation system depends primarily on the ratio of the volume of air to the mass of solids (A/S) required to achieve a given degree of clarification. This ratio will vary with each type of suspension and must be determined experimentally using a laboratory flotation cell. A typical laboratory flotation cell is shown in Fig. 6-22. Procedures for conducting the necessary tests may be found in Ref. 10. Typical A/S ratios encountered in the thickening of sludge in wastewater treatment plants vary from about 0.005 to 0.060.

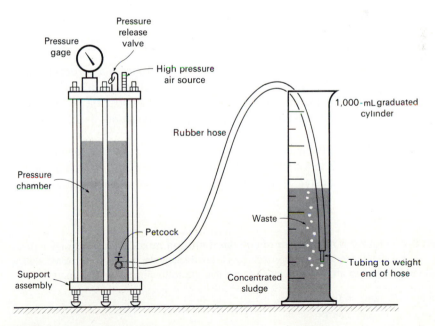

FIGURE 6-22
Schematic of dissolved-air flotation test apparatus.

The relationship between the A/S ratio and the solubility of air, the operating pressure, and the concentration of sludge solids for a system in which all the flow is pressurized is given in Eq. 6-36.

$$\frac{A}{S} = \frac{1.3s_a(fP - 1)}{S_a} \qquad (6\text{-}36)$$

Temp., °C	0	10	20	30
s_a, mL/L	29.2	22.8	18.7	15.7

where A/S = air to solids ratio, mL (air)/mg(solids)

s_a = air solubility, mL/L

f = fraction of air dissolved at pressure P, usually 0.5

P = pressure, atm

$= \dfrac{p + 14.7}{14.7}$ (U.S. customary units)

$= \dfrac{p + 101.35}{101.35}$ (SI units)

p = gage pressure, lb/in^2 gage (kPa)

S_a = sludge solids, mg/L

The corresponding equation for a system with only pressurized recycle is

$$\frac{A}{S} = \frac{1.3s_a(fP - 1)R}{S_aQ} \qquad (6\text{-}37)$$

where R = pressurized recycle, Mgal/d (m^3/d)

Q = mixed-liquor flow, Mgal/d (m^3/d)

In both equations, the numerator represents the weight of air and the denominator the weight of the solids. The factor 1.3 is the weight in milligrams of 1 mL of air, and the term (-1) within the brackets accounts for the fact that the system is to be operated at atmospheric conditions. The use of these equations is illustrated in Example 6-7.

The required area of the thickener is determined from a consideration of the rise velocity of the solids, 0.2 to 4.0 gal/m · ft^2 (8 to 160 L/m^2 · min), depending on the solids concentration, the degree of thickening to be achieved, and the solids-loading rate (see Table 12-14)

Example 6-7 Flotation thickening of activated-sludge mixed liquor. Design a flotation thickener without and with pressurized recycle to thicken the solids in activated-sludge mixed liquor from 0.3 to about 4 percent. Assume that the following conditions apply.

1. Optimum A/S ratio = 0.008 mL/mg
2. Temperature = 20°C, 68°F
3. Air solubility = 18.7 mL/L

4. Recycle-system pressure = 40 lb$_f$/in^2 gage (275 kPa)
5. Fraction of saturation = 0.5
6. Surface-loading rate = 0.2 gal/min · ft^2 (8 L/m^2 · min)
7. Sludge flowrate = 0.1 Mgal/d (400 m^3/d)

Solution (without recycle)

1. Compute the required pressure using Eq. 6-36.

$$\frac{A}{S} = \frac{1.3s_a(fP - 1)}{S_a}$$

$$0.008 = \frac{1.3(18.7 \text{ mL/L})(0.5P - 1)}{3000 \text{ mg/L}}$$

$$0.5P = 0.99 + 1$$

$$P = 3.98 \text{ atm} = \frac{p + 14.7}{14.7}$$

$$P = 43.8 \text{ lb/in}^2 \text{ gage } (302 \text{ } kPa)$$

2. Determine the required surface area.

$$A = \frac{100,000 \text{ gal/d}}{0.2 \text{ gal/min} \cdot \text{ft}^2(24 \text{ h/d})(60 \text{ min/h})}$$

$$= 347 \text{ ft}^2(34.7 \text{ m}^2)$$

3. Check the solids-loading rate.

$$\text{lb/ft}^2 \cdot \text{d} = \frac{(0.1 \text{ Mgal/d}) (3000 \text{ mg/L})[8.34 \text{ Mgal/d} \cdot (\text{mb/L})]}{347 \text{ ft}^2}$$

$$= 7.2 \text{ lb/ft}^2 \cdot \text{d}(34.6 \text{ kg/m}^2 \cdot \text{d})$$

Solution (with recycle)

1. Determine pressure in atmospheres.

$$p = \frac{40 + 14.7}{14.7} = 3.72 \text{ atm}$$

2. Determine the required recycle rate using Eq. 6-37.

$$\frac{A}{S} = \frac{1.3s_a(fP - 1)R}{S_aQ}$$

$$0.008 = \frac{1.3(18.7 \text{ mL/L})[0.5(3.72) - 1] R}{3000 \text{ mg/L} (0.1 \text{ Mgal/d})}$$

$$R = 0.115 \text{ Mgal/d} (435.4 \text{ m}^3/\text{d})$$

Alternatively, the recycle flowrate could have been set and the pressure determined. In an actual design, the costs associated with the recycle pumping, pressurizing systems, and tank construction can be evaluated to find the most economical combination.

3. Determine the required surface area.

$$A = \frac{215,000 \text{ gal}}{0.2 \text{ gal/min} \cdot \text{ft}^2 \ (60 \text{ min/h})(24 \text{ h/d})} = 747 \text{ ft}^2 \ (69.3 \text{ m}^2)$$

6-8 GRANULAR-MEDIUM FILTRATION

Although filtration is one of the principal unit operations used in the treatment of potable water, the filtration of effluents from wastewater treatment processes is a relatively recent practice. Filtration is now used extensively for achieving supplemental removals of suspended solids (including particulate BOD) from wastewater effluents of biological and chemical treatment processes. Filtration is also used to remove chemically precipitated phosphorus.

The ability to design filters and to predict their performance must be based on (1) an understanding of the variables that control the process and (2) a knowledge of the pertinent mechanism or mechanisms responsible for the removal of particulate matter from a wastewater. The discussion in this section therefore covers the following topics: (1) description of the filtration operation, (2) classifications of filtration systems, (3) filtration-process variables, (4) particle-removal mechanisms, (5) general analysis of filtration operation, (6) analysis of wastewater filtration, and (7) need for pilot plant studies. The literature dealing with filtration is so voluminous that the information presented in this section can serve only as an introduction to the subject. For additional details, the references in the text should be consulted. The practical implementation of filtration facilities is detailed in Chap. 11.

Description of the Filtration Operation

The complete filtration operation is comprised of two phases: filtration and cleaning or regeneration (commonly called backwashing). While the description of the phenomena occurring during the filtration phase is essentially the same for all of the filters used for wastewater filtration, the cleaning phase is quite different depending on whether the filter operation is of the semicontinuous or continuous type. As the name implies, in semicontinuous filtration the filtering and cleaning phases occur sequentially, whereas in continuous filtration the filtering and cleaning phases occur simultaneously. The physical and operational characteristics of the granular-medium filters commonly used for wastewater filtration are reported in Tables 6-9 and 6-10, respectively. Definition sketches for the principal types of filters used are presented in Fig. 6-23.

Semicontinuous Filtration Operations. Both the filtration and cleaning phases for conventional semicontinuous filtration are identified in the definition sketch shown in Fig. 6-24. The filtration phase in which particulate material is removed is accomplished by passing the wastewater to be filtered through a filter bed composed of granular material without or with the addition of chemicals. Within the granular-medium filter bed, the removal of the suspended solids contained in the wastewater is accomplished by a complex process involving one or more removal mechanisms such as straining, interception, impaction, sedimentation, flocculation, and adsorption.

The end of the filter run (filtration phase) is reached when the suspended solids in the effluent start to increase (break through) beyond an acceptable level, or when a limiting headloss occurs across the filter bed (see Fig. 6-25). Ideally, both these events should occur at the same time. Once either of these conditions is reached, the filtration phase is terminated, and the filter must be cleaned (backwashed) to remove the material (suspended solids) that has accumulated within the granular filter bed. Usually, this is done by reversing the flow through the filter (see Fig. 6-24). A sufficient flow of washwater is applied until the granular filtering medium is fluidized (expanded). The material that has accumulated within the bed is then washed away. Air is often used in conjunction with the water to enhance the cleaning of the filter bed. In most wastewater treatment plant flow diagrams, the washwater containing the suspended solids that are removed from the filter is returned either to the primary settling facilities or to the biological treatment process.

Continuous Filtration Operations. In filters that operate continuously such as the traveling-bridge filter (see Fig 6-23f) and upflow filter (see Fig. 6-23g), the filtering and cleaning (backwashing) phases take place simultaneously. It should be noted that with filters that operate continuously there is no turbidity breakthrough or terminal headloss.

In the traveling-bridge filter, the incoming wastewater floods the filter bed, flows through the medium by gravity, and exits to the clearwell via effluent ports located under each cell. During the backwash cycle, the carriage and the attached hood (see Fig. 6-23f) move slowly over the filter bed, consecutively isolating and backwashing each cell. The backwash pump, located in the clearwell, draws filtered wastewater from the effluent chamber and pumps it through the effluent port of each cell, forcing water to flow up through the cell and backwashing the filter medium of the cell. The washwater pump located above the hood draws water with suspended matter collected under the hood and transfers it to the backwash water trough. During the backwash cycle, wastewater is filtered continuously through the cells not being backwashed.

In the upflow filter (see Fig. 6-23g), the liquid to be filtered flows upward through the filter bed. At the same time the sand bed, moving in the counter-current direction, is being cleaned continuously. An airlift is used to pump the sand from the bottom of the filter up through a central pipe to a washer assembly located at the top of the filter. As the sand is being pumped up to the top of the filter, the individual sand grains are cleaned of accumulated material by abrasion (sand against sand) and fluid shear forces. In the sand washer, the accumulated material removed from the sand is removed over a weir. Additional washing of the sand occurs as it passes through the zig-zag flow channel in the lower portion of the sand washer and before it falls back on the surface of the sand bed. Because the effluent water level is higher than the water level in the sand washer, there is a positive upward flow of filtered effluent through the sand washer.

TABLE 6-9
Physical characteristics of commonly used granular-medium filters

Type of filter operation	Type of filter (common name)	Filter bed details			Typical direction of fluid flow	Backwash operation	Flowrate through filter	Remarks
		Type of filter bed	Filtering medium	Typical bed depth, in				
Semicontinuous	Conventional	Mono-medium	Sand or anthracite	30	Downward	Batch	Constant/variable	Most commonly used type
Semicontinuous	Conventional	Dual-medium	Sand and anthracite	36	Downward	Batch	Constant/variable	Dual-medium design used to extend filter run length
Semicontinuous	Conventional	Multimedium	Sand, anthracite, and garnet	36	Downward	Batch	Constant/variable	Multimedium design used to extend filter run length
Semicontinuous	Deep bed	Mono-medium	Sand or anthracite	48–72	Downward	Batch	Constant/variable	
Semicontinuous	Deep bed	Mono-medium	Sand	48–72	Upward	Batch	Constant	
Semicontinuous	Pulsed-bed	Mono-medium	Sand	11	Downward	Batch	Constant	Air pulses used to breakup surface mat and increase run length
Continuous	Deep bed	Mono-medium	Sand	48–72	Upward	Continuous	Constant	Sand bed moves in countercurrent direction to fluid flow
Continuous	Traveling-bridge	Mono-medium	Sand	11	Downward	Semicontinuous	Constant	Individual filter cells backwashed sequentially
Continuous	Traveling-bridge	Dual-medium	Sand	16	Downward	Semicontinuous	Constant	Individual filter cells backwashed sequentially

Note: mm × 0.03937 = in.

TABLE 6-10
Operational characteristics of commonly used granular-medium filters

Type of filter	Operation during the filtration phase	Operation during the cleaning phase
Conventional Semicontinuous; mono-, dual-, and multimedium, downflow	Liquid to be filtered is passed downward through the filter bed. Rate of flow through the filter can be constant or variable depending on the flow control method.	When the effluent turbidity starts to increase or the allowable headloss is reached, the filter is backwashed by reversing the direction of flow through the filter. Both air and water are used in the backwash operation.
Deep bed Semicontinuous, mono-medium, downflow	Liquid to be filtered is passed downward through the filter bed. Rate of flow through the filter can be constant or variable depending on the flow control method.	When the effluent turbidity starts to increase or the allowable headloss is reached, the filter is backwashed by reversing the direction of flow through the filter. Both air and water are used in the backwash operation.
Deep bed Semicontinuous, mono-medium, upflow	Liquid to be filtered is passed upward through the filter bed. Rate of flow through the filter is usually constant.	When the effluent turbidity starts to increase or the allowable headloss is reached, the filter is backwashed by increasing the flow rate through the bottom of the filter. Both air and water are used in the backwash operation.
Pulsed-bed Semicontinuous, mono-medium, downflow	Liquid to be filtered is passed downward through the filter bed. As the headloss builds up, air is pulsed up through the bed to break up the surface mat and to redistribute the solids. Rate of flow through the filter is usually constant.	When the effluent turbidity starts to increase or the allowable headloss is reached, the filter is backwashed by reversing the direction of flow through the filter. Liquid to be filtered continues to enter the filter during the backwash operation. Chemical cleaning is also used.
Deep bed Continuous, mono-medium, upflow	Liquid to be filtered is passed downward through the filter bed, which is moving downward in the countercurrent direction. Rate of flow through the filter is usually constant.	The filter medium is backwashed continuously by pumping sand from the bottom of the filter with an air lift to a sand washer located in the top of the filter. After passing through the washer, the clean sand is distributed on the top of the filter bed.
Traveling-bridge Continuous, mono- and dual-medium, downflow	Liquid to be filtered is passed downward through the filter bed. Liquid continues to be filtered while the individual cells are backwashed. Rate of flow through the filter is usually constant.	When the allowable headloss is reached, the individual cells of the filter are backwashed by reversing the direction of flow through each of the cells successively. The backwash water is removed through the backwash hood.

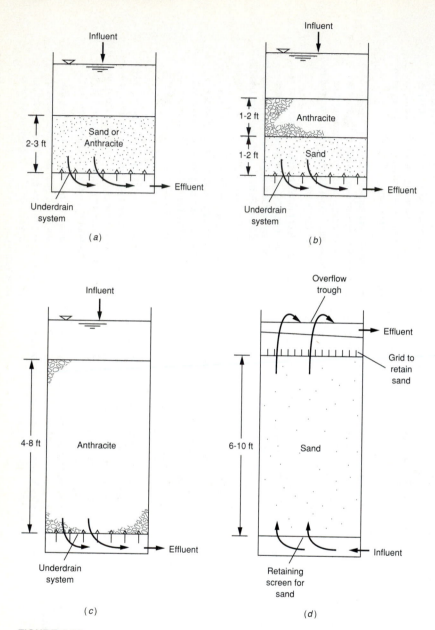

FIGURE 6-23
Types of filters used for the filtration of treated wastewater: (*a*) conventional mono-medium downflow, (*b*) conventional dual-medium downflow, (*c*) conventional mono-medium deep-bed downflow, (*d*) deep-bed upflow (continued on following page).

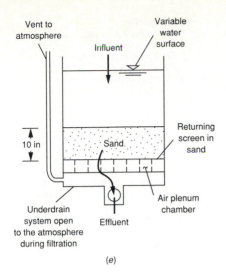

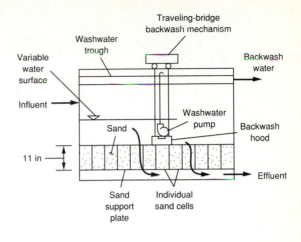

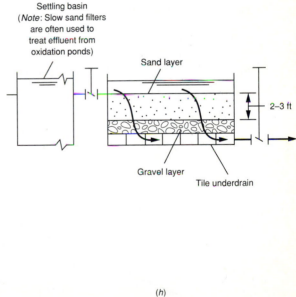

FIGURE 6-23 (*continued*)
(*e*) Pulsed-bed filter, (*f*) traveling bridge filter, (*g*) continuous backwash deep-bed upflow filter, and (*h*) slow sand filter.

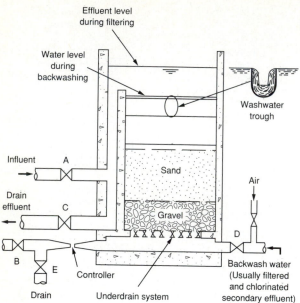

How filter operates:
1. Open valve A. (This allows effluent to flow to filter.)
2. Open valve B. (This allows effluent to flow through filter.)
3. During filter operation, all other valves are closed.

How filter is backwashed:
1. Close valve A.
2. Close valve B when water in filter drops down to top of overflow.
3. Open valves C and D. (This allows water from wash water tank to flow up through the filtering medium, loosening up the sand and washing the accumulated solids out of the filter. Filter backwash water is returned to head end of treatment plant.)

How to filter to waste (if used):
1. Open valves A and E. All other valves closed. Effluent is sometimes filtered to waste for a few minutes after filter has been washed to condition the filter before it is put into service.

FIGURE 6-24
Definition sketch for operation of conventional downflow, granular-medium, gravity-flow filter.

Classification of Filtration Systems

A number of individual filtration-system designs have been proposed and built. The principal types of granular-medium filters may be classified according to (1) the type of operation, (2) the type of filtering medium used, (3) the direction of flow during filtration, (4) the backwashing process, and (5) the method of flowrate control. The

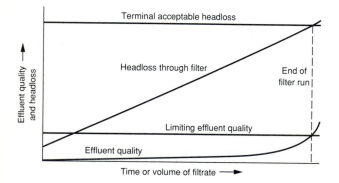

FIGURE 6-25
Definition sketch for the length of filtration run based on effluent quality and headloss.

physical characteristics of commonly used granular-medium filters are reported in Table 6-9.

Type of Operation. With respect to their mode of operation, granular-medium filters can be classified as either semicontinuous or continuous (see Table 6-10). In semicontinuous operation, the filter is operated until the effluent quality starts to deteriorate or the headloss becomes excessive, at which point the filter is taken out of service and backwashed to remove the accumulated solids. In continuous operation, filtration and backwashing occur simultaneously.

Direction of Fluid Flow During Filtration. The principal types of filters used for the filtration of wastewater effluent may be classified according to the direction of flow as downflow or upflow. The downflow filter is by far the most common type used. Its operation was described at the beginning of this section.

Types of Filtering Materials and Filter Bed Configurations. The principal types of filter bed configurations now used for wastewater filtration may be classified according to the number of filtering media used as mono-medium, dual-medium, or multimedium beds (see Fig. 6-26). In conventional downflow filters, the distribution of grain sizes for each medium after backwashing is from small to large. The degree of intermixing in the dual-medium and tri-medium beds depends on the density and size differences of the various media.

Dual- and multimedium and deep bed mono-medium filter beds were developed to allow the suspended solids in the liquid to be filtered so as to penetrate farther into the filter bed, thus using more of the solids-storage capacity available within

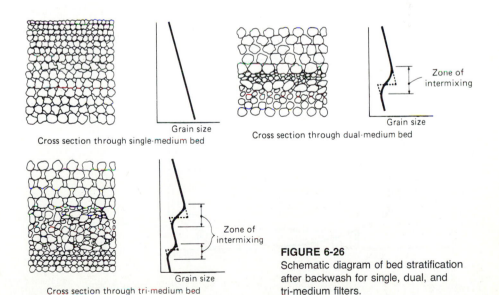

Cross section through single-medium bed

Cross section through dual-medium bed

Zone of intermixing

Grain size

Zone of intermixing

Cross section through tri-medium bed

Grain size

FIGURE 6-26
Schematic diagram of bed stratification after backwash for single, dual, and tri-medium filters.

the filter. By comparison, in shallow mono-medium beds, most of the removal has been observed to occur in the upper few millimeters of the bed. The penetration of the solids farther into the bed also permits longer filter runs because the buildup of headloss is reduced. Data and design information on the various types of media and the characteristic sizes and depths that have been used are presented in Chap. 11.

Filtration Driving Force. Either the force of gravity or an applied pressure force can be used to overcome the frictional resistance to flow offered by the filter bed. Gravity filters of the type shown in Fig. 6-23 are most commonly used for the filtration of treated effluent at large plants. Pressure filters of the type shown in Fig. 6-27 operate in the same manner as gravity filters and are used at smaller plants. The only difference is that, in pressure filters, the filtration operation is carried out in a closed vessel under pressurized conditions achieved by pumping. Pressure filters normally are operated at higher terminal headlosses. This generally results in longer filter runs and reduced backwash requirements.

Flow Control. The rate of flow through a filter may be expressed as follows [24]:

$$\text{Rate of flow} = \frac{\text{driving force}}{\text{filter resistance}} \tag{6-38}$$

In this equation, the driving force represents the pressure drop across the filter. At the start of the filter run, the driving force must overcome only the resistance offered by the clean filter bed and the underdrain system. As solids start to accumulate within

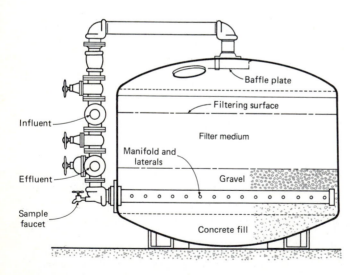

FIGURE 6-27
Cross section through a typical pressure filter.

the filter, the driving force must overcome the resistance offered by the clogged filter bed and the underdrain system. The principal methods now used to control the rate of flow through gravity filters may be classified as (1) constant-rate filtration and (2) variable-declining-rate filtration.

Constant-rate filtration. In constant-rate filtration (see Figs. 6-28a,b,c), the flow through the filter is maintained at a constant rate. Constant-rate filtration systems are either influent controlled or effluent controlled. Pumps or weirs are used for influent control whereas an effluent valve that can be operated manually or mechanically is used for effluent control. In effluent control systems, at the beginning of the run, a large portion of the available driving force is dissipated at the valve, which is almost closed. The valve is opened as the headloss builds up within the filter during the run. Because the required control valves are expensive and because they have malfunctioned on a number of occasions, alternative methods of flowrate control involving pumps and weirs have been developed and are coming into wider use (see Chap. 9).

Variable-rate filtration. In variable-declining-rate filtration (see Fig. 6-28d), the rate of flow through the filter is allowed to decline as the rate of headloss builds up with time. Declining-rate filtration systems are either influent controlled or effluent controlled. When the rate of flow is reduced to the minimum design rate, the filter is removed from service and backwashed. Additional details on this method of control as well as others may be found in Refs. 24–25.

Filtration-Process Variables

The principal variables that must be considered in the design of filters are identified in Table 6-11. In the application of filtration for the removal of residual suspended solids, it has been found that the nature of the particulate matter in the influent to be filtered, the size of the filter material or materials, and the filtration flowrate are perhaps the most important of the process variables (Table 6-11, items 6, 1, and 4).

Influent Characteristics. The most important influent characteristics are the suspended-solids concentration, particle size and distribution, and floc strength. Typically, the suspended-solids concentration in the effluent from activated-sludge and trickling-filter plants varies between 6 and 30 mg/L. Because this concentration usually is the principal parameter of concern, turbidity is often used as a means of monitoring the filtration process. Within limits, it has been shown that the suspended-solids concentrations found in treated wastewater can be correlated to turbidity measurements. A typical relationship for the effluent from a complete-mix, activated-sludge process is [22]

$$\text{Suspended solids, SS, mg/L} = (2.3 \text{ to } 2.4) \times (\text{turbidity, NTU}) \tag{6-39}$$

Typical data on the particle size and distribution in the effluent from a pilot-scale activated-sludge plant operated at a mean cell-residence time of 10 days are

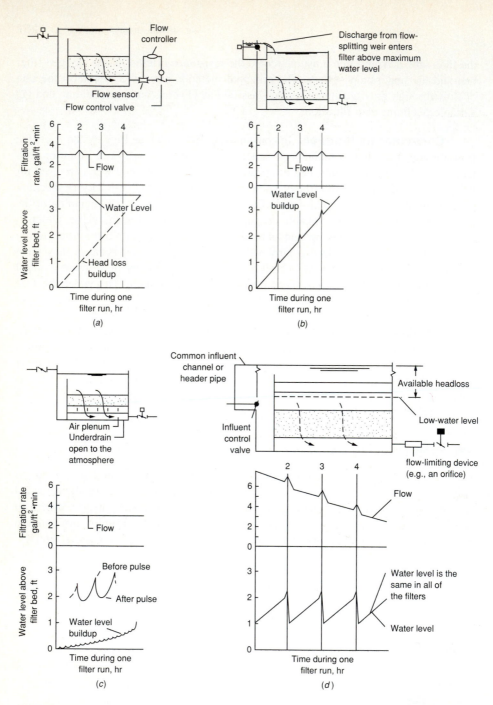

FIGURE 6-28
Definition sketch for constant-rate filtration: (a) fixed head, (b) variable head, (c) variable head with pulsed-bed filter and variable-rate filtration, and (d) variable-flow variable-head. *Note:* Curves for filters in (a), (b), and (d) are for the operation of one filter in a bank of four filters. The numbers represent the filter that is backwashing during the filter run. In practice, the time before backwashing will not be the same for all of the filters [19].

TABLE 6-11
Principal variables in the design of granular-medium filters[a]

Variable	Significance

1. Filter-medium characteristics[b]
   a. Grain size
   b. Grain-size distribution
   c. Grain shape, density
       and composition
   d. Medium charge
	Affect particle removal efficiency and headloss buildup

2. Filter-bed porosity	Determines the amount of solids that can be stored in the filter
3. Filter-bed depth	Affects headloss, length of run
4. Filtration rate[b]	Used in conjunction with variables 1, 2, 3, and 6 to compute clear water headloss
5. Allowable headloss	Design variable

6. Influent wastewater characteristics[b]
   a. Suspended solids
       concentration
   b. Floc or particle size and
       distribution
   c. Floc strength
   d. Floc or particle charge
   e. Fluid properties
	Affect the removal characteristics of a given filter-bed configuration. To a limited extent the listed influent characteristics can be controlled by the designer

[a] Adapted in part from Refs. 18 and 19.
[b] See text for additional discussion of specific variables.

shown in Fig. 6-29. Similar observations have also been made at full-scale plants. As illustrated, the particles fell into two distinct size ranges, small particles varying in areal size (equivalent circular diameter) from 1 to 15 μm and large particles varying in size from 50 to 150 μm. In addition, a few particles larger than about 500 μm are almost always found in settled treated effluent. These particles are

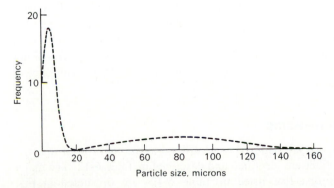

FIGURE 6-29
Typical particle size distribution found in settled wastewater effluent [24].

light and amorphous and do not settle readily (see discussion of hindered settling in Sec. 6-5). From the distribution analysis, the mean size for the smaller particles was estimated to be about 3 to 5 μm and that for the larger particles about 80 to 90 μm. The weight fraction of the smaller particles was estimated to be approximately 40 to 60 percent of the total. This percentage will vary, however, depending on the operating conditions of the biological process and the degree of flocculation achieved in the secondary settling facilities.

The most significant observation relating to particle size is that the distribution of sizes was found to be bimodal. This is important because it will influence the removal mechanisms that may be operative during the filtration process. For example, it seems reasonable to assume that the removal mechanism for particles 1.0 μm in size would be different from that for particles 80 μm in size or larger. The bimodal particle-size distribution has also been observed in water treatment plants [9].

Floc strength, which will vary not only with the type of process but also with the mode of operation, is also important. For example, the residual floc from the chemical precipitation of biologically processed wastewater may be considerably weaker than the residual biological floc before precipitation. Further, the strength of the biological floc will vary with the mean cell-residence time, increasing with longer mean cell-residence time (see Chaps. 8 and 10). The increased strength derives in part from the production of extracellular polymers as the mean cell-residence time is lengthened. At extremely long mean cell-residence times (15 days and longer), it has been observed that the floc strength will decrease.

Filter-Medium Characteristics. Grain size is the principal filter-medium characteristic that affects the filtration operation. Grain size affects both the clear-water headloss and the buildup of headloss during the filter run. If too small a filtering medium is selected, much of the driving force will be wasted in overcoming the frictional resistance of the filter bed. On the other hand, if the size of the medium is too large, many of the small particles in the influent will pass directly through the bed.

Filtration Rate. The rate of filtration is important because it affects the real size of the filters that will be required. For a given filter application, the rate of filtration will depend primarily on the strength of the floc and the size of the filtering medium. For example, if the strength of the floc is weak, high filtration rates will tend to shear the floc particles and carry much of the material through the filter. It has been observed that filtration rates in the range of 2 to 8 gal/ft^2 · min (80 to 320 L/m^2 · min) will not affect the quality of the filter effluents because biological floc is strong. This subject is considered further in Chap. 11.

Particle-Removal Mechanisms

The principal mechanisms that are believed to contribute to the removal of material within a granular-medium filter are identified and described in Table 6-12. The major removal mechanisms (the first five listed in Table 6-12) are illustrated in Fig. 6-30. Straining has been identified as the principal mechanism that is operative in the

TABLE 6-12

Mechanisms operative within a granular-medium filter that contribute to the removal of suspended materials[a]

Mechanism	Description
1. Straining[b]	
a. Mechanical	Particles larger than the pore space of the filtering medium are strained out mechanically
b. Chance contact	Particles smaller than the pore space are trapped within the filter by chance contact
2. Sedimentation[b]	Particles settle on the filtering medium within the filter
3. Impaction[b]	Heavy particles will not follow the flow streamlines
4. Interception[b]	Many particles that move along in the streamline are removed when they come in contact with the surface of the filtering medium
5. Adhesion[b]	Flocculant particles become attached to the surface of the filtering medium as they pass by. Because of the force of the flowing water, some material is sheared away before it becomes firmly attached and is pushed deeper into the filter bed. As the bed becomes clogged, the surface shear force increases to a point at which no additional material can be removed. Some material may break through the bottom of the filter, causing the sudden appearance of turbidity in the effluent
6. Chemical adsorption	
a. Bonding	
b. Chemical interaction	Once a particle has been brought in contact with the surface of the filtering medium or with other particles, either one of these mechanisms, or both, may be responsible for holding it there
7. Physical adsorption	
a. Electrostatic forces	
b. Electrokinetic forces	
c. van der Waals forces	
8. Flocculation	Large particles overtake smaller particles, join them, and form still larger particles. These particles are then removed by one or more of the above removal mechanisms (1 through 5)
9. Biological growth	Biological growth within the filter will reduce the pore volume and may enhance the removal of particles with any of the above removal mechanisms (1 through 5)

[a] Adapted from Ref. 18.

[b] Usually identified in the literature as removal mechanisms.

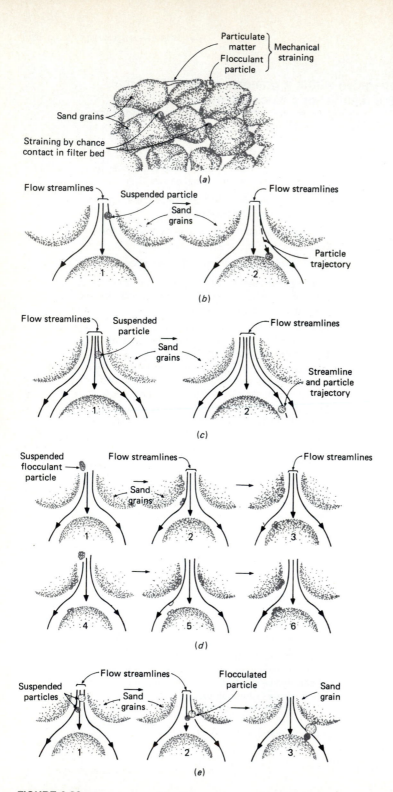

FIGURE 6-30
Removal of suspended particulate matter within a granular filter: (*a*) by straining, (*b*) by sedimentation or inertial impaction, (*c*) by interception, (*d*) by adhesion, or (*e*) by flocculation.

removal of suspended solids during the filtration of settled secondary effluent from biological treatment processes [18,22].

Other mechanisms are probably also operative even though their effects are small and, for the most part, masked by the straining action. These other mechanisms include interception, impaction, and adhesion. In fact, it is reasonable to assume that the removal of some of the smaller particles shown in Fig. 6-29 must be accomplished in two steps involving (1) the transport of the particles to the surface where they will be removed and (2) the removal of particles by one or more of the operative removal mechanisms. O'Melia and Stumm have identified these two steps as transport and attachment [14].

The removal of suspended material by straining can be identified by noting (1) the variation in the normalized concentration-removal curves through the filter as a function of time and (2) the shape of the headloss curve for the entire filter or an individual layer within the filter. If straining is the principal removal mechanism, the shape of the normalized removal curve will not vary significantly with time (see Fig. 6-31), and the headloss curves will be curvilinear.

General Analysis of Filtration Operation

In general, the mathematical characterization of the time-space removal of particulate matter within the filter is based on a consideration of the equation of continuity together with an auxiliary rate equation.

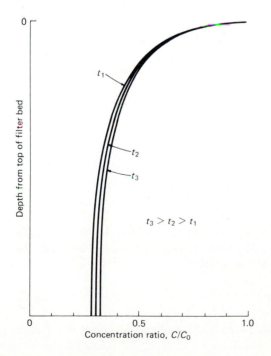

FIGURE 6-31
Concentration-ratio curves for granular-medium filter where straining is the principal particle-removal mechanism.

Equation of Continuity. The equation of continuity for the filtration operation may be developed by considering a suspended-solids mass balance for a section of filter of cross-sectional area A, and of thickness dx, measured in the direction of flow. The mass balance is as follows:

1. General word statement:

$$\begin{array}{c} \text{Rate of accumulation} \\ \text{of solids within the} \\ \text{volume element} \end{array} = \begin{array}{c} \text{Rate of flow of solids into} \\ \text{the volume element} \end{array} - \begin{array}{c} \text{Rate of flow of solids out} \\ \text{of the volume element} \end{array} \quad (6\text{-}40)$$

2. Simplified word statement:

$$\text{Accumulation} = \text{Inflow} - \text{Outflow}$$

3. Symbolic representation:

$$\left(\frac{\partial q}{\partial t} + \alpha(t)\frac{\partial \overline{C}}{\partial t} \right) dV = Q(C) - Q\left(C + \frac{\partial C}{\partial x}dx \right) \quad (6\text{-}41)$$

where $\partial q/\partial t$ = change in quantity of solids deposited within the filter time, mg/in$^3 \cdot$ min

$\alpha(t)$ = average porosity as a function of time

$\partial C/\partial t$ = change in average concentration of solids in pore space with time, mg/in$^3 \cdot$ min

dV = differential volume, in^3

Q = volumetric flowrate, L/min (the unit L/min is used for convenience)

C = concentration of suspended solids, mg/L

$\partial C/\partial x$ = change in concentration of suspended solids in fluid stream with distance, mg/L $\cdot$ in

Substituting $A\ dx$ for dV and Av for Q where v is the filtration velocity (L/in$^2 \cdot$ min) and simplifying, Eq. 6-41 yields

$$-v\frac{\partial C}{\partial x} = \frac{\partial q}{\partial t} + \alpha(t)\frac{\partial \overline{C}}{\partial t} \quad (6\text{-}42)$$

In Eq. 6-42, the first term represents the difference between the mass of suspended solids entering and leaving the section; the second term represents the time rate of change in the mass of suspended solids accumulated within the interstices of the filter medium; and the third term represents the time rate of change in the suspended solids concentration in the pore space within the filter volume.

In a flowing process, the quantity of fluid contained within the bed is usually small compared with the volume of liquid passing through the bed. In this case, the materials balance equation can be written as follows:

$$-v\frac{\partial C}{\partial x} = \frac{\partial q}{\partial t} \quad (6\text{-}43)$$

This equation is the one most commonly found in the literature dealing with filtration theory.

Rate Equation. To solve Eq. 6-43, an additional independent equation is required. The most direct approach is to derive a relationship that can be used to describe the change in concentration of suspended matter with distance, such as

$$\frac{\partial C}{\partial x} = \phi(V_1, V_2, V_3, \dots) \tag{6-44}$$

in which V_1, V_2, and V_3 are the variables governing the removal of suspended matter from solution.

An alternative approach is to develop a complementary equation in which the pertinent process variables are related to the amount of material retained (accumulated) within the filter at various depths. In equation form, this may be written as

$$\frac{\partial q}{\partial t} = \phi(V_1, V_2, V_3, \dots) \tag{6-45}$$

Analysis of Wastewater Filtration

The following analysis, also adapted from Ref. 18, is also based on the assumption that straining is the operative removal mechanism.

Equation of Continuity. Because the shape of the removal curve within the filter does not vary with time, the equation of continuity (Eq. 6-43) may be written as an ordinary differential equation:

$$-v\frac{dC}{dx} = \frac{dq}{dt} \tag{6-46}$$

Rate Equation. From the size and distribution of the influent particles (see Fig. 6-29) and the shape of the normalized curves (see Fig. 6-31), it can be concluded that the rate of change of concentration with distance must be proportional to some removal coefficient that is changing with the degree of treatment or removal achieved in the filter. For example, the entire particle-size distribution in the influent is passed through the first layer. The probability of removing particles from the waste stream is p_1. In the second layer, the probability of removing particles is p_2; p_2 is less than p_1, assuming that some of the larger particles will be removed by the first layer. Continuing this argument, it can be reasoned that the rate of removal must always be changing as a function of the degree of treatment. This phenomenon can be expressed mathematically using the following equation:

$$\frac{dC}{dx} = \left[\frac{1}{(1 + ax)^n}\right] r_o C \tag{6-47}$$

where $C =$ concentration, mg/L
   $x =$ distance, in
   $r_o =$ initial removal rate, in^{-1}
   $a, n =$ constants

In Eq. 6-47, the term within brackets is sometimes called a *retardation factor*. When the exponent n is equal to zero, the term within the brackets is equal to one; under these conditions, Eq. 6-47 represents a logarithmic curve. When n equals one, the value of the term within brackets drops off rapidly in the first 5 in (125 mm) and then more gradually as a function of distance. Therefore, it appears that the exponent n may be related to the distribution of particle sizes in the influent. For example, when dealing with a uniform filter medium and filtering particles of one size, it would be expected that the value of the exponent n would be equal to zero and that the initial removal could be described as a first-order removal function. It should be noted that this equation was verified only for filtration rates up to 10 gal/ft^2 · min (400 L/m^2 · min) [18].

The value of r_o is obtained by computing the slope of the removal curve at near zero depth, since $[1/(1 + ax)^n] \approx 1$. The constants a and n must be determined using an iterative procedure. The easiest way to do this is to rewrite Eq. 6-47 as follows:

$$\left(\frac{Cr_o}{dC/dx} \right)^{1/n} = 1 + ax \qquad (6\text{-}48)$$

If Eq. 6-48 is plotted functionally, the valve of n is equal to the value that results in a straight-line plot. The slope of the line describing the experimental data will be equal to the constant a.

Generalized Rate Equation. On the basis of experimental results derived from this study and data reported in the literature, there appear to be five major factors that affect the time-space removal of the residual suspended matter from a flocculation-sedimentation process within a granular filter for a given temperature. These factors are the size of the filter medium, the rate of filtration, the influent particle size and size distribution, the floc strength, and the amount of material removed within the filter. Therefore, a generalized rate equation accounts for the effect of these factors.

Although a number of different formulations are possible, a generalized rate equation in which all five factors are considered can be developed by multiplying Eq. 6-47 by a factor that takes into account the effect of the material accumulated in the filter. The proposed equation is

$$\frac{dC}{dx} = - \frac{1}{(1 + ax)^n} r_o \, C \left(1 - \frac{q}{q_u} \right)^m \qquad (6\text{-}49)$$

where q = quantity of suspended solids deposited in the filter
q_u = ultimate quantity of solids that can be deposited in the filter
m = a constant related to floc strength

Initially, when the amount of material removed by the filter is low, $q = 0$; $(1 - q/q_u)^m = 1$, and Eq. 6-49 is equivalent to Eq. 6-47. As the upper layers begin to clog, the term $(1 - q/q_u)^m$ approaches zero, and the rate of change in concentration with distance is equal to zero. At the lower depths, the amount of material removed is essentially zero, and the previous analysis applies.

Headloss Development. In the past, the most commonly used approach to determine headloss in a clogged filter was to compute it with a modified form of the equations used to evaluate the clear-water headloss (see Table 6-13 and Example 6-8). In all cases, the difficulty encountered in using these equations is that the porosity must be estimated for various degrees of clogging. Unfortunately, the complexity of this approach renders most of these formulations useless or, at best, extremely difficult to use.

An alternative approach is to relate the development of headloss to the amount of material removed by the filter. The headloss would then be computed using the expression

$$H_t = H_o + \sum_{i=i}^{n} (h_i)_t \tag{6-50}$$

where H_t = total headloss at time t, ft (m)
H_o = total initial clear-water headloss, ft (m)
$(h_i)_t$ = headloss in the ith layer of the filter at time t, ft (m)

From an evaluation of the incremental headloss curves for uniform sand and anthracite, the buildup of headloss in an individual layer of the filter was found to be

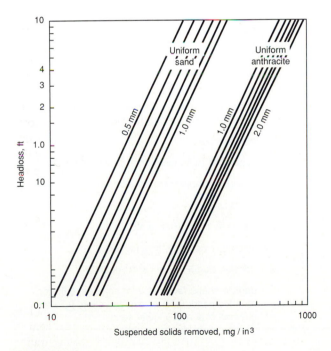

FIGURE 6-32
Headloss versus suspended solids removed for various sizes of uniform sand and anthracite. (Adapted from Refs. 18, 19.)

TABLE 6-13
Formulas used to compute the clear water headloss through a granular medium[a]

Equation	Definition of terms
Carmen-Kozeny: $$h = \frac{f}{\phi}\frac{1-\alpha}{\alpha^3}\frac{L}{d}\frac{V_s^2}{g}$$ $$h_s = \frac{1}{\phi}\frac{1-\alpha}{\alpha^3}\frac{LV_s^2}{g}\sum f\frac{p}{d_g}$$ $$f = 150\frac{1-\alpha}{N_r} + 1.75$$ $$N_R = \frac{\phi d V_s \rho}{\mu}$$ **Fair-Hatch:** $$h = kv\,S^2\frac{(1-\alpha)^2}{\alpha^3}\frac{L}{d^2}\frac{V_s}{g}$$ $$h_s = kv\frac{(1-\alpha)^2}{\alpha^3}\frac{LV_s}{g}\left(\frac{6}{\phi}\right)^2\sum\frac{p}{d_g^2}$$ **Rose:** $$h = \frac{1.067}{\phi}C_d\frac{1}{\alpha^4}\frac{L}{d}\frac{V_s^2}{g}$$ $$h_s = \frac{1.067}{\phi}\frac{LV_s^2}{\alpha^4 g}\sum C_d\frac{p}{d_g}$$ $$C_d = \frac{24}{N_R} + \frac{3}{\sqrt{N_R}} + 0.34$$ **Hazen:** $$h = \frac{1}{CT+10}\frac{60}{d_{10}^2}\frac{L}{}V_h$$	C = coefficient of compactness (varies from 600 for very closely packed sands that are not quite clean to 1200 for very uniform clean sand)
	C_d = coefficient of drag
	d = grain size diameter, ft (m)
	d_g = geometric mean diameter between sieve sizes d_1 and d_2, $\sqrt{d_1 d_2}$, ft (m)
	d_{10} = effective size grain diameter, mm
	f = friction factor
	g = acceleration due to gravity, 32.2 ft/s^2 (9.8 m/s^2)
	h = headloss, ft (m)
	h_s = headloss through stratified filter bed, ft(m)
	k = filtration constant, 5 based on sieve openings, 6 based on size of separation
	L = depth of filter bed or layer, ft (m)
	N_R = Reynolds number
	p = fraction of particles (based on mass) within adjacent sieve sizes
	S = shape factor (varies between 6.0 for spherical particles to 8.5 for crushed materials)
	T = temperature, °F
	V_h = superficial (approach) filtration velocity, m/d
	V_s = superficial (approach) filtration velocity, ft/s (m/s)
	α = porosity
	μ = viscosity, lb · s/ft^2(N · s/m^2)
	ν = kinematic viscosity, ft^2/s (m^2/s)
	ρ = density, slug/ft^3 = lb · s^2/ft^4 (kg/m^3)
	ϕ = particle shape factor (1.0 for spheres, 0.82 for rounded sand, 0.75 for average sand, 0.73 for crushed coal and angular sand)

[a] Adapted from Ref. 19.

related to the amount of material contained within the layer. The form of the resulting equation for headloss in the ith layer is

$$(h_i)_t = a(q_i)_t^b \qquad (6\text{-}51)$$

where $(q_i)_t$ = amount of material deposited in the ith layer at time t, mg/in^3
a, b = constants

In this equation, it is assumed that the buildup of headloss is only a function of the amount of material removed. The buildup of headloss as a function of the amount of material removed within the filter is given in Fig 6-32. Computation of the clear-water headloss through a filter is illustrated in Example 6-8. The determination of the buildup of headloss during the filtration process using the data presented in Fig. 6-31 is illustrated in Example 6-9.

Example 6-8 Computation of clear-water headloss in a granular-medium filter.
Determine the clear-water headloss in a filter bed composed of 12 in of uniform anthracite with an average size of 1.6 mm and 12 in of uniform sand with an average size of 0.5 mm for a filtration rate of 4 gal/ft^2 · min (160 L/m^2 · min). Assume that the operating temperature is 20°C. Use the Rose equation given in Table 6-13 for computing the headloss. Use a ϕ value of 0.73 and 0.82 for the anthracite and sand, respectively.

Solution

1. Determine the Reynolds number for the anthracite and sand layers.

(a) Anthracite layer

$$N_R = \frac{\phi d V_s}{\nu}$$

$$d = 1.6 \text{ mm} = 5.25 \times 10^{-3} \text{ ft}$$

$$V_s = \frac{4 \text{ gal/ft}^2 \cdot \text{min}}{7.48 \text{ gal/ft}^3} = 0.534 \text{ ft/min}$$

(Note that the filtration rate is converted to an equivalent linear velocity by converting the volume expressed in gallons to cubic feet.)

$$\nu \text{ at } 20°C = 1.091 \times 10^{-5} \text{ ft}^2/\text{s (see Appendix C)}$$

$$N_R = \frac{(0.73)(5.25 \times 10^{-3} \text{ ft})[(0.534 \text{ ft/min})/(60 \text{ ft/s})]}{1.091 \times 10^{-5} \text{ ft}^2/\text{s}}$$

$$= 3.12$$

(b) Sand layer

$$N_R = \frac{(0.82)(1.64 \times 10^{-3} \text{ ft})[(0.534 \text{ ft/min})/(60 \text{ ft/s})]}{1.091 \times 10^{-5} \text{ ft}^2/\text{s}}$$

$$= 1.10$$

2. Determine the coefficient of drag C_D.

(a) Anthracite layer

$$C_D = \frac{24}{N_R} + \frac{3}{\sqrt{N_R}} + 0.34$$

$$= \frac{24}{3.12} + \frac{3}{\sqrt{3.12}} + 0.34$$

$$= 9.73$$

(b) Sand layer

$$C_D = \frac{24}{1.10} + \frac{3}{\sqrt{1.10}} + 0.34$$

$$= 25.02$$

3. Determine the headloss through the anthracite and sand layers.
    (a) Anthracite layer

$$h = \frac{1.067}{\phi} C_D \frac{1}{\alpha^4} \frac{L}{d} \frac{V_s^2}{g}$$

$\phi = 0.73$
$C_D = 9.73$
$\alpha = 0.4$ (assumed), $\alpha^4 = 0.0256$
$L = 1$ ft
$d = 5.25 \times 10^{-3}$ ft
$V = 0.534$ ft/min $= 8.9 \times 10^{-3}$ ft/s
$g = 32.2$ ft/s^2

$$h = \frac{1.067}{0.73}(9.73)\left(\frac{1}{0.0256}\right)\left(\frac{1.00}{5.25 \times 10^{-3}}\right) \times \frac{(8.90 \times 10^{-3})^2}{32.2}$$

$$= 0.26 \text{ ft}$$

(b) Sand layer

$\phi = 0.82$
$C_D = 25.02$
$\alpha = 0.4$ (assumed), $\alpha^4 = 0.0256$
$L = 1$ ft
$d = 1.64 \times 10^{-3}$ ft
$V = 0.534$ ft/min $= 8.9 \times 10^{-3}$ ft/s
$g = 32.2$ ft/s^2

$$h = \frac{1.067}{0.82}(25.02)\left(\frac{1}{0.0256}\right)\left(\frac{1.00}{1.64 \times 10^{-3}}\right) \times \frac{(8.90 \times 10^{-3})^2}{32.2}$$

$$= 1.91 \text{ ft}$$

4. Determine the total headloss H_T.

$H_T =$ headloss through anthracite layer + headloss through sand layer
$H_T = 0.26$ ft + 1.91 ft
$\qquad = 2.17$ ft (0.66 m)

Comment. The headloss computations in this example were simplified by assuming that the anthracite and sand layers were of uniform thickness. The same computation procedure can be used for stratified filter beds by considering the total headloss to be the sum of the headlosses in successive layers. The second form of the Carmen-Kozeny, Fair-Hatch, and Rose equations, given in Table 6-13, is used for stratified filter beds.

Need for Pilot Plant Studies

Although the information presented earlier in this section will help the reader understand the nature of the filtration operation as it is applied to the filtration of treated wastewater, it must be stressed that there is no generalized approach to the design of full-scale filters. The principal reason is the inherent variability in the characteristicsu of the influent suspended solids to be filtered. For example, changes in the degree of flocculation of the suspended solids in the secondary settling facilities will significantly affect the particle sizes and their distribution in the effluent. This, in turn, will affect the performance of the filter. Further, because the characteristics of the effluent suspended solids will also vary with the organic loading on the process as well as with the time of day, filters must be designed to function under a rather wide range of operating conditions. The best way to ensure that the filter configuration selected for a given application will function properly is to conduct pilot plant studies.

Because of the many variables that can be analyzed, care must be taken not to change more than one variable at a time so as to confound the results in a statistical sense. Testing should be carried out at several intervals, ideally throughout a full year, to assess seasonal variations in the characteristics of the effluent to be filtered. All test results should be summarized and evaluated in several different ways to ensure their proper analysis. Because the specific details of each test program will be different, no generalization on the best method of analysis can be given. The analysis of some typical pilot plant data is illustrated in Example 6-9.

Example 6-9 Analysis of filtration data from a pilot plant. The normalized suspended-solids-removal-ratio curves shown at top of following page were derived from a filtration pilot plant study conducted at an activated-sludge wastewater treatment plant. Using these curves and the following data, develop curves that can be used to estimate (1) the headloss buildup as a function of the length of run and (2) the length of run to a terminal headloss of 10 ft (3.0 m) as a function of the filtration rate.

Biological treatment process:

1. Mean cell-residence time $\theta_c = 10d$
2. Average suspended-solids concentration in effluent from secondary settling tank = 20 mg/L
3. Particle-size distribution in effluent = similar to that shown in Fig. 6-29

Pilot plant:

1. Type of filter bed = dual-medium
2. Filter media = uniform anthracite and sand

3. Filter-medium characteristics
   (a) Anthracite, $d = 1.6$ mm, $UC \approx 1.4$
   (b) Sand, $d = 0.5$ mm, $UC \approx 1.4$

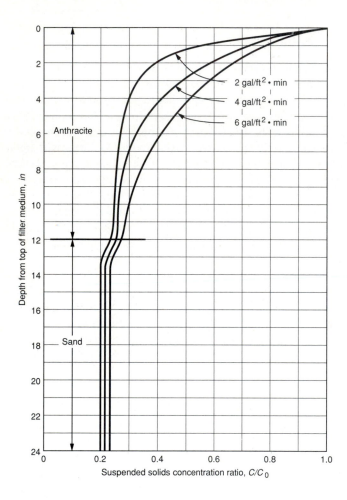

4. Filter-bed depth $= 2$ ft (0.6 m)
   (a) Anthracite $= 1$ ft (0.3 m)
   (b) Sand $= 1$ ft (0.3 m)
5. Filtration rates $= 2, 4,$ and 6 gal/ft² · min (80, 160, and 240 L/m² · min)
6. Temperature $= 20°C$
7. General observation: average concentration-ratio curves plotted in the following figure did not vary significantly with time.

Solution

1. To analyze the concentration-ratio curves, rewrite Eq. 6-46 in a form suitable for numerical analysis:

$$-v\frac{\Delta C}{\Delta x} = \frac{\Delta q}{\Delta t}$$

$$-v\frac{C_{x-1} - C_x}{X_{x-1} - X_x} = \frac{q_2 - q_1}{t_2 - t_1}$$

2. Set up a computation table and determine the value of ΔC $(C_{x-1} - C_x)$ for various depths throughout the filter. The required computations are summarized in the following table. As shown, values of C/C_0 from a normalized removal-ratio curve corresponding to the depths given in column 1 are entered in columns 1, 5, and 8 for each filtration rate. The value of the concentration at each depth is entered in columns 3, 6, and 9 for each filtration rate. The concentration difference ΔC $(C_{x-1} - C_x)$ between the depths given in column 1 is entered in columns 4, 7, and 10 for each filtration rate.

	Filtration rate, gal/ft² · min								
	2.0			4.0			6.0		
Depth, in (1)	C_x/C_0 (2)	C_x (3)	ΔC (4)	C_x/C_0 (5)	C_x (6)	ΔC (7)	C_x/C_0 (8)	C_x (9)	ΔC (10)
0	1.0	20.0		1.0	20.0		1.0	20.0	
			9.0			6.4			4.2
1	0.55	11.0		0.73	14.6		0.79	15.8	
			3.4			3.0			2.2
2	0.38	7.6		0.58	11.6		0.68	13.6	
			1.2			2.0			2.0
3	0.32	6.4		0.48	9.6		0.58	11.6	
			0.6			1.6			1.2
4	0.29	5.8		0.40	8.0		0.52	10.4	
			0.2			1.0			1.2
5	0.28	5.6		0.35	7.0		0.46	9.2	
			0.2			0.6			0.8
6	0.27	5.4		0.32	6.4		0.42	8.4	
			0.2			0.6			0.8
7	0.26	5.2		0.29	5.8		0.38	7.6	
			0.1			0.2			0.8
8	0.255	5.1		0.28	5.6		0.34	6.8	
			0.1			0.2			0.5
9	0.25	5.0		0.27	5.4		0.315	6.3	
			0.1			0.2			0.4
10	0.245	4.9		0.26	5.2		0.295	5.9	
			0.1			0.1			0.3
11	0.24	4.8		0.26	5.1		0.28	5.6	
			0.1			0.3			0.4
12	0.235	4.7		0.24	4.8		0.26	5.2	
			0.6			0.4			0.4
13	0.205	4.1		0.22	4.4		0.24	4.8	
			0.1			0.0			0.0
14	0.20	4.0		0.22	4.4		0.24	4.8	
			0.0			0.0			0.0
15	0.20	4.0		0.22	4.4		0.24	4.8	

3. Set up a computation table and determine the buildup of suspended solids and headloss within each layer of the filter for filter runs of various lengths. The necessary computations for a filtration rate of 4 gal/ft² · min (160 L/m² · min) are summarized in the following table.

Depth in (1)	ΔC, mg/L (2)	Run length, h					
		10		15		20	
		Δq, mg/in³ (3)	Δh, ft (4)	Δq, mg/in³ (5)	Δh, ft (6)	Δq, mg/in³ (7)	Δh, ft (8)
0							
1	6.4	404	1.7	605	4.5	807	8.2
2	3.0	189	0.3	284	0.8	378	1.5
3	2.0	126	0.1	189	0.3	252	0.6
4	1.6	101	0.1	151	0.2	202	0.3
5	1.0	63	—	95	0.1	126	0.1
6	0.6	38	—	57	—	76	—
7	0.6	38	—	57	—	76	—
8	0.2	13	—	19	—	25	—
9	0.2	13	—	19	—	25	—
10	0.2	13	—	19	—	25	—
11	0.1	6	—	9	—	13	—
12	0.3	19	—	28	—	38	—
13	0.4	25	0.4	38	1.1	50	2.0
14	0.0	0		0		0	—
15	0.0	0	—	0	—	0	—
$\Sigma \, \Delta h$, m			2.6		7.0		12.7

Although the required computations for the other filtration rates are not shown, they are the same. The values of ΔC given in column 2 are taken from column 7 of the table prepared in step 2. The values of q shown in columns 3, 5, and 7 are determined by using the difference equation given in step 1. To illustrate for the anthracite layer between 1 and 2 in from the top of the column, the value of Δq after 20 h is as follows:

$$-v\frac{\Delta C}{\Delta x} = \frac{\Delta q}{\Delta t}$$

where $v = 4$ gal/ft² · min $= 0.1051$ L/in² · min
$\Delta C = 3.2$ mg/L
$\Delta X = 1$ in $- 2$ in $= -1$ in
$\Delta t = (20$ h $\times 60$ min/h $- 0) = 1200$ min

$$\Delta q = -0.1051 \text{ L/in}^2 \cdot \text{min} \, \frac{3.0 \text{ mg/L}}{-1 \text{ in}} \, 1200 \text{ min}$$

$$= 379 \text{ mg/in}^3$$

The value of incremental headloss buildup Δh (column 8) for the anthracite layer between 1 and 2 in is obtained from Fig. 6-32 by entering with the value of Δq for this layer. The value of the headloss in the sand layers is determined in a similar manner. To simplify the computations, it is assumed that no intermixing occurs between the anthracite and sand.

Once all of the Δh headloss values are entered, the entire column is summed to obtain the total headloss in the filter bed. The total headloss for other time periods and filtration rates is determined in exactly the same manner. Summary data for the other flowrates are as follows:

Time, h	Headloss, m	
	2.0 gal/ft² · min	8.0 gal/ft² · min
10	1.0	3.7
15	2.6	9.4
20	5.0	18.1

4. Plot curves of headloss versus run length for the three flowrates. The required curves, which are plotted using the data given in step 3, are shown in the following figure.

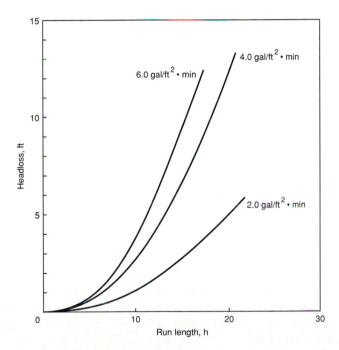

5. Plot the curve of run length to reach a headloss of 10ft versus filtration rate. The required curve is shown below. The data needed to plot this curve are obtained from the headloss curve developed in step 4 by finding the time required to reach a headloss of 3 m for each filtration rate.

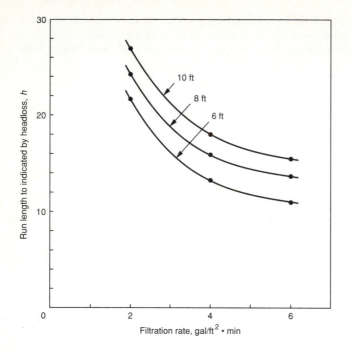

The y-axis is labeled "Run length to indicated by headloss, h" with values 10, 20, 30. The x-axis is labeled "Filtration rate, gal/ft^2 · min" with values 0, 2, 4, 6. Curves are labeled "10 ft", "8 ft", and "6 ft".

Comment. The use of unstratified filter beds for the filtration of treated effluents has been studied by Dahab and Young [6]. They found that unstratified filter beds with the same effective size as that used in the top layer of a dual-medium filter were essentially equivalent to dual-medium filters in terms of effluent quality and length of run. The use of unstratified filter beds is considered further in Chap. 11.

6-9 GAS TRANSFER

Gas transfer may be defined as the process by which gas is transferred from one phase to another, usually from the gaseous to the liquid phase. It is a vital part of a number of wastewater treatment processes. For example, the functioning of aerobic processes, such as activated-sludge biological filtration and aerobic digestion, depends on the availability of sufficient quantities of oxygen. Chlorine, when used as a gas, must be transferred to solution in the water for disinfection purposes. Oxygen is often added to treated effluent after chlorination (postaeration). One process for removing nitrogen compounds consists of converting the nitrogen to ammonia and transferring the ammonia gas from the water to air.

Description

The most common application of gas transfer in the field of wastewater treatment is in the transfer of oxygen in the biological treatment of wastewater. Because of the low solubility of oxygen and the consequent low rate of oxygen transfer, sufficient oxygen to meet the requirements of aerobic waste treatment does not enter water through normal surface air-water interfaces. To transfer the large quantities of oxygen that are needed, additional interfaces must be formed. Either air or oxygen can be introduced into the liquid, or the liquid in the form of droplets can be exposed to the atmosphere. The most commonly used aeration devices are described in Table 6-14 and illustrated in Fig. 6-33. The design and application of many of these devices are considered in Chap. 10 in connection with the design of biological treatment processes.

Oxygen can be supplied by means of air or pure-oxygen bubbles introduced to the water to create additional gas-water interfaces. In wastewater-treatment plants, submerged-bubble aeration is most frequently accomplished by dispersing air bubbles in the liquid at depths up to 30 ft (10 m). Depths up to 100 ft (30 m) have been used in some European designs. As summarized in Table 6-14 and shown in Fig. 6-33, aerating devices include porous plates and tubes, perforated pipes, and various configurations of metal and plastic diffusers. Hydraulic shear devices may also be used to create small bubbles by introducing a flow of liquid at an orifice to break up the air bubbles into smaller sizes. Turbine mixers may be used to disperse air bubbles introduced below the center of the turbine.

In the alternative method of introducing large quantities of oxygen into the liquid, surface aerators generally consist of either low-or high-speed turbines or high-speed floating units operating at the surface of the liquid, partially submerged. They are designed both to mix the liquid in the basin and to expose it to the atmosphere in the form of small liquid droplets.

Analysis of Gas Transfer

Over the past 50 years, a number of mass-transfer theories have been proposed to explain the mechanism of gas transfer. The simplest and the one most commonly used is the two-film theory proposed by Lewis and Whitman in 1924 [13]. The penetration model proposed by Higbie [10] and the surface-renewal model proposed by Danckwertz [4] are more theoretical and take into account more of the physical phenomena involved. The two-film theory remains popular because, in more than 95 percent of the situations encountered, the results obtained are essentially the same as those obtained with the more complex theories. Even in the 5 percent where there is disagreement between the two-film theory and other theories, it is not clear which approach is more correct. For these reasons the two-film theory will be described in the following discussion.

The Two-Film Theory. The two-film theory is based on a physical model in which two films exist at the gas-liquid interface, as shown in Fig. 6-34. The two films, one liquid and one gas, provide the resistance to the passage of gas molecules between the bulk-liquid and the bulk-gaseous phases. For the transfer of gas molecules from the gas phase to the liquid phase, slightly soluble gases encounter the primary resistance to transfer from the liquid film, and very soluble gases encounter the primary resistance

TABLE 6-14
Description of commonly used devices for wastewater aeration

Classification	Description	Use or application
Submerged:		
Diffused air		
Porous (fine bubble)	Bubbles generated with ceramic, vitreous, or resin-bonded porous plates, domes, and tubes	All types of activated-sludge processes
Porous (medium bubble)	Bubbles generated with perforated membrane or plastic tubes	All types of activated-sludge processes
Nonporous (coarse bubble)	Bubbles generated with orifices, nozzles or injectors	All types of activated-sludge processes
Static-tube mixer	Short tubes with internal baffles designed to retain air injected at bottom of tube in contact with liquid	Aerated lagoons and activated-sludge processes
Sparger turbine	Consists of low-speed turbine and compressed-air injection system	All types of activated-sludge processes
Jet	Compresses air injected into mixed liquor as it is pumped under pressure through jet device	All types of activated-sludge processes
Surface:		
Low-speed turbine aerator	Large-diameter turbine used to expose liquid droplets to the atmosphere	Conventional activated-sludge processes and aerated lagoons
High-speed floating aerator	Small-diameter propeller used to expose liquid droplets to the atmosphere	Aerated lagoons
Rotor-brush aerator	Blades mounted on central shaft are rotated through liquid. Oxygen is induced into the liquid by the splashing action of blades and by exposure of liquid droplets to the atmosphere	Oxidation ditch, channel aeration, and aerated lagoons
Cascade	Wastewater flows over a cascade in sheet flow	Post aeration

to transfer from the gaseous film. Gases of intermediate solubility encounter significant resistance from both films.

Addition of Gases. In the systems used in the field of wastewater treatment, the rate of gas transfer is generally proportional to the difference between the existing concentration and the equilibrium concentration of the gas in solution. An equation from this relationship can be expressed as

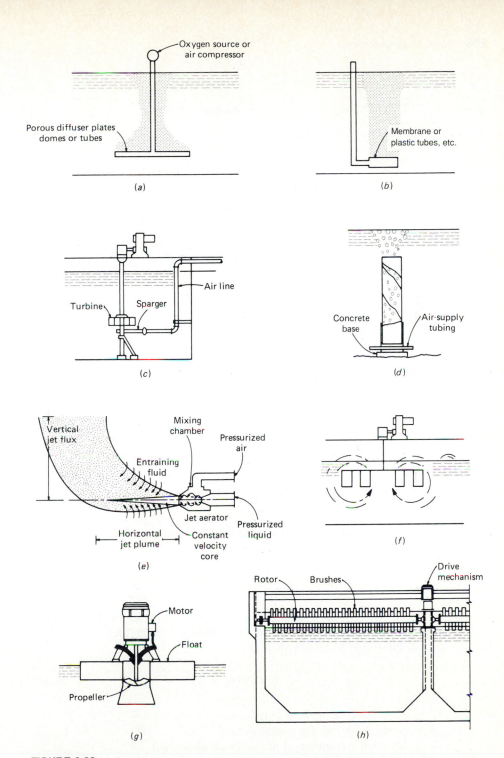

FIGURE 6-33

Typical devices used for the transfer of oxygen: (a) fine bubble diffused-air, (b) medium bubble diffused-air, (c) Sparger turbine, (d) static tube mixer, (e) jet reactor, (f) low-speed turbine, (g) high-speed floating aerator, and (h) rotor-brush aerator.

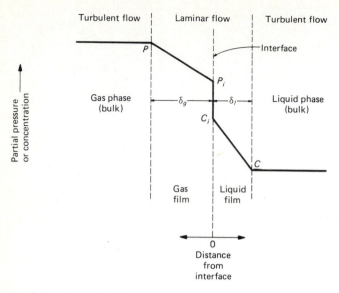

FIGURE 6-34
Definition sketch for two-film theory of gas transfer.

$$r_m = K_g A (C_s - C) \tag{6-52}$$

where r_m = rate of mass transfer
K_g = coefficient of diffusion for gas
A = area through which gas is diffusing
C_s = saturation concentration of gas in solution
C = concentration of gas in solution

Noting that under the conditions of mass transfer encountered in the field $r_m = VdC/dt$, Eq. 6-52 can be written as

$$r_C = \frac{dC}{dt} = K_g \frac{A}{V}(C_s - C) \tag{6-53}$$

In practice, the term $K_g(A/V)$ is replaced by a proportionality factor that is related to existing conditions of exposure. This factor is identified in the literature as $K_L a$. If $K_L a$ is used, Eq. 6-53 can be rewritten as

$$r_C = \frac{dC}{dt} = K_L a(C_s - C) \tag{6-54}$$

where r_C = change in concentration, mg/L · s
$K_L a$ = overall mass-transfer coefficient, s^{-1}
C_s = saturation concentration of gas in solution, mg/L
C = concentration of gas in solution, mg/L

The integrated form of Eq. 6-54 is obtained by integrating between the limits of $C = C_o$ and $C = C$ and $t = 0$ and $t = t$ as follows:

$$\int_{C_o}^{C} \frac{dC}{C_s - C} K_L a \int_0^t dt \tag{6-55}$$

which, when solved yields

$$\frac{C_s - C_t}{C_s - C_o} = e^{-(K_L a)t} \tag{6-56}$$

In Eq. 6-56, the terms $(C_s - C_t)$ and $(C_s - C_o)$ represents the final and initial oxygen-saturation deficits.

Removal of Gases. Where a supersaturated solution is to be degassed, the following alternative form of Eq. 6-56 is used:

$$\frac{C_t - C_s}{C_o - C_s} = e^{-(K_L a)t} \tag{6-57}$$

The derivation of an expression similar to Eq. 6-56 for estimating the amount of oxygen required for the postaeration of treated wastewater is illustrated in Example 6-10.

Example 6-10 Derivation of equation for estimating diffused-air requirements.
Develop an expression that can be used to estimate the diffused-air requirement for the postaeration of effluent following chlorination. Assume that aeration will be accomplished in a plug-flow reactor [8].

Solution

1. The appropriate expression for the oxygen-solution rate is

$$r_m = \frac{dm}{dt} = K_T'(C_s - C)$$

where K_T' = overall mass-transfer coefficient for the given conditions
$$K_T' = K'20 \times (1.024)^{T-20}$$

2. Write an expression for the oxygen-transfer efficiency. The efficiency may be defined as

$$E = \frac{(dm/dt)_{20°C, C=0}}{M}$$

where
E = oxygen-transfer efficiency
$(dm/dt)_{20°C, C=0}$ = oxygen-solution rate at 20°C and zero dissolved oxygen
M = mass rate at which oxygen is introduced

3. Develop a differential expression for the mass rate at which oxygen is introduced. The mass rate at which oxygen is introduced is given by

$$M = \frac{1}{E}\left(\frac{dm}{dt}\right)_{20°C, C=0}$$

$$= \frac{1}{E}\left(\frac{dm}{dt}\right)_T \frac{(dm/dt)_{20°C, C=0}}{(dm/dt)_T}$$

Substituting for $(dm/dt)_{20, C=0}$ and $(dm/dt)_T$ yields

$$M = \frac{1}{E}\left(\frac{dm}{dt}\right)_T \frac{(C_s)_{20°C}}{(C_s - C)_T(1.024)^{T-20}}$$

If the expression is applied to an infinitesimal transverse segment of the tank and $Q\,dC$ is substituted for dm/dt [note that $V(dC/dt) = dm/dt$ and $Q = V/dt$], then the differential form of the above expression can be rewritten as

$$dM = \frac{Q(C_s)_{20°C}}{E(1.024)^{T-2}}\left(\frac{dC}{C_s - C}\right)_T$$

4. Derive the integrated form of the differential expression that was derived in step 3. The integrated form of the equation can be obtained by integrating the expression from the inlet of the tank where $C = C_i$ to the outlet of the tank where $C = C_0$:

$$\int_0^M dM = \frac{Q(C_s)_{20°C}}{E(1.024)^{T-20}}\int_{C_i}^{C_0}\frac{dc}{C_s - C}$$

$$M = \frac{Q(C_s)_{20°C}}{E(1.024)^{T-20}}\left(\ln\frac{C_s - C_i}{C_s - C_0}\right)_T$$

5. Rewrite the equation derived in step 4 in a more practical format. This can be done by noting that the density of air at 68°F is 0.0752 lb/ft³ (see Appendix B) and that air contains about 23 percent oxygen by weight. Using these values and the conversion factor [8.34 lb/Mgal · (mg/L)], the rate of oxygen input expressed in ft³/min, is equal to

$$Q_a = 0.335\frac{Q(C_s)_{20°C}}{E(1.024)^{T-20}}\left(\ln\frac{C_s - C_i}{C_s - C_0}\right)_T$$

where Q_a = required air flowrate, ft³/min
Q = wastewater flowrate, Mgal/d
C_s = saturation concentration of oxygen at 20°C, mg/L

Comment. The value of Q_a is usually multiplied by a factor of 1.1 to account for the fact that the saturation value of oxygen in wastewater is about 95 percent of that in distilled water and to account for the difference in the transfer rates.

Evaluation of Oxygen-Transfer Coefficient

For a given volume of water being aerated, aerators are evaluated on the basis of the quantity of oxygen transferred per unit of air introduced to the water for equivalent conditions (temperature and chemical composition of the water, depth at which the air is introduced, etc.). (See Fig. 6-35). The evaluation of the oxygen transfer coefficient in clean water and wastewater is considered in the following discussion.

Oxygen Transfer in Clean Water. The accepted procedure for determining the overall oxygen transfer coefficient in clean water, as detailed in Ref. 1, may be outlined as follows. The accepted test method involves the removal of dissolved oxygen (DO) from a known volume of water by the addition of sodium sulfite followed by reoxygenation to near the saturation level. The DO of the water volume is monitored during the reaeration period by measuring DO concentrations at several different points selected to best represent the contents of the tank. The minimum number of points, their distribution, and range of DO measurements made at each determination point are specified in the procedure [1].

The data obtained at each determination point are then analyzed by a simplified mass-transfer model (Eq. 6-56) to estimate the apparent volumetric mass-transfer coefficient, $K_L a$, and the equilibrium concentration C_{x*} obtained as the aeration period approaches infinity. The term C_{x*} is substituted for the term C_s in Eq. 6-56. A nonlinear regression analysis is employed to fit Eq. 6-56 to the DO profile measured at each determination point during the reoxygenation test period. In this way, estimates of $K_L a$ and C_{x*} are obtained at each determination point. These estimates are adjusted to standard conditions and the standard oxygen-transfer rate (mass of oxygen dissolved per unit time at a hypothetical concentration of zero DO) is obtained as the average of the products of the adjusted point $K_L a$ values, the corresponding adjusted point C_{x*} values, and the tank volume [1].

Oxygen Transfer in Wastewater. In an activated-sludge system, the $K_L a$ value can be determined by considering the uptake of oxygen by microorganisms. Typically, oxygen is maintained at a level of 1 to 3 mg/L, and the oxygen is used by the microorganisms as rapidly as it is supplied. In equation form,

$$\frac{dC}{dt} = K_L a(C_s - C) - r_M \tag{6-58}$$

FIGURE 6-35
Typical test tank used to test the performance of surface aerators (Aqua Aerobic Systems).

where r_M is the rate of oxygen used by the microorganisms. Typical values of r_M vary from 2 to 7 g/d per gram of mixed-liquor volatile suspended solids (MLVSS). If the oxygen level is maintained at a constant level, dC/dt is zero and

$$r_M = K_L a (C_s - C) \tag{6-59}$$

C in this case is constant also. Values of r_M can be determined in a laboratory by means of the Warburg apparatus. In this case, $K_L a$ can easily be determined as follows:

$$K_L a = \frac{r_M}{C_s - C} \tag{6-60}$$

Example 6-11   Determination of approximate $K_L a$ value from aerator test data. The following field data have been obtained from an aeration test conducted with a surface aerator. Using the field data, estimate the approximate $K_L a$ value at 20 °C by means of a linear regression analysis. The temperature of the water was 15 °C.

Time, min	DO conc., mg/L
4	0.8
7	1.8
10	3.3
13	4.5
16	5.5
19	5.2
22	7.3

Solution

1. To analyze the field data, rewrite Eq. 6-56 in a linear form.

$$\log(C_s - C_t) = \log(C_s - C_o) - \frac{K_L a}{2.303} t$$

2. Determine $C_s - C_t$, and plot $C_s - C_t$ versus t on semilog paper.
   (a) $C_{s(15\,°C)} = 10.15$ (see Appendix E)
   (b) Plot $C_s - C_t$ versus t. See following plot.

Time, min	$C_s - C_t$, mg/L
4	9.35
7	8.55
10	6.85
13	5.65
16	4.65
19	4.95
22	2.85

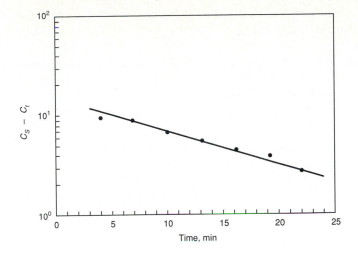

3. Determine the value of $K_L a$ at 20°C.

   (a) From the plot, the value of $K_L a$ at 15°C is

$$K_L a = 2.303 \frac{\log C_{t_1} - \log C_{t_2}}{t_2 - t_1} (60)$$

$$K_L a = 2.303 \frac{\left(\log 8.55 - \log 2.85 \right)}{22 - 7} \quad (60)$$

$$K_L a = 1.91 \text{ h}^{-1}$$

   (b) The approximate value of $K_L a$ at 20°C is

$$= (1.91) 1.024^{15-20}$$

$$= 1.71 \text{ h}^{-1}$$

Comment. The value of $K_L a$ determined in this example is approximate because a linear regression analysis was used. To obtain a more accurate value of $K_L a$ the nonlinear method outlined in Ref. 1 should be used.

Factors Affecting Oxygen Transfer

Prediction of oxygen-transfer rates in aeration systems is nearly always based on an oxygen rate model such as the one given in Eq. 6-54. The overall oxygen mass-transfer coefficient $K_L a$ is usually determined in full-scale facilities or in test facilities such as those shown in Fig. 6-35. If pilot-scale facilities are used to determine $K_L a$ values, scale-up must be considered. The mass-transfer coefficient $K_L a$ is also a function of temperature, intensity of mixing (and hence the type of aeration device used and the geometry of the mixing chamber), and constituents in the water [19]. These factors are considered in the following discussion.

Effect of Temperature. Temperature effects are treated in the same manner here as they were treated in establishing the BOD rate coefficient (i.e., by using an exponential function to approximate the van't Hoff-Arrhenius relationship):

$$K_L a_{(T)} = K_L a_{(20°C)} \theta^{T-20} \tag{6-61}$$

where $K_L a(t)$ = oxygen mass-transfer coefficient at temperature T, s^{-1}
$K_L a_{(20°C)}$ = oxygen mass-transfer coefficient at 20°C, s^{-1}

Reported values for θ vary with the test conditions. Typical θ values are in the range of 1.015 to 1.040. A θ value of 1.024 is typical for both diffused- and mechanical-aeration devices.

Effects of Mixing Intensity and Tank Geometry. Effects of mixing intensity and tank geometry are difficult to deal with on a theoretical basis but must be considered in the design process because aeration devices are often chosen on the basis of efficiency. Efficiency is strongly related to the $K_L a$ value associated with a given aeration unit. In most cases an aeration device is rated for a range of operating conditions using tap water and having a low total-dissolved-solids concentration. A correction factor α is used to estimate the $K_L a$ value in the actual system:

$$\alpha = \frac{K_L a(\text{wastewater})}{K_L a(\text{tap water})} \tag{6-62}$$

Values of α vary with the type of aeration device, the basin geometry, the degree of mixing, and the wastewater characteristics. Values of α vary from about 0.3 to 1.2. Typical values for diffused and mechanical aeration equipment are in the range of 0.4 to 0.8 and 0.6 to 1.2, respectively. If the basin geometry in which the aeration device is to be used is significantly different from that used to test the device, great care must be exercised in selecting an appropriate α value. The selection of α values is considered further in Chap. 10.

Effect of Wastewater Characteristics. A third correction factor, β, is used to correct the test-system oxygen-transfer rate for differences in oxygen solubility due to constituents in the water such as salts, particulates, and surface active substances:

$$\beta = \frac{C_S \text{ (wastewater)}}{C_S \text{ (tap water)}} \tag{6-63}$$

Values of β vary from about 0.7 to 0.98. A β value of 0.95 is commonly used for wastewater. Because the determination of β is within the capability of most wastewater treatment plant laboratories, experimental verification of assumed values is recommended.

Application of Correction Factors. The application of the correction factors cited above can be illustrated by considering the equation used to predict field oxygen-transfer rates for mechanical surface aerators based on measurements made in experimental test facilities (see Fig. 6-35) [1]:

$$\text{OTR}_f = \text{SOTR}\left(\frac{\beta C_S - C_w}{C_{S_{20}}}\right)\theta^{T-20}(\alpha) \tag{6-64}$$

where OTR_f = actual oxygen-transfer rate under field-operating conditions in a respiring system, lb O_2/kW · h

SOTR = standardized oxygen-transfer rate under test conditions at 20°C and zero dissolved oxygen, lb O_2/kW · h

C_s = oxygen saturation concentration for tap water at field-operating conditions, mg/L

C_w = operating oxygen concentration in wastewater, mg/L

$C_{S_{20}}$ = oxygen saturation concentration for tap water at 20°C, mg/L

Other terms are as defined previously.

With diffused-air aeration systems, the C_s values in Eq. 6-64 must be corrected to account for the higher than atmospheric oxygen saturation concentrations achieved in the reactor due to the release of air at the reactor bottom. To use Eq. 6-64 for diffused-air aeration systems, the value of C_s is taken to be the average dissolved oxygen concentration attained at infinite time [1]. The accepted method for determining the appropriate value of C_s for such systems is discussed in detail in Ref. 1.

6-10 VOLATILIZATION AND GAS STRIPPING OF VOLATILE ORGANIC COMPOUNDS (VOCs) FROM WASTEWATER MANAGEMENT FACILITIES

In the past few years a number of volatile organic compounds (VOCs) such as trichloroethylene (TCE) and 1,2-dibromo-3-chloropropane (DBCP) have been detected in wastewater. The uncontrolled release of such compounds that now occurs in wastewater collection systems and wastewater treatment plants is an area of growing concern. It is the purpose of this section to consider the mechanisms governing the release of these compounds and the locations where their release is most prevalent. Methods of controlling the discharge of these compounds to the atmosphere are considered in Chap. 9.

Emission of VOCs

The principal mechanisms governing the release of VOCs in wastewater collection and treatment facilities are (1) volatilization and (2) gas stripping. These mechanisms and the principal locations where VOCs are released are considered in the following discussion.

Volatilization. The release of VOCs from wastewater surfaces to the atmosphere is termed *volatilization*. Volatile organic compounds are released because they partition between the gas and water phase until equilibrium concentrations are reached [15]. The mass transfer (movement) of a constituent between these two phases is a function of the constituent concentration in each phase relative to the equilibrium concentration. Thus, the transfer of a constituent between phases is greatest when the concentration in one of the phases is far from equilibrium. Because the concentration of VOCs in the atmosphere is extremely low, the transfer of VOCs usually occurs from wastewater to the atmosphere.

Gas Stripping. Gas stripping of VOCs occurs when a gas (usually air) is temporarily entrained in wastewater or is introduced purposefully to achieve a treatment objective. When gas is introduced into a wastewater, VOCs are transferred from the wastewater to the gas. The forces governing the transfer between phases are the same as described above. For this reason, gas (air) stripping is most effective when contaminated wastewater is exposed to contaminant free air. In wastewater treatment, air stripping occurs most commonly in aerated grit chambers, aerated biological treatment processes, and in aerated transfer channels.

Locations of Where VOCs Are Emitted. The principal locations where VOCs are emitted from wastewater collection and treatment facilities are summarized in Table 6-15. The degree of VOC removal at an given location will depend on local conditions. Mass transfer is considered in the following section.

Mass-Transfer Rates for VOCs

The mass transfer of VOCs can, for practical purposes, be modeled using the following equation [15,22].

$$r_{\text{VOC}} = -K_L a_{\text{VOC}}(C - C_s) \tag{6-65}$$

where r_{VOC} = rate of VOC mass transfer, $\mu g/ft^3 \cdot h$ ($\mu g/m^3 \cdot h$)

$(K_L a)_{\text{VOC}}$ = overall VOC mass-transfer coefficient, 1/h

C = concentration of VOC in liquid, $\mu g/ft^3$ ($\mu g/m^3$)

C_s = saturation concentration of VOC in liquid, $\mu g/ft^3$ ($\mu g/m^3$)

Before Eq. 6-65 can be applied the terms in Eq. 6-65 must be clarified.
     Based on experimental studies [15], it has been found that the mass-transfer coefficient for VOCs is proportional to the mass-transfer coefficient for oxygen. Thus,

$$(K_L a)_{\text{VOC}} = \Psi (K_L a)_{O_2} \tag{6-66}$$

where $K_L a_{\text{VOC}}$ = overall VOC mass-transfer coefficient, $\mu g/ft^3 \cdot h$ ($\mu g/m^3 \cdot h$)

Ψ = coefficient of proportionality

$K_L a_{O_2}$ = overall oxygen mass-transfer coefficient, $\mu g/ft^3 \cdot h$ ($\mu g/m^3 \cdot h$)

TABLE 6-15
Sources and methods of release of VOCs from wastewater to the atmosphere

Source	Method of release
Domestic, commercial, and industrial discharges	Discharge of small amounts of VOCs in liquid wastes
Wastewater sewers	Volatilization from the surface enhanced by flow induced turbulence
Sewer appurtenances	Volatilization due to turbulence at junctions, etc; Volatilization and air stripping at drop manholes and junction chambers
Pump stations	Volatilization and air stripping at influent wet-well inlets
Bar racks	Volatilization due to turbulence
Comminutors	Volatilization due to turbulence
Parshall flume	Volatilization due to turbulence
Grit chamber	Volatilization due to turbulence in conventional horizontal-flow grit chambers; Volatilization and air stripping in aerated grit chambers
Equalization basins	Volatilization from surface enhanced by local turbulence; Air stripping where diffused air is used
Primary and secondary sedimentation tanks	Volatilization from surface; Volatilization and air stripping at overflow weirs, in effluent channel, and at other discharge points
Biological treatment	Air stripping in diffused-air activated sludge; Volatilization in activated-sludge processes with surface aerators; Volatilization from surface enhanced by local turbulence
Transfer channels	Volatilization from surface enhanced by local turbulence; Volatilization and air stripping in aerated transfer channels
Digester gas	Uncontrolled release of digester gas; Discharge of incompletely combusted or incinerated digester gas

The reported range of values for the coefficient of proportionality Ψ is from 0.55 to 0.65 [15]. The values of Ψ were also found to be essentially the same for clear water as they are for wastewater.

The saturation concentration of a VOC in wastewater is a function of the partial pressure of the VOC in the atmosphere in contact with the wastewater. This relationship is given by Henry's law, as follows:

$$\frac{C_g}{C_s} = H_c \qquad (6\text{-}67)$$

where C_g = concentration of VOC in gas phase, $\mu g/ft^3$ $(\mu g/m^3)$

$\quad\quad C_s$ = saturation concentration of VOC in liquid, $\mu g/ft^3$ $(\mu g/m^3)$

$\quad\quad H_c$ = Henry's law constant, unitless

Values of Henry's law constant for various volatile and semivolatile compounds are reported in Table 6-16. Assuming atmospheric conditions prevail, the following equation is used to convert the values of Henry's constant given in Table 6-16 to the unitless form of Henry's law used in Eq. 6-67.

$$H_c = \frac{H}{RT} \tag{6-68}$$

where H_c = Henry's law constant, unitless

$\quad\quad H$ = Henry's law constant values from Table 6-16, $m^3 \cdot$ atm/g-mole

$\quad\quad R$ = universal gas law constant, 0.000082057 $m^3 \cdot$ atm/g-mole $\cdot$ °K

$\quad\quad T$ = temperature, °K (273 + °C)

Mass Transfer of VOCs from Surface and Diffused-Air Aeration Processes

The amount of VOCs released from a complete-mix reactor used for the activated-sludge process will depend on the method of aeration (e.g., surface aeration or diffused aeration).

Complete-Mix Reactor with Surface Aeration. A materials balance for the stripping of a VOC written around a complete-mix reactor is as follows:

$$\text{Accumulation} = \text{inflow} - \text{outflow} + \text{generation}$$

$$V\frac{dC}{dt} = Q_L C_{L,i} - Q_L C_{L,e} + r_{\text{VOC}} V \tag{6-69}$$

where $\quad\quad V$ = volume of complete-mix reactor, ft^3 (m^3)

$\quad dC/dt$ = rate of change in VOC concentration in reactor

$\quad\quad Q_L$ = flowrate, ft^3/s (m^3/s)

$\quad\quad C_{L,i}$ = concentration of VOC in influent to reactor, $\mu g/ft^3$ $(\mu g/m^3)$

$\quad\quad C_{L,e}$ = concentration of VOC in effluent from reactor, $\mu g/ft^3$ $(\mu g/m^3)$

$\quad\quad r_{\text{VOC}}$ = rate of VOC mass transfer, $\mu g/ft^3 \cdot h$ $(\mu g/m^3 \cdot h)$

Substituting for r_{VOC} from Eq. 6-65 and θ_H for V/Q_L yields

$$\frac{dC}{dt} = \frac{C_{L,i} - C_{L,e}}{\theta_H} - (K_L a)_{\text{VOC}}(C_{L,e} - C_s) \tag{6-70}$$

If steady-state conditions are assumed and it is further assumed that C_s is equal to zero, then the amount of VOC that can be removed by surface aeration is given by the following expression:

TABLE 6-16
Physical properties of selected volatile and semivolatile organic compounds[a,b]

Compounds	mw	mp, °C	bp, °C	vp, mm Hg	vd	sg	sol, mg/L	C_s, g/m³	K_H, m³·atm/mol	log K_{ow}
Benzene	78.11	5.5	80.1	76	2.77	.8786	1780	319	5.49×10^{-3}	2.1206
Chlorobenzene	112.56	-45	132	8.8	3.88	1.1066	500	54	3.70×10^{-3}	2.18–3.79
O-Dichlorobenzene	147.01	18	180.5	1.60	5.07	1.036	150	N/A	1.7×10^{-3}	3.3997
Ethylbenzene	106.17	-94.97	136.2	7	3.66	0.867	152	40	8.43×10^{-3}	3.13
1,2-Dibromoethane	187.87	9.8	131.3	10.25	0.105	2.18	2699	93.61	6.29×10^{-4}	N/A
1,1-Dichloroethane	98.96	-97.4	57.3	297	3.42	1.176	7840	160.93	5.1×10^{-3}	N/A
1,2-Dichloroethane	98.96	-35.4	83.5	61	3.4	1.25	8690	350	1.14×10^{-3}	1.4502
1,1,2,2-Tetrachloroethane	167.85	-36	146.2	14.74	5.79	1.595	2800	13.10	4.2×10^{-4}	2.389
1,1,1-Trichloroethane	133.41	-32	74	100	4.63	1.35	4400	715.9	3.6×10^{-3}	2.17
1,1,2-Trichloroethane	133.4	-36.5	133.8	19	N/A	N/A	4400	13.89	7.69×10^{-4}	N/A
Chloroethene	62.5	-153	-13.9	2548	2.15	0.912	6000	8521	6.4×10^{-2}	N/A
1,1-Dichloroethene	96.94	-122.1	31.9	500	3.3	1.21	5000	2640	1.51×10^{-2}	N/A
c-1,2-Dichloroethene	96.95	-80.5	60.3	200	3.34	1.284	800	104.39	4.08×10^{-3}	N/A
t-1,2-Dichloroethene	96.95	-50	48	269	3.34	1.26	6300	1428	4.05×10^{-3}	N/A
Tetrachloroethene	165.83	-22.5	121	15.6	N/A	1.63	160	126	2.85×10^{-2}	2.5289
Trichloroethene	131.5	-87	86.7	60	4.54	1.46	1100	415	1.17×10^{-2}	2.4200
Bromodichloromethane	163.8	-57.1	90	N/A	N/A	1.971	N/A	N/A	2.12×10^{-3}	N/A
Chlorodibromomethane	208.29	<-20	120	50	N/A	2.451	N/A	N/A	8.4×10^{-4}	N/A
Dichloromethane	84.93	-97	39.8	349	2.93	1.327	20000	1702	3.04×10^{-3}	N/A
Tetrachloromethane	153.82	-23	76.7	90	5.3	1.59	800	754	2.86×10^{-2}	2.7300
Tribromomethane	252.77	8.3	149	5.6	8.7	2.89	800	7.62	5.84×10^{-4}	N/A
Trichloromethane	119.38	-64	62	160	4.12	1.49	7840	1027	3.10×10^{-3}	1.8998
1,2-Dichloropropane	112.99	-100.5	96.4	41.2	3.5	1.156	2600	25.49	2.75×10^{-3}	N/A
2,3-Dichloropropene	110.98	-81.7	94	135	3.8	1.211	insol.	110	N/A	N/A
t-1,3-Dichloropropene	110.97	N/A	112	99.6	N/A	1.224	515	110	N/A	N/A
Toluene	92.1	-95.1	110.8	22	3.14	0.867	515	110	6.44×10^{-3}	2.2095

[a] Data were adapted from Ref. 12.

[b] All values are reported at 20°C.

Note: mw = molecular weight, mp = melting point, bp = boiling point, vp = vapor pressure, vd = vapor density, sg = specific gravity, sol = solubility, C_s = saturation concentration, K_H = Henry's Law Constant, log K_{ow} = logarithm of the octanol-water partition coefficient.

$$1 - \frac{C_{L,e}}{C_{L,i}} = 1 - [1 + (K_L a)\theta_H]^{-1} \tag{6-71}$$

If a significant amount of the VOC is adsorbed or biodegraded, the results obtained with the above equation will be over-estimated. The above analysis can also be used to estimate the release of VOCs at weirs and drops by assuming the time period is about 30 s.

Complete-Mix Reactor with Diffused Aeration. The corresponding expression to Eq. 6-71 for a complete mix reactor with diffused aeration is given by

$$1 - \frac{C_{L,e}}{C_{L,i}} = 1 - \left[1 + \frac{Q_g}{Q_L}(H_c)\left(1 - e^{-\phi}\right)\right]^{-1} \tag{6-72}$$

where Q_g = gas flowrate, ft³/s (m³/s)
$\quad Q_L$ = liquid (wastewater) flowrate, ft³/s (m³/s)
$\quad \phi$ = saturation parameter defined as

$$\phi = \frac{(K_L a)_{VOC} V}{H_c Q_g} \tag{6-73}$$

These above equations are applied in Example 6-12.

Example 6-12 Determination of amount of benzene that can be stripped in a complete-mix reactor equipped with a diffused-air aeration system. Assume that the following conditions apply:

1. Wastewater flowrate = 1.0 Mgal/d
2. Aeration tank volume = 0.25 Mgal
3. Depth of aeration tank = 20 ft
4. Air flowrate = 1,750 ft³/min at standard conditions
5. Oxygen-transfer rate = 6.2/h
6. Influent concentration of benzene = 100 μg/ft³
7. $H = 5.49 \times 10^{-3}$ m³ · atm/mol (see Table 6-16)
8. $\Psi = 0.6$ (assumed)
9. Temperature = 20 °C

Solution

1. Determine the quantity of air referenced to the mid-depth of the aeration tank.

$$Q_g = 1750 \times \frac{14.7}{14.7 + (10/2.31)}$$
$$Q_g = 1351 \text{ ft}^3/\text{min}$$

2. Determine the air/liquid ratio.

$$Q_L = \frac{1.0 \times 10^6 \text{ gal/d}}{7.48 \text{ gal/ft}^3 \times 1440 \text{ min/d}} = 92.8 \text{t}^3/\text{min}$$

$$\frac{Q_g}{Q_L} = \frac{1351}{92.8} = 14.6$$

3. Estimate the mass-transfer coefficient for benzene using Eq. 6-66.

$$(K_L a)_{VOC} = 0.6 \times 6.2/h \times 1/60 \text{ min/h} = 0.062/\text{min}$$

4. Determine the dimensionless value of the Henry's constant using Eq. 6-68.

$$H_c = \frac{H}{RT}$$

$$H_c = \frac{0.00549}{0.000082057 \times (273 + 20)} = 0.228$$

5. Determine the saturation parameter ϕ using Eq. 6-73.

$$\phi = \frac{(K_L a)_{VOC} V}{H_c Q_g}$$

$$\phi = \frac{0.062/\text{min} \times 33{,}422 \text{ ft}^3}{0.228 \times 1351 \text{ ft}^3/\text{min}} = 6.7$$

6. Determine the fraction of benzene removed from the liquid phase using Eq. 6-72.

$$1 - \frac{C_{L,e}}{C_{L,i}} = 1 - \left[1 + \frac{Q_g}{Q_L}(H_c)\left(1 - e^{-\phi}\right)\right]^{-1}$$

$$1 - \frac{C_{L,e}}{C_{L,i}} = 1 - \left[1 + 14.6(0.228)\left(1 - e^{-6.7}\right)\right]^{-1}$$

$$1 - \frac{C_{L,e}}{C_{L,i}} = 1 - 0.23 = 0.77$$

Comment. The computations presented in this example problem are based on the assumption that the concentration of benzene in the influent is not being reduced by adsorption or biological degradation.

DISCUSSION TOPICS AND PROBLEMS

6-1. If the accuracy of a metering flume is ± 5 percent of the maximum flow, the transmitter is ± 1 percent of the maximum flow, and the indicator is ± 3 percent of the maximum flow, what is the overall system accuracy?

6-2. A bar rack is inclined at a 50° angle with the horizontal. The circular bars have a diameter of 0.75 in and a clear spacing of 1.0 in. Determine the headloss when the bars are cleaned and the velocity approaching the rack is 3 ft/s. Is this a very realistic computation in terms of what actually happens at a treatment plant?

6-3. Using the information given in the data table presented in Example 6-1, determine (a) the off-line storage volume needed to equalize the flowrate and (b) the effect of flow equalization on the BOD$_5$ mass-loading rate. How does the BOD$_5$ mass-loading rate curve determined in this problem compare with the curve shown in Fig. 6-7? In your estimation, does the difference in the mass-loading rate justify the cost of the larger basin required for in-line storage?

6-4. Using the information given in the data table presented in Example 6-1, determine the in-line volume required to reduce the variation in the BOD_5 mass-loading rate between the maximum and minimum from the existing ratio of 25.1 : 1 (967 : 37) to a peak value of 5 : 1.

6-5. The contents of a tank are to be mixed with a turbine impeller that has six flat blades. The diameter of the impeller is 6 ft, and the impeller is installed 4 ft above the bottom of the 20 ft tank. If the temperature is 30°C and the impeller is rotated at 30 r/min, what will be the power consumption? Find the Reynolds number using Eq. 6.5.

6-6. It is desired to flash-mix some chemicals with incoming wastewater that is to be treated. Mixing is to be accomplished using a flat paddle mixer 20 in. in diameter with six blades. If the temperature of the incoming wastewater is 10°C and the mixing chamber power number is 1.70, determine
(a) The speed of rotation when the Reynolds number is approximately 100,000
(b) Why it is desirable to have a high Reynolds number in most mixing operations
(c) The required mixer motor size, assuming an efficiency factor of 20%
(d) The Froude number $(F = n^2D/g)$.

6-7. What is the significance of the Froude number in mixing operations? What typical Froude numbers are used for mixing operations in wastewater treatment? Cite three references.

6-8. Assuming that a given flocculation process can be defined by a first-order reaction $(r_N = -kN)$, complete the following table assuming that the process is occurring in a plug-flow reactor with a detention time of 10 min. What would the value be after 5 min if a batch reactor were used instead, assuming that the rate constant is the same?

Time, t	0	5	10
Particles, no/unit volume	10	(?)	3

6-9. If the steady-state effluent from a complete-mix reactor used as a flocculator contained 3 particles/unit volume, determine the concentration of particles in the effluent 5 min after the process started and before steady-state conditions are reached. Assume that the influent contains 10 particles/unit volume, the detention time in the complete-mix reactor is equal to 10 min, and that the first-order kinetics apply $(rN = -kN)$.

6-10. An air flocculation system is to be designed. If a G value of 60 s^{-1} is to be used, estimate the air flowrate that will be necessary for a 6200 ft^3 flocculation chamber. Assume that the depth of the flocculation basin is to be 12 ft and the wastewater temperature is 60° F.

6-11. Determine the required air flow rate to accomplish the flocculation operation in Example 6-2 pneumatically. Assume that the air will be released at a depth of 9 ft.

6-12. Derive Stoke's law by equating Eq. 6-15 to the effective particle mass.

6-13. Determine the settling velocity in feet per second of a sand particle with a specific gravity of 2.65 and a diameter of 0.04in. Assume that the Reynolds number is 175.

6-14. Determine the settling velocity in meters per second of a grit particle with a specific gravity of 2.26 and a diameter of 1 mm. Assume that the Reynolds number is 175.

6-15. Determine the removal efficiency for a sedimentation basin with a critical velocity V_o of 6.5 ft/h in treating wastewater containing particles whose settling velocities are distributed as given in the table at the top of the next page. Plot the particle histogram for the influent and effluent wastewater.

Velocity, ft/h	Number particles
0.0–1.5	20
1.5–3.0	40
3.0–4.5	80
4.5–6.0	120
6.0–7.5	100
7.5–9.0	70
9.0–10.5	20
10.5–12.0	10

6-16. The rate of flow through an ideal clarifier is 2.0 Mgal/d, the detention time is 1 h, the depth is 10 ft. If a full-length moveable horizontal tray is set 3 ft below the surface of the water, determine the percent removal of particles having a settling velocity of 3 ft/h. Could the removal efficiency of the clarifier be improved by moving the tray? If so, where should the tray be located and what would be the maximum removal efficiency? What effect would moving the tray have if the particle-settling velocity were equal to 1.0 ft/h?

6-17. A hydraulic study of the flow-through characteristics of a model sedimentation basin was made using NaCl as a tracer by injecting a slug of salt at the inlet and measuring the salt concentration at the outlet. The results of this study are shown below:

Time, min	NaCl concentration at outlet, mg/L
0	0
5	trace
10	40
15	130
20	110
25	90
30	70
40	50
50	40
60	30
70	20
80	10
90	5

(a) Plot a curve of the ratio of concentration of salt at the outlet to the inlet concentration (C/C_0, as the ordinate) against the ratio of the actual time over the theoretical detention time (t/t_0). Assume that $C_0 = 100$ mg/L and $t_0 = 40$ min.

(b) Calculate the t/t_0 ratios for the mean, median, mode, and minimum times.

(c) From part b, what can you say about the tank as to short circuiting and dead spaces?

(d) If a basin has marked short circuiting and/or dead spaces, does this necessarily mean that it will be less efficient in removing particles that one without short circuiting?

6-18. Prepare a one-page abstract of the following article: Morrill, A. B.: "Sedimentation Basin Research and Design," *Journal AWWA,* vol. 24, pp. 1442, 1932. Using the data from Prob. 6-17, determine the dispersion index and volumetric efficiency of the basin as defined by Morrill.

6-19. Using the data from Prob. 6-17, prepare a cummulative plot of the tracer leaving the basin. Using a cascade of complete-mix reactors, determine the number of reactors needed to model the cummulative dye tracer curve (refer to Appendix G).

6-20. Using the settling test curves shown below, determine the efficiency of a settling tank in removing flocculant particles if the depth is 6.5 ft and the detention time is 20 min.

6-21. For a flocculant suspension, determine the removal efficiency for a basin 10 ft deep with an overflow rate V_0 equal to 10 ft/h, using the laboratory settling data presented in the following table.

Time, min	Percent suspended solids removed at indicated depth (in ft)				
	1.5	3.0	4.5	6.0	7.5
20	61				
30	71	63	55		
40	81	72	63	61	57
50	90	81	73	67	63
60	—	90	80	74	68
70	—	—	86	80	75
80	—	—	—	86	81

6-22. The curve shown below was obtained from a settling test in a 6 ft cylinder. The initial solids concentration was 3600 mg/L. Determine the thickener area required for a concentration Cu of 12,000 mg/L with a sludge flow of 0.4 Mgal/d.

6-23. Given the settling data in the following table from an activated-sludge pilot plant (see the definition sketch figure for Example 6-6), determine the percent recycle rate if the sedimentation tank application rate is 490 gal/ft^2 · d and the concentration of the recycled solids is 10,500 mg/L. What will the recycle rate be if the concentration of the recycled solids is 15,000 mg/L?

Time, min	Sludge concentration, mg/L					
	1,000	2,000	3,000	5,000	10,000	15,000
0	0	0	0	0	0	0
10	3.84	2.97	1.35	0.56	0.16	0.10
20	6.20	5.48	2.76	1.12	0.33	0.20
30	6.30	6.00	4.19	1.67	0.49	0.30
40	6.33	6.17	5.12	2.23	0.66	0.39
50	6.33	6.20	5.45	2.79	0.85	0.46
60	6.36	6.23	5.64	3.35	1.02	0.52
80	6.40	6.27	5.91	4.17	1.35	0.75
100	6.43	6.30	6.04	4.50	1.64	0.95
120	6.46	6.36	6.17	4.82	1.90	1.12

Note: Data in the table correspond to the distance from the top of the settling column to the sludge interface at indicated times, ft.

6-24. Using the data from Prob. 6-20, determine the percent recycle rate if the sedimentation tank design is based on an application rate of 350 gal/ft^2 · d and the concentration of the recycled solids is 9,500 mg/L. What will the recycle rate be if the concentration of the recycled solids is 12,000 mg/L?

6-25. A stock filter sand has the following sieve analysis:

US sieve size designation[a]	Size of opening, mm	Cumulative weight, %
140	0.105	0.4
100	0.149	1.5
70	0.210	4.0
50	0.297	9.5
40	0.420	18.5
30	0.590	31.0
20	0.840	49.0
16	1.190	63.2
12	1.680	82.8
8	2.380	89.0
6	3.360	98.0
4	4.340	100.0

[a] *Note*: Sieve size number 18 has an opening size of 1.0 mm.

(a) Determine the geometric mean size, the geometric standard deviation, the effective size, and the uniformity coefficient for the stock sand.

(b) It is desired to produce from the stock sand a filter sand with an effective size of 0.45 mm and a uniformity coefficient of 1.6. Estimate the amount of stock sand required to procure one ton of filter sand.

(c) What U.S. standard sieve size should be used to eliminate the excess coarse material?

(d) If the material remaining after sieving part c is placed in a filter, what backwash rise rate would be needed to eliminate the excess fine material?

(e) What depth of sieved material would have to be placed in the filter to produce 24 in of usable filter sand?

(f) On log-probability paper, plot the size distribution of the modified sand. Check against the required distribution and sizes.

(g) Determine the headloss through 24 in of the filter sand specified in part b for a filtration rate of 3 gal/ft^2/min. Assume that the maximum and minimum sizes are 1.68 mm and 0.297 mm, respectively, and that the sand is stratified. Assume also that $T = 60°F$ and α (all strats) $= 0.4$.

(Note: Ref. 7 contains an excellent discussion of the procedures involved in developing a usable filter sand from a stock filter sand.)

6-26. Using the equations developed by Fair and Hatch and Rose, determine the headloss through a 30 in sand bed. Assume that the sand bed is composed of spherical unsized sand with a diameter of 0.6 mm, the porosity for the sand is 0.40, and the filtration velocity is 6 gal/ft^2 · min. The temperature is 18°C.

6-27. If a 12 in layer of anthracite is placed on top of the sand bed in Prob. 6-19, determine the ratio of the headloss through the anthracite to that of the sand. Assume that the grain-size diameter of the anthracite is 2.0 mm and that the porosity of the anthracite is 0.50.

6-28. For a given filtration operation, it has been found that straining is the operative particulate-matter-removal mechanism and that the change in concentration with distance can be approximated with a first-order equation $(dC/dx = -rC)$. If the initial concentration of particulate matter is 10 mg/L, the removal-rate constant is equal to 8 in^{-1}, and the filtration velocity is equal to 2.5 gal/ft^2 · min. Determine the amount of material arrested within the filter in the layer between 1 and 2 in over a 1 hr period. Express your answer in mg/in^3. Estimate the headloss in the layer at the end of 6 h.

6-29. The data in the following table were obtained from a pilot plant study on the filtration-settled secondary effluent from an activated-sludge treatment plant. Using these data, estimate the length of run that is possible with and without the addition of polymer if the maximum allowable headloss is 10 ft, the filtration rate is 4.0 gal/ft^2 · min, and the influent suspended-solids concentration is 15 mg/L. Uniform sand with a diameter of 0.55 mm and a depth of 2 ft was used in the pilot filters.

| | Concentration ratio, C/C_o | | | Concentration ratio, C/C_o | |
Depth, in	With polymer addition	Without polymer addition	Depth, in	With polymer addition	Without polymer addition
0	1.00	1.00	14	0.10	0.33
2	0.46	0.70	16	0.10	0.32
4	0.29	0.57	18	0.10	0.31
6	0.20	0.49	20	0.10	0.31
8	0.15	0.44	22	0.10	0.31
10	0.13	0.39	24	0.10	0.31
12	0.11	0.36			

6-30. The data in the following table were obtained from a test program designed to evaluate a new diffused-air aeration system. Using these data, determine the value of $K_L a$ at 20°C and the equilibrium dissolved-oxygen concentration in the test tank. The test program was conducted using tap water at a temperature of 24°C.

C, mg/L	1.5	2.7	3.9	4.8	6.0	7.0	8.2
dC/dt, mg/L · h	8.4	7.5	5.3	4.9	4.2	2.8	2.0

6-31. If the volume of the test tank used to evaluate the aeration system in Prob. 6-30 was equal to 26,400 gal and the air flowrate was equal to 70 ft^3/min, determine the maximum oxygen-transfer efficiency at 20°C and 1.0 atmosphere.

6-32. Using the equation developed in Example 6-10, estimate the air flowrate in ft^3/min required to increase the oxygen content of chlorinated effluent from zero to 4 mg/L. The effluent flowrate is equal to 5.7 Mgal/d. Assume that the transfer efficiency is 6 percent and the temperature is 15°C. What is the air requirement when the temperature is 25°C?

REFERENCES

1. ASCE: *A Standard For The Measurement Of Oxygen Transfer In Clean Water,* New York, July, 1984.
2. Camp, T. R., and P. C. Stein: "Velocity Gradients and Internal Work in Fluid Motion," *J. Boston Soc. Civ. Eng.,* vol. 30, p. 209, 1943.

3. Coe, H. S. and G. H. Clevenger: "Determining Thickener Unit Areas," *Trans. AIME,* vol. 55, no. 3, 1916.

4. Danckwertz, P. V.: "Significance of Liquid Film Coefficients in Gas Absorption," *J. Ind. Eng. Chem.,* vol. 43, p. 1460, 1951.

5. Dick, R. I.: "Folklore in the Design of Final Settling Tanks," *Journal WPCF,* vol. 48, no. 4, 1976.

6. Dick, R. I., and B. B. Ewing: "Evaluation of Activated Sludge Thickening Theories," *J. Sanit. Eng. Div., ASCE,* vol. 93, no. SA-4, 1967.

7. Fair, G. M., J. C. Geyer, and D. A. Okun: *Water and Wastewater Engineering,* vol. 2, Wiley, New York, 1966.

8. Graber, S. D.: "Discussion/Communication by V. Kothandaraman and R. L. Evans on Hydraulic Model Studies of Chlorine Contact Tanks," *Journal WPCF,* vol. 44, no. 10, 1972.

9. Harris, H. S., W. S. Kaufman, and R. B. Krone: "Othokinetic Flocculation in Water Purification," *J. Sanit. Eng. Div., ASCE,* vol. 92, no. SA6, proc. paper 5027, pp. 95-111, December 1966.

10. Higbie, R.: "The Rate of Absorption of Pure Gas into a Still Liquid during Short Periods of Exposure," *Trans. Am. Inst. Chem. Eng.,* vol. 31, p. 365, 1935.

11. Keinath, T. M.: "Operational Dynamics And Control Of Secondary Clarifiers," *Journal WPCF,* vol. 57, no. 7, p. 770, 1989.

12. Lang, R., et al.: *Trace Organic Constituents In Landfill Gas,* prepared by the Department of Civil Engineering, University of California, Davis, CA, California Waste Management Board, Sacramento, CA, 1987.

13. Lewis, W. K., and W. C. Whitman: "Principles of Gas Adsorption," *Ind. Eng. Chem.,* vol. 16, p. 1215, 1924.

14. O'Melia, C. R., and W. Stumm: "Theory of Water Filtration," *Journal AWWA,* vol. 59, no. 11, 1967.

15. Roberts, P. V., C. Munz, P. Dandiker, and C. Matter-Muller: *Volatilization of Organic Pollutants in Wastewater Treatment-Model Studies,* EPA-600/S2-84-047, 1984.

16. Rushton, J. H.: "Mixing of Liquids in Chemical Processing," *Ind. Eng. Chem.,* vol. 44, no. 12, 1952.

17. Talmadge, W. P., and E. B. Fitch: "Determining Thickener Unit Areas," *Ind. Eng. Chem.,* vol. 47, no.1, 1955.

18. Tchobanoglous, G., and R. Eliassen: "Filtration of Treated Sewage Effluent," *J. Sanit. Eng. Div., ASCE,* vol. 96, no. SA2, 1970.

19. Tchobanoglous, G., and E. D. Schroeder, *Water Quality: Characteristics, Modeling, Modification,* Addison-Wesley, Reading, MA, 1985.

20. Tchobanoglous, G., F. Maitski, K. Thompson, and T. H. Chadwick: "Evolution And Performance of City of San Diego Pilot Plant Aquatic Wastewater Treatment System Using Water Hyacinths," *Journal WPCF,* vol. 61, no. 11/12, 1989.

21. Tchobanoglous, G.: "Filtration Treated Wastewater Effluent," presented at the 61st Annual Conference of the WPCF, Dallas, TX, October 1988.

22. Thibodeaux, L. J.: *Chemodynamics: Environmental Movement Of Chemicals In Air,Water, and Soil,* John Wiley & Sons, New York, 1979.

23. U.S. Environmental Protection Agency: Process Design Manual for Upgrading Existing Wastewater Treatment Plants, U.S. Environmental Protection Agency, Technology Transfer, October 1974.

24. U.S. Environmental Protection Agency: Wastewater Filtration-Design Considerations, U.S. Environmental Protection Agency, Technology Transfer Seminar Publication, 1974.

25. U.S. Environmental Protection Agency: Wastewater Filtration-Design Considerations, U.S. Environmental Agency, Technology Transfer Report, 1977.

26. U.S. Environmental Protection Agency: *Report to Congress on the Discharge of Hazardous Wastes to Publicly Owned Treatment Works,* EPA/530-SW-86-004, 1986.

27. Wilson, G. E.: "Is There Grit In Your Sludge," *Civil Engineering,* vol. 55, no. 4, 1985.

28. Yoshika, N., et al.: "Continuous Thickening of Homogeneous Flocculated Slurries," *Kagaku Kogaku,* vol. 26, 1957 (also in *Chem. Eng.,* vol. 21, Tokyo, 1957).

CHAPTER
7

CHEMICAL
UNIT
PROCESSES

Those processes used for the treatment of wastewater in which change is brought about by means of or through chemical reaction are known as chemical unit processes. In the field of wastewater treatment, chemical unit processes are usually used in conjunction with the physical unit operations, discussed in Chap. 6, and the biological unit processes, to be discussed in Chap. 8, to meet treatment objectives.

The chemical processes considered in this chapter and their principal applications are reported in Table 7-1. The use of various chemicals to improve the results of other operations and processes is also noted briefly. Here, as in Chap. 6, each unit process will be described and the fundamentals involved in the engineering analysis of each unit process will be discussed. In these discussions, knowledge of the fundamentals of chemistry is assumed. The practical application of these processes, including such matters as facility design and dosage requirements, is considered in Chap. 9. Some of the unit processes, such as precipitation for phosphorus removal, activated-carbon adsorption for the removal of organic compounds, and breakpoint chlorination for nitrogen removal, are also considered in Chap. 11, which deals with advanced wastewater treatment.

In considering the application of the following chemical unit processes, it is important to remember that one of the inherent disadvantages associated with most chemical unit processes (activated-carbon adsorption is an exception), as compared with the physical unit operations, is that they are additive processes. In most cases, something is added to the wastewater to achieve the removal of something else. As a result, there is usually a net increase in the dissolved constituents in the wastewater. For example, where chemicals are added to enhance the removal efficiency of plain sedimentation, the total dissolved-solids concentration of the wastewater is always

TABLE 7-1
Applications of chemical unit processes in wastewater treatment

Process	Application	See section
Chemical precipitation	Removal of phosphorus and enhancement of suspended-solids removal in primary sedimentation facilities used for physical-chemical treatment	7-1
Adsorption	Removal of organics not removed by conventional chemical and biological treatment methods. Also used for dechlorination of wastewater before final discharge for treated effluent	7-2
Disinfection	Selective destruction of disease-causing organisms (can be accomplished in various ways)	7-3
Disinfection with chlorine	Selective destruction of disease-causing organisms. Chlorine is the most used chemical	7-4
Dechlorination	Removal of total combined chlorine residual that exists after chlorination (can be accomplished in various ways)	7-5
Disinfection with chlorine dioxide	Selective destruction of disease-causing organisms	7-6
Disinfection with bromine chloride	Selective destruction of disease-causing organisms	7-7
Disinfection with ozone	Selective destruction of disease-causing organisms	7-8
Disinfection with ultraviolet light	Selective destruction of disease-causing organisms	7-9
Other chemical applications	Various other chemicals can be used to achieve specific objectives in wastewater treatment	7-10

increased. If the treated wastewater is to be reused, this can be a significant factor. This additive aspect contrasts to the physical unit operations (Chap. 6) and the biological unit processes (Chap. 8), which may be described as being subtractive, in that material is removed from the wastewater. Another disadvantage of chemical unit processes is that they are all intensive in operating costs. The costs of some of these chemicals are tied to the costs of energy and can be expected to increase similarly.

7-1 CHEMICAL PRECIPITATION

Chemical precipitation in wastewater treatment involves the addition of chemicals to alter the physical state of dissolved and suspended solids and to facilitate their removal by sedimentation. In some cases the alteration is slight, and removal is effected by entrapment within a voluminous precipitate consisting primarily of the coagulant itself. Another result of chemical addition is a net increase in the dissolved constituents in the wastewater. Chemical processes, in conjunction with various physical operations, have been developed for the complete secondary treatment of untreated wastewater, including the removal of either nitrogen or phosphorus, or both [4,19]. Other chemical processes have also been developed to remove phosphorus by chemical precipitation and are designed to be used in conjunction with biological treatment.

The purpose in this section is to identify and discuss (1) the precipitation reactions that occur when various chemicals are added to improve the performance of wastewater treatment facilities, (2) the chemical reactions involved in the precipitation of phosphorus from wastewater, and (3) some of the more important theoretical aspects of chemical precipitation. The computations used to determine the quantities of sludge produced as a result of the addition of various chemicals are illustrated in Chap. 9. The removal of phosphorus is considered further in Chap. 11, which deals with advanced wastewater treatment.

Chemical Precipitation
For Improving Plant Performance

In the past, chemical precipitation was used to enhance the degree of suspended solids and BOD removal (1) where there were seasonal variations in the concentration of the wastewater (such as in cannery wastewater), (2) where an intermediate degree of treatment was required, and (3) as an aid to the sedimentation process. Since about 1970, the need to provide more complete removal of the organic compounds and nutrients (nitrogen and phosphorus) contained in wastewater has brought about renewed interest in chemical precipitation.

Over the years a number of different substances have been used as precipitants. The most common chemicals are listed in Table 7-2. The degree of clarification obtained depends on the quantity of chemicals used and the care with which the process is controlled. It is possible by chemical precipitation to obtain a clear effluent, substantially free from matter in suspension or in the colloidal state. From 80 to 90 percent of the total suspended matter, 40 to 70 percent of the BOD_5, 30 to 60 percent of the COD, and 80 to 90 percent of the bacteria can be removed by chemical precipitation. In comparison, when plain sedimentation is used, only 50 to 70 percent of the total suspended matter and 30 to 40 percent of the organic matter settles out.

TABLE 7-2
Chemicals used in wastewater treatment

Chemical	Formula	Molecular weight	Density, lb/ft³	
			Dry	Liquid
Alum	$Al_2(SO_4)_3 \cdot 18H_2O^a$	666.7	60–75	78–80 (49%)
	$Al_2(SO_4)_3 \cdot 14H_2O^a$	594.3	60–75	83–85 (49%)
Ferric chloride	$FeCl_3$	162.1		84–93
Ferric sulfate	$Fe_2(SO_4)_3$	400		
	$Fe_2(SO_4)_3 \cdot 3H_2O$	454		70–72
Ferrous sulfate (copperas)	$FeSO_4 \cdot 7H_2O$	278.0	62–66	
Lime	$Ca(OH)_2$	56 as CaO	35–50	

[a] Number of bound water molecules will vary from 13 to 18.

Note: lb/ft³ × 16.0185 = kg/m³

The chemicals added to wastewater interact with substances that are either normally present in the wastewater or added for this purpose. The reactions involved with (1) alum, (2) lime, (3) ferrous sulfate (copperas) and lime, (4) ferric chloride, (5) ferric chloride and lime, and (6) ferric sulfate and lime are considered in the following discussion [9].

Alum. When alum is added to wastewater containing calcium and magnesium bicarbonate alkalinity, the reaction that occurs may be illustrated as follows:

$$
\underset{\substack{\text{Aluminum}\\\text{sulfate}}}{\overset{666.7}{Al_2(SO_4)_3 \cdot 18\,H_2O}} + \underset{\substack{\text{Calcium}\\\text{bicarbonate}}}{\overset{3 \times 100 \text{ as } CaCO_3}{3\,Ca(HCO_3)_2}} \Leftrightarrow \underset{\substack{\text{Calcium}\\\text{sulfate}}}{\overset{3 \times 136}{3\,CaSO_4}} + \underset{\substack{\text{Aluminum}\\\text{hydroxide}}}{\overset{2 \times 78}{2\,Al(OH)_3}} + \underset{\substack{\text{Carbon}\\\text{dioxide}}}{\overset{6 \times 44}{6\,CO_2}} + \overset{18 \times 18}{18\,H_2O} \quad (7\text{-}1)
$$

The numbers above the chemical formulas are the combining molecular weights of the different substances and, therefore, denote the quantity of each one involved. The insoluble aluminum hydroxide is a gelatinous floc that settles slowly through the wastewater, sweeping out suspended material and producing other changes. The reaction is exactly analogous when magnesium bicarbonate is substituted for the calcium salt.

Because alkalinity in Eq. 7-1 is reported in terms of calcium carbonate ($CaCO_3$), the molecular weight of which is 100, the quantity of alkalinity required to react with 10 mg/L of alum is

$$
10.0 \text{ mg/L} \times \frac{3 \times 100 \text{ g/mol}}{666.7 \text{ g/mol}} = 4.5 \text{ mg/L}
$$

If less than this amount of alkalinity is available, it must be added. Lime is commonly used for this purpose when necessary, but it is seldom required in the chemical treatment of wastewater.

Lime. When lime alone is added as a precipitant, the principles of clarification are explained by the following reactions:

$$
\underset{\substack{\text{Calcium}\\\text{hydroxide}}}{\overset{56 \text{ as } CaO}{Ca(OH)_2}} + \underset{\substack{\text{Carbonic}\\\text{acid}}}{\overset{44 \text{ as } CO_2}{H_2CO_3}} \Leftrightarrow \underset{\substack{\text{Calcium}\\\text{carbonate}}}{\overset{100}{CaCO_3}} + \overset{2 \times 18}{2H_2O} \quad (7\text{-}2)
$$

$$
\underset{\substack{\text{Calcium}\\\text{hydroxide}}}{\overset{56 \text{ as } CaO}{Ca(OH)_2}} + \underset{\substack{\text{Calcium}\\\text{bicarbonate}}}{\overset{100 \text{ as } CaCO_3}{Ca(HCO_3)_2}} \Leftrightarrow \underset{\substack{\text{Calcium}\\\text{carbonate}}}{\overset{2 \times 100}{2CaCO_3}} + \overset{2 \times 18}{2H_2O} \quad (7\text{-}3)
$$

A sufficient quantity of lime must, therefore, be added to combine with all the free carbonic acid and with the carbonic acid of the bicarbonates (half-bound carbonic acid) to produce calcium carbonate, which acts as the coagulant. Much more lime is generally required when it is used alone than when sulfate of iron is also used (see the following discussion). Where industrial wastes introduce mineral acids or acid salts into the wastewater, these must be neutralized before precipitation can take place.

Ferrous Sulfate and Lime. In most cases, ferrous sulfate cannot be used alone as a precipitant because lime must be added at the same time to form a precipitate. The reaction with ferrous sulfate alone is illustrated in Eq. 7-4.

$$
\underset{\substack{\text{Ferrous}\\\text{sulfate}}}{\overset{278}{FeSO_4 \cdot 7H_2O}} + \underset{\substack{\text{Calcium}\\\text{bicarbonate}}}{\overset{100 \text{ as } CaCO_3}{Ca(HCO_3)_2}} \Leftrightarrow \underset{\substack{\text{Ferrous}\\\text{bicarbonate}}}{\overset{178}{Fe(HCO_3)_2}} + \underset{\substack{\text{Calcium}\\\text{sulfate}}}{\overset{136}{CaSO_4}} + \overset{7 \times 18}{7H_2O} \qquad (7\text{-}4)
$$

If lime in the form $Ca(OH)_2$ is now added, the reaction that takes place is

$$
\underset{\substack{\text{Ferrous}\\\text{bicarbonate}}}{\overset{178}{Fe(HCO_3)_2}} + \underset{\substack{\text{Calcium}\\\text{hydroxide}}}{\overset{2 \times 56 \text{ as } CaO}{2Ca(OH)_2}} \Leftrightarrow \underset{\substack{\text{Ferrous}\\\text{hydroxide}}}{\overset{89.9}{Fe(OH)_2}} + \underset{\substack{\text{Calcium}\\\text{carbonate}}}{\overset{2 \times 100}{2CaCO_3}} + \overset{2 \times 18}{2H_2O} \qquad (7\text{-}5)
$$

The ferrous hydroxide is next oxidized to ferric hydroxide, the final form desired, by oxygen dissolved in the wastewater:

$$
\underset{\substack{\text{Ferrous}\\\text{hydroxide}}}{\overset{4 \times 89.9}{4Fe(OH)_2}} + \underset{\text{Oxygen}}{\overset{32}{O_2}} + \overset{2 \times 18}{2H_2O} \Leftrightarrow \underset{\substack{\text{Ferric}\\\text{hydroxide}}}{\overset{4 \times 106.9}{4Fe(OH)_3}} \qquad (7\text{-}6)
$$

The insoluble ferric hydroxide is formed as a bulky, gelatinous floc similar to the alum floc. The alkalinity required for a 10 mg/L dosage of ferrous sulfate (see Eq. 7-4) is

$$
10.0 \text{ mg/L} \times \frac{100 \text{ g/mol}}{278 \text{ g/mol}} = 3.6 \text{ mg/L}
$$

The lime required is

$$
10.0 \text{ mg/L} \times \frac{2 \times 56 \text{ g/mol}}{278 \text{ g/mol}} = 4.0 \text{ mg/L}
$$

The oxygen required is

$$
10.0 \text{ mg/L} \times \frac{32 \text{ g/mol}}{4 \times 278 \text{ g/mol}} = 0.29 \text{ mg/L}
$$

Because the formation of ferric hydroxide is dependent on the presence of dissolved oxygen, the reaction given in Eq. 7-6 cannot be completed with septic wastewater or industrial wastes devoid of oxygen. Ferric sulfate may take the place of ferrous sulfate, and its use often avoids the addition of lime and the requirement of dissolved oxygen.

Ferric Chloride. The reactions for ferric chloride are

$$
\begin{array}{ccccc}
162.1 & & 3 \times 18 & & 106.9 \\
FeCl_3 & + & 3H_2O & \Leftrightarrow & Fe(OH)_3 & + & 3H^+ & +3Cl^- \\
\text{Ferric} & & \text{Water} & & \text{Ferric} \\
\text{chloride} & & & & \text{hydroxide}
\end{array}
\tag{7-7}
$$

$$
\begin{array}{ccc}
3H^+ & + & 3HCO_3^- & \Leftrightarrow & 3H_2CO_3 \\
& & \text{Bicarbonate} & & \text{Carbonic} \\
& & & & \text{acid}
\end{array}
\tag{7-8}
$$

Ferric Chloride and Lime. The reactions for ferric chloride and lime are

$$
\begin{array}{cccc}
2 \times 162 & 3 \times 56 \text{ as CaO} & 3 \times 111 & 2 \times 106.9 \\
2FeCl_3 & + \quad 3Ca(OH)_2 & \Leftrightarrow \quad 3CaCl_2 & + \quad 2Fe(OH)_3 \\
\text{Ferric} & \text{Calcium} & \text{Calcium} & \text{Ferric} \\
\text{chloride} & \text{hydroxide} & \text{chloride} & \text{hydroxide}
\end{array}
\tag{7-9}
$$

Ferric Sulfate and Lime. The reactions for ferric sulfate and lime are

$$
\begin{array}{cccc}
400 & 3 \times 56 \text{ as CaO} & 408 & 2 \times 106.9 \\
Fe_2(SO_4)_3 & + \quad 3Ca(OH)_2 & \Leftrightarrow \quad 3CaSO_4 & + \quad 2Fe(OH)_3 \\
\text{Ferric} & \text{Calcium} & \text{Calcium} & \text{Ferric} \\
\text{sulfate} & \text{hydroxide} & \text{sulfate} & \text{hydroxide}
\end{array}
\tag{7-10}
$$

Chemical Precipitation for Phosphate Removal

The removal of phosphorus from wastewater involves the incorporation of phosphate into suspended solids and the subsequent removal of those solids. Phosphorus can be incorporated into either biological solids (e.g., microorganisms) or chemical precipitates. Biological phosphorus removal is considered in Sec. 8-11 in Chap. 8 and in Chap.11. The removal of phosphorus in chemical precipitates is introduced in this section. The topics to be considered include (1) the strategies for phosphorus removal and (2) the chemistry of phosphate precipitation.

Strategies for Phosphorus Removal. Chemicals that have been used for the removal of phosphorus include metal salts and lime. The most common metal salts used are ferric chloride and aluminum sulfate (alum). Ferrous sulfate and ferrous chloride, which are available as by-products of steel making operations (pickle liquor), are also used. Polymers have also been used effectively in conjunction with iron salts and alum. Lime is used less frequently because of the substantial increase in the mass of sludge as compared to metal salts and the operating and maintenance problems associated with the handling, storage, and feeding of lime. Typical chemical dosages used for phosphorus removal are given in Sec. 11-8. The precipitation of phosphorus from wastewater can occur in a number of different locations within a process flow diagram (see Fig. 7-1). The general locations where phosphorus can be removed may be classified as (1) pre-precipitation, (2) coprecipitation, and (3) post-precipitation [11].

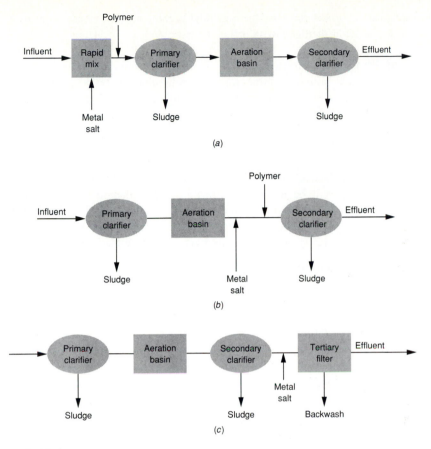

FIGURE 7-1
Flow diagrams for the removal of phosphorus: (a) pre-precipitation, (b) co-precipitation, and (c) post-precipitation [11].

Pre-precipitation. The addition of chemicals to raw wastewater for the precipitation of phosphorus in primary sedimentation facilities is termed "pre-precipitation." The precipitated phosphate is removed with the primary sludge.

Co-precipitation. The addition of chemicals to form precipitates that are removed along with waste biological sludge is defined as "co-precipitation." Chemicals can be added to (1) the effluent from primary sedimentation facilities, (2) to the mixed liquor (in the activated-sludge process), or (3) to the effluent from a biological treatment process before secondary sedimentation.

Post-precipitation. Post-precipitation involves the addition of chemicals to the effluent from secondary sedimentation facilities and the subsequent removal of chemical precipitates. In this process, the chemical precipitates are usually removed in separate sedimentation facilities or in effluent filters.

Chemistry of Phosphate Removal. The chemical precipitation of phosphorus is brought about by the addition of the salts of multivalent metal ions that form precipitates of sparingly soluble phosphates. The multivalent metal ions used most commonly are calcium [CA(II)], aluminum [Al(III)], and iron [Fe(III)]. Because the chemistry of phosphate precipitation with calcium is quite different from that with aluminum and iron, the two different types of precipitation are considered separately in the following discussion.

Calcium is usually added in the form of lime, $Ca(OH)_2$. From the equations presented previously, it will be noted that when lime is added to water it reacts with the natural bicarbonate alkalinity to precipitate $CaCO_3$. As the pH value of the wastewater increases beyond about 10, excess calcium ions will then react with the phosphate, as shown in Eq. 7-11, to precipitate hydroxylapatite $Ca_{10}(PO_4)_6(OH)_2$.

Phosphate precipitation with calcium:

$$10Ca^{+2} \; + \; 6PO_4^{-3} \; + \; 2OH^- \; \Leftrightarrow \; \underset{\text{Hydroxylapatite}}{Ca_{10}(PO_4)_6(OH)_2} \tag{7-11}$$

Because of the reaction of lime with the alkalinity of the wastewater, the quantity of lime required will, in general, be independent of the amount of phosphate present and will depend primarily on the alkalinity of the wastewater. The quantity of lime required to precipitate the phosphorus in wastewater is typically about 1.4 to 1.5 times the total alkalinity expressed as $CaCO_3$. Because a high pH value is required to precipitate phosphate, coprecipitation is usually not feasible. When lime is added to raw wastewater or to secondary effluent, pH adjustment is usually required before subsequent treatment or disposal. Recarbonation with carbon dioxide (CO_2) is used to lower the pH value.

The basic reactions involved in the precipitation of phosphorus with aluminum and iron are as follows.

Phosphate precipitation with aluminum:

$$Al^{+3} \; + \; H_n PO_4^{3-n} \; \Leftrightarrow \; AlPO_4 \; + \; nH^+ \tag{7-12}$$

Phosphate precipitation with iron:

$$Fe^{+3} \; + \; H_n PO_4^{3-n} \; \Leftrightarrow \; FePO_4 \; + \; nH^+ \tag{7-13}$$

In the case of alum and iron, 1 mole will precipitate 1 mole of phosphate; however, these reactions are deceptively simple and must be considered in light of the many competing reactions and their associated equilibrium constants and the effects of alkalinity, pH, trace elements, and ligands found in wastewater. Because of the many competing reactions, Eqs. 7-12 and 7-13 cannot be used to estimate the required chemical dosages directly. Therefore, dosages are generally established on the basis of bench-scale tests and occasionally by full-scale tests, especially if polymers are

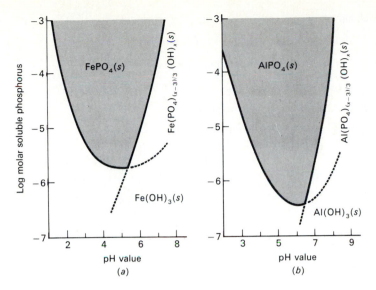

FIGURE 7-2
Concentration of ferric and aluminum phosphate in equilibrium with soluble phosphorus:
(a) Fe(III)-phosphate and (b) Al(III)-phosphate [7].

used (see Example 7-1). For example, for equimolar initial concentrations of Al(III), Fe(III), and phosphate, the total concentration of soluble phosphate in equilibrium with both insoluble $FePO_4$ and $AlPO_4$ is shown in Fig. 7-2. The solid lines trace the concentration of residual soluble phosphate after precipitation. Pure metal phosphates are precipitated within the shaded area, and mixed complex precipitates are formed outside toward the higher pH values. Practical details on the removal of phosphate, including process-flow diagrams and required chemical dosages, are presented in Chap. 11.

Example 7-1 Determination of alum dosage for phosphorus removal. Determine the amount of liquid alum required to precipitate phosphorus in a wastewater which contains 8 mg P/L. Also determine the required storage capacity if a 30 day supply is to be stored at the treatment facility. Based on laboratory testing 1.5 mole of Al will be required per mole of P. The flowrate is 3 Mgal/d (10,455 m³/d). The following data are for the liquid alum supply.

1. Formula for liquid alum $Al_2(SO_4)_3 \cdot 18H_2O$

2. Alum strength = 48 percent

3. Density of liquid alum solution = 80 lb/ft³ (see Table 7-2)
   = 10.7 lb/gal

Solution

1. Determine the weight of aluminum (Al) available per gallon of liquid alum

   (a) The weight of alum per gal is

   $$\text{Alum/gal} = 0.48 \times 10.7 \text{ lb/gal} = 5.14 \text{ lb/gal}$$

(b) The weight of aluminum per gal is

$$\text{Molecular weight of alum} = 666.7 \text{ (see Table 7-2)}$$

$$\text{Molecular weight of aluminum} = 26.98$$

$$\text{Aluminum/gal} = 5.14 \text{ lb/gal} \times (2 \times 26.98/666.7) = 0.416 \text{ lb/gal}$$

2. Determine the weight of Al required per unit weight of P
    (a) Theoretical dosage = 1 mole Al per 1 mole P (see Eq. 7-12)
    (b) Aluminum required = 1.0 lb × (mw Al/mw P)
    $$= 1.0 \text{ lb} \times (26.98/30.97) = 0.87 \text{ lb Al/lb P}$$
3. Determine the amount of alum solution required per lb P

$$\text{Alum dosage} = 1.5 \times \left(\frac{0.87 \text{ lb Al}}{\text{lb P}} \right) \left(\frac{1 \text{ gal alum sol.}}{0.416 \text{ lb Al}} \right)$$

$$= 3.13 \text{ gal alum sol./lb P}$$

4. Determine the amount of alum solution required per day

$$\text{Alum} = (3.0 \text{ Mgal/d})(8 \text{ mg P/L})[8.34 \text{ lb/Mgal} \cdot (\text{mg/L})](3.13 \text{ gal alum sol./lb P})$$
$$= 626.5 \text{ gal alum solution/d}$$

5. Determine the required alum solution storage capacity based on average flow

$$\text{Storage capacity} = (626.5 \text{ gal alum sol./d})(30d)$$
$$= 18,795 \text{ gal}$$

Theoretical Aspects of Chemical Precipitation

The theory of chemical precipitation reactions is very complex. The reactions that have been presented explain it only in part, and even they do not necessarily proceed as indicated. They are often incomplete, and numerous side reactions with other substances in wastewater may take place. Therefore, the following discussion is necessarily incomplete but will serve as an introduction to the nature of the phenomena involved.

Nature of Particles in Wastewater. There are two general types of colloidal solid particle dispersions in liquids. When water is the solvent, these are called the hydrophobic, or "water-hating," and the hydrophilic, or "water-loving," colloids. These two types are based on the attraction of the particle surface for water. Hydrophobic particles have relatively little attraction for water; hydrophilic particles have a great attraction for water. It should be noted, however, that water can interact to some extent with hydrophobic particles. Some water molecules will generally adsorb on the typical hydrophobic surface, but the reaction between water and hydrophilic colloids occurs to a much greater extent.

Surface Charge. An important factor in the stability of colloids is the presence of surface charge. It develops in a number of different ways, depending on the chemical composition of the medium (wastewater in this case) and the colloid. Regardless of how it is developed, this stability must be overcome if these particles are to be aggregated (flocculated) into larger particles with enough mass to settle easily.

Surface charge develops most commonly through preferential adsorption, ionization, and isomorphous replacement. For example, oil droplets, gas bubbles, or other chemically inert substances dispersed in water will acquire a negative charge through the preferential adsorption of anions (particularly hydroxyl ions). In the case of substances such as proteins or microorganisms, surface charge is acquired through the ionization of carboxyl and amino groups [13]. This can be represented as $R_{NH_2}^{COO^-}$ at high pH, $R_{NH_3^+}^{COOH}$ at low pH, and $R_{NH_3^+}^{COO^-}$ at the isoelectric point where R represents the bulk of the solid [7]. Charge development through isomorphous replacement occurs in clay and other soil particles, in which ions in the lattice structure are replaced with ions from solution (e.g., the replacement of Si with Al).

When the colloid or particle surface becomes charged, some ions of the opposite charge (known as *counter ions*) become attached to the surface. They are held there through electrostatic and van der Waals forces strongly enough to overcome thermal agitation. Surrounding this fixed layer of ions is a diffuse layer of ions, which is prevented from forming a compact double layer by thermal agitation. This is illustrated schematically in Fig. 7-3. As shown, the double layer consists of a compact layer (Stern) in which the potential drops from ψ_0 to ψ_s and a diffuse layer in which the potential drops from ψ_s to 0 in the bulk solution.

If a particle such as shown in Fig. 7-3 is placed in an electrolyte solution and an electric current is passed through the solution, the particle, depending on its surface charge, will be attracted to one or the other of the electrodes, dragging with it a cloud of ions.

The potential at the surface of the cloud (called the *surface of shear*) is sometimes measured in wastewater treatment operations. The measured value is often called the *zeta potential*. Theoretically, however, the zeta potential should correspond to the potential measured at the surface enclosing the fixed layer of ions attached to the particle, as shown in Fig. 7-3. The use of the measured zeta potential value is limited because it will vary with the nature of the solution components, and it therefore is not a repeatable measurement.

Particle Aggregation. To bring about particle aggregation, steps must be taken to reduce particle charge or to overcome the effect of this charge. The effect of the charge can be overcome by (1) the addition of potential-determining ions, which will be taken up by or will react with the colloid surface to lessen the surface charge, or the addition of electrolytes, which have the effect of reducing the thickness of the diffuse electric layer and thereby reduce the zeta potential; (2) the addition of long-chained organic molecules (polymers), whose subunits are ionizable and are therefore called *polyelectrolytes*, that bring about the removal of particles through adsorption and bridging; and (3) the addition of chemicals that form hydrolyzed metal ions.

Addition of potential-determining ions to promote coagulation can be illustrated by the addition of strong acids or bases to reduce the charge of metal oxides or

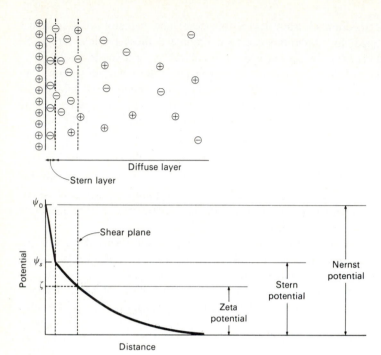

FIGURE 7-3
Stern model of electrical double layer [13].

hydroxides to near zero so that coagulation can occur. Electrolytes can also be added to coagulate colloidal suspensions. Increased concentration of a given electrolyte will cause a decrease in zeta potential and a corresponding decrease in repulsive forces. Similar effects are observed if the electrolyte charge is increased.

Polyelectrolytes may be divided into two categories: natural and synthetic. Important natural polyelectrolytes include polymers of biological origin and those derived from starch products, cellulose derivatives, and alginates. Synthetic polyelectrolytes consist of simple monomers that are polymerized into high-molecular-weight substances. Depending on whether their charge when placed in water is negative, positive, or neutral, these polyelectrolytes are classified as anionic, cationic, and nonionic, respectively.

The action of polyelectrolytes may be divided into three general categories. In the first category, polyelectrolytes act as coagulants that lower the charge of the wastewater particles. Because wastewater particles normally are charged negatively, cationic polyelectrolytes are used for this purpose. In this application, the cationic polyelectrolytes are considered to be primary coagulants.

The second mode of action of polyelectrolytes is interparticle bridging (see Fig. 7-4). In this case, polymers that are anionic and nonionic (usually anionic to a slight extent when placed in water) become attached at a number of adsorption sites to the surface of the particles found in the settled effluent. A bridge is formed when two or

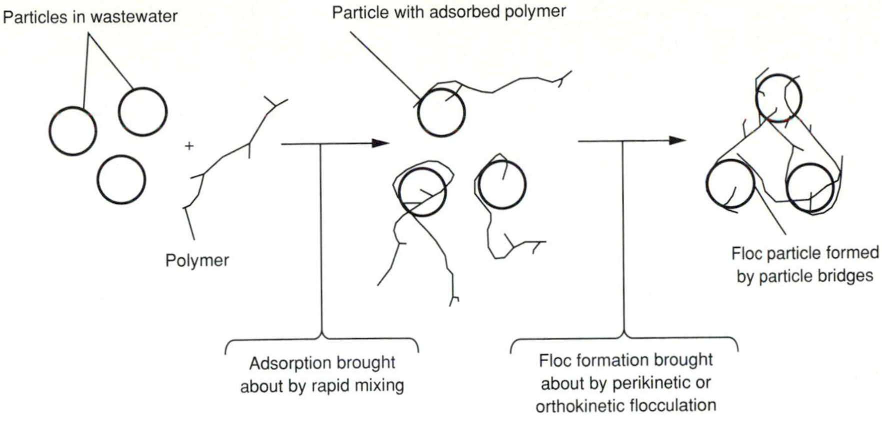

FIGURE 7-4
Definition sketch for interparticle bridging with organic polymers [19].

more particles become adsorbed along the length of the polymer. Bridged particles become intertwined with other bridged particles during the flocculation process. The size of the resulting three-dimensional particles grows until they can be removed easily by sedimentation.

The third type of polyelectrolyte action may be classified as a coagulation-bridging phenomenon, which results from using cationic polyelectrolytes of extremely high molecular weight. Besides lowering the charge, these polyelectrolytes also form particle bridges.

Metal Salt Polymer Formation. In contrast with the aggregation brought about by the addition of chemicals acting as electrolytes and polymers, aggregation brought about by the addition of alum or ferric sulfate is a more complex process. In the past, it was thought that free Al^{+3} and Fe^{+3} were responsible for the effects observed during particle aggregation; however, it is now known that their hydrolysis products are responsible [15,16]. Although the effect of these hydrolysis products is only now appreciated, it is interesting to note that their chemistry was first elucidated in the early 1900s by Pfeiffer (1902–1907), Bjerrum (1907–1920), and Werner (1907) [18]. For example, Pfeiffer proposed that the hydrolysis of trivalent metal salts, such as chromium, aluminum, and iron, could be represented as

$$
\begin{bmatrix}
 & H_2O & & OH_2 & \\
 & \backslash & / & \\
H_2O & - & Me & - & OH_2 \\
 & / & \backslash & \\
 & H_2O & & OH_2 &
\end{bmatrix}^{3+}
\Leftrightarrow
\begin{bmatrix}
 & H_2O & & OH_2 & \\
 & \backslash & / & \\
H_2O & - & Me & - & OH_2 \\
 & / & \backslash & \\
 & H_2O & & OH_2 &
\end{bmatrix}^{++}
+ \quad H^+ \quad (7\text{-}14)
$$

with the extent of the dissociation depending on the anion associated with the metal and on the physical and chemical characteristics of the solution. Further, it was proposed that, upon the addition of sufficient base, the dissociation can proceed to produce a negative ion [18] such as

$$
\begin{bmatrix} & H_2O & OH & \\ & \backslash & / & \\ H_2O & - Me & - OH \\ & / & \backslash & \\ & HO & OH & \end{bmatrix}^{-}
$$

Recently, however, it has been observed that the intermediate hydrolysis reactions of Al(III) are much more complex than would be predicted on the basis of a model in which a base is added to the solution. A hypothetical model proposed by Stumm [7] for Al(III) is shown in Eq. 7-15.

$$[Al(H_2O)_6]^{3+} \xrightarrow{\quad OH^- \quad} [Al(H_2O)_5OH]^{++} \xrightarrow{\quad OH^- \quad} [Al(H_2O)_4(OH)_2]^{+}$$

$$\text{———— } OH^- \text{————}$$

$$\text{———— } OH^- \text{————————————}$$

$$[Al_6(OH)_{15}]^{3+}(aq) \qquad \text{or} \qquad [Al_8(OH)_{20}]^{4+}(aq) \xrightarrow{\quad OH^- \quad}$$

$$[Al(OH)_3(H_2O)_3]\ (s) \xrightarrow{\quad OH^- \quad} [Al(OH)_4(H_2O)_2]^{-} \qquad (7\text{-}15)$$

Before the reaction proceeds to the point where a negative ion is produced, polymerization as depicted in the following formula will usually take place [18].

$$
2\begin{bmatrix} H_2O & OH \\ \backslash & / \\ H_2O - Me & - OH_2 \\ / & \backslash \\ H_2O & OH_2 \end{bmatrix}^{++} \Leftrightarrow \begin{bmatrix} & H & \\ & / O \backslash & \\ (H_2O_4 Me & & Me\ (H_2O)_4 \\ & \backslash O / & \\ & H & \end{bmatrix}^{4+} + 2\ H_2O \qquad (7\text{-}16)
$$

The possible combinations of the various hydrolysis products is endless, and their enumeration is not the purpose here. What is important, however, is the realization that one or more of the hydrolysis products may be responsible for the observed action of aluminum or iron. Further, because the hydrolysis reactions follow a stepwise process, the effectiveness of aluminum and iron will vary with time. For example, an alum slurry that has been prepared and stored will behave differently from a freshly prepared solution when it is added to a wastewater. For a more detailed review of the chemistry involved, the excellent articles on this subject by Stumm and Morgan [15] and Stumm and O'Melia [16] are recommended.

7-2 ADSORPTION

Adsorption, in general, is the process of collecting soluble substances that are in solution on a suitable interface. The interface can be between the liquid and a gas, a solid, or another liquid. Although adsorption is used at the air-liquid interface in the flotation process, only the case of adsorption at the liquid-solid interface will be considered in this discussion. Gas phase adsorption of the volatile odorous compounds and trace organic pollutants that may be emitted from various wastewater operations and processes is considered in Sec. 9-12 in Chap. 9.

In the past, the adsorption process has not been used extensively in wastewater treatment, but demands for a better quality of treated wastewater effluent have led to an intensive examination and use of the process of adsorption on activated carbon. Activated-carbon treatment of wastewater is usually thought of as a polishing process for water that has already received normal biological treatment. The carbon in this case is used to remove a portion of the remaining dissolved organic matter. Depending on the means of contacting the carbon with the water, the particulate matter that is present may also be removed.

Activated-Carbon and Its Use

The nature of activated carbon, the use of granular carbon and powdered carbon for wastewater treatment, and carbon regeneration are discussed below.

Activated-Carbon Production. Activated carbon is prepared by first making a char from materials such as almond, coconut, and walnut hulls, other woods, and coal. The char is produced by heating the material to a red heat in a retort to drive off the hydrocarbons but with an insufficient supply of air to sustain combustion. The char particle is then activated by exposure to an oxidizing gas at a high temperature. This gas develops a porous structure in the char and thus creates a large internal surface area (see Fig. 7-5). The surface properties that result are a function of both the initial material used and the exact preparation procedure, so that many variations are possible. The type of base material from which the activated carbon is derived may also affect the pore-size distribution and the regeneration characteristics. After activation, the carbon can be separated into or prepared in different sizes with different adsorption capacities. The two size classifications are powdered, which has a diameter of less than 200 mesh, and granular, which has a diameter greater than 0.1 mm.

Treatment with Granular Activated Carbon (GAC). A fixed-bed column is often used as a means of contacting wastewater with GAC. A schematic of a typical

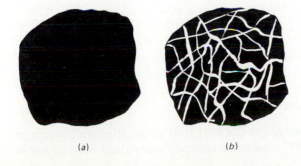

(a)   (b)

FIGURE 7-5
Sketch of activated carbon before and after activation: (a) before activation and (b) after activation.

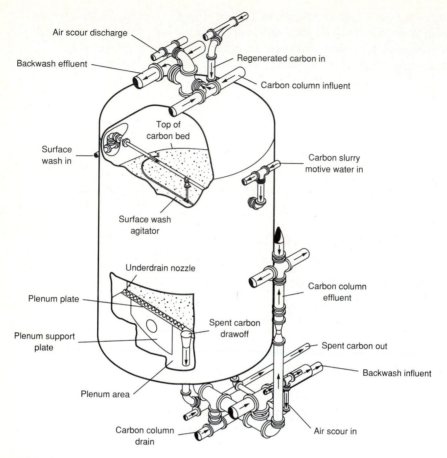

FIGURE 7-6
Typical granular activated-carbon contactor.

activated-carbon contactor used in the treatment of wastewater is shown in Fig. 7-6. The water is applied to the top of the column and withdrawn at the bottom. The carbon is held in place with an underdrain system at the bottom of the column. Provision for backwashing and surface washing is usually necessary to limit the headloss buildup due to the removal of particulate material within the carbon column. Fixed-bed columns can be operated singly, in series, or in parallel (see Fig. 7-7).

Expanded-bed and moving-bed carbon contactors have also been developed to overcome the problems associated with headloss buildup. In the expanded-bed system, the influent is introduced at the bottom of the column and is allowed to expand, much as a filter bed expands during backwash. In the moving-bed system, spent carbon is displaced continuously with fresh carbon. In such a system, headloss does not build up with time after the operating point has been reached.

Treatment with Powdered Activated Carbon (PAC). An alternative means of application is that of adding PAC. Powdered activated carbon has been added to

FIGURE 7-7
Granular actived-carbon contactors used for the treatment of filtered secondary effluent.

the effluent from biological treatment processes, directly to the various biological treatment processes, and in physical-chemical treatment process-flow diagrams. In the case of biological-treatment-plant effluent, PAC is added to the effluent in a contacting basin. After a certain amount of time for contact, the carbon is allowed to settle to the bottom of the tank, and the treated water is then removed from the tank. Because carbon is very fine, a coagulant such as a polyelectrolyte may be needed to aid the removal of the carbon particles, or filtration through granular-medium filters may be required. The addition of PAC directly in the aeration basin of an activated-sludge treatment process has proved to be effective in the removal of a number of soluble refractory organics (see Fig. 11-32 and discussion in Chap. 11).

Carbon Regeneration. Economical application of carbon depends on an efficient means of regenerating the carbon after its adsorptive capacity has been reached. Granular carbon can be regenerated easily in a furnace by oxidizing the organic matter and thus removing it from the carbon surface. Some of the carbon (about 5 to 10 percent) is also destroyed in the regeneration process and must be replaced with new or virgin carbon. The capacity of regenerated carbon is slightly less than that of virgin carbon. A major problem with the use of powdered activated carbon is that the methodology for its regeneration is not well-defined. The use of powdered activated carbon produced from recycled solid wastes may obviate the need to regenerate the spent carbon.

Analysis of the Adsorption Process

The adsorption process takes place in three steps: macrotransport, microtransport, and sorption. Macrotransport involves the movement of the organic material through the water to the liquid-solid interface by advection and diffusion. Microtransport involves the diffusion of the organic material through the macropore system of the GAC to the adsorption sites in the micropores and submicropores of the GAC granule. Adsorption occurs on the surface of the granule and in the macropores and mesopores, but the surface area of these parts of the GAC granule are so small compared with the surface area of the micropores and submicropores that the amount of material adsorbed there is usually considered negligible. Sorption is the term used to describe the attachment of the organic material to the GAC. The term *sorption* is used because it is difficult to differentiate between chemical and physical adsorption. When the rate of sorption equals the rate of desorption, equilibrium has been achieved and the capacity of the carbon has been reached. The theoretical adsorption capacity of the carbon for a particular contaminant can be determined by calculating its adsorption isotherm.

The quantity of adsorbate that can be taken up by an adsorbent is a function of both the characteristics and concentration of adsorbate and the temperature. Generally, the amount of material adsorbed is determined as a function of the concentration at a constant temperature, and the resulting function is called an *adsorption isotherm*. Equations that are often used to describe the experimental isotherm data were developed by Freundlich, by Langmuir, and by Brunauer, Emmet, and Teller (BET isotherm) [13,20]. Of the three, the Freundlich isotherm is used most commonly to describe the adsorption characteristics of the activated carbon used in water and wastewater treatment.

Freundlich Isotherm. The empirically derived Freundlich isotherm is defined as follows.

$$\frac{x}{m} = K_f C_e^{1/n} \tag{7-17}$$

where x/m = amount adsorbate adsorbed per unit weight of absorbent (carbon)
  C_e = equilibrium concentration of adsorbate in solution after adsorption
  K_f, n = empirical constants

The constants in the Freundlich isotherm can be determined by plotting (x/m) versus C and making use of Eq. 7-17 rewritten as

$$\log\left(\frac{x}{m}\right) = \log K_f + \frac{1}{n}\log C_e \tag{7-18}$$

Application of the Freundlich adsorption isotherm is illustrated in Example 7-2.

Langmuir Isotherm. Derived from rational considerations, the Langmuir adsorption isotherm is defined as

$$\frac{x}{m} = \frac{abC_e}{1 + bC_e} \tag{7-19}$$

where x/m = amount adsorbed per unit weight of adsorbent (carbon)

a, b = empirical constants

C_e = equilibrium concentration of adsorbate in solution after adsorption

The Langmuir adsorption isotherm was developed by assuming that (1) a fixed number of accessible sites are available on the adsorbent surface, all of which have the same energy and that (2) adsorption is reversible. Equilibrium is reached when the rate of adsorption of molecules onto the surface is the same as the rate of desorption of molecules from the surface. The rate at which adsorption proceeds, then, is proportional to the driving force, which is the difference between the amount adsorbed at a particular concentration and the amount that can be adsorbed at that concentration. At the equilibrium concentration, this difference is zero.

Correspondence of experimental data to the Langmuir equation does not mean that the stated assumptions are valid for the particular system being studied because departures from the assumptions can have a canceling effect. The constants in the Langmuir isotherm can be determined by plotting $C/(x/m)$ versus C and making use of Eq. 7-19 rewritten as

$$\frac{C_e}{(x/m)} = \frac{1}{ab} + \frac{1}{a}C_e \qquad (7\text{-}20)$$

Application of the Langmuir adsorption isotherm is illustrated in Example 7-2.

Example 7-2 Analysis of activated-carbon adsorption data. Determine the Freundlich and Langmuir isotherm coefficients for the following GAC adsorption test data. The liquid volume used in the batch adsorption tests was 1 L.

Mass of GAC in solution, m (g)	Equilibrium concentration of adsorbate in solution, C_e (mg/L)
0.0	3.37
0.001	3.27
0.010	2.77
0.100	1.86
0.500	1.33

Solution

1. Derive the values needed to plot the Freundlich and Langmuir adsorption isotherm using the batch adsorption test data.

Contaminant mass, mg				x/m,	
C_0	C_e	x	m, g	mg/mg	$C_e/(x/m)$
3.37	3.37	0.00	0.000	—	—
3.37	3.27	0.10	0.001	0.1000	32.7
3.37	2.77	0.60	0.010	0.0600	41.2
3.37	1.86	1.51	0.100	0.0151	123.2
3.37	1.33	2.04	0.500	0.0041	324.4

2. Plot the Freundlich and Langmuir adsorption isotherms using the data developed in Step 1. See following figures.

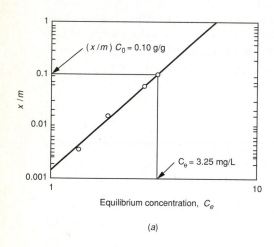

(a)

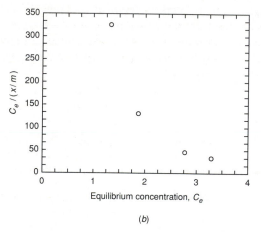

(b)

3. Determine the adsorption isotherms' coefficients.

   (a) Freundlich

   When $C_e = 1.0$, $(x/m) = 0.0015$, and $K_f = 0.0015$
   When $(x/m) = 1.0$, $C_e = 6.2$, and $1/n = 3.56$. Thus,
   $(x/m) = 0.0015 C_e^{3.56}$

   (b) Langmuir

   Because the plot for the Langmuir isotherm is curvalinear, use of the Langmuir adsorption isotherm is inappropriate.

Adsorption of Mixtures. In the application of adsorption to wastewater treatment, mixtures of organic compounds are always encountered. Typically, there is a depression of the adsorptive capacity of any individual compound in a solution of many compounds, but the total adsorptive capacity of the adsorbent may be larger than the adsorptive capacity with a single compound. The amount of inhibition due to competing adsorbates is related to the size of the molecules being adsorbed, their adsorptive affinities, and their relative concentrations. The adsorption from mixtures is considered further in Refs. 2 and 3.

Process Analysis

As noted previously, both granular carbon (in downflow and upflow columns) and powdered activated carbon are used for wastewater treatment. The analysis procedures for both types are described briefly in the following discussion.

Mass-Transfer Zone. The area of the GAC bed in which sorption occurs is called the *mass transfer zone* (MTZ) (see Fig. 7-8). After the contaminated water passes through a region of the bed whose depth is equal to the MTZ, the concentration of the contaminant in the water will have been reduced to its minimum value. No further adsorption will occur within the bed below the MTZ. As the top layers of carbon granules become saturated with organic material, the MTZ will move down in the bed until breakthrough occurs. A certain minimum empty bed contact time is required for the MTZ to be developed fully within the GAC bed. If the empty bed contact time of the column is too short (i.e., the hydraulic loading rate is too great), the length of the MTZ will be larger than the GAC bed depth, and the adsorbable contaminant will not be completely removed by the carbon.

   The thickness of the MTZ varies with the flowrate because dispersion, diffusion, and channeling in a granular medium are directly related to the flowrate. The only way to use the capacity at the bottom of the column is to have two columns in series and switch them as they are exhausted, or to use multiple columns in parallel. The

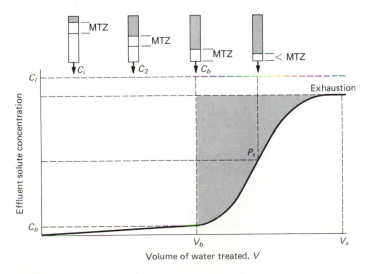

FIGURE 7-8
Typical breakthrough curve for activated carbon showing movement of mass-transfer zone (MTZ) with throughput volume.

optimum flowrate and bed depth, as well as the operating capacity of the carbon, must be established to determine the dimensions and the number of columns necessary for continuous treatment. Because these parameters can be determined only from dynamic column tests, pilot plant studies are recommended.

Carbon Adsorption Capacity. The adsorptive capacity of a given carbon is estimated from isotherm data as follows. If isotherm data are plotted, the resulting isotherm will be as shown in Fig. 7-9. Using this figure, the adsorptive capacity of the carbon can be estimated by extending a vertical line from the point on the horizontal scale corresponding to the initial concentration C_o and extrapolating the isotherm to intersect this line. The $(x/m)C_o$ value at the point of intersection can be read from the vertical scale. This $(x/m)C_o$ value represents the amount of constituent adsorbed per unit weight of carbon when the carbon is at equilibrium with the initial concentration of constituent. This condition should exist in the upper section of a carbon bed during column treatment, and it therefore represents the ultimate capacity of the carbon for a particular waste.

Breakthrough Adsorption Capacity. In the field, the breakthrough adsorption capacity, $(x/m)_b$, of the GAC in a full-scale column is some percentage of the theoretical adsorption capacity found from the isotherm. The $(x/m)_b$ of a single column can be assumed to be approximately 25 to 50 percent of the theoretical capacity $(x/m)_o$[12]. Once $(x/m)_b$ is known, the time to breakthrough can be calculated by solving the following equation for t_b [12].

$$\left(\frac{x}{m}\right)_b = \frac{X_b}{M_c} = Q\left(C_i - \frac{C_b}{2}\right)\frac{t_b}{M_c}[8.34 \text{ lb/Mgal} \cdot (\text{mg/L})] \qquad (7\text{-}21)$$

where $(x/m)_b$ = field breakthrough adsorption capacity, lb/lb or g/g

X_b = mass of organic material adsorbed in the GAC column at breakthrough, lb or g

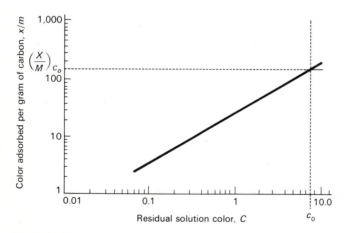

FIGURE 7-9
Typical decolorization isotherm.

M_c = mass of carbon in the column, lb or g
Q = flow rate, Mgal/d
C_i = influent organic concentration, mg/L
C_b = breakthrough organic concentration, mg/L
t_b = time to breakthrough, d

Equation 7-21 was developed assuming that C_i is constant and that the effluent concentration increases linearly with time from 0 to C_b (see Fig. 7-9). Rearranging Eq. 7-21, the time to breakthrough can be calculated using the following relationship.

$$t_b = \frac{(x/m)_b M_c}{Q\left[C_i - (C_b/2)\right]\left[8.34 \text{ lb/Mgal} \cdot (\text{mg/L})\right]} \qquad (7\text{-}22)$$

The application of Eq. 7-22 to the design of a carbon column is illustrated in Example 7-3.

As noted previously, because of the breakthrough phenomenon (see Fig. 7-8), the usual practice is either to use two columns in series and rotate them as they become exhausted, or to use multiple columns in parallel so that breakthrough in a single column will not significantly affect the effluent quality. With proper sampling from points within the column, TOC breakthrough can be anticipated.

Powdered Activated Carbon. For a powdered-carbon application, the isotherm adsorption data can be used in conjunction with a materials-balance analysis to obtain an estimate of the amount of carbon that must be added [20]. Here again, because of the many unknown factors involved, pilot plant tests to develop the optimum design data are recommended.

Example 7-3 Estimation of activated-carbon adsorption breakthrough time. Determine the breakthrough time for a GAC filter column when operated at a filtration rate of 5.0 gal/ft^2 · min. Assume that the surface area of the filter column is 10 ft^2, that the depth of the filter column is 5.0 ft, and that the carbon adsorption data given in Example 7-2 are applicable. The influent TOC concentration is 3.25 mg/L and the breakthrough TOC concentration has been set at 0.75 mg/L. The density of the GAC to be used in the filter columns is 38 lb/ft^3 (600 kg/m^3).

Solution

1. Using the Freundlich adsorption isotherm plotted in Example 7-2, find the value of the theoretical capacity $(x/m)_0$ at an influent TOC concentration of 3.25 mg/L. From Example 7-3,

$$\frac{x}{m} = 0.0015 \, C_e^{3.56}$$

$$(x/m)_0 = 0.0015(3.25)^{3.56} = 0.0996 \text{ mg/mg, say } 0.10 \text{ mg/mg} = 0.10 \text{ lb/lb}$$

2. Determine the breakthrough time using Eq. 7-22.

$$t_b = \frac{(x/m)_b M_c}{Q(C_i - C_b/2)\left[8.34 \text{ lb/Mgal} \cdot (\text{mg/L})\right]}$$

(a) Assume that the following conditions apply:

$(x/m)_b$ = 50 percent of $(x/m)_o$ = 0.5(0.10 lb/lb) = 0.050 lb/lb

Surface area = 10.0 ft^2

M_c = (10.0 ft^2) × (5.0 ft) × 38 lb/ft^3 = 1,900 lb

Q = 5.0 gal/ft^2 · min × 1440 min/d × 10 ft^2 = 72,000 gal/d = 0.072 Mgal/d

C_i = 3.2 mg/L

C_b = 0.75 mg/L

(b) The time to breakthough is

$$t_b = \frac{(0.050 \text{ lb/lb})(1,900 \text{ lb})}{(0.072 \text{ Mgal/d})(3.2 \text{ mg/L} - 0.375 \text{ mg/L})[8.34 \text{ lb/Mgal} \cdot (\text{mg/L})]} = 56.0 \text{ d}$$

Comment. The use of multiple carbon columns that can be operated in parallel and rotated as they become exhausted will improve the effectiveness of the process.

7-3 DISINFECTION

Disinfection refers to the selective destruction of disease-causing organisms. All the organisms are not destroyed during the process. This differentiates disinfection from sterilization, which is the destruction of all organisms. In the field of wastewater treatment, the three categories of human enteric organisms of the greatest consequence in producing disease are bacteria, viruses, and amoebic cysts. Diseases caused by waterborne bacteria include typhoid, cholera, paratyphoid, and bacillary dysentery; diseases caused by waterborne viruses include poliomyelitis and infectious hepatitis (see Table 3-12). The purpose in this section is to introduce the reader to the general concepts involved in the disinfection of microorganisms. The remaining sections of this chapter deal with disinfection using chlorine (Sec. 7-4), dechlorination (Sec. 7-5), chlorine dioxide (Sec. 7-6), bromine chloride (Sec. 7-7), ozone (Sec. 7-8), and UV radiation (Sec. 7-9).

Description of Disinfection Methods and Means

The requirements for an ideal chemical disinfectant are reported in Table 7-3. As shown, an ideal disinfectant would have to possess a wide range of characteristics. Although such a compound may not exist, the requirements set forth in Table 7-3 should be considered in evaluating proposed or recommended disinfectants. It is also important that the disinfectant be safe to handle and apply and that its strength or concentration in treated waters be measurable. Disinfection is most commonly accomplished by the use of (1) chemical agents, (2) physical agents, (3) mechanical means, and (4) radiation. Each of these techniques is considered in the the following discussion.

Chemical Agents. Chemical agents that have been used as disinfectants include (1) chlorine and its compounds, (2) bromine, (3) iodine, (4) ozone, (5) phenol and phenolic compounds, (6) alcohols, (7) heavy metals and related compounds, (8) dyes, (9) soaps and synthetic detergents, (10) quaternary ammonium compounds, (11) hydrogen peroxide, and (12) various alkalies and acids.

Of these, the most common disinfectants are the oxidizing chemicals, and chlorine is the one most universally used. Bromine and iodine have also been used for wastewater disinfection. Ozone is a highly effective disinfectant, and its use is increasing even though it leaves no residual (see Sec. 7-8). Highly acidic or alkaline water can also be used to destroy pathogenic bacteria because water with a pH greater than 11 or less than 3 is relatively toxic to most bacteria.

Physical Agents. Physical disinfectants that can be used are heat and light. Heating water to the boiling point, for example, will destroy the major disease-producing nonspore-forming bacteria. Heat is commonly used in the beverage and dairy industry, but it is not a feasible means of disinfecting large quantities of wastewater because of the high cost. However, pasteurization of sludge is used extensively in Europe.

Sunlight is also a good disinfectant. In particular, ultraviolet radiation can be used. Special lamps that emit ultraviolet rays have been used successfully to sterilize small quantities of water. The efficiency of the process depends on the penetration of the rays into water. The contact geometry between the ultraviolet-light source and the water is extremely important because suspended matter, dissolved organic molecules, and water itself, as well as the microorganisms, will absorb the radiation. It is therefore difficult to use ultraviolet radiation in aqueous systems, especially when large amounts of particulate matter are present.

Mechanical Means. Bacteria and other organisms are also removed by mechanical means during wastewater treatment. Typical removal efficiencies for various treatment operations and processes are reported in Table 7-4. The first four operations listed may be considered to be physical. The removals accomplished are a by-product of the primary function of the process.

Radiation. The major types of radiation are electromagnetic, acoustic, and particle. Gamma rays are emitted from radioisotopes such as cobalt 60. Because of their penetration power, gamma rays have been used to disinfect (sterilize) both water and wastewater. A schematic diagram of a high-energy electron-beam device for the irradiation of wastewater or sludge is shown in Fig. 7-10 [5].

Mechanisms of Disinfectants

Four mechanisms that have been proposed to explain the action of disinfectants are (1) damage to the cell wall, (2) alteration of cell permeability, (3) alteration of the colloidal nature of the protoplasm, and (4) inhibition of enzyme activity [10].

Damage or destruction of the cell wall will result in cell lysis and death. Some agents, such as penicillin, inhibit the synthesis of the bacterial cell wall.

Agents such as phenolic compounds and detergents alter the permeability of the cytoplasmic membrane. These substances destroy the selective permeability of the membrane and allow vital nutrients, such as nitrogen and phosphorus, to escape.

TABLE 7-3
Comparison of ideal and actual characteristics of commonly used disinfectants[a]

Characteristic	Properties/response	Chlorine	Sodium hypochlorite
Toxicity to microorganisms	Should be highly toxic at high dilutions	High	High
Solubility	Must be soluble in water or cell tissue	Slight	High
Stability	Loss of germicidal action on standing should be low	Stable	Slightly unstable
Nontoxic to higher forms of life	Should be toxic to microorganisms and nontoxic to man and other animals	Highly toxic to higher life forms	Toxic
Homogeneity	Solution must be uniform in composition	Homogeneous	Homogeneous
Interaction with extraneous material	Should not be absorbed by organic material other than bacterial cells	Oxidizes organic matter	Active oxidizer
Toxicity at ambient temperatures	Chould be effective in ambient temperature range	High	High
Penetration	Should have the capacity to penetrate through surfaces	High	High
Noncorrosive and nonstaining	Should not disfigure metals or stain clothing	Highly corrosive	Corrosive
Deodorizing ability	Should deodorize while disinfecting	High	Moderate
Availability	Should be available in large quantities and reasonably priced	Low cost	Moderately low cost

[a] Adapted from Refs. 10, 24, and 25.

Heat, radiation, and highly acidic or alkaline agents alter the colloidal nature of the protoplasm. Heat will coagulate the cell protein and acids or bases will denature proteins, producing a lethal effect.

Another mode of disinfection is the inhibition of enzyme activity. Oxidizing agents, such as chlorine, can alter the chemical arrangement of enzymes and deactivate the enzymes.

TABLE 7-3
(*continued*)

Calcium hypochlorite	Chlorine dioxide	Bromine chloride	Ozone	UV radiation
High	High	High	High	High
High	High	Slight	High	N/A
Relatively stable	Unstable, must be generated as used	Slightly unstable	Unstable, must be generated as used	Must be generated as used
Toxic	Toxic	Toxic	Toxic	Toxic
Homogeneous	Homogeneous	Homogeneous	Homogeneous	N/A
Active oxidizer	High	Oxidizes organic matter	Oxidizes organic matter	
High	High	High	High	High
High	High	High	High	Moderate
Corrosive	Highly corrosive	Corrosive	Highly corrosive	N/A
Moderate	High	Moderate	High	
Moderately low cost	Moderately low cost	Moderately low cost	Moderately high cost	Moderately high cost

Analysis of Factors Influencing the Action of Disinfectants

In applying the disinfection agents or means that have been described, the following factors must be considered: (1) contact time, (2) concentration and type of chemical agent, (3) intensity and nature of physical agent, (4) temperature, (5) number of organisms, (6) types of organisms, and (7) nature of suspending liquid [10].

TABLE 7-4
Removal or destruction of bacteria by different treatment processes

Process	Percent removal
Coarse screens	0–5
Fine screens	10–20
Grit chambers	10–25
Plain sedimentation	25–75
Chemical sedimentation	40–80
Trickling filters	90–95
Activated sludge	90–98
Chlorination of treated wastewater	98–99

Contact Time. Perhaps one of the most important variables in the disinfection process is contact time. In general, as shown in Fig. 7-11, it has been observed that for a given concentration of disinfectant, the longer the contact time, the greater the kill. This observation was first formalized in the literature by Chick [1]. In differential form, Chick's law is

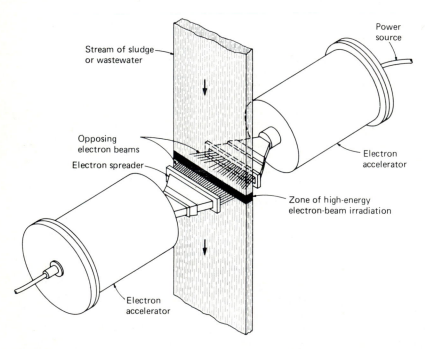

FIGURE 7-10
Schematic diagram of high-energy electron-beam device for the irradiation of wastewater or sludge [15].

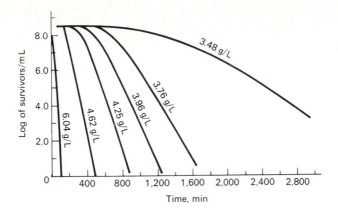

FIGURE 7-11
Effect of time and concentration on survival of E. coli using phenol as disinfectant at 35°C [1].

$$\frac{dN}{dt} = -kN_t \qquad (7\text{-}23)$$

where N_t = number of organisms at time t
t = time
k = constant, time^{-1}

If N_0 is the number of organisms when t equals 0, Eq. 7-23 can be integrated to

$$\frac{N_t}{N_o} = e^{-kt} \qquad (7\text{-}24)$$

or

$$\ln\frac{N_t}{N_o} = -kt \qquad (7\text{-}25)$$

Departures from this rate law are common. Rates of kill have been found to increase with time in some cases and to decrease with time in other cases. To formulate a valid relationship for the kill of organisms under a variety of conditions, an assumption often made is that

$$\ln\frac{N_t}{N_o} = -kt^m \qquad (7\text{-}26)$$

where m is a constant. If m is less than 1, the rate of kill decreases with time, and, if m is greater than 1, the rate of kill increases with time. The constants in Eq. 7-26 can be obtained by plotting $-\ln (N/N_0)$ versus the contact time t on log-log paper. The straight-line form of the equation is

$$\log\left(-\ln\frac{N_t}{N_o}\right) = \log k + m \log t \qquad (7\text{-}27)$$

Another formulation that has been used to describe the observed effects of contact time is

$$\frac{N_t}{N_o} = kt^m \qquad (7\text{-}28)$$

Equation 7-28 results from the analysis of chlorination data that have been found to plot as straight lines on log-log paper.

Concentration and Type of Chemical Agent. Depending on the type of chemical agent, it has been observed that, within limits, disinfection effectiveness is related to concentration. The effect of concentration has been formulated empirically [6]:

$$C^n t_p = \text{constant} \qquad (7\text{-}29)$$

where C = concentration of disinfectant
   n = constant
   t_p = time required to effect a constant percentage kill

The constants in Eq. 7-29 can be evaluated by plotting on log-log paper the concentration versus the time required to effect a given percentage kill. The slope of the line then corresponds to the value of $-1/n$. In general, if n is greater than 1, contact time is more important than the dosage; if n equals 1, the effects of time and dosage are about the same [6].

Intensity and Nature of Physical Agent. As noted earlier, heat and light are physical agents that have been used from time to time in the disinfection of wastewater. It has been found that their effectiveness is a function of intensity. For example, if the decay of organisms can be described with a first-order reaction such as

$$\frac{dN}{dt} = -kN \qquad (7\text{-}30)$$

where N = number of organisms
   t = time, min
   k = reaction velocity of constant, 1/min

then the effect of the intensity of the physical disinfectant is reflected in the constant k through some functional relationship.

Temperature. The effect of temperature on rate of kill can be represented by a form of the van't Hoff-Arrhenius relationship. Increasing the temperature results in a more rapid kill. In terms of the time t required to effect a given percentage kill, the relationship is

$$\ln\frac{t_1}{t_2} = \frac{E(T_2 - T_1)}{RT_1 T_2} \qquad (7\text{-}31)$$

where t_1, t_2 = time for given percentage kill at temperatures T_1 and T_2, °K, respectively

E = activation energy, J/mol (cal/mol)

R = gas constant, 8.314 J/mol · °K (1.99 cal/°K · mol)

Some typical values for the activation energy for various chlorine compounds at different pH values are reported in Table 7-5.

Number of Organisms. In a dilute system such as wastewater, the concentration of organisms is seldom a major consideration. However, it can be concluded from Eq. 7-29 that the larger the organism concentration, the longer the time required for a given kill. An empirical relationship that has been proposed to describe the effect of organism concentration is [6]

$$C^q N_p = \text{constant} \tag{7-32}$$

where C = concentration of disinfectant

N_p = concentration of organisms reduced by a given percentage in given time

q = constant related to strength of disinfectant

Types of Organisms. The effectiveness of various disinfectants will be influenced by the nature and condition of the microorganisms. For example, viable growing bacteria cells are killed easily. In contrast, bacterial spores are extremely resistant, and many of the chemical disinfectants normally used will have little or no effect. Other disinfecting agents, such as heat, may have to be used. This subject is considered further in Sec. 7-4.

Nature of Suspending Liquid. In addition to the foregoing factors, the nature of the suspending liquid must be evaluated carefully. For example, extraneous organic material will react with most oxidizing disinfectants and reduce their effectiveness.

TABLE 7-5
Activation energies for aqueous chlorine and chloramines at normal temperatures[a]

Compound	pH	E, cal/mol[b]
Aqueous chlorine	7.0	8,200
	8.5	6,400
	9.8	12,000
	10.7	15,000
Chloramines	7.0	12,000
	8.5	14,000
	9.5	20,000

[a] From Ref. 6.
[b] Cal × 4.1876 = J.

Turbidity will reduce the effectiveness of disinfectants by absorption and by protecting entrapped bacteria.

7-4   DISINFECTION WITH CHLORINE

As noted earlier, of all the chemical disinfectants, chlorine is perhaps the one most commonly used throughout the world. The reason is that it satisfies most of the requirements specified in Table 7-3. Because the practical aspects of chlorination are discussed in Chap. 9, the following discussion is limited to a brief description of chlorine chemistry and breakpoint chlorination, and an analysis of the performance of chlorine as a disinfectant and the factors that may influence the effectiveness of the chlorination process.

Chlorine Chemistry

The most common chlorine compounds used in wastewater treatment plants are chlorine gas (Cl_2), calcium hypochlorite [$Ca(OCl)_2$], sodium hypochlorite ($NaOCl$), and chlorine dioxide (ClO_2). Calcium and sodium hypochlorite are most often used in very small treatment plants, such as package plants, where simplicity and safety are far more important than cost. Sodium hypochlorite is often used at large facilities, primarily for reasons of safety as influenced by local conditions. Because chlorine dioxide has some unusual properties (it does not react with ammonia), it is also used in a number of treatment facilities. Even though a variety of other chlorine compounds are used, the discussion in this section is based primarily on the use of chlorine in the form of a gas because it is the most commonly used form.

Reactions in Water. When chlorine in the form of Cl_2 gas is added to water, two reactions take place: hydrolysis and ionization.

Hydrolysis may be defined as

$$Cl_2 + H_2O \Leftrightarrow HOCl + H^+ + Cl^- \tag{7-33}$$

The stability constant for this reaction is

$$K = \frac{[HOCl][H^+][Cl^-]}{[Cl_2]} \approx 4.5 \times 10^{-4} \text{ at } 25°C \tag{7-34}$$

Because of the magnitude of this coefficient, large quantities of chlorine can be dissolved in water.

Ionization may be defined as

$$HOCl \Leftrightarrow H^+ + OCl^- \tag{7-35}$$

The ionization constant for this reaction is

$$K_i = \frac{[H^+][OCl^-]}{[HOCl]} = 2.9 \times 10^{-8} \text{ at } 25°C \tag{7-36}$$

The variation in the value of K_i with temperature is reported in Table 7-6.

TABLE 7-6
Values of the ionization constant of hypochlorous acid at different temperatures[a]

Temperature, °C	$K_i \times 10^8$, mol/L
0	1.49
5	1.75
10	2.03
15	2.32
20	2.62
25	2.90

[a] From Ref. 25.

Note: 1.8 (°C) + 32 = °F

The quantity of HOCl and OCl⁻ that is present in water is called the *free available chlorine*. The relative distribution of these two species (see Fig. 7-12) is very important because the killing efficiency of HOCl is about 40 to 80 times that of OCl⁻. The percentage distribution of HOCl at various temperatures can be computed using Eq. 7-37 and the data in Table 7-6.

$$\frac{[\text{HOCl}]}{[\text{HOCl}] + [\text{OCl}^-]} = \frac{1}{1 + [\text{OCl}^-]/[\text{HOCl}]} = \frac{1}{1 + K_i/[\text{H}^+]} \qquad (7\text{-}37)$$

Free chlorine can also be added to water in the form of hypochlorite salts. The pertinent reactions are as follows:

$$\text{Ca(OCl)}_2 + 2\text{H}_2\text{O} \rightarrow 2\text{HOCl} + \text{Ca(OH)}_2 \qquad (7\text{-}38)$$

$$\text{NaOCl} + \text{H}_2\text{O} \rightarrow \text{HOCl} + \text{NaOH} \qquad (7\text{-}39)$$

Reactions with Ammonia. As noted in Chap. 3, untreated wastewater contains nitrogen in the form of ammonia and various combined organic forms. The effluent from most treatment plants also contains significant amounts of nitrogen, usually in

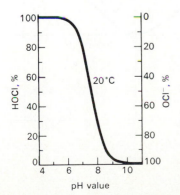

FIGURE 7-12
Distribution of hypochlorous acid and hypochlorite in water at different pH values.

the form of ammonia, or nitrate if the plant is designed to achieve nitrification (see Chaps. 8 and 11). Because hypochlorous acid is a very active oxidizing agent, it will react readily with ammonia in the wastewater to form three types of chloramines in the successive reactions:

$$NH_3 + HOCl \rightarrow NH_2Cl \text{ (monochloramine)} + H_2O \tag{7-40}$$

$$NH_2Cl + HOCl \rightarrow NHCl_2 \text{ (dichloramine)} + H_2O \tag{7-41}$$

$$NHCl_2 + HOCl \rightarrow NCl_3 \text{ (nitrogen trichloride)} + H_2O \tag{7-42}$$

These reactions are very dependent on the pH, temperature, contact time, and the ratio of chlorine to ammonia [25]. The two species that predominate, in most cases, are monochloramine (NH_2Cl) and dichloramine ($NHCl_2$). The chlorine in these compounds is called *combined available chlorine*. As will be discussed subsequently, these chloramines also serve as disinfectants, although they are extremely slow-reacting.

Breakpoint Reaction

The maintenance of a residual (combined or free) for the purpose of wastewater disinfection is complicated by the fact that free chlorine not only reacts with ammonia, as noted previously, but also is a strong oxidizing agent. The stepwise phenomena that result when chlorine is added to wastewater containing ammonia can be explained by referring to Fig. 7-13.

As chlorine is added, readily oxidizable substances, such as Fe^{+2}, Mn^{+2}, H_2S, and organic matter, react with the chlorine and reduce most of it to the chloride ion

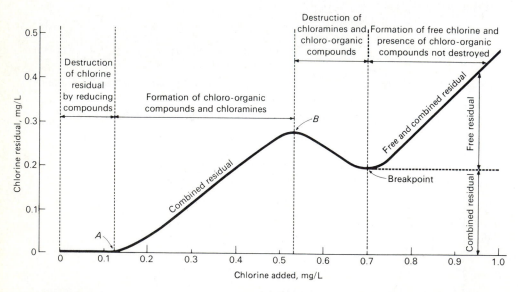

FIGURE 7-13
Generalized curve obtained during breakpoint chlorination.

(point A in Fig. 7-13). After meeting this immediate demand, the chlorine continues to react with the ammonia to form chloramines between points A and B. For mole ratios of chlorine to ammonia less than one, monochloramine and dichloramine will be formed. The distribution of these two forms is governed by their rates of formation, which are dependent on the pH and temperature. Between point B and the breakpoint, some chloramines will be converted to nitrogen trichloride (see Eq. 7-42), the remaining chloramines will be oxidized to nitrous oxide (N_2O) and nitrogen (N_2), and the chlorine will be reduced to the chloride ion. With continued addition of chlorine, most of the chloramines will be oxidized at the breakpoint. Theoretically, as determined in Example 7-4, the weight ratio of chlorine to ammonia nitrogen at the breakpoint is 7.6 : 1.

Possible reactions to account for the appearance of the aforementioned gases and the disappearance of chloramines are as follows (see also Eq. 7-42):

$$NH_2Cl + NHCl_2 + HOCl \rightarrow 4HCl \tag{7-43}$$

$$4NH_2Cl + 3Cl_2 + H_2O \rightarrow N_2 + N_2O + 10HCl \tag{7-44}$$

$$2NH_2Cl + HOCl \rightarrow N_2 + H_2O + 3HCl \tag{7-45}$$

$$NH_2Cl + NHCl \rightarrow N_2 + 3HCl \tag{7-46}$$

Continued addition of chlorine past the breakpoint, as shown in Fig. 7-14a, will result in a directly porportional increase in the free available chlorine (unreacted hypochlorite).

The main reason for adding enough chlorine to obtain a free chlorine residual is that usually disinfection can then be ensured. Occasionally, serious odor problems have developed during breakpoint-chlorination operations because of the formation of nitrogen trichloride and related compounds. The presence of additional compounds during chlorination will react with the alkalinity of the wastewater, and under most circumstances, the pH drop will be slight. The presence of additional compounds that will react with chlorine, such as organic nitrogen, may greatly alter the shape of the breakpoint curve, as shown in Fig. 7-14b. The amount of chlorine that must be added to reach a desired level of residual is called the *chlorine demand*.

Example 7-4 Breakpoint chlorination. Determine the stoichiometric weight ratio of chlorine to ammonia nitrogen at the breakpoint.

Solution

1. Write an overall reaction to describe the breakpoint phenomenon. This can be done using Eqs. 7-40 and 7-45.

$$2NH_3 + 2HOCl \rightarrow 2NH_2Cl + 2H_2O$$

$$\frac{2NH_2Cl + HOCl \rightarrow N_2 + H_2O + 3HCl}{2NH_3 + 3HOCl \rightarrow N_2 + 3H_2O + 3HCl}$$

2. Determine the molecular weight of the ammonia (NH_3) expressed as N and the hypochlorous acid (HOCl) expressed as Cl_2.

$$\text{Molecular weight of NH}_3 \text{ expressed as N} \quad = \frac{17}{17}(14 \text{ g/mol})$$
$$= 14$$

$$\text{Molecular weight of HOCl expressed as Cl}_2 \quad = \frac{52.45}{52.45}(70.9 \text{ g/mol})$$
$$= 70.9$$

3. Determine the weight ratio of chlorine to ammonia nitrogen.

$$\frac{Cl_2}{NH_3 - N} = \frac{3(70.9)}{2(14)} = \frac{7.6}{1}$$

Comment. The ratio computed in step 3 will vary somewhat, depending on the actual reactions involved, which at present are unknown. In practice, the actual ratio has been found to vary from 8:1 to 10:1.

Acid Generation. In practice, the hydrochloric acid formed during chlorination (see reaction given in Step 1 of Example 7-4) will react with the alkalinity of the wastewater, and under most circumstances, the pH drop will be slight. Stoichiometrically, 14.3 mg/L of alkalinity, expressed as $CaCO_3$, will be required for each 1.0 mg/L of ammonia nitrogen that is oxidized in the breakpoint-chlorination process. In practice, it has been found that about 15 mg/L of alkalinity are actually required because of the hydrolysis of chlorine [22].

Buildup of Total Dissolved Solids. In addition to the formation of hydrochloric acid, the chemicals added to achieve the breakpoint reaction will also contribute an incremental increase to the total dissolved solids of the wastewater. In situations where the level of total dissolved solids may be critical with respect to reuse applications, this incremental buildup from breakpoint chlorination should always be checked. The total dissolved-solids contribution for each of several chemicals that may be used in the breakpoint reaction is summarized in Table 7-7. The magnitude of the possible

TABLE 7-7
Effects of chemical addition on total dissolved solids in breakpoint chlorination

Chemical addition	Increase in total dissolved solids per unit of NH_4^+ consumed
Breakpoint with chlorine gas	6.2:1
Breakpoint with sodium hypochlorite	7.1:1
Breakpoint with chlorine gas-neutralization of all acidity with lime (CaO)	12.2:1
Breakpoint with chlorine gas-neutralization of all acidity with sodium hydroxide (NaOH)	14.8:1

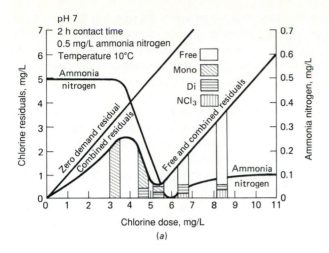

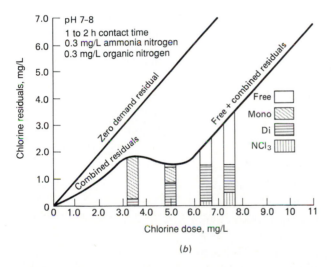

FIGURE 7-14

Curves of chlorine residual versus chlorine dosage for wastewater [adapted from Ref. 25]: (a) for wastewater containing nitrogen in the form of ammonia (NH_3) and (b) for wastewater containing nitrogen in the form of ammonia and organic nitrogen. *Note:* $1.8(°C) + 32 = °F$.

buildup of total dissolved solids is illustrated in Example 11-4, in which the use of breakpoint chlorination is considered for the seasonal control of nitrogen.

Factors That Affect Disinfection Efficiency of Chlorine

The purpose of the following discussion is to explore the important factors that affect the disinfection efficiency of chlorine to the extent that they are now known. These include (1) the germicidal efficiency of chlorine, (2) the germicidal efficiency of the various clorine compounds, (3) the importance of initial mixing, (4) the breakpoint reaction, (5) the contact time, (6) the characteristics of the wastewater, and (7) the characteristics of the microorganisms. To provide a framework in which to view these factors, it will be appropriate to consider first how the effectiveness of the chlorination process is now assessed and how the results are analyzed.

Germicidal Efficiency of Chlorine. When using chlorine for the disinfection of wastewater, the principal parameters that can be measured, apart from environmental variables such as pH and temperature, are the number of organisms and the chlorine residual remaining after a specified period of time. The coliform group of organisms can be determined using the most probable number (MPN) procedure or the plate count procedure as discussed in Chap. 3.

The chlorine residual (free and combined) should be measured using the amperometric method, which has proved to be the most consistently reliable method now available. Also, because almost all the commercial analyzers of residual chlorine use it, the adoption of this method will allow the results of independent studies to be compared directly. Numerous tests have shown that when all the physical parameters controlling the chlorination process are held constant, the germicidal efficiency of disinfection, as measured by bacterial survival, depends primarily on the residual bactericidal chlorine present R and the contact time t. It has also been found that by increasing either one of the two variables R or t and simultaneously decreasing the other one, it is possible to achieve approximately the same degree of disinfection. Thus, the efficiency of disinfection may be expressed as a function of the product $(R \times t)$.

Using the batch reactor whose contents were well-stirred, it has been found that the reductions of coliform organisms in a chlorinated primary treated effluent can be defined by the following relationship [25]:

$$\frac{N_t}{N_0} = (1 + 0.23 C_t t)^{-3} \tag{7-47}$$

where N_t = number of coliform organisms at time t
N_0 = number of coliform organisms at time t_0
C_t = total amperometric chlorine residual at time t, mg/L
t = residence time, min

The data from which this relationship was developed are shown in Fig. 7-15. The application of Eq. 7-47 is considered in Probs. 7-10 and 7-11.

Germicidal Efficiency of Various Chlorine Compounds. A comparison of the germicidal efficiency of hypochlorous acid (HOCl), hypochlorite ion (OCl), and monochloramine (NH_2Cl) is presented in Fig. 7-16. For a given contact time or residual, the germicidal efficiency of hypochlorous acid, in terms of either time or residual, is significantly greater than that of either the hypochlorite ion or monochloramine. However, it should be noted that, given an adequate contact time, monochloramine is nearly as effective as chlorine in achieving disinfection.

Referring to Fig. 7-16, it is clear that hypochlorous acid offers the most positive way of achieving disinfection. For this reason, with proper mixing, the formation of hypochlorous acid following breakpoint is most effective in achieving wastewater

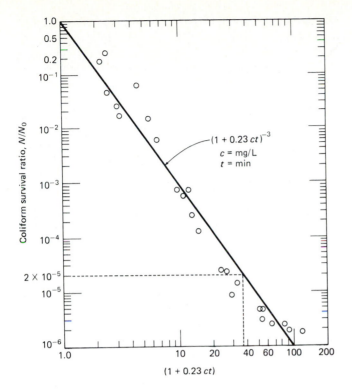

FIGURE 7-15

Coliform survival in a batch reactor as a function of amperomatic chlorine residual and contact time (temperature range 15–18°C) [25]. *Note:* 1.8 (°C) + 32 = °F.

chlorination. If sufficient chlorine cannot be added to achieve the breakpoint reaction, great care must be taken to ensure that the proper contact time is maintained. Because of the equilibrium between hypochlorous acid and the hypochloric ion, maintenance of the proper pH is also important if effective disinfection is to be achieved.

Initial Mixing. The importance of initial mixing on the disinfection process can not be overstressed. It has been shown that the application of chlorine in a highly turbulent regime ($N_R \geq 10^4$) will result in kills two orders of magnitude greater than when chlorine is added separately to a complete-mix reactor under similar conditions. Although the importance of initial mixing is well delineated, the optimum level of turbulence is not known. Mixing times on the order of one second are desirable. The design of mixing facilities is considered in Chap. 9.

Breakpoint Reaction. The basic aspects of the breakpoint reaction and its effects on the disinfection process have been discussed previously. The discussion here is concerned with the practice of using chlorinated wastewater for the chlorine injection water (see Chap. 9). The contention is that, if nitrogenous compounds are present in the wastewater, a portion of the chlorine that is added will react with these compounds

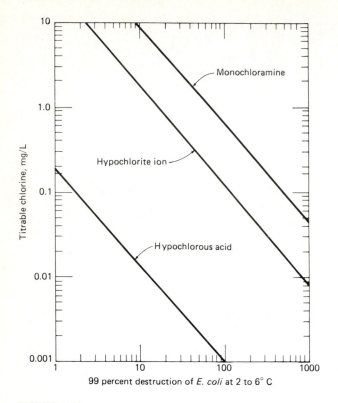

Titrable chlorine, mg/L

99 percent destruction of *E. coli* at 2 to 6° C

FIGURE 7-16
Comparison of the germicidal efficiency of hypochlorous acid, hypochlorite ion, and monochloramine for 99 percent destruction of *E. coli* at 2 to 6°C [25]. *Note:* 1.8(°C) + 32 = °F.

and that, by the time it is injected, it will be in the form of monochloramine or dichloramine. This can be a problem in small installations or where the chlorine solution lines from the chlorinator to the point of injection are quite long. It has been shown, however, that with proper initial mixing bacterial kills are the same whether untreated or treated effluent is used for the injector water supply [25]. From the evidence to date, proper initial mixing appears to be more important in achieving effective disinfections than the form in which the chlorine is injected. Again, it should be remembered that hypochlorous acid (HOCl) and monochloramine (NH_2Cl) are equally effective as disinfecting compounds; only the contact time required is different (see Fig. 7-16).

Contact Time. Because of the reaction of chlorine with the nitrogenous compounds found in untreated and treated wastewater and because chlorination beyond the break-point to obtain free hypochlorous acid is not economically feasible in many situations, the fundamental importance of contact time in the disinfection of wastewater cannot be overemphasized.

In Sec. 7-3, it was noted that the effect of contact time has at one time or another been described with each of the following relationships:

$$\ln\frac{N_t}{N_0} = -kt \tag{7-25}$$

$$\ln\frac{N_t}{N_0} = -kt^m \tag{7-26}$$

$$\frac{N_t}{N_0} = -kt^m \tag{7-28}$$

Of these relationships, Eq. 7-28 appears to provide the best fit for the data obtained from the chlorination of wastewater. The probable reason that Eq. 7-28 applies to wastewater data, as opposed to Eq. 7-25, is that in most cases the chlorine residual is made up of chloramines.

Because of the importance of contact time, either a batch or plug-flow reactor should be used to achieve effective disinfection; further, because a batch reactor for chlorination is impractical, plug-flow reactors are used at most treatment plants. The proper design of a plug-flow chlorination basin is considered in Chap. 9.

Characteristics of the Wastewater. It has often been observed that, for treatment plants of similar design with exactly the same effluent characteristics measured in terms of BOD, COD, and nitrogen, the effectiveness of the chlorination process varies significantly from plant to plant. To investigate the reasons for this observed phenomenon and to assess the effects of the compounds present on the chlorination process, Sung [17] studied the characteristics of the compounds in untreated and treated wastewater. Among the more important conclusions derived from Sung's study are the following:

1. In the presence of interfering organic compounds, the total chlorine residual cannot be used as a reliable measure for assessing the bactericidal efficiency of chlorine.
2. The degree of interference of the compounds studied depended on their functional groups and their chemical structure.
3. Saturated compounds and carbohydrates exert little or no chlorine demand and do not appear to interfere with the chlorination process.
4. Organic compounds with unsaturated bonds may exert an immediate chlorine demand, depending on their functional groups. In some cases, the resulting compounds may titrate as chlorine residual and yet may possess little or no disinfection potential.
5. Compounds with polycyclic rings containing hydroxyl groups and compounds containing sulfur groups react readily with chlorine to form compounds that have little or no bactericidal potential, but that still titrate as chlorine residual.
6. To achieve low bacterial counts in the presence of interfering organic compounds, additional chlorine and longer contact times will be required.

From the results of this work, it is easy to see why the efficiency of chlorination at plants with the same effluent characteristics can be quite different. Clearly, it is not the value of the BOD or COD that is significant but the nature of the compounds

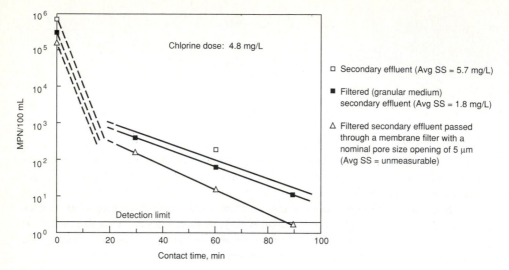

FIGURE 7-17
Typical disinfection results obtained when suspended solids are present [8].

that make up the measured values. Thus, the nature of the treatment process used in any plant will also have an effect on the chlorination process.

Another factor that must be considered is the presence of suspended solids in the wastewater to be chlorinated. As shown in Fig. 7-17, when suspended solids are present, the disinfection process is controlled by two different mechanisms. The large bacterial kill that is observed initially is of individual bacteria and bacteria in small clumps. The initial bacterial kill can be described using Eq. 7-24. The subsequent rate of bacterial kill is controlled by the presence of suspended solids (see Fig. 7-17). Thus, when suspended solids are present, a single equation can not be used to describe the chorination process.

Characteristics of the Microorganisms. Another important variable in the chlorination process is the age of the microorganisms [17]. For a young bacterial culture (1 d old or less) with a chlorine dosage of 2 mg/L, only 1 min was needed to reach a low bacterial number. When the bacterial culture was 10 d old or more, approximately 30 min was required to achieve a comparable reduction for the same applied chlorine dosage. It is likely that the resistance offered by the polysaccharide sheath, which the microorganisms develop as they age, accounts for this observation. In the activated-sludge treatment process, the operating mean cell residence time, which to some extent is related to the age of the bacterial cells in the system, will thus affect the performance of the chlorination process (see Chap. 10).

In view of the renewed interest in wastewater reclamation, the viricidal efficiency of the chlorination process is of great concern. Unfortunately, definitive data on this subject are not available at present. Some representative data on the effectiveness of chlorine in killing *E. coli* and three enteric viruses are reported in Fig. 7-18. From the evidence available, it appears that chlorination beyond the breakpoint to obtain free

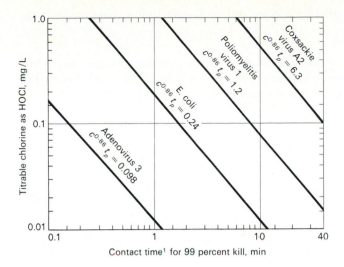

FIGURE 7-18
Concentration of chlorine as HOCl required for 99 percent kill of *E. coli* and three enteric viruses at 0 to 6°C [2].

chlorine will be required to kill many of the viruses of concern. Where breakpoint chlorination is used, it will be necessary to dechlorinate the treated wastewater before reuse in order to reduce any residual toxicity that may remain after chlorination.

7-5 DECHLORINATION

Dechlorination is the practice of removing the total combined chlorine residual that exists after chlorination to reduce the toxic effects of chlorinated effluents discharged to receiving waters or to be used for reuse applications.

Toxicity of Chlorine Residuals

Chlorination is one of the most commonly used methods for the destruction of pathogenic and other harmful organisms that may endanger human health. As noted in the previous discussion, however, certain organic constituents in wastewater interfere with the chlorination process. Many of these organic compounds may react with the chlorine to form toxic compounds that can have long-term adverse effects on the beneficial uses of the waters to which they are discharged. To minimize the effects of these potentially toxic chlorine residuals on the environment, it has been found necessary to dechlorinate wastewater treated with chlorine.

Analysis of Dechlorination

Where effluent toxicity requirements are applicable or where dechlorination is used as a polishing step following the breakpoint chlorination process for the removal of ammonia nitrogen, sulfur dioxide is used most commonly for dechlorination. Activated carbon has also been used. Both of these means are discussed below. Other chemicals that have been used are sodium sulfite (Na_2SO_3) and sodium metabisulfite ($Na_2S_2O_5$).

Sulfur Dioxide. Sulfur dioxide gas successively removes free chlorine, monochloramine, dichloramine, nitrogen trichloride, and poly-n-chlor compounds. When sulfur dioxide is added to wastewater, the following reactions occur [22]:

Reactions with chlorine:

$$SO_2 + H_2O \rightarrow HSO_3^- + H^+ \qquad (7\text{-}48)$$

$$HOCl + HSO_3^- \rightarrow Cl^- + SO_4^{-2} + 2H^+ \qquad (7\text{-}49)$$

$$SO_2 + HOCl + H_2O \rightarrow Cl^- + SO_4^{-2} + 3H^+ \qquad (7\text{-}50)$$

Reactions with chloramines:

$$SO_2 + H_2O \rightarrow HSO_3^- + H^+ \qquad (7\text{-}51)$$

$$NH_2Cl + HSO_3^- + H_2O \rightarrow Cl^- + SO_4^{-2} + NH_4^+ + H^+ \qquad (7\text{-}52)$$

$$SO_2 + NH_2Cl + 2H_2O \rightarrow Cl^- + SO_4^{-2} + NH_4^+ + 2H^+ \qquad (7\text{-}53)$$

For the overall reaction between sulfur dioxide and chlorine (Eq. 7-50), the stoichiometric weight ratio of sulfur dioxide to chlorine is 0.9:1. In practice, it has been found that about 1.0 mg/L of sulfur dioxide will be required for the dechlorination of 1.0 mg/L of chlorine residue (expressed as Cl_2). Because the reactions of sulfur dioxide with chlorine and chloramines are nearly instantaneous, contact time is not usually a factor and contact chambers are not used; however, rapid and positive mixing at the point of application is an absolute requirement.

The ratio of free chlorine to the total combined chlorine residual before dechlorination determines whether the dechlorination process is partial or proceeds to completion. If the ratio is less than 85 percent, it can be assumed that significant organic nitrogen is present and that it will interfere with the free residual chlorine process.

In most situations, sulfur dioxide dechlorination is a very reliable unit process in wastewater treatment, provided that the precision of the combined chlorine residual monitoring service is adequate. Excess sulfur dioxide dosages should be avoided not only because of the chemical wastage but also because of the oxygen demand exerted by the excess sulfur dioxide. The relatively slow reaction between excess sulfur dioxide and dissolved oxygen is given by the following expression:

$$HSO_3^- + 0.5O_2 \rightarrow SO_4^{-2} + H^+ \qquad (7\text{-}54)$$

The result of this reaction is a reduction in the dissolved oxygen contained in the wastewater, a corresponding increase in the measured BOD and COD, and a possible drop in the pH. All these effects can be eliminated by proper control of the dechlorination system.

Sulfur dioxide dechlorination systems are similar to chlorination systems because sulfur dioxide equipment is interchangeable with chlorination equipment. The components of these systems are discussed in Chap. 9. The key control parameters of this process are (1) proper dosage based on precise (amperometric) monitoring of the combined chlorine residual and (2) adequate mixing at the point of application of sulfur dioxide.

Activated Carbon. Carbon adsorption for dechlorination provides complete removal of both combined and free residual chlorine [22]. When activated carbon is used for dechlorination, the following reactions occur.

Reactions with chlorine:

$$C + 2Cl_2 + 2H_2O \rightarrow 4HCl + CO_2 \tag{7-55}$$

Reactions with chloramines:

$$C + 2NH_2Cl + 2H_2O \rightarrow CO_2 + 2NH_4^+ + 2Cl^- \tag{7-56}$$

$$C + 4NHCl_2 + 2H_2O \rightarrow CO_2 + 2N_2 + 8H^+ + 8Cl^- \tag{7-57}$$

Granular activated carbon is used in either a gravity or pressure filter bed. If carbon is to be used solely for dechlorination, it must be preceded by an activated-carbon process for the removal of other constituents susceptible to removal by activated carbon. In treatment plants where granular activated carbon is used to remove organics, either the same or separate beds can be used for dechlorination, and regeneration will be feasible.

Because granular carbon in column applications has proved to be very effective and reliable, activated carbon should be considered where dechlorination is required. However, this method is quite expensive. It is expected that the primary application of activated carbon for dechlorination will be in situations where high levels of organic removal are also required.

7-6 DISINFECTION WITH CHLORINE DIOXIDE

Chlorine dioxide is another bacteriocide, equal to or greater than chlorine in disinfecting power. Chlorine dioxide has proven to be more effective in achieving inactivation of viruses than chlorine. A possible explanation is that, because chlorine dioxide is adsorbed by peptone (a protein) and because viruses have a protein coat, adsorption of chlorine dioxide onto this coating could cause inactivation of the virus. In the past, it did not receive much consideration as a wastewater disinfectant due to its high costs.

Chlorine Dioxide Generation

Chlorine dioxide is an unstable and explosive gas and for this reason it must be generated on site. Generation of chlorine dioxide involves reacting sodium chlorite ($NaClO_2$) with chlorine to produce gaseous chlorine dioxide according to the following reaction:

$$2NaClO_2 + Cl_2 \rightarrow 2ClO_2 + 2NaCl \tag{7-58}$$

Based on Eq. 7-58, 1.34 mg sodium chlorite reacts with 0.5 mg chlorine to yield 1.0 mg chlorine dioxide. Because technical grade sodium chlorite is only about 80 percent pure, about 1.68 mg of the technical grade chlorite would be required to produce 1.0 mg of chlorine dioxide.

Effectiveness of Chlorine Dioxide

The active disinfecting agent in a chlorine dioxide system is free dissolved chlorine dioxide (ClO_2). The complete chemistry of chlorine dioxide in an aqueous environment is not clearly understood. Chlorine dioxide has an extremely high oxidation potential, which probably accounts for its potent germicidal powers. Because of its extreme high oxidizing potential, possible bacteriocidal mechanisms may include inactivation of critical enzyme systems or disruption of protein synthesis.

By-Product Formation. Certain potentially toxic end products, chlorite and chlorate, can be formed with chlorine dioxide and found as components of the total residual. The chlorine dioxide residuals and end products are believed to degrade quicker than chlorine residuals, and therefore may not pose as serious a threat to aquatic life as the chlorine residuals. An advantage in using chlorine dioxide is that it does not react with ammonia to form the potentially toxic chloramines. It has also been reported that halogenated organic compounds are not produced to any appreciable extent. This is especially true with respect to the formation of chloroform, which is a suspected carcinogenic substance.

Environmental Impacts. The environmental impacts associated with the use of chlorine dioxide as a wastewater disinfectant are not well known. It has been reported that the impacts are less adverse than those associated with chlorination. Chlorine dioxide does not dissociate or react with water as does chlorine. However, because chlorine dioxide is normally produced from chlorine and sodium chlorite, free chlorine may remain in the resultant chlorine dioxide solution (depending on the process) and impact the aquatic environment, as do chlorine residuals. A free chlorine dioxide residual will also remain, but it has been found to be less harmful to aquatic life than chlorine.

Dechlorination of Chlorine Dioxide

Dechlorinating wastewater disinfected with chlorine dioxide can be achieved using sulfur dioxide. The reaction that takes place in the chlorine dioxide solution can be expressed as:

$$SO_2 + H_2O \rightarrow H_2SO_3 \tag{7-59}$$

$$H_2SO_3 + 2ClO_2 + H_2O \rightarrow 5H_2SO_4 + 2HCl \tag{7-60}$$

Based on Eq. 7-60, it can be seen that 2.5 mg of sulfur dioxide will be required for each mg of chlorine dioxide residual (expressed as ClO_2). In practice, 2.7 mg/mg would normally be used.

7-7 DISINFECTION WITH BROMINE CHLORIDE

Because the practical aspects of disinfection with bromine chloride are discussed in Chap. 9, the following discussion is limited to a brief description of the chemistry of bromine chloride, an analysis of the performance of bromine chloride as a disinfectant, and the factors that may influence the effectiveness of the bromine disinfection process.

Bromine Chloride Chemistry

The reactions of bromine chloride with water and ammonia are considered in the following discussion.

Reactions in Water. Bromine chloride gas hydrolyzes in water to form hypobromous acid, the most potent germicide of all the bromine compounds, according to the following reaction:

$$BrCl + H_2O \rightarrow HOBr + HCl \qquad (7\text{-}61)$$

As shown, bromine chloride hydrolyzes to form hypobromous acid (HOBr) and hydrochloric acid. The hydrolysis constant for BrCl in water is

$$\frac{[HOBr][H^+][Cl^-]}{[BrCl]} = 2.97 \times 10^{-4} \text{ at } 0°C \qquad (7\text{-}62)$$

Because hypobromous acid (HOBr) is a weak acid, it dissociates according to the following relationship.

$$HOBr \rightarrow H^+ + OBr^- \qquad (7\text{-}63)$$

The following expression, derived from thermodynamic considerations, can be used to define the dissociation of hypobromous acid [22].

$$\frac{[H^+][OBr^-]}{[HOBr]} = 5.24 \times 10^{-6} \exp(-2265/T) \qquad (7\text{-}64)$$

where T = temperature, °K

Reactions with Ammonia. As with chlorination, chlorobromination also generates halogenated amines. Bromine chloride reacts with ammonia to form abromanines as follows:

$$NH_3 + HOBr \rightarrow NH_2Br + H_2O \qquad (7\text{-}65)$$

$$NH_2Br + HOBr \rightarrow NHBr_2 + H_2O \qquad (7\text{-}66)$$

$$NHBr_2 + HOBr \rightarrow NBr_3 + H_2O \qquad (7\text{-}67)$$

The bromamines are typically less stable than chloramines and break down to harmless chloride and bromide salts in less than an hour.

Effectiveness of Bromine Chloride

While bromine chloride cannot be classified from the data available as a fully proven disinfectant as compared to chlorine, bromine chloride appears to be as reliable, flexible, and effective as chlorine. Although additional research is needed to establish the actual cellular disinfection mechanism (because of the similarity of hypobromous acid to hypochlorous acid), it appears reasonable to assume that it adsorbs into the bacterial cell and disrupts critical enzymatic activity. Bromamines have been shown to be more effective germicides than chloramines and also degrade quicker. It has been reported that bromine chloride inactivates equal amounts of poliovirus at about one-half the dose of chlorine. Contact time required for adequate bacterial kill is generally less than that required for chlorine, but no standards have yet been established. A contact time equal to that used in chlorination would appear to be more than adequate for this disinfectant.

Additional studies are needed to verify the bromine chloride dosage for a given wastewater effluent quality, to determine the most effective and efficient bromine chloride application method, to determine effectiveness of bromine chloride for the ancillary uses, and to obtain additional field data on bromine chloride's short- and long-term impacts on aquatic life in the receiving water.

By-Product Formation.   Other brominated organic substances are also believed to be formed as a result of bromine chloride disinfection. It is believed that these brominated organic substances are susceptible to hydrolytic and photochemical degradation. Thus, appreciable quantities should not persist in the receiving waters. In studies conducted by the EPA Environmental Research Laboratory-Duluth, it was found that brominated organic chemicals were bio-accumulated in fish exposed to wastewater disinfected with bromine chloride. However, the organobromine residues in the fish were at concentrations less than other known toxic chemicals (i.e., PCB and chlordane). Due to the limited data that does exist on the long-term environmental impacts associated with bromine chloride disinfection and because conflicting data have been reported, additional studies are indeed warranted.

Environmental Impacts of Using Bromine Chloride.   Because bromine chloride is similar in many respects to chlorine, it is expected that the environmental impacts associated with its use as a disinfectant would be similar to those associated with chlorination. However, investigators have shown that the environmental impacts associated with bromine chloride disinfection are less adverse than those associated with chlorination. Conflicting results have been cited in the literature concerning the toxic effects of bromine residuals on aquatic life. However, it is generally agreed that bromine residuals are less toxic than chlorine residuals, and, as such, less stringent total bromine residuals have been proposed.

7-8 DISINFECTION WITH OZONE

Ozone was first used to disinfect water supplies in France in the early 1900s. Its use there increased and eventually spread into several Western European countries. Today nearly 1,000 ozone disinfection installations exist (primarily in Europe), almost entirely for treating water supplies. A common use for ozone at these installations is to control taste-, odor-, and color-producing agents. Although historically used primarily for the disinfection of water, recent advances in ozone generation and solution technology have made the use of ozone economically more competitive for wastewater disinfection. Ozone can also be used in wastewater treatment for odor control and in advanced wastewater treatment for the removal of soluble refractory organics, in lieu of the carbon-adsorption process. The generation of ozone, the chemistry of ozone, an analysis of the performance of ozone as a disinfectant, and the application of the ozonation process are considered in the following discussion. The practical aspects of disinfection with ozone are discussed in Chap. 9.

Ozone Generation

Because ozone is chemically unstable, it decomposes to oxygen very rapidly after generation, and thus must be generated on-site. The most efficient method of producing ozone today is by electrical discharge (see Fig. 7-19). Ozone is generated either from air or pure oxygen when a high voltage is applied across the gap of narrowly spaced electrodes. The high-energy corona created by this arrangement dissociates one oxygen molecule, which re-forms with two other oxygen molecules to create two ozone molecules. The gas stream generated by this process from air will contain about 0.5 to 3 percent ozone by weight and from pure oxygen about twice that amount, or 1 to 6 percent ozone.

Ozone Chemistry

Some of the chemical properties displayed by ozone may be described by its decomposition reactions which are thought to proceed as follows:

$$O_3 + H_2O \rightarrow HO_3^+ + OH^- \tag{7-68}$$

$$HO_3^+ + OH^- \rightarrow 2HO_2 \tag{7-69}$$

$$O_3 + HO_2 \rightarrow HO + 2O_2 \tag{7-70}$$

$$HO + HO_2 \rightarrow H_2O + O_2 \tag{7-71}$$

The free radicals formed, HO_2 and HO, have great oxidizing powers and are probably the active form in the disinfection process. These free radicals also possess the oxidizing power to react with other impurities in aqueous solutions.

Effectiveness of Ozone

Ozone is an extremely reactive oxidant, and it is generally believed that bacterial kill through ozonation occurs directly because of cell wall disintegration (cell lysis).

FIGURE 7-19
Typical ozone generator (from Emery Chemicals, Inc.).

Ozone is also a very effective virucide and is generally believed to be more effective than chlorine. Ozonation does not produce dissolved solids and is not affected by the ammonium ion or pH influent to the process. For these reasons, ozonation is considered a viable alternative to either chlorination or hypochlorination, especially where dechlorination may be required.

Environmental Impacts of Using Ozone. Unlike the other chemical disinfecting agents previously discussed, ozone will mainly exert beneficial impacts on the environment. It has been reported that ozone residuals can be acutely toxic to aquatic life. However, because ozone dissipates rapidly, ozone residuals will normally not be found by the time the effluent is discharged into the receiving water. Several investigators have reported that ozonation can produce some toxic mutagenic and/or carcinogenic compounds. These compounds are usually unstable, however, and are present only for a matter of minutes in the ozonated water. Thus, these compounds will normally not be present by the time the effluent reaches the receiving water. White [25] has reported that ozonation destroys certain harmful refractory organic substances such as humic acid (precursor of trihalomethane formation) and malathion. Whether toxic intermediates are formed during ozonation depends on the ozone dose, the contact time, and the precursor compounds. It has been reported that ozone treatment before chlorination for disinfection purposes reduces the likelihood of trihalomethane formation [25].

Other Benefits of Using Ozone. An additional benefit associated with the use of ozone for disinfection is that the dissolved oxygen concentration of the effluent will be elevated to near saturation levels as ozone rapidly decomposes after application to oxygen. This may eliminate the need for reaeration of the effluent to meet required dissolved oxygen water quality standards. Further, because ozone decomposes rapidly, no chemical residual that may require removal, as is the case with chlorine residuals, persists in the treated effluent.

7-9 DISINFECTION WITH ULTRAVIOLET LIGHT

Radiation emitted from ultraviolet (UV) light sources has been used to a limited extent since the early 1900s for disinfection of water supplies. Primarily used on high-quality water supplies at first, new interest has recently focused on the use of ultraviolet light as a wastewater disinfectant. A proper dosage of ultraviolet radiation has shown to be an effective bacteriocide and virucide while not contributing to the formation of toxic compounds.

Source of UV Radiation

At present, the low-pressure mercury arc lamp is the principal means of generating UV energy used for disinfection. The mercury lamp is favored because about 85 percent of the light output is monochromatic at a wavelength of 253.7 nm, which is within the optimum range (250 to 270 nm) for germicidal effects. The lamps are typically about 2.5 to 5.0 ft (0.75 to 1.5 m) in length and about 0.6 to 0.8 in (15 to 20 mm) in diameter (see Fig. 7-20). To produce UV energy, the lamp, which contains mercury vapor, is charged by striking an electric arc. The energy generated by the

(a)

(b)

FIGURE 7-20
Typical UV installation: (a) UV lamps installed in the contact channel and (b) UV lamps removed from contact basin for cleaning (from Trojan Technologies, Inc.).

excitation of the mercury vapor contained in the lamp results in the emission of UV light. Operationally, the lamps are either suspended outside of the liquid to be treated or submerged in the liquid. Where the lamps are submerged, the lamps are encased in quartz tubes to prevent cooling effects on the lamps.

Effectiveness of UV Radiation

Ultraviolet light is a physical rather than a chemical disinfecting agent. Radiation with a wavelength of around 254 nm penetrates the cell wall of the microorganism and is absorbed by cellular materials including DNA and RNA, which either prevents replication or causes death of the cell to occur. Because the only ultraviolet radiation effective in destroying bacteria is that which reaches the bacteria, the water must be relatively free from turbidity that would absorb the ultraviolet energy and shield the bacteria. It has also been reported that ultraviolet light is not an effective disinfectant on wastewaters that contain high solids concentrations. For practical purposes, the inactivation of bacteria by UV radiation can be described using first-order kinetics.

Optimizing the Performance of UV Radiation. Because the distance over which ultraviolet light is effective is very limited, most effective disinfection occurs when the *thin film* approach is used. To limit the liquid thickness the ultraviolet light must penetrate, most ultraviolet units are constructed with an array of ultraviolet lamps through which the wastewater is passed. Typically, these units would be installed in the effluent channel, eliminating the need for a contact tank or channel. They should normally be enclosed in a structure to protect the electrical equipment used to power the ultraviolet lamps.

Environmental Impacts of Using UV Radiation. Because ultraviolet light is not a chemical agent, no toxic residuals are produced. However, certain chemical compounds may be altered by the ultraviolet radiation. It is generally believed that the compounds are broken down into a more innocuous form, but additional investigation into this occurrence is still warranted. Therefore, at present, disinfection with ultraviolet light must be considered to have no adverse or beneficial environmental impacts.

7-10   OTHER CHEMICAL APPLICATIONS

In addition to the major applications of chemicals discussed in this chapter, a number of other applications are occasionally encountered in the collection, treatment, and disposal of wastewater. The more important of these applications and the chemicals used are identified in Table 7-8. As shown, chlorine is by far the most commonly used chemical, although hydrogen peroxide is gaining in popularity. The effectiveness of the various chemical additions is site-specific, so optimum dosage requirements are unavailable. Because chlorine has been used extensively, however, some representative dosage ranges have been established, and these are given in Chap. 9.

TABLE 7-8
Additional chemical applications in wastewater collection, treatment, and disposal

Application	Chemicals used[a]	Remarks
Collection		
Slime-growth control	Cl_2, H_2O_2	Control of fungi and slime-producing bacteria
Corrosion control (H_2S)	Cl_2, H_2O_2, O_3	Control brought about by destruction of H_2S in sewers
Corrosion control (H_2S)	$FeCl_3$	Control brought about by precipitation of H_2
Odor control	Cl_2, H_2O_2, O_3	Especially in pumping stations and long, flat sewers
Treatment		
Grease removal	Cl_2	Added before preaeration
BOD reduction	Cl_2, O_3	Oxidation of organic substances
pH control	KOH, $Ca(OH)_2$, NaOH	
Ferrous sulfate oxidation	Cl_2[b]	Production of ferric sulfate and ferric chloride
Filter-ponding control	Cl_2	Residual at filter nozzles
Filter-fly control	Cl_2	Residual at filter nozzles, used during fly season
Sludge-bulking control	Cl_2, H_2O_2, O_3	Temporary control measure
Digester supernatant oxidation	Cl_2	
Digester and Imhoff tank foaming control	Cl_2	
Ammonia oxidation	Cl_2	Conversion of ammonia to nitrogen gas
Odor control	Cl_2, H_2O_2, O_3	
Oxidation of refractory organic compounds	O_3	
Disposal		
Bacterial reduction	Cl_2, H_2O_2, O_3	Plant effluent, overflows, and stormwater
Odor control	Cl_2, H_2O_2, O_3	

[a] Cl_2 = chlorine, H_2O_2 = hydrogen peroxide, O_3 = ozone, KOH = potassium hydroxide, $Ca(OH)_2$ = calcium hydroxide, NaOH = sodium hydroxide.
[b] $6(FeSO_4 \cdot 7H_2O) + 3CL_2 \rightarrow 2FeCl_3 + Fe_2(SO_4)_3 + 42H_2O$.

DISCUSSION TOPICS AND PROBLEMS

7-1. To aid sedimentation in the primary settling tank, 25 mg/L of ferrous sulfate ($FeSO_4 \cdot 7H_2O$) is added to the wastewater. Determine the minimum alkalinity required to react initially with the ferrous sulfate. How many grams of lime should be added as CaO to react with $Fe(HCO_3)_2$ and the dissolved oxygen in the wastewater to form insoluble $Fe(OH)_3$?

7-2. Copperas ($FeSO_4 \cdot 7H_2O$) is to be added at a rate of 150 lb/Mgal to a wastewater to improve the efficiency of an existing primary sedimentation tank. Assuming that sufficient alkalinity is present as $Ca(HCO_3)_2$ determine the following:
  (*a*) How many lbs of lime should be added as CaO to complete the reaction?
  (*b*) What must the concentration of oxygen be in the wastewater to oxidize the ferrous hydroxide formed?
  (*c*) How many lbs of sludge will result per Mgal? and
  (*d*) Compute the amount (lb) of alum needed to obtain the same quantity of sludge as in part (c), assuming that $Al(OH)_3$ is the precipitate formed.

7-3. Assume that 110 lb of (a) alum (mol wt 666.7) and (b) ferrous sulfate and lime as $Ca(OH)_2$ is added per 1.0 Mgal of wastewater. Also assume that all insoluble and very slightly soluble products of the reactions, with the exception of 15 mg/L $CaCO_3$, are precipitated as sludge. How many lb of sludge/Mgal will result in each case?

7-4. Raw wastewater is to be treated chemically for suspended-solids and phosphorus removal through coagulation and sedimentation. The wastewater characteristics are as follows: $Q = 17$ Mgal/d; orthophosphorus $= 8$ mg/L as P; alkalinity $= 200$ mg/L expressed as $CaCO_3$ [essentially all due to the presence of $Ca(HCO_3)_2$]; total suspended solids $= 220$ mg/L.
  (*a*) Determine the sludge production in lb dry wt/d and Mgal/d under the following conditions: (1) Alum ($Al_2(SO_4)_3 \cdot 14.3 H_2O$) dosage of 150 mg/L; (2) 100 percent removal of orthophosphorus as insoluble $AlPO_4$; (3) 95 percent removal of original TSS; (4) all alum that is not required for reaction with phosphate reacts with alkalinity to form $Al(OH)_3$, which is 100 percent removed; (5) wet sludge has a water content of 93 percent and a specific gravity of 1.04.
  (*b*) Determine the sludge production in a lb dry wt/day and Mgal/d under the following conditions: (1) Lime ($Ca(OH)_2$) dosage of 450 mg/L to give pH of approximately 11.2; (2) 100 percent removal of orthophosphorus as insoluble hydroxylapatite ($Ca_{10}(PO_4)_6(OH)_2$); (3) 95 percent removal of original TSS; (4) added lime (i) reacts with phosphate, (ii) reacts with all alkalinity to form $CaCO_3$; 20 mg/L of $CaCO_3$ is soluble and remains in solution and the rest is 100 percent removed, and (iii) remainder stays in solution; (5) wet sludge has a water content of 92 percent and a specific gravity of 1.05.
  (*c*) Determine the net increase in calcium hardness in mg/L as $CaCO_3$ for the treatment specified in part *b*.

7-5. Laboratory tests were conducted on a waste containing 50 mg/L phenol. Four jars containing 1 liter of the waste were dosed with powdered activated carbon. When equilibrium was reached, the contents of each jar were analyzed for phenol. The results are shown in the following table. Determine the constants *a* and *b* in the Langmuir equation and the dosage required to yield an effluent with a phenol concentration of 0.10 mg/L.

Jar	Carbon added, g	Equilibrium conc. of phenol, mg/L
1	0.5	6.0
2	0.64	1.0
3	1.0	0.25
4	2.0	0.08

7-6. A treated and filtered wastewater to be used for golf course irrigation has an initial threshold odor of 10. When activated carbon is used to absorb the odor, the following test results are obtained.

Carbon added, mg/L	0	0.4	1.0	6.0
Odor number	10	6.9	4.5	1.5

Using the Freundlich adsorption isotherm (Eq. 7-17), determine the minimum dosage of activated carbon required to reduce the odor to a residual value of 0.20.

7-7. The following disinfection test data were obtained in a series of laboratory tests performed on an effluent from a secondary wastewater treatment process.

Chlorine dosage, mg/L	Residual fecal coliform count, no./100 ml		
	Contact time, min.		
	15	30	60
1	10,000	2,000	500
2	3,000	350	90
4	400	65	20
6	110	30	12
8	54	19	6
10	30	10	1

(a) Plot the number of organisms remaining versus the dosage of chlorine on log-log paper. Using the plotted data, determine the value of the exponent n and the constant in Eq. 7-29 for residual coliform counts of 200/100 ml and 1000/100 ml.

(b) The following data apply to the wastewater treatment plant.

Item	May–October	November–April
Average flow, Mgal/d	5.3	6.9
Peak daily flow, Mgal/d	10.6	13.7
Maximum permissable fecal coliform count in effluent, no./100ml	200	1,000

Determine the required volume in ft^3 of a chlorine contact chamber designed to provide 30-min contact at the average winter flow. Using the equations developed in part a, determine the minimum dosage required in mg/L to give the required kill under each of the four flow conditions given above. Assuming that the yearly chlorine requirement can be computed on the basis of the average flow for each of the two 6-month periods, determine the minimum yearly chlorine requirement in pounds. (Courtesy E. Foree.)

7-8. Use the following chlorination test survival data for *E.coli*, expressed as percentage to solve the problems below.

Free available Cl. mg/L	Contact time, min[a]				
	1	3	5	10	20
0.05	97	82	63	21	0.3
0.07	93	60	28	0.5	—
0.14	67	11	0.7	—	—

[a] Test conditions: Determine pH = 8.5; temp = 5°C.

(a) Determine the values of m and K for the various concentrations of the modified form of Chick's law (Eq. 7-26).

(b) Using Eq. 7-29, determine the values of the constant and exponent for a 99 percent kill of *E.coli*.

(c) If the temperature of the wastewater was 20°C, estimate the time required for 99 percent kill, using a chlorine dosage of 0.05 mg/L.

(d) What is the significance of the exponents m and n with respect to disinfection kinetics?

7-9. The chlorine residuals measured when various dosages of chlorine were added to a wastewater are given below. Determine (a) the breakpoint dosage and (b) the design dosage to obtain a residual of 0.75 mg/L free available chlorine.

Dosage, mg/L	0.1	0.5	1.0	1.5	2.0	2.5	3.0
Residual, mg/L	0.0	0.4	0.8	0.4	0.4	0.9	1.4

7-10. Derive a rate expression from Eq. 7-47 that can be used to assess the efficiency of a complete mix continuous-flow stirred-tank reactor as a chlorine contact basin.

7-11. Using Eq. 7-47 and the rate expression derived in Prob. 7-10, compare the volume required for a continuous-flow stirred-tank reactor to that for a plug-flow reactor to achieve a 10^4 reduction in the coliform count of a treated effluent. Assume that in both cases the chlorine residual to be maintained is 5 mg/L.

7-12. Determine the amount of activated carbon that would be required per year to dechlorinate treated effluent containing a chlorine residual of 5 mg/L (as Cl_2) from a plant with an average flowrate of 1.0 Mgal/d. What dosage of sulfur dioxide would be required?

7-13. Discuss the advantages and disadvantages of using ozone as a disinfectant. Cite a minimum of four recent references (after 1985) in your discussion.

7-14. Discuss the advantages and disadvantages of using UV radiation as a disinfectant. Cite a minimum of four recent references (after 1985) in your discussion.

REFERENCES

1. Chick, H.: "Investigation of the Laws of Disinfection," *J. Hygiene,* vol. 8, p. 92, 1908.
2. Crittenden, J. C., P. Luft, D. W. Hand, J. L. Oravitz, S. W. Loper, and M. Ari: "Prediction Of Multicomponent Adsorption Equilibria Using Ideal Solution Theory," *Environmental Science and Technology,* vol. 19, p. 1037, 1985.
3. Crittenden, J. C., and W. J. Weber: "Model For Design of Multicomponent Adsorption Systems," *Journal Env. Eng. Div., ASCE,* vol. 104, p. 1175, 1978.
4. Culp, G. L.: "Chemical Treatment of Raw Sewage/1 and 2," *Water Wastes Eng.,* vol. 4, nos. 7, 10, 1967.
5. Eliassen, R., and J. Trump: "High-Energy Electrons Offer Alternative to Chlorine," *Calif. Water Pollut. Control Assoc. Bull.,* vol. 10, no. 3, January 1974.
6. Fair, G. M., et al.: "The Behavior of Chlorine as a Water Disinfectant," *Journal AWWA,* vol. 40, no. 10, 1948.
7. Fair, G. M., J. C. Geyer, and D. A. Okun: *Water and Wastewater Engineering,* vol. 2, Wiley, New York, 1968.
8. Manglik, P. K., J. R. Johnston, T. Asano, and G. Tchobanoglous: "Effect of Particles On Chlorine Disinfection of Wastewater," *Proceedings of Water Reuse Symposium IV Implementing Water Reuse,* AWWA Research Foundation, Denver, CO, 1988.
9. Metcalf, L., and H. P. Eddy: *American Sewerage Practice,* vol. 3, 3rd ed., McGraw-Hill, New York, 1935.
10. Pelczar, M. J., Jr., and E. C. S. Chan: *Microbiology,* 5th ed., McGraw-Hill, New York, 1986.
11. *Principles and Practice of Nutrient Removal From Municipal Wastewater,* The Soap and Detergent Association, New York, 1988.
12. Schroeder, E. D., and G. Tchobanoglous: *Water and Wastewater Treatment,* Addison Wesley, Reading, MA, 1990.
13. Shaw, D. J.: *Introduction to Colloid and Surface Chemistry,* Butterworth, London, 1966.
14. Stumm, W., and J. J. Morgan: *Aquatic Chemistry,* 2nd ed., Wiley-Interscience, New York, 1980.
15. Stumm, W., and J. J. Morgan: "Chemical Aspects of Coagulation," *Journal AWWA,* vol. 54, no. 8, 1962.
16. Stumm, W., and C. R. O'Melia: "Stoichiometry of Coagulation," *Journal AWWA,* vol. 60, no. 5, 1968.
17. Sung, R. D.: Effects of Organic Constituents in Wastewater on the Chlorination Process, Ph.D. thesis, Department of Civil Engineering, University of California, Davis, CA 1974.
18. Thomas, A. W.: *Colloid Chemistry,* McGraw-Hill, New York, 1934.
19. U.S. Environmental Protection Agency: *Physical-Chemical Wastewater Treatment Plant Design,* U.S. Environmental Protection Agency, Technology Transfer Seminar Publication, 1973.
20. U.S. Environmental Protection Agency: *Process Design Manual For Carbon Adsorption,* U.S. Environmental Protection Agency, Technology Transfer, National Environmental Research Center, Cincinnati, OH, 1973.
21. U.S. Environmental Protection Agency: *Design Manual, Odor and Corrosion Control in Sanitary Sewerage Systems and Treatment Plants,* EPA/625/1-85/018, October 1985.
22. U.S. Environmental Protection Agency: *Design Manual, Municipal Wastewater Disinfection,* EPA/625/1-86/021, October 1986.
23. U.S. Environmental Protection Agency: *Phosphorus Removal, Design Manual,* EPA/625/1-87/001, Cincinnati, OH, September 1987.
24. Water Pollution Control Federation: *Wastewater Disinfection,* Manual of Practice FD-10, Alexandria, VA, 1986.
25. White, G. C.: *Handbook of Chlorination,* 2nd. ed, Van Nostrand-Reinhold, New York, 1985.

CHAPTER 8

BIOLOGICAL UNIT PROCESSES

With proper analysis and environmental control, almost all wastewaters can be treated biologically. Therefore, it is essential that the environmental engineer understand the characteristics of each biological process to ensure that the proper environment is produced and controlled effectively. In view of the importance of biological treatment, it is the purpose of this chapter (1) to present an overview of biological wastewater treatment, (2) to introduce the important aspects involved in microbial metabolism, (3) to introduce the principal organisms responsible for wastewater treatment, (4) to discuss the key factors governing biological growth and waste treatment kinetics, and (5) to illustrate the application of fundamentals and kinetics to the analysis of the biological processes used most commonly for wastewater treatment. The biological removal of nutrients and pond processes are considered in separate sections. The information presented in this chapter provides the background for the design of biological treatment processes discussed in Chaps. 10 through 12.

8-1   OVERVIEW OF BIOLOGICAL WASTEWATER TREATMENT

The objectives of biological treatment and the role of microorganisms in the biological treatment of wastewater are considered first to provide a perspective for the material to be presented in this chapter.

Objectives of Biological Treatment

The objectives of the biological treatment of wastewater are to coagulate and remove the nonsettleable colloidal solids and to stabilize the organic matter. For domestic wastewater, the major objective is to reduce the organic content and, in many cases, the nutrients such as nitrogen and phosphorus. In many locations, the removal of trace organic compounds that may be toxic is also an important treatment objective. For agricultural return wastewater, the objective is to remove the nutrients, specifically nitrogen and phosphorus, that are capable of stimulating the growth of aquatic plants. For industrial wastewater, the objective is to remove or reduce the concentration of organic and inorganic compounds. Because many of these compounds are toxic to microorganisms, pretreatment may be required.

Role of Microorganisms

The removal of carbonaceous BOD, the coagulation of nonsettleable colloidal solids, and the stabilization of organic matter are accomplished biologically using a variety of microorganisms, principally bacteria. The microorganisms are used to convert the colloidal and dissolved carbonaceous organic matter into various gases and into cell tissue. Because cell tissue has a specific gravity slightly greater than that of water, the resulting cells can be removed from the treated liquid by gravity settling.

It is important to note that, unless the cell tissue produced from the organic matter is removed from the solution, complete treatment has not been accomplished because the cell tissue, which itself is organic, will be measured as BOD in the effluent. If the cell tissue is not removed, the only treatment that has been achieved is that associated with the bacterial conversion of a portion of the organic matter originally present to various gaseous end products.

8-2 INTRODUCTION TO MICROBIAL METABOLISM

Basic to the design of a biological treatment process or to the selection of the type of process to be used is an understanding of the biochemical activities of the important microorganisms. The two major topics considered in this section are (1) the general nutritional requirements of the microorganisms commonly encountered in wastewater treatment, and (2) the nature of microbial metabolism based on the need for molecular oxygen.

Nutritional Requirements for Microbial Growth

To continue to reproduce and function properly, an organism must have (1) a source of energy, (2) carbon for the synthesis of new cellular material, and (3) inorganic elements (nutrients) such as nitrogen, phosphorus, sulfur, potassium, calcium, and magnesium. Organic nutrients (growth factors) may also be required for cell synthesis. Carbon and energy sources, usually referred to as *substrates*, and nutrient and growth factor requirements for various types of organisms are considered in the following discussion.

Carbon and Energy Sources. Two of the most common sources of cell carbon for microorganisms are organic matter and carbon dioxide. Organisms that use organic carbon for the formation of cell tissue are called *heterotrophs*. Organisms that derive cell carbon from carbon dioxide are called *autotrophs*. The conversion of carbon dioxide to organic cell tissue is a reductive process that requires a net input of energy. Autotrophic organisms must therefore spend more of their energy for synthesis than do heterotrophs, resulting in generally lower growth rates among the autotrophs.

The energy needed for cell synthesis may be supplied by light or by a chemical oxidation reaction. Those organisms that are able to use light as an energy source are called *phototrophs*. Phototrophic organisms may be either heterotrophic (certain sulfur bacteria) or autotrophic (algae and photosynthetic bacteria). Organisms that derive their energy from chemical reactions are known as *chemotrophs*. As with the phototrophs, chemotrophs may be either heterotrophic (protozoa, fungi, and most bacteria) or autotrophic (nitrifying bacteria). Chemoautotrophs obtain energy from the oxidation of reduced inorganic compounds such as ammonia, nitrite, and sulfide. Chemoheterotrophs usually derive their energy from the oxidation of organic compounds. The classification of microorganisms by sources of energy and cell carbon is summarized in Table 8-1. Schematic representations of the common types of bacterial metabolism are given in Figs. 8-1 to 8-3.

Nutrient and Growth Factor Requirements. Nutrients, rather than carbon or energy source, may at times be the limiting material for microbial cell synthesis and growth. The principal inorganic nutrients needed by microorganisms are N, S, P, K, Mg, Ca, Fe, Na, and Cl. Minor nutrients of importance include Zn, Mn, Mo, Se, Co, Cu, Ni, V, and W [34].

In addition to the inorganic nutrients cited above, organic nutrients may also be needed by some organisms. Required organic nutrients, known as "growth factors," are compounds needed by an organism as precursors or constituents of organic cell material that cannot be synthesized from other carbon sources. Although growth factor requirements differ from one organism to another, the major growth factors fall into the following three classes: (1) amino acids, (2) purines and pyrimidines, and (3) vitamins [34].

TABLE 8-1
General classification of microorganisms by sources of energy and carbon[a]

Classification	Energy source	Carbon source
Autotrophic:		
Photoautotrophic	Light	CO_2
Chemoautotrophic	Inorganic oxidation-reduction reaction	CO_2
Heterotrophic:		
Chemoheterotrophic	Organic oxidation-reduction reaction	Organic carbon
Photoheterotrophic	Light	Organic carbon

[a] Adapted from Ref. 34.

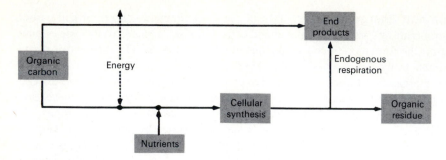

FIGURE 8-1
Schematic representation of chemoheterotrophic bacterial metabolism.

Microbial Nutrition and Biological Treatment Processes. The major objective in most biological treatment processes is the reduction of organic content (carbonaceous BOD) in the wastewater. In accomplishing this type of treatment, the chemoheterotrophic organisms are of primary importance because of their requirement for organic compounds in addition to both carbon and energy source. When treatment objectives include the conversion of ammonia to nitrate, the chemoautotrophic nitrifying bacteria are significant.

Municipal wastewater typically contains adequate amounts of nutrients (both inorganic and organic) to support biological treatment for the removal of carbonaceous BOD. In industrial wastewaters, however, nutrients may not be present in sufficient quantities. In these cases, nutrient addition is necessary for the proper growth of the bacteria and the subsequent degradation of the organic waste.

Types of Microbial Metabolism

Chemoheterotrophic organisms may be further grouped according to their metabolic type and their requirement for molecular oxygen. Organisms that generate energy

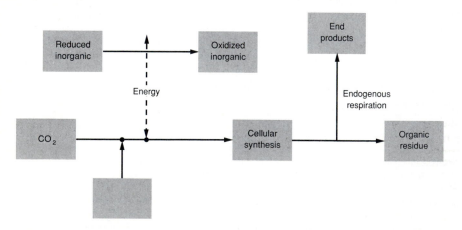

FIGURE 8-2
Schematic representation of chemoautotrophic bacterial metabolism.

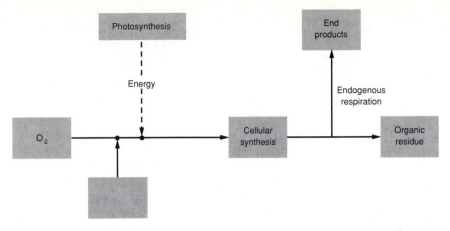

FIGURE 8-3
Schematic representation of photoautotrophic bacterial metabolism.

by enzyme-mediated electron transport from an electron donor to an external electron acceptor are said to have a *respiratory metabolism.* In contrast, *fermentative metabolism* does not involve the participation of an external electron acceptor. Fermentation is a less efficient energy-yielding process than respiration; as a consequence, heterotrophic organisms that are strictly fermentative are characterized by lower growth rates and cell yields than respiratory heterotrophs.

When molecular oxygen is used as the electron acceptor in respiratory metabolism, the process is known as *aerobic respiration.* Organisms that are dependent on aerobic respiration to meet their energetic needs can exist only when there is a supply of molecular oxygen. These organisms are called *obligately aerobic.* Oxidized inorganic compounds such as nitrate and nitrite can function as electron acceptors for some respiratory organisms in the absence of molecular oxygen (see Table 8-2). In environmental engineering, processes that make use of these organisms are often referred to as *anoxic.*

Organisms that generate energy by fermentation and that can exist only in an environment that is devoid of oxygen are *obligately anaerobic. Facultative anaerobes* have the ability to grow in either the presence or absence of molecular oxygen. The

TABLE 8-2
Typical electron acceptors in bacterial reactions commonly encountered in the management of wastewaters

Environment	Electron acceptor	Process
Aerobic	Oxygen, O_2	Aerobic metabolism
Anaerobic	Nitrate, NO_3^-	Denitrification[a]
	Sulfate, SO_4^{2}	Sulfate reduction
	Carbon dioxide, CO_2	Methanogenesis

[a] Also known as anoxic dentrification.

facultative organisms fall into two subgroups, based on their metabolic abilities. True facultative anaerobes can shift from fermentative to aerobic respiratory metabolism, depending upon the presence or absence of molecular oxygen. *Aerotolerant anaerobes* have a strictly fermentative metabolism but are relatively insensitive to the presence of molecular oxygen.

8-3 IMPORTANT MICROORGANISMS IN BIOLOGICAL TREATMENT

On the basis of cell structure and function, microorganisms are commonly classified as eucaryotes, eubacteria, and archaebacteria, as shown in Table 3-11. The procaryotic groups (eubacteria and archaebacteria) are of primary importance in biological treatment and are generally referred to simply as bacteria. The eucaryotic group includes plants, animals, and protists. Eucaryotes important in biological treatment include (1) fungi, (2) protozoa and rotifers, and (3) algae.

Bacteria

Bacteria are single-celled procaryotic organisms. Their usual mode of reproduction is by binary fission, although some species reproduce sexually or by budding. Even though there are thousands of different species of bacteria, their general form falls into one of three categories: spherical, cylindrical, and helical. Bacteria vary widely in size. Representative sizes are 0.5 to 1.0 μm in diameter for the spherical, 0.5 to 1.0 μm in width by 1.5 to 3.0 μm in length for the cylindrical (rods), and 0.5 to 5 μm in width by 6 to 15 μm in length for the helical (spiral).

Cell Structure. In general, most bacterial cells are quite similar (see Fig. 8-4). As shown in Fig. 8-4, the interior of the cell, called the cytoplasm, contains a colloidal suspension of proteins, carbohydrates, and other complex organic compounds. The

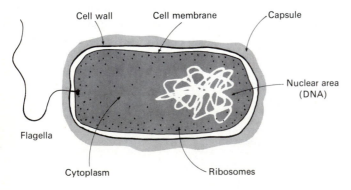

FIGURE 8-4
Generalized schematic of a bacterial cell [31].

cytoplasmic area contains ribonucleic acid (RNA), whose major role is in the synthesis of proteins. Also within the cytoplasm is the area of the nucleus, which is rich in deoxyribonucleic acid (DNA). DNA contains all the information necessary for the reproduction of all the cell components and may be considered the blueprint of the cell.

Cell Composition. Tests on a number of different bacteria indicate that they are about 80 percent water and 20 percent dry material, of which 90 percent is organic and 10 percent inorganic. Typical values for the composition of bacterial cells are reported in Table 8-3. An approximate formula for the organic fraction is $C_5H_7O_2N$ [16]. As indicated by the formula, about 53 percent by weight of the organic fraction is carbon. The formulation $C_{60}H_{87}O_{23}N_{12}P$ can be used when phosphorus is also considered. Compounds comprising the inorganic portion include P_2O_5 (50 percent), SO_3 (15 percent), Na_2O (11 percent), CaO (9 percent), MgO (8 percent), K_2O (6 percent), and Fe_2O_3 (1 percent). Because all these elements and compounds must be derived from the environment, a shortage of any of these substances would limit and, in some cases, alter growth.

Environmental Requirements. Environmental conditions of temperature and pH have an important effect on the survival and growth of bacteria. In general, optimal growth occurs within a fairly narrow range of temperature and pH, although the bacteria may be able to survive within much broader limits. Temperatures below the optimum typically have a more significant effect on growth rate than temperatures above the optimum; it has been observed that growth rates double with approximately

TABLE 8-3
Typical composition of bacterial cells[a]

Element	Percentage of dry mass	
	Range	Typical
Carbon	45–55	50
Oxygen	16–22	20
Nitrogen	12–16	14
Hydrogen	7–10	8
Phosphorus	2–5	3
Sulfur	0.8–1.5	1
Potassium	0.8–1.5	1
Sodium	0.5–2.0	1
Calcium	0.4–0.7	0.5
Magnesium	0.4–0.7	0.5
Chlorine	0.4–0.7	0.5
Iron	0.1–0.4	0.2
All others	0.2–0.5	0.3

[a] Adapted from Refs. 12, 34, and 35.

every 10°C increase in temperature until the optimum temperature is reached. According to the temperature range in which they function best, bacteria may be classified as *psychrophilic, mesophilic, or thermophilic*. Typical temperature ranges for bacteria in each of these categories are presented in Table 8-4. For a more detailed discussion of the organisms in the various temperature ranges, see Refs. 13–15, 34.

The pH of the environment is also a key factor in the growth of organisms. Most bacteria cannot tolerate pH levels above 9.5 or below 4.0. Generally, the optimum pH for bacterial growth lies between 6.5 and 7.5.

Fungi

Fungi of importance in environmental engineering are considered to be multicellular, non-photosynthetic, heterotrophic protists. Fungi are usually classified by their mode of reproduction. They reproduce sexually or asexually, by fission, budding, or spore formation. Molds, or "true fungi," produce microscopic units (hyphae) that collectively form a filamentous mass called the *mycelium*. Yeasts are fungi that cannot form a mycelium and are therefore unicellular.

Most fungi are strict aerobes. They have the ability to grow under low-moisture conditions and can tolerate an environment with a relatively low pH. The optimum pH for most species is 5.6; the range is 2 to 9. Fungi also have a low nitrogen requirement, needing approximately one-half as much as bacteria. The ability of the fungi to survive under low pH and nitrogen-limiting conditions, coupled with their ability to degrade cellulose, makes them very important in the biological treatment of some industrial wastes and in the composting of solid organic wastes.

Protozoa and Rotifers

Protozoa are motile, microscopic protists that are usually single cells. The majority of protozoa are aerobic heterotrophs, although a few are anaerobic. Protozoa are generally larger than bacteria and often consume bacteria as an energy source. In effect, the protozoa act as polishers of the effluents from biological waste-treatment processes by consuming bacteria and particulate organic matter.

TABLE 8-4
Some typical temperature ranges for various bacteria

Type	Temperature, °C	
	Range	Optimum
Psychrophilic[a]	−10–30	12–18
Mesophilic	20–50	25–40
Thermophilic	35–75	55–65

[a] Also called *Cryophilic*.

Note: 1.8(°C) + 32 = °F.

The rotifer is an aerobic, heterotrophic, and multicellular animal. Its name is derived from the fact that it has two sets of rotating cilia on its head, which are used for motility and capturing food. Rotifers are very effective in consuming dispersed and flocculated bacteria and small particles of organic matter. Their presence in an effluent indicates a highly efficient aerobic biological purification process.

Algae

Algae are unicellular or multicellular, autotrophic, photosynthetic protists. They are of importance in biological treatment processes for two reasons. In ponds, the ability of algae to produce oxygen by photosynthesis is vital to the ecology of the water environment. For an aerobic or facultative oxidation pond to operate effectively, algae are needed to supply oxygen to aerobic, heterotrophic bacteria. This symbiotic relationship between algae and bacteria will be expanded upon in Sec. 8-12, which deals with aerobic and facultative oxidation ponds.

Algae are also important in biological treatment processes because the problem of preventing excessive algal growth in receiving waters has, to date, centered around nutrient removal in the treatment process. Some scientists advocate the removal of nitrogen from treatment plant effluents, others recommend the removal of phosphorus, and still others recommend removal of both. The choice of treatment objectives influences the type of biological process selected.

8-4 BACTERIAL GROWTH

Effective environmental control in biological waste treatment is based on an understanding of the basic principles governing the growth of microorganisms. The following discussion is concerned with the growth of bacteria, the microorganisms of primary importance in biological treatment.

General Growth Patterns in Pure Cultures

As mentioned earlier, bacteria can reproduce by binary fission, by a sexual mode, or by budding. Generally, they reproduce by binary fission (i.e., by dividing, the original cell becomes two new organisms). The time required for each fission, which is termed the generation time, can vary from days to less than 20 min. For example, if the generation time is 30 min, one bacterium would yield 16,777,216 bacteria after a period of 12 h. This computed value is a hypothetical figure, for bacteria would not continue to divide indefinitely because of various environmental limitations such as substrate concentration, nutrient concentration, or even system size.

Growth in Terms of Bacterial Numbers. The general growth pattern of bacteria in a batch culture is shown in Fig. 8-5. Initially, a small number of organisms are inoculated into a fixed volume of culture medium, and the number of viable organisms is recorded as a function of time. The growth pattern based on the number of cells has four more or less distinct phases.

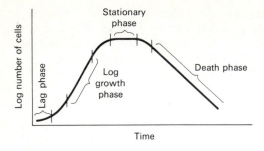

FIGURE 8-5
Typical bacterial growth curve in terms of numbers.

1. *The lag phase.* Upon addition of an inoculum to a culture medium, the lag phase represents the time required for the organisms to acclimate to their new environment and begin to divide.
2. *The log-growth phase.* During this period the cells divide at a rate determined by their generation time and their ability to process food (constant percentage growth rate).
3. *The stationary phase.* Here the population remains stationary. Reasons advanced for this phenomenon are (a) that the cells have exhausted the substrate or nutrients necessary for growth, and (b) that the growth of new cells is offset by the death of old cells.
4. *The log-death phase.* During this phase, the bacteria death rate exceeds the production of new cells. The death rate is usually a function of the viable population and environmental characteristics. In some cases, the log-death phase is the inverse of the log-growth phase.

Growth in Terms of Bacterial Mass. The growth pattern can also be discussed in terms of the variation of the mass of microorganisms with time. This growth pattern consists of the following four phases:

1. *The lag phase.* Again, bacteria require time to acclimate to their nutritional environment. The lag phase in terms of bacterial mass is not as long as the corresponding lag phase in terms of numbers because mass begins to increase before cell division takes place.
2. *The log-growth phase.* There is always an excess amount of food surrounding the microorganisms, and the rate of metabolism and growth is only a function of the ability of the microorganism to process the substrate.
3. *Declining growth phase.* The rate of increase of bacterial mass decreases because of limitations in the food supply.
4. *Endogenous phase.* The microorganisms are forced to metabolize their own protoplasm without replacement because the concentration of available food is at a minimum. During this phase, a phenomenon known as *lysis* can occur in which the nutrients remaining in the dead cells diffuse out to furnish the remaining cells with food (known as "cryptic growth").

Growth in Mixed Cultures

It is important to note that the preceding discussions concerned a single population of microorganisms. Most biological treatment processes are comprised of complex, interrelated, mixed biological populations, with each particular microorganism in the system having its own growth curve. The position and shape of a particular growth curve in the system, on a time scale, depend on the food and nutrients available and on environmental factors such as temperature, pH, and whether the system is aerobic or anaerobic. The variation of microorganism predominance with time in the aerobic stabilization of liquid organic waste is given in Fig. 8-6. While the bacteria are of primary importance, many other microorganisms take part in the stabilization of the organic waste. When designing or analyzing a biological treatment process, the engineer should think in terms of an ecosystem or community, such as the one shown in Fig. 8-6, and not in terms of a "black box" that contains mysterious microorganisms.

8-5 KINETICS OF BIOLOGICAL GROWTH

The need for a controlled environment and biological community in the design of biological waste-treatment units is stressed throughout this chapter. The classes of microorganisms of importance in wastewater treatment have been discussed, along with their metabolic characteristics and their growth patterns. Although the characteristics of the environment needed for their growth have been described, nothing has been said about how to control the environment of the microorganisms. Environmental conditions can be controlled by pH regulation, temperature regulation, nutrient or trace-element addition, oxygen addition or exclusion, and proper mixing. Control

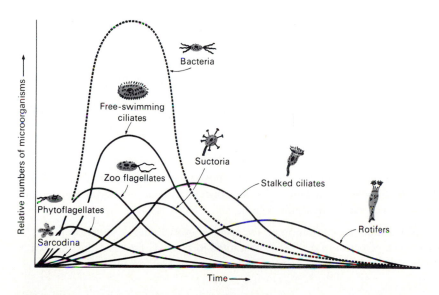

FIGURE 8-6
Relative growth of microorganisms stabilizing organic waste in a liquid environment [24].

of the environmental conditions will ensure that the microorganisms have a proper medium in which to grow.

To ensure that the microorganisms will grow, they must be allowed to remain in the system long enough to reproduce. This period depends on their growth rate, which is related directly to the rate at which they metabolize or utilize the waste. Assuming that the environmental conditions are controlled properly, effective waste stabilization can be ensured by controlling the growth rate of the microorganisms. The purpose of this section is to consider the kinetics of biological growth.

Cell Growth

In both batch and continuous culture systems the rate of growth of bacterial cells can be defined by the following relationship.

$$r_g = \mu X \tag{8-1}$$

where r_g = rate of bacterial growth, mass/unit volume · time
μ = specific growth rate, time^{-1}
X = concentration of microorganism, mass/ unit volume

Because $dX/dt = r_g$ for batch culture (see Appendix G), the following relationship is also valid for a batch reactor:

$$\frac{dX}{dt} = \mu X \tag{8-2}$$

Substrate Limited Growth

In a batch culture, if one of the essential requirements (substrate and nutrients) for growth were present in only limited amounts, it would be depleted first and growth would cease (see Fig. 8-5). In a continuous culture, growth is limited. Experimentally, it has been found that the effect of a limiting substrate or nutrient can often be defined adequately using the following expression proposed by Monod [25,26]:

$$\mu = \mu_m \frac{S}{K_s + S} \tag{8-3}$$

where μ = specific growth rate, time^{-1}
μ_m = maximum specific growth rate, time^{-1}
S = concentration of growth-limiting substrate in solution, mass/unit volume
K_s = half-velocity constant, substrate concentration at one-half the maximum growth rate, mass/ unit volume

The effect of substrate concentration on the specific growth rate is shown in Fig. 8-7.

If the value of μ from Eq. 8-3 is substituted in Eq. 8-1, the resulting expression for the rate of growth is

$$r_g = \frac{\mu_m X S}{K_s + S} \tag{8-4}$$

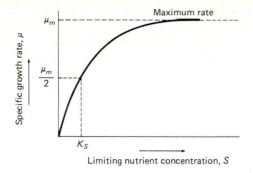

FIGURE 8-7
Plot showing the effects of a limiting nutrient on the specific growth rate.

Cell Growth and Substrate Utilization

In both batch- and continuous-growth culture systems, a portion of the substrate is converted to new cells and a portion is oxidized to inorganic and organic end products. Because the quantity of new cells produced has been observed to be reproducible for a given substrate, the following relationship has been developed between the rate of substrate utilization and the rate of growth.

$$r_g = -Y r_{su} \tag{8-5}$$

where r_g = rate of bacterial growth, mass/unit volume · time
 Y = maximum yield coefficient, mg/mg (defined as the ratio of the mass of cells formed to the mass of substrate consumed, measured during any finite period of logarithmic growth)
 r_{su} = substrate utilization rate, mass/unit volume · time

On the basis of laboratory studies, it has been concluded that yield depends on (1) the oxidation state of the carbon source and nutrient elements, (2) the degree of polymerization of the substrate, (3) pathways of metabolism, (4) the growth rate, and (5) various physical parameters of cultivation.

If the value of r_g from Eq. 8-4 is substituted in Eq. 8-5, the rate of substrate utilization can be defined as follows:

$$r_{su} = -\frac{\mu_m X S}{Y(K_s + S)} \tag{8-6}$$

In Eq. 8-6, the term μ_m / Y is often replaced by the term k, defined as the maximum rate of substrate utilization per unit mass of microorganisms:

$$k = \frac{\mu_m}{Y} \tag{8-7}$$

If the term k is substituted for the term (μ_m / Y) in Eq. 8-6, the resulting expression is

$$r_{su} = -\frac{kX S}{K_s + S} \tag{8-8}$$

Effects of Endogenous Metabolism

In bacterial systems used for wastewater treatment, the distribution of cell ages is such that not all the cells in the system are in the log-growth phase. Consequently, the expression for the rate of growth must be corrected to account for the energy required for cell maintenance. Other factors, such as death and predation, must also be considered. Usually, these factors are lumped together, and it is assumed that the decrease in cell mass caused by them is proportional to the concentration of organisms present. This decrease is often identified in the literature as the endogenous decay. The endogenous decay term can be formulated as follows:

$$r_d \text{ (endogenous decay)} = -k_d X \tag{8-9}$$

where k_d = endogenous decay coefficient, time^{-1}

$\quad\quad X$ = concentration of cells, mass/unit volume

When Eq. 8-9 is combined with Eqs. 8-4 and 8-5, the following expressions are obtained for the net rate of growth:

$$r'_g = \frac{\mu_m X S}{K_s + S} - k_d X \tag{8-10}$$

$$r'_g = -Y r_{su} - k_d X \tag{8-11}$$

where r'_g = net rate of bacterial growth, mass/unit volume · time.

The corresponding expression for the net specific growth rate is given by Eq. 8-12, which is the same as the expression proposed by Van Uden [39]:

$$\mu' = \mu_m \frac{S}{K_s + S} - k_d \tag{8-12}$$

where μ' = net specific growth rate, time^{-1}.

The effects of endogenous respiration on the net bacterial yield are accounted for by defining an observed yield as follows [29,39]:

$$Y_{obs} = -\frac{r'_g}{r_{su}} \tag{8-13}$$

Effects of Temperature

The temperature dependence of the biological reaction-rate constants is very important in assessing the overall efficiency of a biological treatment process. Temperature not only influences the metabolic activities of the microbial population but also has a profound effect on such factors as gas-transfer rates and the settling characteristics of the biological solids. The effect of temperature on the reaction rate of a biological process is usually expressed in the following form:

$$r_T = r_{20} \theta^{(T-20)} \tag{8-14}$$

where r_T = reaction rate at $T°C$

r_{20} = reaction rate at 20°C

θ = temperature-activity coefficient

T = temperature, °C

Some typical values of θ for some commonly used biological processes are presented in Table 8-5. These values should not be confused with values given previously in Chap. 3 for the BOD determination.

Other Rate Expressions

In reviewing the kinetic expressions used to describe the growth of microorganisms and the removal of substrate, it is very important to remember that the expressions presented are empirical and were used for the purpose of illustration and that they are not the only expressions available. Other expressions which have been used to describe the rate of substrate utilization include the following:

$$r_{su} = -k \tag{8-15}$$

$$r_{su} = -kS \tag{8-16}$$

$$r_{su} = -kXS \tag{8-17}$$

$$r_{su} = -kX\frac{S}{S_0} \tag{8-18}$$

Expressions for the specific growth rate (see Eq. 8-3) have been proposed by a number of persons including Monod, Teissier, Contois, and Moser [26,34].

What is fundamental in the use of any rate expression is its application in a mass-balance analysis. In this connection, it does not matter if the rate expression selected has no relationship to those used commonly in the literature, so long as it describes the observed phenomenon. It is equally important to remember that specific rate expressions should not be generalized to cover a broad range of situations on the basis of limited data or experience.

TABLE 8-5
Temperature activity coefficients for various biological treatment processes

Process	θ value	
	Range	Typical
Activated sludge	1.00–1.08	1.04
Aerated lagoons	1.04–1.10	1.08
Trickling filters	1.02–1.08	1.035

Application of Growth and Substrate Removal Kinetics to Biological Treatment

Before discussing the individual biological processes used for the treatment of waste-water, the general application of the kinetics of biological growth and substrate removal will be explained. The purpose here is to illustrate (1) the development of microorganism and substrate balances and (2) the prediction of effluent microorganism and substrate concentrations. In this discussion, an aerobic treatment process carried out in a complete-mix reactor without recycle will be considered (see Fig. 8-8). The schematic shown is the same as that for the activated-sludge process without recycle to be considered in Sec. 8-7. It is interesting to note that the complete-mix reactor is also essentially the same as a chemostat used in laboratory studies (see Fig. 8-9).

Microorganism and Substrate Mass Balances. A mass balance for the mass of microorganisms in the complete-mix reactor shown in Fig. 8-8 can be written as follows:

1. General word statement:

$$\begin{array}{l}\text{Rate of accumulation} \\ \text{of microorganism} \\ \text{within the system} \\ \text{boundary}\end{array} = \begin{array}{l}\text{Rate of flow of} \\ \text{microorganism} \\ \text{into the system} \\ \text{boundary}\end{array} - \begin{array}{l}\text{Rate of flow of} \\ \text{microorganism} \\ \text{out of the system} \\ \text{boundary}\end{array} + \begin{array}{l}\text{Net growth of} \\ \text{microorganism} \\ \text{within the} \\ \text{system boundary}\end{array} \qquad (8\text{-}19)$$

2. Simplified word statement:

$$\text{Accumulation} = \text{Inflow} - \text{Outflow} + \text{Net growth} \qquad (8\text{-}20)$$

3. Symbolic representation:

$$\frac{dX}{dt} V_r = QX_0 - QX + V_r r_g' \qquad (8\text{-}21)$$

where dX/dt = rate of change of microorganism concentration in the reactor measured in terms of mass (volatile suspended solids), mass VSS/unit volume · time

V_r = reactor volume

Q = flowrate, volume/time

X_0 = concentration of microorganisms in influent, mass VSS/unit volume

X = concentration of microorganisms in reactor, mass VSS/unit volume

r_g' = net rate of microorganism growth, mass VSS/unit volume · time

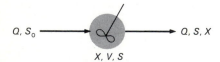

Q, S_0 Q, S, X

X, V, S

FIGURE 8-8
Schematic of a complete-mix reactor without recycle.

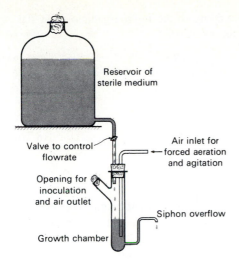

FIGURE 8-9
A schematic diagram of a laboratory chemostat [34].

In Eq. 8-21 and subsequent expressions derived from it, the volatile fraction of the total biological suspended solids is used as an approximation of active biological mass. The assumption is made that the volatile fraction is proportional to the activity of the microbial mass in question. Although a number of other measures, such as nitrogen, protein, DNA, and ATP content, have been used, the volatile suspended solids test is used principally because of its simplicity.

If the value of r'_g from Eq. 8-10 is substituted into Eq. 8-21, the result is

$$\frac{dX}{dt} V_r = QX_0 - QX + V_r\left(\frac{\mu_m X S}{K_s + S} - k_d X\right) \tag{8-22}$$

where S = substrate concentration in effluent from reactor, mg/L.

If it is assumed that the concentration of microorganisms in the influent can be neglected and that steady-state conditions prevail $(dX/dt = 0)$, Eq. 8-22 can be simplified to yield

$$\frac{Q}{V_r} = \frac{1}{\theta} = \frac{\mu_m S}{K_s + S} - k_d \tag{8-23}$$

where θ = hydraulic detention time, V/Q.

In Eq. 8-23, the term $1/\theta$ corresponds to the net specific growth rate (see Eq. 8-12). The term $1/\theta$ also corresponds to $1/\theta_c$ where θ_c is the mean cell residence time. In the field of wastewater treatment, θ_c may be defined as the mass of organisms in the reactor divided by the mass of organisms removed from the system each day. (A second commonly used definition is given in Sec. 8-7.) For the reactor shown in Fig. 8-8, θ_c is given by the following expression.

$$\theta_c = \frac{V_r X}{QX} = \frac{V_r}{Q} \tag{8-24}$$

Performing a substrate balance corresponding to the microorganism mass balance given in Eq. 8-22 results in the following expression.

$$\frac{dS}{dt}V_r = QS_o - QS + V_r\left(\frac{kXS}{K_s + S}\right) \tag{8-25}$$

At steady state $(dS/dt = 0)$, the resulting equation is

$$(S_o - S) - \theta\left(\frac{kXS}{K_s + S}\right) = 0 \tag{8-26}$$

where $\theta = V_r/Q$.

Effluent Microorganism and Substrate Concentrations. The effluent microorganism and substrate concentrations may be obtained as follows. If Eq. 8-23 is solved for the term $S/(K_s + S)$ and the resulting expression is substituted into Eq. 8-26 and simplified using Eq. 8-7, then effluent steady-state concentration is found to be given as

$$X = \frac{\mu_m(S_o - S)}{k(1 + k_d\theta)} = \frac{Y(S_o - S)}{(1 + k_d\theta)} \tag{8-27}$$

Similarly, the effluent substrate concentration is found to be equal to

$$S = \frac{K_s(1 + \theta k_d)}{\theta(Yk - k_d) - 1} \tag{8-28}$$

Thus, if the kinetic coefficients are known, Eqs. 8-27 and 8-28 can be used to predict effluent microorganism and substrate concentrations (see Fig. 8-10). It is important to note that the effluent concentrations predicted using the above equations are based on a soluble waste and do not take into account any influent suspended solids that may be present. Actual effluent substrate and suspended-solids concentrations from the treatment process are dependent on the performance of the sedimentation tanks.

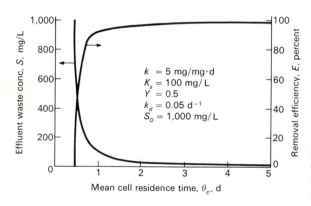

$k = 5$ mg/mg·d
$K_s = 100$ mg/L
$Y = 0.5$
$k_d = 0.05$ d^{-1}
$S_o = 1,000$ mg/L

FIGURE 8-10
Effluent waste concentration and removal efficiency versus mean cell-residence time for a complete-mix reactor without recycle $(\theta = \theta_c)$.

The observed yield, Y_{obs}, is given by the following expression.

$$Y_{obs} = \frac{Y}{1 + k_d \theta} \tag{8-29}$$

Equation 8-29 is derived by substituting the value of X given by Eq. 8-27 for r_g' in Eq. 8-13 and by dividing by the term $(S_0 - S)$, which corresponds to the value of r_{su} expressed as a concentration value.

8-6 BIOLOGICAL TREATMENT PROCESSES

The purpose of this section is to introduce the reader to the principal types of biological treatment processes that have been developed for the treatment of wastewaters and to identify their applications. The individual treatment processes used most commonly for the treatment of wastewater are discussed in the remainder of this chapter.

Some Useful Definitions

To understand the concepts of biological treatment, it will be helpful to know the following terms:

Aerobic processes are biological treatment processes that occur in the presence of oxygen.

Anaerobic processes are biological treatment processes that occur in the absence of oxygen.

Anoxic denitrification is the process by which nitrate nitrogen is converted biologically to nitrogen gas in the absence of oxygen. This process is also known as anaerobic denitrification.

Biological nutrient removal is the term applied to the removal of nitrogen and phosphorus in biological treatment processes.

Facultative processes are biological treatment processes in which the organisms can function in the presence or absence of molecular oxygen.

Carbonaceous BOD removal is the biological conversion of the carbonaceous organic matter in wastewater to cell tissue and various gaseous end products. In the conversion, it is assumed that the nitrogen present in the various compounds is converted to ammonia.

Nitrification is the biological process by which ammonia is converted first to nitrite and then to nitrate.

Denitrification is the biological process by which nitrate is converted to nitrogen and other gaseous end products.

Substrate is the term used to denote the organic matter or nutrients that are converted during biological treatment or that may be limiting in biological treatment. For example, the carbonaceous organic matter in wastewater is referred to as the substrate that is converted during biological treatment.

Suspended-growth processes are the biological treatment processes in which the microorganisms responsible for the conversion of the organic matter or other constituents in the wastewater to gases and cell tissue are maintained in suspension within the liquid.

Attached-growth processes are the biological treatment processes in which the microorganisms responsible for the conversion of the organic matter or other constituents in the wastewater to gases and cell tissue are attached to some inert medium such as rocks, slag, or specially designed ceramic or plastic materials. Attached-growth treatment processes are also known as fixed-film processes.

Biological Treatment Processes

The major biological processes used for wastewater treatment are identified in Table 8-6. There are five major groups: aerobic processes, anoxic processes, anaerobic processes, combined aerobic, anoxic, and anaerobic processes, and pond processes. The individual processes are further subdivided, depending on whether treatment is accomplished in suspended-growth systems, attached-growth systems, or combinations thereof.

It should be noted that all of the biological processes used for the treatment of wastewater, as reported in Table 8-6, are derived from processes occurring in nature. The aerobic and anaerobic cycles, shown in Figs. 8-11 and 8-12, respectively, are typical examples. By controlling the environment of the microorganisms, the decomposition of wastes is speeded up. Regardless of the type of waste, the biological treatment process consists of controlling the environment required for optimum growth of the microorganisms involved.

Application of Biological Treatment Processes

The principal applications of these processes, also identified in Table 8-6, are for (1) the removal of the carbonaceous organic matter in wastewater, usually measured as BOD, total organic carbon (TOC), or chemical oxygen demand (COD); (2) nitrification; (3) denitrification; (4) phosphorus removal; and (5) waste stabilization. In the remainder of this chapter, the emphasis will be on the removal of carbonaceous material, both aerobically and anaerobically. For clarity, biological nutrient removal and pond processes are considered in separate sections. Nitrification, denitrification, and phosphorus removal are considered in greater detail in Chap. 11. Sludge stabilization is discussed in Chap 12.

8-7 AEROBIC SUSPENDED-GROWTH TREATMENT PROCESSES

The principal suspended-growth biological treatment processes used for the removal of carbonaceous organic matter are (1) the activated-sludge process, (2) aerated lagoons,

(3) a sequencing batch reactor, and (4) the aerobic digestion process. Of these, the activated-sludge process is by far the one most commonly used for the secondary treatment of domestic wastewater, and for this reason it will be stressed in this section. Suspended-growth nitrification is considered in Sec. 8-11, which deals with biological nutrient removal.

Activated-Sludge Process

The activated-sludge process was developed in England in 1914 by Ardern and Lockett [3] and was so named because it involved the production of an activated mass of microorganisms capable of stabilizing a waste aerobically. Many versions of the original process are in use today, but fundamentally they are all similar. The system shown in Fig. 8-13 is the complete-mix activated-sludge system. Other activated-sludge systems are listed in Table 8-6 and discussed in Chap. 10.

Process Description. Operationally, biological waste treatment with the activated-sludge process is typically accomplished using a flow diagram such as that shown in Fig. 8-13. Organic waste is introduced into a reactor where an aerobic bacterial culture is maintained in suspension. The reactor contents are referred to as the "mixed liquor." In the reactor, the bacterial culture carries out the conversion in general accordance with the stoichiometry shown in Eqs. 8-30 and 8-31.

Oxidation and synthesis:

$$COHNS + O_2 + nutrients \xrightarrow{\text{bacteria}} CO_2 + NH_3 + C_5H_7NO_2 + \text{other end products} \tag{8-30}$$

(organic matter)　　　　　　　　　　　　　　　　　　　(new bacterial cells)

Endogenous respiration:

$$C_5H_7NO_2 + 5O_2 \xrightarrow{\text{bacteria}} 5CO_2 + 2H_2O + NH_3 + \text{energy} \tag{8-31}$$

(cells)

$$\begin{array}{cc} 113 & 160 \\ 1 & 1.42 \end{array}$$

In these equations, COHNS represents the organic matter in wastewater. Although the endogenous respiration reaction results in relatively simple end products and energy, stable organic end products are also formed. From Eq. 8-31, it can be seen that, if all of the cells can be oxidized completely, the ultimate BOD of the cells is equal to 1.42 times the concentration of cells.

The aerobic environment in the reactor is achieved by the use of diffused or mechanical aeration, which also serves to maintain the mixed liquor in a completely mixed regime. After a specified period of time, the mixture of new cells and old cells is passed into a settling tank, where the cells are separated from the treated wastewater. A portion of the settled cells is recycled to maintain the desired concentration of organisms in the reactor, and a portion is wasted (see Fig. 8-13b). The portion wasted corresponds to the new growth of cell tissue, r'_g (see Eq. 8-11), associated with a

TABLE 8-6
Major biological treatment processes used for wastewater treatment

Type	Common name	Use[a]	See section
Aerobic processes:			
Suspended-growth	Activated-sludge process Conventional (plug-flow) Complete-mix Step aeration Pure oxygen Sequencing batch reactor Contact stabilization Extended aeration Oxidation ditch Deep tank (90 ft) Deep shaft	Carbonaceous BOD removal (nitrification)	8-7, 10-1, 10-2, 10-3
	Suspended-growth nitrification	Nitrification	8-11, 11-6
	Aerated lagoons	Carbonaceous BOD removal (nitrification)	8-7
	Aerobic digestion Conventional air Pure oxygen	Stabilization, carbonaceous BOD removal	8-7, 12-9
Attached-growth	Trickling filters Low-rate High-rate	Carbonaceous BOD removal, nitrification	8-8, 10-5
	Roughing filters	Carbonaceous BOD removal	8-8, 10-5
	Rotating biological contactors	Carbonaceous BOD removal (nitrification)	8-8, 10-6
	Packed-bed reactors	Carbonaceous BOD removal (nitrification)	8-8
Combined suspended- and attached-growth processes	Activated biofilter process, trickling-filter solids-contact process, biofilter activated-sludge process, series trickling-filter activated-sludge process	Carbonaceous BOD removal (nitrification)	10-7

	Use[a]	
Anoxic processes:		
Suspended-growth	Suspended-growth denitrification	Denitrification	8-11, 11-7
Attached-growth	Fixed-film denitrification	Denitrification	8-11
Anaerobic processes:		
Suspended-growth	Anaerobic digestion		8-9, 12-8
	Standard rate, single-stage	Stabilization, carbonaceous BOD removal	
	High-rate, single-stage	Stabilization, carbonaceous BOD removal	
	Two-stage	Stabilization, carbonaceous BOD removal	
	Anaerobic contact process	Carbonaceous BOD removal	8-9
	Upflow anaerobic sludge-blanket	Carbonaceous BOD removal	8-9
Attached-growth	Anaerobic filter process	Carbonaceous BOD removal, waste stabilization (denitrification)	8-10
	Expanded bed	Carbonaceous BOD removal, waste stabilization	8-10
Combined aerobic, anoxic, and anaerobic processes:		
Suspended-growth	Single- or multi-stage processes, Various proprietary processes	Carbonaceous BOD removal, nitrification, denitrification, phosphorus removal	8-11, 11-9
Combined suspended- and attached-growth	Single- or multi-stage processes	Carbonaceous BOD removal, nitrification, denitrification, and phosphorus removal	8-11
Pond processes	Aerobic ponds	Carbonaceous BOD removal	8-12, 10-8
	Maturation (tertiary) ponds	Carbonaceous BOD removal (nitrification)	8-12, 10-8
	Facultative ponds	Carbonaceous BOD removal	8-12, 10-8
	Anaerobic-ponds	Carbonaceous BOD removal (waste stabilization)	8-12, 10-8

[a] Major uses are presented first; other uses are identified in parentheses.

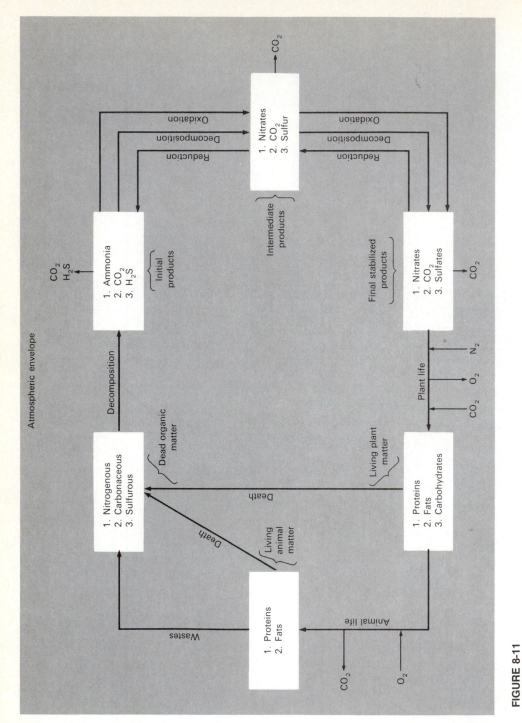

FIGURE 8-11
The aerobic cycle in nature.

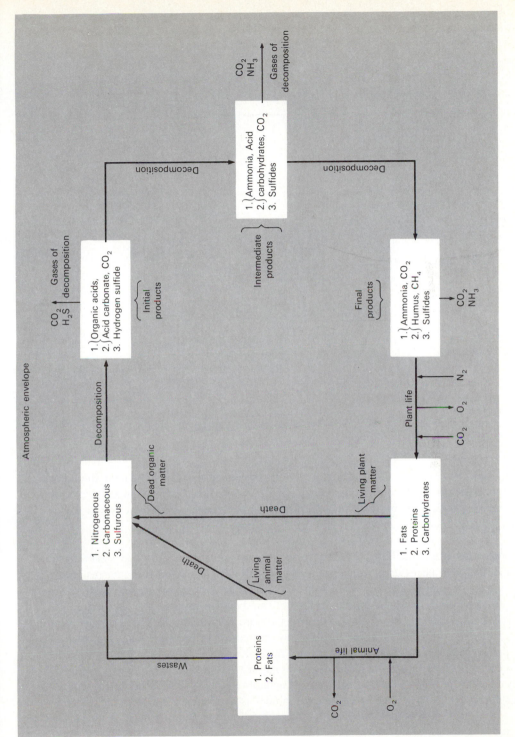

FIGURE 8-12
The anaerobic cycle in nature.

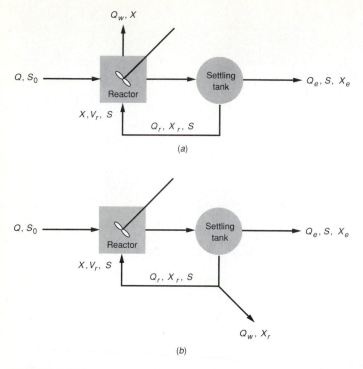

FIGURE 8-13
Schematic of complete-mix reactor with cellular recycle and wasting: (*a*) from the reactor and (*b*) from the recycle line.

particular wastewater. The level at which the biological mass in the reactor should be kept depends on the desired treatment efficiency and other considerations related to growth kinetics. Microorganism concentrations maintained in various activated-sludge treatment systems are listed in Table 10-5 in Chap. 10.

Process Microbiology. To design and operate an activated-sludge system efficiently, it is necessary to understand the importance of the microorganisms in the system. In nature, the key role of the bacteria is to decompose organic matter produced by other living organisms. In the activated-sludge process, the bacteria are the most important microorganisms because they are responsible for the decomposition of the organic material in the influent. In the reactor or aeration tank, a portion of the organic waste is used by aerobic and facultative bacteria to obtain energy for the synthesis of the remainder of the organic material into new cells, as shown in Fig. 8-1. Only a portion of the original waste is actually oxidized to low-energy compounds such as NO_3^-, SO_4^{-2}, and CO_2; the remainder is synthesized into cellular material. Also, many intermediate products are formed before the end products, shown in the right-hand side of Eq. 8-30, are produced.

In general, the bacteria in the activated-sludge process include members of the genera *Pseudomonas*, *Zoogloea*, *Achromobacter*, *Flavobacterium*, *Nocardia*,

Bdellovibrio, Mycobacterium, and the two nitrifying bacteria, *Nitrosomonas* and *Nitrobacter* [13,14]. Additionally, various filamentous forms, such as *Sphaerotilus, Beggiatoa, Thiothrix, Lecicothrix*, and *Geotrichum*, may also be present [13,14]. While the bacteria are the microorganisms that actually degrade the organic waste in the influent, the metabolic activities of other microorganisms are also important in the activated-sludge system. For example, protozoa and rotifers act as effluent polishers. Protozoa consume dispersed bacteria that have not flocculated, and rotifers consume small biological floc particles that have not settled.

Further, although it is important that bacteria decompose the organic waste as quickly as possible, it is also important that they form a satisfactory floc, which is a prerequisite for the effective separation of the biological solids in the settling unit. It has been observed that as the mean cell-residence time of the cells in the system is increased, the settling characteristics of the biological floc are enhanced. For domestic wastes, mean cell-residence times on the order of 3 to 4 d are required to achieve effective settling. Typical values of mean cell-residence times used in the design and operation of various activated-sludge processes are shown in Table 10-5.

Even though excellent floc formation is obtained, the effluent from the system could still be high in biological solids as a result of poor design of the secondary settling unit, poor operation of the aeration units, or the presence of filamentous microorganisms such as *Sphaerotilus, E. coli,* and fungi [14,17,42]. These subjects are considered later in this section and in detail in Chap. 10.

Process Analysis: Complete Mix with Recycle. In the complete-mix system, shown schematically in Fig. 8-13 and pictorially in Fig. 8-14, the contents of the

FIGURE 8-14
Typical complete-mix activated sludge reactor with surface aeration.

reactor are mixed completely, and it is assumed that there are no microorganisms in the wastewater influent. As shown in Fig. 8-13, an integral part of the activated-sludge process is a solids separation unit (sedimentation tank) in which the cells from the reactor are separated (settled) and then returned to the reactor. Because of the presence of this solids separation unit, two additional assumptions must be made in the development of the kinetic model for this system:

1. Waste stabilization by the microorganisms occurs only in the reactor unit. This assumption leads to a conservative model (in some systems there may be some waste stabilization in the settling unit).
2. The volume used in calculating the mean cell-residence time (discussed below) for the system includes only the volume of the reactor unit.

In effect, it is assumed that the sedimentation tank serves as a reservoir from which solids are returned to maintain a given solids level in the aeration tank. If the system is such that these assumptions do not hold true, then the model should be modified. For example, in high-purity-oxygen activated-sludge systems, it has been found that up to 50 percent of the total solids in the system may be present in the secondary settling tank. This subject is considered further in the following discussion and in Chap. 10.

The mean hydraulic retention time for the system θ_s is defined as

$$\theta_s = \frac{V_T}{Q} = \frac{V_r + V_s}{Q} \tag{8-32}$$

where V_T = volume of reactor plus volume of settling tank
Q = influent flowrate
V_r = volume of reactor
V_s = volume of settling tank

The mean hydraulic retention time for the reactor θ is defined as

$$\theta = \frac{V_r}{Q} \tag{8-33}$$

where V_r is the volume of the reactor.

For the system shown in Fig. 8-13a, the mean cell-residence time θ_c, defined as the mass of organisms in the reactor divided by the the mass of organisms removed from the system each day, is given by the following expression.

$$\theta_c = \frac{V_r X}{Q_w X + Q_e X_e} \tag{8-34}$$

where Q_w = flowrate of liquid containing the biological cells to be removed (wasted) from the system (in this case from the reactor)
Q_e = flowrate of liquid from the separation unit
X_e = microorganism concentration in effluent from solids separation unit

For the system shown in Fig. 8-13b, the mean cell residence time θ_c is given by the following expression.

$$\theta_c = \frac{V_r X}{Q'_w X_r + Q_e X_e} \tag{8-35}$$

where X_r = microorganism concentration in return sludge line
Q'_w = cell wastage rate from recycle line

It should be noted that in the literature the value of θ_c is often computed by considering the mass of organisms in both the reactor and the sedimentation tank. Either method is acceptable as long as the basis for computation is noted clearly. Comparing Eq. 8-34 or 8-35 with Eqs. 8-32 and 8-33, it can be seen that a given reactor volume θ_c is theoretically independent of both θ and θ_s. Practically speaking, however, θ_c cannot be completely independent of θ and θ_s. The factors relating θ_c to θ and θ_s are discussed later.

Referring to Fig. 8-13a, a mass balance for the microorganisms in the entire system can be written as follows:

1. General word statement:

Rate of accumulation of microorganism within the system boundary		Rate of flow of microorganism into the system boundary	Rate of flow of microorganism out of the system boundary	Net growth of microorganism within the system boundary	
	=		−	+	(8-36)

2. Simplified word statement:

$$\text{Accumulation} = \text{Inflow} - \text{Outflow} + \text{Net growth} \tag{8-37}$$

3. Symbolic representation:

$$\frac{dX}{dt} V_r = QX_0 - [Q_w X + Q_e X_e] + V_r(r'_g) \tag{8-38}$$

Substituting Eq. 8-11 for the rate of growth and assuming that the cell concentration in the influent is zero and steady-state conditions prevail $(dX/dt = 0)$ yields

$$\frac{Q_w X + Q_e X_e}{V_r X} = -Y \frac{r_{su}}{X} - k_d \tag{8-39}$$

The left-hand side of Eq. 8-39 represents the inverse of the mean cell-residence time as defined previously (see Eq. 8-34). Making use of Eq. 8-35, Eq. 8-39 can be simplified and rearranged to yield

$$\frac{1}{\theta_c} = -Y \frac{r_{su}}{X} - k_d \tag{8-40}$$

The term r_{su} is determined using the following expression:

$$r_{su} = -\frac{Q}{V_r}(S_0 - S) = -\frac{S_0 - S}{\theta} \tag{8-41}$$

where $(S_o - S)$ = mass concentration of substrate utilized, mg/L
S_o = substrate concentration in influent, mg/L
S = substrate concentration in effluent, mg/L
θ = hydraulic detention time, d

The mass concentration of microorganisms X in the reactor can be obtained by substituting Eq. 8-41 into Eq. 8-40 and solving for X.

$$X = \frac{\theta_c \, Y(S_o - S)}{\theta \, (1 + k_d\theta_c)} \tag{8-42}$$

Performing a substrate balance, the effluent substrate concentration is found to be equal to

$$S = \frac{K_s(1 + \theta_c k_d)}{\theta_c(Yk - k_d) - 1} \tag{8-43}$$

It should be noted that Eq. 8-43 is the same as Eq. 8-28, which was developed for a complete-mix reactor without recycle. The corresponding equation for the observed yield in a system with recycle is the same as Eq. 8-29, given previously, with θ_c or θ_{ct} substituted for θ as given below.

$$Y_{obs} = \frac{Y}{1 + k_d\theta_c \text{ or } \theta_{ct}} \tag{8-44}$$

Process Design and Control Relationships. Although Eqs. 8-42 and 8-43 can be useful in predicting the effects of various system changes, they are somewhat difficult to use from a design standpoint because of the many constants involved. For this reason, more usable process design relationships have been developed. The relationships to be considered in the following discussion include the specific utilization rate, mean cell-residence time, and the food-microorganism ratio. The relationship between the specific utilization rate and the mean cell-residence time is also examined.

In Eq. 8-40, the term $(-r_{su}/X)$ is known as the specific substrate utilization rate, U. Using the definition of r_{su} given in Eq. 8-41, the specific utilization rate is calculated as follows:

$$U = -\frac{r_{su}}{X} = \frac{S_o - S}{\theta X} = \frac{Q}{V_r}\frac{S_o - S}{X} \tag{8-45}$$

If the term U is substituted for the term $(-r_{su}/X)$ in Eq. 8-35, the resulting equation is

$$\frac{1}{\theta_c} = YU - k_d \tag{8-46}$$

From Eq. 8-45, it can be seen that $1/\theta_c$, the net specific growth rate, and U, the specific utilization ratio, are related directly. To determine the specific utilization ratio U, the substrate utilized and the mass of microorganisms effective in this utilization must be known. The substrate utilized can be evaluated by determining the difference between the influent and the effluent COD or BOD_5. The evaluation of the active mass of microorganisms is usually what makes the use of U impractical as a control parameter.

Using θ_c as a treatment control parameter, there is neither the need to determine the amount of active biological solids in the system nor the need to evaluate the amount of food utilized. The use of θ_c is simply based on the fact that, to control the growth rate of microorganisms and hence their degree of waste stabilization, a specified percentage of the cell mass in the system must be wasted each day. Thus, if it is determined that a θ_c of 10 days is needed for a desired treatment efficiency, then 10 percent of the total cell mass is wasted from the system per day.

In the complete-mix system with recycle, cell wastage can be accomplished by wasting from the reactor or mixed-liquor return line. If wasting is directly from the reactor and the solids in the effluent X_e are negligible then, referring to Eq. 8-34, only Q_w and V_r need to be known to determine θ_c. Wasting cells in this manner provides for a direct method of controlling and measuring θ_c. In practice, to obtain a thicker sludge, wasting is accomplished by drawing off sludge from the recycle line. Assuming that X_e is very small, Eq. 8-35 can be rewritten as

$$\theta_c \approx \frac{V_r X}{Q'_w X_r} \tag{8-47}$$

Thus, wasting from the recycle line requires that the microorganism concentrations in both the mixed liquor and return sludge be known.

A term closely related to the specific utilization rate U and commonly used in practice as a design and control parameter is known as the food-microorganism ratio (F/M), which is defined as follows:

$$F/M = \frac{S_o}{\theta X} \tag{8-48}$$

The terms U and F/M are related by the process efficiency as follows:

$$U = \frac{(F/M)E}{100} \tag{8-49}$$

where E is the process efficiency as defined by Eq. 8-49:

$$E = \frac{S_o - S}{S_o} \times 100 \tag{8-50}$$

where E = process efficiency, percent
S_o = influent substrate concentration
S = effluent substrate concentration

The application of these process design relationships is illustrated in Example 8-1.

Example 8-1 Activated-sludge process analysis. An organic waste having a soluble BOD_5 of 250 mg/L is to be treated with a complete-mix activated-sludge process. The effluent

BOD_5 is to be equal to or less than 20 mg/L. Assume that the temperature is 20°C, the flowrate is 5.0 Mgal/d, and that the following conditions are applicable.

1. Influent volatile suspended solids to reactor are negligible.
2. Return sludge concentration = 10,000 mg/L of suspended solids = 8,000 mg/L volatile suspended solids.
3. Mixed-liquor volatile suspended solids (MLVSS) = 3,500 mg/L = 0.80 × total MLSS.
4. Mean cell-residence time θ_c = 10 days.
5. Hydraulic regime of reactor = complete mix.
6. Kinetic coefficients, $Y = \dfrac{0.65 \text{ lb cells}}{\text{lb } BOD_5 \text{ utilized}}$, $k_d = 0.06 \ d^{-1}$
7. It is estimated that the effluent will contain about 20 mg/L of biological solids, of which 80 percent is volatile and 65 percent is biodegradable. Assume that the biodegradable biological solids can be converted from ultimate BOD demand to a BOD_5 demand using the factor 0.68 [e.g., BOD K value = $0.1d^{-1}$ (base 10)].
8. Waste contains adequate nitrogen, phosphorus, and other trace nutrients for biological growth.

Solution

1. Estimate the soluble BOD_5 in the effluent.
   Effluent BOD_5 = influent soluble BOD_5 escaping treatment + BOD_5 of effluent biological solids.
   $20 = S + 20(0.65)(1.42)(0.68)$
   $S = 7.4$ mg/L soluble BOD_5
   The biological treatment efficiency based on soluble BOD_5 would be

   $$E_s = \frac{250 - 7.4}{250}(100) = 97\%$$

   The overall plant efficiency would be

   $$E_{overall} = \frac{250 - 20}{250}(100) = 92\%$$

2. Compute the reactor volume. The volume of the reactor can be determined using Eq. 8-42 by substituting V/Q for θ and rearranging the equation as follows:

   $$XV = \frac{YQ\theta_c(S_o - S)}{1 + k_d\theta_c}$$

   $$3{,}500 \text{ mg/L } (V \text{ Mgal}) = \frac{0.65 \ (5 \text{ Mgal/d})(10 \text{ d})(250 \text{ mg/L} - 7.4 \text{ mg/L})}{1 + (0.06/d)(10d)}$$

   $$V = 1.4 \text{ Mgal}$$

3. Compute the sludge-production rate on a mass basis.
   (a) The observed yield is

   $$Y_{obs} = \frac{Y}{(1 + k_d\theta_c)} = \frac{0.65}{1 + 0.06(10)} = 0.406$$

(b) The biomass production rate is

Biomass production, lb VSS/d $= Y$ lb/lb$[(S_o - S)$ mg/L$][(Q$ Mgal/d)$[8.34$ lb/Mgal$\cdot$(mg/L)$]$

$$= 0.406(250 - 7.4)(5)(8.34) = 4,107 \text{ lb SS/d}$$

4. Compute the biomass-wasting rate if wasting is accomplished from the reactor, as shown in Fig. 8-13a, or from the recycle line, as shown in Fig. 8-13b. Take into account the solids lost in the plant effluent (see also the comment at the end of the problem). Also assume that $Q_e = Q$ and the VSS in the effluent is equal to 16.0 mg/L (0.80 $\times$ 20 mg/L).

(a) Determine the wasting rate from the reactor using Eq. 8-34.

$$\theta_c = \frac{V_r X}{Q_w X + Q_e X_e}$$

$$10 = \frac{(1.4 \text{ Mgal})(3,500 \text{ mg/L})}{(Q_w \text{ Mgal/d})(3,500 \text{ mg/L}) + (5 \text{ Mgal/d})(16 \text{ mg/L})}$$

$$Q_w = 0.114 \text{ Mgal/d}$$

(b) Determine the wasting rate from the recycle line using Eq. 8-35.

$$\theta_c = \frac{V_r X}{Q'_w X_r + Q_e X_e}$$

$$10 = \frac{(1.4 \text{ Mgal})(3,500 \text{ mg/L})}{(Q'_w \text{ Mgal/d})(8,000 \text{ mg/L}) + (5 \text{ Mgal/d})(16 \text{ mg/L})}$$

$$Q'_w = 0.050 \text{ Mgal/d}$$

Note that in either case, the weight of sludge wasted is the same (4,107 lb VSS/d), and that either wasting method will achieve a θ_c of 10 days for the system.

5. Compute the recirculation ratio using a suspended solids mass balance around the reactor neglecting the suspended solids in the influent.

$$\text{Aerator VSS conc} = 3,500 \text{ mg/L}$$

$$\text{Return VSS conc} = 8,000 \text{ mg/L}$$

$$3,500(Q + Q_r) = 8,000(Q_r)$$

$$\frac{Q_r}{Q} = R = 0.78$$

6. Compute the hydraulic retention time for the reactor.

$$\text{HRT} = \frac{V}{Q} = \frac{1.4 \text{ Mgal}}{5 \text{ d}} = 0.28 \text{ d} = 6.7 \text{ hr}$$

7. Check the specific substrate utilization rate, the food-to-microorganism ratio, and the volumetric loading rate

(a) The specific substrate utilization rate is

$$U = \frac{S_o - S}{\theta X} = \frac{(250 - 7.4) \text{ mg/L}}{0.28 \text{ d } (3,500 \text{ mg/L})} = 0.25 \frac{\text{mg BOD}_5 \text{ utilized}}{\text{mg MLVSS} \cdot \text{d}}$$

(b) The food-to-microorganism ratio is

$$F/M = \frac{S_o}{\theta X} = \frac{250 \text{ mg/L}}{0.28 \text{ d } (3,500 \text{ mg/L})} = 0.255 \frac{\text{mg BOD}_5 \text{ applied}}{\text{mg MLVSS} \cdot \text{d}}$$

(c) The volumetric loading rate expressed as lb $BOD_5/10^3$ ft^3 is

$$VLR = \frac{(S_o \text{ mg/L})(Q \text{ Mgal/d})[8.34 \text{ lb/Mgal} \cdot (\text{mg/L})](1,000 \text{ ft}^3/10^3 \text{ft}^3)}{(V \text{ Mgal})(10^6 \text{gal}/1.0 \text{ Mgal})/(7.48 \text{ gal/ft}^3)}$$

$$= \frac{250(5)8.34(1,000)}{1.4 \times 10^6/7.48} = \frac{56 \text{ lb BOD}_5 \text{ applied}}{10^3 \text{ ft}^3}$$

Comment. If the solids in the effluent are not considered when the wasting rate is determined, the actual value of the mean cell-residence time will be less than the assumed design value. In this present example, if the volatile solids in the effluent were neglected, the actual mean cell-residence would be about 8.4 days.

Process Performance and Stability. The effects of the kinetics, considered above, on the performance and stability of the system shown in Fig. 8-13 will now be examined further. It was shown in Eq. 8-46 that $1/\theta_c$, the net microorganism growth rate, and U, the specific utilization ratio, are related directly. Combining Eqs. 8-45 with Eq. 8-26, it can be shown that

$$U = \frac{kS}{K_s + S} \tag{8-51}$$

from which the following equation is obtained:

$$S = \frac{UK_s}{k - U} \tag{8-52}$$

For a specified waste, a given biological community, and a particular set of environmental conditions, the kinetic coefficients Y, k, K_s, and k_d are fixed. (It is important to note that domestic wastewater may have significant variability in its composition and may not always be treated as a single waste type in evaluating the kinetic coefficients.) For given values of the coefficients, the effluent-waste concentration from the reactor is a direct function of either θ_c or U, as shown in Eq. 8-51. Setting one of these three parameters not only fixes the other two but also specifies the efficiency of biological waste stabilization. Equations 8-42 and 8-43 are plotted in Fig. 8-15 for a growth-specified complete-mix system with recycle. As shown, the effluent concentration S and the treatment efficiency E are related directly to θ_c.

It can also be seen from Fig. 8-15 that there is a certain value of θ_c below which waste stabilization does not occur. This critical value of θ_c is called the minimum mean cell-residence time θ_c^M. Physically, θ_c^M is the residence time at which the cells are washed out or wasted from the system faster than they can reproduce. The minimum mean cell-residence time can be calculated using Eq. 8-53, which is derived

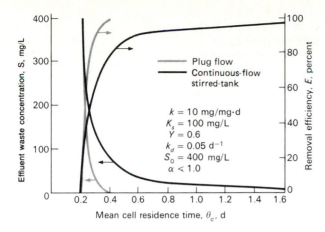

FIGURE 8-15
Effluent waste concentration and removal efficiency for complete-mix and plug-flow reactors with recycle versus mean cell residence time [19].

from Eqs. 8-39, 8-6, and 8-7 . It should be noted that, when washout occurs, the influent concentration S_o is equal to the effluent waste concentration S.

$$\frac{1}{\theta_c^M} = Y\frac{kS_o}{K_s + S_o} - k_d \tag{8-53}$$

In many situations encountered in waste treatment, S_o is much greater than K_s so that Eq. 8-46 can be rewritten to yield

$$\frac{1}{\theta_c^M} \approx Yk - k_d \tag{8-54}$$

Equations 8-53 and 8-54 can be used to determine the minimum mean cell-residence time θ_c^M. Typical kinetic coefficients that can be used to solve for θ_c^M are given in Table 8-7. Obviously, biological treatment systems should not be designed with θ_c values equal to θ_c^M. To ensure adequate waste treatment, biological treatment systems are usually designed and operated with a θ_c value from 2 to 20 times θ_c^M. In effect, the ratio of θ_c to θ_c^M can be considered to be a process safety factor, SF [19]:

$$SF = \frac{\theta_c}{\theta_c^M} \tag{8-55}$$

Plug Flow with Recycle. The plug-flow system with cellular recycle, shown schematically in Fig. 8-16 and pictorially in Fig. 8-17, can be used to model certain forms of the activated-sludge process. The distinguishing feature of this recycle system is that the hydraulic regime of the reactor is of a plug-flow nature. In a true plug-flow model, all the particles entering the reactor stay in the reactor an equal amount of time. Some particles may make more passes through the reactor because of recycle, but, while they are in the tank, they all pass through in the same amount of time.

A kinetic model of the plug-flow system is mathematically difficult, but Lawrence and McCarty [18] have made two simplifying assumptions that lead to a useful kinetic model of the plug-flow reactor:

TABLE 8-7
Typical kinetic coefficients for the activated-sludge process for domestic wastewater[a]

Coefficient	Basis[c]	Value[b] Range	Value[b] Typical
k	d^{-1}	2–10	5
K_s	mg/L BOD_5	25–100	60
	mg/L COD	15–70	40
Y	mg VSS/mg BOD_5	0.4–0.8	0.6
k_d	d^{-1}	0.025–0.075	0.06

[a] Derived in part from Refs. 12, 19, and 42.
[b] Values reported are for 20°C.
[c] VSS = volatile suspended solids.
Note: 1.8(°C) + 32 = °F

1. The concentration of microorganisms in the influent to the reactor is approximately the same as that in the effluent from the reactor. This assumption applies only if $\theta_c/\theta > 5$. The resulting average concentration of microorganisms in the reactor is symbolized as $\overline{X}$.

2. The rate of substrate utilization as the waste passes through the reactor is given by the following expression:

$$r_{su} = -\frac{kS\overline{X}}{K_s + S} \tag{8-56}$$

Integrating Eq. 8-54 over the retention time of the waste in the tank and simplifying gives the following expression:

$$\frac{1}{\theta_c} = \frac{Yk(S_o - S)}{(S_o - S) + (1 + \alpha)K_s \ln(S_i/S)} - k_d \tag{8-57}$$

where S_o = influent concentration
S = effluent concentration

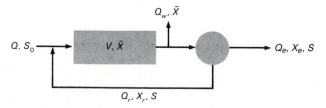

FIGURE 8-16
Plug-flow reactor with cellular recycle.

(a)

(b)

FIGURE 8-17
Typical plug-flow reactors: (a) with dome fine-bubble diffusers and (b) with coarse-bubble diffusers.

S_i = influent concentration to reactor after dilution with recycle flow

$$= \frac{S_o + \alpha S}{1 + \alpha}$$

α = recycle ratio

Other terms are as defined previously.

Equation 8-57 is quite similar to Eq. 8-40, which applied to complete-mix systems, with or without recycle. The main difference in the two equations is that in Eq. 8-57, θ_c is also a function of the influent waste concentration S_o.

The true plug-flow-recycle system is theoretically more efficient in the stabilization of most soluble wastes than the complete-mix recycle system. This is shown graphically in Fig. 8-15. In practice, a true plug-flow regime is difficult to obtain because of longitudinal dispersion. This difficulty, plus the fact that the plug-flow system cannot handle shock loads as well as the complete-mix system, tends to reduce differences in treatment efficiency in the two models. By dividing the aeration tank into a series of reactors, it has been shown that treatment performance can be improved without a major loss in the ability of the system to handle shock loads. Reactor selection is discussed further in Chap. 10.

Sedimentation Facilities for the Activated-Sludge Process. It is important to note and remember that the sedimentation tank is an *integral* part of the activated-sludge process. The design of the reactor cannot be considered independently of the design of the associated settling facilities. To meet discharge requirements for suspended solids and BOD associated with the volatile suspended solids in the effluent and to maintain θ_c independent of θ, it must be possible to separate the mixed-liquor solids and to return a portion to the reactor.

Because of the variable process microbiology that is possible, it has been found that the settling characteristics of the biological solids in the mixed liquor will differ with each plant, depending on the characteristics of the wastewater and the many variables associated with process design and operation. For this reason, when settling facilities are designed for an existing or a proposed new treatment facility, column-settling tests should be performed, and the design should be based on the results of these tests. If it is not possible to perform settling tests, the design should be based on an approach in which both the hydraulic and the solids loadings are considered. Both approaches are considered in Chap. 10.

Bulking in the Activated-Sludge Process. "Bulking" is the term applied to a condition in which an overabundance of filamentous organisms is present in the mixed liquor in the activated-sludge process (see Fig. 8-18). The presence of filamentous organisms causes the biological flocs in the reactor to be bulky and loosely packed. The bulky flocs do not settle well and are often carried over in great quantities in the effluent from the sedimentation tank. The filamentous organisms found in the activated-sludge process include a variety of filamentous bacteria, actinomycetes, and fungi [17,42]. Conditions favoring the growth of filamentous organisms are numerous and vary from plant to plant.

Control of filamentous organisms has been accomplished in a number of ways including the addition of chlorine or hydrogen peroxide to the return waste-activated sludge, alteration of the dissolved oxygen concentration in the aeration tank, alteration of the points of waste addition to the aeration tank to increase the F/M ratio, the addition of major nutrients (i.e., nitrogen and phosphorus), the addition of trace nutrients and growth factors, and more recently the use of selectors [2,17,42,44]. Control of the growth of filamentous organisms in the complete-mix process has been achieved by mixing the return sludge with the incoming wastewater in a small anoxic contact tank known as a "selector" [2,17].

From practical experience, it has been found that that the mixed liquor from plug-flow activated-sludge processes settles better than that from complete-mix processes and tends to have fewer filaments. Improved settling has also been observed in the sequencing batch reactor (see subsequent discussion). Experimentally, as shown in Fig. 8-19, it has been found that the relative abundance of filamentous and nonfilamentous organisms is related to their relative growth rates when exposed to varying concentrations of substrate (e.g., high concentration in the plug-flow process and low concentration in the complete-mix process). Referring to Fig. 8-19, it can be concluded that nonfilamentous floc-formers have a high μ_{max} but

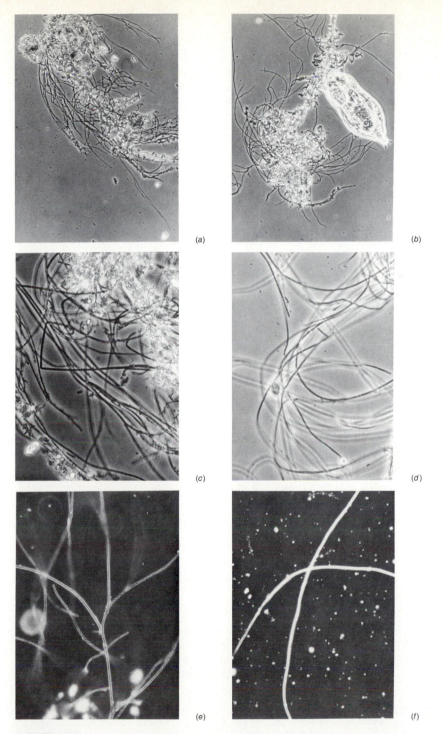

FIGURE 8-18
Typical filamentous organisms found in bulking sludge: (*a* and *b*) phase contrast, 100X, (*c*) phase contrast, 400X, (*d* and *e*) filaments of *Sphaerotilus,* phase contrast and dark field, 400X, and (*f*) filaments of *Thiothrix,* dark field, 400X.

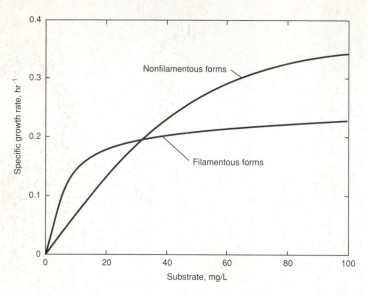

FIGURE 8-19
Typical growth curves for filamentous and nonfilamentous organisms.

a low affinity for the substrate (high K_s), whereas the filamentous forms have a low μ_{max} but a high affinity for the substrate (low K_s). Thus, the low substrate concentrations found in the complete-mix reactor favor the growth of filamentous microorganisms. Additional details on the use of selectors in the activated-sludge process may be found in Ref. 17. The design of selectors is considered further in Chap. 10.

Aerated Lagoons

Aerated lagoons (sometimes called "aerated ponds") evolved from facultative stabilization ponds when surface aerators were installed to overcome the odors from organically overloaded ponds (see Fig. 8-20). Although a number of definitions of aerobic aerated-lagoon processes will be found in the literature, the following process description will be used in this text.

Process Description. The aerated-lagoon process is essentially the same as the conventional extended-aeration activated-sludge process ($\theta_c = 10$ days), except that an earthen basin is used for the reactor, and the oxygen required by the process is supplied by surface or diffused aerators. In an aerobic lagoon, all the solids are maintained in suspension. In the past, aerated lagoons were operated as flow-through activated-sludge systems without recycle, usually followed by large settling ponds. To meet secondary treatment standards of the U.S. Environmental Protection Agency (see Table 4-1), many aerated lagoons are now used in conjunction with settling facilities and incorporate the recycle of biological solids.

FIGURE 8-20
Typical aerated lagoon.

Process Microbiology. Because the aerated-lagoon process is essentially the same as the activated-sludge process, the microbiology is also similar. Some differences occur because the large surface area associated with aerated lagoons can cause more significant temperature effects than are normally encountered in the conventional activated-sludge process.

Seasonal and continuous nitrification may be achieved in aerated-lagoon systems. The degree of nitrification depends on the design and operating conditions within the system and on the wastewater temperature. Generally, with higher wastewater temperatures and lower loadings (increased sludge-retention time), higher degrees of nitrification can be achieved.

Process Analysis. The analysis of an aerated lagoon can be carried out using either the approach described in Sec. 8-5 for a complete-mix aerobic treatment system without recycle or the approach described previously in this section for the activated-sludge process with recycle, depending on the method of operation to be used.

Another approach is to assume that the observed BOD_5 removal (either overall, including soluble and suspended-solids contribution, or soluble only) can be described in terms of a first-order ($r_{su} = -kS$) or a quasi-second-order ($r_{su} = -kSX$) removal function. The required analysis for a complete-mix reactor without recycle has been outlined previously in this chapter (see Sec. 8-5) and in Appendix G. The pertinent equations for a single aerated lagoon are as follows:

For first-order kinetics,

$$\frac{S}{S_o} = \frac{1}{1 + k_1(V/Q)} \tag{8-58}$$

For quasi-second-order kinetics,

$$\frac{S}{S_o} = \frac{1}{1 + k_2 X (V/Q)} \tag{8-59}$$

where
S = effluent BOD_5 concentration, mg/L
S_o = influent BOD_5 concentration, mg/L
k_1, k_2 = observed overall BOD_5 removal rate constant, L/mg · d
V = volume, Mgal
Q = flowrate, Mgal/d
X = mixed-liquor volatile suspended solids, mg/L

The corresponding equation derived from a consideration of soluble substrate-removal kinetics, as given by Eq. 8-8, is

$$\frac{S}{S_o} = \frac{1}{1 + [kX/(K_s + S)](V/Q)} \tag{8-60}$$

The terms in Eq. 8-60 are as defined previously. Application of Eqs. 8-58, 8-59, and 8-60 is considered in Prob. 8-16 and in Example 10-5 in Chap. 10.

Sequencing Batch Reactor

A sequencing batch reactor (SBR) is a fill-and-draw activated-sludge treatment system. The unit processes involved in the SBR and conventional activated-sludge systems are identical. Aeration and sedimentation/clarification are carried out in both systems. However, there is one important difference. In conventional plants, the processes are carried out simultaneously in separate tanks, whereas in SBR operation the processes are carried out sequentially in the same tank.

Process Description. As currently used, all SBR systems have five steps in common that are carried out in sequence as follows: (1) fill, (2) react (aeration), (3) settle (sedimention/clarification), (4) draw (decant), and (5) idle. Each of these steps is illustrated in Fig. 8-21 and described in Table 8-8. A number of process modifications have been made in the times associated with each step to achieve specific treatment objectives [37].

Sludge wasting is another important step in the SBR operation that greatly affects performance. Wasting is not included as one of the five basic process steps because there is no set time period within the cycle dedicated to wasting. The amount and frequency of sludge wasting is determined by performance requirements, as with a conventional continuous-flow system. In an SBR operation, sludge wasting usually occurs during the settle or idle phases. A unique feature of the SBR system is that there is no need for a return activated-sludge (RAS) system. Because both aeration and settling occur in the same chamber, no sludge is lost in the react step, and none has to be returned from the clarifier to maintain the sludge content in the aeration chamber [37]. Some modifications of the SBR process also include continuous flow modes of operation.

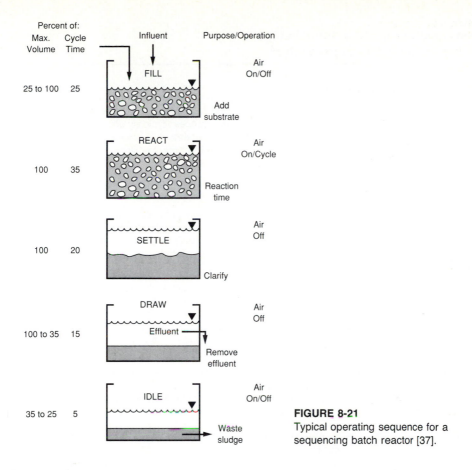

FIGURE 8-21
Typical operating sequence for a sequencing batch reactor [37].

Process Application. In the early 1960s, with the development of the new technology and equipment, interest was revived in the fill-and-draw systems. Improvements in aeration devices and control systems have allowed the development of fill-and-draw systems to achieve their present level of efficiency, which now enables SBR technology to compete successfully with conventional systems. All wastewaters commonly treated by conventional activated-sludge plants can be treated with SBRs.

Aerobic Digestion

Aerobic digestion is an alternative method of treating the organic sludges produced from various treatment operations. Aerobic digesters may be used to treat (1) only waste-activated or trickling-filter sludge, (2) mixtures of waste-activated or trickling-filter sludge and primary sludge, or (3) waste sludge from activated-sludge treatment plants designed without primary settling. Today, two variations of the aerobic digestion process are in common use: conventional and pure oxygen. Thermophilic aerobic

TABLE 8-8
**Description of the operational steps
for the sequencing batch reactor[a]**

Operational step	Description
Fill	The purpose of the fill operation is to add substrate (raw wastewater or primary effluent) to the reactor. The fill process typically allows the liquid level in the reactor to rise from 25 percent of capacity (at the end of idle) to 100 percent. If controlled by time, the fill process normally lasts approximately 25 percent of the full cycle time.
React	The purpose of react is to complete the reactions that were initiated during fill. Typically, react takes up 35 percent of the total cycle time.
Settle	The purpose of settle is to allow solids separation to occur, providing a clarified supernatant to be discharged as effluent. In an SBR, this process is normally much more efficient than in a continuous-flow system because in the settle mode the reactor contents are completely quiescent.
Draw[b]	The purpose of draw is to remove clarified treated water from the reactor. Many types of decant mechanisms are in current use, with the most popular being floating or adjustable weirs. The time dedicated to draw can range from 5 to 30 percent of the total cycle time (15 minutes to 2 hours), with 45 minutes being a typical draw period.
Idle[b]	The purpose of idle in a multitank system is to provide time for one reactor to complete its fill cycle before switching to another unit. Because idle is not a necessary phase, it is sometimes omitted.

[a] Adapted from Ref. 37.

[b] Sludge wasting usually occurs during the settle or idle phases, but wasting can occur in the other phases depending on the mode of operation.

digestion has also been used. Additional details on all these processes are presented in Chap. 12.

Process Description. In conventional aerobic digestion, the sludge is aerated for an extended period of time in an open, unheated tank using conventional air diffusers or surface aeration equipment. The process may be operated in a continuous or batch mode. Smaller plants use the batch system in which sludge is aerated and completely mixed for an extended period of time, followed by quiescent settling and decantation [37]. In continuous systems, a separate tank is used for decantation and concentration. High-purity oxygen aerobic digestion is a modification of the aerobic digestion process in which high-purity oxygen is used in lieu of air. The resultant sludge is similar to conventional aerobically digested sludge.

Thermophilic aerobic digestion represents still another refinement of the aerobic digestion process. Carried out with thermophilic bacteria at temperatures ranging from 77 to 122°F (25 to 50°C) above the ambient air temperature, this process can achieve high removals of the biodegradable fraction (up to 80 percent) at very short detention times (3 to 4 days).

Process Microbiology. Aerobic digestion, as mentioned, is similar to the activated-sludge process. As the supply of available substrate food is depleted, the microorganisms begin to consume their own protoplasm to obtain energy for cell-maintenance reactions. When this occurs, the microorganisms are said to be in the endogenous phase. As shown in Eq. 8-31, cell tissue is aerobically oxidized to carbon dioxide, water, and ammonia. Actually, only about 75 to 80 percent of the cell tissue can be oxidized; the remainder is composed of inert components and organic compounds that are not biodegradable. The ammonia from this oxidation is subsequently oxidized to nitrate as digestion proceeds.

If activated or trickling-filter sludge is mixed with primary sludge and the combination is to be aerobically digested, there will be both direct oxidation of the organic matter in the primary sludge and endogenous oxidation of the cell tissue. Operationally, most aerobic digesters can be considered to be arbitrary-flow reactors without recycle.

Process Analysis. Factors that must be considered in the analysis of aerobic digesters include hydraulic residence time, process loading criteria, oxygen requirements, energy requirements for mixing, environmental conditions, and process operation. The design of aerobic digesters is considered in Chap. 12.

8-8   AEROBIC ATTACHED-GROWTH TREATMENT PROCESSES

Aerobic attached-growth biological treatment processes are usually used to remove organic matter found in wastewater. They are also used to achieve nitrification (the conversion of ammonia to nitrate). The attached-growth processes include the trickling filter, the roughing filter, rotating biological contactor, and fixed-film nitrification reactor. Because the trickling-filter process is used most commonly, it will be considered in greater detail than the other processes. Fixed-film nitrification is considered in Sec. 8-11 in which biological nutrient removal is considered.

Trickling Filter

The first trickling filter was placed in operation in England in 1893. The concept of a trickling filter grew from the use of contact filters, which were watertight basins filled with broken stones. In operation, the contact bed was filled with wastewater from the top, and the wastewater was allowed to contact the media for a short time. The bed was then drained and allowed to rest before the cycle was repeated. A typical cycle required 12 hours (6 hours for operation and 6 hours of resting). The limitations of the contact filter included a relatively high incidence of clogging, the long rest period required, and the relatively low loading that could be used.

Process Description. The modern trickling filter (see Fig. 8-22) consists of a bed of a highly permeable medium to which microorganisms are attached and through which wastewater is percolated or trickled—hence the name. The filter media usually consist of either rock (slag is also used) or a variety of plastic packing materials. In rock-filled trickling filters, the size of the rock typically varies from 1 to 4 in (25 to 100 mm) in diameter. The depth of the rock varies with each particular design but usually ranges from 3 to 8 ft (0.9 to 2.5 m) and averages 6 ft (1.8 m). Rock filter beds are usually circular, and the liquid wastewater is distributed over the top of the bed by a rotary distributor.

Trickling filters that use plastic media have been built in round, square, and other shapes with depths varying from 14 to 40 ft (4 to 12 m). Three types of plastic media are commonly used: (1) vertical-flow packing (see Fig. 8-23), (2) cross-flow packing (see Fig. 10-33), and (3) a variety of random packings (see Fig. 10-33).

Filters are constructed with an underdrain system for collecting the treated wastewater and any biological solids that have become detached from the media. This underdrain system is important both as a collection unit and as a porous structure through which air can circulate (see Fig. 8-22). The collected liquid is passed to a settling tank where the solids are separated from the treated wastewater. In practice, a portion of the liquid collected in the underdrain system or the settled effluent is recycled, usually to dilute the strength of the incoming wastewater and to maintain the biological slime layer in a moist condition.

The organic material present in the wastewater is degraded by a population of microorganisms attached to the filter media (see Fig. 8-24). Organic material from the liquid is adsorbed onto the biological film or slime layer. In the outer portions of the biological slime layer (0.1 to 0.2 mm), the organic material is degraded by aerobic microorganisms. As the microorganisms grow, the thickness of the slime layer increases, and the diffused oxygen is consumed before it can penetrate the full depth of the slime layer. Thus, an anaerobic environment is established near the surface of the media.

As the slime layer increases in thickness, the adsorbed organic matter is metabolized before it can reach the microorganisms near the media face. As a result of having no external organic source available for cell carbon, the microorganisms near the media face enter into an endogenous phase of growth and lose their ability to cling to the media surface. The liquid then washes the slime off the media, and a new slime layer starts to grow. This phenomenon of losing the slime layer is called "sloughing" and is primarily a function of the organic and hydraulic loading on the filter. The hydraulic loading accounts for shear velocities, and the organic loading accounts for the rate of metabolism in the slime layer. In modern trickling filters, the hydraulic loading rate is adjusted to maintain a slime layer of uniform thickness.

Process Microbiology. The biological community in the filter includes aerobic, anaerobic, and facultative bacteria, fungi, algae, and protozoans. Higher animals, such as worms, insect larvae, and snails, are also present. Facultative bacteria are the predominating microorganisms in the trickling filter. Along with the aerobic and anaerobic bacteria, their role is to decompose the organic material in the wastewater.

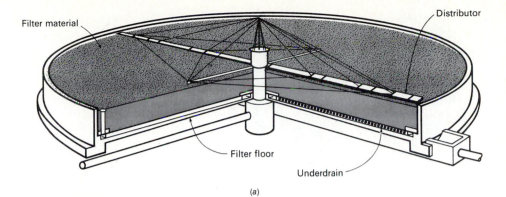

Filter material

Distributor

Filter floor

Underdrain

(a)

(b)

(c)

FIGURE 8-22
Typical trickling filters: (a) cutaway view of a trickling filter (from Dorr-Oliver), (b) conventional rock-filled type filter, and (c) tower trickling filters.

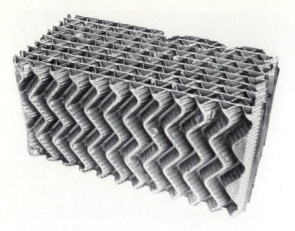

FIGURE 8-23
Typical vertical-flow plastic module used in tower trickling filters.

Achromobacter, *Flavobacterium*, *Pseudomonas*, and *Alcaligenes* are among the bacterial species commonly associated with the trickling filter. Within the slime layer, where adverse conditions prevail with respect to growth, the filamentous forms *Sphaerotilus natans* and *Beggiatoa* will be found. In the lower reaches of the filter, the nitrifying bacteria *Nitrosomonas* and *Nitrobacter* will be present [13,14].

The fungi present are also responsible for waste stabilization, but their contribution is usually important only under low-pH conditions or with certain industrial wastes. At times, their growth can be so rapid that the filter clogs and ventilation becomes restricted. Among the fungi species that have been identified are *Fusazium*, *Mucor*, *Pencillium*, *Geotrichum*, *Sporatichum*, and various yeasts [13,14].

Algae can grow only in the upper reaches of the filter where sunlight is available. *Phormidium*, *Chlorella*, and *Ulothrix* are among the algae species commonly found in trickling filters [13,14]. Generally, algae do not take a direct part in waste degradation, but during the daylight hours they add oxygen to the percolating wastewater. From an operational standpoint, the algae are troublesome because they can cause clogging of the filter surface, which produces odors.

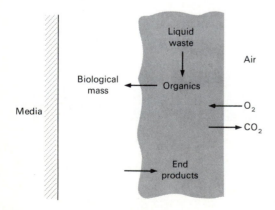

FIGURE 8-24
Schematic representation of the cross section of biological slime in a trickling filter.

The protozoa in the filter are predominantly of the ciliata group, including *Vorticella*, *Opercularia*, and *Epistylis* [13,14]. As in the activated-sludge process, their function is not to stabilize the waste but to control the bacterial population. The higher animals, such as snails, worms, and insects, feed on the biological films in the filter and, as a result, help to keep the bacterial population in a state of high growth or rapid food utilization. The higher animal forms are not as common in high-rate tower trickling filters. Snails are especially troublesome in nitrifying filters, where they have been known to consume most of the growth of nitrifying bacteria.

Variations in the individual population of the biological community occur throughout the filter depth with changes in organic loading, hydraulic loading, influent wastewater composition, pH, temperature, air availability, and other factors, as described in the following discussion.

Process Analysis. In predicting the performance of trickling filters, the organic and hydraulic loadings and the degree of treatment required are among the important factors that must be considered. Over the years, a number of investigators have proposed equations to describe the removals observed, including Atkinson [4], Bruce and Merkens [6], Eckenfelder [7], Fairall [9], Galler and Gotass [10], Germain [11], Logen, et. al. [20,21], the National Research Council [27], Schultz [32], and Velz [40]. In the following discussion, both the theoretical mass-balance approach used in the modeling of the trickling-filter process and a practical approach of developing models from an analysis of field data are presented and discussed.

Mass-balance approach. Atkinson and his coworkers [4] have proposed the following model to describe the rate of flux of organic material into the slime layer, assuming that diffusion into the slime layer controls the rate of reaction and that there is no concentration gradient across the liquid film (see Fig. 8-25).

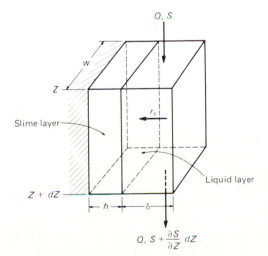

FIGURE 8-25
Definition sketch for the analysis
of the trickling-filter process [31].

$$r_s = -\frac{E\,hk_o\overline{S}}{K_m + \overline{S}} \tag{8-61}$$

where r_s = rate of flux of organic material into the slime layer, ft/d
  E = effectiveness factor $(0 \le E \le 1)$
  h = thickness of slime layer, ft
  k_o = maximum reaction rate, d^{-1}
  $\overline{S}$ = average substrate (e.g., BOD) concentration in the bulk liquid in the volume element, mg/L
  K_m = half-velocity constant, mg/L

Because the effectiveness factor E is approximately proportional to the BOD concentration in the liquid, Eq. 8-61 can be rewritten as follows:

$$r_s = -\frac{f\,hk_o\overline{S}^2}{K_m + \overline{S}} \tag{8-62}$$

where f = proportionality factor.

This model can be applied to the analysis of a trickling filter by performing a mass-balance analysis for the organic material contained in the liquid volume (see Fig. 8-25).

1. General word statement:

Rate of accumulation of substrate within the volume element	=	Rate of flow of substrate into the volume element	−	Rate of flow of substrate out of the volume element	+	Rate of substrate flux into the slime layer from the volume element

$$\tag{8-63}$$

2. Simplified word statement:

$$\text{Accumulation} = \text{Inflow} - \text{Outflow} + \text{Utilization} \tag{8-64}$$

3. Symbolic representation:

$$\frac{\partial \overline{S}}{\partial t}dV = QS - Q\left(S + \frac{\partial S}{\partial D}dD\right) + dDw\left(-\frac{f\,hk_o\overline{S}^2}{K_m + \overline{S}}\right) \tag{8-65}$$

where Q = volumetric flowrate, ft^3/d
  w = width of section under consideration, ft
  D = filter depth, ft

Assuming that steady-state conditions prevail $(\partial \overline{S}/\partial t = 0)$, Eq. 8-65 can be simplified to yield

$$Q\frac{dS}{dD} = -f k_o hw \frac{\overline{S}^2}{K_m + \overline{S}} \tag{8-66}$$

If it is now assumed that the value of the saturation coefficient K_m is small relative to the value of BOD, then Eq. 8-66 can be written as

$$\frac{dS}{dZ} = -\frac{f h k_o w \overline{S}}{Q} \tag{8-67}$$

Equation 8-67 can now be integrated between the limits of S_e and S_i and 0 and D to yield

$$\frac{S_e}{S_i} = \exp\left[-(f h k_o)\frac{wD}{Q}\right] \tag{8-68}$$

where S_e = effluent concentration, mg/L
   Si = influent concentration resulting after the untreated incoming wastewater is mixed with recycled effluent, mg/L

The use of Eq. 8-68 involves the determination of the coefficients f, h, and k_0 for a given set of operating conditions.

The above analysis has been presented to illustrate the general approach followed in preparing a mass-balance analysis of an attached-growth biological process. It should be noted, however, that many of the models developed from theoretical considerations have not worked especially well in terms of modeling the actual performance of trickling filters. Some of the practical models that have been developed to describe the performance of trickling filters are presented below.

NRC equations. Because of the irregular nature of the rock, river stone, and slag used in rock filters, it has been difficult to develop meaningful theoretical relationships that can be used to predict the performance of rock filters. The NRC equations for trickling-filter performance are empirical expressions developed from an extensive study of the operating records of trickling-filter plants serving World War II military installations [27]. The formulas are primarily applicable to single-stage and multistage rock systems, with varying recirculation rates (see Fig. 8-26). For a single-stage or first-stage rock filter, the equation is

$$E_1 = \frac{100}{1 + 0.0561\sqrt{\dfrac{W}{VF}}} \tag{8-69}$$

where E_1 = efficiency of BOD removal for process at 20°C, including recirculation and sedimentation, percent
   W = BOD loading to filter, lb/day
   V = volume of filter media, 10^3 ft^3
   F = recirculation factor

The recirculation factor is calculated using Eq. 8-70

$$F = \frac{1 + R}{(1 + R/10)^2} \tag{8-70}$$

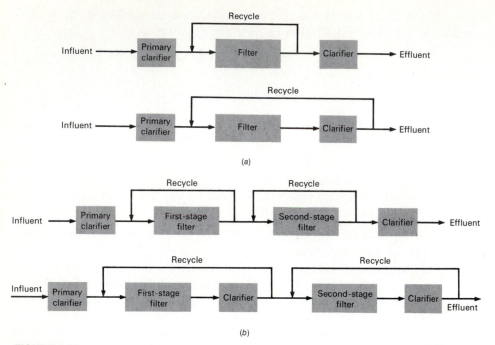

FIGURE 8-26
Typical trickling-filter flow diagrams with various recirculation patterns: (a) single-stage filters and (b) two-stage filters (see also Fig. 10-31).

where R = recirculation ratio Q_r/Q
$\quad\quad Q_r$ = recirculation flow
$\quad\quad Q$ = wastewater flow

The recirculation factor represents the average number of passes of the influent organic matter through the filter. The term $R/10$ takes into account the experimental observation that the removability of organics appears to decrease as the number of passes increases [27]. Typical recycle ratios for various types of filters are reported in Table 10-13. For the second-stage filter (see Fig. 8-26), the equation is

$$E_2 = \frac{100}{1 + \dfrac{0.0561}{1-E_1}\sqrt{\dfrac{W'}{VF}}} \tag{8-71}$$

where E_2 = efficiency of BOD removal for second-stage filter at 20°C, including recirculation and settling, percent
$\quad\quad E_1$ = fraction of BOD removal in first-stage filter
$\quad\quad W'$ = BOD loading applied to second-stage filter, lb/day

The effect of wastewater temperature on the efficiency of the process can be approximated using Eq. 8-14 by substituting E_T for r_T and E_{20} for r_{20}. A value of 1.035

is commonly used for θ, the temperature-activity coefficient. Application of the NRC equations is illustrated in Example 8-2.

Example 8-2 Trickling-filter sizing using NRC equations. A municipal waste having a BOD_5 of 250 mg/L is to be treated by a two-stage trickling filter. The desired effluent quality is 25 mg/L of BOD_5. If both of the filter depths are to be 6 ft and the recirculation ratio is 2:1, find the required filter diameters. Assume $Q = 2$ Mgal/d, wastewater temperature $= 20°C$, and that $E_1 = E_2$.

Solution

1. Compute E_1 and E_2.

$$\text{Overall efficiency} = \frac{200 - 25}{200}(100) = 87.5\%$$

$$E_1 + E_2(1 - E_1) = 0.875$$

$$E_1 = E_2 = 0.646$$

2. Compute the recirculation factor.

$$F = \frac{1 + R}{(1 + R/10)^2} = \frac{1 + 2}{(1.2)^2} = 2.25$$

3. Compute the BOD_5 loading for the first filter.

$$W = (C \text{ mg/L})(Q \text{ Mgal/d})[8.34 \text{ lb/Mgal} \cdot (\text{mg/L})]$$

$$W = 200(8.34)(2) = 3{,}336 \text{ lb } BOD_5/d$$

4. Compute the volume for the first stage.

$$E_1 = \frac{100}{1 + 0.0561\sqrt{\dfrac{W}{VF}}}$$

$$64.6 = \frac{100}{1 + 0.0561\sqrt{\dfrac{3{,}336}{V(2.08)}}}$$

$$V = 16.81 \times 10^3 \text{ ft}^3$$

5. Compute the diameter of the first filter.

$$A = \frac{V}{d} = \frac{16{,}810 \text{ ft}^3}{6} = 2{,}802 \text{ ft}^2$$

$$d = 59.7 \text{ ft}$$

6. Compute the BOD_5 loading for the second-stage filter.

$$W' = (1 - E_1)W = 0.354(3{,}336) = 1{,}181 \text{ lb } BOD_5/d$$

7. Compute the volume of the second-stage filter.

$$E_2 = \cfrac{100}{1 + \cfrac{0.0561}{1 - E_1}\sqrt{\cfrac{W'}{VF}}}$$

$$64.6 = \cfrac{100}{1 + \cfrac{0.0561}{1 - 0.646}\sqrt{\cfrac{1,181}{V2.08}}}$$

$$V = 47.49 \times 10^3 \text{ ft}^3$$

8. Compute the diameter of the second filter.

$$A = \frac{V}{d} = \frac{47,490 \text{ ft}^3}{6} = 7,915 \text{ ft}^2$$

$$d = 100.4 \text{ ft}$$

9. Compute the BOD_5 loading to each filter.
   (a) First-stage filter.

$$BOD_5 \text{ loading} = \frac{3,336 \text{ lb/d}}{16.81 \times 10^3 \text{ ft}^3} = 198.5 \text{ lb/}10^3 \text{ ft}^3$$

   (b) Second-stage filter.

$$BOD_5 \text{ loading} = \frac{1,181 \text{ lb/d}}{47.49 \times 10^3 \text{ ft}^3} = 24.9 \text{ lb/}10^3 \text{ ft}^3$$

10. Compute the hydraulic loading to each filter.
    (a) First-stage filter.

$$\text{Hydraulic loading} = \frac{(1 + 2)(2 \times 10^6 \text{ gal/d})/1440 \text{ min/d}}{2,802 \text{ ft}^2} = 1.49 \text{ gal/ft}^2 \cdot \text{min}$$

    (b) Second-stage filter

$$\text{Hydraulic loading} = \frac{(1 + 2)(2 \times 10^6 \text{ gal/d})/1440 \text{ min/d}}{7,915 \text{ ft}^2} = 0.53 \text{ gal/ft}^2 \cdot \text{min}$$

Comment. To accomodate standard rotary distributor mechanisms, the diameters of the two filters should be rounded to the nearest 5 ft. To reduce construction costs, the two trickling filters are often made the same size. Where two filters of equal diameter are used the efficiencies will be unequal. In many cases, the hydraulic-loading rate will be limited by state standards.

Formulations for plastic media. Because of the predictable properties of the plastic media, a number of more-or-less empirical relationships have been developed to predict the performance of trickling filters packed with plastic media. Two of the expressions used most commonly to describe the observed performance of plastic-packed trickling filters are those proposed by Eckenfelder [7] and by Germain [11] and Schultz [32]. The expression proposed by Eckenfelder is given below:

$$\frac{S_e}{S_i} = \exp[-KS_a^m D(Q_v)^{-n}] \tag{8-72}$$

where K = observed reaction-rate constant for a given depth of filter (value usually obtained from pilot plant studies), ft/d
   D = filter depth, ft
   S_a = specific surface area of filter = $\dfrac{\text{surface area } A_s, \text{ ft}^2}{\text{unit volume } V, \text{ ft}^3}$
   Q_v = volumetric flowrate applied to filter ft^3/ft$^2 \cdot$ d
   Q_v = (Q/A)
   Q = flowrate applied to filter, ft^3/d
   A = cross-sectional area of filter, ft^2
   m, n = empirical constants

The general form of the equation proposed by Germain [11] and Schultz [32] is as follows:

$$\frac{S_e}{S_i} = \exp[-k_{20}D(Q_v)^{-n}] \tag{8-73}$$

where S_e = Total BOD$_5$ of settled effluent from filter, mg/L
   S_i = Total BOD$_5$ of wastewater applied to the filter, mg/L
   k_{20} = treatability constant corresponding to a filter of depth D at 20°C, (gal/min)^n ft
   D = depth of filter, ft
   Q_v = volumetric flowrate applied per unit volume of filter, gal/ft$^2 \cdot$ min
   Q_v = (Q/A)
   Q = flowrate applied to filter without recirculation, gal/min
   A = cross-sectional area of filter, ft^2
   n = experimental constant, usually 0.5

The treatability constant, k_{20}, in Eq. 8-73 takes into account the rate constant K_T and the specific surface area, A_s, of the filter medium as given in Eq. 8-72. The effect of wastewater temperature on the process efficiency is accounted for by adjusting the k_{20} using a θ value of 1.035. The range of θ values found in the field is reported in Table 8-5.

Based on the analysis of data from a variety of operating filters, Albertson [1] has found that the treatability constant must be corrected for depth when a k_{20} value determined at one depth is to be applied to the design of a filter at another depth. The relationship proposed by Alberson [1] is as follows:

$$k_2 = k_1 \left(\frac{D_1}{D_2}\right)^x \tag{8-74}$$

where k_2 = treatability constant corresponding to a filter of depth D_2
k_1 = treatability constant corresponding to a filter of depth D_1
D_1 = depth of filter one, ft
D_2 = depth of filter two, ft
x = 0.5 for vertical and rock media filters
0.3 for cross flow plastic medium filters

In using Eq. 8-68, 8-72, or 8-73 for the design of trickling filters, it should be remembered that the value of the term $f h k_0$, K, or k_{20} will vary with so many local factors that coefficient values derived from the literature must be used with great caution. Further, because most of the k values given in the literature have not been normalized with respect to depth, it is difficult, if not impossible, to compare reported k values directly. Application of Eq. 8-73 for the sizing of trickling filters is illustrated in Example 8-3.

Example 8-3   Trickling-filter sizing using first-order equation. Determine the surface area required for a 20 and 30 ft deep plastic media filter to treat wastewater with a BOD_5 of 300 mg/L after primary sedimentation. The final effluent BOD_5 is to be 25.0 mg/L or less. Assume that the treatability constant determined using a 20 ft deep test filter is 0.085 (gal/min)$^{0.5}$ft at 20°C.

Solution

1. Determine the surface area required for a 20 ft deep filter using Eq. 8-73.

$$\frac{S_e}{S_i} = \exp[-k_{20}D(Q_v)^{-n}]$$

(a) Substituting Q/A for Q_v in the above equation and rearranging yields

$$A = Q\left(\frac{-\ln S_e/S_i}{k_{20}D}\right)^{1/n}$$

(b) Substitute known values and solve for the area A.

$$S_e = 25 \text{ mg/L}$$

$$S_i = 300 \text{ mg/L}$$

$$n = 0.5$$

$$k_{20} = 0.085 \text{ (gal/min)}^{0.5} \text{ ft for } n = 0.5$$

$$D = 20 \text{ ft}$$

$$Q = (2 \times 10^6 \text{ gal/d})/(1,440 \text{ min/d}) = 1,389 \text{ gal/min}$$

$$A = (1,389)\left(\frac{-\ln 25/300}{0.085(20)}\right)^2 = 2,968 \text{ ft}^2$$

2. Determine the surface area required for a 30 ft deep filter using Eq. 8-73.

(a) Determine the k_{20} value for a filter depth of 30 ft using Eq. 8-74.

$$k_{30} = k_{20}\left(\frac{D_{20}}{D_{30}}\right)^x$$

$$k_{30} = 0.085\left(\frac{20}{30}\right)^{0.5} = 0.069$$

(b) Substitute known values and solve for the area A.

$$S_e = 25 \text{ mg/L}$$
$$S_i = 300 \text{ mg/L}$$
$$n = 0.5$$
$$k_{30} = 0.069 \text{ (gal/min)}^{0.5} \text{ ft for } n = 0.5$$
$$D = 30 \text{ ft}$$
$$Q = (2 \times 10^6 \text{ gal/d})/(1,440 \text{ min/d}) = 1,389 \text{ gal/min}$$
$$A = (1,389)\left(\frac{-\ln 25/300}{0.069 (30)}\right)^2 = 2,002 \text{ ft}^2$$

Recirculation. Another factor over which there is a considerable amount of misunderstanding is the effect of recirculation on filter performance. In the past, recirculation has been reported to improve the efficiency (performance) of rock filters. However, based on a more recent assessment [1,35], it appears that the benefits of recirculations are due primarily to improved wetting and flushing of filter media. By properly managing the hydraulic loading rate, it has been possible to maintain a thinner biomass layer consistently, with a concomitant improvement in performance, and to avoid the periodic sloughing phenomenon often observed in most rock-type trickling filters.

Recirculation as applied to synthetic filter media involves a somewhat different concept than is applied to rock filters. Typically synthetic filter media require a higher minimum wetting rate (flow per unit area) to induce a biological slime to develop throughout the depth of the medium. Thus, recirculation in synthetic filter media is required to maintain the required degree of wetting for a given medium. The proper dosage rates for rock and synthetic filter media are considered in Chap. 10. The effect of recirculation on performance of a tower trickling filter packed with a plastic medium is considered in Example 8-4.

Example 8-4 Evaluation of the effects of recirculation in a trickling-filter process. Examine the effect of recirculation ratios varying from 0 to 4 for a trickling-filter process. Use the data and information from Example 8-3 for the 30 ft tower trickling filter.

Solution

1. Determine the fraction of the applied organic load removed using Eq. 8-73, when the recycle ratio is equal to zero.

$$\frac{S_e}{S_i} = \exp\left[-k_{20} D \left(\frac{A}{Q}\right)^n\right]$$

(a) The pertinent data from Example 8-3 are:

$$S_e = 25 \text{ mg/L}$$

$$S_i = 300 \text{ mg/L}$$

$$n = 0.5$$

$$k_{20} = 0.069 \text{ (gal/min)}^{0.5} \text{ ft for } n = 0.5$$

$$D = 30 \text{ ft}$$

$$Q = (2 \times 10^6 \text{ gal/d})/(1,440 \text{ min/d}) = 1,389 \text{ gal/min}$$

$$A = 2,002 \text{ ft}^2$$

(b) The fraction of the applied organic load removed is

$$\frac{S_e}{S_i} = \frac{25}{300} = 0.083$$

$$\text{Fraction removed} = 1 - \frac{S_e}{S_i}$$

$$= 1 - 0.083 = 0.917$$

(Note that, when the recycle ratio is equal to zero, the fraction removed as computed above also corresponds to the fraction of the influent BOD removed.)

2. Determine the fraction of the applied load removed for various recycle ratios.

   (a) Use Eq. 8-73 in the following form, where S_i is the influent to the filter including recirculation.

$$\frac{S_e}{S_i} = \exp\left[-0.069(30)\left(\frac{2,002}{(1 + \alpha)1,389}\right)^{0.5}\right]$$

   (b) Set up a computation table.

α	0	1	2	3	4
S_e/S_i	0.083	0.173	0.238	0.289	0.329
$1 - S_e/S_i$	0.917	0.827	0.762	0.711	0.671

3. Determine the fraction of the influent load removed from the data determined in step 2. This can be accomplished by performing a materials balance on the influent to the filter, taking into account the influent and recycle flows as follows:

$$QS_o + \alpha QS_e = (1 + \alpha)QS_i$$

where S_o = BOD in the influent wastewater before recirculation
   S_e = BOD in the recycle flow
   α = recycle ratio
   S_i = BOD in the influent applied to the filter

Assume that $Q = 1$, and set up the previous expression in a format for computing the term S_o / S_e using the data from step 2:

$$\frac{S_o}{S_e} = \left[(1 + \alpha) \frac{S_i}{S_e} - \alpha \right]$$

Compute the fraction of the influent removed using the data from step 2:

α	$1 + \alpha$	$(1 + \alpha)\dfrac{S_i}{S_e}$	$(1 + \alpha)\dfrac{S_i}{S_e} - \alpha$	$\dfrac{S_e}{S_o}$	$1 - \dfrac{S_e}{S_o}$
0	1	12.048	12.048	0.083	0.917
1	2	11.561	10.561	0.095	0.905
2	3	12.605	10.605	0.094	0.906
3	4	13.841	10.841	0.092	0.908
4	5	15.198	11.198	0.089	0.911

Comment. From the preceding computation, it can be seen that this model predicts that, as the recycle ratio is increased, the degree of treatment is somewhat reduced, whether it is measured in terms of the load applied to the filter $(1 - S_e/S_i)$ or the incoming load $(1 - S_e/S_o)$. The use of recycle to reduce the strength of the applied load, to maintain optimum wetting of the filter, or for hydraulic control is of critical importance even if process efficiency is not improved.

Mass-Transfer Limitations. One of the problems encountered in the design of trickling filters is the determination of the maximum organic material that can be applied to the filter before oxygen becomes a limiting variable. Recognizing the limitations of any analytical approach because of the many variables involved, this problem can nevertheless be approached by equating the transfer of organic material from the liquid film, as defined by Eq. 8-62, to the rate of oxygen transfer, using Eq. 6-54. A factor to account for the yield must be included. This can be done by multiplying the substrate transfer-rate term by a factor such as $(1 - y)$, where y is the expected yield expressed as a decimal. From data reported in the literature, it appears that when the BOD concentrations are in the range of 400 to 500 mg/L, oxygen transfer may become a limiting factor [31]. The airflow through filters is considered in Chap. 10.

Solids Separation Facilities for Trickling Filters. As in the activated-sludge process, solids separation is an important part of the trickling-filter process. It is needed for removal of suspended solids sloughed off during periods of unloading with low-rate filters and for removal of lesser amounts of solids sloughed off continuously by high-rate filters. If recirculation is used, some of the settled solids may be recycled and some may be wasted, but the recycle of the settled biological solids is not as important as in the activated-sludge process. In the trickling-filter process, the majority of the active microorganisms are attached to the filter media and do not pass out of the reactor as in the activated-sludge process. Although recirculation can help in seeding the filter, the primary purposes of recirculation are to dilute strong influent wastewater and to bring the filter effluent back in contact with the biological population for further treatment. Recirculation is almost always included in high-rate trickling-filter systems.

Roughing Filters

Roughing filters are specially designed trickling filters operated at high hydraulic loading rates. Roughing filters are used principally to reduce the organic loading on downstream processes and in seasonal nitrification applications where the purpose is to reduce the organic load so that a downstream biological process will dependably nitrify the wastewater during the summer months.

Process Description. Although the earliest roughing filters were shallow, stone-media systems, the present trend is toward use of synthetic media or redwood at depths of 12 to 40 ft. (3.7 to 12 m). As with other biological processes, roughing-filter performance is temperature-sensitive. When roughing filters are used for the removal of a portion of the organic material present, or to enhance downstream nitrification, a drop in efficiency is not critical.

Roughing filters are typically operated at high hydraulic loadings, necessitating the use of high recycle rates. The higher hydraulic loadings cause nearly continuous sloughing of the slime layer. If unsettled filter effluent is used for recycle, the sloughed biological solids in the recycle stream may contribute to organic removal within the filter as in a suspended-growth process. If this mechanism is significant, process efficiency may be greater than predicted by an attached-growth model.

Process Microbiology. The biological activity in a roughing filter is essentially the same as that described for the trickling filter. Some differences will be noted in the organisms present because of the higher shearing action resulting from the higher hydraulic flowrates applied to those units.

The biological growth is susceptible to the same heavy metals and organic substances as conventional suspended-growth systems, but the process has shown greater resistance to shock loading than suspended-growth systems. Because of the relatively short hydraulic retention time available, organics that are not readily biodegradable are not affected.

Process Analysis and Design. Roughing filters are normally designed using loading factors developed from pilot plant studies and data derived from full-scale installations, although the analysis presented for the trickling filter can be used. Appropriate design values will be found in Chap.10.

Rotating Biological Contactors

A rotating biological contactor consists of a series of closely spaced circular disks of polystyrene or polyvinyl chloride. The disks are submerged in wastewater and rotated slowly through it (see Fig. 8-27).

FIGURE 8-27
Rotating biological contractor equipped with air capture cups (from Envirex, Inc.).

Process Description. In operation, biological growths become attached to the surfaces of the disks and eventually form a slime layer over the entire wetted surface area of the disks. The rotation of the disks alternately contacts the biomass with the organic material in the wastewater and then with the atmosphere for adsorption of oxygen. The disk rotation affects oxygen transfer and maintains the biomass in an aerobic condition. The rotation is also the mechanism for removing excess solids from the disks by shearing forces it creates and maintaining the sloughed solids in suspension so they can be carried from the unit to a clarifier. Rotating biological contactors can be used for secondary treatment, and they can also be operated in the seasonal and continuous-nitrification and denitrification modes.

Process Performance Analysis. Rotating biological contactors are usually designed on the basis of loading factors derived from pilot plant and full-scale installations, although their performance can be analyzed using an approach similar to that for trickling filters. Both hydraulic and organic loading-rate criteria are used in sizing units for secondary treatment. The loading rates for warm weather and year-round nitrification will be considerably lower than the corresponding rates for secondary treatment. Typical design values are presented in Chap. 10.

Properly designed, rotating biological contactors generally are quite reliable because of the large amount of biological mass present (low-operating F/M). This large biomass also permits them to withstand hydraulic and organic surges more effectively. The effect of staging in this plug-flow system eliminates short circuiting and dampens shock loadings.

Packed-Bed Reactors

Still another attached-growth process is the packed-bed reactor, used for both the removal of carbonaceous BOD and nitrification. Typically, a packed-bed reactor consists of a container (reactor) that is packed with a medium to which the microorganisms can become attached. Wastewater is introduced from the bottom of the container through an appropriate underdrain system or inlet chamber. Air or pure oxygen necessary for the process is also introduced with the wastewater.

8-9 ANAEROBIC SUSPENDED-GROWTH TREATMENT PROCESSES

In the past ten years a number of different anaerobic processes have been developed for the treatment of sludges and high-strength organic wastes. The more common processes now in use are shown schematically in Fig. 8-28. Of the processes shown in Fig. 8-28, the most common anaerobic suspended-growth processes used for the treatment of wastewater is the complete-mix anaerobic digestion process. Because the complete-mix anaerobic digestion process is of such fundamental importance in the stabilization of organic material and biological solids, it will be emphasized in the following discussion. The anaerobic contact process and the upflow anaerobic sludge-blanket process are also described.

Anaerobic Digestion

Anaerobic digestion is one of the oldest processes used for the stabilization of sludges. It involves the decomposition of organic and inorganic matter in the absence of molecular oxygen. The major applications have been, and remain today, in the stabilization of concentrated sludges produced from the treatment of wastewater and in the treatment of some industrial wastes. More recently, it has been demonstrated that dilute organic wastes can also be treated anaerobically.

Process Description. In the anaerobic digestion process, the organic material in mixtures of primary settled and biological sludges is converted biologically, under anaerobic conditions, to a variety of end products including methane (CH_4) and carbon dioxide (CO_2). The process is carried out in an airtight reactor. Sludge, introduced continuously or intermittently, is retained in the reactor for varying periods of time. The stabilized sludge, withdrawn continuously or intermittently from the reactor, is reduced in organic and pathogen content and is nonputrescible.

The two types of commonly used anaerobic digesters are identified as standard-rate and high-rate. In the standard-rate digestion process (see Fig. 8-29a), the contents of the digester are usually unheated and unmixed. Detention times for the standard-rate process vary from 30 to 60 days. In a high-rate digestion process (see Fig. 8-29b), the contents of the digester are heated and mixed completely. The required detention

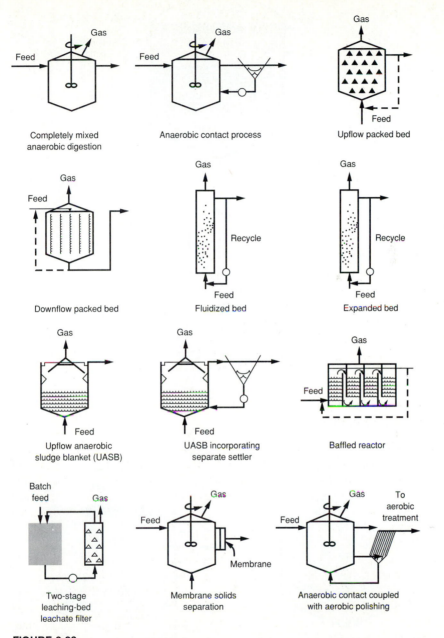

FIGURE 8-28
Typical reactor configurations used in anaerobic wastewater treatment [33].

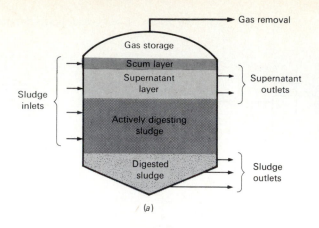

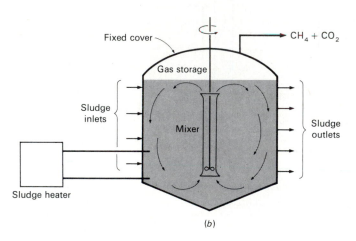

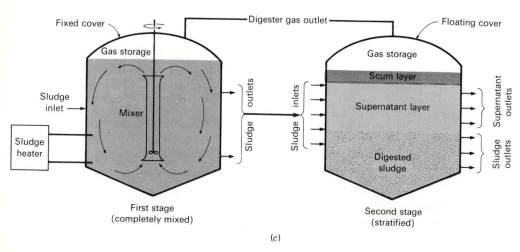

FIGURE 8-29
Typical anaerobic digesters: (*a*) conventional standard-rate single-stage process, (*b*) high-rate, complete-mix, single-stage process, and (*c*) two-stage process.

time for high-rate digestion is typically 15 days or less. A combination of these two basic processes is known as the "two-stage process" (see Fig. 8-29c). The primary function of the second stage is to separate the digested solids from the supernatant liquor; however, additional digestion and gas production may occur.

Process Microbiology. The biological conversion of the organic matter in treatment plant sludges is thought to occur in three steps (see Fig. 8-30). Referring to Fig. 8-30, the first step in the process involves the enzyme-mediated transformation (hydrolysis) of higher-molecular-mass compounds into compounds suitable for use as a source of energy and cell carbon. The second step (acidogenesis) involves the bacterial conversion of the compounds resulting from the first step into identifiable lower-molecular-mass intermediate compounds. The third step (methanogenesis) involves the bacterial conversion of the intermediate compounds into simpler end products, principally methane and carbon dioxide [15,22,23].

In a digester, a consortium of anaerobic organisms work together to bring about the conversion of organic sludges and wastes. One group of organisms is responsible for hydrolyzing organic polymers and lipids to basic structural building blocks such as monosaccharides, amino acids, and related compounds (see Fig. 8-30). A second group of anaerobic bacteria ferments the breakdown products to simple organic acids, the most common of which in an anaerobic digester is acetic acid. This group of

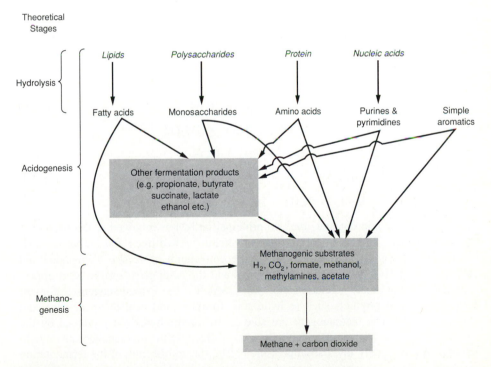

FIGURE 8-30
Schematic diagram of the patterns of carbon flow in anaerobic digestion [15].

microorganisms, described as nonmethanogenic, consists of facultative and obligate anaerobic bacteria. Collectively, these microorganisms are often identified in the literature as "acidogens," or "acid formers." Among the nonmethanogenic bacteria that have been isolated from anaerobic digesters are *Clostridium spp.*, *Peptococcus anaerobus*, *Bifidobacterium spp.*, *Desulphovibrio spp.*, *Corynebacterium spp.*, *Lactobacillus*, *Actinomyces*, *Staphylococcus*, and *Escherichia coli*. Other physiological groups present include those producing proteolytic, lipolytic, ureolytic, or cellulytic enzymes [14,15].

A third group of microorganisms converts the hydrogen and acetic acid formed by the acid formers to methane gas and carbon dioxide. The bacteria responsible for this conversion are strict anaerobes and are called methanogenic. Collectively, they are identified in the literature as "methanogens," or "methane formers." Many of the methanogenic organisms identified in anaerobic digesters are similar to those found in the stomachs of ruminant animals and in organic sediments taken from lakes and rivers. The principal genera of microorganisms that have been identified include the rods (*Methanobacterium*, *Methanobacillus*) and spheres (*Methanococcus*, *Methanosarcina*) [14,15]. The most important bacteria of the methanogenic group are the ones that utilize hydrogen and acetic acid. They have very slow growth rates; as a result, their metabolism is usually considered rate-limiting in the anaerobic treatment of an organic waste. Waste stabilization in anaerobic digestion is accomplished when methane and carbon dioxide are produced. Methane gas is highly insoluble, and its departure from solution represents actual waste stabilization.

It is important to note that methane bacteria can only use a limited number of substrates for the formation of methane. Currently, it is known that methanogens use the following substrates: $CO_2 + H_2$, formate, acetate, methanol, methylamines, and carbon monoxide. Typical energy-yielding conversion reactions involving these compounds are as follows:

$$4H_2 + CO_2 \rightarrow CH_4 + 2H_2O \tag{8-75}$$

$$4HCOOH \rightarrow CH_4 + 3CO_2 + 2H_2O \tag{8-76}$$

$$CH_3COOH \rightarrow CH_4 + CO_2 \tag{8-77}$$

$$4CH_3OH \rightarrow 3CH_4 + CO_2 + 2H_2O \tag{8-78}$$

$$4(CH_3)_3N + H_2O \rightarrow 9CH_4 + 3CO_2 + 6H_2O + 4NH_3 \tag{8-79}$$

In an anaerobic digester, the two principal pathways involved in the formation of methane (see Fig. 8-31) are (1) the conversion of hydrogen and carbon dioxide to methane and water (Eq. 8-75) and (2) the conversion of acetate to methane and carbon dioxide (Eq. 8-77). The methanogens and the acidogens form a "syntrophic" (mutually beneficial) relationship in which the methanogens convert fermentation end products such as hydrogen, formate, and acetate to methane and carbon dioxide. The methanogens are able to utilize the hydrogen produced by the acidogens because of their efficient hydrogenase. Because the methanogens are able to maintain an extremely low partial pressure of H_2, the equilibrium of the fermentation reactions is shifted towards the formation of more oxidized end products (e.g., formate and acetate). The utilization of the hydrogen, produced by the acidogens and other

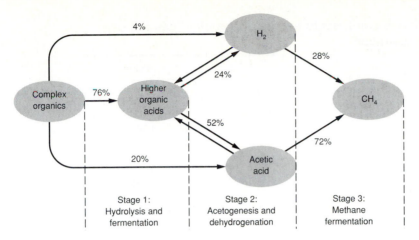

FIGURE 8-31
Steps in the anaerobic digestion process with energy flow [33].

anaerobes, by the methanogens is termed *interspecies hydrogen transfer*. In effect, the methanogenic bacteria remove compounds that would inhibit the growth of acidogens.

To maintain an anaerobic treatment system that will stabilize an organic waste efficiently, the nonmethanogenic and methanogenic bacteria must be in a state of dynamic equilibrium. To establish and maintain such a state, the reactor contents should be void of dissolved oxygen and free from inhibitory concentrations of such constituents as heavy metals and sulfides. Also, the pH of the aqueous environment should range from 6.6 to 7.6. Sufficient alkalinity should be present to ensure that the pH will not drop below 6.2 because the methane bacteria cannot function below this point. When digestion is proceeding satisfactorily, the alkalinity will normally range from 1000 to 5000 mg/L, and the volatile fatty acids will be less than 250 mg/L. A sufficient amount of nutrients, such as nitrogen and phosphorus, must also be available to ensure the proper growth of the biological community. Depending on the nature of the sludges or waste to be digested, growth factors (see Sec. 8-2) may also be required. Temperature is another important environmental parameter. The optimum temperature ranges are the mesophilic, 85 to 100°F (30 to 38°C), and the thermophilic, 120 to 135°F (49 to 57°C).

Process Analysis. The disadvantages and advantages of the anaerobic treatment of an organic waste, as compared to aerobic treatment, stem directly from the slow growth rate of the methanogenic bacteria. Slow growth rates require a relatively long detention time in the digester for adequate waste stabilization to occur. However, the low growth yield signifies that only a small portion of the degradable organic waste is being synthesized into new cells. Typical kinetic coefficients for anaerobic digestion are reported in Table 8-9. With the methanogenic bacteria, most of the organic waste is converted to methane gas, which is combustible and therefore a useful end product. If sufficient quantities are produced, as is customary with municipal wastewater sludge, the methane gas can be used to operate dual-fuel engines to produce electricity and

TABLE 8-9
Typical kinetic coefficients for the anaerobic digestion of various substrates[a]

	Coefficient	Basis	Range	Typical
			Value[b]	
Domestic sludge	Y	mg VSS/mg BOD_5	0.040–0.100	0.06
	k_d	d^{-1}	0.020–0.040	0.03
Fatty acid	Y		0.040–0.070	0.050
	k_d	d^{-1}	0.030–0.050	0.040
Carbohydrate	Y		0.020–0.040	0.024
	k_d	d^{-1}	0.025–0.035	0.03
Protein	Y		0.050–0.090	0.075
	k_d	d^{-1}	0.010–0.020	0.014

[a] Derived in part from Refs. 12, 18, and 43.
[b] Values reported are for 20°C.
Note: 1.8(°C) + 32 = °F

to provide building heat. The amount of gas produced from the anaerobic conversion of organic matter can be estimated as illustrated in Example 8-5.

Example 8-5   Conversion of BOD_L to methane gas. Determine the amount of methane produced per pound of ultimate BOD_L stabilized. Assume that the starting compound is glucose ($C_6H_{12}O_6$).

Solution

1. Write a balanced equation for the conversion of glucose to CO_2 and CH_4 under anaerobic conditions.

$$C_6H_{12}O_6 \rightarrow 3CO_2 + 3CH_4$$
$$180 \qquad\qquad 132 \quad\ 48$$

Note that although the glucose has been converted, the methane has an oxygen requirement for complete conversion to carbon dioxide and water.

2. Write a balanced equation for the oxidation of methane to CO_2 and H_2O, and determine the kilograms of methane formed per kilogram of BOD_L.

$$3CH_4 + 6O_2 \rightarrow 3CO_2 + 6H_2O$$
$$48 \qquad 192$$

Using this and the previous equation, the ultimate BOD_L per pound of glucose is (192/180) lb, and 1.0 lb of glucose yields (48/180) lb of methane, so that the ratio of the amount of methane produced per pound of BOD_L converted is

$$\frac{\text{lb } CH_4}{\text{lb } BOD_L} = \frac{48/180}{192/180} = 0.25$$

Therefore, for each pound of BOD_L converted, 0.25 lb of methane is formed.

3. Determine the volume equivalent of the 0.25 lb of methane produced from the stabilization of 1.0 lb of BOD_L.

$$V_{CH_4} = (0.25 \text{ lb})\left(\frac{454 \text{ g}}{\text{lb}}\right)\left(\frac{1 \text{ mol}}{16 \text{ g}}\right)\left(\frac{22.4 \text{ L}}{\text{mol}}\right)\left(\frac{\text{ft}^3}{28.32 \text{ L}}\right)$$

$$= 5.62 \text{ ft}^3 \text{ of } CH_4 \text{ at standard conditions (32°F and 1 atm)}$$

Therefore, 5.62 ft^3 of methane is produced per pound of ultimate BOD_L converted.

Because of the low cellular growth rate and the conversion of organic matter to methane gas and carbon dioxide, the resulting solid matter is reasonably well-stabilized. After drying or dewatering, the digested sludge may be suitable for disposal in sanitary landfills, for composting, or for application on land. Because of the large proportion of cellular organic material, the sludge solids resulting from aerobic processes are most commonly digested, usually anaerobically.

The high temperatures necessary to achieve adequate treatment are often listed as disadvantages of the anaerobic treatment process; however, high temperatures are necessary only when sufficiently long mean cell-residence time cannot be obtained at nominal temperatures. In the anaerobic treatment systems shown in Fig. 8-29, the mean cell-residence time of the microorganisms in the reactor is equivalent to the hydraulic detention time of the liquid in the reactor. As the operation temperature is increased, the minimum mean cell-residence time is reduced significantly. Thus, heating of the reactor contents lowers not only the mean cell-residence time necessary to achieve adequate treatment but also the hydraulic detention time, so a smaller reactor volume can be used.

Anaerobic Contact Process

Some industrial wastes that are high in BOD can be stabilized very efficiently by anaerobic treatment. In the anaerobic contact process (see Fig. 8-28), untreated wastes are mixed with recycled sludge solids and then digested in a reactor sealed off from the entry of air [33]. The contents of the digester are mixed completely. After digestion, the mixture is separated in a clarifier or vacuum flotation unit, and the supernatant is discharged as effluent, usually for further treatment. Settled anaerobic sludge is then recycled to seed the incoming wastewater. Because of the low synthesis rate of anaerobic microorganisms, the excess sludge that must be disposed of is minimal. This process has been used successfully for the stabilization of meat-packing and other high-strength soluble wastes. Typical process loading and performance data for the anaerobic contact processes are reported in Table 8-10.

Upflow Anaerobic Sludge-Blanket Process

In the upflow anaerobic sludge-blanket (UASB) process (see Fig. 8-28), the waste to be treated is introduced in the bottom of the reactor. The wastewater flows

TABLE 8-10
Typical process and performance data for anaerobic processes used for the treatment of industrial wastes

Process	Input COD, mg/L	Hydraulic detention time, h	Organic loading, lb COD/ft³ · d	COD removal, %
Anaerobic contact process	1,500–5,000	2–10	0.03–0.15	75–90
Upflow anaerobic sludge-blanket (UASB)	5,000–15,000	4–12	0.25–0.75	75–85
Fixed-bed	10,000–20,000	24–48	0.06–0.30	75–85
Expanded-bed	5,000–10,000	5–10	0.30–0.60	80–85

Note: lb/COD/ft³ · d × 16.0185 = kg COD/m³ · d

upward through a sludge blanket composed of biologically formed granules or particles. Treatment occurs as the wastewater comes in contact with the granules. The gases produced under anaerobic conditions (principally methane and carbon dioxide) cause internal circulation, which helps in the formation and maintenance of the biological granules. Some of the gas produced within the sludge blanket becomes attached to the biological granules. The free gas and the particles with the attached gas rise to the top of the reactor. The particles that rise to the surface strike the bottom of the degassing baffles, which causes the attached gas bubbles to be released. The degassed granules typically drop back to the surface of the sudge blanket. The free gas and the gas released from the granules is captured in the gas collection domes located in the top of the reactor. Liquid containing some residual solids and biological granules passes into a settling chamber, where the residual solids are separated from the liquid. The separated solids fall back through the baffle system to the top of the sludge blanket. To keep the sludge blanket in suspension, upflow velocities in the range of 2 to 3 ft/h (0.6 to 0.9 m/h) have been used. Typical process loading and performance data for the UASB process are reported in Table 8-10.

8-10 ANAEROBIC ATTACHED-GROWTH TREATMENT PROCESSES

The two most common anaerobic attached-growth treatment processes are the anaerobic filter and the expanded-bed processes used for the treatment of carbonaceous organic wastes. Attached-growth treatment processes used for denitrification are considered in Sec. 8-11 and in Chap. 11. Typical process loading and performance data for the anaerobic filter and expanded-bed processes are reported in Table 8-10.

Anaerobic Filter Process

The anaerobic filter is a column filled with various types of solid media used for the treatment of the carbonaceous organic matter in wastewater. The waste flows upward through the column, contacting the media on which anaerobic bacteria grow and are retained. Because the bacteria are retained on the media and not washed off in the effluent, mean cell-residence times on the order of 100 days can be obtained. Large values of θ_c can be achieved with short hydraulic retention times, so the anaerobic filter can be used for the treatment of low-strength wastes at ambient temperature.

Expanded-Bed Process

In the expanded-bed process (see Fig. 8-28), the wastewater to be treated is pumped upward through a bed of an appropriate medium (e.g., sand, coal, expanded aggregate) on which a biological growth has been developed. Effluent is recycled to dilute the incoming waste and to provide an adequate flow to maintain the bed in an expanded condition. Biomass concentrations exceeding 15,000 to 40,000 mg/L have been reported. Because a large biomass can be maintained, the expanded-bed process can also be used for the treatment of municipal wastewater at very short hydraulic detention times. When treating municipal wastewater, the presence of sulfate can lead to the formation of hydrogen sulfide. A number of methods have been proposed for the capture of the hydrogen sulfide in the solution phase. As the quantity of sludge produced in the expanded-bed process is considerably less than that produced in aerobic systems, such as the activated-sludge process, it is anticipated that greater use will be made of this and other attached-growth anaerobic processes for the treatment of municipal wastewater. The recovery of methane, a usable gas, is another important advantage of the anaerobic processes.

8-11   BIOLOGICAL NUTRIENT REMOVAL

Removal of nutrients from wastewater prior to disposal is being required more frequently. Because both nitrogen and phosphorus can impact receiving water quality, the discharge of one or both of these constituents may have to be controlled [41]. Nutrient removal options that need to be considered include the following:

1. Nitrogen removal without phosphorus removal
2. Nitrogen and phosphorus removal
3. Phosphorus removal with or without nitrification
4. Year-round removal of phosphorus with seasonal removal of nitrogen

The information presented in this section is intended to serve as a brief introduction to biological nutrient removal. A more comprehensive examination of biological nutrient removal processes can be found in Chap. 11.

Nutrient Removal Processes

Biological nutrient removal is a relatively low-cost means of removing nitrogen and phosphorus from wastewater. Recent experience has shown that biological processes are reliable and effective in removing nitrogen and phosphorus.

Nitrogen Removal. Nitrogen can occur in many forms in wastewater and undergo numerous transformations in wastewater treatment (see Fig. 8-32). These transformations allow the conversion of ammonia-nitrogen to products that can easily be removed from the wastewater. The two principal mechanisms for the removal of nitrogen are assimilation and nitrification-denitrification. Because nitrogen is a nutrient, microbes present in the treatment processes will assimilate ammonia-nitrogen and incorporate it into cell mass. A portion of this ammonia-nitrogen will be returned to the wastewater on the death and lysis of the cells. In nitrification-denitrification, the removal of nitrogen is accomplished in two conversion steps. In the first step, nitrification, the oxygen demand of ammonia is reduced by converting it to nitrate. However, the nitrogen has merely changed forms and not been removed. In the second step, deni-

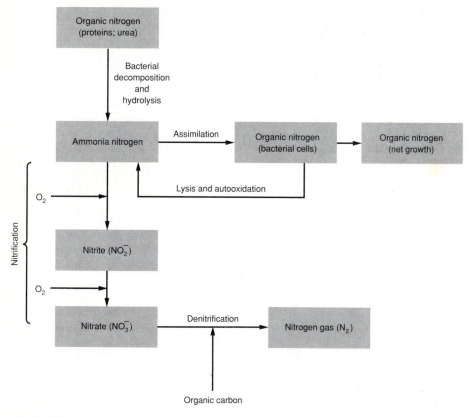

FIGURE 8-32
Nitrogen transformations in biological treatment processes [28].

trification, nitrate is converted to a gaseous product for removal. These two processes are considered separately in this section.

Phosphorus Removal. Phosphorus uptake by a microorganism occurs in staged reactors. By controlling the environmental conditions properly, microorganisms can be made to take up excess phosphorus. The removal of phosphorus is accomplished by wasting or by microbial leaching [8,28]. Both methods of phosphorus removal are considered further in this section and in greater detail in Chap. 11.

Biological Nitrification

Nitrification is the first step in the removal of nitrogen by the nitrification-denitrification process. A description of the nitrification process and its application are considered in the following discussion.

Process Description. Two bacteria genera are responsible for nitrification, *Nitrosomonas* and *Nitrobacter*. *Nitrosomonas* oxidizes ammonia to the intermediate product nitrite. Nitrite is converted to nitrate by *Nitrobacter*. The conversion from ammonia to nitrite involves a complex series of reactions that control the overall conversion process as evidenced by the lack of nitrite build-up in the system. Approximate equations for the reactions that occur can be written as follows.

For *Nitrosomonas* the equation is

$$55\,NH_4^+ + 76\,O_2 + 109\,HCO_3^- \rightarrow C_5H_7O_2N + 54\,NO_2^- + 57\,H_2O + 104\,H_2CO_3 \quad (8\text{-}80)$$

For Nitrobacter the equation is

$$400\,NO_2^- + NH_4^+ + 4H_2CO_3 + HCO_3^- + 195\,O_2 \rightarrow C_5H_7O_2N + 3H_2O + 400\,NO_3^- \quad (8\text{-}81)$$

These equations allow the amount of chemicals required for the processes to be calculated. Approximately 4.3 mg O_2 per mg of ammonia-nitrogen oxidized to nitrate-nitrogen is needed. In the conversion process, a large amount of alkalinity is consumed: 8.64 mg HCO_3^- per mg of ammonia-nitrogen oxidized. It should be noted that changing ammonia-nitrogen to nitrate-nitrogen does not facilitate nitrogen removal but does eliminate its oxygen demand.

Nitrifying bacteria are sensitive organisms and extremely susceptible to a wide variety of inhibitors. A variety of organic and inorganic agents can inhibit the growth and action of these organisms. High concentrations of ammonia and nitrous acid can be inhibitory. The effect of pH is also significant. A narrow optimal range between pH 7.5 to 8.6 exists, but systems acclimated to lower pH conditions have successfully nitrified. Temperature also exerts a tremendous influence on the growth of nitrifying bacteria. However, quantification of this effect has been difficult. Dissolved oxygen concentrations above 1 mg/L are essential for nitrification to occur. If DO levels drop below this value, oxygen becomes the limiting nutrient and nitrification slows or ceases.

Process Application. The principal nitrification processes may be classified as suspended-growth and attached-growth processes. In the suspended-growth process, nitrification can be achieved either in the same reactor used in the treatment of the carbonaceous organic matter or in a separate suspended-growth reactor following a conventional activated-sludge treatment process. When carbonaceous removal and nitrification are achieved in the same reactor, the process is often identified as a single-stage nitrification. When a separate facility is used for nitrification, it normally includes a reactor and settling tank of the same general design configuration used for the activated-sludge process. The oxidation of ammonia to nitrate can be carried out with either air or high-purity oxygen. The details of the nitrification process are considered in Chap. 11.

As with suspended-growth reactors, nitrification can be achieved in the same attached-growth reactor used for carbonaceous organic matter removal or a separate reactor. Trickling filters, rotating biological contactors, and packed towers can be used for nitrifying systems. These systems are resistant to shock loads, but may be susceptible to breakthrough of ammonia at peak flows. In combined carbon oxidation-nitrification systems, the biological films are thicker than those in nitrifying reactors. Low soluble carbonaceous BOD loadings, necessary to promote the growth of nitrifying cultures, account for the difference in film thickness. Higher loading in the combined systems may lead to excessive film growth and sloughing.

Biological Denitrification

Denitrification is the second step in the removal of nitrogen by the nitrification-denitrification process. A description of the denitrification process and its application are considered in the following discussion.

Process Description. The removal of nitrogen in the form of nitrate by conversion to nitrogen gas can be accomplished biologically under "anoxic" (without oxygen) conditions. The process is known as *denitrification*. In the past, the conversion process was often identified as anaerobic denitrification. However, the principal biochemical pathways are not anaerobic but rather a modification of aerobic pathways; therefore, the use of the term anoxic in place of anaerobic is considered appropriate [36]. Conversion of nitrate-nitrogen to a readily removable form can be accomplished by several genera of bacteria. Included in this list are *Achromobacter*, *Aerobacter*, *Alcaligenes*, *Bacillus*, *Brevibacterium*, *Flavobacterium*, *Lactobacillus*, *Micrococcus*, *Proteus*, *Pseudomonas*, and *Spirillum*. These bacteria are heterotrophs capable of dissimilatory nitrate reduction, a two-step process. The first step is conversion of nitrate to nitrite. This stage is followed by production of nitric oxide, nitrous oxide, and nitrogen gas.

The reactions for nitrate reduction are

$$NO_3^- \rightarrow NO_2^- \rightarrow NO \rightarrow N_2O \rightarrow N_2 \tag{8-82}$$

The last three compounds are gaseous products that can be released to the atmosphere.

In denitrifying systems, dissolved oxygen concentration is the critical parameter. The presence of DO will suppress the enzyme system needed for denitrification. Alkalinity is produced during the conversion of nitrate to nitrogen gas resulting in an increase in pH. The optimal pH lies between 7 and 8 with different optimums for different bacterial populations. Temperature affects the removal rate of nitrate and the microbial growth rate. The organisms are sensitive to changes in temperature.

Process Application. As with nitrification, the principal denitrification processes may also be classified as suspended-growth and attached-growth. Suspended-growth denitrification is usually carried out in a plug-flow type of activated-sludge system (i.e., following any process that converts ammonia and organic nitrogen to nitrates [nitrification]). The anaerobic bacteria obtain energy for growth from the conversion of nitrate to nitrogen gas but require a source of carbon for cell synthesis. Because nitrified effluents are usually low in carbonaceous matter, an external source of carbon is required. In some biological denitrification systems, the incoming wastewater or cell tissue is used to provide the needed carbon. In the treatment of agricultural wastewaters that are deficient in organic carbon, methanol has been used as a carbon source. Industrial wastes that are poor in nutrients but contain organic carbon have also been used. Because the nitrogen gas formed in the denitrification reaction hinders the settling of the mixed liquor, a nitrogen-gas stripped reactor should precede the denitrification clarifier.

Attached-growth (fixed-film) denitrification is carried out in a column reactor containing stone or one of several synthetic media upon which the bacteria grow. Depending on the size of the media, this process may or may not need to be followed by a clarifier. Adequate wasting of solids occurs through the low-level suspended-solids carryover in the effluent. Periodic backwashing and/or air scour is necessary to prevent solids buildup in the column that can cause excessive headloss. As in the suspended-growth denitrification process, an external carbon source is usually necessary. Most applications of this process involve the downflow mode (either gravity or pressure), but expanded-bed (upflow) techniques are also used. Additional process modifications in terms of the physical systems used are considered in Chap. 11.

Phosphorus Removal

Phosphorus appears in wastewater as orthophosphate (PO_4^{-3}), polyphosphate (P_2O_7), and organically bound phosphorus. The last two components may account for up to 70 percent of the influent phosphorus. Microbes utilize phosphorus during cell synthesis and energy transport. As a result, 10 to 30 percent of the influent phosphorus is removed during secondary biological treatment. Additional uptake beyond that needed for normal cell maintenance and synthesis is required to achieve low effluent concentration levels. Under certain aerobic conditions more phosphorus than is needed may be taken up by the microorganisms. Phosphorus may be released from cells under anoxic conditions. Biological phosphorus removal is accomplished by sequencing and producing the appropriate environmental conditions in the reactor(s).

Process Description. *Acinetobacter* are one of the primary organisms responsible for removal of phosphorus. These organisms respond to volatile fatty acids (VFAs) in the influent wastewater under anaerobic conditions by releasing stored phosphorus. The VFAs are an important substrate for the *Acinetobacter* during competition with heterotrophs. When an anaerobic zone is followed by an aerobic (oxic) zone, the microorganisms exhibit phosphorus uptake above normal levels. Phosphorus is not only utilized for cell maintenance, synthesis, and energy transport but also stored for subsequent use by the microorganisms. The sludge containing the excess phosphorus is either wasted or removed and treated in a side stream to release the excess phosphorus. Release of phosphorus occurs under anoxic conditions. Thus, biological phosphorus removal requires both anaerobic and aerobic reactors or zones within a reactor.

As noted above, two mechanisms exist for the removal of phosphorus: sludge wasting and treatment of side streams. Currently, a number of proprietary processes take advantage of one of these mechanisms. Two of these processes are the PhoStrip and the Bardenpho (see Fig. 8-33). Both of these processes, as shown in Fig. 8-33, feature anaerobic-aerobic contacting sequences with slightly different modifications. In the PhoStrip process, the biological release of phosphorus, under anoxic conditions, is used to concentrate the nutrient in a side stream for chemical treatment. Generally, lime is applied for phosphorus precipitation. In the Bardenpho process, a sequence of anaerobic, anoxic, and aerobic steps is used to achieve both nitrogen and phosphorus removal. Phosphorus is removed by wasting sludge from the system.

Combined Nitrogen and Phosphorus Removal

Combination processes such as the Bardenpho process (see Fig. 8-33b) are used where both nitrogen and phosphorus are to be removed. Combination processes for nutrient removal are considered further in Chap. 11.

8-12 POND TREATMENT PROCESSES

Ponds systems can be classified as (1) aerobic, (2) maturation, (3), facultative, and (4) anaerobic with respect to the presence of oxygen.

Aerobic Stabilization Ponds

In their simplest form, aerobic stabilization ponds are large, shallow earthen basins that are used for the treatment of wastewater by natural processes involving the use of both algae and bacteria. Although it is common to group all pond systems together when discussing them, in this chapter they are discussed according to the classification presented in Table 8-6. In Chap. 10, their design is considered collectively.

Process Description. An aerobic stabilization pond contains bacteria and algae in suspension, and aerobic conditions prevail throughout its depth. There are two basic types of aerobic ponds. In the first type, the objective is to maximize the production of algae. These ponds are usually limited to a depth of about 0.5 to 1.5 ft (150 to 450 mm).

In the second type, the objective is to maximize the amount of oxygen produced, and pond depths of up to 5 ft (1.5 m) are used. In both types, oxygen, in addition to

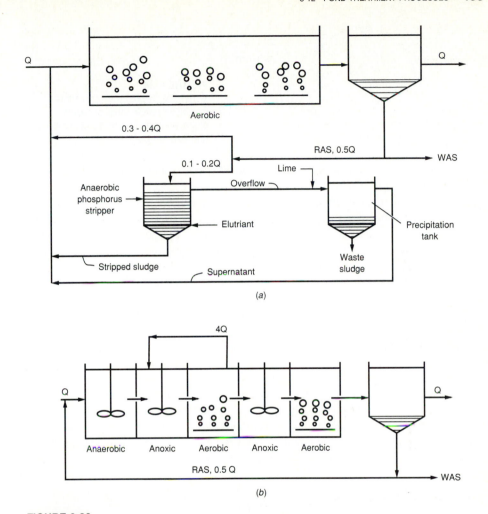

FIGURE 8-33
Typical treatment processes used for the biological removal of phosphorus: (a) PhoStrip process
and (b) five-stage Bardenpho (adapted from Ref. 28).

that produced by algae, enters the liquid through atmospheric diffusion. To achieve
best results with aerobic ponds, their contents must be mixed periodically using pumps
or surface aerators.

Process Microbiology. In aerobic photosynthetic ponds, the oxygen is supplied by
natural surface reaeration and by algal photosynthesis. Except for the algal population,
the biological community present in stabilization ponds is similar to that present in
an activated-sludge system. The oxygen released by the algae through the process of
photosynthesis is used by the bacteria in the aerobic degradation of organic matter.
The nutrients and carbon dioxide released in this degradation are, in turn, used by
the algae. This cyclic-symbiotic relationship is shown in Fig. 8-34. Higher animals,

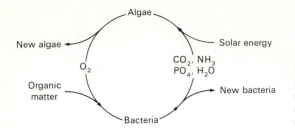

FIGURE 8-34
Schematic representation of the symbiotic relationship between algae and bacteria in a high-rate stabilization pond.

such as rotifers and protozoa, are also present in the pond, and their main function is to polish the effluent.

The particular algal group, animal group, or bacterial species present in any section of an aerobic pond depends on such factors as organic loading, degree of pond mixing, pH, nutrients, sunlight, and temperature. Temperature has a profound effect on the operation of aerobic ponds, particularly in regions with cold winters.

Process Analysis. The efficiency of BOD_5 conversion in aerobic ponds is high, ranging up to 95 percent; however, it must be remembered that, although the soluble BOD_5 has been removed from the influent wastewater, the pond effluent will contain an equivalent or larger concentration of algae and bacteria that may ultimately exert a higher BOD_5 than the original waste. Various means of removing the algae from the treated wastewater are discussed in Chap. 10.

A number of theoretical approaches have been proposed for the analysis of aerobic stabilization ponds. Because of the many uncontrollable variables involved, however, the ponds are still usually designed by using appropriate loading factors derived from pilot plant studies and observations of operating systems. The pond loading is adjusted to reflect the amount of oxygen available from photosynthesis and atmospheric reaeration.

Facultative Ponds

Ponds in which the stabilization of wastes is brought about by a combination of aerobic, anaerobic, and facultative bacteria are known as facultative (aerobic-anaerobic) stabilization ponds.

Process Description. As shown in Fig. 8-35, three zones exist in a facultative pond: (1) a surface zone where aerobic bacteria and algae exist in a symbiotic relationship, as previously discussed; (2) an anaerobic bottom zone in which accumulated solids are decomposed by anaerobic bacteria; and (3) an intermediate zone that is partly aerobic and partly anaerobic, in which the decomposition of organic wastes is carried out by facultative bacteria. Conventional facultative ponds are earthen basins filled with screened and, in some cases, comminuted raw wastewater or primary effluent. Large solids settle out to form an anaerobic sludge layer. Soluble and colloidal organic materials are oxidized by aerobic and facultative bacteria using oxygen produced by algae growing abundantly near the surface. Carbon dioxide produced in

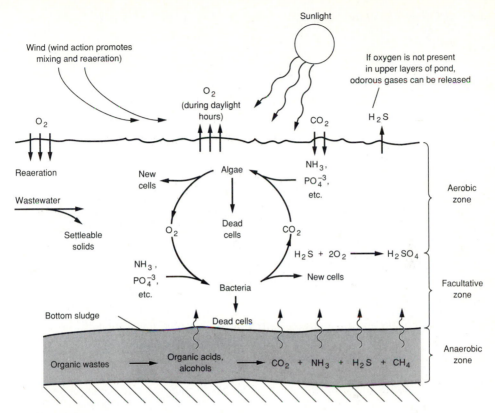

FIGURE 8-35
Schematic representation of a waste-stabilization pond [35].

the organic oxidation serves as a carbon source for the algae. Anaerobic breakdown of the solids in the sludge layer results in the production of dissolved organics and gases such as CO_2, H_2S, and CH_4, which are either oxidized by the aerobic bacteria or vented to the atmosphere.

In practice, oxygen is maintained in the upper layer of facultative lagoons by the presence of algae and by surface reaeration. In some cases, surface aerators have been used (see Chap. 10). If surface aerators are used, algae are not required. The advantage of using surface aerators is that a higher organic load can be applied. However, the organic load must not exceed the amount of oxygen that can be supplied by the aerators without completely mixing the pond contents, or the benefits to be derived from anaerobic decomposition will be lost.

Process Microbiology. The biological community in the upper or aerobic layer is similar to that of an aerobic pond. The microorganisms in the lower and bottom layers of the pond are facultative and anaerobic. Respiration also occurs in the presence of sunlight; however, the net reaction is the production of oxygen. Equations

8-83 and 8-84 represent simplified biochemical reactions for photosynthesis and respiration.

Photosynthesis:

$$CO_2 + 2H_2O \xrightarrow{\text{light}} (CH_2O) + O_2 + H_2O \qquad (8\text{-}83)$$

New
algae
cells

Respiration:

$$CH_2O + O_2 \longrightarrow CO_2 + H_2O \qquad (8\text{-}84)$$

Because algae use carbon dioxide in photosynthetic activity, high pH conditions can result, especially in wastewaters with low alkalinity. In many cases, algae in facultative ponds will obtain the carbon needed for cell growth from the bicarbonate ion. When the bicarbonate ion is used as a source of cell carbon, high diurnal variations in the pH may be observed. In addition, as the pH increases, the alkalinity components change, and carbonate and hydroxide alkalinity tend to predominate. If the wastewater has a high concentration of calcium, calcium carbonate will precipitate when the carbonate and calcium-ion concentrations become great enough to exceed the solubility product. This removal of the carbonate ion by precipitation will keep the pH from continuing to increase.

Process Analysis. The amount of effort that has been devoted to the characterization of facultative ponds is staggering, and an equal amount has probably been spent trying to develop appropriate design equations. Although many design equations have been published, there is no universal equation. Part of the explanation for this is that, to a large extent, the process is undefined because of the vagaries of nature. For example, all predictive equations for effluent quality are essentially meaningless when windy conditions prevail. Under such conditions, the effluent quality will be a function of the degree of wind mixing and the quantity of the deposited solids that have been suspended. For this reason, facultative ponds are usually designed on the basis of loading factors developed from field experience.

Tertiary-Maturation Ponds

Tertiary-maturation low-rate stabilization ponds are designed to provide for secondary effluent polishing and seasonal nitrification. The biological mechanisms involved are similar to other aerobic suspended-growth processes. Operationally, the residual biological solids are endogenously respired, and ammonia is converted to nitrate using the oxygen supplied from surface reaeration and from algae. A detention time of 18 to 20 days has been suggested as the minimum periods required to provide for complete endogenous respiration of the residual solids. To maintain aerobic conditions, the applied loadings are quite low.

Anaerobic Ponds

Anaerobic ponds are used for the treatment of high-strength organic wastewater that also contains a high concentration of solids. Typically, an anaerobic pond is a deep earthen pond with appropriate inlet and outlet piping. To conserve heat energy and to maintain anaerobic conditions, anaerobic ponds have been constructed with depths up to 30 ft (9.1 m). The wastes that are added to the pond settle to the bottom. The partially clarified effluent is usually discharged to another treatment process for further treatment.

Usually, these ponds are anaerobic throughout their depth, except for an extremely shallow surface zone. Stabilization is brought about by a combination of precipitation and the anaerobic conversion of organic wastes to CO_2, CH_4, other gaseous end products, organic acids, and cell tissues. Conversion efficiencies for BOD_5 of up to 70 percent are obtainable routinely. Under optimum operating conditions, removal efficiencies up to 85 percent are possible.

DISCUSSION TOPICS AND PROBLEMS

8-1. A 1 L sample contains 0.55 lb of casein ($C_8H_{12}O_3N_2$). If 0.5 lb of bacterial cell tissue ($C_5H_7NO_2$) is synthesized per lb of casein consumed, determine the amount of oxygen required to complete the oxidation of casein to end products and cell tissue. The end products of the oxidation are carbon dioxide (CO_2), ammonia (NH_3), and water. Assume that the nitrogen not incorporated in cell-tissue production will be converted to ammonia.

8-2. Assuming that the endogenous coefficient k_d can be neglected, develop expressions that can be used to determine the substrate and cell concentration as a function of time for a batch reactor. If the initial concentration of substrate and cell is 100 and 200 mg/L, respectively, determine the amount of substrate remaining after 1 h. If the endogenous coefficient is equal to $0.04\ d^{-1}$, estimate the error made by neglecting this factor. Assume that the following constants apply: $k = 2.0\ h^{-1}$; $K_s = 80$ mg/L; $Y = 0.4$ lb/lb.

8-3. If the dilution rate D is defined as Q/V and the endogenous coefficient is neglected, develop expressions that can be used to estimate the effluent substrate and cell concentration from a complete-mix reactor without recycle as a function of the dilution rate. If $Y = 0.5$ lb/lb, $\mu_m = 1.0\ h^{-1}$, $K_s = 200$ mg/L, and $S_o = 10,000$ mg/L, prepare a plot of the substrate and cell concentration versus the dilution rate. Plot the dilution rate going from zero to $1.0\ h^{-1}$. Use 1/2 in (1 cm) divisions for each 1000 mg/L of substrate and in (2cm) divisions for each 1000 mg/L of cells.

8-4. Derive Eq. 8-29, which is used to determine the observed cell yield.

8-5. A wastewater is to be treated aerobically in a complete-mix reactor with no recycle. Determine θ_c^M using the following constants: $K_s = 50$ mg/L; $k = 5.0\ d^{-1}$; $k_d = 0.06\ d^{-1}$; and $Y = 0.60$. The initial wastewater substrate concentration is 200 mg/L.

8-6. Using a design value of $\theta_c = 2$ d and the constants given in Prob. 8-5, determine the effluent substrate concentration, the specific utilization rate $(-r_{su}/X)$, the food-to-microorganism ratio, $(S_o/X\theta)$, and the concentration of microorganisms in the reactor.

8-7. The following data were obtained using four bench-scale complete-mix activated-sludge units to treat a food-processing waste. Using these data, determine Y and k_d.

		Parameter	
Unit	X, gMLVSS	r'_g, gMLVSS/d	U, gBOD$_5$/gMLVSS · d
1	18.81	0.88	0.17
2	7.35	1.19	0.41
3	7.65	1.42	0.40
4	2.89	1.56	1.09

8-8. Derive Eq. 8-57 for a plug-flow reactor.

8-9. It has been proposed that a complete-mix reactor with recycle should be used for the treatment of a medium-strength wastewater (see Table 3-16). Determine the amount of oxygen required for the carbonaceous oxidation of the wastewater (assume nitrification does not occur) at a mean cell-residence time of 6 d. Use the kinetic coefficients, given in Table 8-7, and assume that the organic compounds in wastewater can be represented as $C_6H_{12}O_6$, the nitrogen as NH_4^+, and the phosphorus as $H_2PO_4^-$. Represent the cell tissue produced in the process as $C_{60}H_{87}O_{23}N_{12}P$. What percentage of the influent nitrogen and phosphorus will be present in the effluent?

8-10. Assuming that the waste specified in Prob. 8-9 can be nitrified completely at a mean cell-residence time of 15 d, estimate the total amount of oxygen required. How does the amount of oxygen required for carbonaceous oxidation compare?

8-11. If 75 percent of the cell tissue produced during biological treatment is biodegradable, estimate the ultimate carbonaceous production of cell tissue using Eq. 8-6. If the K value (base 10) is equal to 0.1, determine the BOD$_5$ of the cell tissues. Express your answer in terms of mg BOD$_5$/mg cell tissue.

8-12. Determine the kinetic coefficients k, K_s, μ_m, Y, and k_d from the following data derived using a laboratory scale complete-mix reactor with solids recycle (see Fig. 8-13b). Refer to Appendix H for a general discussion of the determination of kinetic coefficients from laboratory data.

Unit no.	S_o, mg/L	S, mg/L	θ, d	X, mgVSS/L	θ_c, d
1	400	10	0.167	3,950	3.1
2	400	14.3	0.167	2,865	2.1
3	400	21.0	0.167	2,100	1.6
4	400	49.5	0.167	1,050	0.8
5	400	101.6	0.167	660	0.6

8-13. Prepare a one-page abstract of Reference 19.

8-14. An activated-sludge process with a short aeration time is to be used following a tower trickling filter for treatment of domestic wastewater. Using the following information and data, determine the cell concentration (MLVSS) that must be maintained in the aeration tank if both effluent BOD$_5$ and suspended solids must be less than 25 mg/L. What recycle rate will be required under typical and best operating characteristics for the secondary settling tank? Assume that the BOD$_5$ of the effluent solids is equal to 0.65 times the concentration of the solids. (1) Average flowrate $= 1.0$ Mgal/d; (2) peak flowrate $= 2.0$ Mgal/d; (3) detention time in aeration basin at peak flow $= 0.5$ h; (4) trickling-filter effluent: $BOD_5 = 60$ mg/L, suspended solids $= 60$ mg MLVSS/d (see table presented below); (5) settling data for the MLVSS derived at a nearby location with a similar plant; (6) settled biological solids are to be recycled to head end of the aerator; (7) kinetic coefficients for the aeration process; $k = 5.0$ d^{-1}, $K_s = 60$ mg/L, $Y = 0.5$ lb/lb, $k_d = 0.06$ d^{-1}; (8) aeration tank type $=$ complete-mix.

X, mg MLVSS/L	X_r, mg MLVSS/L	
	Typical	Best
1,000	3,200	6,000
2,000	5,200	8,000
3,000	6,600	9,400
4,000	8,000	10,200

8-15. Prepare a one-page abstract of the following article: Garrett, M. T., Jr.: "Hydraulic Control of Activated Sludge Growth Rate," *Sewage and Industrial Wastes,* vol. 39, no. 3, 1958.

8-16. A conventional complete-mix activated-sludge treatment process is to be used to treat 1.0 Mgal/d of a wastewater with a BOD_5 of 250 mg/L after primary settling. The process loading is to be 0.30 lb BOD_5/lb MLVSS · d. If the detention time is 6 hr and the recirculation ratio is 0.33, determine the value of the MLVSS.

8-17. A complete-mix aerated lagoon is to be designed with a detention time of 5 d. Using Eq. 8-60 and the data given below, determine the effluent soluble BOD_5. Estimate the total BOD_5 by considering the BOD_5 of the biological solids. What would the value of k_1 in Eq. 8-58 and k_2 in Eq. 8-59 have to be to yield the same results as obtained with Eq. 8-60. How do the values you computed compare to values reported in the literature? Cite at least three recent references (after 1980).

1. Influent characteristics:

   Flowrate $= 1.0$ Mgal/d

   Total BOD_5 $= 200$ mg/L

   Filtered BOD_5 $= 150$ mg/L

   Suspended solids $= 200$ mg/L

2. Kinetic coefficients:

   $k = 4.0$ d^{-1}

   $Ks = 80$ mg/L

   $Y = 0.45$ lb/lb

   $k_d = 0.05$ d^{-1}

3. BOD_5 of effluent solids $= 0.65$ (suspended solids)

4. Assume that the BOD_5 associated with suspended solids is totally converted in the process.

8-18. Prepare a one-page abstract of the following article: Thirumurthi, D.: "Design Principles of Waste Stabilization Ponds," *Journal of the Sanitary Division, ASCE,* vol. 95, no. SA2, 1969.

8-19. Find the theoretical diameters of the two trickling filters in a two-stage trickling-filter process (such as shown in Fig. 8-26b) for an installation with the following characteristics and requirements.

1. A flow of 6.0 Mgal/d with a BOD_5 of 300 mg/L.
2. To maintain stream standards, the effluent BOD_5 must be equal to or less than 21 mg/L.
3. The filters are to have equal diameters and a depth of 5 ft.
4. The recirculation ratio chosen shall result in a hydraulic loading of 690 gal/ft² · d.
5. The primary sedimentation tank provides a BOD_5 removal of 30 percent.
6. Use the NRC two-stage trickling-filter loading criteria.

8-20. The data below were obtained from a pilot plant study involving the treatment of a combined domestic-industrial wastewater with a tower trickling filter filled with a plastic medium. The BOD_5 applied to the filter after primary settling was equal to 350 mg/L. The area of the pilot filter was 10 ft². The wastewater temperature at the time the tests were run was 16°C. Using these data, determine the value of the terms k_{20} and n in Eq. 8-73.

	Removal efficiency, %			
	Flowrate, ft/d			
Depth, ft	20	40	60	80
6	52	34	22	18
12	77	55	40	32
18	89	70	52	43
24	95	79	62	50
30	96	81	67	51

8-21. Using the kinetic coefficient derived in Prob. 8-20, determine the maximum rate of flow that can be applied to a 20 ft tower filter that is to be designed to remove 50 percent of applied BOD_5 under winter conditions. The applied BOD_5 is equal to 350 mg/L and the critical sustained winter wastewater temperature is 8°C. If the average summer wastewater temperature is equal to 22°C, what degree of removal can be expected during the summer?

8-22. Estimate the amount of methane that can be produced from the fermentation of tricarboxylic acid having the empirical formula $C_3H_8O_6$. Ignore cell growth.

8-23. What is the approximate molecular weight and specific weight at standard conditions of a digester gas that contains 70 percent methane and 28 percent carbon dioxide?

8-24. An anaerobic digester is designed to remove 85 percent of the BOD_5 of an industrial organic waste with an ultimate BOD equal to 2000 mg/L. If the mean cell-residence time is 12 days, estimate the quantity of sludge to be wasted daily and the quantity of gas produced each day. Assume that the flow is equal to 0.1 Mgal/d, Y equals 0.1, and k_d equals 0.01d⁻¹. How much gas and sludge will be produced if the mean cell-residence time is increased to 20d? Is the added cost associated with increasing the size of the digester worthwhile?

REFERENCES

1. Albertson, O. E., and G. Davis: "Analysis of Process Factors Controlling Performance of Plastic Bio-Media," Presented at the 57th Annual Meeting of the Water Pollution Control Federation, New Orleans, LA, October 1984.

2. Albertson, O. E.: "The Control of Bulking Sludges: From the Early Innovators to Current Practice," *Journal WPCF,* vol. 59, p. 172, April 1987.

3. Ardern, E., and W. T. Lockett: "Experiments on the Oxidation of Sewage without the Aid of Filters," *J. Soc. Chem, Ind.,* vol. 33, pp. 523, 1122, 1914.

4. Atkinson, B., I. J. Davies, and S. Y. How: "The Overall Rate of Substrate Uptake by Microbial Films, parts I and II," *Trans. Inst. Chem. Eng.,* 1974.

5. Benefield, L. D., and C. W. Randall: *Biological Process Design for Wastewater Treatment,* Prentice-Hall, Englewood Cliffs, NJ, 1980.

6. Bruce, A. M., and J. C. Merkens: "Further Studies of Partial Treatment of Sewage by High-Rate Biological Filtration," *J. Inst. Water Pollut. Contr.,* vol. 72, no. 5, London, 1973.

7. Eckenfelder, W. W., Jr.: "Trickling Filtration Design and Performance," *Trans. ASCE,* vol. 128, 1963.

8. Eckenfelder, W. W., Jr.: "Biological Phosphorus Removal: State of the Art Review," *Pollution Engineering,* p. 88, September 1987.

9. Fairall, J. M.: "Correlation of Trickling Filter Data," *Sewage and Ind. Wastes,* vol. 28, no. 9, 1956.

10. Galler, W. S., and H. B. Gotass: "Optimization Analysis for Biological Filter Design," *J. Sanit. Eng. Div., ASCE,* vol. 92, no. SA1, 1966.

11. Germain, J. E.: "Economic Treatment of Domestic Waste By Plastic-Medium Tricking Filter," *Journal WPCF,* vol. 38, no. 2, 1966.

12. Grady, C. P. L., Jr., and H. C. Lim: *Biological Wastewater Treatment: Theory and Application,* Marcel Dekker, New York, 1980.

13. Hawkes, H. A.: *The Ecology of Waste Water Treatment,* Macmillan, New York, NY, 1963.

14. Higgins, I. J., and R. G. Burns: *The Chemistry and Microbiology of Pollution,* Academic, London, 1975.

15. Holland, K. T., J. S Knapp, and J. G. Shoesmith: *Anaerobic Bacteria,* Chapman and Hall, New York, 1987.

16. Hoover, S. R., and N. Porges: "Assimilation of Dairy Wastes by Activated Sludge, II: The Equation of Synthesis and Oxygen Utilization," *Sewage and Ind. Wastes,* vol. 24, 1952.

17. Jenkins, D., M. G. Richard, and G. T. Daigger: *Manual On The Causes And Control of Activated Sludge Bulking And Foaming,* prepared for Water Research Commission, Republic of South Africa and U.S. EPA, Cincinnati, OH, Water Research Commission, Pretoria, South Africa, April 1986.

18. Lawrence, A. W., and P. L. McCarty: "Kinetics of Methane Fermentation in Anaerobic Treatment," *J. Water Pollut. Control Fed.,* vol. 41, no. 2, part 2, 1969.

19. Lawrence, A. W., and P. L. McCarty: "A Unified Basis for Biological Treatment Design and Operation," *J. Sanit. Eng. Div., ASCE,* vol. 96, no. SA3, 1970.

20. Logan, B. E., S. W. Hermanowicz, and D. S. Parker: "Engineering Implications of a New Trickling Filter Model," *Journal WPCF,* vol. 59, no. 12, 1987.

21. Logan, B. E., S. W. Hermanowicz, and D. S. Parker: "A Fundamental Model For Trickling Filter Process Design," *Journal WPCF,* vol. 59, no. 12, 1987.

22. McCarty, P. L.: "Anaerobic Waste Treatment Fundamentals," *Public Works,* vol. 95, nos. 8-12, 1964.

23. McCarty, P. L.: "Kinetics of Waste Assimilation in Anaerobic Treatment," *Developments in Industrial Microbiology,* vol. 7, American Institute of Biological Sciences, Washington, DC, 1966.

24. McKinney, R. E.: *Microbiology for Sanitary Engineers,* McGraw-Hill, New York, 1962.

25. Monod, J.: *Recherches sur la croissance des cultures bacteriennes,* Herman et Cie., Paris, 1942.

26. Monod, J.: "The Growth of Bacterial Cultures," *Ann. Rev. Microbiol.,* vol. 3, 1949.

27. National Research Council: "Trickling Filters (in Sewage Treatment at Military Installations)," *Sewage Works J.,* vol. 18, no. 5, 1946.

28. *Principles and Practice of Nutrient Removal From Municipal Wastewater,* The Soap and Detergent Assoc., New York, 1988.

29. Ribbons, D. W.: "Quantitative Relationships between Growth Media Constituents and Cellular Yields and Composition," in J. W. Norris and D. W. Ribbons (eds.), *Methods in Microbiology,* vol. 3A, Academic, London, 1970.

30. Sarner, E.: *Plastic Packed Trickling Filters,* Ann Arbor Science Publishers, Ann Arbor, MI, 1980.

31. Schroeder, E. D., and G. Tchobanoglous: "Mass Transfer Limitations on Trickling Filter Design," *Journal WPCF,* vol. 48, no. 4, 1976.

32. Schultz, K. L.: "Load and Efficiency of Trickling Filters," *Journal WPCF,* vol. 33, no. 3, pp. 245-260, 1960.

33. Speece, R. E.: "Anaerobic Biotechnology for Industrial Wastewater Treatment," *Environmental Science and Technology,* vol. 17, no. 9, 1983.

34. Stanier, R. Y., J. L. Ingraham, M. L. Wheelis, and P. R. Painter: *The Microbial World,* 5th ed., Prentice-Hall, Englewood Cliffs, NJ, 1986.

35. Tchobanoglous, G. and E. D. Schroeder, *Water Quality: Characteristics, Modeling, Modification,* Addison-Wesley, Reading, MA, 1985.

36. U.S. Environmental Protection Agency: *Process Design Manual for Nitrogen Control,* Office of Technology Transfer, Washington, DC., October 1975.

37. U.S. Environmental Protection Agency: *Sequencing Batch Reactors,* EPA/625/8-86/011, Cincinnati, OH, October 1986.

38. U.S. Environmental Protection Agency: *Phosphorus Removal, Design Manual,* EPA/625/1-87/001, Cincinnati, OH, September 1987.

39. Van Uden, N.: "Transport-Limited Growth in the Chemostat and Its Competitive Inhibition; A Theoretical Treatment," *Arch. Mikrobiol.,* vol. 58, 1967.

40. Velz, C. J.: "A Basic Law for the Performance of Biological Beds," *Sewage Works J.,* vol. 20, no. 4, 1948.

41. Water Pollution Control Federation: *Nutrient Control,* Manual of Practice FD-7, Washington, DC, 1983.

42. Water Pollution Control Federation: *Activated Sludge,* Manual of Practice OM-9, Alexandria, VA, 1987.

43. Water Pollution Control Federation: *Anaerobic Sludge Digestion,* Manual of Practice no. 16, 2nd ed., Alexandria, VA, 1987.

44. Wood, D. K., and G. Tchobanoglous: "Trace Elements in Biological Waste Treatment with Specific References to the Activated Sludge Process," *Proc. 29th Ind. Waste Conf.,* 1974.

CHAPTER
9

DESIGN
OF FACILITIES
FOR PHYSICAL
AND CHEMICAL
TREATMENT
OF WASTEWATER

The purpose of this chapter is to discuss the design of the unit operations and processes that were described in Chaps. 6 and 7. The principal unit operations and processes and their functions as applied to the treatment of wastewater are reported in Table 9-1 (see also Fig. 6-1). As shown, physical operations are used for the removal of coarse solids, suspended and floating solids, grease, and volatile organic compounds. Chemical processes are used for the precipitation of suspended and colloidal solids, disinfection of the wastewater, and control of odors. Although each operation and process identified in Table 9-1 will be discussed in this chapter, those most commonly encountered in the design of wastewater treatment facilities will be considered in greater detail. The unit operations and processes considered in detail in this chapter include (1) screening, (2) communition, (3) grit removal, (4) primary sedimentation, and (5) disinfection with chlorine. Filtration and microscreening of wastewater is discussed in Chap. 11; processing of sludge produced from primary and secondary treatment operations and processes is discussed separately in Chap. 12.

9-1 BAR RACKS AND SCREENS

The first step in wastewater treatment is the removal or reduction of coarse solids. The usual procedure is to pass the untreated wastewater through bar racks or screens. Bar

445

TABLE 9-1

Functions of various physical and chemical operations and processes used for wastewater treatment

Operation or process	Function	See section
Coarse screening	Removal of coarse solids by interception. Considered a preliminary treatment operation.	9-1
Communition	In-channel grinding of solids. Considered a preliminary treatment operation.	9-2
Grit removal	Removal of grit, sand, and gravel, usually following screening and comminution. Considered a preliminary treatment operation.	9-3
Flow equalization	Equalization of flow and mass loadings of BOD and suspended solids on subsequent treatment facilities.	9-4
Other preliminary treatment operations:		9-5
Preaeration	Replenishment of dissolved oxygen. Improvement of hydraulic distribution.	
Flocculation	Improvement of settling characteristics of suspended solids.	
Sedimentation	Removal of settleable solids and floating material. Principal operation used in the primary treatment of wastewater.	9-6
Other solids-removal operations and units:		9-7
Flotation	Used as a replacement for gravity sedimentation or as a pretreatment unit before primary sedimentation to achieve improved suspended-and floatable-solids removal.	
Fine screening	Used as a replacement for gravity sedimentation; may also be used for grit removal in preliminary treatment.	
Chemical precipitation	Removal of settleable and colloidal solids and phosphorus. Used as a first unit process in the independent physical-chemical treatment of wastewater.	9-8
Disinfection with chlorine compounds:		9-9
Chlorination	Used principally for the disinfection of wastewater; also used for odor control.	
Dechlorination	Dechlorination of treated, chlorinated effluent.	
Other means of disinfection	Disinfection by bromine chloride, ozone, or UV radiation	9-10
Post-aeration	Addition of dissolved oxygen to treated effluent.	9-11
Odor control[a]	Various operations and processes used for the removal and elimination of odors emanating from various treatment facilities.	9-12
VOC control[a]	Used for the treatment, destruction, or disposal of off gases containing VOCs.	9-13

[a] Not strictly defined as an operation or process.

racks normally have clear openings between bars of 5/8 in (15 mm) or larger. Screens have clear openings smaller than 5/8 in (15 mm) and are used in facilities (usually small plants) where smaller solids are removed from the incoming wastewater. Before considering the design of bar racks and screens, the characteristics of screenings are described.

Characteristics of Screenings

Screenings are the material retained on bar racks and screens. The smaller the screen opening, the greater the quantity of collected screenings will be. Although no precise definition of screenable material exists and no recognized method of measuring quantities of screenings is available, screenings exhibit some common properties.

Screenings Retained on Bar Racks. Coarse screenings (collected on racks or bars of about 5/8 in or greater spacing) consist of debris such as rocks, branches, pieces of lumber, leaves, paper, tree roots, plastics, and rags. Organic matter can collect as well. The rag content can be substantial and has been visually estimated to comprise from 60 to 70 percent of the total screenings volume on 1 and 4 in (25 and 100 mm) screens, respectively. Coarse screenings are highly volatile (80 to 90 percent or more volatile solids content), and have a dry solids content of 15 to 25 percent and density of 40 to 60 lb/ft^3 (640 to 960 kg/m^3).

Screenings Retained on Screens. Fine screenings consist of materials that are retained on screens with openings less than 5/8 in (15 mm). Screens with 0.09 to 0.25 in openings remove 5 to 10 percent of influent SS, whereas those with openings of 0.03 to 0.06 in can remove 10 to 15 percent, although greater removals have been reported. Fine screenings have been reported to have volatile solids contents varying from 65 to 95 percent. Compared to coarse screenings, their bulk densities are slightly lower and moisture contents are somewhat greater. Because putrescible matter, including pathogenic fecal material, is contained within screenings, they must be properly handled and disposed. Fine screenings contain substantial grease and scum, which requires similar care.

Bar Racks

Bar racks may be hand cleaned or mechanically cleaned. Characteristics of these two types are compared in Table 9-2. Details about each type of bar rack and some of the factors that must be considered in the design of bar rack installations are presented in the following discussion.

Hand-cleaned Bar Racks. Hand-cleaned bar racks are frequently used ahead of pumps in small wastewater pumping stations. In the past they have been used at the headworks for small wastewater treatment plants. The recent practice has been to provide mechanically cleaned racks, even for small installations, not only to minimize

TABLE 9-2
Typical design information for hand cleaned and mechanically cleaned bar racks

Item	Hand-cleaned	Mechanically cleaned
Bar size:		
Width, in	0.2–0.6	0.2–0.6
Depth, in	1.0–1.5	1.0–1.5
Clear spacing between bars, in	1.0–2.0	0.6–3.0
Slope from vertical, degree	30–45	0–30
Approach velocity, ft/s	1.0–2.0	2.0–3.25
Allowable headloss, in	6	6

Note: in × 25.4 = mm
     ft/s × 0.3048 = m/s

the manual labor required to clean the racks and remove and dispose of the rakings but also to reduce flooding and overflows due to clogging.

Where used, the length of the hand-cleaned rack should not exceed the distance that can be conveniently raked by hand, approximately 10 ft (3 m). The rack bars are usually not less than 3/8 in (10 mm) thick by 2 in (50 mm) deep. They are welded to spacing bars located at the rear face, out of the way of the tines of the rake. A perforated drainage plate should be provided at the top of the rack, where the rakings may be stored temporarily for drainage. A typical hand-cleaned rack is shown in Fig. 9-1.

The rack channel should be designed to prevent the accumulation of grit and other heavy materials in the channel ahead of the rack and following it. The channel floor should be level or should slope downward through the screen without pockets to trap solids. Fillets may be desirable at the base of the sidewalls. The channel should preferably have a straight approach, perpendicular to the bar rack, to promote uniform distribution of screenable solids throughout the flow and on the rack.

To provide adequate rack area for accumulation of screenings between raking operations, it is essential that the velocity of approach be limited to approximately 1.5 ft/s (0.45 m/s) at average flow. Additional area to limit the velocity may be obtained by widening the channel at the rack and by placing the rack at a flatter angle to increase the submerged area. As screenings accumulate, partially plugging the rack, the upstream head will increase, submerging new areas for the flow to pass through. The structural design of the screen should be adequate to prevent collapse if it becomes completely plugged.

Mechanically Cleaned Bar Racks. Mechanically cleaned bar racks have been used in wastewater treatment plants for over 50 years. The design of the racks has evolved over the years to reduce the operating and maintenance problems and to improve the screenings removal capabilities. Many of the newer designs include extensive use of corrosion-resistant materials including stainless steel and plastics. Mechanically cleaned bar racks are divided into four principal types: (1) chain-operated, the most

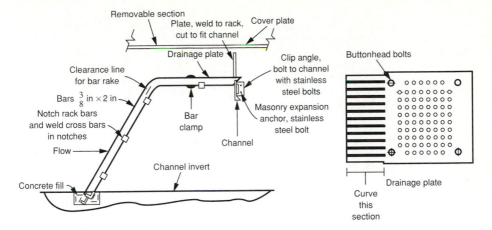

FIGURE 9-1
Typical hand-cleaned bar rack.

prevalent type; (2) reciprocating rake; (3) catenary; and (4) cable. Examples of the different types of mechanically cleaned bar racks are shown in Fig. 9-2.

Chain-operated mechanically cleaned bar racks can be divided into categories based on whether the rack is raked clean from the front (upstream) or the back (downstream) and whether the rakes return to the bottom of the bar rack from the front or back. Each type has its advantages and disadvantages, although the general mode of operation is similar. In general, front-cleaned front-return racks (see Fig. 9-2a) are newer and more efficient in retaining captured solids, but they are less rugged and susceptible to jamming by solids that collect at the base of the rake. Front-cleaned front-return bar racks are used to serve collection systems that largely consist of separate sanitary sewers. This type of bar rack is seldom used for plants serving combined sewers where large objects can jam the rakes. In front-cleaned back-return racks, the cleaning rakes return to the bottom of the bar rack on the downstream side of the rack, pass under the bottom of the rack, and clean the bar rack as the rake rises. The potential for jamming is minimized, but a hinged plate is required to seal the pocket under the rack. The hinged plate, however, is also subject to jamming.

In back-cleaned racks, the bars protect the rake from damage by the debris. However, this type of rack is more susceptible to solids carryover to the downstream side, particularly as rake wipers wear out. The bar rack of the back-cleaned back-return racks is less rugged than the other types because the top of the bar rack is unsupported to allow the rake tines to pass through. Most of the chain-operated racks share the disadvantage of submerged sprockets, which require frequent operator attention and are difficult to maintain. Additional disadvantages include adjustment and repair of the heavy chains and the need to dewater the channels to inspect and repair the submerged parts.

The reciprocating rake type bar rack (see Fig. 9-2b) imitates the movements of a person raking the rack. The rake moves to the base of the rack, engages the bars,

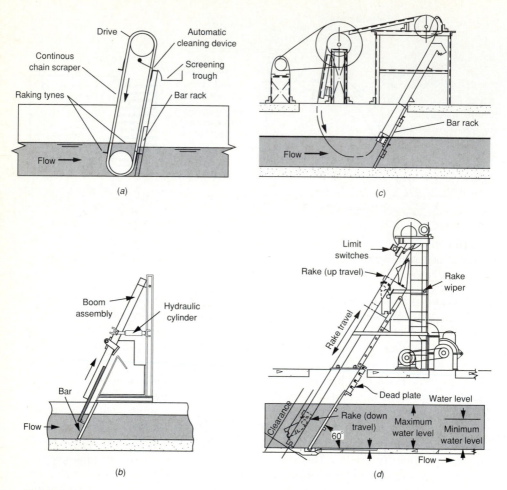

FIGURE 9-2
Typical mechanically cleaned bar racks: (a) chain operated, (b) reciprocating rake (from Franklin Miller), (c) catenary (from Dresser Industries), and (d) cable driven.

and rakes the screenings to the top of the rack, where they are removed. A major advantage is that all parts requiring maintenance are above the water and can be easily inspected and maintained without dewatering the channel. The front-clean front-return minimizes solids carryover. A disadvantage is that this type of rack uses only one rake instead of multiple rakes on the chain-operated racks. As a result the reciprocating rake type bar rack has limited capacity in handling heavy screenings loads, particularly in deep channels where a long "reach" is necessary. The high overhead clearance required to accommodate the mechanism can limit its use in retrofit applications.

In the front-cleaned front-return catenary rack (see Fig. 9-2c), the rake is held against the rack by the weight of the chain. An advantage is that the driving mechanism has no submerged sprockets. A disadvantage is the relatively large amount of space required for installation.

Cable-driven mechanically cleaned bar racks (see Fig. 9-2*d*) are front-cleaned front-return devices that use a pivoting rake that is raised and lowered on tracks by a cable and drum drive. The rake is lowered by gravity, pivots to engage the bar rack, and is raised by the cable drive. An advantage is that the rake itself is the only mechanical part entering the wastewater. Disadvantages include limited raking capacity and maintenance problems related to slack cables, fouled cable reels, and improperly operating brake mechanisms.

Design of Mechanically Cleaned Bar Rack Installations. For most installations, two or more units should be installed so that one unit may be taken out of service for maintenance. Slide gates or stop-log grooves should be provided ahead of and behind each rack so that the unit can be dewatered for chain or cable replacement, replacement of teeth, removal of obstructions, and straightening of bent bars. If only one unit is installed, it is absolutely essential that a bypass channel with a manually cleaned bar rack be provided for emergency use. Sometimes the manual rack is arranged as an overflow in case the mechanical screen becomes inoperative, especially during unattended hours. Flow through the bypass channel normally would be prevented by a closed slide or sluice gate.

The rack channel should be designed to prevent the settling and accumulation of grit and other heavy materials. An approach velocity of at least 1.25 ft/s (0.4 m/s) is recommended to minimize solids deposition in the channel. To prevent the pass through of debris at peak flowrates, the velocity through the bars should not exceed 3 ft/s (0.9 m/s).

The velocity through the bar rack can be controlled by installation of a downstream head control device such as a Parshall flume, or, for screens located upstream of a pumping station, by controlling the wetwell operating levels. If the channel velocities are controlled by wetwell levels, lower velocities can be tolerated provided flushing velocities occur during normal operating conditions.

The headloss through the bar racks is typically limited to about 6 in (150 mm) by operational controls. The raking mechanisms are normally provided with "hand"-"off"-"automatic" controls. On "hand" position, the rakes operate continuously. On "automatic" position, they may be operated when the differential headloss increases above a certain minimum value or by a time clock. Operation by a time clock for a period adjustable by the operator and having a cycle length of approximately 15 minutes is recommended with either a high-water or high-differential contact that will place the rack in continuous operation when needed.

Screenings discharged from the rake mechanism are usually discharged directly into a hopper or container, onto a sorting table, or into a screenings press. For installations with multiple units, the screenings may be discharged onto a conveyor or into a pneumatic ejector system and transported to a common screenings storage hopper, screenings press, or to an incinerator. As an alternative, screenings grinders may be used to grind and shred the screenings. Ground screenings are then returned to the wastewater. Sheet metal enclosures with access doors should be provided for the headworks of the racks above the operating flow level.

Screens

Early coarse screens were of the circular or disk type and were equipped with a perforated bronze screen plate with slotted openings 1/8 in (3 mm) wide or less. At present, comparatively few treatment plants use screens of this type. For a description of these early units, including more information on the quantity and character of screenings and data on removal efficiencies, the reader is referred to Ref. 10. Modern coarse screens are of the static (fixed) or rotary drum type, with stainless steel or nonferrous wire mesh screen materials. Typically, the openings vary from 0.01 to 0.25 in (0.2 to 6 mm). Examples of both types of coarse screens are illustrated in Fig. 9-3. Application of these types of screens is limited to small plants or plants where headloss through the screens is not a problem.

Static wedgewire screens (see Fig. 9-3a) with 0.01 to 0.06 in (0.2 to 1.2 mm) clear openings are designed for flowrates of about 10 to 30 gal/ft^2 · min (400 to 1200 L/m^2 · min) of screen area and require 4 to 7 ft (1.2 to 2.1 m) of headloss. The wedge-wire medium consists of small, stainless steel wedge-shaped bars with the flat part

(a)

FIGURE 9-3
Examples of coarse screens: (a) static screen and (b) rotary drum screen.

of the wedge facing the flow (see Fig. 6-3a). The screens require appreciable floor area for installation and must be cleaned once or twice daily with high-pressure hot water, steam, or degreaser to remove grease buildup.

For the drum type screen, the screening or straining medium is mounted on a cylinder that rotates in a flow channel. The construction varies, principally with regard to the direction of flow through the screening medium. The wastewater flows either into one end of the drum and outward through the screen with the solids collection on the interior surface (see Fig. 9-3b) or from the top of the unit and outward through the interior with the solids collection on the exterior. Stainless steel mesh or wedge wire is used as the screening medium. Provision is made for the continuous removal of the collected solids, supplemented by water sprays to keep the screening medium clean. Headloss through the screens may range from about 2.5 to 4.5 ft (0.8 to 1.4 m). Drum screens are available in various sizes, from 3 to 5 ft (0.9 to 1.5 m) in diameter and from 4 to 12 ft (1.2 to 3.7 m) in length.

Quantities of Screenings

The quantity and characteristics of screenings collected for disposal varies depending on the type of rack or screen used, size of the rack or screen opening, type of sewer system, and geographic location. If screenings data are not readily available,

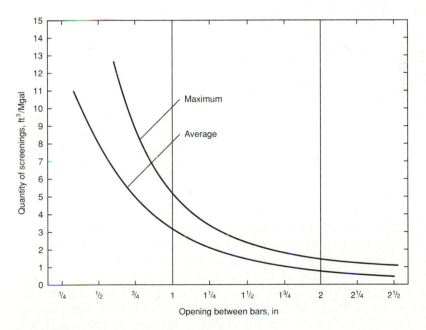

FIGURE 9-4
Average and maximum volume of coarse screenings collected per unit wastewater volume as a function of the size of the openings between bars [20].

the quantity of screenings may be estimated from Fig. 9-4. In plants that serve combined sewers, the quantity of screenings has been observed to increase greatly during periods of storm flow. The screenings removed by bar racks have typically amounted to approximately 0.5 to 5.0 ft^3/Mgal of wastewater treated. The quantities of fine screenings vary considerably, ranging from 5 to 30 ft^3/Mgal (0.0375 to 0.225 m^3/10^3 m^3) or more, equivalent to 5 to 15 percent of the suspended matter [10].

Disposal of Screenings

Means of disposal of screenings include (1) removal by hauling to disposal areas (landfill), (2) disposal by burial on the plant site (small installations only), (3) incineration either alone or in combination with sludge and grit (large installations only), (4) disposal with municipal solid wastes, or (5) discharge to grinders or mascerators where they are ground and returned to the wastewater. The first method of disposal is most commonly used. In some states, screenings are required to be lime stabilized before disposal in landfills. This practice may become more widespread in the future. Disposal on site should be done only in conformance with environmental regulations. Grinding the screenings and returning them to the wastewater flow shares many of the disadvantages cited under comminution, discussed below. Screenings grinders themselves require high maintenance.

9-2 COMMINUTION

As an alternative to racks or coarse screens, comminutors can be used to grind up the coarse solids without removing them from the flow. Comminutors function to cut up (comminute) coarse solids to improve the downstream operations and processes and to eliminate problems caused by the varied sizes of solids present in wastewater. The solids are cut up into a smaller, more uniform size for return to the flow stream for removal in the subsequent downstream treatment operations and processes. Comminutors can theoretically eliminate the messy and offensive task of screenings handling and disposal. Their use is particularly advantageous in a pumping station to protect the pumps against clogging by rags and large objects and to eliminate the need to handle and dispose of screenings. In cold climates, the use of comminutors precludes the need to prevent collected screenings from freezing.

There is a wide divergence of views, however, on the suitability of using comminution devices at wastewater treatment plants. One school of thought maintains that once material has been removed from wastewater it should not be returned, regardless of the form. The other school of thought maintains that once cut up, the solids are more easily handled in the downstream processes.

A disadvantage of comminutors is that the comminuted solids often present downstream problems. The problems are particularly bad with rags, which tend to recombine after comminution into ropelike strands, if agitated. At treatment plants, this agitation is provided in grit chambers and aerated channels. These recombined rags can have a number of negative impacts such as clogging pump impellers, sludge pipelines, and heat exchangers, and accumulating on air diffusers.

Description

Different types of comminutors are available from a number of manufacturers. One type of comminutor (see Fig. 9-5) consists of a vertical revolving-drum screen with 1/4 in (6 mm) slots in small machines and 3/8 in (10 mm) slots in large machines. Coarse material is cut by the cutting teeth and the shear bars on the revolving drum as the solids are carried past a stationary comb. The small sheared particles pass through the drum slots, out of a bottom opening, through an inverted siphon, and into the downstream channel.

Other types of comminuting devices consist of (1) a stationary semicircular screen grid mounted in a rectangular channel with rotating or oscillating circular cutting disks, (2) a unit containing two large-diameter vertical rotating shafts equipped with cutting blades, and (3) a unit containing a conical-shaped screen grid, the axis of

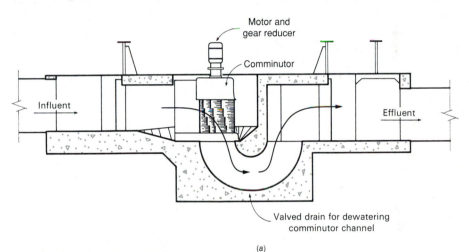

(a)

(b)

FIGURE 9-5
Typical comminutor installation: (a) cross-sectional view (from FMC, Chicago Pump) and (b) pictorial view of comminutor.

which is located parallel to the channel flow. This unit is also equipped with cutting blades. In all of these types, the screen grid intercepts the larger solids, while smaller solids pass through the space between the grid and cutting blades. The method of cutting or shredding the solids is the principal difference in each type.

Grinders are also used for in-line pipeline installations to shred solids, particularly ahead of wastewater and sludge pumps. For these applications, sizes range from 4 to 16 in (100 to 400 mm) in diameter. Grinders for sludge applications are discussed in Chap. 12.

Application and Design

Comminuting devices may be preceded by grit chambers to prolong the life of the equipment and to reduce the wear on the cutting surfaces and on portions of the mechanism where there is a small clearance between moving and stationary parts. They are used especially in smaller communities that are served by separate sanitary sewers carrying a minimum of grit. Comminutors should be constructed with a bypass arrangement so that a manual bar screen is used in case flowrates exceed the capacity of the comminutor or in case there is a power or mechanical failure. Stop gates and provisions for draining should also be included to facilitate maintenance. Headloss through a comminutor usually ranges from several inches to 1 ft (0.3 m) and can approach 3 ft (0.9 m) in large units at maximum flowrates.

In cases where a comminutor precedes grit chambers, the cutting teeth are subject to high wear and require frequent sharpening or replacement. Units that use cutting mechanisms ahead of the screen grid should be provided with rock traps in the channel upstream of the comminutor to collect material that could jam the cutting blade.

Because these units are complete in themselves, no detailed design is necessary. Manufacturers' data and rating tables for these units should be consulted for recommended channel dimensions, capacity ranges, upstream and downstream submergence, and power requirements. Because manufacturers' capacity ratings are usually based on clean water, the ratings should be decreased by approximately 20 to 25 percent to account for partial clogging of the screen.

9-3 GRIT REMOVAL

Grit removal may be accomplished in grit chambers or by the centrifugal separation of sludge. Grit chambers are designed to remove grit, consisting of sand, gravel, cinders, or other heavy solid materials that have subsiding velocities or specific gravities substantially greater than those of the organic putrescible solids in wastewater. The characteristics of grit and the design of grit chambers is considered in the following discussion.

Characteristics of Grit

Grit consists of sand, gravel, cinders, or other heavy materials that have specific gravities or settling velocities considerably greater than those of organic putrescible solids. In addition to these materials, grit includes eggshells, bone chips, seeds, coffee grounds, and large organic particles such as food wastes. Generally, what is removed as grit is predominantly inert and relatively dry. However, grit composition can be highly variable, with moisture content ranging from 13 to 65 percent, and volatile content from 1 to 56 percent. The specific gravity of clean grit particles reaches 2.7 for inerts, but can be as low as 1.3 when substantial organic material is agglomerated with inerts. A bulk density of 100 lb/ft^3 (1600 kg/m^3) is commonly used for grit. Often, enough organics are present in the grit so that it quickly putrifies if not properly handled after removal from the wastewater. Grit particles larger than 65 mesh (0.2 mm) have been cited as the cause of most downstream problems.

The actual size distribution of retained grit exhibits variation due to differences in collection system characteristics, as well as variations in grit removal efficiency. Generally, most grit particles are retained on a No. 100 mesh (0.15 mm) sieve, reaching nearly 100 percent retention in some instances. However, grit can be much finer. In the southeast, where fine sand known as "sugar sand" constitutes a portion of the grit, less than 60 percent of one city's grit was retained on a No. 100 mesh screen.

Grit Chambers

Grit chambers are provided to (1) protect moving mechanical equipment from abrasion and accompanying abnormal wear; (2) reduce formation of heavy deposits in pipelines, channels, and conduits; and (3) reduce the frequency of digester cleaning caused by excessive accumulations of grit. The removal of grit is essential ahead of centrifuges, heat exchangers, and high-pressure diaphragm pumps.

Grit chambers are most commonly located after the bar racks and before the primary sedimentation tanks. In some installations, grit chambers precede the screening facilities. Generally, the installation of screening facilities ahead of the grit chambers makes the operation and maintenance of the grit removal facilities easier.

Locating grit chambers ahead of wastewater pumps, when it is desirable to do so, would normally involve placing them at considerable depth at added expense. It is therefore usually deemed more economical to pump the wastewater, including the grit, to grit chambers located at a convenient position ahead of the treatment plant units, recognizing that the pumps may require greater maintenance.

There are three general types of grit chambers: horizontal-flow, either of a rectangular or square configuration; aerated; or vortex-type. In the horizontal-flow type, the flow passes through the chamber in a horizontal direction and the straight-line velocity of flow is controlled by the dimensions of the unit, special influent distribution gates, and the use of special weir sections at the effluent end. The aerated type consists of a spiral-flow aeration tank where the spiral velocity is induced and controlled by the tank dimensions and quantity of air supplied to the unit. The vortex-type consists of a cylindrical tank in which the flow enters tangentially creating a vortex-flow pattern; centrifugal and gravitational forces cause the grit to separate.

Design of grit chambers is commonly based on the removal of grit particles having a specific gravity of 2.65 and a wastewater temperature of 60°F (15.5°C).

However, analysis of grit removal data indicates that the specific gravity ranges from 1.3 to 2.7 [21].

Rectangular Horizontal-flow Grit Chambers. The oldest type of grit chamber used is the horizontal-flow velocity-controlled type. These units were designed to maintain a velocity as close to 1.0 ft/s (0.3 m/s) as practical and to provide sufficient time for grit particles to settle to the bottom of the channel. The design velocity will carry most organic particles through the chamber and will tend to resuspend any organic particles that settle but will permit the heavier grit to settle out.

The design of horizontal-flow grit chambers should be such that, under the most adverse conditions, the lightest particle of grit will reach the bed of the channel prior to its outlet end. Normally, grit chambers are designed to remove all grit particles that will be retained on a 65-mesh screen (0.21 mm diameter), although many chambers have been designed to remove grit particles retained on a 100-mesh screen (0.15 mm diameter). The length of channel will be governed by the depth required by the settling velocity and the control section, and the cross-sectional area will be governed by the rate of flow and by the number of channels. Allowance should be made for inlet and outlet turbulence; at least a 50 percent increase in the theoretical length is recommended. Representative design data for horizontal-flow grit chambers are presented in Table 9-3. The detailed design of horizontal-flow grit chambers is illustrated in the first edition of this text [11] and in Refs. 4, 10, and 20.

TABLE 9-3
Typical design information for horizontal-flow grit chambers

Item	Value	
	Range	Typical
Detention time, s	45–90	60
Horizontal velocity, ft/s	0.8–1.3	1.0
Settling velocity for removal of:		
65-mesh material, ft/min[a]	3.2–4.2	3.8
100-mesh material, ft/min[a]	2.0–3.0	2.5
Headloss in a control section as percent of depth in channel, %	30–40	36[b]
Allowance for inlet and outlet turbulence	$2D_m - 0.5L$[c]	

[a] If the specific gravity of the grit is significantly less than 2.65, lower velocities should be used.

[b] For Parshall flume control.

[c] D_m = maximum depth in grit chamber;
    L = theoretical length of grit chamber.

Note: ft/s $\times 0.3048$ = m/s
       ft/min $\times 0.3048$ = m/min

Grit removal from horizontal-flow grit chambers is accomplished usually by a conveyor with scrapers, buckets, or plows. Screw conveyors or bucket elevators are used to elevate the removed grit for washing or disposal. In small plants, grit chambers are sometimes cleaned manually.

Square Horizontal-flow Grit Chambers. Square horizontal-flow grit chambers, such as that shown in Fig. 9-6, have been in use for over 50 years. Influent to the unit is distributed over the cross section of the tank by a series of vanes or gates, and the distributed wastewater flows in straight lines across the tank and overflows a weir in a free discharge. It is also generally advisable to use at least two units or provide a temporary bypass. These types of grit chambers are designed on the basis of overflow rates dependent on particle size and the temperature of the wastewater. They are nominally designed to remove 95 percent of the 100-mesh particles at peak flow. A typical set of design curves is shown in Fig. 9-7.

In square grit chambers, the solids are raked by a rotating mechanism to a sump at the side of the tank. Settled grit may be moved up an incline by a reciprocating rake mechanism (see Fig. 9-6), or grit may be pumped from the tank through a cyclone degritter to separate the remaining organic material and concentrated grit. The concentrated grit then may be washed again in a classifier using a submerged reciprocating rake or an inclined screw conveyor. By either method, organic solids are separated from the grit and flow back into the basin, resulting in a cleaner, dryer grit.

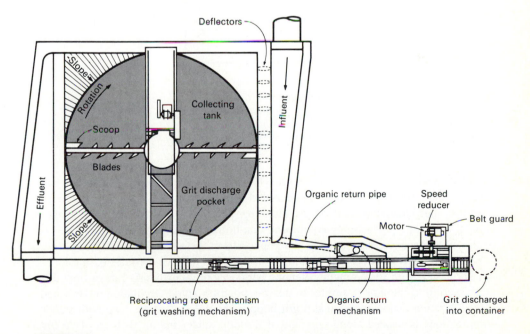

FIGURE 9-6
Typical square horizontal-flow grit chamber (from Dorr-Oliver).

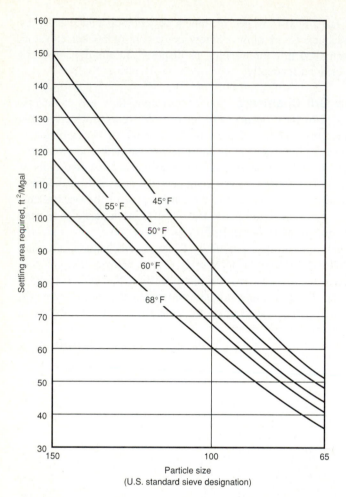

FIGURE 9-7
Area required for settling grit particles with a specific gravity of 2.65 in wastewater at indicated temperatures (from Dorr-Oliver). *Note*: 0.555 (°F − 32) = °C.

Aerated Grit Chambers. The discovery of grit accumulations in spiral-flow aeration tanks preceded by grit chambers led to the development of the aerated grit chamber. The excessive wear on grit-handling equipment and the necessity in most cases for separate grit-washing equipment with horizontal-flow grit chambers are two of the major factors contributing to the popularity of the aerated grit chamber.

Aerated grit chambers are nominally designed to remove particles 65 mesh (0.2 mm) or larger, with 2 to 5 minute detention periods at the peak hourly rate of flow. The cross section of the tank is similar to that provided for spiral circulation in activated-sludge aeration tanks, except that a grit hopper about 3 ft (0.9 m) deep with steeply sloping sides is located along one side of the tank under the air diffusers (see Fig. 9-8). The diffusers are located about 1.5 to 2 ft (0.45 to 0.6 m) above the normal plane of the bottom. Influent and effluent baffles are used frequently for hydraulic

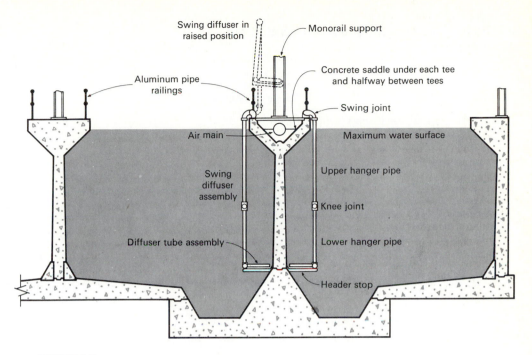

FIGURE 9-8
Typical section through an aerated grit chamber.

control and improved grit removal effectiveness. Basic design data for aerated grit chambers are presented in Table 9-4. The design of aerated grit chambers is illustrated in Example 9-1.

The velocity of roll or agitation governs the size of particles of a given specific gravity that will be removed. If the velocity is too great, grit will be carried out of the chamber; if it is too small, organic material will be removed with the grit. Fortunately, the quantity of air is easily adjusted. With proper adjustment, almost 100 percent removal will be obtained, and the grit will be well-washed. (Grit that is not well-washed and contains organic matter is an odor nuisance and attracts insects.) Wastewater will move through the tank in a spiral path (see Fig. 9-9) and will make two to three passes across the bottom of the tank at maximum flow and more at lesser flows. Wastewater should be introduced in the direction of the roll. To determine the required headloss through the chamber, the expansion in volume caused by the air must be considered.

For grit removal, aerated grit chambers are often provided with grab buckets traveling on monorails and centered over the grit collection and storage trough (see Fig. 9-10). An added advantage of a grab bucket grit removal system is that the grit can be further washed by dropping the grit from the bucket through the tank contents. Other installations are equipped with chain-and-bucket conveyors, running the full length of the storage troughs, which move the grit to one end of the trough and elevate it above the wastewater level in a continuous operation. Screw conveyors, tubular

TABLE 9-4
Typical design information for aerated grit chambers

Item	Value	
	Range	Typical
Detention time at peak flowrate, min	2–5	3
Dimensions:		
Depth, ft	7–16	
Length, ft	25–65	
Width, ft	8–23	
Width-depth ratio	1:1–5:1	1.5:1
Length-width ratio	3:1–5:1	4:1
Air supply, $ft^3/min \cdot ft$ of length	2.0–5.0	
Grit quantities, $ft^3/Mgal$	0.5–27	2.0

Note: $ft \times 0.3048 = m$
     $ft^3/min \cdot ft \times 0.0929 = m^3/min \cdot m$
     $ft^3/Mgal \times 0.00748 = m^3/10^3 m^3$

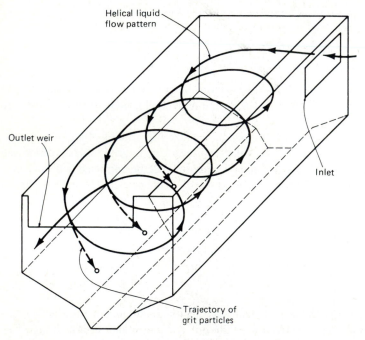

Helical liquid flow pattern

Outlet weir

Inlet

Trajectory of grit particles

FIGURE 9-9
Helical flow pattern in an aerated grit chamber.

FIGURE 9-10
Grab bucket used to remove grit from aerated grit chamber.

conveyors, jet pumps, and air lifts have also been used. Grit removal equipment for aerated grit chambers is subject to the same wear as experienced in the horizontal-flow units.

In areas where industrial wastewater is discharged to the collection system, the release of VOCs by the air agitation in aerated grit chambers needs to be considered. As discussed in Chap. 3, the release of significant amounts of VOCs can be a health risk to the treatment plant operators. In cases where release of VOCs is an important consideration, covers may be required, or nonaerated type grit chambers may be used.

Example 9-1   Design of aerated grit chamber. Design an aerated grit chamber for the treatment of municipal wastewater. The average flowrate is 11.4 Mgal/d (0.5 m³/s). Assume that the peaking factor curve given in Fig. 2-5 is applicable.

Solution

1. Establish the peak hourly flowrate for design. Assume that the aerated grit chamber will be designed for the peak hourly flowrate. The peaking factor is 2.75, and the peak design flowrate is

$$\text{Peak flowrate} = 11.4 \text{ Mgal/d} \times 2.75 = 31.35 \text{ Mgal/d}$$

2. Determine the grit chamber volume. Because it will be necessary to drain the chamber periodically for routine maintenance, use two chambers. Assume that the detention time at the peak flowrate is 3 min.

$$\text{Grit chamber volume, ft}^3 = \frac{(1/2) \ (31.35 \text{ Mgal/d}) \times 10^6}{(7.48 \text{ ft}^3/\text{gal}) \ (24 \text{ hr/d}) \ (60 \text{ min/hr})} \times 3 \text{ min}$$

$$= 4,366 \text{ ft}^3$$

3. Determine the dimensions of the grit chamber. Use a width-to-depth ratio of 1.2:1 and assume that the depth is 10 ft.

(a) Width = 1.2(10 ft) = 12 ft

$$(b) \text{ Length} = \frac{\text{volume}}{\text{width} \times \text{depth}} = \frac{4,366 \text{ ft}^3}{10 \text{ ft} \times 12 \text{ ft}} = 36.4 \text{ ft}$$

4. Determine the air supply requirement. Assume that 5 ft^3/min · ft of length will be adequate.

$$\text{Air required (length basis)} = 36.4 \times 5 \text{ ft}^3/\text{min} \cdot \text{ft} = 182 \text{ ft}^3/\text{ft} \cdot \text{min}$$

5. Estimate the average quantity of grit that must be handled. Assume a value of 7 ft^3/ Mgal.

$$\text{Volume grit} = 11.4 \text{ Mgal/d} \times 7 \text{ ft}^3/\text{Mgal} = 79.8 \text{ ft}^3/\text{d}$$

Comment. In designing aerated grit chambers, means should be provided to vary the air flowrate to control grit removal rates and the cleanliness of the grit.

Vortex-type Grit Chambers. Grit is also removed in devices that use a vortex-flow pattern. Two types of devices are shown in Fig. 9-11. The teacup separator was also considered previously in Chapter 6. In one type, illustrated in Fig. 9-11a, wastewater enters and exits tangentially. The rotating turbine maintains constant flow velocity, and its adjustable pitch blades promote separation of organics from the grit. The action of the rotating turbine produces a toroidal-flow path for grit particles. The grit settles by gravity into the hopper in one revolution of the basin's contents. Solids are removed from the hopper by a grit pump or an air lift pump. Grit removed by a grit pump can be discharged to a hydroclone for removal of the remaining organic material. Grit removed by an air lift may be dewatered on a wedge wire screen (see "Screens" for description). Typical design data are presented in Table 9-5. If more than two units are installed, special arrangements for flow splitting are required.

In the second type, illustrated in Fig. 9-11b, a free vortex is generated by the flow entering tangentially at the top of the unit. Effluent exits the center of the top of the unit from a rotating cylinder, or "eye" of the fluid. Gravitational forces within this cylinder minimize the release of particles with densities greater than water. Grit settles by gravity to the bottom of the unit, while organics, including those separated from grit particles by centrifugal forces, exit principally with the effluent. Organics remaining with the settled grit are separated as the grit particles move along the unit floor. Headloss in the unit is a function of the size particle to be removed and increases significantly for very fine particles. This type of grit removal unit is a relatively recent development and is sized to handle peak flowrates up to 7 Mgal/d (0.3 m^3/s) per unit. Grit is removed from the unit by a cleated belt conveyor. Because

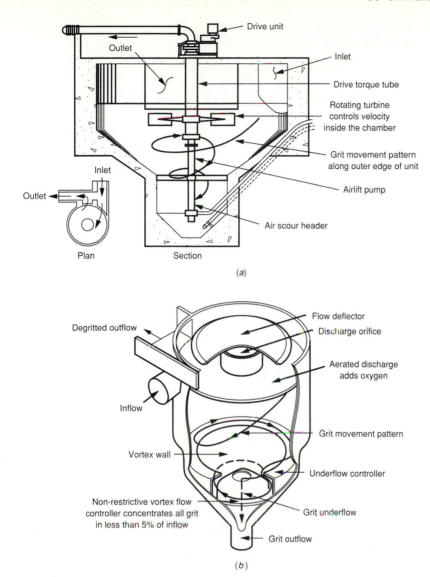

FIGURE 9-11
Vortex-type grit chambers: (a) PISTA unit (from Smith & Loveless) and (b) teacup unit (from Eutek).

of its overall height, this type of grit system requires a deep basement or a lift station if the grit removal unit is installed above grade.

Sludge Degritting

In some treatment facilities, grit chambers are not used and the grit is allowed to settle in the primary settling tanks. Grit removal is accomplished by pumping dilute

TABLE 9-5
Typical design information for vortex-type grit chambers

Item	Value	
	Range	Typical
Detention time at average flowrate, s		30
Dimensions:		
Diameter		
Upper chamber, ft	4.0–24.0	
Lower chamber, ft	3.0–6.0	
Height, ft	9.0–16.0	
Removal rates, percent		
50 mesh (0.30 mm)		95+
70 mesh (0.24 mm)		85+
100 mesh (0.15 mm)		65+

Note: ft ×.3048 = m

primary sludge to a cyclone degritter. The cyclone degritter acts as a centrifugal separator in which the heavy particles of grit and solids are separated by the action of a vortex and discharged separately from the lighter particles and the bulk of the liquid. The advantage of this system is the elimination of the cost of grit chambers. The disadvantages of this system are (1) pumping of dilute quantities of sludge usually requires adding sludge thickeners, and (2) pumping of grit with the primary sludge causes increased maintenance of the sludge collectors and the primary sludge pumps as well as increasing the sludge-pumping costs. Sludge degritting is also discussed in Chap. 12.

Quantities of Grit

The quantities of grit will vary greatly from one location to another, depending on the type of sewer system, the characteristics of the drainage area, the condition of the sewers, the frequency of street sanding to counteract icing conditions, the types of industrial wastes, the number of household garbage grinders served, and the amount of sandy soil in the area. Typical values for aerated grit chambers are reported in Table 9-4.

It is difficult to interpret grit removal data because grit itself is poorly characterized and almost no data exist on relative removal efficiencies. The information on grit characteristics derives from what has been removed as grit. Sieve analyses are not normally performed on grit chamber influents and effluents. For these reasons, the efficiencies of grit removal systems cannot be compared.

Disposal of Grit

The most common method of grit disposal is as fill, covered to prevent objectionable conditions. In some large plants, grit is incinerated with sludge. As with screenings, some states require grit to be lime stabilized before disposal in a landfill. Disposal in all cases should be done in conformance with the appropriate environmental regulations.

Grit Separation and Washing. The character of grit normally collected in horizontal-flow grit chambers and from cyclone degritters varies widely from what might be normally considered as clean grit to grit that includes a large proportion of putrescible organic material. Unwashed grit may contain 50 percent or more of organic material. Unless promptly disposed of, this material may attract insects and rodents. It has a distinctly disagreeable odor.

Removal of a major part of the organic material may be accomplished by grit separators and grit washers. When some of the heavier organic matter remains with the grit, grit washers are commonly used to provide a second stage of volatile solids separation. An example of a grit separation and washing unit is shown in Fig. 9-12.

Two principal types of grit washers are available. One type relies on an inclined, submerged rake that provides the necessary agitation for separation of the grit from the organic materials and, at the same time, raises the washed grit to a point of discharge above water level (similar to the unit shown in Fig. 9-6). Another type (see Fig. 9-12) uses an inclined screw and moves the grit up the ramp. Both types can be equipped with water sprays to assist in the cleansing action.

Removal from Plant. Grit is normally hauled to the disposal areas in trucks. In larger plants, elevated grit storage facilities may be provided with bottom-loading gates. Difficulties experienced in getting the grit to flow freely from the storage hoppers have been minimized by using steep slopes on the storage hoppers, by applying air beneath the grit, and by the use of hopper vibrators. Drainage facilities

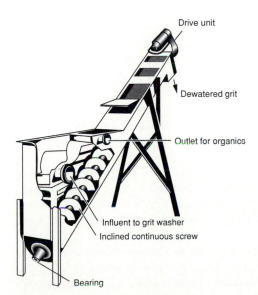

Drive unit

Dewatered grit

Outlet for organics

Influent to grit washer

Inclined continuous screw

Bearing

FIGURE 9-12
Example of grit separation and washing unit (from Wemco).

for collection and disposal of drippings from the bottom-loading gates are desirable. Grab buckets operating on a monorail system may also be used to load trucks directly from the grit chambers (see Fig. 9-10).

In some plants, grit is successfully conveyed to grit-disposal areas by pneumatic conveyors. This system requires no elevated storage hoppers and eliminates problems in storage and trucking. But the wear on piping, especially bends, is considerable.

9-4 FLOW EQUALIZATION

As noted in Chap. 6, both in-line and off-line flow equalization can be used to equalize the flowrate to subsequent treatment operations and processes. Where equalization of the plant loadings is also desired, in-line equalization must be used. Off-line equalization is sometimes used to capture the "first flush" from combined sewers. From a design standpoint, the principal factors that must be considered are (1) basin construction including cleaning, access, and safety; (2) mixing and air requirements; and (3) pump and pump control systems.

Basin Construction

Important considerations in the design of new equalization basins are the materials of construction, basin geometry, and operational appurtenances. If existing tanks are to be converted to equalization basins, the principal concern is with the necessary modifications. Piping and structural changes are usually required.

Construction Materials. New basins may be of earthen, concrete, or steel construction; earthen basins are generally the least expensive. Depending on local conditions, the side slopes may vary between 3:1 and 2:1. A section through a typical earthen basin is shown in Fig. 9-13. In many installations, a liner is required to prevent groundwater contamination. Either diffused-air aeration or floating aerators may be used to prevent septicity. If a floating aerator is used, a concrete pad should be provided below the aerator to minimize erosion. With floating aerators, some minimum operating level is needed to protect the aerator; typically, the depth will vary from 5 to 6 ft (1.5 to 2 m). The freeboard required depends on the surface area of the basin and local wind conditions. To prevent wind-induced erosion in the upper portions of the basin, it may be necessary to protect the slopes with riprap, soil cement, or a partial gunite layer. Fencing should also be provided to prevent public access to the basins.

In areas of high groundwater, drainage facilities should be provided to prevent embankment failure. To further ensure a stable embankment, the tops of the dikes should be of adequate width. The use of an adequate dike width will also reduce construction costs, especially where mechanical compaction equipment is used.

Basin Geometry. The importance of basin geometry varies somewhat, depending on whether in-line or off-line equalization is used. If in-line equalization is used to dampen both the flowrate and the mass loadings, it is important to use a geometry

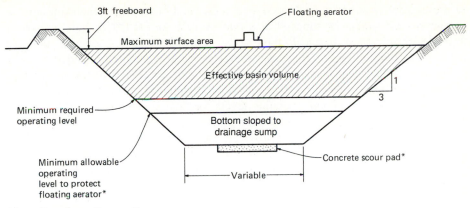

FIGURE 9-13
Section through typical flow equalization basin.

that allows the basin to function as a complete-mix reactor insofar as possible. Therefore, elongated designs should be avoided, and the inlet and outlet configurations should be arranged to minimize short circuiting. Discharging the influent near the mixing equipment usually minimizes short circuiting. If the geometry of the basins is controlled by the available land area and an elongated geometry must be used, it may be necessary to use multiple inlets and outlets. Provisions should be included in the basin design for access by cleaning equipment such as front-end loaders. Multiple compartments are also desirable to reduce cleaning costs and to control odors.

Operational Appurtenances. Among the appurtenances that should be included in the design of equalization basins are (1) facilities for flushing any solids and grease that may tend to accumulate on the basin walls; (2) an emergency overflow in case of pump failure; (3) a high-water takeoff for the removal of floating material and foam; and (4) water sprays to prevent the accumulation of foam on the sides of the basin if foam could be a problem. Solids removed from equalization basins should be returned to the head of the plant for processing.

Mixing and Air Requirements

The proper operation of both in-line and off-line equalization basins generally requires proper mixing and aeration. Mixing equipment should be sized to blend the contents of the tank and to prevent deposition of solids in the basin. To minimize mixing requirements, grit removal facilities should precede equalization basins where possible. Mixing requirements for blending a medium-strength municipal wastewater having a suspended-solids concentration of approximately 220 mg/L range from 0.02 to 0.04 hp/10^3 gal (0.004 to 0.008 kW/m^3) of storage. Aeration is required to prevent the wastewater from becoming septic and odorous. To maintain aerobic conditions, air should be supplied at a rate of 1.25 to 2.0 $ft^3/10^3$ gal $\cdot$ min (0.01 to 0.015 $m^3/m^3 \cdot$ min). In equalization basins that follow primary sedimentation and have short detention times (less than two hours), aeration may not be required.

One method of providing for both mixing and aeration is to use mechanical aerators. Baffling may be necessary to ensure proper mixing, particularly with circular tank configuration. Minimum operating levels for floating aerators generally exceed 5 ft (1.5 m) and vary with the horsepower and design of the unit. To protect the unit, low-level shutoff controls should be provided. Because it may be necessary to dewater the equalization basins periodically, the aerators should be equipped with legs or draft tubes that allow them to come to rest on the bottom of the basin without damage. Various types of diffused-air systems including static tube and aspirating aerators may also be used for mixing and aeration (see Chap. 10).

Pumps and Pump Control

Because flow equalization imposes an additional head requirement within the treatment plant, pumping facilities are frequently required. Pumping may precede or follow equalization, but pumping into the basin is generally preferred for reliability of treatment operation. As a minimum, the required pumping head is equal to the sum of the dynamic losses and the normal surface-level variation. Additional head may be required if the basin is to be dewatered by gravity. In some cases, the pumping of both basin influent and equalized flows will be required.

An automatically controlled flow-regulating device will be required where gravity discharge from the basin is used. Where basin effluent pumps are used, instrumentation should be provided to control the preselected equalization rate. Regardless of the discharge method used, a flow-measuring device should be provided on the outlet of the basin to monitor the equalized flow.

9-5 OTHER PRELIMINARY TREATMENT OPERATIONS

Other preliminary treatment operations have been used to improve the treatability of wastewater and to remove grease and scum from wastewater prior to primary sedimentation. Preaeration and flocculation have been used for this purpose. Typical design information for these operations is presented in Table 9-6. Although these preliminary treatment operations were commonly used in the past, their use is limited today.

Preaeration

The objectives that are often given for aerating wastewater prior to primary sedimentation are (1) to improve its treatability, (2) to provide grease separation, odor control, grit removal, and flocculation, (3) to promote uniform distribution of suspended and floating solids to treatment units, and (4) to increase BOD removals. Of these objectives, the promotion of a more uniform distribution of suspended and floating solids

TABLE 9-6
Typical design information for preaeration tanks and flocculation tanks

Item	Value	
	Range	Typical
Preaeration tanks:		
Detention time, min	10–45	30
Tank depth, ft	10–20	15
Air requirement, ft^3/gal	0.1–0.4	0.25
Flocculation tanks:		
Detention time, min	20–60	30
Paddle-induced flocculation, maximum paddle peripheral speed with turndown adjustment to 30% of maximum speed, ft/s	1.3–3.3	2.0
Air agitation flocculation, with porous tube diffusers, ft^3/Mgal	80–160	100

Note: ft $\times$ 0.3048 = m
      ft^3/Mgal $\times$ 0.00748 = m^3/10^{3}m^3
        ft/s $\times$ 0.3048 = m/s
      ft^3/gal $\times$ 7.4805 = m^3/m^3

is probably its best application. It has been shown that short-period preaeration of 3 to 5 minutes formerly used does not significantly improve BOD or grease removal [8]. Current practice, when preaeration is used, frequently consists of increasing the detention period in aerated grit chambers. In this case, provisions for grit removal may be needed in only the first portion of the tanks. Preaeration periods of 10 to 15 minutes have been suggested if odor control and prevention of septicity are the primary objectives [20]. As discussed under aerated grit chambers, the release of VOCs by air agitation may need to be considered if preaeration is used.

    Aerated channels are used for distributing wastewater to primary sedimentation tanks in large plants to keep the solids in suspension at all rates of flow. Although aerated channels add dissolved oxygen to the wastewater, some odors and volatile organic compounds may be released. The amount of air required ranges from 2 to 5 ft^3/lin ft $\cdot$ min (0.2 to 0.5 m^3/lin m $\cdot$ min) of channel. Aerated channels are often used for distributing mixed liquor to activated-sludge final settling tanks.

Flocculation

The purpose of wastewater flocculation is to form aggregates, or flocs, from the finely divided matter. Although not used routinely, the flocculation of wastewater by mechanical or air agitation may be worthy of consideration when it is desired to

(1) increase the removal of suspended solids and BOD in primary settling facilities, (2) condition wastewater containing certain industrial wastes, and (3) improve the performance of secondary settling tanks following the activated-sludge process.

When used, flocculation can be accomplished (1) in separate tanks or basins specifically designed for the purpose, (2) in in-line facilities such as the conduits and pipes connecting the treatment units, and (3) in combination flocculator-clarifiers (see Fig. 9-14). Paddles for mechanical agitation should have variable-speed drives permitting the adjustment of the top paddle speed downward to 30 percent of the top value. Similarly, where air flocculation is employed, the air supply system should be adjustable so that the flocculation energy level can be varied throughout the tank. In both mechanical- and air-agitation flocculation systems, it is common practice to taper the energy input so that the flocs initially formed will not be broken as they leave the flocculation facilities (whether separate or in-line).

9-6 PRIMARY SEDIMENTATION TANKS

When a liquid containing solids in suspension is placed in a relatively quiescent state, those solids having a higher specific gravity than the liquid will tend to settle, and those with a lower specific gravity will tend to rise. These principles are used in the design of sedimentation tanks for treatment of wastewaters. The objective of treatment by sedimentation is to remove readily settleable solids and floating material and thus reduce the suspended-solids content.

Primary sedimentation tanks may provide the principal degree of wastewater treatment, or they may be used as a preliminary step in the further processing of the wastewater. When these tanks are used as the only means of treatment, they

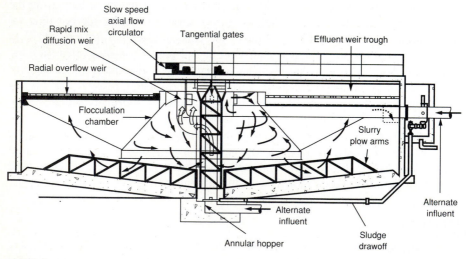

FIGURE 9-14
Typical flocculator-clarifier used in wastewater treatment.

provide for the removal of (1) settleable solids capable of forming sludge deposits in the receiving waters, (2) free oil and grease and other floating material, and (3) a portion of the organic load discharged to the receiving waters. As discussed in Chap. 1, primary sedimentation is being phased out as the only means of treatment. When primary sedimentation tanks are used ahead of biological treatment, their function is to reduce the load on the biological treatment units. Efficiently designed and operated primary sedimentation tanks should remove from 50 to 70 percent of the suspended solids and from 25 to 40 percent of the BOD_5.

Primary sedimentation tanks that precede biological treatment processes may be designed to provide shorter detention periods and a higher rate of surface loading than tanks serving as the only method of treatment, except when waste activated sludge is returned to the primary sedimentation tanks for cosettling with primary sludge.

Sedimentation tanks have also been used as stormwater retention tanks, which are designed to provide a moderate detention period (10 to 30 minutes) for overflows from either combined sewers or storm sewers. The purpose is to remove a substantial portion of the organic solids that otherwise would be discharged directly to the receiving water and that could form offensive sludge deposits. Sedimentation tanks have also been used to provide sufficient detention periods for effective chlorination of such overflows. The discussion of the treatment of combined sewer overflows is included in Chap. 15.

Basis of Design

If all solids in wastewater were discrete particles of uniform size, uniform density, reasonably uniform specific gravity, and fairly uniform shape, the removal efficiency of these solids would be dependent on the surface area of the tank and time of detention. The depth of the tank would have little influence, provided that horizontal velocities would be maintained below the scouring velocity. However, the solids in most wastewaters are not of such regular character but are heterogeneous in nature; the conditions under which they are present range from total dispersion to complete flocculation. The bulk of the finely divided solids reaching primary sedimentation tanks are incompletely flocculated but are susceptible to flocculation.

Flocculation is aided by eddying motion of the fluid within the tanks and proceeds through the coalescence of fine particles at a rate that is a function of their concentration and of the natural ability of the particles to coalesce upon collision. As a general rule, therefore, coalescence of a suspension of solids becomes more complete as time elapses. For this reason, detention time is also a consideration in the design of sedimentation tanks. The mechanics of flocculation are such, however, that as the time of sedimentation increases, less and less coalescence of remaining particles occurs.

Detention Time. Normally, primary sedimentation tanks are designed to provide 1-1/2 to 2-1/2 hours of detention based on the average rate of wastewater flow. Tanks that provide shorter detention periods (1/2 to 1 h), with less removal of suspended solids, are sometimes used for primary treatment ahead of biological treatment units.

Temperature effects are normally not an important consideration in primary clarifier design. However, in cold climates, increases in water viscosity at lower temperatures retard particle settling in clarifiers and reduce performance at wastewater temperatures below 68°F (20°C). A curve showing the increase in detention time necessary to equal the detention time at 68°F is presented in Fig. 9-15 [21]. For wastewater having a temperature of 50°F, for example, the detention period is 1.38 times that required at 68°F to achieve the same efficiency. Therefore, when cold wastewater temperatures are expected, safety factors should be considered in clarifier design to ensure adequate performance.

Surface-loading Rates. Sedimentation tanks are normally designed on the basis of a surface-loading rate (commonly termed "overflow rate") expressed as gallons per square foot of surface area per day (cubic meters per square meter of surface area per day). The selection of a suitable loading rate depends on the type of suspension to be separated. Typical values for various suspensions are reported in Table 9-7. Designs for municipal plants must also meet the approval of state regulatory agencies, many of which have adopted standards for surface-loading rates that must be followed.

The effect of the surface-loading rate and detention time on suspended-solids removal varies widely depending on the character of the wastewater, proportion of settleable solids, concentration of solids, and other factors. It should be emphasized

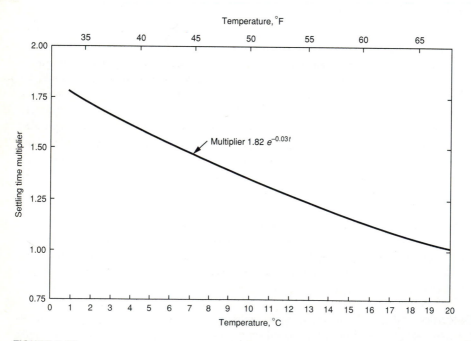

FIGURE 9-15
Settling detention time versus temperature for primary clarification [21].

TABLE 9-7
Typical design information for primary sedimentation tanks[a]

	Value	
Item	Range	Typical
Primary settling followed by secondary treatment:		
Detention time, hr	1.5–2.5	2.0
Overflow rate, gal/ft^2 · d		
Average flow	800–1,200	
Peak hourly flow	2,000–3,000	2,500
Weir loading, gal/ft · d	10,000–40,000	20,000
Primary settling with waste activated-sludge return:		
Detention time, h	1.5–2.5	2.0
Overflow rate, gal/ft^2 · d		
Average flow	600–800	
Peak hourly flow	1,200–1,700	1,500
Weir loading, gal/ft · d	10,000–40,000	20,000

[a] Comparable data for secondary clarifiers are presented in Chap. 10.

Note: gal/ft^2 · d × 0.0407 = m^3/m^2 · d
     gal/ft · d × 0.0124 = m^3/m · d

that overflow rates must be set low enough to ensure satisfactory performance at peak rates of flow, which may vary from over 3 times the average flow in small plants to 2 times the average flow in large plants (see discussion of peak flowrates in Chap. 5).

When the area of the tank has been established, the detention period in the tank is governed by water depth. Overflow rates in current use result in nominal detention periods of 2 to 2.5 h, based on average design flowrate. As design flowrates in all cases are usually based on some future condition, the actual detention periods during the early years of operation are somewhat longer.

Weir Rates. In general, weir-loading rates have little effect on the efficiency of primary sedimentation tanks and should not be considered when reviewing the appropriateness of clarifier design. The placement of the weirs (see following discussion) and the design of the tanks are more important. For general information purposes only, typical weir-loading rates are given in Table 9-7.

Scour Velocity. Scour velocity is important in sedimentation operations. Forces on settled particles are caused by the friction of water flowing over the particles. In sewers, velocities should be maintained high enough that solid particles will be kept from settling. In sedimentation basins, horizontal velocities should be kept low so that settled particles are not scoured from the bottom of the basin. The critical velocity is given by Eq. 9-1, which was developed by Camp [5] using the results from studies by Shields:

$$V_H = \left[\frac{8k(s-1)gd}{f} \right]^{1/2} \tag{9-1}$$

where V_H = horizontal velocity that will just produce scour
    k = constant which depends on type of material being scoured
    s = specific gravity of particles
    g = acceleration due to gravity
    d = diameter of particles
    f = Darcy-Weisbach friction factor

Typical values of k are 0.04 for unigranular sand and 0.06 for more sticky, interlocking matter. The term f (the Darcy-Weisbach friction factor) depends on the characteristics of the surface over which flow is taking place and the Reynolds number. Typical values of f are 0.02 to 0.03. Either U.S. customary or SI units may be used in Eq. 9-1, so long as they are consistent, because k and f are dimensionless. Information on the Darcy-Weisbach equation may be found in Ref. 12.

Tank Type, Size, and Shape

Almost all treatment plants of any size, except for those with Imhoff tanks, now use mechanically cleaned sedimentation tanks of standardized circular or rectangular design. The selection of the type of sedimentation unit for a given application is governed by the size of the installation, by rules and regulations of local control authorities, by local site conditions, and by the experience and judgment of the engineer. Two or more tanks should be provided so that the process may remain in operation while one tank is out of service for maintenance and repair work. At large plants, the number of tanks is determined largely by size limitations. Typical dimensions and other data for rectangular and circular sedimentation tanks are presented in Table 9-8.

Rectangular Tanks. Rectangular sedimentation tanks may use either chain-and-flight sludge collectors or traveling-bridge type collectors. A rectangular tank that uses a chain-and-flight type collector is shown in Fig. 9-16. Sludge removal equipment for this type of tank is available from a number of manufacturers and usually consists of a pair of endless conveyor chains, manufactured of either alloy steel, cast iron, or thermoplastic. Attached to the chains at approximately 10 ft (3 m) intervals are sludge removal flights made of wood or fiberglass, extending the full width of the tank or bay (see Fig. 9-17). The solids settling in the tank are scraped to sludge hoppers in small tanks and to transverse troughs in large tanks. The transverse troughs are equipped with collecting mechanisms (cross collectors), usually either chain-and-flight or screw-type collectors, which convey solids to one or more sludge hoppers. In very long units (over 150 ft), two collection mechanisms can be used to scrape sludge to collection points near the middle of the tank length.

TABLE 9-8
Typical design information for rectangular and circular sedimentation tanks used for primary treatment of wastewater

Item	Value	
	Range	Typical
Rectangular		
Depth, ft	10–15	12
Length, ft	50–300	80–130
Width, ft[a]	10–80	16–32
Flight speed, ft/min	2–4	3
Circular		
Depth, ft	10–15	12
Diameter, ft	10–200	40–150
Bottom slope, in/ft	$\frac{3}{4}$–2	1
Flight travel speed, r/min	0.02–0.05	0.03

[a] If widths of rectangular mechanically cleaned tanks are greater than 20 ft, multiple bays with individual cleaning equipment may be used, thus permitting tank widths up to 80 ft or more.

Note: ft × 0.3048 = m
      in/ft × 83.333 = mm/m

Rectangular tanks may also be cleaned by a bridge-type mechanism, which travels up and down the tank on rubber wheels or on rails supported on the sidewalls (see Fig. 9-18). One or more scraper blades are suspended from the bridge. Some of the bridge mechanisms are designed so that the scraper blades can be lifted clear of the sludge blanket on the return travel.

Where cross collectors are not provided, multiple sludge hoppers must be installed. Sludge hoppers have operating difficulties, notably sludge accumulation on the slopes and in the corners and even arching over the sludge-drawoff piping. Wastewater may also be drawn through the sludge hopper, bypassing some of the accumulated sludge, resulting in a "rathole" effect. A cross collector is more advisable, except possibly in small plants, because a more uniform and concentrated sludge can be withdrawn and many of the problems associated with sludge hoppers can be eliminated.

Influent channels should be provided across the inlet end of the tanks, and effluent channels should be provided across the effluent end of the tanks. It is also desirable to locate sludge-pumping facilities close to the hoppers where sludge is collected at the influent end of the tanks. One sludge-pumping station can conveniently serve two or more tanks.

In rectangular tanks, flow distribution into the tank is critical. Possible approaches to inlet design include (1) full width inlet channels with inlet weirs, (2) inlet channels with submerged ports or orifices, or (3) inlet channels with wide gates and slotted baffles. Inlet weirs, although effective in spreading flow across the tank width, introduce a vertical velocity component into the sludge hopper that may resuspend the sludge particles. Inlet ports can provide good distribution across the tank width if the

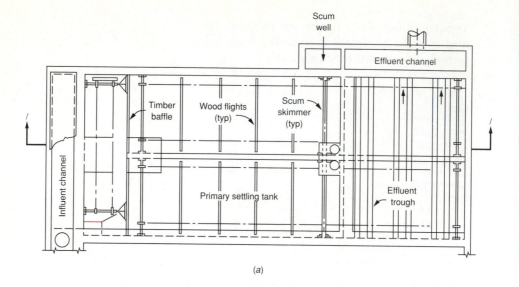

(a)

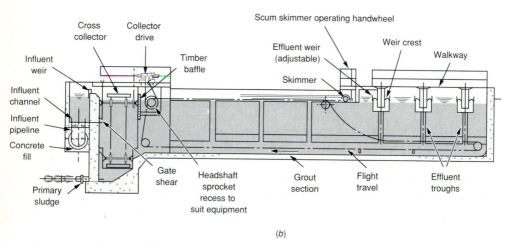

(b)

FIGURE 9-16
Typical rectangular primary sedimentation tank.

velocities are maintained in the 10 to 30 ft/min (3 to 9 m/min) range. Inlet baffles are effective in reducing the high initial velocities and distribute flow over the widest possible cross-sectional area. Where full-width baffles are used, they should extend from 6 in (150 mm) below the surface to 12 in (300 mm) below the entrance opening.

For large multiple installations of rectangular tanks, a pipe and operating gallery can be constructed integrally with the tanks along the influent end to contain the sludge pumps. This gallery can be connected to service tunnels for access to other plant units.

FIGURE 9-17
Empty rectangular sedimentation tank showing sludge removal flights.

Scum is usually collected at the effluent end of rectangular tanks with the flights returning at the liquid surface. The scum is moved by the flights to a point where it is trapped by baffles before removal. The scum can also be moved by water sprays. The scum can be scraped manually up an inclined apron, or it can be removed hydraulically or mechanically, and for this process a number of means have been developed. For small installations, the most common scum-drawoff facility consists of a horizontal, slotted pipe that can be rotated by a lever or a screw. Except when drawing scum, the open slot is above the normal tank water level. When drawing scum, the pipe is rotated so that the open slot is submerged just below the water level, permitting the scum accumulation to flow into the pipe. Use of this equipment results in a relatively large volume of scum liquor.

Another method for removing scum by mechanical means is a transverse rotating helical wiper attached to a shaft. This apparatus makes it possible to draw the scum from the water surface over a short inclined apron for discharge to a cross-collecting scum trough. The scum may then be flushed to a scum ejector or hopper ahead of a scum pump. Another method of scum removal consists of a chain-and-flight type of collector that collects the scum at one side of the tank and scrapes it up a short incline for deposit in scum hoppers, whence it can be pumped to disposal units. Scum is also collected by special scum rakes in those rectangular tanks that are equipped with

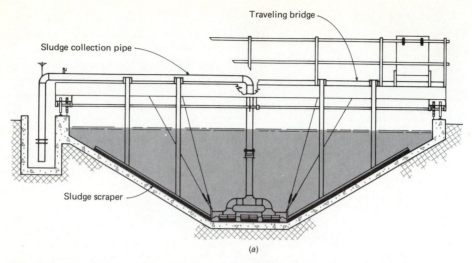

(a)

(b)

FIGURE 9-18
Sedimentation tanks equipped with traveling-bridge sludge removal mechanism: (a) sedimentation tank with sloping sidewalls. Solids accumulated on the bottom are removed with a pump system (from Aqua Aerobic Systems, Inc.) and (b) conventional rectangular sedimentation tank.

the carriage or bridge type of sedimentation tank equipment. In installations where appreciable amounts of scum are collected, the scum hoppers are usually equipped with mixers to provide a homogeneous mixture prior to pumping. Scum is usually disposed of with the sludge produced at the plant; however, separate scum disposal is used by many plants.

Multiple rectangular tanks require less land area than multiple circular tanks and for this reason are used where ground area is at a premium. Rectangular tanks also lend themselves to nesting with preaeration tanks and aeration tanks in activated-sludge plants thus permitting common wall construction and reducing construction costs. They are also generally used where tank roofs or covers are required. On sites with limited space, rectangular sedimentation tanks may also be constructed in a stacked or two-story configuration (see Chap. 10).

Circular Tanks. In circular tanks, the flow pattern is radial (as opposed to horizontal in rectangular tanks). To achieve a radial-flow pattern, the wastewater to be settled can be introduced in the center or around the periphery of the tank, as shown in Fig. 9-19. Both flow configurations have proved to be satisfactory generally, although the center-feed type is more commonly used. Some problems have been experienced with flow distribution and scum removal with the peripheral feed units.

In the center-feed design (see Fig. 9-19a), the wastewater is carried to the center of the tank in a pipe suspended from the bridge or encased in concrete beneath the tank floor. At the center of the tank, the wastewater enters a circular well designed to distribute the flow equally in all directions. The center well has a diameter typically between 15 and 20 percent of the total tank diameter and ranges from 3 to 8 ft (1 to 2.5 m) in depth. The sludge removal mechanism revolves slowly and may have two or four arms equipped with scrapers. The arms also support blades for scum removal. A typical center-feed circular clarifier equipped with a scraper mechanism for sludge removal is shown in Fig. 9-20.

In the peripheral-feed design (see Fig. 9-19b), a suspended circular baffle a short distance from the tank wall forms an annular space into which the wastewater is discharged in a tangential direction. The wastewater flows spirally around the tank and underneath the baffle, and the clarified liquid is skimmed off over weirs on both sides of a centrally located weir trough. Grease and scum are confined to the surface of the annular space.

Circular tanks 12 to 30 ft (3.6 to 9 m) in diameter have the sludge removal equipment supported on beams spanning the tank. Tanks 35 ft (10.5 m) in diameter and larger have a central pier that supports the mechanism and is reached by a walkway or bridge (see Fig. 9-20). The bottom of the tank is sloped at about 1 in/ft (1 in 12) to form an inverted cone, and the sludge is scraped to a relatively small hopper located near the center of the tank.

Multiple tanks are customarily arranged in groups of two or four. The flow is divided among the tanks by a flow split structure, commonly located between the tanks. Sludge is usually withdrawn by sludge pumps for discharge to the sludge-disposal units.

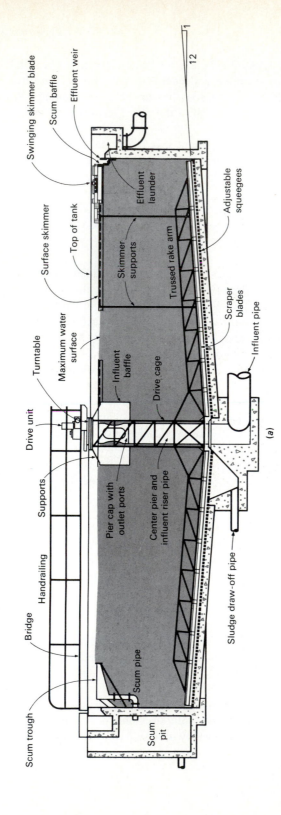

FIGURE 9-19a

Typical circular primary sedimentation tanks: (*a*) center feed (from Infilco Degremont).

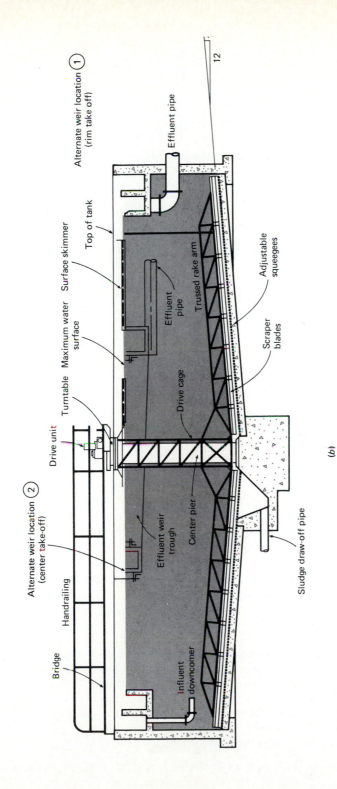

FIGURE 9-19b

Typical circular primary sedimentation tanks: (b) rim feed (from Ecodyne and Clow-Yeomans).

(b)

483

FIGURE 9-20
Empty center-feed sedimentation basin equipped with sludge scrapers.

Quantities of Sludge

The volume of sludge produced in primary settling tanks must be known or estimated so that these tanks and subsequent sludge-pumping, -processing, and disposal facilities can be properly designed. The sludge volume will depend on (1) the characteristics of the untreated wastewater, including strength and freshness; (2) the period of sedimentation and the degree of purification to be effected in the tanks; (3) the condition of the deposited solids, including specific gravity, water content, and changes in volume under the influence of tank depth or mechanical-sludge-removal devices; and (4) the period between sludge-removal operations. Data on the specific gravity and solids content of the sludge removed from primary sedimentation tanks are reported in Table 9-9. Example 9-2 and the subsequent discussion illustrate how these factors enter into the calculation of the required storage capacity.

Example 9-2   Sludge volume estimation. Estimate the volume of primary sludge produced per Mgal from a typical medium-strength wastewater. Assume that the detention time in the primary tank is 2 h and that removal efficiency of suspended solids is 60 percent.

1. Estimate the suspended-solids concentration. From Table 3-16, a medium-strength wastewater is found to contain 220 mg/L suspended solids.

2. Determine the mass of dry solids removed per Mgal.

$$\text{Dry solids} = 0.6(220 \text{ mg/L})[8.34 \text{ lb/Mgal} \cdot (\text{mg/L})](1.0 \text{ Mgal}) = 1100 \text{ lb}$$

3. Determine the volume of sludge, using the data in Table 9-9 for primary sludge and Eq. 12-2. If the specific gravity of the sludge is 1.03 and it contains 6 percent solids (94 percent moisture), the volume is

$$\text{Volume, gal} = \frac{1100 \text{ lb}}{(1.03)(0.06)(62.4 \text{lb/ft}^3) / (7.48 \text{ gal/ft}^3)} = 2134 \text{ gal}$$

Comment. Because of the many problems associated with pumping, treatment, and disposal of sludge, it is important to produce a sludge that is as thick as possible (i.e., minimize volume to be handled) consistent with the processing facilities. The determination of the volume of sludge when chemicals are used is illustrated in Example 9-3.

The calculation in Example 9-2 is directly applicable to the design of sludge-pumping facilities for primary sedimentation tanks. Sludge should be removed by pumping at least once per shift and more frequently in hot weather to avoid deterioration of the effluent. In large plants, sludge-pumping may be (1) continuous; or (2) intermittent, controlled by a time clock to provide on-off operation. Alternatively, sludge pumping may be controlled by time clock to initiate pumping and by measurement of sludge densities for pump shutdown.

In primary sedimentation tanks used in activated-sludge plants, provision may be required for handling the excess activated sludge that may be discharged into the influent of the primary tanks for settlement and consolidation with the primary sludge. For treatment plants where waste activated sludge is returned to the primary sedimentation tanks, the primary sedimentation tanks should include provisions for light, flocculant sludge of 98 to 99.5 percent moisture, and for concentrations ranging from 1500 to 10,000 mg/L in the influent mixed liquor.

9-7 OTHER SOLIDS-REMOVAL OPERATIONS

Flotation and fine screening are unit operations that may be used in place of primary sedimentation for removal of suspended and floating solids.

TABLE 9-9
Typical information on the specific gravity and concentration of sludge from primary sedimentation tanks

Type of sludge	Specific gravity	Solids concentration, %[a]	
		Range	Typical
Primary only:			
Medium-strength wastewater[b]	1.03	4–12	6
From combined sewer system	1.05	4–12	6.5
Primary and waste activated sludge	1.03	2–6	3
Primary and trickling-filter humus sludge	1.03	4–10	5

[a] Percent dry solids.
[b] See Table 3-16.

Flotation

Flotation has been used for untreated wastewater, settled wastewater, and stormwater overflows. The process has the advantage of high surface-loading rates and high removals of grease and floatable material. For these applications, design air-solids ratios have not been well defined. From practical experience, it appears that air quantities of 2 to 3 percent by volume of the wastewater flowrate yield satisfactory results. Flotation is also used extensively for waste activated-sludge thickening and is discussed in detail in Chap. 12.

A typical air-flotation process schematic is shown in Fig. 6-21b, which uses pressurization of the recycle stream. The design provides for injecting air into the retention tank and mixing the air and recycle wastewater in the tank. Such designs enable 80 to 95 percent saturation compared to 50 percent for static designs. The semisaturated recycle stream is then piped to the flotation tanks. A backpressure valve maintains the retention tank pressure within 4 to 5 lb_f/in^2 (28 to 35 kN/m^2). Turbulence or energy dissipation should be avoided in the inlet design to prevent reduction in flotation efficiency.

Fine Screens

With the development of better screening materials and equipment, the use of fine screens for grit removal and as a replacement for (and a means of upgrading the performance of) primary sedimentation tanks is increasing. The three most common types of screens used for this purpose are the inclined self-cleaning type (see Fig. 9-3a), the rotary drum type (see Fig. 9-3b), and the rotary disk screen (see Fig. 9-21). Typical design information on these screens is presented in Table 9-10. From information on a number of full-scale installations, it appears that grit removals of 80 to 90 percent, BOD_5 removals of 15 to 25 percent, and suspended-solids removals of 15 to 30 percent can be achieved with the inclined and rotary drum screens. Suspended solids and BOD removals of 40–50 and 25–35 percent, respectively have been achieved with the rotary disk screen. It also has been found that if the solids in the wastewater are ground up using comminutors, the BOD removals will not be as high [18].

Where fine screens are used as alternatives to primary sedimentation basins, the following (secondary) facilities must be sized appropriately to handle the solids and BOD_5 not removed by the screens as compared to the use of primary sedimentation facilities.

9-8 CHEMICAL PRECIPITATION

Chemical precipitation was a well-established method of wastewater treatment in England as early as 1870. Chemical treatment was used extensively in the United States in the 1890s and early 1900s, but, with the development of biological treatment, the use of chemicals was abandoned and biological treatment was adopted. In the early 1930s, attempts were made to develop new methods of chemical treatment, and

FIGURE 9-21
Rotary disk screen used for primary treatment (shown with cover removed) (see also Fig. 13-24).

TABLE 9-10
Typical design information on screening devices used for the primary treatment of wastewater

Item	Type of screen		
	Inclined	Rotary drum	Rotary disk
Screening surface			
Size classification	Medium	Medium	Fine
Size range, μm	0.01–0.06	0.01–0.06	0.001–0.01
Screen material	Stainless steel wedge wire	Stainless steel wedge wire	Stainless steel woven wire
Hydraulic capacity, gal/ft² · min	15–60	0.12–1.0	0.10–1.0
Composition of waste solids, % solids by weight	10–15	10–15	6–12
Suspended-solids removal, %	15–30	15–30	40–50

Note: gal/ft² · min × 0.0407 = m³/m² · min

a number of plants were installed. Details on these early processes may be found in Refs. 7 and 10.

In current practice, chemical precipitation is used (1) as a means of improving the performance of primary settling facilities, (2) as a basic step in the independent physical-chemical treatment of wastewater, and (3) for the removal of phosphorus. The first two applications are considered in the following discussion. Phosphorus removal is considered in Chap. 11. Aside from the determination of the required chemical dosages, the principal design considerations related to the use of chemical precipitation involve the analysis and design of the necessary sludge-processing facilities and the selection and design of the chemical storage, feeding, piping, and control systems.

Enhanced Removal of Suspended Solids

The degree of clarification obtained when chemicals are added to untreated wastewater depends on the quantity of chemicals used and the care with which the process is monitored and controlled. With chemical precipitation, it is possible to remove 80 to 90 percent of the suspended solids, 50 to 80 percent of the BOD_5, and 80 to 90 percent of the bacteria. Comparable removal values for well-designed and well-operated primary sedimentation tanks without the addition of chemicals are 50 to 70 percent of the suspended solids, 25 to 40 percent of the BOD_5, and 25 to 75 percent of the bacteria. Because of the variable characteristics of wastewater, the required chemical dosages should be determined from bench- or pilot-scale tests. Recommended surface-loading rates for various chemical suspensions to be used in the design of the sedimentation facilities are given in Table 9-11.

Independent Physical-Chemical Treatment

In some localities, industrial wastes have rendered municipal wastewater difficult to treat by biological means. In such situations, physical-chemical treatment may be an

TABLE 9-11
Recommended surface-loading rates for sedimentation tanks for various chemical suspensions

	Loading rate, gal/ft² · d	
Suspension	Range	Peak flow
Alum floc[a]	600–1,200	1,200
Iron floc[a]	600–1,200	1,200
Lime floc[a]	750–1,500	1,500
Untreated wastewater	600–1,200	1,200

[a] Mixed with the settleable suspended solids in the untreated wastewater and colloidal or other suspended solids swept out by the floc.

Note: gal/ft² · d × 0.0407 = m³/m² · d

alternative approach. This method of treatment has met with limited success because of its lack of consistency in meeting discharge requirements, high costs for chemicals, handling and disposal of the great volumes of sludge resulting from the addition of chemicals, and numerous operating problems. Based on typical performance results of full-scale plants using activated carbon, the activated-carbon columns removed only 50 to 60 percent of the applied total BOD_5, and the plants did not consistently meet the effluent standards for secondary treatment. In some instances, substantial process modifications have been required to reduce the operating problems and meet performance requirements, or the process has been replaced by biological treatment. Because of these reasons, new applications of physical-chemical treatment for municipal wastewater are rare. Physical-chemical treatment is used more extensively for the treatment of industrial wastewater. Depending on the treatment objectives, the required chemical dosages and application rates should be determined from bench- or pilot-scale tests.

A flow diagram for the physical-chemical treatment of untreated wastewater is presented in Fig. 9-22. As shown, after first-stage precipitation and pH adjustment by recarbonation (if required), the wastewater is passed through a granular medium filter to remove any residual floc and then through carbon columns to remove dissolved organic compounds. The filter is shown as optional, but its use is recommended to reduce the blinding and headloss buildup in the carbon columns. The treated effluent from the carbon column is usually chlorinated before discharge to the receiving waters.

Estimation of Sludge Quantities

The handling and disposal of the sludge resulting from chemical precipitation is one of the greatest difficulties associated with chemical treatment. Sludge is produced in great volume from most chemical precipitation operations, often reaching 0.5 percent of the volume of wastewater treated when lime is used. The computational procedures involved in estimating the quantity of sludge resulting from chemical precipitation with ferrous sulfate and lime are illustrated in Example 9-3.

Example 9-3 Estimation of sludge volume from chemical precipitation of untreated wastewater. Estimate the mass and volume of sludge produced from untreated wastewater without and with the use of ferrous sulfate and lime for the enhanced removal of SS. Assume that 60 percent of the suspended solids is removed in the primary settling tank without the addition of chemicals and that the addition of ferrous sulfate and lime results in an increased removal of SS to 85 percent. Also, assume that the following data apply to this situation:

1. Wastewater flowrate $= 1.0$ Mgal/d
2. Wastewater suspended solids $= 220$ mg/L
3. Ferrous sulfate $(FeSO_4 \cdot 7H_2O)$ added $= 70$ lb/Mgal
4. Lime added $= 600$ lb/Mgal
5. Calcium carbonate solubility $= 15$ mg/L

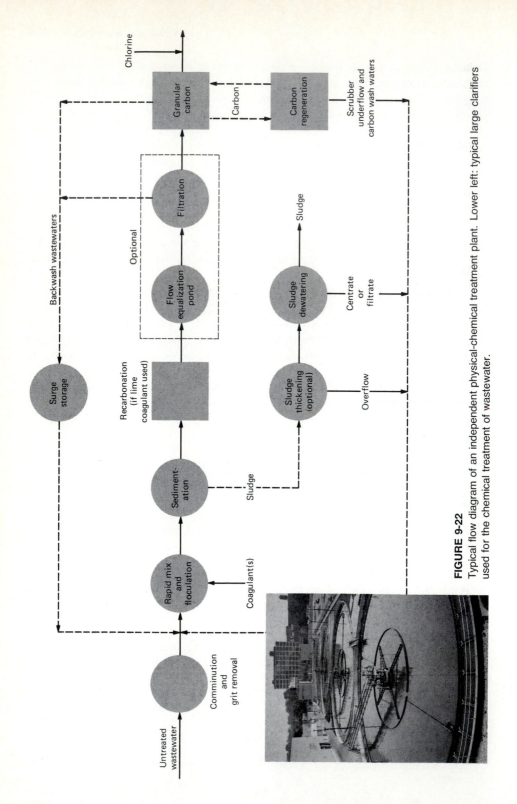

FIGURE 9-22

Typical flow diagram of an independent physical-chemical treatment plant. Lower left: typical large clarifiers used for the chemical treatment of wastewater.

490

Solution

1. Compute the mass of SS removed without and with chemicals.

   (a) Determine the mass of suspended solids removed without chemicals.

   $$M_{ss} = (0.6)(220 \text{ mg/L})[8.34 \text{ lb/Mgal} \cdot (\text{mg/L})](1.0 \text{ Mgal/d}) = 1,100 \text{ lb/d}$$

   (b) Determine the mass of suspended solids removed with chemicals.

   $$M_{ss} = (0.85)(220 \text{ mg/L})[8.34 \text{ lb/Mgal} \cdot (\text{mg/L})](1.0 \text{ Mgal/d}) = 1,560 \text{ lb/d}$$

2. Using Eqs. 7-4 through 7-6, determine the mass of ferric hydroxide ($Fe(OH)_3$) produced from the addition of 70 lb/Mgal of ferrous sulfate.

   $$\text{Ferric hydroxide formed} = 70 \times \frac{106.9}{278} = 27 \text{ lb/Mgal}$$

3. Using Eqs. 7-4 and 7-5, determine the mass of calcium carbonate ($CaCO_3$) produced from the addition of 70 lb/Mgal of ferrous sulfate.

   $$CaCO_3 \text{ formed} = 70 \times \frac{112}{278} \times \frac{100}{56} = 50 \text{ lb/Mgal}$$

4. Using Eqs. 7-4 and 7-5, determine the mass of calcium carbonate ($CaCO_3$) produced from the addition of 600 lb/Mgal of lime. Note the lime will react with CO_2 and the bicarbonates.

   $$CaCO_3 \text{ formed} = \frac{300}{112} \times \left(600 - 50 \times \frac{560}{100}\right) = 1,530 \text{ lb/Mgal}$$

5. Determine the total amount of calcium carbonate precipitated, taking into account the solubility of calcium carbonate.

   (a) Calcium carbonate not precipitated.

   $$CaCO_3 \text{ not precipitated} = (15 \text{ mg/L})[8.34 \text{ lb/Mgal} \cdot (\text{mg/L})](1.0 \text{ Mgal/d}) = 125 \text{ lb/Mgal}$$

   (b) Calcium carbonate precipitated.

   $$CaCO_3 \text{ precipitated} = (50 + 1530) - 125 = 1,455 \text{ lb/Mgal}$$

6. Determine the total amount of sludge on a dry basis.

   $$\text{Total dry solids} = 1,560 + 27 + 1,455 = 3,042 \text{ lb/d}$$

7. Determine the total volume of sludge resulting from chemical precipitation, assuming that the sludge has a specific gravity of 1.05 and a moisture content of 92.5 percent (see Chap. 12).

   $$V_s = \frac{3,042 \text{ lb/d}}{1.05 \times 62.4 \text{ lb/ft}^3 (0.075)} = 619 \text{ ft}^3/\text{d}$$

8. Determine the total volume of sludge without chemical precipitation, assuming that the sludge has a specific gravity of 1.03 and a moisture content of 94 percent (see Chap. 12).

   $$V_s = \frac{1,100 \text{ lb/d}}{1.03 \times 62.4 \text{ lb/ft}^3 (0.06)} = 285 \text{ ft}^3/\text{d}$$

9. Prepare a summary table of sludge masses and volumes without and with chemical precipitation.

Treatment	Sludge	
	Mass, lb/d	Volume, ft³/d
Without chemical precipitation	1,100	285
With chemical precipitation	3,042	619

Comment. The magnitude of the sludge-disposal problem when chemicals are used is evident from a review of the data presented in the summary table given in step 9.

Chemical Storage, Feeding, Piping, and Control Systems

The design of chemical precipitation operations involves not only the sizing of the various unit operations and processes but also the necessary appurtenances. Because of the corrosive nature of many of the chemicals used and the different forms in which they are available, special attention must be given to the design of chemical storage, feeding, piping, and control systems. A brief discussion of these topics is included in this section; for more detailed information, Refs. 19 and 20 may be consulted.

In domestic wastewater treatment systems, the chemicals employed are generally in the solid or liquid form. Coagulants in the solid form generally convert to solution or slurry form prior to introduction into the wastewater. Coagulants in the liquid form are usually delivered to the plant in a concentrated form and have to be diluted prior to introduction into the wastewater. The types of chemical-feed systems are termed dry and liquid feed.

Dry Chemical-Feed Systems. A dry chemical-feed system generally consists of a storage hopper, dry chemical feeder, a dissolving tank, and a pumped or gravity distribution system (see Fig. 9-23). The units are sized according to the volume of wastewater, treatment rate, and optimum length of time for chemical feeding and dissolving. Hoppers used with compressable and archable powder such as lime are equipped with positive hopper agitators and a dust collection system. Dry chemical feeders are either of the volumetric or gravimetric type. The volumetric type measures the volume of the dry chemical fed; the gravimetric type weighs the amount of chemical fed. With a dry feed system, the dissolving operation is critical. The capacity of the dissolving tank is based on the detention time, which is directly related to the wettability or rate of solution of the chemical. When the water supply is controlled for the purpose of forming a constant strength solution, mechanical mixers should be used. Solutions or slurries are often stored after dissolving and discharged to the application point at metered rates by chemical-feed pumps.

Liquid Chemical-Feed Systems. Liquid chemical-feed systems typically include a solution storage tank, transfer pump, day tank for diluting the concentrated solution, and chemical-feed pump for distribution to the application point (see Fig. 9-24). In systems where the liquid chemical does not require dilution, the chemical-feed pumps draw liquid directly from the solution storage tank. The storage tank is sized based on

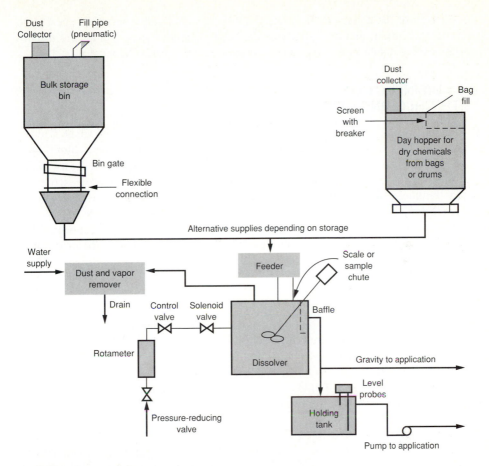

FIGURE 9-23
Typical dry chemical-feed system [19].

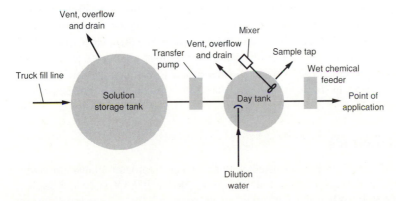

FIGURE 9-24
Typical liquid chemical-feed system [20].

the stability of the chemical, feed-rate requirements, delivery constraints (cost, size of tank truck, etc.), and availability of the supply. Solution-feed pumps are usually of the positive displacement type for accurate metering of the chemical feed.

9-9 DISINFECTION WITH CHLORINE COMPOUNDS

The chemistry of chlorine in water and wastewater has been discussed in Chap. 7 along with an analysis of how chlorine functions as a disinfectant. However, chlorine has been applied for a wide variety of objectives other than disinfection in the wastewater treatment field, including prechlorination for hydrogen sulfide control, activated-sludge bulking control, and odor control. Therefore, the purpose of this section is to discuss briefly (1) the various uses and required dosages, (2) the chlorine compounds most commonly used, (3) the equipment and methods used in its application, (4) the design of mixing and chlorine contact facilities for disinfection, and (5) methods of dechlorination.

Application

To aid in the design and selection of the required chlorination facilities and equipment, it is important to know the uses, including dosage ranges, to which chlorine and its compounds have been applied.

Uses. The principal uses of chlorine and its compounds in the collection, treatment, and disposal of wastewater were reported in Table 7-8. Of the many different applications of chlorine, disinfection of wastewater effluents is still the most important.

Dosages. Ranges of dosages for various applications of chlorine are reported in Table 9-12. A range of dosage values is given because they will vary depending on the characteristics of the wastewater. It is for this reason that laboratory chlorination studies should be conducted to determine optimum chlorine dosages.

Chlorination capacities for disinfection are generally selected to meet the specific design criteria of the state or other regulatory agencies controlling the receiving body of water (see Example 9-4). In any case, where the residual in the effluent is specified or the final number of coliform bacteria is limited, onsite testing is preferred to determine the dosage of chlorine required. However, in the absence of more specific data, the maximum values given in Table 9-12 can be used as a guide in sizing chorination equipment.

Chlorine Compounds

The principal chlorine compounds used at wastewater treatment plants are chlorine (Cl_2), chlorine dioxide (ClO_2), calcium hypochlorite [$Ca(OCl)_2$], and sodium hypochlorite (NaOCl). When the latter two forms are used, the chlorination process is known as "hypochlorination."

TABLE 9-12
Typical dosages for various chlorination applications in wastewater collection, treatment, and disposal

Application	Dosage range, mg/L
Collection	
Corrosion control (H_2S)	2–9[a]
Odor control	2–9[a]
Slime growth control	1–10
Treatment	
BOD reduction	0.5–2[b]
Digester- and Imhoff tank-foaming control	2–15
Digester supernatant oxidation	20–140
Ferrous sulfate oxidation	–[c]
Filter fly control	0.1–0.5
Filter-ponding control	1–10
Grease removal	2–10
Sludge-bulking control	1–10
Disposal (disinfection)	
Untreated wastewater (prechlorination)	6–25
Primary effluent	5–20
Chemical precipitation effluent	2–6
Trickling-filter plant effluent	3–15
Activated-sludge plant effluent	2–8
Filtered effluent (after activated-sludge treatment)	1–5

[a] Per mg/L of H_2S

[b] Per mg/L of BOD_5 destroyed.

[c] $6(Fe\ SO_4 \cdot 7H_2O) + 3Cl_2 \rightarrow 2FeCl_3 + 2Fe_2(SO_4) + 42H_2O.$

Chlorine. Chlorine is supplied as a liquefied gas under high pressure in containers varying in size from 150 lb (68 kg) cylinders to 1 ton (0.907 Mg) containers, multiunit tank cars containing fifteen 1 ton (0.907 Mg) containers, and tank cars with capacities of 16, 30, and 55 tons (14.5, 27.2, and 49.9 Mg). Selection of the size of the chlorine pressure vessel should depend on an analysis of the rate of chlorine usage, cost of chlorine, facility's requirements, and dependability of supply. Storage and handling facilities can be designed with the aid of information developed by the Chlorine Institute and in Ref. 16 fire code requirements must also be considered. Although all the safety devices and precautions that must be designed into the chlorine-handling facilities are too numerous to mention, the following are fundamental:

1. Chlorine gas is toxic and very corrosive. Adequate exhaust ventilation at floor level should be provided because chlorine gas is heavier than air. The ventilation system should be capable of at least 60 air changes per hour. Emergency caustic scrubbing systems may also be required to neutralize leaking chlorine.

2. Chlorine storage and chlorinator equipment rooms should be walled off from the rest of the plant and should be accessible only from the outdoors. A fixed glass viewing window should be included in an inside wall. Fan controls should be located at the room entrance. Air masks should also be located nearby in protected but readily accessible locations.

3. Temperatures in the scale and chlorinator areas should be controlled to avoid freezing.

4. Dry chlorine liquid and gas can be handled in black steel piping, but chlorine solution is highly corrosive and should be handled in Schedule 80 polyvinylchloride (PVC) piping.

5. Adequate storage of standby cylinders should be provided. The amount of storage should be based on the availability and dependability of the supply and the quantities used. Cylinders in use are set on platform scales and the loss of weight is used as a positive record of chlorine dosage.

6. Chlorine cylinders should be protected from direct sunlight in warm climates to prevent overheating of the full cylinders.

7. In larger systems, chlorine residual analyzers should be provided for monitoring and control purposes to prevent the under- or over-dosing of chlorine.

8. The chlorine storage and feed facilities should be protected from fire hazards. In addition, chlorine leak detection equipment should be provided and connected to an alarm system and to the emergency scrubbing system, if provided.

Chlorine Dioxide. In the generation of chlorine dioxide (refer to Fig. 9-26, presented later in the chapter), liquid chlorine is vaporized and metered through standard evaporators and chlorinators, and then converted into a chlorine solution using an injector. Sodium chlorite may be purchased and stored as a liquid (generally a 25 percent solution) and metered directly into the reaction tower, or it may be purchased in the salt form, with the liquid solution prepared onsite. The chlorine and sodium chlorite solutions are brought together at the base of a porcelain ring filled reaction tower. As this combined solution flows upward, chlorine dioxide is generated. A contact time of about 1 minute is generally adequate. To increase the reaction rate and obtain the highest yield of chlorine dioxide, a slight excess of chlorine is recommended. Because sodium chlorite is about ten times as expensive as chlorine on a weight basis, economical considerations must be taken into account. The solution discharged from the tower is only partly chlorine dioxide, with the remaining portion being chlorine in solution as hypochlorous acid.

Calcium Hypochlorite. Calcium hypochlorite is available commercially in either a dry or a wet form. High-test calcium hypochlorite contains at least 70 percent available chlorine. In dry form, it is available as a powder or as granules, compressed tablets, or pellets. A wide variety of container sizes is available depending on the source. Calcium hypochlorite granules or pellets are readily soluble in water and, under proper storage conditions, are relatively stable. Because of its oxidizing potential, calcium hypochlorite should be stored in a cool, dry location away from other chemicals in corrosion resistant containers. Many of the safety concerns related to the transport, storage, and feeding of liquid-gaseous chlorine are eliminated by the use of either calcium or sodium hypochlorite. Hypochlorite is more expensive than liquid chlorine, loses its available strength on storage, and may be difficult to handle. Because it tends to crystallize, calcium hypochlorite may clog metering pumps, piping, and valves. Calcium hypochlorite is used mainly at small installations.

Sodium Hypochlorite. Many large cities including New York, Chicago, and San Francisco use sodium hypochlorite because of the safety concerns related to liquid chlorine. Sodium hypochlorite solution can be either purchased in bulk lots of 12 to 15 percent of available chlorine or manufactured on site. The solution decomposes more readily at high concentrations and is affected by exposure to light and heat. A 16.7 percent solution stored at 80°F (26.7°C) will lose 10 percent of its strength in 10 days, 20 percent in 25 days, and 30 percent in 43 days. It must therefore be stored in a cool location in a corrosion-resistant tank. Another disadvantage of sodium hypochlorite is the chemical cost. The purchase price may range from 150 to 200 percent of the cost of liquid chlorine. The handling of sodium hypochlorite requires special design considerations because of its corrosiveness and the presence of chlorine fumes.

Several proprietary systems are available for the generation of sodium hypochlorite from sodium chloride (NaCl) or seawater. These systems are electric power intensive and, in the case of seawater, result in a very dilute solution, a maximum of 0.8 percent hypochlorite. The onsite generation systems have been used only on a limited basis due to their complexity and high power cost.

Application Flow Diagrams and Dosage Control

In this section, the equipment used to inject (feed) chlorine or its related compounds into the wastewater and the methods used to control the required dosages are discussed.

Flow Diagram for Chlorine. Chlorine may be applied directly as a gas or in an aqueous solution. A typical chlorine-feed system is shown in Fig. 9-25. Chlorine can be withdrawn from storage containers either in liquid or gas form. If withdrawn as a gas, the evaporation of the liquid in the container results in frost formation that restricts gas withdrawal rates to 40 lb/d (18 kg/d) for 150-lb (68-kg) cylinders and 450 lb/d (205 kg/d) for 1-ton (0.907-Mg) containers at 70°F (21°C). Evaporators are normally used where the maximum rate of chlorine-gas withdrawal from a 1-ton (0.907-Mg) container must exceed approximately 400 lb/d (180 kg/d). Although multiple-ton cylinders can be connected to provide more than 400 lb/d, the use of an evaporator conserves space. Evaporators are almost always used when the total dosage exceeds 1500 lb/d (680 kg/d) [24]. Chlorine evaporators are available in sizes ranging from 4,000 to 10,000 lb/d (1818 to 4545 kg/d) capacities; chlorinators are available normally in sizes ranging from 500 to 10,000 lb/d (227 to 4545 kg/d). The sizing of chlorinators is considered in Example 9-4.

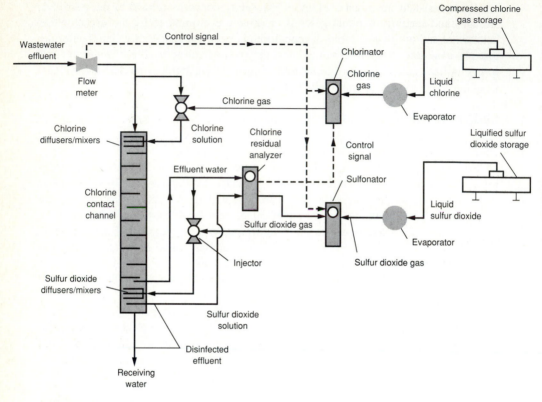

FIGURE 9-25
Chlorination/sulfur dioxide dechlorination schematic flow diagram [9].

Example 9-4 Chlorinator selection. Determine the capacity of a chlorinator for a treatment plant with an average wastewater flow of 0.26 Mgal/d. The peak hourly factor for the treatment plant is 3.0 and the maximum required chlorine dosage (set by state regulations) is to be 20 mg/L.

Solution

1. Determine the capacity of the chlorinator at peak flow.

$$Cl_2, lb/d = (20 mg/L)(0.26 \ Mgal/d)(3) \ [8.34 \ lb/Mgal \cdot (mg/L)] = 130 \ lb/d$$

Use the next largest standard size chlorinator: two 200 lb (90 kg/d) units with one unit serving as a spare. Although the peak capacity will not be required during most of the day, it must be available to meet the chlorine requirements at peak flow. The best design practice calls for the availability of a standby chlorinator.

2. Estimate the daily consumption of chlorine. Assume an average dosage of 10 mg/L.

$$Cl_2, lb/d = (10 \ mg/L)(0.26 \ Mgal/d) \ [8.34 \ lb/Mgal \cdot (mg/L)] = 21.7 \ lb/d$$

Comment. In sizing and designing chlorination systems, it is also important to consider the low-flow/dosage requirements. The chlorination system should have sufficient turndown capability for these conditions so that excessive chlorine is not applied.

Flow Diagram for Chlorine Dioxide. The chlorine dioxide produced by these processes is generated and remains in an aqueous solution. Application of this solution to the water or wastewater stream is done in the same manner as that used for typical chlorination systems. In this regard, a more efficient diffusing or mixing arrangement would increase the effectiveness of disinfection by chlorine dioxide just as for chlorine/hypochlorite. A schematic process flow diagram of a typical chlorine dioxide installation is shown in Fig. 9-26.

Flow Diagram for Hypochlorite Solutions. The most satisfactory means of feeding sodium or calcium hypochlorite is through the use of low-capacity proportioning pumps (see Fig. 9-27). Generally, the pumps are available in capacities up to 120 gal/d (450 L/d), with adjustable stroke for any value below this. Large capacities

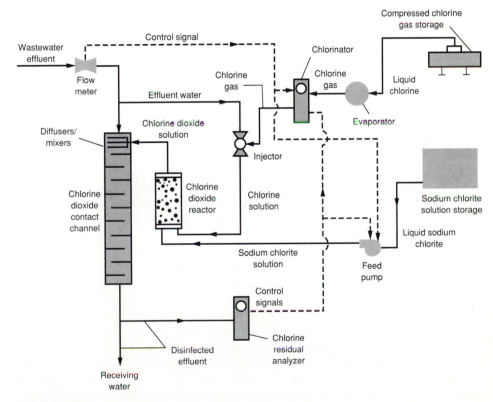

FIGURE 9-26
Chlorine dioxide dechlorination schematic flow diagram [9].

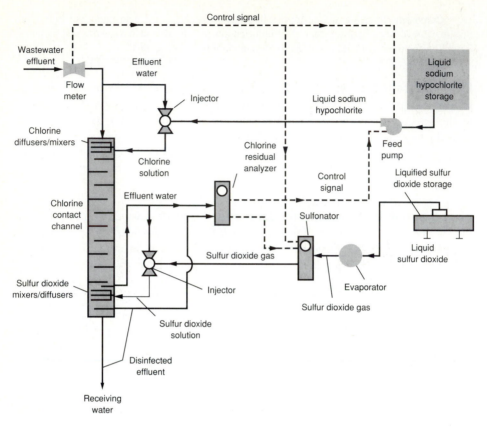

FIGURE 9-27
Hypochlorination/dechlorination schematic flow diagram [9].

or multiple units are available from some of the manufacturers. The pumps can be arranged to feed at a constant rate, or they can be provided with variable speed and with analog signals for varying the feed rate. The stroke length can also be controlled.

Dosage Control. Dosage may be controlled in several ways. The simplest method is manual control; the operator changes the feed rate to suit conditions. The required dosage is usually determined by measuring the chlorine residual after 15 minutes of contact time and adjusting the dosage to obtain a residual of 0.5 mg/L. A second method is to pace the chlorine flowrate to the wastewater flowrate as measured by a primary meter such as a magnetic meter, Parshall flume, or flow tube. A third method is to control the chlorine dosage by automatic measurement of the chlorine residual. An automatic analyzer with signal transmitter and recorder is required. Finally, a compound system that incorporates both the second and third methods may be used. In a compound system, the control signals obtained from the wastewater flowmeter and from the residual recorder provide more precise control of chlorine dosage and residual. Additional details on these systems may be found in Ref. 24.

Chlorine Mixing and Contact

As pointed out in Chap. 7, other things being equal, effective mixing of the chlorine solution with the wastewater, the contact time, and the chlorine residual are the principal factors involved in achieving effective bacterial kill. The contact time is usually specified by the regulatory agency and may range from 15 to 45 min; periods of 15 min at peak flow are common. The appropriate chlorine residual that needs to be maintained, if not specified in the regulations, should be determined from actual plant studies. In the absence of any other information, the chlorine residual may be estimated by using Eq. 7-47. The necessary computations are illustrated in Example 9-5.

Important practical factors that must be considered in the design of chlorine mixing and contact facilities include (1) method of chlorine addition and provision for mixing, (2) design of the chlorine contact basin, (3) maintenance of solids transport velocity, and (4) outlet control and chlorine residual measurement. These topics are considered in the following discussion.

Example 9-5   Estimation of required chlorine residuals. Estimate the chlorine residual that must be maintained to achieve a coliform count equal to or less than 200/100 mL in an effluent from an activated-sludge treatment facility, assuming that the effluent contains a coliform count of 10^7/100 mL. The specified contact time is 30 min. What will be the required residual to meet the specified effluent coliform count for a flowrate corresponding to the 1-d sustained value as given in Fig. 2-5?

Solution

1. Determine the chlorine residual needed to meet the effluent discharge requirement using Eq. 7-47.

$$\frac{N_t}{N_0} = (1 + 0.23 \, C_t t)^{-3}$$

$$\frac{2 \times 10^2}{10^7} = (1 + 0.23 \, C_t t)^{-3}$$

$$2 \times 10^{-5} = (1 + 0.23 \, C_t t)^{-3}$$

$$1 + 0.23 \, C_t t = (0.5 \times 10^5)^{1/3} = 36.84$$

$$C_t t = (36.84 - 1)/0.23 = 155.8$$

For a value of equal to 30 min,

$$C_t = 155.8/30 = 5.2 \text{ mg/L}$$

2. Determine the residual for the peak hourly flowrate. From Fig. 2-5, the ratio of the peak hourly flowrate to the average flowrate is 2.75. Because the chlorine contact time will be reduced by this value, the corresponding residual is

$$C_t = 155.8/(30/2.75) = 14.3 \text{ mg/L}$$

Comment. The chlorination system should be designed to provide chlorine residuals over a range of operating conditions and should include an adequate margin of safety. Further, in applying Eq. 7-47, it has been assumed that the chlorine contact was either a batch reactor or an ideal plug-flow reactor and that ideal initial mixing was achieved. Because batch reactors are seldom used in other than very small plants, a plug-flow reactor would normally be used. Therefore, to account for the effects of the inherent axial dispersion in a plug-flow reactor, it will usually be necessary to increase the value of the residual computed in step 1.

Injection and Initial Mixing. The design of any chlorine contact system should provide for the injection and mixing of chlorine. Addition of chlorine solution is commonly done through a diffuser, which may be a plastic pipe with drilled holes through which the chlorine solution can be uniformly distributed into the path of wastewater flow, or the solution can flow directly to the propeller of a rapid mixer for instantaneous and complete diffusion. Typical diffusers are shown in Fig. 9-28. In most cases, the type of injection diffuser will depend on the means to be used to accomplish the initial mixing of the chlorine solution and the wastewater.

The effective initial mixing of the chlorine solution with the wastewater can be accomplished by use of a turbulent-flow regime or mechanical means. Turbulent mixing (described in Chap. 6) can be accomplished by (1) hydraulic jumps in open channels, (2) Venturi flumes, (3) pipelines, (4) pumps, (5) static mixers, or (6) vessels (chambers) with the aid of mechanical-mixing devices. Current practice favors the use of mechanical mixing to ensure rapid and complete mixing (see Fig. 9-29). A preferred design would be to achieve mixing times on the order of one second or less with a mixer capable of providing velocity gradients (G) ranging from 1500 to 3000 s^{-1}.

As alternatives to injecting and mixing chlorine solution, jet pumps or aspirating-type mixers can be used with gaseous chlorine. In these devices, a vacuum is created and the chlorine gas is drawn into the mixing device (see Fig. 9-30). Chlorine is rapidly dispersed into the wastewater. Advantages of this type of system are the following: (1) breakout of molecular chlorine that may occur in conventional chlorine solution mixing systems is minimized; (2) more efficient disinfection may be achieved [23]; and (3) separate chlorine injector water pumps are not required.

The particular method of achieving effective initial mixing will vary with each situation and may be dictated by local or state regulations. Additional details on initial mixing may be found in Refs. 6, 23, and 24.

Chlorine-contact Basin Design. Because of the importance of contact time, careful attention should be given to design of the contact chamber so that at least 80 to 90 percent of the wastewater is retained in the basin for the specified contact time. The best way to achieve this is by using a plug-flow around-the-end type of contact chamber or a series of interconnected basins or compartments. Plug-flow chlorine-contact basins that are built in a serpentine fashion (i.e., folded back and forth) to conserve space require special attention in their design. The reason for this is the development of dead zones with respect to flow that will reduce the hydraulic detention times. Length-to-width ratios (L/W) of at least 10 to 1 and preferably 40 to 1 will minimize short circuiting. Short circuiting may also be minimized by reducing the velocity of the wastewater entering contact tanks. Baffles similar to those used on rectangular sedimentation tanks may be used for inlet velocity control. The placement

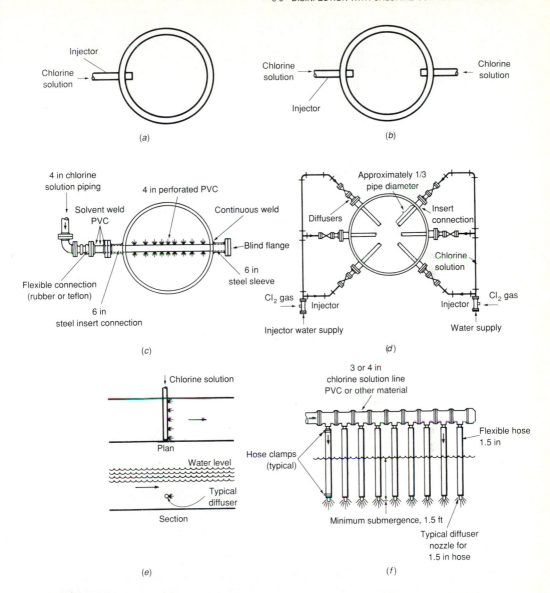

FIGURE 9-28
Typical diffusers used to inject chlorine solution: (a) single injector for small pipe, (b) dual injector for small pipe, (c) across-the-pipe diffuser for pipes larger than 3 ft in diameter, (d) diffuser system for large conduits, (e) single across-the-channel diffuser, and (f) typical hanging-nozzle-type chlorine diffuser for open channels. (Adapted in part from Ref. 23.)

of longitudinal baffles and turning vanes can reduce short circuiting and improve the actual detention time. A chlorine contact basin with deflection baffles at the channel bends is shown in Fig. 9-31.

For most treatment plants, two or more contact basins should be used to facilitate maintenance and accumulated-sludge removal. Provisions should also be included

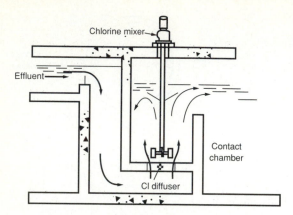

FIGURE 9-29
Typical mechanical mixing chamber for wastewater chlorination.

for draining and scum removal. Vacuum-type cleaning equipment may be used as an alternative to draining the basin for sludge removal. Bypassing the contact basin for maintenance should be practiced on rare occasions, only with the approval of regulatory agencies.

If the time of travel in the outfall sewer at the maximum design flow is sufficient to equal or exceed the required contact time, it may be possible to eliminate the chlorine-contact chambers, provided regulatory authorities agree. In some small plants, chlorine-contact basins have been constructed of large-diameter sewer pipe.

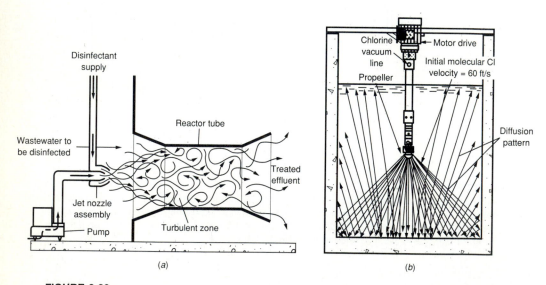

FIGURE 9-30
Typical chlorine injector mixers: (*a*) injector pump type (from Pentech-Houdaille) and (*b*) aspirating-type (from Gardiner Equipment Co.).

FIGURE 9-31
Plug-flow chlorine-contact basin with deflection baffles at the channel bends.

Maintenance of Solids-transport Velocity. The horizontal velocity at minimum flow in a chlorine-contact basin should be sufficient to scour the bottom or at least to give a minimum deposition of sludge solids that may have passed through the settling tank. Horizontal velocities should be at least 6.5 to 15 ft/min (2 to 4.5 m/min).

Outlet Control and Chlorine Residual Measurement. The flow at the end of the contact chamber may be metered by means of a V-notch or rectangular weir or a Parshall flume. Control devices for chlorination in direct proportion to the flowrate may be operated from these meters or from the main plant flowmeter. Final determination of the success of a chlorine-contact chamber must be based on samples taken and analyzed to correlate chlorine residual and the MPN of coliform organisms. When the chlorine residual is used for chlorinator control, chlorine residual sample pumps should be located at the front end of the first pass of the contact basin immediately after rapid mixing. More precise control of the chlorine feed can be maintained as compared to monitoring the chlorine residual at the chlorine basin effluent. Chlorine residual measurements should also be taken at the contact tank outlet to ensure compliance with the regulatory agency requirements. In the event that no chlorine contact chamber is provided and the outfall sewer is used for contact, the sample can be obtained at the point of chlorination and held for the theoretical detention time, and the residual can be determined. The sample is then dechlorinated and subsequently analyzed for bacteria by normal laboratory procedures.

Dechlorination

In cases where low-level chlorine residuals may have potential toxic effects on aquatic organisms, dechlorination of treated effluent is practiced. Dechlorination may be accomplished by reaction with a reducing agent such as sulfur dioxide or sodium metabisulfite or by adsorption on activated carbon. Sulfur dioxide is the most common substance used, particularly for treatment plants over 1 Mgal/d (3800 m^3/d) capacity. The purpose of this section is to discuss briefly the design considerations for sulfur dioxide and activated-carbon systems.

Sulfur Dioxide. Sulfur dioxide (SO_2) is available commercially as a liquified gas under pressure in steel containers with capacities of 100, 150, and 2000 lbs (45, 68, and 907 kg). Sulfur dioxide is handled in equipment very similar to standard chlorine systems. When added to water, sulfur dioxide reacts to form sulfurous acid (H_2SO_3), a strong reducing agent. The sulfurous acid dissociates to form HSO_3^-, which will react with free and combined chlorine, resulting in formation of chloride and sulfate ions. The reaction with the total chlorine residual is accomplished in less than two minutes.

The principal elements of a sulfur dioxide system are the sulfur dioxide containers, scales, sulfur dioxide feeders (sulfonators), solution injectors, diffuser, mixing chamber, and interconnecting piping. For facilities requiring large withdrawal rates of SO_2, evaporators are used because of the low vaporization pressure of 35 lb_f/in^2 at 70°F (241 kN/m^2 at 21°C). Common sulfonator sizes are 475, 1900, and 7500 lb/d (216, 864, and 3409 kg/d). Typical design information is presented in Table 9-13.

Activated Carbon. The common method of activated-carbon treatment used for dechlorination is downflow through either an open or enclosed vessel. Typical loading rates and contact times are presented in Table 9-13. The activated-carbon system, while significantly more costly than other dechlorination approaches, may be appropriate when activated carbon is being used as an advanced wastewater treatment process. Further discussion about carbon adsorption is presented in Chap. 11.

9-10 OTHER MEANS OF DISINFECTION

Other means of disinfection that have been used include (1) bromine chloride, (2) ozone, and (3) UV radiation. Each of these alternative means of disinfection is considered in the following discussion.

Disinfection with Bromine Chloride

Bromine chloride is a hazardous and corrosive chemical, and thus special transportation, storage, and handling precautions are required. However, bromine chloride is less hazardous than chlorine due to its lower vaporization rate. Unlike liquid bromine, bromine chloride exhibits relatively low corrosivity to steel, which permits its use with piping and containers usually associated with chlorine. Bromine chloride is normally shipped in the liquid form in cylinders, tank cars, or 3000 lb containers. (Bromine chloride has a higher density than chlorine; its specific gravity is 2.34 as compared

TABLE 9-13
Typical design information for sulfur dioxide and activated-carbon dechlorination facilities

Item	Value Range	Value Typical
Sulfur dioxide		
Dosage, mg/L per mg/L of chlorine residual		
Average	1.0–1.6	1.3
Peak flowrates	2–5	4
Rapid-mix contact time at peak flowrate, s	30–60	45
Gas withdrawal rate, lb/d		
From 150 lb containers		30
From 2000 lb containers		370
Activated carbon		
Loading rate, gal/ft$^2 \cdot$ d	3,000–4,000	3,700
Contact time, min	15–25	20

Note: lb/d $\times$ 0.4536 = kg/d
gal/ft$^2 \cdot$ d $\times$ .04075 = m^3/m$^2 \cdot$ d

to 1.47 for chlorine). As a wastewater disinfectant, bromine chloride has had only limited use.

In wastewater disinfection applications, bromine chloride is dispensed as liquified gas. The bromine chloride supply is artificially pressurized with nitrogen (or "dry air") to discharge the liquid at a constant pressure to the feed module. The liquid feeder module adds the liquid bromine chloride to a stream of dilution water producing bromine chlorine solution for application to the wastewater. A schematic process flow diagram of a typical bromine chloride system is shown in Fig. 9-32. Bromine chloride residual decreases quickly in the contact tank; therefore, good mixing of the bromine chloride solution with the wastewater is required at the application point. If bromine chloride residual is used for feed rate control, the sample line should be positioned at a location that represents about five minutes of contact time.

Disinfection with Ozone

The concentration of ozone generated from either air or pure oxygen is so low that the transfer efficiency to the liquid phase is an extremely important economic consideration. For this reason, very deep and covered contact chambers are normally used. The ozone is generally diffused from the bottom of the chamber in fine bubbles that provide mixing of the wastewater as well as achieving maximum ozone transfer and utilization. A properly designed diffuser system should normally achieve a 90

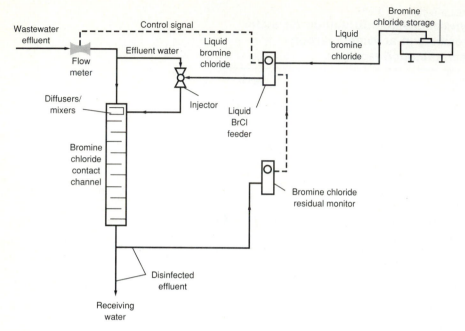

FIGURE 9-32
Bromine chloride disinfection schematic flow diagram [9].

percent transfer of ozone. The off-gasses from the contact chamber must be treated to destroy any remaining ozone as it is an extremely irritating and toxic gas. The product formed by destruction of the remaining ozone is pure oxygen, which can be recycled if pure oxygen is being used to generate the ozone. A schematic process flow diagram of a typical ozone system is shown in Fig. 9-33.

Disinfection with UV Radiation

Because no chemical agent is employed for ultraviolet disinfection, it must be considered the safest alternative disinfection system of those reviewed in Table 7-3. At present, the use of UV radiation for wastewater disinfection cannot be considered fully-proven. In some installations, scale has formed on the quartz tubes that enclose the ultraviolet lamps. The scale that builds up tends to reduce the effectiveness and reliability of the system. Unfortunately, the present mechanical wiper or sonic cleaning systems are not as effective as they might be in restoring the effectiveness of the UV tubes. A schematic process flow diagram of a typical UV disinfection system is shown in Fig. 9-34.

9-11 POST AERATION

The requirement for post-aeration systems has developed in recent years with the introduction of effluent standards and permits that include high dissolved-oxygen

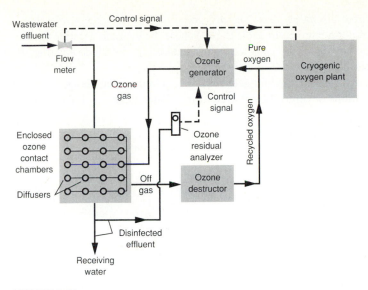

FIGURE 9-33
Ozone disinfection schematic flow diagram [9].

levels (4 to 8 mg/L). Dissolved-oxygen levels have become standard for discharge to water quality limited stream sections. The regulatory intent is to ensure that low dissolved-oxygen levels in the treated effluent do not cause immediate depression after mixture with the waters of the receiving stream. To meet post-aeration requirements, three methods are most commonly used: (1) cascade aeration, (2) mechanical aeration, and (3) diffused air.

Cascade Aeration

If site constraints and hydraulic conditions permit gravity flow, the least costly method to raise dissolved-oxygen levels is to use cascade aeration. Cascade aeration consists of using the available discharge head to create turbulence as the wastewater falls in a thin film over a series of concrete steps. Performance depends on the initial dissolved-

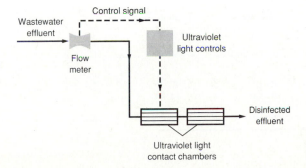

FIGURE 9-34
Ultraviolet radiation disinfection schematic flow diagram [9].

oxygen level, required discharge dissolved oxygen, and wastewater temperature. Typical design information is given in Table 9-14. Where the cascade aeration facility joins the chlorine-contact basin, the post-aeration structure may be made equal to the chlorine-contact basin width for ease of construction.

The most common method used for determining required cascade height is based on the following equations developed by Barrett at the Water Research Laboratory in England [3].

$$H = \frac{R - 1}{0.11 \, ab \, (1 + 0.46 \, T)} \qquad \text{U.S. customary units} \qquad (9\text{-}2)$$

$$H = \frac{R - 1}{0.361 \, ab \, (1 + 0.46 \, T)} \qquad \text{SI units} \qquad (9\text{-}2a)$$

where R = Deficit ratio = $\dfrac{C_s - C_o}{C_s - C}$

C_s = Dissolved-oxygen saturation concentration of the wastewater at temperature T, mg/L

C_o = Dissolved-oxygen concentration of the post-aeration influent, mg/L

C = Required final dissolved-oxygen level after post-aeration, mg/L

a = Water quality parameter equal to 0.8 for a wastewater treatment plant effluent

b = Weir geometry parameter. (For a weir, b = 1.0; for steps, b = 1.1; for step weir, b = 1.3)

T = Water temperature in °C

H = Height through which water falls, ft (m)

A key element in the use of this method is the proper selection of the critical wastewater temperature that affects the dissolved-oxygen saturation concentration, C_s. This effect is illustrated in Example 9-6.

TABLE 9-14
Typical design information for a cascade type post-aeration system

Item	Value	
	Range	Typical
Hydraulic-loading rate at average design flow, gal/ft of width · d	100,000–500,000	240,000
Step dimension, in		
Height	6–12	8
Length	12–24	18
Cascade height, ft	6–16	

Note: gal/d · d × 0.0124 = m³/m · d
ft × 0.3048 = m
in × 25.4 = mm

Example 9-6 Calculation of cascade aeration height. Calculate the height of a cascade aeration system for a wastewater treatment plant in a warm climate where the wastewater temperature averages 20°C in the winter and 25°C in the summer. The dissolved oxygen in the influent to the post-aeration system, C_o, is 1.0 mg/L and the required final dissolved-oxygen concentration, C, is 6.0 mg/L.

Solution

1. Determine the dissolved-oxygen saturation concentration, C_s, at the wastewater temperatures.
   (a) From Appendix D, the dissolved-oxygen solubilities are for 20°C = 9.17 mg/L; for 25°C = 8.38 mg/L.

2. Calculate the cascade height for T = 20°C using Eq. 9-2.
   (a) Calculate the deficit ratio.

$$R = \text{Deficit ratio} = \frac{C_s - C_o}{C_s - C} = \frac{9.17 - 1.0}{9.17 - 6.0} = 2.58$$

   (b) Calculate the cascade height, assuming steps.

$$H = \frac{R - 1}{0.11\, ab\, (1 + 0.46\, T)}$$

$$H = \frac{2.58 - 1}{0.11\, (0.8)\, (1.1)\, (1 + 0.46 \times 20)} = \frac{1.58}{0.186}$$

$$H = 8.5 \text{ ft (2.59 m)}$$

3. Calculate the cascade height for T = 25°C using Eq. 9-2.
   (a) Calculate the deficit ratio.

$$R = \text{Deficit ratio} = \frac{C_s - C_o}{C_s - C} = \frac{8.38 - 1.0}{8.38 - 6.0} = 3.10$$

   (b) Calculate the cascade height, assuming steps and using the same computation procedure as 2b above.

$$H = 10.0 \text{ ft (3.05 m)}$$

Comment. The increased wastewater temperature increases the dissolved-oxygen deficit ratio and consequently affects the height of the cascade. Therefore, the maximum wastewater temperatures should be checked so that the cascade height is not underdesigned.

Mechanical Aeration

Two major types of mechanical aeration equipment are commonly used for post-aeration systems: low-speed surface aerators and submerged turbine aerators. Low-speed surface aerators are preferred because they are usually the most economical, except where high oxygen-transfer rates are required. For high oxygen-transfer rates, submerged turbine units are preferred. For the calculation of oxygen requirements for surface aerators, refer to Chap. 10. Most installations consist of two or more aerators in rectangular basins. Detention times for post aeration using either mechanical or diffused-air aeration usually range from 10 to 20 minutes at peak flowrates.

Diffused-Air Aeration

In larger treatment plants, diffused aeration systems may be more appropriate. Nonporous and porous diffusers may be used. Depending on the depth of submergence, transfer efficiencies of 5 to 8 percent may be attained with nonporous (coarse bubble) diffusers and 15 to 25 percent with porous (fine bubble) diffusers. For the calculation of oxygen requirements for diffused-air systems, refer to Chap. 10. After secondary treatment, the alpha factors should be from 0.85 to 0.95 for coarse bubble systems and from 0.70 to 0.85 for fine bubble systems.

9-12 ODOR CONTROL

In wastewater treatment plants, the principal sources of odors are from (1) septic wastewater containing hydrogen sulfide and other odorous compounds, (2) industrial wastes discharged to the collection system, (3) screenings and unwashed grit, (4) septage-handling facilities, (5) scum on primary settling tanks, (6) organically overloaded biological treatment processes, (7) sludge-thickening tanks, (8) waste gas-burning operations where lower-than-optimum temperatures are used, (9) sludge-conditioning and dewatering facilities, (10) sludge incinerators, (11) digested sludge in drying beds or sludge-holding basins, and (12) sludge-composting operations. This section will describe some of the general approaches to odor control and provide an overview of some of the methods used to treat odors in the gaseous form. Additional information on odor control methods may be found in Refs. 1 and 15.

Approach to Odor Control

With the proper attention to design details, such as the use of submerged inlets and weirs, proper process loadings, containment of odor sources, the combustion of off-gases at proper temperatures, and good housekeeping, the routine development of odors at treatment plants can be minimized. It must also be recognized, however, that odors will occasionally develop. When they do, it is important that immediate steps be taken to control them. Often, this will involve operational changes or the addition of chemicals such as chlorine, hydrogen peroxide, lime, or ozone.

In cases where the treatment facilities are close to developed areas, it may be necessary to cover some of the treatment units such as the headworks, primary clarifiers, and sludge thickeners. Where covers are used, the trapped gases must be collected and treated. The specific method of treatment will depend on the characteristics of the odorous compounds. Buffer zones may also be effective in isolating odors from developed areas; examples of buffer distances used by New York state are presented in Table 9-15. If buffer zones are used, odor studies should be conducted that identify the type and magnitude of the odor source, meteorological conditions, dispersion characteristics, and type of adjacent development.

Where there are chronic odor problems at treatment facilities, approaches to solving these problems may include (1) operational changes to the treatment process or plant upgrading to eliminate odor sources, (2) the control of wastewater discharged to the collection system and treatment plant that creates odor problems, and (3) the

TABLE 9-15
Suggested minimum buffer distances from treatment units for odor containment[a,b]

Treatment process unit	Buffer distance, ft
Sedimentation tank	400
Trickling filter	400
Aeration tank	500
Aerated lagoon	1,000
Sludge digester (aerobic or anaerobic)	500
Sludge-handling units	
Open drying beds	500
Covered drying beds	400
Sludge-holding tank	1,000
Sludge-thickening tank	1,000
Vacuum filter	500
Wet-air oxidation	1,500
Effluent recharge bed	800
Secondary effluent filters	
Open	500
Enclosed	200
Advanced wastewater treatment	
Tertiary effluent filters	
Open	300
Enclosed	200
Denitrification	300
Polishing lagoon	500
Land disposal	500

[a] *Source:* New York State Department of Environmental Conservation.

[b] Actual buffer distance requirements depend upon a number of conditions. See text.

Note: ft × 0.3048 = m

application of chemical control to the liquid (wastewater) phase. Applying chemical or physical control to the gas phase (odor-bearing air or gas streams) is discussed in the following section.

Operational Changes. Operational changes that can be instituted can include the following: (1) reduce overloading on plant processes, (2) increase the aeration rate in biological treatment processes, (3) increase the plant treatment capacity by operating standby process units, (4) reduce the solids inventory and sludge backlog, (5) increase the frequency of pumping of sludge and scum, (6) add chlorinated dilution water to sludge thickeners, (7) reduce free-fall turbulence by controlling water levels, (7) control the release of aerosols, (8) increase the frequency of disposal of grit and screenings, and (9) clean odorous accumulations more frequently.

Control of Discharges to Collection System. Control of wastewater discharges to the collection system can be accomplished by (1) adopting more stringent waste discharge ordinances and enforcement of their requirements, (2) requiring pretreatment of industrial wastewater, and (3) providing flow equalization at the source.

Control of Odor in the Liquid Phase. The release of odors from the liquid phase may be accomplished by (1) maintaining aerobic conditions by increasing the aeration rate to add oxygen or to improve mixing or by adding hydrogen peroxide or air to long force mains; (2) controlling anaerobic microbial growth by disinfection or pH control; (3) oxidizing odorous compounds by chemical addition; and (4) controlling turbulence. For detailed information on the occurrence, effect, and control of biological transformations, the companion volume to this text may be consulted [12].

Control of Odorous Gases

The principal methods for controlling odorous gases may be classified as physical, biological, and chemical. The major methods within each category are summarized in Table 9-16. Two of the most common methods of odor treatment, chemical scrubbers and activated carbon, are illustrated in Figs. 9-35 and 9-36.

TABLE 9-16
Methods to control odorous gases found in wastewater systems[a]

Method	Description and/or application
Physical methods:	
Containment	Installation of covers, collection hoods, and air-handling equipment for containing and directing odorous gases to disposal or treatment system.
Dilution with odor-free air	Gases can be mixed with fresh air sources to reduce the odor unit values. Alternatively, gases can be discharged through tall stacks to achieve atmospheric dilution and dispersion.
Combustion	Gaseous odors can be eliminated by combustion at temperatures varying from 1200 to 1500°F (650 to 815°C). Gases can be combusted in conjunction with treatment plant solids or separately in a fume incinerator.
Adsorption, activated carbon	Odorous gases can be passed through beds of activated carbon to remove odors. Carbon regeneration can be used to reduce costs. Additional details may be found in Chap. 7.
Adsorption on sand, soil, or compost beds	Odorous gases can be passed through sand, soil, or compost beds. Odorous gases from pumping stations may be vented to the surrounding soils or to specially designed beds containing sand or soils. Odorous gases collected from treatment units may be passed through compost beds.
Oxygen injection	The injection of oxygen (either air or pure oxygen) into the wastewater to control the development of anaerobic conditions has proven to be effective.

[a] Developed in part from Ref. 15.

Improvements have been made in the design of chemical scrubbers in order to increase the efficiency of odor removal and to reduce the odor levels at discharge. Wet scrubber types include counter-current packed towers, spray chamber absorbers, and crossflow scrubbers (see Fig. 9-35). The basic objective of each type is to provide contact between air, water, and chemicals (if used) to provide oxidation or entrainment of the odorous compounds. The commonly used oxidizing scrubbing liquids are chlorine (particularly sodium hypochlorite) and potassium permanganate solutions. Sodium hydroxide is also used in systems where H_2S concentrations are high. Hypochlorite scrubbers can be expected to remove oxidizable odorous gases when other gas concentrations are minimal. Typical removal efficiencies are reported in Table 9-17. In cases where the concentrations of odorous components in the exhaust gas from the scrubbers are still above desirable levels, multistage scrubbers are often used. The steps in designing a wet scrubber system should include (1) determining the characteristics and volumes of gas to be treated, (2) defining the exhaust requirements for the treated gas, (3) selecting a scrubbing liquid based on the chemical nature and concentration of the odorous compounds to be removed, and (4) conducting pilot tests to determine design criteria and performance.

TABLE 9-16
(continued)

Method	Description and/or application
Masking agents	Perfume scents can be sprayed in fine mists near offending process units to overpower or mask objectionable odors. In some cases, the odor of the masking agent is worse than the original odor. Effectiveness of masking agents is limited.
Scrubbing towers	Odorous gases can be passed through specially designed scrubbing towers to remove odors. Some type of chemical or biological agent is usually used in conjunction with the tower.
Chemical methods:	
Scrubbing with various alkalies	Odorous gases can be passed through specially designed scrubbing towers to remove odors. If the level of carbon dioxide is high, costs may be prohibitive.
Chemical oxidation	Oxidizing the odor compounds in wastewater is one of the most common methods used to achieve odor control. Chlorine, ozone, hydrogen peroxide, and potassium permanganate are among the oxidants that have been used. Chlorine also limits the development of a slime layer.
Chemical precipitation	Chemical precipitation refers to the precipitation of sulfide with metallic salts, especially iron.
Biological methods:	
Trickling filters or activated-sludge aeration tanks	Odorous gases can be passed through trickling filters or used as process air for activated-sludge aeration tanks to remove odorous compounds.
Special biological stripping towers	Specially designed towers can be used to strip odorous compounds. Typically, the towers are filled with plastic media of various types on which biological growths can be maintained.

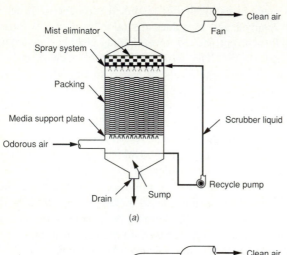

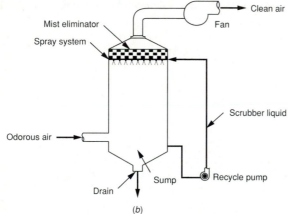

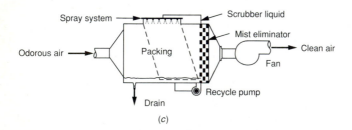

FIGURE 9-35
Typical wet scrubber systems for odor control: (a) countercurrent packed tower, (b) spray chamber absorber, and (c) crossflow scrubber [15].

TABLE 9-17
Effectiveness of hypochlorite wet scrubbers for removal of several odorous gases[a]

Gas	Expected removal efficiency, %
Hydrogen sulfide	98
Ammonia	98
Sulfur dioxide	95
Mercaptans	90
Other oxidizable compounds	70–90

[a] Ref. 15.

Activated-carbon adsorbers are commonly used for odor control (see Fig.9-36). Activated carbon has different rates of adsorption for different substances. Activated carbon may be effective in removing hydrogen sulfide and will work on reducing organic odors. It has also been found that the removal of odors depends on the concentration of the hydrocarbons in the odorous gas. It appears that the hydrocarbons are adsorbed preferentially before compounds such as H_2S are removed. The composition of the odorous gases to be treated must be defined if activated carbon is to be used.

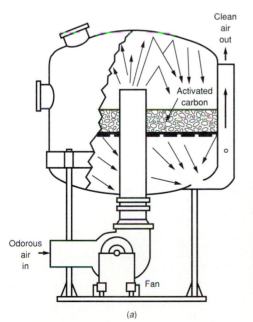

(a)

(b)

FIGURE 9-36
Typical activated-carbon system for odor control: (a) schematic and (b) typical installation.

The life of the carbon bed is limited; carbon must be regenerated or replaced regularly for continued odor removal.

Sometimes two-stage systems are used with the first stage being a wet scrubber followed by activated carbon. In selected applications, a system of this type prolongs the life of the carbon.

A method of biological odor control is the use of a soil or compost filter (see Fig. 9-37). In this system, a moistened-bulk solid medium, such as soil or composted sludge, provides the contact surfaces for microbiological reactions to oxidize odorants. Moisture content and temperature are important environmental conditions for microorganism activity. Foul air residence times are often 15 to 30 seconds or longer in these systems. Soil depths of up to 10 ft (3 m) have been used and bed-loading factors have ranged up to 2 ft^3/min per square foot of bed surface area (0.61 m^3/m^2 · min) for an H$_2$S concentration of 20 mg/L [1].

The specific method of odor control to be applied will vary with local conditions. However, because odor control measures are expensive, the cost of making process changes or modifications to the facilities to eliminate odor development should always be evaluated and compared to the cost of various alternative odor control measures before their adoption is suggested.

9-13   CONTROL OF VOCs RELEASED FROM WASTEWATER MANAGEMENT FACILITIES

The release of VOCs from wastewater management facilities was considered in Chap. 6. The purpose of this section is to consider strategies that can be used to control the release and discharge of VOCs to the atmosphere.

Control Strategies for VOCs

Volatilization and gas stripping are, as noted in Chap. 6, the principal means by which VOCs are released from wastewater treatment facilities. In general, it can be shown

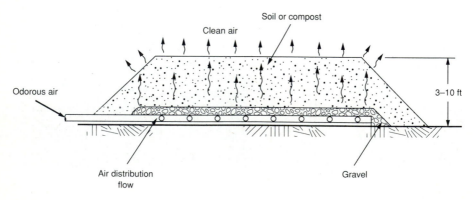

FIGURE 9-37
Soil/compost filter for odor control [1].

that the release of VOCs from open surfaces is quite low compared to the release of VOCs at points of liquid turbulence and by gas stripping. Thus, the principal strategies for controlling the release of VOCs, as reported in Table 9-18, are (1) source control, (2) elimination of points of turbulence, and (3) the covering of various treatment facilities. Two serious problems associated with the covering of treatment facilities are (1) the treatment of the off gases containing VOCs and (2) corrosion of mechanical parts. The treatment of off gases is considered in the following discussion. At this time, little information is available on how covering treatment facilities will affect the corrosion of the enclosed equipment.

Treatment of Off Gas Containing VOCs

The off gases containing VOCs from covered treatment facilities will have to be treated before they can be discharged to the atmosphere. Some options for the treatment of the off gases include (1) vapor-phase adsorption on granular activated carbon or other VOC selective resins, (2) thermal incineration, (3) catalytic incineration, (4) combustion in a flare, and (5) combustion in a boiler or process heater [17]. The

TABLE 9-18
Strategies for the control of VOCs released from wastewater management facilities

Source	Suggested control strategies
Domestic, commercial, and industrial discharges	Institute active source control program to limit the discharge of VOCs to municipal sewers.
Wastewater sewers	Seal existing manholes. Eliminate the use of structures that create turbulence and enhance volatilization.
Sewer appurtenances	Isolate and cover existing appurtenances.
Pump stations	Vent gases from wet-well to VOC treatment unit. Use variable-speed pumps to reduce the size of the wet-well.
Bar racks	Cover existing units. Reduce headloss through bar racks.
Comminutors	Cover existing units. Use in-line enclosed comminutors.
Parshall flume	Cover existing units. Use alternative measuring device.
Grit chamber	Cover existing aerated grit chambers. Reduce turbulence in conventional horizontal-flow grit chambers; cover if necessary. Avoid the use of aerated grit chambers.
Equalization basins	Cover existing units. Use submerged mixers, and reduce air flow.
Primary and secondary sedimentation tanks	Cover existing units. Replace conventional weirs with drops with submerged weirs.
Biological treatment	Cover existing units. Use submerged mixers and reduce aeration rate.
Transfer channels	Use enclosed transfer channels.
Digester gas	Controlled thermal incineration, combustion, or flaring of digester gas.

application of these processes will depend primarily on the volume of air to be treated and the types and concentrations of the VOCs contained in the air stream. The first four of these off-gas treatment processes are considered in greater detail in the following discussion.

Vapor-phase Adsorption. Adsorption is the process whereby hydrocarbons and other compounds are adsorbed selectively on the surface of such materials as activated carbon, silica gel, or alumina. Of the available adsorbants, activated carbon is used most widely. The adsorption capacity of an adsorbent for a given VOC is often represented by adsorption isotherms that relate the amount of VOC adsorbed (adsorbate) to the equilibrium pressure (or concentration) at constant temperature. Typically, the adsorption capacity increases with the molecular weight of the VOC adsorbed. In addition, unsaturated compounds are generally more completely adsorbed than saturated compounds, and cyclical compounds are more easily adsorbed than linearly structured materials. Also, the adsorption capacity is enhanced by lower operating termperatures and higher concentrations. VOCs characterized by low vapor pressures are more easily adsorbed than those with high vapor pressures [17].

Carbon adsorption is usually carried out as a semi-continuous operation involving multiple beds (see Fig. 9-38). The two main steps in the adsorption operation include adsorption and regeneration, usually performed in sequence. For control of continuous emission streams, at least one bed remains on line in the adsorption mode,

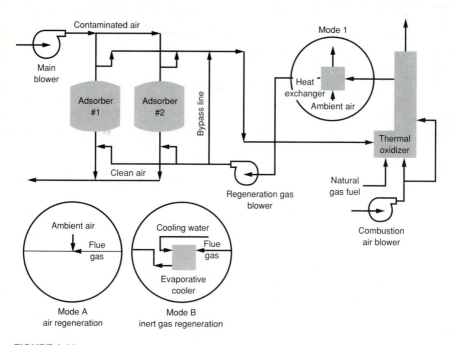

FIGURE 9-38
Gas phase carbon adsorption and regeneration system for the treatment of VOCs in off gas (from Calgon Carbon Corp.).

while the other is being regenerated. In a typical semi-continuous operation, the off gas containing VOCs is passed through the carbon bed where the VOCs are adsorbed on the bed surface. As the adsorption capacity of the bed is approached, traces of VOCs appear in the exit stream, indicating that the breakthrough point of the bed has been attained. The off gas is then directed to a parallel bed containing regenerated adsorbent, and the process continues. Concurrently, the saturated bed is regenerated by the passage of hot air (see Fig. 9-38, Mode A), hot inert gases (see Fig. 9-38, Mode B), low-pressure steam, or a combination of vacuum and hot gas. Because adsorption is a reversible process, the VOCs adsorbed on the bed can be desorbed by supplying heat (equivalent to the amount of heat released during adsorption). Small residual amounts of VOCs are always left on the carbon bed because complete desorption is technically difficult to achieve and economically impractical. Regeneration with hot air and a hot inert gas is considered in the following discussion.

Hot air regeneration is used when the VOCs are either nonflammable or have a high ignition temperature and, thus, do not pose a risk of carbon fires. A portion of the hot flue gas in the oxidizer is mixed with ambient air to cool the gas to below 350°F. The regeneration gas is driven upflow (or counter-current to adsorption flow) through the GAC adsorber. As the temperature of the carbon bed rises, the desorbed organics are transferred to the regeneration gas stream. The regeneration gas containing the desorbed VOCs is sent directly to the thermal oxidizer, where the VOCs are destroyed. After the bed has been maintained at the desired regeneration temperature for a sufficient period of time, regeneration is ended. The bed is then cooled to approximately ambient temperature by shutting off the hot regeneration gas and continuing to pass ambient air through the carbon bed. The regeneration and cooling times are predetermined based on the amount of carbon in the adsorber and the expected loading on the carbon [17].

Where the VOCs contained in the off gas include compounds, such as ketones and aldehydes, that may pose fire risks at elevated temperatures in the presence of oxygen, inert gas regeneration is used. A relatively inert gas can be obtained by passing a portion of the hot flue gas from the thermal oxidizer through an evaporative cooler. Using this technique, it is possible to keep the oxygen concentration in the regeneration gas at a low of 2 to 5 percent by volume. The desorbed VOCs are transferred along with the regeneration gas to the thermal oxidizer. A controlled amount of secondary air is added to the oxidizer. This addition of air ensures complete combustion of the VOCs but limits the excess oxygen level in the oxidizer to an acceptable range (e.g., 2 to 5 percent by volume). Regeneration is complete when the carbon bed has reached the necessary temperature for a given period of time, and VOCs are no longer being desorbed from the bed. Cooling of the bed is accomplished by increasing the water-flow rate to the evaporative cooler and reducing the regeneration gas temperature to between 220 and 250°F.

Thermal Incineration. Thermal incineration (see Fig. 9-39) is used to oxidize VOCs at high temperatures. The most important variables to consider in thermal incinerator design are the combustion temperature and residence time because these design variables determine the VOC destruction efficiency of the incinerator. Further,

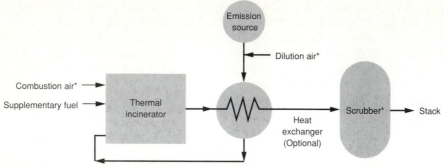

*Required for specific situations

FIGURE 9-39
Schematic diagram of a thermal incinerator system for VOCs in off gas released from treatment facilities [17].

at a given combustion temperature and residence time, destruction efficiency is also affected by the degree of turbulence, or mixing of the emission stream and hot combustion gases, in the incinerator. In addition, halogenated organics are more difficult to oxidize than unsubstituted organics; hence, the presence of halogenated compounds in the emission stream requires higher temperature and longer residence times for complete oxidation. When emission streams treated by thermal incineration are dilute (i.e., low heat content), supplementary fuel is required to maintain the desired combustion temperatures. Supplementary fuel requirements may be reduced by recovering the energy contained in the hot flue gases from the incinerator.

Catalytic Incineration. In catalytic incineration (see Fig. 9-40), VOCs in an emission stream are oxidized with the help of a catalyst. A catalyst is a substance that accelerates the rate of a reaction at a given temperature without being appreciably changed during the reaction. Catalysts typically used for VOC incineration include platinum and palladium; other formulations are also used, including metal oxides for emission streams containing chlorinated compounds. The catalyst bed (or matrix) in the incinerator is generally a metal mesh-mat, ceramic honeycomb, or other ceramic matrix structure designed to maximize catalyst surface area. The catalysts may also be in the form of spheres or pellets. Before passing through the catalyst bed, the emission stream is preheated, if necessary, in a natural gas-fired preheater [17].

The performance of a catalytic incinerator is affected by several factors including (1) operating temperature, (2) space velocity (reciprocal of residence time), (3) VOC composition and concentration, (4) catalyst properties, and (5) presence of catalyst poisons or inhibitors in the emission stream. In catalytic incinerator design, the important variables are the operating temperature at the catalyst bed inlet and the space velocity. The operating temperature for a particular destruction efficiency is dependent on the concentration and composition of the VOC in the emission stream and the type of catalyst used [17].

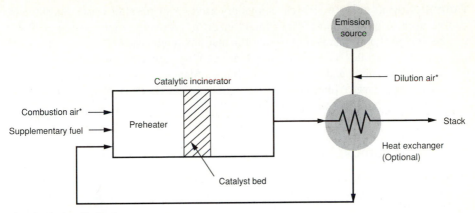

*Required for specific situations

FIGURE 9-40
Schematic diagram of a catalytic incinerator system for VOCs in off gas released from treatment facilities [17].

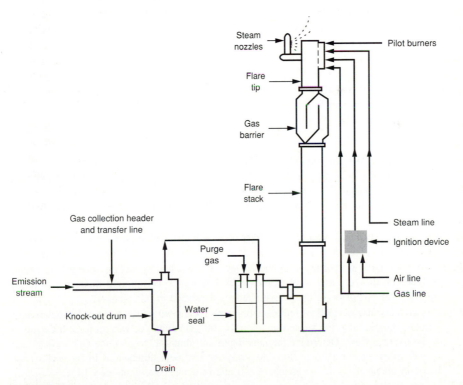

FIGURE 9-41
A typical steam-assisted flare system for VOCs in off gas released from treatment facilities [17].

Combustion in a Flare. Flares, commonly used for disposal of waste digester gas, can be used to destroy most VOCs found in off-gas streams. Flares (see Fig. 9-41) can be designed and operated to handle fluctuations in emission VOC content, inerts content, and flowrate. Several different types of flares are available including steam-assisted, air-assisted, and pressure head flares. Steam-assisted flares are employed in cases where large volumes of waste gases are released. Air-assisted flares are generally used for moderate off-gas flows. Pressure head flares are used for small gas flows.

Release to Atmosphere

A recent practice for the ultimate disposal of treated off gases is to discharge them through high stacks. Stacks as high as 100 to 130 ft have been constructed and found effective. The dispersion results in lowering the concentration of any residuals left in the gas stream.

DISCUSSION TOPICS AND PROBLEMS

9-1. A vertical bar rack with 1-in openings is used to screen wastewater arriving at the treatment plant in a circular sewer with $d = 4$ ft, $n = 0.013$, $s = 0.00064$. The maximum carrying capacity is four times the average dry-weather flow. Find the size of the steel bars that make up the rack, the number of bars in the rack, and the headloss for dry-weather flow conditions. Use rectangular bars.

9-2. Design an aerated grit chamber for a plant with an average flowrate of 4.0 Mgal/d and a peak hourly flowrate of 10.0 Mgal/d. Determine the amount of air required and the pressure at the discharge of the blowers. Allow a 10 in loss in the diffusers, and add the submergence plus 30 percent for loss in piping and valves. Determine the power required using an appropriate blower formula. Use a blower efficiency of 60 percent. Determine the monthly power bill, assuming a motor efficiency of 90 percent and a power cost of $0.08/kWh.

9-3. Design an aerated grit chamber installation for an average wastewater flowrate of 0.3 m³/s and a peak flowrate of 1.0 m³/s. The average depth is 3 m, the width-to-depth ratio is 1.5 :1, and the detention time at peak flow is 3.5 min. The aeration rate is 0.4 m³/min per m of tank length. Determine the dimensions of the grit chambers and the total air required.

9-4. Discuss the advantages and disadvantages of aerated grit chambers versus vortex-type grit chambers.

9-5. Visit your local treatment plant and review the grit and screenings operations. What methods do they use and what problems do they have? How might their operations be improved as compared to alternative methods described in this chapter?

9-6. Design a circular radial-flow sedimentation tank for a town with a projected population of 45,000. Assume that the wastewater flow is 100 gal/capita · d. Design for 2 h detention at the average flow. Determine the tank depth and diameter to produce an overflow rate of 900 gal/ft² · d for average flow. Assume standard tank dimensions to fit mechanisms made in diameters with increments of 5 ft and in depth increments of 1 ft.

9-7. A rectangular settling tank has an overflow rate of 750 gal/ft² · d and dimensions of 8 ft deep by 20 ft wide by 50 ft long. Determine whether or not particles with a diameter of 0.1 mm and a specific gravity of 2.5 will be scoured from the bottom. Use $f = 0.03$ and $k = 0.04$.

9-8. Determine the percentage increase in the hydraulic and solids-loading rates of the primary settling facilities of a treatment plant when 55,000 gal/d of settled waste activated sludge containing 2000 mg/L of suspended solids is discharged to the existing primary facilities for thickening. The average plant flowrate is 5.7 Mgal/d, and the influent suspended-solids concentration is about 350 mg/L. The design overflow rate for the primary settling tanks without the waste sludge is 800 gal/ft$^2 \cdot$ d and the detention time is 2.8 h. Do you believe that the added incremental loadings will affect the performance of the primary settling facilities? Document the basis for your answer.

9-9. Prepare a table and compare the data from a minimum of six references with regard to the following primary sedimentation tank design parameters: (1) detention time (with and without preaeration); (2) expected BOD removal; (3) expected suspended-solids removal; (4) mean horizontal velocity; (5) surface loading rate in gal/ft$^2 \cdot$ d; (6) effluent weir overflow rate per unit length; (7) Froude number; (8) size of organic particle removed; (9) length-to-width ratio (rectangular tanks); (10) average depth. List all references.

9-10. A medium-size treatment plant is being designed, and circular and rectangular primary sedimentation tanks are being considered. What factors should be considered in the evaluation and selection of the type of tank? List the advantages and disadvantages for each type. Cite at least three recent references (since 1980).

9-11. Contrast dissolved-air flotation with sedimentation discussing the following parameters:
   (*a*) Detention time
   (*b*) Surface-loading rate
   (*c*) Power input
   (*d*) Efficiency
   (*e*) Most favorable application for each type

9-12. Determine the quantity of chlorine, in lbs per day, necessary to disinfect a daily average primary effluent flow of 10 Mgal/d, and determine the size of the contact tank. Use a dosage of 16 mg/L, and size the contact chamber for a contact time of 15 min at a peak hourly flow, which is assumed to be 2.5 times the average flow.

9-13. You are called in as a consultant by the community of Rolling Hills to improve the performance of the chlorination facilities at their treatment plant. Their problem is that it has not been possible to achieve the bacterial kills called for in the discharge permit. When you arrive at the treatment facilities, the city manager proudly shows you the chlorine-contact basin designed as a complete-mix reactor. The first thing the city manager says is, "Isn't it beautiful?" What is your answer to this statement, if any, and what long-term remedies might you propose? Assume that the disinfection process can be described adequately with first-order kinetics ($r_N = -kN$).

9-14. A disinfection system needs to be designed for a large (50 Mgal/d) secondary treatment plant located near developed residential areas. Discharge from the plant is to a river, and no chlorine residual is allowed in the discharge for toxicity reasons. Three alternative disinfection systems are being considered: chlorine, sodium hypochlorite, and ozone. Describe the facilities that are required for each system, and compare the advantages and disadvantages of each. Based on technical merit, which system would you select? Justify your answer. Review of Refs. 22, 23, and 24 is suggested.

9-15. The total sulfur concentration ($H_2S = HS^- + S^=$) in a wastewater is 6 mg/L as S. Using the following expressions and data, determine the pH at which 99 percent of the total sulfur will remain in solution, assuming equilibrium conditions. If the concentration of hydrogen sulfide in the sewer atmosphere is not to exceed 2.0 ppm by volume, what pH must be maintained? In solving this problem, assume that the gas volume is equal to the liquid volume.

$$H_2S \text{ (gas)} \leftrightarrow H_2S \text{ (aq)} \qquad \frac{[H_2S]}{P_{H_2S}} = 0.1$$

$$H_2S \text{ (aq)} \leftrightarrow H^+ + HS^- \qquad \frac{[H][HS]}{[H_2S]} = 10^{-7}$$

$$HS^- \leftrightarrow H^+ + S^= \qquad \frac{[H][S]}{[HS]} = 10^{-15}$$

9-16. Based on the results of pilot plant studies, it has been found that the H_2S saturation value for activated carbon is about 0.2 lb H_2S/lb activated carbon. The same saturation value has been found to apply to the gaseous hydrocarbons found in sewer off gases.

(a) If the density of activated carbon is 34 lb/ft^3, determine the number of cubic feet of gas containing 10 ppm of H_2S by volume that can be processed per cubic foot of activated carbon.

(b) How much activated carbon would be required on an annual basis if the H_2S in the air from within a pumping station is to be removed before being discharged to the atmosphere? The pump station dry-well volume below grade is 3500 ft^3 and the air in the pump station contains 5 ppm of H_2S by volume and 100 ppm of hydrocarbons by volume (molecular weight = 100). Assume that 30 air changes per hour will be required.

REFERENCES

1. American Society of Civil Engineers: *Sulfide in Wastewater Collection and Treatment Systems,* ASCE Manuals and Reports on Engineering Practice No. 69, 1989.
2. Ball, W. P., M. D. Jones, and M. C. Kavanaugh: "Mass Transfer of Volatile Organic Compounds in Packed Tower Aeration," *Journal WPCF,* vol. 56, no. 2, 1984.
3. Barrett, M. J.: "Aeration Studies of Four Weir Systems" *Water and Wastes Engineering,* vol. 64, no. 9, 1960.
4. Camp, T. R.: "Grit Chamber Design," *J. Sewage Works,* vol. 14, no. 3, 1942.
5. Camp, T. R.: "Sedimentation and the Design of Settling Tanks," *Trans. ASCE,* vol. 111, 1946.
6. Collins, H. F.: "Effects of Initial Mixing and Residence Time Distribution on the Efficiency of the Wastewater Chlorination Process," paper presented at the California State Department of Health Annual Symposium, Berkeley and Los Angeles, CA, May 1970.
7. Culp, G. L.: "Chemical Treatment of Raw Sewage/1 and 2," *Water Wastes Eng.,* vol. 4, nos. 7 and 10, 1967.
8. Eliassen, R., and D. F. Coburn: "Pretreatment—Versatility and Expandability," presented at the ASCE Environmental Engineering Conference, Chattanooga, TN, 1968.
9. Krause, T. L., C. T. Anderson, D. R. Martenson, and J. D. Seyfert, "Disinfection: Is Chlorination Still The Best Answer," presented at the 53rd Annual Conference of the WPCF, Las Vegas, NV, September 1980.
10. Metcalf, L., and H. P. Eddy: *American Sewerage Practice,* vol. 3, 3rd ed., McGraw-Hill, New York, 1935.
11. Metcalf & Eddy, Inc.: *Wastewater Engineering: Collection, Treatment, Disposal,* McGraw-Hill, New York, 1972.
12. Metcalf & Eddy, Inc.: *Wastewater Engineering: Collection and Pumping of Wastewater,* McGraw-Hill, New York, 1981.
13. U.S. Environmental Protection Agency, *Physical-Chemical Wastewater Treatment Plant Design,* Technology Transfer Seminar Publication, 1973.
14. U.S. Environmental Protection Agency, *Process Design Manual for Upgrading Existing Wastewater Treatment Plants,* Technology Transfer, October 1974.

15. U.S. Environmental Protection Agency: *Design Manual, Odor and Corrosion Control in Sanitary Sewerage Systems and Treatment Plants,* EPA/625/1-85/018, October 1985.

16. U.S. Environmental Protection Agency: *Design Manual, Municipal Wastewater Disinfection,* EPA/625/1-86/021, October 1986.

17. U.S. Environmental Protection Agency: *Handbook: Control Technologies For Hazardous Air Pollutants,* EPA/625/6-86/014, September, 1986.

18. U.S. Environmental Protection Agency: *Design and Operational Considerations—Preliminary Treatment,* EPA 430/09-87-007, September 1987.

19. U.S. Environmental Protection Agency: *Design Manual, Phosphorus Removal,* Office of Research and Development, September 1987.

20. Water Pollution Control Federation: *Sewage Treatment Plant Design,* Manual of Practice 8, Washington, DC, 1977.

21. Water Pollution Control Federation: *Clarifier Design,* Manual of Practice FD-8, 1985.

22. Water Pollution Control Federation: *Wastewater Disinfection,* Manual of Practice FD-10, Alexandria, VA, 1986.

23. White, G. C.: *Disinfection of Wastewater and Water for Reuse,* Van Nostrand Reinhold, New York, 1978.

24. White, G. C.: *Handbook of Chlorination,* 2nd ed., Van Nostrand Reinhold, New York, 1986.

DESIGN OF FACILITIES FOR THE BIOLOGICAL TREATMENT OF WASTEWATER

Biological processes are used to convert the finely divided and dissolved organic matter in wastewater into flocculant settleable biological and inorganic solids that can be removed in sedimentation tanks. In many cases, these processes (also called "secondary processes") are employed in conjunction with the physical and chemical processes that are used for the preliminary and primary treatment of wastewater, discussed in Chap. 9. Primary sedimentation is most efficient in removing settleable solids, whereas the biological processes are most efficient in removing organic substances that are either in the colloidal size range or soluble. Some processes, however, such as aerated lagoons, stabilization ponds, and extended aeration systems, are designed to operate without primary sedimentation.

The most commonly used biological processes are (1) the activated-sludge process, (2) aerated lagoons, (3) trickling filters, (4) rotating biological contactors, and (5) stabilization ponds. The activated-sludge process, or one of its many modifications, is most often used for large installations; stabilization ponds are most often used for small installations. Typical treatment plant process flow diagrams are illustrated in Fig. 10-1. The physical facilities and the process design required for the implementation of these important processes are discussed in detail in this chapter. The use of combined aerobic biological treatment systems is also discussed briefly. The design of biological nutrient removal processes is covered in Chap. 11. The treatment and processing of sludge is considered in Chap. 12.

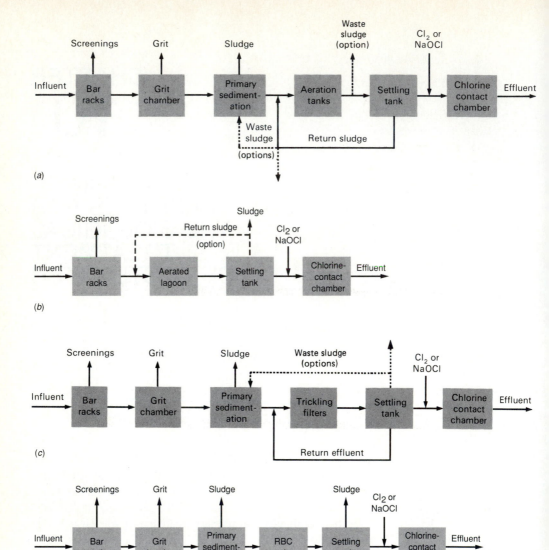

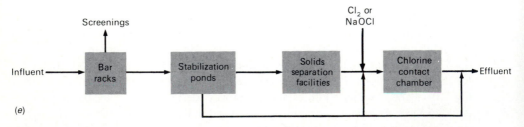

FIGURE 10-1
Typical (simplified) flow diagrams for biological processes used for wastewater treatment: (*a*) activated-sludge process, (*b*) aerated lagoons, (*c*) trickling filters, (*d*) rotating biological contactors, and (*e*) stabilization ponds.

10-1 THE ACTIVATED-SLUDGE PROCESS

The activated-sludge process has been used extensively in its original form as well as in many modified forms. Theoretical aspects of the process, including the microbiology, reaction kinetics, and to some extent operation, were discussed in Chap. 8. The practical application of this process is considered in this section and in Sec. 10-2 and 10-3.

Process Design Considerations

In the design of the activated-sludge process, consideration must be given to (1) selection of the reactor type, (2) loading criteria, (3) sludge production, (4) oxygen requirements and transfer, (5) nutrient requirements, (6) control of filamentous organisms, and (7) effluent characteristics. Because solids separation is one of the most important aspects of biological wastewater treatment, a separate discussion of this subject is provided in the next section.

Selection of Reactor Type. One of the main steps in the design of any biological process is the selection of the type of reactor or reactors (see Chap. 5) to be used in the treatment process. Operational factors that are involved include (1) reaction kinetics governing the treatment process, (2) oxygen-transfer requirements, (3) nature of the wastewater to be treated, (4) local environmental conditions, and (5) construction, operation, and maintenance costs, considered in conjunction with the secondary settling facilities. Because the relative importance of these factors will vary with each application, they should be considered separately when the type of reactor is to be selected. Their importance to the activated-sludge process is described briefly in the following discussion.

The first factor, the effect of reaction kinetics on the selection of the reactor, was illustrated in detail in Chap. 8. The two types of reactors commonly used are the complete-mix (continuous-flow stirred-tank) reactor and the plug-flow reactor. From a practical standpoint, it is interesting to note that the hydraulic detention times of many of the complete-mix and plug-flow reactors in actual use are about the same. The reason is that the combined substrate (soluble and nonsoluble) removal rate for domestic wastes is approximately zero order with respect to the concentration of the substrate. It is quasi first-order with respect to the concentration of cells.

The second factor that must be considered in the selection of reactors for the activated-sludge process is oxygen-transfer requirements. In conventional plug-flow aeration systems, it was often found that supplying sufficient oxygen to meet the requirements of the head end of the reactor was not possible. This condition led to development of the following modifications of the activated-sludge process: (1) the tapered aeration process in which an attempt is made to match the air supplied to the oxygen demand, (2) the step-feed process, where the incoming wastewater is distributed along the length of the reactor (usually at quarter points), and (3) the complete-mix process, where the air supplied uniformly matches or exceeds the oxygen demand. Most of the past oxygen-transfer limitations have been overcome by better selection of process operational parameters and improvements in the design and application of aeration equipment.

The third factor that can influence the type of reactor selected is the nature of the wastewater. For example, because the incoming wastewater is more-or-less dispersed uniformly in a complete-mix reactor, the biological solids in the reactor can, as compared to a plug-flow reactor, more easily withstand shock loads resulting from the slug discharge of organic and toxic materials to the collection system. The complete-mix process has been used in a number of installations for this reason.

The fourth factor is local environmental conditions. Of these, temperature, pH, and alkalinity are perhaps the most important. Temperature is significant because changes in the wastewater temperature can affect the biological reaction rate. For example, a decrease of 18°F (10°C) will reduce the reaction rate by about half. In most cases, temperature changes occur gradually so that modifications in the process operation can be adjusted accordingly. Where significant changes in wastewater temperatures are expected, a series of complete-mix reactors, or a plug-flow reactor whose length could be adjusted by stop gates, can be used effectively. Alkalinity and pH are also important, particularly in the operation of nitrification processes (see Chap. 11). Low pH values may inhibit the growth of nitrifying organisms (and encourage the growth of filamentous organisms) and pH adjustment may be required. Low-alkalinity wastewaters have little buffering capacity and the mixed-liquor pH may drop because of the production of carbon dioxide by bacterial respiration. Industrial waste discharges may also affect the pH in low-alkalinity wastewaters.

The fifth factor, process costs (both capital and operating and maintenance), is extremely important in selecting the type and size of reactor. It is often cost-effective to spend more on physical facilities (capital cost) to reduce ongoing operation and maintenance costs.

Loading Criteria. Over the years, a number of both empirical and rational parameters have been proposed for the design and control of the activated-sludge process. The two most commonly used parameters are (1) the food-to-microorganism ratio (F/M), and (2) the mean cell-residence time θ_c (see Chap. 8).

The food-to-microorganism ratio is defined as

$$F/M = \frac{S_o}{\theta X} \qquad (8\text{-}48)$$

where F/M = food-to-microorganism ratio, d^{-1}
S_o = influent BOD or COD concentration, mg/L (g/m^3)
θ = hydraulic detention time of the aeration tank = V/Q, d
V = aeration tank volume, Mgal (m^3)
Q = influent wastewater flowrate, Mgal/d (m^3/d)
X = concentration of volatile suspended solids in the aeration tank, mg/L (g/m^3)

The relationship of the food-to-microorganism ratio to the specific utilization rate U is

$$U = \frac{(F/M)E}{100} \qquad (8\text{-}49)$$

where E = process efficiency, %

Substituting Eq. 8-48 for the food-to-microorganism ratio and $[(S_o - S)/S_o)](100)$ for the efficiency yields

$$U = \frac{S_o - S}{\theta X} \qquad (8\text{-}45)$$

where S = effluent BOD or COD concentration, mg/L (g/m^3)

The mean cell-residence time can be defined with either of the following two general relationships, depending on the volume used:

Definition based on aeration tank volume:

$$\theta_c = \frac{V_r X}{Q_w X_w + Q_e X_e} \qquad (10\text{-}1)$$

where θ_c = mean cell-residence time based on the aeration tank volume, d
  V_r = aeration tank volume, Mgal (m^3)
  X = concentration of volatile suspended solids in the aeration tank, mg/L (g/m)^3
  Q_w = waste sludge flowrate, Mgal/d (m^3/d)
  X_w = concentration of volatile suspended solids in the waste sludge, mg/L (g/m^3)
  Q_e = treated effluent flowrate, Mgal/d (m^3/d)
  X_e = concentration of volatile suspended solids in the treated effluent, mg/L (g/m^3)

Definition based on total system volume:

$$\theta_{ct} = \left[\frac{X_t}{(Q_w X_w + Q_e X_e)}\right]\left[\frac{1}{8.34\text{lb/Mgal} \cdot (\text{mg/L})}\right] \quad \text{U.S. customary units} \qquad (10\text{-}2)$$

$$\theta_{ct} = \frac{X_t}{Q_w X_w + Q_e X_e} \qquad \text{SI units} \qquad (10\text{-}2a)$$

where θ_{ct} = mean cell-residence time based on the total system
  X_t = total mass of volatile suspended solids in the system, including the solids in the aeration tank, in the settling tank, and in the sludge-return facilities, lb (g)
Other terms are as defined in Eq. 10-1.

It is recommended that the design of the reactor be based on θ_c (Eq. 10-1) on the assumption that substantially all the substrate conversion occurs in the aeration tank. In systems where a large portion of the total solids may be present in the settling tank and sludge-return facilities, Eq. 10-2 can be used to compute the amount of solids to be wasted. The amount of solids contained in the settling tank may be determined by measuring the sludge-blanket depth and the solids concentration in the return sludge. The use of Eq. 10-2 is based on the assumption that the biological solids will undergo endogenous respiration regardless of where they are in the system under either aerobic or anaerobic conditions.

Comparing these parameters, the specific utilization rate U (F/M ratio multiplied by the efficiency) can be considered a measure of the rate at which substrate (BOD) is utilized by a unit mass of organisms, and θ_c can be considered a measure of the average residence time of the organisms in the system. The relationships between mean cell-residence time θ_c, the food-to-microorganism ratio F/M, and the specific utilization rate U is

$$\frac{1}{\theta_c} = Y\frac{F}{M}\frac{E}{100} - k_d = YU - k_d \qquad (8\text{-}46)$$

where Y = cell yield coefficient, lb cell produced per lb organic matter removed
E = process efficiency, %
k_d = endogenous decay coefficient, time^{-1}

Typical values for the food-to-microorganism ratio reported in the literature vary from 0.05 to 1.0. On the basis of laboratory studies and actual operating data from a number of different treatment plants throughout the United States, it has been found that mean cell-residence times of about 3 to 15 d result in the production of a stable, high-quality effluent and a sludge with excellent settling characteristics.

Empirical relationships based on detention time and organic loading factors have also been used. The detention time is usually based on the influent wastewater flowrate. Typically, detention times in the aeration tank range from 4 to 8 h. Organic loadings, expressed in terms of pounds of BOD$_5$ applied daily per thousand cubic feet of aeration tank volume, may vary from 20 to more than 200 (0.3 to more than 3 kg/m^3 ·d). Although the mixed-liquor concentration, the food-to-microorganism ratio, and the mean cell-residence time (which may be considered operating variables as well as design parameters) are ignored when such empirical relationships are used, these relationships do have the merit of requiring a minimum aeration tank volume that has proved adequate for the treatment of domestic wastewater. Problems have developed, however, when such relationships are used to design facilities for the treatment of wastewater containing industrial wastes.

Sludge Production. It is important to know the quantity of sludge to be produced per day because it will affect the design of the sludge-handling and disposal facilities necessary for the excess (waste) sludge. The quantity of sludge that is produced (and must be wasted) on a daily basis can be estimated by Eq. 10-3:

$$P_x = Y_{obs}Q(S_o - S) \times [8.34 \text{ lb/Mgal} \cdot (\text{mg/L})] \qquad \text{U.S. customary units} \qquad (10\text{-}3)$$

$$P_x = Y_{obs}Q(S_o - S) \times (10^3 \text{g/kg})^{-1} \qquad \text{SI units} \qquad (10\text{-}3a)$$

where P_x = net waste activated sludge produced each day, measured in terms of volatile suspended solids, lb/d (kg/d)
Y_{obs} = observed yield, lb/lb (g/g)
Q, S_o, S = as defined previously

The observed yield can be computed using Eq. 8-44:

$$Y_{obs} = \frac{Y}{1 + k_d(\theta_c \text{ or } \theta_{ct})} \qquad (8\text{-}44)$$

The use of θ_c or θ_{ct} in Eq. 8-44 depends on whether the solids in the aeration tank or the solids in the total system are considered in the analysis. If a high percentage of the solids is retained in the settling tank and sludge return facilities, the use of θ_{ct} is reasonable, especially if it is assumed that endogenous respiration goes on regardless of whether the bacterial culture is in an aerobic or anaerobic environment. However, it should be noted that the value of the constant would be different from the values reported in the literature. Because no suitable value exists at the present time for a combined-aerobic-anaerobic k_d, the aerobic value can be used as an estimate.

Oxygen Requirements and Transfer. The theoretical oxygen requirements can be determined from the BOD_5 of the waste and the amount of organisms wasted from the system per day. The reasoning is as follows. If all the BOD_5 were converted to end products, the total oxygen demand would be computed by converting BOD_5 to BOD_L, using an appropriate conversion factor. It is known that a portion of the waste is converted to new cells subsequently wasted from the system; therefore, if the BOD_L of the wasted cells is subtracted from the total, the remaining amount represents the amount of oxygen that must be supplied to the system. From Eq. 8-31, given below, it is known that the BOD_L of one mole of cells is equal to 1.42 times the concentration of cells.

$$C_5H_7NO_2 + 5O_2 \rightarrow 5CO_2 + 2H_2O + NH_3 + \text{ energy} \qquad (8\text{-}31)$$

$$\begin{array}{cc} 113 & 5(32) \\ \text{cells} & \\ 1 & 1.42 \end{array}$$

Therefore, the theoretical oxygen requirements for the removal of the carbonaceous organic matter in wastewater for an activated-sludge system can be computed as

$$\text{lb } O_2/d = \left(\begin{array}{c}\text{total mass of } BOD_L \\ \text{utilized, lb/d}\end{array}\right) - 1.42\left(\begin{array}{c}\text{mass of organisms} \\ \text{wasted, lb/d}\end{array}\right) \qquad (10\text{-}4)$$

In terms that have been defined previously,

$$\text{lb } O_2/d = \frac{Q(S_o - S) \times 8.34}{f} - 1.42(P_x) \qquad \text{U.S. customary units} \quad (10\text{-}5)$$

$$\text{kg, } O_2/d = \frac{Q(S_o - S) \times (10^3 \text{g/kg})^{-1}}{f} - 1.42(P_x) \qquad \text{SI units} \qquad (10\text{-}5a)$$

where f = conversion factor for converting BOD_5 to BOD_L
  8.34 = conversion factor, $[\text{lb/Mgal} \cdot (\text{mg/L})]$
Other terms are as defined previously.

When nitrification has to be considered, the total oxygen requirements can be computed as the lb O_2/d for removal of carbonaceous organic matter plus the lb O_2/d required for nitrogen conversion (from ammonia to nitrate), as follows:

$$\text{lb } O_2/d = \frac{Q(S_o - S) \times 8.34}{f} - 1.42(P_x)$$
$$+ 4.57Q(N_o - N) \times 8.34 \qquad \text{U.S. customary units} \qquad (10\text{-}6)$$

$$\text{kg, } O_2/d = \frac{Q(S_o - S) \times (10^3 \text{g/kg})^{-1}}{f} - 1.42(P_x)$$
$$+ 4.57Q(N_o - N) \times (10^3 \text{g/kg})^{-1} \qquad \text{SI units} \qquad (10\text{-}6\text{a})$$

where N_o = influent TKN, mg/L (g/m^3)
$\quad\quad\quad N$ = effluent TKN, mg/L (g/m^3)
$\quad\quad 4.57$ = conversion factor for amount of oxygen required for complete oxidation of TKN

Then, if the oxygen-transfer efficiency of the aeration system is known or can be estimated, the actual air requirements may be determined. The air supply must be adequate to (1) satisfy the BOD of the waste, (2) satisfy the endogenous respiration by the sludge organisms, (3) provide adequate mixing, and (4) maintain a minimum dissolved-oxygen concentration of 1 to 2 mg/L throughout the aeration tank.

For food-to-microorganism ratios greater than 0.3, the air requirements for the conventional process amount to 500 to 900 ft^3/lb (30 to 55 m^3/kg) of BOD$_5$ removed for coarse bubble (nonporous) diffusers and 400 to 600 ft^3/lb (24 to 36 m^3/kg) for fine bubble (porous) diffusers. The characteristics of air diffusers are described in Sec. 10-2. At lower food-to-microorganism ratios, endogenous respiration, nitrification, and prolonged aeration periods increase air use to 1200 to 1800 ft^3/lb (75 to 115 m^3/kg) of BOD$_5$ removed. In the *Ten States Standards* [14], the normal air requirements for all activated-sludge processes, except extended aeration, are 1500 ft^3/lb BOD$_5$ (93.5 m^3/kg BOD$_5$) for peak aeration tank loading. For the extended aeration process, normal air requirements are 2000 ft^3/lb BOD$_5$ (125 m^3/kg BOD$_5$).

For diffused-air aeration, the amount of air used has commonly ranged from 0.5 to 2.0 ft^3/gal (3.75 to 15.0 m^3/m^3) at different plants, with 1.0 ft^3/gal (7.5 m^3/m^3) an early rule-of-thumb design factor. Because the air use depends on the strength of the wastewater, the air-to-wastewater ratio has become a quantity derived for recordkeeping purposes and is not used as a basic design criterion. A similar early rule of thumb for mechanical aeration systems is 1.0 to 1.2 lb O$_2$/lb BOD$_5$ removed [63].

To meet the sustained peak organic loadings discussed in Chap. 5, it is recommended that the aeration equipment be designed with a safety factor of at least two times the average BOD load. Aeration equipment should also be sized based on a residual dissolved oxygen of 2 mg/L at the average load and 0.5 mg/L at peak load. The *Ten States Standards* [14] require the air diffusion system to be capable of providing oxygen to meet the diurnal peak oxygen demand or 200 percent of the design average, whichever is larger.

Nutrient Requirements. If a biological system is to function properly, nutrients must be available in adequate amounts. As discussed in Chaps. 3 and 8, the principal nutrients are nitrogen and phosphorus. Based on an average composition of cell tissue of $C_5H_7NO_2$, about 12.4 percent by weight of nitrogen will be required. The phosphorus requirement is usually assumed to be about one-fifth of this value. These are typical values, not fixed quantities, because it has been shown that the percentage distribution of nitrogen and phosphorus in cell tissue varies with the age of the cell and environmental conditions.

Other nutrients required by most biological systems are reported in Table 10-1. The inorganic composition of *E. coli* is shown in Table 10-2. The data in Table 10-2 can be used to estimate the concentration of trace elements required for the maintenance of proper biological growth. Because the total amount of nutrients required will depend on the net mass of organisms produced, nutrient quantities will be reduced for processes operated with long mean cell-residence times. This fact can often be used to explain why two similar activated-sludge plants operated at different θ_c may not perform the same way when treating the same waste. The role of trace elements is discussed in greater detail in Ref. 68.

Control of Filamentous Organisms. The growth of filamentous microorganisms is the most common operational problem in the activated-sludge process. A proliferation of filamentous organisms in the mixed liquor results in poorly settling sludge, commonly termed "bulking sludge." The single-stage complete-mix system in particular tends to promote the growth of filamentous organisms because of the low-substrate levels uniformly present in the reactor. In some plug-flow reactors where significant back-mixing occurs, a similar phenomenon takes place. Recent research has focused on the factors influencing the growth of filamentous organisms and practical methods of control. One concept that has gained recognition for the prevention and control of filamentous organism growth is the use of a separate compartment, or

TABLE 10-1
Inorganic ions necessary for most organisms[a]

Substantial quantities	Trace quantities
Sodium (except for plants)	Iron
Potassium	Copper
Calcium	Manganese
Phosphate	Boron (required by plants and certain protists)
Chloride	
Sufate	Molybdenum (required by plants, certain protists, and animals)
Bicarbonate	
	Vanadium (required by certain protists and animals)
	Cobalt (required by certain animals, protists, and plants)
	Iodine (required by certain animals)
	Selenium (required by certain animals)

[a] Ref. 20.

TABLE 10-2
Inorganic composition of E. coli[a]

Element	Percentage of dry-cell weight
Potassium	1.5
Calcium	1.4
Sodium	1.3
Magnesium	0.54
Chloride	0.41
Iron	0.2
Manganese	0.01
Copper	0.01
Aluminum	0.01
Zinc	0.01

[a] Ref. 23

"selector," as the initial contact zone of a biological reactor where the primary effluent and return activated sludge are combined. The selector may be used in combination with the complete-mix activated-sludge process or in a plug-flow reactor and may consist of a separate tank or a sectionalized compartment.

The selector concept entails the selective growth of floc-forming organisms at the initial stage of the biological process by providing a high food-to-microorganism (F/M) ratio at controlled dissolved-oxygen levels. An F/M ratio of at least 2.27 lb BOD_5/lb MLSS · d is suggested [2]. Initial F/M ratio ranging as high as 20–25 lb COD/lb MLVSS · d have also been reported [64]. The high substrate-driving force permits the rapid adsorption of the soluble organics into the floc-forming organisms. The rapid removal of the soluble organics leaves very little available for subsequent assimilation by the filamentous organisms. Good results have been obtained in the selector zones that are aerated or unaerated, anoxic or anaerobic, or in an alternating environment [2]. Sufficient air should be provided to ensure adequate mixing of the selector compartment contents, or mechanical mixers should be used.

The contact time in the selector is relatively short, commonly ranging from 10 to 30 min. Pilot plant testing is highly recommended to define the design parameters. In a selector that is too small, a significant amount of the influent soluble substrate will pass into the main aeration basin. In a selector that is too large, the influent soluble substrate is diluted, resulting in too low an F/M ratio [58]. Examples of bench and pilot plant testing may be found in Refs. 11 and 22. Additional discussion about bulking sludge is provided later in this section.

Effluent Characteristics. Organic content is a major parameters of effluent quality. The organic content of effluent from biological treatment processes is usually composed of the following three constituents:

1. Soluble biodegradable organics
   a. Organics that escaped biological treatment
   b. Organics formed as intermediate products in the biological degradation of the waste
   c. Cellular components (result of cell death or lysis)
2. Suspended organic material
   a. Biological solids produced during treatment that escaped separation in the final settling tank
   b. Colloidal organic solids in the plant influent that escaped treatment and separation
3. Nonbiodegradable organics
   a. Those originally present in the influent
   b. By-products of biological degradation

The kinetic equations developed in Chap. 8 for the effluent quality theoretically apply only to the soluble organic waste that escaped biological treatment. Clearly, this is only a portion of the organic waste concentration in the effluent. In a well-operating activated-sludge plant that is treating domestic wastes, the soluble carbonaceous BOD_5 in the effluent, determined on a filtered sample, will usually vary from 2 to 10 mg/L. Suspended organic material will range from 5 to 15 mg/L, and nonbiogradable organics will range from 2 to 5 mg/L.

Types of Processes and Modifications

The activated-sludge process is very flexible and can be adapted to almost any type of biological waste treatment problem. Several of the conventional activated-sludge processes and some of the modifications that have become standardized are described in Table 10-3. The operational characteristics, application, and typical removal efficiencies for these processes are listed in Table 10-4; design parameters are shown in Table 10-5.

Process Control

Control of the activated process is important to maintain high levels of treatment performance under a wide range of operating conditions. The principal factors used in process control are (1) maintaining dissolved-oxygen levels in the aeration tanks, (2) regulating the amount of return activated sludge (RAS), and (3) controlling the waste activated sludge (WAS). As discussed previously in "Loading Criteria," the most commonly used parameters for controlling the activated-sludge process are the F/M ratio and the mean cell-residence time, θ_c. The mixed-liquor suspended-solids (MLSS) concentration is also used as a control parameter. Return activated sludge is important in maintaining the MLSS concentration (the "M" in the F/M ratio), and the WAS is important in the controlling θ_c. The use of oxygen uptake rates (OUR) is also receiving recognition as a means of monitoring and controlling the

TABLE 10-3
Description of activated-sludge processes and process modifications

Process or process modification	Description	See Figure
Conventional plug-flow	Settled wastewater and recycled activated sludge enter the head end of the aeration tank and are mixed by diffused-air or mechanical aeration. Air application is generally uniform throughout tank length. During the aeration period, adsorption, flocculation, and oxidation of organic matter occurs. Activated-sludge solids are separated in a secondary settling tank.	10-2
Complete-mix	Process is an application of the flow regime of a continuous-flow stirred-tank reactor. Settled wastewater and recycled activated sludge are introduced typically at several points in the aeration tank. The organic load on the aeration tank and the oxygen demand are uniform throughout the tank length.	10-3
Tapered aeration	Tapered aeration is a modification of the conventional plug-flow process. Varying aeration rates are applied over the tank length depending on the oxygen demand. Greater amounts of air are supplied to the head end of the aeration tank, and the amounts diminish as the mixed liquor approaches the effluent end. Tapered aeration is usually achieved by using different spacing of the air diffusers over the tank length.	
Step-feed aeration	Step feed is a modification of the conventional plug-flow process in which the settled wastewater is introduced at several points in the aeration tank to equalize the F/M ratio, thus lowering peak oxygen demand. Generally three or more parallel channels are used. Flexibility of operation is one of the important features of this process.	10-4
Modified aeration	Modified aeration is similar to the conventional plug-flow process except that shorter aeration times and higher F/M ratios are used. BOD removal efficiency is lower than other activated-sludge processes.	
Contact stabilization	Contact stabilization uses two separate tanks or compartments for the treatment of the wastewater and stabilization of the activated sludge. The stabilized activated sludge is mixed with the influent (either raw or settled) wastewater in a contact tank. The mixed liquor is settled in a secondary settling tank and return sludge is aerated separately in a reaeration basin to stabilize the organic matter. Aeration volume requirements are typically 50 percent less than conventional plug flow.	10-5

TABLE 10-3
(continued)

Process or process modification	Description	See Figure
Extended aeration	Extended aeration process is similar to the conventional plug-flow process except that it operates in the endogenous respiration phase of the growth curve, which requires a low organic loading and long aeration time. Process is used extensively for prefabricated package plants for small communities (see Chap. 14).	
High-rate aeration	High-rate aeration is a process modification in which high MLSS concentrations are combined with high volumetric loadings. This combination allows high F/M ratios and long mean cell-residence times with relatively short hydraulic detention times. Adequate mixing is very important.	
Kraus process	Kraus process is a variation of the step aeration process used to treat wastewater with low nitrogen levels. Digester supernatant is added as a nutrient source to a portion of the return sludge in a separate aeration tank designed to nitrify. The resulting mixed liquor is then added to the main plug-flow aeration system.	
High-purity oxygen	High-purity oxygen is used instead of air in the activated-sludge process. Oxygen is diffused into covered aeration tanks and is recirculated. A portion of the gas is wasted to reduce the concentration of carbon dioxide. pH adjustment may also be required. The amount of oxygen added is about four times greater than the amount that can be added by conventional aeration systems.	10-6
Oxidation ditch	The oxidation ditch consists of a ring- or oval-shaped channel and is equipped with mechanical aeration devices. Screened wastewater enters the ditch, is aerated, and circulates at about 0.8 to 1.2 ft/s (0.25 to 0.35 m/s). Oxidation ditches typically operate in an extended aeration mode with long detention and solids retention times. Secondary sedimentation tanks are used for most applications.	10-7
Sequencing batch reactor	The sequencing batch reactor is a fill-and-draw type reactor system involving a single complete-mix reactor in which all steps of the activated-sludge process occur. Mixed liquor remains in the reactor during all cycles, thereby eliminating the need for separate secondary sedimentation tanks.	8-21

TABLE 10-3
(continued)

Process or process modification	Description	See Figure
Deep shaft reactor	The deep vertical shaft reactor is a form of the activated-sludge process. A vertical shaft about 400 to 500 ft (120 to 150 m) deep replaces the primary clarifiers and aeration basin. The shaft is lined with a steel shell and fitted with a concentric pipe to form an annular reactor. Mixed liquor and air are forced down the center of the shaft and allowed to rise upward through the annulus.	10-8
Single-stage nitirification	In single-stage nitrification, both BOD and ammonia reduction occur in a single biological stage. Reactor configurations can be either a series of complete-mix reactors or plug-flow. More details on single-stage nitrification are given in Chap. 11.	
Separate stage nitrification	In separate stage nitrification, a separate reactor is used for nitrification, operating on a feed waste from a preceding biological treatment unit. The advantage of this system is that operation can be optimized to conform to the nitrification needs. More details are given in Chap. 11.	

activated-sludge process. A brief description of OUR monitoring is also provided in the following discussion.

Dissolved-Oxygen Control. The amount of oxygen transferred in the aeration tanks theoretically equals the amount of oxygen required by the microorganisms in the activated-sludge system (including the secondary clarifiers and return sludge lines) to oxidize the organic material and to maintain residual dissolved-oxygen operating

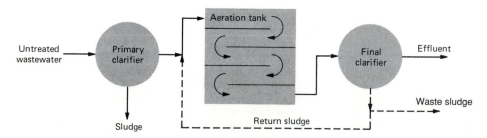

FIGURE 10-2
Typical schematic for a conventional plug-flow activated process.

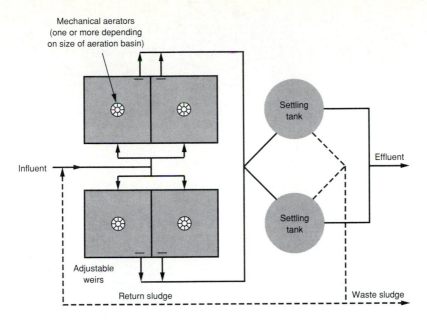

FIGURE 10-3
Complete-mix activated-sludge process (typical schematic for four-cell process).

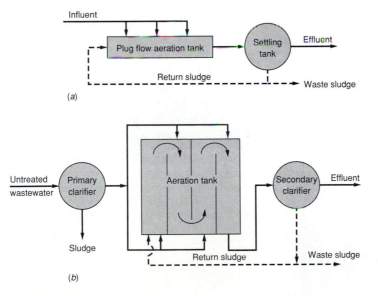

FIGURE 10-4
Flow diagram for step-feed aeration activated-sludge process: (a) simplified schematic and (b) typical physical configuration.

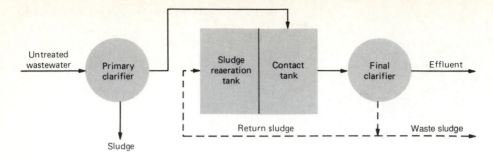

FIGURE 10-5
Flow diagram for contact stabilization activated-sludge process.

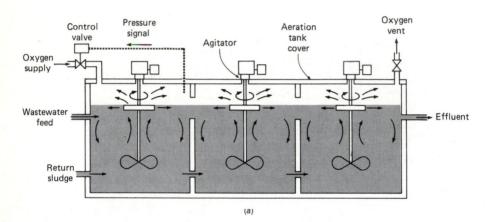

(a)

(b)

(c)

FIGURE 10-6
High-purity oxygen activated-sludge process: (a) schematic of three-stage configuration, (b) view of high-purity oxygen generation unit, and (c) typical mixer drive unit and oxygen injection point.

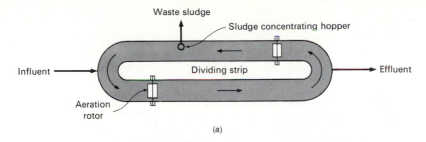

(a)

(b)

FIGURE 10-7
Oxidation ditch activated-sludge process: (a) schematic of oxidation ditch process and (b) aerial view of oxidation ditch process (from Envirex Inc.).

levels. When oxygen limits the growth of microorganisms, filamentous organisms may predominate, and the settleability and quality of the activated sludge may be poor (see further discussion on "Bulking Sludge"). In practice, the dissolved-oxygen concentration in the aeration tank should be maintained at about 1.5 to 4 mg/L in all areas of the aeration tank; 2 mg/L is a commonly used value. Values above 4 mg/L do not improve operations significantly, but increase the aeration costs considerably [61].

Return Activated-Sludge Control. The purpose of the return of activated sludge is to maintain a sufficient concentration of activated sludge in the aeration tank so that the required degree of treatment can be obtained in the time interval desired. The return of activated sludge from the final clarifier to the inlet of the aeration tank is

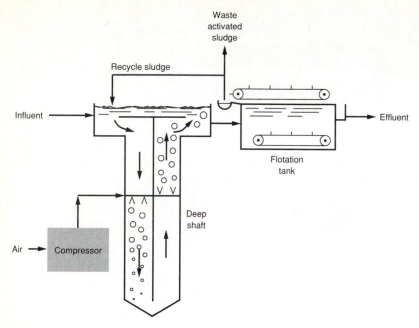

FIGURE 10-8
Schematic diagram of deep shaft activated-sludge reactor [64].

the essential feature of the process. Ample return sludge pump capacity should be provided. It is also important to prevent the loss of sludge solids in the effluent. The solids form a sludge blanket in the bottom of the clarifier. The sludge blanket varies in thickness from time to time and may fill the entire depth of the clarifier at peak flows, if the return sludge-pumping capacity is inadequate. Return sludge-pumping capacities of 50 to 100 percent of the wastewater flowrate are normally provided for large plants and up to 150 percent of the wastewater flowrate for small plants.

Several techniques are used to calculate the desirable return-sludge flowrate. The control stategies are based on either maintaining a target MLSS level in the aeration tanks or a given sludge-blanket depth in the final clarifiers. The most commonly used techniques are (1) settleability, (2) sludge-blanket level control, (3) secondary clarifier mass balance, (4) aeration tank mass balance, and (5) sludge quality [61].

Using the settleability test, the return sludge-pumping rate is set so that the flowrate is approximately equal to the percentage ratio of the volume occupied by the settleable solids from the aeration tank effluent to the volume of the clarified liquid (supernatant) after settling for 30 min in a 1000 mL graduated cylinder. This ratio should not be less than 15 percent at any time. For example, if the settleable solids occupied a volume of 275 mL after 30 min of settling, the percentage volume would be equal to 38 percent [(275 mL/725 mL) × 100]. If the plant flow were 46 Mgal/d (2 m³/s), the return-sludge rate should be 0.38 × 46 Mgal/d = 17.5 Mgal/d (0.76 m³/s).

Another settleability test method often used to control the rate of return-sludge pumping is based on an empirical measurement known as the "sludge-volume index"

(SVI). This index is defined as the volume in milliliters occupied by one gram of activated-sludge mixed-liquor solids, dry weight, after settling for 30 min in a 1000 mL graduated cylinder. In practice, it is taken to be the percentage volume occupied by the sludge in a mixed-liquor sample (taken at the outlet of the aeration tank) after 30 min of settling, O_v, divided by the suspended-solids concentration of the mixed liquor expressed as a percentage, P_w. If the sludge-volume index is known, then the percentage of return sludge, in terms of the recirculation ratio Q_r/Q required to maintain a given percentage of mixed-liquor solids concentration in the aeration tank is $100Q_r/Q = 100/[(100/P_w\text{SVI}) - 1]$. For example, to maintain a mixed-liquor solids concentration of 0.3 percent (3000 mg/L), the percentage of sludge that must be returned when the sludge volume index is 100 is equal to $100/[(100/0.30 \times 100) - 1]$, or 43 percent.

With the sludge-blanket level control method, an optimum sludge-blanket level is maintained in the clarifiers. The optimum level is determined by experience and is a balance between efficient settling depth and sludge storage. The optimum depth of the sludge blanket usually ranges between 1 to 3 ft (0.3 to 0.9 m). This method of control requires considerable operator attention because of the diurnal flow and sludge production variations and changes in the settling characteristics of the sludge. Several methods are available to detect the sludge-blanket levels including air lift pumps, gravity-flow tubes, portable sampling pumps, core samplers, and sludge-supernatant interface detectors. Additional details may be found in Ref. 61.

The return sludge-pumping rate may also be determined by making a mass-balance analysis around either the settling tank or the aeration tank. The appropriate limits for the two mass-balance analyses are illustrated in Fig. 10-9. Assuming that the sludge-blanket level in the settling tank remains constant and that the solids in the effluent from the settling tank are negligible, the mass balance around the settling tank is as follows:

$$\text{accumulation} = \text{inflow} - \text{outflow}$$

$$0 = X(Q + Q_r)(8.34) - X_r Q_r(8.34) + X_r Q'_w(8.34)$$

where X = mixed-liquor suspended solids, mg/L
Q = secondary influent flow, Mgal/d
Q_r = return sludge flow, Mgal/d
X_r = return activated-sludge suspended solids, mg/L
Q'_w = waste sludge flow, Mgal/d
8.34 = conversion factor, [lb/Mgal · (mg/L)]

Solving for Q_r yields

$$Q_r = \frac{XQ - X_r Q'_w}{X_r - X} \qquad (10\text{-}7)$$

The required RAS pumping rate can also be estimated by performing a mass balance around the aeration tank. If new cell growth is considered negligible, then the solids entering the tank will equal the solids leaving the tank. Under conditions such as high organic loadings, this assumption may be incorrect. Solids enter the aeration

TABLE 10-4
Operational characteristics of activated-sludge processes

Process modification	Flow model	Aeration system	BOD removal efficiency, %	Remarks
Conventional	Plug-flow	Diffused-air, mechanical aerators	85–95	Use for low-strength domestic wastes. Process is susceptible to shock loads.
Complete-mix	Continuous-flow stirred-tank reactor	Diffused-air, mechanical aerators	85–95	Use for general application. Process is resistant to shock loads, but is susceptible to filamentous growths.
Step-feed	Plug-flow	Diffused-air	85–95	Use for general application for a wide range of wastes
Modified aeration	Plug-flow	Diffused-air	60–75	Use for intermediate degree of treatment where cell tissue in the effluent is not objectionable.
Contact stabilization	Plug-flow	Diffused-air, mechanical aerators	80–90	Use for expansion of existing systems and package plants.
Extended aeration	Plug-flow	Diffused-air, mechanical aerators	75–95	Use for small communities, package plants, and where nitrified element is required. Process is flexible.
High-rate aeration	Continuous-flow stirred-tank reactor	Mechanical aerators	75–90	Use for general applications with turbine aerators to transfer oxygen and control floc size.
Kraus process	Plug-flow	Diffused-air	85–95	Use for low-nitrogen, high-strength wastes.

			BOD removal efficiency, %	Remarks
High-purity oxygen	Continuous-flow stirred-tank reactors in series	Mechanical aerators (sparger turbines)	85–95	Use for general application with high-strength waste and where limited space is available at site. Process is resistant to slug loads.
Oxidation ditch	Plug-flow	Mechanical aerators (horizontal axis type)	75–95	Use for small communities or where large area of land is available. Process is flexible.
Sequencing batch reactor	Intermittent-flow stirred-tank reactor	Diffused-air	85–95	Use for small communities where land area is limited. Process is flexible and can remove nitrogen and phosphorus.
Deep shaft reactor	Plug-flow	Diffused-air	85–95	Use for general application with high-strength wastes. Process is resistant to slug loads.
Single-stage nitrification	Continuous-flow stirred-tank reactors or plug-flow	Mechanical aerators, diffused-air	85–95	Use for general application for nitrogen control where inhibitory industrial wastes are not present.
Separate stage nitrification	Continuous-flow stirred-tank reactors or plug-flow	Mechanical aerators, diffused-air	85–95	Use for upgrading existing systems, where nitrogen standards are stringent, or where inhibitory industrial wastes are present and can be removed in earlier stages.

TABLE 10-5
Design parameters for activated-sludge processes

Process modification	θ_c, d	F/M, lb BOD5 applied/ lb MLVSS·d	Volumetric loading, lb BOD5/ 10^3ft³·d	MLSS, mg/L	V/Q, h	Q_r/Q
Conventional	5–15	0.2–0.4	20–40	1,500–3,000	4–8	0.25–0.75
Complete-mix	5–15	0.2–0.6	50–120	2,500–4,000	3–5	0.25–1.0
Step-feed	5–15	0.2–0.4	40–60	2,000–3,500	3–5	0.25–0.75
Modified aeration	0.2–0.5	1.5–5.0	75–150	200–1,000	1.5–3	0.05–0.25
Contact stabilization	5–15	0.2–0.6	60–75	(1,000–3,000)[a] (4,000–10,000)[b]	(0.5–1.0)[a] (3–6)[b]	0.5–1.50
Extended aeration	20–30	0.05–0.15	10–25	3,000–6,000	18–36	0.5–1.50
High-rate aeration	5–10	0.4–1.5	100–1,000	4,000–10,000	2–4	1.0–5.0
Kraus process	5–15	0.3–0.8	40–100	2,000–3,000	4–8	0.5–1.0
High-purity oxygen	3–10	0.25–1.0	100–200	2,000–5,000	1–3	0.25–0.5
Oxidation ditch	10–30	0.05–0.30	5–30	3,000–6,000	8–36	0.75–1.50
Sequencing batch reactor	N/A	0.05–0.30	5–15	1,500–5,000[d]	12–50	N/A
Deep shaft reactor	NI	0.5–5.0	NI	NI	0.5–5	NI
Single-stage nitrification	8–20	0.10–0.25 (0.02–0.15)[c]	5–20	2,000–3,500	6–15	0.50–1.50
Separate stage nitrification	15–100	0.05–0.20 (0.04–0.15)[c]	3–9	2,000–3,500	3–6	0.50–2.00

[a] Contact unit.

[b] Solids stabilization unit.

[c] TKN/MLVSS.

[d] MLSS varies depending on the portion of the operating cycle.

Note: lb/10^3 ft³·d × 0.0160 = kg/m³·d

lb/lb·d = kg/kg·d

N/A = not applicable

NI = no information

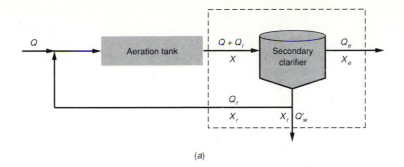

(a)

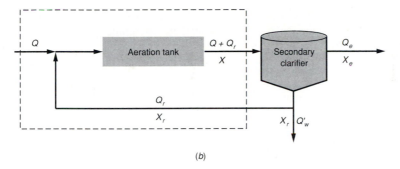

(b)

FIGURE 10-9
Typical suspended solids mass balances for return-sludge control: (a) secondary clarifier mass balance and (b) aeration tank mass balance [61].

tank in the return sludge and in the influent to the secondary process. However, because the influent solids are negligible compared to the MLSS, the mass balance around the aeration tank results in the following expression.

$$\text{accumulation} = \text{inflow} - \text{outflow}$$

$$0 = X_r Q_r (8.34) - X(Q + Q_r)(8.34)$$

Solving for Q_r yields

$$Q_r = Q\frac{X}{X_r - X} \tag{10-8}$$

Still another approach to return sludge control entails determining the sludge settling characteristics. Sludge settleability curves are developed from which return sludge rates are determined [61].

Sludge Wasting. The excess activated sludge produced each day must be wasted to maintain a given food-to-microorganism ratio or mean cell-residence time. The most common practice is to waste sludge from the return sludge line because it is more concentrated and requires smaller waste sludge pumps. The waste sludge is discharged to the primary tanks, to thickening tanks, or to other sludge-thickening facilities. An alternative method can be used to withdraw mixed liquor directly from the aeration tank or from the aeration tank effluent pipe, where the concentration of solids is uniform. The waste mixed liquor can then be discharged to a sludge-thickening tank or to the primary tanks, where it mixes and settles with the untreated primary sludge.

The actual amount of liquid that must be pumped to achieve process control depends on the method used and the location from which the wasting is to be accomplished. (Also, because the solids capture of the sludge-processing facilities is not 100 percent and some solids are returned, the actual wasting rate will be higher than the theoretically determined value. This subject is considered further in Chap. 12.) For example, if the mean cell-residence time is used for process control and wasting is from the sludge return line, the wasting rate can be computed using Eq. 8-35.

$$\theta_c = \frac{V_r X}{(Q'_w X_r + Q_e X_e)} \tag{8-35}$$

where Q'_w = waste sludge flowrate from the return line, Mgal/d
$\quad\; X_r$ = concentration of sludge in the return line, mg/L
Other terms are as defined in Eq. 10-1.

If it is assumed that the concentration of solids in the effluent from the settling tank is low, then Eq. 8-35 reduces to

$$\theta_c \approx \frac{V_r X}{Q'_w X_r} \tag{8-47}$$

and

$$Q'_w \approx \frac{V_r X}{\theta_c X_r} \tag{10-9}$$

To determine the waste sludge flowrate using Eq. 10-9, the solids concentration in both the aeration tank and return line must be known.

If the mean cell-residence time is used for process control, wasting is from the aeration tank, and the solids in the plant effluent are again neglected, then the rate of pumping can be estimated using the following relationship:

$$\theta_c \approx \frac{V_r}{Q_w} \tag{10-10}$$

or

$$Q_w \approx \frac{V_r}{\theta_c} \tag{10-11}$$

where Q_w = waste sludge flowrate from the aeration tank, Mgal/d.

Thus, the process may be controlled by daily wasting of a quantity of flow equal to the volume of the aeration tank divided by the mean cell-residence time.

If the food-to-microorganism method of control is adopted, the wasting flowrate from the return line can be determined using the following relationship:

$$P_x = Q_w X_r (8.34) \tag{10-12}$$

where P_x = waste activated sludge, lb/d
Q_w = waste sludge flowrate, Mgal/d
X_r = solids concentration in the return line, mg/L
8.34 = conversion factor, [lb/Mgal · (mg/L)]

In this case, the concentration of solids in the sludge return line must be known.

If process control is based on one of the other loading criteria, the quantity of solids to be wasted must be established by successive trials.

Oxygen-Uptake Rates. Microorganisms in the activated-sludge process use oxygen as they consume food. The rate at which they use oxygen, the oxygen-uptake rate (OUR), can be taken as a measure of the biological activity. High OURs indicate high biological activity; low OURs indicate low biological activity. The value of OUR is obtained by taking a sample of mixed liquor, saturated with DO, and with a DO probe measuring the decrease in DO with time. The results are typically reported as mg O_2/L · min or mg O_2/L · h[61].

Oxygen uptake is most valuable for plant operations when combined with VSS data. The combination of the OUR with the concentration of MLVSS yields a value termed "specific oxygen-uptake rate" (SOUR), or respiration rate. SOURs indicate the amount of oxygen used by microorganisms and are reported as mg O_2/g MLVSS · h. Based on recent research, it appears that the mixed-liquor SOUR and the final effluent COD can be correlated, thereby allowing predictions of final effluent quality to be made during transient loading conditions [16].

Operational Problems

The most common problems encountered in the operation of an activated-sludge plant are bulking sludge, rising sludge, and Nocardia foam. Because few plants have escaped these problems, it is appropriate to discuss their nature and methods for their control. For descriptions of other operating problems that occur at activated-sludge plants, Ref. 61 may be consulted.

Bulking Sludge. A bulking sludge is one that has poor settling characteristics and poor compactability. Two principal types of sludge-bulking problems have been identified. One is caused by the growth of filamentous organisms or organisms that can grow in a filamentous form under adverse conditions. This is the predominant form of bulking. The other is caused by bound water, in which the bacterial cells composing the floc swell through the addition of water to the extent that their density is reduced and they will not settle. The causes of sludge-bulking that are most commonly cited in the literature are related to (1) the physical and chemical characteristics of the wastewater, (2) treatment plant design limitations, and (3) plant operation.

Wastewater characteristics that can affect sludge bulking include fluctuations in flow and strength, pH, temperature, staleness, nutrient content, and the nature of the waste components. Design limitations include air supply capacity, clarifier design, return sludge-pumping capacity limitations, short circuiting, or poor mixing. Operational causes of filamentous bulking include low dissolved oxygen in the aeration tank, insufficient nutrients, widely varying organic waste loading, low food-to-microorganism (*F/M*) ratio, and insufficient soluble BOD_5 gradient. Operating causes of nonfilamentous bulking are improper organic loading, overaeration, or the presence of toxics [61]. In almost all cases, all of the aforementioned conditions represent some sort of adverse operating condition.

In the control of bulking, where a number of variables are possible causes, a checklist of things to investigate is valuable. The following items are recommended: (1) wastewater characteristics, (2) dissolved-oxygen content, (3) process loading, (4) return sludge-pumping rate, (5) process microbiology, (6) internal plant overloading, and (7) clarifier operation.

The nature of the components found in wastewater or the absence of certain components, such as trace elements, can lead to the development of a bulked sludge [68]. If it is known that industrial wastes are being introduced into the system either intermittently or continuously, the quantity of nitrogen and phosphorus in the wastewater should be checked first because limitations of both or either are known to favor bulking. Wide fluctuations in pH are also known to be deterimental in plants of conventional design. Wide fluctuations in organic waste loads due to batch-type operations can also lead to bulking and should be checked.

Limited dissolved oxygen has been noted more frequently than any other cause of bulking. If the problem is due to limited oxygen, it can usually be confirmed by operating the aeration equipment at full capacity. Under these conditions, the aeration equipment should have adequate capacity to maintain at least 2 mg/L of dissolved oxygen in the aeration tank under normal loading conditions. If this level of oxygen cannot be maintained, the solution to the problem may require the installation of improvements to the existing aeration system.

The *F/M* ratio should be checked to make sure that it is within the range of generally accepted values (see Table 10-5). Low *F/M* ratios, particularly in complete-mix systems, may encourage the growth of certain types of filamentous organisms. High *F/M* may result in the presence of small dispersed floc, a condition that can be remedied by increasing the sludge-wasting rate. When plant operation is controlled based on mean cell-residence time, the *F/M* ratio need not be checked. The mean cell-residence time should be checked to make sure it is within the range normally found to provide efficient treatment (Table 10-5). If it is not within the range given in this table, the sludge-wasting rate should be adjusted as discussed previously.

If bulking is caused by filamentous microorganisms, the types of organisms present should be identified so that a proper solution can be undertaken. More than 20 different morphological types of filamentous organisms have been found in activated sludge [12,13]. Typical characteristics of filamentous types of organisms and their frequency of occurrence in U.S. plants may be found in Ref. 42. Depending upon the environmental conditions, different types of filamentous organisms may proliferate.

The use of indicator organisms associated with a specific bulking problem is also suggested in Ref. 42. Identification of organisms should be done by microbiologists or technicians skilled in wastewater examination. The prevention and control of the growth of filamentous organisms in complete-mix systems has been accomplished effectively by the addition of a selector compartment, as discussed previously.

To avoid internal plant overloading, recycle loads should be controlled so they are not returned to the plant flow during times of peak hydraulic and organic loading. Examples of recycle loads are centrate or filtrate from sludge-dewatering operations and supernatant from sludge digesters.

The operating characteristics of the clarifier may also be a cause of sludge bulking. Bulking is often a problem in center-feed circular tanks where sludge is removed from the tank directly under the point where the mixed liquor enters the tank. Examination of the sludge blanket may show that a large part of the sludge is actually retained in the tank for many hours rather than the desired 30 min. If this is the case, then the design is at fault, and changes must be made in the sludge withdrawal equipment.

In an emergency situation or while the aforementioned factors are being investigated, chlorine and hydrogen peroxide may be used to provide temporary help. Chlorination of return sludge has been practiced quite extensively as a means of controlling bulking. Although chlorination is effective in controlling bulking caused by filamentous growths, it is ineffective when bulking is due to light floc containing bound water. Chlorination of return sludge in the range of 2 to 3 mg/L of Cl_2 per 1000 mg/L of MLVSS is suggested, with dosages of 8 to 10 mg/L per 1000 mg/L in severe cases [61]. Chlorination normally results in the production of a turbid effluent until such time as the sludge is free of the filamentous forms. Chlorination of a nitrifying sludge will also produce a turbid effluent because of the death of the nitrifying organisms. Hydrogen peroxide has also been used in the control of filamentous organisms in bulking sludge. Dosage of hydrogen peroxide and treatment time depend on the extent of the filamentous development.

Rising Sludge. Occasionally sludge that has good settling characteristics will be observed to rise or float to the surface after a relatively short settling period. The cause of this phenomenon is denitrification, in which the nitrites and nitrates in the wastewater are converted to nitrogen gas (see Chap. 11). As nitrogen gas is formed in the sludge layer, much of it is trapped in the sludge mass. If enough gas is formed, the sludge mass becomes buoyant and rises or floats to the surface. Rising sludge can be differentiated from bulking sludge by noting the presence of small gas bubbles attached to the floating solids.

Rising sludge problems may be overcome by (1) increasing the return activated-sludge withdrawal rate from the clarifier to reduce the detention time of the sludge in the clarifier, (2) decreasing the rate of flow of aeration tank mixed liquor into the offending clarifier if the sludge depth cannot be reduced by increasing the return activated-sludge withdrawal rate, (3) where possible, increasing the speed of the sludge-collecting mechanism in the settling tanks, and (4) decreasing the mean cell-residence time by increasing the sludge-wasting rate.

Nocardia Foam. A viscous, brown foam that covers the aeration basins and secondary clarifiers has produced many problems in activated-sludge plants, including safety hazards, deteriorated effluents, and odors. The foam is associated with a slow-growing filamentous organism of the actinomycete group, usually of the *Nocardia* genus. Some of the probable causes of the foaming problem are (1) low *F/M* in the aeration tanks, (2) buildup of a high mixed-liquor suspended-solids concentration (thereby increasing the sludge age) due to insufficient sludge wasting, and (3) operation in the sludge reaeration mode [61]. Higher air flowrates necessary to meet high MLSS concentrations will tend to expand the foam and worsen the foaming problem. Measures for *Nocardia* control include (1) reducing sludge age, (2) reducing the air flowrate to lower the depth of foam accumulation, (3) adding a selector compartment to control the growth of filamentous organisms, (4) injecting a mutant bacterial additive, (5) chlorinating the return sludge, (6) spraying chlorine solution or sprinkling powdered calcium hypochlorite directly onto the foam, and (7) reducing the pH in the mixed liquor by chemical addition or by initiating nitrification [39]. Reducing the sludge age is the method that has been used most commonly for *Nocardia* control.

10-2 SELECTION AND DESIGN OF PHYSICAL FACILITIES FOR ACTIVATED-SLUDGE PROCESS

The physical facilities used in the design of activated-sludge treatment systems are discussed in this section. These facilities include (1) diffused-air aeration, (2) mechanical aerators, (3) high-purity oxygen, (4) aeration tanks and appurtenances, and (5) solids-separation facilities.

Diffused-Air Aeration

The two basic methods of aerating wastewater are (1) to introduce air or pure oxygen into the wastewater with submerged diffusers or other aeration devices or (2) to agitate the wastewater mechanically so as to promote solution of air from the atmosphere. A diffused-air system consists of diffusers submerged in the wastewater, header pipes, air mains, and the blowers and appurtenances through which the air passes. The selection of diffusers and the design of blowers and air-piping are considered in the following discussion. For an extensive review of recent (1989) information on fine pore aeration systems, Ref. 57 may be consulted.

Diffusers. In the past, the various diffusion devices have been classified as either fine-bubble or coarse-bubble, with the connotation that fine bubbles were more efficient in transferring oxygen. The definition of terms and the demarcation between fine and coarse bubbles, however, have not been clear. Therefore, the current preference is to categorize the diffused-aeration systems by the physical characteristics of the equipment. Three categories are defined: (1) porous or fine pore diffusers, (2) nonporous diffusers, and (3) other diffusion devices such as jet aerators, aspirating aerators, and U-tube aerators. The various types of diffused-air devices are described in Table 10-6 and shown schematically in Fig. 10-10.

TABLE 10-6
Description of air diffusion devices[a]

Type of diffuser or device	Transfer efficiency	Description	See Figure
Porous			
Plate	High	Square ceramic plates installed in fixed holders on tank floor.	
Dome	High	Dome-shaped ceramic diffusers mounted on air distribution pipes near tank floor.	10-10a
Disc	High	Rigid ceramic discs or flexible porous membrane mounted on air distribution pipes near tank floor.	10-10b
Tube	Moderate to high	Tubular-shaped diffuser that uses rigid ceramic media or flexible plastic or synthetic rubber sheath mounted on air distribution pipes.	10-10c
Nonporous			
Fixed orifice			
Perforated piping	Low	Air distribution piping with small holes drilled along the length.	
Spargers	Low	Devices usually constructed of molded plastic and mounted on air distribution pipes.	10-10d
Slotted tube	Low	Stainless steel tubing containing perforations and slots to provide a wide band of diffused air.	
Valved orifice	Low	Device that contains a check valve to prevent backflow when air is shut off. Mounts on air distribution piping.	10-10e
Static tubes	Low	Stationary vertical tube mounted on basin bottom that functions like an air lift pump.	10-10f
Perforated hose	Low	Perforated hose that runs lengthwise along basin and is anchored to the floor.	
Other devices			
Jet aeration	Moderate to high	Device that discharges a mixture of pumped liquid and compressed air through a nozzle assembly located near the tank bottom.	10-10g
Aspirating	Low	Inclined propeller pump assembly mounted at basin surface that draws in air and discharges air/water mixture below water surface.	10-10h
U-tube	High	Compressed air is discharged into the down leg of a deep vertical shaft.	10-10i

[a] Adapted from Ref. 63.

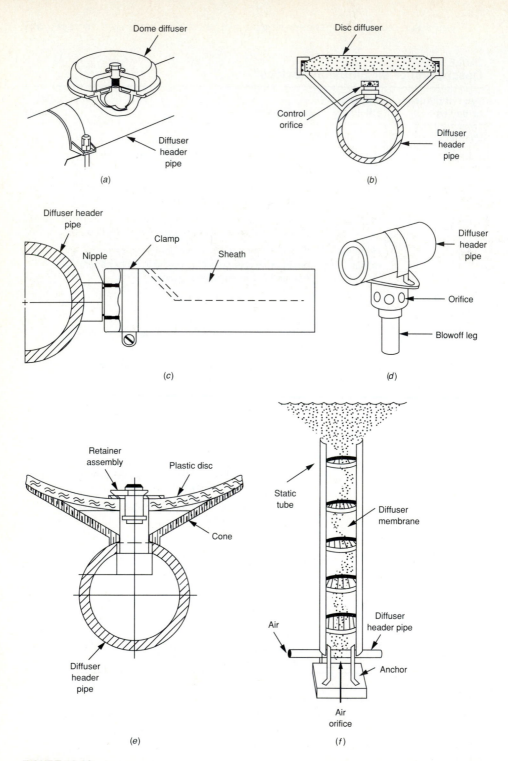

FIGURE 10-10
Typical diffused-air aeration devices: (a) dome diffuser, (b) disc diffuser, (c) tubular diffuser, (d) sparger, (e) valved orifice diffuser, and (f) static tube aerator.

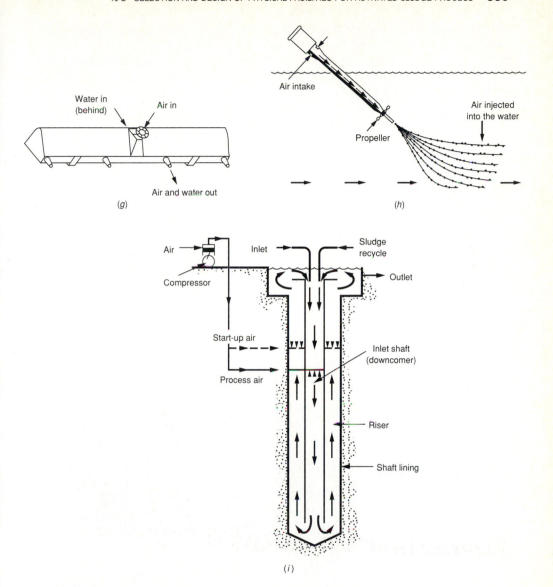

FIGURE 10-10 *(continued)*
(*g*) Jet aerator, (*h*) aspirating device, and (*i*) U-tube aerator.

Porous diffusers. Porous diffusers are made in many shapes, the most common being plates, domes, discs, and tubes (Fig. 10-10*a*, *b*, *c*). Plates were once the most popular, but are costly to install and difficult to maintain. Plate diffusers are installed in concrete or aluminum plate holders, supporting six or more plates, which may be set either in recesses or on the bottom of the aeration tank. Groups of plate holders are connected to the air supply piping at intervals along the tank length, and each group is controlled by a valve. Porous domes, discs, and tubes have largely supplanted plates in newer installations. Domes, discs, or tube diffusers are mounted on or screwed into air manifolds, which may run the length of the tank close to the bottom and along one side; or short manifold headers may be mounted on movable drop pipes on one side of the tank. With the movable drop pipes, it is possible to raise a header out of the water without interrupting the process and without dewatering the tank. The diffusers can then be removed for cleaning or replacement. Dome and disc diffusers may also be installed in a grid pattern on the bottom of the aeration tank to provide uniform aeration throughout the tank (see Fig. 10-11).

A number of materials have been used in the manufacturing of porous diffusers. These materials generally fall into the categories of rigid ceramic and plastic materials, flexible plastic, and flexible plastic, rubber, or cloth sheaths. The ceramic materials consist of rounded or irregular shaped mineral particles bonded together to produce a network of interconnecting passsageways through which compressed air flows. As the air emerges from the surface pores, pore size, surface tension, and air flowrate interact to produce a bubble of given size. Porous plastic materials are more recent developments. Similar to the ceramic materials, the plastics contain a number of interconnecting channels or pores through which the compressed air can pass. Thin, flexible sheaths made from soft plastic or synthetic rubber have also been developed

FIGURE 10-11
Plug-flow aeration tank equipped with dome aeration devices (from Aerocor Co).

and adapted to discs and tubes. Air passages are created by punching minute holes in the sheath material. When the air is turned on, the sheath expands and each slot acts as a variable aperture opening; the higher the air flowrate, the greater the opening.

It is essential that the air supplied be clean and free of dust particles that might clog the porous diffusers. Air filters, often consisting of viscous impingement and dry-barrier types, are commonly used. Precoated bag filters and electrostatic filters have also been used. The filters should be installed on the blower inlet.

Nonporous diffusers. Several types of nonporous diffusers are available (see Fig. 10-10*d*, *e*, *f*). The fixed and valved orifice diffusers produce larger bubbles than porous diffusers and consequently have somewhat lower aeration efficiency; but the advantages of lower cost, less maintenance, and the absence of stringent air-purity requirements offset the slightly lower efficiency. Typical system layouts for the fixed and valved orifice diffusers closely parallel the layouts for porous dome and disc diffusers.

In the static tube aerator (Fig. 10-10*f*), air is introduced at the bottom of a circular tube that can vary in height from 1.5 to 4.0 ft (0.5 to 1.25 m). Internally, the tubes are fitted with alternately placed deflection plates to increase the contact of the air with the wastewater. Mixing is accomplished because the tube aerator acts as an airlift pump. Static tubes are normally installed in a grid-type floor coverage pattern.

Other diffuser types. Other types of diffused-aeration systems include jet, aspirating, and U-tube aeration. Jet aeration (see Fig. 10-10*g*) combines liquid pumping with air diffusion. The pumping system recirculates liquid in the aeration basin, ejecting it with compressed air through a nozzle assembly. This system is particularly suited for deep (25 ft) tanks. Aspirating aeration (Fig. 10-10*h*) consists of a motor-driven aspirator pump. The pump draws air in through a hollow tube and injects it underwater where both high velocity and propeller action create turbulence and diffuse the air bubbles. The aspirating device can be mounted on a fixed structure or on pontoons. U-tube aeration consists of a deep shaft divided into two zones (Fig. 10-10*i*). Air is added to the influent wastewater in the downcomer under high pressure; the mixture travels to the bottom of the tube and then back to the surface. The great depth to which the air-water mixture is subjected results in high oxygen-transfer efficiencies because the high pressure forces all the oxygen into solution. U-tube aeration has particular application for high-strength wastes.

Diffuser Performance. The efficiency of oxygen transfer depends on many factors including the type, size, and shape of the diffuser; the air flowrate; the depth of submersion; tank geometry including the header and diffuser location; and wastewater characteristics. Aeration devices are conventionally evaluated in clean water and the results adjusted to process operating conditions through widely used conversion factors. Typical clean water transfer efficiencies and air flowrates for various diffused-air devices are reported in Table 10-7. Typically, the standard oxygen-transfer efficiency (SOTE) increases with depth; the transfer efficiencies shown in Table 10-7 are shown for the 15 ft (4.57 m) depth, the most common depth of submergence. Data on the variation of SOTE with water depth for various diffuser types can be found in Ref. 63. The variation of oxygen-transfer efficiencies with the type of diffuser and diffuser arrangement are illustrated in Table 10-7. Additional data on the effects of diffuser arrangement on transfer efficiency are reported in Refs. 55 and 57.

Oxygen-transfer efficiency (OTE) of porous diffusers may also decrease with use due to internal clogging or exterior fouling. Internal clogging may be due to impurities in the compressed air that have not been removed by the air filters. External fouling may be due to the formation of biological slimes or inorganic precipitants. The effect of fouling on OTE is described by the term F. The rate at which F decreases with time is designated as f_F, which is expressed as the decimal fraction of OTE lost per unit time. The rate of fouling will depend on the operating conditions, changes in

TABLE 10-7
Typical information on the clean water oxygen-transfer efficiency of various diffuser systems[a]

Diffuser type and placement	Air flowrate, ft³/min · diffuser	SOTE (%) at 15 ft submergence[b]
Ceramic discs—grid	0.4–3.4	25–40
Ceramic domes—grid	0.5–2.5	27–39
Ceramic plates—grid	2.0–5.0[c]	26–33
Rigid porous plastic tubes		
Grid	2.4–4.0	28–32
Dual spiral roll	3.0–11.0	17–28
Single spiral roll	2.0–12.0	13–25
Nonrigid porous plastic tubes		
Grid	1.0–7.0	26–36
Single spiral roll	2.0–7.0	19–37
Perforated membrane tubes		
Grid	1.0–4.0	22–29
Quarter points	2.0–6.0	19–24
Single spiral roll	2.0–6.0	15–19
Jet aeration		
Side header	54.0–300	15–24
Nonporous diffusers		
Dual spiral roll	3.3–10.0	12–13
Mid-width	4.2–45.0	10–13
Single spiral roll	10.0–35.0	9–12

[a] Adapted from Refs. 57 and 63.

[b] SOTE = standard oxygen-transfer efficiency. Standard conditions: tap water, 68°F, at 14.7 lb$_f$/in² and initial dissolved oxygen = 0 mg/L.

[c] Units are ft³/ft² of diffuser · min.

Note: lb$_f$/in² × 6.8948 = kN/m²
      ft³/min × 0.0283 = m³/min
        ft × 0.3048 = m
        $\frac{5}{9}$ (°F − 32) = °C

wastewater characteristics, and the time in service. The fouling rates are important in determining the loss of OTE and the expected frequency of diffuser cleaning. Fouling and the rate of fouling can be estimated by (1) conducting full-scale OTE tests over a period of time, (2) monitoring aeration system efficiency, and (3) conducting OTE tests on fouled and new diffusers [57].

The factors commonly used to convert the oxygen transfer required for clean water to wastewater are the alpha, beta, and theta factors as described in Chap. 6. The alpha factor, the ratio of the $K_L a$ of wastewater to the $K_L a$ of clean water (see Chap. 6), is especially important because the alpha factor varies with the physical features of the diffuser system, the geometry of the reactor, and the characteristics of the wastewater. Wastewater constituents may affect porous diffuser oxygen-transfer efficiencies to a greater extent than other aeration devices, resulting in lower alpha factors [17]. The presence of constituents such as detergents, dissolved solids, and suspended solids can affect the bubble shape and size and result in diminished oxygen-transfer capability. Values of alpha varying from 0.4 to 0.9 have been reported for fine-bubble diffuser systems [18]. Therefore, considerable care must be exercised in the selection of the appropriate alpha factors.

Another measure of the performance of porous diffusers is the combination of the alpha and fouling factors, designated by the term αF. It has been found from a number of in-process studies that αF values have ranged widely, from 0.11 to 0.79 with a mean of < 0.5, and were significantly lower than anticipated [57]. The variability of αF was found to be site specific, and demonstrated the need for the designer to investigate and evaluate carefully the environmental factors that may affect porous diffuser performance in selecting an appropriate α or αF factor.

Because the amount of air used per pound of BOD removed varies greatly from one plant to another, there is risk in comparing the air use at different plants, not only because of the factors mentioned above but also because of different loading rates, control criteria, and operating procedures. Extra-high air flowrates applied along one side of a tank reduce the efficiency of oxygen transfer and may even reduce the net oxygen transfer by increasing circulating velocities. The result is a shorter residence time of air bubbles as well as larger bubbles with less transfer surface.

Methods of cleaning porous diffusers may consist of the refiring of ceramic plates, high-pressure water sprays, brushing, or chemical treatment in acid or caustic baths. Additional details on cleaning methods may be found in Refs. 57 and 63.

Blowers. There are two types of blowers in common use: centrifugal and rotary-lobe positive displacement (see Fig. 10-12). Centrifugal turbines have also been used, especially in Europe. Centrifugal blowers are commonly used where the unit capacity is greater than 3,000 ft³/min (85 m³/min) of free air. At the lower capacities, turndown of the blower should be checked to ensure air requirements can be met at low flow conditions. Rated discharge pressures range normally from 7 to 9 lb$_f$/in² (48 to 62 kN/m²). Centrifugal blowers emit a high-pitched whine unless inlet and outlet silencers are installed.

In wastewater treatment plants, the blowers must supply a wide range of air flows with a relatively narrow pressure range under varied environmental conditions.

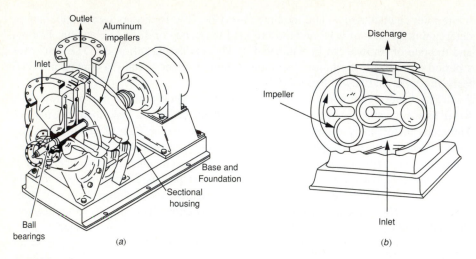

FIGURE 10-12
Typical blowers used in diffused-air systems: (a) centrifugal blower (from Hoffman) and (b) positive displacement (from Roots-Connersville).

Usually, a blower can only meet one particular set of operating conditions efficiently. Because it is necessary to meet a wide range of air flows and pressures at a wastewater treatment plant, provisions have to be included in the blower system design to regulate or turndown the blowers. Methods to achieve regulation or turndown are (1) flow blowoff or bypassing, (2) inlet throttling, (3) adjustable discharge diffusers, (4) variable speed drivers, and (5) parallel operation of multiple units. Inlet throttling and an adjustable discharge diffuser are only applicable to centrifugal blowers; variable speed drivers are more commonly used on positive displacement blowers. Flow blowoff and bypassing is also an effective method of controlling the surging of a centrifugal blower, a phenomenon that occurs when the blower alternately operates at zero capacity and full capacity, resulting in vibration and overheating. Surging occurs when the blower operates in a low volumetric range.

Centrifugal blowers have operating characteristics similar to a low specific-speed centrifugal pump. The discharge pressure rises from shutoff to a maximum at about 50 percent of capacity and then drops off. The operating point of the blower is determined, similar to a centrifugal pump, by the intersection of the head-capacity curve and the system curve. Blowers are rated at standard air conditions, defined as a temperature of 20°C (68°F), a pressure of 14.7 lb_f/in^2 (760 mm Hg), and a relative humidity of 36 percent. Standard air has a specific weight of 0.0750 lb/ft^3 (1.20 kg/m^3). The air density affects the performance of a centrifugal blower, and any change in the inlet air temperature or barometric pressure will change the density of the compressed air. The greater the gas density, the higher the pressure will rise. As a result, greater power is needed for the compression process (see Fig. 10-13). (Typical values for the specific weight of ambient air are presented in Appendix B.) Blowers

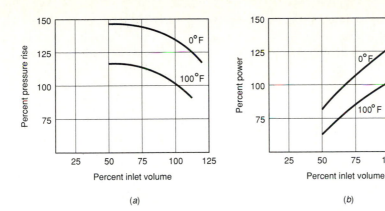

FIGURE 10-13

Characteristic curves for a centrifugal blower at various inlet air temperatures: (a) percent pressure rise versus percent inlet volume and (b) percent power increase versus percent inlet volume.

must be selected for adequate capacity for a hot summer day and be provided with a driver with adequate power for the coldest winter weather. The power requirement for adiabatic compression is given in Eq. 10-13.

$$P_w = \frac{wRT_1}{550ne}\left[\left(\frac{p_2}{p_1}\right)^{0.283} - 1\right] \qquad \text{U.S. customary units} \qquad (10\text{-}13)$$

$$P_w = \frac{wRT_1}{29.7ne}\left[\left(\frac{p_2}{p_1}\right)^{0.283} - 1\right] \qquad \text{SI units} \qquad (10\text{-}13a)$$

where P_w = power requirement of each blower, hp (kW)
  w = weight of flow of air, lb/s (kg/s)
  R = engineering gas constant for air, 53.3 ft·lb/(lb air)·°R
        (U.S. customary units)(Note: °R = °F + 460)
     = 8.314 kJ/k mol °K (SI units)
  T_1 = absolute inlet temperature, °R (°K)
  p_1 = absolute inlet pressure, lb$_f$/in^2 (atm)
  p_2 = absolute outlet pressure, lb$_f$/in^2 (atm)
  n = $(k - 1)/k$ = 0.283 for air
  k = 1.395 for air
  550 = ft·lb/s·hp
  29.7 = constant for SI units conversion
  e = efficiency (usual range for compressors is 0.70 to 0.90

For higher discharge pressure applications (> 8 lb$_f$/in^2) and for capacities smaller than 3,000 ft^3/min (85 m^3/min) of free air per unit, rotary-lobe positive displacement blowers are commonly used. Positive displacement blowers are also used when significant water level variations are expected. The positive displacement blower is a machine of constant capacity with variable pressure. The units cannot be throttled, but capacity control can be obtained by the use of multiple units or a variable speed drive. Rugged inlet and discharge silencers are essential.

Air Piping. Air piping consists of mains, valves, meters, and other fittings used to transport compressed air from the blowers to the air diffusers. Because the pressures are low (less than 10 lb_f/in^2), lightweight piping can be used. Piping is usually sized on a velocity basis. The piping should be sized so that losses in tank headers and diffuser manifolds are small in comparison to the losses in the diffusers. Valves should be provided for flow regulation. Typical air velocities are given in Table 10-8.

Because of the high temperature of the air discharged by blowers (140 to 180°F), condensation in the air piping is not a problem except where piping is submerged in the wastewater. It is essential, however, that provisions be made for pipe expansion and contraction. Where porous diffusers are used, pipes must be made of nonscaling materials or must be lined with noncorrosive materials. Pipe materials are often stainless steel, fiberglass, or plastics suitable for higher temperatures. Other materials used include mild steel or cast iron with external coatings (e.g., coal-tar epoxy or vinyl). Interior surfaces include cement lining or coal-tar or vinyl coatings.

Piping losses should be computed for maximum summer temperatures. The theoretical adiabatic temperature rise during compression is

$$\Delta T_{ad} = T_1 \left[\left(\frac{p_2}{p_1} \right)^n - 1 \right] \tag{10-14}$$

where ΔT_{ad} = adiabatic temperature rise °R (°K)
Other terms are as defined in Eq. 10-13.

The actual temperature rise is approximated by dividing ΔT_{ad} by the blower efficiency. Between the blowers and the aeration tanks, air temperatures will probably drop not more than 10 to 20°F but will quickly approach the temperature of the wastewater in the submerged piping.

TABLE 10-8
Typical air velocities in header pipes

Pipe diameter, in	Velocity, ft/min[a]
1–3	1,200–1,800
4–10	1,800–3,000
12–24	2,700–4,000
30–60	3,800–6,500

[a] At standard conditions.

Note: in × 25.4 = mm
ft/min × 0.3048 = m/min

Friction losses in air piping can be calculated using the Darcy-Weisbach equation written in the following form:

$$h_L = f \frac{L}{D} h_i \qquad (10\text{-}15)$$

where h_L = friction loss, in of water

f = dimensionless friction factor obtained from Moody diagram (Fig. I-1 in Appendix I) based on relative roughness from Fig. I-2. (It is recommended that f be increased by at least 10 percent to allow for an increase in the friction factor as the pipe ages.)

L/D = length of pipe in diameters

h_i = velocity head of air, in of water

In determining the value of the friction factor f using Fig. I-1, the Reynolds number N_R may be computed using the following relationship:

$$N_R = \frac{28.4 \, q_s}{d \mu} \qquad (10\text{-}16)$$

where q_s = air flow in pipe, ft^3/min under the prevailing pressure and temperature conditions

d = inside diameter, in

μ = viscosity of air, centipoises (Note: centipoise $\times$ 0.000672 = lb-ft-sec)

In the range of 0 to 200°F, the viscosity μ can be approximated by using

$$\mu, \text{ centipoises} = (161 + 0.28t) \times 10^{-4} \qquad (10\text{-}17)$$

where t = temperature °F.

The velocity head h_i in inches of water at 70°F and 14.7 lb$_f$/in^2 can be computed by

$$h_i = \frac{(v, \text{ft/min})^2}{2(32.17 \text{ ft/sec}^2)} \left[\frac{1}{(60 \text{ s/min})^2} \right] (\gamma_a, \text{lb air/ft}^3) \frac{1}{62.3 \text{ lb/ft}^3} (12 \text{ in/ft})$$

or

$$h_i = \left(\frac{v}{1,096} \right)^2 \gamma_a \qquad (10\text{-}18)$$

where v = air velocity, ft/min

γ_a = specific weight of air at 70°F and 14.7 lb$_f$/in^2, lb/ft^3 (see Table B-1)

Equation 10-18 can be used to estimate the headloss at other temperatures, but the value of γ_a must be corrected for other temperatures and pressures. Application of the above equations is illustrated in the following example.

Example 10-1 Computation of headloss in air piping. Determine the headloss in 1,000 ft of 15-in diameter commercial steel pipe designed to carry 3,400 ft^3/min of air at standard conditions. The ambient air temperature is 86°F (30°C), and the plant is located at sea level, 14.7 lb$_f$/in^2 (760 mm Hg). Assume that the blower efficiency is 70 percent and the discharge pressure is 8 lb$_f$/in^2 (gage).

Solution

1. Determine the temperature rise during compression using the modified form of Eq. 10-14 given below where n is the efficiency of the blower expressed as a decimal.

$$\Delta T = \frac{T_1}{n}\left[\left(\frac{p_2}{p_1}\right)^n - 1\right]$$

Solving for ΔT yields

$$\Delta T = \frac{460 + 86}{0.70}\left[\left(\frac{22.7}{14.7}\right)^{0.283} - 1\right] = 102°F$$

Therefore, the air temperature at the blower discharge is 188°F (86 + 102°F).

2. Compute the Reynolds number using Eqs. 10-16 and 10-17. Given that the air temperature at the blower is 188 °F, assume that the average temperature in the pipe is 160°F.

$$\mu = [161 + 0.28(160)] \times 10^{-4}$$

$$= 205.8 \times 10^{-4}$$

$$N_R = \frac{28.4\ (3,400)}{15(205.8) \times 10^{-4}} = 3.13 \times 10^5$$

3. Determine the friction factor f from Fig. I-2 using the curve for commercial steel ($e = 0.00015$). The value of e/D is 0.00012. Entering Fig. I-1 with an e/D value of 0.00012 and N_R of 3.13× 10^5, the value of f is 0.0155. Add 10 percent and use an f value of 0.017 for design.

4. Determine the air flowrate in the pipe using the following relationship by substituting the volumetric flowrate for V.

$$\frac{P_1V_1}{T_1} = \frac{P_2V_2}{T_2}$$

$$\frac{(14.7\text{lb}_f/\text{in}^2)(3400\ \text{ft}^3/\text{min})}{460 + 68} = \frac{(14.7 + 8\text{lb}_f/\text{in}^2)V_2}{(460 + 160)}$$

$$V_2 = \text{flowrate} = 3,400\left(\frac{14.7}{22.7}\right)\left(\frac{460 + 160}{460 + 68}\right) = 2,585\text{ft/min}$$

5. Determine the velocity in the pipe.

$$v = \frac{2,585\ \text{ft}^3/\text{min}}{3.14/4(1.25\ \text{ft})^2} = 2,107\ \text{ft/min}$$

6. Determine the specific weight of air at a pressure of 22.7 lb$_f$/in^2 and a temperature of 160°F using the following expression:

$$\gamma_a = \frac{p}{RT}$$

$$\gamma_a = \frac{(22.7 \text{ lb/in}^2)(144 \text{ in}^2/\text{ft}^2)}{[53.3 \text{ ft} \cdot \text{lb/(lb air)}^\circ R][(460 + 160)^\circ R]} = 0.0989 \text{ lb/ft}^3$$

7. Determine the velocity head using Eq. 10-18.

$$h_i = \left(\frac{2,107}{1,096}\right)^2 0.0989 = 0.366 \text{ in of water}$$

8. Determine the headloss using Eq. 10-15.

$$h_L = 0.017\left(\frac{1,000}{1.25}\right)0.366 = 4.98 \text{ in of water}$$

Comment. Losses in elbows, tees, valves, etc., can be computed as a fraction of velocity head using the K values given in the companion volume to this text [29] or in standard hydraulic texts. Meter losses can be estimated as a fraction of the differential head, depending on the type of meter. Losses in air filters, blower silencers, and check valves should be obtained from equipment manufacturers. The discharge pressure at the blowers will be the sum of the above losses, the depth of water over the air diffusers, and the loss through the diffusers.

Mechanical Aerators

Mechanical aerators are commonly divided into two groups based on major design and operating features: aerators with a vertical axis and aerators with a horizontal axis. Both groups are further subdivided into surface and submerged aerators. In surface aerators, oxygen is entrained from the atmosphere; in submerged aerators, oxygen is entrained from the atmosphere and, for some types, from air or pure oxygen introduced in the tank bottom. In either case, the pumping or agitating action of the aerators helps to keep the contents of the aeration tank or basin mixed. In the following discussion, the various types of aerators will be described, along with aerator performance and the energy requirement for mixing.

Surface Mechanical Aerators with a Vertical Axis.
Surface mechanical aerators with a vertical axis are designed to induce either updraft or downdraft flows through a pumping action (Fig. 10-14). They consist of submerged or partially submerged impellers attached to motors mounted on floats or on fixed structures. The impellers are fabricated from steel, cast iron, noncorrosive alloys, and fiberglass-reinforced plastic and are used to agitate the wastewater vigorously, entraining air in the wastewater and causing a rapid change in the air-water interface to facilitate dissolution of the air. Surface aerators may be classified according to the type of impeller used (centrifugal, radial-axial or axial) or the speed of rotation of the impeller (low or high speed). Centrifugal impellers belong to the low-speed category; the axial-flow impeller type aerators operate at high speed. In low-speed aerators, the impeller is driven through a reduction gear by an electric motor. The motor and gearbox are usually mounted on a platform supported either by piers extending to the bottom of the tank or by beams that span the tank. Low-speed aerators may also be mounted on floats. In high-speed aerators, the impeller is coupled directly to the rotating element

(a)   (b)

FIGURE 10-14
Typical surface mechanical aerators: (a) floating high-speed aerator and (b) fixed platform low-speed aerator.

of the electric motor. High-speed aerators are almost always mounted on floats. These units were originally developed for use in ponds or lagoons where the water surface elevation fluctuates or where a rigid support would be impractical. Surface aerators may be obtained in sizes from 1 to 150 hp (0.75 to 100 kW).

Submerged Mechanical Aerators with a Vertical Axis. Most surface mechanical aerators are upflow types that rely on violent agitation of the surface and air entrainment to achieve oxygen transfer. With submerged mechanical aerators, however, air or pure oxygen may also be introduced by diffusion into the wastewater beneath the impeller or downflow of radial aerators. The impeller is used to disperse the air bubbles and mix the contents of the tank (Fig. 10-15). A draft tube may be used

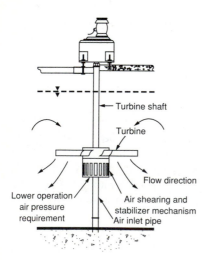

FIGURE 10-15
Typical submerged mechanical aerator.

with either upflow or downflow models to control the flow pattern of the circulating liquid within the aeration tank. The draft tube is a cylinder with flared ends mounted concentrically with the impeller and extends from just above the floor of the aeration tank to just beneath the impeller. Submerged mechanical aerators may be obtained in sizes from 1 to 150 hp (0.75 to 100 kW).

Mechanical Aerators with a Horizontal Axis. Mechanical aerators with a horizontal axis are divided into two groups: surface and submerged aerators. The surface aerator is patterned after the original Kessener brush aerator, a device used to provide both aeration and circulation in oxidation ditches. The brush-type aerator had a horizontal cylinder with bristles mounted just above the water surface. The bristles were submerged in the water, and the cylinder was rotated rapidly by an electric motor drive, spraying wastewater across the tank, promoting circulation, and entraining air in the wastewater. Now, angle steel, steel of other shapes, or plastic bars or blades are used instead of bristles. A typical horizontal axis surface aerator is shown in Fig. 10-16a.

Submerged horizontal axis aerators are similar in principle to the surface type except that they use disks or paddles attached to rotating shafts to agitate the water. The disk aerator (Fig.10-16b) has been used in numerous applications for channel and oxidation ditch aeration. The disks are submerged in the wastewater for approximately one-eighth to three-eighths of the diameter and enter the water in a continuous, nonpulsating manner. Recesses in the disks introduce entrapped air beneath the surface as the disks turn. Spacings of the disks can vary depending on the oxygen and mixing requirements of the process. Typical power requirements are reported as 0.15 to 1.00 hp/disk (0.1 to 0.75 kW/disk) [63].

Aerator Performance. Mechanical aerators are rated in terms of their oxygen-transfer rate, expressed as pounds of oxygen per horsepower-hour (kilograms of oxygen per kilowatt-hour) at standard conditions. Standard conditions exist when

(a) (b)

FIGURE 10-16
Typical horizontal axis mechanical aerators: (a) brush aerator and (b) disk aerator.

the temperature is 20°C, the dissolved oxygen is 0.0 mg/L, and the test liquid is tap water. Testing and rating are normally done under nonsteady-state conditions using fresh water, deaerated with sodium sulfite. Commercial-size surface aerators range in efficiency from 2 to 4 lb O_2/hp · h (1.20 to 2.4 kg O_2/kW · h). Oxygen-transfer data for various types of mechanical aerators are reported in Table 10-9. Efficiency claims for aerator performance should be accepted by the design engineer only when they are supported by test data for the actual model and size of the aerator under consideration. For design purposes, the standard performance data must be adjusted to reflect anticipated field conditions. This adjustment is accomplished using the following equation. The term within the brackets represents the correction factor.

$$N = N_o \left(\frac{\beta C_{W_{alt}} - C_L}{C_{s_{20}}} \right) 1.024^{T-20} \alpha \qquad (10\text{-}19)$$

where
N = lb O_2/hp · h transferred under field conditions
N_o = lb O_2/hp · h transferred in water at 20°C and zero dissolved oxygen
β = salinity-surface tension correction factor, usually 1
$C_{W_{alt}}$ = oxygen saturation concentration for tap water at given temperature and altitude (see Appendix E and Fig. 10-17), mg/L
$C_{s_{20}}$ = oxygen saturation concentration in tap water 20°C, mg/L
C_L = operating oxygen concentration, mg/L
T = temperature, °C
α = oxygen-transfer correction factor for waste (see Table 10-10)

TABLE 10-9
Typical ranges of oxygen-transfer capabilities for various types of mechanical aerators[a]

Aerator type	Transfer rate, lb O_2/hp · h	
	Standard[b]	Field[c]
Surface low-speed	2.0–5.0	1.2–2.4
Surface low-speed with draft tube	2.0–4.6	1.2–2.1
Surface high-speed	2.0–3.6	1.2–2.0
Surface downdraft turbine	2.0–4.0	1.0–2.0[d]
Submerged turbine with sparger	2.0–3.3	1.2–1.8[d]
Submerged impeller	2.0–4.0	1.2–1.8
Surface brush and blade	1.5–3.6	0.8–1.8

[a] Derived in part from Refs. 47, 48, and 63.
[b] Standard conditions: tap water 20°C; at 14.7 lb$_f$/in^2 and initial dissolved oxygen = 0 mg/L.
[c] Field conditions: wastewater, 15°C; altitude 500ft, α = 0.85, β = 0.9; operating dissolved oxygen = 2 mg/L.
[d] Recent research suggests that α values may be lower than 0.85.
Note: lb/hp · h × 0.6083 = kg/kW · h
lb$_f$/in^2 × 6.8948 = kN/m^2
1.8(°C) + 32 = °F

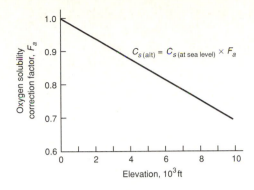

FIGURE 10-17
Oxygen solubility correction factor versus
elevation. *Note:* ft × 0.3048 = m.

The application of this equation is illustrated in Sec. 10-4, which deals with the design
of aerated lagoons.

Energy Requirement for Mixing. As with diffused-air systems, the size and shape
of the aeration tank is very important if effective mixing is to be achieved. Aeration
tanks may be square or rectangular and may contain one or more aerators. The depth
and width of the aeration tanks for mechanical surface aerators are dependent on
aerator size; typical values are given in Table 10-11. Depths up to 35 ft (10.7 m)
have been used with submerged draft tube mixers.

In diffused-air systems, the air requirement to ensure good mixing varies from
20 to 30 ft^3/min $\cdot 10^3$ ft^3 of tank volume (20 to 30 m^3/min $\cdot 10^3$ m^3) for a spiral
roll aeration pattern. For a grid aeration system in which the diffusers are installed
uniformly along the aeration basin bottom, mixing rates of 10 to 15 ft^3/min $\cdot 10^3$ ft^3
(10 to 15 m^3/min $\cdot 10^3$ m^3) have been suggested [63]. Typical power requirements
for maintaining a completely mixed flow regime with mechanical aerators vary from
0.75 to 1.50 hp/10^3 ft^3 (19 to 39 kW/10^3 m^3), depending on the design of the aerator
and the geometry of the tank, lagoon, or basin. In the design of aerated lagoons for

TABLE 10-10
**Typical values of alpha factor for low-speed surface
aerators and selected wastewater types[a]**

Wastewater type	BOD$_5$, mg/L		α factor[b]	
	Influent	Effluent	Influent	Effluent
Municipal wastewater	180	3	0.82	0.98
Pulp and paper	187	50	0.68	0.77
Kraft paper	150–300	37–48	0.48–0.68	0.7–1.1
Bleached paper	250	30	0.83–1.98	0.86–1.0
Pharmaceutical plant	4,500	380	1.65–2.15	0.75–0.83
Synthetic fiber plant	5,400	585	1.88–3.25	1.04–2.65

[a] Ref. 63.
[b] Recent research suggests that α values may be lower and more variable than values
listed in table.

TABLE 10-11
Typical aeration tank dimensions for mechanical surface aerators

Aerator size, hp	Tank dimensions, ft	
	Depth	Width
10	10–12	30–40
20	12–14	35–50
30	13–15	40–60
40	12–17	45–65
50	15–18	45–75
75	15–20	50–85
100	15–20	60–90

Note: hp × 0.7457 = kW
ft × 0.3048 = m

the treatment of domestic wastewater, it is extremely important that the mixing power requirement be checked because, in most instances, it will be the controlling factor.

Generation and Dissolution of High-Purity Oxygen

After the quantity of oxygen required is determined, it is necessary, where high-purity oxygen is to be used, to specify the type of oxygen generator that will best serve the needs of the plant. There are two basic oxygen generator designs: (1) a pressure swing adsorption (PSA) system for smaller and more common plant sizes (less than 40 Mgal/d) and (2) the traditional cryogenic air-separation process for large applications. Liquid oxygen can also be trucked in and stored onsite.

Pressure Swing Adsorption. The pressure swing adsorption system uses a multibed adsorption process to provide a continuous flow of oxygen gas [47]. A schematic diagram of the four-bed system is shown in Fig. 10-18a. The operating principle of the pressure swing adsorption generator is that the oxygen is separated from the feed-air by adsorption at high pressure, and the adsorbent is regenerated by blowdown to low pressure. The process operates on a repeated cycle with two basic steps, adsorption and regeneration. During the adsorption step, feed air flows through one of the adsorber vessels until the adsorbent is partially loaded with impurity. At that time the feed-air flow is switched to another adsorber, and the first adsorber is regenerated. During regeneration, the impurities are cleaned from the adsorbent so that the bed will be available again for the adsorption step. Regeneration is carried out by depressurizing to atmospheric pressure, purging with some of the oxygen, and repressurizing back to the pressure of the feed air.

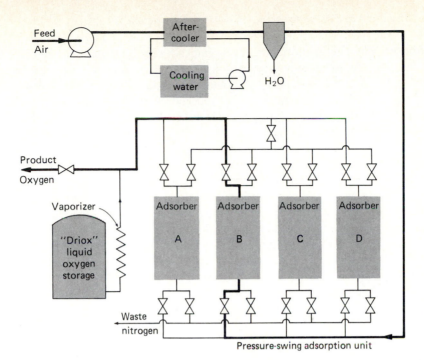

FIGURE 10-18
Schematic diagrams of systems for the generation of oxygen used in the high-purity oxygen activated-sludge process: (a) pressure swing adsorption system and (b) cryogenic generation system [47].

Cryogenic Air Separation. The cryogenic air separation process involves the liquefaction of air, followed by fractional distillation to separate it into its components (mainly nitrogen and oxygen) [47]. A schematic diagram of this process is shown in Fig. 10-18b. First, the entering air is filtered and compressed. The compressed air is fed to the reversing heat exchangers, which perform the dual function of cooling and removing the water vapor and carbon dioxide by freezing these mixtures out into the exchanger surfaces. This process is accomplished by periodically switching or reversing the feed air and the waste nitrogen streams through identical passes of the exchangers to regenerate their water vapor and carbon dioxide removal capacity.

Next, the air is processed through "cold and gel traps," adsorbent beds that are used to remove the final traces of carbon dioxide as well as most hydrocarbons from the feed air. The processed air is then divided into two streams. The first stream is fed directly to the lower column of the distillation unit. The second stream is returned to the reversing heat exchangers and partially warmed to provide the required temperature difference across the exchanger. This stream is then passed through an expansion turbine and fed into the upper column of the distillation unit. An oxygen-rich liquid exits from the bottom of the lower column, and the liquid nitrogen exits from the top. Both streams are then subcooled and transferred to the upper column. In this column, the descending-liquid phase becomes progressively richer in oxygen, and the liquid that subsequently collects in the condenser reboiler is the oxygen product stream. The liquid oxygen is recirculated continually through an adsorption trap to remove all possible residual traces of hydrocarbons. The waste nitrogen exits from the top portion of the upper column and is heat exchanged along with the oxygen product to recover all available refrigeration and to regenerate the reversing heat exchangers.

Dissolution of Commercial Oxygen. Oxygen is very insoluble in water—even pure oxygen—and requires special considerations to ensure high absorption efficiency. Oxygen dissolution equipment designed for air only optimizes energy consumption because the air is free and efficient oxygen absorption is not relevant. However, because of the cost of commercial oxygen, the facilities used for its dissolution must be designed both to efficiently absorb the commercial oxygen as well as to minimize the unit energy consumption. These requirements rule out the more common aeration equipment alternatives [40].

Dissolution time. A key feature that must be incorporated into a commercial oxygen dissolution system is oxygen retention time. To optimize the absorption of pure oxygen, it has been found that a detention time of about 100 seconds is required [40]. Further, two-phase flows must be maintained to avoid the coalescence of the oxygen bubbles to maintain absorption efficiency. Unfortunately, some pure oxygen dissolution systems consume as much energy to dissolve a ton of pure oxygen as standard surface aerators consume in dissolving a ton of oxygen from air.

Downflow bubble contactor. One system that incorporates prolonged oxygen bubble contact time and high rates of oxygen transfer is a cone-shaped chamber, downflow bubble contact aerator—DBCA (see Fig. 10-19a). Wastewater enters the chamber at the apex with a velocity of approximately 10 ft/s. This inlet velocity provides the energy to maintain a two-phase bubble swarm in the cone, ensuring a very high bubble-water interface and resulting in a proportionately high gas-transfer rate. The expanding horizontal cross section of the cone reduces the downward flow velocity of the wastewater to less than 1 ft/s. Because the bubbles have a nominal buoyant velocity of about 1 ft/s, if the downflow velocity of the wastewater is reduced below the buoyant velocity of the bubbles, they will remain indefinitely in the cone, thus satisfying the required bubble residence time. The wastewater, however, has a residence time of about 10 seconds, reflecting the relatively small volume of reactor cone. This system incorporates the desired features of relatively small size, high rate of oxygen transfer and more than adequate bubble residence time. The energy consumption is about 500 kW · h/ton O_2 if the cone is at ambient pressure and drops to about 100 kW · h/ton O_2 if the cone is at 75 lb_f/in^2 gage [40].

U-Tube contactor. Another oxygen-transfer system that incorporates desirable features for efficient dissolution of commercial oxygen with low unit energy consumption is the U-Tube (see Fig. 10-19b). At a depth of 100 ft and a throughput velocity of 8 ft/s, the residence time is 25 seconds. Because a contact time of 25 seconds is low, off-gas recycle back can be used to increase the contact time to about 100 seconds where efficient absorption occurs. The energy requirements are low because the bubble-water mixture is pumped through a filled U-Tube pipe that is hydrostatically pressurized by its vertical configuration. Use of the U-Tube enhances gas transfer

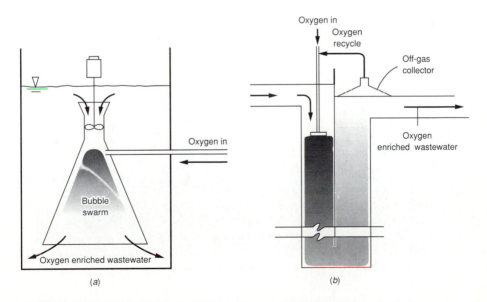

FIGURE 10-19
Pure oxygen dissolution systems: (a) downflow bubble contactor and (b) U-tube contactor [40].

significantly. Energy consumption is 100 kW · h/ton O_2 while producing an effluent dissolved oxygen of 60 mg/L[40].

Conventional diffused aeration. Conventional diffused aeration or surface aerators must operate in a closed headspace to absorb commercial oxygen efficiently. A concrete cover usually is placed over the aeration tank to enclose the headspace. The oxygen activated-sludge systems with surface aerators operating under the cover in an enriched oxygen atmosphere have an energy consumption of 500 to 650 kW · h/ton O_2 [40].

Design of Aeration Tanks and Appurtenances

After the activated-sludge process and the aeration system have been selected and a preliminary design has been prepared, the next step is to design the aeration tanks and support facilities. The following discussion covers (1) aeration tanks, (2) flow distribution, and (3) froth control systems.

Aeration Tanks. Aeration tanks are usually constructed of reinforced concrete and left open to the atmosphere. A cross section of a typical aeration tank using porous tube diffusers is shown in Fig. 10-20. The rectangular shape permits common-wall construction for multiple tanks. The total tank capacity required should be determined from the biological process design. For plants in a capacity range of 0.5 to 10 Mgal/d, at least two tanks should be provided (a minimum of two tanks is preferred for smaller

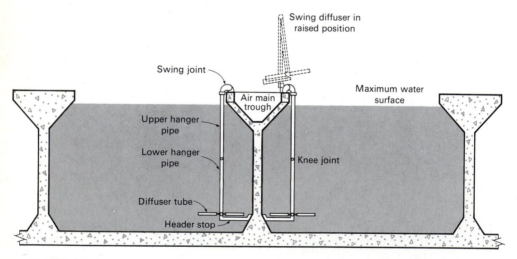

FIGURE 10-20
Cross section of a typical activated-sludge aeration tank using a porous tube diffused-air system.

plants as well). In the range of 10 to 50 Mgal/d (0.44 to 2.2 m³/s), four tanks are often provided to allow operational flexibility and ease of maintenance. Large plants, over 50 Mgal/d (2.2 m³/s) in capacity, should contain six or more tanks. Some of the largest plants contain from 30 to 40 tanks arranged in several groups or batteries. Although the air bubbles dispersed in the wastewater occupy perhaps one percent of the total volume, no allowance is made for this in tank sizing. The volume occupied by submerged piping is usually negligible.

If the wastewater is to be aerated with diffused air, the geometry of the tank may significantly affect the aeration efficiency (α factor) and the amount of mixing obtained. The depth of wastewater in the tank should be between 15 and 25 ft (4.57 and 7.62 m) so that the diffusers can work efficiently. Freeboard from 1 to 2 ft (0.3 to 0.6 m) above the waterline should be provided. The width of the tank in relation to its depth is important if spiral-flow mixing is used in the plug-flow configuration. The width-to-depth ratio for such tanks may vary from 1.0:1 to 2.2:1, with 1.5:1 being the most common. In large plants, the channels become quite long and sometimes exceed 500 ft (150 m) per tank. Tanks may consist of one to four channels with round-the-end flow in multiple-channel tanks. The length-to-width ratio of each channel should be at least 5:1. Where complete-mix diffused-air systems are used, the length-to-width ratio may be reduced to save construction cost.

For tanks with diffusers on both sides or in the center of the tank, greater widths are permissible. The important point is to restrict the width of the tank so that "dead spots," or zones of inadequate mixing, are avoided. The dimensions and proportions of each independent unit should be such as to maintain adequate velocities so that deposition of solids will not occur. In spiral-flow tanks, triangular baffles or fillets may be placed longitudinally in the corners of the channels to eliminate dead spots and to deflect the spiral flow.

For mechanical aeration systems, the most efficient arrangement is one aerator per tank. Where multiple aerators are installed in the same tank for best efficiency, the length-to-width ratio of the tank should be in even multiples with the aerator centered in a square configuration to avoid interference at the hydraulic boundaries. The width and depth should be sized in accordance with the power rating of the aerator, as illustrated in Table 10-11. Two-speed aerators are desirable to provide operating flexibility to cover a wide range of oxygen demand conditions. Freeboard of about 3.5 to 5 ft (1 to 1.5 m) should be provided for mechanical aeration systems.

Individual tanks should have inlet and outlet gates or valves so that they may be removed from service for inspection and repair. The common walls of multiple tanks must therefore be able to withstand the full hydrostatic pressure from either side. Aeration tanks must have adequate foundations to prevent settlement, and, in saturated soil, they must be designed to prevent flotation when the tanks are dewatered. Methods of preventing flotation include thickening the floor slab or installing hydrostatic pressure relief valves. Drains or sumps for aeration tanks are desirable for dewatering. In large plants where tank dewatering might be more common, it may be desirable to install mud valves in the bottoms of all tanks. The mud valves should be connected to a central dewatering pump or to a plant drain discharging to the wet well of the plant pumping station. For small plants, portable pumps are suitable for dewatering service. Dewatering systems are commonly designed to empty a tank in 16 h.

Flow Distribution. For wastewater treatment plants containing multiple units of primary sedimentation basins and aeration tanks, consideration has to be given to equalizing the distribution of flow to the aeration tanks. In many designs, the wastewater from the primary sedimentation basins is collected in a common conduit or channel for transport to the aeration tanks. For efficient use of the aeration tanks, a method of splitting or controlling the flowrate to each of the individual tanks should be used. Methods commonly used are splitter boxes equipped with weirs or control valves or aeration tank influent control gates. Hydraulic balancing of the flow by equalizing the headloss from the primary sedimentation basins to the individual aeration tanks is also practiced. In flow regimes where step feed is used, a positive means of flow control is especially important. Where channels are used for aeration tank influent or effluent transport, they should be equipped with channel aeration diffusers to prevent deposition of solids.

Froth Control Systems. Wastewater normally contains soap, detergents, and other surfactants that produce foam when the wastewater is aerated. If the concentration of mixed-liquor suspended solids is high, the foaming tendency is minimized. Large quantities of foam may be produced during startup of the process, when surfactants are present in the wastewater. The foaming action produces a froth that contains sludge solids, grease, and large numbers of wastewater bacteria. The wind may lift the froth off the tank surface and blow it about, contaminating whatever it touches. The froth, besides being unsightly, is a hazard to those working with it because it is very slippery, even after it collapses. In addition, once the froth has dried, it is difficult to clean off.

It is important, therefore, to consider some method for controlling froth formation, particularly in spiral-flow tanks where the froth collects along the side of the tank. A commonly used system for spiral-flow tanks consists of a series of spray nozzles mounted along the top edge of the aeration tank, opposite the air diffusers. Screened effluent or clear water is sprayed through these nozzles and physically breaks down the froth as it forms. Another approach is to meter a small quantity of antifoaming chemical additive into the inlet of the aeration tank or preferably into the spray water.

Design of Solids-Separation Facilities

The function of the activated-sludge settling tank is to separate the activated-sludge solids from the mixed liquor. Solids separation is the final step in the production of a well-clarified, stable effluent low in BOD and suspended solids and, as such, represents a critical link in the operation of an activated-sludge treatment process.

Although much of the information presented in Chaps. 6 and 9 in connection with the design of primary sedimentation tanks is applicable, the presence of a large volume of flocculant solids in the mixed liquor requires that special consideration be given to the design of activated-sludge settling tanks. As mentioned previously, these solids tend to form a sludge blanket that will vary in thickness. This blanket may fill the entire depth of the tank and overflow the weirs at peak flowrates if the return-sludge pump capacity or the size of the settling tank is inadequate. Further, the mixed liquor, on entering the tank, has a tendency to flow as a density current, interfering with the separation of the solids and the thickening of the sludge. To cope successfully with these characteristics, the engineer must consider the following factors in the design of these tanks: (1) tank types, (2) settling characteristics of the sludge as related to the thickening requirements for proper plant operation, (3) surface- and solids-loading rates, (4) sidewater depth, (5) flow distribution, (6) inlet design, (7) weir placement and loading rates, and (8) scum removal.

Tank Types. The most commonly used types of activated-sludge settling tanks are either circular or rectangular (Fig. 10-21). Square tanks are used on occasion, but are not as effective in retaining separated solids as circular or rectangular tanks. Solids accumulate in the corners of the square tanks and are frequently swept over the weirs by the agitation of the sludge collectors. Circular tanks have been constructed with diameters ranging from 10 to 200 ft (3 to 60 m), although the more common range is from 30 to 140 ft ($\approx$10 to 40 m). The tank radius should preferably not exceed five times the sidewater depth. There are two basic types of circular tanks to choose from: the center-feed and the rim-feed clarifier. Both types use a revolving mechanism to transport and remove the sludge from the bottom of the clarifier. Mechanisms are of two types: those that scrape or plow the sludge to a center hopper similar to the types used in primary sedimentation tanks and those that remove the sludge directly from the tank bottom through suction orifices that serve the entire bottom of the tank

(a)

(b)

FIGURE 10-21
Typical secondary settling tanks: (a) circular and (b) partially covered rectangular.

in each revolution. Of the latter, in one type the suction is maintained by reduced static head on the individual suction pipes (Fig. 10-22a). In another patented suction system, sludge is removed through a manifold either hydrostatically or by pumping (Fig. 10-22b).

Rectangular tanks must be proportioned to achieve proper distribution of incoming flow so that horizontal velocities are not excessive. It is recommended that the maximum length of rectangular tanks not exceed 10 to 15 times the depth, but lengths up to 300 ft (90 m) have been used successfully in large plants. Where widths of rectangular tanks exceed 20 ft (6 m), multiple sludge collection mechanisms may be used to permit tank widths up to 80 ft (24 m). Regardless of tank shape, the sludge collector selected should be able to meet the following two operational conditions: (1) the collector should have a high capacity so that when a high sludge recirculation rate is desired, channeling of the overlying liquid through the sludge will not result; and (2) the mechanism should be sufficiently rugged to be able to transport and remove the very dense sludges that could accumulate in the settling tank during periods of mechanical breakdown or power failure.

Two types of sludge collectors are commonly used in rectangular tanks: (1) traveling flights, and (2) traveling bridges (see Fig. 10-23). Traveling flights are similar to those used for the removal of sludge in primary settling tanks. For very long tanks, it is desirable to use two sets of chains and flights in tandem with a central hopper to receive the sludge. Sludge is usually collected at the influent end of the tank although some designs provide mechanisms that move the sludge to the effluent end for collection. The traveling bridge, which is similar to a traveling overhead crane, travels along the sides of the sedimentation tank or a support structure if several bridges are used. The bridge serves as the support for the sludge removal system, which usually consists of a scraper or a suction manifold from which the sludge is pumped. The sludge is discharged to a collection trough that runs the length of the tank.

Other types of settling tanks include tray clarifiers, tube and lamella (parallel plate) settlers, and intrachannel clarifiers. Tray clarifiers (see Fig. 10-24) are used in installations where limited land area is available for clarifiers. Two general types are used: series-flow (Fig. 10-24a) and parallel-flow (Fig. 10-24b). The parallel-flow tray clarifier has been used extensively in Japan and is now being considered in the United States [19].

The efficiency of conventional or shallow clarifiers may be improved by the installation of tubes or parallel plates to establish laminar flow (Fig. 10-25). In the United States, the tube settler clarifier has been used to a limited extent in the expansion and retrofitting of existing plants. Constructed of bundles of tubes or plates set at selected angles (usually 60°) from the horizontal, these settlers have a very short settling distance, and circulation is dampened because of the small size of the tubes. Sludge that collects in the tubes or on the plates tends to slide out due to gravitational forces. The major drawback in wastewater treatment is a tendency of these tubes to clog because of the accumulation of biological growths and grease.

Intrachannel clarifiers (Fig. 10-26) have been developed to improve the performance of the oxidation ditch activated-sludge process. These devices permit liquid and solids separation and sludge return to occur within the aeration channel. Sludge

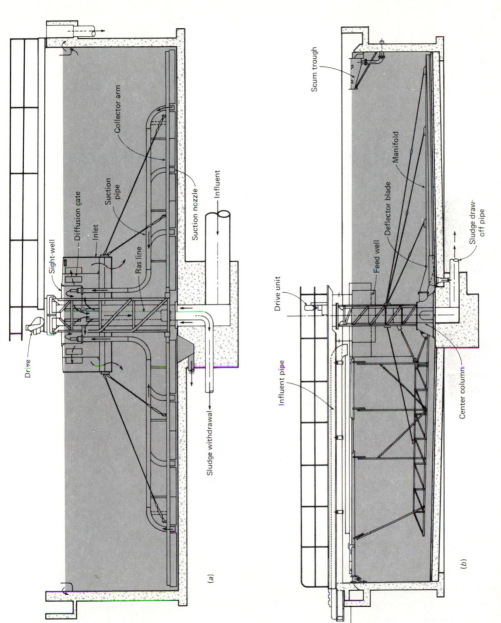

FIGURE 10-22

Typical circular secondary settling tanks designed for rapid sludge removal: (a) sludge removed through suction pipes (from Walker Process Equipment Division, Chicago Bridge & Iron Company) and (b) sludge removed through manifold (from Envirex).

(a)

Drive

Sight-well

Diffusion gate

Inlet

Suction pipe

Collector arm

Ras line

Suction nozzle

Influent

Sludge withdrawal

(b)

Drive unit

Influent pipe

Feed well

Deflector blade

Manifold

Scum trough

Sludge draw-off pipe

Center column

583

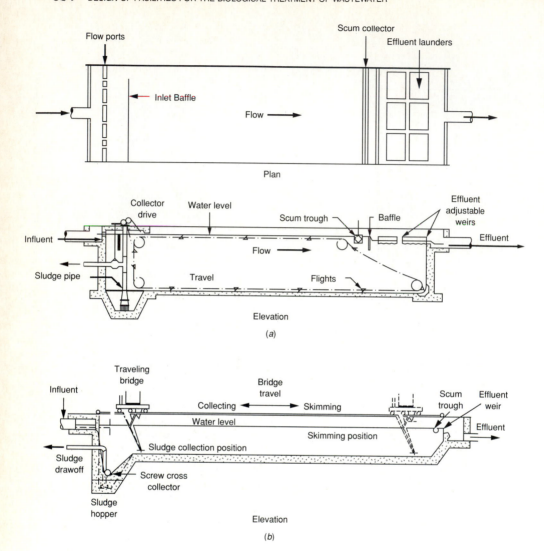

FIGURE 10-23
Typical rectangular secondary settling tanks: (a) chain-and-flight collector and (b) traveling-bridge collector.

wasting is done either from the aeration channel or the intrachannel clarifier. Because these units are relatively new, long-term performance data are not available. For more information on the various types of intrachannel clarifiers that are available, Ref. 9 may be consulted.

Settling Characteristics of Sludge. Operationally, secondary settling facilities must perform two functions: (1) separation of the mixed-liquor suspended solids from the treated wastewater, which results in a clarified effluent, and (2) thickening of the

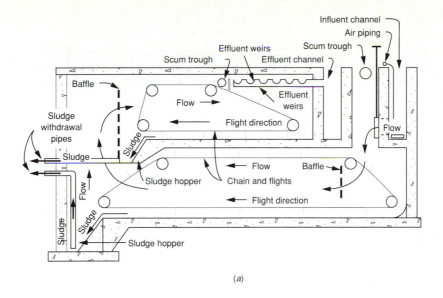

(a)

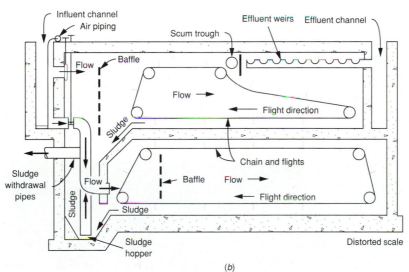

(b)

FIGURE 10-24
Typical tray-type clarifiers: (a) series flow type and (b) parallel flow type [19]. *Note:* In parallel flow type, the upper effluent weirs serve both the upper and lower clarifiers. Channels for the discharge of effluent from the lower to the upper clarifier are located on either side of the sludge collection mechanism in the upper clarifier.

return sludge. Both functions must be taken into consideration if secondary settling facilities are to be designed properly. Because both functions will be affected by clarifier depth, adequate consideration must be given to selection of a depth that will provide the necessary volume for both functions. For example, ample volume must be provided for storage of the solids during periods in which sustained peak plant

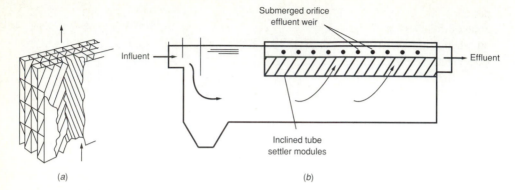

(a)   (b)

FIGURE 10-25
Tube settler clarifier: (a) module of inclined tubes and (b) tubes installed in a rectangular-sedimentation tank.

loadings are experienced (see Chap. 5). Also, peak daily flowrate variations must be considered because they affect sludge removal requirements.

In general, the area required for clarification is based on the overflow rate equivalent of the rise rate of the smallest particle to be removed from the clarified liquid in the upper portions of the settling tank. Often, the design settling rate for clarification is taken as the interface settling rate, derived from settling-column tests for the sludge concentration at which the plant is to operate. Unfortunately, this velocity is usually much greater than the rate required to remove the light, fluffy particles usually found in the effluent from activated-sludge plants. If these light

FIGURE 10-26
Typical intrachannel clarifier installed in an oxidation ditch (from United Industries, Inc.).

particles are to be removed, adequate volume must be provided in the clarification zone of the clarifier. The time required for adequate settling of these particles depends on whether the settling phenomenon for the particles can be described as discrete or flocculant.

The area required for the thickening of the applied mixed liquor depends on the limiting solids flux that can be transported to the bottom of the sedimentation basin. Because the solids flux varies with the characteristics of the sludge, settling-column tests should be conducted to determine the relationship between the sludge concentration and the settling rate. The required area can then be determined using the solids-flux analysis procedure described in Chap. 6. The depth of the thickening portion of the sedimentation tank must be adequate enough to (1) ensure maintenance of an adequate sludge-blanket depth so that unthickened solids are not recycled, and (2) temporarily store the solids that are periodically applied in excess of the transmitting capacity of the given suspension.

Activated-sludge solids have a specific gravity so near to that of water that the increased density and viscosity of the wastewater under winter conditions affect the settling properties of the sludge adversely. In addition, the settling properties of the sludge may vary from time to time because of changes in the amount and specific gravity of the suspended solids passing through the primary settling tanks, the character and amount of industrial wastes contained in the wastewater, and the composition of the microbial life of the floc. For these reasons, it is necessary to use conservative design criteria to avoid occasional loss of sludge solids.

The sludge volume index has also been used as an indication of the settling characteristics of the sludge. However, the index value that is characteristic of a good settling sludge varies with the characteristics and concentration of the mixed-liquor solids, so observed values at a given plant are not comparable to other reported values. For example, if the solids did not settle at all but occupied the entire 1000 mL at the end of 30 min, the maximum index value would be obtained and would vary from 1000 for a mixed-liquor solids concentration of 1000 mg/L to 100 for a mixed-liquor solids concentration of 10,000 mg/L. For such conditions, the computation has no meaning other than the determination of limiting values.

When industrial wastes are to be treated by the activated-sludge process, it is recommended that pilot plant studies be conducted to evaluate the settling character-istics of the mixed liquor. These studies are also desirable in the case of the more familiar municipal wastes where the process variables, such as the concentration of the mixed-liquor suspended solids and the mean cell-residence time, are outside the range of common experience. It is important that such studies be conducted over a temperature range that is representative of both the average and the coldest tempera-tures to be encountered.

Surface and Solids Loading Rates. It is often necessary to design settling facil-ities without the benefit of settling tests. When this situation develops, published val-ues for surface- and solids-loading rates must be used. Because of the large amount of solids that may be lost in the effluent if design criteria are exceeded, effluent overflow rates should be based on peak flow conditions. The overflow rates given in Table 10-12 are typical values used for the design of biological systems. These values are based on wastewater flowrates instead of on the mixed-liquor flowrates because the overflow rate is equivalent to an upward flow velocity. The return-sludge flow is drawn off the bottom of the tank and does not contribute to the upward flow velocity.

The solids-loading rate on an activated-sludge settling tank may be computed by dividing the total solids applied by the surface area of the tank. The preferred units, which are the same as those used to compute the solids flux discussed previously, are pounds per square foot per hour, although units of pounds per square foot per day are common in the literature. The former is favored because the solids-loading factor should be evaluated at both peak and average flow conditions. If peaks are of short duration, average 24 h values may govern; if peaks are of long duration, peak values should be assumed to govern to prevent the solids from overflowing the tank.

In effect, the solids-loading rate represents a characteristic value for the suspension under consideration. In a settling tank of fixed surface area, the effluent quality will deteriorate if the solids loading is increased beyond the characteristic value for the suspension. Typical solids-loading values used for the design of biological systems are given in Table 10-12. Without extensive experimental work covering all seasons and operating variables, higher rates should not be used for design.

Sidewater Depth. Liquid depth in a secondary clarifier is normally measured at the sidewall in circular tanks and at the effluent end wall for rectangular tanks. The liquid depth is a factor in the effectiveness of suspended-solids removal and in the concentration of the return sludge. In recent years, the trend has been toward

TABLE 10-12
Typical design information for secondary clarifiers[a]

Type of treatment	Overflow rate, gal/ft² · d		Solids loading, lb/ft² · h		Depth, ft
	Average	Peak	Average	Peak	
Settling following air activated-sludge (excluding extended aeration)	400–800	1,000–1,200	0.8–1.2	2.0	12–20
Settling following oxygen activated-sludge	400–800	1,000–1,200	1.0–1.4	2.0	12–20
Settling following extended aeration	200–400	600–800	0.2–1.0	1.4	12–20
Settling following trickling filtration	400–600	1,000–1,200	0.6–1.0	1.6	10–15
Settling following rotating biological contractors:					
Secondary effluent	400–800	1,000–1,200	0.8–1.2	2.0	10–15
Nitrified effluent	400–600	800–1,000	0.6–1.0	1.6	10–15

[a] Adapted in part from Ref. 60.

Note: gal/ft² · d × 0.0407 = m³/m² · d
     lb/ft² · h × 4.8824 = kg/m² · h
        ft × 0.3048 = m

increasing liquid depths to improve overall performance, particularly in plants that have low-density activated sludge. It should be noted, however, that in some cases tanks with relatively shallow sidewater depths have been used successfully. Current practice favors a minimum sidewater depth of 12 ft (3.7 m) for large secondary clarifiers, and depths ranging up to 20 ft (6.1 m) have been used [60]. Typical sidewater depths are presented in Table 10-12. The advantages of deeper tanks include greater flexibility of operation and a larger margin of safety when changes in the activated-sludge system occur. The cost of tank construction has to be considered in selecting a sidewater depth, especially in areas of high groundwater levels. Other factors such as inlet design, type of sludge removal equipment, sludge-blanket depth, and weir type and location also affect clarifier performance [10,34,43].

Flow Distribution. Flow imbalance between multiple process units can cause under- or overloading of the individual units and affect overall system performance. In plants where parallel tanks of the same size are used, flow between the tanks should be equalized. In cases where the tanks are not of equal capacity, flows should be distributed in proportion to surface area. Methods of flow distribution to the secondary sedimentation tanks include weirs, flow control valves, hydraulic distribution using hydraulic symmetry,and feed gate or inlet port control (see Fig. 10-27). Effluent weir control, although frequently used to effect flow splitting, is usually ineffective and should be used only where there are two tanks of equal size.

Tank Inlet Design. Poor distribution or jetting of the tank influent can increase the formation of density currents and scouring of settled sludge, resulting in unsatisfactory tank performance. Tank inlets should dissipate influent energy, distribute the flow evenly in horizontal and vertical directions, mitigate density currents, minimize sludge-blanket disturbance, and promote flocculation. In circular center-feed tanks,

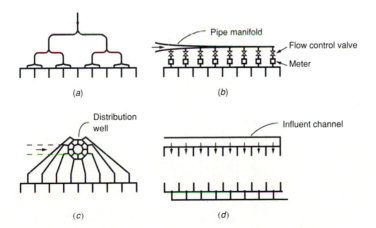

FIGURE 10-27
Alternative methods of flow splitting: (a) hydraulic symmetry, (b) flow measurement and feedback control, (c) hydraulic weir split, and (d) inlet feed gate control [60].

the most common design used, small, solid skirted, cylindrical baffles are provided to dissipate the influent energy and distribute flow. It has been observed in full-scale studies that a density current waterfall can be created using skirted baffles resulting in poor vertical flow distribution [10]. Methods to overcome these problems include the use of a large-center diffusion well or a flocculating-type clarifier. The large-center diffusion well, with a minimum diameter of 25 percent of the tank diameter, provides a greater area for the dissipation of the influent energy and the distribution of the incoming mixed liquor. The bottom of the feed well should end well above the sludge-blanket interface to minimize turbulence and resuspension of the solids. In flocculating center-feed clarifiers (see Fig. 10-28), a flocculating mechanism is incorporated in the center-feed well. Typical flocculation-feed wells have diameters of 30 to 35 percent of the tank diameter. It has been found that the settling characteristics of poorly flocculated mixed liquor may be enhanced significantly by slow flocculation in the feed well [34, 43]. In rectangular tanks, inlet ports or baffles should be provided to achieve flow distribution. Inlet port velocities are typically 15 to 30 ft/min (75 to 150 mm/s) [60]. For addtional information on inlet design, Refs. 44 and 60 may be consulted.

Weir Placement and Loading. When density currents occur in a secondary clarifier, mixed liquor entering the tank flows along the tank bottom until it encounters a counter-current pattern or an end wall. When the mixed liquor encounters an end wall, it tends to mound up and may be discharged over the effluent weirs, especially those located at the end of the tank. The presence of density currents is considered in the design of sedimentation facilities. In experimental work conducted in Chicago on tanks 126 ft (38.4 m) in diameter, it was found that a circular weir trough placed at two-thirds to three-fourths of the radial distance from the center was in the optimum position to intercept well-clarified effluent [5]. With low surface loadings and weir

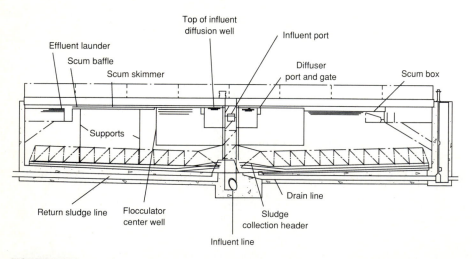

FIGURE 10-28
Typical secondary clarifier with flocculating center well.

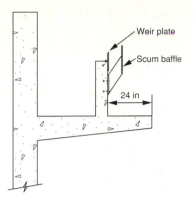

FIGURE 10-29
Horizontal baffle at clarifier weir trough for reducing solids carryover [43].

rates, the placement of the weirs in small tanks does not significantly affect the performance of the clarifier. Circular clarifiers are manufactured with overflow weirs located near both the center and the perimeter of the tank. If weirs are located at the tank perimeter or at end walls in rectangular tanks, a horizontal baffle should be provided to deflect the density currents toward the center of the tank and away from the effluent weir (see Fig. 10-29) [43].

Weir-loading rates are commonly used in the design of clarifiers, although they are less critical in clarifier design than hydraulic overflow rates. Weir-loading rates used in large tanks should preferably not exceed 30,000 gal/lin ft ·d (375 m³/lin m ·d) of weir at maximum flow when located away from the upturn zone of the density current, or 20,000 gal/lin ft ·d (250 m³/lin m ·d) when located within the upturn zone. In small tanks, the weir-loading rate should not exceed 10,000 gal/lin ft ·d (125 m³/lin m ·d) at average flow or 20,000 gal/lin ft ·d at maximum flow. The upflow velocity in the immediate vicinity of the weir should be limited to about 12 to 24 ft/h (3.7 to 7.3 m/h).

Scum Removal. In many well-operating secondary plants, very little scum is formed in the secondary clarifiers. However, occasions arise (see section on "Operating Problems") when some floating material is present, necessitating its removal. Where primary settling tanks are not used, skimming of the final tanks is essential. It has become standard practice in recent years to provide scum removal on all secondary circular clarifiers. Typical scum removal equipment includes beach- and scraper-type, rotating pipe-through skimmer, and slotted pipes. For additional information on skimming devices, Ref. 60 may be consulted.

10-3 ACTIVATED-SLUDGE PROCESS DESIGN

Application of the aforementioned factors to the design of an activated-sludge treatment process is illustrated in Examples 10-2 and 10-3. For purposes of Example 10-2, a complete-mix (continuous-flow stirred-tank) system has been selected. Schematically, the complete-mix system is depicted as shown in Fig. 10-3. Its distinguishing features are (1) uniform distribution of the inflow and return solids to the reactor (aeration tank) and (2) uniform withdrawal of mixed liquor from the reactor. The design features of a sequencing batch reactor are considered in Example 10-3. Application of the principles discussed in this chapter and in Chap. 8 to other types of systems is covered in the problems at the end of this chapter.

Example 10-2 Design of complete-mix activated-sludge system. Design a complete-mix activated-sludge process and secondary settling facilities to treat 5.71 Mgal/d (0.25 m³/s) of settled wastewater with 250 mg/L of BOD_5. The effluent is to have 20 mg/L of BOD_5 or less. Assume that the temperature is 20°C and that the following conditions are applicable:

1. Influent volatile suspended solids to reactor are negligible.
2. Ratio of mixed-liquor volatile suspended solids (MLVSS) to mixed-liquor suspended solids (MLSS) = 0.8.
3. Return-sludge concentration = 10,000 mg/L of suspended solids (SS).
4. Mixed-liquor volatile suspended solids (MLVSS) = 3500 mg/L.
5. Design mean cell-residence time θ_c = 10 d.
6. Effluent contains 22 mg/L of biological solids, of which 65 percent is biodegradable.
7. $BOD_5 = 0.68 \times BOD_L$.
8. Wastewater contains adequate nitrogen, phosphorus, and other trace nutrients for biological growth.
9. The peak hourly flowrate is 2.5 times the average flowrate.
10. The following MLSS settling data was derived from a pilot plant study:

MLSS, mg/L	1,600	2,500	2,600	4,000	5,000	8,000
Initial settling, velocity, ft/h	11.0	8.0	5.0	2.0	1.0	0.3

Solution

1. Estimate the concentration of soluble BOD_5 in the effluent using the following relationship:

$$\text{Effluent } BOD_5 = \text{influent soluble } BOD_5 \text{ escaping treatment} + BOD_5 \text{ of effluent suspended solids}$$

   (a) Determine the BOD_5 of the effluent suspended solids.

       i. Biodegradable portion of effluent biological solids is 0.65(22 mg/L) = 14.3 mg/L

       ii. Ultimate BOD_L of the biodegradable effluent solids is [0.65(22 mg/L)](1.42 mg O_2 consumed/mg cell oxidized) = 20.3 mg/L

       iii. BOD_5 of effluent suspended solids = 20.3 mg/L (0.68) = 13.8 mg/L

   (b) Solve for the influent soluble BOD_5 escaping treatment.

$$20 \text{ mg/L} = S + 13.8 \text{ mg/L}$$

$$S = 6.2 \text{ mg/L}$$

2. Determine the treatment efficiency E using Eq. 8-50.

$$E = \frac{S_o - S}{S_o} 100$$

   (a) The efficiency based on soluble BOD_5 is

$$E_s = \frac{(250 - 6.2) \text{ mg/L}}{250 \text{ mg/L}} 100 = 97.5\%$$

(b) The overall plant efficiency is

$$E_{overall} = \frac{(250 - 20) \text{ mg/L}}{250 \text{ mg/L}} 100 = 92\%$$

3. Compute the reactor volume. The volume of the reactor can be determined using Eq. 8-42,

$$X = \frac{\theta_c Y(S_o - S)}{\theta(1 + k_d\theta_c)}$$

and Eq. 8-33,

$$\theta = \frac{V_r}{Q}$$

(a) Substituting for θ in Eq. 8-42 and solving for V yields

$$V_r = \frac{\theta_c QY(S_o - S)}{X(1 + k_d\theta_c)}$$

(b) Compute the reactor volume using the following data:

$$\theta_c = 10 \text{ d}$$

$$Q = 5.71 \text{ Mgal/d}$$

$$Y = 0.50 \text{ lb/lb (assumed, see Table 8-7)}$$

$$So = 250 \text{ mg/L}$$

$$S = 6.2 \text{ mg/L}$$

$$X = 3500 \text{ mg/L}$$

$$k_d = 0.06 \text{ d}^{-1} \text{ (assumed, see Table 8-7)}$$

$$V = \frac{(10 \text{ d})(5.71 \text{ Mgal/d})(0.50)[(250 - 6.2)\text{mg/L}]}{(3500 \text{ mg/L})(1 + 0.06 \times 10)}$$

$$= 1.24 \text{ Mgal } (4694 \text{ m}^3)$$

4. Compute the quantity of sludge that must be wasted each day.
   (a) Determine Y_{obs} using Eq. 8-44.

$$Y_{obs} = \frac{Y}{1 + k_d\theta_c} = \frac{0.5}{(1 + 0.06 \times 10)} = 0.3125$$

(b) Determine the increase in the mass of mixed-liquor volatile suspended solids (MLVSS) using Eq. 10-3.

$$P_x = Y_{obs}Q(S_o - S)(8.34)$$

$$= 0.3125(5.71 \text{ Mgal/d})(250 - 6.2 \text{ mg/L})[8.34 \text{ lb/Mgal·(mg/L)}]$$

$$= 3628 \text{ lb/d } (1646 \text{ kg/d})$$

(c) Determine the increase in the total mass of mixed-liquor suspended solids (MLSS).

$$P_{x(SS)} = 3628/0.8$$
$$= 4535 \text{ lb/d } (2057 \text{ kg/d})$$

(d) Determine the amount of sludge to be wasted.

Mass to be wasted = increase in MLSS − SS lost in effluent

$$= 4535 \text{ lb/d} - 5.71 \text{ Mgal/d} \times 22 \text{ mg/L} \times [8.34 \text{ lb/Mgal} \cdot (\text{mg/L})]$$

$$= 3487 \text{ lb/d } (1582 \text{ kg/d})$$

Note: If in step 4a it had been assumed that the additional amount of sludge in the settling tanks and sludge return lines was equal to 30 percent of the amount of sludge in the aerator, then, assuming that the values of Y and k_d are applicable, the computed value of Y_{obs} would have been equal to 0.281. The mass of sludge computed in step 4c would then have been equal to 4078 lb/d (1850 kg/d) instead of 4535 lb/d (2057 kg/d).

5. Compute the sludge-wasting rate if wasting is accomplished from the reactor. Assume that $Q_e = Q$ and that the VSS in the effluent is equal to 80 percent of the SS (see Comment at end of example).
   Using Eq. 8-34,

$$\theta_c = \frac{V_r X}{Q_w X + Q_e X_e}$$

$$10 \text{ d} = \frac{(1.24 \text{ Mgal})(3500 \text{ mg/L})}{(Q_w \text{ Mgal/d})(3,500 \text{ mg/L}) + (5.71 \text{ Mgal/d})(22 \text{ mg/L} \times 0.8)}$$

$$Q_w = 0.095 \text{ Mgal/d } (360 \text{ m}^3/\text{d})$$

6. Estimate the recirculation ratio by writing a mass balance around the reactor.
   Note: For a more accurate estimate, the net cell growth within the reactor must be considered in computing the recirculation ratio.

Aerator VSS concentration = 3500 mg/L

Return VSS concentration = 8000 mg/L

$$3500(Q + Q_r) = 8000(Q_r)$$

$$\frac{Q_r}{Q} = \alpha = 0.78$$

7. Compute the hydraulic retention time for the reactor.

$$\theta = \frac{V_r}{Q} = \frac{1.24 \text{ Mgal}}{(5.71 \text{ Mgal/d})} = 0.217 \text{ d } = 5.2 \text{ h}$$

8. Compute the oxygen requirements based on ultimate carbonaceous demand, BOD_L.
   Note: Although O_2 requirements for nitrification are neglected in this example, they must be considered in the design of systems operating at mean cell-residence times sufficiently high to allow nitrification to occur (see discussion in Chap. 11).

   (a) Compute the mass of ultimate BOD_L of the incoming wastewater that is converted in the process, assuming that the BOD_5 is equal to 0.68 BOD_L.

$$\text{Mass of BOD}_L \text{ utilized} = \frac{Q(S - S_o)}{0.68} \times 8.34$$

$$= \frac{5.71 \text{ Mgal/d } (250 \text{ mg/L} - 6.2 \text{ mg/L})}{0.68} \times \frac{8.34 \text{ lb}}{\text{Mgal} \cdot (\text{mg/L})}$$

$$= 17,074 \text{ lb/d } (7744 \text{ kg/d})$$

(b) Compute the oxygen requirement using Eq. 10-6.

$$\text{lb, } O_2/d = 17,074 \text{ lb/d} - 1.42 (3628) \text{ lb/d}$$
$$= 11,922 \text{ lb/d } (5408 \text{ kg/d})$$

9. Check the *F/M* ratio and the volumetric loading factor.

(a) Determine the *F/M* ratio using Eq. 8-48.

$$F/M = \frac{S_o}{\theta X} = \frac{250 \text{ mg/L}}{(0.217 \text{ d})(3500 \text{ mg/L})} = 0.33 \text{ d}^{-1}$$

(b) Determine the volumetric loading.

$$\text{Volumetric loading, lb/}10^3\text{ft}^3\cdot\text{d} = \frac{S_o Q}{V_r} \times 8.34 \times (1000/10^3)$$

$$= \frac{(250 \text{ mg/L})(5.71 \text{ Mgal/d})}{1,240,000 \text{ gal/}(7.48 \text{ gal/ft}^3)}$$
$$\times [8.34 \text{ lb/Mgal} \cdot (\text{mg/L})](1000 \text{ ft}^3/10^3 \text{ ft}^3)$$

$$= 71.8 \text{ lb BOD}_5/10^3 \text{ ft}^3\cdot\text{d } (1.15 \text{ kg BOD}_5/\text{m}^3 \cdot \text{d})$$

10. Compute the volume of air required, assuming that the oxygen-transfer efficiency for the aeration equipment to be used is 8 percent. A safety factor of 2 should be used to determine the actual design volume for sizing the blowers.

(a) The theoretical air requirement, assuming that air contains 23.2 percent oxygen by weight, is

$$\frac{11,922 \text{ lb/d}}{(0.075 \text{ lb/ft}^3)(0.232)} = 685,200 \text{ ft}^3/d \text{ } (19,400 \text{ m}^3/d)$$

(b) Determine the actual air requirement at an 8 percent transfer efficiency.

$$\frac{685,000 \text{ ft}^3/d}{0.08} = 8,565,000 \text{ ft}^3/d \text{ } (242,535 \text{ m}^3/d)$$

or

$$\frac{8,565,000 \text{ ft}^3/d}{1440 \text{ min/d}} = 5948 \text{ ft}^3/\text{min } (166 \text{ m}^3/\text{min})$$

(c) Determine the design air requirement.

$$2(5948) = 11,896 \text{ ft}^3/\text{min } (337 \text{ m}^3/\text{min})$$

11. Check the air volume using the actual value determined in step 10b.

(a) Air requirement per unit volume:

$$\frac{8,565,000 \text{ ft}^3/d}{5,710,000 \text{ gal/d}} = 1.50 \text{ ft}^3/\text{gal } (11.2 \text{ m}^3/\text{m}^3)$$

(*b*) Air requirement per pound of BOD_5 removed:

$$\frac{8,565,000 \text{ ft}^3/\text{d}}{(250\text{mg/L} - 6.2\text{mg/L})(5.71 \text{ Mgal/d}) \times [8.34 \text{ lb/Mgal} \cdot (\text{mg/L})]}$$

$$= 738 \text{ ft}^3/\text{lb of } BOD_5 \text{ removed } (46.1 \text{ m}^3/\text{kg})$$

12. Develop the gravity solids-flux curve using the settling data for design of the required settling facilities.

   (*a*) Plot the given settling-column test data on log-log paper (see following figure).

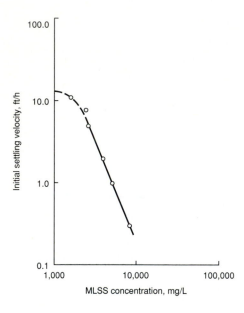

   (*b*) Using the curve plotted in the figure, obtain the data necessary to develop the solids-flux curve.

Solids concentration X, mg/L	1,000	1,500	2,000	2,500	3,000	4,000	5,000	6,000	7,000	8,000	9,000
Initial settling velocity V_i, ft/hr	13.2	11.5	9.2	5.9	3.7	1.8	1.0	0.66	0.43	0.31	0.23
Solids flux lb/ft^2 · d[a]	0.82	1.08	1.15	0.92	0.69	0.45	0.31	0.25	0.19	0.15	0.13

[a] $SF_g = \dfrac{XY_i}{16,030}$ (see Eq. 6-28)

(c) Plot the solids-flux values determined in step b versus the concentration (see following figure).

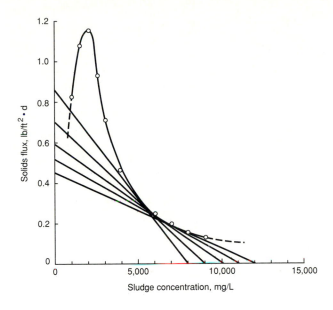

13. Using the solids-flux curve developed in step 12(c), determine the limiting solids-flux values for underflow concentrations varying from 8000 to 12000 mg/L.

(a) Using the alternative geometric construction procedure outlined in Chap. 6, draw straight lines tangent to the solids-flux curve passing through the desired underflow concentration (see figure).

(b) Prepare a summary table of the limiting solids-flux values (y intercept) for the various underflow concentrations.

Underflow concentration, mg/L	8,000	9,000	10,000	11,000	12,000
Limiting solids flux SF_L, lb/ft^2 · d	0.86	0.70	0.58	0.51	0.43

14. Determine the recycle ratio necessary to maintain mixed-liquor suspended-solids concentration at 4375 mg/L (3500 mg/L/0.8).

(a) The required recycle ratio can be determined by performing a materials balance on the influent to the reactor. The resulting expression is

$$Q(X_o) + Q_r(X_u) = (Q + Q_r) \times 4375 \text{ mg/L}$$

where Q = influent flow rate, Mgal/d
       Q_r = recycle flow rate, Mgal/d
       X_o = influent suspended solids, mg/L
       X_u = underflow suspended solids, mg/L

Assuming that $X_o = 0$ and $Q_r = \alpha Q$, the above expression can be written as

$$\alpha Q X_u - \alpha(4375 \text{ mg/L})Q = Q(4375 \text{ mg/L})$$

$$\alpha = \frac{4375 \text{ mg/L}}{X_u \text{ mg/L} - 4375 \text{ mg/L}}$$

where α = recycle ratio, Q_r/Q

(b) Determine the required recycle ratios for the various underflow concentrations.

X_u, mg/L	8,000	9,000	10,000	11,000	12,000
$X_u - 4{,}375$, mg/L	3,625	4,625	5,625	6,625	7,625
α	1.21	0.95	0.78	0.66	0.57

15. Determine the required thickening area of the clarifier for the various underflow concentrations and recycle ratios using the following modified form of Eq. 6-33.

$$SF_a = \frac{(1 + \alpha)(Q)(X)(8.34)}{24A}$$

where SF_a = average applied solids flux, lb/ft^2·h
   α = recycle ratio
   Q = flowrate, Mgal/d
   X = mixed liquor suspended solids concentration, mg/L
   A = cross-sectional area of clarifier, ft^2
   24 = h/d

(a) Assume that $SF_a = SF_L$, the limiting solids flux determined in step 13.
(b) Set up a computation table to determine the required area.

X_u, mg/L	8,000	9,000	10,000	11,000	12,000
SF_L, lb/ft^2·h	0.86	0.70	0.58	0.51	0.43
α	1.21	0.95	0.78	0.66	0.57
A, ft^2	22,300	24,300	26,500	28,100	31,700

16. Determine the overflow rates corresponding to the solids loading computed in step 15b.

X_u, mg/L	8,000	9,000	10,000	11,000	12,000
Solids loading, lb/ft^2·h[a]	0.86	0.70	0.58	0.51	0.43
OF[b], gal/ft^2·d	256	235	215	203	180

[a] Corresponds to limiting solids flux.
[b] OF = overflow rate based on plant flow and area computed in step 15b.

17. Check the clarification requirement, assuming that the final design will be based on an underflow concentration of 10,000 mg/L.

   (a) As shown in step 16, the overflow rate for an underflow concentration of 10,000 mg/L is equal to 215 gal/ft$^2 \cdot$ d. This is equivalent to a settling rate of 1.2 ft/h (0.37 m/h).

   (b) Referring to the settling curve, a settling rate of 1.2 ft/h (0.37 m/h) would correspond to a solids concentration of 4700 mg/L. Because the concentration of the solids at the interface will be below this value, the area for clarification appears to be conservative.

18. Estimate the required depth for thickening. Assume that the minimum allowable depth for the clarified zone in the sedimentation tank is to be set at 5 ft (1.5 m).

   (a) Estimate the required depth of the thickening zone. Assume that under normal conditions the mass of sludge retained in the secondary settling tank is equal to 30 percent of the mass in the aeration tank and that the average concentration of solids in the sludge zone is equal to approximately 7000 mg/L [(4000 + 10,000) mg/L/2].

   i. Determine the mass of solids in the aeration basin.

   $$\text{Aeration tank solids} = 1.24 \text{ Mgal}(4375 \text{ mg/L})[8.34 \text{ lb/Mgal} \cdot (\text{mg/L})] = 45,245 \text{ lb}$$

   ii. Determine the mass of solids in the sedimentation basin.

   $$\text{Sedimentation basin solids} = 0.3(45,245) = 13,573 \text{ lb}$$

   iii. Determine the depth of the sludge zone in the sedimentation basin using the following relationship:

   $$(A, \text{ft}^2)(d, \text{ft})(7000 \text{ mg/L})(k) = 13,573 \text{ lb}$$

   where k (conversion factor) $= 1/16{,}030 \text{ lb/ft}^3 \cdot (\text{mg/L})$ (see Eq. 6-28)

   $$d = \frac{(13,573 \text{ lb})}{(7000 \text{ mg/L})(26,500 \text{ ft}^2)}\left(\frac{16,030 \text{ ft}^3(\text{mg/L})}{\text{lb}}\right)$$

   $$= 1.17 \text{ ft } (0.36 \text{ m})$$

   (b) Estimate the required storage capacity in the sludge zone assuming that excess solids must be stored in the secondary sedimentation tank at peak flow conditions because of limitations in the sludge-processing facilities. Assume that the 2 d sustained peak flowrate of 2.5 Q_{avg} and the 7 d sustained peak BOD loading of 1.5 BOD_{avg} occur simultaneously (see Chap. 5).

   i. Estimate the solids produced under the given conditions using Eq. 10-3.

   $$P_x = Y_{obs}Q(S_o - S) \times (8.34)$$

   $$Y_{obs} = 0.3125$$

   $$Q = 2.5(5.71 \text{ Mgal/d}) = 14.28 \text{ Mgal/d}$$

   $$S_o = 1.5(250 \text{ mg/L}) = 375 \text{ mg/L}$$

   $$S = 15 \text{ mg/L (assumed under increased loading conditions)}$$

   $$(P_x)_{sp} = 0.3125(14.28)(375 - 15)(8.34)$$

   $$= 13,398 \text{ lb } (6077 \text{ kg})$$

ii. Because the peak flowrate is sustained for 2 d, the total solids for the 2 d period are equal to 26,796 lb.

iii. Compute the required depth for sludge storage in the sedimentation tank. Assume that the total solids in the sedimentation tank are now equal to 40,369 lb (26,796 + 13,573).

$$d = \frac{(40,369 \text{ lb})}{(7000 \text{ mg/L})(26,500 \text{ ft}^2)} \left(\frac{16,030 \text{ ft}^3(\text{mg/L})}{\text{lb}} \right)$$

$$= 3.49 \text{ ft } (1.06 \text{ m})$$

(c) Estimate the total required depth.

$$\text{Depth} = (5.0 + 1.17 + 3.49)\text{ft}$$

$$= 9.66 \text{ ft (use minimum of 12 ft, see Table 10-12)}$$

19. Check the surface overflow rate at peak flow.

(a) The peak flow is

$$Q_p = 2.5(5,710,000 \text{ gal/d})$$

$$= 14,275,000 \text{ gal/d}$$

(b) The surface overflow rate at peak flow is

$$\text{Peak overflow rate} = \frac{14,275,000 \text{ gal/d}}{26,500 \text{ ft}^2}$$

$$= 539 \text{ gal/ft}^2 \cdot \text{d } (21.9 \text{ m}^3/\text{m}^2 \cdot \text{d})$$

This value is well below the peak value in Table 10-12.

20. Prepare a summary table of the sedimentation tank design data.

	Value	
Item	**U.S. customary units**	**SI units**
Surface area	26,500 ft²	2,462 m²
Sidewater depth[a]	12 ft	3.7 m
Detention time (avg)	10.1 h	10.1 h
Mixed-liquor suspended solids	4,375 mg/L	4,375 mg/L
Limiting solids flux	14.0 lb/ft² · d	2.85 kg/m² · h
Overflow rate		
At average flow	215 gal/ft² · d	8.8 m³/m² · d
At peak flow	539 gal/ft² · d	21.9 m³/m² · d

[a] Does not include freeboard.

Comment. The volatile fraction of the suspended solids discharged in the plant effluent will vary with the type of process and mode of operation. The water lost with the dewatered sludge and by evaporation is neglected by assuming that $Q_e = Q$.

Example 10-3 Sequencing batch reactor design. A sequencing batch reactor activated-sludge process is to be used to treat wastewater with the characteristics given below. Determine the mass of suspended solids in the reactor over a 7-day operating period. The effluent is to have 20 mg/L of BOD_5 or less. Determine also the depth of clear liquid measured from the top of the settled sludge to the lowest liquid level reached during the decant cycle. Use the following design criteria and constraints.

1. Wastewater characteristics
   (a) Influent flowrate $= 1.0$ Mgal/d ($3,800$ m³/d).
   (b) Influent suspended solids $= 200$ mg/L
   (c) Influent VSS $= 150$ mg/L
   (d) Wastewater temperature is 20°C
   (e) Influent BOD_5 (see table below)

Day	Average BOD, mg/L
1	250
2	400[a]
3	400[a]
4	400[a]
5	400[a]
6	250
7	250

[a] Assume that increase over 250 mg/L is soluble BOD.

2. Design criteria and constraints
   (a) Hydraulic detention time $= 24$ h
   (b) Design F/M $= 0.1$ lb BOD_5 applied/lb MLVSS·d (see Table 10-5)
   (c) Ratio of MLVSS/MLSS produced from the conversion of organic matter in the influent $= 0.8$
   (d) Kinetic coefficients: $Y = 0.65$ lb/lb and $k_d = 0.05$ d^{-1} (Table 8-7)
   (e) Average concentration of settled sludge $= 8,000$ mg/L
   (f) Settled sludge specific gravity $= 1.02$
   (g) Assume 60 percent of the reactor volume will be decanted each day
   (h) Liquid depth of SBR $= 22$ ft
   (i) Sludge wasting is done once a week
   (j) Effluent is estimated to contain 20 mg/L of biological solids, of which 65 percent are biodegradable
   (k) $BOD_5 = 0.68 BOD_L$
   (l) BOD_L of one mole of cells $= 1.42$ times the concentration of cells
   (m) Wastewater contains adequate nitrogen, phosphorus, and other trace nutrients for biological growth

Solution

1. Estimate the concentration of soluble BOD_5 in the effluent using the following relationship:

$$\text{Effluent } BOD_5 = \text{influent soluble } BOD_5 \text{ escaping treatment}$$
$$+ BOD_5 \text{ of effluent suspended solids}$$

(a) Determine the BOD_5 of the effluent suspended solids.

  i. Biodegradable portion of effluent biological solids is 0.65(20 mg/L) = 13.0 mg/L
  ii. Ultimate BOD_L of the biodegradable effluent solids is (13.0 mg/L) (1.42 mg/mg) = 18.5 mg/L
  iii. BOD_5 of effluent suspended solids = 18.5 mg/L (0.68) = 12.6 mg/L

(b) Solve for the influent soluble BOD_5 escaping treatment.

$$20 \text{ mg/L} = S + 12.6 \text{ mg/L}$$

$$S = 7.4 \text{ mg/L}$$

2. Compute the mixed-liquor volatile and total suspended solids concentration and the mass of VSS in the reactor

(a) The volatile suspended solids can be estimated using Eq. 8-42 as given below.

$$X = \frac{QS_o}{V(F/M)}$$

Because the hydraulic detention time is equal to 1.0 d and 60 percent of the total reactor volume is to be decanted each day, the required tank volume is

$$V = \frac{1.0 \text{ Mgal/d} \times 1.0 \text{ d}}{0.60} = 1.67 \text{ Mgal}$$

Using this tank volume, the required MLVSS concentration is

$$X = \frac{1.0 \text{ Mgal/d } (250 \text{ mg/L})}{1.67 \text{ Mgal}(0.1 \text{ lb BOD/lb MLVSS·d})} = 1497 \text{ mg/L}$$

(b) The total SS concentration in the reactor can be estimated as follows:

$$SS_T = \text{Average inert influent SS} + VSS/0.8$$
$$= (200 - 150)\text{mg/L} + (1497 \text{ mg/L})/0.8$$
$$= 50 \text{ mg/L} + 1871 = 1921 \text{ mg/L}$$

(c) The mass of VSS in the reactor is

$$\text{Mass of VSS} = (1.67 \text{ Mgal})(1497 \text{ mg/L})[8.34 \text{ lb/Mgal} \cdot (\text{mg/L})] = 20,850 \text{ lb}$$

(d) The total mass of SS in the reactor is

$$\text{Mass of SS} = (1.67 \text{ Mgal})(1921 \text{ mg/L})[8.34 \text{ lb/Mgal} \cdot (\text{mg/L})] = 26,755 \text{ lb}$$

3. Estimate the volume occupied by the settled sludge at the end of seven days before the waste sludge is to be removed from the reactor.

(a) Determine the mass of SS in the reactor at the end of each day using the following relationship:

$$X_n = X_o + \sum_{n=1}^{n=7} [(P_{x_n})/0.8 + SS_{i_n}]$$

FIGURE 10-30
Typical aerated lagoon with large floating slow-speed aerators.

in Chap. 8. The mean cell-residence time should be selected to ensure (1) that the suspended microorganisms will bioflocculate for easy removal by sedimentation and (2) that an adequate safety factor is provided when compared to the mean cell-residence time of washout. Typical design values of θ_c for aerated lagoons used for treating domestic wastes vary from about 3 to 6 d. Once the value of θ_c has been selected, the soluble substrate concentration of the effluent can be estimated, and the removal efficiency can then be computed using the equations given in Chap. 8.

An alternative approach is to assume that the observed BOD_5 removal (either overall, including soluble and suspended-solids contribution, or soluble only) can be described in terms of a first-order removal function. The BOD_5 removal is measured between the influent and lagoon outlet (not the outlet of the sedimentation facilities following the lagoon). The pertinent equation for a single aerated lagoon (see Appendix G for derivation) is

$$\frac{S}{S_0} = \frac{1}{1 + k(V/Q)}$$ (10-20)

where S = effluent BOD_5 concentration, mg/L
 S_0 = influent BOD_5 concentration, mg/L
 k = overall first-order BOD_5 removal-rate constant, d^{-1}
 V = volume, Mgal (m^3)
 Q = flowrate, Mgal/d (m^3/d)

Reported overall k values vary from 0.25 to 1.0. Removal rates for soluble BOD_5 would be higher. Application of this equation is illustrated in Example 10-5 presented later in this section.

Effluent Characteristics. The important characteristics of the effluent from an aerated lagoon include the BOD_5 and the suspended-solids concentration. The effluent BOD_5 will be made up of those components previously discussed in connection with the activated-sludge process and occasionally may contain the contribution of small amounts of algae. The solids in the effluent are composed of a portion of the incoming suspended solids, the biological solids produced from waste conversion, and occasionally small amounts of algae. The solids produced from the conversion of soluble organic wastes can be estimated using Eq. 8-27.

Oxygen Requirement. The oxygen requirement is computed as previously outlined in Sec. 10-1, which deals with the activated-sludge process design. Based on operating results obtained from a number of industrial and domestic installations, the amount of oxygen required has been found to vary from 0.7 to 1.4 times the amount of BOD_5 removed.

Temperature. Because aerated lagoons are installed and operated in locations with widely varying climatic conditions, the effects of temperature change must be considered in their design. The two most important effects of temperature are (1) reduced biological activity and treatment efficiency and (2) the formation of ice.

The effect of temperature on biological activity is described in Chap. 8. From a consideration of the influent wastewater temperature, air temperature, surface area of the pond, and wastewater flowrate, the resulting temperature in the aerated lagoon can be estimated using the following equation developed by Mancini and Barnhart [24]:

$$(T_i - T_w) = \frac{(T_w - T_a)fA}{Q} \tag{10-21}$$

where T_i = influent waste temperature, °F (°C)
       T_w = lagoon water temperature, °F (°C)
       T_a = ambient air temperature, °F (°C)
        f = proportionality factor
        A = surface area, ft^2 (m^2)
        Q = wastewater flowrate, Mgal/d (m^3/d)

The proportionality factor incorporates the appropriate heat transfer coefficients and includes the effect of surface area increase due to aeration, wind, and humidity. A typical value for the eastern United States is 12×10^{-6} in U.S. customary units (0.5 in SI units). To compute the lagoon temperature, Eq. 10-21 is rewritten as

$$T_w = \frac{AfT_a + QT_i}{Af + Q} \tag{10-22}$$

Alternatively, if climatological data are available, the average temperature of the lagoon may be determined from a heat budget analysis by assuming that the lagoon is mixed completely.

Where icing may be a problem, its effects on the operation of lagoons may be minimized by increasing the depth of the lagoon or by altering the method of operation.

where X_o = initial mass of SS after decanting

P_{x_n} = net mass of solids produced in day n from the conversion of the organic matter in the wastewater

SS_{i_n} = mass of inert solids added each day

The value of P_{x_n} based on VSS can be computed for any day using the following expression:

$$P_{x_n} = Y(S_o - S)Q(8.34) - k_d X_{n-1}$$

where X_{n-1} = mass of VSS in the system at the beginning of day n. To be more precise, the value of X used in the above equation should be an average value for day n. However, by using the value of X at the beginning of day n, the value of P_{x_n} is over-estimated by a small amount (see subsequent computations). The slight over-estimation of P_{x_n}, in turn results in a more conservative estimate of the total mass in the reactor, which is acceptable.

For example, the net mass of VSS produced in day one is equal to

$$P_{x_1} = 0.65(250 - 7.4)(1.0)(8.34) - (0.05 \times 20,850) = 273 \text{ lb}$$

The mass of inert SS added in day one is equal to

$$P_{x_1} = (50)(1.0)(8.34) = 417 \text{ lb}$$

The mass of SS in the reactor at the end of day one is equal to

$$X_1 = 26,755 + (273)/0.8 + 417 = 27,513 \text{ lb}$$

Similarly, the net mass of VSS produced in day two is equal to

$$P_{x_2} + 0.65(400 - 7.4)(1.0)(8.34) - [0.05 \times (20,850 + 273)] = 1072 \text{ lb}$$

The mass of inert SS added in day two is equal to

$$P_{x_2} = (50)(1.0)(8.34) = 417 \text{ lb}$$

The mass of SS in the reactor at the end of day two is equal to

$$X_2 = 27,513 + (1072)/0.8 + 417 = 29,270 \text{ lb}$$

Following the same line of reasoning, the total mass of VSS and SS in the reactor at the end of each day is summarized in the following table.

Day	BOD, mg/L	P_x, lb/d	SS_i, lb/d	VSS_T, lb	SS_T, lb
1	250	273	417	21,123	27,513
2	400	1,072	417	22,195	29,270
3	400	1,018	417	23,213	30,959
4	400	967	417	24,180	32,585
5	400	919	417	25,099	34,151
6	250	60	417	25,159	34,643
7	250	57	417	25,216	35,131

4. Determine the volume required for sludge storage. For an average concentration of the settled sludge of 8,000 mg/L and the specific gravity of 1.02, the approximate volume for settled sludge is

$$V_s = \frac{35,131 \text{ lb}}{1.02(62.4 \text{ lb/ft}^3)(8,000/10^6)} = 68,995 \text{ ft}^3 = 516,080 \text{ gal}$$

The available volume is 670,000 gal. Because the required volume for sludge storage is less than the available volume, the decant system should function acceptably (see "Comment").

5. Determine the depth of clear liquid above the top of the sludge layer:

$$\text{Total liquid depth after decanting} = (1 - 0.60) \times 22 \text{ ft} = 8.80 \text{ ft}$$

$$\text{Sludge depth} = (8.80 \text{ ft})\frac{516,080 \text{ gal}}{670,000 \text{ gal}} = 6.77 \text{ ft}$$

$$\text{Clear liquid depth} = 8.80 - 6.77 = 2.03 \text{ ft}$$

Comment. The decant volume in SBRs is often limited to about 50 percent of the total volume. At the lowest point in the decant cycle, the liquid level should be an adequate distance above the top of the settled sludge to avoid the discharge of settled solids.

10-4 AERATED LAGOONS

An aerated lagoon is a basin in which wastewater is treated either on a flow-through basis or with solids recycle. The essential function of this treatment process is waste conversion. Oxygen is usually supplied by means of surface aerators or diffused air units. As with other suspended-growth systems, the turbulence created by the aeration devices is used to maintain the contents of the basin in suspension.

Depending on the detention time, the effluent from an aerated lagoon contains about one-third to one-half the value of the incoming BOD in the form of cell tissue. Most of these solids must be removed by settling prior to discharge (a settling tank or basin is a normal component of most lagoon systems). If the solids are returned to the lagoon, there is no difference between this process and a modified activated-sludge process. A typical aerated lagoon is shown in Fig. 10-30.

Process Design Considerations

Factors that must be considered in the process design of aerated lagoons include (1) BOD removal, (2) effluent characteristics, (3) oxygen requirements, (4) temperature effects, (5) energy requirement for mixing, and (6) solids separation. The first four factors are considered in the following discussion, and their application is illustrated in Example 10-4. The energy required for mixing was discussed previously (see "Mechanical Aerators"). Solids separation is discussed at the end of this section.

BOD Removal. Because an aerated lagoon can be considered a complete-mix reactor without recycle, the basis of design can be the mean cell-residence time, as outlined

FIGURE 10-30
Typical aerated lagoon with large floating slow-speed aerators.

in Chap. 8. The mean cell-residence time should be selected to ensure (1) that the suspended microorganisms will bioflocculate for easy removal by sedimentation and (2) that an adequate safety factor is provided when compared to the mean cell-residence time of washout. Typical design values of θ_c for aerated lagoons used for treating domestic wastes vary from about 3 to 6 d. Once the value of θ_c has been selected, the soluble substrate concentration of the effluent can be estimated, and the removal efficiency can then be computed using the equations given in Chap. 8.

An alternative approach is to assume that the observed BOD_5 removal (either overall, including soluble and suspended-solids contribution, or soluble only) can be described in terms of a first-order removal function. The BOD_5 removal is measured between the influent and lagoon outlet (not the outlet of the sedimentation facilities following the lagoon). The pertinent equation for a single aerated lagoon (see Appendix G for derivation) is

$$\frac{S}{S_o} = \frac{1}{1 + k(V/Q)} \qquad (10\text{-}20)$$

where S = effluent BOD_5 concentration, mg/L
S_o = influent BOD_5 concentration, mg/L
k = overall first-order BOD_5 removal-rate constant, d^{-1}
V = volume, Mgal (m^3)
Q = flowrate, Mgal/d (m^3/d)

Reported overall k values vary from 0.25 to 1.0. Removal rates for soluble BOD_5 would be higher. Application of this equation is illustrated in Example 10-5 presented later in this section.

Effluent Characteristics. The important characteristics of the effluent from an aerated lagoon include the BOD_5 and the suspended-solids concentration. The effluent BOD_5 will be made up of those components previously discussed in connection with the activated-sludge process and occasionally may contain the contribution of small amounts of algae. The solids in the effluent are composed of a portion of the incoming suspended solids, the biological solids produced from waste conversion, and occasionally small amounts of algae. The solids produced from the conversion of soluble organic wastes can be estimated using Eq. 8-27.

Oxygen Requirement. The oxygen requirement is computed as previously outlined in Sec. 10-1, which deals with the activated-sludge process design. Based on operating results obtained from a number of industrial and domestic installations, the amount of oxygen required has been found to vary from 0.7 to 1.4 times the amount of BOD_5 removed.

Temperature. Because aerated lagoons are installed and operated in locations with widely varying climatic conditions, the effects of temperature change must be considered in their design. The two most important effects of temperature are (1) reduced biological activity and treatment efficiency and (2) the formation of ice.

The effect of temperature on biological activity is described in Chap. 8. From a consideration of the influent wastewater temperature, air temperature, surface area of the pond, and wastewater flowrate, the resulting temperature in the aerated lagoon can be estimated using the following equation developed by Mancini and Barnhart [24]:

$$(T_i - T_w) = \frac{(T_w - T_a)fA}{Q} \tag{10-21}$$

where T_i = influent waste temperature, °F (°C)
T_w = lagoon water temperature, °F (°C)
T_a = ambient air temperature, °F (°C)
f = proportionality factor
A = surface area, ft² (m²)
Q = wastewater flowrate, Mgal/d (m³/d)

The proportionality factor incorporates the appropriate heat transfer coefficients and includes the effect of surface area increase due to aeration, wind, and humidity. A typical value for the eastern United States is 12×10^{-6} in U.S. customary units (0.5 in SI units). To compute the lagoon temperature, Eq. 10-21 is rewritten as

$$T_w = \frac{AfT_a + QT_i}{Af + Q} \tag{10-22}$$

Alternatively, if climatological data are available, the average temperature of the lagoon may be determined from a heat budget analysis by assuming that the lagoon is mixed completely.

Where icing may be a problem, its effects on the operation of lagoons may be minimized by increasing the depth of the lagoon or by altering the method of operation.

The effect of reducing the surface area is illustrated in Example 10-4. As computed, reducing the area by one-half increases the temperature about 6.8°F (3.8°C), which corresponds roughly to about a 50 percent increase in the rate of biological activity. As the depth of the lagoon is increased, maintenance of a completely mixed flow regime becomes difficult. If the depth is increased much beyond 12 ft (3.7 m), draft tube aerators or diffused aeration must be used.

Example 10-4 Effect of pond surface area on liquid temperature. Determine the effect of reducing the surface area of an aerated lagoon from 100,000 to 50,000 ft² (9,290 to 4,645 m²) by doubling the depth for the following conditions:

1. Wastewater flowrate, $Q = 1$ Mgal/d (3800 m³/d)
2. Wastewater temperature, $T_i = 60$ °F (15.6 °C)
3. Air temperature, $T_a = 20$ °F (−6.7°C)
4. Proportionality constant, $f = 12 \times 10^{-6}$

Solution

1. Determine the lagoon water temperature for a surface area of 100,000 ft² using Eq. 10-22:

$$T_w = \frac{Af T_a + QT_i}{Af + Q}$$

$$T_w = \frac{100,000(12 \times 10^{-6})(20) + 1(60)}{100,000(12 \times 10^{-6}) + 1} = 38.2°F \ (3.4°C)$$

2. Determine the lagoon water temperature for a surface of 50,000 ft².

$$T_w = \frac{50,000(12 \times 10^{-6})(20) + 1(60)}{50,000(12 \times 10^{-6}) + 1} = 45°F \ (7.2°C)$$

In multiple lagoon systems, cold weather effects can be mitigated by seasonal changes in the method of operation. During the warmer months, the lagoons would be operated in parallel. In the winter, they would be operated in series. In the winter operating mode, the downstream aerators could be turned off and removed, and the lagoon surface allowed to freeze. In spring when the ice melts, the parallel method of operation is again adopted. With this method of operation, it is possible to achieve a 60 to 70 percent removal of BOD$_5$ even during the coldest winter months. Still another method that can be used to improve performance during the winter months is to recycle a portion of the solids removed by settling.

Aerated Lagoon Process Design

The design of an aerated lagoon is illustrated in Example 10-5.

Example 10-5 Design of an aerated lagoon. Design a flow-through aerated lagoon to treat a wastewater flow of 1 Mgal/d (3800 m³/d), including the number of surface aerators and their horsepower rating. The treated liquid is to be held in a settling basin with a 2 d detention time before discharge. Assume that the following conditions and requirements apply:

1. Influent suspended solids = 200 mg/L.
2. Influent suspended solids are not biologically degraded.
3. Influent soluble BOD_5 = 200 mg/L.
4. Effluent soluble BOD_5 = 20 mg/L.
5. Effluent suspended solids after settling = 20 mg/L.
6. Kinetic coefficients: Y = 0.65, K_s = 100 mg/L, k = 6.0 d^{-1}, k_d = 0.07 d^{-1}.
7. Total biological solids produced are equal to computed volatile suspended solids divided by 0.80.
8. First-order soluble BOD_5 removal-rate constant k_{20} = 2.5 d^{-1} at 20°C.
9. Summer air temperature = 86°F (30°C).
10. Winter air temperature = 50°F (10°C).
11. Wastewater temperature = 60°F (15.6°C).
12. Temperature coefficient: θ = 1.06.
13. Aeration constants: α = 0.85, β = 1.0.
14. Elevation = 2000 ft (610 m).
15. Oxygen concentration to be maintained in liquid = 1.5 mg/L.
16. Lagoon depth = 10 ft (3 m).
17. Design mean cell-residence time θ_c = 4 d.

Solution

1. On the basis of a mean cell-residence time of 4 d, determine the surface area of the lagoon.

$$\text{Volume } V = Q\theta_c = (1 \text{ Mgal/d}) 4d = 4,000,000 \text{ gal} \times \frac{1 \text{ ft}^3}{7.48 \text{ gal}}$$

$$= 535,000 \text{ ft}^3 \ (15,150 \text{ m}^3)$$

$$\text{Surface area} = \frac{535,000 \text{ ft}^3}{10 \text{ ft}} = 53,500 \text{ ft}^2 \ (0.5 \text{ ha})$$

2. Estimate the summer and winter liquid temperatures using Eq. 10-22.

Summer:

$$T_w = \frac{53,500 \, (12 \times 10^{-6})(86) + 1(60)}{53,500 \, (12 \times 10^{-6}) + 1}$$
$$= 70.2°F \ (21.2°C)$$

Winter:

$$T_w = \frac{53,500 \, (12 \times 10^{-6})(50) + 1(60)}{53,500 \, (12 \times 10^{-6}) + 1}$$
$$= 56.1°F \ (13.4°C)$$

3. Estimate the soluble effluent BOD_5 measured at the lagoon outlet during the summer using Eq. 8-28.

$$S = \frac{K_s(1 + \theta k_d)}{\theta(Yk - k_d) - 1}$$

$$= \frac{100[1 + 4(0.07)]}{4[0.65(6) - 0.07] - 1}$$

$$= 8.9 \text{ mg/L}$$

(*Note:* The value in the effluent from the settling facilities will be essentially the same.) This value was computed using kinetic-growth constants derived for the temperature in the range from 20 to 25°C. Thus, during the summer months, the effluent requirement of 20 mg/L or less will be met easily. Because there is no reliable information on how to correct these constants for the winter temperature of 13.4°C (56.1°F), an estimate of the effect of temperature can be obtained using the first-order soluble BOD_5 removal-rate constant.

4. Estimate the effluent BOD_5.

(*a*) Correct the removal-rate constant for temperature effects using Eq. 8-14:

$$\frac{k_T}{k_{20}} = \theta^{T-20}$$

Summer (70.2°F) (21.2°C):

$$k_{21.2} = 2.5(1.06)^{21.2-20} = 2.71$$

Winter (56.1°F) (13.4°C):

$$k_{13.4} = 2.5(1.06)^{13.4-20} = 1.7$$

(*b*) Determine the effluent BOD_5 using Eq. 10-20, substituting θ for $= V/Q$:

$$\frac{S}{S_o} = \frac{1}{1 + k\theta}$$

Summer (70.2°F):

$$\frac{S}{200} = \frac{1}{1 + 2.71(4)}$$

$$S = 16.9 \text{ mg/L}$$

Winter (56.1°F):

$$\frac{S}{200} = \frac{1}{1 + 1.7(4)}$$

$$S = 25.6 \text{ mg/L}$$

$$\text{Ratio of } \frac{S_{winter}}{S_{summer}} = \frac{25.6}{16.9} = 1.5$$

Applying the ratio to the soluble effluent BOD_5 computed using the kinetic growth constants yields a value of about 13 mg/L. Using the ratio of the removal-rate constants yields approximately the same value.

Note: The foregoing calculations were presented only to illustrate the method. The value of the removal-rate constant must be evaluated for the wastewater in question, in a bench or pilot-scale test program as outlined in Chap. 8 and Appendix H.

5. Estimate the concentration of biological solids produced using Eq. 8-27.

$$X = \frac{Y(S_o - S)}{1 + k_d \theta} = \frac{0.65(200 - 8.9)}{1 + 0.07(4)} = 97 \text{ mg/L VSS}$$

An approximate estimate of the biological solids produced can be obtained by multiplying the assumed growth-yield constant (BOD_5 basis) by the BOD_5 removed.

6. Estimate the suspended solids in the lagoon effluent before settling.

$$SS = 200 \text{ mg/L} + \frac{97 \text{ mg/L}}{0.80} = 321 \text{ mg/L}$$

With the extremely low overflow rate provided in a holding basin with a detention time of 2 d, an effluent containing less than 20 mg/L of suspended solids should be attainable.

7. Estimate the oxygen requirement using Eq. 10-5.

$$\text{lb } O_2/d = \frac{Q(S_o - S) \times 8.34}{f} - 1.42 P_x$$

(a) Determine P_x, the amount of biological solids wasted per day.

$$P_x = (97 \text{ mg/L})(1.0 \text{ Mgal/d}) [8.34 \text{ lb/Mgal} \cdot (\text{mg/L})] = 809 \text{ lb/d}$$

(b) Assuming that the conversion factor for BOD_5 to BOD_L is 0.68, determine the oxygen requirements.

$$\text{lb } O_2/d = \frac{(1.0 \text{ Mgal/d})[(200 - 8.9) \text{ mg/L} \times [8.34 \text{ lb/Mgal} \cdot (\text{mg/L})]}{0.68} - 1.42(809 \text{ lb/d})$$

$$= 1195 \text{ lb/d } (543 \text{ kg/d})$$

8. Compute the ratio of oxygen required to BOD_5 removed.

$$\frac{O_2 \text{ required}}{BOD_5 \text{ removed}} = \frac{1195 \text{ lb/d}}{[(200 - 8.9) \text{ mg/L}](1.0 \text{ Mgal/d})[8.34 \text{ lb/Mgal} \cdot (\text{mg/L})]}$$

$$= 0.75$$

9. Determine the surface aerator power requirements, assuming that the aerators used are rated at 3.0 lb $O_2/hp \cdot h(1.8 \text{ kg } O_2/kW \cdot h)$.

(a) Determine the correction factor for surface aerators for summer conditions using Eq. 10-19. (Note: Correction factor $= N/N_o$)

   i. Oxygen saturation concentration at 21.2°C $= 8.87$ mg/L (see Appendix E)
   ii. Oxygen saturation concentration at 21.2°C corrected for altitude $= 8.87 \times 0.94 = 8.34$ mg/L (see Fig. 10-17)
   iii. $C_{s_{20}}$ from Appendix E $= 9.08$

$$\text{Correction factor} = \left[\frac{\beta C_{walt} - C_L}{C_{s_{20}}} 1.024^{T-20} \alpha \right]$$

$$= \frac{8.34 - 1.5}{9.08}(1.024^{21.2-20})0.85$$

$$= 0.67$$

(b) The field-transfer rate N is equal to

$$N = N_o(0.67) = 3.0(0.67) = 2.01 \text{ O}_2/\text{hp} \cdot \text{hr}$$

The amount of O_2 transferred per day per unit is equal to 48.2 lb $O_2/\text{hp} \cdot \text{d}$. The total power required to meet the oxygen requirements is

$$\text{hp} = \frac{1195 \text{ lb O}_2/\text{d}}{48.2 \text{ lb O}_2/\text{hp} \cdot \text{d}} = 24.8$$

10. Check the energy requirements for mixing. Assume that, for a completely mixed-flow regime, the power requirement is 0.6 hp/1000 ft^3.

   (a) Lagoon volume $= 535,000$ ft^3
   (b) Power required $= 0.6 \times 535 = 321$ hp (239 kW)
   (c) Use eight 40 hp (30 kW) surface aerators.

   Comment. For installations designed to treat domestic wastewater, the energy requirement for mixing is usually the controlling factor in sizing the aerators. The energy needed to meet the oxygen required is often the controlling factor in sizing the aerators where industrial wastes are to be treated. It should be noted that in some cases when the power requirements for mixing are significantly greater than the power required for oxygen transfer, aerated lagoons have not been operated in the complete-mix mode.

Solids Separation

If the effluent from aerated lagoons must meet the minimum standards for secondary treatment as defined by the U.S. Environmental Protection Agency (see Table 4-1), it will be necessary to provide some type of settling facility. Usually, sedimentation is accomplished in a large, shallow earthen basin used expressly for the purpose or in more conventional settling facilities. Where large earthen basins are used, the following requirements must be considered carefully: (1) the detention time must be adequate to achieve the desired degree of suspended-solids removal; (2) sufficient volume must be provided for sludge storage; (3) algal growth must be minimized; (4) odors that may develop as a result of the anaerobic decomposition of the accumulated sludge must be controlled; and (5) the need for a lining must be assessed. In some cases, because of local conditions, these requirements may be in conflict with each other.

In most cases, a minimum detention time of 6 to 12 h is required to achieve solids separation [1]. If a 6 to 12 h detention time is used, adequate provision must be made for sludge storage so that the accumulated solids will not reduce the actual liquid detention time. Further, if all the solids become deposited in localized patterns, it may be necessary to increase the detention time to counteract the effects of poor hydraulic distribution. Under anaerobic conditions, about 40 to 60 percent of the deposited volatile suspended solids will be degraded each year. Assuming that first-order removal kinetics apply, the following expression can be used to estimate the decay of volatile suspended solids [1].

$$W_t = W_o e^{-k_d t} \tag{10-23}$$

where W_t = mass of volatile suspended solids that have not degraded after time t, lb (kg)

W_o = mass of solids deposited initially, lb (kg)

k_d = decay of coefficient, d^{-1} or yr^{-1}

t = time, d or yr

Two problems that are often encountered with the use of settling basins are the growth of algae and the production of odors. Algal growths can usually be controlled by limiting the hydraulic detention time to 2 d or less. If longer detention times must be used, the algal content may be reduced by using either a rock filter (see Sec. 10-8) or a microstrainer. Odors arising from anaerobic decomposition can generally be controlled by maintaining a minimum water depth of 3 ft (1 m). In extremely warm areas, depths up to 6 ft (1.8 m) have been needed to eliminate odors, especially those of hydrogen sulfide.

If space for large settling basins is unavailable, conventional settling facilities can be used. To reduce the construction costs associated with conventional concrete and steel settling tanks, lined earthen basins can be used. The design of a large earthen sedimentation basin for an aerated lagoon is illustrated in Example 10-6.

Example 10-6 Design of a large earthen sedimentation basin for an aerated lagoon. Design an earthen sedimentation basin for the aerated lagoon designed in Example 10-5. Assume that the hydraulic detention time is to be 2 d and that the liquid level above the sludge layer at its maximum level of accumulation is to be 5 ft (1.5 m). For the purposes of this example, assume that 70 percent of the total solids discharged to the sedimentation basin are volatile. Also assume that the sedimentation pond is cleaned after 4 years.

Solution

1. Determine the mass of sludge that must be accumulated in the basin each year without anaerobic decomposition.

$$\text{Mass} = (SS_i - SS_e)Q[8.34 \text{ lb/Mgal} \cdot (\text{mg/L})](365 \text{ d/yr})$$

where SS_i = suspended solids in the influent to the sedimentation basin, mg/L

SS_e = suspended solids in the effluent from the sedimentation basin, mg/L

Q = flowrate, Mgal/d

(*a*) Compute the total mass of solids added per year.

$$\text{Mass} = [(321 - 20) \text{ mg/L}](1.0 \text{ Mgal/d})[8.34 \text{ lb/Mgal} \cdot (\text{mg/L})](365 \text{ d/yr})$$

$$= 916,300 \text{ lb/yr}$$

(*b*) Compute the mass of volatile and fixed solids added per year, assuming that VSS = $0.70 \times$ SS.

i. Volatile solids:

$$(\text{Mass})_{VSS} = (916,300 \text{ lb/yr})(0.7)$$
$$= 641,400 \text{ lb/yr}$$

ii. Fixed solids:

$$(\text{Mass})_{FS} = (916,300 - 641,400) \text{ lb/yr}$$
$$= 274,900 \text{ lb/yr}$$

2. Determine the amount of sludge that will accumulate at the end of 4 years.
   Assume that the maximum volatile solids reduction that will occur is equal to 75 percent and that it will occur within 1 year. To simplify the problem, assume that the deposited volatile suspended solids undergo a linear decomposition. Because the volatile solids will decompose to the maximum extent within 1 year, the following relationship can be used to determine the maximum amount of volatile solids available at the end of each year of operation:

$$(\text{VSS})_t = [0.7 + 0.25(t - 1)](641,400 \text{ lb/yr})$$

   where $(\text{VSS})_t$ = mass of volatile suspended solids at the end of t yr, lb
   t = time, yr

   (a) Mass of volatile suspended solids accumulated at the end of 4 yr:

$$\text{VSS}_t = [0.7 + 0.25(4 - 1)](641,400 \text{ lb/yr})$$
$$= 930,030 \text{ lb}$$

   (b) Total mass of solids accumulated at the end of 4 yr:

$$\text{SS}_t = 930,030 \text{ lb} + 4 \text{ yr } (274,900 \text{ lb/yr})$$
$$= 2,029,630 \text{ lb}$$

3. Determine the required liquid volume and the dimensions for the sedimentation basin.
   (a) Volume of sedimentation basin:

$$V = (2 \text{ d})(1.0 \text{ Mgal/d}) = 2 \text{ Mgal}$$
$$= 2,000,000 \text{ gal } (1 \text{ ft}^3/7.48 \text{ gal}) = 267,400 \text{ ft}^3$$

   (b) Surface area of sedimentation basin:

$$A_S = \frac{267,400 \text{ ft}^3}{5} = 53,480 \text{ ft}^2$$

   The aspect ratio for the surface area of the sedimentation basin (ratio of width to length) depends on the geometry of the available site.

4. Determine the depth required for the storage of sludge.
   (a) Determine the mass of accumulated sludge per square foot.

   Accumulated mass of sludge $= 2,029,630$ lb

$$\text{Mass per unit area} = \frac{2,029,630 \text{ lb}}{53,480 \text{ ft}^2} = 38.0 \text{ lb/ft}^2$$

   (b) Determine the required depth, assuming that the deposited solids will compact to an average value of 15 percent and that the density of the accumulated solids is equal to 1.06.

$$\frac{38.0 \text{ lb/ft}^2}{d, \text{ ft}} = (1.06)(0.15)(62.4 \text{ lb/ft}^3)$$

$$d = \frac{38.0 \text{ lb/ft}^2}{(1.06)(0.15)(62.4 \text{ lb/ft}^3)}$$
$$= 3.83 \text{ ft } (1.17 \text{ m})$$

If it is difficult to provide a total depth of 8.83 ft (5.0 ft + 3.83 ft), it may be necessary to increase the detention time or to clean the sedimentation basins more frequently.

10-5 TRICKLING FILTERS

Trickling filters have been used to provide biological wastewater treatment for nearly 100 years. Modern trickling filters have a bed of media over which wastewater is continuously distributed. The process microbiology and theoretical analysis of trickling filters are described in Chap. 8. The discussion in this section covers the filter classification, design of physical facilities, and trickling-filter process design.

Filter Classification

Trickling filters are classified by hydraulic- or organic-loading rates. Classifications are low- or standard-rate, intermediate-rate, high-rate, super high-rate, and roughing. Frequently, two-stage filters are used, in which two trickling filters are connected in series. The range of loadings normally encountered and other operational characteristics are shown in Table 10-13.

Low-Rate Filters. A low-rate filter is a relatively simple, highly dependable device that produces an effluent of consistent quality with an influent of varying strength. The filters may be circular or rectangular in shape. Generally, a constant hydraulic loading is maintained, not by recirculation, but by suction level controlled pumps or a dosing siphon. Dosing tanks are small, usually with only a 2 min detention time based on twice the average design flow so that intermittent dosing is minimized. Even so, at small plants, low nighttime flows may result in intermittent dosing and recirculation may be necessary to keep the media moist [48]. If the interval between dosing is longer than 1 or 2 h, the efficiency of the process deteriorates because the character of the biological slime is altered by a lack of moisture.

 In most low-rate filters, only the top 2 to 4 ft (0.6 to 1.2 m) of the filter medium will have appreciable biological slime. As a result, the lower portions of the filter may be populated by autotrophic nitrifying bacteria that oxidize ammonia nitrogen to nitrite and nitrate forms. If the nitrifying population is sufficiently well-established and if climatic conditions and wastewater characteristics are favorable, a well-operated low-rate filter can provide good BOD removal and a highly nitrified effluent.

 With a favorable hydraulic gradient, the ability to use gravity flow is a distinct advantage. Pumping may be required if the site is too flat to permit gravity flow. Odors are a common problem, especially if the wastewater is stale or septic or if the weather is warm. Filters should not be located where occasional odor events would create a nuisance. Filter flies (*Psychoda*) may breed in the filters unless control measures are used.

TABLE 10-13
Typical design information for trickling filters[a]

Item	Low-rate	Intermediate rate	High-rate	Super high-rate	Roughing	Two-stage
Filter medium[b]	Rock, slag	Rock, slag	Rock	Plastic	Plastic, redwood	Rock, plastic
Hydraulic loading,						
gal/ft$^2 \cdot$ min	0.02–0.06	0.06–0.16	0.16–0.64	0.2–1.20	0.8–3.2	0.16–0.64
Mgal/acre $\cdot$ d	1–4	4–10	10–40	15–90	50–200[c]	10–40[c]
BOD$_5$ loading, lb/10^3 ft$^3 \cdot$ d	5–25	15–30	30–60	30–100	100–500	60–120
Depth, ft	6–8	6–8	3–6	10–40	15–40	6–8
Recirculation ratio	0	0–1	1–2	1–2	1–4	0.5–2
Filter flies	Many	Some	Few	Few or none	Few or none	Few or none
Sloughing	Intermittent	Intermittent	Continuous	Continuous	Continuous	Continuous
BOD$_5$ removal efficiency, %	80–90	50–70	65–85	65–80	40–65	85–95
Effluent	Well-nitrified	Partially nitrified	Little nitrification	Little nitrification	No nitrification	Well-nitrified

[a] Adapted in part from Refs. 36 and 62.
[b] See Table 10-15 for physical characteristics of various filter mediums.
[c] Does not include recirculation.

Note: ft × 0.3048 = m
gal/ft$^2 \cdot$ min × 58.6740 = m^3/m$^2 \cdot$ d
Mgal/acre $\cdot$ d × 0.9354 = m^3/m$^2 \cdot$ d
lb/10^{3}ft$^3 \cdot$ d × 0.0160 = kg/m$^3 \cdot$ d

Intermediate-Rate and High-Rate Filters. In intermediate-rate and high-rate filters, recirculation of the filter effluent or final effluent permits higher organic loadings. Flow diagrams for various intermediate- and high-rate configurations are shown in Fig. 10-31. The intermediate-rate filters are similar to the low-rate filters and may be circular or rectangular. Flow to the filter is usually continuous, although intermittent wetting of the filter medium is permissible.

High-rate filters are designed for loadings substantially higher than low-rate filters. Recirculation of effluent from the trickling filter clarifier permits the high-rate filter to achieve similar removal efficiencies as the low-rate or intermediate-rate filter. Recirculation of filter effluent around the filter (first flow diagram in Fig. 10-31a and b) results in the return of viable organisms and often improves treatment efficiency. Recirculation also helps to prevent ponding in the filter and to reduce the nuisance from odors and flies [48]. High-rate filters use either a rock or a plastic packing medium. The filters are usually circular and flow is continuous.

Super High-Rate Filters. Super high-rate trickling filters are loaded at high hydraulic and organic rates (see Table 10-13). The major differences between super high-rate and high-rate filters are greater hydraulic loadings and greater filter depths (see Fig. 10-32). The greater depths are possible because lighter, plastic media are used. Most of these types of filters are in the form of packed towers [62].

Roughing Filters. Roughing filters are high-rate type filters that treat an organic load of more than 100 lb BOD/10^3 ft$^3 \cdot$ d (1.6 kg/m$^3 \cdot$ d) and hydraulic loadings up to 3.2 gal/ft$^2 \cdot$ min (187 m^3/m$^2 \cdot$ d). In most cases, these types of filters are used to treat wastewater prior to secondary treatment. Most roughing filters are designed to use plastic media [62].

Two-Stage Filters. A two-stage filter system with an intermediate clarifier to remove solids generated by the first filter is most often used with high-strength wastewater (see second flow diagram in Fig. 10-31b). A design example for a two-stage trickling filter is presented in Chap. 8, Example 8-2. Two-stage systems are also used where nitrification is required. The first-stage filter and intermediate clarifier reduce carbonaceous BOD, and nitrification takes place in the second stage.

Design of Physical Facilities

Factors that must be considered in the design of trickling filters include (1) the dosing rate, (2) the type and dosing characteristics of the distribution system, (3) the type and physical characteristics of filter medium to be used, (4) the configuration of the underdrain system, (5) provision for adequate ventilation, either natural or forced air, and (6) the design of the required settling tanks.

Dosing Rate. To optimize the treatment performance of trickling filters, there should be a continual and uniform (1) growth of biomass and (2) sloughing of excess biomass as a function of the organic loading [4]. To achieve uniform growth and sloughing,

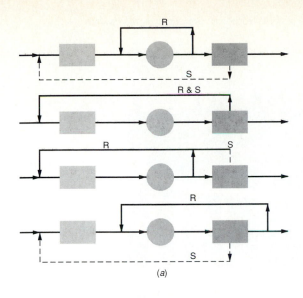

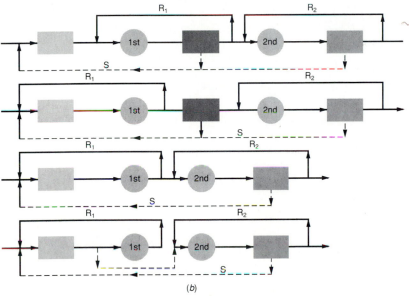

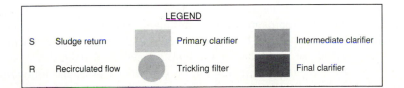

FIGURE 10-31
Intermediate-rate and high-rate trickling-filter flow diagrams with various recirculation patterns:
(a) single-stage filters and (b) two-stage filters.

(a)

(b)

FIGURE 10-32
Typical super high-rate tower trickling filters: (a) uncovered (from American Surfpac Corp.) and (b) view inside of a covered filter.

it has been found that higher periodic dosing rates than commonly used are required [4,66]. Suggested dosing rates are reported in Table 10-14. The required dosing rate in in/pass can also be approximated by multiplying the organic-loading rate expressed in lb $BOD_5/10^3$ ft^3 by 0.12. The instantaneous dosing rate is a function of the rotational speed of the distributor or the on-off times for a fixed distributor. The rotational speed for a rotary distributor can be determined using the following relationship [4,66].

$$n = \frac{1.6(Q_T)}{(A)(DR)} \qquad (10\text{-}24)$$

TABLE 10-14
Typical dosing rates
for trickling filters[a]

Organic loading rate, lb BOD$_5$/10^3 ft^3	Dosing rate, in/pass
< 25	3
50	6
75	9
100	12
150	18
200	24

[a] Adapted from Refs. 8, 66.
Note: lb/10^3 ft^3 × 0.0160 = kg/m^3
in × 2.54 = cm

where n = rotational speed of distributor, rev/min
Q_T = total applied hydraulic loading rate, gal/ft^2 · min
= $Q + Q_R$
Q = influent wastewater hydraulic loading rate, gal/ft^2 · min
Q_R = recycle flow hydraulic loading rate, gal/ft^2 · min
A = number of arms in rotary distributor assembly
DR = dosing rate, in/pass of distributor arm

To achieve the suggested dosing rates, the speed of the rotary distributor can be controlled (1) by reversing the location of some of the existing orifices to the front of the distributor arm,(2) by adding reversed deflectors to the existing orifice discharges, and (3) by converting the rotary distributor to a variable-speed electric drive [4]. At the slowest speed, a dosing rate of at least 4 in/pass should be produced for filters loaded at less than 25 lb BOD$_5$/ 10^3 ft^3.

Distribution Systems. The rotary distributor for trickling filtration has become a standard for the process because it is reliable and easy to maintain. A distributor consists of two or more arms that are mounted on a pivot in the center of the filter and revolve in a horizontal plane. The arms are hollow and contain nozzles through which the wastewater is discharged over the filter bed. The distributor assembly may be driven either by the dynamic reaction of the wastewater discharging from the nozzles or by an electric motor. The speed of rotation, which varies with the flowrate and the organic-loading rate, can be determined using Eq. 10-24. Clearance of 6 to 9 in (150 to 225 mm) should be allowed between the bottom of the distributor arm and the top of the bed. This clearance permits the wastewater streams from the nozzles to spread out and cover the bed uniformly, and it prevents ice accumulations from interfering with the distributor motion during freezing weather.

Distributors are manufactured for trickling filters with diameters up to 200 ft (60 m). Distributor arms may be of constant cross section for small units, or they may be tapered to maintain minimum transport velocity. Nozzles are spaced unevenly so that greater flow per unit of length is achieved near the periphery than at the center. For uniform distribution over the area of the filter, the flowrate per unit of length should be proportional to the radius from the center. Headloss through the distributor is in the range of 2 to 5 ft (0.6 to 1.5 m). Important features that should be considered in selecting a distributor are the ruggedness of construction, ease of cleaning, ability to handle large variations in flowrate while maintaining adequate rotational speed, and corrosion resistance of the material and its coating system.

Fixed nozzle distribution systems consist of a series of spray nozzles located at the points of equilateral triangles covering the filter bed. A system of pipes placed in the filter is used to distribute the wastewater uniformly to the nozzles. Special nozzles with a flat spray pattern are used, and the pressure is varied systematically so that the spray falls first at a maximum distance from the nozzle and then at a decreasing distance as the head slowly drops. In this way, a uniform dose is applied over the whole area of the bed. Half-spray nozzles are used along the sides of the filter. Twin dosing tanks with sloping bottoms that provide more volume at the higher head (required by the greater spray area) supply the nozzles by discharging through automatic siphons and are arranged to fill and dose alternately. The head required, measured from the surface of the filter to the maximum water level in the dosing tank, is normally 8 to 10 ft (2.4 to 3 m).

Filter Media. The ideal filter medium is a material that has a high surface area per unit of volume, is low in cost, has a high durability, and does not clog easily. Typical packing media are shown in Fig. 10-33. The physical characteristics of commonly used filter media, including those shown in Fig. 10-33, are reported in Table 10-15. Until the mid-1960s, the material used most commonly was either high-quality granite or blast furnace slag. Because of cost and such problems as minimal void area and the potential for biomass plugging, rock media have been replaced in later designs by plastic, redwood, or pressure-treated wood.

Where locally available, rock media have the advantage of low cost. The most suitable material is generally available river gravel or crushed stone, graded to a uniform size so that 95 percent is within the range of 3 to 4 in (75 to 100 mm). The specification of size uniformity is a way of ensuring adequate pore space for wastewater flow and air circulation. Other important characteristics of filter media are strength and durability. Durability may be determined by the sodium sulfate test, which is used to test the soundness of concrete aggregates. Because of the weight of the media, the depth of rock filters is usually limited to 5 to 10 ft (1.5 to 3.0 m).

Various forms of plastic media are illustrated in Fig. 10-33. Molded plastic media have the appearance of a honeycomb. Flat and corrugated sheets of polyvinyl-chloride are bonded together in rectangular modules. The sheets usually have a corrugated surface for enhancing slime growth and retention time. Each layer of modules is turned at right angles to the previous layer to further improve wastewater distribution. The two basic types of corrugated plastic sheet media are vertical and crossflow (see Fig. 10-33*b*, *c*, and *d*). Both types of media are reported to be effective in BOD and SS removal over a wide range of loadings [8,15]. Filters as deep as 40 ft (12 m) have been constructed using plastic or wood media. The high hydraulic

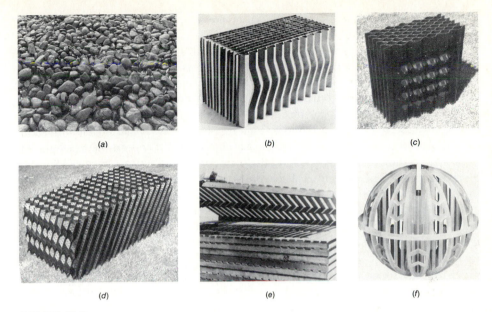

FIGURE 10-33
Typical packing media for trickling filters: (a) rock, (b) and (c) plastic-vertical flow, (d) plastic-cross flow, (e) redwood horizontal, and (f) random pack. (Figs. (c) and (d), from American Surfpac Corp., (e) from Neptune Microfloc, and (f) from Jaeger Products, Inc.)

TABLE 10-15
Physical properties of trickling-filter media[a]

Medium	Nominal size, in	Mass/unit volume, lb/ft^3	Specific surface area, ft^2/ft^3	Void space, %
River rock				
Small	1–2.5	78–90	17–21	40–50
Large	4–5	50–62	12–50	50–60
Blast furnace slag				
Small	2–3	56–75	17–21	40–50
Large	3–5	50–62	14–18	50–60
Plastic				
Conventional	24 × 24 × 48[b]	2–6	24–30	94–97
High-specific surface	24 × 24 × 48[b]	2–6	30–60	94–97
Redwood	48 × 48 × 20[b]	9–11	12–15	70–80
Random pack[c]	1–3.5	3–6	38–85	90–95

[a] Adapted in part from Ref. 50.

[b] Module size.

Note: in×25.4 = mm
    lb/ft^3×16.0185 = kg/m^3
    ft^2/ft^3×3.2808 = m^2/m^3

capacity and resistance to plugging offered by these types of media can best be used in a high-rate type filter.

Underdrains. The wastewater collection system in a trickling filter consists of underdrains that catch the filtered wastewater and solids discharged from the filter medium and convey them to the final sedimentation tank. The underdrain system for a rock media filter usually has precast blocks of vitrified clay or fiberglass grating laid on a reinforced concrete subfloor (Fig. 10-34). The floor and underdrains must have sufficient strength to support the media, slime growth, and the wastewater. The floor and underdrain block slope to central or peripheral drainage channels at a 1 to 5 percent grade. The effluent channels are sized to produce a minimum velocity of 2 ft/s (0.6 m/s) at the average daily flowrate [62]. Underdrains may be open at both ends so that they may be inspected easily and flushed out if they become plugged. The underdrains also ventilate the filter, providing the air for the microorganisms that live in the filter slime, and they should at least be open to a circumferential channel for ventilation at the wall as well as to the central collection channel.

The underdrain and support system for plastic media consists of either a beam and column or grating. A typical underdrain system for a tower filter is shown in Fig. 10-35. The beam and column system typically has precast concrete beams supported by columns or posts. The media are placed over the beams, which have channels in their tops to ensure free flow of wastewater and air. All underdrain systems should be designed so that forced air ventilation can be added at a later date if filter operating conditions should change.

Airflow. An adequate flow of air is of fundamental importance to the successful operation of a trickling filter. The principal factors responsible for airflow in an open top filter are natural draft and wind forces. In the case of natural draft, the driving force for airflow is the temperature difference between the ambient air and the air inside the pores. If the wastewater is colder than the ambient air, the pore air will be cold and the direction of flow will be downward. If the ambient air is colder than the wastewater, the flow will be upward. The latter is less desirable from a mass-transfer

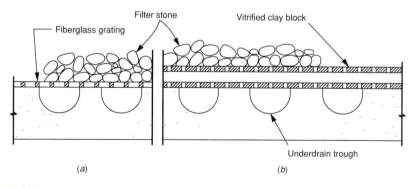

(a)   (b)

FIGURE 10-34
Typical underdrains for rock filter: (a) fiberglass grating and (b) vitrified clay block media.

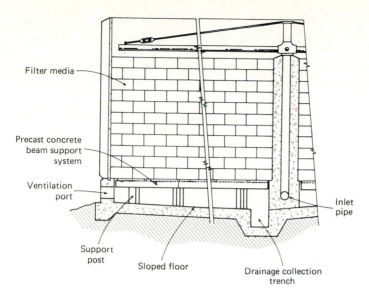

FIGURE 10-35
Typical underdrain system for tower filter.

point of view because the partial pressure of oxygen (and thus the oxygen-transfer rate) is lowest in the region of highest oxygen demand. In many areas of the country, there are periods, especially during the summer, when essentially no airflow occurs through the trickling filter because temperature differentials are negligible.

Draft, which is the pressure head resulting from the temperature difference, may be determined using Eq. 10-25 [38]:

$$D_{air} = 7.64 \left(\frac{1}{T_c} - \frac{1}{T_h} \right) Z \quad \text{U. S. customary units} \tag{10-25}$$

$$D_{air} = 3.53 \left(\frac{1}{T_c} - \frac{1}{T_h} \right) Z \quad \text{SI units} \tag{10-25a}$$

where D_{air} = natural air draft, in of water (mm)
T_c = cold temperature, °R (460 + °F) [°K]
T_h = hot temperature, °R (460 + °F) [°K]
Z = height of the filter, ft (m)

A more conservative estimate of the average pore air temperature is obtained by using the log-mean temperature, T_m:

$$T_m = \frac{T_2 - T_1}{\ln(T_2/T_1)} \tag{10-26}$$

where T_1 = warmer temperature, °R(°K)
T_2 = colder temperature, °R(°K)

The volumetric air flowrate may be estimated by setting the draft equal to the sum of the headlosses that result from the passage of air through the filter and underdrain system [3].

Natural draft has proved adequate for trickling filters, provided that the following precautions are taken [59]:

1. Underdrains and collecting channels should be designed to flow no more than half full to provide a passageway for the air.

2. Ventilating manholes with open grating types of covers should be installed at both ends of the central collection channel.

3. Large diameter filters should have branch-collecting channels with ventilating manholes or vent stacks installed at the filter periphery.

4. The open area of the slots in the top of the underdrain blocks should not be less than 15 percent of the area of the filter.

5. One square foot gross area of open grating in ventilating manholes and vent stacks should be provided for each 250 ft^2 (23 m^2) of filter area.

In extremely deep or heavily loaded filters, there may be some advantage in forced air ventilation if designed, installed, and operated properly. Such a design should provide for a minimum air flow of 1 ft^3/ft$^2 \cdot$ min (0.3 m^3/m$^2 \cdot$ min) of filter area in either direction. It may be necessary during periods of extremely low air temperature to restrict the flow of air through the filter to keep it from freezing.

Settling Tanks. The function of settling tanks that follow trickling filters is to produce a clarified effluent. They differ from activated-sludge settling tanks in that sludge recirculation, which is essential to the activated-sludge process, is lacking. All the sludge from trickling-filter settling tanks is removed to sludge-processing facilities. The design of these tanks is similar to the design of primary settling tanks, except that the surface-loading rate is based on the plant flow plus the recycle flow (see Fig. 10-1) minus the underflow (often neglected). Suggested overflow rates for settling tanks following trickling filters are reported in Table 10-12.

Trickling-filter Process Design

As noted in Chap. 8, a universal equation is not available for the design of trickling filters. However, Eq. 8-73 has proven to be adequate in terms of describing the observed removals in trickling filters packed with plastic material.

$$\frac{S_e}{S_i} = \exp[-k_{20}D(Q_v)^{-n}] \tag{8-73}$$

where S_e = total BOD$_5$ of settled effluent from filter, mg/L
S_i = total BOD$_5$ of wastewater applied to the filter, mg/L
k_{20} = treatability constant corresponding to a filter of depth D at 20°C, (gal/min)^n ft
D = depth of filter, ft

TABLE 10-16
Typical treatability constants for 20-foot tower trickling filter packed with plastic media[a]

Type of wastewater	Treatability constant, k, gal/min$^{0.5}$ft
Domestic	0.065–0.10
Domestic and food waste	0.060–0.08
Fruit-canning wastes	0.020–0.05
Meat packing	0.030–0.05
Paper mill wastes	0.020–0.04
Potato processing	0.035–0.05
Refinery	0.020–0.07

[a] Data are for 20°C.

Q_v = volumetric flowrate applied per unit volume of filter, gal/min · ft^2 (Q/A)
Q = total flowrate applied to filter without recirculation, gal/min
A = cross-sectional area of filter, ft^2
n = experimental constant, usually 0.5

When a treatability constant measured at one depth (see Table 10-16) is used to design a filter of a different depth, the treatability constant must be corrected for the new depth using the following equation:

$$k_2 = k_1 \left(\frac{D_1}{D_2}\right)^x \tag{8-74}$$

where k_2 = treatability constant corresponding to a filter of depth D_2
k_1 = treatability constant corresponding to a filter of depth D_1
D_1 = depth of filter one, ft
D_2 = depth of filter two, ft
x = 0.5 for vertical and rock media filters
= 0.3 for crossflow plastic medium filters

The process design for a trickling filter is illustrated in Example 10-7.

Example 10-7 Trickling filter design. Design a 30 ft deep tower trickling filter using a plastic packing to treat wastewater from a rural community in which a small vegetable cannery is located. Assume that the following information and data, derived from local records and pilot tests, apply. Assuming that a rotary distributor will be used, also determine the rotational speed in rev/min for the summer and winter conditions.

1. Average year-round domestic wastewater flowrate $= 2.5$ Mgal/d$(9,460$ m^3/d$)$
2. Sustained peak seasonal cannery flowrate $= 1.25$ Mgal/d$(4,730$ m^3/d$)$
3. The canning season is May through October.
4. Average year-round domestic BOD$_5$ = 220 mg/L.

5. Sustained peak combined domestic and cannery BOD_5 = 550 mg/L.
6. Effluent BOD_5 requirement = 30 mg/L.
7. Critical wastewater temperature data
   (a) Sustained low for May and October = 20°C.
   (b) Sustained low for January = 10°C.
8. Treatability constant = 0.10 $(gal/min)^{0.5}$ ft. (The treatability constant was derived from pilot plant studies conducted using a 20 ft deep test filter during the summer when the average temperature was 25°C.)

Solution

1. Determine the surface area required for a 30 ft deep filter during the canning season using Eq. 8-73.

$$\frac{S_e}{S_i} = \exp[-k_{20}D(Q_v)^{-n}]$$

(a) Substituting Q/A for Q_v in Eq. 8-73 and rearranging yields

$$A = Q\left(\frac{-\ln S_e/S_i}{k_{20}D}\right)^{\frac{1}{n}}$$

(b) Correct the observed BOD treatability constant for the sustained wastewater temperatures observed during May and October.

$$k_{20} = k_{25}\theta^{T-25}$$
$$k_{20} = 0.10 \ (gal/min)^{0.5} \ ft(1.035^{20-25})$$
$$= 0.084 \ (gal/min)^{0.5} \ ft$$

(c) Correct the observed BOD treatability constant for depth using Eq 8-74.

$$k_{30} = k_{20}\left(\frac{D_{20}}{D_{30}}\right)^x$$

$$k_{30} = 0.084\left(\frac{20}{30}\right)^{0.5} = 0.069$$

(d) Substitute known values and solve for the area A.

$S_e = 550$ mg/L

$S_i = 30$ mg/L

$n = 0.5$

$k_{20} = 0.069 \ (gal/min)^{0.5}$ ft for $n = 0.5$

$D = 30$ ft

$Q = [(2.5 + 1.25) \times 10^6 \ gal/d]/(1440 \ min/d) = 2,604$ gal/min

$$A = (2604)\left(\frac{-\ln 30/550}{0.069(30)}\right)^2 = 5,142 \ ft^2$$

2. Determine the surface area required for a 30 ft deep filter during the winter to meet the effluent requirements.

 (a) Correct the observed BOD treatability constant for the sustained temperatures observed during January.

$$k_{20} = k_{25} \theta^{T-25}$$

$$k_{20} = 0.10 \ (\text{gal/min})^{0.5} \ \text{ft} (1.035^{10-25})$$

$$= 0.060 \ (\text{gal/min})^{0.5} \ \text{ft}$$

 (b) Correct the observed BOD treatability constant for 30 ft depth

$$k_{30} = 0.060 \left(\frac{20}{30}\right)^{0.5} = 0.049$$

 (c) Substitute known values and solve for the area A.

$$S_e = 220 \ \text{mg/L}$$

$$S_i = 30 \ \text{mg/L}$$

$$n = 0.5$$

$$k_{20} = 0.049 \ (\text{gal/min})^{0.5} \ \text{ft for } n = 0.5$$

$$D = 30 \ \text{ft}$$

$$Q = (2.5 \times 10^6 \ \text{gal/d})/1440 = 1736 \ \text{gal/min}$$

$$A = (1,736) \left(\frac{-\ln 30/220}{0.49(30)}\right)^2 = 3189 \ \text{ft}^2$$

Because the area required for the summer condition is larger, the design is controlled by summer conditions.

3. Check the hydraulic loadings.

 (a) Summer Condition:

$$(\text{HLR})_S = \frac{2604 \ \text{gal/min}}{5142 \ \text{ft}^2} = 0.51 \ \text{gal/min} \cdot \text{ft}^2$$

 (b) Winter Condition:

$$(\text{HLR})_W = \frac{1736 \ \text{gal/min}}{5142 \ \text{ft}^2} = 0.34 \ \text{gal/min} \cdot \text{ft}^2$$

4. Check the organic loadings.

 (a) Summer Condition:

$$(\text{OLR})_s = \frac{(3.75 \ \text{Mgal/d})(550 \ \text{mg/L})[8.34 \ \text{lb/Mgal} \cdot (\text{mg/L})]}{(30 \ \text{ft})(5142 \ \text{ft}^2)/(1000 \ \text{ft}^3/10^3 \ \text{ft})} = 112 \ \text{lb BOD}/10^3 \ \text{ft}^3$$

 (b) Winter Condition:

$$(\text{OLR})_w = \frac{(2.5)(220)(8.34)}{(30 \times 5142)/(1000/10^3)} = 30 \ \text{lb BOD}/10^3 \ \text{ft}^3$$

5. Determine rotational speed of rotary distributor using Eq. 10-24.

$$DR = \frac{1.6(Q_T)}{(A)(n)}$$

(a) Summer Condition:
  i. The required dosing rate for the summer condition is 0.12×112 lb BOD/10^3 ft^3 = 13.4 in/pass of arm.
  ii. The required rotational speed is

$$n = \frac{1.6(Q_T)}{(A)(DR)} = \frac{1.6(0.51)}{2 \times 13.4} = 0.03 \text{ rev/min}$$

or one revolution every 33 min.

(a) Winter Condition:
  i. The required dosing rate for the winter condition is 0.12×30 lb BOD/10^3 ft^3 = 3.6 in/pass of arm.
  ii. The required rotational speed is

$$n = \frac{1.6(Q_T)}{(A)(DR)} = \frac{1.6(0.34)}{2 \times 3.6} = 0.076 \text{ rev/min}$$

or one revolution every 13.2 min.

10-6   ROTATING BIOLOGICAL CONTACTORS

Rotating biological contactors (RBCs) were first installed in West Germany in 1960 and were later introduced in the United States (see Fig. 10-36). In the United States and Canada, 70 percent of the RBC systems installed are used for carbonaceous BOD removal only, 25 percent for combined carbonaceous BOD removal and nitrification, and 5 percent for nitrification of secondary effluent [52–54]. A general process description and the theoretical aspects of the process are presented in Chap. 8. The discussion in this section covers process design considerations, the description of the equipment, operating problems, and process design.

Process Design Considerations

Properly designed, the RBC system may be superior in performance to other fixed-film systems due to lower organic loading per mass of biological solids, longer detention time in the biological stage, and better control of short circuiting. A typical flow diagram of an RBC application for secondary treatment is shown in Fig. 10-37. In the design of a rotating biological contactor system, consideration must be given to (1) staging of the RBC units, (2) loading criteria, (3) effluent characteristics, and (4) settling tank requirements.

Staging of RBC Units. The stage configuration of the RBC system is an integral part of the overall design process. Staging is the compartmentalization of the RBC media to form a series of independent cells. Stages can be accomplished by using

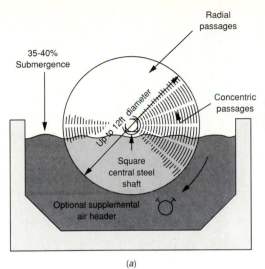

(a)

(b)

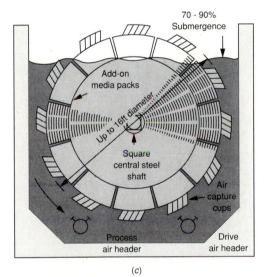

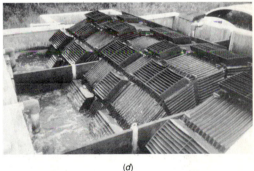

(c)

(d)

FIGURE 10-36
Typical RBC units: (a) conventional RBC with mechanical drive and optional air input, (b) conventional RBC in enclosed reactor, (c) submerged-type RBC equipped with air capture cups (air is used both to rotate and to aerate the biodisks), and (d) typical submerged RBC equipped with air capture cups (from Envirex Inc.).

baffles in a single tank or by use of separate tanks in series. Staging promotes a variety of conditions where different organisms can flourish in varying degrees. The degree of development in any stage depends primarily on the soluble organic concentration in the stage bulk liquid. As the wastewater flows through the system, each subsequent stage receives an influent with a lower organic concentration than

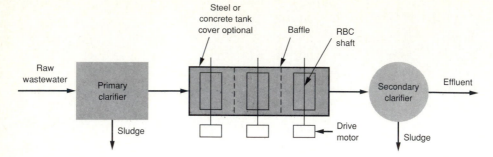

FIGURE 10-37
Typical RBC schematic for secondary treatment.

the previous stage. Typical RBC staging arrangements are illustrated in Fig. 10-38. In secondary treatment applications, three or more stages are generally provided for each flow stream. Additional stages are added for nitrification or for combined BOD and ammonia removal.

For small plants, RBC drive shafts are oriented parallel to the direction of flow with disk clusters separated by baffles (Fig. 10-38a). In larger installations, shafts are mounted perpendicular to flow with several stages in series to form a process train (Fig. 10-38b). To handle the loading on the initial units, a step-feed (Fig. 10-38c) or a tapered system (Fig. 10-38d) may be used. Two or more parallel-flow trains should be installed so that the units can be isolated for turndown or repairs. Tank construction may be reinforced concrete or steel, with steel preferred at smaller plants.

Loading Criteria. When RBC systems were originally introduced into the United States, the process design was based on a hydraulic loading expressed in gal/ft$^2 \cdot$ d, to achieve required removals. Over the last 15 years, the design approach has shifted, first to the use of total BOD per unit of surface area (lb TBOD/10^{3}ft^2) and most recently to soluble BOD per unit of surface area (lb SBOD/10^{3}ft^2), or, in case of nitrification, lb NH$_3$/10^{3}ft^2.

Poor performance has been observed where systems are overloaded resulting in low DO, H$_2$S odors, and poor first-stage removals. Under these conditions, filamentous organisms such as *Beggiatoa*, a sulfate-reducing organism, may develop. Overloading problems can be overcome by removing baffles between first and second stages to reduce surface loading and increase oxygen-transfer capability. Other approaches include supplemental air systems, step feed, or recycle from the last stage. Earlier designs used overly optimistic manufacturer recommendations that did not account for peak loads, sludge recycle flows, and temperature considerations.

Effluent Characteristics. RBC systems can be designed to provide secondary or advanced levels of treatment. Effluent BOD$_5$ characteristics for secondary treatment are comparable to well-operated activated-sludge processes. Where a nitrified effluent is required, RBCs can be used to provide combined treatment for BOD and ammonia

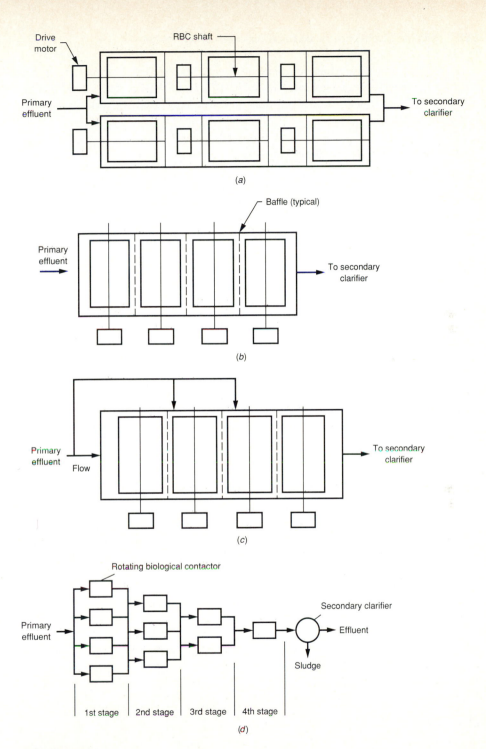

FIGURE 10-38
Typical rotating biological contactor arrangements: (a) flow parallel to shaft, (b) flow perpendicular to shaft, (c) step feed, and (d) tapered feed.

nitrogen or to provide separate nitrification of secondary effluent. Typical ranges of effluent characteristics are indicated in Table 10-17. An RBC process modification in which the media shaft is totally submerged has been used for denitrification of wastewater.

Physical Facilities for RBC Process

The principal elements of an RBC unit and their importance in the process are described in this section. The design of RBC equipment differs between manufacturers in nearly every component. For details of equipment differences, Ref. 54 may be consulted.

Shafts. RBC shafts are used to support and rotate the plastic media. Maximum shaft length is presently limited to 27 ft (8.23 m) with 25 ft (7.62 m) occupied by media. Shorter shaft lengths ranging from 5 to 25 ft (1.52 to 7.62 m) are also available. The shape and design details vary significantly between manufacturers. The structural properties of the shaft and the method of attachment of the media are important design considerations. Additional discussion on structural shaft failure is provided in the section "Operating Problems."

Media. The media used for RBCs are manufactured of high-density polyethylene and are provided in different configurations or corrugation patterns. Corrugations increase the available surface area and enhance structural stability. The types of media are

TABLE 10-17
Typical design information for rotating biological contractors

Item	Treatment level		
	Secondary	Combined nitrification	Separate nitrification
Hydraulic loading, gal/ft$^2 \cdot$ d	2.0–4.0	0.75–2.0	1.0–2.5
Organic loading			
lb SBOD$_5$/10^3 ft$^2 \cdot$ d$^{a, b}$	0.75–2.0	0.5–1.5	0.1–0.3
lb TBOD$_5$/10^3 ft$^2 \cdot$ d$^{a, c}$	2.0–3.5	1.5–3.0	0.2–0.6
Maximum loading on first stage			
lb SBOD$_5$/10^3 ft$^2 \cdot$ d$^{a, b}$	4–6	4–6	
lb TBOD$_5$/10^3 ft$^2 \cdot$ d$^{a, c}$	8–12	8–12	
NH$_3$ loading, lb/10^3 ft$^2 \cdot$ d		0.15–0.3	0.2–0.4
Hydraulic retention time, θ, h	0.7–1.5	1.5–4	1.2–2.9
Effluent BOD$_5$, mg/L	15–30	7–15	7–15
Effluent NH$_3$, mg/L		<2	1–2

[a] Wastewater temperature above 55°F (13°C).
[b] SBOD = Soluble BOD.
[c] TBOD = Total BOD.

Note: gal/ft$^2 \cdot$ d × 0.0407 = m^3/m$^2 \cdot$ d
lb/10^3 ft$^2 \cdot$ d × 0.0049 = kg/m$^2 \cdot$ d

classified based on the area of media on the shaft and are commonly termed low-(or standard) density, medium-density, and high-density. Standard-density media, defined as media with a surface area of 100,000 ft^2 (9290 m^2) per 27 ft (8.23 m) shaft, have larger spaces between media layers and are normally used in the lead stages of an RBC process train. Medium- and high-density media have surface areas of 120,000 to 180,000 ft^2 (11,149 to 16,723 m^2) per 27 ft (8.23 m) shaft and are used typically in the middle and final stages of an RBC system where thinner biological growths occur.

Drive Systems. Most RBC units are rotated by direct mechanical drive units attached directly to the central shaft. Air drive units are also available. The air drive assembly consists of deep plastic cups attached to the perimeter of the media, an air header located beneath the media, and an air compressor. The release of air into the cups creates a buoyant force that causes the shaft to turn. Both systems have proven to be mechanically reliable. Variable speed features can be provided to regulate the speed of rotation of the shaft.

Tankage. Tankage for RBC systems has been optimized at 0.12 gal/ft^2 (0.0049 m^3/m^2) of media, resulting in a stage volume of 12,000 gal (45.42 m^3) for a 100,000 ft^2 (9290 m^2) shaft. Based on this volume, a detention time of 1.44 h is provided for a hydraulic loading of 2 gal/ft^2 · d (0.08 m^3/m^2 · d). A typical sidewater depth is 5 ft (1.52 m) to accomodate a 40 percent submergence of the media.

Enclosures. Segmented fiberglass-reinforced plastic covers are usually provided over each shaft. In some cases, units have been housed in a building for protection against cold weather, for improved access, or for aesthetic reasons. RBCs are enclosed to (1) protect the plastic media from deterioration due to ultraviolet light, (2) protect the process from low temperatures, (3) protect the media and equipment from damage, and (4) control the buildup of algae in the process (see Fig. 10-36b).

Settling Tanks. Settling tanks for RBCs are similar to trickling-filter settling tanks in that all of the sludge from the settling tanks is removed to the sludge-processing facilities. Design overflow rates for settling tanks used with RBCs are given in Table 10-12.

Operating Problems

Many of the early RBC units had operating problems consisting of shaft failures, media breakage, bearing failures, and odor problems. Shaft failures have been the most serious equipment problem because of the loss of a process unit from service and the possible damage to a portion of the media. Causes of shaft breakage may be attributed to inadequate structural design, metal fatigue, and excessive biomass accumulation on the media. Media breakage has been caused by exposure to heat, organic solvents, or ultraviolet radiation, or by inadequate design of the media support systems. Bearing failures have been attributed to inadequate lubrication. Odor

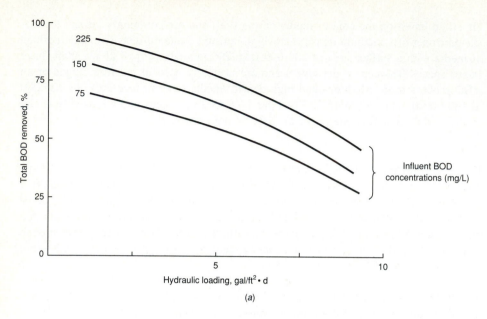

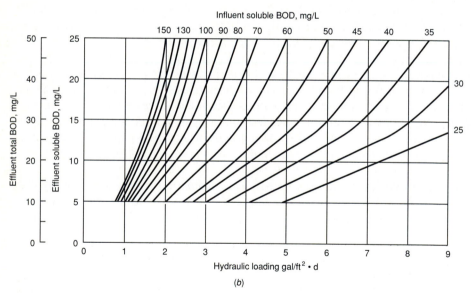

FIGURE 10-39
RBC process design curves for: (a) total BOD removal and (b) total and soluble effluent BOD
(T > 55°F) [53].

problems are most frequently caused by excessive organic loadings, particularly in the first stage. Modifications to the equipment have been made to mitigate many of these problems and to simplify maintenance. Units with increased submergence have been developed recently to reduce shaft and bearing loads and to improve equipment reliability. Descriptions of specific operating problems are included in Refs. 52 and 53.

RBC Process Design

Although several theoretical loadng performance models have been developed by data analysis for a given facility, the usefulness of these models to predict performance at other installations is not well-established. As a consequence, the design of RBCs is largely based on the use of design parameters. The design parameters presented in this section are derived from the experience gained by evaluating the operating records of numerous full-scale RBC installations. Application of the design parameters discussed in this section is illustrated in Example 10-8.

Process Sizing. A range of design parameters associated with major process modifications are presented in Table 10-17 and in Fig. 10-39. The design parameters given in Table 10-17 are for mixtures of domestic wastewater containing minor amounts of industrial process water. Selection of a loading value within the reported range is made based on effluent requirements, temperature range, degree of uncertainty as to waste load, and expected competency of the operating staff. Total media area is normally sized based on annual average design year conditions, unless information is available on significant loading variations occurring during the year. After the total surface area required is determined, the design must be checked to avoid exceeding oxygen-transfer capacity of first-stage units. This loading level has been reported at about 8 to 12 lb $BOD_5/10^3$ ft^2 or 4 to 6 lb soluble $BOD_5/10^3$ ft^2.

Effect of Temperature. When wastewater temperatures less than 55°F are expected, organic removal rates may decrease. To compensate for cold temperature effects, the required surface area of the RBC is increased. Surface area correction curves for temperatures below 55°F are shown in Fig. 10-40.

Example 10-8 RBC Process Design. A municipal wastewater with a soluble and total BOD_5 of 150 and 250 mg/L, respectively, is to be treated with an RBC process. The effluent BOD_5 is to be equal to or less than 25 mg/L. The average design flowrate is 0.75 Mgal/d. Assume that the temperature of the incoming wastewater is is 20°C and that the peaking factor for both the peak hourly flowrate and organic loading is 3.5. Determine the sizes of the RBC unit and the settling facilities.

Solution

1. Determine the required surface area of the RBCs.
   - (a) To achieve an effluent BOD_5 of 25 mg/L or less, an appropriate loading factor is about 1.5 lb $SBOD/10^3$ $ft^2 \cdot$ d (see Table 10-17)
   - (b) The required surface area is

$$A = \frac{0.75 \text{ Mgal/d}(150 \text{ mg/L SBOD})}{1.5 \text{ lb SBOD}/10^3 \text{ ft}^3 \cdot \text{d}}[8.34 \text{ lb/Mgal} \cdot (\text{mg/L})] = 625,500 \text{ ft}^2$$

2. Check the design for organic overloading.

   (a) Organic load peaking factor $= 3.5$

   (b) Determine the loading per unit area

$$\text{OLR}_{\text{Peak}} = \frac{0.75(150)8.34(1000)(3.5)}{625,500 \text{ ft}^2} = 5.25 \text{ lb SBOD}_5/10^3 \text{ ft}^3$$

   Referring to Table 10-17, the computed maximum organic-loading rate on the first stage is acceptable.

3. Determine the required surface area for the settling facilities.

   (a) Determine area based on average flow using an overflow rate of 600 gal/ft$^2 \cdot$ d

$$A_{\text{Avg flow}} = \frac{0.75 \text{ Mgal/d } (10^6 \text{ gal/Mgal})}{600 \text{ gal/ft}^2 \cdot \text{d}} = 1250 \text{ ft}^2$$

   (b) Determine area based on peak flow using an overflow rate of 1,200 gal/ft$^2 \cdot$ d

$$A_{\text{Peak flow}} = \frac{(3.5)0.75 \text{ Mgal/d } (10^6 \text{ gal/Mgal})}{1200 \text{ gal/ft}^2 \cdot \text{d}} = 2186 \text{ ft}^2$$

   (c) Based on the above computations, the size of the settling facilities is controlled by the maximum flowrate.

   Comment. In small plants subject to wide fluctuations in flowrates, the sizing of the settling facilities will almost always be based on the peak hourly flowrate.

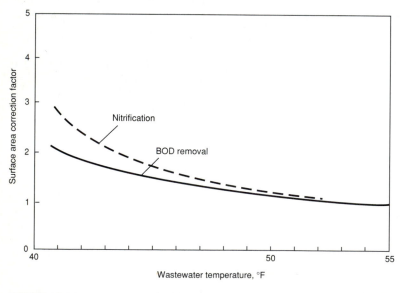

FIGURE 10-40
Surface area correction curves for RBCs for temperatures below 55°F.

10-7 COMBINED AEROBIC TREATMENT PROCESSES

Recent research has considered how aerobic treatment processes might be combined to obtain the best performance and the most economical treatment of wastewater [31]. A large number of treatment systems can be derived by combining the various aerobic processes discussed previously. The principal reason for combining processes is that they provide the stability and resistance to shock loads of attached-growth processes and the high-quality effluent of suspended-growth systems. The use of combined aerobic systems has increased in recent years, stimulated largely by the improvements in trickling-filter media discussed previously. Examples of the more common combined systems now in use will be considered in this section: (1) activated biofilter, (2) trickling filter followed by a solids contactor, (3) roughing filter followed by an activated-sludge process, (4) biofilter followed by an activated-sludge process, and (5) trickling filter followed by an activated-sludge process. Schematic flow diagrams of these systems are shown in Fig. 10-41, and typical design information is presented in Table 10-18. In several of these applications, the first process in the series can be considered a "roughing process" that functions to reduce the loading on the following process to a level that will allow it to function in an optimum manner.

Activated Biofilter Process

The activated biofilter process (ABF) resembles a high-rate trickling filter, except the secondary sludge is recycled to the trickling filter (see Fig. 10-41a). A separate suspended-growth process is generally not used, although one modification incorporates short-term aeration prior to secondary sedimentation. The return sludge is controlled to maintain a high concentration of suspended growth in the filter. The biofilter uses redwood media instead of other types of media. The advantages of this process are as follows: (1) significantly higher levels of BOD removal can be achieved by producing a combination of attached and suspended growth; (2) BOD loadings of four to five times higher than those used in conventional filters can be applied. Design loadings normally range from 200 to 250 lb/10^3 ft$^3 \cdot$ d (3.21 to 4.0 kg/m$^3 \cdot$ d) for 60 to 65 percent BOD removal in the biofilter [7].

The combined BOD removal through both the biofilter and secondary clarifier may be determined using Eq. 10-27 [7]:

$$\frac{L_e}{L_o} = e^{-K_T[1/(\text{TL})]^{0.48}} \qquad \text{U.S. customary units} \quad (10\text{-}27)$$

$$\frac{L_e}{L_o} = e^{-K_T[0.016/(\text{TL})]^{0.48}} \qquad \text{SI units} \quad (10\text{-}27a)$$

where L_e = secondary effluent total BOD, mg/L
L_o = primary effluent total BOD, mg/L
K_T = treatability constant or removal rate at temperature T, °C
   = $K_{20}\theta^{T-20}$

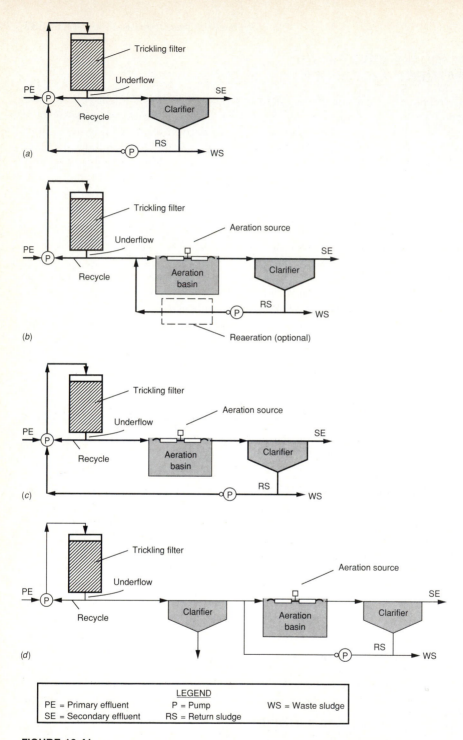

FIGURE 10-41
Typical combined aerobic treatment system flow diagrams: (*a*) activated biofilter, (*b*) trickling filter solids contact and roughing filter/activated sludge, (*c*) biofilter/activated sludge, and (*d*) series trickling filter/activated sludge.

TABLE 10-18
Typical design information for combined aerobic treatment processes[a]

Process combination	Trickling-filter loading	θ_c, d	Aeration basin	
			F/M, lb BOD$_5$ applied/ lb MLVSS · d	MLSS, mg/L
Activated biofilter	Low[b]	N/A	N/A	1,500–4,000
Trickling-filter/solids-contact	Low	0.5–2.0	N/A	1,000–3,000
Roughing filter/activated-sludge	High[c]	2–5	0.5–1.2	1,500–3,000
Biofilter/activated-sludge	High	2–5	0.5–1.2	1,500–4,000
Trickling-filter/activated-sludge	High	4–8	0.2–0.5	1,500–4,000

[a] Adapted from Ref. 62.
[b] Typically less than 40 lb BOD$_5$/10^3 ft^3 · d.
[c] Typically greater than 100 lb BOD$_5$/10^3 ft^3 · d.
Note: lb/10^3 ft^3 · d × 0.0160 = kg BOD$_5$/m^3 · d
N/A = Not applicable

$$K_{20} = 12.16 \text{ for wastewater}$$
$$TL = \text{biofilter organic loading in lb/10}^3 \text{ ft}^3 \cdot \text{d (kg/m}^3 \cdot \text{d)}$$
$$\theta = 1.016 \text{ for domestic wastewater}$$

Removal across the biofilter is reported to be independent of media depth greater than 14 ft (4.27 m) and hydraulic-loading rates above 1.5 gal/ft^2 · min (88 m^3/m^2 · d) [31].

Trickling-Filter Solids-Contact Process

The trickling-filter solids-contact (TF/SC) process consists of a trickling filter, an aerobic contact tank, and a final clarifier (see Fig. 10-41b). Modifications to this system include a return-sludge aeration tank and flocculating center-well clarifiers. The trickling filters are sized to remove the major portion of the BOD, typically 60 to 85 percent [35]. The biological solids formed on the trickling filter are sloughed off and concentrated through sludge recirculation in the contact tank. In the contact tank, the suspended growth is aerated for less than one hour, causing the flocculation of the suspended solids and further removal of soluble BOD. When short solids contact times are used, a sludge reaeration tank is usually required. Because of the high level of dispersed solids in contact tank effluent, flocculating center-well clarifiers have been found to be effective in maximizing solids capture.

Overall removal of BOD in a TF/SC process is determined by computing the removal of soluble BOD$_5$ in the trickling filter and the aerobic contact tank. A model has been developed for predicting soluble BOD$_5$ removal in trickling filters using plastic media, and an example performance curve is shown in Fig. 10-42 [35]. The curve is based on crossflow filter media with module depths of 2 ft (0.61 m). Using Fig. 10-42 for example, a 50 percent soluble BOD$_5$ reduction can be achieved in a tower filter with 8 ft (2.44 m) of media at a hydraulic loading of 0.68 gal/ft^2 · min (40 m^3/m^2 · d). For estimating soluble BOD$_5$ removal in the aerobic contact tank, first-

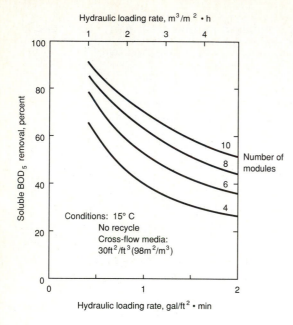

FIGURE 10-42

Effect of hydraulic loading and number of plastic filter media modules on trickling-filter efficiency [35].

order kinetics in a plug-flow reactor configuration can be used in accordance with Eq. 10-28 [26, 35]:

$$\ln\frac{C}{C_o} = \left[-K_{20}\theta^{(T-20)}X_v\right]t \tag{10-28}$$

where C_o = mixed-liquor soluble carbonaceous BOD at the contact tank inlet, mg/L
C = soluble carbonaceous BOD after time t, mg/L
K_{20} = first-order reaction rate coefficient at 20°C, L/mg · min
θ = temperature correction coefficient (assume $\theta = 1.035$)
T = wastewater temperature, °C
X_v = MLVSS, mg/L
t = contact time, min, based on total flow in the tank

C_o at the beginning of the contact tank can be related to the soluble effluent BOD_5 in the trickling filter effluent by the following mass balance:

$$(1 + R)C_o = RC_1 + S_1 \tag{10-29}$$

where R = return-sludge recycle ratio
C_1 = trickling-filter soluble BOD_5
S_1 = aerobic solids-contact tank effluent soluble BOD_5

Roughing-Filter Activated-Sludge Process

The roughing-filter activated-sludge (RF/AS) process configuration is identical to that of the TF/SC system (see Fig. 10-41b). The RF/AS system, however, operates at higher total organic loadings. The trickling filter is used to remove a portion of the BOD and to provide process stabilization, particularly when shock loads occur. The aeration basins are required to treat the organic loading not removed by the trickling filters.

Biofilter Activated-Sludge Process

The biofilter activated-sludge (BF/AS) process is similar to the ABF process, except an aeration tank is used following the trickling filter (see Fig 10-41c). Return activated sludge is recycled over the trickling filter. The average organic loading and aeration tank hydraulic retention times typically are similar to those of the RF/AS system. The concept of system F/M ratio, considering the biofilter and aeration basin as one integral treatment system, is introduced in Ref. 7. The system F/M value typically used for the design of the aeration basin for normal carbonaceous BOD removal is between 1.0 and 1.5, which is three to four times higher than the corresponding value for a conventional activated-sludge aeration basin not preceded by a biofilter. As a result, the aeration basin size is reduced to about one-fourth of that for a conventional activated-sludge system.

Series Trickling-Filter
Activated-Sludge Process

The trickling-filter process using an upstream trickling filter followed by an activated-sludge process (see Fig. 10-41d) is often used to upgrade an existing activated-sludge system. An alternative arrangement involves the addition of an activated-sludge process downstream from an existing trickling filter. This system is also used to reduce the strength of wastewater where industrial and domestic wastewater is treated in common treatment facilities and in applications where nitrification is required. In some systems, particularly those treating high-strength wastes, intermediate clarifiers between the trickling filters and the activated-sludge units are provided.

The microbiology for these combined processes is essentially the same as for the individual processes described previously in Chap. 8. Some microorganism population shifts occur in the trickling filter because of the high hydraulic loadings that are normally used. Also, many of the microorganisms associated with conventional rock or slag trickling filters are not present because tower filters are normally used.

10-8 STABILIZATION PONDS

A stabilization pond (or lagoon) is a relatively shallow body of wastewater contained in an earthen basin. The often used term "oxidation pond" is synonymous. Ponds have become very popular in small communities because their low construction and operating costs offer a significant financial advantage over other treatment methods. Ponds are also used extensively for the treatment of industrial wastewater and mixtures of industrial and domestic wastewater amenable to biological treatment. Pond installations also serve such industries as oil refineries, slaughterhouses, dairies, poultry processing plants, and rendering plants.

The purpose of this section is to describe (1) the various types of ponds and their application, (2) process design, (3) solids separation techniques, and (4) the design of the physical facilities.

Pond Classification and Application

Stabilization ponds are usually classified according to the nature of the biological activity taking place: aerobic, anaerobic, or aerobic-anaerobic. This general scheme is used in Table 8-1 where the various pond processes are also classified according to whether they are suspended-growth, attached-growth, or combination-growth processes. The principal types of stabilization ponds commonly used are identified in Table 10-19. Other classification schemes that have been used are based on the type of influent (untreated, screened, or settled wastewater or activated-sludge effluent), the pond overflow condition (nonexistent, intermittent, or continuous), and the method of oxygenation (photosynthesis, atmospheric surface reaeration, or mechanical aerators).

Stabilization ponds have been used singly or in various combinations to treat both domestic and industrial wastes. Typical applications are also reported in Table 10-19. As shown, aerobic ponds are used primarily for the treatment of soluble organic wastes and effluents from wastewater treatment plants. The aerobic-anaerobic ponds are the most common type and have been used to treat domestic wastewater and a wide variety of industrial wastes (see Fig. 10-43). Anaerobic ponds are especially effective in bringing about rapid stabilization of strong organic wastes. Usually, anaerobic ponds are used in series with aerobic-anaerobic ponds to provide complete treatment.

States where ponds are commonly used have regulations governing their design, installation, and management (operation). A minimum of a 60 d detention time is often required for flow-through facultative ponds receiving untreated wastewater. Higher detention times (90 to 120 d) have been specified frequently. A high degree of coliform removal is assured even with a 30 d detention.

Process Design and Analysis

The design of stabilization ponds is perhaps the least well-defined of all the biological treatment process designs . Numerous methods have been proposed in the literature, yet, when the results are correlated, a wide variance is usually found. A summary of design approaches is provided in Ref. 51. Typical design parameters for the different types of ponds are reported in Table 10-20; data for aerated lagoons are also included for comparison purposes. Most of the data were derived from operating experience with a wide variety of individual ponds and pond systems. Some methods that have been proposed for pond design, including a consideration of sludge buildup, are addressed in the following discussion.

Aerobic Ponds. Process design is usually based on organic-loading rates and hydraulic residence times; ranges in common use are reported in Table 10-20. Large systems are often designed as complete-mix reactors, using two or three reactors in series. A second approach is to use the first-order removal-rate equation developed

TABLE 10-19
Types and applications of stabilization ponds in common use

Type of pond or pond system	Common name	Identifying characteristics	Application
Aerobic	a. Low-rate pond	Designed to maintain aerobic conditions throughought the liquid depth.	Treatment of soluble organic wastes and secondary effluents
	b. High-rate pond	Designed to optimize the production of algae cell tissue and achieve high yields of harvestable protein.	Nutrient removal, treatment of soluble organic wastes, conversion of wastes
	c. Maturation or tertiary pond	Similar to low-rate ponds but very lightly loaded.	Used for polishing effluents from conventional secondary treatment processes such as trickling-filter or activated-sludge
Aerobic-anaerobic (oxygen source: supplemental aeration)	Facultative pond with aeration	Deeper than high-rate pond; aeration and photosynthesis provide oxygen for aerobic stabilization in upper layers. Lower layers are facultative. Bottom layer of solids undergoes anaerobic digestion.	Treatment of screened untreated or primary settled wastewater or industrial wastes
Aerobic-anaerobic (oxygen source: algae)	Facultative pond	As above, except without supplemental aeration. Photosynthesis and surface reaeration provide oxygen for upper layers.	Treatment of screened untreated or primary settled wastewater or industrial wastes
Anaerobic	Anaerobic lagoon, anaerobic pretreatment pond	Anaerobic conditions prevail throughout, usually followed by aerobic or facultative ponds.	Treatment of municipal wastewater and industrial wastes
Anaerobic followed by aerobic-anaerobic	Pond system	Combination of pond types described above. Aerobic-anaerobic ponds may be followed by an aerobic pond. Recirculation frequently used from aerobic to anaerobic ponds.	Complete treatment of municipal wastewater and industrial wastes with high bacterial removal

FIGURE 10-43
Typical facultative stabilization ponds.

by Wehner and Wilhelm [65] for a reactor with an arbitrary flow-through pattern (between a complete-mix pattern and a plug-flow pattern), as follows:

$$\frac{S}{S_o} = \frac{4a \exp (1/2d)}{(1 + a)^2 \exp(a/2d) - (1 - a)^2 \exp(-a/2d)} \tag{10-30}$$

where S = effluent substrate concentration
  S_o = influent substrate concentration
  a = $\sqrt{1 + 4ktd}$
  d = dispersion factor = D/uL
  D = axial dispersion coefficient, ft^2/h (m^2/h)
  u = fluid velocity, ft/h (m/h)
  L = characteristic length, ft (m)
  k = first-order reaction constant, 1/h
  t = detention time, h

To facilitate use of Eq. 10-30 for stabilization ponds, Thirumurthi developed the graph in Fig. 10-44, in which the term kt is plotted against S/S_o for dispersion factors varying from zero for an ideal plug-flow reactor to infinity for a complete-mix reactor [45]. For most stabilization ponds, the dispersion factors are within the range of 0.1 to 2.0. Because the contents of aerobic ponds must be mixed to achieve the best performance, it is estimated that a typical value for the pond dispersion factor would be about 1.0. Typical values for the overall first-order BOD$_5$ removal-rate constant k vary from about 0.05 to 1.0 per day, depending on the operational and hydraulic characteristics of the pond. The use of Fig. 10-44 is illustrated in Example 10-9. The design of an aerobic stabilization pond is illustrated in Example 10-10.

TABLE 10-20
Typical design parameters for stabilization ponds

Parameter	Aerobic low rate[a]	Aerobic high rate	Aerobic maturation	Aerobic-anaerobic facultative[b]	Anaerobic pond	Aerated lagoon
Flow regime	Intermittently mixed	Intermittently mixed	Intermittently mixed	Mixed surface layer		Completely mixed
Pond size, acres	<10 multiples	0.5–2	2–10 multiples	2–10 multiples	0.5–2 multiples	2–10 multiples
Operation[c]	Series or parallel	Series	Series or parallel	Series or parallel	Series	Series or parallel
Detention time,[c] d	10–40	4–6	5–20	5–30	20–50	3–10
Depth, ft	3–4	1–1.5	3–5	4–8	8–16	6–20
pH	6.5–10.5	6.5–10.5	6.5–10.5	6.5–8.5	6.5–7.2	6.5–8.0
Temperature range, °C	0–30	5–30	0–30	0–50	6–50	0–30
Optimum temperature, °C	20	20	20	20	30	20
BOD_5 loading,[d] lb/acre · d	60–120	80–160	≤ 15	50–180	200–500	
BOD_5 conversion, %	80–95	80–95	60–80	80–95	50–85	80–95
Principal conversion	Algae, CO_2, bacterial cell tissue	Algae, CO_2, bacterial cell tissue	Algae, CO_2, bacterial cell tissue NO_3	Algae, CO_2, CH_4, bacterial cell tissue	CO_2, CH_4, bacterial cell tissue	CO_2, bacterial cell tissue
Algal concentration, mg/L	40–100	100–260	5–10	5–20	0–5	
Effluent suspended solids,[e] mg/L	80–140	150–300	10–30	40–60	80–160	80–250

[a] Conventional aerobic ponds designed to maximize the amount of oxygen produced rather than the amount of algae produced.
[b] Pond includes supplemental aeration. For ponds without supplemental aeration, typical BOD_5 loadings are about one-third of those listed.
[c] Depends on climactic conditions.
[d] Typical values. Much higher values have been applied at various locations. Loading values are often specified by state regulatory agencies.
[e] Includes algae, microorganisms, and residual suspended solids. Values are based on an influent soluble BOD_5 of 200 mg/L and, with the exception of the aerobic ponds, an influent suspended solids of 200 mg/L.

Note: acre × 0.4047 = ha
           ft × 0.3048 = m
           lb/acre · d × 1.1209 = kg/ha · d

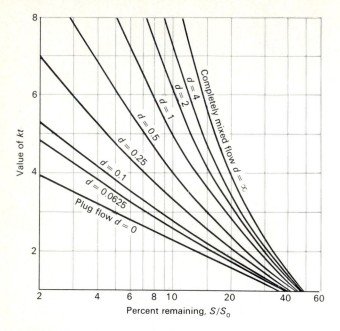

FIGURE 10-44
Values of kt in the Wehner and Wilhelm equation versus percent remaining for various dispersion factors [45].

Example 10-9 Bacterial die-off in series of stabilization ponds. It has been found that the observed die-off coefficient for *E. coli* in biological stabilization ponds can be described adequately with first-order kinetics. Assuming that the value of the specific reaction-rate constant is 1.0 d^{-1}, determine the concentration of *E. coli* in the effluent from a series of three ponds, when the initial concentration, N_o, is 10^6 organisms/mL and the average flowrate is 1.32 Mgal/d (5000 m^3/d). The ponds are rectangular and have an average depth of 5 ft (1.5 m). The surface areas of the ponds are 2.5, 5.0, and 2.5 acres (1, 2 and 1 ha).

Solution

1. Estimate the dispersion factors from the information given. Assume that the dispersion factor is 0.5 for the smaller ponds and 0.25 for the large pond.
2. Determine the kt value for the ponds.

   (*a*) For the smaller ponds:

   $$kt = k\frac{V}{Q} = k\frac{A \times d}{Q}$$
   $$= 1.0\left(\frac{1}{d}\right)\frac{(2.5 \text{ acres})(43,560 \text{ ft}^2/\text{acre})(5 \text{ ft})}{1,320,000 \text{ gal/d}}\left(\frac{7.48 \text{ gal}}{\text{ft}^3}\right)$$
   $$= 3.08$$

   (*b*) For the larger pond:

   $$kt = k\frac{V}{Q} = k\frac{A \times d}{Q}$$

$$= 1.0 \left(\frac{1}{d}\right) \frac{(5 \text{ acres})(43,560 \text{ ft}^2/\text{acre})(5 \text{ ft})}{1,320,000 \text{ gal/d}} \left(\frac{7.48 \text{ gal}}{\text{ft}^3}\right)$$

$$= 6.16$$

3. Determine the corresponding S/S_o values from Fig. 10-44.

   (a) For the smaller ponds:

$$S/S_o = 0.15$$

   (b) For the large pond:

$$S/S_o = 0.03$$

4. Estimate the concentration of organisms in the effluent. The ratio of the number of organisms in the influent to the number in the effluent is equal to the product of the dispersion values in the pond system.

$$\frac{N}{N_o} = (0.15)(0.03)(0.15)$$

$$= 6.75 \times 10^{-4}$$

$$N = 10^6 \text{ organisms/mL } (6.75 \times 10^{-4})$$

$$= 675 \text{ organisms/mL}$$

Example 10-10 Design of an aerobic stabilization pond. Design an aerobic stabilization pond to treat an industrial wastewater flow of 1.0 Mgal/d (3800 m³/d) with a soluble BOD_5 of 100 mg/L. Assume that the following conditions apply:

1. Influent suspended solids = negligible
2. BOD_5 (conversion) = 90 percent
3. First-order soluble BOD_5 removal-rate constant = 0.25 d^{-1} at 20°C
4. Temperature coefficient θ = 1.06 at 20°C
5. Pond temperature in summer = 32°C
6. Pond temperature in winter = 10°C
7. Maximum individual pond area = 10 acres (4 ha)
8. Maximum pond depth = 3.0 ft (0.9 m)
9. Pond dispersion factor = 1.0

Solution

1. From Fig. 10-44, determine the value of kt for the pond for a dispersion factor of 1.0 and a removal efficiency of 90 percent.

$$kt = 5$$

2. Determine the temperature coefficient for summer and winter conditions.

   (a) Winter:

$$k_{10°C} = k_{20°C} \theta^{T-20}$$

$$k_{10} = 0.25(1.06)^{10-20}$$

$$= 0.14 \text{ d}^{-1}$$

(b) Summer:

$$k_{32°C} = k_{20°C}\theta^{T-20}$$
$$k_{32} = 0.25(1.06)^{32-20}$$
$$= 0.5 \text{ d}^{-1}$$

3. Determine the detention time for winter and summer conditions.
   (a) Winter:

$$0.14 \text{ d}^{-1}(t) = 5$$
$$t = 35.7 \text{ d}$$

   (b) Summer:

$$0.5 \text{ d}^{-1}(t) = 5$$
$$t = 10 \text{ d}$$

4. Determine the pond surface area requirements for winter and summer conditions.
   (a) Winter:

$$\text{Surface area} = \frac{1,000,000 \text{ gal/d} \times 37.5 \text{ d}}{3 \text{ ft} \times 43,560 \text{ ft}^2/\text{acre}} \left[1 \text{ ft}^3/7.48 \text{ gal}\right]$$

$$= 36.5 \text{ acres } (14.8 \text{ ha})$$

   (b) Summer:

$$\text{Surface area} = \frac{1,000,000 \text{ gal/d} \times 10 \text{ d}}{3 \text{ ft} \times 43,560 \text{ ft}^2/\text{acre}} \left(\frac{1 \text{ ft}^3}{7.48 \text{ gal}}\right)$$
$$= 10.2 \text{ acres } (4.1 \text{ ha})$$

Therefore, winter conditions govern.

Aerobic-Anaerobic (Facultative) Ponds. The design of aerobic-anaerobic ponds closely follows the method used for the design of aerobic ponds. Because of the method of operation (for example, maintenance of quiescent conditions to promote the removal of suspended solids by sedimentation), it is anticipated that dispersion factors for such ponds will vary from 0.3 to 1.0.

Another factor that must be considered is sludge accumulation, which is important in terms of the oxygen resources and the overall performance of the pond. For example, in cold climates, a portion of the incoming BOD_5 is stored in the accumulated sludge during the winter months. As the temperature increases in the spring and summer, the accumulated BOD_5 is converted anaerobically, and the oxygen demand of gases and acids produced may exceed the oxygen resources of the aerobic surface layer of the pond. When it is anticipated that BOD_5 storage will be a problem, surface aerators are recommended. If the design is based on BOD_5, the aerators should have a capacity adequate to satisfy from 175 to 225 percent of the incoming BOD_5. Another problem caused by the accumulation of sludge is a reduction in performance of the pond, as measured by the suspended-solids content of the effluent. The design of an aerobic-anaerobic pond using surface aerators is illustrated in Example 10-11.

EXAMPLE 10-11 Aerobic-anaerobic stabilization pond design. Design an aerobic-anaerobic stabilization pond to treat a wastewater flow of 1.0 Mgal/d (3800 m^3/d). Because the ponds are to be installed near a residential area, surface aerators will be used to maintain oxygen in the upper layers of the pond. Assume that the following conditions apply:

1. Influent suspended solids = 200 mg/L
2. Influent BOD_5 = 200 mg/L
3. Summer liquid temperature = 25°C (77°F)
4. Winter liquid temperature = 15°C (59°F)
5. Overall first-order BOD_5 removal-rate constant = 0.25 d^{-1} at 20°C
6. Temperature coefficient θ = 1.06
7. Pond depth = 6 ft (1.8 m)
8. Pond dispersion factor = 0.5
9. Overall BOD_5 removal efficiency = 80 percent

Solution

1. From Fig. 10-44, determine the value of kt for a dispersion factor of 0.5 and a BOD_5 removal efficiency of 80 percent.

$$kt = 2.4$$

2. Determine the temperature coefficient for summer and winter conditions.
   (a) Winter:

   $$k_{15} = (0.25 \text{ d}^{-1})[(1.06)^{15-20}] = 0.187 \text{ d}^{-1}$$

   b. Summer:

   $$k_{25} = (0.25 \text{ d}^{-1})[(1.06)^{25-20}] = 0.335 \text{ d}^{-1}$$

3. Determine the detention time for winter and summer conditions.
   (a) Winter:

   $$(0.187 \text{ d}^{-1})(t) = 2.4$$
   $$t = 12.8 \text{ d}$$

   (b) Summer:

   $$(0.335 \text{ d}^{-1})(t) = 2.4$$
   $$t = 7.2 \text{ d}$$

4. Determine the pond volumes and surface requirements.
   (a) Winter:

   $$\text{Volume} = (1,000,000 \text{ gal/d})(12.8 \text{ d})(1 \text{ ft}^3/7.48 \text{ gal})$$
   $$= 1,711,000 \text{ ft}^3$$

   $$\text{Surface area} = \frac{1,711,000 \text{ ft}^3}{6 \text{ ft}} \frac{\text{acre}}{43,560 \text{ ft}^2} = 6.5 \text{ acres (2.6 ha)}$$

(*b*) Summer:

$$\text{Volume} = 962,600 \text{ ft}^3$$

$$\text{Surface area} = 3.7 \text{ acres } (1.5 \text{ ha})$$

Therefore, winter conditions control the design.

5. Determine the surface loading.

$$\text{lb BOD}_5/\text{acre} \cdot \text{d} = \frac{(1.0 \text{ Mgal/d})(200 \text{ mg/L})}{6.5 \text{ acres}} [8.34 \text{ lb/Mgal} \cdot (\text{mg/L})]$$

$$= 257 \text{ lb BOD}_5 /\text{acre} \cdot \text{d}$$

6. Determine the power requirements for the surface aerators. Assume that the oxygen-transfer capacity of the aerators will be twice the value of the BOD$_5$ applied per day and that a typical aerator will transfer about 48 lb O$_2$/hp · d.

$$\text{lb O}_2/\text{d required} = 2(1.0 \text{ Mgal/d})(200 \text{ mg/L})[8.34 \text{ lb/ Mgal} \cdot (\text{mg/L})]$$

$$= 3336 \text{ lb/d}$$

$$\text{hp} = \frac{3336 \text{ lb/d}}{48 \text{ lb O}_2/\text{hp} \cdot \text{d}} = 69.5 \text{ hp } (51.8\text{kW})$$

Use five 15 hp units.

7. Check the power input to determine the degree of mixing.

$$\text{hp/10}^3 \text{ ft}^3 = \frac{75 \text{ hp}}{1711 \times 10^3 \text{ ft}^3/10^3 \text{ ft}^3} = 0.04 \text{ hp/10}^3 \text{ ft}^3 (1 \text{ kW/10}^3 \text{ m}^3)$$

Comment. Regardless of how the ponds are operated (for example, series or parallel), the power required to keep the surface aerated will not be sufficient to mix the pond contents [about 0.6 to 1.15 hp/10^3 ft^3 (15 to 30 kW/10^3 m^3) is about the minimum required].

Anaerobic Ponds. The design of anaerobic stabilization ponds follows the principles presented in Chap. 8 and previously in this chapter. Because anaerobic ponds are similar to anaerobic digesters, with the exception of mixing, the process design methods outlined in Chap. 12 should be reviewed.

Pond Systems. Pond systems, such as those previously discussed, are designed by applying the aforementioned equations sequentially, taking into account recirculation where it is used. Stabilization ponds may be used in parallel or series arrangements to achieve special objectives. Series operation is beneficial where a high level of BOD or coliform removal is important. The effluent from aerobic-anaerobic ponds in series operation has a much lower algal concentration than that obtained in parallel operation, with a resultant decrease in color and turbidity. Many serially operated multiple unit installations have been designed to provide complete treatment or complete retention of the wastewater, with the liquid evaporated into the atmosphere or percolated into the ground. Parallel units provide better distribution of settled solids. Smaller units are conducive to better circulation and have less wave action. The additional cost of equipping units for both series and parallel operation is usually nominal. In some instances, savings can be demonstrated because of the lesser volume of earth moving needed to adapt smaller units to the topography.

Recirculation. Recirculation of pond effluent has been used effectively to improve the performance of pond systems operated in series. Occasionally, internal recirculation is used. If three aerobic-anaerobic ponds are used in series, the normal mode of operation involves recirculating effluent from either the second or the third pond to the first pond. The same situation applies if an anaerobic pond is substituted for the first aerobic-anaerobic pond. Recirculation rates varying from 0.5 to 2.0 Q (plant flow) have been used. If recirculation is considered, it is recommended that the pumps have a capacity of at least Q.

Solids Separation

In the minimum requirements for secondary treatment set forth by the EPA (see Table 4-1), provision is made for the adjustment of effluent quality from waste stabilization ponds. Adjustment may be made provided the ponds are the principal process used and, based on a review of operation and maintenance records, it can be shown that the SS values in the regulations cannot be achieved. Where adjustment of these requirements cannot be made, solids removal facilities may have to be used for algae reduction. The principal means of solids separation are summarized in Table 10-21. Additional details of solids removal processes may be found in Refs. 30, 46 and 51.

Design of Physical Facilities for Ponds

Although the process design for ponds is imprecise, careful attention must be given to the design of the physical facilities to ensure optimum performance. Factors that should be considered include (1) inlet and outlet structure design, (2) transfer lines, (3) dike construction, (4) liquid depth, (5) construction of lagoon bottom, and (6) control of surface runoff.

Inlet and Outlet Structures. Many ponds have been installed with a single inlet, located near the center of the pond. Multiple inlet arrangements are preferred to achieve better hydraulic distribution and pond performance. For large aerobic-anaerobic ponds, multiple inlets are particularly desirable to distribute settleable solids over a larger area. For increased flexibility, movable inlets can be used. The outlet should be located as far as possible from the inlet and should allow for lowering the water level at a rate of less than 1 ft/week (0.3 m/week) while the facility is receiving its normal load. The outlet should be large enough to permit easy access for normal maintenance. During ice-free periods, discharge should be taken from just below the water surface for the release of effluent of the highest quality and retention of floating solids. For flow-through ponds, the maximum rate of effluent discharge is less than the rate of peak wastewater flow, because of pond losses and the leveling out of peak flows. Overflow structures comparable to a sewer manhole are most frequently used,

TABLE 10-21
Types of solids separation facilities used with stabilization ponds

Operation or process	Description or application
Sedimentation ponds or tanks	Earthen lagoons or settling tanks following the stabilization pond. Sedimentation ponds may also serve as maturation ponds for effluent polishing. Provision for sludge removal and storage need to be included.
Chemical precipitation	Chemical addition, flocculation, and sedimentation. May be accomplished by adding chemicals to pond influent for continuous discharge ponds or by adding chemicals by motor boat to the pond surface for controlled discharge ponds (batch treatment). Chemicals may include alum, lime, ferric chloride, or magnesium hydroxide.
Flotation	Flotation of solids in pond effluent by dissolved air flotation or autoflotation. Dissolved air flotation (see Chap. 9) may also include chemical addition and flocculation for algae removal. Autoflotation, the natural removal of algae by gas supersaturation in stabilization ponds, is limited by climatic and biological factors.
Fine screens	Addition of fine screens or microstrainers for removing solids from pond effluent. Results may be highly variable depending on the screening medium and the algal community in the pond effluent.
Intermittent sand filters	Intermittent application of pond effluent to sand filters. Adequate underdrainage is required. Removal and replacement of the top layer of sand is required periodically to restore filtering ability. Application is mainly suited to small systems (see Chap. 14).
Rock filters	Rock or coarse-medium filter constructed in the pond near the outlet, allowing the algae to settle out on the rock surface and into the void space (see Fig. 10-45).
Rapid sand filters	Conventional rapid sand filters for polishing the pond effluent (see Chap. 11). Because of the high effluent solids concentrations, media selection is critical. Dual media are preferred. Filters may require frequent backwashing.
Natural treatment systems	Land treatment methods or aquaculture for effluent polishing (see Chap. 13). Applicable land treatment methods include rapid infiltration and overland flow. Applicable aquaculture methods include water hyacinths and constructed wetlands.

and selected level discharge is facilitated through valved piping or other adjustable overflow devices. Overflow lines should be vented to prevent siphoning. Provision for complete draining of the pond is desirable for maintenance purposes. All inlet, outlet and transfer lines should be provided with seepage collars.

Transfer Lines. Pond transfer structure placement and size will affect the flow patterns within the pond system. Pond transfer lines should be constructed to minimize headloss at peak flowrates and to ensure uniform distribution to all pond areas.

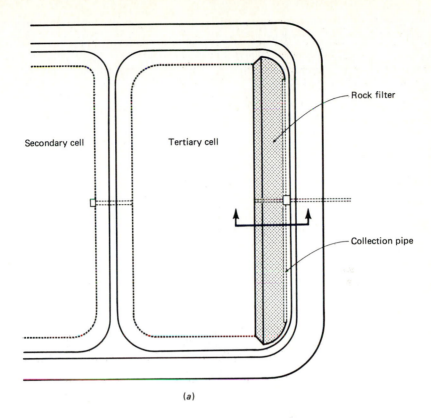

(a)

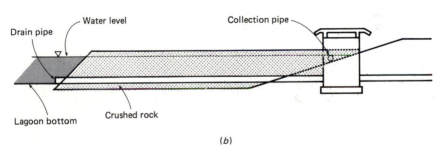

(b)

FIGURE 10-45
Rock filter for the separation of solids from effluent from facultative pond: (a) plan of rock filter and
(b) cross-section of a rock filter.

Several transfer pipes should be provided and should be sufficiently large to limit
peak headloss to 3 to 4 in (70 to 100 mm) with pipes flowing two-thirds to three-
fourths full [46].

Dike Construction. Dikes should be constructed in a way that minimizes seepage.
Compaction afforded by the use of conventional construction equipment is usually

adequate. Vegetation should be removed, and the area on which the embankment is to be placed should be scarified.

The dike should be wide enough to accommodate mowing machines and other maintenance equipment. An all-weather gravel surface will facilitate access for inspection and maintenance. A width of 10 ft (3 m) at the top of the dike is generally preferred, and narrower dikes may be satisfactory for small installations. Slopes are influenced by the nature of the soil and the size of the installation. For outer slopes, a 3 horizontal to 1 vertical is satisfactory. Inner slopes are generally from 1 vertical to 3 to 4 horizontal, although slopes exceeding 1 to 5 for larger installations are sometimes specified. The selected slope will depend upon the dike material and the water erosion protection to be provided.

The freeboard is to some extent influenced by the size and shape of the installation because wave heights are greater on larger bodies of water. Three feet (0.9 m) above maximum liquid level is usually specified as the minimum freeboard, but 2 ft (0.6 m) is considered adequate by some states, particularly for installations of 5 acres (2 ha) or less not exposed to severe winds.

A problem common to many ponds is erosion of the interior slopes. Erosion is caused by surface runoff and wind-induced wave action. The basic approaches for erosion control are to minimize wave energy, to reduce the raindrop impact on embankment soils, and to increase the embankment soil resistance to erosion. Techniques commonly used to control or arrest erosion problems include vegetative cover, revetments, and breakwaters. In applying vegetative cover, the slope, soils, depth of topsoil, and type of vegetation must be considered. A revetment is a heavy facing on a slope to protect it from wave action. A typical rock revetment is illustrated in Fig. 10-46. Breakwaters dissipate the energy of approaching waves and are either fixed or floating. Additional information on slope protection methods for ponds may be found in Ref. 56.

Liquid Depth. Optimum liquid depth for circulation is influenced to some extent by the pond area; greater depth is allowed for large units. Shallow ponds encourage the growth of vegetation and may foster mosquito breeding.

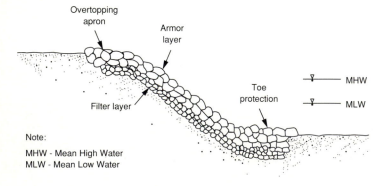

FIGURE 10-46
Typical revetment used for the slope protection of lagoons [56].

There is a distinct advantage for facilities that permit operation at selected depths up to 5 ft (1.5 m), and provision for additional depth may be desirable for large installations. Facilities for adjusting pond levels can be provided at a small cost. For ponds 30 acres (12 ha) or larger, provision for periodic operation at depths greater than 5 ft may be advantageous.

Pond Bottom Construction. The bottom of aerobic and most aerobic-anaerobic ponds should be made as level as possible except around the inlet. The finished elevation should not vary more than 6 in (15 cm) from the average elevation of the bottom, except where the bottom of an aerobic-anaerobic pond is designed specifically to retain the settleable solids in hoppered compartments or cells. The bottom should be well-compacted to avoid excessive seepage. Where excessive percolation may result in pollution of the groundwater, a pond sealer or liner will be required. Types of sealers or liners include (1) synthetic and rubber liners, (2) earthen and cement liners, and (3) natural and chemical treatment sealers. For more detailed information on pond sealers and liners, Ref. 51 may be consulted.

Surface Runoff Control. Ponds should not receive significant amounts of surface runoff. If necessary, provision should be made for diverting surface water around the ponds. For new installations, where maintenance of a satisfactory water depth is a problem, the diversion structure may be designed to admit surface runoff to the lagoon when necessary.

DISCUSSION TOPICS AND PROBLEMS

10-1. Using Example 10-2, compute the required quantities of nitrogen and phosphorus if the nitrogen requirement is $0.12P_x$ and the phosphorus requirement is one-fifth of the nitrogen requirement. In what forms should these nutrients be added?

10-2. A complete-mix activated-sludge process is used to treat 14 Mgal/d of primary effluent containing 175 mg/L of BOD_5 and 125 mg/L of SS. If the effluent concentration for BOD_5 and SS is required to be 20 mg/L, determine the theoretical oxygen requirements. The effluent suspended solids are assumed to be 65 percent biodegradable.

10-3. In Prob. 10-2, alternative diffused-air aeration devices are being considered for installation at a submergence of 15 ft in an aeration tank. Determine the standard oxygen-transfer rate and theoretical air requirements for both ceramic dome diffusers installed in a grid pattern and nonporous diffusers installed for a dual spiral roll. The wastewater temperature is 20°C and the α factors are 0.64 for ceramic domes and 0.75 for the nonporous diffusers, respectively.

10-4. A conventional activated-sludge plant is to treat 1.0 Mgal/d of wastewater with a BOD_5 of 200 mg/L after settling. The process loading is 0.30 lb BOD/d · lb MLVSS. The detention time is 6 h and the recirculation ratio is 0.33. Determine the value of MLVSS.

10-5. A conventional activated-sludge plant is operated at a mean cell-residence time of 10 d. The reactor volume is 2 Mgal, and the MLSS concentration is 3000 mg/L. Determine (a) the sludge production rate, (b) the sludge-wasting flowrate when wasting from the reactor, and (c) the sludge wasting flowrate when wasting from the recycle line. Assume that the concentration of suspended solids in the recycle is equal to 10,000 mg/L.

10-6. The step aeration activated-sludge system shown in Fig. 10-4 is to be analyzed as a series of complete-mix reactors (see following figure). Using the design parameters given below, determine the MLVSS concentration in each tank.

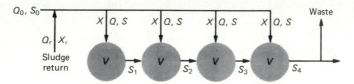

$V = 240 \text{ m}^3$	$S_4 = 10 \text{ mg/L}$	$X_r = 10,000 \text{ mg/L}$
$S_1 = 4 \text{ mg/L}$	$S_o = 250 \text{ mg/L}$	$Y = 0.65$
$S_2 = 6 \text{ mg/L}$	$Q_o = 4000 \text{ m}^3/\text{d}$	$k_d = 0.05$
$S_3 = 8 \text{ mg/L}$	$Q_r = 800 \text{ m}^3/\text{d}$	

10-7. The following data were obtained from an aerator test performed in tap water at a temperature of 7.5°C. Determine the value of $K_L a$ expressed in terms of h^{-1} using both Eqs. 6-56 and Eq. 6-58.

Time, min	C, mg/L	Time, min	C, mg/L	Time, min	C, mg/L
0	0	15	4.4	30	7.2
5	1.8	20	5.5	35	7.9
10	3.2	25	6.4	40	8.4

10-8. Using the $K_L a$ value determined in Prob. 10-7, estimate the maximum strength of waste that could be treated in a complete-mix activated-sludge process if the observed value of the yield is equal to 0.35 lb/lb.

10-9. Using the sequencing batch reactor sequence times of 2 h for fill, 4 h for react, and 1 h for settle-decant and the following flowrate characteristics, determine the number of batch reactors required and the daily cycle of each unit, including the idle time when the reactor is empty. Assume that 30 percent of the reactor contents is retained after decanting.

Time	Flowrate, Mgal/d	Time	Flowrate, Mgal/d
Midnight	1.85	1P.M.	2.64
1A.M.	1.58	2	2.38
2	1.45	3	3.04
3	1.12	4	3.17
4	1.06	5	4.36
5	1.12	6	4.22
6	1.98	7	4.09
7	3.04	8	3.56
8	3.70	9	3.43
9	3.43	10	2.77
10	2.90	11	2.11
Noon	2.80	Midnight	1.85

10-10. Using Example 10-3, determine the average F/M ratios for the 7-day operating period of the sequencing batch reactor.

10-11. Determine the temperature of the wastewater in an 8-acre aerated lagoon. Wastewater is discharged to the lagoon at a rate of 0.5 Mgal/d. Use a typical f value of 60×10^{-6} for the midwestern United States. The temperature of the air is 50°F, and the temperature of the incoming wastewater is 68°F.

10-12. Design an aerated lagoon to treat 10,000 m^3/d of wastewater under the following conditions:

    (a) Influent soluble BOD_5 and suspended solids = 150 mg/L

    (b) Overall first-order BOD_5 removal-rate constant = 2.0 d^{-1} at 20°C

    (c) Summer temperature = 27°C

    (d) Winter temperature = 7°C

    (e) Wastewater temperature = 15°C

    (f) Temperature coefficient = 1.07

    (g) $\alpha = 0.85$, $\beta = 1.0$

    (h) Elevation = 1250 m

    (i) Oxygen concentration to be maintained = 2.0 mg/L

    (j) Lagoon depth = 2 m

    (k) Hydraulic residence time = 10 d

    (l) Temperature proportionality constant $f = 0.5$

Determine the surface area, summer and winter temperatures in the lagoon, and the effluent BOD_5 in summer and winter. If the growth yield is approximately 0.5 (BOD_5 basis), determine the biological solids concentration in the lagoon, the oxygen requirements, and the power requirements for summer and winter conditions. Use surface aerators rated at 1.5 kg $O_2/kW \cdot h$. Perform all of the computations in SI units and show answers in both SI and U.S. customary units.

10-13. Prepare a plot of the natural draft available in a 20 ft tower trickling filter for air flow in the upward and downward direction between the ambient air and the air within the filter. Use (1) the wastewater temperature and (2) the log-mean temperature as estimates of the air temperature within the filter. Assume that the maximum temperature difference between the ambient air and the wastewater is ± 35°F and that the temperature of the wastewater is 80°F.

10-14. A tower trickling filter 20 ft in height is to be used to treat a combined domestic-industrial waste. The temperature of the wastewater is 80°F. Using the following temperature data taken at the plant site and the plot developed in Prob. 10-13, determine the air draft available throughout the day in inches of water. Use the log-mean estimate for the air temperature within the filter.

Time	Air temperature, °F	Time	Air temperature, °F
Midnight	72	2	108
2	65	4	106
4	62	6	101
6	63	8	90
8	72	10	79
10	90	Midnight	72
Noon	105		

If a bulk air flowrate through the filter of 0.3 ft^3/ft$^2 \cdot$ min is needed to meet the oxygen requirements of the waste, determine the number of hours during each day that the air flow will be insufficient. Assume that the area of the vent openings in the bottom of the filter is equal to 5 percent of the filter surface area and that the headloss through the filter can be approximated using the following expression:

$$h_F = 0.017 \frac{V_v^2}{2g} \frac{\rho_a}{\rho_w}$$

where h_F = headloss through the filter, in of H_2O
V_v = air flow velocity through the vent, ft/min
g = acceleration due to gravity = 32.2 ft/s^2
ρ_a = density of air, lb/ft^3
ρ_w = density of water, lb/ft^3

10-15. An industrial waste is to be treated with a tower trickling filter followed by an activated-sludge process. Primary settling will not be used. The packing medium in the tower trickling filter will be plastic, and the operational mean cell-residence time for the activated-sludge process will be 5 d during the critical summer period and vary from 5 to 15 d during the winter. The lowest average sustained winter temperature (at least two weeks) is 5°C and the highest average sustained summer temperature is 26°C. The characteristics of the industrial waste, data derived from pilot plant studies, and related design data are presented below. Using these data, size the units and determine the concentration of mixed-liquor suspended solids to be maintained during summer and winter operation, the recycle rates around the filter and activated-sludge process, the quantity of sludge to be disposed, and the quantity of nutrients that must be added. Assume that the flowrate of 20,000 m^3/d has been equalized. Perform all computations in SI units and show answers in both SI and U.S. customary units.

Wastewater characteristics:

$$BOD_5 = 1200 \text{ mg/L}$$

$$SS = 100 \text{ mg/L}$$

$$VSS = 0 \text{ mg/L}$$

$$\text{Total nitrogen as } N = 10 \text{ mg/L}$$

$$\text{Total phosphorus as } P = 4 \text{ mg/L}$$

$$\text{Total iron as } Fe = 0.15 \text{ mg/L}$$

Trickling-filter pilot plant data:

$$K_{20°C} = 0.075 \text{ m/d}$$

$$Y(BOD_5) = 0.70 \text{ mg/mg}$$

$$\theta = 1.06$$

Activated-sludge pilot plant data:

$$Y(BOD_5) = 0.8 \text{ mg/mg}$$

$$K_d = 0.1 \text{ d}^{-1}$$

$$k = 6.0 \ d^{-1}$$

$$K_s = 90 \ mg/L$$

$$\theta = 1.035$$

Design parameters:

$$w(\text{for trickling filter}) = 1.0 \ m/m^2 \ \text{of trickling-filter cross-sectional area}$$

$$\theta_c = 5 \ d \ (\text{critical summer period})$$

$$\theta_c = 5 \ \text{to} \ 15 \ d \ (\text{winter})$$

10-16. Using the design parameters summarized in Tables 10-12 and 10-17 for secondary level of treatment, design a treatment process using rotating biological contactors to treat a wastewater with the characteristics given in Example 10-2. Compare and contrast your design to that given in Example 10-2.

10-17. Prepare a plot of soluble BOD_5 removal efficiency in percent versus contact time in minutes for a contact tank to be used in a TF/SC process. The operating conditions are

$$T = 15°C$$

$$MLVSS = 2,000 \ mg/L$$

$$K_{20} = 0.08$$

10-18. Using the plot in Prob. 10-17, design two alternative TF/SC processes for treating a primary effluent with a soluble BOD_5 of 100 mg/L. The average wastewater flowrate is 7 Mgal/d and the alternative processes are to be designed based on a trickling-filter hydraulic load of 5 ft/hr for Alt.1 and 8.2 ft/hr for Alt. 2. Compare the advantages and disadvantages of both alternatives and select the best alternative, citing your reasons.

10-19. Design an aerobic stabilization pond to treat 2.5 Mgal/d of wastewater with a BOD_5 removal efficiency of 90 percent under the following conditions:

(a) Influent BOD_5 = 250 mg/L

(b) Overall first-order BOD_5 removal-rate constant = $0.2 \ d^{-1}$ at 20°C

(c) Pond temperature in summer = 30°C

(d) Pond temperature in winter = 12°C

(e) Temperature coefficient = 1.06

(f) Maximum pond area = 10 acres

(g) Maximum pond depth = 5 ft

(h) Pond dispersion factor = 0.5

Determine the detention times and area requirements for summer and winter conditions.

REFERENCES

1. Adams, C. E., Jr., and W. W. Eckenfelder, Jr. (eds.): *Process Design Techniques for Industrial Waste Treatment*, Enviro, Nashville, 1974.
2. Albertson, O. E.: "The Control of Bulking Sludges: From the Early Innovators to Current Practice," *Journal WPCF,* vol. 59, no. 4, 1987.
3. Albertson, O. E., and R. N. Okey: "Trickling Filters Need To Breathe Too," presented at the Iowa Water Pollution Control Federation, Des Moines, June 1988.

4. Albertson, O. E.: "Optimizing Rotary Distributor Speed For Trickling Filters," *WPCF Operators Forum,* vol. 2, no. 1, 1989.

5. Anderson, N. E.: "Design of Final Settling Tanks for Activated Sludge," *Sewage Works Journal,* vol. 17, no. 1, 1945.

6. Arora, M. L., E. F. Barth, and M. B. Umphres: "Technology Evaluation of Sequencing Batch Reactors," *Journal WPCF,* vol. 57, no. 8, 1985.

7. Arora, M. L., and M. B. Umphres: "Evaluation of Activated Biofiltration and Activated Biofiltration/Activated Sludge Technologies," *Journal WPCF,* vol. 59, no. 4, 1987.

8. Aryan, A. F., and S. H. Johnson: "Discussion of: A Comparison of Trickling Filter Media," *Journal WPCF,* vol. 59, no. 10, 1987.

9. Bender, J. H.: "Assessment of Design Tradeoffs When Using Intrachannel Clarification," *Journal WPCF,* vol. 59, no. 10, 1987.

10. Crosby, R. M., and J. H. Bender: "Hydraulic Considerations That Affect Clarifier Performance," *EPA Technology Transfer,* March 1980.

11. Daigger, G. T., M. H. Robbins, Jr., and B. R. Marshall: "The Design of a Selector to Control Low F/M Filamentous Bulking," *Journal WPCF,* vol. 57, no. 3, 1985.

12. Eikelboom, D. H.: "Filamentous Organisms in Activated Sludge," *Water Research,* 9:365, 1975.

13. Eikelboom, D. H.: "Identification of Filamentous Organisms in Bulking Sludge," *Progress in Water Technology,* 8:153, 1977.

14. Great Lakes-Upper Mississippi River Board of State Sanitary Engineers: *Recommended Standards for Sewage Works (Ten State Standards),* 1978.

15. Harrison, J. R., and G. T. Daigger: "A Comparison of Trickling Filter Media," *Journal WPCF,* vol. 59, no. 7, 1987.

16. Huang, J. Y. C., and M. D. Cheng: "Measurement and New Applications of Oxygen Uptake Rates in Activated Sludge Processes," *Journal WPCF,* vol. 56, no. 3, 1984.

17. Huibregtse, G. L., T. C. Rooney, and D. C. Rasmussen: "Factors Affecting Fine Bubble Diffused Aeration," *Journal WPCF,* vol. 55, no. 8, 1983.

18. Hwang, H. J., and M. K. Stenstrom: "Evaluation of Fine-Bubble Alpha Factors in Near Full-Scale Equipment," *Journal WPCF,* vol. 57, no. 12, 1985.

19. Kelly, K.: "New Clarifiers Help Save History," *Civil Engineering,* October 1988.

20. Kimball, J. W.: *Biology,* 2nd ed., Addison-Wesley, Reading, MA, 1968.

21. Krause, T. L., D. E. Schmidtke, and S. Modrick: "Rock Trickling Filter Media Can Still Be Viable," *Operations Forum,* June 1988.

22. Linne, S. R., and S. C. Chiesa: "Operational Variables Affecting Performance of the Selector-Complete-Mix Activated Sludge Process," *Journal WPCF,* vol. 59, no. 7, 1987.

23. Luria, S. E.: "The Bacterial Protoplasm: Composition and Organization," in I. C. Gunsalus and R. Y. Stanier (eds.), *The Bacteria I,* Academic, New York, 1960.

24. Mancini, J. L., and E. L. Barnhart: "Industrial Waste Treatment in Aerated Lagoons," in E. F. Gloyna and W. W. Eckenfelder, Jr., (eds.), *Advances in Water Quality Improvement,* University of Texas Press, Austin, 1968.

25. Mandt, M. G., and B. A. Bell: *Oxidation Ditches in Wastewater Treatment,* Ann Arbor Science Publishers, Ann Arbor, MI, 1982.

26. Matasci, R. N., C. Kaempfer, and J. A. Heidman: "Full-scale Studies of the Trickling Filter/Solids Contact Process," *Journal WPCF,* vol. 58, no. 11, 1986.

27. Mehta, D. S., H. H. Davis, and R. P. Kingsburg: "Oxygen Theory in Biological Treatment Plant Design," *J. San. Eng. Div. ASCE,* vol. 98, no. SA3, 1972.

28. Metcalf & Eddy, Inc.: *Wastewater Engineering: Collection, Treatment, Disposal,* McGraw-Hill, New York, 1972.

29. Metcalf & Eddy, Inc.: *Wastewater Engineering: Collection and Pumping of Wastewater,* McGraw-Hill, New York, 1981.

30. Middlebrooks, E. J., C. H. Middlebrooks, J. H. Reynolds, G. Z. Watters, C. S. Reed, and D. B. George: *Wastewater Stabilization Lagoon Design, Performance, and Upgrading,* Macmillan, New York, 1982.

31. Newbry, B. W., G. T. Daigger, and D. Taniguchi-Dennis: "Unit Process Tradeoffs for Combined Trickling Filter and Activated Sludge Processes," *Journal WPCF,* vol. 60, no. 10, 1988.

32. O'Brien, W. J.: "Algae Removal by Rock Filtration," *Transactions of the Twenty-Fifth Annual Conference on Sanitary Engineering,* University of Kansas, 1975.

33. Parker, D. S., and M. S. Merrill: "Oxygen and Air Activated Sludge: Another View," *Journal WPCF,* vol. 48, no. 11, 1976.

34. Parker, D. S.: "Assessment of Secondary Clarification Design Concepts," *Journal WPCF,* vol. 55, no. 4, 1983.

35. Parker, D. S.: "The TF/SC Process at Eight Years Old: Past, Present, and Future," presented at the 59th Annual Conference of the California Water Pollution Control Association, April 1987.

36. Qasim, S. R.: *Wastewater Treatment Plants: Planning, Design, and Operation,* Holt, Rinehart and Winston, New York, 1985.

37. Schroeder, E. D.: *Water and Wastewater Treatment,* McGraw-Hill, New York, 1977.

38. Schroeder, E. D., and G. Tchobanoglous: "Mass Transfer Limitations on Trickling Filter Design," *Journal WPCF,* vol. 48, no. 4, 1976.

39. Sezgin, M., and P. R. Karr: "Control of Actinomycete Scum on Aeration Basins and Clarifiers," *Journal WPCF,* vol. 58, no. 10, 1986.

40. Speece, R. E., N. N. Khauden, and G. Tchobanoglous: "Commercial Oxygen Utilization In Water Quality Management," *Water Environment & Technology,* vol. 2, no. 7, 1990.

41. Stall, T. R., and J. H. Sherrard: "Evaluation of Control Parameters for the Activated Sludge Process," *Journal WPCF,* vol. 50, no. 3, 1978.

42. Strom, P. F., and D. Jenkins: "Identification and Significance of Filamentous Microorganisms in Activated Sludge," *Journal WPCF,* vol. 56, no. 5, 1984.

43. Stukenberg, J. R., L. C. Rodman, and J. H. Touslee: "Activated Sludge Clarifier Design Improvements," *Journal WPCF,* vol. 55, no. 4, 1983.

44. TeKippe, R. J., and J. H. Bender: "Activated Sludge Clarifiers: Design Requirements and Research Priorities," *Journal WPCF,* vol. 59, no. 10, 1987.

45. Thirumurthi, D.: "Design of Waste Stabilization Ponds," *J. San. Eng. Div., ASCE,* vol. 95, no. SA2, 1969.

46. U.S. Environmental Protection Agency: *Upgrading Lagoons,* EPA Technology Transfer Publication, August 1973.

47. U.S. Environmental Protection Agency: *Oxygen Activated Sludge Wastewater Treatment Systems: Design Criteria and Operating Experience,* rev. ed., Technology Transfer Seminar Publication, January 1974.

48. U.S. Environmental Protection Agency: *Process Design Manual for Upgrading Existing Wastewater Treatment Plants,* Office of Technology Transfer, Washington, DC, October 1974.

49. U.S. Environmental Protection Agency: *Process Design Manual for Suspended Solids Removal,* January 1975.

50. U.S. Environmental Protection Agency: *Process Design Manual for Nitrogen Control,* Office of Technology Transfer, Washington, DC, October 1975.

51. U.S. Environmental Protection Agency: *Design Manual, Municipal Wastewater Stabilization Ponds,* October 1983.

52. U.S. Environmental Protection Agency: *Review of Current RBC Performance and Design Procedures,* EPA-600/2-85-033, 1984.

53. U.S. Environmental Protection Agency: *Design Information on Rotating Biological Contactors,* EPA-600/2-84-106, June 1984.

54. U.S. Environmental Protection Agency: *Summary of Design Information on Rotating Biological Contactors,* EPA-430/9-84-008, September 1984.

55. U.S. Environmental Protection Agency and ASCE: *Summary Report: Fine Pore (Fine Bubble) Aeration Systems,* EPA 625/8-85-010, Water Engineering Research Laboratory, Washington, DC, 1985.

56. U.S. Environmental Protection Agency: "Protection of Wastewater Lagoon Interior Slopes, EPA Design Information Report," *Journal WPCF,* vol. 58, no. 10, 1986.

57. U.S. Environmental Protection Agency: *Design Manual, Fine Pore Aeration Systems,* EPA/625/1-89/023, September 1989.

58. van Niekirk, A. M., D. Jenkins, and M. G. Richard: "A Mathematical Model of the Carbon-Limited Growth of Filamentous and Floc-Forming Organisms in Low F/M Sludge," *Journal WPCF,* vol. 60, no. 1, 1988.

59. Water Pollution Control Federation: *Wastewater Treatment Plant Design,* Manual of Practice no. 8, 1977.

60. Water Pollution Control Federation: *Clarifier Design,* Manual of Practice FD-8, 1985.

61. Water Pollution Control Federation: *Activated Sludge,* Manual of Practice OM-9, 1987.

62. Water Pollution Control Federation: *O&M of Trickling Filters, RBCs, and Related Processes,* Manual of Practice OM-10, 1988.

63. Water Pollution Control Federation: *Aeration,* Manual of Practice FD-13, 1988.

64. Water Pollution Control Federation: *Wastewater Treatment Plant Design,* Manual of Practice no. 8 (draft revisions), 1988.

65. Wehner, J. F., and R. F. Wilhelm: "Boundary Conditions of Flow Reactor," *Chem. Eng. Sci.,* 6:89, 1958.

66. West German Ein Regelwerk der Abwassertechnischen Vereinsgung (ATV), Arbeitblatt A 135, Section 3.2.2, Tropfkorperbemessung, p. 6, April 1983.

67. Weston, R. F.: *Review Of Current RBC Performance And Design Procedures,* EPA-600/2-85/033, 1985.

68. Wood, D. K., and G. Tchobanoglous: "Trace Elements in Biological Waste Treatment," *Journal WPCF,* vol. 47, no. 7, 1975.

CHAPTER 11

ADVANCED WASTEWATER TREATMENT

Advanced wastewater treatment is defined as the additional treatment needed to remove suspended and dissolved substances remaining after conventional secondary treatment. These substances may be organic matter or suspended solids or may range from relatively simple inorganic ions, such as calcium, potassium, sulfate, nitrate, and phosphate, to an ever-increasing number of highly complex synthetic organic compounds. In recent years, the effects of many of these substances on the environment have become understood more clearly. Research on potential toxic substances is continuing to determine their environmental effects and how these substances can be removed by both conventional and advanced wastewater treatment processes. As a result, wastewater treatment requirements are becoming more stringent both in terms of limiting concentrations of many of these substances in the treatment plant effluent and of establishing whole effluent toxicity limits, as outlined in Chap. 3. To meet these new requirements, many of the existing secondary treatment facilities will have to be retrofitted, and new advanced wastewater treatment facilities will have to be constructed.

Since the early 1970s, the number of advanced wastewater treatment facilities has increased significantly, and a great deal of information has been published, especially with respect to the removal of nitrogen and phosphorus. The purpose of this chapter is not to report on all these developments but rather to present an overview of this subject in relation to the removal of specific constituents of concern. The chapter contains a brief summary of the need for advanced wastewater treatment, an overview of the available technologies used for the removal of the contaminants of concern (identified in Chaps. 3 and 4), and a review of the more important technologies as applied to the specific constituents. The ultimate disposal of residuals from advanced wastewater treatment is considered in Chap. 12.

11-1   NEED FOR ADVANCED WASTEWATER TREATMENT

With increased scientific knowledge of the constituents found in wastewater and the availability of an expanded information base, derived from environmental monitoring studies, permit requirements for the discharge of treated effluent are becoming increasingly more strict. Permit requirements in many areas may include the removal of organic matter, suspended solids, nutrients, and specific toxic compounds that cannot be accomplished by conventional secondary treatment processes. In some areas of the United States where water supplies are limited, reuse of wastewater is becoming an important factor in water resources planning (see Chap. 16). The important residual constituents in treated wastewater and their potential impacts are discussed in this section.

Residual Constituents in Treated Wastewater

The typical composition of domestic wastewater was reported in Table 3-16. Most domestic wastewaters also contain a wide variety of trace compounds and elements, although they are not measured routinely. If industrial wastewater is discharged to domestic sewers, the distribution of the constituents will vary considerably from that reported in Table 3-16 and may include some of the priority pollutants described in Table 3-9. Some of the substances found in wastewater that may cause problems when discharged to the environment are reported in Table 11-1. This list is not meant to be exhaustive; rather, it suggests that a wide variety of substances must be considered and that they will vary with each wastewater treatment application.

Impacts of Residual Constituents

The potential effects of residual constituents contained in treated effluents may vary considerably. Some of the effects of specific constituents and their critical concentrations are listed in Table 11-1. Although suspended solids and biodegradable organics are addressed specifically in secondary treatment requirements prescribed by EPA, additional removals may be required in special circumstances (e.g., discharge to small streams and lakes and other environmentally sensitive water bodies).

Compounds containing available nitrogen and phosphorus have received considerable attention since the mid-1960s. Initially, nitrogen and phosphorus in wastewater discharges became important because of their effects in accelerating eutrophication of lakes and promoting aquatic growths. More recently, nutrient control has become a routine part of treating wastewaters used for the recharge of groundwater supplies. Nitrification of wastewater discharges is also required in many cases to reduce ammonia toxicity or to lessen the impact on the oxygen resources in flowing streams or estuaries.

Since the early 1980s, regulatory agencies have focused more attention on priority pollutants and volatile organic compounds (VOCs), many of which have been found to be toxic to humans and the aquatic environment. These constituents are of particular concern where wastewater is discharged to surface water or groundwater, both of which may subsequently be used as a domestic water supply.

TABLE 11-1
Typical constituents that may be found in treated wastewater and their effects

Constituent	Effect	Critical concentration, mg/L
Suspended solids	May cause sludge deposits or interfere with receiving water clarity	Variable
Biodegradable organics	May deplete oxygen resources	Variable
Priority pollutants (see Table 3-9)	Toxic to humans; carcinogenic	Varies by individual constituent
	Toxic to aquatic environment	Varies based on presence in water column, biota, or sediment
Volatile organic compounds	Toxic to humans; carcinogenic; form photochemical oxidants (smog)	Varies by individual constituent
Nutrients		
Ammonia	Increases chlorine demand; can be converted to nitrates and, in the process, can deplete oxygen resources; with phosphorus, can lead to the development of undesirable aquatic growths	Any amount
	Toxic to fish	Variable[a]
Nitrate	Stimulates algal and aquatic growth;	0.3[b]
	Can cause methemoglobinemia in infants (blue babies)	45[c]
Phosphorus	Stimulates algal and aquatic growth;	0.015[b]
	Interferes with coagulation;	0.2–0.4
	Interferes with lime-soda softening	0.3
Other inorganics		
Calcium and magnesium	Increase hardness and total dissolved solids	
Chloride	Imparts salty taste	250
	Interferes with agricultural and industrial processes	75–200
Sulfate	Cathartic action	600–1,000
Other organics		
Surfactants	Cause foaming and may interfere with coagulation	1.0–3.0

[a] Depends on pH and temperature.

[b] For quiescent lakes.

[c] As NO_3, per U.S. Environmental Protection Agency, Primary Drinking Water Standards.

11-2 TREATMENT TECHNOLOGIES USED FOR ADVANCED WASTEWATER TREATMENT

Over the past 20 years, a wide variety of treatment technologies have been studied, developed, and applied for the removal of the constituents of concern reported in Table 3-2, as well as other compounds and substances (see Table 11-1). The classification of these technologies and typical performance data are presented in this section.

Classification of Technologies

Advanced wastewater treatment systems may be classified by the type of unit operation or process or by the principal removal function performed. To facilitate a general comparison of the various operations and processes, information on (1) the principal constituent removal function, (2) the types of operations or processes that can be used to perform this function, (3) the type of wastewater treated, and (4) the section and chapter where each is considered is reported in Table 11-2. As will be noted in reviewing Table 11-2, many of the operations and processes have already been discussed and analyzed in detail in Chaps. 6 through 10. In addition, some of the operations or processes may be capable of more than one principal removal function, as discussed later in this chapter.

Typical Process Performance Data

Selection of a given operation, process, or combination thereof depends on (1) the potential use of the treated effluent, (2) the nature of the wastewater, (3) the compatibility of the various operations and processes, (4) the available means to dispose of the ultimate contaminants, and (5) the environmental and economic feasibility of the various systems. Because of special conditions required for contaminant removal, economic feasibility may not be a controlling factor in the design of an advanced wastewater treatment system.

Several examples of advanced wastewater treatment systems are illustrated in Fig. 11-1. Typical residual pollutant concentrations in treated effluent for the combinations of unit operations and processes shown in Fig. 11-1 are reported in Table 11-3. Many other combinations of operations and processes are possible depending upon the constituents to be removed and the economics of the treatment analysis. In applications where granular-medium filters and activated-carbon contactors are used, flow equalization may be beneficial in reducing the size and number of units and in optimizing performance.

11-3 REMOVAL OF RESIDUAL SUSPENDED SOLIDS BY GRANULAR-MEDIUM FILTRATION

The important filtration process variables and operative particle removal mechanisms were considered in detail in Chap. 6. With that information serving as a background, this section identifies the major factors that should be considered in the design of

TABLE 11-2
Constituent removal by advanced wastewater treatment operations and processes

Principal removal function	Description of operation or process	Type of wastewater treated[a]	See Sec.
Suspended-solids removal	Filtration	EPT, EST	11- 3
	Microstrainers	EST	11- 4
Ammonia oxidation	Biological nitrification	EPT, EBT, EST	11- 6
Nitrogen removal	Biological nitrification/ denitrification	EPT, EST	11- 7
Nitrate removal	Separate-stage biological denitrification	EST + nitrification	11- 7
Biological phosphorus removal	Mainstream phosphorus removal[b]	RW, EPT	11- 8
	Sidestream phosphorus removal	RAS	11- 8
Combined nitrogen and phosphorus removal by biological methods	Biological nitrification/ denitrification and phosphorus removal	RW, EPT	11- 9
Nitrogen removal by physical or chemical methods	Air stripping	EST	11-10
	Breakpoint chlorination	EST + filtration	11-10
	Ion exchange	EST + filtration	11-10
Phosphorus removal by chemical addition	Chemical precipitation with metal salts	RW, EPT, EBT, EST	11-11
	Chemical precipitation with lime	RW, EPT, EBT, EST	11-11
Toxic compounds and refractory organics removal	Carbon adsorption	EST + filtration	11-12
	Activated-sludge-powdered activated carbon	EPT	11-12
	Chemical oxidation	EST + filtration	11-12
Dissolved inorganic solids removal	Chemical precipitation	RW, EPT, EBT, EST	11-11
	Ion exchange	EST + filtration	11-13
	Ultrafiltration	EST + filtration	11-13
	Reverse osmosis	EST + filtration	11-13
	Electrodialysis	EST + filtration + carbon adsorption	11-13
Volatile organic compounds	Volatilization and gas stripping	RW, EPT	6-10, 9-13

[a] EPT = effluent from primary treatment
EBT = effluent from biological treatment (before clarification)
EST = effluent from secondary treatment (after clarification)
RW = raw (untreated) wastewater
RAS = return activated sludge

[b] Removal process occurs in the main flowstream as opposed to sidestream treatment.

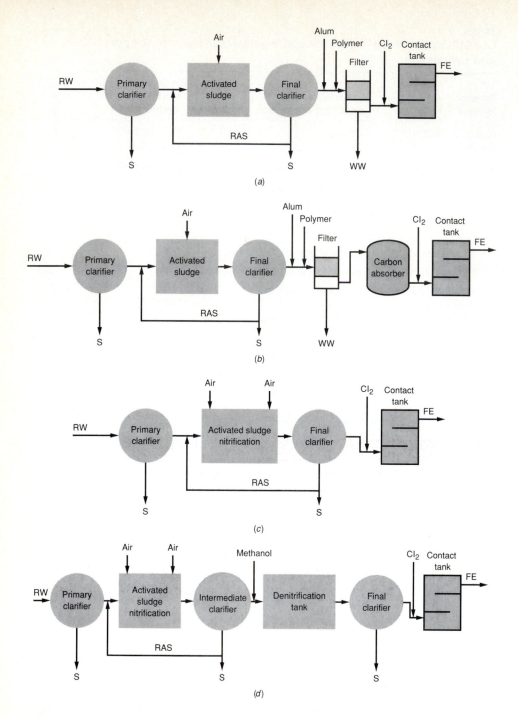

FIGURE 11-1
Examples of advanced wastewater treatment flow diagrams: (*a*) activated sludge + filtration, (*b*) activated sludge + filtration + activated carbon, (*c*) activated sludge nitrification (single stage), and (*d*) activated sludge nitrification/denitrifiction using methanol.

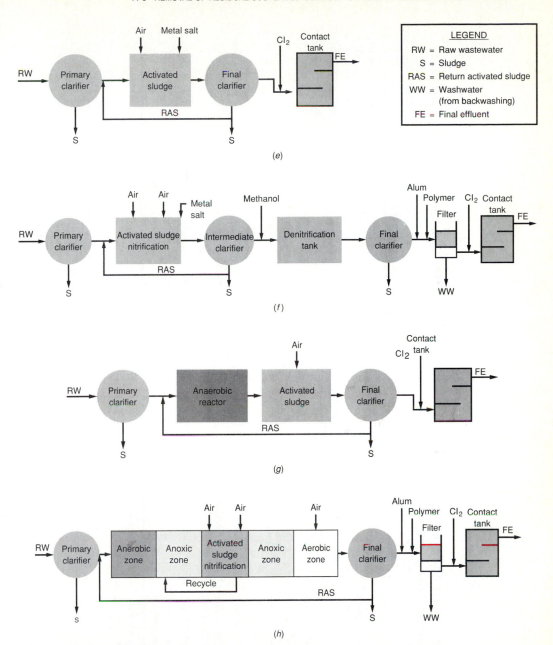

FIGURE 11-1 (*continued*)
(*e*) Metal salt to addition activated sludge for phosphorus removal, (*f*) metal salt addition to activated sludge for phosphorus removal + nitrification/denitrification using methanol, (*g*) mainstream biological phosphorus removal, and (*h*) biological nitrogen and phosphorus removal + filtration.

TABLE 11-3
Treatment levels achievable with various combinations of unit operations and processes used for advanced wastewater treatment[a]

Treatment process	SS, mg/L	BOD$_5$, mg/L	COD, mg/L	Total N, mg/L	NH$_3$-N, mg/L	PO$_4$ as P, mg/L	Turbidity, NTU
			Typical effluent quality				
Activated sludge + granular-medium filtration	4–6	<5–10	30–70	15–35	15–25	4–10	0.3–5
Activated sludge + granular-medium filtration + carbon adsorption	<3	<1	5–15	15–30	15–25	4–10	0.3–3
Activated sludge/nitrification, single stage	10–25	5–15	20–45	20–30	1–5	6–10	5–15
Activated sludge/nitrification-denitrification, separate stages	10–25	5–15	20–35	5–10	1–2	6–10	5–15
Metal salt addition to activated sludge	10–20	10–20	30–70	15–30	15–25	<2	5–10
Metal salt addition to activated sludge + nitrification/denitrification + filtration	<5–10	<5–10	20–30	3–5	1–2	<1	0.3–3
Mainstream biological phosphorus removal[b]	10–20	5–15	20–35	15–25	5–10	<2	5–10
Mainstream biological nitrogen and phosphorus removal[b] + filtration	<10	<5	20–30	<5	<2	<1	0.3–3

[a] Adapted in part from Refs. 7, 38, 39, and 43.
[b] Removal process occurs in the main flowstream as opposed to sidestream treatment.

effluent filtration systems for the removal of residual suspended solids. Topics that will be considered are (1) application of granular-medium filtration; (2) number and size of filter units; (3) selection of a type of filter; (4) filter bed configurations; (5) characteristics of filtering materials; (6) filter backwashing systems; (7) filter appurtenances; (8) filter problems; (9) filter control systems and instrumentation; and (10) effluent filtration with chemical addition. Specific details for piping and physical structures involved are not presented because they vary in each situation. It should be noted that many of the filters to be considered in the following discussion are proprietary and are supplied by the manufacturer as a complete unit. Thus, many of the design details presented in this section apply only to individually designed filters. Intermittent and recirculating sand filters, usually limited to small systems, are discussed in Chap. 14.

Application of Granular-Medium Filtration

Early applications of granular-medium filters for wastewater treatment essentially followed the design procedures developed for the treatment of potable water. Because wastewater is significantly different in physical and chemical characteristics from most natural waters, wastewater filtration entails special design considerations. In general, wastewater filters receive larger, heavier, and more variable particle sizes, as well as uneven solids loadings. The filtration mechanisms are complex and may consist of a combination of factors including straining (mechanical and chance contact), interception within the media, gravity settling, inertial impaction of the particles with adhesion to the filtering medium, and growth of biological solids within the filter bed, which further enhances solids removal [1]. Because the performance of wastewater filters is affected by many factors, pilot studies are recommended in cases where strict effluent quality limits must be met.

Filtration of wastewater is most commonly used for the removal of residual biological floc in settled effluents from secondary treatment before discharge to the receiving waters. Filtration is also used to remove residual precipitates from the metal salt or lime precipitation of phosphates and is used as a pretreatment operation before treated wastewater is discharged to activated-carbon columns. In reuse applications, filtration of treated wastewater is required for application to food crops, park and playground irrigation, and body-contact recreational impoundments (see Chap. 16).

Number and Size of Filter Units

One of the first decisions to be made in the design of a granular-medium filtration system is determining the number and size of required filter units. The surface area required is based on the peak filtration and peak plant flowrates. The allowable peak filtration rate is usually established on the basis of regulatory requirements. Operating ranges for a given filter type are based on past experience, the results of pilot plant studies, and manufacturers' recommendations. The number of units should generally be kept to a minimum to reduce the cost of piping and construction but should be sufficient to ensure (1) that the backwash flowrates do not become excessively large and (2) that, when one filter unit is taken out of service for backwashing, the transient loading on the remaining units will not be high enough to dislodge material contained in the filters [6]. Transient loadings due to backwashing are not an issue with filters that backwash continuously. The sizes of the individual units should be consistent with the sizes of equipment available for use as underdrains, wash-water troughs, and surface washers. Typically, width-to-length ratios for individually designed gravity filters vary from 1:1 to 1:4. For proprietary and pressure filters, it is common practice to use the manufacturer's standard sizes.

Selection of the Type of Filter

The types of filters that have been used for wastewater filtration and their performance characteristics are reviewed in this section.

Types of Filters. The principal types of granular-medium filters are identified in Table 11-4. As shown in Table 11-4, the filters can be classified in terms of their operation as semicontinuous or continuous. Within each of these two classifications, there are a number of different types, depending on bed depth (e.g., conventional, shallow-bed, and deep-bed), the type of filtering medium used (mono-, dual-, and multi-medium), whether the filtering medium is stratified or unstratified, the type of operation (downflow or upflow), and the method used for the management of solids (i.e., surface or internal storage). Mono- and dual-medium semicontinuous filters can be further classified by driving force (e.g., gravity or pressure). Another important distinction that must be noted for the filters reported in Table 11-4 is whether they are proprietary or individually designed. With proprietary filters, the manufacturer is responsible for providing the complete filter unit and its controls, based on basic design criteria and performance specifications. In individually designed filters, the designer is responsible for working with several suppliers in developing the design of the system components. Contractors and suppliers then furnish the materials and equipment in accordance with the engineer's design.

Performance Characteristics of Different Types of Filters. The critical question associated with the selection of any granular-medium filter is whether it will perform as anticipated. Insight into the performance of granular-medium filters can be gained from a review of the data presented in Fig. 11-2. The results of testing six different types of pilot-scale filters on the effluent from the same activated-sludge plant are shown [30]. The principal conclusions to be reached from an analysis of the data presented in Fig. 11-2 are that (1) given a high-quality filter influent (turbidity less than 7 to 9 NTU), all of the filters tested produced an effluent with an average turbidity of 2 NTU or less and, (2) when the influent turbidity was greater than about 7 to 9 NTU, chemical addition was required for all of the filters to achieve an effluent turbidity of 2 NTU or less. Using the relationship between turbidity and suspended solids given in Eq. 6-39, an influent turbidity of 7 to 9 NTU corresponds to a suspended-solids concentration of 16 to 23 mg/L.

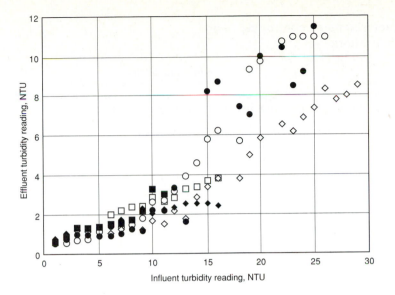

FIGURE 11-2
Performance data for six different types of granular medium filters used for wastewater applications tested on the effluent from the same activated-sludge plant at a filtration rate of 4 gal/ft^2 · min [30].

For new wastewater treatment plants, extra care should be devoted to the design of the secondary settling facilities. With properly designed settling facilities, the decision on what type of filtration system to use is often based on plant-related variables such as the space available, duration of filtration period (seasonal versus year-round), the time available for construction, and costs. For existing plants that do not function well and must be retrofitted with effluent filtration, it may be appropriate to consider the type of a filter that can continue to function even when heavily overloaded. The pulsed-bed filter and both downflow and upflow deep-bed coarse medium filters have been used in such applications.

Filter Bed Options

The principal types of filter beds and their characteristics are considered further in the following discussion.

Stratified and Unstratified Filter Beds. As noted previously, filters may be classified according to the stratification of the filtering medium. Depending on the procedure used to backwash a filter, the filtering materials may become stratified or unstratified. Using water alone for backwashing and fluidizing a single filter medium results in the filter becoming stratified with the smaller particles at the surface. Use of a simultaneous air-water wash backwashing procedure, when fluidizing and washing a single medium, will produce an unstratified bed due to the mixing of the fine

TABLE 11-4
Physical characteristics of commonly used granular-medium filters

Type of filter operation	Type of filter (common name)	Filter bed details		
		Type of filter bed	Filtering medium	Typical bed depth, in
Semicontinuous	Conventional	Mono-medium (stratified or unstratified)	Sand or anthracite	30
Semicontinuous	Conventional	Dual-medium (stratified)	Sand and anthracite	36
Semicontinuous	Conventional	Multimedium (stratified)	Sand, anthracite, and garnet	36
Semicontinuous	Deep-bed	Mono-medium (stratified or unstratified)	Sand or anthracite	72
Semicontinuous	Deep-bed	Mono-medium (stratified)	Sand or anthracite	72
Semicontinuous	Pulsed-bed	Mono-medium (stratified)	Sand	11
Continuous	Deep-bed	Mono-medium (unstratified)	Sand	72
Continuous	Traveling-bridge	Mono-medium (stratified)	Sand	11
Continuous	Traveling-bridge	Dual-medium (stratified)	Sand and anthracite	16

Note: in × 25.4 = mm   *(continued horizontally)*

and coarser particles. Use of a simultaneous air-water wash with dual- or multi-medium requires use of water alone at the end of the backwash cycle to achieve stratification.

Shallow Mono-Medium Stratified Filter Beds. Single-medium beds with depths of less than 1 ft (0.3 m) are being used increasingly for wastewater filtration. Two

TABLE 11-4
(continued)

Typical direction of fluid flow	Backwash operation	Flowrate through filter	Solids storage location	Remarks	Type of design
Downward	Batch	Constant/variable	Surface and upper bed	Rapid headloss buildup	Individual
Downward	Batch	Constant/variable	Internal	Dual-medium design used to extend filter run length	Individual
Downward	Batch	Constant/variable	Internal	Multimedium design used to extend filter run length	Individual
Downward	Batch	Constant/variable	Internal		Individual
Upward	Batch	Constant	Internal		Proprietary
Downward	Batch	Constant	Surface and upper bed	Air pulses used to break up surface mat and increase run length	Proprietary
Upward	Continuous	Constant	Internal	Sand bed moves in countercurrent direction to fluid flow	Proprietary
Downward	Semi-continuous	Constant	Surface and upper bed	Individual filter cells backwashed sequentially	Proprietary
Downward	Semi-continuous	Constant	Surface and upper bed	Individual filter cells backwashed sequentially	Proprietary

principal types are used: the pulsed-bed filter and the traveling-bridge filter. Typical design data are presented in Table 11-5. The pulsed-bed filter, illustrated in Fig. 11-3a, uses a stratified fine sand medium. An air diffuser, located just above the surface of the bed, keeps the solids above the filter bed in suspension. Periodically an air pulse is generated through the backwash/underdrain system, which resuspends the solids retained on the surface of the bed and redistributes the solids trapped within the

TABLE 11-5
Typical design data for mono-medium filters[a]

Characteristic	Value Range	Value Typical
Shallow-bed (stratified)		
Sand:		
Depth, in	10–12	11
Effective size, mm	0.35–0.6	0.45
Uniformity coefficient	1.2–1.6	1.5
Filtration rate, gal/ft$^2 \cdot$ min	2–6	3
Anthracite:		
Depth, in	12–20	16
Effective size, mm	0.8–1.5	1.3
Uniformity coefficient	1.3–1.8	1.6
Filtration rate, gal/ft$^2 \cdot$ min	2–6	3
Conventional (stratified)		
Sand:		
Depth, in	20–30	24
Effective size, mm	0.4–0.8	0.65
Uniformity coefficient	1.2–1.6	1.5
Filtration rate, gal/ft$^2 \cdot$ min	2–6	3
Anthracite:		
Depth, in	24–36	30
Effective size, mm	0.8–2.0	1.3
Uniformity coefficient	1.3–1.8	1.6
Filtration rate, gal/ft$^2 \cdot$ min	2–8	4
Deep-bed (unstratified)		
Sand:		
Depth, in	36–72	48
Effective size, mm	2–3	2.5
Uniformity coefficient	1.2–1.6	1.5
Filtration rate, gal/ft$^2 \cdot$ min	2–10	5
Anthracite:		
Depth, in	36–84	60
Effective size, mm	2–4	2.75
Uniformity coefficient	1.3–1.8	1.6
Filtration rate, gal/ft$^2 \cdot$ min	2–10	5

[a] Developed in part from Refs. 6 and 30.

Note: in $\times$ 25.4 = mm
gal/ft$^2 \cdot$ min $\times$ 40.7458 = L/m$^2 \cdot$ min

upper portion of the filter bed. Either after a set number of pulses or when terminal headloss is reached, the filter is backwashed through the underdrain system. Another unusual feature of the pulsed-bed filter is that underdrain of the filter is open to the atmosphere. The pulsed-bed filter has been used successfully for the filtration of primary and secondary effluent.

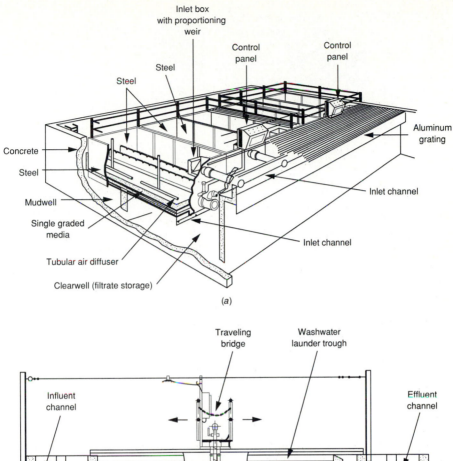

(a)

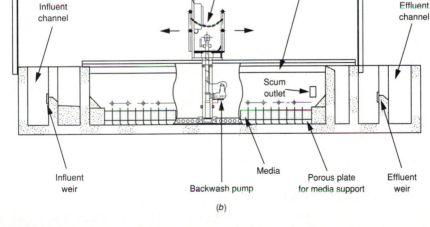

(b)

FIGURE 11-3
Alternative shallow bed filters: (a) pulsed-bed (from Zimpro Passavant) and (b) traveling-bridge (from Infilco Degremont).

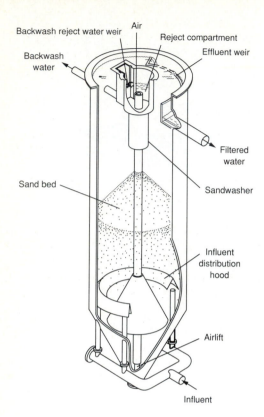

Backwash reject water weir

Air

Reject compartment

Effluent weir

Backwash water

Sand bed

Filtered water

Sandwasher

Influent distribution hood

Airlift

Influent

FIGURE 11-4
Continuous-backwash upflow filter. Because of the method used to backwash the sand, the filter bed is unstratified (from Parkson Corporation).

The second type of shallow depth filter uses a stratified sand medium 11 in (0.28 m) in depth in a series of small filter cells (see Fig. 11-3b). A backwash system mounted on a traveling bridge moves continuously along the length of the filter hydraulically backwashing each cell. This method of backwashing (1) does not require the total filter, comprised of all filter cells, to be taken out of service for backwashing, (2) reduces the headlosses through the filter to a range of 6 to 10 in (150 to 250 mm), and (3) reduces the washwater flowrate, thereby eliminating the need for a washwater collection and equalization basin. This type of filter is used mainly for filtration of effluent from secondary and advanced wastewater treatment plants.

Mono-Medium Unstratified Filter Beds. Two types of single-medium unstratified filter beds are now in use. In the first type, a single, uniform, coarse medium (2 to 3 mm in effective size) is used in beds with depths up to 6 ft (~2 m). These coarse-medium deep-bed filters offer longer filter runs. Depending on the type of treatment process, these filters can also be used for the simultaneous denitrification of the wastewater, although the filtration rate will be significantly lower. The principal disadvantages are (1) the need for a uniform size of medium, (2) the high backwash velocities required to fluidize the bed for effective cleaning, and (3) the added cost for the backwashing facilities and the structure needed to contain the deep beds.

In the second type, a single medium of varying grain sizes is used with a combined air-water backwash. This type has proved to be an effective alternative to the filter with a single medium of uniform size. The combined air-water backwash scours the accumulated material from the filtering medium without the need for fluidizing the entire bed. This backwash system also eliminates the normal stratification that occurs in single-medium and multi-medium beds when only a water backwash or an air-followed-by-water backwash is used. Thus, it is possible to obtain a filter bed with a more or less uniform pore-size distribution through its depth. From the analysis in Chap. 6, it can be concluded that the uniform pore size distributions achieved in unstratified beds will increase the potential for the removal of suspended particles in the lower portions of the filter. By comparison, in a stratified filter bed, the potential decreases with depth because of increasing pore size.

A third type of unstratified filter is the continuous-backwash upflow filter, shown in Fig 11-4. As described in Chap. 6, the flow moves upward through the filtering medium (usually sand), which is moving in the counter-current direction. Additional details may be found in Chap. 6.

Typical design data for single-medium unstratified filter beds are also presented in Table 11-5. Additional details on the performance of unstratified filters may be found in Ref. 10. The appurtenances used in conjunction with unstratified filters are essentially the same as those used for conventional downflow filters.

Dual-Medium and Multimedium Stratified Filter Beds. Some dual-medium filter beds are composed of (1) anthracite and sand, (2) activated carbon and sand, (3) resin beads and sand, and (4) resin beads and anthracite. Multimedium beds may be composed of (1) anthracite, sand, and garnet or ilmenite (see Fig. 11-5), (2) activated carbon, anthracite, and sand, (3) weighted, spherical resin beads (charged and uncharged), anthracite, and sand, and (4) activated carbon, sand, and garnet or ilmenite. Typical data on the depth and characteristics of the filtering materials used most commonly in dual- and multimedium filters are presented in Table 11-6. Because filter performance is related directly to the characteristics of the liquid and the design of the filtering material, it is desirable to conduct pilot plant studies to determine the optimum combination of filter materials. If it is not possible to conduct such studies, the data in Table 11-6 may be used as a guide.

Characterization of Filtering Materials

Once a type of filter has been selected, the next step is to specify the characteristics of the filtering medium or media, if more than one is used. Typically, this involves the selection of the grain size as specified by the effective size and uniformity coefficient, the specific gravity, solubility, hardness, and depth of the various materials used in the filter bed. Sometimes it is advantageous to specify the 99 percent passing size and the 1 percent passing size to define more accurately the gradation curve for each filter medium. In addition, during conceptual design, it is necessary (1) to determine the type of underdrain system required to support the filtering materials, and (2) to determine the submergence requirements of the filter bed to minimize or prevent negative heads from occurring in the filter.

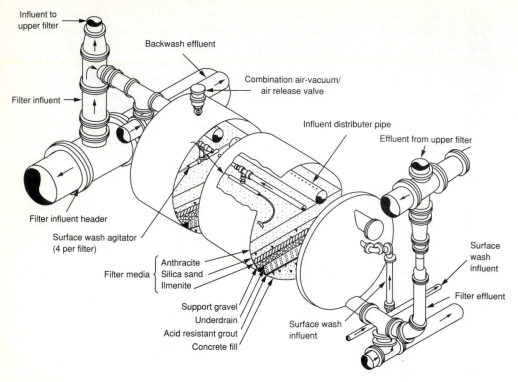

FIGURE 11-5
Typical pressure filter with multi-medium filter bed and surface wash used for the treatment of waste-water.

Filter Backwashing Systems

A filter bed can function properly only if the backwashing system cleans the material removed within the filter effectively. The methods commonly used for backwashing granular-medium filter beds include (1) water backwash with auxiliary surface water-wash agitation, (2) water backwash with auxiliary air scour, and (3) combined air-water backwashing. With the first two methods, fluidization of the granular medium is necessary to achieve effective cleaning of the filter bed at the end of the run. With the third method, fluidization is not necessary. Typical backwash flowrates required to fluidize various filter beds are reported in Table 11-7.

Water Backwash with Auxiliary Surface Wash. Surface washers (see Fig. 11-6; see also Fig. 11-5) are often used to provide the shearing force required to clean the grains of the wastewater filtering medium. Operationally, the surface-washing cycle is started about 1 or 2 min before the water backwashing cycle is started. Both cycles are continued for about 2 min, at which time the surface wash is terminated. Water usage is as follows: for a single-sweep surface backwashing system, from 0.5 to 1.0

TABLE 11-6
Typical design data for dual- and multi-medium filters[a]

Characteristic	Value Range	Typical
Dual-medium		
Anthracite:		
Depth, in	12–30	24
Effective size, mm	0.8–2.0	1.3
Uniformity coefficient	1.3–1.8	1.6
Sand:		
Depth, in	6–12	12
Effective size, mm	0.4–0.8	0.65
Uniformity coefficient	1.2–1.6	1.5
Filtration rate, gal/ft^2 · min	2–10	5
Multimedium		
Anthracite (top layer of quad-media filter):		
Depth, in	8–20	16
Effective size, mm	1.3–2.0	1.6
Uniformity coefficient	1.5–1.8	1.6
Anthracite (second layer of quad-media filter):		
Depth, in	4–16	8
Effective size, mm	1.0–1.6	1.1
Uniformity coefficient	1.5–1.8	1.6
Anthracite (top layer of tri-media filter):		
Depth, in	8–20	16
Effective size, mm	1.0–2.0	1.4
Uniformity coefficient	1.4–1.8	1.6
Sand:		
Depth, in	8–16	10
Effective size, mm	0.4–0.8	0.5
Uniformity coefficient	1.3–1.8	1.6
Garnet or ilmenite:		
Depth, in	2–6	4
Effective size, mm	0.2–0.6	0.3
Uniformity coefficient	1.5–1.8	1.6
Filtration rate, gal/ft^2 · min	2–10	5

[a] Developed in part from Refs. 6 and 30.

Note: in × 25.4 = mm
gal/ft^2 · min × 40.7458 = L/m^2 · min

gal/ft^2 · min (20 to 40 L/m^2 · min); for a dual-sweep surface backwashing system, from 1.5 to 2.0 gal/ft^2 · min (60 to 80 L/m^2 · min) [6].

Water Backwash with Auxiliary Air Scour. Air scouring the filter provides a more vigorous washing action than water alone. Operationally, air is usually applied for 3 to 4 min before the water backwashing cycle begins. In some systems, air is

TABLE 11-7
Typical backwash flowrates required to fluidize various filter beds[a]

Type of filter	Size of critical granular medium	Minimum backwash velocity needed to fluidize bed[b]	
		gal/ft^2 · min	ft/min
Single-medium (sand)	2 mm	44–48	6–6.5
Dual-media (anthracite and sand)	See Table 11-6	20–30	2.5–4
Tri-media (anthracite, sand, and garnet or ilmenite)	See Table 11-6	20–30	2.5–4

[a] Adapted in part from Refs. 10, 32, and 35.
[b] Varies with size, shape, and specific gravity of the medium and the temperature of the backwash water.
Note: gal/ft^2 · min × 0.04075 = m^3/m^2 · min
         ft/min × 0.3048 = m/min

also injected during the first part of the water-washing cycle. Typical air flowrates range from 3 to 5 ft^3/ft^2 · min (10 to 16 m^3/m^2 · min) [6].

Combined Air-Water Backwash. The combined air-water backwash system is used in conjunction with the single-medium unstratified filter bed. Operationally, air and water are applied simultaneously for several minutes. The specific duration of the combined backwash varies with the design of the filter bed. Ideally, during the backwash operation, the filter bed should be agitated sufficiently so that the grains of the filter medium move in a circular pattern from the top to the bottom of the filter as

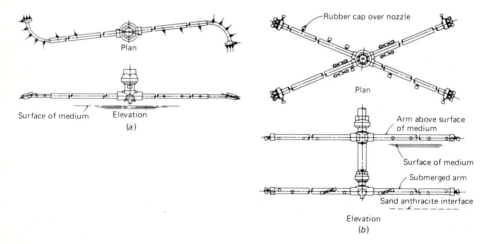

FIGURE 11-6
Typical surface-wash agitators: (a) single-arm agitator and (b) dual-arm agitator (from F.B. Leopold Co.).

the air and water rise up through the bed. Some typical data on the quantity of water and air required are reported in Table 11-8. The reduced washwater requirements for the air-water backwash system can be appreciated by comparing the values given in Table 11-8 with those given in Table 11-7. At the end of the combined air-water backwash, a 2 to 3 min water backwash at subfluidization velocities is used to remove any air bubbles that may remain in the filter bed [10]. This last step is required to eliminate the possibility of air binding within the filter.

Filter Appurtenances

The principal filter appurtenances are as follows: (1) the underdrain system used to support the filtering materials, collect the filtered effluent, and distribute the backwash water and air (where used), (2) the washwater troughs used to remove the spent backwash water from the filter, and (3) the surface-washing systems used to help remove attached material from the filter medium.

Underdrain Systems. The choice of an underdrain system depends on the type of backwash system. In conventional water backwashed filters without air scour, it is common practice to place the filtering medium on a support consisting of several layers of graded gravel. The design of a gravel support for a granular medium is delineated in the *AWWA Standard for Filtering Material B100-89.* Typical underdrain systems are shown in Fig. 11-7.

Washwater Troughs. Washwater troughs are constructed of fiberglass, plastic, sheet metal or concrete with adjustable weir plates. The particular design of the trough will depend to some extent on the other design equipment and construction of the

TABLE 11-8
Air and water backwash rates used with single-medium sand and anthracite filters[a]

Medium	Medium characteristics		Backwash rate	
	Effective size, mm	Uniformity coefficient	Water, $gal/ft^2 \cdot min$	Air, $ft^3/ft^2 \cdot min$[b]
Sand	1.00	1.40	10	43
	1.49	1.40	15	65
	2.19	1.30	20	86
Anthracite	1.10	1.73	7	22
	1.34	1.49	10	43
	2.00	1.53	15	65

[a] Adapted in part from Ref. 10.
[b] Air at 70°F (21°C) and 1.0 atm.
Note: $gal/ft^2 \cdot min \times 0.04075 = m^3/m^2 \cdot min$
$ft^3/ft^2 \cdot min \times 0.3048 = m^3/m^2 \cdot min$

684

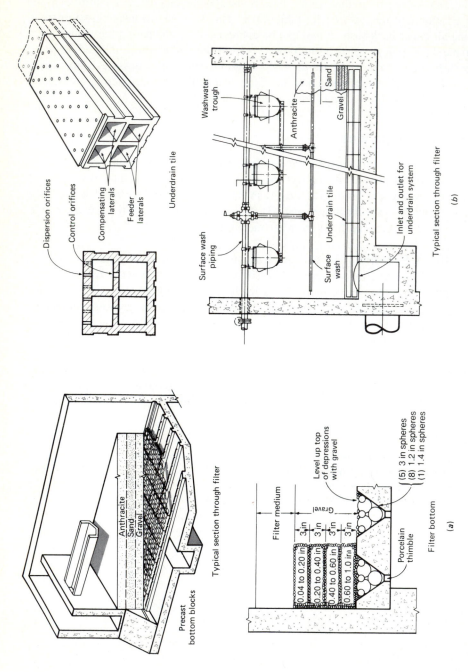

Dispersion orifices

Control orifices

Compensating laterals

Feeder laterals

Underdrain tile

Washwater trough

Anthracite

Sand

Gravel

Surface wash piping

Surface wash

Underdrain tile

Inlet and outlet for underdrain system

Typical section through filter

(b)

Anthracite
Sand
Gravel

Typical section through filter

Precast bottom blocks

Level up top of depressions with gravel

Filter medium

Gravel

3 in
3 in
3 in
3 in

0.04 to 0.20 in
0.20 to 0.40 in
0.40 to 0.60 in
0.60 to 1.0 in

(5) 3 in spheres
(8) 1.2 in spheres
(1) 1.4 in spheres

Porcelain thimble

Filter bottom

(a)

FIGURE 11-7

Typical underdrain systems used with gravel support layer (*a*) and (*b*) and without gravel support layer (*c*): (*a*) Wheeler underdrain system (from BIF) and (*b*) Leopold underdrain system (from F.B. Leopold Co.).

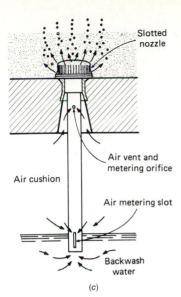

FIGURE 11-7 (*continued*)
(*c*) Air-water nozzle underdrain system
(from Infilco-Degremont).

filter. Loss of filter material during backwashing is a common operating problem. To reduce this problem, baffles can be placed on the underside of the washwater troughs [23].

Surface Washers. Surface washers for filters can be fixed or mounted on rotary sweeps, as shown in Fig. 11-6. According to data on a number of systems, rotary-sweep washers appear to be the most effective.

Filter Problems

The principal problems encountered in wastewater filtration and effective control measures are reported in Table 11-9. Because these problems can affect both the performance and operation of a filter system, care should be taken in the design phase to provide the necessary facilities that will minimize their impact. When filtering secondary effluent containing residual biological floc, semicontinuous filters should be backwashed at least once every 24 hours to avoid the formation of mudballs and the buildup of grease. In most cases, the frequency of backwashing will be more often.

Filter Instrumentation and Control Systems

The supervisory control facilities for wastewater filtration include instrumentation systems for the control and monitoring of the filters. The control systems are similar to those used for water treatment; however, full automation of gravity wastewater filters is generally not required. Although not required, a number of fully automatic control systems are available for wastewater filtration.

TABLE 11-9
Summary of commonly encountered problems in the filtration of wastewater and control measures for those problems[a]

Problem	Description/control
Turbidity breakthrough[b]	Unacceptable levels of turbidity are recorded in the effluent from the filter, even though the terminal headloss has not been reached. To control the buildup of effluent turbidity levels, chemicals and polymers have been added to the filter. The point of chemical or polymer addition must be determined by testing.
Mudball formation	Mudballs are an agglomeration of biological floc, dirt, and the filtering medium or media. If the mudballs are not removed, they will grow into large masses that often sink into the filter bed and ultimately reduce the effectiveness of the filtering and backwashing operations. The formation of mudballs can be controlled by auxiliary washing processes such as air scour or water-surface wash concurrent with, or followed by, water wash.
Buildup of emulsified grease	The buildup of emulsified grease within the filter bed increases the headloss and thus reduces the length of filter run. Both air scour and water-surface wash systems help control the buildup of grease. In extreme cases, it may be necessary to steam clean the bed or to install a special washing system.
Development of cracks and contraction of filter bed	If the filter bed is not cleaned properly, the grains of the filtering medium become coated. As the filter compresses, cracks develop, especially at the sidewalls of the filter. Ultimately, mudballs may develop. This problem can be controlled by adequately backwashing and scouring.
Loss of filter medium or media (mechanical)	In time, some of the filter material may be lost during backwashing and through the underdrain system (where the gravel support has been upset or the underdrain system has been installed improperly). The loss of the filter material can be minimized through the proper placement of washwater troughs and underdrain system. Special baffles are also effective.
Loss of filter medium or media (operational)	Depending on the characteristics of the biological floc, grains of the filter material can become attached to it, forming aggregates light enough to be floated away during the backwashing operations. The problem can be minimized by the addition of an auxiliary air and/or water-scouring system.
Gravel mounding	Gravel mounding occurs when the various layers of the support gravel are disrupted by the application of excessive rates of flow during the backwashing operation. A gravel support with an additional 2 to 3 in (50 to 75 mm) layer of high-density material, such as ilmenite or garnet, can be used to overcome this problem.

[a] Adapted in part from Ref. 6.

[b] Turbidity breakthrough does not occur with filters that operate continuously.

Flow through the filters may be controlled from a water level upstream of the filters or from the water level in each filter. These water levels are used in conjunction with rate-of-flow controllers or a control valve to limit or regulate the flowrate through a filter. Filter hydraulic operating parameters requiring monitoring include filtered water flowrate, total headloss across each filter, surface-wash and backwash water flowrates, and air flowrate if an air-water backwash system is employed. Water quality parameters in filtered water that are usually monitored include BOD, suspended solids, phosphorus, and nitrogen. Turbidity may also be monitored in systems where chemical addition is used. Signals from effluent trubidity monitors and effluent flowrate are often used to pace the chemical-feed system. All filter operating data should be logged to provide records of performance.

The sequencing of the backwash cycle for a conventional gravity filter should preferably be semiautomatic, incorporating a manual start and followed by automatic operation to carry the backwash cycle through its various steps. The design of backwash systems must recognize the impact of maximum wastewater temperatures experienced at treatment plants. Local control units should be provided at the filters to allow for local operation and backwashing by plant operators. Additional details for filter instrumentation and control systems may be found in Refs. 32 and 35.

Effluent Filtration with Chemical Addition

Depending on the quality of the settled secondary effluent, chemical addition has been used to improve the performance of effluent filters. Chemical addition has also been used to achieve specific treatment objectives including the removal of specific contaminants such as phosphorus, metal ions, and humic substances. The removal of phosphorus by chemical addition is considered in Sec. 11-11. In Switzerland, to control eutrophication, the contact filtration process is used to remove phosphorus from wastewater treatment plant effluents that are discharged to lakes. Chemicals commonly used in effluent filtration include a variety of organic polymers, alum, and ferric chloride. Use of organic polymers and the effects of the chemical characteristics of the wastewater on alum addition are considered in the following discussion.

Use of Organic Polymers. Organic polymers are typically classified as long-chain organic molecules with molecular weights varying from 10^4 to 10^6. With respect to charge, organic polymers can be cationic (positively charged), anionic (negatively charged), or nonionic (no charge). Polymers are added to settled effluent to bring about the formation of larger particles by bridging, as described in Chap. 7. Because the chemistry of the wastewater has a significant effect on the performance of a polymer, the selection of a type of polymer for use as a filter aid generally requires experimental testing. Common test procedures for polymers involve adding an initial dosage (usually 1.0 mg/L) of a given polymer and observing the effects. Depending upon the effects observed, the dosage should be increased by 0.5 mg/L increments or decreased by 0.25 mg/L increments (with accompanying observation of effects) to obtain an operating range. After the operating range is established, additional testing can be done to establish the optimum dosage.

A recent development is the use of lower molecular weight polymers that are intended to serve as alum substitutes. When these polymers are used, the dosage is considerably higher (≥ 10 mg/L) than with higher molecular weight polymers (0.25 to 1.25 mg/L). As with the mixing of alum, the initial mixing step is critical in achieving maximum effectiveness of a given polymer. In general, mixing times of less than 1 second with G values of > 1500 s^{-1} are recommended (see Table 6-6).

Effects of Chemical Characteristics of Wastewater on Alum Addition. As with polymers, the chemical characteristics of the treated wastewater effluent can have a significant impact on the effectiveness of aluminum sulfate (alum) when it is used as an aid to filtration. For example, the effectiveness of alum is dependent on pH (see Fig. 11-8). Although Fig. 11-8 was developed for water treatment applications, it has been found to apply to most wastewater effluent filtration uses with minor variations. As shown in Fig 11-8, the approximate regions in which the different phenomenon associated with particle removal in conventional sedimentation and filtration processes are operative are plotted as a function of the alum dose and the pH of the treated effluent after alum has been added. For example, optimum particle removal by sweep

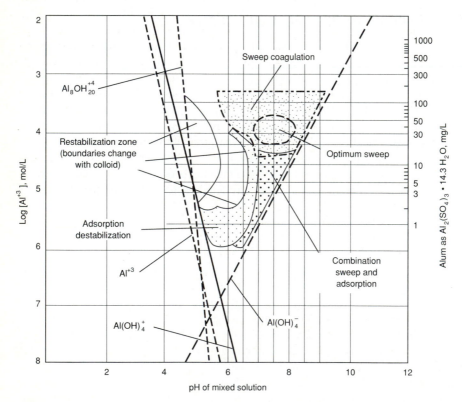

FIGURE 11-8
Typical operating ranges for alum coagulation (adapted from Ref. 2).

floc occurs in the pH range of 7 to 8 with an alum dose of 20 to 60 mg/L. Generally, for many wastewater effluents that have high pH values (e.g., 7.3 to 8.5), low alum dosages in the range of 5 to 10 mg/L will not be effective. Although it is possible to operate with low alum dosages without proper pH control, the most common practice in wastewater filtration applications is to operate in the sweep floc region.

11-4 REMOVAL OF RESIDUAL SUSPENDED SOLIDS BY MICROSCREENING

The microscreen is a surface filtration device used to remove a portion of the residual suspended solids from secondary effluents and from stabilization pond effluents.

Description

Microscreening involves the use of variable low-speed (up to 4 r/min), continuously backwashed, rotating drum filters operating under gravity conditions (see Fig. 11-9). The principal filtering fabrics have openings of 23 or 35 μm and are fitted on the drum periphery. The wastewater enters the open end of the drum and flows outward through the rotating screening cloth. The collected solids are backwashed by high-pressure jets into a trough located within the drum at the highest point of the drum.

The typical suspended-solids removal achieved with these units is about 55 percent. The range is from about 10 to 80 percent. Problems encountered with micro-

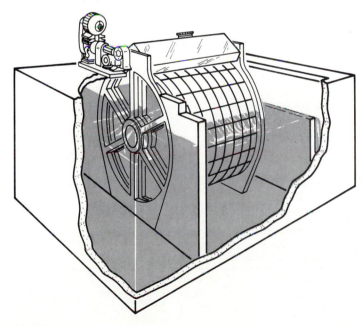

FIGURE 11-9
Microscreen for secondary effluent.

screens include incomplete solids removals and inability to handle solids fluctuations. Reducing the rotating speed of the drum and less frequent flushing of the screen have resulted in increased removal efficiencies but reduced capacity.

Functional Design

The functional design of a microscreen unit involves the following considerations: (1) the characterization of the suspended solids with respect to the concentration and degree of flocculation; (2) the selection of unit design parameter values that will not only ensure capacity to meet maximum hydraulic loadings with critical solids characteristics but also provide desired design performance over the expected range of hydraulic and solids loadings; and (3) the provision of backwash and cleaning facilities to maintain the capacity of the screen. Typical design information for microscreens is presented in Table 11-10. Because of the variable performance of these units, it is recommended that pilot plant studies be conducted, especially if the units are to be used to remove solids from stabilization pond effluents.

TABLE 11-10
Typical design information for microscreens used for screening secondary settled effluent[a]

Item	Typical value	Remarks
Screen size	$20 - 35$ μm	Stainless steel or polyester screen cloths are available in sizes ranging from 15 to 60 μm
Hydraulic loading rate	$75 - 150$ gal/ft^2 · min	Based on submerged surface area of drum
Headloss through screen	$3 - 6$ in	Bypass should be provided when headloss exceeds 8 in
Drum submergence	$70 - 75\%$ of height; $60 - 70\%$ of area	Varies depending on screen design
Drum diameter	$8 - 16$ ft	10 ft (3 m) is most commonly used size; smaller sizes increase back-wash requirements
Drum speed	15 ft/min at 3 in headloss; $115 - 150$ ft/min at 6 in headloss	Maximum rotational speed is limited to 150 ft/min
Backwash requirements	2% of throughput at 50 lb$_f$/in^2; 5% of throughput at 15 lb$_f$/in^2	

[a] Adapted in part from Ref. 30.

Note: gal/ft^2 · min × 0.04075 = m^3/m^2 · min
in × 25.4 = mm
ft × 0.3048 = m
lb$_f$/in^2 × 6.8948 = kPa
ft/min × 0.3048 = m/min

11-5 CONTROL OF NUTRIENTS

Nitrogen and phosphorus are the principal nutrients of concern in treated wastewater discharges. Discharges containing nitrogen and phosphorus may accelerate the eutrophication of lakes and reservoirs and may stimulate the growth of algae and rooted aquatic plants in shallow streams. In addition to being aesthetically unsightly, the presence of algae and aquatic plants may interfere with beneficial uses of the water resources, particularly when they are used for water supplies, fish propagation, and recreation. Significant concentrations of nitrogen in treated effluents may also have other adverse effects including depleting dissolved oxygen in receiving waters, exhibiting toxicity toward aquatic life, affecting chlorine disinfection efficiency, presenting a public health hazard, and affecting the suitability of wastewater for reuse. Therefore, the control of nitrogen and phosphorus is becoming increasingly important in water quality management and in the design of wastewater treatment plants.

Nutrient control strategies, control and removal of nitrogen, and removal of phosphorus are considered in the following discussion. More detailed information on specific nutrient control processes is presented in Secs. 11-6 through 11-11.

Nutrient Control Strategies

In selecting a nutrient control strategy, it is important to assess the characteristics of the untreated wastewater, the type of existing wastewater facility, and the level of nutrient control required. The need for seasonal versus year-round nutrient removal must also be considered. The approaches used for nutrient control may involve the addition of a single process for control of a specific nutrient (e.g., adding alum for the precipitation of phosphorus) or may involve the integration of nutrient removal with the main biological treatment system. The approach used and the process flow diagram selected will depend upon the required reliability in meeting the effluent quality objectives, flexibility of operation, and cost.

Various treatment methods have been used employing chemical, physical, and biological systems to limit or control the amount and form of nutrients discharged by the treatment system. The processes most used initially were biological nitrification for ammonia oxidation and control, biological denitrification using methanol for nitrogen removal, and chemical precipitation for phosphorus removal. In recent years, a number of biological treatment processes have been developed for removal of phosphorus alone or in combination with nitrogen. These processes have considerable appeal to designers and operators because the use of chemicals has been eliminated or reduced substantially.

Control and Removal of Nitrogen

Nitrogen in untreated wastewater is principally in the form of ammonia or organic nitrogen, both soluble and particulate. Soluble organic nitrogen is mainly in the form of urea and amino acids. Untreated wastewater usually contains little or no nitrite or nitrate. A portion of the organic particulate matter is removed by primary sedimentation. During biological treatment, most of the particulate organic nitrogen

TABLE 11-11
Effect of various treatment operations and processes on nitrogen compounds[a]

Treatment operation or process	Nitrogen compound			Removal of total nitrogen entering process, %[b]
	Organic nitrogen	$NH_3-NH_4^+$	NO_3^-	
Conventional treatment				
Primary	10–20% removed	No effect	No effect	5–10
Secondary	15–50% removed[c] urea → $NH_3-NH_4^{+d}$	< 10% removed	Slight effect	10–30
Biological processes				
Bacterial assimilation	No effect	40–70% removed	Slight	30–70
Denitrification	No effect	No effect	80–90% removed	70–95
Harvesting algae	Partial transformation to $NH_3-NH_4^+$	→ Cells	→ Cells	50–80
Nitrification	Limited effect	→ NO_3^-	No effect	5–20
Oxidation ponds	Partial transformation to $NH_3-NH_4^+$	Partial removal by stripping	Partial removal by nitrification/denitrification	20–90

Chemical processes

Breakpoint chlorination	Uncertain	90–100% removed	No effect	80–95
Chemical coagulation	50–70% removed	Slight effect	Slight effect	20–30
Carbon adsorption	30–50% removed	Slight effect	Slight effect	10–20
Selective ion exchange for ammonium	Slight, uncertain	80–97% removed	No effect	70–95
Selective ion exchange for nitrate	Slight effect	Slight effect	75–90% removed	70–90
Physical operations				
Filtration	30–95% of suspended organic N removed	Slight effect	Slight effect	20–40
Air stripping	No effect	60–95% removed	No effect	50–90
Electrodialysis	100% of suspended organic N removed	30–50% removed	30–50% removed	40–50
Reverse osmosis	60–90% removed	60–90% removed	60–90% removed	80–90

[a] Adapted from Ref. 33.
[b] Depends on the fraction of influent nitrogen for which the process is effective and other processes in treatment plant.
[c] Soluble organic nitrogen, in the form of urea and amino acids, is reduced substantially by secondary treatment.
[d] Arrow denotes "conversion to."

is transformed to ammonium and other inorganic forms. A portion of the ammonium is assimilated into the cell material of the biomass. Most of the nitrogen in treated secondary effluent is in the ammonium form. Less than 30 percent of the total nitrogen is removed by conventional secondary treatment. Unit operations and processes used for the conversion and removal of nitrogen from wastewater are reported in Table 11-11. Four major treatment categories and their effects on organic, ammonia, and nitrate nitrogen in wastewater are listed. Typical removals of total nitrogen are also reported.

Conventional biological treatment processes are discussed in Chaps. 8 and 10. The principal nitrogen conversion and removal processes for the remaining three categories are considered in this chapter: conversion of ammonia nitrogen to nitrate by biological nitrification in Sec. 11-6, removal of nitrogen by biological nitrification/denitrification in Sec. 11-7, combined removal of nitrogen and phosphorus by biological methods in Sec. 11-9, and the removal of nitrogen by physical and chemical systems in Sec. 11-10. These operations and processes were selected for detailed discussion because they have been used frequently for the control of nitrogen. The removal of nitrogen in natural treatment systems is discussed in Chap. 13.

Removal of Phosphorus

With most wastewaters, approximately 10 percent of the phosphorus corresponding to the insoluble portion is normally removed by primary settling. Except for the amount incorporated into cell tissue, the additional removal achieved in conventional biological treatment is minimal because almost all the phosphorus present after primary sedimentation is soluble. The effects of conventional and other treatment processes on phosphorus removal are listed in Table 11-12.

Removal of phosphorus can be accomplished by chemical, biological, and physical methods. Chemical precipitation using iron and aluminum salts or lime has commonly been employed for phosphorus removal. Biological treatment methods are based on stressing the microorganisms so that they will take up more phosphorus than is required for normal cell growth. A number of biological processes have been developed in recent years as alternatives to chemical precipitation. Filtration is used in combination with either chemical or biological processes where low levels of phosphorus (usually less than 1 mg/L as P) are required. Other physical processes such as ultrafiltration and reverse osmosis are effective in phosphorus reduction but are used primarily for overall dissolved inorganic solids reduction. Biological phosphorus removal methods are discussed in Sec. 11-8, combined removal of nitrogen and phosphorus by biological methods in Sec. 11-9, and removal of phosphorus by chemical addition in Sec. 11-11. Ultrafiltration and reverse osmosis are discussed as a part of Sec. 11-13.

11-6 CONVERSION OF AMMONIA BY BIOLOGICAL NITRIFICATION

The process in which the nitrogen in the untreated or settled wastewater is substantially converted to nitrate is known as "biological nitrification." The discharge of nitrified

TABLE 11-12
Effect of various treatment operations and processes on phosphorus removal[a]

Treatment operation or process	Removal of phosphorus entering system, %
Conventional treatment	
Primary	10–20
Activated-sludge	10–25
Trickling-filter	8–12
Rotating biological contactors	8–12
Biological phosphorus removal only	
Mainstream treatment	70–90
Sidestream treatment	70–90
Combined biological nitrogen and phosphorus removal	70–90
Chemical removal	
Precipitation with metal salt	70–90
Precipitation with lime	70–90
Physical removal	
Filtration	20–50
Reverse osmosis	90–100
Carbon adsorption	10–30

[a] Adapted in part from Ref. 24.

wastewater will generally satisfy receiving water requirements where reduction of the nitrogen oxygen demand is required or where reduction of ammonia toxicity is necessary. In this section, the biological nitrification process is described, types of processes used are classified and described, operating considerations are discussed, and the various nitrification alternatives are compared.

Process Description

Nitrification is an autotrophic process (i.e., energy for bacterial growth is derived by the oxidation of nitrogen compounds, primarily ammonia). In contrast to heterotrophs, nitrifiers use carbon dioxide (inorganic carbon) rather than organic carbon for synthesis of new cells. Nitrifier cell yield per unit of substrate metabolized is many times smaller than the cell yield for heterotrophs.

As described in Sec. 8-11, nitrification of ammonium nitrogen is a two-step process involving two genera of microorganisms, *Nitrosomonas* and *Nitrobacter*. In the first step, ammonium is converted to nitrite; in the second step, nitrite is converted to nitrate. The conversion process is described as follows:

First step,

$$NH_4^+ + \frac{3}{2}O_2 \xrightarrow{\text{\textit{Nitrosomonas}}} NO_2^- + 2H^+ + H_2O \qquad (11\text{-}1)$$

Second step,

$$NO_2^- + \frac{1}{2}O_2 \xrightarrow{Nitrobacter} NO_3^- \qquad (11\text{-}2)$$

Equations 11-1 and 11-2 are energy yielding reactions. *Nitrosomonas* and *Nitrobacter* use the energy derived from these reactions for cell growth and maintenance. The overall energy reaction is represented in Eq. 11-3:

$$NH_4^+ + 2O_2 \rightarrow NO_3^- + 2H^+ + H_2O \qquad (11\text{-}3)$$

Along with obtaining energy, some of the ammonium ion is assimilated into cell tissue. The biomass synthesis reaction can be represented as follows:

$$4CO_2 + HCO_3^- + NH_4^+ + H_2O \rightarrow C_5H_7O_2N + 5O_2 \qquad (11\text{-}4)$$

As noted in Chap. 8, the chemical formula $C_5H_7O_2N$ is used to represent the synthesized bacterial cells.

The overall oxidation and synthesis reaction can be represented as follows [33]:

$$NH_4^+ + 1.83O_2 + 1.98HCO_3^- \rightarrow$$
$$0.021C_5H_7O_2N + 0.98NO_3^- + 1.041H_2O + 1.88H_2CO_3 \quad (11\text{-}5)$$

The oxygen required to oxidize ammonia to nitrate in Eq. 11-5 (4.3 mg O_2/mg ammonium nitrogen) is in close agreement with the value of 4.57, which is often recommended for design calculations. The value of 4.57 is derived from Eq. 11-3, in which cell synthesis is not considered [33].

Classification of Nitrification Processes

Nitrification processes may be classified based on the degree of separation of the carbon oxidation and nitrification functions. Carbon oxidation and nitrification may occur in a single reactor, termed "single stage." In separate-stage nitrification, carbon oxidation and nitrification occur in different reactors. Suspended or attached-growth reactors may be used for either single-stage or separate-stage systems. Examples of single-stage and separate-stage nitrification are illustrated in Fig. 11-10.

Nitrifying organisms are present in almost all aerobic biological treatment processes, but usually their numbers are limited. The ability of various activated-sludge processes to nitrify has been correlated to the BOD_5/TKN (total Kjeldahl nitrogen) ratio [33]. For BOD_5/TKN ratios between 1 and 3, which roughly correspond to the values encountered in separate-stage nitrification systems, the fraction of nitrifying organisms is estimated to vary from 0.21 at a BOD_5/TKN ratio of 1, to 0.083 at a ratio of 3 (see Table 11-13) [33]. In most conventional activated-sludge processes, the fraction of nitrifying organisms would therefore be considerably less than the 0.083 value. It has been found that when the BOD_5/TKN ratio is greater than about 5, the process can be classified as a combined carbon oxidation and nitrification process, and, when the ratio is less than 3, it can be classified as a separate-stage nitrification process [33].

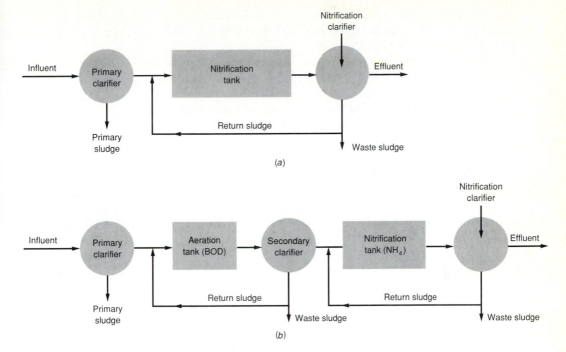

FIGURE 11-10
Typical suspended-growth carbon oxidation and nitrification processes: (a) single-stage and
(b) separate-stage.

Single-Stage Carbon Oxidation and Nitrification

Nitrification can be accomplished in any of the suspended-growth activated-sludge
processes identified in Table 8-6. The processes most commonly used are conventional plug-flow, complete-mix, extended aeration, and various modifications of the
oxidation ditch. As discussed in Sec. 8-11, to achieve nitrification, all that is required
is the maintenance of conditions suitable for the growth of nitrifying organisms. For

TABLE 11-13
**Relationship between the fraction of nitrifying organisms
and the BOD$_5$/TKN ratio**[a]

BOD$_5$/TKN ratio	Nitrifier fraction	BOD$_5$/TKN ratio	Nitrifier fraction
0.5	0.35	5	0.054
1	0.21	6	0.043
2	0.12	7	0.037
3	0.083	8	0.033
4	0.064	9	0.029

[a] Ref. 33.

example, in most warm climates, increased nitrification can be brought about by increasing the mean cell-residence time and the air supply. This technique is often used to achieve seasonal nitrification.

The two attached-growth processes that can be used for combined carbon oxidation nitrification processes are the trickling filter and the rotating biological contactor (see Table 8-6). As with the suspended-growth processes, nitrification in the attached-growth process can be brought about or encouraged by adjustment of operating parameters. Nitrification can usually be accomplished by reducing the applied loading rate.

Suspended-Growth Process. The kinetics of the nitrification process are reasonably well-defined for suspended-growth systems. In general, it has been found that the kinetic expressions developed in Chap. 8 for aerobic suspended-growth processes are applicable to the nitrification process, subject to environmental constraints. To avoid the necessity of referring to Chap. 8, the pertinent kinetic expressions used in the following analysis are summarized in Table 11-14. Details on the development of these equations may be found in Chap. 8 (Secs. 8-5 and 8-7).

From experience and laboratory studies [33,39], it has been found that the following factors have a significant effect on the nitrification process: ammonia and nitrite concentrations, BOD_5/TKN ratio, dissolved-oxygen concentration, temperature, and pH. The impact of these variables on the nitrification process and the approach developed to account for them is reported in Table 11-15. Representative kinetic coefficients for the suspended-growth nitrification process are given in Table 11-16.

Application of the kinetic approach to the analysis of the suspended-growth nitrification process in a complete-mix reactor involves the following steps:

1. Select an appropriate safety factor to handle peak, diurnal, and transient loadings. (A minimum safety factor of 2.0 applied to θ_c is recommended to ensure that ammonia breakthrough does not occur during diurnal peaks in loads.)
2. Select the minimum mixed-liquor dissolved-oxygen concentration, DO. A minimum DO level of 2.0 mg/L is recommended to avoid depressing effects of low DO on the rate of nitrification.
3. Determine the process operating pH. A pH range of 7.2 to 9.0 is recommended. Every mg/L of $NH_4^+ -N$ oxidized causes the destruction of 7.14 mg/L of alkalinity expressed as $CaCO_3$.
4. Estimate the maximum growth rate of nitrifying organisms consistent with the most adverse temperature, dissolved oxygen, and pH conditions.
5. Determine the minimum mean cell-residence time based on the adjusted growth rate determined in step 4.
6. Determine the design mean cell-residence time using the safety factor determined in step 1.
7. Determine the effluent nitrogen concentration.
8. Determine the hydraulic retention time to achieve the necessary effluent nitrogen concentration.

TABLE 11-14
Summary of kinetic expressions used for the analysis of suspended-growth nitrification and denitrification processes

Equation	Number	Definition of terms
$\mu = \mu_m \dfrac{S}{K_S + S}$	8-3	μ = specific growth rate, time^{-1}
		r_{su} = substrate utilization rate, mass/unit volume
$r_{su} = -\dfrac{\mu_m XS}{Y(K_S + S)}$	8-6	μ_m = maximum specific growth rate, time^{-1}
		S = concentration of growth limiting substrate in solution, mass/unit volume
$k = \dfrac{\mu_m}{Y}$	8-7	X = concentration of microorganisms, mass/unit volume
$r_{su} = \dfrac{kSX}{K_S + S}$	8-56	Y = maximum yield coefficient measured during a finite period of logarithmic growth, mass of cell formed per mass of substrate consumed
$U = -\dfrac{r_{su}}{X}$	8-45	K_S = half velocity constant, mass/unit volume
$U = \dfrac{S_o - S}{\theta X}$	8-45	k = maximum rate of substrate utilization, time^{-1}
		k_d = endogenous decay coefficient, time^{-1}
$U = \dfrac{kS}{K_S + S}$	8-51	U = substrate utilization rate, time^{-1}
$\dfrac{1}{\theta_c} = YU - k_d$	8-46	θ = hydraulic detention time, time
		θ_c = design mean cell-residence time, time
$\dfrac{1}{\theta_c^M} \approx Yk - k_d$	8-54	θ_c^M = minimum mean cell-residence time, time
$SF = \dfrac{\theta_c}{\theta_c^M}$	8-55	SF = safety factor
		S_o = influent concentration, mass/unit volume

9. Determine the organic substrate utilization rate U where a single-stage oxidation-nitrification process is to be used.

The application of these steps is illustrated in Example 11-1. The key concept involved in this analysis is the determination of the minimum mean cell-residence time subject

TABLE 11-15
Effects of the major operational and environmental variables on the suspended-growth nitrification process[a]

Factor	Description of effect
Ammonia and nitrite concentration	It has been observed that the concentration of ammonia and nitrite will affect the maximum growth rate of *Nitrosomonas* and *Nitrobacter*. The effect of either constituent can be made using a Monad-type kinetic expression: $$\mu = \mu_m \frac{S}{K_S + S}$$ Because it has been found that the growth rate of *Nitrobacter* is considerably greater than *Nitrosomonas*, the rate of nitrification is usually modeled using the conversion of ammonia to nitrite as the rate-limiting step.
BOD$_5$/TKN	The fraction of nitrifying organisms present in the mixed liquor of a single-state carbon oxidation nitrification process has been found to be reasonably well-related to the BOD$_5$/TKN ratio. For ratios greater than 5, the fraction of nitrifying organisms decreases from a value of 0.054 (see Table 11-13).
Dissolved-oxygen concentration	The DO level has been found to affect the maximum specific growth rate μ_m of the nitrifying organisms. The effect has been modeled with the following relationship: $$\mu'_{m_n} = \mu_{m_n} \frac{DO}{K_{O_2} + DO}$$ Based on limited information, a value of 1.3 can be used for K_{O_2}.
Temperature	Temperature has a significant effect on nitrification-rate constants. The overall nitrification rate decreases with decreasing temperature and is accounted for with the following two relationships: $$\mu'_{m_n} = \mu_m e^{0.098(T - 15)}$$ $$K_n = 10^{0.051T - 1.158}$$ where $T = °C$
pH	It has been observed that the maximum rate of nitrification occurs between pH values of about 7.2 and 9.0. For combined carbon oxidation nitrification systems, the effect of pH can be accounted for using the following relationship: $$\mu'_{m_n} = \mu[1 - 0.833(7.2 - pH)]$$

[a] Developed from information in Ref. 33.

to the most critical environmental constraints and the use of an appropriate safety factor. This approach is essentially the same as that used in the design of the suspended-growth activated-sludge process in a complete-mix reactor. The analysis of the plug-flow suspended-growth process is outlined in Sec. 8-7.

TABLE 11-16
Typical kinetic coefficients for the suspended-growth nitrification process (pure culture values)[a, b]

Coefficient	Basis	Value	
		Range	Typical[c]
Nitrosomonas			
μ_m	d^{-1}	0.3–2.0	0.7
K_S	NH_4^+ –N, mg/L	0.2–2.0	0.6
Nitrobacter			
μ_m	d^{-1}	0.4–3.0	1.0
K_S	NO_2^- –N, mg/L	0.2–5.0	1.4
Overall			
μ_m	d^{-1}	0.3–3.0	1.0
K_S	NH_4^+ –N, mg/L	0.2–5.0	1.4
Y	NH_4^+ –N, mg VSS/mg	0.1–0.3	0.2
k_d	d^{-1}	0.03–0.06	0.05

[a] Derived in part from Refs. 26 and 33.

[b] Values for nitrifying organisms in activated sludge will be considerably lower than the values reported in this table.

[c] Values reported are for 20°C.

Note: 1.8(°C) + 32 = °F

Example 11-1 Design of single-stage suspended-growth carbon oxidation-nitrification process. Determine the design criteria for an activated-sludge process to achieve essentially complete nitrification when treating domestic wastewater. Assume that the following conditions apply for this example:

1. Influent flowrate = 0.90 Mgal/d (3400 m³/d)
2. BOD$_5$ after primary settling = 200 mg/L
3. TKN after primary settling = 40 mg/L
4. Minimum sustained temperature = 15°C
5. Dissolved oxygen to be maintained in the reactor = 2.5 mg/L
6. Buffer capacity of the wastewater is adequate to maintain the pH at or above a value of 7.2
7. Use the kinetic coefficients given in Table 11-16 except let $\mu_m = 0.5$ d^{-1}

Solution

1. Estimate the safety factor to be used in the design based on the peak nitrogen loading. From a review of Fig. 5-6c, it appears that a safety factor of 2.5 should be adequate.
2. Determine the maximum growth rate for the nitrifying organisms under the stated operating conditions.

(a) The following expression developed from Table 11-15 can be used:

$$\mu'_m = \mu_m e^{0.098(T-15)} \quad \times \quad \frac{DO}{K_{o_2} + DO} \quad \times \quad [1 - 0.833(7.2 - pH)]$$

temperature	dissolved-	pH
correction	oxygen	correction
factor	factor	factor

where μ'_m = growth rate under the stated conditions of temperature, dissolved oxygen, and pH

μ_m = maximum specific growth rate

T = temperature

DO = dissolved oxygen

K_{o_2} = dissolved-oxygen half-velocity constant = 1.3

pH = operating pH, the numerical value of the pH term is taken as 1 for the above values

(b) Substitute the known values and determine μ'_m.

$$\mu_m = 0.5 \text{ d}^{-1}$$

$$T = 15°C$$

$$DO = 2.5 \text{ mg/L}$$

$$K_{o_2} = 1.3$$

$$pH = 7.2$$

$$\mu'_m = (0.5 \text{ d}^{-1}) \left[e^{0.098(T-15)} \right] \left(\frac{2.5}{1.3 + 2.5} \right) \left[1 - 0.833(7.2 - 7.2) \right]$$

$$= 0.5 \text{ d}^{-1} \frac{2.5}{1.3 + 2.5}$$

$$= 0.33 \text{ d}^{-1}$$

3. Determine the maximum rate of substrate utilization k using Eq. 8-7 (see Table 11-14).

$$k' = \frac{\mu'_m}{Y}$$

$$\mu'_m = 0.33 \text{ d}^{-1} \quad \text{(from step 2b above)}$$

$$Y = 0.2 \quad \text{(from Table 11-16)}$$

$$k' = \frac{0.33 \text{ d}^{-1}}{0.2} = 1.65 \text{ d}^{-1}$$

4. Determine the minimum and design mean cell-residence times.
   (a) Minimum θ_c^M:

$$\frac{1}{\theta_c^M} \sim Yk' - k_d$$

$$Y = 0.2$$

$$k' = 1.65 \text{ d}^{-1} \quad \text{(from step 3)}$$

$$k_d = 0.05 \quad \text{(from Table 11-16)}$$

$$\frac{1}{\theta_c^M} = 0.2\left(1.65 \text{ d}^{-1}\right) - 0.05 \text{ d}^{-1}$$

$$= 0.28 \text{ d}^{-1}$$

$$\theta_c^M = \frac{1}{0.28 \text{ d}^{-1}} = 3.57 \text{ d}$$

(b) Design θ_c (using a safety factor of 2.5):

$$\theta_c = \text{SF}\left(\theta_c^M\right) = 2.5 \left(3.57 \text{ d}\right) = 8.93 \text{ d}$$

5. Determine the design substrate-utilization factor U for the oxidation of ammonia.

$$\frac{1}{\theta_c} = YU - k_d$$

$$U = \left(\frac{1}{\theta_c} + k_d\right)\frac{1}{Y}$$

$$= \left(\frac{1}{8.93 \text{ d}} + 0.05 \text{ d}^{-1}\right)\frac{1}{0.2} = 0.81 \text{ d}^{-1}$$

6. Determine the concentration of ammonia in the effluent using Eq. 8-51.

$$U = \frac{kN}{K_N + N}$$

$$= 0.81 \text{ d}^{-1}$$

$$k = 1.65 \text{ d}^{-1}$$

$$T = 15°C$$

$$N = \text{effluent NH}_4^+ - \text{N concentration, mg/L}$$

$$K_n = 10^{0.051T - 1.158} = 0.40 \text{ mg/L (see Table 11-15)}$$

$$0.81 = \frac{1.65N}{0.40 + N}$$

$$N = \frac{1.65N}{0.810} - 0.4$$

$$N\left(1 - \frac{1.65}{0.81}\right) = -0.40$$

$$N = 0.39 \text{ mg/L}$$

7. Determine the BOD removal rate for the activated-sludge process using Eq. 8-46.

$$\frac{1}{\theta_c} = YU - k_d$$

$$\theta_c = 8.93 \text{ d} \qquad \text{(from step 4)}$$

$$Y = 0.5 \text{ lb VSS/lb BOD}_5 \qquad \text{(from Table 8-7)}$$

$$k_d = 0.06 \text{ d}^{-1} \qquad \text{(from Table 8-7)}$$

$$U = \left(\frac{1}{8.93 \text{ d}} - 0.06 \text{ d}^{-1} \right) \frac{1}{0.5}$$

$$U = 0.34 \text{ lb BOD}_5 \text{ removed/lb MLVSS} \cdot \text{d}$$

If it is assumed that the process efficiency is 90 percent, the corresponding value of the food-to-microorganism ratio is equal to 0.38 lb BOD_5 applied per lb MLVSS $\cdot$ d.

8. Determine the required hydraulic detention time for BOD oxidation and nitrification using Eq. 8-45.

$$U = \frac{S_o - S}{\theta X}$$

(a) BOD$_5$ oxidation:

$$\theta = \frac{S_o - S}{UX}$$

$$S_o = 200 \text{ mg/L} \qquad \text{(from problem specification)}$$

$$S = 20 \text{ mg/L} \qquad \text{(assumed value)}$$

$$U = 0.34 \text{ d}^{-1} \qquad \text{(from step 7)}$$

$$X = \text{MLVSS, mg/L (assume that } X = 2000 \text{ mg/L)}$$

$$\theta = \frac{(200 - 20) \text{ mg/L}}{0.34 \text{ d}^{-1} \, (2000 \text{ mg/L})} = 0.26 \text{ d} = 6.4 \text{ h}$$

(b) Ammonia oxidation (nitrification):

$$\theta = \frac{N_o - N}{UX}$$

$$N_o = 40 \text{ mg/L} \qquad \text{(from problem specification)}$$

$$N = 0.39 \text{ mg/L} \qquad \text{(from step 6)}$$

$$U = 0.81 \text{ d}^{-1}$$

$$X = 2000 \text{ mg/L} \times 0.08 \qquad \text{(assumed fraction of nitrifiers)}$$

$$= 160 \text{ mg/L}$$

$$\theta = \frac{(40 - 0.39) \text{ mg/L}}{0.81 \text{ d}^{-1} \, (2000 \text{ mg/L} \times 0.08)} = 0.31 \text{ d} = 7.3 \text{ h}$$

Conclusion: Nitrification process controls the required hydraulic detention time.

9. Determine the required aeration tank volume.

$$V = Q\theta = (0.90 \text{ Mgal/d}) (0.31 \text{ d}) = 0.279 \text{ Mgal}$$

10. Determine the total amount of oxygen required.

(a) The total amount of oxygen required based on average conditions can be estimated using the following expression:

$$\text{lb } O_2/d = \frac{Q(S_o - S) \times 8.34}{f} - 1.42\,(P_x) + 4.57\,Q\,(N_o - N) \times 8.34 \qquad (10\text{-}6)$$

where Q = flowrate, Mgal/d
S_o = influent BOD_5, mg/L (g/m³)
S = effluent BOD, mg/L (g/m³)
8.34 = conversion factor, [lb/Mgal · (mg/L)]
f = factor to convert BOD_5 value to BOD_L, 0.68
P_x = net mass of volatile solids (cells) produced
1.42 = conversion factor for cell tissue to BOD_L
N_o = influent TKN, mg/L
N = effluent TKN, mg/L
4.57 = conversion factor for amount of oxygen required for complete oxidation of TKN

(b) Alternatively, the following expression can be used as a rough estimate:

$$O_2 \text{ lb/d} = Q(kS_o + 4.57 \text{ TKN}) \times 8.34$$

where k = conversion factor for BOD for low loadings on nitrification systems. The range for k is from 1.1 to 1.25.

(c) Using the expression given in step 10b, the total oxygen required per day, with a k value of 1.15, and a factor of safety of 2.5, is equal to

$$O_2 \text{ lb/d} = (0.90 \text{ Mgal/d}) \,[1.15(200 \text{ mg/L}) + 4.57\,(40 \text{ mg/L})]$$

$$\times [8.34 \text{lb/Mgal·(mg/L)}] \times (2.5)$$

$$= 7746 \text{ lb/d } (3521 \text{ kg/d})$$

Comment. In addition to these computations, the alkalinity requirements should be checked. If the natural alkalinity of the wastewater is insufficient, it may be necessary to install a pH control system.

Attached-Growth Processes. The principal attached-growth processes are trickling filters and rotating biological contactors (RBCs). To date, the most common approach to describe the performance of the attached-growth processes has been to use loading factors. Typical loading data to achieve nitrification with these processes are reported in Table 11-17.

For trickling filters using a rock medium, organic loadings will affect nitrification efficiency because the bacterial film is dominated by heterotrophic bacteria at high organic loadings. To attain high nitrification efficiency, organic loadings have to be maintained in the ranges indicated in Table 11-17. Because filters using a plastic packing have greater surface contact areas (and larger quantities of active microorganisms), higher organic loadings can be applied while still achieving good nitrification. Another factor favoring plastic medium filters is better ventilation, which permits higher oxygen transfer. Performance comparisons between rock and plastic medium indicate that plastic medium with 80 percent greater surface area will be able

TABLE 11-17
Typical loading rates for attached-growth processes to achieve nitrification[a]

Process	Percent nitrification	Loading rate, lb $BOD_5/10^3 ft^3 \cdot d$
Trickling filter, rock medium	75–85	10–6
	85–95	6–3
Tower trickling-filter, plastic medium	75–85	18–12
	85–95	12–6
Rotating biological contactor	(see Fig. 11-11)	(see Fig. 10-39)

[a] Developed in part from Refs. 26 and 33.
Note: $lb/10^3 ft^3 \cdot d = 0.0160 = kg/m^3 \cdot d$

to nitrify about 60 percent more ammonium nitrogen per unit volume in a combined carbon oxidation nitrification system [39].

In RBCs, the amount of ammonia that can be oxidized depends on the surface area of the process units. A two-step procedure for determining surface area requirements for RBCs is described in Ref. 39. Significant nitrification will not occur in the RBC process until the soluble BOD_5 concentration is reduced to 15 mg/L or less. Using the process design information in Chap. 10 for BOD_5 reduction (Fig. 10-39), the first step is to determine the surface area of medium required to reduce the soluble BOD_5 concentration. For influent ammonium nitrogen concentrations of 15 mg/L and above, it is necessary to reduce the soluble BOD_5 concentration to the same value as the ammonium nitrogen. The second step involves the use of a nitrification design curve, Fig. 11-11, to determine the total media area necessary to reduce the influent ammonium concentration to the required effluent concentration. The sum of the two surface areas represents the total surface area required for the combined BOD_5 oxidation and nitrification. If low-temperature conditions exist, surface area calculations must be adjusted using temperature correction factors. Temperature correction curves for wastewater temperatures below 55°F are shown in Fig. 10-40 in Chap. 10.

Separate-Stage Nitrification

Both suspended-growth and attached-growth processes are used to achieve separate-stage nitrification. Nitrification in a separate reactor allows greater process flexibility and reliability, and each process (carbonaceous oxidation and nitrification) can be operated independently to achieve optimum performance. Potential toxic effects may also be reduced because biodegradable organic materials, which may be toxic to nitrifying bacteria, are removed in the carbon oxidation stage. A flow diagram and an aerial view of a typical treatment plant with separate stages for suspended-growth carbon oxidation and nitrification are shown in Fig. 11-12.

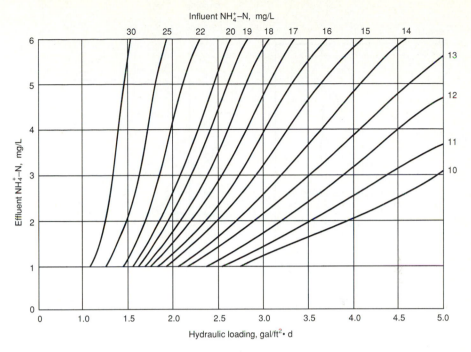

FIGURE 11-11
Nitrification of domestic wastewater using a rotating biological contactor ($T > 55°F$) [39].
Note: gal/ft$^2 \cdot$ d $\times$ 40.74 = L/m$^2 \cdot$ d.

The degree of organic carbon removal in the carbon oxidation stage will affect the selection and operation of the nitrification process. Low levels of organic carbon in the influent to the nitrification process may be advantageous to attached-growth reactors because these low levels may eliminate the need for clarification following nitrification. In suspended-growth nitrification reactors, low organic carbon in the influent may cause an imbalance between the solids lost from the sedimentation basins and the solids synthesized in the reactor. This imbalance often necessitates continuous wasting or increasing the BOD in the nitrification reactor influent to maintain the inventory of biological solids in the nitrification system.

Suspended-Growth. In most details, separate-stage suspended-growth nitrification processes are similar in design to the activated-sludge process. When very low ammonia concentrations are desired, complete-mix staged-flow or plug-flow reactors are favored.

For separate-stage nitrification, nitrification rate determination is the approach often used. Experimentally measured rates are considered more appropriate for use than theoretical rates because of the difficulty of assessing the nitrifier fraction of the mixed liquor [39]. Nitrification rates increase as the temperature increases. The value

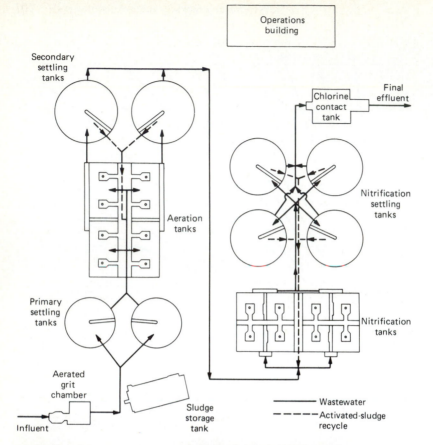

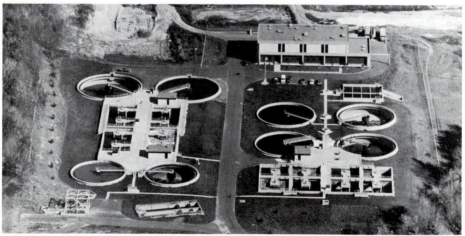

FIGURE 11-12
Flow diagram and aerial view of treatment plant with separate stages for carbon oxidation and nitrification (Marlborough, MA; design average flowrate = 5.5 Mgal/d).

of the BOD_5/TKN ratio is important in nitrification with nitrification rates increasing as the ratio decreases. Nitrification rates are also affected by the pH of the mixed liquor and how far it deviates from the optimum pH for nitrification. Nitrification rates have been observed to vary from 0.05 to 0.6 lb NH_4^+–N oxidized/lb MLVSS · d, depending on the nature and temperature of the wastewater [33]. Because of this wide variation, pilot plant studies are required to determine the appropriate rates for design. For additional information on separate-stage nitrification, Refs. 5 and 39 may be consulted.

Attached-Growth. Two different types of attached-growth processes have been used frequently for separate-stage nitrification: trickling filters and rotating biological contactors (RBCs). The packed-bed reactor has also been used, but only in a few applications. The packed-bed reactor is similar to an upflow sand filter except air or high-purity oxygen is added at the bottom of the reactor to sustain nitrification.

Trickling filters may be used for separate-stage stage nitrification following a suspended-growth process for carbon oxidation. The more common process combination is the use of a two-stage trickling-filter system; the first stage is used for carbon oxidation and the second stage for nitrification. Tower filters using plastic media are particularly well-suited for nitrification because of the large surface area available. Tower filters should always be designed so that forced air ventilation can be used, if required (see Chap. 10).

Data for separate-stage nitrification using a trickling filter are limited. A comprehensive study of nitrification in tower trickling filters using plastic media is reported in Ref. 12. Wastewater was applied to a 21.5 ft (6.6 m) tower and the performance characteristics are shown in Fig. 11-13. As shown in Fig. 11-13, nitrogen efficiency

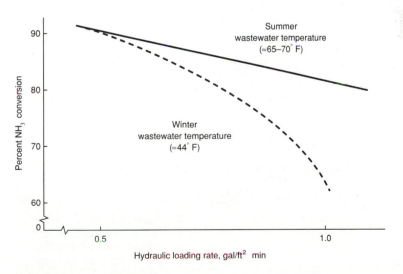

FIGURE 11-13
Effect of raw wastewater hydraulic-loading rate on ammonia conversion in a trickling filter [12].
Note: gal/ft² · min × 40.74 = L/m² · min.

decreases with increasing surface loading and decreasing wastewater temperature. At surface-loading values below 0.5 gal/ft^2 · min, a high degree of nitrification can be achieved year-round. Typical design graphs developed from the data in Ref. 12 are reported in Fig. 11-14 [15].

When used for biological nitrification, rotating biological contactors are often based on influent ammonia concentration rather than influent unfiltered or soluble TKN concentration. For nitrification of biologically treated effluents, this approach should not result in a sizing problem. In combined carbon oxidation nitrification applications, however, such an approach could lead to gross undersizing of the required media surface area in the nitrification portion of the reactor [13]. The design data presented on RBCs in the section "Single-Stage Carbon Oxidation and Nitrification" can be used to size the nitrification stage.

Operating Considerations

Assuming that sufficient air can be supplied, nitrification generally can be assured at moderate temperatures in a conventional activated-sludge process. If nitrification is to be accomplished in a single-stage activated-sludge process, certain operational adjustments must be made beyond those necessary for stabilization of the organic matter:

1. Additional oxygen must be provided for the nitrification process.
2. A longer mean cell-residence time must be used. Because the bacteria that are responsible for nitrification are strict autotrophs, they are distinctly different from the heterotrophic bacteria responsible for the degradation of the organic matter. Nitrifying bacteria have a slower growth rate than that of the heterotrophic bacteria,

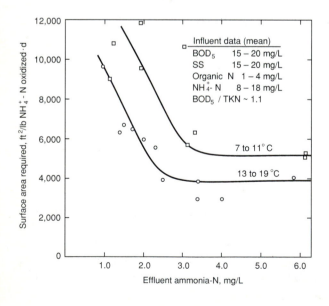

FIGURE 11-14
Typical design graphs for a packed tower trickling filter performing separate-stage nitrification [15].
Note: ft^2 /lb NH$_4^+$−N · d × 0.2 = m^2/kg·d.

so they require a longer mean cell-residence time to be effective. Typical kinetic coefficients for the nitrification process are presented in Table 11-16.

3. Because the microbial conversion causes a drop in pH, provision should be made for the addition of lime, soda ash, or caustic in low-alkalinity wastewaters.

In applications using the RBC process for nitrification, a pH range of 7.2 to 8.6 should be maintained within the nitrification section of the reactor. When pH values fall below 7.2, the process efficiency is retarded. Dissolved-oxygen concentrations also should be maintained above 1.5 mg/L [39].

Comparison of Nitrification Processes

The advantages and disadvantages of the various nitrification processes are reported in Table 11-18. The selection of a particular process flow diagram depends on a number of factors, including (1) whether nitrification is being incorporated into an existing treatment plant or a new treatment plant is being designed, (2) whether seasonal or year-round standards are to be met, (3) range of operating temperatures, (4) desired effluent ammonia concentration, (5) effluent standards for other parameters, and (6) costs.

11-7 REMOVAL OF NITROGEN BY BIOLOGICAL NITRIFICATION/DENITRIFICATION

Of the methods proposed for the removal of nitrogen, biological nitrification/denitrification is often the best for the following reasons: (1) high potential removal efficiency, (2) high process stability and reliability, (3) relatively easy process control, (4) low land area requirements, and (5) moderate cost. The removal of nitrogen by biological nitrification/denitrification is a two-step process. In the first step, ammonia is converted aerobically to nitrate (NO_3^-) (nitrification). In the second step, nitrates are converted to nitrogen gas (denitrification).

Process Description

As described in Chap. 8, the removal of nitrate by conversion to nitrogen gas can be accomplished biologically under anoxic (without oxygen) conditions. Two types of enzyme systems are involved in the reduction of $NO_3^- - N$: assimilatory and dissimilatory. In the assimilatory nitrate reduction process, $NO_3^- - N$ is converted to ammonia nitrogen for use by the cells in biosynthesis and occurs when $NO_3^- - N$ is the only form of nitrogen available. In the dissimilatory nitrate reduction process, nitrogen gas is formed from $NO_3^- - N$; this process results in the denitrification of wastewater. In most biological nitrification/denitrification systems, the wastewater to be denitrified must contain sufficient carbon (organic matter) to provide the energy source for the conversion of nitrate to nitrogen gas by the bacteria. The carbon requirements may be provided by internal sources, such as wastewater and cell material, or by an external source (e.g., methanol).

TABLE 11-18
Comparison of nitrification alternatives

System type	Advantages	Disadvantages
Combined carbon oxidation nitrification		
Suspended-growth	Combined treatment of carbon and ammonia in a single stage; low effluent ammonia is possible; inventory control of mixed-liquor stable due to high BOD$_5$/TKN ratio	No protection against toxicants; only moderate stability of operation; stability linked to operation of secondary clarifier for biomass return; large reactors required in cold weather
Attached-growth	Combined treatment of carbon and ammonia in a single stage; stability not linked to secondary clarifier as organisms are attached to media	No protection against toxicants; only moderate stability of operation; effluent ammonia is normally 1–3 mg/L (except RBC); cold weather operation impractical in most cases
Separate-stage nitrification		
Suspended-growth	Good protection against most toxicants; stable operation; low effluent ammonia possible	Sludge inventory requires careful control when BOD$_5$/TKN ratio is low; stability of operation linked to operation of secondary clarifier for biomass return; greater number of unit processes required than for combined carbon oxidation nitrification
Attached-growth	Good protection against most toxicants; stable operation; stability not linked to secondary clarifier as organisms are attached to media	Effluent ammonia normally 1–3 mg/L; greater number of unit processes required than for combined carbon oxidation nitrification

The rate of denitrification can be described by the following equation:

$$U'_{DN} = U_{DN} \times 1.09^{(T-20)} (1 - DO) \qquad (11\text{-}6)$$

where U'_{DN} = overall denitrification rate
U_{DN} = specific denitrification rate, lb NO_3-N/lb MLVSS · d
T = wastewater temperature, °C
DO = dissolved oxygen in the wastewater, mg/L

The DO term in Eq. 11-6 indicates that the denitrification rate decreases linearly to zero when the dissolved-oxygen concentration reaches 1.0 mg/L. Specific denitrification rates for various carbon sources are given in Table 11-19. A method of calculating denitrification rates is developed in Ref. 13. The application of Eq. 11-6 is illustrated in Example 11-2.

TABLE 11-19
Typical denitrification rates for various carbon sources[a]

Carbon source	Denitrification rate, U_{DN}, lb NO_3^- –N/lb VSS $\cdot$ d	Temperature, °C
Methanol	0.21–0.32	25
Methanol	0.12–0.90	20
Wastewater	0.03–0.11	15–27
Endogenous metabolism	0.017–0.048	12–20

[a] Ref. 27.

Note: 1.8(°C) + 32 = °F

Example 11-2 Calculation of denitrification basin residence time. Calculate the residence time for an anoxic basin used for denitrification for the following conditions:

1. Influent nitrate to basin = 22 mg/L
2. Effluent nitrate from basin = 3 mg/L
3. MLVSS = 2000 mg/L
4. Temperature = 10°C
5. Dissolved oxygen = 0.1 mg/L
6. $U_{DN\,(20°)}$ = 0.10 day^{-1}

Solution

1. Calculate the denitrification rate for 10°C using Eq. 11-6.

$$U'_{DN} = (0.10) \times 1.09^{(10-20)} (1 - 0.1)$$

$$= (0.10)\,(0.42)\,(0.9)$$

$$= 0.038 \text{ day}^{-1}$$

2. Calculate the residence time using Equation 8-45.

$$U = \frac{S_o - S}{\theta X}$$

$$\theta = \frac{S_o - S}{UX}$$

$$\theta = \frac{22 - 3}{0.038 \times 2,000}$$

$$= 0.237 \text{ d}$$

$$= 5.7 \text{ hr}$$

Classification of Nitrification/Denitrification Processes

The denitrification processes in Chap. 8 were identified as being anoxic suspended-growth and anoxic attached-growth. (It was noted in Chap. 8 that the term "anoxic" is used in preference to the term "anaerobic" when describing the denitrification

l split and corrs Fri May 18 1990 -clyde

process because the principal biochemical pathways are not anaerobic but only modifications of aerobic pathways [33]). In the following discussion, classification will be based on whether denitrification is accomplished (1) in combined carbon oxidation nitrification/denitrification systems using internal and endogenous carbon sources or (2) in separate reactors using methanol or another suitable external source of organic carbon. As noted previously, combined systems are commonly termed "single-sludge systems," and nitrification/denitrification systems using separate reactors are often termed "separate or two-sludge systems." It should be noted that the sludges generated in the separate sludge system are of different character.

Combined Nitrification/Denitrification (Single-Sludge) System

Because of the high cost of external organic carbon sources, processes have been developed in which the carbon oxidation nitrification/denitrification steps are combined into a single process, using carbon naturally developed in the wastewater. Specific advantages of these processes include (1) reduction in the volume of air needed to achieve nitrification and BOD_5 removal, (2) elimination of the need for supplemental organic carbon sources (e.g., methanol) required for denitrification, and (3) elimination of intermediate clarifiers and return-sludge systems required in a staged nitrification/denitrification system. Most of these systems are capable of removing from 60 to 80 percent of total nitrogen; removal rates ranging from 85 to 95 percent have also been reported [27]. Example flow diagrams of two of these combined-stage processes are presented in Fig. 11-15.

In these combined processes, the carbon in the wastewater and the carbon remaining in the bacterial cell tissue fragments after endogenous decay of the organisms are used to achieve denitrification. For denitrification, a series of alternating aerobic and anoxic stages without intermediate settling have been used (Fig. 11-15a). Anoxic zones can be accomplished in oxidation ditches by controlling the oxygenation levels (Fig. 11-15b). The sequencing batch reactor process is also particularly adaptable to providing aerobic and anoxic periods during the operating cycle. Use of the sequencing batch reactor for biological nutrient removal is discussed in Sec. 11-8.

The maximum denitrification rates for wastewater in a single-sludge denitrification system range from 0.075 to 0.115 lb NO_3^- −N/lb MLVSS · d at 20°C in an anoxic reactor under noncarbon-limiting conditions [39]. Denitrification rates in single-sludge systems are approximately one-half of the rates of a separate-sludge system. Using endogenous carbon sources, the denitrification rates range from 0.017 to 0.048 lb NO_3^- −N/lb MLVSS · d (see Table 11-19). The lower denitrification rates occur in systems with a higher θ_c.

Bardenpho Process (Four-Stage). The four-stage proprietary *Bardenpho process*, shown in Fig. 11-15a, uses both the carbon in the untreated wastewater and carbon from endogenous decay to achieve denitrification. Separate reaction zones

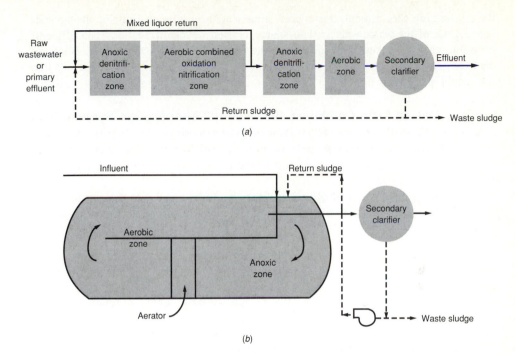

FIGURE 11-15
Combined-stage nitrification/denitrification systems: (a) four-stage Bardenpho and (b) oxidation ditch.

are used for carbon oxidation nitrification and anoxic denitrification. The wastewater initially enters an anoxic denitrification zone to which nitrified mixed liquor is recycled from a subsequent combined carbon oxidation nitrification compartment. The carbon present in the wastewater is used to denitrify the recycled nitrate. Because the organic loading is high, denitrification proceeds rapidly. The ammonia in the wastewater passes unchanged through the first anoxic basin to be nitrified in the first aeration basin. The nitrified mixed liquor from the first aeration basin passes into a second anoxic zone, where additional denitrification occurs using the endogenous carbon source. The second aerobic zone is relatively small and is used mainly to strip entrained nitrogen gas prior to clarification. Ammonia released from the sludge in the second anoxic zone is also nitrified in the last aerobic zone. A modification of the Bardenpho process (the addition of a fifth stage) is also used for combined nitrogen and phosphorus removal (see Sec. 11-9).

Oxidation Ditch. The oxidation ditch has been used to achieve nitrification and denitrification (see Fig. 11-15b). In an oxidation ditch, mixed liquor flows around a loop-type channel, driven and aerated by mechanical aeration devices (see Chap. 10). For nitrification/denitrification applications, an aerobic zone is established immediately downstream of the aerator, and an anoxic zone is created upstream of the aerator.

By discharging the influent wastewater stream at the upstream limit of the anoxic zone, some of the wastewater carbon source is used for denitrification. The effluent from the reactor is taken from the end of the aerobic zone for clarification. Because the system has only one anoxic zone, nitrogen removals are lower than those of the Bardenpho process.

Process Design for a Combined Nitrification/Denitrification System. The design procedures for a combined suspended-growth nitrification/denitrification system vary based on the type of process used. A simplified method for determining the required aerobic and anoxic residence times and recycle ratio, based on Refs. 13 and 27, is presented below.

Assuming complete denitrification of the $NO_3^- - N$ recycled to the anoxic stage, and neglecting nitrogen assimilation, the required recycle ratio (mixed liquor + return sludge) is given by

$$R = \frac{(NH_4^+ - N)_o - (NH_4^+ - N)_e}{(NO_3^- - N)_e} - 1 \qquad (11\text{-}7)$$

where
$$R = \text{overall recycle (mixed liquor + return sludge) ratio}$$
$$(NH_4^+ - N)_o, (NH_4^+ - N)_e = \text{influent and effluent ammonium nitrogen, respectively, mg/L}$$
$$(NO_3^- - N)_e = \text{effluent nitrate nitrogen, mg/L}$$

Because nitrifiers can only grow in the aerobic zone, the solids retention time required for nitrification can be expressed by

$$\theta_c' = \frac{\theta_c}{V_{\text{aerobic}}} \qquad (11\text{-}8)$$

where
$$\theta_c' = \text{solids retention time required for nitrification in a combined-stage (single-sludge) system, d}$$
$$\theta_c = \text{solids retention time required for nitrification in a conventional system (obtained by using Eqs. 8-46 and 8-55 and the equations listed in Table 11-15), d}$$
$$V_{\text{aerobic}} = \text{aerobic volume fraction}$$

In Chap. 8, the mass concentrations in the reactor are expressed by

$$X = \frac{\theta_c}{\theta} \frac{Y(S_o - S)}{(1 + k_d \theta_c)} \qquad (8\text{-}42)$$

The overall system aerobic residence time can be calculated by using the following modification of Eq. 8-42:

$$\theta_a = \frac{\theta_c' Y_h (S_o - S)}{X_a [1 + k_{df} \text{vss} \theta_c']} \qquad (11\text{-}9)$$

where θ_a = overall aerobic hydraulic residence time, d

Y_h = heterotrophic yield coefficient, mg VSS/mg BOD_5 (a value used commonly is 0.55)

$S_o - S$, mg/L (BOD removed in the system and, in some cases, it is approximately equal to influent BOD, S_o)

k_d = endogenous decay rate coefficient, d^{-1}

X_a = MLVSS, mg/L

f VSS = degradable fraction of MLVSS under aeration (a term that is often added to account for the fraction of the degradable MLVSS)

Because the degradable fraction varies with the solids retention time and the endogenous rate coefficient, the degradable fraction of the MLVSS, f VSS, can be expressed as [25]

$$f \text{ VSS} = \frac{f'\text{VSS}}{[1 + (1 - f'\text{VSS})k_d\theta_c']} \tag{11-10}$$

where $f'\text{VSS}$ = degradable fraction of VSS at generation (typically the maximum degradable portion ranges from 0.75 to 0.8)

Other terms are as defined previously.

The anoxic residence time is given by

$$\theta_{DN} = (1 - V_{\text{aerobic}})\,\theta_a \tag{11-11}$$

The anoxic residence time for denitrification θ'_{DN} is determined by:

$$\theta'_{DN} = \frac{N_{\text{Denit}}}{U_{DN}X_a} \tag{11-12}$$

where N_{Denit} = the amount of nitrate to be denitrified, mg/L

U_{DN} = denitrification rate, d^{-1} (see typical values in Table 11-19)

If $\theta_{DN} = \theta'_{DN}$, the calculation is completed. If $\theta_{DN} \neq \theta'_{DN}$ a different V_{aerobic} is assumed and the calculation is repeated. Application of the above calculation procedure is illustrated in Example 11-3.

Example 11-3   Calculation of aerobic and anoxic residence times in a combined nitrification/denitrification reactor. Calculate the required aerobic and anoxic residence times and the recycle ratio for a combined nitrification/denitrification reactor, assuming the following conditions:

1. Influent BOD = 200 mg/L
2. Influent ammonia = 25 mg/L as N
3. Effluent ammonia = 1.5 mg/L as N
4. Effluent nitrate = 5 mg/L as N

5. Temperature $= 15°C$
6. $Y_h = 0.55$ mg VSS/mg BOD
7. $k_{d\ (15°C)} = 0.04\ d^{-1}$
8. $U_{DN\ (15°C)} = 0.042$ mg NO_3-N/mg VSS · d
9. DO in aeration basin $= 2.0$ mg/L
10. $X_a = 2500$ mg/L MLVSS
11. $\theta_c = 8.9$ d for nitrification (from Example 11-1)
12. $V_{aerobic} = 0.71$
13. $f'_{VSS} = 0.8$

Solution

1. Calculate the total recycle ratio using Eq. 11-7.

$$R = \frac{25 - 1.5}{5} - 1$$
$$= 3.7$$

2. Calculate the overall sludge age using Eq. 11-8.

$$\theta'_c = \frac{8.9}{0.71}$$
$$= 12.5\ d$$

3. Calculate the degradable fraction of the MLVSS using Eq. 11-10.

$$f_{VSS} = \frac{0.8}{1 + [(1 - 0.8)(0.04\ d^{-1})(12.5\ d)}$$
$$= 0.73$$

4. Calculate the total aerobic residence time using Eq. 11-9.

$$\theta_a = \frac{(0.55\ \text{mg VSS/mg BOD})(200\ \text{mg/L})(12.5\ d)}{2,500\ \text{mg VSS/L}[1 + (0.04\ d^{-1})(0.73)(12.5\ d)]}$$
$$= 0.40\ d = 9.6\ h$$

5. The anoxic residence time using Eq. 11-11 is

$$\theta_{DN} = (1 - 0.71)(0.40)$$
$$= 0.12\ d = 2.9\ h$$

6. The required anoxic residence time for denitrification is (from Eq. 11-12)

$$\theta'_{DN} = \frac{(25 - 1.5 - 5)\ \text{mg/L}}{(0.042\ d^{-1})(2500\ \text{mg/L})}$$
$$= 0.18\ d = 4.3\ h$$

Because $\theta_{DN} \neq \theta'_{DN}$, additional computations are left as a student homework assignment (see Prob. 11-10).

Separate-Stage
Denitrification (Separate-Sludge) System

In the early 1970s, the most generally accepted approach to biological denitrification was the addition of a separate biological system using methanol as the carbon source to remove the nitrate. Several alternative systems are illustrated in Fig. 11-16. Because the carbon oxidation nitrification/denitrification occurs in separate reactors, sludge is generated separately in each reactor, hence the name "separate-sludge system" is

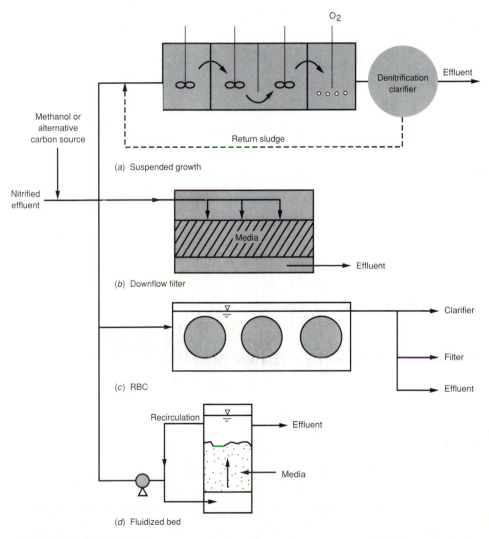

FIGURE 11-16
Alternative separate-stage denitrification processes using a separate carbon source.

used frequently. Because any excess carbon in the form of methanol added over that required for the process will be measured in the effluent BOD, careful attention must be devoted to the design and operation of this aspect of the system.

Stoichiometry. Using methanol as the carbon source, the stoichiometry of separate-stage denitrification can be described as follows. The energy reaction may be represented by the following equations:

Energy reaction, step 1:

$$6NO_3^- + 2CH_3OH \rightarrow 6NO_2^- + 2CO_2 + 4H_2O \qquad (11\text{-}13)$$

Energy reaction, step 2:

$$6NO_2^- + 3CH_3OH \rightarrow 3N_2 + 3CO_2 + 3H_2O + 6OH^- \qquad (11\text{-}14)$$

Overall energy reaction:

$$6NO_3^- + 5CH_3OH \rightarrow 5CO_2 + 3N_2 + 7H_2O + 6OH^- \qquad (11\text{-}15)$$

A typical synthesis reaction as given by McCarty [22] is as follows:

Synthesis:

$$3NO_3^- + 14CH_3OH + CO_2 + 3H^+ \rightarrow 3C_5H_7O_2N + H_2O \qquad (11\text{-}16)$$

In practice, 25 to 30 percent of the amount of methanol required for energy is required for synthesis. On the basis of laboratory studies, the following empirical equation was developed to describe the overall nitrate-removal reaction [22].

Overall nitrate removal:

$$NO_3^- + 1.08CH_3\,OH + H^+ \rightarrow$$
$$0.065C_5H_7O_2N + 0.47N_2 + 0.76CO_2 + 2.44H_2O \quad (11\text{-}17)$$

If all the nitrogen is in the form of nitrate, the overall methanol requirement can be determined using Eq. 11-17. However, biologically processed wastewater that is to be denitrified may contain some nitrite and dissolved oxygen. Where nitrate, nitrite, and dissolved oxygen are present, the methanol requirement can be computed using the following empirically derived equation [22]:

$$C_m = 2.47N_o + 1.53N_1 + 0.87D_o \qquad (11\text{-}18)$$

where C_m = required methanol concentration, mg/L
N_o = initial nitrate-nitrogen concentration, mg/L
N_1 = initial nitrite-nitrogen concentration, mg/L
D_o = initial dissolved-oxygen concentration, mg/L

Kinetic coefficients for the denitrification process are summarized in Table 11-20.

Suspended-Growth. The design of suspended-growth denitrification systems is similar in many respects to the design of the activated-sludge systems used to remove organic carbon. Both complete-mix and plug-flow reactors have been used. Because

TABLE 11-20
Typical kinetic coefficients for the denitrification process[a]

Coefficient	Basis	Value[b] Range	Value[b] Typical
μ_m	d^{-1}	0.3–0.9	0.3
K_S	$NO_3^- -N$, mg/L	0.06–0.20	0.1
Y	$NO_3^- -N$, mg VSS/mg	0.4–0.9	0.8
k_d	d^{-1}	0.04–0.08	0.04

[a] Derived in part from Refs. 26 and 33.
[b] Values reported are for 20°C.
Note: 1.8(°C) + 32 = °F

the nitrogen gas released during the denitrification process often becomes attached to the biological solids, a nitrogen release step is included between the reactor and the sedimentation facilities used to separate the biological solids. The removal of the attached nitrogen gas bubbles can be accomplished either in aerated channels that can be used to connect the biological reactor and the settling facilities or in a separate tank in which the solids are aerated for a short period of time (5 to 10 min).

One of the flow diagrams that has been used for the separate-stage removal of nitrogen from domestic wastewater is illustrated in Fig. 11-17. Typical design parameters for each of the processes shown in Fig. 11-17 are given in Table 11-21. This flow diagram can also be used for the removal of phosphorus by adding alum to precipitate the phosphorus in the nitrification clarifier. In addition to removing phosphorus, this technique also can be used to overcome the difficulties of separating organisms growing in the dispersed growth phase. Effluent from the denitrification step can be filtered, or alum can be added before filtration for the removal of residual phosphorus and suspended solids. Additional details on facilities design may be found in Ref. 33.

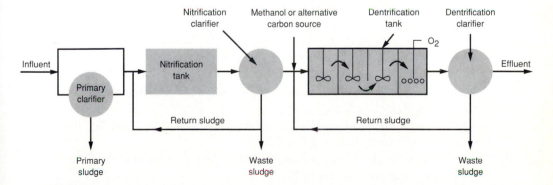

FIGURE 11-17
Flow diagram for a two-stage biological treatment process for nitrogen removal.

TABLE 11-21
Design parameters for a two-stage biological treatment process to remove nitrogen from domestic wastewater

Treatment process	Type of reactor	θ_c, d^a	θ, h^a	Design parameter MLVSS, mg/L	pH	Temperature coefficient[b]
Single-stage nitrification	Plug-flow	8–20	6–15	2,000–3,500	7.0–8.0[c]	1.08–1.10
Denitrification[d]	Plug-flow	1–5	0.2–2	1,000–2,000[e]	6.5–7	1.14–1.16

[a] Indicated values for θ_c and θ are for 20°C.
[b] Temperature coefficient to be used in the equation $K_T = K_{20}\, \theta^{T-20}$
[c] Lower values have been reported.
[d] Methanol requirement can be computed using Eq. 11-18.
[e] Higher values may be observed depending on the degree of solids carryover.
Note: mg/L = g/m³

The mixed-liquor volatile suspended solids in the nitrification reactors are composed of those organisms responsible for the conversion of organic carbon (BOD) and those responsible for nitrification. The distribution of the two groups of organisms varies with each installation. The total mixed-liquor suspended solids in the nitrification reactor are normally 50 to 100 percent higher than the mixed-liquor volatile suspended solids and may include residual chemical precipitates if alum precipitation is used for phosphorus removal. In denitrification reactors, the mixed-liquor volatile suspended solids have been observed to be about 40 to 70 percent of the mixed-liquor suspended solids.

The effects of the major operational and environmental variables on the separate-stage denitrification process are reported in Table 11-22. The kinetic expressions used to analyze the denitrification process for a complete-mix reactor were reported in Table 11-14. The application of the kinetic approach, using Table 11-22 and Table 11-14 to analyze the suspended-growth denitrification process, is as follows:

1. Using the kinetic data given in Table 11-20 and Eq. 8-7, determine the minimum mean cell-residence time θ_c^M for denitrification. The kinetic coefficients must be corrected for the operating temperature using the expression given in Table 11-22.

2. Using Eq. 8-55 and an assumed safety factor, determine the design mean cell-residence time, θ_c.

3. Using the design mean cell-residence time determined in step 2 and Eq. 8-46, determine the substrate utilization rate U.

4. Using the substrate utilization rate determined in step 3, determine the effluent substrate concentration using Eq. 8-51.

5. Determine the hydraulic retention time using Eq. 8-45.

6. Determine the sludge-wasting rate using the standard definition given in Chap. 8.

TABLE 11-22
Effect of the major operational and environmental variables on the denitrification process[a]

Factor	Description of effect
Nitrate concentration	It has been observed that the concentration of nitrate will affect the maximum growth of the organisms responsible for denitrification. The effect of the nitrate concentration has been modeled using the following expression: $$\mu'_D = \mu_{mD} \frac{M}{K_{SN} + C_N}$$
Carbon concentration	The effect of the carbon concentration has also been modeled using a Monod type expression. The relationship using methanol as a carbon source is: $$\mu'_D = \mu_{mD} \frac{M}{K_M + M}$$ where M = methanol concentration, mg/L $\quad\;\; K_M$ = half saturation constant for methanol, mg/L
Temperature	The effect of temperature is significant. It can be estimated using the following expression: $$P = 0.25\,T^2$$ where P = percent of denitrification growth rate at 20°C $\quad\;\; T$ = temperature, °C
pH	From available evidence, it appears that the optimum pH range is between about 6.5 and 7.5, and the optimum condition is around 7.0

[a] Developed from information in Ref. 33.
Note: 1.8(°C) + 32 = °F

Temperature will affect performance significantly and will need to be taken into account. It cannot be overstressed that unless temperature is considered properly in the design of both nitrification and denitrification systems, effluent quality (measured in terms of the amount of ammonia or nitrate in the effluent) will deteriorate at low temperatures. For example, using a temperature coefficient of 1.12, the reactor volume at 10°C (50°F) would be approximately three times the volume required at 20°C (68°F). From a design standpoint, providing flexibility in the selection of the reactor volume is an important consideration. Additional reactor volume could be provided by using a plug-flow reactor whose length could be changed using stop gates. Alternatively, the solids in the system could be increased to accommodate cold weather operation, or covered reactors could be used.

Attached-Growth. A number of attached-growth denitrification processes, many of them proprietary, have been developed. The principal attached-growth processes, shown in Fig. 11-16b, c, and d, are described in Table 11-23. Fluidized-bed reac-

TABLE 11-23
Description of attached-growth denitrification systems

Classification	Description	Typical removal rates at 20°C, lb/10^3ft^3 · d[a]
Packed-bed reactor		
Gas-filled	The reactor is covered and filled with nitrogen gas, which eliminates the necessity of having to submerge the medium to maintain anoxic conditions.	100–112
Liquid-filled	With both high- and low-porosity liquid-filled packed-bed reactors, backwashing of packing medium is usually required to control the biomass.	6–8
Fluidized-bed reactor		
High-porosity medium, fine sand	Porosity is varied by adjusting the density of the medium and the flowrate.	750–1,000
High-porosity medium, activated carbon		300–375
Rotating biological contactors	Contactors are similar to aerobic process except media are submerged.	(see Fig. 11-20)

[a] Data are reported for comparative purposes only. If any of these processes are to be applied, pilot plant testing is recommended to verify reported removal rates.

Note: lb/10^3ft^3 · d = 0.0160 = kg/m^3 · d
1.8 (°C) + 32 = °F

tors and rotating biological contactors (RBCs) are the most commonly used. In the fluidized-bed reactor (see Fig. 11-18), the wastewater to be treated passes upward through a bed of fine-grained material, such as sand, at sufficient velocity to suspend or fluidize the media. Fluidization significantly increases the specific surface and allows high biomass concentrations in the reactor. The reactor requires a relatively small space and is relatively simple to operate. A typical design loading curve showing the effect of temperature is presented in Fig. 11-19 [19]. Additional information on fluidized-bed performance for denitrification may be found in Ref. 39.

Operation of RBCs for denitrification is similar to that for aerobic processes, except the medium is totally submerged to avoid oxygenating the liquid. Clarification is required following biological treatment to remove the excess sloughed biomass. Typical design curves for RBCs, developed by Antonie, are shown in Fig. 11-20 [3].

The approach most commonly used in assessing the performance of the attached-growth denitrification processes has involved the use of removal-rate parameters. Although some application data are reported in Table 11-23, it is recommended that

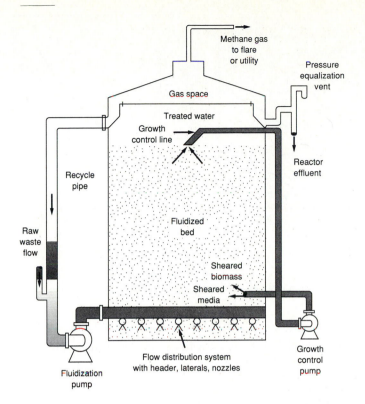

FIGURE 11-18
Fluidized-bed reactor used for the denitrification of wastewater (from Envirex).

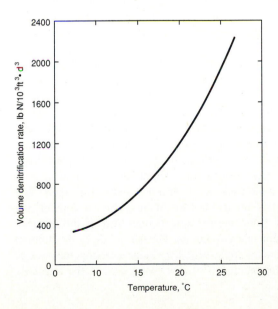

FIGURE 11-19
Design loading curve for the denitrification of municipal wastewater in a fluidized-bed reactor [19].
Note: lb/10^3 ft^3· d · × 16 = glm^3 · d.

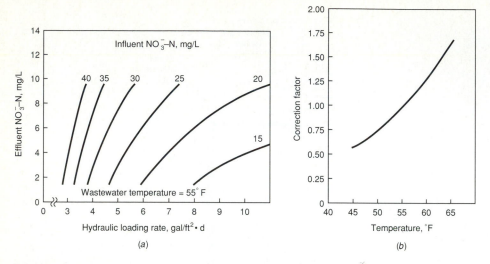

FIGURE 11-20
Design curves for denitrification using a rotating biological contactor: (*a*) loading curve and
(*b*) temperature correction curve [3].

pilot plant tests be conducted to define process kinetics when any of these processes
are considered for use.

Comparison of Denitrification Processes

A general comparison of the various denitrification processes is presented in Table
11-24. Because many of the processes have been in limited use, caution should be
exercised in making recommendations. In almost all cases, pilot plant studies are
recommended. Where such studies cannot be conducted, the use of conservative
design criteria is advised.

11-8 REMOVAL OF PHOSPHORUS BY BIOLOGICAL METHODS

In recent years, a number of biological phosphorus removal processes have been
developed as alternatives to chemical treatment. As described in Chap. 8, phosphorus is removed in biological treatment by means of incorporating orthophosphate,
polyphosphate, and organically bound phosphorus into cell tissue. The total amount
removed depends on the net solids produced, as determined using Eq. 10-3. The
phosphorus content of the cell tissue is about one-fifth of the nitrogen content; the
actual phosphorus content may vary from one-seventh to one-third of the nitrogen
value, depending on specific environmental conditions. On the average, the amount
of phosphorus removed during secondary treatment by sludge wasting may range
from 10 to 30 percent of the influent amount [36]. By using one of the specially

TABLE 11-24
Comparison of denitrification alternatives

System type	Advantages	Disadvantages
Combined carbon oxidation nitrification/ denitrification in suspended-growth reactor using endogenous carbon source	No methanol required; lesser number of unit processes required; better control of filamentous organisms in activated-sludge process possible; single basin can be used; adaptable to sequencing batch reactor; process can be adapted to include biological phosphorus removal.	Denitrification occurs at very slow rates; longer detention time and much larger structures required than methanol-based system; stability of operation linked to clarifier for biomass return; difficult to optimize nitrification and denitrification separately; biomass requires sufficient dissolved-oxygen level for nitrification to occur; less nitrogen removal than methanol-based system.
Combined carbon oxidation nitrification/ denitrification in suspended-growth reactor using wastewater carbon source	No methanol required; lesser number of unit processes required; better control of filamentous organisms in activated-sludge process possible; single basin can be used; adaptable to sequencing batch reactor; process can be adapted to include biological phosphorus removal.	Denitrification occurs at slow rates; longer detention time and larger structures required than methanol-based system; stability of operation linked to clarifier for biomass return; difficult to optimize nitrification and denitrification separately; biomass requires sufficient dissolved-oxygen level for nitrification to occur; less nitrogen removal than methanol-based system.
Suspended-growth using methanol following a nitrification stage	Denitrification rapid; small structures required; demonstrated stability of operation; few limitations in treatment sequence options; excess methanol oxidation step can be easily incorporated; each process in system can be separately optimized; high degree of nitrogen removal possible.	Methanol required; stability of operation linked to clarifier for biomass return; greater number of unit processes required for nitrification/denitrification than in combined systems.
Attached-growth (column) using methanol following a nitrification stage	Denitrification rapid; small structures required; demonstrated stability of operation; stability not linked to clarifier as organisms on media; few limitations in treatment sequence options; high degree of nitrogen removal possible; each process in the system can be separately optimized.	Methanol required; excess methanol oxidation process not easily incorporated; greater number of unit processes required for nitrification/ denitrification than in combined systems.

developed biological phosphorus removal processes, removals significantly in excess of this range may be achieved.

The key to the biological phosphorus removal is the exposure of the microorganisms to alternating anaerobic and aerobic conditions. As discussed in Chap. 8, exposure to alternating conditions stresses the microorganisms so that their uptake of phosphorus is above normal levels. Phosphorus is not only used for cell maintenance, synthesis, and energy transport but is also stored for subsequent use by the microorganisms. The sludge containing the excess phosphorus is either wasted or removed through a sidestream to release the excess. The alternating exposure to anaerobic and aerobic conditions can be accomplished in the main biological treatment process, or "mainstream," or in the return-sludge stream, or "sidestream" (see Fig. 11-21). Several typical biological treatment processes used for phosphorus removal are described in this section. These processes are (1) the proprietary A/O process for mainstream phosphorus removal, (2) the proprietary PhoStrip process used for sidestream phosphorus removal, and (3) the sequencing batch reactor (SBR). The SBR is used for smaller wastewater flows and also provides the flexibility of operation

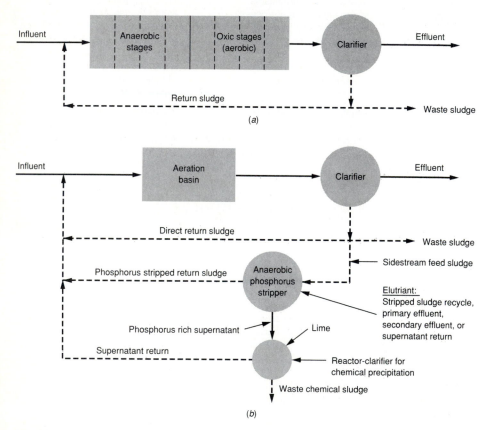

FIGURE 11-21
Alternative biological phosphorus removal systems: (a) A/O process and (b) PhoStrip process.

that allows for the removal of nitrogen in addition to phosphorus. Typical design information for these processes is given in Table 11-25.

A/O Process
(Mainstream Phosphorus Removal)

The proprietary A/O process is used for combined carbon oxidation and phosphorus removal from wastewater. The A/O process is a single-sludge suspended-growth system that combines anaerobic and aerobic sections in sequence (Fig. 11-21a). Provision may be made for nitrification by supplying the necessary detention time in the aerobic stage. Settled sludge is returned to the influent end of the reactor and mixed with the incoming wastewater. Under anaerobic conditions, the phosphorus contained in the wastewater and the recycled cell mass is released as soluble phosphates. Some BOD reduction also occurs in this stage. The phosphorus is then taken up by the cell mass in the aerobic zone. Phosphorus is removed from the liquid stream in the waste activated sludge. The concentration of phosphorus in the effluent is dependent mainly on the ratio of BOD to phosphorus of the wastewater treated. It has been reported that when this ratio exceeds 10 to 1, effluent soluble phosphorus values of 1 mg/L or less can be achieved [43]. In cases where the BOD to phosphorus ratios are less than 10 to 1, metal salts can be added to the process to achieve low effluent phosphorus concentrations.

TABLE 11-25
Typical design information for biological phosphorus removal processes[a]

		Process		
Design parameter	Units	A/O	PhoStrip	Sequencing batch reactor
Food-to-microorganism ratio (F/M)	lb BOD/ lb MLVSS · d	0.2 – 0.7	0.1 – 0.5	0.15 – 0.5
Solids retention time, θ_c	d	2–25	10–30	
MLSS	mg/L	2,000–4,000	600–5,000	2,000–3,000
Hydraulic retention time, θ	h			
Anaerobic zone		0.5–1.5	8–12	1.8–3
Aerobic zone		1–3	4–10	1.0–4
Return activated sludge	% of influent	25–40	20–50	
Internal recycle	% of influent		10–20[b]	

[a] Adapted from Ref. 41.

[b] Stripper underflow.

PhoStrip Process
(Sidestream Phosphorus Removal)

In the proprietary PhoStrip process, a portion of the return activated sludge from the biological treatment process is diverted to an anaerobic phosphorus stripping tank (see Fig. 11-21*b*). The retention time in the stripping tank typically ranges from 8 to 12 hours. The phosphorus released in the stripping tank passes out of the tank in the supernatant, and the phosphorus-poor activated sludge is returned to the aeration tank. The phosphorus-rich supernatant is treated with lime or another coagulant in a separate tank and discharged to the primary sedimentation tanks or to a separate flocculation/clarification tank for solids separation. Phosphorus is removed from the system in the chemical precipitant. Conservatively designed PhoStrip and associated activated-sludge systems are capable of consistently producing an effluent with a total phosphorus content of less than 1.5 mg/L before filtration [38].

Sequencing Batch Reactor

The SBR (described in Sec. 8-7) can be operated to achieve any combination of carbon oxidation, nitrogen reduction, and phosphorus removal (see Fig. 11-22). Reduction of these constituents can be accomplished with or without chemical addition by changing the operation of the reactor. Phosphorus can be removed by coagulant addition or biologically without coagulant addition. In the configuration shown in Fig. 11-22, phosphorus release and BOD uptake will occur in the anaerobic stir phase, with subsequent phosphorus uptake in the aerobic stir phase. By modifying the reaction times as illustrated in Fig. 11-23, nitrification or nitrogen removal can also be accomplished. Overall cycle times may vary from 3 to 24 h [4]. A carbon source in the anoxic phase is required to support denitrification—either an external source or endogenous respiration of the existing biomass [41].

Comparison of Biological
Phosphorus Removal Processes

A general comparison of the various biological phosphorus removal processes is presented in Table 11-26. Based on the preceding discussion, biological processes

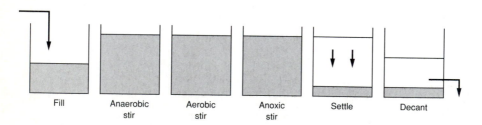

| Fill | Anaerobic stir | Aerobic stir | Anoxic stir | Settle | Decant |

FIGURE 11-22
Sequencing batch reactor operation for carbon, nitrogen, and phosphorus removal.

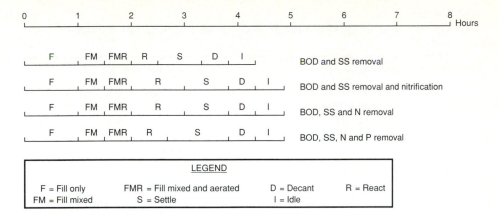

FIGURE 11-23
Suggested SBR operating strategies for the removal of carbon, nitrogen, and phosphorus [4].

offer many advantages for integrating nutrient removal into the treatment process. As the need for nutrient removal increases, modifications to these processes will continue to develop and more of these types of processes will be used. Because successful performance of many of these processes depends upon the specific local conditions, pilot plant testing is recommended to develop operating data and design criteria.

11-9 COMBINED REMOVAL OF NITROGEN AND PHOSPHORUS BY BIOLOGICAL METHODS

A number of biological processes have been developed for the combined removal of nitrogen and phosphorus. Many of these are proprietary and use a form of the activated-sludge process but employ combinations of anaerobic, anoxic, and aerobic zones or compartments to accomplish nitrogen and phosphorus removal. Some of these processes were developed originally for phosphorus removal and later evolved into combined phosphorus and nitrogen removal systems. The most commonly used processes for combined nitrogen and phosphorus removal are (1) the A^2/O process, (2) the five-stage Bardenpho process, (3) the UCT process, and (4) the VIP process, which are described in this section. These four processes are shown schematically in Fig. 11-24. Typical design information is presented in Table 11-27. The sequencing batch reactor, described in the previous section, is also used for the combined removal of nitrogen and phosphorus.

A^2/O Process

The proprietary A^2/O process is a modification of the A/O process and provides an anoxic zone for denitrification (Fig. 11-24a). The detention period in the anoxic zone is approximately one hour. The anoxic zone is deficient in dissolved oxygen, but

TABLE 11-26
Advantages and disadvantages of biological phosphorus removal processes[a]

Process	Advantages	Disadvantages
A/O	Operation is relatively simple compared to other processes.	Is not capable of achieving high levels of nitrogen and phosphorus removal simultaneously.
	Waste sludge has a relatively high phosphorus content (3–5%) and has fertilizer value.	Performance under cold weather operating conditions uncertain.
	Relatively short hydraulic retention time.	High BOD/P ratios are required.
	Where reduced levels of phosphorus removal efficiency are acceptable, process may achieve complete nitrification.	With reduced aerobic cell detention time, very high-rate oxygen-transfer devices may be necessary.
		Limited process control flexibility is available.
PhoStrip	Can be incorporated easily into existing activated-sludge plants.	Requires lime addition for phosphorus precipitation.
	Process is flexible; phosphorus removal process is not controlled by BOD/phosphorus ratio.	Requires higher mixed-liquor dissolved oxygen to prevent phosphorus release in final clarifier.
	Several installations in U.S.	Additional tankage required for stripping.
	Significantly less chemical usage than mainstream chemical precipitation.	Lime scaling may be a maintenance problem.
	Can achieve reliably effluent orthophosphate concentrations of less than 1.5 mg/L.	
Sequencing batch reactor	Process is very flexible for combining nitrogen and phosphorus removal.	Suitable only for smaller flows.
	Process is simple to operate.	Redundant units are required.
	Mixed-liquor solids cannot be washed out by hydraulic surges.	Effluent quality depends upon reliable decanting facility.
		Limited design data available.

[a] Adapted from Refs. 31, 36, 38, and 39.

chemically bound oxygen in the form of nitrate or nitrite is introduced by recycling nitrified mixed liquor from the aerobic section. Effluent phosphorus concentrations of less than 2 mg/L can be expected without effluent filtration; with effluent filtration, effluent phosphorus concentrations may be less than 1.5 mg/L [38].

Bardenpho Process (Five-Stage)

The proprietary Bardenpho process, described in Sec. 11-7, can be modified for combined nitrogen and phosphorus removal. The Phoredox modification of the Bardenpho

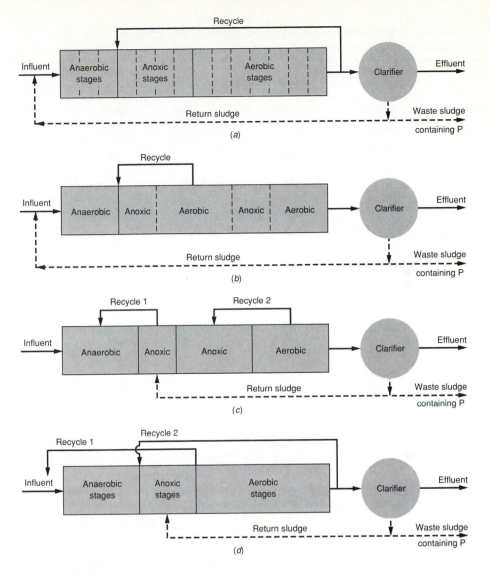

FIGURE 11-24
Combined biological nitrogen and phosphorus removal processes: (a) A^2/O process, (b) five-stage Bardenpho process, (c) UCT process, and (d) VIP process. *Note*: Nitrogen is released to the atmosphere in the anoxic stages.

process incorporates a fifth (anaerobic) stage for phosphorus removal (see Fig. 11-24b). The staging sequence and recycle method are different from the A^2/O process. The five-stage system provides anaerobic, anoxic, and aerobic stages for phosphorus, nitrogen, and carbon removal. A second anoxic stage is provided for additional denitrification using nitrate produced in the aerobic stage as the electron acceptor and the endogenous organic carbon as the electron donor. The final aerobic stage is used to strip

TABLE 11-27
Typical design information for combined removal of nitrogen and phosphorus by biological processes[a]

Design parameter	Units	A²/O	Bardenpho (5-stage)	UCT	VIP
			Process		
Food-to-microorganism ratio (F/M)	lb BOD/ lb MLVSS · d	0.15–0.25	0.1–0.2	0.1–0.2	0.1–0.2
Solids retention time, θ_c	d	4–27	10–40	10–30	5–10
MLSS	mg/L	3,000–5,000	2,000–4,000	2,000–4,000	1,500–3,000
Hydraulic retention time, θ	h				
Anaerobic zone		0.5–1.5	1–2	1–2	1–2
Anoxic zone – 1		0.5–1.0	2–4	2–4	1–2
Aerobic zone – 1		3.5–6.0	4–12	4–12	2.5–4
Anoxic zone – 2			2–4	2–4	
Aerobic zone – 2			0.5–1		
Return activated sludge	% of influent	20–50	50–100	50–100	50–100
Internal recycle	% of influent	100–300	400	100–600	200–400

[a] Adapted from Refs. 11, 27, and 41.

residual nitrogen gas from solution and to minimize the release of phosphorus in the final clarifier. Mixed liquor from the first aerobic zone is recycled to the anoxic zone. As shown in Table 11-27, the process uses a longer θ_c (10 to 40 days) than the A²/O process, which increases the carbon oxidation capability.

UCT Process

The UCT process, developed at the University of Cape Town, is similar to the A²/O process, with two exceptions (see Fig. 11-24c). The return activated sludge is recycled to the anoxic stage instead of the aeration stage, and the internal recycle is from the anoxic stage to the anaerobic stage. By returning the activated sludge to, the anoxic stage, the introduction of nitrate to the anaerobic stage is eliminated, thereby improving the release of phosphorus in the anaerobic stage. The internal recycle feature provides for increased organic utilization in the anaerobic stage. The mixed liquor from the anoxic stage contains substantial soluble BOD but little nitrate. The recycle of the anoxic mixed liquor provides for optimal conditions for fermentation uptake in the anaerobic stage. At present (1989), there are no facilities in the United States known to be using this process.

VIP Process

The VIP process (named for the Virginia Initiative Plant in Norfolk, Virginia) is similar to the A^2/O and UCT processes except for the methods used for recycle systems (Fig. 11-24d). The return activated sludge is discharged to the inlet of the anoxic zone along with nitrified recycle from the aerobic zone. The mixed liquor from the anoxic zone is returned to the head end of the anaerobic zone. Based on test data, it appears that some of the organic matter in the process influent is stabilized through anaerobic mechanisms in the anaerobic zone, which further reduces process oxygen requirements [11]. A full sized plant using this process is under construction at the time of writing of this text (1989).

Comparison of Combined Biological Nitrogen and Phosphorus Removal Processes

A comparison of the various combined biological nitrogen and phosphorus removal processes is presented in Table 11-28. Advantages shared by all of these processes are that the sludge quantities generated are comparable to sludge production from conventional activated-sludge systems, and little or no chemicals are required for phosphorus removal. Some of the processes in modified form may also be used for nitrogen or phosphorus removal alone.

11-10 REMOVAL OF NITROGEN BY PHYSICAL AND CHEMICAL PROCESSES

The principal physical and chemical processes used for nitrogen removal, identified in Table 11-11, are air stripping, breakpoint chlorination, and selective ion exchange. In a survey of advanced wastewater treatment facilities, only six out of over 1200 unit operations or processes used air stripping, eight used breakpoint chlorination, and one used ion exchange [42]. The reasons for the limited use of these processes are cost, inconsistent performance, and operating and maintenance problems. Because air stripping and ion exchange have limited use, they will only be described briefly here. For the theory of air stripping of ammonia and ion exchange, the previous edition of this text may be consulted [23]. The advantages and disadvantages of each process are summarized in Table 11-29.

Air Stripping of Ammonia

Ammonia nitrogen can be removed from wastewater by volatilization of gaseous ammonia. The process is simple in concept, but it has serious drawbacks that make it expensive to operate and maintain. The rate of ammonia transfer is enhanced by converting most of the ammonia to a gaseous form at a high pH, usually in the range of 10.5 to 11.5, by the addition of lime (see Fig. 11-25). Because of the high operating and maintenance costs associated with air stripping, the practical application of air stripping of ammonia is limited to special cases such as the need for a high pH for other reasons.

TABLE 11-28
Advantages and disadvantages of combined nitrogen and phosphorus removal processes[a]

Process	Advantages	Disadvantages
A²/O	Waste sludge has a relatively high phosphorus content (3–5%) and has fertilizer value.	Performance under cold weather operating conditions uncertain.
	Provides better denitrification capability than A/O.	More complex than A/O.
Bardenpho	Produces least sludge of all biological phosphorus removal systems.	Large internal cycle increases pumping energy and maintenance requirements.
	Waste sludge has relatively high phosphorus content and has fertilizer value.	Limited experience in U.S.
		Requirements for chemical addition uncertain.
	Total nitrogen is reduced to levels lower than most processes.	Requires more reactor volume than A²/O process.
	Alkalinity is returned to the system, thereby reducing or eliminating the need for chemical addition.	Primary setting reduces ability of process to remove nitrogen and phosphorus.
	Has been widely used in South Africa and substantial data are available.	High BOD/P ratios are required.
		Temperature effects on process performance are not well-known.
UCT	Recycle to anoxic zone eliminates nitrate recycle and provides better phosphorus removal environment in the anaerobic zone.	No installations in U.S.
		Large internal cycle increases pumping energy and maintenance requirements.
	Has slightly less reactor volume than Bardenpho process.	Requirements for chemical addition uncertain.
		High BOD/P ratios are required.
		Temperature effects on process performance are not well-known.
VIP	Recycle of nitrate to anoxic zone reduces oxygen requirements and alkalinity consumption.	Large internal recycle increases pumping energy and maintenance requirements.
	Recycle of anoxic zone effluent to anaerobic zone reduces nitrate loading on aerobic zone.	Few operating installations in U.S.
		Low temperatures reduce nitrogen removal capabilities.
	Adaptable to year-round phosphorus removal and seasonal nitrogen removal.	

[a] Adapted from Refs. 11, 31, 36, 38, and 39.

TABLE 11-29
Advantages and disadvantages of physical and chemical nitrogen removal processes[a]

Process	Advantages	Disadvantages
Air stripping	Process can be controlled for selected ammonia removals. Most applicable if required seasonally in combination with lime system for phosphorus removal. Process may be able to meet total nitrogen standards. Not sensitive to toxic substances.	Process is temperature sensitive. Ammonia solubility increases with lower temperatures. Air requirements also vary. Fogging and icing occur in cold weather. Ammonia reaction with sulfur dioxide may cause air pollution problems. Process usually requires lime for pH control, thereby increasing treatment cost and lime-related operating and maintenance problems. Carbonate scaling of packing and piping. Potential noise and aesthetic problems.
Breakpoint chlorination	With proper control, all ammonia nitrogen can be oxidized. Process can be used following other nitrogen removal processes for fine-tuning of nitrogen removal. Concurrent effluent disinfection. Limited space requirement. Not sensitive to toxic substances and temperature. Low capital costs. Adaptable to existing facility.	May produce high chlorine residuals that are toxic to aquatic organisms. Wastewater contains a variety of chlorine demanding substances which increase cost of treatment. Process is sensitive to pH, which affects dosage requirements. High operating cost due to chemical requirements. Trihalomethane formation may impact quality of water supplies. Addition of chlorine raises effluent TDS. Process may not be able to meet total nitrogen standards. Requires careful control of pH to avoid formation of nitrogen trichloride gas. Requires highly skilled operator.
Ion exchange	Can be used where climatic conditions inhibit biological nitrification and where stringent effluent standards are required. Produces a relatively low TDS effluent. Produces a reclaimable product (aqueous ammonia). Process may be able to meet total nitrogen standards. Ease of product quality control.	Organic matter in effluent from biological treatment can cause resin binding. Pretreatment by filtration is usually required to prevent the buildup of excessive headloss due to suspended-solids accumulation. High concentrations of other cations will reduce ammonia removal capability. Regeneration recovery may require the addition of another unit progress (e.g., gas stripping). High capital and operating costs. Regeneration products must be disposed of. Requires highly skilled operator.

[a] Adapted in part from Refs. 39 and 42.

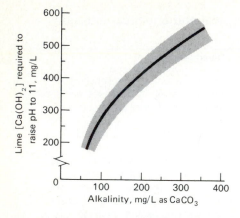

FIGURE 11-25
Lime dosage required to raise the pH to 11 as a function of untreated wastewater alkalinity [29].

In most cases where air stripping has been applied, a number of problems have developed such as calcium carbonate scaling within the tower and feed lines and poor performance during cold weather operation. The high pH range results in absorption of carbon dioxide from the air and the development of carbonate scaling. The amount and nature (soft to extremely hard) of the calcium carbonate scale varies with the characteristics of the wastewater and local environmental conditions. As the temperature decreases, the amount of air required increases significantly for the same degree of removal (Fig. 11-26). When icing occurs, the liquid-air contact geometry in the tower is altered and the overall efficiency is further reduced.

Breakpoint Chlorination

As described in Chap. 7, breakpoint chlorination involves the addition of chlorine to wastewater to oxidize the ammonia nitrogen in solution to nitrogen gas and other

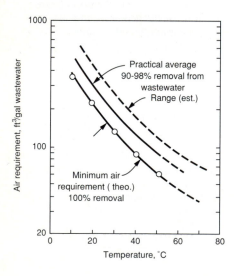

FIGURE 11-26
Effect of temperature on air requirements for ammonia stripping [29].
Note: ft^3/gal × 7.4805 = m^{3}air/m^3 wastewater.

stable compounds. Breakpoint chlorination is also an alternative method of achieving nitrogen control (see Table 11-11). Perhaps the most important advantage of this process is that, with proper control, all the ammonia nitrogen in the wastewater can be oxidized. The process has a number of disadvantages, as cited in Table 11-29, that have limited its application. Because of its effectiveness in ammonia removal, further research in breakpoint chlorination has been recommended to determine if the operating problems can be resolved [42].

Theory. The theory of breakpoint chlorination was described in Sec. 7-4. From that discussion, a representative equation, based on ammonia expressed as NH_3, that can be used to describe the overall reaction is

$$2NH_3 + 3HOCl \rightarrow N_2 + 3H_2O + 3HCl \tag{11-19}$$

The stoichiometric mass ratio of chlorine as Cl_2 to ammonia as N, as computed in Example 7-4, is 7.6:1. In practice, the ratio has been found to vary from 8:1 to 10:1.

From both laboratory and full-scale testing programs, it has been found that the optimum operating pH range for breakpoint chlorination is between 6 and 7. If breakpoint chlorination is accomplished outside this range, it has been observed that the chlorine dosage required to reach the breakpoint increases significantly and that the rate of reaction is slower. Temperature does not appear to have a major effect on the process in the ranges normally encountered in wastewater treatment. The effect of interfering substances on the process is discussed in Sec. 7-4.

Application. The breakpoint chlorination process can be used for the removal of ammonia nitrogen from treatment plant effluents, either alone or in combination with other processes. To avoid the large chlorine dosages required when used alone, breakpoint chlorination can be used following biological nitrification to achieve low levels of ammonia in the effluent.

To optimize the performance of this process and to minimize equipment and facility costs, flow equalization is usually required. Also, because of the potential toxicity problems that may develop if chlorinated compounds are discharged to the environment (see Sec. 7-5), it is usually necessary to dechlorinate the effluent. The use of the breakpoint chlorination process for seasonal nitrogen control is considered in Example 11-4.

Example 11-4 Analysis of breakpoint chlorination process used for seasonal control of nitrogen. Estimate the daily required chlorine dosage and the resulting buildup of total dissolved solids when breakpoint chlorination is used for the seasonal control of nitrogen. Assume that the following data apply to this problem:

1. Plant flowrate = 1.0 Mgal/d (3800 m^3/d)
2. Effluent characteristics
   (a) BOD_5 = 20 mg/L
   (b) Suspended solids = 25 mg/L
   (c) $NH_3 - N$ concentration = 23 mg/L
3. Required effluent $NH_3 - N$ concentration = 1.0 mg/L

Solution

1. Estimate the required Cl_2 dosage. Assume that the required mass ratio of chlorine to ammonia is 9:1.

$$lb\ Cl_2/d = (1.0\ Mgal/d)[(23 - 1)\ mg/L](9.0)[8.34\ lb/Mgal \cdot (mg/L)]$$
$$= 1651\ lb/d$$

2. Determine the increment of total dissolved solids added to the wastewater. Using the data reported in Table 7-7, the total dissolved solids increase per mg/L of ammonia consumed is equal to 6.2.

$$Total\ dissolved\ solids\ increment = 6.2\ (23 - 1)\ mg/L = 136\ mg/L$$

Comment. In this example, it was assumed that the acid produced from the break-point reaction would not require the addition of a neutralizing agent such as NaOH (sodium hydroxide). If the addition of NaOH were required, the total dissolved solids increase would have been significantly large. Although breakpoint chlorination can be used to control nitrogen, it may be counterproductive if in the process the treated effluent is rendered unusable for reuse applications because of the buildup of total dissolved solids.

Ion Exchange

Ion exchange is a unit process in which ions of a given species are displaced from an insoluble exchange material by ions of a different species in solution. It may be operated in either a batch or a continuous mode. In a batch process, the resin is stirred with the water to be treated in a reactor until the reaction is complete. The spent resin is removed by settling and subsequently is regenerated and reused. In a continuous process, the exchange material is placed in a bed or a packed column, and the water to be treated is passed through it. Theoretical and operational aspects of the ion exchange process may be found in Ref. 39.

For nitrogen control, the ion typically removed from the waste stream is ammonium, NH_4^+. The ion that the ammonium displaces varies with the nature of the solution used to regenerate the bed. (Regeneration is the process of removing the accumulated NH_4^+ from the ion exchange media so that the media can be reused.)

Although both natural and synthetic ion exchange resins are available, synthetic resins are used more widely because of their durability. Some natural resins (zeolites) have been applied in the removal of ammonia from wastewater. For the removal of ammonium ions from wastewater, clinoptilolite, a naturally occurring zeolite, is one of the best natural exchange resins. In addition to having a greater affinity for ammonium ions than other exchange media, it is relatively inexpensive when compared to synthetic media. One of the novel features of this zeolite is the regeneration system employed. Upon exhaustion, the zeolite is regenerated with lime $Ca(OH)_2$, and the ammonium ion removed from the zeolite is converted to ammonia. A flow diagram for this process is shown in Fig. 11-27.

To make ion exchange economical for advanced wastewater treatment, it would be desirable to use regenerants and restorants that would remove both the inorganic anions and the organic material from the spent resin. Chemical and physical restorants found to be successful in the removal of organic material from resins include sodium

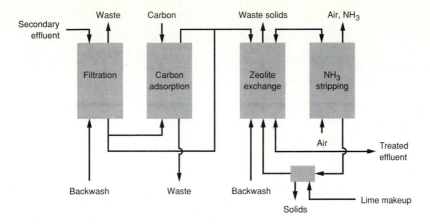

FIGURE 11-27
Flow diagram for the removal of ammonia by zeolite exchange.

hydroxide, hydrochloric acid, methanol, and bentonite [13,14]. Ion exchange has had limited application because of the extensive pretreatment required, concerns about the life of the ion exchange media, and the complex regeneration system required.

11-11 REMOVAL OF PHOSPHORUS BY CHEMICAL ADDITION

The addition of certain chemicals to wastewater produces insoluble or low-solubility salts when combined with phosphate. The principal chemicals used for this purpose are alum, sodium aluminate, ferric chloride or sulfate, and lime. Ferrous sulfate and ferrous chloride, available as byproducts of steel-making operations (pickle liquor), are also used. Polymers have been used effectively in conjunction with alum and lime as flocculant aids. The chemistry of the precipitation reactions involved was described in Sec. 7-1. Factors affecting the choice of chemical for phosphorus removal are reported in Table 11-30. A summary of the pertinent reactions required for

TABLE 11-30
Factors affecting the choice of chemical for phosphorus removal[a]

1. Influent phosphorus level
2. Wastewater suspended solids
3. Alkalinity
4. Chemical cost (including transportation)
5. Reliability of chemical supply
6. Sludge handling facilities
7. Ultimate disposal methods
8. Compatibility with other treatment processes

[a] Adapted in part from Ref. 21.

determining the quantity of sludge when using alum, iron, or lime for the precipitation of phosphorus is given in Table 11-31.

Phosphorus Removal
Using Metal Salts and Polymers

Iron or aluminum salts can be added at a variety of different points in the treatment process (see Fig. 11-28), but, because polyphosphates and organic phosphorus are less easily removed than orthophosphorus, adding aluminum or iron salts after secondary treatment (where organic phosphorus and polyphosphorus are transformed into orthophosphorus) usually results in the best removal. Some additional nitrogen removal occurs because of better settling due to chemical addition, but essentially no ammonia is removed unless chemical additions to primary treatment reduce BOD loadings to the point where nitrification can occur. An increase in total dissolved solids can be expected because of the added chemicals. A number of the important features of adding chemicals at different points in the treatment process are discussed in this section.

Metal Salt Addition to Primary Sedimentation Facilities. When aluminum or iron salts are added to untreated wastewater, they react with the soluble orthophosphate to produce a precipitate. Organic phosphorus and polyphosphate are removed by more complex reactions and by adsorption onto floc particles. The insolubilized phosphorus, as well as considerable quantities of BOD and suspended solids, are removed from the system as primary sludge. Adequate mixing and flocculation are necessarily done upstream of primary facilities, whether separate basins are provided or existing facilities are modified to provide these functions. Polymer addition may be required to aid in settling. In low-alkalinity waters, the addition of a base is sometimes necessary to keep pH in the 5 to 7 range. Alum and ferric chloride are generally applied in a

TABLE 11-31
Summary of pertinent reactions required to determine quantities of sludge produced during the precipitation of phosphorus with lime, alum, and iron Fe (III)

Reaction	Chemical species in sludge
Lime	
1. $5 Ca^{+2} + 3 PO_4^{-3} + OH^- \leftrightarrow Ca_5(PO_4)_3(OH)$	$Ca_5 (PO_4)_3(OH)$
2. $Mg^{+2} + 2 OH^- \leftrightarrow Mg (OH)_2$	$Mg(OH)_2$
3. $Ca^{+2} + CO_3^{-2} \leftrightarrow CaCO_3$	$CaCO_3$
Alum	
1. $Al^{+3} + PO_4^{-3} \leftrightarrow AlPO_4$	$AlPO_4$
2. $Al^{+3} + 3 OH^- \leftrightarrow Al(OH)_3$	$Al(OH)_3$
Iron Fe(III)	
1. $Fe^{+3} + PO_4^{-3} \leftrightarrow FePO_4$	$FePO_4$
2. $Fe^{+3} + 3 OH^- \leftrightarrow Fe(OH)_3$	$Fe (OH)_3$

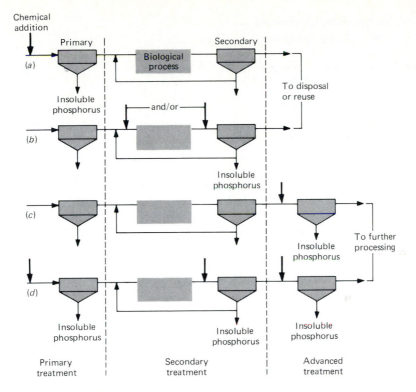

FIGURE 11-28
Alternative points of chemical addition for phosphorus removal: (*a*) before primary sedimentation, (*b*) before and/or following biological treatment, (*c*) following secondary treatment, and (*d*) at several locations in a process (known as "split treatment").

molar ratio in the range of a 1 to 3 metal ions to 1 phosphorus ion (see Table 11-32 and Fig. 11-29). The exact application rate is determined by onsite testing and varies with the characteristics of the wastewater and the desired phosphorus removal.

Metal Salt Addition to Secondary Treatment. Metal salts can be added to the untreated wastewater in the activated-sludge aeration tank or the final clarifier influent channel. In trickling-filter systems, the salts are added to the untreated wastewater or to the filter effluent. Multipoint additions have also been used. Phosphorus is removed from the liquid phase through a combination of precipitation, adsorption, exchange, and agglomeration and removed from the process with either the primary or secondary sludges, or both. Theoretically, the minimum solubility of $AlPO_4$ occurs at pH 6.3, and that of $FePO_4$ occurs at pH 5.3; however, practical applications have yielded good phosphorus removal anywhere in the range of pH 5.5 to 7.0, which is compatible with most biological treatment processes.

The use of ferrous salts is limited because they produce low phosphorus levels only at high pH values. In low-alkalinity waters, either sodium aluminate and alum or ferric plus lime, or both, can be used to maintain the pH higher than 5.5. Improved

TABLE 11-32
Typical alum dosage requirements for various levels of phosphorus removal[a]

Phosphorus reduction,%	Mole ratio, Al:P	
	Range	Typical
75	1.25:1–1.5:1	1.4:1
85	1.6:1–1.9:1	1.7:1
95	2.1:1–2.6:1	2.3:1

[a] Developed in part from Ref. 34.

settling and lower effluent BOD result from chemical addition, particularly if polymer is also added to the final clarifier. Dosages generally fall in the range of a 1 to 3 metal ion-phosphorus molar ratio.

Metal Salt and Polymer Addition to Secondary Clarifiers. In certain cases, such as trickling filtration and extended aeration activated-sludge processes, solids may not flocculate and settle well in the secondary clarifier. This settling problem may become acute in overloaded plants. The addition of aluminum or iron salts will cause the precipitation of metallic hydroxides or phosphates, or both. Aluminum and iron salts, along with certain organic polymers, can also be used to coagulate colloidal particles and to improve removals on filters. The resultant coagulated colloids and precipitates will settle readily in the secondary clarifier, reducing the suspended solids in the effluent and effecting phosphorus removal. Dosages of aluminum and iron

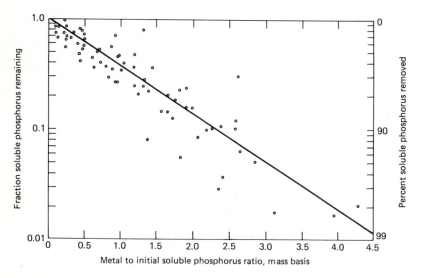

FIGURE 11-29
Soluble phosphorus removal by ferric chloride addition [34].

salts usually fall in the range of 1 to 3 metal ion/phosphorus on a molar ratio basis if the residual phosphorus in the secondary effluent is greater than 0.5 mg/L. To achieve phosphorus levels below 0.5 mg/L, significantly higher metal salt dosages and filtration will be required.

Polymers may be added (1) to the mixing zone of a highly mixed or internally recirculated clarifier, (2) preceding a static or dynamic mixer, or (3) to an aerated channel. Although mixing times of 10 to 30 seconds have been used for polymers, shorter mixing times are favored. Polymers should not be subjected to insufficient or excessive mixing because the process efficiency will diminish, resulting in poor settling and thickening characteristics.

Phosphorus Removal Using Lime

The use of lime for phosphorus removal is declining because of (1) the substantial increase in the mass of sludge to be handled compared to metal salts and (2) the operation and maintenance problems associated with the handling, storage, and feeding of lime [36]. When lime is used, the principal variables controlling the dosage are the degree of removal required and the alkalinity of the wastewater. The operating dosage must usually be determined by onsite testing. Lime has been used customarily either as a precipitant in the primary sedimentation tanks or following secondary treatment clarification.

Although lime recalcination lowers chemical costs, it is only a feasible alternative for large plants. Where a lime recovery system is required for a cost-effective operation, the system includes a thermal regeneration facility, which converts the calcium carbonate in the sludge to lime by heating at 1800°F (980°C). The carbon dioxide from this process or other onsite stack gas (containing 10 to 15 percent carbon dioxide) is generally used as the source of recarbonation for pH adjustment of the wastewater.

Lime Addition to Primary Sedimentation Tanks. Both low- and high-lime treatment can be used to precipitate a portion of the phosphorus (usually about 65 to 80 percent). When lime is used, both the calcium and the hydroxide react with the orthophosphorus to form an insoluble hydroxyapatite $[Ca_5(PO_4)_3OH]$. A residual phosphorus level of 1.0 mg/L can be achieved with the addition of effluent filtration facilities to which chemicals can be added. In the high-lime system, sufficient lime is added to raise the pH to about 11 (see Fig. 11-25). After precipitation, the effluent must be recarbonated before biological treatment. In activated-sludge systems, the pH of the primary effluent should not exceed 9.5 or 10; higher pH values can result in biological process upsets. In the trickling-filter process, the carbon dioxide generated during treatment is usually sufficient to lower the pH without recarbonation. The dosage for low-lime treatment is usually in the range of 75 to 250 mg/L as $Ca(OH)_2$ at pH values of 8.5 to 9.5. In low-lime systems, however, the conditions required for precipitation are more specialized; the Ca^{+2}/Mg^{+2} mol ratio is $\leq 5/1$ [27].

The additional BOD and suspended-solids removals afforded by chemical addition to primary treatment may also solve overloading problems on downstream biological systems, or may allow seasonal or year-round nitrification, depending on biological system designs. The BOD removal in the primary sedimentation operation is in the order of 50 to 60 percent at a pH of 9.5 [42]. The amount of primary sludge will also increase significantly. The computational procedures involved in estimating the quantity of sludge resulting from the chemical precipitation of phosphorus with lime are illustrated in Example 11-5.

Example 11-5 Estimation of sludge volume from the chemical precipitation of phosphorus with lime in a primary sedimentation tank. Estimate the mass and volume of sludge produced in a primary sedimentation tank from the precipitation of phosphorus with lime. Assume that 60 percent of the suspended solids is removed without the addition of lime and that the addition of 400 mg/L of $Ca(OH)_2$ increases the removal of suspended solids to 85 percent. Assume that the flowrate and suspended solids are the same as Example 9-3 and that the following data apply:

1. Wastewater flowrate = 1.0 Mgal/d
2. Wastewater suspended solids = 220 mg/L
3. Wastewater volatile suspended solids = 150 mg/L
4. Wastewater PO_4^{-3} as P = 10 mg/L
5. Wastewater total hardness as $CaCO_3$ = 241.3 mg/L
6. Wastewater Ca^{+2} = 80 mg/L
7. Wastewater Mg^{+2} = 10 mg/L
8. Effluent PO_4^{-3} as P = 0.5 mg/L
9. Effluent Ca = 60 mg/L
10. Effluent Mg = 0 mg/L

Solution

1. Compute the mass and volume of solids removed without chemicals, assuming that the sludge contains 94 percent moisture and has a gravity of 1.03.

   (*a*) Determine the mass of suspended solids removed.

   $$M_{ss} = 1100 \text{ lb/d (see Example 9-3)}$$

   (*b*) Determine the volume of sludge produced at a specific gravity of 1.03 and a moisture content of 94 percent.

   $$V_s = 285 \text{ ft}^3/\text{d (see Example 9-3)}$$

2. Using the equations summarized in Table 11-31, determine the mass of $Ca_5(PO_4)_3OH$, $Mg(OH)_2$, and $CaCO_3$ produced from the addition of 400 mg/L of lime.

   (*a*) Determine the mass of $Ca_5(PO_4)_3OH$ formed.

   i. Determine the moles of P removed.

   $$\text{mol P removed} = \frac{10 \text{ mg/L} - 0.5 \text{ mg/L}}{30.97 \text{ g/mol} \times 10^3 \text{ mg/g}}$$
   $$= 0.307 \times 10^{-3} \text{ mol/L}$$

   ii. Determine the moles of $Ca_5(PO_4)_3OH$ formed.

$$\text{mol } Ca_5(PO_4)_3OH \text{ formed} = 1/3 \times 0.307 \times 10^{-3} \text{ mol/L}$$
$$= 0.102 \times 10^{-3} \text{ mol/L}$$

iii. Determine the mass of $Ca_5(PO_4)_3OH$ formed.

$$\text{Mass } Ca_5 (PO_4)_3OH = 0.102 \times 10^{-3} \text{ mol/L} \times 502 \text{ g/mol} \times 103 \text{ mg/g}$$
$$= 51.3 \text{ mg/L}$$

(b) Determine the mass of $Mg(OH)_2$ formed.

i. Determine the moles of Mg removed.

$$\text{mol } Mg^{+2} \text{ removed} = \frac{10 \text{ mg/L}}{24.31 \text{ g/mol} \times 10^3 \text{ mg/g}}$$
$$= 0.411 \times 10^{-3} \text{ mol/L}$$

ii. Determine the mass of $Mg(OH)^2$ formed.

$$\text{mol } Mg(OH)_2 = 0.411 \times 10^{-3} \text{ mol/L} \times 58.3 \text{ g/mol} \times 10^3 \text{ mg/g}$$
$$= 24.0 \text{ mg/L}$$

(c) Determine the mass of $CaCO_3$ formed.

i. Determine the mass of Ca^{+2} in $Ca_5(PO_4)_3(OH)$.

$$\text{Mass Ca in } Ca_5(PO_4)_3(OH) = 5(40 \text{ g/mol}) \times 0.102 \times 10^{-3} \text{ mol/L} \times 10^3 \text{ mg/g}$$
$$= 20.4 \text{ mg/L}$$

ii. Determine the mass of Ca added in the original dosage.

$$\text{Mass Ca in } Ca(OH)_2 = \frac{40 \text{ g/mol} \times 400 \text{ mg/L}}{74 \text{ g/mol}}$$
$$= 216.2 \text{ mg/L}$$

iii. Determine the mass of Ca present as $CaCO_3$.

$$Ca \text{ in } CaCO_3 = Ca \text{ in } Ca(HO)_2 + Ca \text{ in influent wastewater}$$
$$- Ca \text{ in } Ca_5(PO_4)_3OH - Ca \text{ in effluent wastewater}$$
$$= 216.2 + 80 - 20.4 - 60$$
$$= 215.8 \text{ mg/L}$$

iv. Determine the mass of $CaCO_3$.

$$\text{Mass } CaCO_3 = \frac{100 \text{ g/mol} \times 215.8 \text{ mg/L}}{40 \text{ g/mol}}$$
$$= 540 \text{ mg/L}$$

3. Determine the total mass of solids removed as a result of the lime dosage.

(a) Suspended solids in wastewater:

$$M_{ss} = 0.85 \ (220 \text{ mg/L}) \ (1.0 \text{ Mgal/d}) \ [8.34 \text{ lb/Mgal} \cdot (\text{mg/L})] = 1560 \text{ lb/d}$$

(b) Chemical solids:

$$M_{Ca5(PO4)3(OH)} = (51.2 \text{ mg/L}) \ (1.0 \text{ Mgal/d}) \ [8.34 \text{ lb/Mgal} \cdot (\text{mg/L})] = 427 \text{ lb/d}$$

$$M_{Mg(OH)2} = (24 \text{ mg/L}) \ (1.0 \text{ Mgal/d}) \ [8.34 \text{ lb/Mgal} \cdot (\text{mg/L})] = 200 \text{ lb/d}$$

$$M_{CaCO3} = (540 \text{ mg/L}) \ (1.0 \text{ Mgal/d}) \ [8.34 \text{ lb/Mgal} \cdot (\text{mg/L})] = 4504 \text{ lb/d}$$

(c) Total mass of solids removed:

$$M_T = (1560 + 427 + 200 + 4504) \text{ lb/d}$$
$$= 6691 \text{ lb/d}$$

4. Determine the total volume of sludge resulting from chemical precipitation, assuming that the sludge has a specific gravity of 1.07 and a moisture content of 92.5 percent (see Chap. 12).

$$V_s = \frac{6691 \text{ lb/d}}{1.07 \times 62.4 \text{ lb/ft}^3 \, (0.075)} = 1336 \text{ ft}^3/\text{d}$$

5. Prepare a summary table of sludge masses and volumes without and with chemical precipitation.

	Sludge	
Treatment	Mass, lb/d	Volume, ft³/d
Without chemical precipitation	1,100	285
With chemical precipitation	6,691	1,336

Comment. The sludge disposal problem associated with high-lime treatment for phosphorus removal as compared to biological phosphorus removal is illustrated in this example.

Lime Addition Following Secondary Treatment. Lime can be added to the waste stream after biological treatment to reduce the level of phosphorus and suspended solids. Single-stage and two-stage process flow diagrams are shown in Fig. 11-30. In the first-stage clarifier of the two-stage process (see Fig. 11-30b), sufficient lime is added to raise the pH above 11 to precipitate the soluble phosphorus as basic calcium phosphate (apatite). The calcium carbonate precipitate formed in the process acts as a coagulant for suspended-solids removal. The excess soluble calcium is removed in the second-stage clarifier as a calcium carbonate precipitate by adding carbon dioxide gas to reduce the pH to about 10. Generally, there is a second injection of carbon dioxide to the second-stage effluent to reduce the formation of scale. To remove the residual levels of suspended solids and phosphorus, the secondary clarifier effluent is passed through a multimedia filter. Care should be taken to limit excess calcium in the filter feed to ensure that cementing of the filter media will not occur.

Comparison of Chemical Phosphorus Removal Processes

The advantages and disadvantages of the removal of phosphorus by the addition of chemicals at various points in a treatment system are summarized in Table 11-33. It is recommended that each alternative point of application be evaluated carefully.

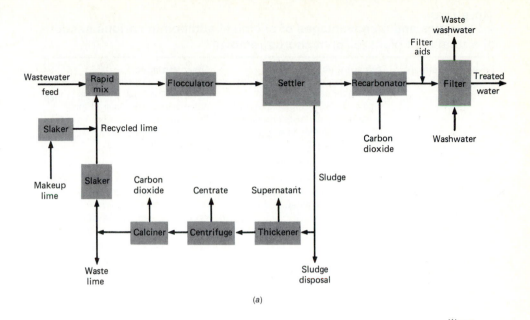

(a)

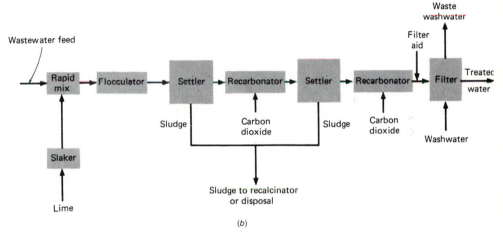

(b)

FIGURE 11-30
Typical lime treatment process flow diagrams for phosphorus removal: (a) single-stage system and (b) two-stage system.

11-12 REMOVAL OF TOXIC COMPOUNDS AND REFRACTORY ORGANICS

As stated earlier in this chapter, increasing attention has been given in recent years to the definition of toxic substances, their effects on public health and the environment, and applicable treatment methods for their removal. Special attention is being given the priority pollutants and refractory organic compounds described in Chap. 3. Refractory

TABLE 11-33
Advantages and disadvantages of chemical addition in various sections of a treatment plant for phosphorus removal[a]

Level of treatment	Advantages	Disadvantages
Primary	Applicable to most plants; increased BOD and suspended-solids removal; lowest degree of metal leakage; lime recovery demonstrated.	Least efficient use of metal; polymer may be required for flocculation; sludge more difficult to dewater than primary sludge.
Secondary	Lowest cost; lower chemical dosage than primary; improved stability of activated sludge; polymer not required.	Overdose of metal may cause low pH toxicity; with low-alkalinity wastewaters, a pH control system may be necessary; cannot use lime because of excessive pH; inert solids added to activated-sludge mixed liquor, reducing the percentage of volatile solids.
Advanced	Lowest phosphorus effluent; most efficient metal use; lime recovery demonstrated.	Highest capital cost; highest metal leakage.

[a] Adapted from Ref. 21.

organics are compounds resistant to microbial degradation in conventional biological treatment processes and the natural environment. Processes commonly used for the treatment of toxic compounds and the removal of refractory organics are reviewed in this section. Removal of toxic substances by land application is discussed in Chap. 13.

Treatment Methods Used for the Removal of Toxic Compounds

Many treatment methods can be used for the treatment of toxic compounds. Because of the complex nature of toxicity, the treatment methods must consider the specific characteristics of the wastewater and the nature of the toxic compounds. Treatment processes used to remove some of the specific compounds or groups of compounds are summarized in Table 11-34. Three of the processes, activated-carbon adsorption, activated-sludge powdered activated carbon, and chemical oxidation are discussed in this section. Air stripping is discussed in Chap. 6, chemical coagulation is discussed in Chaps. 7 and 9, and conventional biological treatment is discussed in Chaps. 8 and 10. Carbon adsorption for the removal of volatile organic compounds is also discussed in Chap. 9. For additional information on the removal performance of the various processes, including conventional biological processes, on specific toxic compounds, Refs. 16, 17, and 40 may be consulted. In process selection, pilot plant testing

TABLE 11-34
Treatment processes used for the removal of toxic compounds[a]

Process	Removal application
Activated-carbon adsorption	Natural and synthetic organic compounds including VOCs; pesticides; PCBs; heavy metals
Activated-sludge-powdered activated carbon	Heavy metals; ammonia; selected refractory priority pollutants
Air stripping	Volatile organic compounds (VOCs) and ammonia
Chemical coagulation, sedimentation, and filtration	Heavy metals and PCBs
Chemical oxidation	Ammonia; refractory and toxic halogenated aliphatic and aromatic compounds
Conventional biological treatment (activated-sludge, trickling filter)	Phenols; PCBs; selected hydrogenated hydrocarbons

[a] Adapted from Refs. 13 and 40.

is recommended for the development of treatment performance data and design criteria.

Carbon Adsorption

Carbon adsorption, discussed in Sec. 7-2, is an advanced wastewater treatment method used for the removal of the refractory organic compounds as well as residual amounts of inorganic compounds such as nitrogen, sulfides, and heavy metals. Granular-medium filters are commonly used upstream of the activated-carbon contactors to remove the soluble organics associated with the suspended solids present in secondary effluent. High influent suspended-solids concentrations (more than 20 mg/L) will form deposits on the carbon granules resulting in pressure loss, flow channeling or blockages, and loss of adsorption capacity. If soluble organic removal is not maintained at a high level, more frequent regeneration of the carbon may be required. Lack of consistency in pH, temperature, and flowrate may also affect performance of carbon contactors [41].

Both granular and powdered carbon are used and appear to have a low adsorption affinity for low molecular weight polar organic species. If biological activity is low in the carbon contactor or in other biological unit processes, these species are difficult to remove with activated carbon. Under normal conditions, after treatment with carbon, the effluent BOD ranges from 2 to 7 mg/L, and the effluent COD ranges from 10 to 20 mg/L. Under optimum conditions, it appears that the effluent COD can be reduced to about 10 mg/L.

Types of Carbon Contactors. Several types of activated-carbon contactors are used for advanced wastewater treatment. Typical systems may be either pressure or gravity type and may be upflow-countercurrent type with packed or expanded carbon beds or may be used as upflow or downflow fixed-bed units with two or three columns in series. Typical schematic diagrams are shown in Fig. 11-31.

Upflow columns. Upflow columns are arranged so that the liquid moves from the base of the column upward. As the carbon adsorbs organics, the apparent density of the carbon particles increases and encourages migration of the heavier or more spent carbon downward. Upflow columns may have more carbon fines in the effluent than downflow columns because upflow tends to expand, not compress, the carbon. Bed expansion creates fines (because the carbon particles collide) and allows the fines to escape through passageways created by the expanded bed.

Downflow columns. Downflow columns usually consist of two or three columns operated in series. The advantage of a downflow design is that adsorption of organics and filtration of suspended solids are accomplished in a single step. Downflow filters may require more frequent backwashing because of the accumulation of suspended material on the surface of the contactor. Plugging of the carbon pores may require premature removal of the carbon for regeneration, thereby decreasing the useful life of the carbon. Sand and gravel resting on a filter block form the supporting media for downflow contactors.

Fixed beds. In fixed-bed contactors, the carbon remains fixed as in the downflow mode. Fixed beds remove particulates and require backwashing to dispose of the accumulated particulate matter. Usually fixed beds employ downward flow to lessen the chance of accumulating particulate material in the bottom of the bed where the particulate material would be difficult to remove by backwashing.

Expanded beds. The upflow column led to the development of the moving- or pulsed-bed. In this method, wastewater flows upward through a descending fixed bed of carbon. When the adsorptive capacity of the carbon at the bottom of the carbon

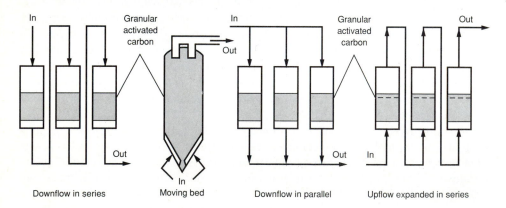

FIGURE 11-31
Types of activated-carbon contactors (from Calgon Carbon Corp.).

is exhausted, the bottom portion of carbon is removed, and an equivalent amount of regenerated or virgin carbon is added to the top of the column. Because this type of contactor cannot be backwashed, residual organic content in the contactor influent should be very low to avoid plugging.

Sizing of Carbon Contactors. The sizing of carbon contactors is based on four factors: contact time, hydraulic loading rate, carbon depth, and the number of contactors. Typical design information for the first three factors is presented in Table 11-35. A minimum of two parallel carbon contactors is recommended for design. Multiple units permit one or more units to remain in operation while one unit is taken out of service for removal and regeneration of spent carbon or for maintenance. For additional information on the prediction of performance of carbon contactors, Ref. 13 may be consulted.

Activated-Sludge-Powdered Activated-Carbon Treatment

A proprietary process, "PACT," combines the use of powdered activated carbon with the activated-sludge process (see Fig. 11-32). In this process, when the activated carbon is added directly to the aeration tank, biological oxidation and physical adsorption occur simultaneously. A feature of this process is that it can be integrated into existing activated-sludge systems at nominal capital cost. The addition of powdered activated carbon has several process advantages including (1) system stability during shock loads, (2) reduction of refractory priority pollutants, (3) color and ammonia

TABLE 11-35
Typical design information for activated-carbon contactors used for treating secondary effluent[a]

Design parameter	Units	Range	Typical
Contact time for an effluent COD of			
10–20 mg/L	min	15–20	
5–15 mg/L	min	30–35	
Hydraulic loading rate			
Upflow columns	$gal/ft^2 \cdot min$	4–10	
Downflow columns	$gal/ft^2 \cdot min$	3–5	
Carbon depth	ft	10–40[b]	15–20[b]
Operating pressure	$lb_f/in^2 \cdot ft$ of depth		<1

[a] Adapted in part from Ref. 41.
[b] A freeboard allowance of 10 to 50 percent should be added for backwash or expanded bed operation.
Note: $gal/ft^2 \cdot min \times 40.7458 = L/m^2 \cdot min$
$ft \times 0.3048 = m$
$lb_f/in^2 \times 6.8948 = kPa$

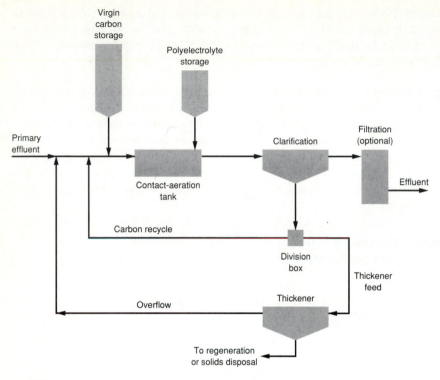

FIGURE 11-32
Flow diagram of powdered activated-carbon activated-sludge process (from Zimpro).

removal, and (4) improved sludge settleability. In some industrial waste applications where nitrification is inhibited by toxic organics, the application of powdered activated carbon may reduce or limit this inhibition.

The dosage of powdered activated carbon and the mixed-liquor-powdered activated-carbon suspended-solids concentration are related to the sludge age as follows [13]:

$$X_p = \frac{X_i \, \theta_c}{\theta} \tag{11-20}$$

where X_p = equilibrium powdered activated-carbon MLSS content, mg/L
X_i = powdered activated-carbon dosage, mg/L
θ_c = solids retention time, d
θ = hydraulic retention time, d

Carbon dosages range typically from 20 to 200 mg/L. With higher sludge ages, the organic removal per unit of carbon is enhanced, thereby improving the process efficiency. Reasons cited for this phenomena include (1) additional biodegradation due to decreased toxicity, (2) degradation of normally nondegradable substances due to increased exposure time to the biomass through adsorption on the carbon, and (3) replacement of low molecular weight compounds with high molecular weight compounds, resulting in improved adsorption efficiency and lower toxicity [13].

Chemical Oxidation

In advanced wastewater treatment applications, chemical oxidation can be used to remove ammonia, to reduce the concentration of residual organics, and to reduce the bacterial and viral content of wastewaters. The use of chlorine for the oxidation of ammonia is discussed in Sec. 11-10. Because chlorine forms trihalomethanes (THMs) when added to wastewater, alternatives to chlorine have been investigated where THMs are of principal concern in drinking water supplies. Alternative oxidants include chlorine dioxide and ozone. When these chemicals are used for this purpose, disinfection of the wastewater is usually an added benefit. A further benefit of using ozone is the removal of color.

Typical chemical dosages for both chlorine and ozone for the oxidation of the organics in wastewater are reported in Table 11-36. The dosages increase with the degree of treatment, which is reasonable when it is considered that the organic compounds that remain after biological treatment are typically composed of low molecular weight polar organic compounds and complex organic compounds built around the benzene ring structure.

It is recommended that pilot plant studies be conducted when either chlorine, chlorine dioxide, or ozone is to be used for the oxidation of organics. Because ozone can be generated conveniently at treatment plants that use the high-purity oxygen activated-sludge process, it is anticipated that its use may become more common at these locations in the future.

TABLE 11-36
Typical chemical dosages for the oxidation of organics in wastewater[a]

Chemical	Use	Dosage, lb/lb destroyed	
		Range	Typical
Chlorine	BOD$_5$ reduction	0.5–2.5	1.75[b]
		1.0–3.0	2.0[c]
Ozone	COD reduction	2.0–4.0	3.0[b]
		3.0–8.0	6.0[c]

[a] Derived in part from Ref. 44.

[b] For settled wastewater.

[c] In secondary effluent.

11-13 REMOVAL OF DISSOLVED INORGANIC SUBSTANCES

As reported in Table 11-2, a number of different unit operations and processes have been investigated in various advanced wastewater treatment applications. Although many of them are technically feasible, other factors, such as cost, operational requirements, and aesthetic considerations, are not favorable in some cases. Nevertheless, it is important that environmental engineers be familiar with the more important operations and processes so that in any given situation they can consider all treatment possibilities. These important operations and processes are chemical precipitation, ion exchange, ultrafiltration, reverse osmosis, and electrodialysis.

Chemical Precipitation

As discussed in Chaps. 7 and 9 and in Sec. 11-11, precipitation of phosphorus in wastewater is usually accomplished by the addition of coagulants such as alum, lime or iron salts, and organic polymers. Coincidentally, the addition of these chemicals for the removal of phosphorus removes various inorganic ions, principally some of the heavy metals. Where both industrial and domestic wastes are treated together, it may be necessary to add chemicals to the primary settling facilities, especially if onsite pretreatment measures prove to be ineffective. When chemical precipitation is used, anaerobic digestion for sludge stabilization may not be possible because of the toxicity of the precipitated heavy metals. As noted in Chap. 7, one of the disadvantages of chemical precipitation is that it usually results in a net increase in the total dissolved solids of the wastewater being treated. Other disadvantages include the large amount of sludge requiring treatment, which, in turn, may contain toxic compounds that may be difficult to treat and dispose of.

Ion Exchange

Ion exchange is a unit process by which ions of a given species are displaced from an insoluble exchange material by ions of a different species in solution. The most widespread use of this process is in domestic water softening, where sodium ions from a cationic exchange resin replace the calcium and magnesium ions in the treated water, thus reducing the hardness. For the reduction of the total dissolved solids, both anionic and cationic exchange resins must be used. The wastewater is first passed through a cation exchanger where the positively charged ions are replaced by hydrogen ions. The cation exchanger effluent is then passed over an anionic exchange resin where the anions are replaced by hydroxide ions. Thus, the dissolved solids are replaced by hydrogen and hydroxide ions that react to form water molecules.

Ion exchangers are usually of the downflow, packed-bed column type. Wastewater enters the top of the column under pressure, passes downward through the resin bed, and is removed at the bottom. When the resin capacity is exhausted, the column is backwashed to remove trapped solids and then regenerated. The cationic exchange resin is regenerated with a strong acid such as sulfuric or hydrochloric. Sodium hydroxide is the commonly used regenerant for the anion exchange resin.

Ion exchange demineralization can take place in separate exchange columns arranged in series, or both resins can be mixed in a single reactor. Wastewater application rates range from 5 to 10 gal/ft^2 · min (0.20 to 0.40 m^3/m^2 · min). Typical bed depths are 2 to 6.5 ft (0.75 to 2.0 m).

High concentrations of influent suspended solids can plug the ion exchange beds, causing high headlosses and inefficient operation. Resin binding can be caused by residual organics found in biological treatment effluents. Some form of chemical treatment and clarification is required before ion exchange demineralization. Not all dissolved ions are removed equally; each ion exchange resin is characterized by a selectivity series, and some dissolved ions at the end of the series are only partially removed.

In reuse applications, treatment of a portion of the wastewater by ion exchange, followed by blending with wastewater not treated by ion exchange, would possibly reduce the dissolved solids to acceptable levels.

Ultrafiltration

Ultrafiltration systems are pressure-driven membrane operations that use porous membranes for the removal of dissolved and colloidal material. These systems differ from reverse osmosis systems by the relatively low driving pressures, usually under 150 lb$_f$/in^2 (1034 kN/m^2). Ultrafiltration is normally used to remove colloidal material and large molecules with molecular weights in excess of 5000. Applications for ultrafiltration include removal of oil from aqueous streams and the removal of turbidity from color colloids. Recent research indicates that effluent from ultrafiltration using spiral wound elements is suitable as a feed source for reverse osmosis. A system flow diagram using ultrafiltration for pretreatment for reverse osmosis is illustrated in Fig. 11-33 [28]. Ultrafiltration has also been suggested as a unit operation for the removal of phosphorus [39].

Reverse Osmosis (Hyperfiltration)

Reverse osmosis is a process in which water is separated from dissolved salts in solution by filtering through a semipermeable membrane at a pressure greater than the osmotic pressure caused by the dissolved salts in the wastewater (see Fig. 11-34). With existing membranes and equipment, operating pressures vary from atmospheric to 1000 lb$_f$/in^2 (6900 kN/m^2). Reverse osmosis has the advantage of removing dissolved organics that are less selectively removed by other demineralization techniques. The primary limitations of reverse osmosis are its high cost and the limited operating experience in the treatment of domestic wastewaters.

The basic components of a reverse osmosis unit are the membrane, a membrane support structure, a containing vessel, and a high-pressure pump. Cellulose acetate and nylon have been used as membrane materials. Four types of membrane support

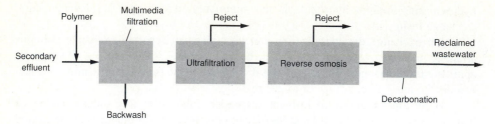

FIGURE 11-33
Flow diagram for ultrafiltration and reverse osmosis used for municipal wastewater reclamation [28].

configurations have been used: spiral wound, tubular, and hollow fiber configurations. The spiral wound configuration is the most successful for use with domestic wastewater effluents [28]. Reverse osmosis units can be arranged either in parallel to provide adequate hydraulic capacity or in series to effect the desired degree of demineralization.

A very high-quality feed is required for efficient operation of a reverse osmosis unit. Membrane elements in the reverse osmosis unit can be fouled by colloidal matter in the feed stream. Pretreatment of a secondary effluent by chemical clarification and multimedia filtration or by multimedia filtration and ultrafiltration is usually necessary (see Fig. 11-33). Also, the removal of iron and manganese is sometimes necessary to

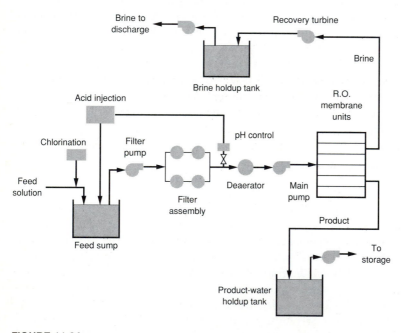

FIGURE 11-34
Typical flow diagram for a single-stage reverse osmosis process [39].

decrease scaling potential. The pH of the feed should be adjusted to a range of 4.0 to 7.5 to inhibit scale formation. Regular chemical cleaning of the membrane elements (about once a month) is necessary to restore the membrane flux [28].

Electrodialysis

In the electrodialysis process, ionic components of a solution are separated through the use of semipermeable ion-selective membranes. Application of an electrical potential between the two electrodes causes an electric current to pass through the solution, which, in turn causes a migration of cations toward the negative electrode and a migration of anions toward the positive electrode. Because of the alternate spacing of cation- and anion-permeable membranes, cells of concentrated and dilute salts are formed.

Wastewater is pumped through the membranes, which are separated by spacers and assembled into stacks. The wastewater is usually retained for about 10 to 20 s in a single stack or stage. Dissolved solids removals vary with the (1) wastewater temperature, (2) amounts of electrical current passed, (3) type and amount of ions, (4) permselectivity of the membrane, (5) fouling and scaling potential of the wastewater, (6) wastewater flowrates, and (7) number and configuration of stages.

This process may be operated in either a continuous or a batch mode. The units can be arranged either in parallel to provide the necessary hydraulic capacity or in series to effect the desired degree of demineralization. Makeup water, usually about 10 percent of the feed volume, is required to wash the membranes continuously. A portion of the concentrate stream is recycled to maintain nearly equal flowrates and pressures on both sides of each membrane. Sulfuric acid is fed to the concentrate stream to maintain a low pH and thus minimize scaling.

Problems associated with the electrodialysis process for wastewater renovation include chemical precipitation of salts with low solubility on the membrane surface and clogging of the membrane by the residual colloidal organic matter in wastewater treatment plant effluents. To reduce membrane fouling, activated-carbon pretreatment, possibly preceded by chemical precipitation and some form of multimedia filtration, may be necessary.

DISCUSSION TOPICS AND PROBLEMS

11-1. A wastewater contains 10 mg/L of ammonia nitrogen and no organic carbon. If the plant flowrate is 2.5 Mgal/d, estimate the methanol requirement and cell production in pounds per day for the complete bacterial assimilation of ammonia.

11-2. Estimate the methanol requirement and cell production in Prob. 11-1 in SI units using a plant flowrate of 10,000 m³/d.

11-3. A conventional activated-sludge plant treating 1 Mgal/d of wastewater is to be operated to produce a nitrified effluent. How would this be done? Assuming that a nitrified effluent is produced containing 15 mg/L of nitrate nitrogen, 1.5 mg/L of nitrite nitrogen, and 2.0 mg/L of dissolved oxygen, compute the methanol requirement for denitrification. How will the activated-sludge effluent BOD affect the methanol requirement?

11-4. Compute the methanol requirements in Prob. 11-3 in SI units using a plant flowrate of 4000 m³/d.

11-5. A wastewater contains 40 mg/L of nitrate nitrogen (177 mg/L as NO_3^-) and has a flowrate of 2.5 Mgal/d. Effluent requirements have been set at 2 mg/L total nitrogen. Using a mean cell-residence time of 15 d and a mixed-liquor concentration of 1500 mg/L, determine the volume of a complete-mix reactor that will be required to provide the necessary treatment. Use the kinetic coefficients reported in Table 11-20. Also determine the rate of cell production and the methanol utilization rate, assuming that the influent DO is equal to 5 mg/L. Assume that the final clarifier will produce an effluent with 10 mg/L suspended solids.

11-6. Determine the volume of a complete-mix reactor, the rate of cell production, and methanol utilization rate in Prob. 11-5 in SI units using a plant flowrate of 10,000 m³/d.

11-7. Design an aerobic-anoxic process for nitrification/denitrification of wastewater. After preliminary treatment, 5 Mgal/d of wastewater containing 150 mg/L of BOD_5 and 30 mg/L of ammonia nitrogen is to be treated. The following conditions are to be used in the design: denitrification rate = 0.1 mg NO_3-N/mg VSS · d at 20°C and MLVSS = 3500 mg/L. Calculate the anoxic volume requirement for denitrification and the BOD removal in the anoxic and aerobic zones.

11-8. Design the aerobic-anoxic process in SI units using the conditions described in Prob. 11-7 and a plant flowrate of 20,000 m³/d.

11-9. Based on a review of at least four articles dealing with the use of trickling filters for nitrification, recommend an appropriate loading factor or design approach to achieve complete nitrification using a trickling filter following an activated-sludge process. Assume that the activated-sludge process is designed to remove only the carbonaceous organic matter (after secondary treatment soluble BOD_5 = 2.0 mg/L and NH_3 as N = 40 mg/L). Cite the literature references reviewed.

11-10. What $V_{aerobic}$ is required to satisfy the anoxic residence time requirements in Example 11-3? Show your calculation methodology.

11-11. Prepare a distribution diagram of the relative amounts of NH_3 and NH_4^+ (expressed as a percent) that would be present in a water sample at 25°C as a function of pH.

11-12. Phosphorus is to be removed from a secondary effluent. The plant discharge requirements have been set at 1.0 mg/L. If the soluble phosphorus in the effluent is equal to 10 mg/L, estimate the alum dosage that will be required to achieve the desired degree of removal. If the concentration of the settled alum sludge is 6 percent and the specific gravity is 1.05, estimate the volume of sludge that must be disposed of per day if the plant flowrate is 10 Mgal/d.

11-13. Using the same plant design data as Prob. 11-12 and a primary effluent BOD_5 of 130 mg/L, compute and compare the reactor sizes using the A^2/O and Bardenpho biological phosphorus removal processes. Select one process and justify your selection.

11-14. Based on a review of articles on sequencing batch reactors in actual operation, prepare a summary of design data for three operating plants, at least one of which removes either nitrogen or phosphorus. Summarize the operating performance of each plant and cite any operating problems noted. Discuss how the plants might be modified to achieve both nitrogen and phosphorus removal.

11-15. A wastewater is to be treated with activated carbon to remove residual COD. The following data were obtained from a laboratory adsorption study in which 1 g of activated carbon was added to a beaker containing 1 L of wastewater at selected COD values. Using these data, determine the more suitable isotherm (Langmuir or Fruendlich) to describe the data.

Initial COD, mg/L	Equilibrium COD, mg/L
140	10
250	30
300	50
340	70
370	90
400	110
450	150

11-16. Using the results from Prob. 11-15, determine the amount of activated carbon that would be required to treat a flow of 5000 m^3/d to a final COD concentration of 20 mg/L if the COD concentration after secondary treatment is equal to 120 mg/L.

11-17. A quantity of sodium-form ion exchange resin (5 g) is added to a water containing 2 meq of potassium chloride and 0.5 meq of sodium chloride. Calculate the residual concentration of potassium if the exchange capacity of the resin is 4.0 meq/g of dry weight and the selectivity coefficient is equal to 1.46.

11-18. Gravity filters are to be used to treat 6 Mgal/d of settled effluent at a filtration rate of 5 gal/ft^2 · min. The filtration rate with one filter taken out of service for backwashing is not to exceed 6 gal/ft^2 · min. Determine the number of units and the area of each unit to satify these conditions. If each filter is backwashed for 5 min every 24 h at a wash rate of 24 gal/ft^2 · min, determine the percentage of filter output used for washing if the filter is out of operation for a total of 30 min/d. What would be the total percentage of filter output used for backwashing if a surface-washing system that requires 0.75 gal/ft^2 · min of filtered effluent is to be installed?

REFERENCES

1. American Society of Civil Engineers, Task Committee on Design of Wastewater Filtration Facilities: "Tertiary Filtration of Wastewaters," *J. Env. Eng. Div.,* vol. 112, p. 1008, December 1986.
2. Amirtharajah, A., and K. M. Mills: "Rapid Mix Design for Mechanisms of Alum Coagulation," *Journal AWWA,* vol. 74, no. 4, 1982.
3. Antonie, R. L.: "Nitrogen Control with the Rotating Biological Contactor," in M. Wanichita and W. W. Eckenfelder, Jr., eds., *Advances in Water and Wastewater Treatment-Biological Nutrient Removal,* Ann Arbor Science Publishers, Inc., 1978.
4. Arora, M. L., E. F. Barth, and M. B. Umphres: "Technology Evaluation of Sequencing Batch Reactors," *Journal WPCF,* vol. 57, no. 8, 1985.
5. Benefield, L. D., and C. W. Randall: *Biological Process Design for Wastewater Treatment,* Prentice-Hall, Inc., Englewood Cliffs, NJ, 1980.
6. Bishop, S. L., and B. W. Behrman: "Filtration of Wastewater Using Granular Media," paper presented at the 1976 Thomas R. Camp Lecture Series on Wastewater Treatment and Disposal, Boston Society of Civil Engineers, Boston, 1976.
7. Burdick, C. R., D. R. Refling, and H. D. Stensel: "Advanced Biological Treatment to Achieve Nutrient Removal," *Journal WPCF,* vol. 54, no. 7, 1982.

8. Chapman, P. M., G. P. Romberg, and G. A. Vigers: "Design of Monitoring Studies for Priority Pollutants," *Journal WPCF,* vol. 54, no. 3, 1982.

9. Chapman, R. F. (ed.): *Separation Processes in Practice,* Reinhold, New York, 1961.

10. Dahab, M. F., and J. C. Young: "Unstratified-Bed Filtration of Wastewater," *J. Env. Eng. Div., ASCE,* vol. 103, EE 12714, 1977.

11. Daigger, G. T., G. D. Waltrip, E. D. Romm, and L. M. Morales: "Enhanced Secondary Treatment Incorporating Biological Nutrient Removal," *Journal WPCF,* vol. 60, no. 10, 1988.

12. Duddles, G. A., S. E. Richardson, and E. F. Barth: "Plastic Medium Trickling Filters for Biological Nitrogen Control," *Journal WPCF,* vol. 46, no. 5, 1974.

13. Eckenfelder, W. W., Jr.: *Industrial Water Pollution Control,* 2nd ed., McGraw-Hill, New York, 1989.

14. Eliassen, R., and G. E. Bennett: "Anion Exchange and Filtration Techniques for Wastewater Renovation," *Journal WPCF,* vol. 39, no. 10, part 2, 1967.

15. Grady, C. P. L., Jr., and H. C. Lim: *Biological Wastewater Treatment, Theory and Application,* Marcel Decker, Inc., New York, 1980.

16. Hannah, S. A., B. M. Austern, A. E. Eralp, and R. H. Wise: "Comparative Removal of Toxic Pollutants by Six Wastewater Treatment Processes," *Journal WPCF,* vol. 58, no. 1, 1986.

17. Hannah, S. A., B. M. Austern, A. E. Eralp, and R. A. Dobbs: "Removal of Organic Toxic Pollutants by Trickling Filter and Activated Sludge," *Journal WPCF,* vol. 60, no. 7, 1988.

18. Irvine, R. L., L. H. Ketchum, Jr., M. L. Arora, and E. F. Barth: "An Organic Loading Study of Full-Scale Sequencing Batch Reactors," *Journal WPCF,* vol. 57, no. 8, 1985.

19. Jeris, J. S., and R. W. Owens: "Pilot Scale High Rate Biological Denitrification," *Journal WPCF,* vol. 47, no. 8, 1975.

20. Ketchum, L. H., Jr., R. L. Irvine, R. E. Breyfogle, and J. F. Manning, Jr.: "A Comparison of Biological and Chemical Phosphorus Removals in Continuous and Sequencing Batch Reactors," *Journal WPCF,* vol. 59, no. 1, 1987.

21. Kugelman, I. J.: *Status of Advanced Waste Treatment,* in H. W. Gehm and J. I. Bregman (eds.), *Handbook of Water Resources and Pollution Control,* Van Nostrand, New York, 1976.

22. McCarty, P. L., L. Beck, and P. St. Amant: "Biological Denitrification of Wastewaters by Addition of Organic Materials," *Proceedings of the 24th Purdue Industrial Waste Conference,* Lafayette, IN, 1969.

23. Metcalf & Eddy, Inc.: *Wastewater Engineering: Treatment, Disposal, Reuse,* 2nd ed., McGraw-Hill, New York, 1979.

24. Qasim, S. R.: *Wastewater Treatment Plants Planning, Design, and Operation,* Holt, Rinehart, and Winston, New York, 1985.

25. Quirk, T., and W. W. Eckenfelder, Jr.: "Active Mass in Activated Sludge Analysis and Design," *Journal WPCF,* vol. 58, no. 9, 1986.

26. Schroeder, E. D.: *Water and Wastewater Treatment,* McGraw-Hill, New York, 1977.

27. Soap and Detergent Association: *Principles and Practice of Nutrient Removal from Municipal Wastewater,* October 1988.

28. Sudak, R. G.: *A Summary Report on Municipal Wastewater Reclamation with Ultrafiltration,* Separation Processes, Inc., San Marcos, CA, 1989.

29. Tchobanoglous, G.: "Physical and Chemical Processes for Nitrogen Removal—Theory and Application," *Proceedings of the Twelfth Sanitary Engineering Conference,* University of Illinois, Urbana, 1970.

30. Tchobanoglous, G.: "Filtration of Secondary Effluent for Reuse Applications," presented at the 61st Annual Conference of the WPCF, Dallas, TX, October 1988.

31. Tetreault, M. J., A. H. Benedict, C. Kaempfer, and E. F. Barth: "Biological Phosphorus Removal: A Technology Evaluation," *Journal WPCF,* vol. 58, no. 7, 1986.

32. U.S. Environmental Protection Agency: *Wastewater Filtration—Design Considerations,* Technology Transfer Seminar Publication, 1974.

33. U.S. Environmental Protection Agency: *Process Design Manual for Nitrogen Control,* Office of Technology Transfer, Washington, D.C., October 1975.

34. U.S. Environmental Protection Agency: *Process Design Manual for Phosphorus Removal*, Office of Technology Transfer, Washington, D.C., April 1976.

35. U.S. Environmental Protection Agency: *Wastewater Filtration—Design Considerations*, Technology Transfer Report, 1977.

36. U.S. Environmental Protection Agency: *Phosphorus Removal Design Manual*, EPA/625/1-87/001, September 1987.

37. U.S. Environmental Protection Agency: *Retrofitting POTWs for Phosphorus Removal in the Chesapeake Bay Drainage Basin*, EPA/625/6-87/017, September 1987.

38. Walsh, T. K., B. W. Behrman, G. W. Weil, and E. R. Jones: "A Review of Biological Phosphorus Removal Technology," presented at the Water Pollution Control Federation Annual Conference, October 1983.

39. Water Pollution Control Federation: *Nutrient Control*, Manual of Practice FD-7, 1983.

40. Water Pollution Control Federation: *Removal of Hazardous Wastes in Wastewater Facilities: Halogenated Organics*, Manual of Practice FD-11, 1986.

41. Water Pollution Control Federation: *Wastewater Treatment Plant Design*, Draft Manual of Practice no. 8, October 1988.

42. Weston, Roy F., Inc.: *Advanced Waste Treatment Performance Evaluation Summary Report*, USEPA Contract No. 68-03-3019, 1984.

43. Weston, Roy F., Inc.: *Emerging Technology Assessment of PhoStrip, A/O, and Bardenpho Processes for Biological Phosphorus Removal*, USEPA Contract No. 68-03-3055, February 1985.

44. White, G. C.: *Handbook of Chlorination*, 2nd ed., Van Nostrand Reinhold, New York, 1986.

45. Young, J. C., E. R. Baumann, and D. J. Wall: "Packed-Bed Reactors for Secondary Effluent BOD and Ammonia Removal," *Journal WPCF*, vol. 47, no. 1, 1975.

CHAPTER 12

DESIGN
OF FACILITIES
FOR THE
TREATMENT
AND DISPOSAL
OF SLUDGE

The constituents removed in wastewater treatment plants include screenings, grit, scum, and sludge. The sludge resulting from wastewater treatment operations and processes is usually in the form of a liquid or semisolid liquid that typically contains from 0.25 to 12 percent solids by weight, depending on the operations and processes used. Of the constituents removed by treatment, sludge is by far the largest in volume, and its processing and disposal is perhaps the most complex problem facing the engineer in the field of wastewater treatment. For this reason, a separate chapter has been devoted to this subject. The disposal of grit and screenings is discussed in Chap. 9. The problems of dealing with sludge are complex because (1) it is composed largely of the substances responsible for the offensive character of untreated wastewater; (2) the portion of sludge produced from biological treatment requiring disposal is composed of the organic matter contained in the wastewater but in another form, which can also decompose and become offensive; and (3) only a small part of the sludge is solid matter.

The main purpose of this chapter is to describe the operations and the processes that (1) are used to reduce the water and organic content of sludge and (2) are used to render it suitable for final disposal or reuse. The principal methods used to process and dispose of sludge are listed in Table 12-1. Thickening (concentration), conditioning, dewatering, and drying are used primarily to remove moisture from sludge; digestion, composting, incineration, wet-air oxidation, and vertical tube reactors are used primarily to treat or stabilize the organic material in the sludge. To make the study of these operations and processes more meaningful, the first three sections of this chapter are devoted to a discussion of the sources, characteristics, and quantities of sludge, the current regulatory environment, and a presentation of representative sludge treatment flow diagrams. Because sludge pumping is a fundamental part of wastewater treatment plant design, a separate discussion (Sec. 12-4) is devoted to sludge and scum pumping. The various methods used in the processing of sludge are discussed in Secs. 12-5 through 12-15. The preparation of solids balances for treatment facilities is described and illustrated in Sec. 12-16. Land application of sludge, other beneficial uses of sludge, and the conveyance and ultimate disposal of the sludge and residuals after processing are discussed in Secs. 12-17 through 12-19, respectively.

12-1 SOLIDS AND SLUDGE SOURCES, CHARACTERISTICS, AND QUANTITIES

To design sludge-processing, treatment, and disposal facilities properly, the sources, characteristics, and quantities of the solids and sludge to be handled must be known. Therefore, the purpose of this section is to present background data and information on these topics, which will serve as a basis for the material to be presented in the subsequent sections of this chapter.

Sources

The sources of solids in a treatment plant vary according to the type of plant and its method of operation. The principal sources of solids and sludge and the types generated are reported in Table 12-2. For example, in a complete-mix activated-sludge process, if sludge wasting is accomplished from the mixed-liquor line or aeration chamber, the activated-sludge settling tank is not a source of sludge. On the other hand, if wasting is accomplished from the solids return line, the activated-sludge settling tank constitutes a sludge source. If the sludge from the mixed-liquor line or aeration chamber is returned to the primary settling tank for thickening, this may obviate the need for a thickener, reducing by one the number of independent sludge sources in the treatment plant. Processes used for thickening, digesting, conditioning, and dewatering the sludge produced from primary and secondary settling tanks also constitute sludge sources.

TABLE 12-1
Sludge-processing and disposal methods

Unit operation, unit process, or treatment method	Function	See Sec.
Preliminary operations		
Sludge grinding	Size reduction	12-5
Sludge degritting	Grit removal	12-5
Sludge blending	Blending	12-5
Sludge storage	Storage	12-5
Thickening		
Gravity thickening	Volume reduction	12-6
Flotation thickening	Volume reduction	12-6
Centrifugation	Volume reduction	12-6
Gravity belt thickening	Volume reduction	12-6
Rotary drum thickening	Volume reduction	12-6
Stabilization		
Lime stabilization	Stabilization	12-7
Heat treatment	Stabilization	12-7
Anaerobic digestion	Stabilization, mass reduction	12-8
Aerobic digestion	Stabilization, mass reduction	12-9
Composting	Stabilization, product recovery	12-10
Conditioning		
Chemical conditioning	Sludge conditioning	12-11
Heat treatment	Sludge conditioning	12-11
Disinfection		
Pasteurization	Disinfection	12-12
Long-term storage	Disinfection	12-12
Dewatering		
Vacuum filter	Volume reduction	12-13
Centrifuge	Volume reduction	12-13
Belt filter press	Volume reduction	12-13
Filter press	Volume reduction	12-13
Sludge drying beds	Volume reduction	12-13
Lagoons	Storage, volume reduction	12-13
Heat Drying		
Flash dryer	Weight and volume reduction	12-14
Spray dryer	Weight and volume reduction	12-14
Rotary dryer	Weight and volume reduction	12-14
Multiple hearth dryer	Weight and volume reduction	12-14
Multiple-effect evaporator	Weight and volume reduction	12-14
Thermal reduction		
Multiple-hearth incineration	Volume reduction, resource recovery	12-15
Fluidized-bed incineration	Volume reduction	12-15
Co-incineration with solid wastes	Volume reduction	12-15
Wet-air oxidation	Stabilization, volume reduction	12-15
Vertical, deep-well reactor	Stabilization, volume reduction	12-15
Ultimate disposal		
Land application	Final disposal	12-17
Distribution and marketing	Beneficial use	12-18
Chemical fixation	Beneficial use, final disposal	12-18
Landfill	Final disposal	12-19
Lagooning	Volume reduction, final disposal	12-19

TABLE 12-2
Sources of solids and sludge from a conventional wastewater treatment plant

Unit operation or process	Types of solids or sludge	Remarks
Screening	Coarse solids	Coarse solids are removed by mechanical and hand-cleaned bar screens. In small plants, screenings are often comminuted for removal in subsequent treatment units.
Grit removal	Grit and scum	Scum removal facilities are often omitted in grit removal facilities.
Preaeration	Grit and scum	In some plants, scum removal facilities are not provided in preaeration tanks. If the preaeration tanks are not preceded by grit removal facilities, grit deposition may occur in preaeration tanks.
Primary sedimentation	Primary sludge and scum	Quantities of sludge and scum depend upon the nature of the collection system and whether industrial wastes are discharged to the system.
Biological treatment	Suspended solids	Suspended solids are produced by the biological conversion of BOD. Some form of thickening may be required to concentrate the waste sludge stream from biological treatment.
Secondary sedimentation	Secondary sludge and scum	Provision for scum removal from secondary settling tanks is a requirement of the U.S. Environmental Protection Agency.
Sludge-processing facilities	Sludge, compost, and ashes	The characteristics of the end products depend on the characteristics of the sludge being treated and the operations and processes used. Regulations for the disposal of residuals are becoming increasingly stringent.

Characteristics

To treat and dispose of the sludge produced from wastewater treatment plants in the most effective manner, it is important to know the characteristics of the solids and sludge that will be processed. The characteristics vary depending on the origin of the solids and sludge, the amount of aging that has taken place, and the type of processing to which they have been subjected. Some of the physical characteristics of sludges are summarized in Table 12-3.

General Composition. Typical data on the chemical composition of untreated and digested sludges are reported in Table 12-4. Many of the chemical constituents, including nutrients, are important in considering the ultimate disposal of the processed sludge and the liquid removed from the sludge during processing. The measurement

TABLE 12-3
Characteristics of solids and sludge produced during wastewater treatment

Solids or sludge	Description
Screenings	Screenings include all types of organic and inorganic materials large enough to be removed on bar racks. The organic content varies, depending on the nature of the system and the season of the year.
Grit	Grit is usually made of the heavier inorganic solids that settle with relatively high velocities. Depending on the operating conditions, grit may also contain significant amounts of organic matter, especially fats and grease.
Scum/grease	Scum consists of the floatable materials skimmed from the surface of primary and secondary settling tanks. Scum may contain grease, vegetable and mineral oils, animal fats, waxes, soaps, food wastes, vegetable and fruit skins, hair, paper and cotton, cigarette tips, plastic materials, condoms, grit particles, and similar materials. The specific gravity of scum is less than 1.0 and usually around 0.95.
Primary sludge	Sludge from primary settling tanks is usually gray and slimy and, in most cases, has an extremely offensive odor. Primary sludge can be readily digested under suitable conditions of operation.
Sludge from chemical precipitation	Sludge from chemical precipitation with metal salts is usually dark in color, though its surface may be red if it contains much iron. Lime sludge is grayish brown. The odor of chemical sludge may be objectionable, but is not as bad as primary sludge. While chemical sludge is somewhat slimy, the hydrate of iron or aluminum in it makes it gelatinous. If the sludge is left in the tank, it undergoes decomposition similar to primary sludge, but at a slower rate. Substantial quantities of gas may be given off and the sludge density increased by long residence times in storage.
Activated sludge	Activated sludge generally has a brownish, flocculant appearance. If the color is dark, the sludge may be approaching a septic condition. If the color is lighter than usual, there may have been underaeration with a tendency for the solids to settle slowly. Sludge in good condition has an inoffensive "earthy" odor. The sludge tends to become septic rapidly and then has a disagreeable odor of putrefaction. Activated sludge will digest readily alone or when mixed with primary sludge.

(*continued*)

of pH, alkalinity, and organic acid content is important in process control of anaerobic digestion. The content of heavy metals, pesticides, and hydrocarbons has to be determined when incineration and land application methods are considered. The energy (thermal) content of sludge is important where a thermal reduction process such as incineration is considered.

TABLE 12-3
(continued)

Solids or sludge	Description
Trickling-filter sludge	Humus sludge from trickling filters is brownish, flocculant, and relatively inoffensive when fresh. It generally undergoes decomposition more slowly than other undigested sludges. When trickling-filter sludge contains many worms, it may become inoffensive quickly. Trickling-filter sludge digests readily.
Digested sludge (aerobic)	Aerobically digested sludge is brown to dark brown and has a flocculant appearance. The odor of aerobically digested sludge is not offensive; it is often characterized as musty. Well-digested aerobic sludge dewaters easily on drying beds.
Digested sludge (anaerobic)	Anaerobically digested sludge is dark brown to black and contains an exceptionally large quantity of gas. When thoroughly digested, it is not offensive, its odor being relatively faint and like that of hot tar, burnt rubber, or sealing wax. When drawn off onto porous beds in thin layers, the solids first are carried to the surface by the entrained gases, leaving a sheet of comparatively clear water. The water drains off rapidly and allows the solids to sink down slowly on to the bed. As the sludge dries, the gases escape, leaving a well-cracked surface with an odor resembling that of garden loam.
Composted sludge	Composted sludge is usually dark brown to black, but the color may vary if bulking agents such as recycled compost or wood chips have been used in the composting process. The odor of well-composted sludge is inoffensive and resembles that of commercial garden-type soil conditioners.
Septage	Sludge from septic tanks is black. Unless the sludge is well digested by long storage, it is offensive because of the hydrogen sulfide and other gases that it gives off. The sludge can be dried on porous beds if spread out in thin layers, but objectionable odors can be expected while it is draining unless it is well-digested.

Specific Constituents. Characteristics of sludge that affect its suitability for land application and beneficial use include organic content (usually measured as volatile solids), nutrients, pathogens, metals, and toxic organics. The fertilizer value of sludge, which should be evaluated where the sludge is to be used as a soil conditioner, is based primarily on the content of nitrogen, phosphorus, and potassium (potash). Typical nutrient values of sludge as compared to commercial fertilizers are reported in Table 12-5. In most land application systems, sludge provides sufficient nutrients for good plant growth. In some applications, the phosphorus and potassium content of wastewater sludge may be too low to satisfy specific plant uptake requirements.

Trace elements in sludge are those inorganic chemical elements that, in very small quantities, can be essential or detrimental to plants and animals. The term

TABLE 12-4
Typical chemical composition and properties of untreated and digested sludge [a]

Item	Untreated primary sludge		Digested primary sludge		Activated sludge, range
	Range	Typical	Range	Typical	
Total dry solids (TS), %	2.0–8.0	5.0	6.0–12.0	10.0	0.83–1.16
Volatile solids (% of TS)	60–80	65	30–60	40	59–88
Grease and fats (% of TS)					
Ether soluble	6–30	—	5–20	18	—
Ether extract	7–35	—	—	—	5–12
Protein (% of TS)	20–30	25	15–20	18	32–41
Nitrogen (N, % of TS)	1.5–4	2.5	1.6–6.0	3.0	2.4–5.0
Phosphorus (P_2O_5, % of TS)	0.8–2.8	1.6	1.5–4.0	2.5	2.8–11.0
Potash (K_2O, % of TS)	0–1	0.4	0.0–3.0	1.0	0.5–0.7
Cellulose (% of TS)	8.0–15.0	10.0	8.0–15.0	10.0	—
Iron (not as sulfide)	2.0–4.0	2.5	3.0–8.0	4.0	—
Silica (SiO_2, % of TS)	15.0–20.0	—	10.0–20.0	—	—
pH	5.0–8.0	6.0	6.5–7.5	7.0	6.5–8.0
Alkalinity (mg/L as $CaCO_3$)	500–1,500	600	2,500–3,500	3,000	580–1,100
Organic acids (mg/L as HAc)	200–2,000	500	100–600	200	1,100–1,700
Energy content, Btu/lb	10,000–12,500	11,000	4,000–6,000	5,000	8,000–10,000

[a] Adapted from Ref. 42.

Note: Btu/lb × 2.3241 = kJ/kg

TABLE 12-5
Comparison of nutrient levels in commercial fertilizers and wastewater sludge

	Nutrients, %		
	Nitrogen	Phosphorus	Potassium
Fertilizers for typical agricultural use [a]	5	10	10
Typical values for stabilized wastewater sludge	3.3	2.3	0.3

[a] The concentrations of nutrients may vary widely depending upon the soil and crop needs.

"heavy metals" is used to denote several of the trace elements present in sludge. Concentrations of heavy metals may vary widely, as indicated in Table 12-6. For land application of sludge, concentrations of heavy metals may limit the sludge application rate and the useful life of the application site (see Sec. 12-17).

Quantities

Data on the quantities of sludge produced from various processes and operations are presented in Table 12-7. Corresponding data on the sludge concentrations to be expected from various processes are given in Table 12-8. Although the data in Table 12-7 are useful as presented, it should be noted that the quantity of sludge produced will vary widely.

TABLE 12-6
Typical metal content in wastewater sludge [a]

Metal	Dry sludge, mg/kg	
	Range	Median
Arsenic	1.1–230	10
Cadmium	1–3,410	10
Chromium	10–99,000	500
Cobalt	11.3–2,490	30
Copper	84–17,000	800
Iron	1,000–154,000	17,000
Lead	13–26,000	500
Manganese	32–9,870	260
Mercury	0.6–56	6
Molybdenum	0.1–214	4
Nickel	2–5,300	80
Selenium	1.7–17.2	5
Tin	2.6–329	14
Zinc	101–49,000	1700

[a] Ref. 45.

TABLE 12-7
Typical data for the physical characteristics and quantities of sludge produced from various wastewater treatment operations and processes

Treatment operation or process	Specific gravity of sludge solids	Specific gravity of sludge	Dry solids, lb/10^3 gal	
			Range	Typical
Primary sedimentation	1.4	1.02	0.9–1.4	1.25
Activated sludge (waste sludge)	1.25	1.005	0.6–0.8	0.7
Trickling filtration (waste sludge)	1.45	1.025	0.5–0.8	0.6
Extended aeration (waste sludge)	1.30	1.015	0.7–1.0	0.8 [a]
Aerated lagoon (waste sludge)	1.30	1.01	0.7–1.0	0.8 [a]
Filtration	1.20	1.005	0.1–0.2	0.15
Algae removal	1.20	1.005	0.1–0.2	0.15
Chemical addition to primary sedimentation tanks for phosphorus removal				
Low lime (350–500 mg/L)	1.9	1.04	2.0–3.3	2.5 [b]
High lime (800–1,600 mg/L)	2.2	1.05	5.0–11.0	6.6 [b]
Suspended-growth nitrification	—	—	—	— [c]
Suspended-growth denitrification	1.20	1.005	0.1–0.25	0.15
Roughing filters	1.28	1.02	—	— [d]

[a] Assuming no primary treatment.
[b] Sludge in addition to that normally removed by primary sedimentation.
[c] Negligible.
[d] Included in sludge production from biological secondary treatment processes.
Note: lb/10^3 gal × 120.48 = kg/10^3 m^3

Quantity Variations. The quantity of solids entering the wastewater treatment plant daily may be expected to fluctuate over a wide range. To ensure capacity capable of handling these variations, the designer of sludge-processing and disposal facilities should consider (1) the average and maximum rates of sludge production and (2) the potential storage capacity of the treatment units within the plant. The variation in daily quantity that may be expected in large cities is shown in Fig. 12-1. The curve is characteristic of large cities having a number of large sewers laid on flat slopes; even greater variations may be expected at small plants.

A limited quantity of solids may be stored temporarily in the sedimentation and aeration tanks. The storage capacity can be used to equalize short-term peak loads. Where digestion tanks with varying levels are used, their large storage capacity provides a substantial dampening effect on peak digested sludge loads. In sludge treatment systems where digestion is used, the design is usually based on maximum monthly loadings. Where digestion is not used, the sludge treatment process should be

TABLE 12-8

Expected sludge concentrations from various treatment operations and processes

Operation or process application	Sludge solids concentration, % dry solids	
	Range	Typical
Primary settling tank		
Primary sludge	4.0–10.0	5.0
Primary sludge to a cyclone	0.5–3.0	1.5
Primary and waste activated sludge	3.0–8.0	4.0
Primary sludge and trickling-filter humus	4.0–10.0	5.0
Primary sludge with iron addition for phosphorus removal	0.5–3.0	2.0
Primary sludge with low lime addition for phosphorus removal	2.0–8.0	4.0
Primary sludge with high lime addition for phosphorus removal	4.0–16.0	10.0
Scum	3.0–10.0	5.0
Secondary settling tank		
Waste activated sludge		
With primary settling	0.5–1.5	0.8
Without primary settling	0.8–2.5	1.3
High-purity oxygen activated sludge		
With primary settling	1.3–3.0	2.0
Without primary settling	1.4–4.0	2.5
Trickling-filter humus sludge	1.0–3.0	1.5
Rotating biological contractor waste sludge	1.0–3.0	1.5
Gravity thickener		
Primary sludge only	5.0–10.0	8.0
Primary and waste activated sludge	2.0–8.0	4.0
Primary sludge and trickling-filter humus	4.0–9.0	5.0
Dissolved-air flotation thickener		
Waste activated sludge only		
With chemical addition	4.0–6.0	5.0
Without chemical addition	3.0–5.0	4.0
Centrifuge thickener		
Waste activated sludge only	4.0–8.0 [a]	5.0
Gravity belt thickener		
Waste activated sludge only with chemical addition	3.0–6.0 [b]	5.0
Anaerobic digester		
Primary sludge only	5.0–10.0	7.0
Primary and waste activated sludge	2.5–7.0	3.5
Primary sludge and trickling-filter humus	3.0–8.0	4.0
Aerobic digester		
Primary sludge only	2.5–7.0	3.5
Primary and waste activated sludge	1.5–4.0	2.5
Waste activated sludge only	0.8–2.5	1.3

[a] Adapted from Ref. 57.

[b] Adapted from Ref. 18.

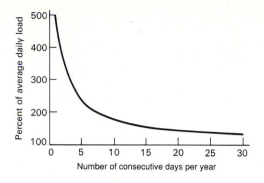

FIGURE 12-1
Peak sludge load as a function of the average daily load.

capable of handling the solids production of the maximum week. Certain components of the sludge system, such as sludge pumping and thickening, may need to be sized to handle the maximum day conditions. The total quantities of sludge that must be processed are determined by preparing a series of solids balances for the treatment process. The preparation of solids balances is considered in Sec. 12-16.

Volume-Weight Relationships. The volume of sludge depends mainly on its water content and only slightly on the character of the solid matter. A 10 percent sludge, for example, contains 90 percent water by weight. If the solid matter is composed of fixed (mineral) solids and volatile (organic) solids, the specific gravity of all of the solid matter can be computed using Eq. 12-1.

$$\frac{W_s}{S_s \rho_w} = \frac{W_f}{S_f \rho_w} + \frac{W_v}{S_v \rho_w} \qquad (12\text{-}1)$$

where W_s = weight of solids
S_s = specific gravity of solids
ρ_w = density of water
W_f = weight of fixed solids (mineral matter)
S_f = specific gravity of fixed solids
W_v = weight of volatile solids
S_v = specific gravity of volatile solids

Therefore, if one-third of the solid matter in a sludge containing 90 percent water is composed of fixed mineral solids with a specific gravity of 2.5 and two-thirds is composed of volatile solids with a specific gravity of 1.0, then the specific gravity of all of the solids S_s would be equal to 1.25, as follows:

$$\frac{1}{S_s} = \frac{0.33}{2.5} + \frac{0.67}{1} = 0.802$$

$$S_s = \frac{1}{0.802} = 1.25$$

If the specific gravity of the water is taken to be 1.0, the specific gravity of the sludge is 1.02, as follows:

$$\frac{1}{S} = \frac{0.1}{1.25} + \frac{0.9}{1.0} = 0.98$$

$$S = \frac{1}{0.98} = 1.02$$

The volume of a sludge may be computed with the following expression:

$$V = \frac{W_s}{\rho_w S_{sl} P_s} \tag{12-2}$$

where W_s = weight of dry solids, lb
ρ_w = density of water, lb/ft^3
S_{sl} = specific gravity of the sludge
P_s = percent solids expressed as a decimal

For approximate calculations for a given solids content, simply remember that the volume varies inversely with the percent of solid matter contained in the sludge as given by

$$\frac{V_1}{V_2} = \frac{P_2}{P_1} \qquad \text{(approximate)}$$

where V_1, V_2 = sludge volumes
P_1, P_2 = percent of solid matter

The application of these volume and weight relationships is illustrated in Example 12-1.

Example 12-1 Volume of untreated and digested sludge. Determine the liquid volume before and after digestion and the percent reduction for 1000 lb (dry basis) of primary sludge with the following characteristics:

	Primary	Digested
Solids, %	5	10
Volatile matter, %	60	60 (destroyed)
Specific gravity of fixed solids	2.5	2.5
Specific gravity of volatile solids	≈1.0	≈1.0

Solution

1. Compute the average specific gravity of all the solids in the primary sludge using Eq. 12-1.

$$\frac{1}{S_s} = \frac{0.4}{2.5} + \frac{0.6}{1.0} = 0.76$$

$$S_s = \frac{1}{0.76} = 1.32 \qquad \text{(primary solids)}$$

2. Compute the specific gravity of the primary sludge.

$$\frac{1}{S_{sl}} = \frac{0.05}{1.32} + \frac{0.95}{1}$$

$$S_{sl} = \frac{1}{0.99} = 1.01$$

3. Compute the volume of the primary sludge using Eq. 12-2.

$$V = \frac{1000 \text{ lb}}{(62.4 \text{ lb/ft}^3)(1.01)(0.05)}$$
$$= 317 \text{ ft}^3 \ (9.0 \text{ m}^3)$$

4. Compute the percentage of volatile matter after digestion.

$$\% \text{ volatile matter} = \frac{\text{total volatile solids after digestion}}{\text{total solids after digestion}} \times 100$$
$$= \frac{0.4(0.6 \times 1000)}{400 + 0.4(600)}(100) = 37.5$$

5. Compute the average specific gravity of all the solids in the digested sludge using Eq. 12-1.

$$\frac{1}{S_s} = \frac{0.625}{2.5} + \frac{0.375}{1} = 0.625$$

$$S_s = \frac{1}{0.625} = 1.6 \quad \text{(digested solids)}$$

6. Compute the specific gravity of the digested sludge.

$$\frac{1}{S_{ds}} = \frac{0.1}{1.6} + \frac{0.90}{1} = 0.96$$

$$S_{ds} = \frac{1}{0.96} = 1.04$$

7. Compute the volume of digested sludge using Eq. 12-2.

$$V = \frac{400 \text{ lb} + 0.4(600 \text{ lb})}{(62.4 \text{ lb/ft}^3)(1.04)(0.10)}$$
$$= 98.6 \text{ ft}^3 \ (2.8 \text{ m}^3)$$

8. Determine the percentage reduction in the sludge volume after digestion.

$$\text{Reduction, } \% = \frac{(317 - 98.6)}{317}100 = 68.9$$

12-2 REGULATIONS FOR THE REUSE AND DISPOSAL OF SLUDGE

In selecting the appropriate methods of sludge processing, reuse, and disposal, consideration must be given to the regulations controlling the disposal of sludge from wastewater treatment plants. As discussed in Chap. 4, new standards have been proposed by the EPA that establish pollutant numerical limits and management practices for (1) application of sludge to agricultural and nonagricultural land, (2) distribution and marketing, (3) monofilling, (4) surface disposal, and (5) incineration. The proposed regulations are under review at the time of writing this text (1989) and may change substantially when they are finally promulgated. The new regulations may have a direct impact on the methods of sludge processing, reuse, and disposal being used or considered.

The regulations for the disposal of sludge from the treatment of domestic wastewater, as currently proposed in 40CFR Part 503, prescribe limits for certain metals and organic compounds [12]. A listing is provided in Table 12-9 of the reuse and disposal options and those metals and organic compounds for which specific numerical limits are being proposed. The list of regulated pollutants is also required to be updated whenever additional constituents of concern are identified. It is also proposed that for treatment plants with flowrates greater than 10 Mgal/d, the concentrations of the regulated pollutants should be monitored on a monthly basis. As a

TABLE 12-9
Regulated pollutants in wastewater sludge [a]

	Type of disposal or reuse				
Pollutant	Land application	Distribution & marketing	Monofilling	Surface disposal	Incineration
Aldrin	✓	✓			
Arsenic	✓	✓	✓	✓	✓
Benzene			✓	✓	
Benzo(a)pyrene	✓	✓	✓	✓	
Beryllium					✓
Bis(2-ethylhexyl)phthalate			✓	✓	
Cadmium	✓	✓	✓	✓	✓
Chlordane	✓	✓	✓	✓	
Chromium	✓	✓			✓
Copper	✓	✓	✓	✓	
DDD/DDE/DDT	✓	✓	✓	✓	
Dieldrin	✓	✓			
Dimethyl nitrosamine	✓		✓	✓	
Heptachlor	✓	✓			
Hexachlorobenzene	✓	✓			
Hexachlorobutadiene	✓	✓			
Lead	✓	✓	✓	✓	✓
Lindane	✓	✓	✓	✓	
Mercury	✓	✓	✓	✓	✓
Molybdenum	✓				
Nickel	✓	✓	✓	✓	✓
PCB	✓	✓	✓	✓	
Selenium	✓	✓			
Toxaphene	✓	✓	✓	✓	
Trichloroethylene	✓		✓	✓	
Total hydrocarbons					✓
Zinc	✓	✓			

[a] From Ref. 12.

result of the uncertainties concerning the possible significant changes in sludge regulations, the impact on the planning and design of new facilities and the operation and retrofitting of existing facilities is difficult to assess at the present time. Therefore, it is imperative that the design engineer become familiar with the current and proposed sludge regulations early in the planning stage of a project. The future direction in sludge regulations will most likely result in improved source control of heavy metals and toxic organics, such as PCBs and pesticides, that are not affected by biological treatment processes.

12-3 SLUDGE TREATMENT FLOW DIAGRAMS

A generalized flow diagram incorporating the unit operations and processes to be discussed in this chapter is presented in Fig. 12-2. As shown, an almost infinite number of combinations are possible. In practice, the most commonly used process flow diagrams for sludge treatment may be divided into two general categories, depending on whether or not biological treatment is involved. Typical flow diagrams incorporating biological processing are presented in Fig. 12-3. Depending on the source of the sludge, thickeners may be used contingent upon the method of sludge stabilization, dewatering, and disposal. Following biological digestion, any of the several methods shown may be used to dewater the sludge, the choice influenced by economic evaluations and local conditions.

Because the presence of industrial and other toxic wastes has presented problems in the operation of biological digesters, a number of plants have been designed with other means for sludge treatment. Three representative process flow diagrams without biological treatment are shown in Fig. 12-4.

12-4 SLUDGE AND SCUM PUMPING

Sludge produced in wastewater treatment plants must be conveyed from one plant point to another in conditions ranging from a watery sludge or scum to a thick sludge. Sludge may also be pumped offsite for long distances for treatment and disposal. For each type of sludge and pumping application, a different type of pump may be needed.

Pumps

Pumps used most frequently to convey sludge include the plunger, progressive cavity, centrifugal, torque-flow, diaphragm, high-pressure piston, and rotary-lobe types. Other types such as peristaltic (hose or rotor) pumps and concrete slurry pumps have also been used to pump sludge. Diaphragm and centrifugal pumps also are used extensively for pumping scum.

Plunger Pumps. Plunger pumps (see Fig. 12-5a) have been used frequently and, if rugged enough for the service, have proved to be quite satisfactory. The advantages of plunger pumps are as follows:

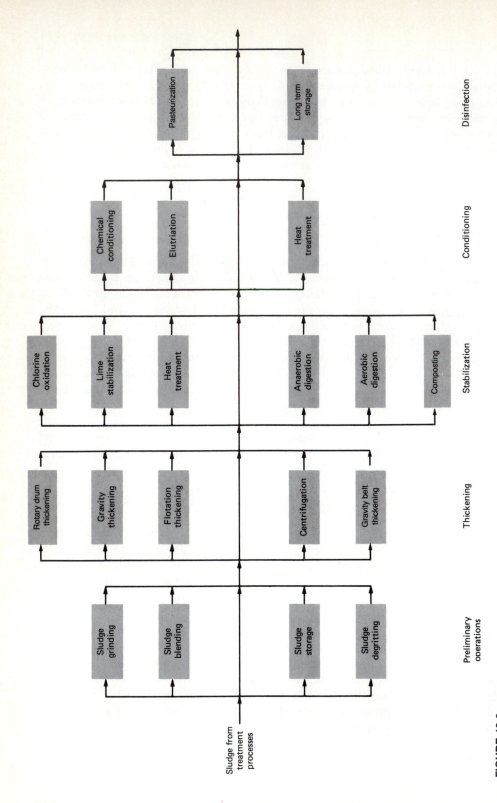

FIGURE 12-2
Generalized sludge processing and disposal flow diagram.

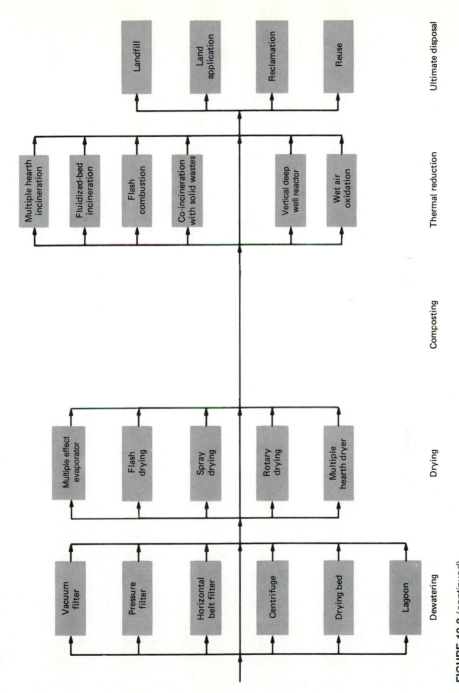

FIGURE 12-2 (*continued*)
Generalized sludge processing and disposal flow diagram.

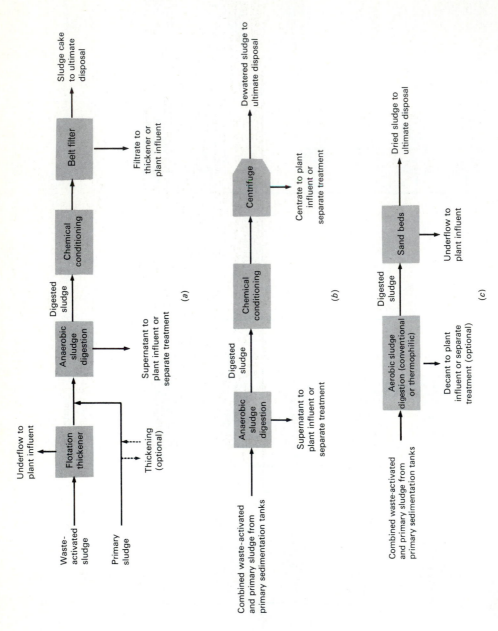

FIGURE 12-3
Typical sludge treatment flow diagrams with biological digestion and three different sludge dewatering processes: (a) belt filter press, (b) centrifugation, and (c) drying beds.

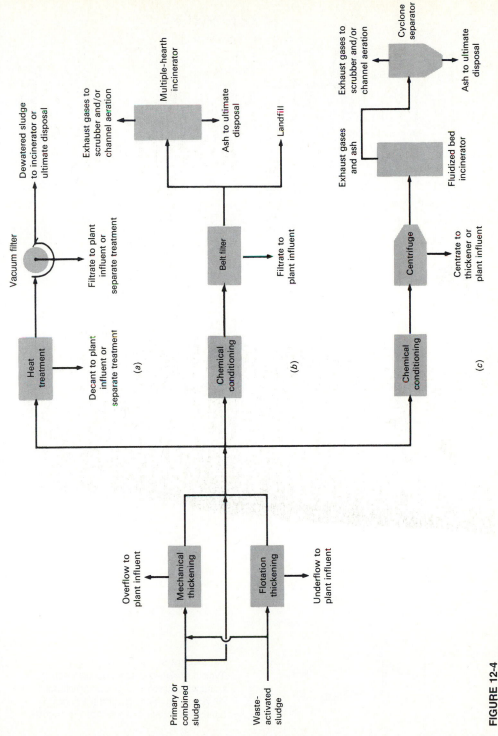

FIGURE 12-4
Typical nonbiological sludge treatment flow diagrams: (a) heat treatment with vacuum filter dewatering, (b) multiple-hearth incineration, and (c) fluidized bed incineration.

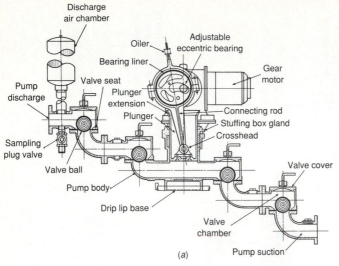

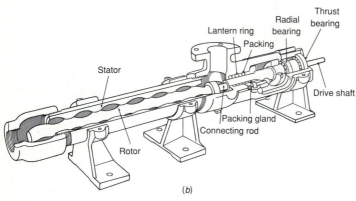

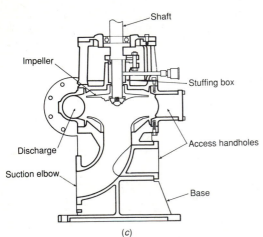

FIGURE 12-5
Typical sludge and scum pumps used in wastewater treatment plants: (*a*) plunger, (*b*) progressive cavity, and (*c*) nonclog centrifugal (continued on following page).

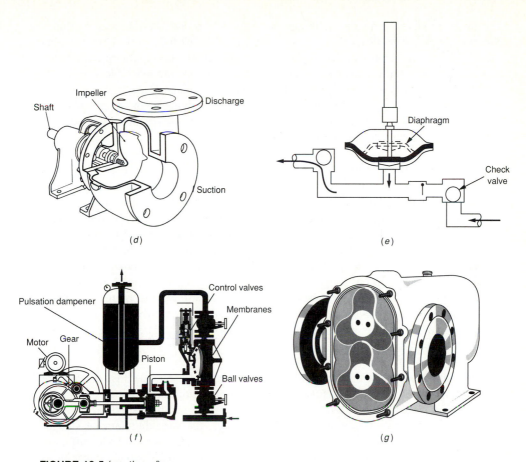

FIGURE 12-5 (*continued*)
(*d*) Torque flow, (*e*) diaphragm, (*f*) high-pressure piston, and (*g*) rotary lobe.

1. Pulsating action of simplex and duplex pumps tends to concentrate the sludge in the hoppers ahead of the pumps and resuspend solids in pipelines when pumping at low velocities.

2. They are suitable for suction lifts up to 10 ft (3 m) and are self-priming.

3. Low pumping rates can be used with large port openings.

4. Positive delivery is provided unless some object prevents the ball check valves from seating.

5. They have constant but adjustable capacity, regardless of large variations in pumping head.

6. High discharge heads may be provided for.

7. Heavy solids concentrations may be pumped if the equipment is designed for the load conditions.

Plunger pumps come with one, two, or three plungers (called simplex, duplex, or triplex units), with capacities of 40 to 60 gal/min (2.5 to 3.8 L/s) per plunger, and larger models are available. Pump speeds should be between 40 and 50 r/min. The pumps should be designed for a minimum head of 80 ft (24 m) in small plants and 115 ft (35 m) or more in large plants because grease accumulations in sludge lines cause a progressive increase in head with use. Pumps are available with heads up to 230 ft (70 m) and should be considered for large plants. Capacity is decreased by shortening the stroke of the plunger; however, the pumps seem to operate more satisfactorily at or near full stroke. For this reason, many pumps are provided with variable-pitch, V-belt drives for speed control of capacity.

Progressive Cavity Pumps. The progressive cavity pump (see Fig. 12-5b) has been used successfully on almost all types of sludges. The pump is composed of a single-threaded rotor that operates with a minimum of clearance in a double-threaded helix stator made of rubber. A volume, or "cavity," moves progressively from suction to discharge as the rotor turns. The pump is self-priming at suction lifts up to 28 ft (8.5 m), but it must not be operated dry because it will burn out the rubber stator. It is available in capacities up to 1200 gal/min (75 L/s) and may be operated at discharge heads of 450 ft (137 m) on sludge. For primary sludges, a grinder normally precedes these pumps. The pumps are expensive to maintain because of wear on the rotors and the stators, particularly in primary sludge-pumping applications where grit is present. Advantages of the pumps are (1) easily controlled flowrates, (2) minimal pulsation, and (3) relatively simple operation.

Centrifugal Pumps. Centrifugal pumps of nonclog design (see Fig. 12-5c) are commonly used. In using centrifugal pumps to pump sludge, a problem arises over choosing the proper size. At any given speed, centrifugal pumps operate well only if the pumping head is within a relatively narrow range; the variable nature of sludge, however, causes pumping heads to change. The selected pumps must have sufficient clearance to pass the solids without clogging and have a small enough capacity to avoid pumping a sludge diluted by large quantities of wastewater overlying the sludge blanket. Throttling the discharge to reduce the capacity is impractical because of frequent stoppages; hence, it is absolutely essential that these pumps be equipped with variable-speed drives. Centrifugal pumps of special design—torque-flow, screw-feed, and bladeless—have been used for pumping primary sludge in large plants. Screw-feed and bladeless pumps have not been used very much in recent applications because of the successful use of torque-flow pumps.

Torque-flow pumps (see Fig. 12-5d) have fully recessed impellers and are very effective in conveying sludge. The size of particles that can be handled is limited only by the diameter of the suction or discharge openings. The rotating impeller develops a vortex in the sludge so that the main propulsive force is the liquid itself. Most of the fluid does not actually pass through the vanes of the impeller, thereby minimizing abrasive contact; however, pumps used in sludge service should have nickel or chrome abrasion resistant volute and impellers. The pumps can operate only over a narrow head range at a given speed, so the system operating conditions must be evaluated carefully. Variable speed control is recommended where the pumps are expected to operate over a wide range of head conditions.

For high-pressure applications, multiple pumps may be used and connected together in series.

Slow-speed centrifugal and mixed-flow pumps are commonly used for returning activated sludge to the aeration tanks. Screw pumps are also being used for this service.

Diaphragm Pumps. Diaphragm pumps use a flexible membrane that is pushed and pulled to contract and enlarge an enclosed cavity (see Fig. 12-5e). Flow is directed through this cavity by check valves, which may be either ball or flap type. The capacity of a diaphragm pump is altered by changing either the length of the diaphragm stroke or the number of strokes per minute. Pump capacity can be increased and flow pulsations smoothed out by providing two pump chambers and using both strokes of the diaphragm for pumping. Diaphragm pumps are relatively low capacity and low head; the largest available air-diaphragm pump delivers 220 gal/min (14 L/s) against 50 ft (15 m) of head.

High-Pressure Piston Pumps. High-pressure piston pumps are used in high-pressure applications such as pumping sludge long distances. Several types of piston pumps have been developed for high-pressure applications and are similar in action to plunger pumps. The high-pressure piston pumps use separate power pistons or membranes to separate the drive mechanisms from contacting the sludge. A piston membrane type pump is shown in Fig. 12-5f. Advantages of these types of pumps are that (1) they can pump relatively small flowrates at high pressures, up to 2000 lb_f/in^2 (13,800 kN/m^2), (2) large solids up to the discharge pipe diameter can be passed, (3) a range of solids concentrations can be handled, and (4) the pumping can be accomplished in a single stage. The pumps, however, are very expensive.

Rotary-Lobe Pumps. Rotary-lobe pumps (see Fig. 12-5g) are positive displacement pumps in which two rotating, synchronous lobes push the fluid through the pump. Rotational speed and shearing stresses are low. For sludge pumping, lobes are made of hard metal or hard rubber. An advantage cited for the rotary-lobe pump is that lobe replacement is less costly than rotor and stator replacement for progressive cavity pumps. Rotary-lobe pumps, like other positive displacement pumps, must be protected against pipeline obstructions.

Application of Pumps to Types of Sludge

Types of sludge that are pumped include primary, chemical, and trickling-filter sludges and activated, thickened, and digested sludges. Scum that accumulates at various points in a treatment plant must also be pumped. The application of pumps to types of sludge is summarized in Table 12-10.

Headloss Determination

The headloss encountered in the pumping of sludge depends on the flow properties (rheology) of sludge, the pipe diameter, and the flow velocity. It has been observed

TABLE 12-10
Application of pumps to types of sludge [a]

Type of sludge or solids	Applicable pump	Comment
Ground screenings	Pumping screenings should be avoided	Pneumatic ejectors may be used.
Grit	Torque-flow centrifugal	The abrasive character of grit and the presence of rags make grit difficult to handle. Hardened casings and impellers should be used for torque-flow pumps. Pneumatic ejectors may also be used.
Scum	Plunger; progressive cavity; diaphragm; centrifugal	Scum is often pumped by the sludge pumps; valves are manipulated in the scum and sludge lines to permit this. In larger plants, separate scum pumps are used. Scum mixers are often used to ensure homogeneity prior to pumping. Pneumatic ejectors may also be used.
Primary sludge	Plunger; torque-flow centrifugal; diaphragm; progressive cavity; rotary-lobe	In most cases, it is desirable to obtain as concentrated a sludge as practicable from primary sedimentation tanks, usually by collecting the sludge in hoppers and pumping intermittently, allowing the sludge to collect and consolidate between pumping periods. The character of untreated primary sludge will vary considerably, depending on the characteristics of the solids in the wastewater and the types of treatment units and their efficiency. Where biological treatment follows, the quantity of solids from (1) waste activated sludge, (2) humus sludge from settling tanks following trickling filters, (3) overflow liquors from digestion tanks, and (4) centrate or filtrate return from dewatering operations will also affect the sludge characteristics. In many cases, the character of the sludge is not suitable for the use of conventional nonclog centrifugal pumps.
Chemical precipitation	Same as for primary sludge	

(continued)

that headlosses increase with increased solids content, increased volatile content, and lower temperatures. When the percent of volatile matter multiplied by the percent of solids exceeds 600, difficulties may be encountered in pumping sludge.

Water, oil, and most other fluids are "Newtonian," which means that the pressure drop is proportional to the velocity and viscosity under laminar-flow conditions. As the velocity increases past a critical value, the flow becomes turbulent. Dilute sludges such as unconcentrated activated and trickling-filter sludges behave similar to water. Concentrated wastewater sludges, however, are non-Newtonian fluids. The pressure drop under laminar conditions for non-Newtonian fluids is not proportional to flow, so the viscosity is not a constant. Special procedures may be used to determine headloss under laminar-flow conditions and the velocity at which turbulent flow

TABLE 12-10
(continued)

Type of sludge or solids	Applicable pump	Comment
Digested sludge	Plunger; torque-flow centrifugal; progressive cavity; diaphragm; high-pressure piston; rotary-lobe	Well-digested sludge is homogeneous, containing 5 to 8% solids and a quantity of gas bubbles, but may contain up to 12% solids. Poorly digested sludge may be difficult to handle. If good screening and grit removal is provided, nonclog centrifugal pumps may be considered.
Trickling-filter humus sludge	Nonclog and torque-flow centrifugal; progressive cavity; plunger; diaphragm	Sludge is usually of homogeneous character and can be easily pumped.
Return or waste activated sludge	Nonclog and torque-flow centrifugal; progressive cavity; diaphragm	Sludge is dilute and contains only fine solids so that nonclog pumps may commonly be used. For nonclog pumps, slow speeds are recommended to minimize the breakup of flocculant particles.
Thickened or concentrated sludge	Plunger; progressive cavity; diaphragm; high-pressure piston; rotary-lobe	Positive displacement pumps are most applicable for concentrated sludge because of their ability to generate movement of the sludge mass. Torque-flow pumps may be used but may require the addition of flushing or dilution facilities.

[a] Adapted in part from Ref. 42.

begins. In this section, both the simplified approach of calculating headloss and a method using the sludge rheology will be discussed.

The headloss in pumping unconcentrated activated and trickling-filter sludges may be from 10 to 25 percent greater than for water. Primary, digested, and concentrated sludges at low velocities may exhibit a plastic flow phenomenon in which a definite pressure is required to overcome resistance and start flow. The resistance then increases approximately with the first power of the velocity throughout the laminar range of flow, which extends to about 3.5 ft/s (1.1 m/s), the lower critical velocity. Above the higher critical velocity at about 4.5 ft/s (1.4 m/s), the flow may be considered turbulent. In the turbulent range, the losses for well-digested sludge may be from two to three times the losses for water. The losses for primary and concentrated sludges may be considerably greater.

Simplified Headloss Computations. Relatively simple procedures are used to compute headloss for short sludge pipelines. The accuracy of these procedures is often quite adequate, especially at solids concentrations less than 3 percent by weight. To determine the headloss, the factor k is obtained from empirical curves for a given solids content and type of sludge. The headloss when pumping sludge is computed by multiplying the headloss of water, determined by using the Darcy-Weisbach, Hazen-Williams, or Manning equations by k. Because sludge lines at treatment plants are usually of short length, simplified computational methods are practical and convenient where friction losses are conservatively estimated. Long sludge lines, however, require careful study of methods of estimating friction losses from engineering, economic, and operating considerations. A method of determining friction losses in long pipelines using sludge rheology concepts is presented in the following section.

Approximate estimates of headloss can be obtained using Fig. 12-6a. This figure should be used under laminar-flow conditions when (1) velocities are at least 2.5 ft/s (0.8 m/s), (2) thixotropic behavior is not considered, and (3) the pipe is not obstructed by grease or other materials. Another approximate method makes use of multiplication factor charts (see Fig. 12-6b) developed from the work of various researchers [7]. This method involves only velocity and percent solids concentration.

Usually, the consistency of untreated primary sludge changes during pumping. At first, the most concentrated sludge is pumped. When most of the sludge has been

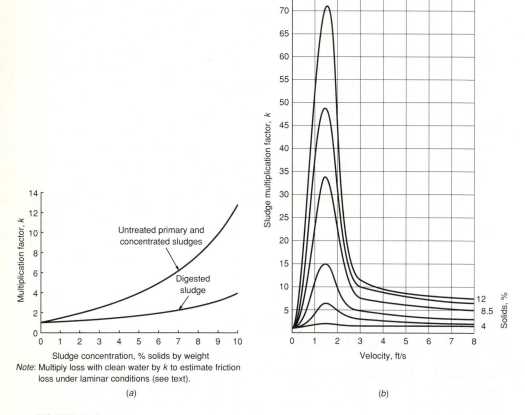

Note: Multiply loss with clean water by k to estimate friction loss under laminar conditions (see text).

(a)

(b)

FIGURE 12-6
Head loss multiplication factors: (a) for different sludge types and concentrations and (b) for different pipeline velocities and sludge concentrations.

pumped, the pump must handle a dilute sludge that has essentially the same hydraulic characteristics as water. This change in characteristics causes a centrifugal pump to operate farther out on its curve. The pump motor should be sized for the additional load, and a variable-speed drive should be supplied to reduce the flow under these conditions. If the pump motor is not sized for the maximum load obtainable when pumping water at top speed, it is likely to go out on overload or be damaged if the overload devices do not function or are set too high.

To determine the operating speeds and motor power required for a centrifugal pump handling sludge, system curves should be computed (1) for the most dense sludge anticipated, (2) for average conditions, and (3) for water. These system curves should be plotted on a graph of the pump curves for a range of available speeds. The maximum and minimum speeds required of a particular pump are obtained from the intersection of the pump head-capacity curves with the system curves at the desired capacity. Where the maximum speed head-capacity curve intersects the system curve for water determines the power required. In constructing the system curves for sludge for velocities from 0 to 3.5 ft/s (1.1 m/s), the headloss can be considered constant at the figure computed for 3.5 ft/s. The intersection of the pump curves with the system curve for average conditions can be used to estimate hours of operations, average speed, and power costs.

Because the usual flow formulas cannot be used in the plastic and laminar range, the engineer must rely on judgment and experience. In this range, capacities will be small, and plunger, progressive cavity, or rotary-lobe pumps should be used with ample head and capacity, as recommended previously.

Application of Rheology to Headloss Computations. For pumping sludge over long distances, an alternative method of computing headloss characteristics has been developed based on the flow properties of the sludge. A method of computing headloss for laminar-flow conditions was derived originally by Babbitt and Caldwell, based on the results of experimental and theoretical studies [3]. Additional studies have been performed for the transition from laminar to turbulent flow and are reported in Refs. 32 and 42 and are summarized in Ref. 38. Long-distance pumping of mixtures of untreated (raw) primary and secondary sludge is discussed in Ref. 6. The approach used in those studies for turbulent flow, which is of critical importance for long pipelines, is described below. For laminar and transitional flow, computational procedures described in Ref. 38 are recommended.

Water, oil, and most other common fluids are "Newtonian," which means the pressure drop is directly proportional to the velocity and viscosity under laminar-flow conditions. As the velocity increases past a critical value, the flow becomes turbulent. The transition from laminar to turbulent flow depends on the Reynolds number, which is inversely proportional to the fluid viscosity. Wastewater sludge, however, is a non-Newtonian fluid. The pressure drop under laminar conditions is not proportional to flow, so the viscosity is not a constant. The precise Reynolds number at which turbulent flow characteristics are encountered is uncertain for sludges.

Sludge has been found to behave much like a Bingham plastic, a substance with a straight-line relationship between shear stress and flow only after flow begins. A Bingham plastic is described by two constants: (1) the yield stress, s_y,

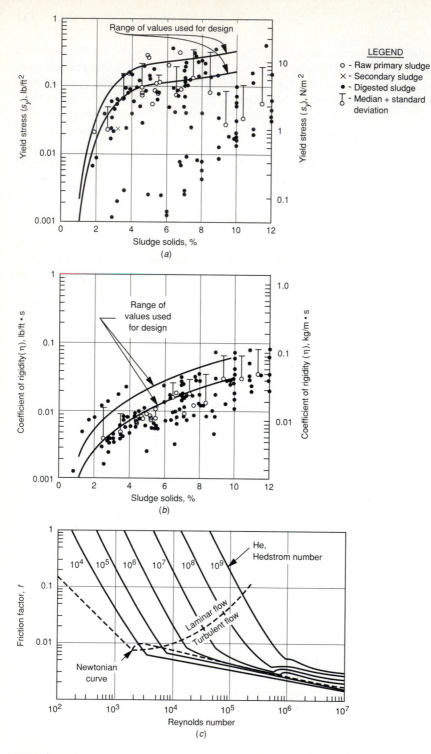

FIGURE 12-7
Curves for computing pipeline headloss by the sludge rheology method: (a) yield stress versus % sludge solids [32], (b) coefficient of rigidity versus % sludge solids [32], and (c) friction factor for sludge analyzed as a Bingham plastic [42].

and the coefficient of rigidity, η. Typical ranges of values for yield stress and coefficient of rigidity are shown in Figs. 12-7a and 12-7b [32]. If the two constants can be determined, the pressure drop over a wide range of velocities can be obtained using ordinary equations for water and the use of Fig. 12-7c [42]. As observed in Figs. 12-7a and 12-7b, published data quantifying yield stress and the coefficient of rigidity values for wastewater sludges are highly variable.
Pilot studies should be conducted to determine the rheological data for specific applications. Procedures for developing yield stress and the coefficient of rigidity are also given in Ref. 6.

Two dimensionless numbers can be used to determine the pressure drop due to friction for sludge: Reynolds number and Hedstrom number. Reynolds number is calculated by Eq. 12-3[42]:

$$Re = \frac{\gamma V D}{\eta} \qquad \text{U.S. customary units} \qquad (12\text{-}3)$$

$$Re = \frac{\rho V D}{\eta} \qquad \text{SI units} \qquad (12\text{-}3a)$$

where Re = Reynolds number, dimensionless
$\quad \gamma$ = specific weight of sludge, lb/ft^3
$\quad \rho$ = specific mass of sludge, kg/m^3
$\quad V$ = average velocity, ft/s (m/s)
$\quad D$ = diameter of pipe, ft (m)
$\quad \eta$ = coefficient of rigidity, lb/ft · s (kg/m · s)

Hedstrom number is calculated as follows [42]:

$$He = \frac{D^2 s_y g \gamma}{\eta^2} \qquad \text{U.S. customary units} \qquad (12\text{-}4)$$

$$He = \frac{D^2 s_y \rho}{\eta^2} \qquad \text{SI units} \qquad (12\text{-}4a)$$

where He = Hedstrom number, dimensionless
$\quad s_y$ = yield stress, lb$_f$/ft^2 (N/m^2)
$\quad g$ = 32.2 ft/s^2
Other terms are as defined previously.

Using the calculated Reynolds number and the Hedstrom number, the friction factor, f, can be determined from Fig. 12-7c. The pressure drop for turbulent conditions can then be calculated from the following relationship [42]:

$$\Delta p = \frac{2f \gamma L V^2}{g D} \qquad \text{U.S. customary units} \qquad (12\text{-}5)$$

$$\Delta p = \frac{2f \rho L V^2}{D} \qquad \text{SI units} \qquad (12\text{-}5a)$$

where Δp = pressure drop due to friction, lb_f/ft^2 (N/m^2)
f = friction factor (from Fig. 12-7c)
L = length of pipeline, ft (m)

In using Eqs. 12-3, 12-4, and 12-5, it should be noted that the Reynolds number is not the same as the Reynolds number based on viscosity. In plastic flow, an effective viscosity may be defined, but it is variable and can be much greater than the coefficient of rigidity. Consequently, the two Reynolds numbers can differ greatly. The friction factor f will usually differ significantly from the f values reported in standard hydraulic texts for clear water, which may be four times the values used in Fig. 12-7c. These equations apply to the entire range of laminar and turbulent flows, except that Fig. 12-7c does not allow for pipe roughness. To allow for pipe roughness, if ordinary water formulas give a higher pressure drop than Eq. 12-5, then roughness is dominant, the flow is fully turbulent, and the pressure drop given by the ordinary water formula will be reasonably accurate. A safety factor on the order of 1.5 is recommended for worst-case design conditions [32]. The use of Eq. 12-3, 12-4, and 12-5 is illustrated in Example 12-2.

Example 12-2   Computation of headloss using sludge rheology. Calculate the headloss in an 8 in pipeline, 33,000 ft long conveying untreated (raw) sludge at an average flowrate of 640 gal/min. Determine also if the flow is turbulent. By testing, the following sludge rheology data were found:

Yield stress, s_y, = 0.0325 lb/ft^2
Coefficient of rigidity, η, = 0.025 lb/ft · s
Specific gravity = 1.01

Solution

1. Calculate the pipeflow velocity
   (a) Determine the pipe cross-sectional area.

$$A = \pi \frac{D^2}{4} = 3.14 \times \frac{(0.667)^2}{4} = 0.349 \ ft^2$$

   (b) Determine velocity.

$$Q = 640 \ gal/min \ (1 \ ft^3/7.48 \ gal)(1 \ min/60 \ s) = 1.42 \ ft^3/s$$

$$V = \frac{Q}{A} = \frac{1.42 \ ft^3/s}{0.349 \ ft^2} = 4.07 \ ft/s$$

2. Compute sludge specific weight.

$$\gamma = 62.4 \ lb/ft^3 \times 1.01 = 63.0 \ lb/ft^3$$

3. Compute Reynolds number using Eq. 12-3.

$$Re = \frac{\gamma V D}{\eta} = \frac{(63.0 \ lb/ft^3)(4.07 \ ft/s)(0.667 \ ft)}{0.025 \ lb/ft \cdot s} = 6.8 \times 10^3$$

4. Compute Hedstrom number using Eq. 12-4.

$$He = \frac{D^2 s_y g \gamma}{\eta^2} = \frac{(0.667 \text{ ft})^2 (0.0325 \text{ lb/ft}^2)(32.2 \text{ ft/s}^2)(63.0 \text{ lb/ft}^3)}{(0.025 \text{ lb/ft} \cdot \text{s})^2}$$

$$He = 4.7 \times 10^4$$

5. Determine friction factor f from Fig. 12-7c using the computed Reynolds and Hedstrom numbers.

$$f = 0.007$$

Note, on Fig. 12-7c, that the flow is in the turbulent zone.

6. Compute pressure drop using Eq. 12-5.

$$\Delta p = \frac{2f \gamma L V^2}{gD} = \frac{2(0.007)(63.0 \text{ lb/ft}^3)(33,000 \text{ ft})(4.07 \text{ ft/s})^2}{(32.2 \text{ ft/s}^2)(0.667 \text{ ft})}$$

$$= 22,449 \text{ lb/ft}^2$$

Convert to feet of water:

$$\Delta p = \frac{22,449 \text{ lb/ft}^2}{62.4 \text{ lb/ft}^3} = 360 \text{ ft}$$

Comment. In this example, only one set of rheology data was used. In actual design, test data should be used for a range of probable conditions so that a family of headloss curves can be developed for the range of operating conditions. In addition, appropriate safety factors should be used for worst-case conditions. Comparison of the headloss to the headloss for water using the Hazen-Williams formula is left as a homework problem.

Sludge Piping

In treatment plants, conventional sludge piping should not be smaller than 6 in (150 mm) in diameter, although smaller diameter glass-lined pipes have been used successfully. Pipe sizes need not be larger than 8 in (200 mm), unless the velocity exceeds 5 to 6 ft/s (1.5 to 1.8 m/s), in which case, the pipe is sized to maintain that velocity. Gravity sludge withdrawal lines should not be less than 8 in (200 mm) in diameter. It is common practice to install a number of cleanouts in the form of plugged tees or crosses instead of elbows so that the lines can be rodded if necessary. Pump connections should not be smaller than 4 in (100 mm) in diameter.

Grease has a tendency to coat the inside of piping used for transporting primary sludge and scum. Grease accumulation is more of a problem in large plants than in small ones. The coating results in a decrease in the effective diameter and a large increase in pumping head. For this reason, low-capacity positive-displacement pumps are designed for heads greatly in excess of the theoretical head. Centrifugal pumps, with their larger capacity, are used to pump a more dilute sludge, often containing some wastewater. Buildup of head due to grease accumulations appears to occur more slowly in systems where more dilute sludges are pumped. In some plants, provisions have been made for melting the grease by circulating hot water, steam, or digester supernatant through the main sludge lines.

In treatment plants, friction losses are usually low because of short pipe runs, and there is little difficulty in providing an ample safety factor. In the design of long sludge lines, special design features should be considered including (1) providing two pipes unless a single pipe can be shutdown for several days without causing problems, (2) providing for external corrosion and pipe loads, (3) adding facilities for applying dilution water for flushing the line, (4) providing means to insert a pipe cleaner at the treatment plant, (5) including provisions for steam injection, (6) providing air relief and blowoff valves for the high and low points, respectively, and (7) considering the potential effects of waterhammer. A discussion of waterhammer in force mains is provided in the companion volume to this text [26].

12-5   PRELIMINARY OPERATIONS

Sludge grinding, degritting, blending, and storage are necessary to provide a relatively constant, homogeneous feed to sludge-processing facilities. Blending and storage can be accomplished either in a single unit designed to do both or separately in other plant components.

Sludge Grinding

Sludge grinding is a process in which large and stringy material contained in sludge is cut or sheared into small particles to prevent the clogging of or wrapping around rotating equipment (see Fig. 12-8). Some of the processes that must be preceded by sludge grinders and the purposes of grinding are reported in Table 12-11. Grinders historically have required high maintenance, but newer designs of slow-speed grinders have been more durable and reliable. These designs include improved bearings and seals, hardened steel cutters, overload sensors, and mechanisms that reverse the cutter rotation to clear obstructions or shut down the unit if the obstruction cannot be cleared.

Sludge Degritting

In some plants where separate grit removal facilities are not used ahead of the primary sedimentation tanks or where the grit removal facilities are not adequate to handle peak flows and peak grit loads, it may be necessary to remove the grit before further processing of the sludge. Where further thickening of the primary sludge is desired, a practical consideration is sludge degritting. The most effective method of degritting sludge is through the application of centrifugal forces in a flowing system to achieve separation of the grit particles from the organic sludge. Such separation is achieved through the use of cyclone degritters, which have no moving parts. The sludge is applied tangentially to a cylindrical-feed section, thus imparting a centrifugal force. The heavier grit particles move to the outside of the cylinder section and are discharged through a conical-feed section. The organic sludge is discharged through a separate outlet.

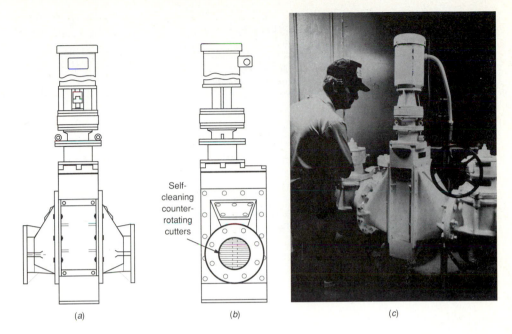

(a)   (b)   (c)

FIGURE 12-8
In-line sludge grinder: (a) side view, (b) end view, and (c) typical installation (from Disposable Waste Systems, Inc.).

The efficiency of the cyclone degritter is affected by pressure and by the concentration of the organics in the sludge. To obtain effective grit separation, the sludge must be relatively dilute. As the sludge concentration increases, the particle size that can be removed decreases. The general relationship between sludge concentration

TABLE 12-11
Operations or processes requiring the grinding of sludge

Operation or process	Purpose of grinding
Pumping with progressive cavity pumps	To prevent clogging and reduce wear.
Solid bowl centrifuges	To prevent clogging. Large solid bowl units generally can handle larger particles and may not require sludge grinding.
Belt filter press	To prevent clogging of the sludge distribution system, to prevent warping of rollers, and to provide more uniform dewatering.
Heat treatment	To prevent clogging of high-pressure pumps and heat exchangers.
Chlorine oxidation	To enhance chlorine contact with sludge particles.

TABLE 12-12
Grit removal efficiency using cyclone degritters for primary sludges[a, b]

Primary sludge concentration, % total solids	Mesh size of material removed[c]
1	150
2	100
3	65
4	28–35

[a] Ref. 57.

[b] For a 12 in (0.3 m) hydroclone at 6 lb_f/in^2 gage (42 kN/m^2) at 205 gal/min (12.9 L/s).

[c] About 95 percent or more of indicated particle size is removed.

Note: Normal design range is for 1–1.5% feed solids.

and the effectiveness of removal for primary sludges is shown in Table 12-12. When a cyclone degritter is used, the degritted sludge is usually discharged to a thickener to increase the solids concentration.

Sludge Blending

Sludge is generated in primary, secondary, and advanced wastewater treatment processes. Primary sludge consists of settleable solids carried in the raw wastewater. Secondary sludge consists of biological solids as well as additional settleable solids. Sludge produced in the advanced wastewater may consist of biological and chemical solids. Sludge is blended to produce a uniform mixture to downstream operations and processes. Uniform mixtures are most important in short detention time systems, such as sludge dewatering, heat treatment, and incineration. Provision of a well-blended sludge with consistent characteristics to these treatment units will greatly enhance plant operability and performance.

Sludge from primary, secondary, and advanced processes can be blended in several ways:

1. *In primary settling tanks.* Secondary or advanced wastewater treatment sludges can be returned to the primary settling tanks, where they will settle and mix with the primary sludge.
2. *In pipes.* This procedure requires careful control of sludge sources and feed rates to ensure the proper blend. Without careful control, wide variations in sludge consistency may be expected.
3. *In sludge processing facilities with long detention times.* Aerobic and anaerobic digesters (complete-mix type) can blend the feed sludges uniformly.
4. *In a separate blending tank.* This practice provides the best opportunity to control the quality of the blended sludges.

In treatment plants of less than 1 Mgal/d (0.044 m³/s) capacity, blending is usually accomplished in the primary settling tanks. In large facilities, optimum efficiency is achieved by separately thickening sludges before blending. Blending tanks are usually equipped with mechanical mixers and baffles to ensure good mixing.

Sludge Storage

Sludge storage must be provided to smooth out fluctuations in the rate of sludge production and to allow sludge to accumulate during periods when subsequent sludge-processing facilities are not operating (e.g., night shifts, weekends, and periods of unscheduled equipment downtime). Sludge storage is particularly important in providing a uniform feed rate ahead of the following processes: lime stabilization, heat treatment, mechanical dewatering, drying, and thermal reduction.

Short-term sludge storage may be accomplished in wastewater-settling tanks or in sludge-thickening tanks. Long-term sludge storage may be accomplished in sludge stabilization processes with long detention times (e.g., aerobic and anaerobic digestion) or in specially designed separate tanks. In small installations, sludge is usually stored in the settling tanks and digesters. In large installations that do not use aerobic and anaerobic digestion, sludge is often stored in separate blending and storage tanks. Such tanks may be sized to retain the sludge for a period of several hours to a few days. If sludge is stored longer than two to three days, it will deteriorate and will be more difficult to dewater. The determination of the required storage volume is illustrated in Example 12-3. Sludge is often aerated to prevent septicity and to promote mixing. Mechanical mixing may be necessary to assure complete blending of the sludge. Chlorine and hydrogen peroxide have been used with limited success to arrest septicity and to control the odors from sludge storage and blending tanks. Sodium hydroxide or lime may also be used for odor control by raising the pH and keeping the hydrogen sulfide in solution.

Example 12-3 Determination of volume required for sludge storage. Assume that the yearly average rate of sludge production from an activated-sludge treatment plant is 26,500 lb/d (12,000 kg/d). Develop a curve of sustained sludge mass-loading rates that can be used to determine the size of sludge storage facilities required with various downstream sludge-processing units. Then, using this curve, determine the volume required for sludge storage, assuming that sludge accumulated for 7 d is to be processed in 5 working days and that sludge accumulated for 14 d is to be processed in 10 working days. Note that the 5 and 10 d work periods correspond to 1 and 2 weeks, respectively, assuming that certain sludge-processing facilities, such as belt filter presses, will not be operated on the weekends.

Solution

1. Develop a curve of sustained sludge mass loadings.
   (a) Because no information is specified, it will be assumed that the sustained sludge production will mirror the sustained BOD plant loadings given in Fig. 5-6a and used in Example 5-3.
   (b) Set up an appropriate computation table and compute the values necessary to plot the curve.

(1) Length of sustained peak, d	(2) Peaking factor[a]	(3) Peak solids mass loading, lb/d	(4) Total sustained loading, lb[b]
1	2.4	63,600	63,600
2	2.1	55,650	111,300
3	1.9	50,350	151,050
4	1.8	47,700	190,800
5	1.7	45,050	225,250
10	1.4	37,100	371,000
15	1.3	34,450	516,750
365	1.0	26,500	

[a] From Fig. 5-6a.
[b] Total mass produced for the corresponding sustained period given in column 1.

(c) Plot the sustained solids loading curve (see following figure).

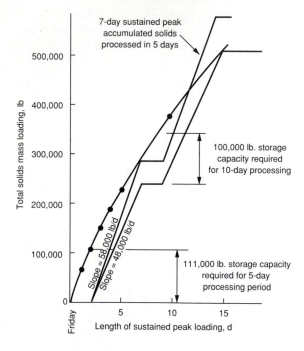

2. Determine the sludge storage volume required for the stated operating conditions.

(a) Determine the daily rate at which sludge must be processed to handle the 7 d sustained peak (from figure) in 5 working days.

$$\text{lb/d} = \frac{290,000 \text{ lb}}{5 \text{ d}} = 58,000 \text{ lb/d}$$

(b) Determine the daily rate at which sludge must be processed to handle the 14 d sustained peak (from figure) in 10 working days.

$$\text{lb/d} = \frac{480,000 \text{ lb}}{10 \text{ d}} = 48,000 \text{ lb/d}$$

(c) Assuming that the sludge storage facilities are empty on Friday, plot on the figure the average daily rate at which sludge must be processed during the 5 and 10 d periods.

(d) From the figure, the required storage capacity in pounds of solids is

    i. Capacity based on 5 working days = 111,000 lb

    ii. Capacity based on 10 working days = 100,000 lb

Comment. The downstream processing equipment can now be sized using the daily rate at which sludge must be processed. For example, if the number of pounds per hour that can be processed with a belt filter press is known, then the size and number of units can be computed from the number of shifts to be used per day and the assumed value of the actual working hours per shift. In sizing equipment, a tradeoff analysis should always be performed between the cost of storage and processing facilities versus labor costs (for both one shift and two shifts) to determine the most cost-effective combination.

12-6 THICKENING (CONCENTRATION)

The solids content of primary, activated, trickling-filter, or mixed sludge (i.e., primary plus activated) varies considerably, depending on the characteristics of the sludge, the sludge removal and pumping facilities, and the method of operation. Representative values of percent total solids from various treatment operations or processes are shown in Table 12-8. Thickening is a procedure used to increase the solids content of sludge by removing a portion of the liquid fraction. To illustrate, if waste activated sludge, which is typically pumped from secondary settling tanks with a content of 0.8 percent solids, can be thickened to a content of 4 percent solids, then a five-fold decrease in sludge volume is achieved. Thickening is generally accomplished by physical means, including gravity settling, flotation, centrifugation, and gravity belts. Typical sludge-thickening methods are described in Table 12-13.

Application

The volume reduction obtained by sludge concentration is beneficial to subsequent treatment processes, such as digestion, dewatering, drying, and combustion, from the following standpoints: (1) capacity of tanks and equipment required, (2) quantity of chemicals required for sludge conditioning, and (3) amount of heat required by digesters and amount of auxiliary fuel required for heat drying or incineration, or both.

On large projects where sludge must be transported a significant distance, such as to a separate plant for processing, a reduction in sludge volume may result in a reduction of pipe size and pumping costs. On small projects, the requirements of a minimum practicable pipe size and minimum velocity may necessitate the pumping of significant volumes of wastewater in addition to sludge, thereby diminishing the value of volume reduction. Volume reduction is very desirable when liquid sludge is transported by tank trucks for direct application to land as a soil conditioner.

Sludge thickening is achieved at all wastewater treatment plants in some manner—in the primary clarifiers, in sludge digestion facilities, or in specially designed

TABLE 12-13
Occurrence of thickening methods in sludge processing

Method	Type of sludge	Frequency of use and relative success
Gravity	Untreated primary	Commonly used with excellent results. Sometimes used with hydroclone degritting of sludge.
Gravity	Untreated primary and waste activated sludge	Often used. For small plants, generally satisfactory results with sludge concentrations in the range of 4 to 6%. For large plants, results are marginal.
Gravity	Waste activated sludge	Seldom used; poor solids concentration (2 to 3%).
Dissolved-air flotation	Untreated primary and waste activated sludge	Some limited use; results similar to gravity thickeners.
Dissolved-air flotation	Waste activated sludge	Commonly used; good results (3.5 to 5% solids concentration).
Imperforate basket centrifuge	Waste activated sludge	Limited use; excellent results (8 to 10% solids concentration).
Solid bowl centrifuge	Waste activated sludge	Increasing; good results (4 to 6% solids concentration).
Gravity belt thickener	Waste activated sludge	Increasing; good results (3 to 6% solids concentration).
Rotary drum thickener	Waste activated sludge	Limited use; good results (5 to 9% solids concentration).

separate units. If separate units are used, the recycled flows are normally returned to the wastewater treatment facilities. In treatment plants with less than 1 Mgal/d (0.044 m³/s) capacity, separate sludge thickening is seldom practiced. In small plants, gravity thickening is accomplished in the primary settling tank or in the sludge digestion units, or both. In larger treatment facilities, the additional costs of separate sludge thickening are often justified by the improved control over the thickening process and the higher concentrations attainable.

Description of Thickening Equipment

The following discussion is intended to introduce the reader to the equipment used for the thickening of sludges. Most of the equipment is mechanical; therefore, the design engineer is usually more concerned with its proper application to meet a given treatment objective than with the theory of mechanical design.

Gravity Thickening. Gravity thickening is accomplished in a tank similar in design to a conventional sedimentation tank. Normally, a circular tank is used. Dilute sludge is fed to a center-feed well. The feed sludge is allowed to settle and compact, and the thickened sludge is withdrawn from the bottom of the tank. Conventional sludge-collecting mechanisms with deep trusses (see Fig. 12-9) or vertical pickets are used to

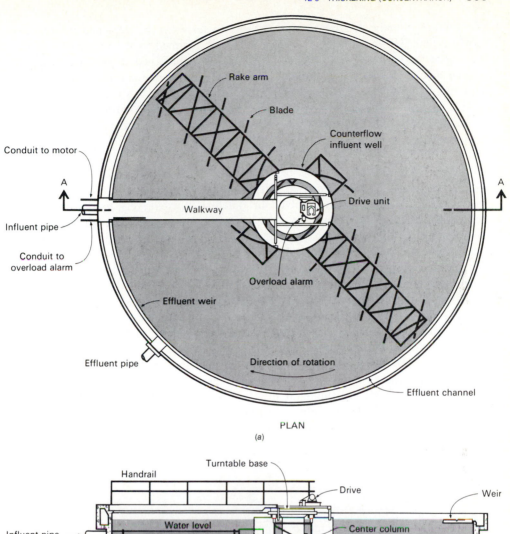

PLAN

(a)

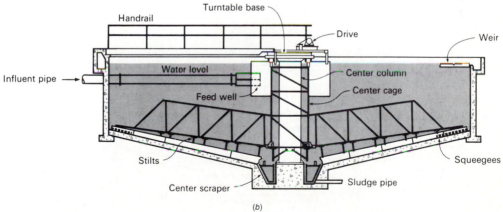

(b)

FIGURE 12-9
Schematic of gravity thickener: (a) plan and (b) section A-A (from Dorr-Oliver).

stir the sludge gently, thereby opening up channels for water to escape and promoting densification. The supernatant flow that results is returned to the primary settling tank or to the headworks of the treatment plant. The thickened sludge that collects on the bottom of the tank is pumped to the digesters or dewatering equipment as required; thus, storage space must be provided for the sludge. As indicated in Table 12-13, gravity thickening is most effective on primary sludge. Provisions for dilution water and occasional chlorine addition are frequently included to improve process performance.

Flotation Thickening. As described in Chap. 6, there are three basic variations of the flotation thickening operation: dissolved-air flotation, vacuum flotation, and dispersed-air flotation. Only dissolved-air flotation is extensively used for sludge thickening in the United States. In dissolved-air flotation, air is introduced into a solution that is being held at an elevated pressure. A typical unit used for thickening waste activated sludge is shown in Fig. 12-10. When the solution is depressurized, the dissolved air is released as finely divided bubbles carrying the sludge to the top, where it is removed. In locations where freezing is a problem or where odor control is of concern, flotation thickeners are normally enclosed in a building.

Flotation thickening is used most efficiently for waste sludges from suspended-growth biological treatment processes, such as the activated sludge process or the suspended-growth nitrification process. Other sludges such as primary sludge, trickling-filter humus, aerobically digested sludge, and sludges containing metal salts from chemical treatment have also been flotation thickened.

The float solids concentration that can be obtained by flotation thickening of waste activated sludge is influenced primarily by the air-to-solids ratio, sludge characteristics (in particular the sludge volume index, SVI), solids-loading rate, and

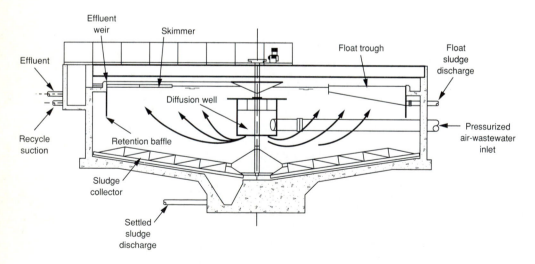

FIGURE 12-10
Dissolved-air flotation unit used for thickening waste-activated sludge (from Envirex, Inc.).

polymer application. Although float solids concentrations have ranged historically between 3 and 6 percent by weight, float solids concentration is difficult to predict during the design stage without bench or pilot plant testing [57]. The air-to-solids ratio is probably the most important factor affecting performance of the flotation thickener and is defined as the weight ratio of air available for flotation to the solids to be floated in the feed stream. The air-to-solids ratio at which the float solids concentration is maximized varies from 2 to 4 percent. The SVI is also important because better thickening performance has been reported when the SVI is less than 200, using nominal polymer dosages. At high SVIs, the float concentration deteriorates and high polymer dosages are required. Performance data also indicate that float concentrations have declined at high solids-loading rates, in excess of 100 $lb/ft^2 \cdot d$ ($\sim$470 $kg/m^2 \cdot d$) [57].

Centrifugal Thickening. Centrifuges are used both to thicken and to dewater sludges. As indicated in Table 12-13, their application in thickening is limited normally to waste activated sludge. Thickening by centrifugation involves the settling of sludge particles under the influence of centrifugal forces. The two basic types of centrifuges currently used for sludge thickening are solid bowl and imperforate basket centrifuges (see Fig. 12-11).

The solid bowl centrifuge consists of a long bowl, normally mounted horizontally and tapered at one end. Sludge is introduced into the unit continuously and the

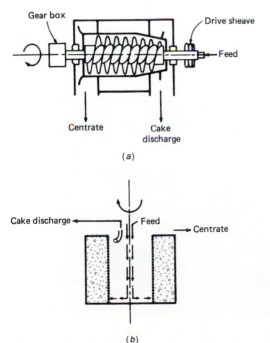

FIGURE 12-11
Centrifuges used for the thickening of sludge: (a) solid bowl and (b) imperforate basket.

solids concentrate on the periphery. A helical scroll, spinning at a slightly different speed, moves the accumulated sludge toward the tapered end where additional solids concentration occurs. The sludge is then discharged.

The imperforate basket centrifuge operates on a batch basis. The liquid sludge is introduced into a vertically mounted spinning bowl. The solids accumulate against the wall of the bowl and the centrate is decanted. When the solids-holding capacity of the machine has been achieved (usually 60 to 85 percent of the maximum depth), the bowl decelerates and a scraper is positioned in the bowl to help remove the accumulated solids. The imperforate basket centrifuge is particularly well-suited for soft or fine solids that are difficult to filter or where the nature of the solids varies widely.

Under normal conditions, thickening can be accomplished by centrifugal thickening without polymer addition. Maintenance and power costs for the centrifugal thickening process, however, can be substantial. Therefore, the process is usually attractive only at facilities larger than 5 Mgal/d (0.2 m³/s), where space is limited and skilled operators are available, or for sludges that are difficult to thicken by more conventional means.

Gravity Belt Thickening. Gravity belt thickeners are a recent development in sludge thickening, stemming from the application of belt presses for sludge dewatering (See Fig. 12-12). In belt press dewatering, particularly for sludges with solids concentrations of less than 2 percent, effective thickening occurs in the gravity drainage

FIGURE 12-12
Gravity belt thickener (from Ashbrook-Simon-Hartley).

section of the press. The equipment developed for thickening consists of a gravity belt that moves over rollers driven by a variable-speed drive unit. The sludge is conditioned with polymer and fed into a feed/distribution box at one end. The box is used to distribute the sludge evenly across the width of the moving belt as the water drains through and the sludge is carried toward the discharge end of the thickener. The sludge is ridged and furrowed by a series of plow blades placed along the travel of the belt, allowing the water released from the sludge to pass through the belt. After the thickened sludge is removed, the belt travels through a wash cycle. The gravity belt thickener has been used for thickening raw and digested sludges; polymer addition is required. Additional information on gravity belt thickeners may be found in Ref. 18.

Rotary Drum Thickening. Rotary media-covered drums are also used to thicken sludges. A rotary drum thickening system consists of a waste activated-sludge conditioning system (including a polymer feed system) and rotating cylindrical screens (see Fig. 12-13). Polymer is mixed with thin sludge in the mixing and conditioning drum. The conditioned sludge is then passed to rotating screen drums, which separate the flocculated solids from water. Thickened sludge rolls out the end of the drums, while separated water decants through the screens. A thickening range of 3 to 4 percent has been reported for waste activated sludge. Polymer addition is required. Advantages of rotary drum thickeners are low maintenance, low energy use, and small space requirements. Recent designs also allow coupling of the rotary drum unit with a belt filter press for combination thickening and dewatering.

FIGURE 12-13
Rotary drum thickener (from Ralph B. Carter Co.).

Design of Thickeners

In designing thickening facilities, it is important to (1) provide adequate capacity to meet peak demands and (2) prevent septicity, with its attendant odor problems, during the thickening process. The design of three types of thickeners commonly used in wastewater treatment plants (gravity, dissolved-air flotation, and centrifugal) is discussed in this section.

Gravity Thickeners. Gravity thickeners are sized on the basis of solids loading. Typical solids loadings are listed in Table 12-14. To maintain aerobic conditions in

TABLE 12-14
Typical concentrations of unthickened and thickened sludges and solids loadings for gravity thickeners [a]

Type of sludge	Sludge concentration, percent		Solids loading for gravity thickeners, $lb/ft^2 \cdot d$
	Unthickened	Thickened	
Separate			
Primary sludge	2–7	5–10	18–28
Trickling-filter humus sludge	1–4	3–6	7–10
Rotating biological contactor	1–3.5	2–5	7–10
Air activated sludge	0.5–1.5	2–3	2.5–7
High-purity oxygen activated sludge	0.5–1.5	2–3	2.5–7
Extended aeration activated sludge	0.2–1.0	2–3	5–7
Anaerobically digested primary sludge from primary digester	8	12	25
Combined			
Primary and trickling-filter humus sludge	2–6	4–9	12–20
Primary and rotating biological contactor	2–6	4–8	10–16
Primary and modified aeration sludge	3–4	5–10	12–20
Primary and air activated sludge	2–5	2–8	8–16
Waste activated sludge and trickling-filter humus sludge	0.5–2.5	2–4	2.5–7
Anaerobically digested primary and waste activated sludge	4	8	14
Thermally conditioned			
Primary sludge	3–6	12–15	40–50
Primary and waste activated sludge	3–6	8–15	28–40
Waste activated sludge	0.5–1.5	6–10	20–28

[a] Adapted in part from Ref. 42.
Note: $lb/ft^2 \cdot d \times 4.8824 = kg/m^2 \cdot d$

gravity thickeners, provisions should be made for adding 600 to 750 gal/ft$^2 \cdot$ d (24 to 30 m^3/m$^2 \cdot$ d) of final effluent to the thickening tank.

In operation, a sludge blanket is maintained on the bottom of the thickener to aid in concentrating the sludge. An operating variable is the sludge volume ratio, which is the volume of the sludge blanket held in the thickener divided by the volume of the thickened sludge removed daily. Values of the sludge volume ratio normally range between 0.5 and 20 d; the lower values are required during warm weather. Alternatively, sludge-blanket depth should be measured. Blanket depths may range from 2 to 8 ft (0.6 to 2.4 m); shallower depths are maintained in the warmer months.

Flotation Thickeners. Higher loadings can be used with flotation thickeners than are permissible with gravity thickeners because of the rapid separation of solids from the wastewater. Flotation thickeners may be operated at the solids loadings given in Table 12-15. For design, the minimum loadings should be used. The higher solids loadings generally result in lower concentrations of thickened sludge. Typical polymer dosages for flotation and other types of thickeners are reported in Table 12-16. The pressure and air-to-solids ratio requirements for dissolved-air flotation thickeners are discussed in Chap. 6.

Primary tank effluent or plant effluent is recommended as the source of air-charged water rather than flotation tank effluent, except when chemical aids are used, because of the possibility of fouling the air-pressure system with solids. The use of polymers as flotation aids is effective in increasing the solids recovery in the floated sludge from 85 to 98 or 99 percent and in reducing the recycle loads.

Centrifuge Thickeners. The performance of a centrifuge is often quantified by the percent capture, which is defined as

$$\text{Percent capture} = \left[1 - \frac{C_r(C_c - C_s)}{C_s(C_c - C_r)}\right]100 \qquad (12\text{-}6)$$

TABLE 12-15
Typical solids loadings for dissolved-air flotation units [a, b]

	Loading, lb/ft$^2 \cdot$ d	
Type of sludge	Without chemical addition	With chemicals
Air activated sludge	10	Up to 45
High-purity oxygen activated sludge	14–20	Up to 55
Trickling-filter humus sludge	14–20	Up to 45
Primary + air activated sludge	14–30	Up to 45
Primary + trickling-filter humus sludge	20–30	Up to 60
Primary only	20–30	Up to 60

[a] Adapted in part from Ref. 42.
[b] Loading rates necessary to produce a minimum 4 percent solids concentration in the float.
Note: lb/ft$^2 \cdot$ d $\times$ 4.8824 = kg/m$^2 \cdot$ d

TABLE 12-16
Typical levels of polymer addition for various sludges and for various methods of thickening[a]

	lb of dry polymer added/ton of dry solids			
Type of sludge	Dissolved-air flotation unit	Solid bowl centrifuge	Basket centrifuge	Gravity belt filter
Waste activated	4–10	0–8	2–6	6–14
Aerobically digested		8–16		
Anaerobically digested		8–16		

[a] Adapted from Refs. 18 and 61.
Note: lb/ton × 0.5 = kg/10^{3}kg

where C_r = concentration of solids in reject wastewater (centrate), mg/L, %
$\quad\quad C_c$ = concentration of solids in the cake, mg/L, %
$\quad\quad C_s$ = concentration of solids in sludge feed, mg/L, %

For a constant feed concentration, the percent capture increases as the concentration of solids in the centrate decreases. In concentrating sludge solids, capture is important if a minimum amount of solids is to be returned to the treatment process. Many systems are designed with standby polymer systems for use when capture rates have to be increased.

The principal operational variables include the following: (1) characteristics of the feed sludge (its water-holding structure and the sludge volume index), (2) rotational speed, (3) hydraulic-loading rate, (4) depth of the liquid pool in the bowl (solid bowl machines), (5) differential speed of the screw conveyor (solid bowl machines), and (6) the need for polymers to improve the performance [57]. Because the interrelationships of these variables will be different in each location, specific design recommendations are not available; in fact, bench scale or pilot plant tests are recommended.

12-7 STABILIZATION

Sludges are stabilized to (1) reduce pathogens, (2) eliminate offensive odors, and (3) inhibit, reduce, or eliminate the potential for putrefaction. The success in achieving these objectives is related to the effects of the stabilization operation or process on the volatile or organic fraction of the sludge. Survival of pathogens, release of odors, and putrefaction occur when microorganisms are allowed to flourish in the organic fraction of the sludge. The means to eliminate these nuisance conditions through stabilization are (1) the biological reduction of volatile content, (2) the chemical oxidation of volatile matter, (3) the addition of chemicals to the sludge to render it unsuitable for the survival of microorganisms, and (4) the application of heat to disinfect or sterilize the sludge.

When designing a sludge stabilization process, it is important to consider the sludge quantity to be treated, the integration of the stabilization process with the other treatment units, and the objectives of the stabilization process. The objectives of the stabilization process are often affected by existing or pending regulations. If sludge is to be applied on land, pathogen reduction by various methods of sludge stabilization has to be considered. The effects of regulations on land application of sludge are discussed in Sec. 12-17.

The technologies for sludge stabilization discussed in this text include (1) lime stabilization, (2) heat treatment, (3) anaerobic digestion, (4) aerobic digestion, and (5) composting. The first two processes are discussed in this section. Because of the importance of the anaerobic and aerobic digestion processes and composting, they are discussed separately in Secs. 12-8, 12-9, and 12-10, respectively. Chemical oxidation with chlorine is rarely used; for information on chlorine oxidation, Ref. 25 may be consulted.

Lime Stabilization

In the lime stabilization process, lime is added to untreated sludge in sufficient quantity to raise the pH to 12 or higher. The high pH creates an environment that is not conducive to the survival of microorganisms. Consequently, the sludge will not putrefy, create odors, or pose a health hazard, so long as the pH is maintained at this level. Two methods of lime stabilization used are (1) addition of lime to sludge prior to dewatering, termed "lime pre-treatment" and (2) the addition of lime to sludge after dewatering, or "lime post-treatment." Either hydrated lime, $Ca(OH)_2$, or quicklime, CaO, may be used for lime stabilization. Fly ash, cement kiln dust, and carbide lime have also been used as a substitute for lime in some cases.

Lime Pre-Treatment. Lime pre-treatment of liquid sludge requires more lime per unit weight of sludge processed than that necessary for dewatering. The higher lime dose is necessary to attain the required higher pH. In addition, sufficient contact time must be provided before dewatering so as to effect a high level of pathogen kill. The recommended design objective is to maintain the pH above 12 for about two hours so as to ensure pathogen destruction (the minimum EPA criteria for lime stabilization) and to provide enough residual alkalinity so that the pH does not drop below 11 for several days [42]. The lime dosage required varies with the type of sludge and solids concentration. Typical dosages are reported in Table 12-17. Generally, as

TABLE 12-17
Typical lime dosages for stabilizing liquid sludge [a]

Type of sludge	Solids concentration, %		Lime dosage, lb $Ca(OH)_2$/lb dry solids [b]	
	Range	Average	Range	Average
Primary	3–6	4.3	120–340	240
Waste activated	1–1.5	1.3	420–860	600
Aerobically digested mixed	6–7	6.5	280–500	380
Septage	1–4.5	2.7	180–1,020	400

[a] Adapted from Ref. 59.
[b] Amount of $Ca(OH)_2$ required to maintain a pH of 12 for 30 minutes.

the percent solids concentration increases, the required lime dose decreases for a constant temperature increase. Testing should be performed for specific applications to determine the actual dosage requirements.

Because lime stabilization does not destroy the organics necessary for bacterial growth, the sludge must be treated with an excess of lime or disposed of before the pH drops significantly. An excess dosage of lime may range up to 1.5 times the amount needed to maintain the initial pH of 12. For additional details about pH decay following lime stabilization, Ref. 59 may be consulted.

Lime Post-Treatment. Although the use of lime to stabilize organic matter is not a new concept, post-treatment of dewatered wastewater treatment plant sludge using lime is a relatively recent development. In this process, hydrated lime or quicklime is mixed with dewatered sludge in a pugmill, paddle mixer, or screw conveyor to raise the pH of the mixture. Quicklime is preferred because the exothermic reaction of quicklime and water can raise the temperature of the mixture above 50°C, sufficient to inactivate worm eggs. The theoretical temperature increase by the addition of quicklime is illustrated in Fig. 12-14.

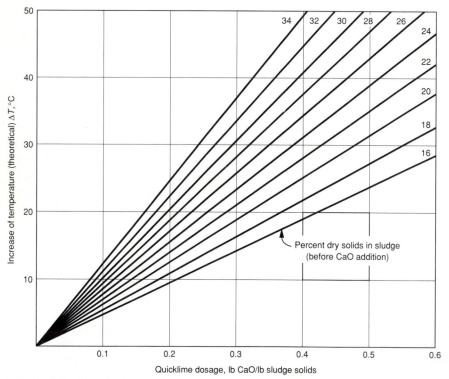

Note: In practice, higher temperature values are expected. 1.8 (°C) + 32 = °F

FIGURE 12-14
Theoretical temperature increase in post-lime stabilized sludge using quicklime [37].

Post-lime stabilization has several significant advantages when compared to pre-lime stabilization: (1) dry lime can be used; therefore, no additional water is added to the dewatered sludge; (2) there are no special requirements for dewatering; and (3) scaling problems and associated maintenance problems of lime sludge dewatering equipment are eliminated. Adequate mixing is critical for a post-lime stabilization system so as to avoid pockets of putrescible matreial. A post-lime stabilization system consists typically of a dry lime feed system, dewatered sludge cake conveyor, and a lime-sludge mixer (see Fig. 12-15). Good mixing is especially important to ensure contact between lime and small particles of sludge. When the lime and sludge are well-mixed, the resulting mixture has a crumbly texture, which allows it to be stored for long periods or easily distributed on land by a conventional manure spreader [37].

Heat Treatment

Heat treatment is a continuous process in which sludge is heated in a pressure vessel to temperatures up to 500°F (260°C) at pressures up to 400 lb_f/in^2 gage (2760 kN/m^2) for short periods of time (approximately 30 minutes). Heat treatment serves essentially as both a stabilization process and a conditioning process; in most cases, it is classified as a conditioning process. Heat treatment conditions the sludge by rendering the solids capable of being dewatered without the use of chemicals. When the sludge is subjected to the high temperatures and pressures, the thermal activity releases bound water and results in the coagulation of solids. In addition, hydrolysis of proteinaceous materials occurs, resulting in cell destruction and release of soluble organic compounds and ammonia nitrogen. This method of treatment is considered in greater detail in Sec. 12-11, where sludge conditioning is discussed.

12-8 ANAEROBIC SLUDGE DIGESTION

Anaerobic digestion is among the oldest forms of biological wastewater treatment, and its history can be traced from the 1850s with the development of the first tank designed

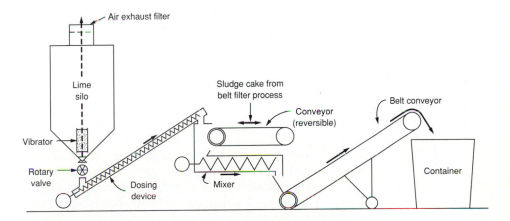

FIGURE 12-15
Typical lime post-treatment system (from Roediger Pittsburgh).

to separate and retain solids. One of the first installations in the United States using separate digestion tanks was the wastewater treatment plant in Baltimore, Maryland. Three rectangular digestion tanks were built as part of the original plant in 1911. In the period from 1920 to 1935, the anaerobic digestion process was studied extensively. Heat was applied to separate digestion tanks, and major improvements were made in the design of the tanks and associated appurtenances. It is interesting to note that the same practice is being followed today, but great progress has been made in the fundamental understanding and control of the process, the sizing of tanks, and the design and application of equipment. Because of the emphasis on energy conservation and recovery and the desirability of obtaining beneficial use of wastewater sludge, anaerobic digestion continues to be the dominant sludge stabilization process.

Process Description

The microbiology of anaerobic digestion and the optimum environmental conditions for the microorganisms involved are discussed in Chap. 8. The operation and physical facilities for anaerobic digestion in standard-rate, single-stage high-rate, two-stage, and separate digesters are described in this section. The processes described below normally operate in the mesophilic range, between 85 to 100°F (30 to 38°C). Thermophilic digestion is discussed at the end of the section

Standard-Rate Digestion. Standard-rate sludge digestion is usually carried out as a single-stage process (see Fig. 8-29*a*). The functions of digestion, sludge thickening, and supernatant formation are carried out simultaneously. A cross section of a typical standard-rate digester is shown in Fig. 12-16. Operationally, in a single-stage process, untreated sludge is added in the zone where the sludge is actively digesting and the gas is being released. The sludge is heated by means of an external heat exchanger. As gas rises to the surface, it lifts sludge particles and other materials, such as grease, oils, and fats, ultimately giving rise to the formation of a scum layer.

As a result of digestion, the sludge stratifies by forming a supernatant layer above the digesting sludge and becomes more mineralized (for example, the percentage of fixed solids increases). The biochemistry of the reactions taking place in the digesting zone is described in Chap. 8. As a result of the stratification and the lack of intimate mixing, not more than 50 percent of the volume of a standard-rate single-stage digester is used. Because of these limitations, the standard-rate process is used principally for small installations.

Single-Stage High-Rate Digestion. The single-stage high-rate digestion process differs from the standard-rate single-stage process in that the solids-loading rate is much greater (see "Process Design"). The sludge is mixed intimately by gas recirculation, mechanical mixers, pumping, or draft tube mixers (separation of scum and supernatant does not take place), and sludge is heated to achieve optimum digestion rates (see Fig. 12-17). With the exception of higher loading rates and improved mixing, there are only a few differences between the primary digester in a conventional two-stage process and a single-stage high-rate digester. The mixing equipment should have greater capacity and should reach to the bottom of the tank; the gas piping will be

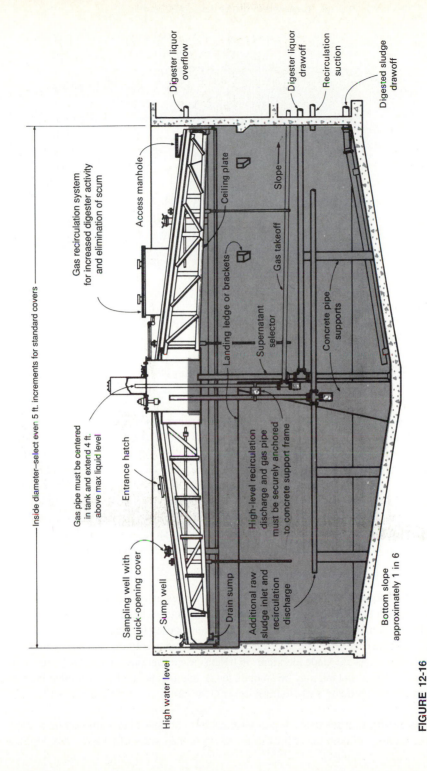

Inside diameter—select even 5 ft. increments for standard covers

Gas recirculation system for increased digester activity and elimination of scum

Gas pipe must be centered in tank and extend 4 ft. above max liquid level

Digester liquor overflow

Access manhole

Ceiling plate

Slope

Landing ledge or brackets

Supernatant selector

Gas takeoff

Concrete pipe supports

Digester liquor drawoff

Recirculation suction

Digested sludge drawoff

Entrance hatch

High-level recirculation discharge and gas pipe must be securely anchored to concrete support frame

Sampling well with quick-opening cover

Sump well

Drain sump

Additional raw sludge inlet and recirculation discharge

High water level

Bottom slope approximately 1 in 6

FIGURE 12-16

Cross section through a typical standard-rate digester.

815

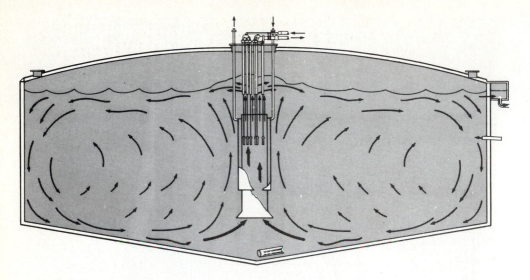

FIGURE 12-17
Section through a gas-mixed high-rate sludge digestion tank (from Walker Process Equipment Division, Chicago Bridge & Iron Co.).

somewhat larger; fewer multiple sludge drawoffs replace the supernatant drawoffs; and the tank should be deeper, if practicable, to aid the mixing process for the high-rate digester.

Sludge should be pumped to the digester continuously or by time clock on a 30 min to 2 h cycle to help maintain constant conditions in the reactor. The incoming sludge displaces digested sludge to a holding tank or sump. Because there is no supernatant separation in the high-rate digester and the total solids are reduced by 45 to 50 percent and given off as gas, the digested sludge is about half as concentrated as the untreated sludge feed. Digestion tanks may have fixed roofs or floating covers. Any or all of the floating covers may be of the gas holder type, which provides excess gas storage capacity. Alternatively, gas may be stored in a separate gas holder or compressed and stored under pressure. A large digester installation is shown in Fig. 12-18.

Two-Stage Digestion. Frequently, a high-rate digester is coupled in series with a second digestion tank. In this process, the first tank is used for digestion and is heated and equipped with mixing facilities. The second tank is used for the storage and concentration of digested sludge and for the formation of a relatively clear supernatant. Frequently, the tanks are made identical, in which case either one may be the primary. In other cases, the second tank may be an open tank, an unheated tank, or a sludge lagoon. Tanks may have fixed roofs or floating covers, the same as single-stage digestion.

Separate Sludge Digestion. Most wastewater treatment plants employing anaerobic digestion use common tanks for the digestion of a mixture of primary and biologi-

FIGURE 12-18
Aerial view of several large anaerobic digesters (Owls Head Wastewater Treatment Plant, New York City; design average flowrate = 120 Mgal/d).

cal sludge. The solid-liquid separation of digested primary sludge, however, is downgraded by even small additions of biological sludge, particularly activated sludge. The rate of reaction under anaerobic conditions is also slowed slightly. Some recent designs have separated the digestion of primary and the biological sludges, in some cases the biological sludge is digested aerobically instead of anaerobically. Reasons cited for separate digestion include (1) the excellent dewatering characteristics of the digested primary sludge are maintained, (2) the digestion process is specifically tailored to the sludge being treated, and (3) optimum process control conditions can be maintained. Design criteria and performance data for the separate anaerobic digestion of biological sludges, however, are very limited.

Process Design

Ideally, the design of anaerobic sludge digestion processes should be based on an understanding of the fundamental principles of biochemistry and microbiology discussed in Chap. 8. Because these principles have not been appreciated fully in the past, a number of empirical methods have also been used in the design of digesters. The purpose of this discussion is to illustrate the various methods that have been used to design digesters in terms of size. These methods are based on (1) the concept of mean cell-residence time, (2) the use of volumetric loading factors, (3) observed volume reduction, and (4) loading factors based on population.

Mean Ce ... Digester design based on mean cell-residence time
involves a ... rinciples discussed in Chap. 8. To review briefly, the
respiration ... nd products of anaerobic digestion are methane gas and
carbon dio ... ntity of methane gas can be calculated using Eq. 12-7 [22]:

$$V_{CH4} = (5.62)\left[(S_o - S)(Q)(8.34) - 1.42 P_x\right] \tag{12-7}$$

where V_{CH4} = olume of methane produced at standard conditions
(32°F and 1 atm), ft³/d

5.62 = theoretical conversion factor for the amount of methane produced
from the complete conversion of one pound of BOD_L to methane
and carbon dioxide, ft³ CH_4/lb BOD_L oxidized

Q = flowrate, Mgal/d

S_o = ultimate BOD_L in influent, mg/L

S = ultimate BOD_L in effluent, mg/L

8.34 = conversion factor, $\dfrac{lb}{Mgal \cdot (mg/L)}$

P_x = net mass of cell tissue produced per day, lb/d

The derivation of the theoretical conversion factor for the amount of methane produced
from the conversion of 1 lb of BOD_L is illustrated in Example 8-5.

The typical reduction in volatile solids achieved in anaerobic digestion for mixed
sludges (primary plus secondary) varies from 45 to 60 percent.

For a complete-mix high-rate digester without recycle, the mass of biological
solids synthesized daily, P_{xj} can be estimated using Eq. 12-8.

$$P_x = \frac{Y\left[(S_o - S)(Q)(8.34)\right]}{1 + k_d \theta_c} \tag{12-8}$$

where Y = yield coefficient, lb/lb

k_d = endogenous coefficient, d^{-1}

θ_c = mean cell-residence time, d

Other terms are as defined previously.

TABLE 12-18
Suggested mean cell-residence times for use in the design of complete-mix digesters [a]

Operating temperature, °C	θ_c^M, d (minimum)	θ_c, d suggested for design
18	11	28
24	8	20
30	6	14
35	4	10
40	4	10

[a] Refs. 22 and 23.

Note: 1.8(°C) + 32 = °F

Note that for a complete-mix flow-through digester, θ_c is the same as the hydraulic retention time θ. Values for Y and k_d, as found for various substrates, are given in Table 8-9. Typical values for θ_c for various temperatures are reported in Table 12-18. The application of Eqs. 12-7 and 12-8 in the process design of a high-rate digester is illustrated in Example 12-4.

Example 12-4 Estimation of digester volume and performance. Estimate the size of digester required to treat the sludge from a primary treatment plant designed to treat 10 Mgal/d (37,800 m³/d) of wastewater. Check the volumeric loading, and estimate the percent stabilization and the amount of gas produced per capita. For the wastewater to be treated, it has been found that the quantity of dry solids and BOD_L removed is 1200 lb/Mgal (~0.15 kg/m³) and 1150 lb/Mgal (~ 0.14 kg/m³), respectively. Assume that the sludge contains about 95 percent moisture and has a specific gravity of 1.02. Other pertinent design assumptions are as follows:

1. The hydraulic regime of the reactor is complete mix.
2. $\theta_c = 10$ days at 35°C (see Table 12-18).
3. Efficiency of waste utilization $E = 0.60$.
4. The sludge contains adequate nitrogen and phosphorus for biological growth.
5. $Y = 0.05$ lb cells/lb BOD_L utilized and $K_d = 0.03$ d⁻¹.
6. Constants are for a temperature of 35°C.

Solution

1. Compute the daily sludge volume and BOD_L loading.

$$\text{Sludge volume} = \frac{(1200 \text{ lb/Mgal})(10 \text{ Mgal/d})}{1.02(62.4 \text{ lb/ft}^3)(0.05 \text{ lb/lb})}$$

$$= 3770 \text{ ft}^3/\text{d} \ (107 \text{ m}^3/\text{d})$$

$$BOD_L \text{ loading} = 1150 \text{ lb/Mgal} \ (10 \text{ Mgal/d})$$

$$= 11,500 \text{ lb/d} \ (5227 \text{ kg/d})$$

2. Compute the digester volume.

$$\frac{V}{Q} = \theta_c$$

where $Q =$ sludge flowrate

$$V = Q\theta_c = (3770 \text{ ft}^3/\text{d})(10 \text{ d})$$

$$= 37,700 \text{ ft}^3 \text{ (also, the volume of first-stage digester in two-stage process)}$$

3. Compute the volumetric loading.

$$\text{lb } BOD_L/\text{ft}^3 \cdot \text{d} = \frac{11,500 \text{ lb/d}}{37,700 \text{ ft}^3} = 0.30 \text{ lb/ft}^3 \cdot \text{d}$$

4. Compute the quantity of volatile solids produced per day using Eq. 12-8.

$$S_o = 11,500 \text{ lb/d}$$

$$S = 11,500 \ (1 - 0.60) = 4600 \text{ lb/d}$$

$$P_x = \frac{0.05\,(11,500 - 4600)\ \text{lb/d}}{1 + 0.03\ \text{d}^{-1}(10\ \text{d})}$$

$$= 265\ \text{lb/d}$$

5. Compute the percent stabilization.

$$\text{Percent stabilization} = \frac{(11,500 - 4600) - 1.42\,(265)\ \text{lb/d}}{11,500\ \text{lb/d}} \times 100$$

$$= 56.7$$

6. Compute the volume of methane produced per day using Eq. 12-7.

$$V_{CH_4} = 5.62\ \text{ft}^3/\text{lb}[(11,500 - 4600)\ \text{lb/d} - 1.42\,(265\ \text{lb/d})]$$

$$= 38,402\ \text{ft}^3\ (1087\ \text{m}^3)$$

7. Estimate the total gas production. Because digester gas is about two-thirds methane, the total volume of gas produced is

$$\text{Total gas volume} = \frac{38,402}{0.67}$$

$$= 57,316\ \text{ft}^3\ (1622\ \text{m}^3)$$

Loading Factors. One of the most common methods used to size digesters is to determine the required volume on the basis of a loading factor. Although a number of different factors have been proposed, the two that seem most favored are based on (1) the pounds (kilograms) of volatile solids added per day per cubic foot (cubic meter) of digester capacity and (2) the pounds (kilograms) of volatile solids added per day per pound (kilogram) of volatile solids in the digester. From the information presented in Chap. 8, the similarity between these loading factors and the food-to-microorganism ratio is apparent. In applying these loading factors, another factor that should also be checked is the hydraulic detention time because of its relationship to organism growth and washout (see Table 12-18) and to the type of digester used (e.g., only 50 percent or less of the capacity of a conventional standard-rate single-stage digester is effective).

Ideally, the conventional single-stage digestion tank is stratified into three layers with the supernatant at the top, the active digestion zone in the middle, and the thickened sludge at the bottom. Because of the storage requirements for the digested sludge and the supernatant and the excess capacity provided for daily fluctuations in sludge loading, the volumetric loading for standard-rate digesters is low. Detention times based on cubic feet (cubic meters) of untreated sludge pumped vary from 30 to more than 90 d for this type of tank. The recommended solids loadings for standard rate digesters are from 0.03 to 0.10 lb/ft³ · d (0.5 to 1.6 kg/m³ · d) of volatile solids.

For high-rate digesters, loading rates of 0.10 to 0.30 lb/ft³ · d (1.6 to 4.8 kg/m³ · d) of volatile solids and hydraulic detention periods of 10 to 20 d are practicable. Mixing has proved to be a problem at sludge-loading rates greater than about 0.30 lb/ft³ · d (4.0 kg/m³ · d). The effect of sludge concentration and hydraulic detention time on the volatile solids loading factor is reported in Table 12-19.

The degree of stabilization obtained is also often measured by the percent reduction in volatile solids. This reduction can be related either to the mean cell-

TABLE 12-19
Effect of sludge concentration and hydraulic detention time on volatile solids loading factor [a]

Sludge concentration, %	Volatile solids loading factor, lb/ft³ · d			
	10d[b]	12d	15d	20d
4	0.19	0.16	0.13	0.10
5	0.24	0.20	0.16	0.12
6	0.28	0.24	0.19	0.14
7	0.33	0.28	0.22	0.17
8	0.38	0.32	0.25	0.19
9	0.43	0.36	0.29	0.21
10	0.48	0.40	0.32	0.24

[a] Based on 75 percent volatile content of sludge and a sludge specific gravity of 1.02 (concentration effects neglected).
[b] Hydraulic detention time, d.
Note: lb/ft³ · d × 16.0185 = kg/m³ · d

residence time or to the detention time based on the untreated sludge feed. Because the untreated sludge feed can be measured easily, this method is more commonly used. In plant operation, calculation of voliatile solids reduction should be made routinely as a matter of record whenever sludge is drawn to processing equipment or drying beds. Alkalinity and volatile acids content should also be checked daily as a measure of the stability of the digestion process.

In calculating the volatile solids reduction, the ash content of the sludge is assumed to be conservative; that is, the number of pounds of ash going into the digester is equal to that being removed. A typical calculation of volatile solids reduction is presented in Example 12-5.

Example 12-5 Determination of volatile solids reduction. From the following analysis of untreated and digested sludge, determine the total volatile solids reduction achieved during digestion. It is assumed that (1) the weight of fixed solids in the digested sludge equals the weight of fixed solids in the untreated sludge and (2) the volatile solids are the only constituent of the untreated sludge lost during digestion.

Solution

1. Determine the weight of the digested solids. Because the quantity of fixed solids remains the same, the weight of the digested solids based on 1.0 lb of dry untreated sludge, as computed below, is 0.6 lb.

	Volatile solids, %	Fixed solids, %
Untreated sludge	70	30
Digested sludge	50	50

$$\text{Fixed-solids untreated sludge, } 30\% = \frac{(0.3 \text{ lb})(100)}{0.3 \text{ lb} + 0.7 \text{ lb}}$$

Let x equal the weight of volatile solids after digestion. Then,

$$\text{Fixed-solids digested sludge, } 50\% = \frac{(0.3 \text{ lb})(100)}{0.3 \text{ lb} + x}$$

$$x = 0.3 \text{ lb volatile solids}$$

$$\text{Weight of digested solids} = 0.3 \text{ lb} + x = 0.6 \text{ lb}$$

2. Determine the percent reduction in total and volatile solids.
   (a) Percent reduction of total solids.

$$R_{\text{TSS}} = \frac{(1.0 - 0.6 \text{ lb})(100)}{1.0 \text{ lb}} = 40\%$$

   (b) Percent reduction in volatile solids.

$$R_{\text{VSS}} = \frac{(0.7 - 0.3 \text{ lb})(100)}{0.7 \text{ lb}} = 57.1\%$$

Volume Reduction. It has been observed that as digestion proceeds, if the supernatant is withdrawn and returned to the head end of the treatment plant, the volume of the remaining sludge decreases approximately exponentially. If a plot is prepared of the remaining volume versus time, the required volume of the digester is represented by the area under the curve and can be computed using Eq. 12-9.

$$V = \left[V_f - \frac{2}{3} \left(V_f - V_d \right) \right] t \tag{12-9}$$

where V = volume of digester, ft^3 (m^3)
 V_f = volume of fresh sludge added per day, ft^3/d (m^3/d)
 V_d = volume of digested sludge removed per day, ft^3/d (m^3/d)
 t = digestion time, d

Population Basis. Digestion tanks are also designed on a volumetric basis by allowing a certain number of cubic feet per capita. Detention times range from 10 to 20 d for high-rate digesters and 30 to 60 d for standard-rate digesters [42]. These detention times are recommended for design based on total tank volume plus additional storage volume, if sludge is dried on beds and weekly sludge drawings are curtailed because of inclement weather.

Typical design criteria for anaerobic digesters are shown in Table 12-20. These requirements are for heated tanks (mesophilic range) and are applied where analyses and volumes of sludge to be digested are not available. For unheated tanks capacities must be increased, depending on local climatic conditions and the storage volume required. The capacities shown in Table 12-20 should be increase 60 percent in a municipality where the use of kitchen food waste grinders is universal

TABLE 12-20
Typical design criteria for sizing mesophilic anaerobic sludge digesters [a]

Parameter	Standard-rate digestion	High-rate digestion
Volume criteria, ft^3/capita		
Primary sludge	2–3	1.3–2.0
Primary sludge + trickling-filter humus	4–5	2.6–3.3
Primary sludge + activated sludge	4–6	2.6–4
Solids-loading rate, lb VSS/10^{3}ft^3 · d	40–100	100–200
Solids retention time, days	30–60	15–20

[a] Adapted from Ref. 42.
Note: ft^3 × 0.028317 = m^3
     lb VSS/10^{3}ft^3 · d × 0.0160 = kg/m^3 · d

and should be increased on a population-equivalent basis to allow for the effect of industrial wastes.

Tank Design

Anaerobic digestion tanks are either cylindrical, rectangular, or egg-shaped. The most common design is a low, vertical cylinder. Rectangular tanks are used infrequently because of the greater difficulty of mixing the tank contents uniformly. Egg-shaped tanks are used extensively in Europe and have been introduced in recent years in the United States.

Cylindrical sludge digestion tanks are seldom less than 20 ft (6 m) or more than 125 ft (38 m) in diameter. They should have a water depth of not less than 25 ft (7.5 m) at the sidewall and may be as deep as 45 ft (14 m) or more. The floor of the digester is usually conical with the bottom sloping to the sludge drawoff in the center, with a minimum slope of 1 vertical to 4 horizontal (see Fig. 12-16). Alternative bottom designs using a "waffle" shape have been employed to minimize grit accumulation and to reduce the need for digester cleaning (see Fig. 12-19).

The purpose of the egg-shaped design is to eliminate the need for cleaning. The digester sides form a cone so steep at the bottom that grit cannot accumulate (see Fig. 12-20). Other advantages cited for the egg-shaped design include better mixing, better control of the scum layer, and smaller land area requirements. Egg-shaped tanks may be constructed of steel or reinforced concrete.

Gas Production, Collection, and Use

Gas from anaerobic digestion contains about 65 to 70 percent CH_4 by volume, 25 to 30 percent CO_2, and small amounts of N_2, H_2, H_2S, water vapor, and other gases.

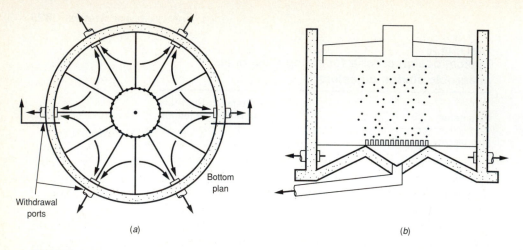

FIGURE 12-19
Typical waffle bottom anaerobic digester: (a) plan view and (b) section.

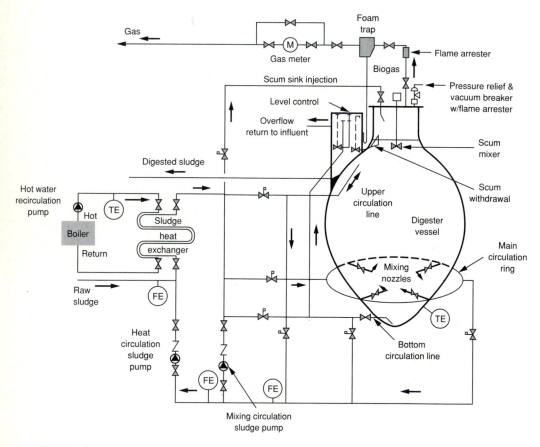

FIGURE 12-20
Schematic diagram of an egg-shaped anaerobic digester installation (from Walker Process).

Digester gas has a specific gravity of approximately 0.86 relative to air. Because production of gas is one of the best measures of the progress of digestion and because digester gas can be used as fuel, the designer should be familiar with its production, collection, and use.

Gas Production. The volume of methane gas produced during the digestion process can be estimated using Eq. 12-7, as discussed previously. Total gas production is estimated usually from the percentage of volatile solids reduction. Typical values vary from 12 to 18 ft³/lb (0.75 to 1.12 m³/kg) of volatile solids destroyed. Gas production can fluctuate over a wide range, depending on the volatile solids content of the sludge feed and the biological activity in the digester. Excessive gas production rates sometimes occur during startup and may cause foaming and escape of foam and gas from around the edges of floating digester covers. If stable operating conditions have been achieved and the foregoing gas production rates are being maintained, the operator can be assured that the result will be a well-digested sludge.

Gas production can also be crudely estimated on a per capita basis. The normal yield is 0.6 to 0.8 ft³/person · d (15 to 22 m³/10³ persons · d) in primary plants treating normal domestic wastewater. In secondary treatment plants, the gas production is increased to about 1.0 ft³/person · d (28 m³/10³ persons · d).

Gas Collection. Digester gas is collected under the cover of the digester; two principal types of covers are used: (1) floating and (2) fixed [60]. Floating covers fit on the surface of the digester contents and allow the volume of the digester to

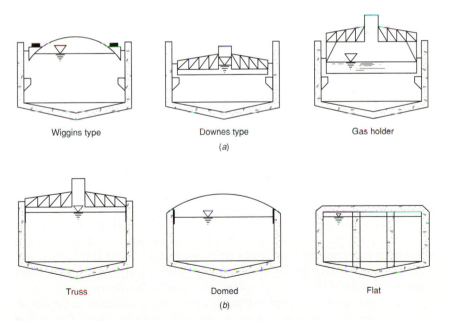

Wiggins type	Downes type	Gas holder
	(a)	

Truss	Domed	Flat
	(b)	

FIGURE 12-21
Types of anaerobic digester covers: (a) floating covers and (b) fixed covers.

change without allowing air to enter the digester (see Fig. 12-21a). Gas and air must not be allowed to mix, or an explosive mixture may result. Explosions have occurred in wastewater treatment plants. Gas piping and pressure relief valves must include adequate flame traps. The covers may also be installed to act as gas holders that store a small quantity of gas under pressure and serve as reservoirs. Floating covers can be used for single-stage digesters or in the second stage of two-stage digesters.

Fixed covers provide a free space between the roof of the digester and the liquid surface (see Fig. 12-21b). Gas storage must be provided so that (1) when the liquid volume is changed, gas, not air, will be drawn into the digester, and (2) gas will not be lost by displacement. Gas can be stored either at low pressure in external gas holders that use floating covers or at high pressure in pressure vessels if gas compressors are used. Gas not used should be burned in a flare. Gas meters should be installed to measure gas produced and gas used or wasted.

Use of Digester Gas. Methane gas at standard temperature and pressure has a net heating value of 960 Btu/ft^3 (35,800 kJ/m^3). Because digester gas is typically about 65 percent methane, the low heating value of digester gas is approximately 600 Btu/ft^3 (22,400 kJm^3). By comparison, natural gas, which is a mixture of methane, propane, and butane, has a low heating value of approximately 1000 Btu/ft^3 (37,300 kJ/m^3).

In large plants, digester gas may be used as fuel for boiler and internal combustion engines, which are in turn used for pumping wastewater, operating blowers, and generating electricity. Hot water from heating boilers or from engine jackets and exhaust-heat boilers may be used for sludge heating and for building heating, or gas-fired sludge heating boilers may be used. Because digester gas contains hydrogen sulfide, particulates, and water vapor, the gas frequently has to be cleaned in dry or wet scrubbers before it is used in internal combustion engines.

Digester Mixing

Proper mixing is one of the most important considerations in achieving optimum process performance. Various systems for mixing the contents of the digester have been employed (see Fig. 12-22). The most common types involve the use of (1) gas injection, (2) mechanical stirring, and (3) mechanical pumping [2,27]. Some treatment plants use a combination of gas mixing and recirculation by pumping. The advantages and disadvantages of the various mixing systems are summarized in Table 12-21, and typical design parameters are given in Table 12-22.

Gas injection systems are classified as unconfined or confined (Figs. 12-22a and 12-22b). Unconfined gas systems are designed to collect gas at the top of the digesters, compress the gas, and then discharge the gas through a pattern of bottom diffusers or through a series of radially placed top-mounted lances. Unconfined gas systems mix the digester contents by releasing gas bubbles that rise to the surface, carrying and moving the sludge. These systems are suitable for digesters with fixed, floating, or gas holder covers. In confined gas systems, gas is collected at the top of the digesters, compressed, and then discharged through confined tubes. Two major types of confined systems are the gas lifter system and the gas piston system. The gas lifter system consists of submerged gas pipes or lances inserted into an eductor

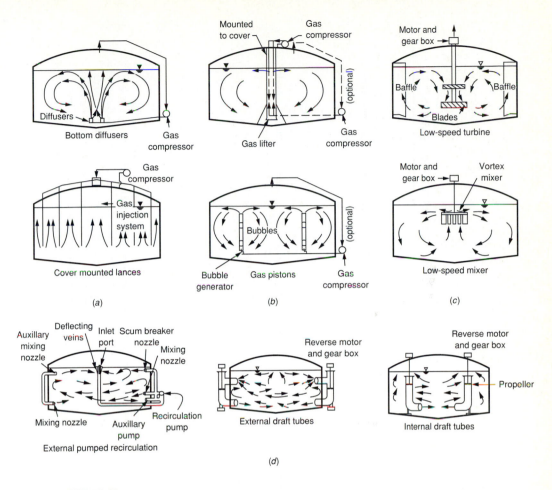

FIGURE 12-22
Devices used for mixing contents of anaerobic high-rate digesters: (a) unconfined gas injection systems, (b) confined gas injections systems, (c) mechanical stirring systems, and (d) mechanical pumping systems [2,27].

tube or gas lifter. Compressed gas is released from the lances or pipes, and the gas bubbles rise, creating an air lift effect. In the gas piston system, gas bubbles are released intermittently at the bottom of a cylindrical tube or piston. The bubbles rise and act like a piston, pushing the sludge to the surface. These systems are suitable for fixed, floating, or gas holder covers.

Mechanical mixing systems commonly use low-speed turbines or mixers (see Fig. 12-22c). In both systems, the rotating impeller(s) displaces the sludge, mixing the digester contents. Low-speed turbine systems usually have one cover-mounted motor with two turbine impellers located at different sludge depths. A low-speed mixer system usually has one cover-mounted mixer. Mechanical stirring systems are suitable for digesters with fixed or floating covers.

TABLE 12-21
Summary of advantages and disadvantages of various anaerobic digester mixing systems [a]

Type mixer	Advantages	Disadvantages
All systems	Increased rate of sludge stabilization.	Corrosion and wear of ferrous metal piping and supports. Equipment wear by grit. Equipment plugging and operational interference by rags.
Gas injection		
Unconfined		
Cover-mounted lances	Lower maintenance and less hindrance to cleaning than bottom-mounted diffusers. Effective against scum buildup.	Corrosion of gas piping and equipment. High maintenance for compressor. Potential gas seal problem. Compressor problems if foam gets inside. Solids deposition. Plugging of gas lances.
Bottom-mounted diffusers	Better movement of bottom deposits than cover-mounted lances.	Corrosion of gas piping and equipment. High maintenance for compressor. Potential gas seal problem. Foam problem. Does not completely mix digester contents. Scum formation. Plugging of diffusers. Bottom deposits can alter mixing patterns. Breakage of bottom-mounted gas piping. Requires digester dewatering for maintenance.
Confined		
Gas lifters	Better mixing and gas production, and better movement of bottom deposits than cover-mounted lances. Lower power requirements.	Corrosion of gas piping and equipment. High maintenance for compressor. Potential gas seal problem. Corrosion of gas lifter. Lifter interferes with digester cleaning. Scum buildup. Does not provide good top mixing. Variable pumping rates. Requires digester dewatering for maintenance if bottom-mounted. Plugging of lances.

Gas Pistons	Good mixing efficiency.	Corrosion of gas piping and equipment. High maintenance for compressor. Potential gas seal problem. Equipment internally mounted. Breakage of bottom-mounted gas piping. Plugging of piston and piping. Requires digester dewatering for maintenance. Pistons interfere with digester cleaning.
Mechanical stirring		
Low-speed turbines	Good mixing efficiency.	Wear of impellers and shafts. Bearing failures. Long overhung loads. Interferences of impellers by rags. Requires oversized gear boxes. Gas leaks at shaft seal.
Low-speed mixers	Break up scum layers.	Not designed to mix entire tank contents. Bearing and gear box failures. Wear of impellers. Interference of impellers by rags.
Mechanical pumping		
Internal draft tubes	Good top to bottom mixing. Minimal scum buildup.	Sensitive to liquid level. Corrosion and wear of impeller. Bearing and gear box failures. Requires oversized gear box. Plugging of draft tube by rags.
External draft tubes	Same as internal draft tube.	Same as internal draft tube.
Pumps	Positive known quantity of mixing. Scum layer recirculated. Sludge deposits can be recirculated. Pumps easier to maintain than compressor.	Nozzle maintenance requires dewatering. Wear of impellers. Plugging of impellers and volutes by rags. Bearing failures.

[a] From Refs. 2 and 27.

TABLE 12-22
Typical design parameters for anaerobic digester mixing systems[a]

Parameter	Definition	Typical value [b]
Unit power	Motor power of mixing equipment in hp divided by digester volume in 10^3 gal, hp/10^3 gal	Mechanical systems: 0.025–0.04 hp/10^3 gal
Unit gas flow	Quantity of gas delivered by gas injection system in ft/min^3 divided by the digester gas volume in 10^3 ft^3, ft^3/10^3 ft^3 · min	Unconfined gas systems: 4.5–5 ft^3/10^3 ft^3 · min Confined gas systems: 5–7 ft^3/10^3 ft^3 · min
Velocity gradient	The square root of the ratio of the power used per unit volume divided by the absolute viscosity of the sludge, G, s^{-1} (see Eq. 6-3)	All mixing systems, 50–80 s^{-1}
Turnover time	The digester volume divided by the sludge flowrate, min	Confined gas mixing systems and mechanical systems: 20–30 min

[a] From Ref. 2.
[b] Actual design values may differ depending on the type of mixing system, manufacturer, and digestion process or function.

Note: hp/10^3 gal × 0.1970 = kW/m^3
Ft^3/10^3 ft^3 · min × 0.001 = m^3/m^3 · min

Most mechanical pumping systems consist of propeller-type pumps mounted in internal or external draft tubes, or axial flow or centrifugal pumps and piping installed externally (see Fig. 12-22*d*). Mixing is promoted by the circulation of sludge. Mechanical pumping systems are suitable for digesters with fixed covers.

Digester Heating

The heat requirements of digesters consist of the amount needed (1) to raise the incoming sludge to digestion tank temperatures, (2) to compensate for the heat losses through the walls, floor, and roof of the digester, and (3) to make up the losses that might occur in the piping between the source of heat and the tank. The sludge in digestion tanks is heated by pumping the sludge and supernatant through external heat exchangers and back to the tank (see Fig. 12-23) or by internal heat exchangers.

Analysis of Heat Requirements. In computing the energy required to heat the incoming sludge to the temperature of the digester, it is assumed that the specific heat of most sludges is essentially the same as that of water. This assumption has proved to be quite acceptable for engineering computations. The heat loss through the digester sides, top, and bottom is computed using the following expression:

$$q = UA\Delta T \tag{12-10}$$

where q = heat loss, Btu/h (W)
U = overall coefficient of heat transfer, Btu/ft^2 · h · °F (W/m^2 · °C)
A = cross-sectional area through which the heat loss is occurring, ft^2 (m^2)
ΔT = temperature drop across the surface in question, °F (°C)

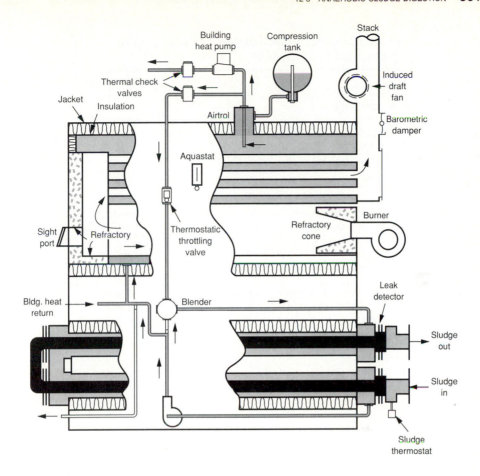

FIGURE 12-23
Typical heat exchanger used for heating digesting sludge (from Ralph B. Carter Co.).

In computing the heat losses from a digester using Eq. 12-10, it is common practice to consider the characteristics of the various heat transfer surfaces separately and to develop transfer coefficients for each one. The application of Eq. 12-10 in the computation of digester heating requirements is illustrated in Example 12-6. (Note, when using SI units, W is equal to J/s and, in the heat loss computation, conversion to J/d should be included.)

Heat Transfer Coefficients. Typical overall heat transfer coefficients are reported in Table 12-23. As shown, separate entries are included for the walls, bottom, and top of the digester.

Digestion tank walls may be surrounded by earth embankments that serve as insulation, or they may be made of compound construction consisting of approximately 12 in (300 mm) of concrete, corkboard insulation, or an insulating air space, plus brick facing or corrugated aluminum facing over rigid insulation. The heat transfer from plain concrete walls below ground level and from floors depends on whether

primary sludge, waste-activated sludge, and digested sludge are good indicators of the range of probable conditioner doses required. Solids concentrations will affect the dosage and the dispersal of the conditioning agent. The pH and alkalinity may affect the performance of the conditioning agents, in particular the inorganic conditioners. When lime is used to maintain a high pH for dewatering, strong ammonia odor and lime-scaling problems may occur. The method of dewatering may also affect the selection of the conditioning chemical because of the differences in mixing equipment used by various vendors and the characteristics of particular methods of dewatering. For example, polymers are used commonly in centrifuge and belt press dewatering but are used less frequently for vacuum and pressure filtration. Laboratory or pilot-scale testing is recommended to determine the types of chemical-conditioning agents required, particularly for sludge that may be difficult to dewater.

Dosage. The chemical dosage required to condition a given sludge is determined in the laboratory. Tests used for selecting chemical dosage include the Buchner Funnel test for the determination of specific resistance of sludge, Capillary Suction Time Test (CST), and the standard jar test. The Buchner Funnel test is a method of testing sludge drainability or dewatering characteristics using various conditioning agents. The Capillary Suction Test relies on gravity and the capillary suction of a piece of thick filter paper to draw out water from a small sample of conditioned sludge. The standard jar test, the easiest method to use, consists of testing standard volumes of sludge samples (usually 1 L) with different conditioner concentrations, followed by rapid mixing, flocculation, and settling using standard jar test apparatus. Detailed descriptions of testing procedures are provided in Ref. 61.

In general, it has been observed that the type of sludge has the greatest impact on the quantity of chemical required. Difficult-to-dewater sludges require larger doses of chemicals, generally do not yield as dry a cake, and have poorer quality of filtrate or centrate. Sludge types, listed in the approximate order of increasing conditioning chemical requirements, are as follows:

1. Untreated (raw) primary sludge
2. Untreated mixed primary and trickling-filter sludge
3. Untreated mixed primary and waste activated sludge
4. Anaerobically digested primary sludge
5. Anaerobically digested mixed primary and waste activated sludge
6. Untreated waste activated sludge
7. Aerobically digested sludge

Typical levels of polymer addition for various types of sludge for diverse methods of dewatering are shown in Table 12-26. Actual dosages in any given case may vary considerably from the indicated values. Polymer dosages will also vary greatly depending on the molecular weight, ionic strength, and activity levels of the polymers used. Manufacturers should be consulted for applicability and dosage information. Dosages of ferric chloride and lime, two of the chemicals used most commonly to condition sludge for vacuum filter dewatering, also vary widely. Factors

TABLE 12-26
Typical levels of polymer addition for various sludges and for various methods of dewatering[a]

	lb of dry polymer added/ton of dry solids		
Type of sludge	Vacuum filter	Belt filter press	Solid bowl centrifuge
Primary	2–10	2–8	1–5
Primary and waste activated	10–20	4–16	4–10
Primary and trickling filter	2.5–5	4–16	—
Waste activated	15–30	8–20	10–16
Anaerobically digested primary	7–14	4–10	6–10
Anaerobically digested primary and air waste activated	3–17	3–17	4–10
Aerobically digested primary and air waste activated	15–20	4–16	—

[a] Adapted from Refs. 58 and 61.
Note: lb/ton × 0.5 = kg/10^3 kg

that affect the dosages of ferric chloride and lime include the type of sludge (primary, secondary, or a mixture) and the type of stabilization process, if any, used prior to dewatering [61].

Sludge Mixing. Intimate admixing of sludge and coagulant is essential for proper conditioning. The mixing must not break the floc after it has formed, and the detention should be kept to a minimum so that sludge reaches the dewatering unit as soon after conditioning as possible. Mixing requirements vary depending on the dewatering method used. A separate mixing and flocculation tank is usually provided ahead of vacuum and pressure filters; a separate flocculation tank may be provided for a belt filter press, or the conditioner may be added directly to the sludge feed line of the belt press unit; and in-line mixers are usually used with a centrifuge. It is generally desirable in design to provide at least two locations for the addition of conditioning chemicals.

Heat Treatment

Heat treatment is both a stabilization and a conditioning process that involves heating the sludge for short periods of time under pressure. Heat treatment is used to coagulate solids, to break down the gel structure, and to reduce the water affinity of sludge solids. As a result, the sludge is sterilized and dewatered readily. The heat treatment process is most applicable to biological sludges that may be difficult to stabilize or condition by other means. The high capital costs of equipment generally limit its use to large plants (more than 5 Mgal/d or 0.2 m³/s) or facilities where space may be limited. Supernatant from the heat treatment unit is high in BOD and may require special sidestream treatment before it is introduced into the mainstream wastewater treatment process. Several types of heat treatment processes have been developed, but many are no longer in operation. A schematic diagram of the system used most commonly, the low-pressure Zimpro system, is shown in Fig. 12-30. The low-pressure Zimpro system uses a sludge-to-sludge heat exchanger, air injection, and live steam injection into the reactor.

The partially oxidized sludge from the heat treatment unit may be dewatered by vacuum filtration, centrifugation, belt presses, or on drainage beds. Advantages cited for heat treatment are as follows: (1) the solids content of the dewatered sludge can range from 30 to 50 percent, depending on the degree of oxidation achieved; (2) the processed sludge does not normally require chemical conditioning; (3) the process stabilizes sludge and will destroy most pathogenic organisms; (4) the processed sludge will have a heating value of 12,000 to 13,000 Btu/lb (28 to 30 kJ/g) of volatile solids; and (5) the process is relatively insensitive to changes in sludge composition. Essentially complete oxidation of volatile solids (approximately 90 percent reduction) can be accomplished with higher pressures and temperatures (see "Wet-Air Oxidation" in Sec. 12-15).

The major disadvantages associated with heat treatment are (1) high capital cost due to its mechanical complexity and the use of corrosion resistant materials, (2) close supervision, skilled operators, and a strong preventative maintenance program are required, (3) the process produces sidestreams with high concentrations of organics, ammonia nitrogen, and color, (4) significant odorous gases are produced that require

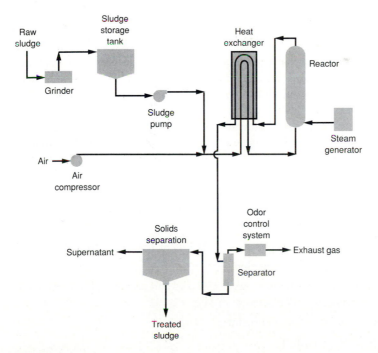

FIGURE 12-30
Schematic diagram for wet-air oxidation heat treatment system (from Zimpro Passavant).

extensive containment, treatment, and /or destruction, and (5) scale formation in the heat exchangers, pipes, and reactor requires acid washing or high-pressure water jets [30]. Few new heat treatment facilities have been constructed in the U.S. in recent years because of these disadvantages.

Other Processes

Other sludge-conditioning methods that have been investigated include (1) chemical conditioning using a combination of organic polymers and inorganic chemicals, (2) use of bulking materials such as newspaper pulp and fly ash, (3) sludge acidification to improve dewatering, (4) freeze-thaw of sludge (see also "Sludge Drying Beds" in Sec. 12-13), (5) solvent extraction of oils, fats, and greases, and (6) irradiation. Most of these methods have been limited to laboratory or pilot-scale demonstrations. For more detailed information, Ref. 61 may be consulted.

A unit operation used in the past for conditioning is elutriation. In elutriation, a solid or a solid-liquid mixture is intimately mixed with a liquid for the purpose of transferring certain components to the liquid. A typical example is the washing of digested sludge before chemical conditioning to remove certain soluble inorganic and organic components that would consume large amounts of chemicals. Elutriation is seldom used because the finely divided solids washed out of the sludge may not be fully captured in the main wastewater treatment facilities. For additional information on elutriation, Ref. 25 may be consulted.

12-12 DISINFECTION

Sludge disinfection is becoming an important consideration as an add-on process because of stricter regulations for the reuse and application of sludge on land. When sludge is applied to the land, protection of public health requires that contact with pathogenic organisms be controlled (see Sec. 12-2).

There are many ways to destroy pathogens in liquid and dewatered sludges. The following methods have been used to achieve pathogen reduction beyond that attained by stabilization [42]:

1. Pasteurization
2. Other thermal processes such as heat conditioning, heat drying, incineration, pyrolysis, or starved air combustion
3. High pH treatment, typically with lime, at a pH higher than 12.0 for 3 h
4. Long-term storage of liquid digested sludge
5. Complete composting at temperatures above 55°C (131°F) and curing in a stockpile for at least 30 d (composting is discussed in Sec. 12-10)
6. Addition of chlorine to stabilize and disinfect sludge
7. Disinfection with other chemicals
8. Disinfection by high-energy irradiation

As indicated in Sec. 12-7, some stabilization processes will also provide disinfection. These processes include lime stabilization, heat treatment, thermophilic anaerobic digestion, and thermophilic aerobic digestion.

Anaerobic and aerobic digestion (excluding thermophilic anaerobic and aerobic digestion) will not disinfect the sludge, but will greatly reduce the number of pathogenic organisms. Disinfection of liquid aerobic and anaerobic digested sludges is best accomplished by pasteurization or long-term storage. Long-term storage and composting are probably the most effective means of disinfecting dewatered aerobic and anaerobic digested sludges.

Pasteurization

There is only one municipal sludge pasteurization facility reported to be operating in the United States. Pasteurization is used in Europe and is required in Germany and Switzerland to disinfect sludges spread on pastures during the spring and summer growing season. For the disinfection of wet sludges, pasteurization at 70°C (158°F) for 30 minutes will inactivate parasitic ova and cysts [42]. Based on European experience, heat pasteurization is a proven technology, requiring skills such as boiler operation and the understanding of high-temperature and pressure processes.

The two methods that are used for pasteurizing liquid sludges involve (1) the direct injection of steam, and (2) indirect heat exchange. Because heat exchangers tend to scale or become fouled with organic matter, it appears that direct steam injection is the most feasible method. A schematic diagram for sludge pasteurization using direct steam injection is shown in Fig. 12-31. Equipment presently used for sludge pasteurization may not be cost effective for plants with capacities of less than 5 Mgal/d ($0.2 \text{ m}^3/\text{s}$) because of the high capital costs. Thermophilic aerobic digestion coupled with anaerobic digestion (dual digestion) may also be used for the pasteurization of sludge (see Sec. 12-9).

Long-Term Storage

Liquid digested sludge is normally stored in earthen lagoons. Storage requires that sufficient land be available. Storage is often necessary in land application systems to retain sludge during periods when it cannot be applied because of weather or crop considerations. In this case, the storage facilities can perform a dual function by providing disinfection as well as storage. Typical detention times for disinfection are 60 days at 68°F (20°C) and 120 days at 39°F (4°C). Because of the potential contamination effects of the stored sludge, special attention must be devoted to the design of these lagoons with respect to limiting percolation and the development of odors. Additional information on sludge storage is provided in Sec. 12-19.

12-13 DEWATERING

Dewatering is a physical (mechanical) unit operation used to reduce the moisture content of sludge for one or more of the following reasons:

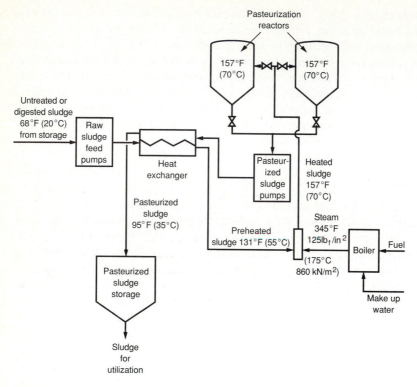

FIGURE 12-31
Schematic diagram of a sludge pasteurization system [42].

1. The costs for trucking sludge to the ultimate disposal site become substantially lower when sludge volume is reduced by dewatering.

2. Dewatered sludge is generally easier to handle than thickened or liquid sludge. In most cases, dewatered sludge may be shoveled, moved about with tractors fitted with buckets and blades, and transported by belt conveyors.

3. Dewatering is required normally prior to the incineration of the sludge to increase the energy content by removal of excess moisture.

4. Dewatering is required before composting to reduce the requirements for supplemental bulking agents or admendments.

5. In some cases, removal of the excess moisture may be required to render the sludge odorless and nonputrescible.

6. Sludge dewatering is required prior to landfilling in monofills to reduce leachate production at the landfill site.

A number of techniques are used in dewatering devices for removing moisture. Some rely on natural evaporation and percolation to dewater the solids. In mechanical dewatering devices, mechanically assisted physical means are used to dewater the sludge more quickly. The physical means include filtration, squeezing, capillary action, vacuum withdrawal, and centrifugal separation and compaction.

The selection of the dewatering device is determined by the type of sludge to be dewatered, characteristics of the dewatered product, and the space available. For smaller plants where land availability is not a problem, drying beds or lagoons are generally used. Conversely, for facilities situated on constricted sites, mechanical dewatering devices are often chosen.

Some sludges, particularly aerobically digested sludges, are not amenable to mechanical dewatering. These sludges can be dewatered on sand beds with good results. When a particular sludge must be dewatered mechanically, it is often difficult or impossible to select the optimum dewatering device without conducting bench-scale or pilot studies. Trailer-mounted full-size equipment is available from several manufacturers for field-testing purposes.

The available dewatering processes include vacuum filters, centrifuges, belt filter presses, recessed plate filter presses, drying beds, and lagoons. The advantages and disadvantages of the various methods of sludge dewatering are summarized in Table 12-27. For additional descriptive material for dewatering devices, Refs. 43, 51, and 58 may be consulted.

Vacuum Filtration

Vacuum filtration has been used for municipal sludge dewatering for over 60 years, but its use has declined in the past ten years because of the development of and improvements to alternative mechanical dewatering equipment. Some of the reasons for its decline in popularity are (1) system complexity, (2) need for conditioning chemicals, and (3) high operating and maintenance costs.

Principles of Operation. In vacuum filtration, atmospheric pressure, due to a vacuum applied downstream of the filter media, is the driving force on the liquid phase that causes it to move through the porous media. The vacuum filter consists of a horizontal cylindrical drum that rotates partially submerged in a vat of conditioned sludge. The surface of the drum is covered with a porous medium, the selection of which is based on the sludge-dewatering characteristics. Types of filter media commonly used are cloth belts or coiled springs. The drum surface is divided into sections around its circumference. Each section is sealed from its adjacent section and the ends of the drum. A separate vacuum/drain line connects each section to a rotary valve at the axis of the drum. The rotary valve controls the various phases of the filtering cycle and channels filtrate away from the drum. As the drum rotates, the valve allows each segment to function in sequence as one of three distinct zones: cake formation, cake dewatering, and cake discharge.

System Operation and Performance. A vacuum filter system usually consists of sludge-feed pumps, chemical feed equipment, sludge conditioning tank, filter drum, sludge cake conveyor or hopper, vacuum system, and filtrate removal system. The schematic diagram of a typical vacuum filtration system is shown in Fig. 12-32.

TABLE 12-27
Comparison of alternative sludge-dewatering methods[a]

Dewatering method	Advantages	Disadvantages
Vacuum filter	Skilled personnel not required Maintenance requirements are low for continuously operating equipment	Highest energy consumer per unit of sludge dewatered Continuous operator attention required Vacuum pumps are noisy Filtrate may have high suspended-solids content, depending on filter medium
Solid bowl centrifuge	Clean appearance, minimal odor problems, fast startup and shutdown capabilities Easy to install Produces relatively dry sludge cake Low capital cost-to-capacity ratio	Scroll wear potentially a high-maintenance problem Requires grit removal and possibly a sludge grinder in the feed stream Skilled maintenance personnel required Moderately high suspended-solids content in centrate
Imperforate basket centrifuge	Same machines can be used for both thickening and dewatering Chemical conditioning may not be required Clean appearance, minimal odor problems, fast startup and shutdown capabilities Very flexible in meeting process requirements Not affected by grit Excellent results for difficult sludges	Limited size capacity Except for vacuum filters, consumes more energy per unit of sludge dewatered Skimming stream may produce significant recycle load For easily dewatered sludges, has highest capital cost-to-capacity ratio For most sludges, produces lowest cake solids concentration Vibration
Belt filter press	Low energy requirements Relatively low capital and operating costs Less complex mechanically and easier to maintain High-pressure machines are capable of producing very dry cake Minimal effort required for system shutdown	Hydraulically limited in throughput Requires sludge grinder in feed stream Very sensitive to incoming sludge feed characteristics Short media life as compared to other devices using cloth media Automatic operation generally not advised

TABLE 12-27
(*continued*)

Dewatering method	Advantages	Disadvantages
Recessed plate filter press	Highest cake solids concentration	Batch operation
	Low suspended solids in filtrate	High equipment cost
		High labor cost
		Special support structure requirements
		Large floor area required for equipment
		Skilled maintenance personnel required
		Additional solids due to large chemical addition require disposal
Sludge drying beds	Lowest capital cost method where land is readily available	Requires large area of land
	Small amount of operator attention and skill required	Requires stabilized sludge
	Low energy consumption	Design requires consideration of climatic effects
	Low to no chemical consumption	Sludge removal is labor intensive
	Less sensitive to sludge variability	
	Higher solids content than mechanical methods	
Sludge lagoons	Low energy consumption	Potential for odor and vector problems
	No chemical consumption	Potential for groundwater pollution
	Organic matter is further stabilized	More land intensive than mechanical methods
	Low capital cost where land is available	Appearance may be unsightly
	Least amount of skill required for operation	Design requires consideration of climatic effects

[a] Adapted in part from Refs. 42 and 58.

The results obtained from a vacuum filter system vary greatly with the characteristics of the sludge being filtered. The solids content of the sludge, among other parameters, is very important. Chemical conditioning of the sludge prior to filtration is practiced to increase the solids content, to reduce filtrate solids, and to improve the dewatering characteristics. The optimum solids content for filtration is about 6 to 8 percent. Higher solids contents make the sludge difficult to distribute and to condition for dewatering; lower solids contents require the use of larger-than-necessary

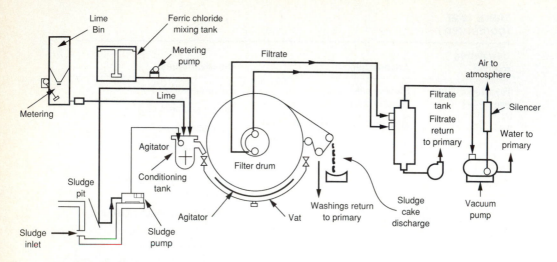

FIGURE 12-32
Typical vacuum filtration system.

vacuum filters. Chemicals that are commonly used for conditioning sludge are lime, ferric chloride, and polymers. Sludge from primary settling tanks, in general, requires smaller amounts of conditioning chemical than sludge from biological waste treatment processes.

The performance of a vacuum filter is measured in terms of the yield of solids on a dry weight basis expressed as pounds per square foot per hour (kilograms per square meter per hour). The quality of the filter cake is measured by its solids content on a wet-weight basis expressed as a percent. Typical yields and cake solids contents are shown in Table 12-28. A design rate of 3.5 lb/ft$^2 \cdot$ h (17.0 kg/m$^2 \cdot$ h) is used frequently when the quality of the sludge must be estimated. For additional information on the calculation of vacuum filter yield, Ref. 25 may be consulted.

Centrifugation

For separating liquids of different density, thickening slurries, or removing solids, the centrifugation process is widely used in the industry. The process is applicable to the dewatering of wastewater sludges and has been used with varying degrees of success in both the United States and Europe. The centrifugal devices used for thickening sludge, solid bowl and imperforate basket centrifuges (discussed in Sec. 12-6), may also be used for sludge dewatering.

Solid Bowl Centrifuge. In the solid bowl machine (see Fig. 12-33), sludge is fed at a constant flowrate into the rotating bowl, where it separates into a dense cake containing the solids and a dilute stream called "centrate." The centrate contains fine, low-density solids and is returned to the wastewater treatment system. The sludge cake, which contains approximately 70 to 80 percent moisture, is discharged from the bowl by a screw feeder into a hopper or onto a conveyor belt. Depending on the type

TABLE 12-28
Typical dewatering performance for vacuum filters[a]

Type of sludge	Cloth media		Coil media	
	Yield, lb/ft^2 · h	Cake solids, %	Yield, lb/ft^2 · h	Cake solids, %
Primary	4–8	27–35	6–8	28–32
Primary and air activated	3–6	18–25	2.5–4	23–27
Primary and oxygen activated	5–6	20–30		
Primary and trickling filter	3–7	20–30		
Waste air activated	2–2.5	13–20		
Waste oxygen activated	3–4	15–25		
Anaerobically digested				
Primary	4–7	25–35		
Primary and waste activated	2–5	18–25	3.5–4.5	20–25
Primary and trickling filter	3.5–8	20–27	4–6	27–33
Thermally conditioned				
Primary and waste activated	4–8	35–45		

[a] Adapted from Ref. 42.

Note: lb/ft^2 · h × 4.8828 = kg/m^2 · h

FIGURE 12-33
Typical solid bowl centrifuge for dewatering sludge (from Bird Machine Co.).

of sludge, solids concentration in the cake generally varies from 10 to 35 percent; newer designs can achieve solids concentrations in the 30 to 35 percent range. Sludge cake concentrations above 25 percent are desirable for disposal by incinerating or by hauling to a sanitary landfill.

Solid bowl centrifuges are suitable generally for a variety of sludge dewatering applications. The units can be used to dewater sludges with no prior chemical conditioning, but the solids capture and centrate quality are improved considerably when solids are conditioned with polymers. Chemicals for conditioning are added to the sludge-feed line or to the sludge within the bowl of the centrifuge. Dosage rates for conditioning with polymers vary from 2 to 15 lb/ton (1.0 to 7.5 kg/10^3 kg) of sludge (dry solids). Typical performance data for solid bowl centrifuges are reported in Table 12-29.

Imperforate Basket Centrifuge. Imperforate basket centrifuges are particularly suitable for small plants. For these applications, basket centrifuges can be used to concentrate and dewater waste activated sludge, with no chemical conditioning, at solids capture rates up to 90 percent. They have also been used at large plants; the County Sanitation Districts of Los Angeles use 48 imperforate basket centrifuges.

The operation of the imperforate basket centrifuge is described in Sec. 12-6 for thickening applications. The dewatering operation is similar to thickening with one additional operation. After the centrifuge is filled with solids, the unit starts to decelerate. In the dewatering mode, a "skimming" step takes place before the initiation of plowing. Skimming is the removal of soft sludge from the inner wall of sludge in the basket. The skimming volume is normally 5 to 15 percent of the bowl volume. The skimming stream is returned to the wastewater treatment system. Typical performance data for the imperforate basket centrifuge are also included in Table 12-29.

Design Considerations. The major difficulty encountered in the operation of centrifuges has been the disposal of the centrate, which is relatively high in suspended, nonsettling solids. The return of these solids to the influent of the wastewater treatment plant has resulted in the passage of these fine solids through the treatment system, thereby reducing effluent quality. Two methods can be used to control the fine solids discharge and to increase the capture—increased residence time or chemical conditioning. Longer residence of the liquid stream is accomplished by reducing the feed rate or by using a centrifuge with a larger bowl volume. Particle size can be increased by coagulating the sludge prior to centrifugation. Solids capture (measured in percent of influent solids) may be increased from a range of 50 to 80 percent to a range of 80 to 95 percent by longer residence time and chemical conditioning.

The addition of lime will also aid in the control of odors that may develop when centrifuging untreated sludge. Untreated primary sludge can usually be dewatered to a lower moisture content than digested sludge because it has not been subjected to the liquefying action of the digestion process, which reduces particle size. Chemical con-

TABLE 12-29
Typical dewatering performance for solid bowl and imperforate basket centrifuges[a]

Type of sludge	Solid bowl			Imperforate basket		
	Cake solids, %	Solids capture, %		Cake solids, %	Solids capture, %	
		Without chemicals	With chemicals		Without chemicals	With chemicals
Untreated						
Primary	25–35	75–90	90+	25–30	90–95	95+
Primary and trickling filter	20–25	60–80	90+	7–11	90+	90+
Primary and air activated	12–20	55–65	90+	12–14	—	90+
Primary and rotating biological contactor				17–24	85–90	95+
Waste sludge						
Trickling filter	10–20	60–80	90+	9–12	90+	95+
Air activated	5–15	60–80	90+	8–14	85–90	90+
Oxygen activated	10–20	60–80	90+			
Anaerobically digested						
Primary	25–35	65–80	85+			
Primary and trickling filter	18–25	60–75	85+			
Primary and air activated	15–20	50–65	85+	8–14	75–80	85+
Aerobically digested						
Waste activated	8–10	60–75	90+			
Thermally conditioned						
Primary and trickling filter	30–40	60–70	90+			
Primary and waste activated	30–40	75–85	90+			

[a] Adapted in part from Ref. 42.

ditioning is usually desirable when dewatering combined primary and waste activated sludge, regardless of whether it has been digested.

Selection of units for plant design is dependent on a manufacturer's rating and performance data. Several manufacturers have portable pilot plant units, which can be used for field testing if sludge is available. Wastewater sludges from supposedly similar treatment processes but in different localities may differ markedly from each other. For this reason, pilot plant tests should be run, whenever possible, before final design decisions are made.

The area required for a centrifuge installation is less than that required for other dewatering devices of equal capacity, and the initial cost is lower. Higher power costs will partially offset the lower initial cost. Special consideration must also be given to providing sturdy foundations and soundproofing because of the vibration and noise that result from centrifuge operation. Adequate electric power is required because large motors may be used.

Belt Filter Press

Belt filter presses are continuous-feed sludge-dewatering devices that involve the application of chemical conditioning, gravity drainage, and mechanically applied pressure to dewater sludge (see Fig. 12-34). The belt filter press was introduced in the U.S in the early 1970s and has become one of the predominant sludge-dewatering devices. Belt filter presses have proven to be effective for almost all types of municipal wastewater sludge [1].

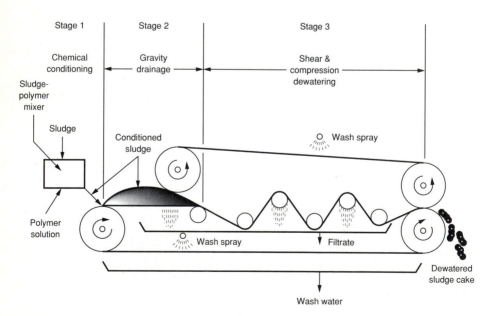

FIGURE 12-34
Three basic stages of belt press dewatering.

Process Description. In most types of belt filter presses, conditioned sludge is first introduced on a gravity drainage section where it is allowed to thicken. In this section, a majority of the free water is removed from the sludge by gravity. On some units, this section is provided with a vacuum assist, which enhances drainage and may help to reduce odors. Following gravity drainage, pressure is applied in a low-pressure section, where the sludge is squeezed between opposing porous cloth belts. On some units, the low-pressure section is followed by a high-pressure section, where the sludge is subjected to shearing forces as the belts pass through a series of rollers. The squeezing and shearing forces thus induce the release of additional quantities of water from the sludge. The final dewatered sludge cake is removed from the belts by scraper blades.

System Operation and Performance. A typical belt filter press system consists of sludge-feed pumps, polymer-feed equipment, a sludge-conditioning tank (flocculator), a belt filter press, a sludge cake conveyor, and support systems (sludge-feed pumps, washwater pumps, and compressed air). Some units do not use a sludge-conditioning tank. A schematic diagram of a typical belt filter press installation is shown in Fig. 12-35.

Many variables affect the performance of the belt filter press: sludge characteristics, method and type of chemical conditioning, pressures developed, machine configuration (including gravity drainage), belt porosity, belt speed, and belt width. The belt filter press is sensitive to wide variations in sludge characteristics, resulting in improper conditioning and reduced dewatering efficiency. Sludge-blending facilities should be included in the system design where the sludge characteristics are likely to

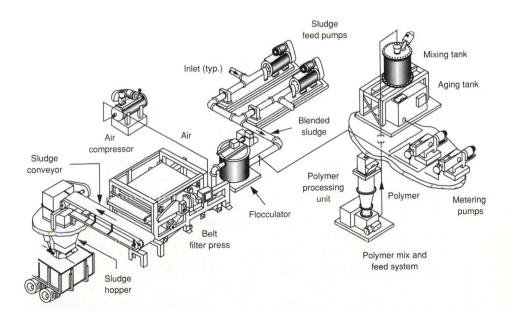

FIGURE 12-35
Schematic diagram of a belt press dewatering system [28].

TABLE 12-30
Typical dewatering performance for belt filter presses[a]

Type of sludge	Feed solids, %	Cake solids, %
Primary	3–7	28–44
Primary and waste activated	3–6	20–35
Primary and trickling filter	3–6	20–35
Waste activated	1–4	12–20
Anaerobically digested		
Primary	3–7	25–35
Primary and waste activated	3–6	20–25
Waste activated	3–4	12–20
Aerobically digested primary		
and waste activated	1–3	12–20
	4–8	12–30
Thermally conditioned primary		
and waste activated	4–8	25–50

[a] From Ref. 58.

vary widely. Based on actual operating experience, it has been found that the solids throughput is greater and the cake dryness is improved with higher solids concentrations in the feed sludge. Typical belt filter press performance data for various types of sludge are reported in Table 12-30.

Design Considerations. Belt filter presses are available in metric sizes from 0.5 to 3.5 m in belt width. The most common size used for municipal sludge applications is 2.0 m. Sludge-loading rates vary from 200 to 1500 lb/m · h (90 to 680 kg/m · h), depending on the sludge type and feed concentrations. Hydraulic throughput based on belt width ranges from 25 to 100 gal/m · min (1.6 to 6.3 L/m · s). Design of a belt filter press is illustrated in Example 12-8.

    Safety considerations in design should include adequate ventilation to remove hydrogen sulfide or other gases and equipment guards to prevent loose clothing from being caught between the rollers.

Example 12-8 Belt filter press design. A wastewater treatment plant produces 19,000 gal/d of thickened sludge containing 3 percent solids. A belt filter press installation is to be designed based on a normal operation of 8 h/d and 5 d/wk, a belt filter press loading rate of 600 lb/m · h, and the following data. Compute the number and size of belt filter presses required and the expected solids capture, in percent. Determine the daily hours of operation required if a sustained 3d peak solids load occurs.

1. Total solids in dewatered sludge = 25 percent
2. Suspended-solids concentration in filtrate = 900 mg/L = 0.09 percent
3. Washwater flowrate = 24 gal/min per m of belt width
4. Specific gravities of sludge feed, dewatered cake, and filtrate are 1.02, 1.07, and 1.01, respectively

Solution

1. Compute the average weekly sludge production rate.

$$\text{Wet sludge} = (19{,}000 \text{ gal/d})(7 \text{ d/wk})(8.34 \text{ lb/gal})(1.02) = 1{,}131{,}404 \text{ lb/wk}$$

$$\text{Dry solids} = 1{,}131{,}404 \times 0.03 = 33{,}942 \text{ lb/wk}$$

2. Compute daily and hourly dry solids processing requirements based on an operating schedule of 5 days per week and 8 hours per day.

$$\text{Daily rate} = (33{,}942 \text{ lb/wk})/(5 \text{ d/wk}) = 6788 \text{ lb/d}$$

$$\text{Hourly rate} = 6788/8 = 849 \text{ lb/h}$$

3. Compute belt filter press size.

$$\text{Belt Width} = \frac{849 \text{ lb/h}}{600 \text{ lb/m} \cdot \text{h}} = 1.42 \text{ m}$$

Use one 1.5 m belt filter press and provide one identical size for standby.

4. Compute filtrate flowrate by developing solids balance and flow balance equations.
   (a) Develop a solids balance equation.

$$\text{Solids in sludge feed} = \text{solids in sludge cake} + \text{solids in filtrate}$$

$$6788 = (S \text{ gal/d})(8.34 \text{ lb/gal})(1.07)(0.25)$$

$$+ (F \text{ gal/d})(8.34 \text{ lb/gal})(1.01)(0.0009)$$

$$6788 = 2.231S + 0.0075F$$

where S = sludge cake flowrate, gal/d
F = filtrate flowrate, gal/d

   (b) Develop a flowrate equation

$$\text{Sludge flowrate} + \text{washwater flowrate} = \text{filtrate flowrate} + \text{cake flowrate}$$

$$\text{Daily sludge flowrate} = 19{,}000 \text{ gal/d} \times 7/5 = 26{,}600 \text{ gal/d}$$

$$\text{Washwater flowrate} = 24 \text{ gal/min} \cdot \text{m} \ (1.5 \text{ m})(60 \text{ min/h})(8 \text{ h/d})$$

$$= 17{,}280 \text{ gal/d}$$

$$26{,}600 + 17{,}280 = 43{,}880 = F + S$$

   (c) Solve the mass balance and flowrate equations simultaneously.

$$F = 40{,}975 \text{ gal/d}$$

5. Determine solids capture.

$$\text{Solids capture} = \frac{\text{solids in feed} - \text{solids in filtrate}}{\text{solids in feed}} \times 100\%$$

$$= \frac{6788 \text{ lb/d} - [(40{,}975 \text{ gal/d})(8.34 \text{ lb/gal})(1.01)(0.0009)]}{6788 \text{ lb/d}} \times 100\%$$

$$= 95.4 \text{ percent}$$

6. Determine operating requirements for the sustained peak sludge load.
   (a) Determine peak 3d sludge load.
      From Fig. 5-6b, the ratio of peak to average mass loading for 3 consecutive days is 2. The peak sludge load is 19,000(2) = 38,000 gal/d.
   (b) Determine daily operating time requirements, neglecting sludge in storage.

$$\text{Dry solids/d} = 38{,}000 \text{ gal/d}(8.34 \text{ lb/gal})(1.02)(0.03)$$

$$= 9698 \text{ lb/d}$$

$$\text{Operating time} = \frac{9698 \text{ lb/d}}{600 \text{ lb/m} \cdot \text{h} \times 1.5 \text{ m}} = 10.8 \text{ h}$$

The operating time can be accomplished by running the standby belt filter press in addition to the duty press or by operating the duty press for an extended shift.

Comment. The value of sludge storage is important in dewatering applications because of the ability to schedule operations to suit labor availability most efficiently. Scheduling sludge-dewatering operations during the day shift is also desirable if sludge has to be hauled offsite.

Filter Presses

In a filter press, dewatering is achieved by forcing the water from the sludge under high pressure. Advantages cited for the filter press include (1) high concentrations of cake solids, (2) filtrate clarity, and (3) high solids capture. Disadvantages include mechanical complexity, high chemical costs, high labor costs, and limitations on filter cloth life. Various types of filter presses have been used to dewater sludge. The two types used most commonly are the fixed-volume and variable-volume, recessed plate filter presses.

Fixed-Volume, Recessed Plate Filter Press. The fixed-volume, recessed plate filter press consists of a series of rectangular plates, recessed on both sides, that are supported face to face in a vertical position on a frame with a fixed and movable head (see Fig. 12-36). A filter cloth is hung or fitted over each plate. The plates are held together with sufficient force to seal them so as to withstand the pressure applied during the filtration process. Hydraulic rams or powered screws are used to hold the plates together.

In operation, chemically conditioned sludge is pumped into the space between the plates, and pressure of 100 to 225 lb_f/in^2 (690 to 1550 kN/m^2) is applied and maintained for 1 to 3 h, forcing the liquid through the filter cloth and plate outlet ports. The plates are then separated and the sludge is removed. The filtrate is normally returned to the headworks of the treatment plant. The sludge cake thickness varies from about 1 to 1.5 in (25 to 38 mm), and the moisture content varies from 48 to 70 percent. The filtration cycle time varies from 2 to 5 h and includes the time required to (1) fill the press, (2) maintain the press under pressure, (3) open the press, (4) wash and discharge the cake, and (5) close the press. Depending on the degree of automation incorporated into the machine, operator attention must be devoted to the filter press during feed, discharge, and wash intervals.

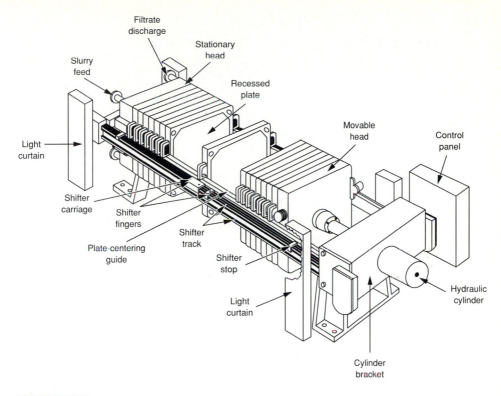

FIGURE 12-36
Fixed-volume, recessed plate filter press used for dewatering sludge (from Eimco Process Equipment Co.).

Variable-Volume, Recessed Plate Filter Press. Another type of filter press used for wastewater sludge dewatering is the variable-volume recessed plate filter press, commonly called the "diaphragm press." This type of filter press is similar to the fixed-volume press except that a rubber diaphragm is placed behind the filter media, as shown in Fig. 12-37. The rubber diaphragm expands to achieve the final squeeze pressure, thus reducing the cake volume during the compression step. Generally about 10 to 20 minutes are required to fill the press and 15 to 30 minutes of constant pressure are required to dewater the cake to the desired solids content. Variable-volume presses are generally designed for 100 to 125 lb_f/in^2 (690 to 860 kN/m^2) for the initial stage of dewatering, followed by 200 to 300 lb_f/in^2 (1380 to 2070 kN/m^2) for final compression [30]. Variable-volume presses can handle a wide variety of sludges with good performance results, but require considerable maintenance.

Design Considerations. Several operating and maintenance problems have been identified for recessed plate filter presses, ranging from difficulties in the chemical-feed and sludge-conditioning system to excessive downtime for equipment maintenance. Features that should be considered in the design of a filter press instal-

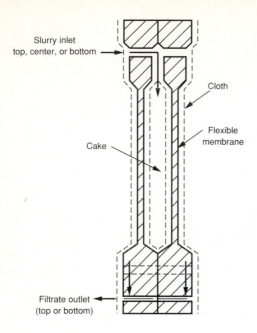

Slurry inlet
top, center, or bottom

Cloth

Cake

Flexible
membrane

Filtrate outlet
(top or bottom)

FIGURE 12-37
Cross section of a variable-volume, recessed
plate filter press.

lation include (1) adequate ventilation in the dewatering room (6 to 12 air changes per hour are recommended depending on the ambient temperature), (2) high-pressure washing systems, (3) an acid wash circulation system to remove calcium scale when lime is used, (4) a sludge grinder ahead of the conditioning tank, (5) cake breakers or shredders following the filter press (particularly if the dewatered sludge is incinerated), and (6) equipment to facilitate removal and maintenance of the plates [29].

Sludge Drying Beds

Drying beds are the most widely used method of sludge dewatering in the United States. Sludge drying beds are typically used to dewater digested sludge. After drying, the sludge is removed and either disposed of in a landfill or used as a soil conditioner. The principal advantages of drying beds are low cost, infrequent attention required, and high solids content in the dried product. Four types of drying beds are used: (1) conventional sand, (2) paved, (3) artificial media, and (4) vacuum-assisted. Because conventional sand drying beds are used most extensively, more detailed discussion is provided for this type of drying bed. For additional information on the other types of drying beds, Refs. 42 and 51 may be consulted.

Conventional Sand Drying Beds. Conventional sand drying beds are generally used for small- and medium-sized communities, although some installations are reported for larger facilities [42]. For cities with populations over 20,000, consideration should be given to alternative means of sludge dewatering. In large municipalities, the initial cost, the cost of removing the sludge and replacing sand, and the large area requirements generally preclude the use of sand drying beds.

In a typical sand drying bed, sludge is placed on the bed in a 8 to 12 in (200 to 300 mm) layer and allowed to dry. Sludge dewaters by drainage through the sludge mass and supporting sand and by evaporation from the surface exposed to the air (see Fig. 12-38). Most of the water leaves the sludge by drainage, thus the provision of an adequate underdrainage system is essential. Drying beds are equipped with lateral drainage lines (vitrified clay pipe laid with open joints or perforated plastic pipe), sloped at a minimum of 1 percent and spaced 8 to 20 ft (2.5 to 6 m) apart. The drainage lines should be adequately supported and covered with coarse gravel or crushed stone. The sand layer should be from 9 to 12 in (230 to 300 mm) deep with an allowance for some loss from cleaning operations. Deeper sand layers generally retard the draining process. Sand should have a uniformity coefficient of not over 4.0 and an effective size of 0.3 to 0.75 mm.

The drying area is partitioned into individual beds, 20 ft wide by 20 to 100 ft long (approximately 6 m wide by 6 to 30 m long), or a convenient size so that one or two beds will be filled in a normal loading cycle. The interior partitions commonly consist of two or three creosoted planks, one on top of the other, to a height of 15 to 18 in (380 to 460 mm), stretching between slots in precast concrete posts. The outer boundaries may be of similar construction or may be made of earthen embankments for open drying beds. Concrete foundation walls are required if the beds are to be covered.

Piping to the sludge beds should drain to the beds and should be designed for a velocity of at least 2.5 ft/s (0.75 m/s). Cast iron or plastic pipe is commonly used. Provisions should be included to flush the lines, if necessary, and to prevent their freezing in cold climates. Distribution boxes are required to divert the sludge flow into the bed selected. Splash plates are placed in front of the sludge outlets to spread the sludge over the bed and to prevent erosion of the sand.

Sludge can be removed from the drying bed after it has drained and dried sufficiently to be spadable. Dried sludge has a coarse, cracked surface and is black or dark brown. The moisture content is approximately 60 percent after 10 to 15 d under favorable conditions. Sludge removal is accomplished by manual shoveling into wheelbarrows or trucks or by a scraper or front-end loader. Provisions should be made for driving a truck onto or alongside of the bed to facilitate loading.

Open beds are used where adequate area is available and is sufficiently isolated to avoid complaints caused by occasional odors. Open sludge beds should be located at least 300 ft (about 100 m) from dwellings to avoid odor nuisance. Covered beds with greenhouse types of enclosures are used where it is necessary to dewater sludge continuously throughout the year regardless of the weather and where sufficient isolation does not exist for the installation of open beds.

Sludge bed loadings are computed on a per capita basis or on a unit loading of pounds of dry solids per square foot per year (kilograms of dry solids per square meter per year). Typical data for various types of sludge are shown in Table 12-31. With covered drying beds, more sludge can be applied per year because of the protection from rain and snow.

In cold climates, the effects of freezing and thawing have been observed to improve the dewatering characteristics of sludge. Freezing and thawing convert the

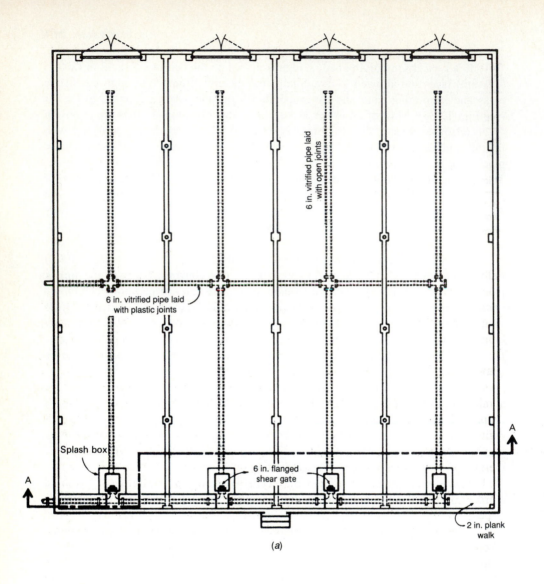

6 in. vitrified pipe laid
with open joints

6 in. vitrified pipe laid
with plastic joints

Splash box

6 in. flanged
shear gate

A

A

2 in. plank
walk

(a)

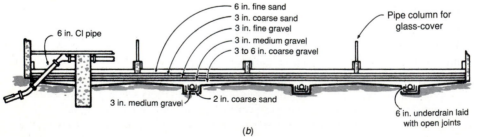

6 in. CI pipe

6 in. fine sand
3 in. coarse sand
3 in. fine gravel
3 in. medium gravel
3 to 6 in. coarse gravel

Pipe column for
glass-cover

3 in. medium gravel
2 in. coarse sand

6 in. underdrain laid
with open joints

(b)

FIGURE 12-38
Plan and section of a typical sand drying bed: (a) plan, and (b) section A–A.

872

TABLE 12-31
Typical area requirements for open sludge drying beds

Type of sludge	Area, ft²/person[a]	Sludge-loading rate, lb dry solids/ft² · yr
Primary digested	1.0–1.5	25–30
Primary and trickling-filter humus digested	1.25–1.75	18–25
Primary and waste activated digested	1.75–2.50	12–20
Primary and chemically precipitated digested	2.0–2.5	20–33

[a] Corresponding area requirements for covered beds vary from about 70 to 75 percent of those for the open beds.

Note: ft² × 0.0929 = m²
lb/ft² · yr × 4.8828 = kg/m² · yr

jelly-like consistency of sludge to a granular-type material that drains readily. Solids concentrations exceeding 20 percent may occur when the material thaws and may increase to 50 to 70 percent with additional drying time. Recent research has indicated that a 3 in (80 mm) layer of sludge is practical for most locations in moderately cold climates. For additional details, Refs. 33 and 34 may be consulted.

Paved Drying Beds. Two types of paved drying beds have been used as an alternate to sand drying beds: a drainage type and a decanting type. The drainage type functions similarly to a conventional bed in that underdrainage is collected, but sludge removal is improved by using a front-end loader. Sludge drying may also be facilitated by frequent agitation with mobile equipment. With this design, the beds are normally rectangular in shape and are 20 to 50 ft (6 to 15 m) wide by 70 to 150 ft (21 to 46 m) long with vertical side walls. Concrete or bituminous concrete linings are used, overlaying a 8 to 12 in (200 to 300 mm) sand or gravel base. The lining should have a minimum 1.5 percent slope to a center unpaved drainage area. For a given amount of sludge, this type of paved drying bed requires more area than conventional sand beds.

The decanting-type paved drying bed is a relatively new design and is advantageous for warm, arid and semi-arid climates. This type of drying bed uses low-cost impermeable paved beds that depend on the decanting of the supernatant and mixing of the drying sludge for enhanced evaporation. Features of this design include (1) a soil cement mixture paving material, (2) drawoff pipes for decanting the supernatant, and (3) a sludge feed pipe at the center of the bed (see Fig. 12-39). Decanting may remove about 20 to 30 percent of the water with a good settling sludge. Solids concentration may range from 40 to 50 percent for a 30 to 40 d drying time in an arid climate for a 12 in (300 mm) sludge layer. The bottom area of the drying bed may be determined by trial using the following equation [51,52]

$$A = \frac{1.04\, S[(1 - s_d)/s_d - (1 - s_e)/s_e] + (62.4)(P)(A)}{(62.4)(k_e)(E_p)} \qquad \text{U.S. customary units}$$

$$\tag{12-14}$$

$$A = \frac{1.04\, S[(1 - s_d)/s_d - (1 - s_e)/s_e] + (1000)(P)(A)}{(10)(k_e)(E_p)} \qquad \text{SI units} \quad (12\text{-}14a)$$

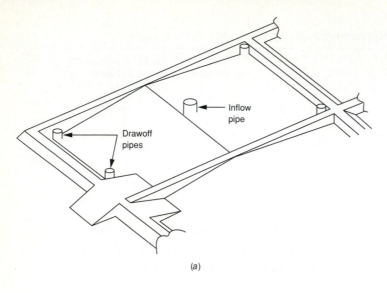

(a)

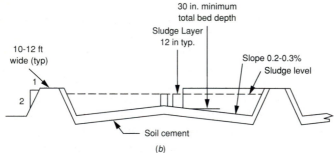

(b)

FIGURE 12-39
Paved sludge drying bed for decantation and evaporation: (a) isometric view and (b) cross section.

where A = bottom area of paved bed, ft^2 (m^2)
$\quad\quad S$ = annual sludge production, dry solids, lb (kg)
$\quad\quad s_d$ = percent dry solids in the sludge after decanting, as a decimal
$\quad\quad s_e$ = percent dry solids required for final disposal, as a decimal
$\quad\quad P$ = annual precipitation, ft (m)
$\quad\quad k_e$ = reduction factor for evaporation from sludge versus a free water surface. Use 0.6 for preliminary estimate; pilot test to determine factor for final design
$\quad\quad E_p$ = free water pan evaporation rate for the area, ft/yr (cm/yr)

Artificial Media Drying Beds. Recent developments in drying bed design include using artificial media such as stainless steel wedge wire or high-density polyurethane formed into panels. Wedge-wire drying beds were developed in England and have been used to a limited extent in the United States. In a wedge-wire drying bed, liquid sludge is introduced onto a horizontal, relatively open drainage medium (see Fig. 12-40). The medium consists of small stainless steel wedge-shaped bars, with the flat part of the wedge on top. The slotted openings between the bars are 0.01 in (0.25 mm) wide. The wedge wire is formed into panels and installed in a false floor. An outlet valve is used to control the flow of drainage. Advantages cited for this method of dewatering are as follows: (1) no clogging of the wedge wire, (2) drainage is constant and rapid, (3) throughput is higher than sand beds, (4) aerobically digested sludges can be dried, and (5) beds are relatively easy to maintain. The principal disadvantage is that the capital costs are higher than conventional drying beds.

In the high-density polyurethane media system, special 12 in (300 mm) square, interlocking panels are formed for installation on a sloped slab or in prefabricated steel self-dumping trays. Each panel has an 8 percent open area for dewatering and contains a built-in underdrain system. The system can be designed for installation in open or covered beds. Advantages cited for this method of dewatering are that (1) dilute sludges can be dewatered including aerobically digested waste activated sludge, (2) filtrate contains low suspended solids, and (3) fixed units can be cleaned easily with a front-end loader.

Vacuum-Assisted Drying Beds. A method used to accelerate dewatering and drying is the vacuum-assisted sludge drying bed (see Fig.12-41). Dewatering and drying is assisted by the application of a vacuum to the underside of porous filter plates. Operation of this method usually consists of the following steps: (1) preconditioning the sludge with a polymer, (2) filling the beds with sludge, (3) dewatering the sludge initially by gravity drainage followed by applying a vacuum, (4) allowing the sludge to air dry for approximately 24 to 48 h, (5) removal of the dewatered sludge by a front-end loader, and (6) washing the surface of the porous plates with a high-pressure hose to remove the remaining sludge residue. Data reported from

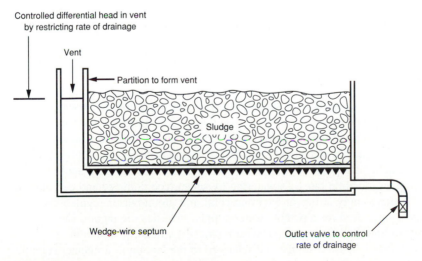

FIGURE 12-40
Cross section of a wedge-wire sludge drying bed [51].

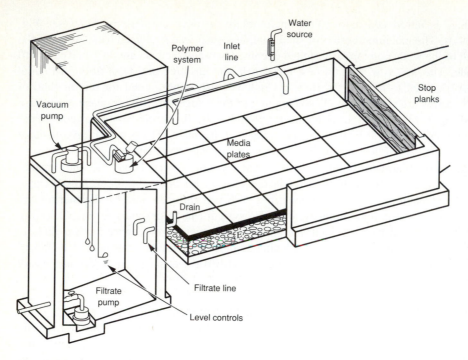

FIGURE 12-41
Sectional view of a vacuum-assisted sludge drying bed [52].

operating systems indicate that solids concentrations ranging from 8 to 23 percent can be achieved with cycle times of 8 to 48 h [48]. The principal advantages cited for this method are the reduced cycle time required for sludge dewatering, thereby reducing the effects of weather on sludge drying, and the smaller area required as compared to other types of drying beds. The principal disadvantage is that further processing may be required for additional moisture reduction.

Lagoons

Drying lagoons may be used as a substitute for drying beds for the dewatering of digested sludge. Lagoons are not suitable for dewatering untreated sludges, limed sludges, or sludges with a high-strength supernatant because of their odor and nuisance potential. The performance of lagoons, like that of drying beds, is affected by climate; precipitation and low temperatures inhibit dewatering. Lagoons are most applicable in areas with high evaporation rates. Dewatering by subsurface drainage and percolation is limited by increasingly stringent environmental and groundwater regulations. If a groundwater aquifer used for a potable water supply underlies the lagoon site, it may be necessary to line the lagoon or otherwise restrict significant percolation.

Unconditioned digested sludge is discharged to the lagoon in a manner suitable to accomplish an even distribution of sludge. Sludge depths usually range from 2.5 to 4 ft (0.75 to 1.25 m). Evaporation is the prime mechanism for dewatering. Facilities for the decanting of supernatant are usually provided, and the liquid is recycled to the treatment facility. Sludge is removed mechanically, usually at a solids content of 25 to 30 percent. The cycle time for lagoons varies from several months to several years. Typically, sludge is pumped to the lagoon for 18 months, and then the lagoon is rested for 6 months. Solids loading criteria range from 2.2 to 2.4 lb/ft$^3 \cdot$ yr (36 to 39 kg/m$^3 \cdot$ yr) of lagoon capacity [51]. A minimum of two cells is essential, even in very small plants, to ensure availability of storage space during cleaning, maintenance or emergency conditions.

12-14 HEAT DRYING

Sludge drying is a unit operation that involves reducing water content by vaporization of water to the air. In conventional sludge drying beds, vapor pressure differences account for evaporation to the atmosphere. In mechanical drying apparatuses, auxiliary heat is provided to increase the vapor-holding capacity of the ambient air and to provide the latent heat necessary for evaporation. The purpose of heat drying is to remove the moisture from the wet sludge so that it can be incinerated efficiently or processed into fertilizer. Drying is necessary in fertilizer manufacturing so as to permit the grinding of the sludge, to reduce its weight, and to prevent continued biological action. The moisture content of the dried sludge is less than 10 percent.

Theory

Under equilibrium conditions of constant rate drying, mass transfer is proportional to (1) the area of wetted surface exposed, (2) the difference between the water content of the drying air and the saturation humidity at the wet-bulb temperature of the sludge-air interface, and (3) other factors such as velocity and turbulence of drying air expressed as a mass-transfer coefficient. The pertinent equation is

$$W = k_y(H_s - H_a)A \tag{12-15}$$

where W = evaporation rate, lb/h (kg/h)
  k_y = mass-transfer coefficient of gas phase, lb mass/ft$^2 \cdot$ h (kg/m$^2 \cdot$ h) per unit of humidity difference (ΔH)
  H_s = saturation humidity of air at sludge-air interface, lb water vapor/lb dry air (kg/kg)
  H_a = humidity of drying air, lb water vapor/lb dry air (kg/kg)
  A = area of wetted surface exposed to drying medium, ft^2 (m^2)

The sludge-air interface temperature may be taken as equal to the wet-bulb temperature of the bulk volume of drying air or hot gases, provided that the temperature of the air and the walls of the container are approximately the same. For an extension of the theory and its application to specific types of drying equipment, Ref. 20 may be consulted.

Drying may be accomplished most rapidly by exposing new areas to the drying air stream. Furthermore, maximum contact between dry air and wet sludge should be obtained to assure a maximum value of ΔH. These factors must be considered in the selection of drying apparatuses for sludge disposal.

Heat Drying Options

Five mechanical processes may be used for drying sludge: (1) flash dryers, (2) spray dryers, (3) rotary dryers, (4) multiple-hearth dryers, and (5) multiple-effect evaporation (the Carver-Greenfield process). Most systems can be made to dry or incinerate. Sludge dryers are normally preceded by dewatering. Flash dryers are the most common type in use at wastewater treatment plants.

Flash Dryers. Flash drying involves pulverizing the sludge in a cage mill or by an atomized suspension technique in the presence of hot gases. The equipment should be designed so that the particles remain in contact with the turbulent hot gases long enough to accomplish mass transfer of moisture from sludge to the gases.

One operation involves a cage mill that receives a mixture of wet sludge or sludge cake and recycled dried sludge. The mixture contains approximately 50 percent moisture. The hot gases and sludge are forced up a duct in which most of the drying takes place and to a cyclone, which separates the vapor and solids. It is possible to achieve a moisture content of 8 percent in this operation. The dried sludge may be used or sold as soil conditioner or it may be incinerated in the furnace in any proportion up to 100 percent of production.

Spray Dryers. A spray dryer uses a high-speed centrifugal bowl into which liquid sludge is fed. Centrifugal force serves to atomize the sludge into fine particles and to spray them into the top of the drying chamber, where steady transfer of moisture to the hot gases takes place. A nozzle may be used in place of the bowl if the design prevents clogging of the nozzle.

Rotary Dryers. Rotary dryers have been used in several plants for the drying of sludge and for the drying and burning of municipal solid and industrial wastes. In direct-heating dryers, the material being dried is in contact with the hot gases (see Fig. 12-42a). In indirect-heating dryers, steam surrounds the central cell and is added to the hollow shaft agitator (see Fig.12-42b). In indirect-direct dryers, the hottest gases surround a central shell containing the material but return through it at reduced temperatures. Coal, oil, gas, municipal solid wastes, or the dried sludge may be used as fuel. Plows or louvers may be installed for lifting and agitating the material as the drum revolves.

Multiple-Hearth Dryers. A multiple-hearth incinerator is frequently used to dry and burn sludges that have been dewatered by mechanical devices. The operation is counter-current in which heated air and products of combustion pass by finely pulverized sludge that is raked continually to expose fresh surfaces. Additional discussion of multiple-hearth incineration is presented in Sec. 12-15.

Multiple-Effect Evaporators. The drying of sludge can also be accomplished using a proprietary multiple-effect process known as the "Carver-Greenfield process" (see Fig. 12-43). The major steps in this process are oil mixing, multiple-effect evaporation, oil-solids separation, and condensate-oil separation. In this process, the oil-sludge mixture, which can be pumped easily and is effective in reducing scaling

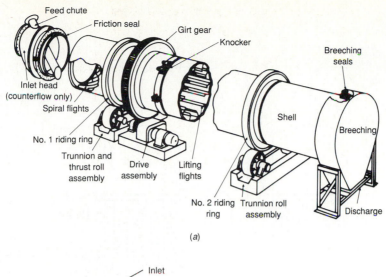

Feed chute
Friction seal
Girt gear
Knocker
Breeching seals
Inlet head (counterflow only)
Spiral flights
Shell
Breeching
No. 1 riding ring
Trunnion and thrust roll assembly
Drive assembly
Lifting flights
No. 2 riding ring
Trunnion roll assembly
Discharge

(a)

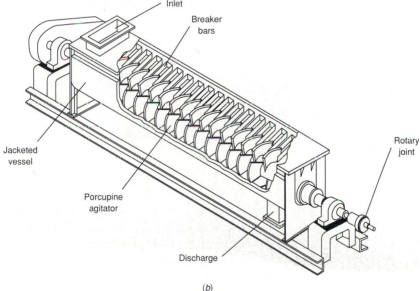

Inlet
Breaker bars
Jacketed vessel
Rotary joint
Porcupine agitator
Discharge

(b)

FIGURE 12-42
Rotary dryers used for drying sludge: (a) direct dryer and (b) indirect dryer (from Bethlehem Corp.).

and corrosion, is passed through falling film evaporators. Water is removed because it has a lower boiling point than the oil carrier. After evaporation, what remains is essentially a mixture of oil and dry sludge. The solids are removed from the oil with a centrifuge. The remaining oil can be separated into a light-oil and heavy-oil residue by exposing it to superheated steam.

The dry solids are suitable for further processing (e.g., pelletizing as a fuel source) or disposal. The recovery of energy and heat from the dried sludge using

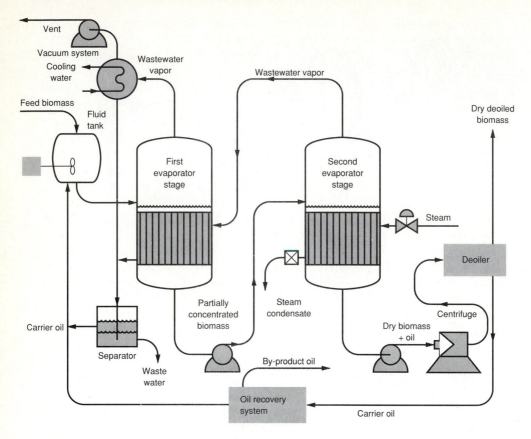

FIGURE 12-43
Schematic diagram of a two-stage Carver-Greenfield sludge drying system.

an incinerator-pyrolysis reactor, or gasifier, is an option that should be investigated when the Carver-Greenfield process is being evaluated. The heat recovered from the dried sludge could be used to supply the energy requirements of the process. Two large Carver-Greenfield facilities are nearing completion of construction in the Los Angeles area at the time of the writing of this text (1989).

Air Pollution and Odor Control

The two most important control measures associated with heat drying of sludge are fly ash collection and odor control. Cyclone separators with efficiencies of 75 to 80 percent are suitable for vent gas temperatures up to 650 or 700°F (340 or 370°C). Wet scrubbers have higher efficiencies and will condense some of the organic matter in the vent gas but may carry over water droplets.

Sludge drying occurs at temperatures of approximately 700°F (370°C), whereas temperatures ranging from 1200 to 1400°F (650 to 760°C) are required for complete incineration. To achieve destruction of odors, the exhaust gases must reach approximately 1350°F (730°C). Thus, if the gases evolved in the drying process are reheated in an incinerator to a minimum of 1350°F, odors will be diminished greatly. At lower temperatures, partial oxidation of odor-producing compounds may occur, resulting in an increase in the intensity of the disagreeable character of odor produced.

12-15 THERMAL REDUCTION

Thermal reduction of sludge involves (1) the total or partial conversion of organic solids to oxidized end products, primarily carbon dioxide and water, by incineration or wet-air oxidation or (2) the partial oxidation and volatilization of organic solids by pyrolysis or starved-air combustion to end products with energy content. The major advantages of thermal reduction are (1) maximum volume reduction, thereby lessening the disposal requirements, (2) destruction of pathogens and toxic compounds, and (3) energy recovery potential [47]. Disadvantages cited include (1) high capital and operating cost, (2) highly skilled operating and maintenance staffs are required, and (3) the residuals produced (air emissions and ash) may have adverse environmental effects, and (4) disposal of residuals, which may be classified as hazardous wastes, may be uncertain and expensive. Thermal reduction processes are used most commonly by medium-to-large sized plants with limited ultimate disposal options.

Sludges processed by thermal reduction are usually dewatered, untreated sludges. It is normally unnecessary to stabilize sludge before incineration. In fact, such practice may be detrimental because stabilization, specifically aerobic and anaerobic digestion, decreases the volatile content of the sludge and consequently increases the requirement for an auxiliary fuel. An exception is the use of heat treatment ahead of incineration. Heat treated sludges dewater extremely well, making the sludge auto-combustible (i.e., no auxiliary fuel is required to sustain the burning process). Sludges may be subjected to thermal reduction separately or in combination with municipal solid wastes.

The thermal reduction processes considered in the following discussion include multiple-hearth incineration, fluidized-bed incineration, co-incineration, wet-air oxidation, and wet oxidation in vertical, deep-well reactors. Before discussing these processes, it will be helpful to review some fundamental aspects of thermal reduction.

Process Fundamentals

Combustion is the rapid exothermic oxidation of combustible elements in fuel. Incineration is complete combustion. Pyrolysis is the destructive distillation, reduction, or thermal cracking and condensation of organic matter under heat and/or pressure in the absence of oxygen. Partial pyrolysis, or more commonly called "starved-air combustion," is incomplete combustion and occurs when insufficient oxygen is provided to satisfy the combustion requirements. Wet oxidation is a form of incomplete combustion that occurs under high temperatures and pressures.

Complete Combustion. The predominant elements in the carbohydrates, fats, and proteins that compose the volatile matter of sludge are carbon, oxygen, hydrogen, and nitrogen (C-O-H-N). The approximate percentages of these may be determined in the laboratory by a technique known as *ultimate analysis*.

Oxygen requirements for complete combustion of a material may be determined from a knowledge of its constituents, assuming that carbon and hydrogen are oxidized to the ultimate end products CO_2 and H_2O. The formula becomes

$$C_aO_bH_cN_d + (a + 0.25c - 0.5b)O_2 \rightarrow aCO_2 + 0.5cH_2O + 0.5dN_2 \quad (12\text{-}16)$$

The theoretical quantity of air required will be 4.35 times the calculated quantity of oxygen because air is composed of 23 percent oxygen on a weight basis. To ensure complete combustion, excess air amounting to about 50 percent of the theoretical amount will be required. A materials balance must be made to include the above compounds and the inorganic substances in the sludge, such as the inert material and moisture, and the moisture in the air. The specific heat of each of these substances and of the products of combustion must be taken into account in determining the heat required for the incineration process.

Heat requirements will include the sensible heat, Q_s, in the ash, plus the sensible heat required to raise the temperature of the flue gases to 1400°F (760°C) or whatever higher temperature of operation is selected for complete oxidation and elimination of odors, less the heat recovered in preheaters or recuperators. Latent heat, Q_e, must also be furnished to evaporate all of the moisture in the sludge. Total heat required, Q, may be expressed as

$$Q = \Sigma Q_s + Q_e = \Sigma C_p W_s(T_2 - T_1) + W_W \lambda \quad (12\text{-}17)$$

where
C_p = specific heat for each category of substance in ash and flue gases
W_s = weight of each substance
T_1, T_2 = initial and final temperatures
W_W = weight of moisture in sludge
λ = latent heat of evaporation per pound (kilogram)

Reduction of moisture content of the sludge is the principal way to lower heat requirements; and the moisture content may determine whether additional fuel will be needed to support combustion.

The heating value of a sludge may be determined by the conventional bomb-calorimeter test. An empirical formula based on a statistical study of fuel values of vacuum filtered sludges of different types and taking into account the amount of coagulant added before filtration, is as follows [11]:

$$Q = a\left[\frac{P_v(100)}{100 - P_c} - b\right]\frac{100 - P_c}{100} \quad (12\text{-}18)$$

where Q = fuel value, Btu/lb dry solids
a = coefficient (131 for primary sludge, untreated or digested; 107 for fresh waste activated)
b = coefficient (10 for primary sludge; 5 for activated sludge)
P_v = percent of volatile solids in sludge
P_c = percent of coagulating solids added to the sludge

The fuel value of sludge ranges widely depending on the type of sludge and the volatile solids content. The fuel value of untreated primary sludge is the highest, especially if it contains appreciable amounts of grease and skimmings. Where kitchen food grinders are used, the volatile and thermal content of the sludge will also be high. Digested sludge has a significantly lower heating value than raw sludge. Typical heating values for various types of sludge are reported in Table 12-32. The heating value for sludge is equivalent to some of the lower grades of coal.

To design an incinerator for sludge volume reduction, a detailed heat balance must be prepared. Such a balance must include heat losses through the walls and pertinent equipment of the incinerator, as well as losses in the stack gases and ash. Approximately 1800 to 2500 Btu (2.0 to 2.5 MJ) are required to evaporate each 1 lb (0.5 kg) of water in the sludge. Heat is obtained from the combustion of the volatile matter in the sludge and from the burning of auxiliary fuels. For untreated primary sludge incineration, the auxiliary fuel is only needed for warming up the incinerator and maintaining the desired temperature when the volatile content of the sludge is low. The design should include provisions for auxiliary heat for startup and for assuring complete oxidation at the desired temperature under all conditions. Fuels such as oil, natural gas, or excess digester gas are suitable.

Pyrolysis. Because most organic substances are thermally unstable, they can, upon heating in an oxygen-free atmosphere, be split through a combination of thermal cracking and condensation reactions into gaseous, liquid, and solid fractions. Pyrolysis is the term used to describe the process. In contrast to the combustion process, which is highly exothermic, the pyrolytic process is highly endothermic. For this reason, the term "destructive distillation" is often used as an alternative for pyrolysis [40].

The characteristics of the three major component fractions resulting from the pyrolysis are as follows:

1. A gas stream containing primarily hydrogen, methane, carbon monoxide, carbon dioxide, and various other gases, depending on the organic characteristics of the material being pyrolyzed
2. A fraction that consists of a tar and/or oil stream that is liquid at room temperatures and has been found to contain chemicals such as acetic acid, acetone, and methanol
3. A char consisting of almost pure carbon plus any inert material that may have entered the process

TABLE 12-32
Typical heating values for various types of sludge[a]

	Heating value, Btu/lb of total dry solids	
Type of sludge	Range	Typical
Raw primary	10,000 –12,500	11,000
Activated	7,000[b]–10,000	9,000
Anaerobically digested primary	4,000 – 6,000	5,000
Raw chemically precipitated primary	6,000 – 8,000	7,000
Biological filter	7,000 –10,000	8,500

[a] Adapted in part from Ref. 62.
[b] Lower value applies to plants operating with long solids retention times.
Note: Btu/lb of dry solids × 2.3241 = kJ/kg

For cellulose ($C_6H_{10}O_5$), the following expression has been suggested as being representative of the pyrolysis reaction:

$$3(C_6H_{10}O_5) \rightarrow 8H_2O + C_6H_8O + 2CO + 2CO_2 + CH_4 + H_2 + 7C$$

$$(12\text{-}19)$$

In Eq. 12-19, the liquid tar and/or oil compounds normally obtained are represented by the expression C_6H_8O. It has been found that distribution of the product fractions varies dramatically with the temperature at which the pyrolysis is carried out. Additional details may be found in Ref. 19.

Starved Air Combustion. Starved air combustion combines some of the features of complete combustion with pyrolysis. The process is easier to control than pyrolysis and provides better control of air emissions than complete combustion. Products of the starved air combustion process are combustible gases, tars, oils, and a solid char that can have appreciable heating value. Some of the advantages in lieu of complete combustion are easier process control, greater solids throughput because of higher hearth loading rates, particulate production per unit weight of solids fed is less than conventional incineration, and lower fuel requirements are required for afterburners used for emissions control [42]. Existing multiple hearth furnaces can also be retrofitted to use starved air combustion. Because starved air combustion is very complex and is not understood completely, pilot plant testing is recommended to determine the yield and composition of the off-gas and residue.

Wet Combustion. Organic substances may be oxidized under high pressures at elevated temperatures with the sludge in a liquid state by feeding compressed air into the pressure vessel. The process, known as wet combustion, was developed in Norway for pulp mill wastes, but has been modified for the oxidation of untreated wastewater sludges pumped directly from the primary settling tank or thickener. Combustion is not complete; the average is 80 to 90 percent completion. Thus, some organic matter, plus ammonia, will be observed in the end products. For this incomplete combustion reaction, the following reaction is characteristic:

$$C_aH_bO_cN_d + 0.5(n_y + 2s + r - c)O_2 \rightarrow$$
$$nC_wH_xO_yN_z + sCO_2 + rH_2O + (d - nz)NH_3 \quad (12\text{-}20)$$

where $r = 0.5[b - nx - 3(d - nz)]$
$s = a - nw$

The results obtained from this equation can also be approximated by the COD of the sludge, which is approximately equal to the oxygen required in combustion. The range of heat released per 1 lb (0.5 kg) of air required has been found to be from 1200 to 1400 Btu (1.3 to 1.5 MJ). Maximum operating temperatures for the system vary from 350 to 600°F (175 to 315°C) with design operating pressures ranging from 150 to 3000 lb_f/in^2 gage (1 to 20 MN/m^2). An application of the wet combustion concept is the Zimpro process, which is also used for heat treatment (see Sec. 12-7). Another adaptation of the wet combustion concept is the vertical, deep-well reactor, where liquid sludge is heated and pressurized in a below-grade pressure vessel. The pressure vessel is an encased, sealed well approximately 4000 to 5000 ft (1200 to 1500 m) deep.

Thermal Reduction Process Applications

Application of various types of thermal reduction processes is described in the discussion that follows: multiple hearth incineration, fluidized-bed incineration, co-incineration, wet-air oxidation, and wet oxidation in a vertical, deep-well reactor.

Multiple-Hearth Incineration. Multiple hearth incineration is used to convert dewatered sludge cake to an inert ash. Because the process is complex and requires specially trained operators, multiple-hearth furnaces are normally used only in large plants. Multiple-hearth incinerators have been used at smaller facilities where land for the disposal of sludge is limited and at chemical treatment plants for the recalcining of lime sludges.

As shown in Fig. 12-44, sludge cake is fed onto the top hearth and is slowly raked to the center. From the center, sludge cake drops to the second hearth where the rakes move it to the periphery. The sludge cake drops to the third hearth and is again raked to the center. The hottest temperatures are on the middle hearths, where the sludge burns and where auxiliary fuel is also burned as necessary to warm up the furnace and to sustain combustion. Preheated air is admitted to the lowest hearth and is further heated by the sludge as the air rises past the middle hearths where combustion is occurring. The air then cools as it gives up its heat to dry the incoming sludge on the top hearths.

The highest moisture content of the flue gas is found on the top hearths, where sludge with the highest moisture is heated and some water is vaporized. Cooling air is initially blown into the central column and hollow rabble arms to keep them from overheating. A large portion of this air, after passing out the central column at the top, is recirculated to the lowest hearth as preheated combustion air.

A multiple-hearth furnace may also be designed as a dryer only. In this case, a furnace is needed to provide hot gases, and the sludge and gases both proceed downward through the furnace in parallel flow. Parallel flow of product and hot gases is frequently used in drying operations to prevent burning or scorching heat-sensitive materials.

Feed sludge must contain more than 15 percent solids because of limitations on the maximum evaporating capacity of the furnace. Auxiliary fuel is usually required when the feed sludge contains between 15 and 30 percent solids. Feed sludge containing more than 50 percent solids may create temperatures in excess of the refractory and metallurgical limits of standard furnaces. Average loading rates of wet cake are approximately 8 lb/ft$^2 \cdot$ h (40 kg/m$^2 \cdot$ h) of effective hearth area but may range from 5 to 15 lb/ft$^2 \cdot$ h (25 to 75 kg/m$^2 \cdot$ h).

In addition to dewatering, required ancillary processes include ash-handling systems and some type of wet or dry scrubber to meet air pollution requirements. In wet scrubbers, scrubber water comes in contact with and removes most of the particulate matter in the exhaust gases. The recycle BOD and COD is nil, and the suspended-solids content is a function of the particulates captured in the scrubber.

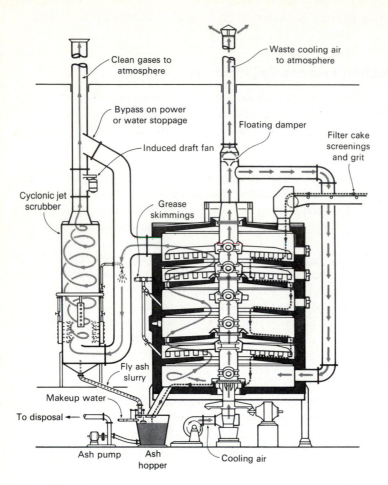

FIGURE 12-44
Typical multiple-hearth incinerator.

Under proper operating conditions, particulate discharges to the air from wet scrubbers are less than 1.3 lb/ton (0.65 kg/10^3 kg) of dry sludge input.

Ash-handling may be either wet or dry. In the wet system, the ash falls into an ash hopper located beneath the furnace, where it is slurried with water from the exhaust gas scrubber. After agitation, the ash slurry is pumped to a lagoon or is dewatered mechanically. In the dry system, the ash is conveyed mechanically to a storage hopper for discharge into a truck for eventual disposal as fill material, if the dry ash is environmentally acceptable. The ash is usually conditioned with water. Ash density is about 0.35 lb/ft^3 (5.6 kg/m^3) dry and 55 lb/ft^3 (880 kg/m^3) wet.

Fluidized-Bed Incineration. The fluidized bed incinerator commonly used for sludge incineration is a vertical, cylindrically shaped refractory-lined steel shell that contains a sand bed (media) and fluidizing air orifices to produce and sustain combustion (see Fig.12-45). The fluidized-bed incinerator ranges in size from 9 to 25 ft (2.7 to 7.6 m) in diameter. The sand bed, when quiescent, is approximately 2.5 ft (0.8 m) thick and rests on a brick dome or refractory-lined grid. The sand bed support area contains orifices, called "tuyeres," through which air is injected into the incinerator at a pressure of 3 to 5 lb_f/in^2 (20 to 35 kN/m^2) to fluidize the bed. At low velocities, combustion gas "bubbles" appear within the fluidized bed. The main bed of suspended particles remains at a certain elevation in the combustion chamber and "boils" in place. Units that function in this manner are called "bubbling bed" incinerators. The mass of suspended solids and gas, when active and at operating temperature, expands to about double the at-rest volume. Sludge is mixed quickly within the fluidized bed by the turbulent action of the bed. Evaporation of the water and combustion of the sludge solids takes place rapidly. Combustion gases and ash leave the bed and are transported through the freeboard area to the gas outlet through the top of the incinerator. No ash exits

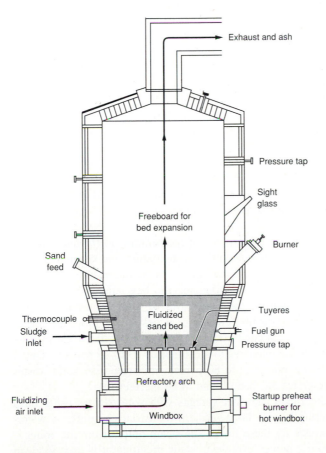

FIGURE 12-45
Fluidized-bed incinerator.

from the bed at the bottom of the incinerator. Combustion gases and entrained ash are scrubbed, normally with a venturi scrubber.

The minimum temperature needed in the sand bed prior to injection of sludge is approximately 1300°F (700°C). The temperature of the sand bed is controlled between 1400 to 1500°F (760 and 820°C).

Recycle flows consist of scrubber water produced at a rate of approximately 3 to 5 gal/lb (25 to 40 L/kg) of dry solids feed to the fluidized bed. Most of the ash (99 percent) is captured in the scrubber water, and the suspended-solids content is approximately 20 to 30 percent of the dry solids feed. Because the BOD and COD are nil, the recycle flow is normally directed to an ash lagoon. Particulates and other air emissions are comparable to those from the multiple-hearth incinerator.

The combustion process is controlled by varying the sludge-feed rate and the air flow to the reactor to oxidize completely all the organic material. If the process is operated continuously or with shutdowns of short duration, there is no need for auxiliary fuel after startup.

A modification of the fluid-bed incineration technology is the "circulating bed" incinerator. In the circulating bed unit, the reactor gas passes through the combustion chamber at much higher velocities, ranging from 10 to 25 ft/s (3 to 8 m/s). At these velocities, the bubbles in the fluidized bed disappear and streamers of solids and gas prevail. The entire mass of entrained particles flow up the reactor shaft to a particle separator, are deposited in storage momentarily, and are recirculated back to the primary combustion zone in the bottom of the reactor. Ash is removed continuously from the bottom of the bed. On turndown, the circulating bed becomes a bubbling bed. A facility using circulating bed technology is under construction in Los Angeles, California at the time of writing of this text (1989).

Like the multiple hearth, the fluidized bed, though very reliable, is complex and requires the use of trained personnel. For this reason, fluidized-bed incinerators are normally used in medium-to-large plants, but they may be used in plants with lower flow ranges where land for the disposal of sludge is limited.

Co-Incineration. Co-incineration is the process of incinerating wastewater sludges with municipal solid wastes. The major objective is to reduce the combined costs of incinerating sludge and solid wastes. At present, co-incineration is not practiced widely. The process has the advantages of producing the heat energy necessary to evaporate water from sludges, supporting the combustion of solid wastes and sludge, and providing an excess of heat for steam generation, if desired, without the use of auxiliary fossil fuels. In properly designed systems, the hot gases from the process can be used to remove moisture from sludges to a content of 10 to 15 percent. Direct feeding of sludge filter cake containing 70 to 80 percent moisture over solid wastes on traveling or reciprocating grates has been found to be ineffective.

Co-incineration of industrial wastewater sludges with solid wastes using flash drying has been done successfully in a large, eastern industrial complex. A water-filled boiler serves as the furnace for these fuels. The steam output from the boiler is used to generate in-plant electric power. An electrostatic precipitator is used to clean the exhaust gases. In Duluth, Minnesota, refuse derived fuel (RDF) and municipal sewage sludge are co-fired in a bubbling bed fluid-bed furnace.

For systems operating without heat recovery, a disposal ratio of 1 lb (0.5 kg) of dry wastewater solids to 5 lb (2.3 kg) of solid wastes is fired in normal operation. In the case of the water-walled boiler with heat recovery, the ratio is approximately 1 lb (0.5 kg) of dry (industrial plant) solids to 8 lb (3.6 kg) of solid wastes.

Based on past experience in municipal solid waste disposal, the application of co-incineration will likely continue to proceed very slowly, despite the advantages to the community in combining the two waste disposal functions. For additional information on co-incineration facilities, Ref. 47 may be consulted.

Wet-Air Oxidation. The proprietary Zimpro process (see Fig. 12-30) involves wet oxidation of untreated sludge at an elevated temperature and pressure. The process is the same as that discussed under heat treatment, except that higher pressures and temperatures are required to oxidize the volatile solids more completely. Untreated sludge is ground and mixed with a specified quantity of compressed air. The mixture is pumped through a series of heat exchangers and then enters a reactor, which is pressurized to keep the water in the liquid phase at the reactor operating temperature of 350 to 600°F (175 to 315°C). High pressure units can be designed to operate at pressures up to 3000 lb_f/in^2 gage (20 MN/m^2). Gases, liquid, and ash leave the reactor.

The liquid and ash are returned through heat exchangers to heat the incoming sludge and then pass out of the system through a pressure-reducing valve. Gases released by the pressure drop are separated in a cyclone and released to the atmosphere. In large installations, it may be economical to expand the gases through a turbine to recover power. The liquid and stabilized solids are cooled by passing through a heat exchanger and are then separated in a lagoon or settling tank or on sand beds. The liquid is returned to the primary settling tank and the solids are disposed of by landfill. The process can be designed to be thermally self-sufficient when untreated sludge is used. When additional heat is needed, steam is injected into the reactor vessel.

A major disadvantage associated with this process is the high-strength recycle liquor produced. The liquors represent a considerable organic load on the treatment system. The BOD content of the liquor may be as high as 40 to 50 percent of that of the unprocessed sludge; the COD typically ranges from 10,000 to 15,000 mg/L [61]. Anaerobic digestion has been used successfully for treatment of the recycle liquor in San Mateo, California.

Wet-air oxidation has been implemented in only a limited number of installations since its introduction in the early 1960s, but many of these units have subsequently been taken out of service because of corrosion, high energy costs, excessive maintenance, or odor problems.

Wet Oxidation in a Vertical, Deep-Well Reactor. Wet oxidation in a vertical, deep-well reactor consists of discharging liquid sludge in the pressure and temperature controlled environment of a tube-and-shell reactor suspended within a deep well (see Fig. 12-46). The concentric tubes of the reactor separate the downflowing and upflowing streams. Oxygen or air is injected into the waste stream, and the mixture flows out of the bottom of the downcomer line in the reactor and rises vertically.

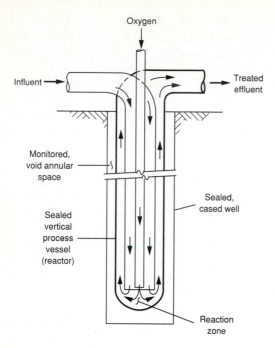

FIGURE 12-46
Cross-section through wet oxidation, deep-well reactor (adapted from Oxidyne).

Pressures in the bottom of the reactor range from 1400 to 2000 lb_f/in^2 (9700 to 13,800 kN/m^2) and temperatures are about 550°F (290°C). The use of natural hydrostatic pressure eliminates the need for high-pressure pumps and high-temperature heat exchangers. Above-ground treatment consists of gas-liquid separation, solids-liquid separation and post treatment of the supernatant stream. Reductions in COD in excess of 80 percent are claimed by manufacturers of this system. A large facility is under construction in Houston, Texas, at the time of the writing of this text (1989). Advantages cited for this process are (1) small space requirements, (2) high removals of suspended solids and organic matter, (3) little odors or objectionable air emissions, and (4) low energy requirements because the process is exothermal. The principal disadvantages are that it does not have a long history of operation and maintenance, and skilled operators are required for process control.

Air Pollution Control

Thermal reduction methods for wastewater sludge have the potential to be significant contributors to air pollution. Air contaminants associated with thermal reduction methods can be divided into two categories: (1) odors and (2) combustion emissions. Odors are particularly offensive to the human senses and special attention is required to minimize nuisance odor emissions. Combustion emissions vary depending upon the type of thermal reduction technology employed and the nature of the sludge and auxiliary fuel used in the combustion process. Combustion emissions of particular concern are particulates, oxides of nitrogen, acid gases, and specific constituents such as hydrocarbons and heavy metals (mercury, beryllium, etc.).

Air pollution control regulations are promulgated by federal, state, and local agencies. The Clean Air Act of 1979 gave the Environmental Protection Agency (EPA) the responsibility and authority to establish a nationwide program of air pollution abatement and air quality enhancement. The state government has primary responsibility for implementing the program under federal supervision. States often delegate the responsibility and authority to local agencies for carrying out the state plan. In addition, the EPA is in the process of promulgating new sludge regulations at the time this text is being written (1989), which include proposed regulations for incinerator operation and emissions control. Designers of thermal reduction facilities must be familiar with the various existing and pending air pollution control regulations. It will be necessary to provide proper odor and emission control methods to meet these regulations and to obtain the necessary permits required. For summaries of regulations for emission control and emission control technologies, Refs. 47 and 63 may be consulted.

12-16 PREPARATION OF SOLIDS MASS BALANCES

Sludge-processing facilities, such as thickening, digestion, and dewatering, produce waste streams that must be recycled to the treatment process or to treatment facilities designed specifically for the purpose. When the flows are recycled to the treatment process, they should be directed to the head of the plant and blended with the plant flow following preliminary treatment. Equalization facilities can be provided for the recycled flows so that their reinjection into the plant flow will not cause a shock loading on the subsequent treatment processes.

The recycled flows impose an incremental solids, hydraulic, and organic load on the wastewater treatment facilities that must be considered in the plant design. To predict these incremental values, it is necessary to perform a mass balance for the treatment system. The preparation of a mass balance is illustrated in Example 12-9. Although all the computational details are given in the example and are, for the most part, self-explanatory, the following discussion is provided to further explain the general methodology involved in preparing mass balances.

Basis for Preparation of Mass Balances

Typically, a mass balance is computed on the basis of average flow and average BOD and suspended-solids concentrations. To size certain facilities properly, such as sludge storage tanks and plant piping, it is also important to perform a mass balance for the maximum expected concentration of BOD and suspended solids in the untreated wastewater. However, the maximum concentrations will not usually result in a proportional increase in the recycled BOD and suspended solids. The principal reason is that the storage capacity in the wastewater and sludge-handling facilities tends to dampen peak solids loads to the plant. For example, for a maximum suspended-solids concentration equal to twice the average value, the resulting peak solids loading to a dewatering unit may be only 1.5 times the average loading. Further, it has been shown that periods of maximum hydraulic loading typically do not correlate with periods of maximum BOD and suspended solids. Therefore, coincident maximum hydraulic loadings should not be used in the preparation of a mass balance for maximum organic loadings (see Chap. 5).

Performance Data for Sludge-Processing Facilities

To prepare a mass balance, it is necessary to have information on the operational performance and efficiency of the various unit operations and processes used for the processing of waste sludge. Some representative data on the solids capture and expected solids concentrations for the most commonly used operations are reported in Tables 12-33 and 12-34. These data were derived from an analysis of the records from a number of installations throughout the United States. However, local conditions have a significant effect on such data, so the reported values should be used as a guide only if specific data are unavailable.

In addition to data on expected solids capture and concentrations, data on the expected concentrations of BOD and suspended solids in the return flows must also be available for the preparation of mass balances. Some representative data for the most commonly used operations and processes are reported in Table 12-35. The wide variation that can occur in the reported values is apparent. It is therefore stressed that the values in Table 12-35 should be used only if specific data are unavailable. The preparation of a solids balance is illustrated in Example 12-9.

TABLE 12-33
Typical solids capture values for various solids-processing facilities

Operation	Solids capture, %	
	Range	Typical
Gravity thickeners		
Primary sludge only	85–92	90
Primary and waste activated	80–90	85
Flotation thickeners		
With chemicals	90–98	95
Without chemicals	80–95	90
Centrifuge thickeners		
With chemicals	90–98	95
Without chemicals	80–90	85
Vacuum filtration		
With chemicals	90–98	95
Belt filter press		
With chemicals	85–98	93
Filter press		
With chemicals	90–98	95
Centrifuge dewatering		
With chemicals	85–98	90
Without chemicals	55–90	80

TABLE 12-34
Typical solids concentration values for the sludge-processing facilities given in Table 12-33

Operation	Solids concentration, %	
	Range	Typical
Gravity thickeners		
Primary sludge only	4–10	6
Primary and waste activated	2–6	4
Flotation thickeners		
With chemicals	3–6	4
Without chemicals	3–6	4
Centrifuge thickeners		
With chemicals	4–8	5
Without chemicals	3–6	4
Vacuum filtration		
With chemicals	15–30	20
Belt filter press filtration		
With chemicals	15–30	22
Filter press		
With chemicals	20–50	36
Centrifuge dewatering		
With chemicals	10–35	22
Without chemicals	10–30	18

TABLE 12-35
Typical BOD and suspended-solids concentrations in the recycle flows from various sludge-processing facilities

Operation	BOD, mg/L		Suspended solids, mg/L	
	Range	Typical	Range	Typical
Gravity thickening				
Primary sludge	100–400	250	80–300	200
Primary sludge and waste activated sludge	60–400	300	100–350	250
Flotation thickening	50–400	250	100–600	300
Centrifuge thickening				
Air activated sludge	400–1,200	800	500–1,500	800
Oxygen activated sludge	1,200–1,600	1,400	1,500–2,000	1,600
Anaerobic digestion				
Standard-rate type	500–1,000	800	1,000–5,000	3,000
High-rate type	2,000–5,000	4,000	1,000–10,000	6,000
Aerobic digestion	200–5,000	500	1,000–10,000	3,400
Heat treatment, top liquor or filtrate	3,000–15,000	7,000	1,000–5,000	2,000
Vacuum filtration				
Undigested sludge	500–5,000	1,000	1,000–5,000	2,000
Digested sludge	500–5,000	2,000	1,000–20,000	4,000
Centrifugation				
Undigested sludge	1,000–10,000	5,000	2,000–10,000	5,000
Digested sludge	1,000–10,000	5,000	2,000–15,000	5,000
Belt filter press				
Undigested sludge	50–500	300	200–2,000	1,000
Digested sludge	50–500	300	200–2,000	1,000

EXAMPLE 12-9   Preparation of solids balance for treatment facility. Prepare a solids balance for the treatment flow diagram shown in the following figure, using an iterative computational procedure. Assume that the design of the biological treatment process is the same as that presented in Example 10-2. Also assume for the purpose of this example that the following data apply:

1. Wastewater flowrates
    (a) Average dry-weather flow = 5.71 Mgal/d
    (b) Peak dry-weather flow = 2.5 × (5.71 Mgal/d) = 14.3 Mgal/d
2. Influent characteristics
    (a) BOD_5 = 375 mg/L
    (b) Suspended solids = 360 mg/L

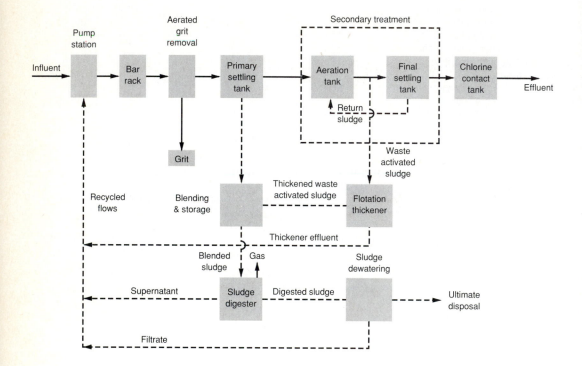

3. Solids characteristics
    (a) Concentration of primary solids = 6%
    (b) Concentration of thickened waste activated sludge = 4%
    (c) Total solids in digested sludge = 5%
    (d) For the purposes of this example, assume that the specific gravity of the solids from the primary sedimentation tank and the flotation thickener is equal to 1.0.
4. Effluent characteristics
    (a) BOD_5 = 20 mg/L
    (b) Suspended solids = 22 mg/L
5. Neglect solids removed in grit chambers in solids balance

Solution

1. Convert the given constituent quantities to daily mass values.
    (a) BOD_5 in influent:

    $$BOD_5, lb/d = 5.71 \text{ Mgal/d } (375 \text{ mg/L})[8.34 \text{ lb/Mgal} \cdot (mg/L)]$$

    $$= 17,858 \text{ lb/d}$$

    (b) Suspended solids in influent:

    $$\text{Suspended solids, lb/d} = 5.71 \text{ Mgal/d } (360 \text{ mg/L})[8.34 \text{ lb/Mgal} \cdot (mg/L)]$$

    $$= 17,143 \text{ lb/d}$$

2. Prepare the first iteration of the solids balance. (In the first iteration, the effluent wastewater suspended solids and the biological solids generated in the process are distributed among the unit operations and processes that make up the treatment system.)
    (a) Primary settling:
        i. Operating parameters:
           BOD_5 removed = 33%
           Suspended solids removed = 70% (typical value)
       ii. BOD_5 removed = 0.33(17,858 lb/d) = 5893 lb/d
      iii. BOD_5 to secondary = (17,858 − 5893) lb/d = 11,965 lb/d
       iv. Suspended solids removed = 0.7(17,143 lb/d) = 12,000 lb/d
        v. Suspended solids to secondary = (17,143 − 12,000) lb/d = 5143 lb/d
    (b) Determine the volatile fraction of primary sludge.
        i. Operating parameters:
           Volatile fraction of suspended solids in influent = 67%
           Volatile fraction of incoming suspended solids discharged to the secondary process
           = 85%
       ii. Volatile suspended solids in influent, lb/d = 0.67(17,143 lb/d) = 11,486 lb/d
      iii. Volatile suspended solids in secondary influent, lb/d = 0.85(5143 lb/d)

           $$= 4372 \text{ lb/d}$$

       iv. Volatile suspended solids in primary sludge, lb/d = (11,486 − 4372) lb/d = 7114 lb/d
        v. Volatile fraction in primary sludge

        $$\text{Volatile fraction, } \% = \frac{7114 \text{ lb/d}}{12,000 \text{ lb/d}}(100\%) = 59.3\%$$

    (c) Secondary process:
        i. Operating parameters (from Example 10-2):
           Mixed-liquor suspended solids = 4375 mg/L
           (The recycled BOD_5 and suspended solids from the sludge-processing facilities were not considered in Example 10-2 in computing the concentration of the mixed-liquor suspended solids.)
           Volatile fraction of mixed-liquor suspended solids = 0.80
           $Y_{obs} = 0.3125$
       ii. Determine the effluent mass quantities.

        $$BOD_5, lb/d = (5.71 \text{ Mgal/d})(20 \text{ mg/L})[8.34 \text{ lb/Mgal} \cdot (mg/L)]$$

        $$= 952 \text{ lb/d}$$

$$\text{Suspended solids, lb/d} = (5.71 \text{ Mgal/d})(22 \text{ mg/L})[8.34 \text{ lb/Mgal} \cdot (\text{mg/L})]$$

$$= 1048 \text{ lb/d}$$

iii. Estimate the mass of volatile solids produced in the activated-sludge process that must be wasted. The required value computed using Eq. 10-3. (Note: BOD_5 to secondary $= 375 \times 0.67 = 250 \text{ mg/L}$).

$$P_{x(VSS)} = Y_{obs}Q(S_o - S)(8.34)$$

$$= 0.3125(5.71 \text{ Mgal/d})(250 - 6.2) \text{ mg/L} [8.34 \text{ lb/Mgal} \cdot (\text{mg/L})]$$

$$= 3628 \text{ lb/d}$$

Note: The actual flowrate will be the primary influent less the flowrate of the primary underflow. However, the primary underflow is normally small and can be neglected. If the underflow is significant, the actual flowrate should be used to determine the volatile solids production.

iv. Estimate the total mass of suspended solids that must be wasted, assuming that the volatile fraction represents 0.80 of the total solids.

$$\text{Suspended solids, lb/d} = 3628/0.80 = 4535 \text{ lb/d}$$

Note: If it is assumed that the fixed solids portion of the influent suspended solids equals 0.15, the mass of fixed solids in the input from the primary settling facilities is equal to $(0.15 \times 5143 = 771 \text{ lb/d})$. This value can then be compared with the fixed solids determined in the above computations, which is equal to $(4535 - 3628 = 907 \text{ lb/d})$. The ratio of these values is 1:18 (907 lb/d to 771 lb/d). Values that have been observed for this ratio vary from about 1.0 to 1.3; a value of 1.15 is considered to be the most representative.

v. Estimate the waste quantities discharged to the thickener. (It is assumed in this example that wasting is from the biological reactor.)

$$\text{Suspended solids, lb/d} = (4535 - 1048) \text{ lb/d}$$

$$= 3487 \text{ lb/d}$$

$$\text{Flowrate} = \frac{3487 \text{ lb/d}}{(4375 \text{ mg/L})[8.34 \text{ lb/Mgal} \cdot (\text{mg/L})]}\left(\frac{10^6 \text{ gal}}{\text{Mgal}}\right) = 95,567 \text{ gal/d}$$

(From Example 10-2, the concentration of mixed-liquor suspended solids in the aerator is 4375 mg/L [(3500 mg/L)/0.8]. This value will increase when the recycle BOD_5 and suspended solids are taken into consideration in the second and subsequent iterations.

(d) Flotation thickeners:

i. Operating parameters:
Concentration of thickened sludge $= 4\%$
Assumed solids recovery $= 90\%$
Assumed specific gravity of feed and thickened sludge $= 1.0$

ii. Determine the flowrate of the thickened sludge.

$$\text{Flowrate, ft}^3/\text{d} = \frac{(3487 \text{ lb/d})(0.9)}{(62.4 \text{ lb/ft}^3)(0.04)} = 1257 \text{ ft}^3/\text{d}$$

$$\text{Flowrate, gal/d} = 1257 \text{ ft}^3/\text{d} \times 7.48 \text{ gal/ft}^3 = 9402 \text{ gal/d}$$

iii. Determine the flowrate recycled to the plant headworks.

$$\text{Recycled flowrate} = (95{,}567 - 9402) \text{ gal/d} = 86{,}165 \text{ gal/d}$$

iv. Determine the suspended solids to the blending tank.

$$\text{Suspended solids, lb/d} = (3487 \text{ lb/d})(0.9) = 3138 \text{ lb/d}$$

v. Determine the suspended solids recycled to the plant headworks.

$$\text{Suspended solids, lb/d} = (3487 - 3138) \text{ lb/d} = 349 \text{ lb/d}$$

vi. Determine the recycled BOD_5.

$$\text{Suspended solids in recycled flow} = \frac{349 \text{ lb/d}}{0.086 \text{ Mgal/d } [8.34 \text{ lb/Mgal} \cdot (\text{mg/L})]}$$

$$= 487 \text{ mg/L}$$

$$BOD_5 \text{ of suspended solids} = (487 \text{ mg/L})(0.65)(1.42)(0.68) = 305.7 \text{ mg/L}$$

Note: The BOD_5 of the suspended solids was estimated using the same procedure applied in Example 10-2.

$$\text{Soluble } BOD_5 \text{ escaping treatment} = 6.2 \text{ mg/L (from Example 10-2)}$$

$$\text{Total } BOD_5 \text{ concentration} = (305.7 + 6.2) \text{ mg/L} = 311.9 \text{ mg/L}$$

$$BOD_5, \text{ lb/d} = (311.9 \text{ mg/L})(0.086 \text{ Mgal/d})[8.34 \text{ lb/Mgal} \cdot (\text{mg/L})] = 224 \text{ lb/d}$$

(e) Sludge digestion:
  i. Operating parameters:
  $\theta = 10$ d
  Volatile solids destruction during digestion $= 50\%$
  Gas production $= 18$ ft^3/lb of volatile solids destroyed
  BOD in digester supernatant $= 5000$ mg/L (0.5%)
  Total solids in digester supernatant $= 5000$ mg/L (0.5%)
  Total solids in digested sludge $= 5\%$
  ii. Determine the total solids fed to the digester and the corresponding flowrate.

$$\text{Total solids} = \text{solids from primary settling plus waste solids from thickener}$$

$$\text{Total solids, lb/d} = 12{,}000 \text{ lb/d} + 3138 \text{ lb/d} = 15{,}138 \text{ lb/d}$$

$$\text{Total flowrate} = \frac{12{,}000 \text{ lb/d}}{0.06(8.34 \text{ lb/gal})} + \frac{3138 \text{ lb/d}}{0.04(8.34 \text{ lb/gal})}$$

$$= (23{,}981 + 9407) \text{ gal/d}$$

$$= 33{,}388 \text{ gal/d}$$

iii. Determine the total volatile solids fed to the digester.

$$\text{Total volatile solids, lb/d} = 7114 \text{ lb/d} + 0.80(3138 \text{ lb/d})$$

$$= (7114 + 2510) \text{ lb/d}$$

$$= 9624 \text{ lb/d}$$

$$\text{Percent volatile solids in sludge mixture fed to digester} = \frac{9624 \text{ lb/d}}{15,138 \text{ lb/d}}(100)$$

$$= 63.6\%$$

iv. Determine the volatile solids destroyed.

$$\text{Volatile solids destroyed, lb/d} = 0.5(9624 \text{ lb/d})$$

$$= 4812 \text{ lb/d}$$

v. Determine the mass flow to the digester.
Primary sludge at 6% solids:

$$\text{Mass flow, lb/d} = \frac{12,000 \text{ lb/d}}{0.06}$$

$$= 200,000 \text{ lb/d}$$

Thickened waste activated sludge at 4% solids:

$$\text{Mass flow, lb/d} = \frac{3138 \text{ lb/d}}{0.04}$$

$$= 78,450 \text{ lb/d}$$

$$\text{Total mass flow} = (200,000 + 78,450) \text{ lb/d} = 278,450 \text{ lb/d}$$

Note: The total mass flow can also be computed by multiplying the total flowrate to the digester by the density of the combined sludge, if known.

vi. Determine the mass quantities of gas and sludge after digestion. Assume that the total mass of fixed solids does not change during digestion and that 50 percent of the volatile solids are destroyed.

$$\text{Fixed solids} = \text{total solids} - \text{volatile solids} = (15,138 - 9624) \text{ lb/d} = 5514 \text{ lb/d}$$

$$\text{Total solids in digested sludge} = 5514 \text{ lb/d} + 0.5(9624)\text{lb/d} = 10,326 \text{ lb/d}$$

Gas production assuming that the density of digester gas is equal to 0.86 times that of air (0.075 lb/ft^3) and 18 ft^3 of gas is produced per lb of volatile solids:

$$\text{Gas, lb/d} = (18 \text{ ft}^3/\text{lb})(0.5)(9624 \text{ lb/d})(0.86)(0.075 \text{ lb/ft}^3) = 5587 \text{ lb/d}$$

Mass balance of digester output:

$$\text{Mass input} = 278,450 \text{ lb/d}$$

$$\text{Less gas} = -5587 \text{ lb/d}$$

$$\text{Mass ouput} = 272,863 \text{ lb/d (solids and liquid)}$$

vii. Determine the flowrate distribution between the supernatant at 5000 mg/L and digested sludge at 5 percent solids. Let S = lb/d of supernatant suspended solids.

$$\frac{S}{0.005} + \frac{10,326 - S}{0.05} = 272,863 \text{ lb/d}$$

$$S + 1033 - 0.1S = 1364 \text{ lb/d}$$

$$0.9S = 331$$

$$S = 368 \text{ lb/d}$$

$$\text{Digested solids} = (10{,}326 - 368) \text{ lb/d} = 9958 \text{ lb/d}$$

$$\text{Supernatant flowrate} = \frac{368 \text{ lb/d}}{[(0.005)8.34 \text{ lb/gal})]} = 8825 \text{ gal/d}$$

$$\text{Digested sludge flowrate} = \frac{9958 \text{ lb/d}}{[(0.05)(8.34 \text{ lb/gal})]} = 23{,}880 \text{ gal/d}$$

viii. Establish the characteristics of the recycled supernatant flow to the plant headworks.

$$\text{Flowrate} = 8825 \text{ gal/d} \left(1 \text{ Mgal/} 10^6 \text{ gal}\right) = 0.0088 \text{ Mgal/d}$$

$$\text{BOD}_5 = 0.0088 \text{ Mgal/d} (5000 \text{ mg/L})[8.34 \text{ lb/Mgal} \cdot (\text{mg/L})]$$

$$= 367 \text{ lb/d}$$

$$\text{Suspended solids} = 0.0088 \text{ Mgal/d} (5000 \text{ mg/L})[8.34 \text{ lb/Mgal} \cdot (\text{mg/L})]$$

$$= 367 \text{ lb/d}$$

(f) Sludge dewatering. (Note: In the analysis that follows, the weight of the polymer or other sludge-conditioning chemicals that may be added were not considered. In some cases, their contribution can be significant and must be considered.)

   i. Operating parameters:

   Sludge cake = 20% solids

   Specific gravity of sludge = 1.06

   Solids capture = 95%

   Filtrate BOD_5 = 1,500 mg/L

   ii. Determine the sludge cake characteristics.

$$\text{Solids} = (9958 \text{ lb/d})(0.95) = 9460 \text{ lb/d}$$

$$\text{Volume} = \frac{9460 \text{ lb/d}}{(1.06)(0.20)(62.4 \text{ lb/ft}^3)} = 715 \text{ ft}^3/\text{d}$$

$$\text{Volume, gal/d} = 715 \text{ ft}^3/\text{d} \times 7.48 \text{ gal/ft}^3 = 5348 \text{ gal/d}$$

   iii. Determine the filtrate characteristics.

$$\text{Flowrate} = (23{,}880 - 5348) \text{ gal/d} = 18{,}532 \text{ gal/d}$$

$$= 18{,}532 \text{ Mgal/} 10^6 \text{ gal} = 0.0185 \text{ Mgal/d}$$

$$\text{BOD}_5 \text{ at 1500 mg/L} = 0.0185 \text{ Mgal/d} (1500 \text{ mg/L})[8.34 \text{ lb/Mgal} \cdot (\text{mg/L})]$$

$$= 231 \text{ lb/d}$$

$$\text{Suspended solids} = 9958 \text{ lb/d} (0.05) = 498 \text{ lb/d}$$

(g) Prepare a summary table of the recycle flows and waste characteristics for the first iteration.

Operation	Flow, gal/d	BOD$_5$ lb/d	Suspended solids, lb/d
Flotation thickener	86,165	224	349
Digester supernatant	8,825	367	367
Dewatering filtrate	18,532	231	498
Total	113,522	822	1,214[a]

[a] The volatile fraction of the returned suspended solids will typically vary from 50 to 75 percent. A value of 60 percent will be used for the computation in the second iteration.

3. Prepare the second iteration of the solids balance.
   (a) Primary settling:
       i. Operating parameters = same as those in the first iteration.
       ii. Total suspended solids and BOD$_5$ entering the primary tanks:

$$\text{Total BOD}_5 = \text{influent BOD}_5 + \text{recycled BOD}_5$$

$$= 17,858 \text{ lb/d} + 822 \text{ lb/d} = 18,680 \text{ lb/d}$$

$$\text{Total suspended solids} = \left(\begin{array}{c} \text{Influent} \\ \text{suspended} \\ \text{solids} \end{array} \right) + \left(\begin{array}{c} \text{Recycled} \\ \text{suspended} \\ \text{solids} \end{array} \right)$$

$$= 17,143 + 1214 \text{ lb/d}$$

$$= 18,357 \text{ lb/d}$$

       iii. BOD$_5$ removed = 0.33(18,680 lb/d) = 6164 lb/d
       iv. BOD$_5$ to secondary = (18,680 − 6164) lb/d = 12,516 lb/d
       v. Suspended solids removed = 0.7(18,357 lb/d) = 12,850 lb/d
       vi. Suspended solids to secondary = (18,357 − 12,850) lb/d = 5507 lb/d
   (b) Determine the volatile fraction of the primary sludge and effluent suspended solids.
       i. Operating parameters:
          Incoming wastewater = same as those for the first iteration
          Volatile fraction of solids in recycle = 60%
       ii. Although the computations are not shown, the computed change in the volatile fractions determined in the first iteration are slight; therefore, the values determined previously are used for the second iteration. If the volatile fraction of the return is less than about 50 percent, the volatile fractions should be recomputed.
   (c) Secondary process:
       i. Operating parameters = same as those for the first iteration
       ii. Determine the BOD$_5$ in the influent to the aeration tank.

$$\text{Flowrate to aeration tank} = \text{influent flowrate} + \text{recycle flowrate}$$

$$= 5.71 + 0.11 = 5.82 \text{ Mgal/d}$$

$$\text{BOD}_5, \text{ mg/L} = \frac{12,516 \text{ lb/d}}{5.82 \text{ Mgal/d}[8.34 \text{ lb/Mgal} \cdot (\text{mg/L})]}$$

$$= 258 \text{ mg/L}$$

       iii. Determine the new concentration of mixed-liquor suspended solids. The volatile suspended solids can be computed using Eq. 8-42, which was also used to determine the volume in Example 10-2. The difference in the following computations is that the volume is now fixed (1.24 Mgal) from Example 10-2.

$$X_{VSS} = \frac{\theta_c Q Y (S_0 - S)}{V(1 + k_d \theta_c)}$$

$$X_{VSS} = \frac{10 \text{ d } (5.82 \text{ Mgal/d})(0.5)(258 - 6.2) \text{ mg/L}}{(1.24 \text{ Mgal})[1 + (0.06 \text{ d}^{-1} \times 10 \text{ d})]}$$

$$= 3693 \text{ mg/L}$$

iv. Determine the mixed-liquor suspended solids.

$$X_{SS} = \frac{X_{VSS}}{0.8}$$

$$= \frac{3693 \text{ mg/L}}{0.8}$$

$$= 4616 \text{ mg/L}$$

v. Determine the cell growth using Eq. 10-3.

$$P_x = Y_{obs} Q (S_0 - S)(8.34)$$

$$= 0.3125(5.82 \text{ Mgal/d})(258 - 6.2) \text{ mg/L}[8.34 \text{ lb/Mgal} \cdot (\text{mg/L})]$$

$$= 3820 \text{ lb/d}$$

$$P_{X(ss)} = \frac{3820 \text{ lb/d}}{0.8} = 4775 \text{ lb/d}$$

vi. Determine the waste quantities discharged to the thickener.

Effluent suspended solids, lb/d = 1048 lb/d (specified in the first iteration)

Total suspended solids to be wasted to the thickener, lb/d = (4775 − 1048) lb/d

$$= 3727 \text{ lb/d}$$

$$\text{Flowrate, gal/d} = \frac{3727 \text{ lb/d}}{4616 \text{ mg/L } [8.34 \text{ lb/Mgal} \cdot (\text{mg/L})]} \left(\frac{10^6 \text{ gal}}{\text{Mgal}} \right) = 96,812 \text{ gal/d}$$

(d) Complete the remainder of the second iteration in the same manner as the first iteration. Computations are not shown, but the resultant values for the recycle flows and characteristics are presented in the following table. The incremental change in the recycle flows and waste characteristics from the previous iteration is also reported.

				Incremental change from previous iteration		
Operation/process	**Flow, gal/d**	**BOD$_5$, lb/d**	**Suspended solids, lb/d**	**Flow, gal/d**	**BOD$_5$, lb/d**	**Suspended solids, lb/d**
Flotation thickener	86,760	239	373	595	15	24
Digester supernatant	9,448	394	394	623	27	27
Dewatering filtrate	19,834	250	533	1,302	19	35
Total	116,042	883	1,300	2,520	61	97

4. Prepare the third iteration of the solids balance. This cycle is computed in the same manner as the second iteration. Computations again are not shown, but the resultant values for the recycle flows, waste characteristics, and incremental values are presented in the following table. This is the final iteration since the incremental change in the return quantities is less than 5 percent. The flow, suspended solids, and BOD_5 values for the various processes are presented in following figure.

				Incremental change from previous iteration		
			Suspended			Suspended
	Flow,	BOD_5,	solids,	Flow,	BOD_5,	solids,
Operation process	gal/d	lb/d	lb/d	gal/d	lb/d	lb/d
Flotation thickener	86,789	240	374	29	1	1
Digester supernatant	9,496	396	396	48	2	2
Dewatering filtrate	19,921	250	535	87	0	2
Total	116,206	886	1,305	164	3	5

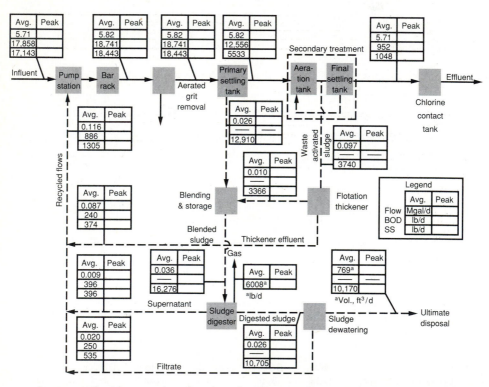

Comment. In this example, after three iterations the incremental change in the return quantities was less than 1 percent. In general, the iterative computational procedure should be carried out until the incremental change in all of the return quantities from the previous iteration is equal to or less than 5 percent.

12-17 LAND APPLICATION OF SLUDGE

Land application of sludge has been practiced successfully for decades. The interest in land application of sludge has increased in recent years as the availability and feasibility of other sludge management options such as landfilling, incineration, and ocean have decreased.

Land application of stabilized municipal wastewater sludge is defined as the spreading of sludge on or just below the soil surface. Land application is currently the most widely employed sludge use and disposal option among small- and medium-sized treatment plants in the United States. Sludge may be applied to (1) agricultural land, (2) forest land, (3) disturbed land, and (4) dedicated land disposal sites. In all four cases, the land application is designed with the objective of providing further sludge treatment. Sunlight, soil microorganisms, and desiccation combine to destroy pathogens and many toxic organic substances found in sludge. Trace metals are trapped in the soil matrix and nutrients are taken up by plants and converted to useful biomass. In the first three cases, sludge is used as a valuable resource to improve the characteristics of the land. Sludge acts as a soil conditioner to facilitate nutrient transport, increase water retention, and improve soil tilth. Sludge also serves as a partial replacement for expensive chemical fertilizers.

The steps involved in the development of a sludge land application system include the following:

1. Characterization of sludge quantity and quality
2. Review of pertinent federal, state, and local regulations
3. Evaluation and selection of site and disposal option
4. Determination of process design parameters—loading rates, land area requirements, application methods and scheduling

These steps are discussed in this section, but a detailed discussion of each step is beyond the scope of this text. The reader should consult Refs. 35, 36, 44, and 45 for further details.

Sludge Characteristics Affecting Land Application

Characteristics of sludge that affect its suitability for land application or affect the design of land application systems include organic content (usually measured as volatile solids), nutrients, pathogens, metals, and toxic organics.

Organic Content and Pathogens. Degradable organic material in unstabilized sludges can lead to odor problems and attract vectors (flies, mosquitoes, and rodents) in a land application setting. Pathogens (bacteria, viruses, protozoa, and eggs of parasitic worms) are concentrated in sludges and can spread diseases if there is human exposure to the sludge. To meet prescribed limits, organic content and pathogens must be reduced significantly prior to land application by means of preapplication treatment processes. Additional discussion on the requirements for pathogen reduction is provided under "Regulatory Requirements."

Nutrients. Major plant nutrients—nitrogen, phosphorus, and potassium—are not removed substantially during sludge processing but are taken up by vegetation after sludge has been applied to the land. Nitrogen is normally the nutrient of concern in land application because of the potential for nitrate contamination of groundwaters. The nitrogen uptake rate of vegetation, therefore, is a key design parameter in determining sludge-loading rates. When nutrient content of wastewater sludge is compared to commercial fertilizers (see Table 12-5), sludge can meet only a portion of the complete nutrient needs of plants in most cases. Nutrient uptake of plants is discussed in Chap. 13, Sec. 13-3.

Metals and Organics. Wastewater sludges contain trace metals and organic compounds that are retained in the soil and pose potential toxic risks to plants, animals, and humans. The principal metal of concern is cadmium because it can accumulate in plants to levels that are toxic to humans and animals but below levels that are toxic to plants (phytotoxic). Because of the potential wide range of constituent concentrations found in various sludges, such as those reported in Table 12-6, a thorough characterization of sludge is necessary when land application of sludge is considered.

Wastewater sludges also contain organic compounds, primarily chlorinated hydrocarbons such as PCBs, which are slow to degrade in the soil profile. The principal concern with such organics is not with plant uptake, which does not occur, but with the direct ingestion of compunds by animals, particularly dairy cattle grazing on sludge treated grasses. There is also evidence that organics can be adsorbed onto the surface of root crops such as carrots [44]. Consequently, loading limits for specific organic compounds are of concern when designing land application systems for sludge.

Regulatory Requirements

As discussed in Sec. 12-2, new regulations are being prepared by the EPA for the use and disposal of sludge. The new regulations may require detailed sampling and analysis of sludge to identify and characterize the sludge constituents so as to determine if the sludge is suitable for land application (see Table 12-9 for constituents of concern). Maximum annual loading rates may be prescribed as well as permissible cumulative loading rates, depending on whether the land is used for agricultural or nonagricultural purposes. Pathogen and vector attraction reduction measures also may need to be addressed. The designer, therefore, will have to be familiar with existing and pending regulations when planning land application facilities.

For land application of sludge, different levels of pathogen control by various stabilization methods will have to be considered. Two levels of pathogen control defined by the EPA are (1) processes to significantly reduce pathogens (PSRP) and (2) processes to further reduce pathogens (PFRP). Sludge applied to the land surface or incorporated into the soil must be treated by a PSRP. Sludge applied to land where crops for human consumption are grown (less than 18 months after application) must be treated by a PFRP. Examples of PSRP stabilization processes are aerobic digestion, air drying, anaerobic digestion, composting, and lime stabilization. Examples of PFRP stabilization processes are composting, heat drying, heat treatment, and thermophilic aerobic digestion. Additional details on pathogen reduction by various stabilization methods may be found in Ref. 59.

Site Evaluation and Selection

A critical step in the land application of sludge is to find a suitable site. The characteristics of the site will determine the actual design and will influence the overall effectiveness of the land application concept. The sites considered potentially suitable will depend on the land application option or options being considered, such as application to agricultural lands, forest lands, etc. The designer should determine which options are potentially feasible for the local situation. The site selection process should include an initial screening on the basis of the factors and criteria described in the following discussion. After the number of potential sites is narrowed, each site should be evaluated in detail, taking into account operational techniques and potential environmental impacts. For screening purposes, it is necessary to have at least a rough estimate of land area requirements for each feasible option. Typical sludge-loading rates given in Table 12-36 may be used to develop preliminary estimates of land area requirements. Final land area requirements must be based on design loading rates plus buffer zones and other land area requirements. Physical site characteristics of concern include topography, soil permeability, site drainage, depth to groundwater, subsurface geology, proximity to critical areas, and accessibility. A summary of typical guidelines used for site evaluation is presented in Table 12-37.

TABLE 12-36
Typical sludge application rates for various land disposal options[a]

Land disposal option	Time period of application	Application rates, tons/acre[b,c]	
		Range	Typical
Agricultural use	Annual	1–30	5
Forest	One time, or at 3- to 5-year intervals	4–100	20
Land reclamation	One time	3–200	50
Dedicated disposal site	Annual	100–400	150

[a] Ref. 44.
[b] Rates are for dry solids.
[c] Rates shown are only for sludge application area and do not include area for buffer zone, sludge storage, and other special requirements.
Note: tons/acre × 2.2417 = Mg/ha

TABLE 12-37
Typical soil limitations for wastewater sludge applied to agricultural land at nitrogen fertilizer rates[a]

Soil features affecting use	Degree of soil limitation		
	Slight	Moderate	Severe
Slope[b]	< 6%	6–12%	> 12%
Depth to seasonal water table	> 4 ft	2–4 ft	< 2 ft
Flooding and ponding	None	None	Occasional to frequent
Depth to bedrock	> 4 ft	2–4 ft	< 2 ft
Permeability of the most restricting layer above a 3 ft depth	0.1–0.3 in/h	0.3–1.0 in/h 0.03–0.1 in/h	< 0.03 in/h > 1.0 in/h
Available water capacity	> 1.0 in/h	0.5–1.0 in/h	< 0.5 in/h

[a] Adapted from Ref. 44.

[b] Slope is an important factor in determining the runoff that is likely to occur. Most soils on 0 to 6% slopes will have slow to very slow runoff; soils on 6 to 12% slopes generally have medium runoff; and soils on steeper slopes generally have a very rapid runoff.
Note: in × 25.4 = mm
ft × 0.3048 = m

Topography. Topography is important as it affects the potential for erosion and runoff of applied sludge for equipment operability. Recommended slope limitations as related to sludge application methods are presented in Table 12-38.

Soils. In general, desirable soils (1) have moderately slow permeabilities, 0.2 to 0.6 in/h (0.5 to 1.5 cm/h), (2) are well-drained to moderately well-drained, (3) are

TABLE 12-38
Typical slope limitations for land application of sludge[a]

Slope, %	Comment
0–3	Ideal; no concern for runoff or erosion of liquid sludge or dewatered sludge.
3–6	Acceptable; slight risk of erosion; surface application of liquid or dewatered sludge is acceptable.
6–12	Injection of liquid sludge required for general cases, except in closed drainage basin and/or when extensive runoff control is provided; surface application of dewatered sludge is usually acceptable.
12–15	No application of liquid sludge should be made without extensive runoff control; surface application of dewatered sludge is acceptable, but immediate incorporation into the soil is recommended.
Over 15	Slopes greater than 15% are suitable only for sites with good permeability where the length of slope is short and where the area with a steep slope is a minor part of the total application area.

[a] Adapted from Ref. 44.

alkaline or neutral (pH > 6.5) so as to control metal solubility, and (4) are deep and relatively fine textured for high moisture and nutrient-holding capacity. With proper design and operation, almost any soil may be suitable for sludge application.

Soil Depth to Groundwater. A basic philosophy inherent in federal and state regulations is to design sludge application systems based on sound agronomic principles so that sludge application poses no greater threat to groundwater than current agricultural practices. Because the groundwater fluctuates on a seasonal basis in many soils, difficulties are encountered in establishing an acceptable minimum depth to groundwater. The quality of the underlying groundwater and the sludge application option have to be considered. Generally, the greater the depth to the water table, the more desirable a site is for sludge application. Typical minimum depths for various sludge application options are listed in Table 12-39. The presence of faults, solution channels, and other similar connections between soil and groundwater is undesirable unless the depth of overlying soil is adequate. When a specific site or sites has been selected for sludge application, a detailed field investigation may be necessary to obtain the required groundwater information.

Proximity to Critical Areas and Accessibility. Buffer zones or setbacks will be required from residential development, inhabited dwellings, surface waters, water wells, and roads. It is important to establish actual distances between such features and the prospective sites. Required setbacks are normally established by local or state regulations. Isolated sites are preferred; however, the site should not be so isolated that it lacks access. Setbacks may range from 50 to over 1,500 ft depending upon the type of sludge application and the type of critical area (residential development, ponds and lakes, high water levels, etc.) [44]. Lack of nearby transport arteries, such as railroads, highways, or navigable waterways, may require construction of access roads or pipelines.

TABLE 12-39
Typical minimum depths to groundwater for land application of sludge[a,b]

Type of site	Drinking water aquifer, ft	Excluded aquifer, ft
Agricultural	3	1.5
Forest	6[c]	2
Drastically disturbed land	3[d]	1.5
Dedicated land disposal	> 3	1.5

[a] Ref. 44.

[b] Clearances are to ensure trafficability of surface, not for groundwater protection.

[c] Seasonal (springtime) high water and/or perched water less than 3 ft is not usually a concern.

[d] Assumes no groundwater contact with leachate from sludge application operation.

Note: ft × 0.3048 = m

Design Loading Rates

Design sludge loading rates for application to agricultural or to nonagricultural lands will be controlled by the pollutant limits set forth in regulatory guidelines or by the nutrient-loading rates necessary to meet vegetation requirements. Generally, nitrogen is the nutrient of principal concern.

Loading Rates Based on Pollutant Loading. The constituents of concern normally will be those listed in Table 12-9. Because pollutant-loading limits are being reviewed by the EPA at the time of the writing of the text, no firm numerical limits have been established. For application to agricultural land (defined as land on which crops are grown for either direct or indirect human consumption or for animal feed), annual pollutant-loading limits and cumulative pollutant-loading limits may be established. In addition, the requirements may also include a maximum annual whole sludge-loading rate limit based on dry tons/acre (dry metric tons/ha). For application to nonagricultural lands (forest land, turf farms, drastically disturbed or reclaimed lands, and dedicated lands), maximum limits for pollutant concentrations may also be established. In general, design loading rates for drastically disturbed lands and dedicated lands are based on pollutant-loading limits only.

Sludge containing significant pollutant concentrations might have an annual sludge application rate limited by one of its constituents such as cadmium. The limits for application of cadmium in this case may result in a lower application rate for nitrogen needed for crop growth, as discussed below. The cumulative amount of sludge that can be applied is based on pollutant limitations, given by the following equation [35,36]:

$$R_m = \frac{L_m}{(C_m)(2000)} \tag{12-21}$$

where R_m = maximum amount of sludge that can be applied over the useful life of the site, ton dry solids/acre

L_m = maximum amount of pollutant that can be applied over the useful life of the site as established by regulatory requirements, lb of pollutant/acre

C_m = percent pollutant content in the sludge, expressed as a decimal (e.g., for sludge with 50 ppm cadmium, C_m = 0.00005)

2000 = lb/ton dry sludge solids

Loading Rates Based on Nutrient Loading. The design approach for sludge application to agricultural lands, turf farms, and forest lands is based on the use of sludge as a plant fertilizer. In most cases, the design sludge-loading rate is based on meeting the annual nitrogen needs of the crop, although a few systems have been designed on the basis of phosphorus loadings. The design loadings must also comply with the pollutant-loading limits described above. Consequently, the final design loading rate will be the lowest rate determined on the basis of nutrient and pollutant loadings.

Nitrogen limitation. Calculation of sludge-loading rates based on nitrogen requirements is complicated because much of the nitrogen in sludge is in the organic form and is slowly mineralized or converted to plant-available forms in the soil over the period of several years. (Nitrogen transformation in soil systems is discussed in Chap. 13.) Nitrogen available during a year from sludge applied the same year may be estimated using the following equation:

$$N_a = (2000)[NO_3 + k_v(NH_4) + f_n(N_o)] \qquad (12\text{-}22)$$

where N_a = plant-available nitrogen in the sludge during the application year, lb nitrogen/ton dry sludge solids · yr
  2000 = lb/ton of dry solids
  NO_3 = percent nitrate in sludge expressed as a decimal
  k_v = volatilization factor for ammonia
    = 0.5 for surface or sprinkler applied liquid sludge
    = 1.0 for incorporated liquid sludge or dewatered sludge
  NH_4 = percent ammonia in sludge expressed as a decimal
  f_n = mineralization factor for year $n = 1$ (see Table 12-40 for values)
  N_o = percent organic nitrogen in sludge expressed as a decimal

Nitrogen available from mineralization of organic nitrogen applied in previous years is calculated using the following equation:

$$N_{ap} = 2000\Sigma f_2(N_o)_2 + f_3(N_o)_3 + \cdots + f_n(N_o)_n \qquad (12\text{-}23)$$

where N_{ap} = plant-available nitrogen from mineralization of organic nitrogen applied in the previous n years, lb nitrogen/ton dry sludge solids · yr
  $(N_o)_n$ = decimal fraction of organic nitrogen remaining in the sludge from year n
  f = mineralization rate from Table 12-40, subscripts refer to the year of concern

TABLE 12-40
Mineralization rates for organic nitrogen in wastewater sludge[a]

Time after sludge application, y	Mineralization rate, %		
	Raw sludge	Anaerobically digested	Composted
1	40	20	10
2	20	10	5
3	10	5	3
4	5	3	3
5	3	3	3
6	3	3	3
7	3	3	3
8	3	3	3
9	3	3	3
10	3	3	3

[a] From Ref. 36.

The total available nitrogen during a given year is the amount available from sludge applied during the year (N_a) plus the amount available from mineralization of sludge applied in previous years (N_{ap}). The annual nitrogen-based sludge loading is then calculated as follows:

$$R_n = \frac{U_n}{N_a + N_{ap}} \tag{12-24}$$

where R_n = annual sludge loading in year n, ton dry solids/acre · yr
  U_n = annual vegetative uptake of nitrogen, lb nitrogen/acre · yr
  (see Table 13-8 in Chap. 13)

Phosphorus limitation. When crop uptake of phosphorus is specified as the limiting parameter, the sludge application rate is calculated using the following equation:

$$R_p = \frac{U_p}{(C_p)(2000)} \tag{12-25}$$

where R_p = phosphorus-limited sludge application rate, ton phosphorus/acre · yr
  U_p = annual crop uptake rate of phosphorus, lb/acre · yr (see Table 13-8)
  C_p = percent phosphorus content in the sludge, expressed as a decimal. Normally, about 50 percent of the total phosphorus in the sludge is assumed to be available.

Land Requirements. Once the design sludge-loading rate is established, land area requirements may be calculated using the following equation [35,36]:

$$A = \frac{Q_s}{R_d} \tag{12-26}$$

where A = application area required, acres
  Q_s = total sludge production, ton dry sludge solids/yr
  R_d = design sludge loading rate, ton dry sludge solids/acre · yr

Application of pollutant and nitrogen loadings on sludge application rates and land area requirements is illustrated in Example 12-10.

Example 12-10   Determination of loading rates and land requirements for land application of sludge. Determine the annual application rate and land area requirements for an agricultural application of digested sludge, based on dry sludge solids. A marketable crop is not intended but the site will be planted with a rye grass mixture. The local regulatory authorities have established a cadmium loading limit of 16 lb cadmium/acre for the useful life of the site and allow a design based on nitrogen fertilization requirements. Assume that the following conditions apply:

   Sludge production rate = 600 dry ton/yr

Sludge characteristics: Cadmium $= 50$ ppm

Organic nitrogen $= 2\%$

Ammonia nitrogen $= 2\%$

Nitrate nitrogen $= 0$

Nitrogen uptake for rye grass $= 180$ lb nitrogen/acre $\cdot$ yr (Table 13-8)

Solution

1. Determine the maximum amount of sludge that can be applied over the lifetime of the site, based on cadmium-loading limitations, using Eq. 12-21:

$$R_m = \frac{16 \text{ lb/acre}}{(0.00005)(2000 \text{ lb/ton})} = 160 \text{ ton/acre}$$

Note: In determining the maximum amount of sludge that can be applied, it is assumed that the sludge characteristics do not change during the period of permissible sludge application. (See discussion of sludge characteristics in Sec. 12-1.)

2. Determine the available nitrogen in the sludge using Eqs. 12-22 and 12-23:

   (a) Determine the nitrogen-loading rate for the first year of application.

   $$N_a = 2000[NO_3 + k_v(NH_4) + f_n(N_o)]$$

   $$N_a = 2000[0 + 0.5(0.02) + 0.2(0.02)]$$

   $$N_a = 28 \text{ lb nitrogen/ton dry solids}$$

   (b) Determine the decimal fraction of organic nitrogen from the first year's application remaining in the soil that is to be mineralized in the second year:

   $$(N_o)_2 = (N_o)_1 - f_1(N_o)$$

   $$(N_o)_2 = 0.02 - 0.2(0.02)$$

   $$(N_o)_2 = 0.016$$

   (c) Determine the amount of nitrogen available in the second year from mineralization of residual organic nitrogen:

   $$(N_a)_2 = 2000[f_2(N_o)_2]$$

   $$(N_a)_2 = 2000[0.1(0.016)]$$

   $$(N_a)_2 = 3.2 \text{ lb nitrogen/ton dry solids}$$

   (d) Determine the total available nitrogen during the second year:

   $$(N_a) = N_a + N_{ap}$$

   $$(N_a)_2 = 28 + 3.2$$

   $$(N_a)_2 = 31.2 \text{ lb nitrogen/ton dry solids}$$

   (e) Similarly, the total available nitrogen in succeeding years is

   $$(N_a)_3 = 32.6 \text{ lb nitrogen/ton dry solids}$$

   $$(N_a)_4 = 33.5 \text{ lb nitrogen/ton dry solids}$$

   $$(N_a)_5 = 34.3 \text{ lb nitrogen/ton dry solids, etc.}$$

3. Determine the annual nitrogen-limited sludge application rate, assuming that 34.3 lb nitrogen/ton dry solids is a steady-state value:

$$R_n = \frac{U_n}{N_a + N_{ap}}$$

$$R_n = \frac{180}{34.3}$$

$$R_n = 5.3 \text{ ton dry solids/acre} \cdot \text{yr}$$

4. Determine the land area required for sludge application using Eq. 12-26:

$$A = \frac{Q_s}{R_d}$$

$$A = \frac{600 \text{ ton/yr}}{5.3 \text{ ton/acre} \cdot \text{yr}} = 113 \text{ acres}$$

5. Determine the useful life of the site for sludge application, based on cadmium loading to ensure that there are no restrictions for future land use restrictions including production of food crops.

$$\text{Useful life} = \frac{160 \text{ ton/acre}}{5.3 \text{ ton/acre} \cdot \text{yr}} = 30.2 \text{ yr}$$

Comment. For agricultural applications, the nutrient-based loading rate must be compared with loading limits for the constituents of concern prescribed by the regulatory agency. The lowest constituent loading rate is then used as a basis for determining the sludge application rate and land area requirements.

Application Methods

The method of sludge application selected will depend on the physical characteristics of the sludge (liquid or dewatered), site topography, and the type of vegetation present (annual field crops, existing forage crops, trees, or preplanted land).

Liquid Sludge Application. Application of sludge in the liquid state is attractive because of its simplicity. Dewatering processes are not required, and the liquid sludge can be transferred by pumping. Typical solids concentrations of liquid sludge applied to land range from 1 to 10 percent. Liquid sludge may be applied to land by vehicle or by irrigation methods similar to those used for wastewater distribution.

Vehicular application may be surface distribution or by subsurface injection or incorporation. Limitations to vehicular application include limited tractability on wet soil and potential reduction in crop yields due to soil compaction from truck traffic. Use of vehicles equipped with high-flotation tires can minimize these problems.

Surface distribution may be accomplished by tank trucks or tank wagons equipped with rear-mounted spreading manifolds or by tank trucks mounted with high-capacity spray nozzles or guns. Specially designed, all-terrain, sludge application vehicles with spray guns are ideally suited for sludge application on forest lands (see Fig. 12-47). Vehicular surface application is the most common method used for field

FIGURE 12-47
Forest land sludge application vehicle (from City of Seattle).

and forage croplands. The procedure used commonly for annual crops is to (1) spread the sludge prior to planting, (2) allow the sludge to dry partially, and (3) incorporate the sludge by disking or plowing. The process is then repeated after harvest.

Liquid sludge can be injected below the soil surface by using tank wagons or tank trucks with injection shanks, or it can be incorporated immediately after the surface application by using plows or discs equipped with sludge distribution manifolds and covering spoons. Important advantages of injection or immediate incorporation methods include minimization of potential odors and vector attraction, minimization of ammonia loss due to volatilization, elimination of surface runoff, and minimum visibility leading to better public acceptance. Injection shanks and plows are very disruptive to perennial forage crops or pastures. To minimize such effects, special grassland sludge injectors have been developed (see Fig. 12-48).

Irrigation methods include sprinkling and furrow irrigation. Flood or graded-border distribution of sludge has generally not been successful and is not recommended. Typically, large-diameter, high-capacity sprinkler guns are used to avoid clogging problems. Sprinkling has been used mainly for application to forested lands and occasionally for application to dedicated disposal sites that are relatively isolated from public view and access. Sprinklers can operate satisfactorily on land too rough or wet for tank trucks or injection equipment and can be used throughout the growing season. Disadvantages to sprinkling include power costs of high-pressure pumps, contact of sludge with all parts of the crop, possible foliage damage to sensitive crops, potential odors and vector attraction problems, and potentially high

(a)   (b)

FIGURE 12-48
Sludge application vehicles: (a) truck equipped with liquid sludge grassland injector (from Ag Chem) and (b) truck equipped with sludge cake spreader.

visibility to the public. Available sprinkling methods are described in more detail in Chap. 13.

Furrow irrigation can be used to apply sludge to row crops during the growing season. Disadvantages associated with furrow irrigation are localized settling of solids and the potential of sludge in the furrows, both of which can result in odor problems.

Dewatered Sludge Application. Application of dewatered sludge to the land is similar to an application of semisolid animal manure. Typical solids concentrations of dewatered sludge applied to land range from 15 to 30 percent. Application of sludge using conventional manure spreaders is an important advantage of dewatered sludge because private farmers can apply sludge on their lands with their own equipment. Other advantages include reduced sludge hauling, storage, and spreading costs. For forest land application where use of dewatered sludge is often impractical, sludge may be dewatered for storage and hauling and reliquified to allow spray application. Dewatered sludge is spread most commonly using tractor-mounted box spreaders or manure spreaders followed by plowing or disking into the soil. For high application rates, bulldozers, loaders, or graders may be used.

12-18   OTHER BENEFICIAL USES OF SLUDGE

The beneficial use of sludge is receiving considerable attention because of the decline in available landfill and the interest in using the beneficial nutrient and soil conditioning properties of sludge [65]. In addition to the benefits of land application, discussed in Sec. 12-17, sludge may be distributed and marketed for residential and commercial uses as a soil amendment and conditioner. Sludge may also be treated chemically to stabilize the sludge for use as landfill cover or for landscaping or land reclamation projects. Distribution and marketing and chemical fixation are discussed briefly in this section.

Distribution and Marketing

The amount of sludge disposed of by distribution and marketing is reported to range from 11 to 19 percent of the sludge generated, depending on treatment plant size (see Table 1-3). Sludge that is distributed and marketed is used as a substitute for topsoil and peat on lawns, golf courses, and parks and in ornamental and vegetable gardens. Usually the sludge used for these purposes is composted. The sludge may be distributed in bulk or in bags. Application rates of sludge may be limited based on whether it is used for food or nonfood crops.

Regulations for the beneficial use of sludge by distribution and marketing varies from state-to-state, but national minimum standards are being proposed by the EPA. The national regulations proposed for the control of pollutants for the distribution and marketing of sludge are similar to the regulations for land application (see Table 12-9). The numerical limits for the pollutants under consideration may vary somewhat [12]. The regulations may also include management practices and other general requirements to reduce the level of pathogenic organisms. Rules may be instituted requiring distributors to provide labels or information sheets identifying the product and to provide instructions on the proper use of the product. When distributing and marketing is considered for the beneficial use of sludge, it is strongly recommended that current and proposed state and federal regulations be reviewed.

Chemical Fixation

The chemical fixation/solidification process has been applied to the treatment of industrial sludge and hazardous wastes to immobilize the undesirable constituents. The process has also been used to stabilize municipal sludge for use as landfill cover and for land reclamation projects. Stabilized sludge may also be disposed of in landfills. The chemical fixation process consists of mixing untreated or treated liquid or dewatered sludge with stabilizing agents such as cement, sodium silicate, pozzolan (fine-grained silicate), and lime so as to chemically react with or encapsulate the sludge [53]. A typical schematic diagram is illustrated in Fig. 12-49. The process may generate a product with a high pH, which inactivates the pathogenic bacteria and viruses. For many chemical treatment processes, the product is similar in consistency to natural clay.

12-19 FINAL SLUDGE AND SOLIDS CONVEYANCE, STORAGE, AND DISPOSAL

The solids removed as sludge from preliminary and biological treatment processes are concentrated and stabilized by biological and thermal means and are reduced in volume in preparation for final disposal. Because the methods of conveyance and final disposal often determine the type of stabilization required and the amount of volume reduction that is needed, they are considered briefly in the following discussion.

Conveyance Methods

Sludge may be transported long distances by (1) pipeline, (2) truck, (3) barge, (4) rail, or (5) any combination of these four modes (see Fig. 12-50). To minimize the danger

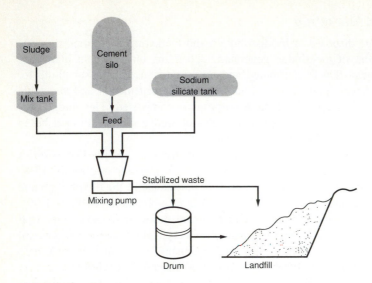

FIGURE 12-49
Schematic diagram of a chemical fixation system [53].

of spills, odors, and dissemination of pathogens to the air, liquid sludges should be transported in closed vessels such as tank trucks, railroad tank cars, or covered or tank barges. Stabilized, dewatered sludges can be transferred in open vessels, such as dump trucks, or in railroad gondolas. If sludge is hauled long distances, the vessels should be covered.

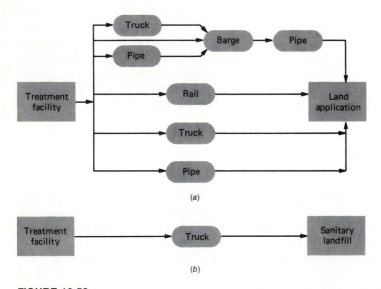

FIGURE 12-50
Conveyance methods for processed sludge: (a) transfer to land application or further treatment and (b) transfer to landfill.

The method of transportation chosen and its costs are dependent on a number of factors, including (1) the nature, consistency, and quantity of sludge to be transported, (2) the distance from origin to destination, (3) the availability and proximity of the transit modes to both origin and destination, (4) the degree of flexibility required in the transportation method chosen, and (5) the estimated useful life of the ultimate disposal facility.

Pipeline. In general, the energy requirements for long distance transportation of untreated sludges with a solids concentration of more than 6 percent are prohibitive. Also, during sludge pumping, grease tends to build up in unlined pipes and corrosion problems may arise. During low-flow conditions, grit tends to accumulate, thereby increasing pipe roughness, and septicity may become a problem. Most of these problems can be reduced or eliminated by maintaining adequate flowrates in large diameter pipes. Transport of digested sludge is somewhat easier because the sludge is more homogeneous and has a lower grease content. The headloss factor for digested sludge is also lower, as previously discussed in Sec. 12-3 and as shown in Fig. 12-6a. For reasons cited, the piping of sludge for long distances cannot be justified economically unless a large and relatively constant volume of liquid sludge is to be transported. Further, because pipelines require a large capital investment and because their routes are fixed, the termination point of the pipeline must be long-lasting to justify the expenditure.

Truck. Trucking is the most flexible and most widely used method for transporting sludge. Either liquid or dewatered sludge may be hauled by truck to diverse destinations. Tank trucks for liquid sludge are available in sizes ranging from 500 to 6000 gal (2000 to 24,000 L). Trucking dewatered sludge is usually the most economical method for small- to medium-sized treatment facilities where the sludge must be disposed of in sanitary landfills or where it must be transported seasonally to several different locations. The capital investment is relatively small and the operation is not complicated. Dump truck sizes range from 8 to 30 yd^3 (6 to 23 m^3). Hopper bottom trucks may be used and are available in sizes from 15 to 25 yd^3 (12 to 19 m^3).

Long-haul trucking of thickened sludges to land application sites may be somewhat more expensive than the other transportation options. In addition, sludge storage facilities may have to be provided offsite in the event sludge cannot be applied to land by the transporting vehicle.

Barge. Many different sizes and types of barges are available for transporting sludge. Generally, double-hulled vessels are used to reduce the possibility of spills in congested harbor areas. Barges may be either towed or self-propelled and may use either gravity or pumped discharges. Barge transport is generally economical only for large facilities treating wastewater flows in excess of 100 Mgal/d (4.38 m^3/s) or in locations where one barge can serve several plants. Barges can also be used for carrying dewatered sludge in containers. Barging sludge for ocean disposal has been prohibited by new federal regulations and is in the process of being phased out.

Rail. Rail transportation of sludge is not widely used in the United States at the present time. Rail may be used to transport sludges of any consistency, but those with high solids content are transported most economically. The use of rail transportation for small quantities of sludges or for transportation of sludges over short distances is not justifiable economically. In the future, rail haul may be used to transport treated sludge to large centralized sludge-processing and -holding facilities.

Environmental Considerations. Each transportation method contributes a minor air pollutant load, either directly or indirectly. A certain amount of air pollution is produced from the facility that generates the electricity necessary for sludge pumping. The engines that move trucks, barges, and railroad cars also produce some air pollutants. On a mass (tonnage) basis, the transportation mode that contributes the lowest pollutant load is piping. Next, in sequence, are barging and unit train rail transportation. The highest pollutant load is from trucking. Other factors of environmental concern include traffic, noise, and construction disturbance.

Sludge Storage

It is often necessary to store sludge that has been digested anaerobically before it is disposed of or used beneficially. Storage of liquid sludge can be accomplished in sludge storage basins, and storage of dewatered sludge can be done on storage pads.

Sludge Storage Basins. Sludge stored in basins becomes more concentrated and is further stabilized by continued anaerobic biological activity. As stated previously in Sec. 12-12, long-term storage is effective in pathogen destruction.

Depth of the sludge storage basins may vary from 10 to 16 ft (3 to 5 m). Solids-loading rates vary from about 20 to 50 lb VSS/10^3 ft^2 · d of surface area (0.1 to 0.25 kg VSS/m^2 · d). If the basins are not loaded too heavily ($\leq$ 20 lb VSS/10^3 ft^2 · d), it is possible to maintain an aerobic surface layer through the growth of algae and by atmospheric reaeration [42]. Alternatively, surface aerators can be used to maintain aerobic conditions in the upper layers (see Fig. 12-51).

The number of basins to be used should be sufficient to allow each basin to be out of service for a period of about six months. Stabilized and thickened sludge can be removed from the basins using a mud pump mounted on a floating platform or by mobile crane using a drag line. Sludge concentrations as high as 35 percent solids have been achieved in the bottom layers of these basins [42].

Sludge Storage Pads. Where dewatered sludge has to be stored prior to land application, sufficient storage area should be provided based on the number of consecutive days that sludge hauling could occur without applying sludge to land. Allowances also have to be made for paved access and for area to maneuver the sludge-hauling trucks, loaders, and application vehicles. The storage pads should be constructed of concrete or bituminous concrete and designed to withstand the truck loadings and sludge piles. Provisions for leachate and stormwater collection and disposal also have to be included.

FIGURE 12-51
Sludge storage basin with floating aerator in foreground and sludge pumping rig (mudcat) in background.

Final Disposal

Final disposal for the sludge and solids that are not beneficially used usually involves some form of land disposal. Ocean disposal of sludge by the major coastal cities of the United States is prohibited and is being phased out because of changes in water pollution control regulations. In addition to spreading sludge on land, other methods of final disposal include landfilling and lagooning; these methods are considered briefly in the following discussion. As in the case of land application of sludge, the regulations for other methods of sludge disposal are becoming increasingly stringent and require close attention and review when planning and designing sludge disposal facilities.

Landfilling. If a suitable site is convenient, a sanitary landfill can be used for disposal of sludge, grease, grit, and other solids. Stabilization may be required depending on state or local regulations. Dewatering of sludge is usually required to reduce the volume to be transported and to control the generation of leachate from the landfill. In many cases, solids concentration is an important factor in determining the acceptability of sludge in landfills. The sanitary landfill method is most suitable if it is also used for disposal of the other solid wastes of the community. In a true sanitary landfill, the wastes are deposited in a designated area, compacted in place with a tractor or roller, and covered with a 12-in (30-cm) layer of clean soil. With daily coverage of the newly deposited wastes, nuisance conditions such as odors and flies are minimized. In some landfills, composted sludge and chemically treated sludge have been used as cover material. Composted sludge also serves to reduce odors that might emanate from the disposal of municipal solid wastes.

In sludge monofills, the regulations may require daily or more frequent covering for vector control and may include limitations on methane gas generation. The pollutants that are regulated for sludge monofills are listed in Table 12-9.

In selecting a land disposal site, consideration must be given to (1) environmentally sensitive areas such as wetlands, flood plains, recharge zones for aquifers, and habitats for endangered species, (2) runoff control to surface water, (3) groundwater protection, (4) air pollution from dust, particulates, and odors, (5) disease vectors, and (6) safety as related to toxic materials, fires, and access. Trucks carrying wet sludge and grit should be able to reach the site without passing through heavily populated areas or business districts. After several years, during which the wastes are decomposed and compacted, the land may be used for recreational or other purposes for which gradual subsidence would not be objectionable. Design details for sludge monofills and landfills for the co-disposal of sludge and refuse may be found in Ref. 41.

Lagooning. Lagooning of sludge is another common disposal method because it is simple and economical if the treatment plant is in a remote location. A lagoon is an earth basin into which untreated or digested sludge is deposited. In untreated sludge lagoons, the organic solids are stabilized by anaerobic and aerobic decomposition, which may give rise to objectionable odors. The stabilized solids settle to the bottom of the lagoon and accumulate. Excess liquid from the lagoon, if there is any, is returned to the plant for treatment. Lagoons should be located away from highways and dwellings to minimize possible nuisance conditions and should be fenced to keep out unauthorized persons. They should be relatively shallow, 4 to 5 ft (1 to 1.5 m), if they are to be cleaned by scraping. If the lagoon is used only for digested sludge, the nuisances mentioned should not be a problem. As stated in Sec. 12-13, subsurface drainage and percolation should be investigated to determine if the underlying groundwater will be affected. If excessive percolation is a problem or if regulations require leachate control, the lagoon may have to be lined. Sludge may be stored indefinitely in a lagoon, or it may be removed periodically after draining and drying.

DISCUSSION TOPICS AND PROBLEMS

12-1. The water content of a sludge is reduced from 98 percent to 95 percent. What is the percent reduction in volume by the approximate method and by the more exact method, assuming that the solids contain 70 percent organic matter of specific gravity 1.00 and 30 percent mineral matter of specific gravity 2.00? What is the specific gravity of the 98 and the 95 percent sludge?

12-2. Consider an activated-sludge treatment plant with a capacity of 10.0 Mgal/d. The untreated wastewater contains 200 mg/L suspended solids. The plant provides 60 percent removal of the suspended solids in the primary settling tank. If the primary sludge alone is pumped, it will contain 5 percent solids. Assume that 0.1 Mgal/d of waste activated sludge containing 0.5 percent solids is to be wasted to the digester. If the waste activated sludge is thickened in the primary settling tank, the resulting mixture will contain 3.5 percent solids. Calculate the reduction in daily volume of sludge pumped to the digester that can be achieved by thickening the waste activated sludge in the primary settling tank, as compared with discharging the primary and waste activated sludge directly to the digester. Assume complete capture of the waste activated sludge in the primary settling tank.

12-3. Sludge is to be withdrawn by gravity from a primary settling tank for heat treatment. The available head is equal to 10 ft, and 300 ft of 6 in pipe is to be used to interconnect the units. Determine the flowrate and velocity, assuming that the solids content of the sludge is 6 percent. Assume that the f value for water in the Darcy Weisbach equation is 0.025 and that the minor losses are equal to 2 ft.

12-4. Sludge is to be pumped from the wastewater treatment plant to an offsite sludge-processing plant located 10 miles away. The treatment plant is located at an elevation of 1200 ft and the sludge-processing plant is located at an elevation of 1500 ft. Using an 8 in pipeline and the sludge flowrate and characteristics in Example 12-2, calculate the pumping head. Assume that minor friction losses due to bends, valves, and fittings are 40 ft. Select two types of pump for the pumping application, and state your reasons for pump selection.

12-5. Determine the required digester volume for the treatment of the sludge quantities specified in Example 12-4 using the (a) volatile solids loading factor, and (b) volume reduction methods. Set up a comparison table to display the results obtained using the three different procedures for sizing digesters (two in this problem and one in Example 12-4). Assume that the following data apply:

1. Volatile solids loading method
   a. Solids concentration = 5%
   b. Detention time = 10 d
   c. Loading factor = 0.24 lb VSS/ft^3 · d (see Table 12-19)
2. Volume reduction method
   a. Initial volatile solids = 75%
   b. Volatile solids destroyed = 60%
   c. Final sludge concentration = 8%
   d. Final sludge specific gravity = 1.04

12-6. A primary wastewater treatment plant providing for separate sludge digestion receives an influent wastewater with the following characteristics:

Average flow = 2.0 Mgal/d
Suspended solids removed by primary sedimentation = 200 mg/L
Volatile matter in settled solids = 75%
Water in untreated sludge = 96%
Specific gravity of mineral solids = 2.60
Specific gravity of organic solids = 1.30

1. Determine the required digester volume using a mean cell-residence time of 12 d.
2. Determine the minimum digester capacity using the recommended loading parameters of pounds of volatile matter per cubic foot per day and cubic feet per 1000 persons.
3. Assuming 90 percent moisture in the digested sludge and a 60 percent reduction in volatile matter during digestion at 90°F, determine the minimum theoretical digester capacity for this plant based on parabolic reduction in sludge volume during digestion and a digestion period of 25 d.

12-7. Consider an industrial waste consisting mainly of carbohydrates in solution. Pilot plant experiments using a complete-mix anaerobic digester without recycle yielded the following data:

Run	BOD$_L$ influent, kg/d	X$_T$ reactor, kg	P$_X$ effluent, kg/d
1	1,000	428	85.7
2	500	115	46

Assuming a waste-utilization efficiency of 80 percent, estimate the percentage of added BOD$_L$ that can be stabilized when treating a waste load of 5000 kg/d. Assume that the design sludge retention time (θ_c) is 10 d.

12-8. A digester is loaded at a rate of 600 lb BOD$_L$/d. Using a waste-utilization efficiency of 75 percent, what is the volume of gas produced when $\theta_c = 40$ d? $Y = 0.10$ and $k_d = 0.02$ d^{-1}.

12-9. Volatile acid concentration, pH, or alkalinity should not be used alone to control a digester. How should they be correlated to predict most effectively how close to failure a digester is at any time?

12-10. Prepare a one-page abstract of each of the following four articles: P. L. McCarty: "Anaerobic Waste Treatment Fundamentals," *Public Works*, vol. 95, nos. 9, 10, 11, and 12, 1964.

12-11. A digester is to be heated by circulation of sludge through an external hot water heat exchanger. Using the following data, find the heat required to maintain the required digester temperature:

(a) U_x = overall heat transfer coefficient, Btu/hr/ft^2 · °F.

(b) $U_{air} = 0.15$, $U_{ground} = 0.12$, $U_{cover} = 0.20$.

(c) Digester is a concrete tank with floating steel cover; diameter = 35 ft m and side-wall depth = 26 ft, 13 ft of which is above the ground surface.

(d) Sludge fed to digester = 4000 gal/d at 58° F.

(e) Outside temperature = -5°F.

(f) Average ground temperature = 40°F.

(g) Sludge in tank is to be maintained at 95°F.

(h) Assume a specific heat of the sludge = 1.0 Btu/lb · °F.

(i) Sludge contains 4 percent solids

(j) Assume a cone-shaped cover with center 2 ft above digester top and a bottom with center 4 ft below bottom edge.

12-12. The ultimate elemental analysis of a dried sludge yields the following data:

Carbon	52.1%
Oxygen	38.3%
Hydrogen	2.7%
Nitrogen	6.9%
Total	100.0%

How many pounds of air will be required per pound of sludge for its complete oxidation?

12-13. Compute the fuel value of the sludge from a primary settling tank (a) if no chemicals are added and (b) if the coagulating solids amount to 10 percent by weight of the dry sludge. The amount of volatile solids is 75 percent.

12-14. Assume that a community of 5,000 persons has asked you to serve as a consultant on their sludge disposal problems. Specifically, you have been asked to determine if it is feasible to compost the sludge from the primary clarifier with the community's solid waste. If this plan is not feasible, you have been asked to recommend a feasible solution.

Currently the waste solids from the communitys' biological process are thickened in the primary clarifier. Assume that the following data are applicable:

Solid waste data:
Waste production = 4.5 lb/person · d
Compostable fraction = 55%
Moisture content of compostable fraction = 22%

Sludge production:
Net sludge production = 0.26 lb/person · d
Concentration of sludge in underflow from primary clarifier = 5%
Specific gravity of underflow solids = 1.08

Compost:
Final moisture content of sludge-solid waste mixture = 55%

12-15. Prepare a solids balance for the peak loading condition for the treatment plant used in Example 12-9. Assume that the following data apply. Enter your final values on the solids balance figure in Example 12-9.

Peak flowrate = 14 Mgal/d
Average BOD_5 at peak flowrate = 340 mg/L
Average suspended solids at peak flowrate = 350 mg/L
Suspended solids after grit removal = 325 mg/L

Use data given in Example 12-9 for other parameters.

12-16. Prepare a solids balance, using the iterative technique delineated in Example 12-9, for the flow diagram shown in the following figure. Also determine the effluent flowrate and suspended-solids concentration. Assume that the following data are applicable:

Influent characteristics:
Flowrate = 4000 m^3/d
Suspended solids = 1000 mg/L

Sedimentation tank:
Removal efficiency = 85%
Concentration of solids in underflow = 7%
Specific gravity of sludge = 1.1

Alum addition:
Dosage = 10 mg/L of filter influent
Chemical solution = 0.5 kg alum/L of solution

Filter:
Removal efficiency = 90%
Washwater solids concentration = 6%
Specific gravity of backwash = 1.08

Thickener:
Effluent solids concentration = 500 mg/L
Concentration of solids in underflow = 12%
Specific gravity of sludge = 1.25
Ferric chloride addition
Dosage = 1% of underflow solids from thickener
Specific gravity of chemical solution = 2.0

Belt filter press:
Concentration of solids in filtrate = 200 mg/L
Concentration of thickened solids = 40%
Specific gravity of thickened sludge = 1.6

In preparing the solids balance, assume that all of the unit operations respond linearly such that the removal efficiency for recycled solids is the same as that for the solids in the influent wastewater. Also assume that the distribution of the chemicals added to improve the performance of the filter and belt filter press is proportional to the total solids in the returns and the effluent solids.

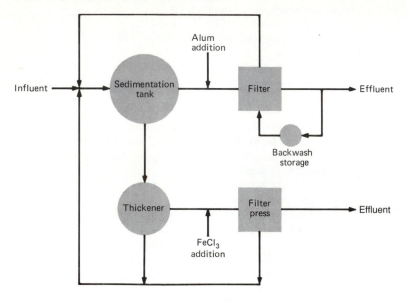

12-17. For Example 12-10, confirm the values for total available nitrogen in the third, fourth, and fifth years. Calculate the total available nitrogen for a 10-year period. Using the total nitrogen in year 10 as a steady-state value, how much more land would be required than the amount calculated in Example 12-10?

12-18. Determine the dry sludge application rate for Reed canary grass on the basis of satisfying crop nitrogen uptake (Table 13-8). Assume that a sludge containing 3 percent nitrogen by weight is applied to a soil that has an initial nitrogen content of zero. Use a decay rate of 30 percent for the first year, 15 percent for the second year, and 5 percent for the third and subsequent years.

12-19. A sludge containing 50 ppm of cadmium on a dry basis is to be applied to the land. If the limiting mass loading to the soil is set at 8 lb/acre, what would be the safe loading rate for 50 yr of application?

REFERENCES

1. American Society of Civil Engineers Task Force on Belt Press Filters: "Belt Filter Press Dewatering of Wastewater Sludge," *ASCE J. Env. Eng. Div.,* vol. 114, no. 5, pp. 991–1006, October 1988.
2. Anonymous: "Water Pollution Control Federation: Anaerobic Digester Mixing Systems," *Journal WPCF,* vol. 59, no. 3, 1987.
3. Babbitt, H. E., and D. H. Caldwell: "Laminar Flow of Sludge in Pipes," *University of Illinois Bulletin,* 319, 1939.
4. Burd, R. S.: *A Study of Sludge Handling and Disposal,* U.S. Department of the Interior Publication WP-20-4, 1968.
5. Carman, P. C.: "A Study of the Mechanism of Filtration, parts I-III," *J. Soc. Chem. Ind.,* vols. 52, 53, London, 1933, 1934.
6. Carthew, G. A., C. A. Goehring, and J. E. van Teylingen: "Development of Dynamic Head Loss Criteria of Raw Sludge Pumping," *Journal WPCF,* vol. 55, no. 5, 1983.

7. Chou, T. L.: "Resistance of Sewage Sludge to Flow in Pipes," *Journal San. Eng. Div., Proceedings ASCE,* 84, SA5, p. 1780, September 1958.

8. Coackley, P., and B. R. S. Jones: "Vacuum Sludge Filtration," *Sewage and Industrial Wastes,* vol. 28, no. 6, 1956.

9. Drier, D. E., and C. A. Obma: "Aerobic Digestion of Solids," *Walker Process Equipment Co. Bulletin,* no. 26-S-18194, Aurora, IL, 1963.

10. Eckenfelder, W. W., Jr.: *Industrial Water Pollution Control,* McGraw-Hill, New York, 1966.

11. Fair, G. M., and E. W. Moore: "Heat and Energy Relations in the Digestion of Sewage Solids," *Sewage Works J.,* vol. 4, pp. 242, 428, 589, and 728, 1932.

12. Federal Register: "40 CFR Parts 257 and 503," *Standards for the Disposal of Sewage Sludge,* February 6, 1989.

13. Finstein, M. S., F. C. Miller, J. A. Hogan, and P. F. Strom: "Analysis of EPA Guidance on Composting Sludge, Part II—Biological Process Control," *Biocycle,* vol. 28, no. 2, 1987.

14. Finstein, M. S., F. C. Miller, J. A. Hogan, and P. F. Strom: "Analysis of EPA Guidance on Composting Sludge, Part IV—Facility Design and Operation," *Biocycle,* vol. 28, no. 4, 1987.

15. Goldstein, N.: "Steady Growth for Sludge Composting," *Biocycle,* vol. 29, no. 10, 1988.

16. Haug, R. T.: *Compost Engineering, Principles and Practices,* Ann Arbor Science Publications, Ann Arbor, MI, 1980.

17. Jewell, W. J., and R. M. Kabrick: "Autoheated Aerobic Thermophilic Digestion with Aeration," *Journal WPCF,* vol. 50, no. 3, 1980.

18. Komanik, R. A., and K. A. Dejewski: "Gravity Belts Thicken Sludge Economically," *Water Engineering and Management,* p. 37, May 1986.

19. Lewis, M. F.: "Sludge Pyrolysis for Energy Recovery on Pollution Control," *Proceedings of the National Conference on Municipal Sludge Management and Disposal,* Information Transfer, Rockville, MD, 1975.

20. McAdams, W. H.: *Heat Transmission,* 2nd ed., McGraw-Hill, New York, 1954.

21. McCabe, B. J., and W. W. Eckenfelder, Jr.: *Biological Treatment of Sewage and Industrial Wastes,* vol. 2, Reinhold, New York, 1958.

22. McCarty, P. L.: "Anaerobic Waste Treatment Fundamentals," *Public Works,* vol. 95, nos. 9-12, 1964.

23. McCarty, P. L.: "Anaerobic Treatment of Soluble Wastes," in E. F. Gloyna and W. W. Eckenfelder, Jr. (eds.), *Advances in Water Quality Improvement,* University of Texas Press, Austin, 1968.

24. McGrew, J. L., G. L. Hartman, J. E. Barnes, and W. G. Purdy: *Wet Oxidation of Municipal Sludge by the Vertical Tube Reactor,* U.S. EPA Contract No. 68-03-2812, February 1986.

25. Metcalf & Eddy, Inc.: *Wastewater Engineering: Treatment, Disposal, Reuse,* 2nd ed., McGraw-Hill, 1979.

26. Metcalf & Eddy, Inc.: *Wastewater Engineering—Collection and Pumping of Wastewater,* McGraw-Hill, 1981.

27. Metcalf & Eddy, Inc.: *Improved Design and Operation of Anaerobic Digester Mixing Systems,* Draft Report to U.S. EPA, Contract 68-03-3208, September 29, 1984.

28. Metcalf & Eddy, Inc.: *Improved Design and Operation of Belt Filter Presses,* Draft Report to U.S. EPA, Contract 68-03-3208, September 29, 1984.

29. Metcalf & Eddy, Inc.: *Improved Design and Operation of Recessed Plate Filter Presses,* Draft Report to U.S. EPA, Contract 68-03-3208, September 29, 1984.

30. Metcalf & Eddy, Inc.: *Achieving Improved Operation of Heat Treatment/Low Pressure Oxidation of Sludge,* Draft Report to U.S. EPA, Contract 68-03-3208, September 29, 1984.

31. Metcalf & Eddy, Inc.: *Improved Design and Operation of Multiple-Hearth and Fluid Bed Sludge Incinerators,* Draft Report to U.S. EPA, Contract 68-03-3208, September 29, 1984.

32. Mulbarger, M.C., S.R. Copas, J.R. Kordic, and F.M. Cash: "Pipeline Friction Losses for Wastewater Sludges," *Journal WPCF,* vol. 51, no. 8, 1981.

33. Reed, S. C.: "Sludge Freezing for Dewatering," *Biocycle,* vol. 28, no. 1, p. 32, January 1987.

34. Reed, S. C., J. Bouzon, and W. Medding: "A Rational Method for Sludge Dewatering Via Freezing," *Journal WPCF,* vol. 58, no. 9, 1986.

35. Reed, S. C., and R. W. Crites: *Handbook of Land Treatment Systems for Industrial and Municipal Wastes,* Noyes Publications, 1984.

36. Reed, S. C., E. J. Middlebrooks, and R. W. Crites: *Natural Systems for Waste Management and Treatment,* McGraw-Hill, New York, 1988.

37. Roediger, H.: "Using Quicklime—Hygienization and Solidification of Dewatered Sludge," *Operations Forum,* April 1987.

38. Sanks, R. L., G. Tchobanoglous, D. Newton, B. E. Bosserman, and G. M. Jones (eds.): Pumping Station Design, Butterworths, Stoneham, MA, 1989.

39. Sorber, C. A., B. E. Moore, D. E. Johnson, H. J. Hardy, and R. E. Thomas: "Microbiological Aerosols from the Application of Liquid Sludge to Land," *Journal WPCF,* vol. 56, no. 6, 1984.

40. Tchobanoglous, G., H. Theisen, and R. Eliassen: *Solid Wastes: Engineering Principles and Management Issues,* McGraw-Hill, New York, 1977.

41. U.S. Environmental Protection Agency: *Process Design Manual for Municipal Sludge Landfills,* EPA-625/1-78-010, October 1978.

42. U.S. Environmental Protection Agency: *Process Design Manual for Sludge Treatment and Disposal,* September 1979.

43. U.S. Environmental Protection Agency: *Design Manual for Dewatering Municipal Wastewater Sludges,* October 1982.

44. U.S. Environmental Protection Agency: *Process Design Manual for Land Application of Municipal Sludge,* EPA 625/1-83-016, September 1983.

45. U.S. Environmental Protection Agency: *Environmental Regulations and Technology, Use and Disposal of Municipal Wastewater Sludge,* EPA 625/10-84-003, September 1984.

46. U.S. Environmental Protection Agency: *Seminar Publication on Composting of Municipal Wastewater Sludges,* EPA/625/4-85/014, August 1985.

47. U.S. Environmental Protection Agency: *Municipal Wastewater Sludge Combustion Technology,* September 1985.

48. U.S. Environmental Protection Agency: *Innovative and Alternative Technology Projects,* 1986 Progress Report, September 1986.

49. U.S. Environmental Protection Agency: *Design Information Report Centrifuges,* Contract 68-03-3208, September 1986.

50. U.S. Environmental Protection Agency: "EPA Design Information Report—Design, Operational, and Cost Considerations for Vacuum Assisted Sludge Dewatering Bed Systems," *Journal WPCF,* 59:228, April 1987.

51. U.S. Environmental Protection Agency: *Design Manual for Dewatering Municipal Wastewater Sludges,* September 1987.

52. U.S. Environmental Protection Agency: *Innovations in Sludge Drying Beds, A Practical Technology,* October 1987.

53. U.S. Environmental Protection Agency: *Seminar Publication, Corrective Action: Technologies and Applications,* EPA/625/4-89/020, September 1989.

54. U.S. Environmental Protection Agency: *Summary Report, In-Vessel Composting of Municipal Wastewater Sludge,* EPA/625/8-89/016, September 1989.

55. Vesilind, P. A., G. C. Hartman, and E. T. Skene: *Sludge Management and Disposal for the Practicing Engineer,* Lewis Publishers, Chelsea, MI, 1986.

56. Wagenhals, H. H., E. J. Theriault, and H. B. Hommon: "Sewage Treatment in the United States," *U.S. Public Health Bulletin,* 132, 1925.

57. Water Pollution Control Federation: *Sludge Thickening,* Manual of Practice FD-1, 1980.

58. Water Pollution Control Federation: *Sludge Dewatering,* Manual of Practice no. 20, 1983.

59. Water Pollution Control Federation: *Sludge Stabilization,* Manual of Practice FD-9, 1985.

60. Water Pollution Control Federation: *Anaerobic Sludge Digestion,* Manual of Practice no. 16, 2nd ed., 1987.

61. Water Pollution Control Federation: *Sludge Conditioning,* Manual of Practice FD-14, 1988.

62. Water Pollution Control Federation: *Incineration,* Manual of Practice OM-11, 1988.

63. Water Pollution Control Federation: *Wastewater Treatment Plant Design,* Draft Manual of Practice no. 8, 1988.

64. Water Pollution Control Federation: *Wastewater Treatment Plant Design,* Manual of Practice no. 8, 1977.

65. Water Pollution Control Federation: *Beneficial Use of Waste Solids,* Manual of Practice FD-15, 1989.

CHAPTER
13

NATURAL
TREATMENT
SYSTEMS

In the natural environment, physical, chemical, and biological processes occur when water, soil, plants, microorganisms, and the atmosphere interact. Natural treatment systems are designed to take advantage of these processes to provide wastewater treatment. The processes involved in natural systems include many of those used in mechanical or in-plant treatment systems—sedimentation, filtration, gas transfer, adsorption, ion exchange, chemical precipitation, chemical oxidation and reduction, and biological conversion and degradation—plus others unique to natural systems such as photosynthesis, photooxidation, and plant uptake. In natural systems, the processes occur at "natural" rates and tend to occur simultaneously in a single "ecosystem reactor," as opposed to mechanical systems in which processes occur sequentially in separate reactors or tanks at accelerated rates as a result of energy input.

The natural treatment systems covered in this chapter include (1) the soil-based or land-treatment systems—slow rate, rapid infiltration, and overland flow—and (2) the aquatic-based systems—constructed and natural wetlands and aquatic plant treatment systems. The specific topics covered in this chapter are (1) the development of land-treatment systems, (2) fundamental considerations in natural treatment systems, (3) slow-rate systems, (4) rapid infiltration systems, (5) overland-flow systems, (6) constructed wetland systems, and (7) aquatic plant systems. Land application of sludge is discussed in Chap. 12.

13-1 DEVELOPMENT OF NATURAL TREATMENT SYSTEMS

An overview of natural treatment systems is provided in this section. The historical practice is traced, and the characteristics and objectives of systems used in current practice are described.

Natural Treatment Systems in the United States

Use of land-based natural treatment systems in the United States dates from the 1880s (see Table 13-1) [8]. As in Europe, sewage farming (the older term used in the early literature) became relatively common as a first attempt to control water pollution. In the first half of the twentieth century, these systems were generally replaced either by in-plant treatment systems or by (1) managed farms where treated wastewater was used for crop production, (2) landscape irrigation sites, or (3) groundwater recharge sites. These newer land-treatment systems tended to predominate in the western United States, where the resource value of wastewater was an added advantage.

The number of U.S. municipalities using natural treatment increased from 304 in 1940 to 571 (serving a population of 6.6 million) in 1972, but this total still represented only a small percentage of the estimated 15,000 total municipal

TABLE 13-1
Selected early land-treatment systems[a]

Location	Date started	Type of system	Area, acre	Flow, Mgal/d
International				
Berlin, Germany	1874	Sewage farm	6,720	N/A
Braunschweig, Germany	1896	Sewage farm	10,870	16.0
Croydon-Beddington, England	1860	Sewage farm	620	4.6
Leamington, England	1870	Sewage farm	395	0.9
Melbourne, Australia	1893	Irrigation	10,280	50.0
Mexico City, Mexico	1900	Irrigation	110,700	570.0
Paris, France	1869	Irrigation	1,580	80.0
Wroclaw, Poland	1882	Sewage farm	1,975	28.0
United States				
Calumet City, MI	1888	Rapid infiltration	12	1.1
Ely, NV	1908	Irrigation	395	1.6
Fresno, CA	1891	Irrigation	3,950	26.0
San Antonio, TX	1895	Irrigation	3,950	20.0
Vineland, NJ	1901	Rapid infiltration	14	0.9
Woodland, CA	1889	Irrigation	170	4.1

[a] Adapted from Ref. 8.

N/A = Not available

Note: acre × 0.4047 = ha
         Mgal/d × 0.0438 = m³/s

treatment facilities. With the passage of the Clean Water Act of 1972, interest in land-based natural treatment systems was revived as a result of the emphasis that was placed on water reuse, nutrient recycling, and the use of wastewater for crop production. Financial support provided by the Act stimulated widespread research and development of natural treatment system technology, leading to its acceptance in the field of wastewater engineering as a management technique that should be considered equally with any others.

The most recent developments in natural treatment system technology have been in the use of constructed wetlands with emergent plants and aquatic systems with floating plants. Interest in the use of constructed wetlands developed as a result of the renovative performance observed in natural wetlands combined with concurrent experience with other aquatic plant and natural treatment systems. Floating plants were used initially to upgrade the performance of conventional lagoon and stabilization ponds, but further development of this application has resulted in the unique technology of aquatic systems.

Characteristics and Objectives of Natural Treatment Systems

The physical features, design objectives, and treatment capabilities of the various types of natural systems are described and compared in this section. Comparisons of major site characteristics, typical design features, and the expected quality of the treated wastewater from the principal types of natural systems are presented in Tables 13-2, 13-3, and 13-4, respectively. All forms of natural treatment systems are preceded by some form of mechanical pretreatment. For wastewater, a minimum of fine screening or primary sedimentation is necessary to remove gross solids that can clog distribution systems and lead to nuisance conditions. The need to provide preapplication treatment beyond some minimum level will depend on the system objectives and regulatory requirements. The capacity of all natural systems to treat wastewater sludge is finite, and systems must be designed and managed to function within that capacity. Details of site evaluation, preapplication treatment, and process design for each type of system are discussed in subsequent sections.

Slow Rate. Slow-rate treatment, the predominant natural treatment process in use today, involves the application of wastewater to vegetated land to provide treatment and to meet the growth needs of the vegetation. The applied water either is consumed through evapotranspiration or percolates vertically and horizontally through the soil profile (see Fig. 13-1). Any surface runoff is usually collected and reapplied to the system. Treatment occurs as the applied water percolates through the soil profile. In most cases, the percolate will enter the underlying groundwater, but in some cases, the percolate may be intercepted by natural surface waters or recovered by means of underdrains or recovery wells. The rate at which water is applied to the land per unit area (hydraulic-loading rate) and the selection and management of the vegetation are functions of the design objectives of the system and the site conditions as described in Sec. 13-3.

TABLE 13-2
Comparison of site characteristics for natural treatment systems

Characteristics	Slow-rate	Rapid infiltration	Overland-flow	Wetland application	Floating aquatic plants
Climatic conditions	Storage often needed for cold weather and during precipitation	None (possibly modify operation in cold weather)	Storage often needed for cold weather and during precipitation	Storage may be needed for cold weather	Storage may be needed for cold weather
Depth to groundwater	2–3 ft (minimum)	10 ft (lesser depths acceptable where underdrainage is provided)	Not critical	Not critical	Not critical
Slope	Less than 15% on cultivated land; less than 40% on forested land	Not critical; excessive slopes require much earthwork	Finish slopes 1–8%	Usually less than 5%	Usually less than 5%
Soil permeability	Moderately slow to moderately rapid	Rapid (sands, loamy sands)	Slow (clays, silts and soils with impermeable barriers)	Slow to moderate	Slow to moderate

Note: ft $\times$ 0.3048 = m

TABLE 13-3
Comparison of design features of alternative natural treatment systems.

Feature	Slow-rate (type 1)	Slow-rate (type 2)	Rapid infiltration	Overland-flow	Wetland application	Floating aquatic plant
Application techniques	Sprinkler or surface[a]	Sprinkler or surface[a]	Usually surface	Sprinkler or surface	Sprinkler or surface	Surface
Annual hydraulic-loading rate, ft/y	5.6–20	2.0–6.7	20–300	24–186	18–60	18–60
Area required, ac/(Mgal/d)[b]	56–200	170–550	3.7–56	6–45	18–62	18–62
Minimum preapplication treatment provided	Primary sedimentation[c]	Primary sedimentation[c]	Primary sedimentation	Screening	Primary sedimentation	Primary sedimentation
Disposition of applied wastewater	Evapo-transpiration and percolation	Evapo-transpiration and percolation	Mainly percolation	Surface runoff and evaporation with some percolation	Evapo-transpiration, percolation, and runoff	Some evapo-transpiration
Need for vegetation	Required	Required	Optional	Required	Required	Required

[a] Includes furrow and graded border.
[b] Field area in acres not including buffer area, roads, or ditches.
[c] Depends on the use of the effluent and the type of crop.

Note: ft/y × 0.3048 = m/y
ac/(Mgal/d) × 0.1069 = ha/(10^{3}m^3/d)

TABLE 13-4
Comparison of expected effluent quality of treated water from slow-rate, rapid infiltration, and overland-flow natural treatment systems.

Constituent	Value, mg/L					
	Slow-rate[a]		Rapid infiltration[b]		Overland-flow[c]	
	Average	Maximum	Average	Maximum	Average	Maximum
BOD	< 2	< 5	2	< 5	10	< 15
Suspended solids	< 1	< 5	2	< 5	15	< 25
Ammonia nitrogen as N	< 0.5	< 2	0.5	< 2	1	< 3
Total nitrogen as N	3	< 8	10	< 20	5	< 8
Total phosphorus as P	< 0.1	< 0.3	1	< 5	4	< 6

[a] Percolation of primary or secondary effluent through 5 ft (1.5 m) of soil.
[b] Percolation of primary or secondary effluent through 15 ft (4.5 m) of soil.
[c] Runoff of continued municipal wastewater over about 150 ft (45 m) of slope.

Slow-rate systems are often classified as type 1 or type 2 depending on design objectives. A slow-rate system is considered to be type 1 when the principal objective is wastewater treatment and the hydraulic-loading rate is not controlled by the water requirements of the vegetation but by the limiting design parameter—soil permeability or constituent loading. Type 2 systems, designed with the objective of water reuse through crop production or landscape irrigation, are often referred to as wastewater irrigation or crop irrigation systems.

Wastewater can be applied to crops or vegetation (including forestland) by a variety of sprinkling methods (see Fig. 13-1) or by surface techniques such as graded-border and furrow irrigation (see Fig. 13-2). Intermittent application cycles, typically every 4 to 10 days, are used to maintain predominantly aerobic conditions in the soil profile. The relatively low application rates combined with the presence of vegetation and the active soil ecosystem provide slow-rate systems with the highest treatment potential of the natural treatment systems (see Table 13-4).

Rapid Infiltration. In rapid-infiltration systems, wastewater that has received some preapplication treatment is applied on an intermittent schedule usually to shallow infiltration or spreading basins, as shown schematically in Fig. 13-3. Application of wastewater by high-rate sprinkling is also practiced. Vegetation is usually not provided in infiltration basins but is necessary for sprinkler application. Because loading rates are relatively high, evaporative losses are a small fraction of the applied water, and most of the applied water percolates through the soil profile where treatment occurs. Design objectives for rapid-infiltration systems include (1) treatment followed by groundwater recharge to augment water supplies or prevent saltwater intrusion, (2) treatment followed by recovery using underdrains or pumped withdrawal (see Fig. 13-3), and (3) treatment followed by groundwater flow and discharge into surface waters. The treatment potential of rapid infiltration systems is somewhat less than slow-rate systems because of the lower retention capacity of permeable soils and the relatively higher hydraulic-loading rates (see Table 13-4).

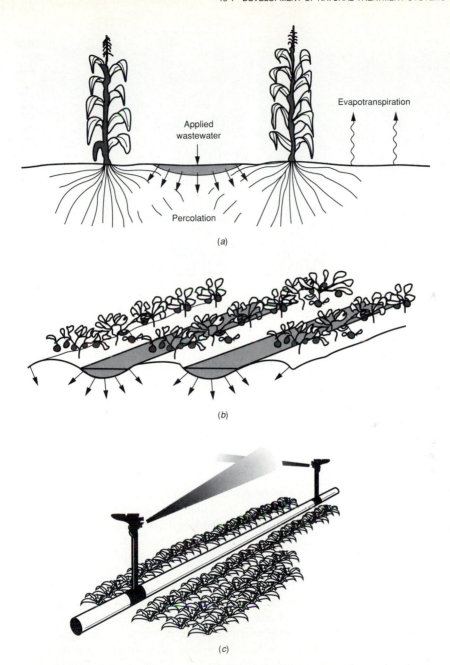

FIGURE 13-1
Slow-rate treatment: (a) hydraulic pathway, (b) surface distribution, and (c) sprinkler distribution.

FIGURE 13-2
Surface distribution using the furrow method with gated pipe.

Overland Flow. In overland flow, pretreated wastewater is distributed across the upper portions of carefully graded, vegetated slopes and allowed to flow over the slope surfaces to runoff collection ditches at the bottom of the slopes. A process schematic is shown in Fig. 13-4. Overland flow is normally used at sites with relatively impermeable surface soils or subsurface layers, although the process has been adapted to a wide range of soil permeabilities because the soil surface tends to seal over time. Percolation through the soil profile is, therefore, a minor hydraulic pathway, and most of the applied water is collected as surface runoff. A portion of the applied water will be lost to evapotranspiration. The percentage of the applied water lost varies with the time of the year and local climate. Systems are operated using alternating application and drying periods, with the lengths of the periods depending on the treatment objectives. Distribution of wastewater may be accomplished by means of high-pressure sprinklers, low-pressure sprays, or surface methods such as gated pipe.

Wetlands. Wetlands are inundated land areas with water depths typically less than 2 ft (0.6 m) that support the growth of emergent plants such as cattail, bulrush, reeds, and sedges (see Fig. 13-5). The vegetation provides surfaces for the attachment of bacteria films, aids in the filtration and adsorption of wastewater constituents, transfers oxygen into the water column, and controls the growth of algae by restricting the penetration of sunlight. Both natural and constructed wetlands have been used for wastewater treatment, although the use of natural wetlands is generally limited to the polishing or further treatment of secondary or advanced treated effluent.

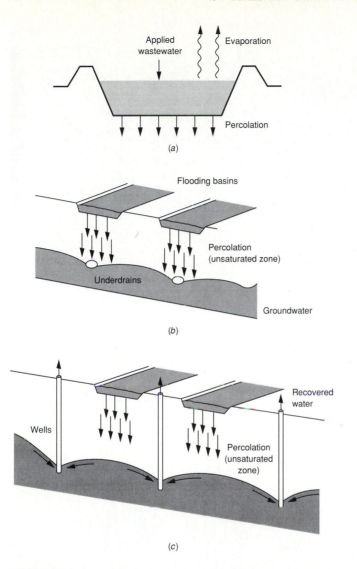

FIGURE 13-3
Rapid infiltration hydraulic pathways: (*a*) hydraulic pathway, (*b*) recovery pathway using underdrains, and (*c*) recovery pathway using wells.

Natural wetlands. From a regulatory standpoint, natural wetlands are usually considered receiving waters. Consequently, discharges to natural wetlands, in most cases, must meet applicable regulatory requirements, which typically stipulate secondary or advanced treatment. Furthermore, the principal objective when discharging to natural wetlands should be enhancement of existing habitat. Modification of

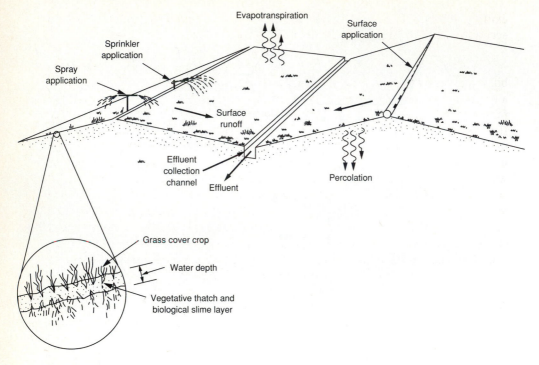

FIGURE 13-4
Overland-flow process schematic.

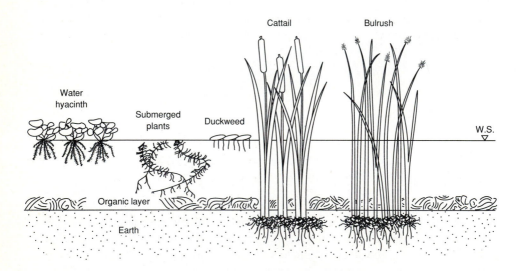

FIGURE 13-5
Common aquatic plants [29].

existing wetlands to improve treatment capability is often very disruptive to the natural ecosystem and, in general, should not be attempted.

Constructed wetlands. Constructed wetlands offer all of the treatment capabilities of natural wetlands but without the constraints associated with discharging to a natural ecosystem. Two types of constructed wetland systems have been developed for wastewater treatment: (1) free water surface (FWS) systems and (2) subsurface flow systems (SFS). When used to provide a secondary level or advanced levels of treatment, FWS systems typically consist of parallel basins or channels with relatively impermeable bottom soil or subsurface barrier, emergent vegetation, and shallow water depths of 0.33 to 2 ft (0.1 to 0.6 m). Pretreated wastewater is normally applied continuously to such systems, and treatment occurs as the water flows slowly through the stems and roots of the emergent vegetation. Free water surface systems may also be designed with the objective of creating new wildlife habitats or enhancing nearby existing natural wetlands. Such systems normally include a combination of vegetated and open water areas and land islands with appropriate vegetation to provide waterfowl with breeding habitats. Subsurface flow systems are designed with an objective of secondary or advanced levels of treatment. These systems have also been called "root zone" or "rock-reed filters" and consist of channels or trenches with relatively impermeable bottoms filled with sand or rock media to support emergent vegetation (see Fig. 13-6).

Floating Aquatic Plants. Floating aquatic plant systems are similar in concept to FWS wetlands systems except that the plants are floating species such as water hyacinth and duckweed (see Fig. 13-5). Water depths are typically deeper than wetlands systems, ranging from 1.6 to 6.0 ft (0.5 to 1.8 m). Supplementary aeration has been used with floating plant systems to increase treatment capacity and to maintain aerobic conditions necessary for the biological control of mosquitoes. Both

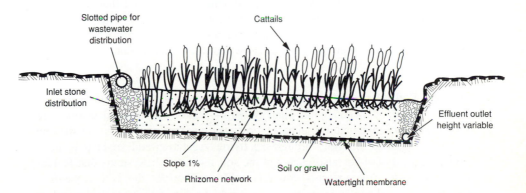

FIGURE 13-6
Cross section of a typical subsurface flow system (SFS) [42].

hyacinth and duckweed systems have been used to remove algae from lagoon and stabilization pond effluents, while hyacinth systems have been designed to provide secondary and advanced levels of treatment. Annual hydraulic loadings and specific area requirements for floating plant systems are similar to wetland systems (see Table 13-3).

Aquaculture. Aquaculture is the growth of fish and other aquatic organisms for the production of food sources. Wastewater has been used in a variety of aquaculture operations around the world. However, in most cases, the production of biomass was the primary objective of the system, and any wastewater treatment was only a side benefit. Most of the treatment achieved in aquaculture systems has been attributed to the bacteria attached to floating aquatic plants. There is little evidence that fish contribute directly to treatment [19]. Combining wastewater treatment and aquaculture into a single operation requires further research. In particular, the health risks associated with use of aquatic organisms grown in wastewater must be defined.

13-2 FUNDAMENTAL CONSIDERATIONS IN THE APPLICATION OF NATURAL TREATMENT SYSTEMS

Knowledge of wastewater characteristics, treatment mechanisms, public health issues, and regulatory requirements is fundamental to the successful design and operation of natural treatment systems.

Wastewater Characteristics and Treatment Mechanisms

As described in the introduction, treatment of wastewater in natural systems is provided by natural physical, chemical, and biological processes that occur in the soil-water-plant ecosystem. Natural systems are capable of removing, at least to some degree, almost all of the major and minor constituents of wastewater that are considered pollutants—suspended solids, organic matter, nitrogen, phosphorus, trace elements, trace organic compounds, and microorganisms (see Table 3-16, Chap. 3). The fundamental processes responsible for the removal of these constituents are described in this section.

Suspended Solids. In systems that feature water flow above the soil surface— overland flow, wetlands, and aquatic plant systems—wastewater suspended solids are removed in part by sedimentation, enhanced by very low flow velocities and shallow depths, and in part by filtration through the living vegetation and vegetative litter. Additional removal of solids also occurs at the soil interface. In systems that feature water flow below the soil surface—slow-rate, rapid infiltration, and SFS wetlands— wastewater suspended solids are removed primarily by filtration through the soil or subsurface media, although sedimentation can be significant in rapid infiltration basins during application. In slow-rate and rapid infiltration systems, most solids are removed at the soil surface. Thus, there is the tendency for wastewater solids to clog or seal the infiltrative surfaces of these systems, so the systems must be designed and operated to minimize loss of infiltrative capacity.

Organic Matter. Degradable organic matter in wastewater, whether soluble or suspended, is removed through microbial degradation. The microbes responsible for the degradation are generally associated with slimes or films that develop on the surfaces of soil particles, vegetation, and litter. In general, natural systems are designed and operated to maintain aerobic conditions so that degradation is performed predominantly by aerobic microorganisms because aerobic decomposition tends to be more rapid and complete than anaerobic decomposition; therefore, potential odors associated with anaerobic degradation are avoided. An exception to the use of aerobic systems occurs when systems are designed to maximize nitrogen removal by denitrification. In such cases, periodic anoxic conditions are imposed on the system to enhance denitrification. The capacity of natural treatment systems to degrade organic matter aerobically is limited by the transfer of oxygen to the system from the atmosphere. Thus, systems must be designed such that the biochemical oxygen demand of the applied organic matter (BOD loading rate) is less than the estimated rate of oxygen transfer to the system.

Nitrogen. The transformation and removal of nitrogen in natural systems involves a complex set of processes and reactions as illustrated in Fig. 13-7. The mechanisms involved in the removal of nitrogen from wastewater (and sludge, see Chap. 12) depend on the form in which the nitrogen is present—nitrate, ammonia, or organic nitrogen. Nitrogen is usually in the form of ammonia or organic nitrogen except in the case of wastewaters that have undergone nitrification as a result of advanced wastewater treatment (see Chap. 11).

Organic nitrogen. Organic nitrogen associated with suspended solids in wastewater is removed by sedimentation and filtration, as described above. Solid-phase organic nitrogen may be incorporated directly into soil humus, which consists of very large, complex organic molecules containing complex carbohydrates, proteins, protein-like substances, and lignins. Some organic nitrogen is hydrolyzed to soluble amino acids that may undergo further breakdown to release ionized ammonia (NH_4^+).

Ammonia nitrogen. Ammonia nitrogen may follow several pathways in natural systems. Soluble ammonia can be removed by volatilization directly into the atmosphere as ammonia gas. This removal pathway is relatively minor (<10 percent) except in the case of stabilization ponds where long detention times and large pH swings combine to produce substantial volatilization of ammonia. Most of the influent and converted ammonia in a natural system is adsorbed temporarily through ion exchange reactions on soil particles and charged organic particles. Adsorbed ammonia is available for uptake by vegetation and microorganisms or for conversion to nitrate nitrogen through biological nitrification under aerobic conditions. Because the ammonia adsorption capacity of natural systems is finite, nitrification is necessary to release adsorbed ammonia and thereby regenerate adsorption sites. This adsorption-release cycle is particularly important in overland-flow systems, where

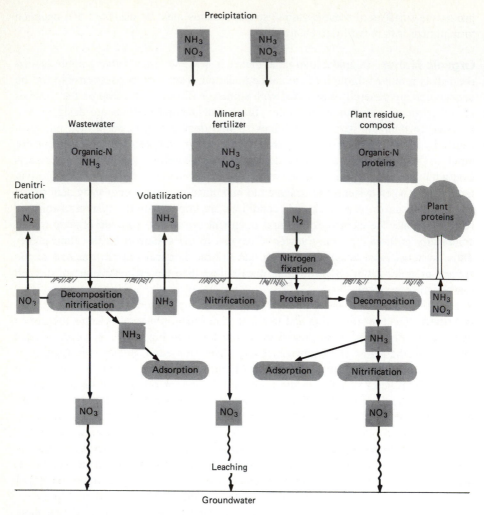

FIGURE 13-7
Nitrogen transformation in natural treatment systems.

adsorption is confined to the slope surface and the capacity for adsorption is more limited.

Nitrate nitrogen. Nitrate nitrogen, being negatively charged, is not held by exchange reactions but remains in solution and is transported in the percolate. If not removed by plant uptake or denitrification, nitrate will leach or percolate into underlying groundwaters. For systems that feature substantial percolation of water, such as slow-rate, rapid infiltration, and sludge application, nitrate in the percolate can pose public health risks (see Public Health Issues). Thus, these systems must be designed and operated to achieve the necessary degree of nitrogen removal to protect groundwaters. Nitrate may be taken up by vegetation, but uptake only occurs in the vicinity of the root zone during active growing periods. To actually achieve nitrogen removal from the system by plant uptake, the vegetation must be harvested and removed from the system. If vegetation is left in the system, the nitrogen contained in the vegetation will be recycled and will reenter the system as organic nitrogen. Plant uptake and harvest is the principal nitrogen removal mechanism in slow-rate systems.

Biological denitrification. Nitrate is also removed by biological denitrification and subsequent release of gaseous nitrous oxide and molecular nitrogen into the atmosphere. Biological denitrification is the major nitrogen removal mechanism in overland flow, rapid infiltration, and aquatic systems. Denitrification is carried out by facultative bacteria under anoxic conditions. It is not necessary for the entire system to be anoxic for denitrification to occur. Denitrification is known to occur in anoxic microsites adjacent to aerobic sites. However, to achieve maximum denitrification, conditions required for denitrification must be optimized. In addition to anoxic conditions, a sufficient carbon/nitrogen ratio is necessary to complete the denitrification reaction. A carbon/nitrogen ratio of at least 2:1 (based on TOC and total N) is necessary to achieve complete denitrification in natural systems [26]. Carbon from decaying vegetation can serve as a partial source of carbon, especially in aquatic systems, but in high-rate systems, such as overland flow and rapid infiltration, the carbon source must be included in the applied wastewater. Thus maximum nitrogen removal cannot be achieved in these systems with secondary effluents that have carbon/nitrogen ratios typically less than 1:1.

Phosphorus. The major phosphorus removal processes in natural treatment systems are chemical precipitation and adsorption, although plants do take up some amounts. The phosphorus, which occurs mainly in the form of orthophosphates, is adsorbed by clay minerals and certain organic soil fractions in the soil matrix. Chemical precipitation with calcium (at neutral to alkaline pH values) and iron or aluminum (at acid pH values) occurs at a slower rate than adsorption, but it is equally important. Adsorbed phosphorus can be held tightly and is generally resistant to leaching.

Although the phosphorus adsorption capacity of soils is finite, it is quite large even for sandy soils. After 88 years of rapid infiltration of untreated municipal wastewater at Calumet, Michigan, concentrations of phosphorus in groundwater are still low (0.1 to 0.4 mg/L) [32]. However, long-term application has caused soil-soluble phosphorus to increase substantially in the top 12 in (0.3 m), indicating that this layer is becoming saturated with phosphorus. The degree of phosphorus removal achievable by a natural treatment system depends on the degree of wastewater contact with the soil matrix. Thus, systems that feature substantial flow of water above the soil surface, such as overland flow and aquatic systems, have a limited potential for phosphorus removal.

Trace Elements. Removal of trace elements (principally metals) occurs mainly through sorption (the term includes adsorption and precipitation reactions) and to a minor extent through plant uptake for some metals. Metals are retained in the soil profile or the sediments of aquatic systems. The retention capacity for most metals in most soils and sediments is generally very high, especially at pH values above 6.5. Under low pH and anaerobic conditions, some metals are more soluble and can be released into solution. The removal of metals varies among systems, depending on the influent concentrations and local site conditions. Reported removal efficiencies for most metals generally range between 80 and 95 percent. Lower removal efficiencies can be expected with FWS wetlands and floating aquatic plant systems due to limited water contact with soils and sediments and anaerobic conditions in sediments.

Trace Organics. Trace organic compounds are removed from wastewater through volatilization and adsorption followed by biological or photochemical breakdown. In general, natural systems are capable of removing large fractions of trace organic compounds; however, the current data base is too small to predict removal efficiencies for individual compounds. Typical results reported in the literature for removal of a few selected organic compounds are presented in Table 13-5.

TABLE 13-5
Removal of trace organic chemicals in natural treatment systems[a]

	Percent removal				
	Slow-rate		Rapid infiltration	Overland-flow	Water hyacinth basins
Organic chemical	Sandy soil	Silty soil			
Chloroform	98.57	99.23	>99.99	96.50	93.61
Toluene	>99.99	>99.99	99.99	99.00	99.99
Benzene	>99.99	>99.99	>99.99	98.09	99.99
Chlorobenzene	99.97	99.98	>99.99	98.99	99.99
Ethylbenzene	[b]	—	—	—	99.99
Bromoform	99.93	99.96	>99.99	97.43	—
Dibromochloromethane	99.72	99.72	>99.99	98.78	99.99
m-Nitrotoluene	>99.99	>99.99	—	94.03	—
PCB 1242	>99.99	>99.99	>99.99	96.46	—
Naphthalene	99.98	99.98	96.15	98.49	85.71
Phenanthrene	>99.99	>99.99	—	99.19	—
Pentachlorophenol	>99.99	>99.99	—	98.06	—
2, 4-Dinitrophenol	—	—	—	93.44	—
Nitrobenzene	>99.99	>99.99	—	88.73	—
m-Dichlorobenzene	>99.99	>99.99	82.27	—	—
Pentane	>99.99	>99.99	—	—	—
Hexane	99.96	99.96	—	—	—
Diethyl phthalate	—	—	90.75	—	75.00
1,1,1-Trichloroethane	—	—	—	—	99.99
Tetrachloroethylene	—	—	—	—	91.49
Phenol	—	—	—	—	80.65
Butylbenzyl phthalate	—	—	—	—	80.95
Isophorone	—	—	—	—	66.67
1, 4-Dichlorobenzene	—	—	—	—	99.99

[a] Adapted from Refs. 19 and 43.

[b] Not reported.

Microorganisms. Removal mechanisms for bacteria and parasites (protozoa and helminths) common to most natural treatment systems include die-off, straining, sedimentation, entrapment, predation, radiation, desiccation, and adsorption. Viruses are removed almost exclusively by adsorption and subsequent die-off. Slow-rate and rapid infiltration systems, which both feature the flow of wastewater through the soil profile, are capable of the complete removal of wastewater microorganisms in the percolate. In medium- to fine-textured soils normally used for slow-rate systems, complete removal can be achieved within 5 ft (1.5 m) of travel. Longer travel distances through the soil are required to achieve removal in rapid infiltration systems, with the distance depending on the soil permeability and the hydraulic-loading rate [40]. All of the other forms of natural treatment systems are capable of reducing wastewater microorganism concentrations by several orders of magnitude but, in general, do not provide sufficient removal to eliminate the need for disinfection where bacterial limits are placed on the system effluent.

Public Health Issues

Aspects of public health related to land treatment include (1) bacteriological agents and the possible transmission of disease to higher biological forms, including humans, (2) chemicals that may reach groundwater and pose risks to health if ingested, and (3) crop quality when crops are irrigated with wastewater effluents.

Bacteriological Agents. The survival of pathogenic bacteria and viruses in sprayed aerosol droplets, on and in the soil, and the effects on workers have received considerable attention [17, 24, 25]. It is important to realize that any connection between pathogens applied to land through wastewater and the contraction of disease in animals or humans would require a long and complex path of epidemiological events. Nevertheless, questions have been raised, concern exists, and precautions should be taken in dealing with possible disease transmission.

Sprinklers, used to apply effluents, produce a mist that may be transported by wind currents. Mist droplets that are extremely small in both dimension and mass are referred to as aerosols. Aerosols are tiny airborne colloidal-like droplets of liquid (0.01 to 50 μ in diameter). Aerosols generated in connection with wastewater that is disinfected inadequately may contain active bacteria and viruses. However, it is reported that aerosolization occurred for only about 0.3 percent of the wastewater being sprinkled, as determined by fluorescein dye tracer tests [25].

Studies of aerosol and mist travel have been conducted using untreated wastewater as well as disinfected secondary effluent [25]. Although bacteria traveled farther in aerosols from undisinfected wastewater, the reported maximum distances ranged from 100 to 600 ft (30 to 200 m). Generally, the wind travel of bacteria increases with increases in relative humidity and wind speed and with decreases in temperature and ultraviolet radiation. An empirical predictive model has been developed to estimate the downwind concentration of aerosol organisms [19].

The need for buffer zones or disinfection to minimize public health risks from aerosols should be assessed on a case-by-case basis considering (1) the degree of public access to the site, (2) the size of the irrigated area, (3) the feasibility of providing buffer zones or plantings of trees or shrubs, and (4) the prevailing climatic conditions. Buffer zone requirements are normally established by regulatory agencies. Setbacks of 50 to 200 ft (15 to 60 m) from roads, property lines, and buildings are typical. Alternatives to buffer zones include plantings of trees, use of sprinklers spraying downward or at low trajectories, and the ceasing of sprinkling, or at least the sprinkling of interior portions of the site, during high winds.

Groundwater Quality. Systems where a portion of the wastewater percolates to groundwaters that serve or potentially could serve as a drinking water supply (principally slow-rate and rapid infiltration) must be designed and managed to maintain receiving groundwater quality above drinking water standards established by the U.S. Environmental Protection Agency [39]. Because nitrate is the causative agent of methemoglobinemia in infants, its concentration in drinking water is limited in the Primary Drinking Water Standards to 10 mg/L as nitrate nitrogen. Sufficient nitrogen removal must be achieved through pretreatment and natural treatment to maintain this standard.

Trace metals applied to natural systems do not pose a threat to groundwater quality because trace metals are usually removed from the percolating water by adsorption or chemical precipitation within the first few feet of soil, even in rapid infiltration systems with high hydraulic-loading rates. In studies on the long-term effects of wastewater application, it has been found that there is no increase in the metals concentration in the soil above the normal range for agricultural soils [19].

Bacterial removal from effluents passing through fine soils is quite complete; it may be extensive in the coarse, sandy soil used for rapid infiltration systems. Fractured rock or limestone cavities may provide a passage for bacteria that can travel several hundred feet from the point of application. This situation can be avoided by proper geological investigations during site selection.

Crop Quality. Trace metals are retained in the soil and sediments of natural treatment systems and are available for uptake by plants. From a public health standpoint, the principal metal of concern is cadmium. Cadmium can accumulate in plants to levels toxic to humans and animals, and these levels are below the concentrations toxic to the plant (phytotoxic). As a result, cadmium is one of the principal limiting constituents in determining sludge-loading rates on agricultural land (see Chap. 12). For most wastewater applications, accumulation of cadmium should not be an issue. The monitoring of a site in Melbourne, Australia, that has been receiving wastewater for 76 years revealed no significant increase in cadmium accumulation in plants as compared to plants grown on a control site that had received no wastewater [17]. Other potential metals of concern are either not taken up by plants (e.g., lead) or are phytotoxic at levels far below concentrations which represent a toxic risk in the food chain (e.g., zinc, copper, and nickel).

13-3 SLOW-RATE SYSTEMS

Slow-rate system design is a two-phase procedure—preliminary and detailed design. After the wastewater characteristics and regulatory requirements have been defined, the preliminary design phase begins. The key steps in preliminary design, summarized in Table 13-6, are discussed in this section. The detailed design phase involves the layout and sizing of individual system components such as pumps, distribution piping or channels, sprinklers, and drainage systems. These detailed design steps can be performed following conventional irrigation system design procedures described in other references [7,14,33–37]. Further details on slow-rate system design may be found in Refs. 16, 18, and 40. The relationships among the key design steps for type 1 and type 2 systems are delineated graphically in Fig. 13-8.

Site Evaluation and Selection

The major site characteristics and general criteria for site selection are listed in Table 13-7. Soil permeability and the depth of soil to groundwater, impermeable layer, or rock are normally the most important characteristics determining the suitability of a site for slow-rate treatment. The vertical permeability or hydraulic conductivity under saturated conditions of the most restrictive layer or horizon in the soil profile will largely determine allowable hydraulic-loading rates for type 1 systems and will affect the type of crops that can be grown and the selection and design of the distribution system.

Soils with permeabilities in the mid-range, 0.2–2.0 in/hr (5 to 50 mm/hr), are best suited for slow-rate systems because they provide the best balance between the retention of wastewater components and drainage. This range of permeabilities is normally associated with medium-textured soils with textural classifications ranging from clay loams to sandy loams. Soils with low permeabilities are associated with fine-textured soils (clays) and soils with cemented sublayers. Wastewater renovation potential of such soils is excellent, but hydraulic-loading rates are restricted and crop management is difficult. Low-permeability soils are better suited to the overland flow process.

Soils with high permeabilities are associated with coarse-textured soils (sands). Such soils can transmit large quantities of water and therefore allow high hyrdaulic-loading rates. However, the capacity of coarse soils to retain moisture in the profile is limited, which makes crop management more difficult. Sites with coarse soils are generally best suited to the rapid infiltration process, although some tree crops and other deep-rooted crops may be grown on sandy soils. The renovation capacity of coarse soils is limited and may restrict the allowable loading rate, based on limiting design factors other than permeability.

Adequate soil depth to groundwater or bedrock is important for retention of wastewater components, bacterial action, and root development. A minimum soil depth of 3 to 4 ft (0.9 to 1.2 m) is required for wastewater treatment, but greater depths are required for deep-rooted crops. For lesser depths, subsurface drainage will generally be required.

Soils with high or low pH (acid or alkaline soils) and soils with a high electrical conductivity (EC) value (saline soils) can limit the growth of many crops, whereas a high exchangeable sodium percentage (ESP) in a soil (sodic soils) can reduce soil permeability. However, it is possible to modify these chemical characteristics through soil reclamation procedures if soil reclamation can be justified economically

TABLE 13-6
Principal steps in the design of natural systems

Slow-rate systems	Rapid infiltration systems	Overland-flow systems	Constructed wetlands and floating aquatic plant systems
1. Site evaluation and selection	1. Site evaluation and selection	1. Site evaluation and selection	1. Site evaluation and selection
2. Determination of pretreatment level	2. Determination of pretreatment level	2. Determination of pretreatment level	2. Determination of pretreatment level
3. Crop selection	3. Selection of distribution method	3. Selection of distribution method	3. Vegetation selection and management
4. Distribution system selection	4. Determination of design hydraulic-loading rate	4. Determination of system design parameters	4. Determination of design parameters
5. Determination of loading rates	5. Determination of design operating cycle	5. Determination of storage requirements	5. Vector control measures
6. Determination of land area requirements	6. Determination of land requirements	6. Calculation of land area requirements	6. Detailed design of system components
7. Determination of storage volume requirements	7. Layout of infiltration area	7. Layout of system components	7. Determination of monitoring requirements
8. Determination of monitoring requirements	8. Layout and sizing of effluent recovery system	8. Selection of cover crop	
	9. Determination of storage requirements	9. Detailed design of system components	
	10. Determination of monitoring requirements	10. Determination of monitoring requirements	

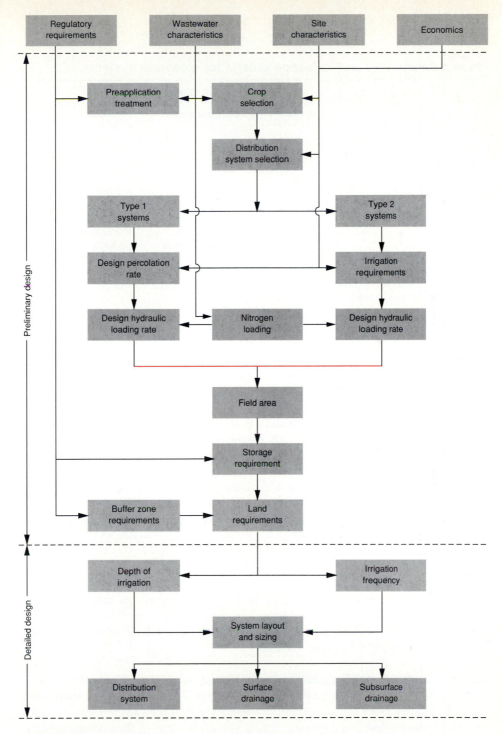

FIGURE 13-8
Typical activity diagram showing key design tasks for slow-rate systems [16].

TABLE 13-7
Site characteristics and selection criteria for slow-rate systems

Characteristic	Acceptability		
	Desirable	Less desirable	Poor
Soil			
pH	5.5–8.4	5.2–5.5	< 5.2; > 8.4
ESP, %	< 5	5–10	> 10[a]
EC, mmhos/cm	< 4	4–8	> 8
Permeability, in/hr	0.2–2.0	0.06–0.2; 2–6	< 0.06; > 6.0
Depth to groundwater, ft	> 5	2–5[b]	< 2
Slope grade, %	0–2	2–15	> 15[c]
Land use	Agricultural	Low-density	Urban/industrial[d]
Hydrology	No flood hazard	Low flood hazard	High flood hazard

[a] > 20 for coarse soils.

[b] Subsurface drainage may be required.

[c] > 30 percent for forested sites.

[d] Landscape and golf course irrigation are possible with high levels of pretreatment.

Note: ft × 0.3048 = m
         in × 25.4 = mm

[16]. It is also possible to select crops tolerant of adverse soil conditions (see "Crop Characteristics").

The ground slope, or grade, should be limited to about 15 percent or less for cultivated crops. Slopes up to 20 percent can be used for noncultivated crops such as pastures, depending on the type of farm equipment to be used. Forested hillsides up to 40 percent in slope have been irrigated successfully with sprinklers.

A suitable site for a slow-rate system would preferably be located in an area not susceptible to flooding and where public access to the site is controlled. For slow-rate systems where public access cannot be restricted, such as irrigation of parks, playgrounds, golf courses, and highway landscaping, high levels of disinfection will be required to protect public health.

Crop Selection

Crop selection is normally the first step in the preliminary design process because most of the other design decisions depend on the type of crop. Advice from local farmers, farm advisors, or agricultural extension specialists will be helpful in making crop selections suitable for local conditions.

Guidelines for Crop Selection. Crops that are most compatible with type 1 systems are those with high nutrient uptake capacity, high consumptive water use, high tolerance for moist soil conditions, low sensitivity to wastewater constituents, and minimum management requirements. Crops with all or most of these characteristics include certain perennial forage grasses, turf grasses, certain tree species, and some field crops. Grasses that have been used successfully include Reed canary grass, tall fescue, perennial rye grass, Italian rye grass, Bermuda grass, orchard grass, and brome grass. The most common tree crops used for type 1 systems have been mixed hardwoods and pines. Potential candidate species include cottonwood, sycamore, green ash, black cherry, black locust, red bud, catalupa, chinese elm, white pine, eucalyptus, willow, and hybrid poplar. Field crops that have been used in type 1 systems when the soil is well-drained and the groundwater is below the rooting depth include corn, milo, sorghum, and barley.

A broader selection of crops may be considered for type 2 systems because excess water is not applied. Thus, in addition to the crops suggested for type 1 systems, candidate crops include all types of forage crops, such as legumes (alfalfa, clover); most field crops including cotton, soybeans, and grains; and some fruit crops including citrus, apples, and grapes.

Crop Characteristics. Crop characteristics of principal interest include nutrient uptake capacity, consumptive water use, and salinity tolerance.

Nutrient uptake. The nutrient uptake capacity of a crop is not a fixed value; it depends on the crop yield and the nutrient content at the time of harvest. Accordingly, design estimates of harvest removals should be based on yield goals and nutrient compositions that can be achieved with effective management on similar soils based on local experience. Typical annual nutrient uptake rates for several commonly selected crops are listed in Table 13-8. To achieve the nitrogen removals given in Table 13-8 for forest crops, it is necessary to practice whole tree harvesting. If only the merchantable stems of trees are removed from the system, the net amount of nitrogen removed by the system will be less than 30 percent of the amount stored in the biomass.

Consumptive water use. Consumptive water use by plants is also termed "evapotranspiration" (ET). Evapotranspiration is an important parameter in the water balance equation used in hydraulic-loading calculations. Consumptive water use varies with the physical characteristics and growth stage of the crop, the soil moisture level, and the local climate. Potential ET or reference ET (ET_0) (defined as the rate of ET from an extended surface of well-watered, full-cover short grass) may be used as a direct estimate of ET for well-managed pasture or full-cover, perennial forage grass crops. Estimated ET_0 values for several different types of climates are given in Table 13-9. Estimates of ET for evergreen trees are 10 to 30 percent greater than ET_0. Estimates of the evaporation (E) from water surfaces and moist bare soil are 5 to 15 percent greater than ET_0, depending on humidity and wind conditions [16]. Values of ET for annual crops vary widely depending on location, and planting and harvest dates. Local agricultural advisors should be consulted to obtain estimates of ET for annual crops.

Salinity tolerance. Reclaimed wastewater often contains higher salinity levels than available irrigation water supplies. Salinity must either be managed through proper leaching and drainage, or crops that are tolerant to expected salinity levels must be selected (see also discussion in Chap. 16). Sensitivity of selected crops to

TABLE 13-8
Nutrient uptake rates for selected crops[a]

Crop	Nutrient uptake, lb/acre · yr		
	Nitrogen	Phosphorus	Potassium
Forage crops			
Alfalfa[b]	200–480	20–30	155–200
Brome grass	116–200	35–50	220
Coastal Bermuda grass	350–600	30–40	200
Kentucky bluegrass	180–240	40	180
Quack grass	210–250	27–41	245
Reed canary grass	300–400	36–40	280
Rye grass	180–250	55–75	240–290
Sweet clover	158	16	90
Tall fescue	135–290	26	267
Orchard grass	230–250	20–50	225–315
Field crops			
Barley	63	15	20
Corn	155–172	17–25	96
Cotton	66–100	12	34
Grain sorghum	120	14	62
Potatoes	205	20	220–288
Soybeans[b]	94–128	11–18	29–48
Wheat	50–81	15	18–4
Forests			
Eastern forests			
Mixed hardwoods	195		
Red pine	100		
Old field with white spruce plantation	250		
Pioneer succession	250		
Southern forests			
Mixed hardwoods	300		
Southern pine[c] with no understory	196		
Southern pine[c] with understory	285		
Lake states forests			
Mixed hardwoods	100		
Hybrid poplar[d]	140		
Western forests			
Hybrid poplar[d]	270–360		
Douglas fir plantation	135–220		

[a] From Ref. 40.

[b] Legumes can fix atmospheric nitrogen.

[c] Principal southern pine is the loblolly pine.

[d] Short-term rotation with harvesting at 4–5 years; represents first growth cycle from planted seedlings.

Note: lb/acre · yr × 1.1209 = kg/ha · yr

TABLE 13-9
Typical monthly potential evapotranspiration rates for selected geographic locations[a]

	Value, in						
Month	Paris, TX	Central MO	Brevard, NC	Jonesboro, GA	Hanover, NH	Seabrook, NJ	Central valley CA	Southern desert CA
Jan	0.6	0.3	0.1	0.5	0.0	0.1	1.1	2.7
Feb	0.6	0.5	0.1	0.5	0.0	0.1	1.8	3.6
Mar	1.4	1.2	1.2	0.8	–	0.8	3.0	5.9
Apr	2.7	2.6	1.8	2.3	1.1	1.6	4.6	7.6
May	3.9	4.3	3.0	4.3	3.2	2.9	5.8	10.1
Jun	5.8	5.7	4.0	5.8	5.1	4.5	7.3	11.4
Jul	6.3	6.7	4.5	6.2	5.4	5.5	7.9	11.6
Aug	6.4	6.0	4.1	5.9	4.7	5.4	6.7	9.6
Sep	3.8	4.1	2.9	4.3	2.9	3.9	5.2	8.5
Oct	2.5	2.5	1.8	2.3	1.6	1.9	3.4	6.3
Nov	1.1	1.0	0.6	1.0	0.8	0.8	1.6	3.5
Dec	0.6	0.4	0.1	0.5	0.0	0.1	1.0	2.0
Annual	35.7	35.3	24.2	34.4	24.8	27.6	49.4	82.8

[a] From Ref. 16.
Note: in × 25.4 = mm

salinity, expressed in terms of electrical conductivity, is reported in Table 13-10. References 12 and 16 may be consulted for more detailed listings of tolerance ratings for various field crops.

Preapplication Treatment

Slow-rate treatment should be considered as a unit process that must be combined with other processes to produce a complete wastewater treatment system. Preapplication treatment is provided for a combination of reasons including protection of public health, nuisance control, distribution system constraints, reduction of limiting wastewater constituents, and soil and crop considerations. The degree of preapplication treatment can range from primary to advanced treatment. The required level of preapplication treatment will depend on the type of slow-rate system. For type 1 systems, the level of preapplication treatment should be the minimum necessary to ensure public health and to avoid nuisance conditions. Screening and primary sedimentation is the minimum level of preapplication treatment recommended for type 1 systems. Type 2 systems are designed to emphasize reuse of wastewater and require greater flexibility in the handling of wastewater; therefore, type 2 systems often require higher levels of preapplication treatment. The required degree of preapplication treatment for crop irrigation will normally be based on a consideration of the state public health regulations or guidelines. Regulations for preapplication treatment differ considerably from state to state and depend on the type of crop grown, the intended

TABLE 13-10
Decrease in yield to be expected for forage and field crops resulting from high electrical conductivity in irrigation water[a]

Location	EC_e values in mmho/cm (saturated paste extract) for a reduction of crop yield of		
	0%	25%	100%
Forage crops			
Alfalfa	2.0	5.4	15.5
Bermuda grass	6.9	10.8	22.5
Clover	1.5	3.6	10.0
Corn (forage)	1.8	5.2	15.5
Orchard grass	1.5	5.5	17.5
Perennial rye grass	5.6	8.9	19.0
Tall fescue	3.9	8.6	23.0
Vetch	3.0	5.3	12.0
Tall wheat grass	7.5	13.3	31.5
Field crops			
Barley	8.0[b]	13.0	28.0
Corn	1.7	3.8	10.0
Cotton	7.7	13.0	27.0
Potato	1.7	3.8	10.0
Soybeans	5.0	6.2	10.0
Sugarbeets	7.0	11.0	24.0
Wheat	6.0[b]	9.5	20.0

[a] From Ref. 1.

[b] Because barley and wheat are less tolerant during germination and seeding stage, the EC value should not exceed 4 or 5 mmho/cm.

use of the crop, the degree of contact by the public with the applied wastewater, and the method of application. For example, the irrigation of certain crops to be eaten raw by humans may require either secondary or advanced wastewater treatment with disinfection, or it may be prohibited altogether.

Distribution Methods

The method of distributing wastewater is selected early in the preliminary design of type 2 systems because the application efficiency of the distribution system is an important design parameter used in the calculation of the total irrigation requirements for type 2 systems. Distribution systems may be classified into three broad categories: sprinkler, surface, and drip. The specific types of sprinkler and surface systems commonly used are listed in Table 13-11 along with conditions suitable for their use and application efficiencies.

Sprinkler Systems. Sprinkler distribution is the most common method used for slow-rate systems because sprinklers can be adapted to a wide range of soil and topographic conditions and used for a variety of crop types. Fixed sprinkling systems, often called "solid-set systems," may be either on the ground surface or buried. Both types usually consist of impact sprinklers mounted on risers spaced along lateral pipelines, which are in turn connected to main pipelines. These systems are adaptable to a wide variety of terrains and may be used for irrigation of either cultivated land or woodlands. Portable aluminum pipe is normally used for above-ground systems (see Fig. 13-9). Sprinkler distribution has the advantage of a relatively low capital cost, but it is easily damaged, has a short expected life because of corrosion, and must be removed during cultivation and harvesting operations. Buried systems generally have the greatest capital cost of any of the irrigation systems. On the other hand, buried systems are probably the most dependable and are well suited to automatic control. There are a number of different moving sprinkling systems, including center-pivot, wheel roll, and traveling gun. The center-pivot system, which consists of a lateral suspended by wheel supports and rotating about a point, is the most widely used of this type for wastewater distribution (see Fig. 13-10).

Surface Application Systems. The two main types of surface application systems are furrow and graded-border irrigation, illustrated in Fig. 13-2. Furrow irrigation is accomplished by gravity flow of reclaimed wastewater down the length of furrows from which it seeps into the soil. Typically, wastewater is distributed into the furrows using gated aluminum pipe, as shown in Fig. 13-11, or using syphon tubes from an open head ditch. Graded-border irrigation consists of low, parallel soil ridges or borders constructed in the direction of slope. Level border or basin irrigation is also practiced. Typical design information for these systems is presented in Table 13-12.

Drip or Trickle Application Systems. Drip or trickle irrigation consists of a distribution piping network with delivery of the water made by small emitters or

FIGURE 13-9
Typical sprinkler irrigation systems.

TABLE 13-11
Types of distribution systems and conditions recommended for their use[a]

Distribution system	Suitability and conditions of use				Application efficiency,[b] %
	Crops	Topography	Soil	Water	
Sprinkler systems					
Portable hand move	Orchards, pasture, grain, alfalfa, vineyards, low-growing vegetable and field crops	Max grade: 20%	Min IR[c]: 0.1 in/h WHC[e]: 3.0 in	Quantity: NR[d] Quality: high TDS water can cause leaf burn	70–80
Wheel roll	All crops less than 3 ft high	Max grade: 15%	Min. IR: 0.1 in/h WHC: 3.0 in	Quantity: NR Quality: see above	70–80
Solid set	NR	NR	Min. IR: 0.05 in/h	Quantity: NR Quality: see above	70–80
Center-pivot or traveling lateral	All crops except trees	Max grade: 15%	Min. IR: 0.3 in/h WHC: 2.0 in	Quantity: large flows required Quality: see above	70–80
Traveling gun	Pasture, grain, alfalfa, field crops, vegetables	Max grade: 15%	Min. IR: 0.3 in/h WHC: 2.0 in	Quantity: 100–1000 gal/min · unit Quality: see above	70–80
Surface systems					
Narrow graded border up to 15 ft wide	Pasture, grain, alfalfa, vineyards	Max grade: 7% Cross slope: 0.2%	Min. IR: 0.3 in/h Max. IR: 6.0 in/h	Quantity: moderate flows required	65–85
Wide graded border up to 100 ft wide	Pasture, grain alfalfa, orchards	Max grade: 0.5–1% Cross slope: 0.2%	Min. IR: 0.3 in/h Max. IR: 6.0 in/h Depth: sufficient for required grading	Quantity: large flows required	65–85

Level border	Grain, field crops, rice, orchards	Max grade: level Cross slope: 0.2%	Min IR: 0.1 in/h Max. IR: 6.0 in/h Depth: sufficient for required grading	Quantity: moderate flows required	75–90
Straight furrows	Vegetables, row crops, orchards, vineyards	Max grade: 3% Cross slope: 10% (erosion hazard)	Min. IR: 0.1 in/h Min. IR: NR if furrow length is adjusted to intake Depth: sufficient for required grading	Quantity: moderate flows required	70–85
Graded contour furrows	Vegetables, row crops, orchards, vineyards	Max grade: 8% Undulating Cross slope: 10% (erosion hazard)	Min. IR: 0.1 in/h Max. IR: NR if furrow length is adjusted to intake Noncracking soils required	Quantity: moderate flows required	70–85
Drip systems	Orchards, vineyards, vegetables, nursery plants	NR	Min. IR: 0.02 in/h	Quantity: NR	70–85

[a] From Ref. 40.
[b] Based on good management and return of runoff water for surface systems.
[c] Infiltration rate.
[d] NR = no restriction.
[e] WHC = water-holding capacity.

Note: in × 25.4 = mm
ft × 0.3048 = m

FIGURE 13-10
Center-pivot irrigation machine.

applicators located near the base of plants to be irrigated. Drip systems are not often used for slow-rate systems because the water supply must be consistently clean to prevent plugging of emitters. The requirement for a high-quality water necessitates a very high level of pretreatment that must be economically justified. Intermittent and recirculating sand filters, as described in Chap. 14, have been used to produce an effluent suitable for disposal by drip irrigation.

(a)   (b)

FIGURE 13-11
Gated aluminum pipe used for wastewater distribution: (a) typical installation and (b) close-up of adjustable discharge port.

TABLE 13-12
Typical design details for surface application methods for slow-rate systems

Item	Value Range	Typical
Ridge-and-furrow system		
Topography[a]	Relatively flat to moderately sloped	
Dimensions		
Furrow length, ft	600–1,400	
Furrow spacing, ft[b]	20–40	
Application[c]		
Pipe type	Gated aluminum	
Pipe length, ft	80–100	
Rest periods	Up to 6 wk	7–14 d
Border-strip flooding		
Strip dimensions[d]		
Border widths, ft	20–100	40–60
Slopes, %	0.2–0.4	0.3
Strip length, ft	600–1,400	
Method of distribution[e]	Concrete-lined ditch, underground pipe, or gated aluminum pipe	
Application rest periods	Up to 6 wk	7–14 d
Application rate per foot width of strip[f]		
Clay, gal/min · ft	9–18	13
Sand, gal/min · ft	25–50	40

[a] Ridge-and-furrow irrigation can be used on relatively flat land (less than 1 percent) with furrows running down the slope or on moderately sloped land with furrows running along the contour.

[b] Furrow spacing depends on the crop.

[c] Short runs of pipe are preferred to minimize pipe diameter and headloss and to provide maximum flexibility. Surface standpipes are used to provide the head of 3 to 4 ft (0.9 to 1.2 m) necessary for even distribution. Application amounts of 3 to 4 in (75 to 100 mm) generally result in a matter of hours with both ridge-and-furrow systems and border-strip flooding.

[d] Strip dimensions vary with types of crop and soil and slope. Relatively permeable soils require the steeper slopes.

[e] Distribution is generally by means of a concrete-lined ditch with slide gates at the head of each strip, underground pipe with risers and alfalfa valves, or gated aluminum pipe.

[f] Application rates at the head of each strip will vary primarily with soil type. The period of application for each strip will vary with strip length and slope.

Note: ft × 0.3048 = m
    gal/min · ft × 12.4193 = L/min · m

Design Hydraulic-Loading Rate

The hydraulic-loading rate is the volume of water applied per unit area of land over a specified time period—typically weekly, monthly, or annually. The corresponding units of expression are in/wk, in/mo, in/yr, and ft/yr. The range of hydraulic-loading rates used are reported in Table 13-3. For type 1 systems, the design hydraulic-loading rate is the hydraulic-loading rate calculated on the basis of the limiting design factor. The factors that must normally be considered for municipal wastewaters are (1) soil permeability and (2) nitrogen loading limits. For industrial wastewaters, other factors such as organic loading, salt loading, or metals loading may require consideration. For type 2 systems, the design hydraulic-loading rate is normally the lesser of the total irrigation water requirement of the crop or the allowed hydraulic-loading rate based on nitrogen loading.

Hydraulic-Loading Rate Based on Soil Permeability—Type 1 Systems. The general water balance equation, with rates based on a monthly time period, is used to determine the monthly hydraulic-loading rate based on soil permeability. The equation, with all runoff assumed to be collected and reapplied, is

$$L_{w(p)} = ET - P + W_p \tag{13-1}$$

where $L_{w(p)}$ = wastewater hydraulic-loading rate based on soil permeability, in/mo
ET = design evapotranspiration rate, in/mo
P = design precipitation rate, in/mo
W_p = design percolation rate, in/mo

The design ET rate is normally the average monthly ET rate of the selected crop. If sufficient historical evaporation data are available (at least 15 consecutive years), it is suggested that a 90 percent exceedance value of $ET - P$ be determined using a frequency distribution analysis [16].

The value for design precipitation rate should be determined from a frequency analysis of wetter-than-normal years. The wettest year in a 10-year period is reasonable in most cases, but it is prudent to check the water balance using the range of precipitation amounts that may be encountered. For purposes of evaluating monthly water balances, the design annual precipitation can be distributed over each month of the year by multiplying the design annual value by the ratio of average monthly precipitation to average annual precipitation for each month.

The design percolation rate is the amount of water allowed to percolate each month beyond the root zone into underlying groundwater or drainage systems. The design value for percolation rate is based on a percentage of the minimum saturated permeability of the upper 8 ft (2.5 m) of the soil profile and the capacity of the down-gradient saturated profile to transmit the added water. A maximum daily value of 2 to 6 percent of the minimum soil profile permeability can be used for preliminary design. The design monthly value is determined by multiplying the maximum daily value by the number of operating days in a given month. Nonoperating days should be allowed for harvesting or cultural procedures and for freezing temperatures. The computation procedure is illustrated in Example 13-1.

Hydraulic-Loading Rate Based on Nitrogen Limits. If percolating water from a slow-rate system will enter a potable groundwater aquifer, then the system should be designed so that the concentration of nitrate nitrogen in the receiving groundwater at the project boundary does not exceed 10 mg/L as nitrogen (Primary Drinking Water Standards). To meet this limiting nitrogen requirement, the allowable hydraulic-loading rate based on an annual nitrogen loading rate ($L_{w(n)}$) must be estimated and compared to the previously calculated ($L_{w(p)}$). The following equation may be used to estimate $L_{w(n)}$:

$$L_{w(n)} = \frac{(C_p \text{ mg/L})(P - \text{ET in/yr}) + (U \text{ lb/acre} \cdot \text{yr})(4.4)}{(1 - f)(C_n \text{ mg/L}) - (C_p \text{ mg/L})} \qquad (13\text{-}2)$$

where $L_{w(n)}$ = allowable hydraulic-loading rate based on annual nitrogen loading rate, in/yr

   C_p = total nitrogen concentration in percolating water, mg/L
   ET = design evapotranspiration rate, in/yr
   P = design precipitation rate, in/yr
   U = nitrogen uptake by crop, lb/acre · yr
   4.4 = combined conversion factor
   C_n = total nitrogen concentration in applied wastewater, mg/L
   f = fraction of applied total nitrogen removed by denitrification and volatilization

The nitrogen uptake of most crops has been determined using fresh water for irrigation, and typical uptake values are given in Table 13-8. Nitrogen uptake values may be higher when wastewater is applied instead of fresh water only because more nitrogen is available. Limited data are available for actual nitrogen uptake by crops in slow-rate systems. Nitrogen uptakes for plants not listed in Table 13-8 can generally be obtained from agricultural extension service agents. When more than one crop per year is grown on one field, the total nitrogen uptake for the entire year should be determined. Nitrogen removal by crop uptake is a function of crop yield and requires the harvesting and physical removal of the crop to be effective.

The extent of denitrification and volatilization depends on the loading rate and characteristics of the wastewater to be applied, as well as the microbiological conditions in the active zones of the soil. Even in aerobic soils, denitrification may account for 15 to 25 percent of the applied nitrogen. Higher values can be expected if the carbon to nitrogen ratio of the applied wastewater exceeds 2.0. Volatilization of ammonia will not be significant for effluents with a pH less than 7 or for nitrified effluents.

If the value of $L_{w(n)}$ calculated using Eq. 13-2 is greater than the value of annual $L_{w(p)}$, then $L_{w(p)}$ is limiting and should be used for design. If the calculated value of annual $L_{w(n)}$ is less than the value of annual $L_{w(p)}$, then $L_{w(n)}$ is limiting and a month-by-month comparison of $L_{w(n)}$ and $L_{w(p)}$ should be made with the lesser of the two values used for design. Monthly values for $L_{w(n)}$ may be calculated using Eq. 13-2 with monthly values for P, ET, and U. Monthly values for U can be estimated by assuming that annual crop uptake of nitrogen is distributed monthly according to the same ratio as the monthly-to-total growing season ET. If actual monthly values of crop nitrogen uptake are available, such values should be used in Eq. 13-2. The procedure for determining nitrogen loading limits is illustrated in Example 13-1.

Example 13-1   Determination of hydraulic-loading rate based on soil permeability and nitrogen loading limits for a type 1 slow-rate system.   Determine the design hydraulic-loading rate for a type 1 slow-rate system, using the following site characteristic data and assumptions:

Month	Precipitation, in	Potential evapo- transpiration, in
January	2.3	0.7
February	2.3	1.5
March	2.1	3.1
April	1.6	3.9
May	0.4	5.2
June	0.2	6.5
July	0.1	7.0
August	Trace	6.5
September	0.2	4.4
October	0.6	3.9
November	1.0	1.5
December	2.2	0.8

1. The design precipitation given above is for the wettest year in 10 years.
2. The average monthly evapotranspiration rates given above are typical for the site.
3. The site is mostly flat and level.
4. The soil is a deep, clay loam with a permeability of 0.35 in/h (9 mm/h).
5. The crop is coastal Bermuda grass. Assume potential ET.
6. Storage will be provided for a portion of the flow during the winter.
7. Runoff, if any, will be collected and stored for reapplication.
8. Total nitrogen in applied water (C_n) is 20 mg/L.
9. Allowable nitrogen concentration in percolate (C_p) is 10 mg/L.
10. Denitrification/volatilization loss fraction (f) is 0.15.
11. The number of nonoperating days is zero.

Solution

1. Determine the design percolation rate (W_p) for each month. Using a design value of 4 percent of the minimum soil permeability, the design monthly percolation rate is

$$W_p = (0.35 \text{ in/h})(24 \text{ h/d})(0.04)(30 \text{ operating d/mo})$$

$$W_p = 10 \text{ in/mo}$$

2. Determine the hydraulic-loading rate based on soil permeability ($L_{w(p)}$). Using Eq. 13-1, the design monthly precipitation is subtracted from the total water losses (ET $+ W_p$) to determine

the amount of wastewater to be applied ($L_{w(p)}$). The required computation table is shown below with the resulting monthly values for $L_{w(p)}$ shown in column 6.

Month (1)	Water losses, in			Water applied, in		
	Evapo-transpiration (2)	Perco-lation (3)	Total (2)+(3)= (4)	Precipi-tation (5)	Wastewater ($L_{w(p)}$) (4)−(5)= (6)	Total (5)+(6)= (7)
Jan	0.7	10.0	10.7	2.3	8.4	10.7
Feb	1.5	10.0	11.5	2.3	9.2	11.5
Mar	3.1	10.0	13.1	2.1	11.0	13.1
Apr	3.9	10.0	13.9	1.6	12.3	13.9
May	5.2	10.0	15.2	0.4	14.8	15.2
June	6.5	10.0	16.5	0.2	16.3	16.5
July	7.0	10.0	17.0	0.1	16.9	17.0
Aug	6.5	10.0	16.5	Trace	16.5	16.5
Sep	4.4	10.0	14.4	0.2	14.2	14.4
Oct	3.9	10.0	13.9	0.6	13.3	13.9
Nov	1.5	10.0	11.5	1.0	10.5	11.5
Dec	0.8	10.0	10.8	2.2	8.6	10.8
Total annual	45.0	120.0	165.0	13.0	152.0	165.0

3. Determine the annual nitrogen uptake rate for coastal Bermuda grass. Use a value of 400 lb/acre · yr (see Table 13-8).

4. Determine the allowable annual nitrogen loading rate, $L_{w(n)}$, using Eq. 13-2.

$$L_{w(n)} = \frac{(10 \text{ mg/L})(13.0\text{-}45.0 \text{ in/yr}) + (400 \text{ lb/acre} \cdot \text{yr})(4.4)}{(1 - 0.15)(20 \text{ mg/L}) - (10 \text{ mg/L})}$$

$$L_{w(n)} = 205.7 \text{ in/yr}$$

5. Compare the value of $L_{w(p)}$ to the value of $L_{w(n)}$. Because $L_{w(n)} > L_{w(p)}$, the design hydraulic-loading rate is based on $L_{w(p)}$ and is 152 in/yr (see column 6). If $L_{w(n)}$ were found to be less than $L_{w(p)}$, monthly values of $L_{w(n)}$ would be determined and compared to the corresponding monthly values of $L_{w(p)}$. The lesser of the two values would be used for the design hydraulic-loading rate for each month. The monthly values then would be summed to yield the annual design hydraulic-loading rate.

Comment. The maximum application of wastewater will be less than 4 in/wk (100 mm/wk) and will occur in July. Storage will be required for a portion of the flow for each month in which the wastewater available exceeds the wastewater applied (see "Storage Requirements").

Hydraulic-Loading Rate Based on Irrigation Water Requirements—Type 2 Systems.
The net irrigation water requirement (R) of a crop over a specified period of time is defined as the amount of water (normally expressed in terms of depth) needed to replace water consumed through evapotranspiration plus that needed for leaching, seed germination, climate control, and fertilizer or chemical application. In

arid climates, leaching is usually required to control the salinity concentration in the root zone. Leaching requirements typically range between 10 and 25 percent of the applied water. The leaching requirement for a particular crop is determined on the basis of the salinity toxicity threshold of the crop and the salinity of the applied wastewater [16]. Considering only ET and leaching requirement (LR), the net irrigation water requirement for any specified period of time is defined by the following equation:

$$R = \frac{ET - P}{1 - LR/100} \qquad (13\text{-}3)$$

where R = net irrigation water requirement, in
       ET = crop evapotranspiration, in
        P = precipitation, in
       LR = leaching requirement, %

Because distribution systems do not apply water uniformly over the irrigated area and some water is lost during application, a depth of water (D) that is greater than the net irrigation water requirement (R) must be applied to ensure that the entire irrigated area receives the net irrigation water requirement. The depth of water (D) is referred to as the total irrigation water requirement and may be determined using the following equation:

$$D = \frac{R}{E_u/100} \qquad (13\text{-}4)$$

where D = total irrigation requirement, in
        R = net irrigation requirement, in
       E_u = unit application efficiency for distribution system, %

The range of unit application efficiencies achieved in practice for the various distribution systems are reported in Table 13-11. An example calculation of monthly hydraulic-loading rates based on irrigation water requirements ($L_{w(r)}$) is presented in Example 13-2 for a double crop of corn and oats and vetch.

Example 13-2   Determination of hydraulic-loading rate based on irrigation water requirement for a type 2 slow-rate system. Determine the design hydraulic-loading rate for a type 2 slow-rate system, using the following data and assumptions:

1. A double crop of corn plus oats and vetch is used. The design ET $-$ P values for each month are listed in column 2 of the computation table given below.
2. The leaching requirement (LR) is 10 percent.
3. A center-pivot sprinkler is used for distribution. Assume an E_u value of 80 percent.

Solution

1. Determine the monthly hydraulic-loading rates based on irrigation water requirements ($L_{w(r)}$) using Eq. 13-4. The required computation table is shown below.

Month (1)	ET − P (2)	100/(100 − LR) (3)	(100/E_u) (4)	L_w(r) (5) = (2)(3)(4)
		Value, in		
January	−3.69	—	—	
February	−2.59	—	—	
March	−1.82	—	—	
April	1.34	1.11	1.25	1.86
May	1.02	1.11	1.25	1.42
June	4.74	1.11	1.25	6.58
July	8.56	1.11	1.25	11.89
August	6.68	1.11	1.25	9.28
September	2.05	1.11	1.25	2.85
October	1.06	1.11	1.25	1.47
November	−2.10	—	—	—
December	−2.98	—	—	—
Total annual				35.35

Comment. The calculated value of $L_{w(r)}$ must be compared with the allowable hydraulic-loading rate based on nitrogen limits $L_{w(n)}$ as described in Example 13-1. If $L_{w(r)}$ exceeds $L_{w(n)}$, then the nitrogen concentration in the applied wastewater must be reduced through pretreatment such that $L_{w(r)}$ equals $L_{w(n)}$, or fresh water must be supplied to supplement the wastewater to meet total irrigation water requirements.

Storage of wastewater will be required during those periods when the wastewater available exceeds the wastewater applied (see "Storage Requirements").

Land Area Requirements

The total land area required for a slow-rate system includes the cropped area, or field area, as well as land for preapplication treatment facilities, buffer zones, service roads, and storage reservoirs. The required field area is determined from the design hydraulic-loading rate using the following equation:

$$A_w = \frac{(Q \text{ ft}^3/\text{d})(365 \text{ d/yr}) + \Delta V_s \text{ ft}^3/\text{yr}}{(L_w \text{ in/yr})(1.0 \text{ ft/12 in})(43,560 \text{ ft}^2/\text{ac})} \tag{13-5}$$

where A_w = field area, acres
Q = average daily wastewater flow, ft^3/d
ΔV_s = net loss or gain in stored water volume due to precipitation, evaporation, and seepage at storage reservoir, ft^3/yr
L_w = design hydraulic-loading rate, in/yr

As previously mentioned, land area requirements for slow-rate systems range from 60 to 200 acres (24 to 80 ha) for type 1 systems and from 175 to 550 acres (70 to 220 ha) for type 2 systems for an average flowrate of 1 Mgal/d.

Storage Requirements

Storage of wastewater, as indicated in Examples 13-1 and 13-2, will be required whenever the quantity of available wastewater exceeds the design hydraulic-loading rate. A monthly water balance calculation procedure may be used to estimate storage volume requirements, as illustrated in Example 13-3. The initial estimated storage volume must be adjusted to account for any net loss or gain in stored water volume due to precipitation, evaporation, and seepage at storage reservoir (ΔV_s).

For slow-rate systems using annual crops, wastewater application is restricted to the growing season, and storage may be required for a period ranging from 1 to 3 months in moderate climates and 4 to 7 months in cold northern states. Irrigation of perennial grasses or double cropping annual crops can extend the period of application. Periods of snow cover and subfreezing conditions may limit the application to perennial grasses and forest land.

With regard to temperature, it has been shown that irrigation systems can usually operate successfully below 32°F (0°C). A forest irrigation system at Dover, Vermont, operates at temperatures down to 10°F (−12.2°C) [32].

Example 13-3 Estimate storage volume requirements for a slow-rate system.
Estimate the required storage volume for a slow-rate system using a water balance calculation and the following data and assumptions:

1. Average wastewater flow is 1.0 Mgal/d and does not vary from month to month.
2. Monthly design hydraulic-loading rates are as given in Example 13-1.
3. Assume that ΔV_s is zero for the first iteration.

Solution

1. Determine the land area requirement using Eq. 13-5.

$$A_w = \frac{(10^6 \text{ gal/d})(0.134 \text{ ft}^3/\text{gal})(365 \text{ d/yr}) + 0}{(152 \text{ in/yr})(1.0 \text{ ft/12 in})(43{,}560 \text{ ft}^2/\text{ac})}$$

$$A_w = 88.5 \text{ (acres)}$$

2. Convert the design monthly hydraulic-loading rates to equivalent volume loading rates (V_w) using the following equation:

$$V_W \text{ (acre} \cdot \text{ft/yr)} = \frac{A_W \text{ (acres)} \times L_W \text{ (in/yr)}}{12 \text{ in/ft}}$$

Tabulate the resulting values in column 2 of the computation table as illustrated in the table given below. The computation table should begin with the first month following the summer growing season in which storage is required. This month is usually October, but it could be earlier or later depending on the climate. In this example, the calculation is started with the month of November.

	Value, acre-ft			
Month (1)	**Wastewater volume loading** (V_w) (2)	**Available wastewater volume,** (Q_m) (3)	**Change in storage,** (ΔS) (4)=(3)−(2)	**Cumulative storage,** $(\Sigma \Delta S)$ (5)
November	77.4	93.4	16.0	−0.2[a]
December	63.4	93.4	30.0	16.0
January	62.0	93.4	31.4	46.0
February	67.8	93.4	25.6	77.4
March	81.0	93.4	12.4	103.0
April	90.7	93.4	2.7	115.4
May	109.2	93.4	−15.8	118.1 (max value)
June	120.2	93.4	−26.8	102.3
July	124.6	93.4	−31.2	75.5
August	121.7	93.4	−28.3	44.3
September	104.5	93.4	−11.1	16.0
October	98.1	93.4	−4.7	4.9
Annual	1,120.8	1,120.8		

[a] Rounding error; assume zero.

3. Tabulate the volume of wastewater available each month (Q_m) in column 3; calculate the difference between Q_m and V_w and tabulate the result as change in storage (ΔS) in column 4.

4. Calculate the cumulative storage volume in column 5 by adding the change in storage during one month to the accumulated total from the previous month. The maximum monthly cumulative volume is the initial estimated required storage volume. In this example, the required cumulative storage volume is 118.1 acre-ft, which occurs at the beginning of the month of May.

5. The final design storage volume is determined by repeating the previous calculation steps with the inclusion of an annual ΔV_s term in Eq. 13–5 and a monthly ΔV_s term in the calculation of monthly change in storage $(\Delta S = \Delta V_s + Q_m - V_w)$ to account for the net gain or loss in storage volume due to precipitation, evaporation, and seepage. To determine ΔV_s, it is necessary to first estimate the surface area (A_s) of the storage reservoir, using the initial estimated storage volume and an assumed water depth. The following equation may be used to calculate ΔV_s:

$$\Delta V_s = (P - E_{pond} - \text{seepage})(A_s)$$

Underdrainage

Underdrain refers to any type of buried conduit with open joints or perforations that is used to collect and convey renovated water that has percolated through the soil. Underdrains should be designed to draw down the water table within a short time after effluent application or a major rainfall event. Underdrains may be required in poorly drained soils or when groundwater levels affect wastewater renovation or crop growth. The topography of the land to be drained and the position, level, and annual fluctuation of the water table are factors to be considered in the preliminary design and layout of a drainage system for a given site. Detailed field investigations will be required before final design because subsoil and groundwater conditions are not always evident from visual inspection of the site.

Underdrain systems normally consist of a network of drainage pipes buried 4 to 10 ft (1.2 to 3 m) below the surface and intercepted at one end of the field by a cutoff ditch. The pipes normally range in diameter from 4 to 8 in (100 to 200 mm). Underdrain spacing will be controlled by soil permeabilities and depth of the water table. Where high loadings occur, as in rapid infiltration, or where permeabilities are low, as for clay soils, underdrain spacing may be as close as 50 to 100 ft (15 to 30 m). For irrigation systems with moderate to rapid surface soil permeabilities, underdrains may be spaced up to 300 ft (90 m) apart. Reference 38 may be consulted for the design of underdrains.

Surface Runoff Control

Requirements for the control of surface runoff resulting from both applied effluent and stormwater depend mainly on the expected quality of the runoff—for which few data exist. In surface runoff control for irrigation systems, consideration must be given to tail water return, storm runoff, and system protection.

Tail Water Return. Surface runoff of applied effluent is usually taken into account in the design of surface application systems, such as ridge-and-furrow irrigation and border-strip flooding, because it is difficult to maintain even distribution across the field with these methods. To improve irrigation efficiency, excess water is applied at the beginning of the field, and the accumulated water at the end of the field is returned for reapplication. Generally, this tail water is collected and returned by means of a series of collection ditches, a small reservoir, a float-actuated pumping station, and a force main to the main storage reservoir or distribution system.

The amount of tail water will vary between 10 and 40 percent of applied flows (depending on the management provided, the type of soil, and the rate of application). In humid climates, the tail water system design may be controlled by stormwater runoff flows.

Storm Runoff. For high-intensity rainfall, some form of stormwater runoff control may be required for slow-rate systems except those with well-drained soils, relatively flat sites, or where the quality of the runoff is acceptable for discharge. Where runoff control is deemed necessary, it generally consists of the collection and treatment or return of the runoff from a storm of specified intensity, with a provision for the overflow of a portion of all larger flows.

The amount of runoff to be expected as a result of precipitation will depend on the infiltration capability of the soil, antecedent moisture condition of the soil, slope, type of vegetation, and temperature of both air and soil. The relationships between runoff and these factors are common to many other hydrologic problems and are covered adequately in standard textbooks dealing with hydrology.

13-4 RAPID INFILTRATION SYSTEMS

As with slow-rate systems, design of rapid infiltration systems is divided into preliminary and detailed phases. The key steps in the preliminary design of rapid infiltration systems, as summarized in Table 13-6, are discussed in this section. Further details on rapid infiltration system design are provided in Refs. 40, 41, and 43. Detailed design involves the sizing, selection, and layout of individual system components such as conveyance piping, valves, and pumping stations, which are covered in the companion volume to this text [13].

Site Evaluation and Selection

Soils with permeabilities of 1.0 in/h (25 mm/h) or more are necessary for successful rapid infiltration. Acceptable soil types include sand, sandy loams, loamy sands, and gravels. Very coarse sand and gravel are not ideal because they allow wastewater to pass too rapidly through the first few feet, where the major biological and chemical action takes place. Uniform soils are preferred because non-uniformity increases the cost and complexity of site investigations.

Other important factors in site selection include depth of soil to groundwater or bedrock, topography, movement and quality of groundwater, and underlying geologic formations. Soil depths greater than 10 ft (3.0 m) are preferred, but depths as shallow as 6.5 ft (2.0 m) are possible with carefully designed underdrain systems. Near level topography is desirable for use of infiltration basins because cut-and-fill construction can adversely affect the permeability of the surface soils and will add substantially to the cost of the project. Sprinkler distribution has been used for rapid infiltration systems on slopes up to 15 percent, but slopes less than 5 percent are recommended [40].

Adequate field investigations of soil and aquifer conditions are critical to the successful design of rapid infiltration systems. Field verification of soil conditions and permeabilities at the actual site are mandatory. Field measurements of infiltration rate or permeability using large basins (10 ft × 20 ft) are favored over standard ring infiltrometer testing because the vertical permeability can be over-estimated using infiltrometers (see also Chap. 14). If the wastewater to be infiltrated contains significant concentrations of suspended solids or BOD, as with primary effluent or industrial wastewaters, it is recommended that the infiltration tests be conducted with the actual wastewater to be treated. Testing should be conducted during the coldest time of the year under conditions of minimum evaporation (calm and cloudy). To control the wastewater after it infiltrates the surface and percolates through the soil matrix, the subsoil and aquifer characteristics must be known. Design of a rapid infiltration system should not be attempted without specific knowledge of the movement of the water in the soil profile and the groundwater aquifer.

Preapplication Treatment

The purpose of preapplication treatment is to reduce soil clogging and to prevent nuisance conditions (particularly odors) from developing during storage or at the application site. The design preapplication treatment level should be based on the objectives of the system. Primary treatment or the equivalent is the minimum recommended preapplication treatment level for all systems and is the recommended level when maximum nitrogen removal is the system objective. Preapplication treatment to secondary or higher levels is suggested if maintaining maximum hydraulic-loading rates is the objective of the system. State regulatory requirements may dictate minimum preapplication treatment levels in some states. Use of oxidation ponds or storage ponds that generate high concentrations of algae should be avoided prior to rapid infiltration systems. Algal solids can quickly cause severe and permanent clogging of the soil surface. Algae clogging is the major cause of rapid infiltration system failure. Disinfection is generally not necessary, as a number of studies have found that rapid infiltration reduces pathogenic bacteria quite effectively [17]. Chlorination of wastewater prior to rapid infiltration should, in general, be avoided to prevent the formation of chlorinated hydrocarbons.

Distribution Methods

Sprinkling and spreading basins are the two distribution methods most suitable for rapid infiltration systems. Factors that should be considered in the selection of the method include soil conditions, topography, climate, and economics.

Sprinkling. Sprinkler distribution is used only when the topography of the site either precludes construction of spreading basins or makes their construction less cost-effective than sprinkling. Normally, vegetation is necessary to protect the surface of the soil and to preclude runoff. Hydrophytic or water-tolerant grasses are usually chosen. Sprinkling on forest land may also be considered for rapid infiltration.

Spreading Basins. Where spreading is used, the infiltration area is divided into an array of shallow spreading or infiltration basins (see Fig. 13-12). The number and dimensions of basins are controlled by topography, hydraulic-loading rates, and soil permeability. Determination of these parameters is discussed under "Layout of Infiltration Area." Water is delivered to individual spreading basins through a network of gravity or low-pressure pipes. Discharge to basins is accomplished by means of multiple inlet structures with splash pads. To ensure uniform distribution of wastewater over the basin surface, the basin bottoms should be level. The basin bottoms are normally bare native soil, but vegetated bottoms have been used with success in some systems. Vegetation can help to intercept suspended solids and maintain infiltration rates through root growth. However, the use of vegetation requires more intensive basin management and can reduce the rate of soil drying. Bare soil surfaces should be scarified or raked when solids accumulate. Use of a gravel layer over native soil is not recommended because solids, which tend to accumulate between the gravel voids, prevent drying of the underlying soil.

FIGURE 13-12
Typical infiltration basins for wastewater disposal.

Design Hydraulic-Loading Rate and Operating Cycle

The design of a rapid infiltration system is defined by two hydraulic-loading rates—the average annual hydraulic-loading rate, expressed in in/yr, and the actual hydraulic-loading rate during the period of wastewater application (termed "average application rate"), expressed in in/d.

Annual Hydraulic-Loading Rate Based on Soil Permeability. The design annual hydraulic-loading rate is based usually on the permeability or effective vertical hydraulic conductivity of the soil profile above the groundwater table or bedrock. In some cases, however, the loading rate of wastewater constituents, such as nitrogen in municipal wastewaters or BOD in industrial wastewaters, may be the limiting factor that controls the design hydraulic-loading rate. The design average annual hydraulic-loading rate based on permeability is determined by multiplying the minimum long-term, field-measured infiltration rate by an application factor, the value of which depends on the method of field measurement and the characteristics of the wastewater to be applied and by the number of operating days per year.

$$L_W = (\text{IR in/h})(1 \text{ ft/12 in})(24 \text{ h/d})(\text{OD d/yr})(F) \qquad (13\text{-}6)$$

where   IR = infiltration rate, in/h
        OD = number of operating days per year, d/yr
        F = application factor (see Table 13-13)

Suggested design application factors for the different methods of field measurement are given in Table 13-13. Hydraulic-loading rates used in practice at operating facilities are compared in Fig. 13-13, with recommended design rates based on measured minimum vertical hydraulic conductivity of the soil profile.

Application Rate and Operating Cycle. Drying periods are necessary in rapid infiltration operation to allow the soil to reaerate between applications and to allow time for decomposition of accumulated organic material and for other biological conversions such as nitrification. The combination of application and drying periods is termed the operating or loading cycle. Operating cycles are selected to maximize either infiltration, nitrogen removal, or nitrification. Suggested operating cycles to meet these objectives are given in Table 13-14. Typical operating cycles at existing systems are reported in Table 13-15.

Because application of wastewater is not continuous, the design average application rate (R_a) is greater than daily equivalent of the average annual hydraulic-loading rate and is calculated from the annual hydraulic-loading rate (L_w) and the operating cycle, according to the following equation:

$$R_a = \left(\frac{L_W \text{ ft/yr}}{365 \text{ d/yr}}\right)\left(\frac{\text{operating cycle time, d}}{\text{application period, d}}\right) \qquad (13\text{-}7)$$

For sprinkler distribution systems, the average application rate should be the design sprinkler application rate, which should be less than the measured infiltration rate or effective vertical hydraulic conductivity of the soil profile to prevent accumulation and runoff. For basin distribution systems, the rate at which water is actually delivered to the basin may exceed the application rate, which may, in turn, exceed the soil infiltration rate. However, the accumulated depth of water in the basin should

TABLE 13-13
Recommended application factors to be used in calculating the average annual hydraulic-loading rates for rapid infiltration systems[a]

Field measurement	Application factor, F
Basin infiltration test	10–15% of the minimum measured infiltration rate
Cylinder infiltrometer and air entry permeameter measurements	2–4% of the minimum measured infiltration rate
Vertical hydraulic conductivity measurements	4–10% of the conductivity of the most restrictive soil layer

[a] From Ref. 40.

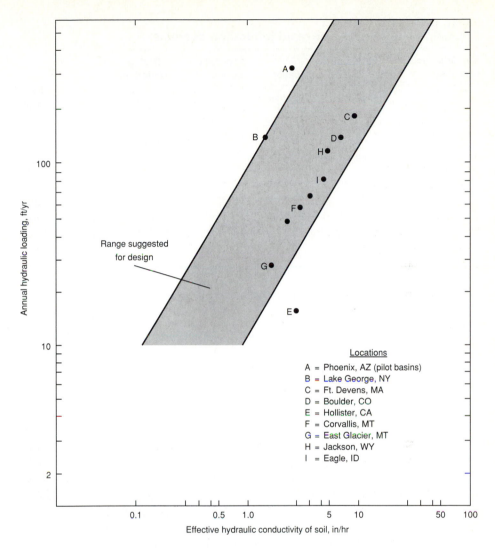

FIGURE 13-13
Suggested design loading rates for rapid infiltration systems [41].

not exceed 12 to 18 in (0.30 to 0.45 m) to minimize compaction of the surface layer and possible secondary effects (algae growth and chemical precipitation).

Hydraulic-Loading Rate Based on Constituent Loading. The wastewater constituents of usual concern are nitrogen in municipal wastewaters and BOD in some industrial wastewaters. Phosphorus may be of concern in some cases where the effluents from rapid infiltration systems enter surface waters sensitive to phosphorus loading. There is increasing concern about the potential effects of residual TOC in the applied wastewaters (humic and fulvic acids and refractory synthetic organics) with

TABLE 13-14
Typical loading cycles for rapid infiltration systems[a]

Loading cycle objective	Applied wastewater	Season	Application period, d[b]	Drying period, d
Maximize infiltration rates	Primary	Summer	1–2	5–7
		Winter	1–2	7–12
	Secondary	Summer	1–3	4–5
		Winter	1–3	5–10
Maximize nitrogen removal	Primary	Summer	1–2	10–14
		Winter	1–2	12–16
	Secondary	Summer	7–9	10–15
		Winter	9–12	12–16
Maximize nitrification	Primary	Summer	1–2	5–7
		Winter	1–2	7–12
	Secondary	Summer	1–3	4–5
		Winter	1–3	5–10

[a] From Ref. 40.
[b] Regardless of the season or cycle objective, application periods for primary effluent should be limited to 1 to 2 days to prevent excessive soil clogging.

regard to direct toxicity in drinking water and THM precursor potential in chlorinated water supplies. Significant reductions of several trace organics through rapid infiltration systems have been reported by a few researchers (see Table 13-5), but the data base on the removal of trace organics is limited.

Because it is generally not known in advance if constituent loading will control design, it is recommended to estimate the removal of the constituent of concern at the proposed design hydraulic-loading rate and operating cycle. In attempting to predict the performance of a proposed rapid infiltration system, a literature review beginning with Refs. 18 and 40 is recommended to search for data on systems with similar conditions (soil texture, soil depth, and hydraulic loadings). If it is not possible to reduce constituent concentrations below prescribed levels under the proposed design conditions, design parameters must be modified or other preapplication treatment provided to meet the performance objectives. A summary of typical rapid infiltration system loading and performance data for BOD, nitrogen, and phosphorus is reported in Table 13-16.

Nitrogen removal. The principal nitrogen removal mechanism in rapid infiltration is denitrification. The maximum amount of nitrogen that can be effectively denitrified during rapid infiltration under optimum operating conditions (ΔN) may be estimated from the wastewater TOC concentration using the following equation.

$$\Delta N = \frac{(TOC - 5)}{2} \tag{13-8}$$

Thus, a carbon to nitrogen ratio (C:N) of at least 2 to 1 is necessary for maximum nitrogen removal. The carbon to nitrogen ratio of secondary effluent is usually less than 2.0. Therefore, primary treatment is the recommended level of preapplication treatment to achieve maximum nitrogen removal.

Nitrogen removal in soils suitable for rapid infiltration is inversely related to soil permeability. Thus, nitrogen removal potential can be increased somewhat by reducing the surface infiltration rate through compaction of the surface. The design hydraulic-loading rate must be reduced accordingly. Pilot testing is recommended to determine the infiltration rate and nitrogen removal following compaction.

BOD removal. Organic material is degraded primarily by aerobic microorganisms in the soil profile. When the BOD loading is high, bacteria multiply rapidly and form slime layers, which can eventually clog the soil pores and reduce not only the infiltration rate but the rate of soil reaeration during drying. Soil clogging coupled with the oxygen consumption by the bacteria during degradation can lead to long-term anaerobic conditions in the soil profile. By-products from the activity of anaerobic bacteria tend to accelerate the rate of soil clogging. The ultimate result of excessive BOD loading will be system failure. Design BOD loadings should be within the range shown in Table 13-16. Higher loadings are reported in the literature, but such systems require intensive management. Design of systems with BOD loadings exceeding 135 lb/acre · d (150 kg/ha · d) should be preceded by long-term pilot studies.

Phosphorus removal. Adsorption and chemical precipitation are the primary phosphorus removal mechanisms in rapid infiltration systems. Although all soil systems have a finite capacity to remove phosphorus, the capacity of many rapid infiltration sites is quite large. Empirical models have been developed to estimate the phosphorus retention capacity of soils and the removal of phosphorus as a function of travel distance in the soil [40].

Land Area Requirements

If flow equalization is provided prior to rapid infiltration, the area required for infiltration only (excluding land required for levees and roads) is determined by dividing the annual average wastewater flowrate by the design annual hydraulic loading as given below.

$$A_i = \frac{(Q \text{ gal/d})(365 \text{ d/yr})}{(L_W \text{ ft/yr})(7.48 \text{ gal/ft}^3)(43,560 \text{ ft}^2/\text{ac})} \qquad (13\text{-}9)$$

If seasonal wastewater flows are not equalized, the highest average seasonal flowrate should be used for design. The initial estimate of required land area computed using Eq. 13-9 may need to be adjusted depending on constraints, as discussed in the section dealing with the layout of the infiltration area. Additional land will be required for access roads, buffer zones, storage or flow equalization, and future expansion. Land area requirements for existing systems range from 2.5 to 55 acres (1 to 22 ha) per 1 Mgal/d (0.044 m^3/s).

TABLE 13-15
Typical hydraulic-loading cycles for rapid infiltration systems using various methods of preapplication treatment[a]

Location	Preapplication treatment	Cycle objective	Application period	Resting period	Bed surface
Boulder, Colorado	Trickling filters	Maximize nitrification and infiltration rates	< 1 d	$< 3\frac{1}{2}$ d	Sand (disked), solids turned into soil
Calumet, Michigan	Untreated	Maximize infiltration rates	1–2 d	7–14 d	Sand (not cleaned)
Flushing Meadows, Arizona	Activated sludge				
Year-round		Maximize nitrification	2 d	5 d	Sand (cleaned)[b]
Summer		Maximize infiltration rates	2 wk	10 d	Sand (cleaned)[b]
Winter		Maximize infiltration rates	2 wk	20 d	Sand (cleaned)[b]
Year-round		Maximize nitrogen removal	9 d	12 d	Sand (cleaned)[b]
Fort Devens, Massachusetts	Primary				
Year-round		Maximize infiltration rates	2 d	14 d	Weeds (not cleaned)
Year-round		Maximize nitrogen removal	7 d[c]	14 d	Weeds (not cleaned)

Location	Pretreatment	Objective	Application period	Drying period	Cover
Hollister, California	Primary				
Summer		Maximize infiltration rates	1 d	14–21 d	Sand
Winter		Maximize infiltration rates	1 d	10–16 d	Sand
Lake George, New York	Trickling filters				
Summer		Maximize infiltration rates	9 h	4–5 d	Sand (cleaned)[b]
Winter		Maximize infiltration rates	9 h	5–10 d	Sand (cleaned)[b]
Tel Aviv, Israel	Ponds, lime precipitation, and ammonia stripping	Maximize polishing	5–6 d	10–12 d	Sand[d]
Vineland, New Jersey	Primary	Maximize infiltration rates	1–2 d	7–10 d	Sand (disked) solids turned into soil
Westby, Wisconsin	Trickling filters	Maximize infiltration rates	2 wk	2 wk	Grassed
Whittier Narrows, California	Activated sludge with filtration[e]	Maximize infiltration rates	9 h	15 h	Pea gravel

[a] From Ref. 40.
[b] Cleaning usually involved physical removal of surface solids.
[c] Caused clogging and reduced long-term hydraulic capacity.
[d] Maintenance of sand cover is unknown.
[e] Treated wastewater blended with surface waters before application.

TABLE 13-16
Typical performance data for rapid infiltration systems[a]

Parameter	Average loading rate, lb/ac · d	Average removals, %	Comments
BOD	40–160	86–98	Higher values are associated with well-designed systems.
Nitrogen	3–37	10–93	Very dependent on preapplication treatment, BOD/N ratio, wet/dry cycle, hydraulic-loading rate.
Phosphorus	1–12	29–99	Removals correlate closely with travel distance through soil.
Fecal coliform	–	2–6 logs	Removals correlate with soil texture, travel distance through soil, and resting time.

[a] From Ref. 43.
Note: lb/acre · d × 1.12 = kg/ha · d

Example 13-4 Determination of hydraulic-loading rate for a rapid infiltration system. Determine the hydraulic-loading rate and area required for a rapid infiltration system, using the following information and data.

1. The average daily flow (Q) is 2.3 Mgal/d.
2. Testing of the site to be used for the infiltration basin yielded a minimum infiltration rate of 1.5 in/h.
3. Use an application factor of 0.1.
4. Use a 10-day operating cycle with a 3-day application period. Assume that all of the applied wastewater infiltrates in 3 days.

Solution

1. Determine the annual hydraulic-loading rate using Eq. 13-6.

$$L_W = \left(1.5 \text{ in/h}\right)\left(1 \text{ ft/12 in}\right)\left(24 \text{ h/d}\right)\left(365 \text{ d/yr}\right)(0.1)$$

$$= 109.5 \text{ ft/yr}$$

2. Determine the average application rate using Eq. 13-7.

$$R_a = \left(\frac{109.5 \text{ ft/yr}}{365 \text{ d/yr}}\right)\left(\frac{10 \text{ d}}{3 \text{d}}\right)$$

$$= 1.0 \text{ ft/d}$$

3. Determine the area in acres required for infiltration using Eq. 13-9.

$$A_i = \frac{2,300,000 \text{ gal/d (365 d/yr)}}{109.5 \text{ ft/yr (7.48 gal/ft}^3) \text{ (43,560 ft}^2 \text{/acre)}}$$

$$= 23.5 \text{ acres}$$

Comment. In reviewing the above computations, the importance of estimating the infiltration rate properly cannot be overstressed.

Layout of Infiltration Area

The layout of infiltration systems with either sprinkler or spreading basin distribution will be controlled by the site geometry and the operating cycle. The infiltration area is divided into several application areas, with one or more areas receiving wastewater while remaining areas undergo drying. The number of application areas provided should be sufficient to allow at least one area to be loading at all times unless storage is provided. The minimum number of application areas required for continuou application is presented in Table 13-17 for several different operating cycles.

TABLE 13-17
Minimum number of infiltration basins required for a rapid infiltration system for the continuous application of wastewater[a]

Loading application period, d	Cycle drying period, d	Minimum number of infiltration basins
1	5–7	6–8
2	5–7	4–5
1	7–12	8–13
2	7–12	5–7
1	4–5	5–6
2	4–5	3–4
3	4–5	3
1	5–10	6–11
2	5–10	4–6
3	5–10	3–5
1	10–14	11–15
2	10–14	6–8
1	12–16	13–17
2	12–16	7–9
7	10–15	3–4
8	10–15	3
9	10–15	3
7	12–16	3–4
8	12–16	3
9	12–15	3

[a] From Ref. 40.

For spreading basins, the number of basins also depends on the total area required for infiltration and the sizing criteria. Generally, sizes of individual basins are greater than 0.5 acre and less than 20 acres. For example, consider a system with a total application area of 62 acres and an operating cycle of 1-day application and 10-day drying. A typical design would provide 22 basins with a unit area of 2.82 acres. With 22 basins, two basins would receive wastewater during each application period.

The geometry or dimensions of individual basins may be controlled by groundwater-mounding considerations. When wastewater is applied to a basin, a groundwater mound will form directly beneath the basin. The extent or height of the groundwater mound will depend on a number of factors including

1. Geometry of the basins

2. Average application rate

3. Minimum depth of existing groundwater table

4. Depth to impermeable layer

5. Slope of groundwater table

6. Horizontal hydraulic conductivity of the aquifer

7. Effective pore space in the soil profile above the water table

8. Elevation and distance to any horizontal controls (stream, river, or lake surface)

9. Determination of monitoring requirements

A groundwater mound analysis must be performed using values of the parameters listed above to estimate the maximum height of the groundwater mound (see Refs. 18, 40, and 43 for analysis procedures). The system must be designed such that the groundwater mound is below the minimum recommended soil depth to groundwater—2.0 ft (0.6 m). The mound rise is strongly dependent on the basin geometry and can be minimized by using long, narrow basin geometries instead of circular or square shapes. Underdrains or recovery wells may be required if the minimum depth to groundwater cannot be maintained through the design of basin geometry.

For the case where natural drainage of groundwater to surface waters is planned, the following equation may be used to determine the required elevation difference between the water level in the stream or lake and the maximum allowable water table level below the infiltration area [2]:

$$WI = \frac{KDH}{L} \tag{13-10}$$

where W = width of infiltration area, ft (m)

$\quad\quad I$ = hydraulic-loading rate, ft/d (m/d)

$\quad\quad K$ = hydraulic conductivity of aquifer, ft/d (m/d)

$\quad\quad D$ = average thickness of zone below water table perpendicular to flow direction, ft (m)

$\quad\quad H$ = elevation difference between water level in stream or lake and maximum allowable water table level below infiltration area, ft (m)

$\quad\quad L$ = distance of lateral flow, ft (m)

The product *WI* represents the amount of the applied water per ft of axial extent for a given section and thereby controls the size of the infiltration basin (see Fig. 13-14). Thus, if the amount of applied water is restricted by the groundwater, relatively high hydraulic-loading rates (*I*) may be used by designing basins of relatively narrow width (*W*).

Effluent Recovery System

If groundwater mounding must be controlled or permanent comingling of percolate with groundwater is not desirable, recovery of effluent by the use of underdrains or recovery wells can be used.

Underdrains. An underdrain system must be provided to recover the percolate and control the development of a mound without interfering with soil detention time and underground travel distance required to achieve the desired quality of renovated water. The quality of applied water, application rate, soil renovation potential and permeability, aquifer conditions, and the use of a cover crop will determine the necessary detention time and travel distance. Optimum depth and spacing of underdrains to recover renovated water from rapid infiltration systems is mainly a matter of opinion. Water table depths of more than 5 ft (1.5 m) would not greatly increase the depth of the aerobic zone during drying of infiltration basins [11].

Proper placement of underdrains for renovated water recovery is more critical than for irrigation systems. An equation has been developed to determine the distance underdrains should be placed away from the infiltration area (see Fig. 13-15) [2]. The height H_c of the water table below the outer edge of the infiltration area can be calculated as follows:

$$H_c^2 = H_d^2 + IW(W + 2L)K \qquad (13\text{-}11)$$

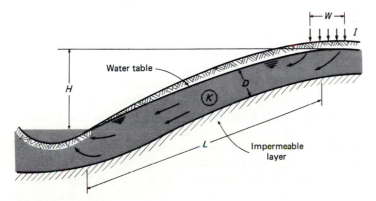

FIGURE 13-14
Natural drainage from rapid infiltration basin into surface water [2].

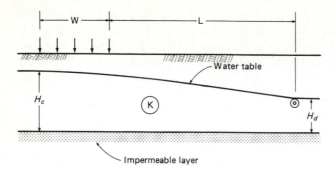

FIGURE 13-15
Collection of renovated water by underdrains [2].

where H_d = drain height above impermeable layer, ft (m)
   I = infiltration rate, ft/d (m/d)
   W = width of infiltration basin, ft (m)
   L = distance measured from centerline of mound to underdrain, ft (m)
   K = hydraulic conductivity of the soil, in/d (m/d)

The location of the drain is selected and H_c is calculated using Eq. 13-11. By adjusting variables L, W, and I, a satisfactory value of H_c is obtained. An L value less than the most desirable distance of underground travel may have to be accepted to obtain a workable system.

   Plastic, concrete, and clay tile lines are used for underdrains. The choice usually depends on price and availability of materials. Most tile drains are laid in a machine-dug trench. Depending on soil conditions, covering the concrete or clay drains with coarse sand may be necessary to keep fine sands and silts out of the tile lines. Plastic drain lines are normally equipped with fiberglass filter socks. In organic soils, loam, and clay loam soils, a filter is not needed. The value of using a filter also depends on the cost of cleaning a plugged drain line versus the cost of the filter materials.

Recovery Wells. The use of wells to recover percolated wastewater is applicable only to rapid infiltration systems. The percolation rates for other methods of natural treatment are generally not high enough to make this process feasible. Recovery may be desired for reuse of the renovated water or to control the water table in order to increase the renovation distance and treatment effectiveness. The potential for percolate recovery at a site depends on several factors such as the depth of the aquifer and the permeability and continuity of an aquiclude. The primary limitations to recovering percolated wastewater are the ability to maintain adequate depth to the groundwater recharge mound and the ability to contain the percolate within a designated area.

   Planning and design considerations for recovery-well systems include configuration of wells relative to infiltration areas, spacing between wells, depth of wells, type of packing, and flowrate. These variables depend on the geology, soil, and groundwater conditions of the site, application rates, and the desired percentage of the renovated water to be recovered. Possible configurations of wells and recharge areas are shown in Fig. 13-16. To select the proper well spacing, the shape and configuration of the cone of depression after pumping should be determined by installing test wells and making pumping tests. Details on these and other aspects of well design may be found in Ref. 3.

Storage Requirements and Climatic Considerations

Because rapid infiltration does not rely on vegetation, it is the natural treatment process most adaptable to cold climates. Also, surface application by flooding basins is less susceptible to freezing than other distribution techniques. At Lake George, New York,

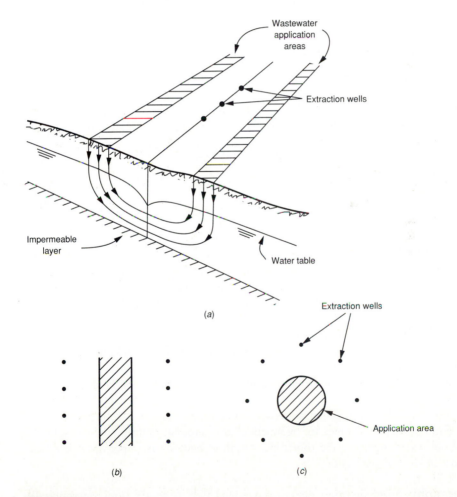

FIGURE 13-16
Extraction well configurations for rapid infiltration systems: (a) wells placed midway between two application areas and (b) and (c) wells (dots) surrounding application areas (cross-hatched areas) [40].

and at Fort Devens, Massachusetts, the systems are operated throughout the winter. When ice forms on the surface of the flooding basins, it is not removed but merely floated by the next application of wastewater. The ice serves to insulate the soil surface from further lowering of the temperature. Rapid infiltration basins have also been operated successfully in the intermountain area of the northwest United States, where air temperatures can be as low as $-35°F$ ($-37°C$). No decline in renovation efficiency, as determined from monitoring wells, has been evident during periods of prolonged cold weather.

Although rapid infiltration systems are usually capable of operating during adverse weather conditions, storage may be needed to regulate application rates to provide flow equalization or for emergencies. Winter storage may be needed in severe climates if the soil permeability is on the low end of the range suitable for rapid infiltration because the water may not drain from the upper profile quickly enough to avoid freezing.

13-5 OVERLAND-FLOW SYSTEMS

The key steps in the design of overland-flow systems are summarized in Table 13-6. Steps 1 through 8, which constitute preliminary system design, are discussed in this section. Further details on overland-flow system design are provided in Refs. 40, 41, and 43. Step 9, detailed design, is performed using standard civil and agricultural engineering design practices covered in Refs 7 and 13.

Site Evaluation and Selection

Site characteristics of importance for overland flow include soil characteristics, topography, and climate.

Soil Characteristics. Although the overland-flow process was originally developed for, and is generally used on, low-permeability soils < 0.06 in/h (< 15 mm/h) or soils with low-permeability sublayers, the process may be used on a variety of soils with permeabilities ranging from slow to moderately rapid, 0.06 to 2.0 in/h (15 to 50 mm/h). Surface soils on overland-flow slopes tend to seal or clog quickly as a result of the growth of biological slimes and deposition of solids in the pore spaces. The result is that percolation losses are small and independent of initial soil permeability. Permeability may also be decreased by compacting the surface during construction. Consequently, field testing for soil permeability is not critical for overland-flow system design.

The depth of soil to groundwater should be a minimum of 1 to 2 ft (0.3 to 0.6 m) to allow sufficient distance for treatment of any percolate entering the groundwater and avoid waterlogging of the root zone. Depth to bedrock is important if it affects the cost of slope construction.

Topography. Ideal site topography for overland flow is gently sloping terrain with a uniform slope in the range of 1 to 8 percent. Sites with level terrain may be adapted for overland flow by constructing slopes with grades greater than 1 percent using a balanced cut-and-fill design. Sites with slopes up to 12 percent have been used for overland flow, but the risk of erosion and channeling increases substantially with slopes above 8 percent. Terraced construction can be used when the natural slope exceeds 8 percent.

Climate. Because overland-flow treatment depends on microbiological activity at or near the surface of the soil, process performance, particularly nitrogen removal, is affected adversely by cold weather. Removal of BOD can continue up to the point of freezing because decreased metabolic activity is compensated for by an increase in microbial population [18], indicating that treatment can continue up to the point of freezing. Thus year-round operation can generally be achieved in the warm climate regions of the United States shown in Fig. 13-17. Winter storage will be required in other regions (see "Storage Requirements").

Preapplication Treatment

Overland flow has been used to treat screened-untreated, primary, secondary, and advanced treated municipal wastewaters as well as high-strength food-processing wastewaters. The minimum level of preapplication treatment required for all systems

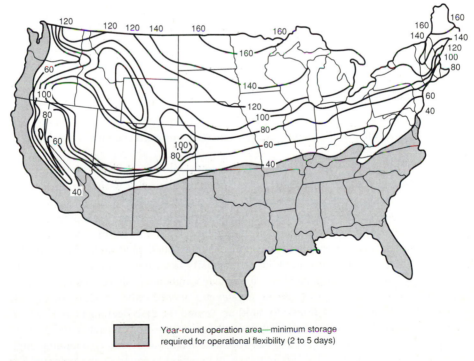

Year-round operation area—minimum storage
required for operational flexibility (2 to 5 days)

FIGURE 13-17
Recommended storage days for overland-flow systems [43].

is fine screening to remove solids that may clog distribution systems. For municipal wastewaters, a minimum screen opening of 1/16 in (1.5 mm) is recommended. The optimum screen size for industrial wastewaters depends on the nature of the solids in the wastewater. It should be noted that screened-untreated municipal wastewater has only been used at research and demonstration facilities and that higher levels of preapplication treatment may be prescribed by regulatory authorities. Primary treatment provides the optimum level of preapplication treatment for municipal wastewaters. Alternatives to primary treatment include aerated lagoons with short detention times (1 to 2 days), and Imhoff tanks for small systems.

Use of oxidation ponds, lagoons, or storage ponds that generate high concentrations of algae should, if possible, be avoided prior to overland-flow treatment. Overland-flow systems do not provide consistent, year-round algae removal to concentrations less than 30 mg/L at conventional application rates because certain types of algae are buoyant or motile and resist removal by settling [44]. Algal solids have been removed consistently to levels below 30 mg/L using very low application rates, but performance is very dependent on the concentration and type of the algae present. At least one full year of pilot testing is recommended prior to design of systems intended for algae removal.

Overland flow has been used for polishing effluents from existing secondary treatment systems to remove nitrogen and trace metals [26]. However, in these cases overland flow is being used to upgrade existing systems. In general, there will be no benefit to process performance to provide secondary treatment prior to overland flow. In the case of nitrogen removal, secondary treatment is actually detrimental to performance because the carbon/nitrogen ratio is reduced below the level required for complete denitrification.

Distribution Methods

Distribution methods for overland-flow systems include gated pipe, fan sprays, and sprinklers. A summary of advantages and limitations of each method is presented in Table 13-18.

Gated Pipe. Gated aluminum pipe, commonly used for furrow irrigation, can be used with municipal wastewaters (see Fig. 13-11). Gates (adjustable plastic slide closures) can be placed on only one side of the pipe or on both sides of the pipe (double-gated) for use with back-to-back slope configurations. A minimum gate spacing of 2 ft is recommended. Wastewater is supplied to the gated pipe under low pressure (2 to 5 lb_f/in^2), and the gates are adjusted manually to achieve uniform distribution. Reference 27 should be consulted for specific recommendations on design of gated pipe systems for overland flow. Because gates are opened only a fraction of the full opening, fibrous material will tend to build up around the gate openings regardless of the preapplication treatment level. Consequently, gates must be inspected and cleaned on a routine basis. Use of gated pipe with industrial wastewaters containing high concentrations of suspended solids is not recommended because of the potential for deposition of solids near the point of discharge.

TABLE 13-18
Summary of overland-flow distribution methods[a]

Method	Advantages	Limitations
Gated pipe	Low energy costs	Less uniform distribution of wastewater then other methods
	Minimum aerosols and wind drift	
	Small buffer zones	Moderate erosion
		Easy to clean
Potential	Easiest of surface methods to balance hydraulically	Potential for freezing and settling
Slotted or perforated pipe	Low energy costs	Same as gated pipe
	Minimum aerosols and wind drift	Small openings clog
	Small buffer zones	Most difficult to balance hydraulically
Low pressure sprays	Better distribution than pipe methods	Nozzles subject to clogging
	Fewer aerosols than sprinkler systems	More aerosols and wind drift than pipe distribution systems
	Relatively low energy costs	
Sprinklers	Most uniform distribution of wastewater	High energy costs
		Aerosol and wind drift potential
		Large buffer zones required

[a] Adapted from Ref. 43.

Fan Sprays. Low-pressure (5 to 15 lb_f/in^2) fan-spray nozzles mounted on vertical risers have been used successfully at municipal systems. Nozzle openings must be sized sufficiently large to prevent clogging with wastewater solids.

Sprinklers. High-pressure (35 to 60 lb_f/in^2) sprinklers distribute wastewater over a much broader area than gated pipe or fan sprays. Normally, full-circle sprinklers are placed approximately one-third of the way down the slope or at the crest of back-to-back slopes. Because wastewater is distributed a considerable distance from the top of the slope, longer slope lengths are recommended with sprinkler systems to provide sufficient slope length for treatment. Typical sprinkler configurations are shown in Fig. 13-18. For industrial wastewaters, sprinklers have been used exclusively because of their ability to distribute wastewater solids and organic load more uniformly over the slope surface. Distribution of the oxygen demand associated with high BOD wastewaters is important because oxygen is supplied to the slope uniformly over the surface through transfer from the atmosphere (see "BOD_5 Loading" under Design Parameters).

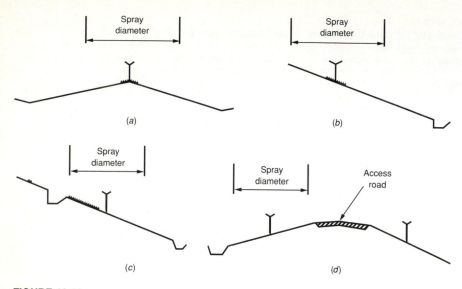

FIGURE 13-18
Typical sprinkler distribution configuration for overland-flow systems [43].

Design Parameters

The principal overland-flow design parameters include application rate, slope length, slope grade, and operating cycle. Suggested guideline values for these parameters are given in Table 13-19 for different levels of preapplication treatment and methods of distribution. These values have been used at existing systems to achieve the effluent quality indicated in Table 13-4. For high-strength industrial wastewaters, BOD loading is also of concern.

Application Rate. The application rate is defined as the volume of wastewater applied per unit time per unit of slope width and is expressed in units of gal/min · ft ($m^3/h \cdot m$). Removal efficiency of most constituents by overland flow increases as the application rate decreases until a lower limit is approached [28]. Many of the early overland-flow systems were designed and operated at or below the lower limit of application rates. Consequently, very conservative design values for application rate have been suggested in the major references dealing with the design of overland-flow systems [40,43]. The suggested design values for application rates given in Table 13-19 reflect more recent experience, in which it has been shown that much higher application rates can be used without sacrificing treatment performance. Values at the low end of the range should be used when the soil temperature drops below 10°C or if maximum removal efficiency for any constituent is desired.

Slope Length. For gated pipe and fan spray distribution systems, slope lengths typically range from 100 to 150 ft (30 to 45 m). In general, overland-flow treatment

TABLE 13-19
Guidelines for overland-flow design parameters for various types of preapplication treatment

Parameter	Units	Screening	Primary/ aerated pond[a]	Oxidation pond[b]	Secondary/ advanced
			Type of preapplication treatment		
Application rate	gal/min · ft	0.25–0.60	0.25–0.60	<0.15	0.33–0.80
Slope length	ft	100–150[c]	100–150[c]	>150[c]	100–150[c]
Application period	h	8–12	8–12	8–12	8–12
Drying period	h	16–12	16–12	16–12	16–12

[a] Detention time of 1 to 2 d.

[b] Not recommended without pilot testing.

[c] For sprinkler distribution systems use a slope length of 150 ft or 65 ft greater than the diameter of the spray diameter, whichever is greater.

Note: gal/min · ft × 12.4193 = L/min · m
ft × 0.3048 = m

efficiency has been shown to be directly related to slope length and inversely related to application rate. Thus, longer slope lengths should be used with higher application rates, and conversely shorter slope lengths should be used with lower application rates to achieve the same degree of treatment. Typical relationships observed for BOD removal efficiency versus downslope distance and application rate are shown in Fig. 13-19 for the treatment of primary effluent using gated pipe distribution. These relationships can be described by a first-order empirical removal model of the following form [28]:

$$\frac{C_z - C}{C_o} = A \exp \frac{-kz}{q^n} \tag{13-12}$$

where C_z = BOD$_5$ concentration of surface flow at a distance (z) downslope, mg/L
C_o = BOD$_5$ concentration of applied wastewater, mg/L
C = background BOD$_5$ level, mg/L
A = empirically determined coefficient dependent on the value of q
k = empirically determined rate constant
z = distance downslope, ft
q = application rate, gal/min · ft
n = empirically determined exponent (<1)

The regression curves shown in Fig. 13–19 may be used to aid design and to check anticipated performance of a given design. Although this model has been verified with data from other systems, caution is advised in the use of the curves due to their empirical derivation. Design values for application rate and slope length should conform to the guidelines presented in Table 13-19. From an analysis of Eq.

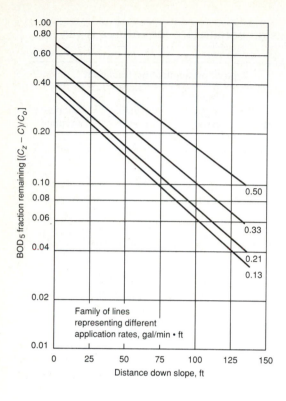

FIGURE 13-19

Typical BOD$_5$ removal performance from overland-flow systems treating primary effluent: fraction of BOD$_5$ remaining versus downslope distance [28].

13-12, it can be concluded that minimum land area will be required for a given level of treatment when higher application rates and longer slope lengths (within the recommended design ranges) are used as opposed to lower application rates and shorter slope lengths.

For sprinkler distribution systems, slope lengths typically range between 150 to 200 ft (45 to 60 m). The slope length should be at least 65 ft (20 m) greater than the diameter of the sprinkler pattern. In some cases, the slope length may be constrained by the geometry of the site.

Slope Grade. Design slope grades should be within the range of 1 to 8 percent. It has been shown through several studies that process performance is not affected by slope grade within this range. Consequently, slope grades should be designed to conform generally to the natural grade of the site to minimize earthwork required to shape the slopes. In almost all cases, rough grading, disking, and final grading will be required to construct the slopes. Final slope grades should be maintained within a tolerance of 0.05 ft (1.5 mm) to ensure sheet flow of the applied wastewater. Individual slopes can vary in grade within the confines of a site, but each should be within the recommended range of 1 to 8 percent. Compound slope grades can be considered for individual slopes when there is an abrupt change in the natural grade down the length of a slope.

Operating Cycle. Almost all overland-flow systems are operated in an intermittent mode with the operating cycle consisting of an application period followed by a drying period. Typically the operating cycle time is one day, or 24 hours, with application periods ranging from 8 to 12 h and corresponding drying periods ranging from 16 to 12 h. With the exception of nitrogen, the removal of most constituents is not affected by variations within these ranges. Ammonia removal from primary effluent was found to vary inversely with the ratio of application period to drying period, as indicated in Fig. 13-20 [9].

There are certain cases where shorter or longer cycle times are used to improve process performance. For example, a system in Davis, California, used to treat high-strength food-processing wastewater (e.g., $BOD_5 = 800$ to 1000 mg/L) is operated with a cycle time of 6 h (3 h application and 3 h drying). This operating cycle allows natural reaeration during drying to "catch up with" and meet the oxygen demand of the wastewater applied during the application period. Longer application periods resulted in reduced BOD_5 removal efficiency. Longer operating cycles (4 d application and 2 d drying) have been used where supplemental carbon addition is employed to denitrify secondary effluent [26]. Limiting the application period to 4 days was found to almost eliminate the propagation of mosquitoes at the site. The extreme operating cycle example is the Melbourne, Australia, system where the operating cycle time is one year (approximately 6 mo application and 6 mo drying).

BOD_5 Loading. For overland-flow treatment of high-strength industrial waste-waters, the BOD_5 loading must be considered. An overland-flow slope can be con-

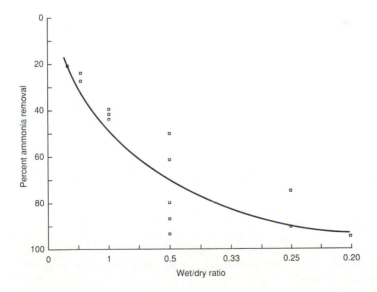

FIGURE 13-20
Typical ammonia removal performance from overland-flow systems: effect of wet/dry ratio on performance using primary effluent [9].

sidered as a film-flow reactor analogous to a trickling filter for purposes of analyzing BOD removal (see Chap. 8). The oxygen required for aerobic microbial decomposition of BOD_5 is supplied from the atmosphere through natural reaeration. To avoid development of anaerobic conditions, the rate of oxygen supply or transfer must be greater than or equal to the rate of oxygen uptake, which is a function of the BOD_5 mass-loading rate. Based on an oxygen-transfer analysis for trickling filters [23], an oxygen-transfer rate of approximately 255 lb/acre · d (285 kg/ha · d) can be estimated for overland-flow systems. Because oxygen is supplied uniformly to the slope area while the BOD_5 load is applied to the upper half of the slope area (for sprinkler distribution), ultimate BOD loadings should be no greater than one-half the rate of oxygen supply. Based on a BOD_u:BOD_5 of 1.5 for industrial wastewater, the recommended maximum daily BOD_5 loading rate is 85 lb/ acre · d (95 kg/ha · d). This limit is consistent with field perfomance data. It has been reported that treatment of food-processing wastewater at BOD_5 loadings greater than 90 lb/acre · d resulted in declining BOD removal efficiency [15].

In addition to the mass loading of BOD_5, the BOD_5 concentration of the applied water must be considered. Based on an oxygen-transfer analysis for trickling filters, a limiting maximum BOD_5 concentration of 400 to 500 mg/L in the applied wastewater is recommended [23]. For wastewaters with higher BOD_5 concentrations, effluent recycling may be practiced to reduce the concentration of the applied wastewater. Alternatively, short operating cycles may be employed to cease application before anaerobic conditions develop (see "Operating Cycle").

Storage Requirements

Storage is recommended when the average daily temperature is below 32°F. Storage requirements may be estimated using Fig. 13-17. For areas below the 40-day storage line shown in Fig. 13-17, winter storage requirements can be estimated from the number of days per year with an average daily temperature less than 32°F with a 20-year return record. Storage is generally not necessary during periods of rainfall. Effluent quality, in terms of BOD and suspended solids, decreases only slightly during rainfall events [5]. A minimum of 2 to 5 days of storage capacity should be provided for operational flexibility. Storage reservoirs should be designed for off-line service so that wastewater that has received preapplication treatment does not pass through storage prior to overland-flow application during nonstorage periods.

Land Requirements

The land required for overland-flow slope area is calculated from the design values for application rate, slope length, operating cycle, application period, and storage data using the following equation:

$$A_s = \frac{[Q + (\Delta V_s/(365 \times 24))] \times [(365 + D_s)/365] \times Z}{(R_a) \times (P_a/O_t)} \qquad (13\text{-}13)$$

where A_s = area required for overland-flow slopes, ft^2
$\quad\quad\quad Q$ = average wastewater flow, ft^3/h
$\quad\quad\quad \Delta V_s$ = net gain or loss in stored volume, ft^3/yr
$\quad\quad\quad D_s$ = number of days of storage
$\quad\quad\quad R_a$ = design application rate, $\text{ft}^3/\text{h} \cdot \text{ft}$
$\quad\quad\quad P_a$ = application period, h or d
$\quad\quad\quad O_t$ = operating cycle time, h or d
$\quad\quad\quad Z$ = slope length, ft

An additional 20 to 25 percent is normally added to this area requirement to allow a portion of the system to be taken out of service each year for slope renovation (rototilling, regrading, and reseeding). Slope area requirements can range from 6 to 45 acres/(Mgal/d) (7 to 50 $\text{m}^2/\text{m}^3/\text{d}$)). Additional land is required for access roads. A value of 10 percent of the slope area is normally allowed for roads in preliminary design. The computation procedure for determining land requirements is illustrated in Example 13-5.

Example 13-5 Determination of land area requirements for an overland-flow site.
Determine the slope area required for a flow of 1 Mgal/d, using a design application rate of 0.33 gal/min · ft , a slope length of 100 ft, an application period of 12 h/d, and an operating cycle time of 24 h. Assume 5 days storage and no net loss or gain from storage. Assume a 20 percent allowance for slope renovation and a 10 percent allowance for roads.

Solution

1. Calculate the overland-flow slope area using Eq. 13-13.
   (a) Convert Q from Mgal/d to ft^3/h

$$Q = \frac{1,000,000 \text{ gal/d}}{(7.48 \text{ gal/ft}^3)(24 \text{ h/d})} = 5,570 \text{ ft}^3/\text{h}$$

   (b) Convert R_a from gal/min · ft to $\text{ft}^3/\text{h} \cdot$ ft.

$$R_a = \frac{(0.33 \text{ gal/min} \cdot \text{ft})(60 \text{ min/h})}{7.48 \text{ gal/ft}^3} = 2.65 \text{ ft}^3/\text{ft} \cdot \text{h}$$

   (c) Determine the slope area.

$$A_s = \frac{(5570 \text{ ft}^3/\text{h})[\,(365 \text{ d/yr} + 5 \text{ d/yr})\,/\,(365 \text{ d/yr})\,](100 \text{ ft})}{2.65 \text{ ft}^3/\text{ft} \cdot \text{h} \times [(12 \text{ h/d})/(24 \text{ h/d})]}$$

$$A_s = 426,136 \text{ ft}^2$$

$$= 9.8 \text{ acres } (4.0 \text{ ha})$$

   (d) Determine the total slope area including the allowance for renovation.

$$\text{Total slope area} = 9.8 \text{ acres} \times 1.2 = 11.8 \text{ acres } (4.8 \text{ ha})$$

2. Determine the total land requirement taking into account the allowance for roads.

$$\text{Total land required} = 11.8 \text{ acres} \times 1.1 = 13.0 \text{ acres } (5.3 \text{ ha})$$

System Layout

The overland-flow site is divided into a network of slopes each having the selected design length. Site geometry may require that slope lengths vary somewhat. The total width of all slopes is determined by dividing the slope area by the design slope length. Slopes should be grouped into a minimum of four or five hydraulically separated, approximately equal application areas or zones to allow operating and harvesting or mowing flexibility. The network of effluent collection channels should be sized to carry effluent flow plus runoff from a storm with a 25-year return frequency without causing flooding of the lower reaches of the slopes. As an example, the overland-flow system at Davis, California, has a total slope area of 170 acres (69 ha) and is divided into 15 application zones, each consisting of two slopes with a length of 150 ft (45 m) and a width of 1640 ft (500 m). Three of the zones are out of service each year for renovation, and two zones at a time are taken out of service for mowing four or five times per year.

Cover Crop Selection and Management

A dense, uniform cover crop is required on overland-flow slopes to prevent erosion and aid in the removal processes. Water tolerant grasses are required for overland-flow vegetation. Suitable types include Reed canary grass, fescue, Italian rye grass, common and coastal Bermuda grass, Dallis, and Bahia grasses, depending on their adaptability to the local climate. Because of the high intensity of water application, grasses from overland-flow systems rarely have any commercial value as feed. Grasses are often mowed with a mulching mower and the cuttings are left on the slopes. Harvesting with green-chopping equipment is also practiced. Mowing and baling is not recommended because of the time required for drying prior to baling.

13-6   CONSTRUCTED WETLANDS

The principal steps in constructed wetland system design are summarized in Table 13-6. Steps 1 through 5, which deal with preliminary design, are discussed in this section. Detailed design involves the sizing, selection, and layout of individual system components such as conveyance piping, valves, and pumping stations, which are covered in the companion volume to this text [13]. Further details on constructed wetland system design are provided in Refs. 19, 42, and 43. It should be noted that although constructed wetlands have been used for a variety of applications including the treatment of septage, acid mine drainage, ash pond seepage, and pulp mill effluents, the discussion in this section is limited to the use of wetlands for the treatment of municipal wastewater. Information on the use of wetlands for other applications may be found in Refs. 10 and 22.

Site Evaluation and Selection

Site characteristics that must be considered in wetland system design include topography, soil characteristics, existing land use, flood hazard, and climate.

Topography. Level to slightly sloping, uniform topography is preferred for wetland sites because free water systems (FWS) are generally designed with level basins or channels, and subsurface flow systems (SFS) are normally designed and constructed with slopes of 1 percent or slightly more. Although basins may be constructed on steeper sloping or uneven sites, the amount of earthwork required will affect the cost of the system. Thus, slope grades for wetland sites are normally less than 5 percent.

Soil. Sites with slowly-permeable ($<$ 0.20 in/h) surface soils or subsurface layers are most desirable for wetland systems because the objective is to treat the wastewater in the water layer above the soil profile. Therefore, percolation losses through the soil profile should be minimized. As with overland-flow systems, the surface soil will tend to seal with time due to deposition of solids and growth of bacterial slimes. Permeabilities of native soils may be purposely reduced by compacting during construction. Sites with rapidly-permeable soils may be used for small systems by constructing basins with clay or artificial liners. Site criteria for soil depth to groundwater or bedrock is the same as that discussed previously for overland-flow systems.

Flood Hazard. In general, wetland sites should be located outside of flood plains, or protection from flooding should be provided. In cases where flooding occurs only in the winter, when the system is not operated, protection from rare flood events may not be required depending on regulatory requirements.

Existing Land Use. Open space or agricultural lands, particularly those near existing natural wetlands, are preferred for wetland sites. Constructed wetlands can enhance existing natural wetlands by providing additional wildlife habitat and, in some cases, by providing a more consistent water supply.

Climate. The use of wetland systems in cold climates is possible. The FWS system in Listowel, Ontario, is operated year-round with wastewater temperatures as low as 3°C [41]. However, the feasibility of operating a system through the winter depends on the temperature of the water in the basin and the treatment objectives. Because the principal treatment mechanisms are biological, treatment performance is strongly temperature sensitive (see Design Parameters). Storage will be required where treatment objectives cannot be met due to low temperatures.

Preapplication Treatment

The minimum level of preapplication treatment for wetlands systems should be primary treatment, short detention-time aerated ponds, or the equivalent. Treatment beyond this level depends on the effluent requirements and the removal capability of the wetlands system. Constructed wetlands have been used in several locations to polish effluent from existing secondary treatment facilities to meet more stringent regulatory requirements. Use of oxidation ponds or lagoons that generate high concentrations of algae should be avoided prior to wetlands treatment because, like overland-flow systems, algae removal through wetlands is inconsistent. Phosphorus removal in the preapplication treatment step is recommended where there are effluent limitations on phosphorus because phosphorus removal in wetlands is minimal.

Vegetation Selection and Management

Vegetation plays an integral role in wetlands treatment by transferring oxygen through their roots and rhizome systems to the bottom of treatment basins and providing a medium beneath the water surface for the attachment of microorganisms that perform most of the biological treatment. Emergent plants, those rooted in the soil or granular support medium that emerge or penetrate the water surface, are used in wetland systems (see Fig. 13-21). The plants used most frequently in constructed wetlands include cattails, reeds, rushes, bulrushes, and sedges. All of these plants are ubiquitous and tolerate freezing conditions. The important characteristics of the plants related to design are the optimum depth of water for FWS systems and the depth of rhizome and root penetration for SFS systems. Cattails tend to dominate in water depths over 6 in (0.15 m). Bulrushes grow well at depths of 2 to 10 in (0.05 m to 0.25 m). Reeds grow along the shoreline and in water up to 5 ft deep (1.5 m), but are poor competitors in shallow waters. Sedges normally occur along the shoreline and in shallower waters than bulrushes. Cattail rhizomes and roots extend to a depth of approximately 12 in (0.3 m), whereas reeds extend to more than 24 in (0.6 m) and

FIGURE 13-21
Typical constructed wetland system with cattails, bulrush, and sedges used to treat the effluent from a series of oxidation ponds, Gustine, CA.

bulrushes to more than 30 in (0.76 m). Reeds and bulrushes are normally selected for SFS systems because the depth of rhizome penetration allows for the use of deeper basins.

Harvesting of wetland vegetation is generally not required, especially for SFS systems. However, dry grasses in FWS systems are burned off periodically to maintain free-flow conditions and to prevent channeling of the flow. Removal of the plant biomass for the purpose of nutrient removal is normally not practical.

Design Parameters

The principal design parameters for constructed wetland systems include hydraulic detention time, basin depth, basin geometry (width and length), BOD_5 loading rate, and hydraulic-loading rate. Typical ranges suggested for design are given in Table 13-20.

Hydraulic Detention Time. For FWS systems designed to achieve BOD removal, required detention time may be estimated using the following first-order removal model [19]:

$$\frac{C_e}{C_o} = A \, \exp\left(-0.7K_T(A_v)^{1.75}t\right) \tag{13-14}$$

where C_e = effluent BOD_5 concentration, mg/L
C_o = influent BOD_5 concentration, mg/L
A = empirically determined coefficient representing the fraction of BOD_5 not removed by settling at the head of the system
0.7 = empirical constant
K_T = temperature-dependent first-order rate constant, d^{-1}
A_v = specific surface area for microbiological activity, ft^2/ft^3
t = hydraulic detention-time, d

TABLE 13-20
Design guidelines for constructed wetlands[a]

Design parameter	Unit	Type of system	
		FWS	SFS
Hydraulic detention time	d	4–15	4–15
Water depth	ft	0.3–2.0	1.0–2.5
BOD_5 loading rate	lb/acre · d	< 60	< 60
Hydraulic-loading rate	Mgal/acre · d	0.015–0.050	0.015–0.050
Specific area	acre/(Mgal/d)	67–20	67–20

[a] Adapted from Ref. 42.

Note: ft × 0.3048 = m
lb/acre · d × 1.1209 = kg/ha · d
Mgal/acre · d × 0.9354 = m^3/m^2 · d
acre/(Mgal/d) × 0.1069 = ha/($10^3 m^3$/d)

Hydraulic detention time is a function of the design flowrate and the system geometry expressed by the following equation:

$$t = \frac{LWnd}{Q} \tag{13-15}$$

where L = basin length, ft
   W = basin width, ft
   n = fraction of cross-sectional area not occupied by plants
   d = depth of basin, ft
   Q = average flowrate through system $[(Q \text{ in} + Q \text{ out})/2]$, ft^3/d

The following values have been estimated for the coefficients in Eqs. 13-14 and 13-15, however, caution is advised in using these values for design because of the limited data base used in their development [19]:

$$A = 0.52$$
$$K_T = K_{20} (1.1)^{(T-20)}, \ T \text{ in } °C$$
$$K_{20} = 0.0057 \ d^{-1}$$
$$A_v = 4.8 \ ft^2/ft^3 \ (15.7 \ m^2/m^3)$$
$$n = 0.75$$

A similar model has been suggested for determining the required detention time for SFS wetland systems used for BOD removal [19]:

$$\frac{C_e}{C_o} = \exp(-K_T t') \tag{13-16}$$

In Eq. 13-16, the detention time, t', is defined as the theoretical detention time based on the porosity of the medium or the pore-space detention time:

$$t' = \frac{LW\alpha d}{Q} \tag{13-17}$$

where t' = pore-space detention time, d
   L = basin length, ft
   W = basin width, ft
   α = porosity of basin medium
   d = depth of basin, ft

The actual detention time t, is a function of media hydraulic conductivity and basin length given by the following relationship:

$$t = \frac{L}{k_s S} \tag{13-18}$$

where L = basin length, ft
   k_s = hydraulic conductivity, ft^3/ft$^2 \cdot$ d
   S = slope of basin, ft/ft

Characteristics of media typically used in SFS systems are given in Table 13-21. Caution is advised in using any of the above equations for wetland system design because the equations are derived from the performance of a limited number of systems. Design values for all parameters should be checked against the recommended range of design values given in Table 13-20, and pilot studies are recommended for large projects.

Nitrogen removal in wetland systems is directly related to detention time, but removal rates generally cannot be predicted by the first-order models used to predict BOD removal. Other factors, such as forms of nitrogen present, C:N ratio, system geometry, and vegetation patterns also strongly affect the removal of nitrogen. Design detention times required for nitrogen removal in wetland systems are currently based on pilot study data or on the performance of existing systems with similar wastewater characteristics and site conditions. Typical TKN removal versus detention time data for an alternating cattail/open water/gravel system are shown in Fig. 13-22. For FWS systems, it appears that a configuration of alternating vegetated and open-water zones may provide the proper combination of environmental conditions necessary to optimize nitrogen removal. Maintaining such a configuration will require periodic (at least annual) harvesting of vegetation that develops in the open-water zones.

Water Depth. For FWS systems, the design water depth depends on the optimum depth for the selected vegetation. In cold climates, the operating depth is normally increased in the winter to allow for ice formation on the surface and to provide the increased detention time required at colder temperatures. Systems should be designed with an outlet structure that allows for varied operating depths. The system in Listowel, Ontario, is operated at a depth of 4 in (0.1 m) in the summer and 12 in (0.3 m) in the winter.

The design depth of SFS systems is controlled by the depth of penetration of the plant rhizomes and roots because the plants supply oxygen to the water through the rhizome/root system.

Basin Area and Geometry. The basin geometry will depend on whether the system is FWS or SFS. Considerations for these two systems are discussed below.

TABLE 13-21
Typical media characteristics for subsurface flow systems[a]

Media type	Maximum 10% grain size, mm	Porosity, α	Hydraulic conductivity, k_s, ft^3/ft$^2 \cdot$ d	K_{20}
Medium sand	1	0.42	1,380	1.84
Coarse sand	2	0.39	1,575	1.35
Gravelly sand	8	0.35	1,640	0.86

[a] From Ref. 42.

Note: ft^3/ft$^2 \cdot$ d × 0.3048 = m^3/m$^2 \cdot$ d

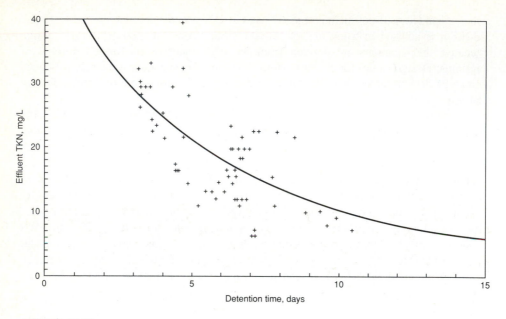

FIGURE 13-22
Typical nitrogen removal performance from constructed wetlands system: effluent TKN versus retention time of an alternating Typha/open-water/gravel system [42].

FWS systems. For FWS systems, the surface area $(L \times W)$ is set by the design detention time and depth according to Eq. 13-15. Guidelines for optimum length to width ratios have not been established firmly, although one study has reported superior performance with long, narrow basins and recommends a length to width ratio of at least 10:1. Use of long, narrow basins reduces the potential for short-circuiting but results in a concentration of the loading at the basin inlet, which can lead to overloaded conditions at the inlet if loading criteria are exceeded (see "BOD Loading Rate"). A serpentine wraparound channel arrangement with step feed (similar to Fig. 13-28d, discussed in Sec. 13-7) can be used to avoid inlet overloading. Typically, existing systems have a total width approximately equal to the length of the basins. The width of the system is divided into multiple (at least two) parallel basins separated by berms to provide better hydraulic control and operating flexibility. With multiple basins, a portion of the system may be taken out of service for vegetation management or basin renovation.

SFS systems. The cross-sectional area (A_c) of subsurface systems is established by the required hydraulic capacity according to the following equation:

$$A_c = \frac{Q}{k_s S} \tag{13-19}$$

where Q, k_s, and S are as defined above. The flow velocity defined by $(k_s\,S)$ should be limited to a value of 22 ft/d (6.8 m/d) to minimize localized shearing of bacterial films [19]. The required width of the system is a function of the cross-sectional area and design depth and is calculated using the following equation:

$$W = \frac{A_c}{d} \qquad (13\text{-}20)$$

The required length may then be calculated using Eq. 13-17. Typically, the length of SFS systems will be substantially less than the width (see Example 13-6).

BOD₅ Loading Rate. As with overland-flow systems, BOD₅ loading must be limited such that the oxygen demand of the the the applied wastewater does not exceed the oxygen-transfer capacity of the wetlands vegetation. Care must be exercised in using area loading criteria (mass/area · time) because the actual load is not applied uniformly but is concentrated at the inlets, whereas oxygen is supplied uniformly over the surface. Estimated oxygen-transfer rates for emergent plants range from 45 to 400 lb/acre · d (5 to 45 g/m² · d) with an average value of 180 lb/ac · d (20 g/m² · d) considered typical for most systems [42]. This oxygen-transfer rate can be compared with an oxygen-transfer rate of 256 lb/acre · d (28.5 g/m² · d) estimated for trickling filters [23]. Oxygen is transferred through the exposed leaves and stems to the rhizomes and roots. For SFS systems in which roots are in contact with the flowing water column, the oxygen transferred to the root system will be available to attached organisms that degrade the soluble BOD in the water column.

The oxygen requirement must be determined on the basis of ultimate oxygen demand. Based on a BOD$_u$:BOD₅ ratio of 1:5, the maximum BOD₅ loading rate for SFS systems should be limited to 120 lb/acre · d (133 kg/ha · d). An upper limit of 100 lb/acre · d (110 kg/ha · d) is typically recommended [43]. Because the BOD load is concentrated at the inlet of the system, it is further recommended that the design ultimate BOD loading rate should not exceed one-half the oxygen-transfer rate [19,42]. Based on this criterion and a BOD$_u$:BOD₅ ratio of 1:5, the maximum BOD₅ loading rate should be limited to 60 lb/acre · d (66.5 kg/ha · d). For systems treating wastewaters with a significant fraction of settleable organic solids, the loading must be even less or distributed along the length of the basin by step feeding to avoid anaerobic conditions at the head of the basins.

For FWS systems, oxygen supply to the water column is limited, as compared to SFS systems, because the root zone is in the soil profile below the water column and any oxygen transported to the root zone will likely be consumed by the large benthic oxygen demand that normally exists in wetlands. Furthermore, oxygen transfer through the water surface by wind-induced reaeration and photosynthesis is minimized when dense vegetation is present. Thus, fully vegetated FWS systems are suitable only for moderate BOD loading rates. In the absence of specific recommendations in current literature, design loadings for such systems should not exceed the value of 60 lb/acre · d (66.5 kg/ha · d) recommended for SFS systems. Successful treatment of oxidation pond effluent using a fully vegetated system has been reported at BOD₅ loading rates up to

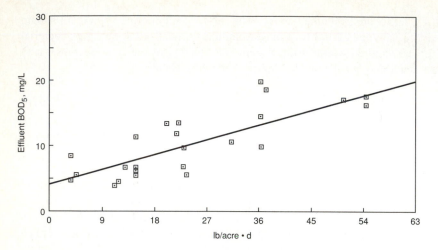

FIGURE 13-23
Typical BOD removal performance from a fully vegetated free water surface (FWS) wetlands system [6].

54 lb/acre · d (60 kg/ha · d), as shown in Fig. 13-23 [6]. Increased oxygen transfer on a systemwide basis can likely be achieved by using alternating vegetated and open-water cells as suggested previously for improving nitrogen removal.

Hydraulic-Loading Rate. The hydraulic-loading rate, L_w, for wetland systems is not usually a primary design parameter, but it is a convenient parameter to use in making comparisons between different systems. Hydraulic-loading rates used in practice range from 15,000 to 55,000 gal/acre · d (150 to 500 m³/ha · d) [42]. The reciprocal of the hydraulic-loading rate, the specific area requirement (A_{sp}), is also used to compare system designs and to make quick preliminary determinations of land area requirements. Specific area requirements used in practice range from about 20 to 65 acres/Mgal · d (2.1 to 6.9 ha/10^3 m³ · d). For wetlands designed to polish secondary or advanced treated effluent and to provide wildlife and aquatic habitat in the central coastal valleys of California, a specific area of 20 acres/Mgal · d (2.1 ha/(10^3 m³ · d)) has been found to provide optimum benefits.

Example 13-6 Determination of basin design for SFS wetlands system. Design an SFS wetlands system using the following information:

1. Influent BOD = 130 mg/L
2. Effluent BOD = 20 mg/L
3. Q = 0.25 Mgal/d = 33,400 ft³/d
4. Vegetation type = cattails
5. Minimum water temperature = 6°C
6. Basin media = coarse sand
7. Basin slope = 0.01

Solution

1. Select basin depth for use with cattail, using 12 in (0.3 m).

2. Select values for α, k_s, and K_{20} from Table 13-21 for coarse sand.

$$\alpha = 0.39$$

$$k_s = 1575 \text{ ft}^3/\text{ft}^2 \cdot \text{d}$$

$$K_{20} = 1.35$$

3. Determine the value of K_T at 6°C.

$$K_T = 1.35(1.1)^{(6-20)}$$

$$K_T = 0.36 \text{ d}^{-1}$$

4. Determine pore-space detention time (t') using Eq. 13-16 rearranged as follows.

$$t' = \frac{-\ln C_e/C_o}{K_T}$$

$$t' = \frac{-\ln 20/130}{0.36}$$

$$t' = 5.2 \text{ d}$$

5. Determine cross-sectional area (A_c) using Eq. 13-19.

$$A_c = \frac{Q}{k_s S}$$

$$A_c = \frac{33,400 \text{ ft}^3/\text{d}}{1575 \text{ ft}^3/\text{ft}^2 \cdot \text{d}(0.01)}$$

$$A_c = 2121 \text{ ft}^2$$

6. Determine basin width (W) using Eq. 13-20.

$$W = \frac{A_c}{d}$$

$$W = \frac{2121}{1.0}$$

$$W = 2121 \text{ ft}$$

7. Determine basin length (L) using Eq. 13-17.

$$L = \frac{t'Q}{Wd\alpha}$$

$$L = \frac{(5.2 \text{ d})(33,400 \text{ ft}^3/\text{d})}{(2121 \text{ ft})(1.0 \text{ ft})(0.39)}$$

$$L = 210 \text{ ft}$$

8. Determine required surface area A_s.

$$A_s = L \times W$$

$$A_s = \frac{(210 \text{ ft})(2,121 \text{ ft})}{43,560 \text{ ft}^2/\text{acre}}$$

$$A_s = 10.2 \text{ acres } (4.1 \text{ ha})$$

9. Check hydraulic-loading rate or specific area requirement.

$$L_w = \frac{Q}{LW}$$

$$L_w = \frac{250,000 \text{ gal/d}}{10.2 \text{ acre}} = 24,510 \text{ gal/d} \cdot \text{acre}$$

$$L_w = 24,510 \text{ gal/d} \cdot \text{acre} \qquad \text{OK} \qquad 16,000 < L_w < 54,000$$

or

$$A_{sp} = \frac{1}{L_w}$$

$$A_{sp} = \frac{1}{0.245 \text{ Mgal/acre}}$$

$$A_{sp} = \frac{40.8 \text{ acres}}{\text{Mgal/d}} \qquad \text{OK} \qquad 20 < A_{sp} < 65$$

10. Check BOD_5 loading rate.

$$LBOD_5 = (0.25 \text{ Mgal/d})(130 \text{ mg/L})[8.34 \text{ lb/Mgal} \cdot (\text{mg/L})]$$

$$LBOD_5 = 271 \text{ lb/d}$$

$$LBOD_5 = 26.6 \text{ lb } BOD_5/\text{acre} \cdot \text{d}, \qquad \text{OK} \qquad LBOD_5 < 60$$

Vector Control

Wetlands, particularly FWS systems, provide ideal breeding habitat for mosquitoes. The issue of vector control may be the critical factor in determining the feasibility of using a constructed wetlands system. Plans for biological control of mosquitoes through the use of mosquito fish (*Gambusia afinis*) plus application of chemical control agents as necessary must be included in the design. Dissolved-oxygen levels above 1 mg/L are necessary to maintain fish populations. Thinning of vegetation may also be necessary to eliminate pockets of water that are inaccessible to fish. Mosquito breeding should not be a problem in SFS systems, provided the system is designed to prevent mosquito access to the subsurface water zone. The surface is normally covered with pea gravel or coarse sand to achieve this purpose.

13-7 FLOATING AQUATIC PLANT TREATMENT SYSTEMS

The steps involved in the design of treatment systems employing floating aquatic plants are essentially the same as those described for constructed wetland systems (see Table 13-6). The principal differences in the design are the type of vegetation used and the physical requirements associated with the plants. Further details on the design of aquatic systems may be found in Refs. 4, 19, 22, 29, 31, 42, and 43.

Site Evaluation and Selection

Site characteristics that must be considered in aquatic system design include topography, soil characteristics, flood hazard, and climate.

Topography. Level to slightly sloping, uniform topography is preferred for the construction of aquatic treatment systems. Although basins and channels may be constructed on steeper-sloping or uneven sites, the amount of earthwork required will affect the cost of the system.

Soil Characteristics. Sites with slowly permeable < 0.2 in/h (< 5 mm/h) surface soils or subsurface layers are most desirable for floating aquatic plant systems because the objective of wetland systems is to treat the wastewater in ponds or basins. Thus, percolation losses through the soil profile should be minimized. As with other pond systems, the pond bottoms will seal with time due to deposition of colloidal and suspended solids and growth of bacterial slimes. Sites with rapidly-permeable soils may be used by constructing basins with clay or artificial liners.

Climate. Because of their sensitivity to cold temperatures, the use of water hyacinths is restricted to the southern portions of California, Arizona, Texas, Mississippi, Alabama, and Georgia, and all of Florida. Water temperatures as low as 10°C can be tolerated if the air temperature does not drop below 5 to 10°C. Duckweed and pennywort are less sensitive to cold temperature and can be applied seasonally throughout most of the United States. and year-round in the southern tier of states. Duckweed can be grown at water temperatures as low as 7°C [19,42,43]. Combined systems of several aquatic plants (e.g., duckweed, pennywort, and water hyacinth) may be suitable for locations with greater climatic variations.

Preapplication Treatment

The minimum level of preapplication treatment should be primary treatment, short detention time aerated ponds, or the equivalent. Preapplication treatment using a rotary disk screen in place of primary sedimentation has also proven to be effective (see Fig. 13-24). Treatment beyond primary depends on the effluent requirements. Aquatic treatment systems have been used in several locations to polish effluent from existing secondary treatment facilities to meet more stringent regulatory requirements. Use of oxidation ponds or lagoons in which high concentrations of algae are generated should be avoided prior to aquatic treatment because, like overland flow, algae removal is inconsistent. When there are effluent limitations on phosphorus, it should be removed in the preapplication treatment step because phosphorus removal in aquatic treatment systems is minimal.

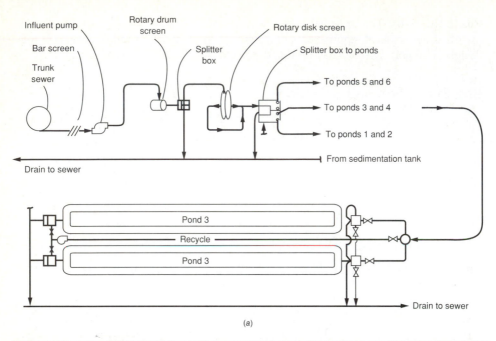

FIGURE 13-24
Aquatic treatment system employing water hyacinth, San Diego, CA: (*a*) schematic and (*b*) preapplication treatment is provided by a rotary drum screen and a rotary disk screen [31].

Plant Selection

The principal floating aquatic plants used in aquatic treatment systems are water hyacinth, duckweed, and pennywort (see Fig. 13-5). These plants are described in greater detail in the following discussion.

Water Hyacinths. Water hyacinth (*Eichhornia crassipes*) is a perennial, freshwater aquatic vascular plant with rounded, upright, shiny green leaves and spikes of lavender flowers. The petioles of the plant are spongy with many air spaces and contribute to the buoyancy of the hyacinth plant. When grown in wastewater, individual plants range from 20 to 48 in (0.5 to 1.2 m) from the top of the flower to the root tips [19]. The plants spread laterally until the water surface is covered, and then the vertical growth increases. The water hyacinth is capable of rapid growth and is ranked eighth among the world's top 10 weeds in growth rate [20]. It reproduces primarily by vegetative propagation, but seeds may be a major source of reinfestation once the parent plants have been removed. The growth of water hyacinth is influenced by (1) efficiency of the plant to use solar energy, (2) nutrient composition of the water, (3) cultural methods, and (4) environmental factors [20].

Plant growth is described in two ways: (1) as the percentage of pond surface covered over a given time period and (2) as the plant density in units of wet plant mass per unit of surface area. Under normal conditions, loosely packed water hyacinths can cover the water surface at relatively low plant densities, about 2 lb/ft^2 (10 kg/m^2) wet weight. Plant densities as high as 16 lb/ft^2 (80 kg/m^2) wet weight can be reached. As in other biological processes, the growth rate of water hyacinths is dependent on temperature. Both air and water temperatures are important in assessing plant vitality.

Duckweed. Duckweeds (*Lemna spp., Spirodela spp., Wolffia spp.*) are small, green freshwater plants with fronds from one to a few millimeters in width. *Lemna* and *Spirodel* have a short root, usually less than $\frac{1}{2}$ in (12 mm) in length. Duckweeds are the smallest and the simplest of the flowering plants and have one of the fastest reproduction rates. A small cell in the frond divides and produces a new frond; each frond is capable of producing at least 10 to 20 times during its life cycle [19,42]. *Lemna spp.* grown in wastewater effluent (at 27°C) doubles in frond numbers, and therefore in area covered, every four days. It is estimated that duckweed can grow 30 percent faster than water hyacinths. The plant is essentially all metabolically active cells with very little structural fiber [19,42].

Small floating plants, particularly duckweed, are sensitive to wind and may be blown in drifts to the leeward side of the pond unless baffles are used. Redistribution of the plants requires manual labor. If drifts are not redistributed, decreased treatment efficiency may result due to incomplete coverage of the pond surface. Odors have also developed where the accumulated plants are allowed to remain and undergo anaerobic decomposition.

Pennywort. Pennywort (*Hydrocotyle umbellata, H. ranunculoides, H. spp.*) is generally a rooted plant. However, under high-nutrient conditions, it may form hydroponic rafts that extend across water bodies. Pennywort tends to intertwine and grows horizontally; at high densities, the plants tend to grow vertically. Unlike water hyacinth, the photosynthetic leaf area of pennywort is small, and, at dense plant stands, yields are significantly reduced as a result of self shading [21,42]. Pennywort exhibits mean growth rates greater than 0.002 lb/ft^2 · d (0.010 kg/m^2 · d) in central Florida [21]. Although rates of nitrogen and phosphorus uptake by water hyacinth drop sharply during the winter, nutrient uptake by pennywort is approximately the same during both warm and cool seasons. Nitrogen and phosphorus uptake during the winter months is greater for pennywort than for water hyacinth. Although annual biomass yields of pennywort are lower than water hyacinth, it is a cool season plant that can be integrated into water hyacinth/water lettuce biomass production systems [21].

Types of Floating Aquatic Plant Treatment Systems

The principal types of floating aquatic plant treatment systems used for wastewater treatment are those employing water hyacinth and duckweed.

Water Hyacinth Systems. Water hyacinth systems represent the majority of aquatic plant systems that have been constructed. Three types of hyacinth systems can be described based on the level of dissolved oxygen and the method of aerating the pond: (1) aerobic nonaerated, (2) aerobic aerated, and (3) facultative anaerobic.

A *nonaerated* aerobic hyacinth system will produce secondary treatment or nutrient (nitrogen) removal depending on the organic-loading rate. This type of system is the most common of the hyacinth systems now in use. The advantages of this type of system include excellent performance with few mosquitoes or odors.

For plant locations in which no mosquitoes or odors can be tolerated, an *aerated* aerobic hyacinth system is required. The added advantages of such a system are that with aeration, higher organic-loading rates are possible, and reduced land area is required (see Fig. 13-25).

The third configuration for a hyacinth system is known as a *facultative* anaerobic hyacinth system. These systems are operated at very high organic-loading rates. Odors and increased mosquito populations are the principal disadvantages of this type of system. Facultative anaerobic hyacinth systems are seldom used because of these problems.

Duckweed Systems. Duckweed and pennywort have been used primarily to improve the effluent quality from facultative lagoons or stabilization ponds by reducing the algae concentration. Conventional lagoon design may be followed for this application (see Chap. 10), except for the need to control the effects of wind. Without controls, duckweed will be blown to the downwind side of the pond, resulting in exposure of large surface areas and defeating the purpose of the duckweed cover. As noted previously, accumulations of decomposing plants can also result in the production of odors. Floating baffles can be used to form cells of limited size to minimize the amount of open surface area exposed to wind action (see Fig. 13-26).

FIGURE 13-25
Views of water hyacinth treatment systems shown schematically in Fig. 13-24a, San Diego, CA.

Design Parameters

The principal design parameters for aquatic treatment systems include hydraulic detention time, water depth, pond geometry, organic-loading rate, and hydraulic-loading
rate. Treatment process kinetics are also considered in the following discussion.
Typical design guidelines for water hyacinth and duckweed systems are summarized in
Table 13-22 for different levels of preapplication treatment. The control of mosquitoes
and plant harvesting and processing are considered subsequently.

Hydraulic Detention Time. Hydraulic detention time depends on the organic-
loading rate, the hydraulic-loading rate, and the depth of the system. In most cases
the organic-loading rate is the controlling factor.

(a)   (b)

FIGURE 13-26
Typical duckweed treatment systems: (a) pond system with floating dividers and (b) duckweed
harvester (from the Lemna Corp.).

TABLE 13-22

Typical design criteria and expected effluent quality from floating aquatic plant treatment systems[a]

Item	Type of water hyacinth treatment system			
	Secondary aerobic (nonaerated)	Secondary aerobic (aerated)	Nutrient removal aerobic (nonaerated)	Duckweed treatment system
Typical design criteria				
Influent wastewater	Screened or settled	Screened or settled	Secondary	Facultative pond effluent
Influent BOD_5, mg/L	130–180	130–180	30	40
BOD_5 loading, lb/acre · d	40–80	150–300	10–40	20–30
Water depth, ft	1.5–3	3–4	2–3	4–6
Detention time, d	10–36	4–8	6–18	20–25
Hydraulic-loading rate, Mgal/ac · d	0.02–0.06	0.10–0.30	0.04–0.16	0.06–0.09
Water temperature, °C	> 10	> 10	> 10	> 7
Harvest schedule	Annually to seasonally	Twice monthly to continuously	Twice monthly to continuously	Monthly for secondary treatment, weekly for nutrient removal
Expected effluent quality				
BOD_5, mg/L	< 20	< 15	< 10	< 30(< 10)[b]
SS, mg/L	< 20	< 15	< 10	< 30(< 10)
TN, mg/L	< 15	< 15	< 5	< 15(< 5)
TP, mg/L	< 6	< 1–2	< 2–4	< 6(< 1–2)

[a] Adapted from Ref. 4.

[b] Values in parentheses are for nutrient removal.

Note: lb/acre · d × 1.1209 = kg/ha · d

ft × 0.3048 = m

Mgal/ac · d × 0.9354 = m³/m² · d

Water Depth. The critical concern with respect to water depth is to control the vertical mixing in the pond so that the wastewater to be treated will come into contact with the plant roots where the bacteria that accomplish the treatment are located (see Fig. 13-27). Typical operating depths for the various types of water hyacinth systems are reported in Table 13-22. A greater depth is sometimes recommended for the final cell in a series of hyacinth ponds because the plant roots will increase in length with decreasing nutrient concentrations. To accommodate variable operating conditions, hyacinth systems should be designed with an outlet structure that allows the operating depth to be varied.

Pond Configuration. Typical pond configurations used for water hyacinth systems are shown in Fig. 13-28. Most of the early hyacinth systems involved rectangular basins operated in series similar to stabilization ponds (see Fig. 13-28a, b). Recycle

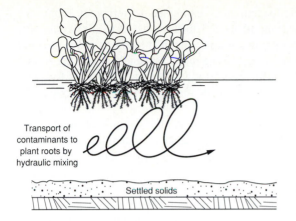

Transport of contaminants to plant roots by hydraulic mixing

Settled solids

FIGURE 13-27
Definition sketch for the transport of the wastewater to be treated to the root zone of the water hyacinth plants.

and step feed (see Fig. 13-28b, c, d) are employed to (1) reduce the concentration of the organic constituent at the plant root zone, (2) improve the transport of wastewater to the root zone, and (3) reduce the formation of odors. The use of a wraparound design (see Fig. 13-28d) shortens the required length of the step feed and recycle lines and reduces recycle pumping costs.

Duckweed systems should be designed as conventional stabilization ponds except for the need to control the effects of wind. As noted previously, floating baffles are used to minimize the amount of surface area exposed to direct wind action. Without this control, duckweed will be blown by the wind and treatment efficiencies cannot be achieved.

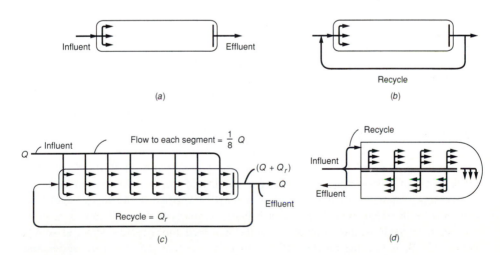

FIGURE 13-28
Alternative flow diagrams for water hyacinth ponds: (a) plug-flow, (b) plug-flow with recycle, (c) step-feed with recycle, and (d) step-feed with recycle in wraparound pond [30,31].

Organic-Loading Rate. Organic-loading rates expressed in terms of BOD_5 for water hyacinth systems can range from 10 to 275 lb/acre · d (10 to 300 kg/ha · d). Without supplemental aeration, odor problems are common at loadings above or about 150 lb/acre · d. Odors can develop at lower loading rates, especially where the sulfate concentration in the wastewater is greater than 50 mg/L. Average loadings for plant systems without aeration should not exceed 90 to 100 lb/acre · d (100 to 110 kg/ha · d).

Hydraulic-Loading Rate. Hydraulic-loading rate is the volume of wastewater applied per day divided by the surface area of the aquatic system. The hydraulic-loading rates applied to water hyacinth facilities have varied from 25,000 to 375,000 gal/acre · d (240 to 3600 m³/ha · d) when treating domestic wastewaters [19,42]. For secondary treatment objectives (BOD_5 and SS ≤ 30 mg/L), the hydraulic-loading rate is typically between 20,000 and 65,000 gal/acre · d (200 and 600 m³/ha · d). For secondary treatment with supplemental aeration, hydraulic-loading rates of 107,000 gal/acre · d (1000 m³/ha · d) have been used successfully. However, organic-loading rates will generally control hydraulic loading [19,42].

Process Kinetics. Based on the results of studies conducted at San Diego, California, and at other locations, it has been found that the BOD_5 removal for a modified step-feed system, such as shown in Fig. 13-28d, and in pond systems where the length to width ratios are not great, can be modeled using first-order kinetics, and the flow regime can be approximated by one or more complete-mix reactors, as shown in Fig. 13-29 [30]. For example, the steady-state materials balance for the first complete-mix reactor in the series of eight reactors, as shown in Fig. 13-29, is given by

$$\text{accumulation} = \text{inflow} - \text{outflow} + \text{generation}$$

$$O = Q_r(C_8) + 0.125Q(C_o) + (Q_r + 0.125Q)(C_1) + k_T(C_1)V_1$$

where
Q_r = recycle flow, Mgal/d
C_8 = concentration of BOD_5 in effluent from reactor 8 in series, mg/L
$0.125Q$ = inflow to each individual cell ($Q/8$), Mgal/d
C_o = concentration of BOD_5 in influent, mg/L
C_1 = concentration BOD_5 in effluent from reactor 1 in series, mg/L
k_T = first-order reaction-rate constant at temperature T, d^{-1}
V_1 = volume of first reactor in series, Mgal

The estimated value of k_T to be used in the above expression for BOD_5 removal is on the order of 1.95 d^{-1} at 20°C [30]. The validity of the modeling approach using one or more complete-mix reactors in series must be verified by testing to determine if the reactor or a segment of the reactor behaves similarly to a complete-mix reactor.

Perhaps the most important aspect of the step-feed and recycle system, as shown in Fig. 13-29, is that the recycle ratio is 16:1 for the first reactor in the series and 23:1 for the last reactor in the series. If the recycle flow had been mixed directly with the influent before being applied to the pond, the recycle ratio would have been 2:1. The difference between these two modes of operation is significant with respect to the performance of the pond [30].

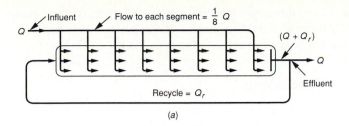

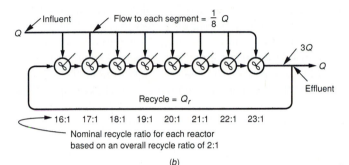

FIGURE 13-29
Definition sketch for the analysis of a water hyacinth pond system with step-feed and recycle:
(a) schematic of actual pond with step-feed and recycle and (b) equivalent system, comprised of a
series of complete-mix reactors, used for process analysis [30].

Mosquitoes and Their Control

In many parts of the United States, the growth of mosquitoes in aquatic treatment
systems may be the critical factor in determining whether or not the use of such
systems will be allowed. The objective of mosquito control is to suppress the mosquito
population below the threshold level required for disease transmission or the nuisance
tolerance level. Strategies that can be used to control mosquito populations include
[30]

1. Stocking ponds with mosquito fish (*Gambusia spp.*)
2. More effective pretreatment to reduce the total organic loading on the aquatic
   system to help maintain aerobic conditions
3. Step feed of influent waste stream with recycle (see Fig. 13-28*d*)
4. More frequent plant harvesting
5. Water spraying in the evening hours
6. Application of chemical control agents (*larvicides*)
7. Diffusion of oxygen (with aeration equipment)
8. Biological control agents (e.g., *BT/israelensis*)

   Fish used for control of mosquitoes (typically *Gambusia spp.*) will die under
the anaerobic conditions that exist in organically overloaded ponds. In addition to

inhibited fish populations, mosquitoes may develop in dense hyacinth systems when plants have been allowed to grow tightly together. As the plants bridge together, pockets of water form that are accessible to the mosquitoes but not the fish.

Plant Harvesting and Processing

The need for plant harvesting depends on the water quality objectives, the growth rates of the plants, and the effects of predators such as weevils. Harvesting of aquatic plants is needed to maintain a crop with high metabolic uptake of nutrients (see Fig. 13-30). For example, frequent harvesting of hyacinths (every three to four weeks) is practiced to achieve nutrient removal. Significant phosphorus removal is achieved only with frequent harvesting. In areas where weevils pose a threat to healthy hyacinth populations, selective harvesting is often used to keep the plants from being infected. Duckweed harvesting for nutrient removal may be required as often as once per week during warm periods.

Harvested water hyacinth plants are typically dried and landfilled or spread on land and tilled into the soil. Water hyacinth can also be composted readily. However, if the plants are not first partially dried or squeezed, the high moisture content tends to reduce the effectiveness of the compost process and results in the production of a

FIGURE 13-30
Harvesting of water hyacinth using a truck equipped with articulated pickup boom. Person in the background is redistributing the water hyacinths in the pond after harvesting.

liquid stream that must be disposed of. Ground duckweed can be used as animal feed without air drying. Continued work on the disposal of harvested water hyacinths and other plants is needed to make the use of aquatic plant systems more feasible.

DISCUSSION TOPICS AND PROBLEMS

13-1. A type 1 slow-rate system designed for an application rate of 2.5 in/wk is to be used at a flowrate of 0.7 Mgal/d (30 L/s). For a year-round operation, what is the required field area? If the system is designed for 36 weeks/yr of application, what is the required field area?

13-2. A sprinkler system is selected for application of the wastewater for a slow-rate system. The sprinklers are spaced in a rectangular grid pattern of 40 ft by 60 ft (12 m by 20 m), and each sprinkler nozzle discharges 30 gal/min. What is the application rate in in/h? Assuming a unit efficiency, E_u, of 80 percent, how many hours must the system be operated in a single area each week to satisfy the application rate of 2.5 in/week?

13-3. A rapid infiltration system is designed for an application rate of 65 ft/yr. The system is operated throughout the year on a cycle of 1 d of application followed by 7 d of drying. If the wastewater has a BOD of 60 mg/L, what is the average annual BOD loading rate in lb/acre? Over the 8 d cycle, what is the average BOD loading rate in lb/acre · d? For the first day of application, what is the loading rate in lb/acre · d?

13-4. Develop a water balance for an overland-flow system. Use the evapotranspiration and precipitation data from Example 13-1. Use a wastewater application rate of 32 in/mo, and assume a percolation rate of 10 percent of the application rate.

13-5. Using the ET data in Example 13-1, determine the allowable monthly nitrogen loading for a slow-rate system. Assume that the crop takes up 270 lb/acre of nitrogen over the year. Distribute the uptake over 12 months in proportion to the percentage of total evapotranspiration occurring in each month. Assume that the applied wastewater contains 20 mg/L of total nitrogen, that the loss to denitrification accounts for 25 percent of the applied nitrogen, and that the allowable concentration of nitrogen in the percolate is 10 mg/L.

13-6. A rapid infiltration system is designed to treat 1.37 Mgal/d of primary effluent at an annual hydraulic-loading rate of 100 ft/yr. What is the required field area? On the basis of a review of current literature, what should the soil permeability be to ensure a successful operation hydraulically? What would be the expected nitrogen removal?

13-7. An overland-flow system is loaded at 8 in/week with a total nitrogen concentration in the applied wastewater of 25 mg/L. If the expected removal of nitrogen on a mass basis is 90 percent, estimate the amount of nitrogen that is removed each year. If the grass is coastal Bermuda grass, what percentage of this nitrogen removal can be accounted for in crop uptake?

13-8. Review at least four current articles on wetlands application and wastewater treatment with aquatic systems. What are the advantages and disadvantages of such treatment systems? What are the advantages and disadvantages of using water hyacinths for the partial treatment of wastewater? Cite the references reviewed.

13-9. Given the following wastewater characteristics and effluent requirements, (a) develop the overland-flow design criteria listed below, and (b) using Fig. 13-19, check the design application rate and the slope length. Use gated-pipe distribution. State all assumptions.

Wastewater characteristics and effluent requirements:

a. Primary effluent flowrate = 1.0 Mgal/d
b. Influent BOD = 120 mg/L
c. Effluent BOD = 20 mg/L
d. Effluent SS = 30 mg/L

Required design criteria:

a. Application rate
b. Slope length
c. Application period
d. Drying period
e. Slope area
f. Total land area

13-10. Using the data and results from Example 13-3, determine the final design storage volume by accounting for the net gain or loss in storage volume due to precipitation in and evaporation and seepage from the reservoir. State all assumptions.

13-11. Using Eqs. 13-12 and 13-3, demonstrate that the minimum slope area required for an overland-flow system occurs when the design application rate is maximized within guideline limits.

REFERENCES

1. Ayers, R. S., and D. W. Westcot: "Water Quality for Agriculture," Food and Agriculture Organization of the United Nations, Irrigation Drainage Paper no. 29, Rome, 1976.

2. Bouwer, H.: "Infiltration-Percolation Systems in Land Application of Wastewater," *Proceedings of a Research Symposium sponsored by the U.S. Environmental Protection Agency, Region III,* Newark, DE, pp. 85–92, November 1974.

3. Campbell, M. D. and J. H. Lehr: *Water Well Technology,* McGraw-Hill, New York, 1973.

4. DeBusk, W. F. and K. R. Reedy: "Wastewater Treatment Using Floating Aquatic Macrophytes: Contaminant Removal Processes and Management Strategies," in K. R. Reedy and W, H. Smith (eds.), *Aquatic Plants for Water Treatment and Resource Recovery,* Magnolia Publishing, pp. 27–48, 1987.

5. de Figueredo, R. F., R. G. Smith, and E. D. Schroeder: "Rainfall and Overland Flow Performance," *Journal of Environmental Engineering Division,* American Society of Civil Engineers, 110:678, 1984.

6. Gearheart, R.: "Constructed Free Surface Wetlands to Treat and Receive Wastewater–Pilot Project to Full Scale," presented at International Conference on Constructed Wetlands for Wastewater Treatment, Chatanooga, TN, 1988.

7. Jensen, M. E. (ed.): *Design and Operation of Farm Irrigation Systems,* American Society of Agricultural Engineers, Monograph no. 3, St. Joseph, MI, 1980.

8. Jewell, W. J., and B. L. Seabrook: *A History of Land Application as a Treatment Alternative,* EPA 430/9-79-012, U.S. EPA, Washington, DC, 1979.

9. Johnston, J., R. G. Smith, and E. D. Schroeder: "Operating Schedule Effects on Nitrogen Removal in Overland Flow Wastewater Treatment Systems," presented at the 61st Annual Water Pollution Control Federation Conference, Dallas, TX, 1988.

10. Hammer, D. A. (ed.): *Constructed Wetlands for Wastewater Treatment: Municipal, Industrial and Agricultural,* Lewis Publishers, Chelsea, MI, 1989.

11. Lance, J. C., F. D. Whisler, and H. Bouwer: "Oxygen Utilization in Soils Flooded with Sewage Water," *Journal of Environ. Qual.,* vol. 2, no. 3, 1973.

12. Mass, E. V.: "Salt Tolerance in Plants," in B. R. Christie (ed.), *The Handbook of Plant Science in Agriculture,* CRC Press, Boca Raton, FL (in press).

13. Metcalf & Eddy, Inc.: *Wastewater Engineering: Collection and Pumping of Wastewater,* McGraw-Hill, New York, 1981.

14. Pair, C. H. (ed.): *Sprinkler Irrigation,* 5th ed., Irrigation Association, Silver Spring, MD, 1983.

15. Perry, L. E., E. J. Reap, and M. Gilliand: "Evaluation of the Overland Flow Process for the Treatment of High-strength Food Processing Wastewaters," *Proceedings of the 14th Mid-Atlantic Industrial Waste Conference,* University of Maryland, June 1982.

16. Pettygrove, G. S., and T. Asano (eds.): *Irrigation with Reclaimed Wastewater—A Guidance Manual,* Lewis Publishers, Chelsea, MI, 1985.

17. Reed S. C.: *Health Aspects of Land Treatment,* GPO 1979-657-093/7086, U.S. EPA, Cincinnati, OH, 1979.

18. Reed, S. C., and R. W. Crites: *Handbook of Land Treatment Systems for Industrial and Municipal Wastes,* Noyes Publications, Park Ridge, NJ, 1984.

19. Reed, S. C., E. J. Middlebrooks, and R. W. Crites: *Natural Systems for Waste Management and Treatment,* McGraw-Hill, New York, 1988.

20. Reedy, K. R. and D. L. Sutton: "Water Hyacinths for Water Quality Improvement and Biomass Production," *Journal of Environmental Quality,* 14:459-462, 1984.

21. Reedy, K. R. and W. F. DeBusk: "Growth Characteristics of Aquatic Macrophytes Cultured in Nutrient Enriched Water: I. Water Hyacinth, Water Lettuce, and Pennywort," *Economic Bot.,* 38:225–235, 1984.

22. Reedy, K. R., and W. H. Smith (eds.): *Aquatic Plants for Water Treatment and Resource Recovery,* Magnolia Publishing, pp. 27–48, 1987.

23. Schroeder, E. D.: *Water and Wastewater Treatment,* McGraw-Hill, New York, 1977.

24. Sepp, E.: *The Use of Sewage for Irrigation—A Literature Review,* Bureau of Sanitary Engineering, California State Department of Public Health, Berkeley, CA, 1971.

25. Sorber, C. A., et al.: "A Study of Bacterial Aerosols at a Wastewater Irrigation Site," *Journal WPCF,* vol. 48, no. 10, 1976.

26. Smith, R. G., G. Hayashi, and R. F. de Figueredo: "Seasonal Denitrification of Secondary Effluent," presented at the 61st Annual Water Pollution Control Federation Conference, Dallas, TX, 1988.

27. Smith, R. G. and E. D. Schroeder: "Physical Design of Overland Flow Systems," *Journal WPCF,* vol. 55, no. 3, 1983.

28. Smith, R. G. and E. D. Schroeder: "Field Studies of the Overland Flow Process for the Treatment of Raw and Primary Treated Municipal Wastewater," *Journal WPCF,* vol. 57, no. 7, 1985.

29. Stowell, R., R. Ludwig, J. Colt, and G. Tchobanoglous: "Concepts in Aquatic Treatment System Design," *Journal of Environmental Engineering Division, Proceedings ASCE,* vol. 107, no. EE5, October 1981.

30. Tchobanoglous, G., F. Maitski, K. Thomson, and T. H. Chadwick: "Evolution and Performance of City of San Diego Pilot Scale Aquatic Wastewater Treatment System Using Water Hyacinths," *Journal WPCF,* vol. 61, no. 11/12, 1989.

31. Tchobanoglous, G.: "Aquatic Plant Systems for Wastewater Treatment: Engineering Considerations," in K. R. Reedy and W. H. Smith (eds.), *Aquatic Plants for Water Treatment and Resource Recovery,* Magnolia Publishing, pp. 27–48, 1987.

32. Uiga, A., and R. S. Shedden: "An Overview of Land Treatment from Case Studies of Existing Systems," presented at the 49th Annual Water Pollution Control Federation Conference, Minneapolis, MN, October 1976.

33. U.S. Department of Agriculture, Soil Conservation Service: "Border Irrigation," Chapter 4, Section 15, in *Irrigation, SCS National Engineering Handbook,* U.S. Government Printing Office, Washington, DC, 1974.

34. U.S. Department of Agriculture, Soil Conservation Service: *Agricultural Waste Management Field Manual,U.S. Government Printing Office,* Washington, DC, 1975.

35. U.S. Department of Agriculture, Soil Conservation Service: "Sprinkler Irrigation," Chapter 11, Section 15 in *Irrigation, SCS National Engineering Handbook,* U.S. Government Printing Office, Washington, DC, 1983.

36. U.S. Department of Agriculture, Soil Conservation Service: "Trickle Irrigation," Chapter 7, Section 15 in *Irrigation, SCS National Engineering Handbook,* U.S. Government Printing Office, Washington, DC, 1986.

37. U.S. Department of Agriculture, Soil Conservation Service: "Furrow Irrigation," Chapter 5, Section 15 in *Irrigation, SCS National Engineering Handbook,* U.S. Government Printing Office, Washington, DC (in press).

38. U.S. Department of the Interior, Bureau of Reclamation: *Drainage Manual,* U.S. Government Printing Office, Washington, DC, 1978.

39. U.S. Environmental Protection Agency: *National Interim Primary Drinking Water Regulations,* EPA 570/9-76-003, Washington, DC, 1976.

40. U.S. Environmental Protection Agency: *Process Design Manual for Land Treatment of Municipal Wastewater,* EPA 625/1-81-013, Cincinnati, OH, October 1981.

41. U.S. Environmental Protection Agency: *Process Design Manual for Land Treatment of Municipal Wastewater: Supplement on Rapid Infiltration and Overland Flow,* EPA 625/1-81-013a, Cincinnati, OH, October 1984.

42. U.S. Environmental Protection Agency: *Design Manual for Constructed Wetlands and Floating Aquatic Plant Systems for Municipal Wastewater Treatment,* EPA 625/1-88-022, Cincinnati, OH, September 1988.

43. Water Pollution Control Federation: *Natural Systems for Wastewater Treatment,* Manual of Practice FD-16, Alexandria, VA, February 1990.

44. Witherow, J. L. and B. E. Bledsoe: "Algae Removal by the Overland Flow Process," *Journal WPCF,* vol. 55, no. 10, 1983.

SMALL WASTEWATER TREATMENT SYSTEMS

A small community is defined, for the purpose of this book, as one with a population of 1000 or less. As noted in Chap. 1, communities of 10,000 or less account for 77 percent of the treatment facilities but only 8 percent of the treatment capacity. Communities with populations of 1000 or less account for about 32 percent of the treatment plants, but only 0.7 percent of the total treatment capacity. Small communities, by their very geography and development, have a number of problems that make the provision of both water and wastewater a difficult undertaking. It is the purpose of this chapter to consider the nature of the problems faced by small communities and to consider designs of wastewater management facilities that are suitable for individual residences, clusters of homes, and small communities.

14-1 SPECIAL PROBLEMS FACED BY SMALL COMMUNITIES

Because of their size, small communities are faced with a variety of problems that make the construction and operation of community-wide managed wastewater facilities a difficult undertaking. The principal problems are related to (1) stringent discharge requirements, (2) high per capita costs, (3) limited finances, and (4) limited operation and maintenance budgets.

Stringent Discharge Requirements

To protect the environment, discharge requirements for treated wastewater are the same for both large and small communities. As a consequence, small communities must provide the same degree of treatment that is now provided by large communities. The challenge is to be able to provide this level of treatment subject to the following economic constraints.

High Per Capita Costs

Because of their size, small communities do not benefit from the economies of scale that are possible with the construction of wastewater management facilities for larger communities. In fact, the return-to-scale curve actually flattens out below some minimum size [20]. As a result, conventional wastewater management facilities for small communities often cost significantly more per capita to construct when compared to those for larger communities. The fact that the population of small communities tends to be spread out also contributes to increased per capita costs. The provision of utilities can cost from two to four times as much per capita in a community of 1000 persons as compared to a community of 100,000 persons.

Limited Finances

In general, small communities have difficulty in financing wastewater management facilities for one or more of the following reasons [5].

1. *Lower household incomes*. In general, incomes of nonmetropolitan households are less than metropolitan households. The percentage of poverty-level households is much higher in nonmetropolitan areas. On the other hand, there are many high income small communities, especially near large cities.
2. *Residential tax base*. Homeowners bear the major brunt of taxes in small communities where there is a smaller commercial and industrial tax base.
3. *Financing*. Small communities have great difficulty entering the bond market. Those that have a bond rating are usually rated low. About 54 percent of small communities have a "C" rating, compared to 6.7 percent of larger communities. Further, small communities are likely to pay a higher interest rate for the same rated bond because of smaller issues.
4. *Impact of recession*. Small communities, because of their general reliance on a small number of major employers, are often harder hit by a poor economic climate.

Limited Operation
and Maintenance Capabilities

In many cases, small communities have limited economic resources and expertise to manage wastewater treatment facilities. Problems are often experienced in design, contracting, inadequate construction supervision, project management, billing, accounting, budgeting, operations, and maintenance [15]. Overcoming these problems makes the implementation of treatment facilities a major undertaking. For example, the salary required for a treatment plant operator may exceed the salary of the mayor or city administrator. In unincorporated areas, such a social problem may be insurmountable. Thus, effective low-maintenance solutions must be developed to provide wastewater treatment for small communities. Methods and techniques that have proven to be successful are considered in this chapter.

14-2 SMALL SYSTEM FLOWRATES AND WASTEWATER CHARACTERISTICS

Small system flowrates and wastewater characteristics differ significantly from those of large systems. Thus, knowledge of the expected wastewater flowrates and characteristics is essential for the effective design of wastewater management facilities for individual residences as well as for clusters of homes and small communities.

Wastewater Flowrates

Per capita flowrates and variations must be considered in the design of both individual systems and systems designed to serve a cluster of homes and small communities.

Per Capita Flowrates. In Chap. 5, average wastewater flowrates, typically measured at treatment facilities, were reported to vary from about 80 to 120 gal/capita · d (300 to 450 L/capita · d). These typical values are higher than would be expected from an individual residence primarily due to contributions from commercial and industrial establishments and infiltration/inflow. Typical per capita flowrates to be expected from various residential units are presented in Table 14-1. Although a range of values

TABLE 14-1
Typical wastewater flowrates from residential dwellings[a,b]

Type of dwelling	Flowrate, gal/capita · d	
	Range	Typical
Single family		
Summer	35–50	42
Low income	40–55	45
Median income	40–80	55
Luxury homes	50–100	65
Apartments	35–50	40
Condominiums	35–50	40

[a] Flow discharged from the residence excluding any extraneous flow contributions.

[b] The average flowrate per capita is based on an average occupancy of about 2.4 to 2.8 residents per home. Flow rates from residences with fewer occupants can be estimated using Eq. 14-1.

Note: gal × 3.7854 = L

is shown, a typical per capita value for residences in unsewered areas is about 55 gal/capita·d (210 L/capita·d) based on an average occupancy of about 2.4 to 2.8 residents per home.

An alternative method that can be used to estimate the flow from individual residences is based on allocating the total water use between household and personal uses. Assuming that the household use consists of 10 gal for dishwashing, 25 gal for laundry and 5 gal for miscellaneous use and that personal use consists of 3 gal for drinking and cooking, 2 gal for oral hygiene, 18 gal for bathing, and 17 gal for toilet flushing, the flow from a residence would be:

Flow, gal/residence·d =
$$40 \text{ gal/residence·d} + 40 \text{ gal/resident·d} \times (\text{Number of residents/home}) \quad (14\text{-}1)$$

Applying Eq. 14-1 to a residence with 2.6 residents results in an average flow per resident of 55 gal. This value correlates well with the values given in Table 14-1. Equation 14-1 can be revised to account for other household uses and the use of low-flush toilets and fixtures. For example, if 1.5 gal/flush toilets are used, the corresponding average flow for a residence with 2.6 occupants, based on five flushes per resident per day, would be 46 gal/capita·d.

Flowrate Variations. The flowrate variations that can be expected from an individual residence are quite variable, ranging from no flow in the early morning hours to peak hourly flowrates as high as 8 to 1 compared to the average daily flowrate. While the flowrate variation from an individual home is quite variable and unpredictable, the flowrate variation for 50 or more homes is quite similar to that given in Fig. 5-2 in Chap. 5. Typical peaking factors for individual residences, small commercial establishments, and small communities are reported in Table 14-2. The peaking factors for individual residences and small commercial establishments are, as shown, considerably greater than those for small communities. Peaking factors are of importance in the design of wastewater management facilities, especially for sizing grease traps for small commercial establishments and secondary settling tanks in package or built-in-place treatment plants.

TABLE 14-2
Peaking factors for wastewater flows from individual residences, small commercial establishments, and small communities[a]

Peaking factor	Individual residence		Small commercial establishment		Small community	
	Range	Typical	Range	Typical	Range	Typical
Peak hour	4–8	6	6–10	8	3–6	4.7
Peak day	2–6	4	4–8	6	2–5	3.6
Peak week	1.25–4	2.0	2–6	3	1.5–3	1.75
Peak month	1.2–3	1.75	1.5–4	2	1.2–2	1.5

[a] The reported peaking factors are exclusive of extreme flow events (i.e., values greater than the 99 percentile value).

Wastewater Characteristics

Typical data on the quantities of feces and urine discharged per person per day are reported in Table 14-3. Using the data reported in Table 5-4 in Chap. 5 and the flowrate data given in Table 14-1, the typical characteristics of the wastewater from individual residences are presented in Table 14-4. It is interesting to note that the values given for the low end of the range [based on a flow of 100 gal/capita · d (380 L/capita · d)] correspond quite closely to those given in Table 3-16 in Chap. 3 for wastewater of medium strength.

14-3 TYPES OF SMALL WASTEWATER MANAGEMENT SYSTEMS

Small wastewater management systems vary in size from systems designed to serve individual residences with a flow of 50 to 500 gal/d (190 to 1900 L/d) to systems designed for wastewater flows of up to 0.1 Mgal/d (380 m³/d). Two types of small systems are considered in this chapter: (1) those systems for individual residences and other community facilities in unsewered areas and (2) those systems for clusters of homes and small communities that are to be sewered or are already sewered. These systems are introduced in this section and are considered in greater detail in the following sections.

Wastewater Management Options for Unsewered Areas

Wastewater from individual dwellings and other community facilities in unsewered locations is usually managed by onsite treatment and disposal systems. Alternative wastewater management options for unsewered areas are reported in Table 14-5. Although a variety of onsite systems have been used, the most common system consists of a septic tank for the partial treatment of the wastewater and a subsurface soil disposal field for final treatment and disposal of the septic tank effluent (see Fig. 14-1). Because conventional disposal fields cannot be used in some locations, many alternative systems have been developed [22]. The most successful of these include

TABLE 14-3
Typical data on the daily quantities of human excrement

Item	Unit	Value Range	Value Typical
Feces	lb/capita · d	0.22–0.30	0.26
	(g/capita · d)	(100–140)	(120)
Urine	gal/capita · d	0.2–0.35	0.3
	(L/capita · d)	(0.8–1.3)	(1.1)

Note: lb × 4.5359 = kg
gal × 3.7854 = L

TABLE 14-4
Typical data on the unit loading factors and expected wastewater constituent concentrations from individual residences

Item	Unit loading factor,[a] lb/capita · d	Unit	Value Range[b]	Typical[c]
BOD_5[d]	0.180	mg/L	216–540	392
SS[d]	0.200	mg/L	240–600	436
NH_3 as N	0.007	mg/L	7–20	14
Org. N as N	0.020	mg/L	24–60	43
TKN as N	0.027	mg/L	31–80	57
Org P as P	0.003	mg/L	4–10	7
Inorg. P as P	0.006	mg/L	6–17	12
Grease		mg/L	45–100	70
Total coliform		Number/100mL	10^7–10^{10}	10^8
Temperature		°F	59–79	70
pH		unitless	5–8	7.2[e]

[a] Data from Table 5-4.

[b] Range of values for constituent concentrations based on 100 and 40 gal/capita · d (380 and 150 L/capita · d).

[c] Based on 55 gal/capita · d (210 L/capita · d).

[d] Values without contribution from ground kitchen wastes. The corresponding values if ground kitchen wastes are included are BOD = 0.22 lb/capita · d and SS = 0.26 lb/capita · d. The values for the nutrients remain about the same.

[e] Median value (average pH value has no meaning).

Note: lb/capita · d $\times$ 0.4538 = kg/capita · d
0.555(°F − 32) = °C

intermittent and recirculating granular-medium filters. Intermittent sand filters have become quite popular in many parts of the country for single family residences because of their excellent performance, reliability, and relatively lower cost. Recirculating granular-medium filters are used for larger flows. Complete recycle systems have been developed for commercial buildings. Holding tanks are used where an acceptable onsite disposal system cannot be installed. All of the above units are described in detail and their application is considered in the Section 14-4.

Wastewater Management Options for Sewered Areas

Often, because the individual lots are too small to accommodate individual onsite systems or the soils and underlying strata are unsuitable, small cluster or community systems are installed. These systems typically consist of (1) a collection system to convey the wastewater away from each residence or establishment, (2) some form of treatment, and (3) an effluent disposal system. The principal wastewater management options available for clusters of homes and small communities are reported in Table 14-6. With the exception of the collection facilities described briefly in Section 14-6,

TABLE 14-5
Wastewater management options for unsewered areas

Source of wastewater[a]	Wastewater treatment and/or containment	Wastewater disposal
Individual residences	Primary treatment	Subsurface disposal
Combined wastewater	Septic tank	Disposal fields
Black water	Imhoff tank	Seepage beds
Gray water	Secondary treatment	Shallow sand-filled disposal trenches
Public facilities	Aerobic/anaerobic unit	Mound systems
Commercial establishments	Aerobic unit	Evapotranspiration/percolation beds
	Intermittent sand filter	Drip application
	Recirculating granular medium filter	Evaporation systems
	Constructed wetlands	Evapotranspiration bed
	Recycle treatment system	Evaporation pond
	Onsite containment	Wetland (marsh)
	Holding tank	Discharge to water bodies
	Privy	Combinations of the above

[a] Many residences, public facilities, and commercial establishments may be equipped with flow reduction devices and appliances.

the other units listed in Table 14-6 have been described elsewhere in this text. The application of these systems is considered in Sections 14-6 through 14-10.

The types of collection systems used include conventional gravity flow sewers, small-diameter variable-slope gravity-flow sewers, small-diameter pressure sewers, and vacuum sewers. The choice of collection system is usually dictated by local topography and cost. The treatment component of cluster and community systems will vary with the size of the installation. Typically, a large septic tank will be used for a cluster of homes. Imhoff tanks, commonly used in the past, are rarely used today because of their relatively high cost. In some communities, septic tanks may be used for the separation of settlable solids and greases and oils. Recirculating granular-medium filters are used in conjunction with septic tanks where a higher level of treatment is required. Pre-engineered and constructed package plants and individually designed plants are used where the flows are larger. Treatment processes and facilities for flows in the range from 0.1 to 1.0 Mgal/d (380 to 3800 m^3/d) that have been described elsewhere in the text are not described again in this chapter. Flow diagrams and design factors for their application in small systems are, however, presented in Section 14-10.

The methods used for effluent disposal will also vary with the size of the system. For small installations serving a cluster of homes, effluent disposal is most commonly

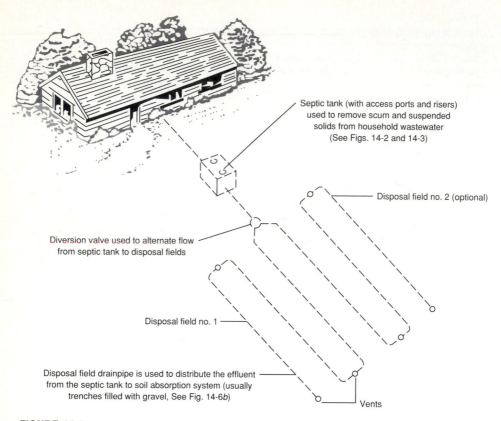

Septic tank (with access ports and risers) used to remove scum and suspended solids from household wastewater (See Figs. 14-2 and 14-3)

Disposal field no. 2 (optional)

Diversion valve used to alternate flow from septic tank to disposal fields

Disposal field no. 1

Disposal field drainpipe is used to distribute the effluent from the septic tank to soil absorption system (usually trenches filled with gravel, See Fig. 14-6*b*)

Vents

FIGURE 14-1
Conventional onsite system consisting of a septic tank and a disposal field served by intermittent gravity flow.

accomplished using disposal fields. As the size of the system increases, the methods used for the disposal of effluent are, as shown in Table 14-6, essentially the same as those used for larger systems as discussed in Chaps. 13, 16, and 17.

14-4   ONSITE SYSTEMS FOR INDIVIDUAL RESIDENCES AND OTHER COMMUNITY FACILITIES IN UNSEWERED AREAS

The purpose of this section is (1) to describe the principal components that are used in onsite systems in unsewered areas and (2) to review the treatment performance of selected onsite systems. The application and design of selected onsite systems is considered in Section 14-5.

Onsite System Components

The principal components of the most common type of onsite wastewater management systems for individual residences and other establishments, as cited above, include:

TABLE 14-6
Wastewater management options for clusters of homes and small community systems

Source of wastewater[a]	Wastewater collection	Wastewater treatment	Wastewater disposal
Individual residences	Conventional gravity flow sewers	Primary treatment	Subsurface soil absorbtion systems
Public facilities		Large septic tank	
Commercial establishments	Small-diameter variable-slope gravity-flow sewers (with septic tanks	Imhoff tank and variations	Drip application
			Surface water discharge
		Secondary treatment	
	Pressure sewers	Aerobic/anaerobic unit	Constructed wetlands
	With septic tanks		Spray irrigation
	Without septic tanks	Activated-sludge system(s)	Reuse
	Vacuum sewers	Sequencing batch reactor	Combinations of the above
		Aerated lagoons	
		Recirculating granular-medium filter	
		Oxidation ditch	
		Oxidation ponds	
		Land treatment	
		Constructed wetlands	
		Trickling filter	

[a] Many residences, public facilities, and commercial establishments may be equipped with flow reduction devices and appliances.

septic tanks, grease interceptor tanks, Imhoff tanks, disposal fields, disposal beds and pits, intermittent sand filters, recirculating granular-medium filters, shallow-trench sand-filled pressure-dosed disposal fields, mound systems, complete recycle units, and graywater systems. Each of these components is considered separately in the following discussions. Typical design criteria for the components considered in the following discussion are presented in Sections 14-5 and 14-7.

Septic Tank. Septic tanks, as shown schematically in Fig. 14-2 and photographically in Fig. 14-3, are prefabricated tanks that serve as a combined settling and skimming tank and as an unheated-unmixed anaerobic digester. The antecedents of the septic tank can be traced back to about 1860 with the early work of Mouras in France [12]. Today, most septic tanks are made of concrete or fiberglass, although other materials such as steel, redwood, and polyethylene have been used. The use of steel and redwood tanks is no longer accepted by most regulatory agencies. Thick-wall polyethylene and fiberglass tanks have been used successfully. Regardless of the material of construction, a septic tank must be water-tight and structurally sound if it is to function properly. Each tank should be tested for water-tightness and structural

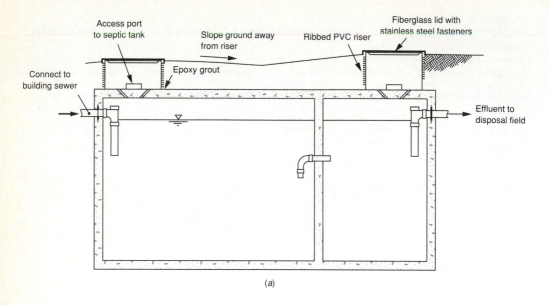

(a)

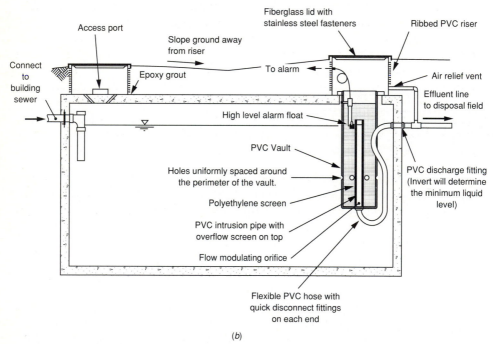

(b)

FIGURE 14-2

Typical septic tanks: (a) conventional two-compartment tank and (b) a single-compartment tank equipped with filter vault (from Orenco Systems, Inc.).

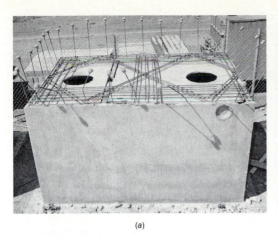

(a)   (b)

FIGURE 14-3
Typical septic tanks: (a) concrete type under construction; note ample use of reinforcing steel (from Pacific Concrete Products) and (b) fiberglass type.

integrity by completely filling the tank with water before and after installation. Water-tight tanks are a necessity for most alternative collection systems.

Sometimes an interior baffle is used to divide the tank, and access ports are provided to permit inspection and cleaning (see Fig. 14-2a). Two compartments have been used to limit the discharge of solids in the effluent from the septic tank. Based on measurements made in both single and double compartments, the benefit of a two-compartment tank appears to depend more on the design of the tank than the use of two compartments. A more effective way to eliminate the discharge of untreated solids involves the use of an effluent filter vault in conjunction with a single compartment tank (see Fig. 14-2b). Operationally, effluent flows into the vault through the inlet holes located in the center of the vault chamber. Before passing into the center of the vault, the effluent must pass through a screen which is located on the inside of the vault. Because of the large surface area of the filter screen, clogging is not excessively rapid. If needed, the screen can be removed and cleaned. It should be noted that the effluent filter vault functions, in effect, as a second chamber. An advantage of the effluent filter vault is that it can be installed in both existing and new septic tanks to limit the discharge of gross untreated solids. In other designs, gas deflection baffles and inclined tubes have been used to limit the discharge of solids (see Fig. 14-4).

Settleable solids in the incoming wastewater settle and form a sludge layer at the bottom of the tank. Greases and other light materials float to the surface where a scum layer is formed as floating materials accumulate. Settled and skimmed wastewater flows from the clear space between the scum and sludge layers to the disposal field or to a treatment unit if one is used. The organic material retained in the bottom of the tank undergoes facultative and anaerobic decomposition and is converted to more stable compounds and gases such as carbon dioxide (CO_2), methane (CH_4), and hydrogen sulfide (H_2S). Although hydrogen sulfide is produced in septic tanks,

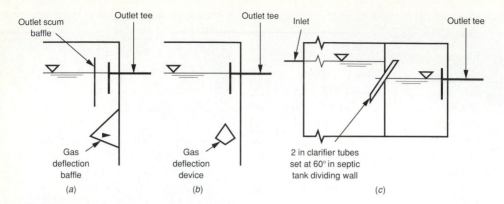

FIGURE 14-4
Typical facilities used in septic tanks to limit the discharge of suspended solids [17,23]: (*a, b*) internal gas deflection baffles and (*c*) inclined clarifier tubes.

odors are not usually a problem because the hydrogen sulfide combines with the metals in the accumulated solids to form insoluble metallic sulfides. Even though the volume of the solid material being deposited is being reduced continually by anaerobic decomposition, there is always a net accumulation of sludge in the tank. Material from the bottom of the tank that is buoyed up by the decomposition gases will often stick to the bottom of the scum layer increasing its thickness. Because the long-term accumulation of scum and sludge can reduce the effective volumetric capacity of the tank, the contents of the tanks should be pumped periodically.

Grease and Oil Interceptor Tanks. Wastewaters from restaurants, laundromats, and service stations typically contain grease, oils, and detergents. If the greases and oils are allowed to enter the septic tank, there is the possibility that they can be discharged along with the septic tank effluent to the soil absorption system. Greases and oils, along with suspended solids, tend to accumulate on the surfaces of the soil absorption system ultimately leading to a reduction in the infiltration capacity. Greases and oils are especially troublesome because of their persistence. Typically, interceptor tanks are used to trap grease by cooling and flotation, and oils by flotation. The tank serves as a heat exchanger by cooling the liquid, which helps to solidify the greases. For flotation to be effective, the interceptor tank must detain the fluid for an adequate period of time (typically greater than 30 minutes).

   Although a number of commercial grease and oil traps are available, they have not proven to be effective because of the limited detention time provided in such units. Also, most commercial units are rated on average flow and not the instantaneous peak flows observed in the field from restaurants and laundries (see Table 14-2). The use of conventional septic tanks as interceptor tanks has proven to be very effective. Depending on the tank configuration some replumbing may be necessary when septic tanks are used as grease traps. The larger volume provided by the septic tank has been beneficial in achieving the maximum possible separation of greases and oils. The presence of lint in wastewaters from laundromats is also a serious concern. The discharge of lint can be limited by using a series of replaceable screens in the effluent channel or a replaceable or cleanable screened outlet in the interceptor tank.

Imhoff Tank. The removal of settleable solids and the anaerobic digestion of these solids in an Imhoff tank is similar to a septic tank. The difference is that the Imhoff tank (see Fig. 14-5) consists of a two-story tank in which sedimentation is accomplished in the upper compartment and digestion of the settled solids is accomplished in the lower compartment. As shown in Fig. 14-5, solids pass through an opening in the bottom of the settling chamber into the unheated lower compartment for digestion. Scum accumulates in the sedimentation compartment. Gas produced in the digestion process in the lower compartment escapes through the vents. Because of the overhanging lip in the bottom of the sedimentation chamber, gases and gas-buoyed sludge particles rising from the sludge layer in the bottom of the tank are not released to the sedimentation compartment.

Disposal Field. Final treatment and disposal of the effluent from a septic tank or other treatment unit is accomplished, most commonly, by means of subsurface-soil absorption. Typically a soil absorption system, commonly known as a disposal field (also known as a leachfield), consists of a series of narrow, relatively shallow [2 to 5 ft (0.6 to 1.5 m)] trenches filled with a porous medium (usually gravel, see Fig. 14-6). The porous medium is used (1) to maintain the structure of the disposal field trenches, (2) to provide partial treatment of the effluent, (3) to distribute the effluent to the infiltrative soil surfaces, and (4) if the trenches are not filled with liquid, to provide temporary storage capacity during peak flows [7]. Effluent from the septic tank is applied to the disposal field by intermittent gravity flow or by periodic dosing using a pump or a dosing siphon.

Effluent from the septic tank discharged to the disposal field infiltrates into the soil primarily through the side walls of the trench. Once the effluent has passed through the soil surface, it enters the vadose zone (unsaturated soil zone between the ground surface and the groundwater or bedrock). Flow in the vadose zone depends on the soil and bedrock conditions. Effluent moves over soil particle surfaces and in capillary pores in response to the force of gravity. Treated effluent travels from the vadose zone to the groundwater or to nearby water courses. Groundwater flow can occur either vertically or horizontally depending on the permeability of the soil and bedrock.

The treatment provided by the disposal field occurs (1) as the effluent flows over and through the porous medium used in the disposal field trenches, (2) as it infiltrates into the soil, and (3) as it percolates through the soil. Treatment on the porous medium in the disposal field occurs through a combination of physical, biological, and chemical mechanisms. The porous medium acts as a submerged anaerobic filter under continuous inundation, and as an aerobic trickling filter under periodic application. Treatment at the soil interface is considered in the following discussion.

Intermittent gravity-flow application. When septic tank effluent is applied to the disposal field by intermittent gravity flow, a biomat develops progressively on the infiltrative surfaces of the disposal field (see Fig. 14-7). Because of the

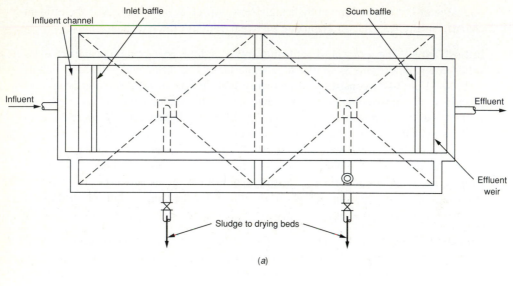

Influent channel
Inlet baffle
Scum baffle
Influent
Effluent
Effluent weir
Sludge to drying beds

(a)

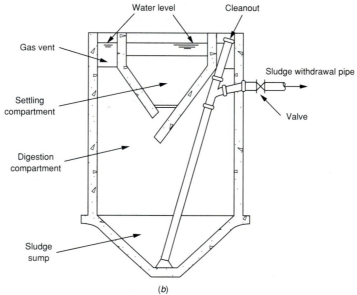

Water level
Cleanout
Gas vent
Sludge withdrawal pipe
Settling compartment
Valve
Digestion compartment
Sludge sump

(b)

FIGURE 14-5
Typical Imhoff tank for a small community: (a) plan view and (b) section through tank.

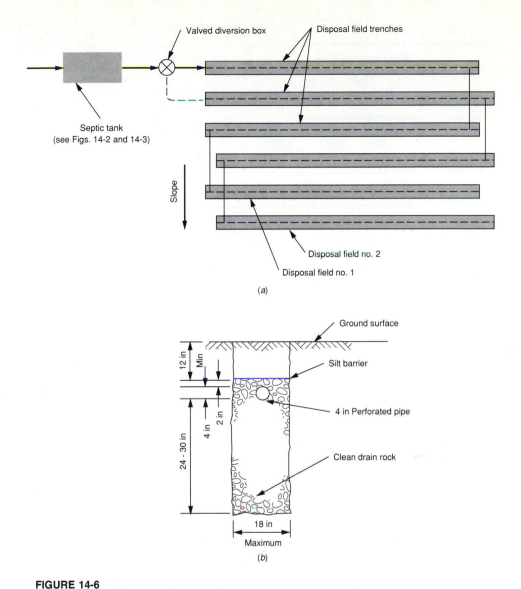

FIGURE 14-6
Typical subsurface soil absorption disposal field: (*a*) valued disposal field system and (*b*) cross section through disposal field trench [26].

relatively high organic and solids loadings, the local environment is usually anaerobic. Particulate materials in the effluent, (e.g., wastewater solids, mineral precipitates, etc.) are strained out on the surface of the soil. A biomat will form at the interface as bacteria and other microorganisms start to colonize and grow on the particulate matter. As the microorganisms metabolize the organic material in the septic-tank effluent, the thickness of the biomat will increase. As biological conversion reactions occur

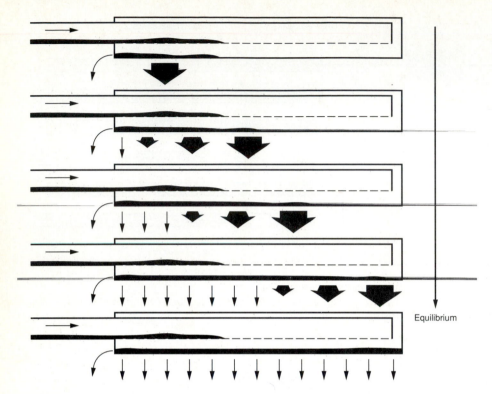

FIGURE 14-7
Definition sketch for the progressive development of a biomat in a disposal field [7].

in the biomat, mineral precipitates such as ferrous sulfide and aluminum, iron, and calcium phosphate will also form [8]. These precipitates may accumulate or leach out of the biomat depending on the environmental conditions (e.g., pH, dissolved oxygen level, etc.). Over a long period of time, the biomat develops a dynamic equilibrium. Effluent solids accumulate and the biomass increases due to growth. Simultaneously, mineralized constituents and particulate material that have been reduced in size are carried away with the percolating water; gases resulting from the biological conversion of the waste are released to the surrounding environment. The biomat, commonly observed in conventional disposal fields, has also been shown to be very effective in the removal of viruses.

In a study conducted in 1955 [13], settled wastewater was spread on five California soils of varying initial permeabilities (about 17-fold). The objective was to determine the factors governing the infiltration and the percolation of wastewater into soil formations and the steady-state infiltration rates that could be expected under continual inundation. The most significant finding from this study, as shown in Fig. 14-8, was that under continual inundation, the long-term infiltration rate was essentially the same for all of the soils. Based on these results, it was concluded that for the soils tested the infiltration capacity of a soil absorption system is controlled primarily

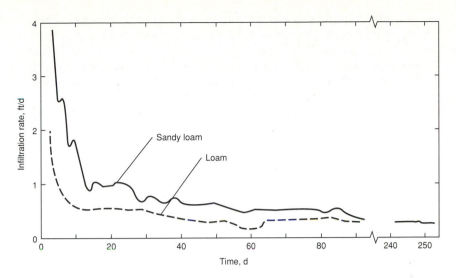

FIGURE 14-8
Effects of continuous application of settled wastewater on infiltration rate in soils of varying initial permeabilities as measured in a soil lysimeter [13].

by the nature of the biomat and not by the permeability of the soil. Another significant finding of this study was that when the soil-wastewater interface was allowed to dry and was exposed to the atmosphere between applications, most of the original infiltrative capacity was restored when wastewater was reapplied. What actually happens is that the biomat which has developed dries out and cracks exposing the soil surface. In soils containing clay, resting is beneficial because the clay particles, which become dispersed under continuous inundation, will reaggregate upon resting and most of the original infiltration capacity will be restored. The restoration of infiltration capacity is also observed in land application systems.

Functionally, the biomat serves as a biological treatment unit and as a mechanical and biological filter. Although the biomat penetrates into the soil surface, the major portion of the biomat is located on the surface of the soil. Because the biomat serves as a mechanical and biological filter, the passage of effluent from the disposal field into the surrounding soil is often controlled by the hydraulic characteristics of the biomat. Exceptions in which the biomat does not control the hydraulic capacity of the disposal field are in those soils composed of very coarse gravel or in soils containing significant amounts of clay. The long-term hydraulic capacity of the biomat is often termed the long-term acceptance rate (LTAR). Typical LTAR values that have been reported are in the range from 0.3 to 0.5 gal/ft$^2 \cdot$d (12 to 20 L/m$^2 \cdot$d) depending on the hydraulic head [8,29,31].

Periodic application (dosing). When the entire disposal field is dosed with septic tank effluent using a pump or dosing siphon, the environment in the disposal field is usually aerobic. Under aerobic conditions, biological treatment of the septic tank effluent occurs more rapidly than under anaerobic conditions. Because the effluent is dispersed over a larger area, the biomat that forms at the gravel-soil interface is not as heavy or uniform as the biomat which forms under intermittent gravity-flow application. Where a continuous biomat does not form (e.g., in very coarse soils), dosing is important because maximum treatment is achieved when the septic-tank effluent flows over the gravel in the disposal field in a thin layer and through the soil vadose zone under unsaturated flow conditions.

While unsaturated flow provides the maximum opportunity for the operative treatment mechanisms to be most effective, it has been observed that even in porous soils, fluid flow tends to occur under conditions of saturated flow in selected flow channels. These selected flow channels are also observed in rapid sand filters. In soils with high permeabilities (e.g., sandy soils), septic-tank effluent should be applied periodically in small doses distributed over the entire disposal field area to achieve effective treatment. Dosing is usually accomplished with a pump or a dosing siphon (see Fig. 14-9). Because the disposal fields do not usually pond when they are dosed, there has been a tendency to reduce the required surface. However, great care should be taken not to reduce the required area to the point where complete saturated flow occurs.

Disposal Beds or Pits. When the bottom width of a disposal field is greater than about 3 to 4 ft it is usually called a diposal bed (also known as a seepage bed). When

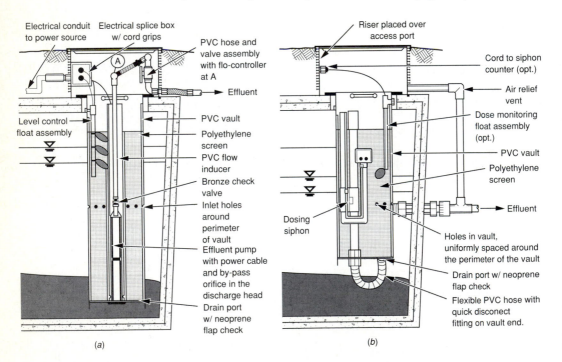

(a)   *(b)*

FIGURE 14-9
Typical facilities used for dosing disposal fields: (a) specially designed pumps and (b) dosing siphon (from Orenco Systems, Inc.).

the geometry is like a vertical cylinder it is known as a disposal pit (also known as a seepage pit) [31]. Disposal beds and pits are still used in some parts of the country. Very deep soil and a great separation from groundwater are needed to use disposal beds or pits. Arid regions of the southwest have areas suitable for disposal beds and pits. In most other locations, the continued use of seepage pits should be discouraged because of the potential to contaminate the underlying groundwater.

Intermittent Sand Filter. Local site conditions that may preclude the use of conventional septic-tank disposal-field systems include shallow soil cover, percolation rates that are considered too slow or too rapid, high groundwater, steepness of slope, and limited area. In locations with limited soil cover and percolation rates that are too rapid, the concern is that partially treated effluent may reach the surface or groundwater. In lots with limited area suitable for disposal fields, the effluent may have to be disposed of after treatment with a sand filter. Drip application at multiple locations has been used with filtered effluent. It should be noted that untreated septic tank effluent is not suitable for drip application because of its tendency to clog the emitters and to produce odor.

Intermittent sand filters are shallow beds of sand [24 to 30 in (600 to 760 mm)] provided with a surface distribution system and an underdrain system (see Fig. 14-10). Septic tank effluent is applied periodically to the surface of the sand bed. The treated liquid is collected in the underdrain system located at the bottom of the filter. The effluent from the filter is commonly discharged to a disposal field or disinfected and discharged to surface waters. Most intermittent sand filters are buried (see Fig. 14-11) although open filters have been used. As the name implies, buried filters are constructed below grade. Open filters are essentially the same as buried filters with the exception that the surface is left open to the atmosphere. Open filters are often provided with a cover for improved maintenance and for increasing the temperature with solar heat in cold climates. It is interesting to note that the intermittent sand filters in use today are essentially the same as those used in 1868 and those used in the 1920s [6].

Treatment of the effluent in an intermittent sand filter is brought about by physical, chemical, and biological transformations. Suspended solids are removed principally by mechanical straining, straining due to chance contact, and sedimentation. Because bacteria colonize within the sand grains, autofiltration caused by the growth of bacteria further enhances the removal of suspended solids. The removal of BOD_5 and the conversion of ammonia to nitrate (nitrification) occurs under aerobic conditions by the microorganisms present in the sand bed. The conversion of nitrate to nitrogen gas (denitrification), routinely occurs resulting in a significant (up to 45 percent) loss of nitrogen. Denitrification is brought about by anaerobic bacteria that coexist in anaerobic microenvironments within the filter bed. Specific constituents are removed by sorption (chemical and physical). To maintain a high performance level, aerobic conditions must be maintained. Intermittent application and venting of the underdrains helps to maintain aerobic conditions within the filter. The design of sand filters is considered in the following section.

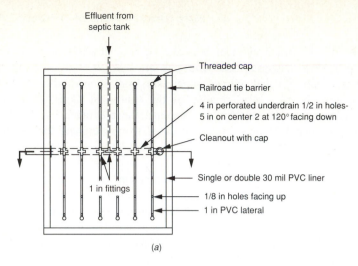

Effluent from
septic tank

Threaded cap

Railroad tie barrier

4 in perforated underdrain 1/2 in holes-
5 in on center 2 at 120° facing down

Cleanout with cap

Single or double 30 mil PVC liner

1 in fittings

1/8 in holes facing up

1 in PVC lateral

(a)

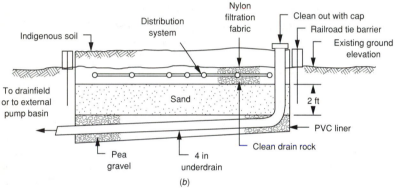

Indigenous soil

Distribution
system

Nylon
filtration
fabric

Clean out with cap

Railroad tie barrier

Existing ground
elevation

To drainfield
or to external
pump basin

Sand

2 ft

PVC liner

Pea
gravel

4 in
underdrain

Clean drain rock

(b)

FIGURE 14-10
Typical schematic for an intermittent sand filter: (a) plan view and (b) section through filter.

FIGURE 14-11
Photographs of an intermittent sand filter and disposal field located under raised planters in front yard of homes in Stinson Beach, CA.

1036

Recirculating Granular-Medium Filter. Functionally, a recirculating granular-medium filter is essentially the same as an intermittent sand filter as described above. The major differences are (1) effluent from a septic tank or another treatment unit is recirculated through the filter as opposed to a single application, (2) the effective filter medium (coarse sand or fine gravel) size is larger, and (3) the loading rate based on the effluent flowrate is greater than that for an intermittent sand filter. Recirculating granular-medium filters, used to provide an improved level of treatment for larger flows such as from an apartment building or small communities, are discussed in greater detail in Section 14-8.

Shallow Sand-Filled Pressure-Dosed Disposal Field. In some locations where the groundwater is high or the underlying strata may not be suitable for a conventional disposal field, shallow sand-filled pressure-dosed disposal fields have been used successfully (see Fig. 14-12). Operationally, the sand-filled pressure-dosed disposal fields function like a combination intermittent sand filter and disposal field. The quality of the effluent after it passes through the sand is very high. Pressure distribution, which serves to distribute the effluent evenly over the sand in the trench is a key factor contributing to the success of this type of disposal field system.

Mound System. The mound system is essentially an intermittent sand filter that is placed above the natural surface of the ground (see Fig. 14-13). Trenches or beds are constructed in sand placed above the natural soil. Septic tank effluent is pumped or dosed through a pressure distribution system placed in a gravel layer. A barrier material (geotextile) and a cap layer are placed above the gravel layer. The mound system is covered with topsoil. Mound systems have been used in locations where: (1) the soils are permeable and the water table is shallow, (2) the underlying strata are highly porous and conventional systems should not be used, (3) slopes are

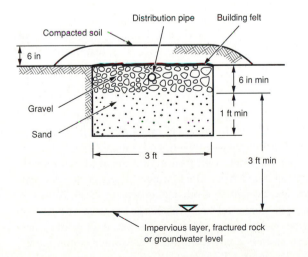

FIGURE 14-12
Typical shallow sand-filled
pressure-disposal field trench.

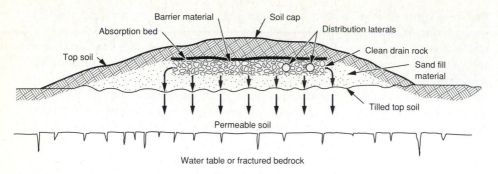

FIGURE 14-13
Typical mound system used for the disposal of septic tank effluent for a site with permeable soil and high groundwater or shallow fractured bedrock (adapted from Ref. 23).

less than 12 percent, and (4) the soils are slowly permeable. While conventional mound systems have been used where the soils are slowly permeable, they have only been partially effective because the applied effluent which accumulates under the mound usually cannot be transported away from under the mound. Many regulatory agencies no longer approve the use of community-sized mound systems because of the high rate of failure and because pressure-dosed sand-filled trenches are far more effective.

Recycle Treatment Systems. Over the past 10 years, a number of self-contained recycle systems have been developed to take sanitary wastewater from buildings, treat it, and return the bulk of the treated effluent for reuse as flushwater in toilets and urinals. One such unit involves three treatment steps: (1) the solids in the wastewater are collected and treated aerobically, (2) the effluent from the biological treatment unit is then passed through a self-cleaning ultrafiltration step where residual organics, microorganisms, and suspended solids are removed, and (3) in the final step the effluent is passed through an activated carbon column for polishing (see Fig. 14-14). The material removed in the ultrafiltration step is returned to the first processing step for further treatment. The effluent from the carbon filters is disinfected with ozone before it is reused for flush water. Although such processes are expensive, they have been used for office buildings located in unsewered areas and where water for domestic use is in short supply.

Gray Water System. Gray water is defined as the water and solids from household fixtures and water-using appliances excluding the water and solids from toilets. It should be noted that the term "black" is often used to describe the water and solids from toilets. Gray water is often separated from black water to reduce the loading on onsite systems. Often, laundry wastewaters and other gray waters are re-routed to wherever the troubled homeowner can more effectively discharge the large volumes of water that cannot be managed in the onsite system. The use of separate gray and black water systems to achieve nitrification and denitrification of wastewater is considered in Section 14-5 under the heading Degree Of Treatment.

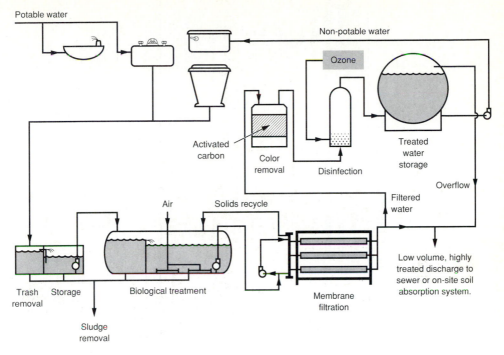

FIGURE 14-14
Typical flow diagram for a complete wastewater recycle system (from Thetford Systems, Inc.).

Treatment Performance of Onsite Systems

One of the important design objectives for individual onsite systems is the effective treatment of the wastewater from individual residences so that it does not cause any nuisance conditions and does not impact any of the beneficial uses of the local groundwater. The principal constituents of concern are BOD_5, SS, nitrogen, phosphorus, bacteria, and viruses. Performance data for various onsite system components are reported in Table 14-7. The constituent concentrations in the effluent from a septic tank are somewhat higher than the corresponding values that would be expected in primary effluent from a wastewater treatment facility because of the lack of dilution from extraneous flows that occurs in most gravity sewers. At a depth of 3 ft (0.9 m) below the bottom of the disposal field trench most of the constituent concentrations are at or below the lower limits of detectability (see Table 14-7, Column 4). Nitrates and phosphorus are exceptions. Priority pollutants and metals, found in septic tank effluent, are also of concern, but the information available on the fate of the constituents in disposal fields is limited.

As reported in columns 5 and 6 of Table 14-7, effluent quality from well-designed intermittent and recirculating filters is excellent. Concentration values for both BOD_5 and SS in the effluent from intermittent sand filters are typically below 10 mg/L and in most cases, below 5 mg/L. The corresponding BOD_5 and SS values for recirculating granular-medium filters are slightly higher. Under normal operating

TABLE 14-7
Treatment performance of onsite system components and intermittent or recirculating sand filters[a]

Parameter	Raw waste (1)	Septic tank effluent (2)	1.0 ft below bottom of leachfield trench (3)	3.0 ft below bottom of leachfield trench (4)	Intermittent sand filter effluent (5)	Recirculating granular-medium filter effluent (6)
BOD_5, mg/L	210–530	140–200	0	0	<10	<15
SS, mg/L	237–600	50–90	0	0	<10	<15
Nitrogen, mg/L						
Total	35–80	25–60	—	—		
NH_4^+	7–40	20–60	20[b]	—	<0.5	<0.5
NO_3^-	<1	<1	40[b]	40[b]	25	25
Total phosphorus, mg/L	10–27	10–30	10[b]	1[b]		
Fecal coliforms, MPN/100 mL	10^6–10^{10}	10^3–10^6	0–10^2	0	10^2–10^4	10^2–10^4
Viruses, PFU/mL[c]	Unknown	10^5–10^7	0–10^3	0		

[a] Adapted in part from Refs 2, 23, 28, and 30.
[b] Value varies from the background level up to the indicated value.
[c] PFU = plaque forming units.
Note: ft × 0.3048 = m

conditions, essentially complete nitrification is achieved as the effluent passes through an intermittent sand filter. In most intermittent sand filters, some denitrification occurs simultaneously as it does in natural soil systems. Typically about 40 to 45 percent of the total nitrogen is lost due to denitrification. Specially designed sand filters can be used to reduce the concentration of nitrates to levels below 10 mg/L [28].

14-5 SELECTION AND DESIGN OF ONSITE SYSTEMS

The purpose of this section is to introduce the reader to the design of individual onsite systems. The subject of septage disposal is also introduced. The formation of onsite wastewater management districts is considered in Section 14-6. The principal considerations in the design of an individual onsite system are related to [8,30]:

1. *Hydraulic assimilation capacity*. Is the proposed site suitable for the disposal of septic tank effluent? Can the expected quantity of effluent be assimilated and transported away from the site, given the conditions of the soil mantle and other local constraints?
2. *Disposal field design*. Is the proposed hydraulic application rate for the disposal field consistent with the characteristics of the wastewater and the properties of the biomat and soil?
3. *Treatment requirements*. Does the proposed onsite system provide enough treatment capacity to protect public health and the environment?

To answer the above questions, the typical design procedure for onsite systems involves (1) a preliminary site assessment, (2) a detailed site evaluation, (3) assessment of the hydraulic assimilative capacity of the site, (4) selection of appropriate onsite systems for evaluation, (5) selection of design criteria for the disposal field, (6) sizing and preliminary layout of the disposal field, and (7) selection of design criteria for physical facilities. These subjects are examined in greater detail in the following paragraphs. Because flow distribution in pressure-dosed disposal fields and intermittent and recirculating granular-medium filters is of paramount importance in the design of these components, this subject is also considered. The designs of an individual onsite system and intermittent sand filter are also illustrated.

Preliminary Site Evaluation

The principal factors that should be considered in the preliminary evaluation of a site for the use of an onsite system are as follows:

1. Geographic features such as gullies, creeks, marshes, etc.
2. Surface slope
3. Flooding potential
4. Existing structures including any water wells
5. Landscaping

The significance of the geographic and topographic features of the site with respect to the use of disposal fields and beds are reviewed in Table 14-8.

Detailed Site Evaluation

The principal factors that should be considered in a detailed evaluation of a site for the use of an onsite system will include: (1) identification of soil characteristics, (2) percolation testing, and (3) hydrogeological characterization. Determination of the acceptance rate and the saturated hydraulic conductivity of the soil mantle, often determined on large projects, is considered following this discussion of site evaluation.

Identification of Site Soil Characteristics. Soil is comprised of solids, water, and air. The typical mineral soil contains about 50 percent solids (by volume) of which about 90 percent is mineral matter and 10 percent is organic matter. The remaining 50 percent consists of variable amounts of water (20 to 30 percent) and air (20 to 30 percent). The success of any waste disposal system requiring land disposal depends on a thorough understanding of the soils. Improper attention to soil conditions has led to failures that have given septic tank systems a reputation of being only a temporary solution for the disposal of wastewater from individual facilities. Properly designed soil absorption systems begin with the soil and work backward to the pretreatment needed.

The properties of the soil that should be considered to assess its hydraulic properties as well as its ability to treat the wastewater include the following:

1. Soil texture
2. Soil structure
3. Soil color
4. Seasonally saturated soils
5. Location of impervious layers
6. Presence of swelling clays
7. Bulk density

Information used for the characterization of soil according to soil texture is summarized in Table 14-9 and in Fig. 14-15. Other soil characteristics (including items 2 through 6 above) are identified in the field on the basis of soil borings or test pits. Test pits, dug with a backhoe (see Fig. 14-16), are favored by many regulatory agencies for characterizing the soils at a site. Unusual conditions such as the presence of impervious layers, soil mottling due to high groundwater, and the presence of swelling clays are noted for consideration in the design of the system [19]. Determination of the coefficient of permeability is discussed in the following section. The significance of the soil characteristics of the site with respect to the use of disposal fields and beds is reviewed in Table 14-8. Additional details on the examination of soils for subsurface wastewater disposal may be found in the excellent booklet prepared by the State of Maine [19].

TABLE 14-8
Typical site criteria for use of disposal fields and beds[a]

Item	Criteria
Landscape position[b]	Level, well-drained areas, crests of slopes, convex slopes more desirable. Avoid depressions, bases of slopes and concave slopes unless suitable surface drainage is provided.
Slope[b]	0 to 25%. Slopes in excess of 25% can be utilized but the use of construction machinery may be limited. Seepage bed systems are limited to 0 to 5%.
Typical horizontal separation distances[c]	
Water supply wells	50–100 ft (15–30 m)
Surface waters, springs	50–100 ft (15–30 m)
Escarpments, manmade cuts	10–20 ft (3–6 m)
Boundary of property	5–10 ft (1.5–3 m)
Building foundations	10–20 ft (3–6 m)
Soil	
Texture	Soils with sandy or loamy textures are best suited. Gravelly and cobbley soils with open pores and slowly permeable clay soils are less desirable.
Structure	Strong granular, blocky or prismatic structures are desirable. Platey or unstructured massive soils should be avoided.
Color	Bright uniform colors indicate well-drained, well aerated soils, Dull, gray or mottled soils indicate continuous or seasonal saturation and are unsuitable.
Layering	Soils exhibiting layers with distinct textural or structural changes should be evaluated carefully to ensure water movement will not be severely restricted.
Unsaturated depth	2 to 4 ft (0.6 to 1.2 m) of unsaturated soil should exist between the bottom of the disposal field and the seasonally high watertable or bedrock.

[a] Adapted from Ref. 23.

[b] Landscape position and slope are more restrictive for seepage beds because of the depths of cut on the upslope side.

[c] Intended only as a guide. Safe distance varies from site to site, based on local codes, topography, soil permeability, groundwater gradients, geology, etc.

Note: ft × 0.3048 = m

TABLE 14-9
Appearance and feeling of various soil textural classes[a]

Soil textural class	Appearance and feeling	
	Dry soil	Moist soil
Sand	Loose, single grains which feel gritty. Squeezed in the hand, the soil mass falls apart when the pressure is released.	Squeezed in the hand it forms a cast which crumbles when lightly touched. Does not form a ribbon between thumb and forefinger.
Loamy sand	Loose, single grains which feel gritty but enough fine particles to stain finger prints in the palm of hand.	Squeezed in the hand, it forms a cast that crumbles when touched and only bears very careful handling.
Sandy loam	Aggregates are easily crushed. Very faint, velvety feeling initially, but as rubbing is continued, the gritty feeling of sand soon dominates.	Forms a cast that bears careful handling without breaking. Doesn't form a ribbon between thumb and forefinger.
Loam	Aggregates are crushed under moderate pressure; clods can be quite firm. When pulverized, loam has a velvety feel that becomes gritty with continued rubbing.	Cast can be handled quite freely without breaking. Slight tendency to ribbon between thumb and forefinger. Rubbed surface is rough.
Silt loam	Aggregates are firm but may be crushed under moderate pressure. Clods are firm to hard. Smooth, flour-like feel dominates when soil is pulverized.	Cast can be freely handled without breaking. Slight tendency to ribbon between thumb and forefinger. Rubbed surface has a broken or rippled appearance.
Silty clay loam	Aggregates are very firm. Clods are hard to very hard.	Cast can be handled very firmly without breaking. Tendency to ribbon between thumb and forefinger with some flaking, greasy feeling, moderately sticky.
Silty clay		Squeezed with proper moisture content into a long ribbon; sticky feel.

[a] From Ref. 19.

Percolation Testing. In many parts of the country, the results of percolation tests are used to determine the required size of the soil absorption system. The allowable hydraulic loading rate for the soil absorption system is determined from a table or curve relating the average percolation rate to the allowable loading rate. Selection of application rates based on the results of percolation tests is considered later in this section.

In the percolation test, test holes varying in diameter from 4 to 12 in (100 to 300 mm) are dug in the general location where the disposal field is to be placed. The bottom of the test hole is placed at the same depth as the proposed bottom of the disposal field or bed. After the formation around the test hole has been soaked

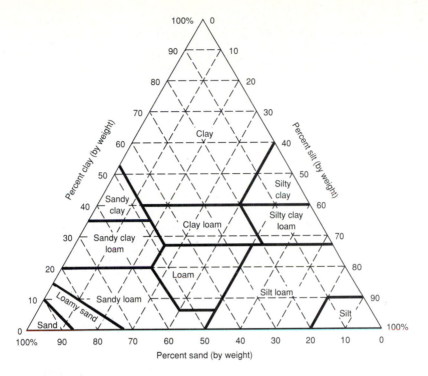

FIGURE 14-15
General soil classification used by the U.S. Department of Agriculture.

(a)   (b)

FIGURE 14-16
Assessing the suitability of a proposed home site for the disposal of septic tank effluent: (a) excavation of test pit with backhoe and (b) examination of exposed soils and soil formations.

for a period of 24 hr, the percolation rate is determined by measuring either (1) the time required for the water surface to drop a specified distance [reported as minutes per inch, min/in (min/100 mm)] or (2) the depth the water surface falls in a specified period of time (see Fig. 14-17). Details for performing the percolation test vary throughout the United States. Additional information on conducting percolation tests and the interpretation of the results may be found in References 3, 8, and 25. Regardless of the test method used, great care must be exercised in both the conduct and interpretation of percolation test results because of the variability that will be observed when multiple tests are conducted. Because of the wide variation in test results that have been observed for a single site, percolation test results can, at best, only be taken as a gross indication of the ability of a soil to accept water.

Hydrogeological Characterization. Important hydrogeological data include the depth to groundwater, the hydraulic gradient, and direction of groundwater flow. Hydrogeological data are used to determine whether the applied effluent can be assimilated and transported from the site without surfacing or developing a groundwater mound, which may also surface when the level of the groundwater becomes high during wet-weather conditions. The slope of the groundwater table is usually determined by measuring the static water level in three wells located in a triangular pattern. The

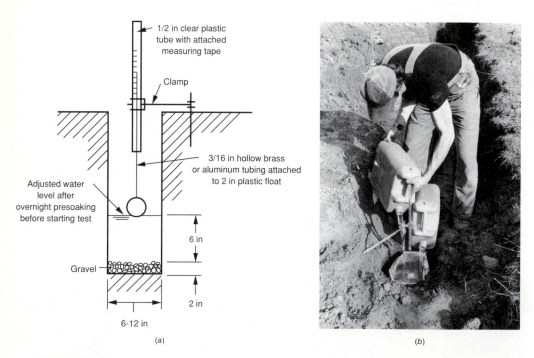

(a) (b)

FIGURE 14-17
Conduct of a percolation test: (a) test and equipment details and (b) field-test setup. Water is being poured into a perforated paper bag to avoid splashing and unnecessary clogging of the sidewall surface area. This test technique was developed by J. T. Winneberger [31].

direction of flow is determined by passing a plane through the three points. If perched water is encountered, additional wells should be drilled. In general, disposal fields should be placed perpendicular to the direction of the groundwater flow to utilize the assimilative capacity of the site effectively.

Analysis of Site Assimilative Capacity

The assimilative capacity of a site is defined as the ability of the soil to accept water. The water may be percolated downward or transported away from the site laterally, taken up by plants, or evapotranspirated as it becomes reintroduced into the natural water cycle [31]. The assimilative capacity of a site that is to be used for the disposal of effluent depends on the permeability of the underlying strata, the location and slope of the groundwater, the slope of the ground surface, and the hydraulic characteristics of the site. Analysis of the assimilation capacity of a site can be done with Darcy's Law and the principles of groundwater flow.

Darcy's Law. In most situations, Darcy's Law can be used to estimate how much water can be transported away from the site. Darcy's Law is

$$v = -ks \tag{14-2}$$

where v = velocity of flow, ft/d (m/d)
k = coefficient of permeability, ft/d (m/d)
s = hydraulic gradient, dh/dL, ft/ft (m/m)

The minus sign in Darcy's equation is used because the headloss dh is negative in the direction of the flow. The coefficient of permeability is also known as the hydraulic conductivity, the effective permeability, or the seepage coefficient. The flowrate is obtained by multiplying the velocity by the cross-sectional area, A, perpendicular to the direction of flow

$$Q = Av = -Aks \tag{14-3}$$

where Q = flowrate, ft^3/d (m^3/d)
A = cross-sectional area, ft^2 (m^2)

In general, the permeability of a soil is influenced by particle size, void ratio, composition, degree of saturation, and temperature. For soil containing clay, the chemical properties of the clay also affect the permeability. From empirical observations, it has been found that the coefficient of permeability can be defined in terms of some characteristic size of the porous medium and the properties of the fluid. The coefficient of permeability can be determined in the laboratory using one of several types of constant or falling head permeameters or in the field using the auger hole method, the two-auger-hole method, the pipe cavity method, shallow well pump-in test, the permeameter method, the pond infiltration method [9], and the shallow trench pump-in test [29]. Typical values for the coefficient of permeability, for the soil types identified in Table 14-9 and in Fig. 14-15, are presented in Table 14-10. As reported in Table 14-10 there is considerable variation in the coefficient of permeability within the various soil classifications. The wide variation in the coefficients of permeability

TABLE 14-10
Approximate coefficients of permeability and the corresponding vertical percolation and assimilation rates for various soils based on soil texture

Soil texture	Approximate coefficient of permeability, K		Vertical percolation rate corresponding to the coefficient of permeability[a]		Vertical assimilation rate corresponding to the coefficient of permeability[a]	
	ft/d	m/d	min/in	min/10²mm	gal/ft² · d	L/m² · d
Gravel, coarse sand	330–3,300	10^2–10^3	0.364–0.0364	1.5–0.15	2,470–24,700	10^5–10^6
Coarse to medium sand	33–330	10^1–10^2	3.64–0.364	15–1.5	247–2,470	10^4–10^5
Fine sand, loamy sand	3.3–33	10^0–10^1	36.4–3.64	150–15	24.7–247	10^3–10^4
Sandy loam, loam	0.33–3.3	10^{-1}–10^0	364–36.4	1,500–150	2.47–24.7	10^2–10^3
Loam, porous silt loam	0.033–0.33	10^{-2}–10^{-0}	3,640–36.4	15,000–150	0.247–24.7	10^1–10^3
Silty clay loam, clay loam	0.0033–0.033	10^{-3}–10^{-1}	36,460–364	15×10^4–15×10^2	0.0247–2.47	10^0–10^2
Clays, colloidal clays	≤0.0033	≤10^{-3}	>36,400	>15×10^4	≤0.0247	≤10^0

[a] Computed values based on approximate coefficients of permeability assuming a unit hydraulic gradient (see Eq. 14-3).

Note: ft × 0.3048 = m

gal/ft² · d × 40.7458 = L/m² · d

is due largely to the presence of varying amounts of fine soil particles within each category. As a consequence, the selection of hydraulic application rates for disposal fields based on soil classification alone has generally proven to be unacceptable.

Determination of Soil Acceptance Rate and Saturated Coefficient of Permeability. A modification of the shallow well pump-in test known as the shallow trench pump-in test can be used to determine the assimilative capacity of a site and the saturated coefficient of permeability [29]. In the trench pump-in method, a shallow trench of adequate length [6 to 10 ft (2 to 3 m)] is dug in the location where the actual disposal field trenches are to be placed (see Fig. 14-18). The bottom and sidewalls of the trench are picked to remove any smearing. A wooden box is placed in the trench and the trench is packed with gravel as shown in Fig. 14-18a. A float is installed in the box to maintain a constant head of water. Water applied to the trench is metered so that an accurate accounting can be made. A series of observation wells is drilled both up- and down-gradient to monitor water levels (see Fig 14-18b). Where the soil is too rocky to drill test wells, long trenches can be excavated to intersect the migrating water.

Once completed, the trench is then filled with water to a given height and the water level is maintained at that height throughout the test period (typically 4 to 8 days). As water is applied to the trench, a portion of the water flows vertically downward into the underlying strata in response to the pull of gravity. Another portion of the applied water moves outwardly in the lateral direction until it can move vertically downward subject to the force of gravity. To determine the acceptance rate of the underlying soil strata, the horizontal extent to which the water spreads under saturated flow conditions is defined. The extent of the spread (or plume) is assumed

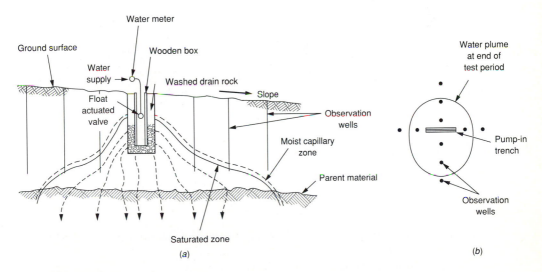

FIGURE 14-18
Field test procedure used to assess the hydraulic capacity of a site: (a) definition sketch for pump-in trench and (b) plan view showing trench and monitoring wells [29].

to correspond to the horizontal cross-sectional area required to transport the applied flow vertically downward under the given hydraulic head in the trench. Often the test is repeated for different heights of water in the trench. Knowing the cross-sectional area through which flow takes place, the total water applied, the water remaining in the trench, in the soil column above the area defined by the plume, and in the capillary fringe, the soil acceptance rate and the saturated coefficient of permeability can be estimated (see Example 14-1).

Example 14-1 Determination of the soil acceptance rate and the saturated coefficient of permeability using data obtained from a trench pump-in test. Using the following data obtained from a trench pump-in test determine the soil acceptance rate and the saturated coefficient of permeability.

1. Length of trench = 10.0 ft (3.0 m), width of trench = 1.5 ft. (0.45 m), depth of trench = 3 ft (0.9 m)

2. The total extent in area of the water plume (see accompanying definition sketch) = 380 ft² (35 m²)

3. Depth of water maintained in the bottom of the trench = 1 ft (0.3 m)

4. Depth below bottom of trench at periphery of plume 2 ft (0.6 m)

5. Total water applied during test = 4500 gal (17,030 L)

6. Height of capillary zone = 12 in (300 mm)

7. Degree of saturation in capillary zone = 30 percent

8. Porosity of soil, α = 0.42

9. Total elapsed time = 153 hr

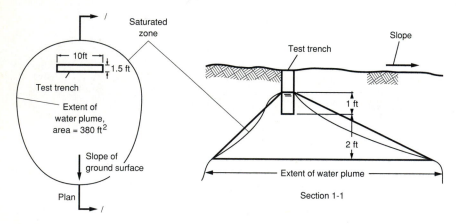

Solution

1. Determine the water remaining in the trench, within the saturated zone (see definition sketch), and within the capillary fringe.

   (*a*) Water remaining in trench

$$w_t = (10.0 \text{ ft} \times 1.5 \text{ ft} \times 1.0 \text{ ft}) = 15 \text{ ft}^3 = 112 \text{ gal}$$

(b) Water remaining in saturated zone

Referring to the definition sketch, the volume of soil that is above the area defined by the plume can be considered to be a truncated pyramid less the volume of the trench. The volume of water in the truncated pyramid is obtained by multiplying the volume by the porosity.

$$w_{sz} = [\frac{1}{2}(380 \text{ ft}^2 + 15 \text{ ft}^2) \times 3.0 \text{ ft} - 15 \text{ ft}^3] \times 0.42$$

$$= 243 \text{ ft}^3 = 1818 \text{ gal}$$

(c) Water remaining in capillary zone

Referring to the definition sketch, the surface area of soil subject to capillary rise is assumed to correspond to the area defined by the plume.

$$w_{cz} = 380 \text{ ft}^2 \times 1.0 \text{ ft} \times 0.42 \times 0.3 = 48 \text{ ft}^3 = 359 \text{ gal}$$

(d) Total water remaining

$$w_{tr} = 112 + 1818 + 359 = 2289 \text{ gal} = 306 \text{ ft}^3$$

2. Determine the water acceptance rate.

(a) Water absorbing into the underlying soil = 4500 gal − 2289 gal = 2211 gal = 296 ft^3

(b) Acceptance rate (AR) =

$$\text{AR} = \frac{\text{total flow absorbing}}{(\text{area})(\text{time})}$$

$$\text{AR} = \frac{2211 \text{ gal}}{(380 \text{ ft}^2)[153 \text{ hr}/(24 \text{ hr/d})]} = 0.91 \text{ gal/ft}^2 \cdot \text{d}$$

3. Determine the saturated coefficient of permeability using Darcy's Law.

(a) Darcy's Law (see Eq. 14-3)

$$Q = -Aks$$

For the situation where the flow below the trench is in the vertical direction, the numerical value of the slope term, s, in the Darcy equation (dh/dL) is approximately equal to minus one. Thus,

$$k = -\frac{Q}{As} = -\frac{296 \text{ ft}^3/[153 \text{ hr}/(24 \text{ hr/d})]}{380 \text{ ft}^2(-1)} = 0.12 \text{ ft/d}$$

Comment. Assuming the long-term acceptance rate through the biomat is on the order of 0.3 gal/ft^2 ·d for a unit head, the acceptance rate of the soil (0.9 gal/ft^2 · d) should not be a limiting factor.

Assessing the Assimilative Capacity of a Site.

Definition sketches for the determination of the assimilative capacity of various types of sites are presented in Fig. 14-19. As shown in Fig. 14-19a, in sloped lots, the transport of effluent from the site is controlled by the characteristics of the site. To maximize the amount of effluent that can be transported from a sloped lot, the disposal fields must be placed perpendicular to the movement of the groundwater. Many soil absorption systems have failed because the disposal fields were placed parallel to the direction of flow of

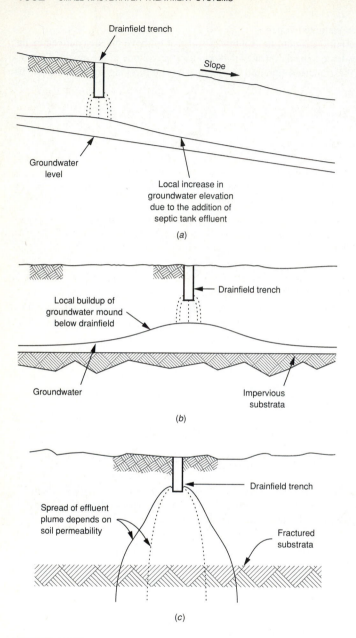

FIGURE 14-19
Definition sketch for the analysis of the assimilative capacity of a site: (*a*) sloped terrain with high groundwater, (*b*) flat area with impervious barrier, and (*c*) flat area with fractured substrata.

the groundwater. In flat areas with limited percolation capacity (see Fig. 14-19b), the flow from the disposal field tends to spread out laterally until it is dissipated. Thus, the placement of the disposal fields in flat areas is often not as critical as it is on sloped lots. For contiguous lots located in flat areas, the analysis is more complex because the subsurface flows from adjacent lots may interfere with each other. In such situations, a more comprehensive modeling effort may be required to determine the flow characteristics. In flat lots underlain with a porous layer where the pressure, for all practical purposes is uniform, the effluent flow is columnar (see Fig. 14-19c). For columnar flow, the hydraulic gradient is approximately equal to negative one. The assessment of the assimilative capacity of an onsite system located on a hillside lot is illustrated in Example 14-2.

Example 14-2 Assessment of the assimilative capacity of a sloped site for the disposal of septic tank effluent. Disposal fields with a depth of 2 ft are to be placed on a sloped lot with a slope of 10 percent as shown in the accompanying figure. The natural groundwater level under wet-weather conditions is located 2.5 ft (0.76 m) below the bottom of the disposal field trenches. For the given conditions, determine the quantity of effluent that can be transported from the site if the assimilated effluent is to remain 3 ft (0.9 m) below the ground surface at the property line. The width of the lot is 100 ft. Assume that the coefficient of permeability for the soil in the direction of the groundwater flow is 2.25 ft/d.

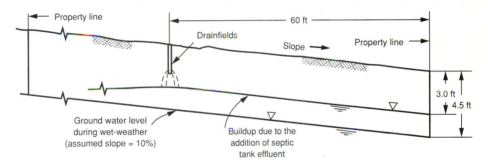

Solution

1. Determine the slope of the hydraulic gradient

   (a) Determine the angle θ

   $$\tan \theta = \frac{-1.0 \text{ ft}}{10.0 \text{ ft}} = -0.10$$

   $$\theta = -5.7$$

   (b) Determine the hydraulic gradient

   $$s = \sin \theta = -0.094$$

2. Determine the total flow that can be transported from the site. Assume that the length of the drainfields is 80 ft and that they are located perpendicular to the direction of groundwater flow.

$$Q = -Aks \text{ (see Eq. 14-3)}$$
$$= -(80 \text{ ft} \times 1.5 \text{ ft}) \times (2.25 \text{ ft/d}) \times (-0.094)$$
$$= 25.4 \text{ ft}^3/\text{d} = 190 \text{ gal/d}$$

In practice, more flow can be transported from the site because of the lateral spreading of the water plume.

Comment. While a more rigorous analysis can be made using a two-dimensional computer flow program, the analysis presented in this example is adequate for most purposes. The critical issues in assessing the assimilative capacity are (1) identification of the maximum level of groundwater under wet-weather conditions, (2) the direction of groundwater flow, and (3) the value of the lateral permeability. The shallow trench pump-in test described previously can also be used to obtain an estimate of the lateral permeability.

Selection of Appropriate
Onsite Systems for Evaluation

The selection of appropriate onsite systems for evaluation depends primarily on the findings from the preliminary and detailed site evaluations as outlined above. The more common types of systems that have been used for normal and difficult site conditions are identified in Table 14-11.

Onsite Systems for Normal Site Conditions. The first system that should be evaluated is the conventional system consisting of a septic tank and a disposal field served by intermittent gravity flow. A conservatively designed conventional onsite system is the favored choice because it is well-known that homeowners are neither trained nor especially interested in maintenance and that a conventional system operates with little attention. If the conventional onsite system is found to be unsuitable, then one or more of the options listed in Table 14-11 should be evaluated.

Depending on the layout of the lot and the site conditions, the disposal field may be pressure dosed using either a pump or a dosing siphon. Occasionally, a regulatory agency has ruled against the use of pumps in individual onsite systems, because proper maintenance cannot be expected. Today's pumps, however, are reliable, rarely need maintenance, and can be found in most locations. Thus, the use of pumps should not be ruled out.

Onsite Systems for Difficult Site Conditions. The principal conditions that necessitate the use of alternative onsite systems include: (1) low soil permeability, (2) shallow impervious substratum, (3) shallow soils over openly fractured bedrock, (4) high soil permeability, (5) steep slopes, (6) small lots, (7) sensitive groundwater areas, and (8) high groundwater. Alternative systems to the conventional septic tank and disposal field system that have been commonly used where the above conditions are encountered are reported in Table 14-11.

In many locations, the design of onsite systems is essentially fixed by local building codes, which have been developed empirically to limit the failure of these systems. As noted in Ref. 30, codes are necessary because local authorities are neither empowered nor funded to provide individual designs for onsite systems. In a number

TABLE 14-11
Applicability of alternative onsite systems for difficult site conditions

System	Normal site conditions	Low soil permeability	Shallow soil over impervious layer	Shallow soil over fractured bedrock	High soil permeability	High groundwater
Septic tank with conventional disposal field	Yes	Yes	No	No	No	No
Septic tank with conventional disposal field with pressure distribution	Yes	Yes	No	No	No	No
Septic tank with shallow sand-filled pressure-dosed disposal field trenches	Yes	Yes	Yes	Yes	Yes	Yes
Septic tank with intermittent sand filter with conventional disposal field	Yes	Yes	Yes	Yes	Yes	Yes
Septic tank with intermittent sand filter with conventional disposal field trenches with pressure distribution	Yes	Yes	Yes	Yes	Yes	Yes
Septic tank with intermittent sand filter with shallow leachfield trenches with pressure distribution	Yes	Yes	Yes	Yes	Yes	Yes
Septic tank with mound system with pressure distribution	Yes	Yes	Yes	Yes	Yes	Yes

of locations, local codes have been modified to allow alternative engineered systems that may function effectively where a conventional system might not work.

Selection of Design Criteria for Disposal Fields

The ability of a disposal field to function properly for an extended period of time will vary with (1) the quality of the applied effluent, (2) the quantity of flow, (3) the hydraulic gradient, (4) the method of application (e.g., gravity or pressure dosed,) (5) the amount of oxygen in the trench and at the soil interface (e.g., aerobic, facultative, anaerobic), and (6) temperature. Although all of these factors must be considered in the selection of an appropriate hydraulic application (loading) rate for the design of disposal fields, little is known quantitatively concerning the individual and collective impact of these factors. Selection of the design criteria for the soil absorption systems is currently based on the use of percolation test results, on soil profile examinations, and on the use of the most conservative criterion. Each of these methods is considered in the following discussion. Suggested design criteria are also presented.

Infiltrative Surfaces Used for Design of Disposal Fields. The infiltrative surfaces in a disposal field trench are the two sidewalls of the trench and the bottom. However, before the sidewalls can become effective, the biomat on the bottom of the trench must develop sufficiently to cause ponding. Ponding occurs when the applied flow exceeds the long-term acceptance rate of the biomat. Because solids tend to settle and accumulate on the bottom, the extent of the biomat formation on the sidewalls is usually not as extensive as it is on the trench bottom. In fact, in a conventional disposal field served by intermittent gravity flow, development of the biomat is progressive, as shown previously in Fig. 14-8. Further, because of the uneven biomat formation, the hydraulic gradient is variable for the sidewalls. Thus, there is no uniform hydraulic application rate that is applicable to all of the infiltrative surfaces. For this reason, the soil absorption area to be considered for design will vary with each local agency. In many locations, only the bottom area is used. In other locations, only the trench sidewall area is considered. In still other locations, both the bottom and the sidewall areas are considered.

Loading Rates Derived from Percolation Test Results. In many parts of the country, as noted previously, current design practice for individual onsite systems typically involves the use of a percolation test to determine the required size of the soil absorption system. Once percolation tests have been performed, the allowable hydraulic loading rate for the soil absorption system is determined from a table or curve relating the percolation rate expressed in min/in ($min/10^2$ mm) to the allowable loading rate, in $gal/ft^2 \cdot d$ ($m^3/m^2 \cdot d$ or $L/m^2 \cdot d$). However, because of the findings reported previously regarding the long-term acceptance rate through the biomat when the soil surface is continuously inundated with septic tank effluent, it can be concluded that there is no direct relationship between the observed short-term percolation rate determined with clean water and the long-term acceptance rate based on the application of septic tank effluent. Therefore, the use of percolation test results as the only criterion for the selection of the hydraulic loading rate is not recommended. The percolation test is, however, useful in identifying problem soils that may be considered too permeable or too slowly permeable.

Loading Rates Based on Soil Characteristics. In many locations, both a detailed soil analysis and the results of percolation tests are used to determine the appropriate hydraulic application rate for soil absorption systems. Some states have abandoned the use of percolation tests in favor of soil profile examinations. Unfortunately, because most soils are so variable within a given classification (see Table 14-10), it is sometimes difficult to characterize the soils on an individual site effectively within economic constraints. For large systems, trench pump-in tests, as described previously, are often conducted to assess the acceptance rate of a soil. Disposal field loading rates recommended by the EPA [23] for the design of soil absorption systems, based on bottom area, for various types of soils and observed percolation rates are reported in Table 14-12. The loading rate data reported in Table 14-12 have also been used to design disposal fields based on side wall area. Based on numerous observations that the biomat which eventually forms in most disposal fields will control the hydraulic capacity of the disposal field, there is little justification for using the variable loading rates given in Table 14-12.

TABLE 14-12
Hydraulic loading rates for disposal field trenches and seepage beds based on bottom area recommended by EPA[a]

Soil texture	Approximate percolation rate		Application rate based on bottom area[b,c]	
	min/in	min/10²mm	gal/ft² · d	L/m² · d
Gravel, coarse sand	<1	<4	not recommended[d]	
Coarse to medium sand	1–5	4–20	1.2	48
Fine sand, loamy sand	6–15	21–60	0.8	32
Sandy loan, loam	16–30	61–120	0.6	24
Loam, porous silt loam	31–60	121–240	0.45	18
Silty clay loam, clay loam[e,f]	61–120	241–480	0.2	8
Clays, colloidal clays	>120	>4800	not recommended	

[a] Adapted from Ref. 23.

[b] Rates based on septic tank effluent from a domestic waste source. A factor of safety may be desirable for wastewaters of significantly different character.

[c] May be suitable for side wall infiltration rates.

[d] Soils with percolation rates < 1 min/in can be used if the soil is replaced with a suitably thick > 2 ft (> 0.6 m) layer of loamy sand or other suitable soil.

[e] Soils without significant amounts of expandable clays.

[f] Soil easily damaged during construction.

Note: gal/ft² · d × 40.7458 = L/m² · d

Loading Rates Based on the Most Conservative Criterion. Recognizing the vagaries associated with the identification and classification of soils based on limited field testing, the variability of most soils, the variability of the percolation test results, the minimal maintenance most onsite systems will receive, and the findings of his own research and those of others, one well-known researcher and practitioner recommends the use of a single hydraulic loading rate regardless of soil permeability, providing onsite conditions of assimilation are satisfactory, the design conforms with assimilative needs, and proper health considerations have been ensured [31]. Some of the evidence that was considered in arriving at the value for the most conservative criterion is presented graphically in Fig. 14-20 in which the hydraulic loading rates for trouble free and troubled soil absorption systems are plotted versus the percolation rate [31]. Referring to Fig. 14-20, if all of the data points corresponding to systems with problems are encompassed, the hydraulic application rates would vary from about 0.5 gal/ft$^2 \cdot$ d (20 L/m$^2 \cdot$ d) at a percolation rate of <1 min/in (<4 min/10^2 mm) to 0.25 gal/ft$^2 \cdot$ d (10 L/m$^2 \cdot$ d) at a percolation rate of 120 min/in (480 min/10^2 mm).

The recommended hydraulic loading rate is 0.125 gal/ft$^2 \cdot$ d (5 L/m$^2 \cdot$ d), based on trench side wall areas only [31]. Where the soils contain significant amounts of clay, it is suggested that the disposal field be divided into two and that the two disposal fields be alternated every year. When two disposal fields are used, the actual loading rate for the field in operation is 0.25 gal/ft$^2 \cdot$ d (10 L/m$^2 \cdot$ d). The proposed loading rate is defined as the most conservative criterion. Several counties in California have adopted the use of the most conservative criterion for sizing disposal fields and have found this approach to the design of disposal fields both practical and satisfactory.

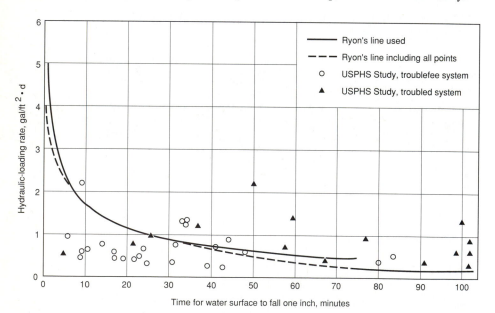

FIGURE 14-20
Permissible hydraulic loading rates for subsurface soil absorption systems for various percolation rates [31].

Suggested Loading Rates. Based on an analysis of the work in the field of onsite waste management [6,8,13,23,27,29,31] and work in the field of land application (see rates are based on the trench side wall areas only. It was also assumed that the onsite system will not receive any maintenance and may otherwise be abused. As shown in Table 14-13, separate hydraulic loading rates are presented for soils not classified as silty clay or clay loam and for soils classified as silty clay loam or clay loam.

For the soils classified as silty clay loam or clay loam, the reported hydraulic loading rates are used to determine the total length of the disposal field trench. The disposal field is then divided in half and each half is used for six months and then rotated. Resting of disposal fields placed in soils containing clay is necessary to restore the infiltration capacity of the soil. Upon resting, the clay particles that have become dispersed under continuous inundation will reaggregate and much of the original infiltration capacity will be restored.

Sizing and Layout of Disposal Fields

Sizing of disposal fields is based on the design per capita flowrates and the recommended hydraulic loading criteria discussed above and presented in Table 14-13. The layout of a disposal field is based on the direction of groundwater flow and other local site constraints. The sizing of a disposal field for an individual residence is illustrated in Example 14-3.

Design Flowrates. Ideally, design flowrates should be estimated on the basis of the expected occupancy of the home and the water-use patterns. Typical per capita flowrate data, as reported in Table 14-1, vary from about 35 to 100 gal/capita · d (130 to 380 L/capita · d). If actual per capita water use data are not available, a value of 55 gal/capita · d (210 L/capita · d) can be used as a guide for the design of disposal fields. As noted previously, the value of 55 gal/capita · d (210 L/capita · d) is based on an average occupancy of 2.4 to 2.8 residents per home. The quantity of flow from a home with one or more occupants may be estimated using Eq. 14-1.

Over a period of years, many regulatory agencies have sought to simplify the design of onsite systems by basing the design of disposal fields on a fixed flow per bedroom. Because there is no quantifiable relationship between the flow from an individual residence and the number of bedrooms [31], the practice of using a fixed flow per bedroom for the design of disposal fields should be discouraged. The design of disposal fields should be based on sound defensible engineering practice, as outlined above, and not on the use of unsubstantiated rules of thumb.

Layout of Soil Disposal Field. The key to laying out disposal fields is to locate the disposal field in the best soil on the site and to distribute the flow as widely as possible so as to maximize the opportunity for treatment and assimilation. Ideally, the disposal fields should be placed perpendicular to the direction of groundwater flow. In locations where site constraints limit the placement of the disposal field(s), the use of an effluent pump should be considered to optimize the placement of the disposal fields.

TABLE 14-13
Recommended hydraulic loading rates for disposal fields based on actual or estimated flow

General soil classification and type of disposal field	Hydraulic loading rate based on trench side wall areas		Remarks
	gal/ft² · d	L/m² · d	
For soils not classified as silty clay loam, clay loam. Conventional single disposal field served by gravity flow or pressure dosed with effluent from			Typical percolation rates less than 60 min/in (e.g., 1–60 min/in)
Septic tank	0.2	8	
Intermittent sand filter	0.4	16	
Recirculating granular-medium filter	0.4	16	
For soils classified as silty clay loam, clay loam. Conventional dual disposal field served by gravity flow or pressure dosed with effluent from			Typical percolation rates greater than 60 min/in (e.g., 60–120 min/in). The reported hydraulic loading rate is used to determine the total length of the disposal field required. The total length of trench is then divided in half and each half is operated for six months and then rotated. Resting in clay soils is necessary to restore the permeability capacity that is partially dependent on soil properties
Septic tank	0.15	6	
Intermittent sand filter	0.35	14	
Recirculating granular-medium filter	0.3	12	
Shallow sand-filled disposal trenches pressure dosed with effluent from septic tank	0.3	12	

Note: gal/ft² · d × 40.7458 = L/m² · d

Example 14-3 Design of onsite wastewater management system. Design an onsite system using a trench-type soil absorption system for an individual residence (median income) in an unsewered area. The maximum occupancy of the residence will be five persons. Use a daily peaking factor of 3 based on average flow. The design for the soil absorption system is to be based on the trench side wall areas using the criteria given in Table 14-13. The average percolation rate based on the results of five tests is 35.5 min/in (126 min/10^2 mm). The soils at the site have been classified as loam. Assume the disposal field will be pressure dosed to achieve a more uniform application of the septic tank effluent and to improve the operation of the soil absorption system.

1. Estimate the average daily flowrate using Eq.14-1
   The design flowrate is:

   $$\text{Flowrate} = 40 \text{ gal/residence} \cdot d + 40 \text{ gal/resident} \cdot d \times (5 \text{ residents/home})$$

   $$= 240 \text{ gal/d}$$

2. Determine the average detention time in the septic tank at peak flow. Based on the local building code, the minimum required septic tank volume is 1200 gal (4.5 m³). Assume 30 percent of the volume is lost because of sludge and scum accumulations.

   (a) The detention time in the septic tank at peak daily flow is

   $$\text{Detention time} = \frac{1200 \text{ gal} \times 0.70}{3 \times 240 \text{ gal/d}}$$

   $$= 1.2 \text{ d}$$

   (b) The minimum acceptable detention time should be about 0.5 d, thus the size of septic tank is acceptable.

3. Determine the required length of the soil absorption system. Assume the maximum trench depth is to be fixed at 5 ft (1.5 m) to conform to local code requirements. Assume the maximum depth in the trenches below the distribution pipe is 4 ft (1.2 m). The trench width is to be 12 in (300 mm)

   (a) From Table 14-13 the allowable hydraulic loading rate is 0.2 gal/ft² · d (8 L/m² · d)

   (b) Determine the percolation capacity per foot of trench length

   $$\text{Trench sidewall capacity} = 2 \ (4.0 \text{ ft}^2/\text{ft} \times 0.2 \text{ gal/ft}^2 \cdot d)$$

   $$= 1.6 \text{ gal/ft of trench}$$

   (c) The required disposal field trench length is

   $$\text{Trench length required} = \frac{240 \text{ gal/d}}{1.6 \text{ gal/ft}} = 150 \text{ ft}$$

   Because the required length of disposal field is quite long, use two 75 ft trenches or three 50 ft trenches.

4. Prepare a typical layout for the proposed system

   (a) A typical layout for the proposed onsite system is shown in the sketch on the following page.

   (b) When locating disposal fields it is extremely important to place the fields perpendicular to the direction of flow of the groundwater

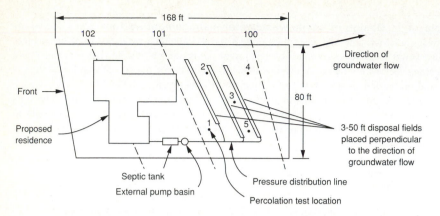

Comment. The allowable loading rate for soil absorption systems will usually be prescribed by local codes. Nevertheless, the prescribed rates should always be checked against data collected for the parcel in question. Whenever possible, the disposal field should be made as large as possible. The cost of constructing longer disposal field trenches is minor compared to the cost and bother of constructing a new system. Also, as noted previously, the soil absorption area to be considered for design purposes will vary. In this example, the side wall areas were used. In many locations, only the trench bottom area is considered.

Degree of Treatment

As noted in Table 14-7, the effluent applied to a soil absorption system is highly treated after it has passed through about 3 ft (0.9 m) of soil. Based on the results reported in a number of studies [8,14,16,30], it can be concluded that a high degree of treatment for BOD_5, SS, and coliform organisms is provided within 3 to 30 ft (0.9 to 9 m) from the disposal field for most of the soils found suitable for soil absorption systems.

Where nitrogen removal is required, an intermittent sand or recirculating granular-medium filter can be used to nitrify the ammonia in the septic tank effluent. The nitrate can be removed by denitrification, by returning the filter effluent to the septic tank. A nitrification/denitrification process employing separate gray and black water systems in which the organic matter in the gray water is used as the carbon source for denitrification is shown in Fig. 14-21.

Selection of Design Criteria for Physical Facilities

Typical design criteria for septic tanks and related septic tank and disposal field appurtenances are presented in Tables 14-14 and 14-15, respectively. Recommended design criteria for disposal fields have been presented previously in Table 14-13. Information on the pumps used for onsite systems may be found in Table 14-16. Typical design criteria for intermittent sand and recirculating granular-medium filters are presented in Table 14-17. The application of these design criteria are illustrated

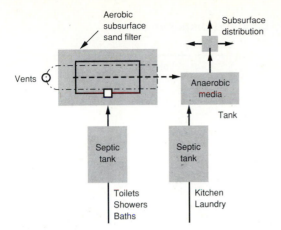

FIGURE 14-21
Onsite system for the removal of
nitrogen from wastewater employing separate black and gray water
systems [from Ref. 8].

TABLE 14-14
Typical design criteria for septic tanks

Design parameter	Unit	Value Range	Value Typical
Liquid volume			
Minimum	gal	750–1,000	750[a]
1–2 bedrooms	gal	750–1,000	750[a]
3 bedrooms	gal	1,000–1,500	1,200[a]
4 bedrooms	gal	1,000–2,000	1,500[a]
5 bedrooms	gal	1,200–2,000	1,500[a]
Additional bedrooms	gal	150–250	250[a]
Compartments			
Number	No.	1–3	2[b]
Volume distribution in multi-compartment tanks			
Two-compart. tank	% 1st, 2nd	67, 33	67, 33
Three-compart. tank	% 1st, 2nd, 3rd	33, 33, 33	33, 33, 33
Length to width	ratio	2:1–4:1	3:1
Depth	ft	1–6	4
Clear space above liquid	in	10–12	10
Depth of water surface below inlet	in	3–4	3
Inspection ports (see Fig. 14-2)	No.	2–3	2
Inlet devices			Inlet tee (see Fig. 4-2)
Outlet devices			Effluent filter vault; scum baffle and outlet tee; scum baffle, outlet tee, and some type of gas defection baffle (see Figs. 14-2 and 14-4)

[a] Most regulatory agencies have minimum size requirements for septic tanks [typically 1200 gal (4550 L)]. If septic tank size is not specified, the values given in this table can be used as a guide.

[b] Two- or three-compartment tanks are used when a screen vault is not used.

Note: gal × 3.7854 = L
ft × 0.3048 = m
in × 25.4 = mm

TABLE 14-15
Typical design criteria for septic tank and disposal field appurtenances[a]

Design parameter	Unit	Value Range	Value Typical
Internal pump vaults			
Type	Screened vault (see Fig. 14-9a)		
Minimum volume between pump on and off levels	gal	50–150	100
Dosing siphon		75–100	90
External pump basin			
Liquid volume below septic tank inlet	factor × Q_{avg}	1–2	1.5
Minimum volume between pump on and off levels	gal	25–150	50
Inspection port opening	in	20–36	≥ 20
Pump setting above bottom of sump	in	≥ 4–8	4
Pumps	See Table 14-16		
Diversion facilities for leachfields	PVC ball valves, three way valves, hydraulic flow splitter, hydro splitter, basin flow splitter, orifice control on main line to lateral		

[a] Most regulatory agencies have minimum size requirements for septic tank appurtenances.
Note: gal × 3.7854 = L
      in × 25.4 = mm

in the example problems. Typical design criteria for Imhoff tanks may be found in Section 14-8.

Flow Distribution in Pressure-Dosed Disposal Fields and Sand Filters

To optimize the performance of pressure-dosed disposal fields and intermittent and recirculating sand filters, a pressurized distribution system should be used. The distribution system piping must be sized so that the discharge from each orifice in the distribution system is nearly the same as possible (see Fig. 14-22). In onsite systems, essentially equal flow from orifices is accomplished by adjusting the size of the distribution pipe so that the headloss in the distribution pipe is low compared to the headloss through the orifices. The difference in discharge between orifices in a distribution system can be assessed as follows for different orifice and pipe sizes.

   Assume the discharge in any orifice is to be held to a value mq_1, where m is a decimal value less than 1 and q_1 is the discharge from the first orifice. The discharge from orifice n can be computed using the following equation.

TABLE 14-16
Typical characteristics of pumps used for onsite systems

Type	Head, ft	Flow, gal/min	Power, hp	Remarks
Small effluent pump	Up to 100	Up to 150	0.25–1.5	Used to pump septic tank effluent to pressure sewer, sand filters, drainfields, etc.
High head effluent pump (multi-stage turbine type)	Up to 500	Up to 60	0.33–1.5	A typical 0.33 hp pump will deliver 5 gal/min at a head of 160 ft. Constructed of stainless steel and thermoplastic. Ideally suited for corrosive environments. Commonly used in STEP systems where a pump with a relatively steep head-capacity curve is desired
Solids handling pump	Up to 55	Up to 250	0.33–2	Used to pump wastewater from basement to septic tanks, from home to gravity sewer line, and in pumping to recirculating granular-medium filter where low head and large flows are required.
Grinder pump	Up to 150	Up to 20+	1–2	Handles solids by grinding into smaller particles. Used in pressure sewers and for pumping wastewater to gravity sewers.

$$q_n = 2.45C(D^2)\sqrt{2gh_n} \qquad (14\text{-}4)$$

where q_n = discharge from orifice n, gal/min
 2.45 = conversion factor used to express the discharge in gal/min when the diameter of the orifice is expressed in in and the velocity is expressed in ft/s
   C = orifice discharge coefficient (usually =0.61 for holes drilled in the field)
   D = diameter of orifice, in
   g = acceleration due to gravity, 32.2 ft/s^2
  h_n = head on orifice n, ft

The head on orifice n is equal to:

$$h_n = \left[\frac{1}{(2.45CD^2)^2 2g}\right]q_n^2 = kq_n^2 = k(mq_1)^2 = m^2 h_1 \qquad (14\text{-}5)$$

where k = constant
  h_1 = head on orifice 1, ft

The headloss between orifice 1 and n, which corresponds to the headloss in the distribution pipe between orifice 1 and n, is:

$$\Delta h_{(1-n)} = h_1 - h_n \qquad (14\text{-}6)$$

TABLE 14-17
Typical design criteria for intermittent sand and recirculating granular-medium filters[a]

| | | Design criteria | | | |
| | | Intermittent | | Recirculating | |
Design factor	Unit	Range	Typical	Range	Typical
Pretreatment		Sedimentation (septic tank or equivalent)			
Filter medium					
Material		Washed durable granular material			
Effective size	mm	0.25–0.5	0.35	1.0–5.0	3.0
Uniformity coefficient	UC	<4	3.5	<2.5	2.0
Depth	in	18–36	24	18–36	24
Underdrains					
Bedding					
Type		Washed durable gravel or crushed stone			
Size	in	$\frac{3}{8}-\frac{3}{4}$		$\frac{3}{8}-\frac{3}{4}$	
Underdrain					
Type		Slotted or perforated drain pipe			
Size	in	3–4	4	3–6	4
Slope	%	0–1.0	Flat	0–1.0	Flat
Venting		Upstream			
Pressure distribution					
Pump types		See Table 14-16			
Pipe size[b]	in	1–2	$1\frac{1}{4}$	1–2	$1\frac{1}{4}$
Orifice size	in	$\frac{1}{8}-\frac{1}{4}$	$\frac{1}{8}$	$\frac{1}{8}-\frac{1}{4}$	$\frac{1}{8}$
Head on orifice	ft H_2O	3–5+	5+	3–5+	5+
Lateral spacing	ft	1.5–4	2	1.5–4	2
Orifice spacing	ft	1.5–4	2	1.5–4	2
Design parameters					
Hydraulic loading[c]	gal/ft^2 · d	0.4–1	0.6	3–5	4
Organic loading	lb BOD$_5$/ft^2 · d	0.0005–0.002	<0.001	0.002–0.008	<0.005
Recirculation ratio		--	--	3:1–5:1	4:1
Dosing frequency	times/d	3–6	4		
Dosing frequency	min/30 min			1–10	4
Dosing tank volume	days flow	0.5–1.0	0.5	0.5–1.0	0.5
Passes through filter	No.	1	1	2–8	4
Filter medium temperature	°F		>41		>41

[a] Adapted from Refs. 2, 23, 28.
[b] Size of distribution pipe depends on the flow rate (see Example 14-4).
[c] Based on estimated flowrate.

Note:
$$\text{in} \times 25.4 = \text{mm}$$
$$\text{ft} \times 3.2808 = \text{m}$$
$$\text{gal/ft}^2 \cdot \text{d} \times 40.7458 = \text{L/m}^2 \cdot \text{d}$$
$$\text{lb BOD}_5/\text{ft}^2 \cdot \text{d} \times 4.8824 = \text{kg BOD}_5/\text{m}^2 \cdot \text{d}$$
$$0.555(°F - 32) = °C$$

FIGURE 14-22
Testing of distribution manifold for intermittent sand filter by observing the height of the individual water jets as water is pumped into the distribution manifold.

Now it can be shown that the headloss between the first and last orifice in a distribution pipe with multiple evenly spaced orifices is approximately equal to one-third of the headloss that would occur if the total flow were to pass through the same length of distribution pipe without orifices [3,4]. Thus:

$$h_{f_{dp}} = \frac{1}{3}h_{f_p} = \Delta h_{(1-n)} \qquad (14\text{-}7)$$

where $h_{f_{dp}}$ = actual headloss through distribution pipe, ft
h_{f_p} = headloss through pipe without orifices, ft

The headloss through the pipe can be computed using the Darcy–Weisbach or the Hazen–Williams equation. The Hazen–Williams equation, commonly used in the field for the determination of the loss of head in closed conduits, is:

$$h_{f_p} = 10.5(L_{1-n})\left(\frac{Q}{C}\right)^{1.85} D^{-4.87} \qquad (14\text{-}8)$$

where h_{f_p} = headloss through the pipe from orifice 1 to orifice n, ft
L_{1-n} = length of pipe between orifice 1 and n, ft
Q = pipe discharge, gal/min
C = Hazen–Williams discharge coefficient, 150 for plastic pipe
D = inside diameter of pipe, in

The difference in discharge between orifice 1 and n, for a given distribution pipe and orifice size, can now be determined using Eqs. 14-4 through 14-8. If the computed value of m is too low (e.g., < 0.98 or some other acceptable value), the size of the distribution pipe can be increased. The use of these equations is illustrated in Example 14-4.

Example 14-4 Design of an onsite wastewater management system employing an intermittent sand filter for a single dwelling. Size and lay out an intermittent sand filter and distribution system for an individual three-bedroom residence in a unsewered area. Determine the difference in the discharge between the orifices at average and peak flow. If the difference in the discharge between the orifices is greater than 2 percent $[(1 - m) \times 100]$ in either case, the distribution system should be resized. Assume the following conditions apply.

1. Average occupancy $= 3.5$ persons/d
2. Assumed peaking factor $= 3$
3. Size of septic tank $= 1200$ gal (4.5 m³)
4. Sand filter application rate $= 0.6$ gal/ft² · d (24 L/m² · d) based on actual flow
5. Sand filter dose rate per day $= 4$ times/d
6. Distribution system orifice size $= 1/8$ in (3.0 mm)
7. Orifice discharge head $= 5$ ft minimum(1.5 m)

Solution

1. Determine the size of the sand filter using the design information given in Tables 14-1 and 14-17.

    (a) Total flow = (3.5 persons) × 55 gal/capita · A d (see Table 14-1) = 193 gal/d

    (b) Area of sand filter = (193 gal/d)/(0.6 gal/ft² · d) = 321 ft²

    (c) Use a filter 16 ft × 20 ft, check area (16 × 20) = 320 ft²

2. Layout sand filter and the effluent distribution system

    (a) The spacing between distribution pipes and orifices is to be 2 ft (0.6 m)

    (b) The layout of the sand filter and distribution system is shown in the sketch shown on the following page. As shown, there are sixteen 9-ft (sixteen 2.75-m) laterals each with 5 orifices equally spaced 2 ft (0.6 m) on centers. The size of the distribution system pipe used as a first try is 1.19 in (nominal 1 in pipe)

3. Determine the flow and rate of discharge in each lateral in the distribution system

    (a) Determine flow discharged per dose
    Flow/dose = (193 gal/d)/4 = 48.3 gal/dose

    (b) Determine the discharge per lateral
    Flow/lateral = (48.3 gal/dose)/16 laterals = 3 gal/lateral · dose

    (c) Determine the flowrate in each lateral
    The flow in the last orifice is

    $$q_n = 2.45C(D^2) \sqrt{2gh_n}$$

    $$q_n = 2.45(0.61)(0.125^2) \sqrt{2(32.2)5} = 0.42 \text{ gal/min}$$

    Total flowrate into each lateral based on 5 orifices per lateral = 5(0.42 gal/min) = 2.1 gal/min · lateral

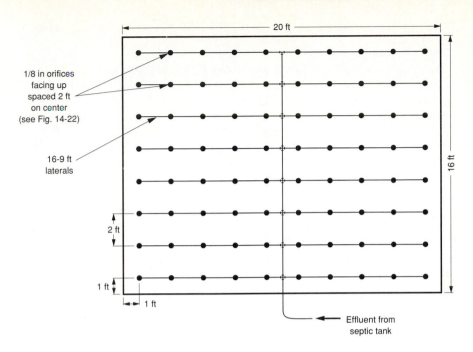

4. Determine the headloss in a lateral distribution pipe using Eqs. 14-7 and 14-8.

   (a) The headloss in pipe is determined using Eq. 14-8.

   $$h_{f_p} = 10.5(L_{1-n}) \left(\frac{Q}{C}\right)^{1.85} D^{-4.87}$$

   $$h_{f_p} = 10.5(9) \left(\frac{2.1}{150}\right)^{1.85} 1.19^{-4.87} = 0.015 \text{ ft}$$

   (b) The headloss in the actual distribution pipe using Eq. 14-7 is

   $$h_{f_{dp}} = \frac{1}{3} h_{f_p} = \Delta h_{(1-n)}$$

   $$h_{f_{dp}} = \frac{1}{3} 0.015 = 0.005 \text{ ft}$$

5. Determine the difference in the discharge between the first and last orifice in each lateral.

   (a) The head on the first orifice is

   $$\Delta h_{(1-n)} = h_1 - h_n$$

   $$h_1 = h_n + \Delta h_{(1-n)}$$

   $$h_1 = 5 + 0.005 = 5.005 \text{ ft}$$

   (b) Determine the value of m using Eq. 14-5.

   $$h_n = m^2 h_1$$

$$m = \sqrt{\frac{5}{5.005}} = 0.9995$$

The difference in the discharge between the first and last orifice in each lateral is about 0.05 percent [$(1 - 0.9995) \times 100$], well below the 2 percent value specified.

6. Check flow distribution at peak flow. It should be noted that the peak flow will have essentially no effect on the computed headloss values. In a pumped system, peak flow will only affect the length of time the dosing pump is on. At average flow, the dosing pump will be operating for about 88 seconds each time the filter is dosed [$(0.4 \text{ ft}^3/\text{lateral} \cdot \text{dose})/(4.57 \times 10^{-3} \text{ ft}^3/\text{lateral} \cdot \text{s})$]. At a peak flow of three times the average, the pump will operate for about 264 seconds each time the filter is dosed.

Comment. The maximum flowrate applied to an intermittent sand filter serving a single-family residence should be kept below 25 gal/min. A pump should be selected to provide the required flowrate with a minimum head of 5 ft on the orifices. Higher pressures (up to 30 ft) are usually not a problem. Some designers include a ball valve to adjust the residual head [typically 5 to 10 ft (1.5 to 3 m)].

Septage Disposal

To maintain performance, the material that accumulates in a septic tank must be pumped out periodically. Septage is the term used to describe the combination of scum, sludge, and liquid pumped from a septic tank. The buildup of sludge and scum in a septic tank will vary depending on usage and whether a kitchen food waste grinder is used. The handling and disposal of septage is one of the most difficult aspects of wastewater management in unsewered areas. The subject of septage management is considered in detail in Section 14-11.

14-6 ONSITE WASTEWATER MANAGEMENT DISTRICTS

Although onsite systems require very little maintenance, they rarely get any. As a result, many onsite systems have failed prematurely. The principal mode of failure has been a premature reduction of the infiltrative capacity of the disposal field below the required capacity for managing the daily flow. In most cases, where premature failure has occurred, it has been found that the disposal fields have been undersized. The discharge of grease from septic tanks that are inadequately designed and constructed has also been a serious problem.

When onsite systems are used on large lots, the failure of an individual system does not cause a serious environmental problem. However, as the density of development increases, and lot sizes become smaller, the failure of one or more onsite systems can pose a nuisance problem and in some cases, a public health problem. To insure that onsite systems will function properly, especially in densely developed areas, it is usually necessary to change the maintenance of onsite systems from private to public responsibility [31]. The formation of onsite wastewater management districts (OSWMD) have proven to be successful for this purpose.

Types of Onsite Wastewater Management Districts

Both private and public onsite wastewater management districts have been formed for the management of onsite systems. The specific type of district used depends on local circumstances. For example, a developer of a residential housing development may form a private management district. In small rural communities the formation of public districts is most common. In each case, the district is the responsible legal entity for the continued long-term performance of the onsite systems. Properly constituted and staffed OSWMDs have proven to be an effective means of ensuring the long-term performance of onsite systems.

The Functions of an OSWMD

The functions of an OSWMD will vary depending on the legal authority under which the district was formed. Typical functions include the following:

1. Approval of design and plans for individual onsite systems, and for construction inspection.
2. Responsibility for design and construction of individual onsite systems.
3. Annual or semi-annual inspection of each onsite system in the district.
4. Issuance of permits to operate, failed-system citations, and abatement orders.
5. Scheduling of routine monitoring and pumping of septic tanks.

Requirements for a Successful OSWMD

The elements required for a successful onsite OSWMD include [17,30]:

1. Regulatory authority.
2. Authority and lawful means to cause a homeowner to correct a failed system.
3. Trained personnel.
4. Economic feasibility.
5. Well-designed and constructed onsite systems.

Although all of the above elements are important, the most important is the authority and the legal means to get a homeowner to correct a failed system. If the district has no hold over the homeowner, getting a failed system corrected has proven, in many cases, to be an impossible task. In Stinson Beach, CA the OSWMD is part of the Stinson Beach County Water District which also has responsibility for providing water service. Ultimately, after all other means have been exhausted, the district has the legal authority to turn off the water supply, which it has done [26]. This method has proven to be effective in getting homeowners to correct failed systems.

14-7   WASTEWATER COLLECTION SYSTEMS FOR SMALL COMMUNITIES

In many rural locations, the density of residential development has increased to the point that continued use of individual onsite systems is no longer feasible. Under these conditions, some form of wastewater collection is needed. The use of conventional gravity flow sewers for the collection and transport of wastewater from residences and commercial establishments has, and continues to be, the accepted norm for sewerage practice in the United States. However, in many areas that are now being developed, the use of gravity-flow sewers may not be economically feasible for reasons of topography, high groundwater, structurally unstable soils, and rocky conditions. Further, in small unsewered communities, the cost of installing conventional gravity-flow sewers is prohibitive especially if the density of development is low. To overcome these difficulties, small-diameter variable-slope, pressure, and vacuum sewers have been developed as alternatives. Because infiltration/inflow is, for all practical purposes, eliminated when alternative sewers are used, a variety of alternative treatment processes can be used. For example, the use of recirculating granular-medium filters is usually not feasible if infiltration/inflow cannot be controlled.

Conventional Gravity Sewers

The use of gravity-flow sewers is accepted because (1) the performance of gravity-flow sewers is well-established and documented and (2) a well-developed body of knowledge is available for their design, construction, and operation. Where the topography is suitable, gravity-flow sanitary sewers have been selected and will probably continue to be selected. Gravity sewers are discussed in detail in the companion text to this volume [11].

Small-Diameter Variable-Slope Sewers

The small-diameter variable-slope (SDVS) sewer system was developed jointly by the Rural Housing Research Unit (RHRU) of USDA-ARS, Tuskegee Institute, and the Farmers Home Administration (FHA). The basic concept involved in an SDVS sewer system is that if one is installed with a net positive slope from the inlet to the outlet, wastewater put in at the upper end or along the SDVS sewer will eventually exit from the lower end. The SDVS sewer is laid at approximately the same depth below the surface of the ground regardless of the grade (see Fig. 14-23). As shown, there will be both downhill and uphill sections. The only requirement is that the outlet be lower than the inlet and lower than any of the house connections to the sewer. Thus, the flow of wastewater through the system will involve delays, surging, and transitions from full to partial pipe flow [17,18]. Under this flow regime, some sections of the sewer will always be full. It is always necessary, however, to plot both the hydraulic grade line and the pipeline profile to locate the proper type of air release valves.

The SDVS sewer is used in conjunction with septic tanks, which are used for solids removal. To ensure that solids from the septic tank do not clog the small-diameter sewer, some positive means of keeping the solids in the septic tank are

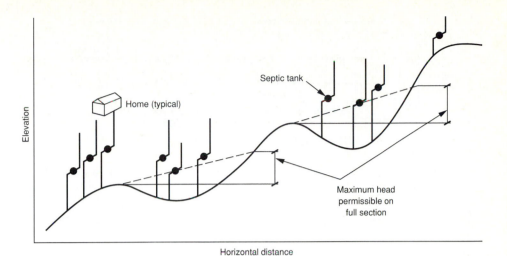

FIGURE 14-23
Definition sketch for a small-diameter variable-slope sewer (adapted from Refs. 17,18).

required. Methods that have been used to reduce the presence of solids in the septic tank effluent include the use of an effluent filter vault (see Fig. 14-2), internal deflection baffles (see Fig. 14-4a,b), and inclined clarifier tubes (see Fig. 14-4c). As discussed previously, the effluent filter has proven to be quite effective in eliminating the discharge of solids.

To date, the experience with the use of SDVS sewers has been favorable. Solids carryover from the septic tank has not been a problem. The use of manholes with these systems is not recommended because they increase the potential for inflow and sediment entry and unnecessary cost. Appropriately spaced cleanout ports should be included for cleaning the lines when necessary. The lines are cleaned using a pigging device in exactly the same manner that water lines are cleaned. As more of these systems are built it is anticipated that they will find greater acceptance. Because these systems are continually being improved, it is advisable that current operating data be obtained when considering the use of SDVS sewers. If possible, visits should be made to existing installations.

Pressure Sewers

In pressure sewer systems, wastewater from individual residences or buildings is collected and discharged into a septic tank or holding tank and then pumped to a pressure or gravity-flow collector sewer. The main components of a pressure sewer system are shown in Fig. 14-24a. Where a septic tank is used to remove the solids from the wastewater before it is pumped, the system is referred to as a *septic tank effluent pumping* (STEP) system. Where a holding tank is used, wastewater is discharged periodically into a pressure main by means of a grinder pump that can reduce the size of the solids in the wastewater. Systems with grinder pumps are usually known as

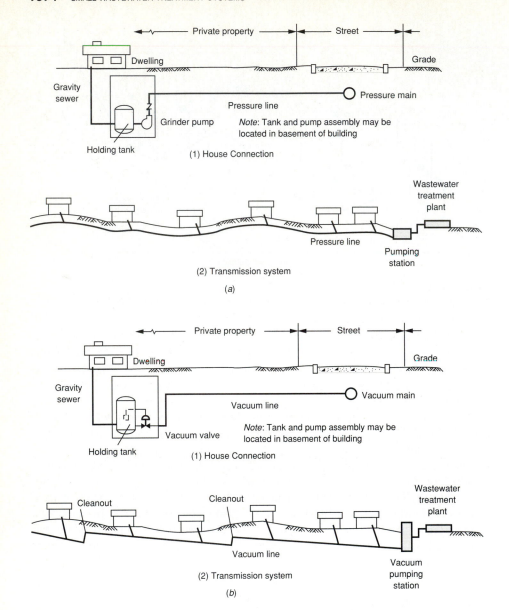

FIGURE 14-24
Principal components of pressure and vacuum sewers: (*a*) pressure sewer and (*b*) vacuum sewer
[11].

grinder pump systems. A septic tank or holding tank and pump are required at each inlet point to the pressure main. To reduce capital and maintenance costs, a single tank or holding tank and pump can be used for several homes. A pressure sewer connected to a residence too low to be served by a gravity sewer is illustrated in Fig.14-25.

Wastewater from the pressure main is discharged either into a gravity line or to the influent facilities of the treatment plant. A pressure sewer system eliminates the need for small pumping stations and makes it possible to substitute a small-diameter plastic pipe placed at shallow depths for a much larger diameter conventional pipe placed at greater depths. However, all of this is accomplished at the expense of having to install an effluent or grinder pump at each inlet to the pressure main. In addition to the initial cost of the pumps, the associated power and maintenance expenses must be considered. Typical design data on pressure sewer systems are reported in Table 14-18. Additional details on pressure sewers may be found in Refs. [1,21].

Vacuum Sewers

The principal features of a vacuum sewer system are shown in Fig. 14-24b; typical operational data are reported in Table 14-18. In these systems, wastewater from an individual building flows by gravity to the location of a vacuum ejector (vacuum valve of special design). The valve seals the line leading to the main so that the required vacuum levels can be maintained in the main. When a given amount of wastewater accumulates behind the valve, the valve is programmed to open and close after the wastewater enters as a liquid plug. Vacuum pumps in a central station maintain the vacuum in the system. The station is usually near the treatment facilities or any other

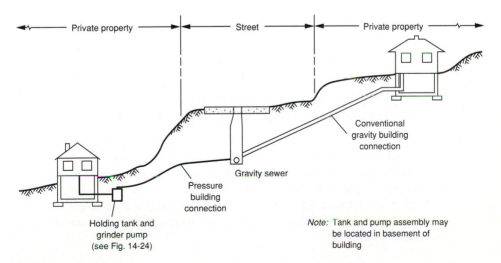

FIGURE 14-25
Typical pressure sewer connection from an isolated low building site to a gravity sewer [11].

TABLE 14-18
Typical design data for pressure and vacuum sewers

Design parameter	Unit	Range	Typical
Pressure sewers			
Grinder pump	hp	1–2	1.5
Grinder pump discharge pressure (gage)	lb_f/in^2	30–40	35
Size of line from pump to pressure main	in	1–2	$1\frac{1}{4}$
Size of pressure main	in	2–12	—[a]
Pump discharge	gal/min	5–25	12
Vacuum sewers			
Height of water level on vacuum discharge valve	in	3–40	30
Size of line from discharge line to vacuum main	in	3–5	4
Air/liquid ratio		1–10	2
Vacuum maintained in collection tank at pumping station	mm Hg	300–500	400

[a] Varies with location in system.

Note: hp × 0.7457 = kW
lb_f/in^2 × 6.8948 = kN/m^2
gal/min × 0.0631 = L/s
in × 25.4 = mm

convenient discharge point. In general, vacuum sewers are not cost-effective when compared to STEP or grinder pump pressure sewer systems.

14-8   SMALL SYSTEMS FOR CLUSTERS OF HOMES AND VERY SMALL COMMUNITIES

The purpose of this section is to introduce the reader to the types of systems that have been used for the treatment of very small wastewater flows. The systems to be considered include (1) those systems in which septic tanks located on the home owners' property are used for solids separation and (2) those systems in which centralized facilities are used for solids processing and additional treatment. The use of prefabricated package and individually designed treatment plants is considered in the following two sections.

Systems with Individual Septic Tanks

In locations where onsite absorption systems can no longer be used, it has been found, in some cases, to be most cost-effective to continue the use of onsite septic tanks for solids separation and to collect the septic tank effluent for further treatment or disposal in a more centralized facility. In most cases, existing septic tanks are replaced with new water-tight tanks. A pressure or small-diameter variable-slope sewer is used to transport the septic tank effluent from several homes to the centralized location for disposal in a large subsurface disposal field. As noted previously, those systems with effluent pumps are known as STEP systems.

The most serious operational problem encountered with systems in which each individual septic tank is retained, has been the carryover of solids to the drainfields due to the lack of proper septic tank maintenance. This problem is most serious where a large central disposal field is to be used for the disposal of the septic tank effluent without any further treatment. The use of an onsite maintenance district (OSWMD) has proven effective in some locations, but as noted earlier, unless the OSWMD has the authority to correct or repair a failed system, little can be done. Recognizing that poor septic tank maintenance will be the order of the day, some regulatory agencies have required the addition of a large septic, or other solids-separation unit before the collected septic tank effluent can be disposed of in subsurface disposal fields.

Centralized Systems Without Individual Septic Tanks

In rural locations where individual onsite absorption systems cannot be used, some form of centralized collection, treatment, and disposal is required. Perhaps the most common system is the one in which a pressure or a conventional gravity sewer is used to transport the wastewater from several homes to a centralized location for treatment and disposal. Treatment options to be considered in the following discussion are large septic and Imhoff tanks and recirculating granular-medium filters. As noted above, package and individually designed treatment plants are considered in the following two sections.

Use of Large Septic and Imhoff Tanks in Centralized System. Although septic tanks are used primarily for individual residences and other community facilities, large septic tanks have been used to serve clusters of homes. In general, large septic tanks are divided into multiple compartments, usually three, and are designed to provide a detention time of one day. Parallel tanks are also used commonly. A typical example of a large septic tank serving a trailer park is shown in Fig. 14-26. Imhoff tanks are used occasionally because they are simple to operate and do not require highly skilled supervision. There is no mechanical equipment to maintain, and operation consists of removing scum daily and discharging it into the nearest gas vent, reversing the flow of wastewater twice a month to even up the solids in the two ends of the digestion compartment, and drawing sludge periodically to the sludge drying beds (see Fig. 14-27). Recent designs developed by manufacturers for a modified form of Imhoff tank provide means for heating the sludge compartment and mechanical removal of sludge. Conventional unheated Imhoff tanks are usually rectangular, although some small circular tanks have been used. Typical design criteria for Imhoff tanks are presented in Table 14-19.

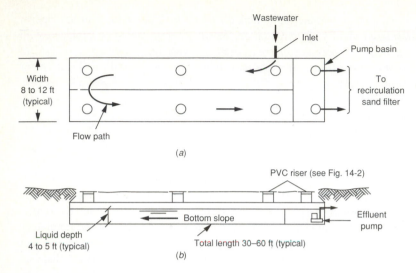

Wastewater

Inlet

Pump basin

Width
8 to 12 ft
(typical)

To
recirculation
sand filter

Flow path

(a)

PVC riser (see Fig. 14-2)

Bottom slope

Effluent
pump

Liquid depth
4 to 5 ft (typical)

Total length 30–60 ft (typical)

(b)

FIGURE 14-26
Typical large septic tank used for a trailer park: (a) plan view and (b) section.

FIGURE 14-27
Typical Imhoff tank used for a small community. Effluent from the Imhoff tank is disposed of in a series of rapid infiltration basins. Note sludge drying beds located adjacent to the Imhoff tank.

TABLE 14-19
Typical design criteria for unheated Imhoff tanks

Design parameter	Unit	Value Range	Value Typical
Settling compartment			
Overflow rate peak hour	gal/ft$^2 \cdot$ d	600–1,000	800
Detention time	h	2–4	3
Length to width	ratio	2:1–5:1	3:1
Slope of settling compartment	ratio	1.25:1 to 1.75:1	1.5:1
Slot opening	in	6–12	10
Slot overhang	in	6–12	10
Scum baffle			
Below surface	in	10–16	12
Above surface	in	12	12
Freeboard	in	18–24	24
Gas vent area			
Surface area	% of total surface area	15–30	20
Width of opening[a]	in	18–30	24
Digestion section			
Volume (unheated)	Storage capacity		6 months of sludge
Volume[b]	ft^3/capita	2–3.5	2.5
Sludge withdrawal pipe	in	8–12	10
Depth below slot to top of sludge	ft	1–3	2
Tank depth			
Water surface to tank bottom	ft	24–32	30

[a] Minimum width of opening must be 18 in to allow a person to enter for cleaning.
[b] Based on a six-month digestion period.
Note: gal/ft$^2 \cdot$ d $\times$ 0.0407 = m^3/m$^2 \cdot$ d
                in $\times$ 25.4 = mm
    ft^3 $\times$ 2.8317 $\times$ 10^{-2} = m^3
        ft $\times$ 0.3048 = m

Use of Recirculating Granular-Medium Filter in Centralized System. Where a higher degree of treatment is required or where discharge to surface water may be possible, the use of a recirculating sand filter should be considered following the removal of solids. A water-tight septic tank and collection are a must in many areas. As noted previously, a recirculating sand filter is similar to an intermittent sand filter with the following exceptions: (1) effluent from a septic tank or other treatment unit is recirculated through the filter, (2) the effective sand size is larger, and (3) the loading rate based on the effluent flowrate is greater than that for an intermittent sand filter.

The principal components of a recirculating granular-medium filter system are shown in Fig. 14-28. Typical design criteria for recirculating granular-medium filters have been presented previously in Table 14.17. As shown in Fig. 14-28, effluent leaves the septic tank and enters a recirculation tank large enough to hold one-half to one-day's flow. A pump located in the recirculation tank is used to pump wastewater from the recirculation tank to the sand filter. Effluent is applied to the filter for approximately 5 minutes every 30 minutes. Treated effluent from the filter

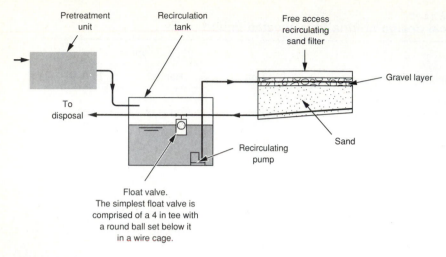

FIGURE 14-28
Principal components of a recirculating sand filter system.

returns to the recirculation tank. The odor level, if any, is low because the septic tank effluent enters a tank containing a treated and oxygenated fluid. Typically, the oxygen concentration in the recirculation tank is greater than 5 mg/L. When the liquid level in the recirculation tank reaches the floating ball valve, effluent from the filter is directed to a sump pump. Treated effluent from the sump pump can be disposed of in a variety of ways depending on local environmental requirements. The high-quality effluent produced, ease of operation, and low maintenance cost are the principal factors contributing to the popularity of recirculating granular-medium filters. The design procedure for a recirculating granular-medium filter is similar to the design procedure for the intermittent-slow sand filter presented previously in Example 14-4. Effluent disposal is considered in Section 14-9.

14-9   SYSTEMS WITH PACKAGE (PRE-ENGINEERED) TREATMENT PLANTS

Commercially available prefabricated treatment plants, known as *package plants*, are often used for the treatment of wastewater for individual properties and small communities. Although package plants are available in capacities up to 1.0 Mgal/d (3800 m³/d), they are used most commonly for wastewater flows in the range from 0.01 to 0.25 Mgal/d (38 to 950 m³/d). Properly sized, operated, and maintained, these plants can usually provide satisfactory treatment for small wastewater flows.

When package plants employing biological treatment first came into use, it was concluded that if they were operated to achieve complete oxidation, no excess biological sludge had to be wasted. As a result of this erroneous conclusion, sludge would build up and periodically discharge from the system. This discharge phenomenon, termed "burping," still occurs in small package plants. In the following discussion the principal operational issues encountered with package plants, the types of package plants most commonly used, and suggested design requirements for package plants are reviewed.

Design and Operational Issues with Package Plants

The major design and operational issues that affect the performance of package plants employing biological treatment (usually some type of activated-sludge process) include [22]:

1. Hydraulic shock loads—the large variations in flow from small communities, accentuated by the use of oversized pumps where wastewater is pumped.
2. Very large fluctuations in both flow and BOD_5 loading.
3. Very small flows that make the design of self-cleansing conduits and channels difficult.
4. Adequate or positive sludge return, requiring provisions for a recirculation rate of up to 3:1, for extended aeration systems to meet all normal conditions.
5. Adequate provision for scum and grease removable from final clarifier.
6. Denitrification in final clarifier, with resultant solids carryover.
7. Inadequate removal and improper provision for handling and disposing of waste sludge.
8. Adequate control of MLSS in the aeration tank.
9. Adequate antifoaming measures.
10. Large and rapid temperature change.
11. Adequate control of air supply rate.
12. Adequate design under organic and solids loadings, which can cause poor treatment performance and odor problems.

Although the above factors are related more specifically to package plants employing biological treatment, many of the factors (e.g., 1, 2, and 3) also apply to package plants employing physical/chemical treatment. Measures that can be taken to address the above issues are discussed below.

Types of Package Plants

The most common types of package plants are (1) extended aeration, (2) contact stabilization, (3) sequencing batch reactors, (4) rotating biological contactor, and (5) physical/chemical. Because these processes have been considered in detail in Chapters 8 and 12, the following discussion is limited to those factors that affect their performance when applied in package plants.

Extended Aeration. A typical example of a commercially available extended-aeration package plant is shown in Fig. 14-29. In general, and as shown in Fig. 14-29, primary clarification is not employed in extended-aeration package plants.

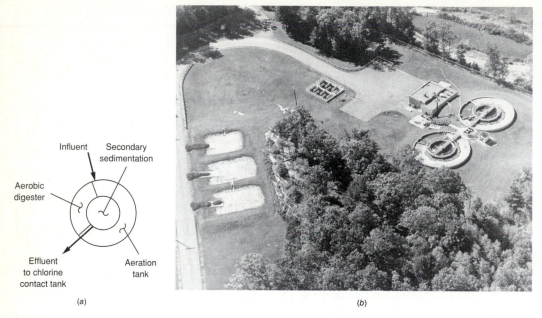

Influent Secondary
sedimentation

Aerobic
digester

Effluent
to chlorine
contact tank

Aeration
tank

(a)

(b)

FIGURE 14-29
Typical example of an extended aeration package plant: (a) schematic-circular configuration and
(b) aerial view of package plant at Sturbridge, MA. (Design average flowrate for each unit = 0.15
Mgal/d.)

To avoid the accumulation of solids, the aeration system should provide sufficient
agitation to keep the solids in suspension. To ensure optimum performance under
field conditions, it is recommended that the maximum organic loading, expressed in
terms of food to microorganism ratio, be in the range from 0.05 to 0.15 lb BOD_5/lb
MLVSS.

    Another critical area of concern is the design of the secondary settling tank and
related facilities. Again, because of the uncertainties of field operation, it is recommended that the overflow rate at the design peak hourly flowrate be limited to 600 to
800 gal/ft$^2 \cdot$ d (24 to 32 m^3/m$^2 \cdot$ d). Positive and effective means should be provided
for returning waste sludge to the aeration chamber. Although air lift pumps have
been used for returning waste sludge, they are undesirable in this application because
the rate of return cannot be adjusted easily or reliably. The secondary settling tank
should also be equipped with scum collection facilities, and an effective system for
the removal of the accumulated scum. Additional design considerations are presented
and discussed at the end of this section.

Contact Stabilization. Contact stabilization (see Fig. 14-30) is most suitable for
the treatment of wastewater in which most of the BOD_5 is in the form colloidal and
suspended particles. The contact stabilization process is used in package plants to
reduce the volume of the aeration tank as compared to the extended aeration process.
Because of the short contact time of 20 to 40 minutes, the contact stabilization process

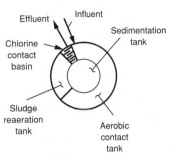

(a)

(b)

FIGURE 14-30
Typical example of a contact stabilization plant: (a) schematic-circular configuration and (b) aerial view of package plant at Penuelas, Puerto Rico. (Design average flowrate = 0.75 Mgal/d.)

should be used in conjunction with flow equalization facilities or for larger flows with less variation in the peak to average flow.

Sequencing Batch Reactor. The sequencing batch reactor (SBR) was pioneered for use in the remote regions of Australia with little or no operator attendance. As with the extended aeration process, to ensure optimum performance under field conditions, it is recommended that the maximum organic loading, expressed in terms of food to microorganism ratio, be in the range from 0.05 to 0.15 lb BOD_5/lb MLVSS. If solids are to be wasted infrequently, adequate volume should be included for the storage of solids (see Example 10-3, Chap. 10).

Rotating Biological Contactor. A typical package plant employing the rotating biological contactor (RBC) process is shown in Fig. 14-31. As shown in Fig. 14-31, a primary settling unit must precede the RBC unit to avoid the deposition of solids in the disk compartment. To ensure effective performance, RBC units should be covered to avoid damage from high winds, to prevent the washing off the biological growth by heavy rains, to avoid freezing problems, to prevent odor migration, and to guard against vandalism.

Physical/Chemical Treatment. A typical flow diagram for a physical/chemical treatment facility is shown in Fig. 14-32. Preliminary treatment usually consists of screening and degritting. The first step of the physical/chemical process is coagulation

TABLE 14-20

Typical design criteria for package treatment plants and other treatment systems for small communities[a]

Design parameter	Unit	Value Range	Value Typical
Plant loadings			
BOD$_5$	lb/capita · d	0.13–0.24	0.18
SS	lb/capita · d	0.13–0.25	0.20
TKN-N	mg/L	15–50	25
NH$_3$	mg/L	5–25	15
PO$_4$-P	mg/L	5–15	10
Extended aeration process			
Pretreatment		Bar screen, communition	
Detention time (aeration tank)	h	18–36	24
BOD$_5$ loading	lb BOD$_5$/lb MLVSS	0.05–0.15	0.10
MLSS (aeration tank)	mg/L	2,500–6,500	3,500
Oxygen required			
Average at 20°C	lb/lb BOD$_5$ applied	2–3	2.5
Peak at 20°C	(value) × avg. flow	1.25–2.0	1.5
Excess sludge	lb/lb BOD$_5$ removed	0.3–0.75	0.4
Settling tank overflow rate based on peak hourly flow	gal/ft^2 · d	600–1,000	800
Aerobic digestion			
Solids detention time	h	10–30	15
VSS loading	lb/ft^3 · d	0.05–0.25	0.15
Sand drying beds	ft^2/capita	1.5–2.5	2.0
Equalization basin volume	% of avg. flow	25–100	50
Rapid sand filters	gal/ft^2 · d (at peak flow)	4–6	5
Chlorination			
Dosage at peak flow	mg/L	15–40	25
Detention time at peak flow	min	15–45	30

followed by sludge blanket clarification. Following clarification, the treated wastewater is passed through a granular-medium filter for the removal of residual solids and an activated-carbon filter for the removal of any residual trace organics. Because of the problems associated with the handling and disposal of sludge and the high operating costs, physical/chemical treatment is not widely used. Physical/chemical

TABLE 14-20
(continued)

Design parameter	Unit	Value Range	Value Typical
Contact stabilization process			
Pretreatment		Bar screen, communition	
Detention time (contact tank)	min	20–40	30
Detention time (reaeration tank)	h	20–36	24
BOD$_5$ loading (aeration tank)	lb/10^3 ft$^3 \cdot$ d		
MLSS (contact tank)	mg/L	1,500–2,500	2,000
MLSS (reaeration tank)	mg/L	2,500–6,500	3,500
Oxygen required			
Average at 20°C	lb/lb BOD$_5$ applied	2–3	2.5
Peak at 20°C	(value) × avg. flow	1.25–2.0	1.5
Excess sludge	lb/lb BOD$_5$ removed	0.3–0.75	0.4
Settling tank overflow rates based on peak hourly flow	gal/ft$^2 \cdot$ d	600–1,000	800
Sequencing batch reactor			
Pretreatment		Bar screen, communition	
Detention time	min	16–36	24
BOD$_5$ loading	lb BOD$_5$/lb MLVSS	0.05–0.15	0.10
MLSS	mg/L	2,500–6,500	3,500
Oxygen required			
Average at 20°C	lb/lb BOD$_5$ applied	2–3	2.5
Peak at 20°C	(value) × avg. flow	1.25–2.0	1.5
Rotating biological contactors			
Pretreatment		Bar screen, communition	
Surface loading	gal/ft$^2 \cdot$ d	1–2.5	1.5
Total BOD$_5$ loading	lb/10^3 ft$^2 \cdot$ d	6–10	8
Other factors		See extended aeration system	

[a] Adapted in part from Refs. 10, 20, 22, and 24.

Note: lb/capita $\cdot$ d × 0.4536 = kg/capita $\cdot$ d
gal/ft$^2 \cdot$ d × 24.5424 = m^3/m$^2 \cdot$ d
lb/ft$^3 \cdot$ d × 16.0185 = kg/m$^3 \cdot$ d
lb/10^{3}ft$^2 \cdot$ d × 0.0049 = kg/m$^2 \cdot$ d

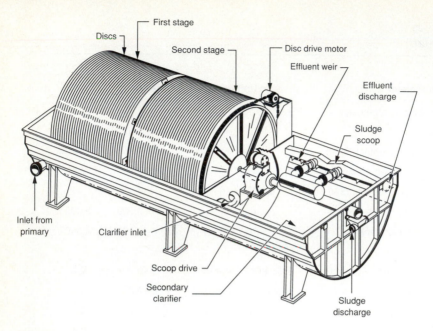

FIGURE 14-31
Typical example of a rotating biological contactor package plant.

package plants are used in cold climate areas because of their small size, their on-off operation, and reliability [22].

Improving the Performance of Package Plants

The performance of most package plants can be improved by sizing the treatment facilities conservatively, especially the secondary settling facilities, and by specifying positive means for handling and pumping the side stream flows. Design criteria for package plants are presented in Table 14-20. Selecting a prefabricated package plant is illustrated in Example 14-5.

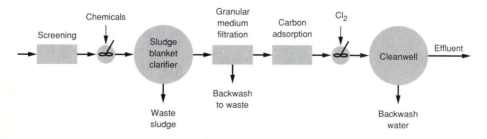

FIGURE 14-32
Flow diagram for a typical physical/chemical package treatment plant.

Example 14-5 Selecting a prefabricated package plant. A prefabricated package plant is to be used to treat the wastewater from a small subdivision consisting of 550 individual family residences. The average occupancy has been estimated to be 2.9 persons per residence. Use a flow of 60 gal/person · d and a daily peaking factor of 2.5 for flow, BOD_5 and SS. Use an hourly peaking factor of 4 for sizing the sedimentation facilities. Select the type of package plant and size the principal components of the plant. The effluent from the package plant must meet EPA secondary standards (see Table 4-1).

Solution

1. Estimate the average and peak wastewater flowrates
    (*a*) The total number of persons is:
$$550 \times 2.9 \text{ persons/home} = 1595 \text{ persons}$$
    (*b*) The corresponding average flowrate based on 60 gal/person · d is:
$$1595 \text{ persons} \times 60 \text{ gal/person} \cdot d = 95,700 \text{ gal/d}$$
    (*c*) The corresponding peak daily flowrate based on factor of 2.5 (see Table 14-2) is:
$$95,700 \text{ gal/d} \times 2.5 = 239,250 \text{ gal/d}$$

2. Estimate the daily BOD_5 and SS mass loading rates to be treated
    (*a*) Using a value of 0.18 lb/capita · d (see Table 14-4), the average BOD_5 mass loading rate is:
$$1595 \times 0.18 \text{ lb/capita} \cdot d = 287 \text{ lb/d}$$
    Based on average flowrate of 95,700 gal/d, the corresponding BOD_5 concentration is:
$$BOD_5, \text{ mg/L} = \frac{287 \text{ lb/d}}{0.0957 \text{ Mgal/d} \times [8.34 \text{ lb/Mgal} \cdot (\text{mg/l})]} = 360 \text{ mg/L}$$
    (*b*) Using a value of 0.20 lb/capita · d (see Table 14-4) the average SS daily mass loading rate is:
$$1595 \times 0.20 \text{ lb/capita} \cdot d = 319 \text{ lb/d}$$
    Based on average flowrate of 95,700 gal/d, the corresponding SS concentration is:
$$SS, \text{ mg/L} = \frac{319 \text{ lb/d}}{0.0957 \text{ Mgal/d} [8.34 \text{ lb/Mgal} \cdot (\text{mg/L})]} = 400 \text{ mg/L}$$
    (*c*) Using a peaking factor of 2.5, the peak daily BOD_5 mass loading rate is:
$$287 \text{ lb/d} \times 2.5 = 718 \text{ lb/d}$$
    (*d*) Using a peaking factor of 2.5, the peak daily SS mass loading rate is:
$$319 \text{ lb/d} \times 2.5 = 798 \text{ lb/d}$$

3. Select the type of treatment process
    (*a*) An extended aeration activated-sludge process package plant is recommended. The principal reasons for selecting an extended aeration activated-sludge process are: (1) excellent effluent quality, (2) relatively low sludge yield, (3) relative simplicity, and (4) relative ease of operation.

(b) Assume a bar rack and rotary drum screen will be used for removing coarse solids, and that primary sedimentation facilities will not be used.

4. Size the principal treatment process components

(a) Using a detention time of one day at average flow (see Table 14-20), the required aeration tank volume is equal to the average daily flow. Thus, the aeration tank volume is equal to:

$$\text{Volume} = (95,700 \text{ gal/d}) \times (1.0 \text{ d}) = 95,700 \text{ gal}$$

(b) The aeration system must be capable of providing the required amount of oxygen to meet the sustained peak demand. Thus, based on the peak organic loading rate and assuming an oxygen transfer efficiency of 6 percent, determine the required capacity of the aeration system.

(1) Assume specific weight of air at standard temperature and pressure is 0.0752 lb/ft^3 and contains 23.2 percent oxygen by weight.

(2) The theoretical air requirement is:

$$\text{Air req.} = \frac{718 \text{ lb/d}}{(0.0752 \text{ lb/ft}^3)(0.232)(0.06)(1440 \text{ min/d})} = 476 \text{ ft}^3/\text{min}$$

(c) Using a peak hour factor of 4 and an overflow rate of $800 \text{ gal/ft}^2 \cdot \text{d}$ (see Table 14-20), the surface area required for the secondary settling tank is:

$$\text{Surface area} = \frac{95,700 \text{ gal/d} \times 4}{800 \text{ gal/ft}^2 \cdot \text{d}} = 479 \text{ ft}^2$$

(d) Using a detention time of 30 min at peak flow, the required size of the chlorine contact tank is:

$$\frac{95,700 \text{ gal/d} \times 4 \times 0.5 \text{ h}}{(24 \text{ h/d})} = 7975 \text{ gal}$$

Comment. To avoid problems with rising sludge due to denitrification, the secondary settling tank must be equipped with positive means for removing the accumulated sludge.

Effluent Disposal Options

The means of disposing of effluent from small community wastewater processing facilities depends on the degree of treatment provided. The principal means of effluent disposal following secondary and advanced treatment are summarized in Table 14-21. Specific information on the various effluent disposal options identified in Table 14-21 may be found in Chaps 13, 16, and 17.

14-10 INDIVIDUALLY DESIGNED TREATMENT FACILITIES

Individually designed built-in-place plants (see Fig. 14-33) can also be designed for communities instead of prefabricated package plants described previously. In the following discussion, the principal types of systems used are identified, some typical process flow diagrams are reviewed, and suggested design requirements for individually designed plants for small communities are presented.

TABLE 14-21
Effluent disposal options for small communities

Disposal option	Following indicated level of treatment		
	Primary	Secondary	Advanced
Subsurface absorption	√	√	√
Rapid infiltration	√	√	√
Spray disposal		√	√
Drip application			√
Irrigation			√
Constructed wetlands		√	√
Surface water discharge		√	√
Indirect reuse			√

Treatment Processes

The most common types of individually designed treatment processes for small communities are

1. Extended aeration activated-sludge
2. Oxidation ditch activated-sludge

FIGURE 14-33
Typical example of an individually designed built in-place oxidation ditch activated-sludge process at Greenville, SC. (Design average flowrate for each unit = 2.0 Mgal/d.) Wrap around design was used to achieve common wall construction. The same type of construction can be used effectively for smaller facilities.

3. Sequencing batch reactor activated-sludge
4. Rotating biological contactor
5. Trickling filter
6. Facultative lagoons
7. Aerated lagoons
8. Land treatment systems
9. Aquatic plant systems

Because these processes have been considered in detail previously in Chapters 8, 10, 11, and 13, the following discussion is limited to a review of some of the alternative types of flow diagrams that have been used and recommended design criteria for these processes.

Process Flow Diagrams

Typical process flow diagrams involving the use of the treatment processes mentioned earlier are shown in Fig. 14-34. The diversity of applications that have been used is apparent from a review of the process flow diagrams shown in Fig. 14-34. As shown, a number of the treatment processes have been combined to achieve specific treatment objectives. The key issue in using these either singly or in various combinations is to select appropriate design criteria.

14-11   SEPTAGE AND SEPTAGE DISPOSAL

Septage, as noted previously, is the combination of sludge, scum, and liquid pumped from a septic tank. To minimize any adverse impacts to the environment, septage must be disposed under controlled conditions. The purpose of the following discussion is to review the characteristics and quantities of septage and methods that can be used for its treatment and disposal. Additional details on the treatment and disposal of sludge may be found in Ref. 24.

Characteristics and Quantities of Septage

Data on the characteristics and quantities of septage must be available to design septage treatment and disposal facilities properly.

Characteristics of Septage. The characteristics of septage will vary depending on usage, whether a kitchen food waste grinder is used, and the frequency of pumping. The degree of digestion the solids in the septic tank have undergone will depend on the frequency of pumping. Because of the changing nature of the input to a septic tank, the characteristics of septage can be quite variable. Typical data on the characteristics of septage are presented in Table 14-22. The application of the data presented in Table 14-22 is illustrated in Example 14-6.

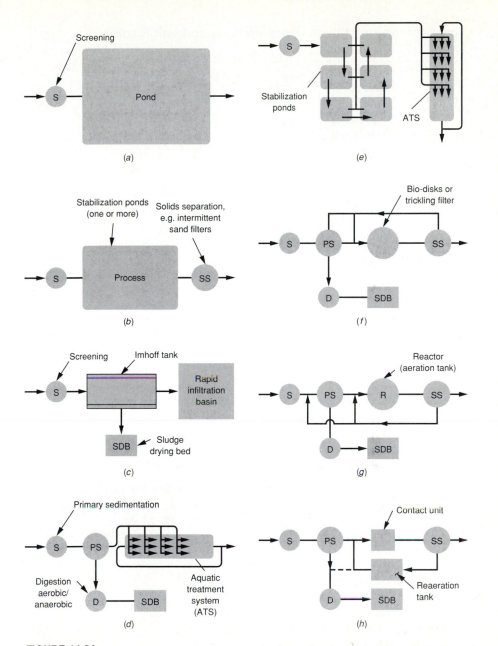

FIGURE 14-34
Typical flow diagrams for individually designed small treatment facilities using: (a) and (b) stabilization ponds, (c) Imhoff tank, (d) aquatic treatment, (e) stabilization ponds and aquatic treatment,
(f) biodisks or trickling filters, and (g) and (h) activated sludge.

TABLE 14-22
Typical septage constituent concentrations and unit loading factors

Constituent	Concentration, mg/L		Unit loading[a], lb/capita · d	
	Range	Typical	Range	Typical
BOD$_5$	2,000–30,000	6,000	0.01–0.03	0.02
TS	4,000–100,000	40,000		
SS	2,000–100,000	15,000	0.02–0.10	0.05
VSS	1,200–14,000	7,000		
TKN	100–1,600	700		
NH$_3$	100–800	400		
TP	50–800	250		
Grease	5,000–10,000	8,000		
Heavy metals[b]	100–1,000	300		

[a] If concentration data are not available, the loading factors given above can be used to estimate the waste loadings that would be expected from the septage of unsewered areas (see Example 14-6).

[b] Primarily iron (Fe), zinc (Zn), and aluminum (Al).

Note: lb × 0.4536 = kg

Quantities of Septage. The buildup of sludge and scum in a septic tank will also vary depending on usage, whether a kitchen food waste grinder is used, and the frequency of pumping. Because the scum and sludge cannot be pumped out of a tank selectively, the usual procedure is to pump out the entire contents of the tank each time the tank is pumped out. Thus the quantity of septage from a residence will depend on the size of the septic tank and the cleaning frequency.

Methods for the Treatment and Disposal of Septage

The principal methods most commonly used for the treatment and disposal of sludge are as follows:

1. Land application
   Surface application
   Subsurface application
2. Co-treatment with wastewater
   Biological treatment
   Chemical treatment
3. Co-disposal with solid wastes
   Landfilled with solid wastes
   Composting
4. Processing at separate facilities
   Biological treatment
   Lime stabilization
   Chemical oxidation
   Composting

Land Application. Land application is one of the methods that has been used most commonly for the treatment and disposal of septage. Both surface and subsurface methods have been used to apply septage to the land.

Surface application. Septage can be applied to the surface of the land with spray guns, trucks equipped with liquid spreaders, and liquid manure spreaders used on farms. Surface spreading should be followed by a short drying period and then disking. The principal concerns associated with the direct application of septage on land are the potential health risks, the possible contamination of groundwater, and the production of nuisance conditions and odors. Because of the many problems associated with the surface application of septage on land, this practice is no longer permitted in a number of locations.

Septage can also be dewatered in lagoons or on drying beds and applied to the land in a solid or semi-solid form. Both surface and subsurface application methods can be used. Dewatered sludge can be applied to the soil surface with specially designed trucks or manure spreaders. Where heavy equipment is required, care must be taken not to damage crops during the growing season.

Subsurface application. Most of the problems associated with the surface application of septage can be eliminated or overcome by subsurface application. The methods used most commonly for the subsurface application of septage are: (1) the furrow cover method in which septage is applied in narrow furrows and covered with soil by a following plow, and (2) the injection method in which septage is injected in a wide band or several narrow bands some 6 to 8 in below the surface of the soil (see Fig. 12-48). Properly managed and subject to constituent loading limitations (see Sec. 12-17, Chap. 12), land application can provide effective treatment and disposal of septage.

Co-Treatment with Wastewater. Co-treatment with wastewater at the local wastewater treatment plant is one of the most effective methods for the treatment and disposal of septage. However, many treatment plants do not have the facilities required to unload the pumper trucks or the excess capacity to process septage.

Septage receiving stations. If septage is to be co-treated with wastewater it will be necessary to construct a septage receiving station. Typically, as shown in Fig. 14-35, such a station will consist of an unloading area, a septage storage tank, and one or more grinder-type transfer pumps. The storage tank is used to store the septage so that it can be discharged to the treatment plant. The storage tank should be covered for odor control. Discharge of septage to a headworks is usually preferred for the removal of grit and screenings. If there are no screening or comminution facilities ahead of the septage discharge facility, the septage should be transferred from the storage tank to the treatment plant with grinder pumps (see Fig. 14-35). In some cases, this transfer can be accomplished by gravity flow. If the septage is especially strong, it can be diluted with treated wastewater. Chemicals such as lime or chlorine can also be added to the septage in the storage tank to neutralize it, to render it more treatable, or to reduce odors. If the capacity limitations do not exist

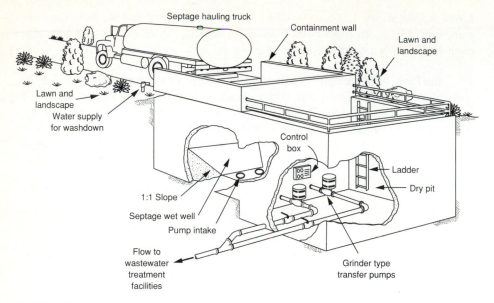

FIGURE 14-35
Typical septage receiving station located at a wastewater treatment plant.

at the treatment plant, the co-treatment and disposal of septage with wastewater is one of the most cost-effective and environmentally sound methods that can be used for the management of septage.

Treatment plant capacity limitations. In some cases septage cannot be discharged directly from the pumper truck into the headworks of a treatment facility, because the plant may become overloaded with organic matter. This is often the situation for small plants with limited capacity (see Example Problem 14-6). Problems of organic overloading can be overcome by collecting and storing the septage during the daytime hours and then discharging it to the treatment plant in the early morning hours or when the incoming organic and solids loadings to the plant are lowest. Most plants have excess treatment capacity for the liquid portion of the septage during the early morning hours. A more serious problem arises when the solids processing capacity of the plant is limited.

Example 14-6   Determination of impact of septage discharges on BOD_5 and SS loadings for small treatment plant. Estimate the added plant BOD_5 and SS loadings resulting from the septage delivered to a small biological treatment plant from a rural unsewered area containing about 2000 homes. Assume that the average volume pumped from each septic tank is 1000 gal (3.8 m^3) and each tank is to be pumped once every four years. Also determine the percent increase in the plant BOD_5 and SS loadings for a wastewater treatment plant having an average flowrate of 1.0 Mgal/d (3800 m^3/d) and BOD_5 and SS concentrations of 220 and 200 mg/L, respectively.

Solution

1. Determine the averge number of septic tanks pumped out per day.
   (a) Number of homes serviced each year is

   $$\text{Homes/year} = \frac{2000}{4} = 500$$

   (b) Assume septic tanks are pumped out during a 250-day period each year.
   (c) The number of septic tanks pumped out per day

   $$\text{Tank/d} = \frac{500 \text{ tanks/yr}}{250 \text{ d/yr}} = 2 \text{ tanks/d}$$

2. Determine the BOD_5 and SS loadings using the concentration values given in Table 14-22.
   (a) The volume of septage pumped out per day

   $$\text{Volume/d} = (1000 \text{ gal/tank}) \times (2 \text{ tanks/d})$$

   $$= 2000 \text{ gal/d}$$

   (b) The BOD_5 loading per day is

   $$BOD_5 \text{ loading} = (0.002 \text{ Mgal/d}) \times (6000 \text{ mg/L}) \times [8.34 \text{ lb/Mgal} \cdot (\text{mg/L})]$$

   $$= 100 \text{ lb/d}$$

   (c) The SS loading per day is

   $$\text{SS loading} = (0.002 \text{ Mgal/d}) \times (15,000 \text{ mg/L}) \times [8.34 \text{ lb/Mgal} \cdot (\text{mg/L})]$$

   $$= 250 \text{ lb/d}$$

3. Determine the increase in the plant BOD_5 and SS loadings
   (a) The plant BOD_5 loading per day without septage is

   $$BOD_5 \text{ loading} = (1.0 \text{ Mgal/d}) \times (220 \text{ mg/L}) \times [8.34 \text{ lb/Mgal} \cdot (\text{mg/L})]$$

   $$= 1834.8 \text{ lb/d}$$

   (b) The plant SS loading per day without septage is

   $$\text{SS loading} = (1.0 \text{ Mgal/d}) \times (200 \text{ mg/L}) \times [8.34 \text{ lb/Mgal} \cdot (\text{mg/L})]$$

   $$= 1668.0 \text{ lb/d}$$

   (c) The increased plant BOD_5 loading per day with septage is

   $$\text{Percent increase} = \frac{100 \text{ lb/d}}{1834.8 \text{ lb/d}} \times 100 = 5.5\%$$

   (d) The increased plant SS loading per day with septage is

   $$\text{Percent increase} = \frac{250 \text{ lb/d}}{1668.0 \text{ lb/d}} \times 100 = 15.0\%$$

Comment. Peak day delivery of up to 3 times the average number of loads can be expected unless controlled by the municipality. With the increased plant BOD_5 and SS loading values computed in this example, it is easy to see why many small plants are overloaded when they accept septage.

Co-Disposal with Solid Wastes. Septage can also be co-disposed with solid wastes. The most common co-disposal methods include landfilling and composting.

Landfilling. In the past, the co-disposal of septage with solid wastes in a landfill has been used extensively by small communities throughout the United States. Because of possible contamination of the underlying groundwater, a number of communities and water pollution control agencies have banned the co-disposal of septage with solid wastes. If the landfill is sealed properly to eliminate the percolation of leachate from the bottom of the landfill, co-disposal of septage with solid wastes has been allowed with proper monitoring.

Where gas (methane) is to be recovered from a sanitary landfill, septage can be used to provide the moisture and microorganisms necessary to accomplish the biological conversion of the solid wastes. If gas recovery is to be part of the landfill operation, special care must be taken to seal the landfill to eliminate leachate contamination.

Composting. Septage solids can be co-composted with solid wastes to produce a humus-like end product. Composting may be defined as the biological stabilization (decomposition) of the organic matter in the septage and waste paper in the presence of oxygen under thermophilic 120 to 135°F (49 to 57°C) dewatered conditions. Composting of septage with solid wastes is usually accomplished by one of three methods: windrow, static aerated pile, and in-vessel. These methods are described in detail in Chap. 12.

Processing at Separate Facilities. If none of the above methods can be used for the treatment and disposal of septage, consideration must be given to the design and construction of facilities specifically for the purpose. Septage processing at specially designed facilities may be accomplished by: (1) biological treatment, (2) combined physical and biological treatment, (3) lime stabilization, and (4) chemical oxidation. One of the major problems associated with the processing of septage at separate facilities is that some method must be found for the disposal of the liquid and solid portions of the septage after treatment.

Biological treatment. The biological treatment of septage is usually accomplished in (1) either aerobic or aerobic/anaerobic (facultative) lagoons (see Fig. 14-36a), (2) conventional biological treatment facilities (see Fig. 14-36b), and (3) combined physical and biological treatment facilities (see Fig. 14-36c). An alternative biological treatment system employing a series of reactors with aquatic plants and a marsh system has also been developed for the treatment of septage (see Fig. 14-36d).

Aerobic lagoons are shallow 1 to 3 ft (0.3 to 0.9 m) impoundments into which the septage is discharged for treatment. Typically, two lagoons are used so that one can be dewatered and dried for solids removal. The dried solids are usually disposed of in a landfill or spread on land. Lagoon effluent can be disposed of (1) in infiltration/percolation beds, (2) by spray disposal on land, (3) by evaporation or (4) by further treatment with a recirculating sand filter and surface water discharge.

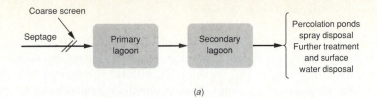

(a)

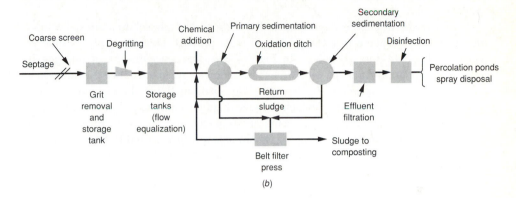

(b)

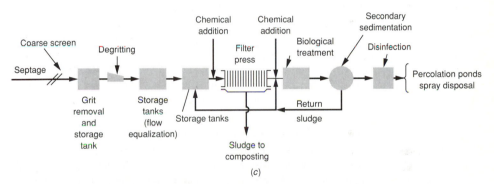

(c)

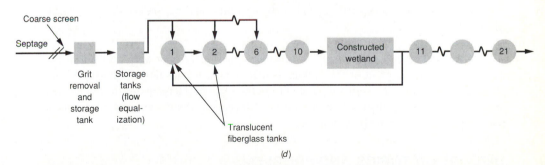

(d)

FIGURE 14-36
Typical flow diagrams for the biological and physical treatment of septage: (a) using a two-stage lagoon system, (b) using chemical addition for phosphorus removal, biological treatment for partial nitrogen removal, and effluent filtration for the removal of residual solids, (c) using chemical addition for phosphorus removal, plate and frame filter press for solids and phosphorus removal, and biological treatment for nitrification, and (d) using a complex ecosystem in a series of complete-mix tanks followed by a wetland (marsh) followed by a second series of complete-mix tanks.

In contrast, facultative lagoons are relatively deep 4 to 10 ft (1.2 to 3.0 m) impoundments. Because the lagoons are deep, they are drained for solids removal only after extended periods (e.g., 5 years). The effluent is disposed of in the same manner as that from aerobic lagoons. Because of the potential nuisance problems that can arise from their use, facultative lagoons are not favored by most regulatory agencies.

Where the discharge requirements for nitrogen and phosphorus are quite low, the treatment of septage is accomplished in more process intensive facilities such as shown in Figs. 14-36b, c, and d. In Fig. 14-36b, the concentration of nitrogen is reduced to about 50 mg/L or below in the biological treatment process. Additional nitrogen removal is accomplished in the spray disposal fields. In Fig. 14-36c, the effluent from the filter press, which is quite high in ammonia, is nitrified in the biological process. Denitrification occurs in the percolation ponds. In Fig. 14-36d, a complex ecosystem of bacteria and higher biological forms is used to bring about the treatment of the septage, including the removal of nitrogen, phosphorus, and heavy metals.

Lime stabilization. In the lime stabilization process, lime is added to stabilize the septage and to destroy pathogenic organisms. For the process to be effective, the pH must be raised to a value of 12 or greater for at least 30 minutes. After lime treatment, the solids must be removed. The liquid and the solids must be disposed of separately. Because of the number of treatment steps involved in the process and the cost of chemicals, this process is not often used on a long-term basis. Lime stabilization can be used to deal with short-term septage disposal problems.

Chemical oxidation. The most common chemical oxidation process involves the use of chlorine gas for the stabilization of the septage. Because of the cost and complexity of this and similar processes, chemical oxidation has not been used extensively for the treatment of septage. As with lime stabilization, it can be used as a temporary treatment method, but both the liquid and separated solids must be disposed of separately.

Composting. Composting, as described previously, can only be used effectively if the solids content of the septage is high. Thus, it is usually used as a further treatment method for the solids separated from the septage. Here again, the composted septage solids require disposal. In a number of locations, the disposal of the compost has proven to be extremely expensive.

DISCUSSION TOPICS AND PROBLEMS

14-1. Obtain the design criteria used for onsite systems in your county or a nearby county where onsite systems are used, and compare them to the criteria given in this chapter with respect to the sizing of septic tanks and the design of the disposal fields. Identify and discuss any major differences and the impact of any differences with respect to the design of onsite systems.

14-2. Size and layout an onsite system with alternating disposal fields for a home with an average occupancy of three persons for the site shown in the accompanying figure. Based on a pump-in test, it has been found that the vertical permeability is 0.67 gal/ft^2 · d. The total depth of the disposal fields is to be 4.5 ft. A soil cover of 1.0 ft will be used and the effluent distribution pipe is to be located 1.5 ft below the ground surface (see Fig. 14-6). Assume that 2.5 ft of the side wall area measured from the bottom of the trench will be useful for the infiltration of septic tank effluent. Use the data given in Tables 14-8 and 14-13.

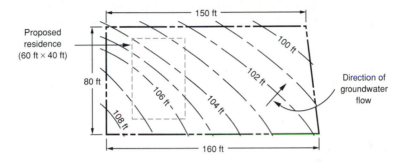

14-3. Assume that the site shown in Problem 14-2 is essentially flat. If the vertical coefficient of permeability is 0.50 gal/ft^2 · d, estimate the maximum amount of septic tank effluent that could be disposed of per day if a factor of safety of 3 is to be used and none of the septic tank effluent is to leave the site laterally. Prepare a preliminary layout for the proposed onsite system, assuming that a single disposal field will be used. Assume that the horizontal separation distances given in Table 14-8 are applicable.

14-4. Size and layout an onsite system with a single disposal field for a home with an average occupancy of five persons for the site shown in the accompanying figure. If the maximum groundwater level is approximately 4.0 ft below the groundwater surface during winter conditions, estimate the distance to the groundwater from the surface of the ground at the edge of the property. The coefficient of permeability in the direction of the groundwater flow is estimated to be 0.65 ft/d. Assume that 10 percent of the applied septic tank effluent will leave the site laterally. Assume pressure-dosed shallow trenches of the type shown in Fig. 14-12 will be used.

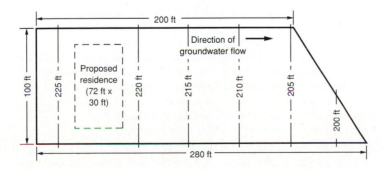

14-5. Size and layout an intermittent sand filter and pressure-dosed alternating disposal fields for the site shown in Prob. 14-2. Assume that the data given in Prob. 14-2 are applicable.

14-6. Size and layout an intermittent sand filter and a single pressure-dosed disposal field for the site shown in Prob. 14-4. Assume that the vertical permeability is 0.67 gal/ft$^2 \cdot$ d and an average home occupancy of 4.0 persons/d.

14-7. Assume that the depth of the soil cover over a relatively impermeable substrata in the site shown in Prob. 14-2 is equal to 1.5 ft. If the groundwater level must remain 1.0 ft below the surface of the ground, how much soil should be added to develop an onsite disposal system? Assume that none of the flow moves vertically downward. Assume that shallow trenches of the type shown in Fig. 14-12 will be used.

14-8. Demonstrate that, for columnar flow, the hydraulic gradient is equal to one.

14-9. Using the following data obtained from a pump-in test, determine the saturated coefficient of permeability.

> Trench dimensions: width = 2.0 ft. length = 20 ft, and depth = 3 ft
> Water depth maintained in test trench = 1.5 ft
> Area of plume formed before applied water moves downward = 423 ft^2 (see Fig. 14-18)
> Depth below trench bottom to the periphery of the plume = 2.7 ft
> Soil porosity = 39%
> Height of capillary zone = 14 in
> Degree of saturation of the capillary zone = 30%
> Total water applied during test = 3750 gal
> Total elapsed time = 189 hr

14-10. Size and prepare a layout sketch of a recirculating sand filter for a community with a population of 1000 persons. Use a flowrate of 55 gal/capita $\cdot$ d.

14-11. Review three recent articles (since 1980) on the travel of bacteria in soil. Compare the conclusions in these articles to the information presented in this chapter.

14-12. It is often stated that groundwater supplies are contaminated by the nitrogen (converted to nitrate) discharged from septic tanks. Estimate the maximum amount of nitrate that might reach the groundwater from a 100 acre development with the following characteristics:

> Average lot size = 0.25 acres
> Area in streets and parks = 15% of total area
> Average residents per home = 3.2 persons/home

Compare the amount of nitrogen added (expressed in lb N/acre $\cdot$ yr) from the residential development to the amount of nitrogen added as result of landscape fertilization and to the amount of nitrogen added in a conventional agricultural operation. Assume that a crop such as alfalfa is grown.

14-13. Size a prefabricated contact stabilization activated-sludge process to treat the waste from 1500 persons. Prepare an outline sketch to scale of the process, using a circular configuration such as the one shown in Fig. 14-30. Include a chlorine-contact basin in the outer ring of the tank. Assume the following conditions:

> Average wastewater flowrate = 60 gal/person $\cdot$ d
> Ratio of peak day flowrate to average day flowrate = 2.6
> Ratio of peak hour flowrate to average day flowrate = 5.5

14-14. Determine the impact on a small wastewater treatment plant, of accepting septage from a rural area containing about 2500 homes. Assume that the following data are applicable:

Average home occupancy in unsewered area = 3.25 persons/home
Average size of septic tank = 1200 gal/tank
Volume pumped = 1000 gal/pumping
Average time between septic tank pumpings = 3 years
Wastewater treatment plant data:

> Average flowrate = 0.65 Mgal/d
> Average BOD_5 = 230 mg/L
> Average SS = 220 mg/L

Use typical septage data given in Table 14-22.

14-15. Prepare a preliminary design for a double-lined, two-stage lagoon system to treat 50,000 gal/d of septage with the characteristics given in Table 14-22. Assume that the first lagoon will be anaerobic and that the second lagoon will be facultative. Use the design criteria given in Chap. 10 to size the lagoons.

14-16. An oxidation ditch activated-sludge process will be used to treat 50,000 gal/d of septage. The septage will be screened and degritted before primary sedimentation and biological treatment. Following secondary sedimentation, the effluent will be filtered before land disposal. Prepare a preliminary design for this process, assuming that the total nitrogen in the influent will be reduced to 40 mg/L or less. Refer to Chap. 11 for criteria that can be used for the design of the process to achieve nitrification and denitrification. Assume that the following data are applicable:

Capacity of flow equalization tank = 150,000 gal
Septage characteristics = (see Table 14-22)

REFERENCES

1. *Alternative Sewer Systems,* Manual of Practice no. FD-12, Facilities Development, Water Pollution Control Federation, Alexandria, VA, 1986.
2. Anderson, D. L., R. L. Siegrist, and R. J. Otis: "Technology Assessment of Intermittent Sand Filters," U.S. Environmental Protection Agency, Office of Municipal Pollution Control (WH-546), Washington, DC, April 1985.
3. Fair, G. M.: "The Hydraulics of Rapid Sand Filters," *Journal of the Institute of Water Engineers,* vol. 5, no. 2, March 1951.
4. Fair, G. M., and J. C. Geyer: *Water Supply and Waste-Water Disposal,* John Wiley & Sons, New York, 1954.
5. Flowers, J. E.: "The Importance of Wastewater Facilities Planning In Small Communities," *Small Alternative Wastewater Systems, Design Workshop Notes,* Developed under Grant #T-901092-010,U.S. Environmental Protection Agency, 1983.
6. Frank, L. C., and C. P. Rhynus: "The Treatment of Sewage From Single Houses And Small Communities," *Public Health Bulletin No. 101,* Treasury Department, United States Public Health Service, Washington, DC, 1920.
7. Kreissl, J. F.: "On-site Wastewater Disposal Research in the United States," A. S. Eikum and R. W. Seabloom (eds.), in *Alternative Wastewater Treatment,* D. Reidel Publishing Company, Boston, MA, 1982.
8. Laak, R.: *Wastewater Engineering Design For Unsewered Areas,* 2nd ed., Technomic Publishing Co., Lancaster, PA, 1986.
9. Luthin, J. N.: *Drainage Engineering,* John Wiley & Sons, New York, 1966.
10. Metcalf & Eddy, Inc.: *Wastewater Engineering: Treatment, Disposal, Reuse,* 2nd ed., McGraw-Hill, New York, 1979.
11. Metcalf & Eddy, Inc.: *Wastewater Engineering: Collection and Pumping of Wastewater,* McGraw-Hill, New York, 1981.

12. Metcalf, L.: "The Antecedents of the Septic Tank," *Transactions of the American Society of Civil Engineers,* vol. XLVI, December 1901.

13. Orlob, G. T., and R. G. Butler: *An Investigation of Sewage Spreading on Five California Soils,* Sanitary Engineering Research Laboratory Technical Bulletin 12, University of California, Berkeley, CA, 1955.

14. Orlob, G. T., and R. B. Krone: *Final Report on Movement of Coliform Bacteria Through Porous Media,* Sanitary Engineering Research Laboratory Technical, University of California, Berkeley, CA, November 1960.

15. Proceedings of National Conference on Less Costly Wastewater Treatment Systems For Small Communities, held in Reston, VA on April 13–15, 1977, U.S. Environmental Protection Agency, Washington, DC, 1977.

16. Romero, J. C.: "The Movement of Bacteria and Viruses Through Porous Media," *Ground Water,* vol. 8, no. 2, 1970.

17. "Small-Diameter, Variable-Grade, Gravity Sewers For Septic Tank Effluent," *Proceedings of the 3rd National Symposium On Individual and Small Community Sewage Treatment,* ASAE Publication no. 1-82, 1982.

18. Simmons, J. D. , and J. O. Newman, et al.: "Design Workbook For Small-Diameter, Variable-Grade, Gravity Sewer," Small Alternative Wastewater Systems, Design Workshop Notes, Developed under Grant no. T-901092-010, U.S. Environmental Protection Agency, 1983.

19. "Site Evaluation For Subsurface Wastewater Disposal Design in Maine," Maine Department of Human Services, Division of Health Engineering, 2nd ed., June 1987.

20. Tchobanoglous, G.: "Wastewater Treatment for Small Communities," *Public Works,* Part One, vol. 105, no. 7, 1974, and Part Two, vol. 105, no. 8, 1974.

21. Thrasher, D.: *Design and Use of Pressure Sewer Systems,* Lewis Publishers, Chelsea, MI, 1987.

22. U.S. Environmental Protection Agency: *Process Design Manual Wastewater Treatment Facilities For Sewered Small Communities,* EPA-625/1-77-009, Environmental Research Information Center, Cincinnati, OH, October 1977.

23. U.S. Environmental Protection Agency: Design Manual: *Onsite Wastewater Treatment and Disposal Systems,* EPA 625/1-80-012, Office of Water Program Operations, Washington, DC, 1980.

24. U.S. Environmental Protection Agency: *Handbook Septage Treatment and Disposal,* UEPA 625/6-84-009, Center for Environmental Research Information, Cincinnati, OH, October 1984.

25. Warshall, P.: *Septic Tank Practices,* Anchor Books, Anchor Press/Doubleday, Garden City, NY, 1979.

26. *Wastewater Management Program Rules and Regulations,* 1989 edition, Stinson Beach County Water District, Stinson Beach, CA, 1989.

27. Weibel, S. R., C. P. Straub, and J. R. Thoman: "Studies On Household Sewage Disposal Systems," Federal Security Agency, Public Health Service, Environmental Health Center, Cincinnati, OH, 1949.

28. Wert, S. R., and R. C. Paeth: "Performance of Disposal Trenches Charged With Recirculating Sand Filter Effluent," *Proceedings of the 5th Northwest On-Site Wastewater Treatment Short Course,* University of Washington, Seattle, WA, September 1985.

29. Wert, S. R.: "Determination of the Assimilative Capacity Using Trench Pump-In Tests," *Proceedings of the 6th Northwest On-Site Wastewater Treatment Short Course,* University of Washington, Seattle, WA, September 1986.

30. Winneberger, J. H. T.: *Nitrogen, Public Health, And The Environment,* Ann Arbor Science Publishers (Now Butterworth Publishers), Boston, MA, 1982.

31. Winneberger, J. H. T.: *Septic Tank Systems: A Consultant's Toolkit,* Butterworth Publishers, Boston, MA, 1984.

CHAPTER

15

MANAGEMENT OF WASTEWATER FROM COMBINED SEWERS

A combined sewer carries both wastewater and stormwater runoff. Although new combined sewer systems are no longer being built in the United States, they are an extensive part of the existing infrastructure in many locations, particularly in older urban areas. Combined sewer overflows (CSOs) discharged to receiving waters have resulted in contamination problems that have often prevented the attainment of water quality standards. Contaminants discharged from CSOs that can cause adverse receiving water effects include bacteria, nutrients, solids, BOD, metals, and other potentially toxic constituents. Some idea of the extent of CSO impacts can be gained from a recent EPA survey that reported the necessity of spending over 16 billion dollars for CSO pollution control [11]. This cost is expected to increase substantially, because the costs for less than one-third of the over 1100 CSO systems in the country were quantified in the survey. Special regulations have been or are being enacted by the federal government and by many states that deal with pollution control from CSO discharges.

Implementation of pollution control measures for CSOs can affect both combined wastewater collection systems and wastewater treatment plants. In response to the need for correction of these problems (and recognizing the special character of combined systems), engineers have utilized a variety of CSO control methods. These include construction of new separate sewers and storm drainage systems (sewer separation), treatment at the combined sewer outlet, and storage followed by treatment at dry-weather facilities. The purpose of this chapter is to provide an introduction to the subject of combined sewers and combined sewer overflows. The topics presented include (1) a brief history of combined sewer systems, (2) description of a combined sewer system, (3) combined sewer flowrates and characteristics, (4) methods for CSO control, (5) treatment of CSOs, and (6) future directions in the management of CSOs.

15-1 HISTORY OF COMBINED SEWER SYSTEMS

Many of the older communities located in the Northeast, mid-Atlantic, and Midwest have combined sewer systems, with a total population served of about 40,000,000. These systems also tend to be concentrated in communities of large population. For example, over 45 percent of communities with populations over 100,000 have combined systems. Current engineering practice for collection systems normally involves design of separate systems for wastewater and stormwater runoff. Wastewater is conveyed to a treatment facility while stormwater is typically discharged without treatment to a receiving water. In the early part of this century, the design of combined systems, where both stormwater and wastewater were collected and transported in a single conduit, was an accepted practice [9]. Some of these combined sewer systems were initially designed to carry storm runoff only.

As communities increased in size, the problem of handling wastewater became more difficult, and homes in some areas were connected to storm drains. These storm drains, as well as those originally designed to carry combined wastewater, transported wastewaters from their source to the nearest surface water body for disposal. As populations and waste quantities increased further, it became apparent that some degree of wastewater treatment would be required to protect public health and to maintain the quality of receiving waters. To collect the discharges from numerous combined sewer outlets in a given community, collection sewers termed *interceptors* were constructed. Interceptors, which often followed watercourse boundaries, were used to intercept combined sewers at a point upstream of their outlet to the receiving water body. The interceptor system was normally designed to discharge the collected combined wastewater to a wastewater treatment plant for treatment. It should be noted that for separate sewer systems the term *interceptor sewer* commonly refers to a major sewer line used to collect and transport flows from smaller lateral sewers.

Interceptor sewers and downstream treatment facilities were designed with sufficient hydraulic capacity to handle the peak dry-weather wastewater flow and, in most instances, a portion of the stormwater flow. The cost of constructing treatment facilities with sufficient capacity to handle an appreciable portion of the stormwater flow was considered prohibitive. Thus, diversion or regulator structures were required to divert flows in excess of the treatment plant capacity to the receiving water via the combined sewer overflow outlets. These structures, constructed at the junction of each combined sewer and the interceptor, served to prevent flooding and hydraulic overloading of the treatment facilities. As development and urbanization continued to increase, the overflows from the combined sewer interceptors resulted in continued degradation of receiving water quality. Recent efforts, therefore, have been focused on the control or elimination of combined sewer overflows.

15-2   COMPONENTS OF COMBINED SEWER SYSTEMS

The principal components of a combined sewer system, as shown in Fig. 15-1, include (1) the contributing drainage area (catchment) and wastewater sources, (2) the combined sewer pipe network and interceptor(s), (3) the regulator and diversion structures, and (4) the CSO outlets. Because effective management of wastewater from CSOs requires an understanding of the combined system, these components are described and illustrated in the following discussion.

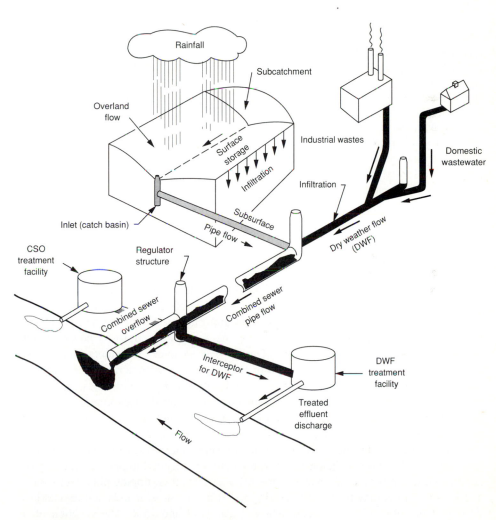

FIGURE 15-1
Schematic diagram of a combined sewer system.

Contributing Drainage Area and Wastewater Sources

As depicted in Fig. 15-1, precipitation falls on the ground surface of the drainage area. Excess water that does not infiltrate or that becomes trapped in surface depressions travels by overland flow to stormwater inlets, where it enters the combined drainage system. Stormwater can also enter the combined sewer system through roof drains, manhole covers, and other inlets. Stormwater enters the combined sewer and mixes with the domestic, commercial, and industrial wastewater that is discharged directly to the combined sewer, resulting in a mixture of diluted raw wastewater.

Combined Sewer Pipe Network and Interceptor

A combined sewer is typically a gravity flow sewer designed to carry both stormwater and wastewater flow. Combined sewers are usually sized to handle stormwater flows corresponding to a given design storm event, as the wastewater portion of the flow is only a small fraction of the design flow [9]. The interceptor sewer leading to the treatment plant receives flow from the individual combined sewers up to an amount that can be safely handled by the treatment facility. Based on a nationwide survey, the wet-to-dry weather capacity ratios for combined interceptor sewer design range from 1:1 to 8:1, with a median ratio of 4:1 [2].

Flow Regulators

The function of flow-regulating devices is to control the flow between the combined sewer and companion facilities. Most frequently, flow regulators are used to control the flow between collection sewers and interceptor(s); however, other facilities such as storage and treatment units and outlets may be involved. During dry weather, the regulator structures allow the wastewater flow to be conveyed to the downstream treatment facility. During wet weather, regulators, known as diversion structures, are used to divert flow above a predetermined or design flowrate away from the interceptor to the CSO outlet or to specially designed CSO storage and treatment facilities. Other types of regulating devices can be designed to limit the flowrate leaving the combined sewer or entering the interceptor or other facility.

Combined wastewater may be diverted by side weirs, transverse weirs, leaping weirs, orifices, and relief siphons. If flow regulation is required to prevent flooding of downstream facilities, such devices as mechanical regulators, tipping plate regulators, and Hydro-Brakes may be used. Several types of diversion structures and regulators, described in Table 15-1, are illustrated in Figs. 15-2 and 15-3. The function of a flow regulator is considered in Example 15-1. Additional information on diversion structures and flow regulators may be found in Refs. 1, 5, and 13.

TABLE 15-1
Description of some typical flow regulators

Regulator	Description
Side weir	Typically consists of a weir parallel to the wastewater flow located in the side of the sewer pipe (Fig. 15-2a). The weir should be high enough to prevent any discharge of dry-weather flows, but low and long enough to discharge the required excess flow during wet weather.
Transverse weir	A weir or small dam placed directly across the sewer, perpendicular to the line of flow, is used to direct dry-weather flow to the interceptor sewer (Fig. 15-2b). Increase of flow during wet weather results in flow overtopping the weir and discharging to the overflow outlet.
Orifice	These diversion structures allow flow from the combined sewer to pass through a circular or rectangular orifice and enter the interceptor. The orifice is sized to allow the dry-weather flow, and possibly some of the wet weather flow, to pass. The orifices can be oriented in a variety of ways, including horizontally at the invert of the sewer ("drop inlet"), and vertically on the side of the sewer (often used in conjunction with a transverse weir as in Fig. 15-2b).
Leaping weir	A leaping weir is formed by an opening in the invert of a sewer of such dimensions as to permit dry-weather wastewater flow to fall through the opening and pass to the interceptor (Fig. 15-2c). During storms, the increased velocity and depth of flow cause most of the flow to leap the opening and enter the overflow outlet. The steel weir plate is normally designed so that it can be adjusted for various flow conditions.
High outlet regulator	A commonly used orifice-type regulator in which the invert of the overflow pipe is typically above the crown of the combined sewer (Fig. 15-2d).
Relief siphon	The relief siphon (Fig. 15-2e) affords a means of regulating the maximum water-surface elevation in a sewer with smaller variations in high-water level than can be obtained with other devices. A siphon works automatically and does not require any auxiliary mechanisms. The siphon inlet is typically set as far below the top water level as possible to minimize the carryover of floating scum and debris to the overflow outlet.
Mechanical regulator	The mechanical regulator (also known as an automatic regulator or reverse taintor gate) responds to the water level in the combined sewer or interceptor sewer (Fig. 15-3a). In either case, the float travel and the corresponding gate travel may be adjusted to regulate closely the flow to the interceptor.
Tipping-plate regulator	In these devices, the plate is pivoted off-center, and its motion is controlled by the difference of water levels above and below the gate (Fig. 15-3b). Multiple gates can be used to increase the capacity of an installation.
Hydro-Brake regulator	The patented configuration of the Hydro-Brake (Fig. 15-3c) imparts a more-or-less centrifugal motion to the entering fluid. This action, which commences when a predetermined liquid head has been reached, effectively reduces the rate of discharge. This device has been used extensively on combined sewers to limit the flow to the interceptor, thus maximizing storage of flows in the combined sewer.

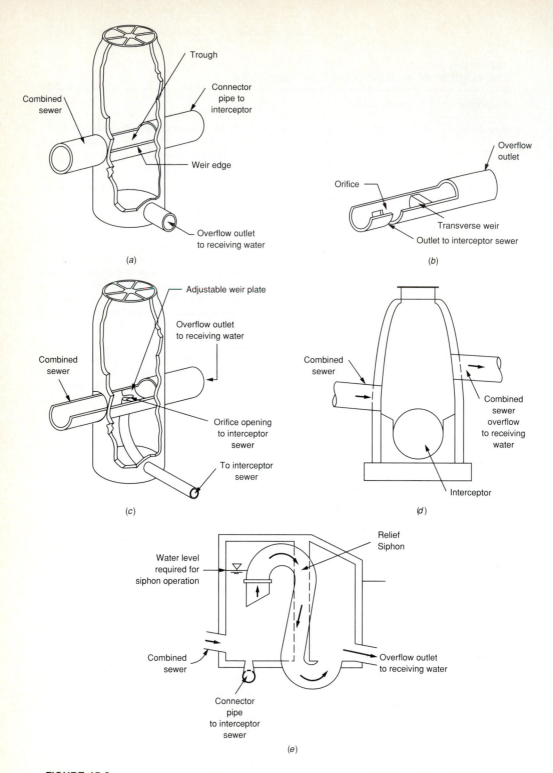

FIGURE 15-2
Typical diversion regulators: (a) side weir, (b) transverse weir with orifice, (c) leaping weir, (d) high outlet regulator, and (e) relief siphon.

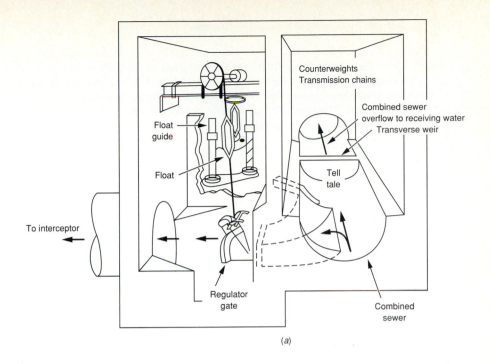

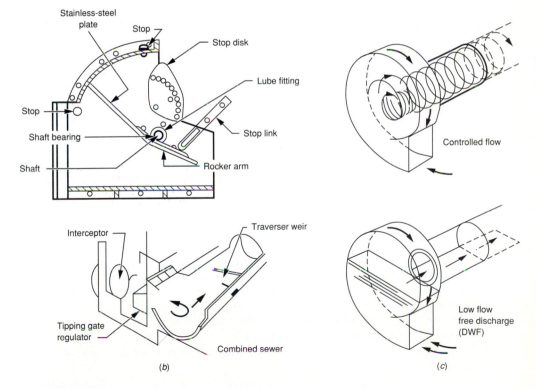

FIGURE 15-3
Typical CSO regulators: (a) mechanized regulator, (b) tipping plate regulator, and (c) Hydro-Brake regulator.

Example 15-1   Regulated flow through a combined sewer. Flow through a 48 in circular combined sewer (see following figure) is diverted by means of a mechanical regulator to a 24 in circular interceptor sewer. The dimensions of the rectangular regulator gate are 9.75 in wide by 7.5 in high. The regulator gate is float operated based on the flow level in the interceptor sewer, with gate closures as follows:

Flow depth in interceptor, in	Gate closure
0–12	None (gate remains fully open)
12–24	Gate closes 1 in for every 2 in rise in flow depth

Determine the maximum flow that will be diverted to the interceptor before overflow occurs, assuming the depth of flow in the interceptor is less than half full. Then determine the flow to the interceptor when the combined sewer is flowing half full and the interceptor is flowing three-quarters full.

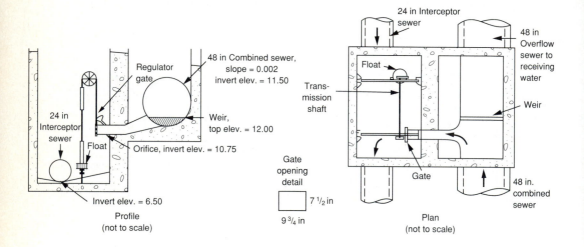

Solution

1. Determine the maximum flow that will be diverted to the interceptor before overflow occurs. Overflow occurs when flow in the 48 in combined sewer tops the diversion weir. Therefore, maximum flow to the interceptor will occur when flow in the combined sewer reaches the top of the weir. Because depth of flow in the interceptor sewer is less than 12 in, the regulator gate is fully open. (Note: the depth of flow in the interceptor sewer would be determined using Manning's equation and the sewer geometry.)

   (a) The orifice equation given below is used to determine the flow through the regulator gate:

   $$Q = CA(2gh)^{1/2}$$

where Q = flow through gate, ft^3/s

$\quad\quad C$ = discharge coefficient; use 0.95 (unitless)

$\quad\quad A$ = cross-sectional area, ft^2

$\quad\quad g$ = gravitational acceleration, 32.2 ft/s^2

$\quad\quad h$ = head on the orifice (measured to the centerline of the orifice), ft

(b) Because the interceptor level is below the regulator gate, a free discharge will occur through the orifice. The head on the orifice, taken to the midpoint of the gate opening, is

$$h = \text{(top of weir elevation)} - \text{(gate midpoint elevation)}$$

$$h = 12.0 \text{ ft} - 10.75 \text{ ft} - \frac{7.5 \text{ in}}{2(12 \text{ in/ft})} = 0.94 \text{ ft}$$

(c) The flow through the orifice is

$$A = (9.75 \text{ in} \times 7.50 \text{ in})/(144 \text{ in}^2/\text{ft}^2) = 0.508 \text{ ft}^2$$

$$Q = 0.95(0.508 \text{ ft}^2)\left[2(32 \text{ ft/s}^2)(0.94 \text{ ft})\right]^{1/2} = 3.75 \text{ ft}^3/\text{s}$$

Thus, overflow will occur when flow in the combined sewer exceeds 3.75 ft^3/s.

2. Determine the flow to the interceptor when the combined sewer is flowing half full and the interceptor is flowing three-quarters full.

(a) The surface elevation of the wastewater in the combined sewer when flowing half full is

$$11.50 \text{ ft (invert elevation)} + 0.5(4 \text{ ft diameter}) = 13.50 \text{ ft}$$

(b) For an 18 in depth of flow in the interceptor, the gate will close 3 in and the head on the orifice is

$$h = 13.5 \text{ ft} - 10.75 \text{ ft} - \frac{4.5 \text{ in}}{2(12 \text{ in/ft})} = 2.56 \text{ ft}$$

(c) The corresponding flow through the orifice is

$$A = (9.75 \text{ in} \times 4.5 \text{ in})/(144 \text{ in}^2/\text{ft}^2) = 0.305 \text{ ft}^2$$

$$Q = 0.95(0.305 \text{ ft}^2)[2(32.2 \text{ ft/s}^2)(2.56 \text{ ft})]^{1/2} = 3.66 \text{ ft}^3/\text{s}$$

Comment. For the purpose of simplicity, a value of 0.95 was assumed for the discharge coefficient. The actual value of the discharge coefficient will decrease as the gate closes.

Outlets

The outlet of the combined wastewater system is at the end of the combined sewer, where the combined wastewater is discharged into a receiving water body. Frequently, there is an outlet at each location where a combined sewer intersects the interceptor sewer. If the CSO outlet discharges at an elevation below the high-water level of the receiving water, a backwater or tide gate is necessary to prevent the receiving water from entering the sewer. A backwater gate (often identified as a flap gate) consists of a flap hung against an inclined seat (see Fig. 15-4a). The hinges may be at the top

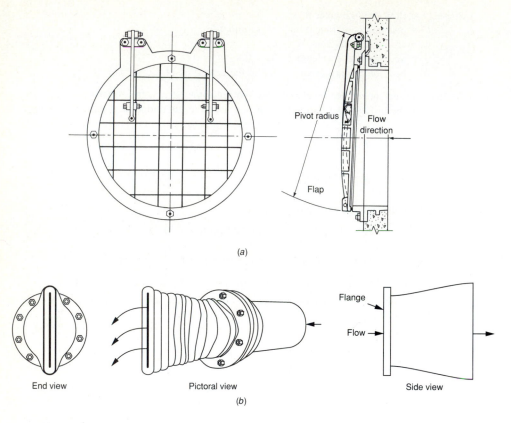

FIGURE 15-4
Typical CSO outlets: (*a*) flap gate and (*b*) elastomeric check valve [13].

when the gate consists of a single leaf, as is usually the case, or they may be at the
side when the gate consists of two leaves. A second type of backwater gate, recently
developed, is an elastomeric check valve, or a "duckbill" type gate (see Fig. 15-4*b*).
This gate consists of a durable rubber sleeve formed in the shape of a duck's bill.
The all-rubber construction enables the gate to discharge at small head differentials
and reduces backwater leakage by sealing around debris caught in the gate during
operation. This type of gate generally requires less maintenance than flap gates.

15-3 COMBINED SEWER FLOWRATES
AND WASTEWATER CHARACTERISTICS

The quantity of flow and quality characteristics of combined wastewater are important
for several reasons. During water quality impact investigations, the volume and
contaminant load overflowing into a receiving water from the combined sewer must

be determined. To select appropriate control facilities and treatment processes, a knowledge of the combined wastewater characteristics is required. Proper operational management of transport and treatment facilities also requires an understanding of these factors. Combined sewer flowrates and the chemical and biological characteristics of combined wastewater are reviewed in this section. Combined sewer flows and characteristics can be determined either by direct measurement or by calculation. Although direct measurement is the most accurate method, collection of such data is time-consuming and expensive. Therefore, calculation techniques are frequently relied upon as a supplement. Both methods are reviewed in this section.

Combined Sewer Flowrates

Flow in the combined sewer system is composed mainly of rainfall runoff and wastewater. Flow enters the combined sewer continuously during both dry and wet weather from the contributing wastewater sources. This flow may include domestic, commercial, and industrial wastewater and infiltration as described in Chap. 2. During a rainfall event, the amount of storm flow is normally much larger than the dry-weather wastewater flow, and the observed flows during wet weather can mask completely the dry weather flow patterns.

As flow proceeds through the combined sewer to the interceptor, it is modified by hydraulic routing effects as well as any surcharged conditions within the system (surcharging results when the pipeline capacity is exceeded). The flow is modified further as additional flow enters the interceptor from regulator structures located further downstream in the system. The flow is then split by the regulator structure, with some entering the interceptor and the remainder exiting through the CSO outlet. In some cases where the combined sewer is undersized, flooding or surcharging may occur at various upstream locations within the system. Typical wet-weather flow variations at a CSO outlet and at a treatment plant serving a combined sewer system are depicted in Fig. 15-5, along with the rainfall measured during the storm event.

The catchment hydrograph (flow vs. time) illustrated in Fig. 15-5, closely resembles that of the variations in rainfall intensity. The short response time between the rainfall event and the increase in the flowrate can be taken as an indication of a short travel time for flow from all points in the upstream combined system. In contrast, the hydrograph at the treatment plant shows less distinct flow peaks and a lag time of several hours for flows to return to normal dry-weather levels following rainfall cessation. The higher flows at this location are due to the larger contributing combined system, and the smoothed peaks result from loss of flow through overflows and hydraulic routing effects.

Direct Measurement of Combined Sewer Flows and Wastewater Characteristics. Combined wastewater flows and characteristics can be monitored at various points in the system, including the combined sewer, regulator structure, interceptor, outlet, and treatment plant. Monitoring points in the sewer or interceptor system

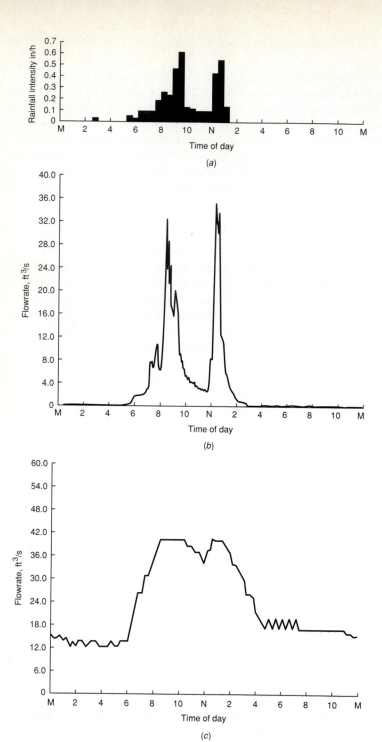

FIGURE 15-5
Flow variations in a combined sewer system during wet weather: (a) rainfall hyetograph, (b) typical catchment flowrate, and (c) observed treatment plant flowrate.

may be needed for a number of other reasons, including defining the flow to be controlled or treated. Frequently, in performing studies of combined sewer systems, temporary flow measurement and wastewater monitoring facilities are installed and left in place for several storm events and then removed. Flow metering in such installations is typically performed using portable, battery-operated depth- and velocity-sensing instrumentation. Similarly, wastewater samples are taken using portable, battery-operated programmable sampling devices. These samplers are preset for desired sampling time intervals and are level or flow actuated. A typical flow meter and automatic sampling installation are shown in Fig. 15-6.

To understand how combined wastewater flowrate and characteristics respond to a storm event, rainfall data must be obtained. Therefore, it is often necessary to install temporary rainfall-monitoring equipment in close proximity to the drainage area tributary to the monitoring location. Continuously recording rain gauges capable of monitoring the rainfall depth over time should be used. The tipping bucket rain gauge is one type that can be used to record rainfall continuously in 0.01 in increments on a clock-driven recorder chart. If the combined system being monitored encompasses a large area, installation of several rain gauges may be necessary to record spatial variation of rainfall characteristics across the entire area.

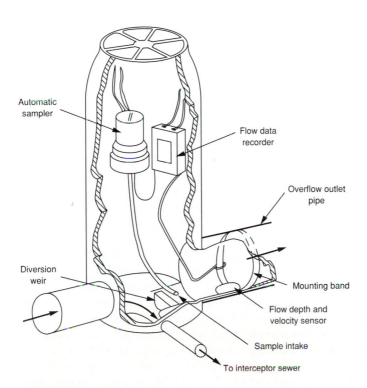

FIGURE 15-6
Typical temporary CSO flow metering and sampling installation.

Although temporary metering installations are designed to operate automatically, it is advisable to have them checked as frequently as possible. Due to the nature of combined wastewater, which tends to contain significant amounts of debris (rags, sticks, paper, cans, etc.), automatic installations are prone to clogging problems. In addition, laboratory quality control requires that sample holding times not be exceeded (for example, the holding time for coliform bacteria should not exceed six hours).

Permanent flow-monitoring installations are used in some systems to allow continuous flow records to be obtained at critical points. Also, these installations can allow centralized control of transport system facilities to maximize storage of combined wastewater in the system, or to control flow to the downstream treatment plant. The flow data recorded at the site may be periodically recovered manually, or the data may be telemetered to a central location for real-time control. A discussion of such metering techniques may be found in Refs. 5 and 14.

Calculation of Combined Sewer Flowrates. Calculation of flowrates in a combined sewer system is a complicated and challenging task. The first step in the process involves quantifying wastewater, rainfall runoff, and other sources of flow such as groundwater infiltration. These sources of flow are then combined and routed through the various components of the system. Finally, the volumes of flow exiting the system through CSO outlets, entering the downstream treatment facility, or being transported to other points in the system are determined.

Computer modeling. Due to the complexity of combined sewer systems, it is normally necessary to use computer models that simulate the combined sewer system, including dry-weather wastewater flows, hydraulic routing through the piping system, flow splitting at the regulators, discharges through the outlets, and flow through the interceptor and treatment plant. Of the many computer models available for the assessment of combined sewer systems, the most widely used model is the Storm Water Management Model (SWMM). Developed originally in 1971 [8], this model has undergone substantial improvements since that time [3,10]. The SWMM is designed to simulate the time-varying processes of precipitation falling onto land of varying characteristics, conversion of rainfall to runoff, and collection and transport of stormwater runoff and wastewater through the combined system.

Both hydraulic and contaminant routing are performed by SWMM. The model can be used for both single-storm simulation and long-term simulation. Single-storm simulation allows detailed assessment of combined sewer system performance during individual storm events using short (minutes) time intervals. Long-term simulations allow development of CSO flow and load statistics based on long-term multiyear rainfall records using long (hours) time intervals. Both types of simulation are important in developing an understanding of the behavior of the CSO system, as well as projecting CSO impacts on the receiving waters. Using the SWMM model, separate calculations are performed for runoff generation and for transport through the piping system. Other computer models, developed for CSO systems, are described in Ref. 13.

Model calibration and verification. The process of calculating flows in a combined system using a computer model such as SWMM normally involves comparison of measured versus model-predicted flows at selected locations in the system (see Fig. 15-7). For using SWMM, the calibration and verification process is recommended. During model calibration the model is run with rainfall data collected from one storm, and the calculated results are compared with the observed field measurements. Estimated input parameters are then adjusted within reasonable bounds to obtain best fit between predictions and measurements. During verification, data sets from other storms are used, and no adjustments of parameters are allowed. This calibration and verification process is used to confirm the predictive capability of the model.

Characteristics of Combined Wastewater

The characteristics of the combined wastewater reflect the combination of wastewater and rainfall runoff that contribute to the system, as well as the resuspension of settled material from the collection system itself. The factors identified in Table 15-2 are among the many that may affect the characteristics. Many of the quantity-related factors, which can be used to determine flow in the system, also affect the resulting quality. The characteristics of municipal wastewater have been discussed in Chap. 3. Rainfall runoff will generally contribute a larger amount of flow, which is of better quality than the wastewater. Because of the variability of precipitation events, drainage area, wastewater factors, and the other contributing factors, combined wastewater characteristics tend to be highly variable from location to location and difficult to predict without obtaining actual measurements in the system.

Comparative Wastewater Quality Data. A comparison of data collected for rainfall, stormwater runoff, combined wastewater, and untreated wastewater is given in Table 15-3. Rainfall has some amounts of oxygen demanding material, nutrients, and metals, in part owing to atmospheric pollution sources. The quality of stormwater runoff becomes worse due to the wash-off of contaminants from the ground surface, including solids, bacteria, oxygen-demanding materials, nutrients and metals. The characteristics of combined wastewater are dependent upon mixing of this stormwater runoff with untreated wastewater, and thus falls somewhere in between.

Variations in Combined Wastewater Quality. Typical variations of BOD, suspended solids, and fecal coliform bacteria measured in a combined sewer are shown in Fig. 15-8, during and after a storm event. As shown, the BOD_5 and fecal coliform bacteria concentrations are low during the storm, when runoff flows are high. After the storm, when runoff subsides and the flow consists primarily of wastewater, concentrations rise significantly. When this rise occurs, it can be concluded that the BOD_5 and fecal coliform concentrations in the stormwater are significantly lower than in the wastewater component. Unlike BOD_5 and fecal coliform bacteria, the suspended solids concentrations rise slightly during the storm and remain unchanged after the storm, indicating that suspended solids concentrations from stormwater runoff and

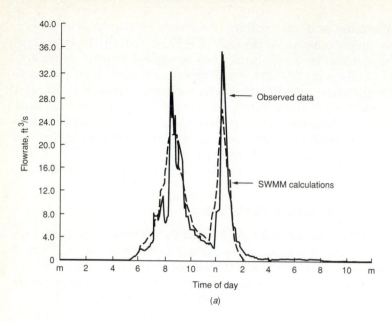

(a)

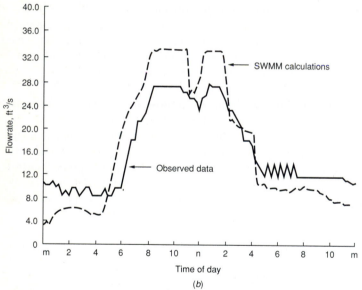

(b)

FIGURE 15-7
Comparison of measured combined wastewater flowrates versus model predicted flowrates using SWMM: (a) at CSO outlet and (b) at wastewater treatment plant.

TABLE 15-2
Typical factors influencing the characteristics of combined wastewater

Parameter	Quantity-related factors	Quality-related factors
Precipitation	Rainfall depth and volume Storm intensity Storm duration	Regional atmospheric quality
Wastewater	Flow rate and variability Type of contributing sources (residential, commercial, etc.)	Type of contributing sources
Drainage basin	Size, time of concentration Land use type Impervious area Soil characteristics Runoff control practices	Pollutant build-up and wash-off Watershed management practices
Sewer system, interceptor	Pipe size, slope, and shape Quantity of infiltration Surcharging or backwater conditions Type of flow regulation or diversion Capacity reduction from sediment build-up	Chemical and biological transformations Quality of infiltration Sediment load resuspended from collection system

wastewater are similar. The slight rise in the suspended solids concentration during the peak flow may be due to a phenomenon, common to many combined sewer systems, known as the "first flush." The first flush has often been observed following the initial phase of a rainfall event during which much of the accumulated surface contaminants are washed into the combined system. In combined sewers, the increased flows may be capable of resuspending material previously deposited during low-flow periods. Together, the resuspended material and contaminants washed off surfaces result in high contaminant concentrations. Factors known to contribute to the magnitude and frequency of the first flush effect include combined sewer slopes, street and catch basin cleaning frequency and design, rainfall intensity and duration, and surface buildup of debris and contaminants.

Calculation of the Characteristics of Combined Wastewater. After flowrates have been quantified, it may be necessary to estimate the characteristics of the combined wastewater in terms of contaminant concentration or loading. This aspect of combined sewer assessment is far less well understood and, therefore, less predictable than for flowrates. Deterministic formulations are used in the SWMM model to estimate contaminant concentrations. Predominant land use in each subcatchment, street-sweeping frequency, number of dry-weather days before a storm, and surface contaminant accumulation rates are defined. These are used in conjunction with street gutter length data for each subcatchment to compute the amount of surface contaminant accumulation or build-up prior to a given storm. For example, dust and dirt build-up has been represented by linear, exponential, and power function equations of the following form [3]:

TABLE 15-3
Comparison of characteristics of combined wastewater with other sources

Parameter	Unit	Range of parameter concentrations			
		Rainfall[a]	Stormwater runoff[b]	Combined wastewater[c]	Municipal wastewater[d]
Suspended solids	mg/L		67–101	270–550	100–350
BOD₅	mg/L	1–13	8–10	60–220	110–400
Chemical oxygen demand	mg/L	9–16	40–73	260–480	250–1,000
Fecal coliform bacteria	MPN/100 ml		1,000–21,000	200,000–1,100,000	10^6–10^7
Nitrogen (total as N)	mg/L			4–17	20–85
total Kjeldahl nitrogen		0.05–1.0	0.43–1.00		20–85
nitrate			0.48–0.91		0
Phosphorus (total as P)	mg/L	0.02–0.15	0.67–1.66	1.2–2.8	4–15
Metals	μg/L				
copper			27–33		
lead		30–70	30–144	140–600	
zinc			135–226		

[a] Adapted from Ref. 3.
[b] Adapted from Ref. 12.
[c] Adapted from Ref. 6.
[d] Adapted from Table 3-16, Chap. 3.

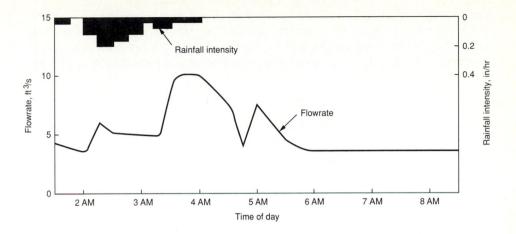

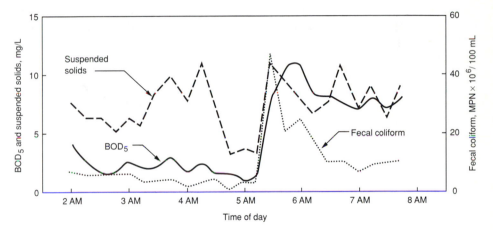

FIGURE 15-8
Typical variations of flowrate and BOD, SS, and fecal coliform in a combined sewer system during a storm event.

$$DD = at + b \qquad (15\text{-}1)$$

$$DD = c(1 - e^{-dt}) \qquad (15\text{-}2)$$

$$DD = et^f \qquad (15\text{-}3)$$

where
DD = dust and dirt build-up (lb per 100 ft of curb length)
t = time
a, b, c, d, e, f = empirical coefficients

A number of individual contaminants, including BOD, solids, bacteria, nutrients, and metals, can be modeled using these equations. Modeling of the individual contaminants can be accomplished by establishing a proportional relationship between the contaminant of concern, based, for example, on laboratory testing. The contaminant accumulation on the land and the contaminant concentration in the rainfall itself are combined based on the amount of rainfall-runoff in a given time period. An empirical wash-off function is used to estimate the amount of contaminant build-up washed off the land surface into the combined sewer system. The quantity of contaminant washed off over a given time period has been estimated by the following expression:

$$P_{off} = P_{shed} \, (1 - e^{-rcoeff \cdot r^y \cdot t}) \qquad (15\text{-}4)$$

where P_{off} = contaminant load washed off, lb
P_{shed} = initial amount of pollutant, lb
rcoeff = wash-off coefficient, in^{-1}
r = runoff rate, in/h
y = wash-off exponent
t = time, h

After estimation of the quantity of contaminants using these empirical equations (see Example 15-2), contaminants from runoff and wastewater in the combined sewer piping system are mixed and routed based on the pipe flows. Other in-pipe processes that affect contaminant load, such as settling, die-off, and decay, are also considered. In addition, the wastewater characteristics entering the piping system are estimated if measured data are not available. The combination of all these factors allows estimation of the contaminant concentrations and loadings that exit the outlets, or that are transported to the treatment plant.

Although models such as SWMM can be used to estimate contaminant loads, it is generally accepted that the processes controlling contaminant concentrations are not well understood at this time. It has been shown that contaminant concentrations are highly variable from location to location and from storm to storm, even at the same location. Comparisons of model predictions and measurements are normally rough at best. It is, therefore, necessary to collect site-specific data to verify that the results obtained from predictive models, even approximately, resemble the measured data. Estimation of the quantities of contaminants to be expected as a result of a storm event is illustrated in Example 15-2.

Example 15-2 Contaminant buildup and wash-off. Dust and dirt build-up data were collected for three different drainage areas and the following relationships were established:

| | | Maximum contaminant accumulation | |
		Contaminant loading, lb/100 ft curb	Time to maximum build-up, d
Area	Build-up equation[a]		
1	$DD = 4.5t^{0.6}$	11.8	5
2	$DD = 1.7t$	11.9	7
3	$DD = 18(1 - e^{-0.4t})$	18.0	—

[a] Equations are valid when a complete contaminant washoff occurs at $t = 0$.
Note: DD = dust and dirt load, t = time in days.

From an analysis of the dust and dirt from the three areas, the following three contaminants were found to be related to the dust and dirt load (DD) as follows:

Biochemical oxygen demand (BOD): 3.4 mg/g DD
Suspended solids (SS): 950 mg/g DD
Fecal coliform bacteria (FC): 38,800/g DD

If the three drainage areas have total curb lengths of 13,700, 22,300, and 9500 ft, respectively, determine the amount of BOD, SS, and FC that would accumulate in each area, assuming that six dry days have elapsed since the last rainstorm in which a complete wash-off occurred.

For the same drainage areas, estimate the rate of contaminant wash-off from Area 2 as a function of amount of contaminant available for wash-off and the runoff rate. A short-duration rainstorm occurred in which runoff lasted approximately 45 minutes at the following rates:

Time, min	Average runoff rate, in/hr
0–10	0.3
10–35	0.5
35–45	0.2

Assume the following coefficients are applicable: rcoeff $= 5.0$ and $y = 2.0$.

Solution: Accumulation of Contaminants

1. Using the appropriate dust and dirt build-up relationship, compute the quantity of dust and dirt accumulated in each area for $t = 6$ days.

   (a) Area 1:
   $$DD = 4.5t^{0.6} = 4.5(6)^{0.6} = 13.2 \text{ lb/100 ft}$$

   Because the maximum dust and dirt accumulation for Area 1 is 11.8, use 11.8 lb/100 ft curb.

   (b) Area 2:
   $$DD = 1.7(t) = 1.7(6) = 10.2 \text{ lb/100 ft curb}$$

   (c) Area 3:
   $$DD = 18(1 - e^{-0.4t}) = 18(1 - e^{-0.4(6)}) = 16.4 \text{ lb/100 ft curb}$$

2. Determine the quantity of BOD, SS, and FC in each area using the total curb length and the ratio of contaminant to dust and dirt.

   (a) Area 1

   $$BOD = \frac{11.8 \text{ lb DD}}{100 \text{ ft curb}} \times \frac{3.4 \text{ mg BOD}}{g \text{ DD}} \times \frac{454 \text{ g}}{lb} \times \frac{1 \text{ lb}}{454,000 \text{ mg}} \times 13,700 \text{ ft curb} = 5.5 \text{ lb}$$

   $$SS = \frac{11.8 \text{ lb DD}}{100 \text{ ft curb}} \times \frac{950 \text{ mg SS}}{g \text{ DD}} \times \frac{454 \text{ g}}{lb} \times \frac{1 \text{ lb}}{454,000 \text{ mg}} \times 13,700 \text{ ft curb} = 1535 \text{ lb}$$

   $$FC = \frac{11.8 \text{ lb DD}}{100 \text{ ft curb}} \times \frac{38,800}{g \text{ DD}} \times \frac{454 \text{ g}}{lb} \times 13,700 \text{ ft curb} = 2.8 \times 10^{10} \text{ organisms}$$

(b) Following a similar procedure for Areas 2 and 3, the following results are obtained.

Drainage area	Dust and dirt build-up, lb/100 ft curb	Total contaminant load for 13,700 ft curb		
		BOD, lb	SS, lb	FC, no.
1	11.8	5.5	1535	2.84×10^{10}
2	10.2	7.7	2161	4.01×10^{10}
3	16.4	5.0	1480	2.74×10^{10}

Solution: Wash-Off of Contaminants

1. Determine the quantity of dust and dirt washed off Area 2 for each runoff rate over the course of the storm using Eq. 15-4.

(a) For $t = 0$ to $t = 10$ min

$$P_{shed} = 10.2 \text{ lb DD} \quad \text{(see Step 2b above)}$$

$$P_{off} = 10.2[1 - e^{-5.0(0.3)^{2.0}(10/60)}] = 0.7 \text{ lb DD}$$

Quantity left after $t = 10$ min: $10.2 - 0.7 = 9.5$ lb DD.

(b) For $t = 10$ to $t = 35$ min (new $P_{shed} = 9.5$ lb DD)

$$P_{off} = 9.5[1 - e^{-5.0(0.5)^{2.0}(25/60)}] = 3.9 \text{ lb DD}$$

Quantity left after $t = 35$ min: $9.5 - 3.9 = 5.6$ lb DD.

(c) For $t = 35$ to $t = 45$ min (new $P_{shed} = 5.6$ lb DD)

$$P_{off} = 5.6[1 - e^{-5.0(0.2)^{2.0}(10/60)}] = 0.2 \text{ lb DD}$$

Quantity left after 45 min: $5.6 - 0.2 = 5.4$ lb DD.

2. Compute total wash-off quantity.

$$10.2 - 5.4 = 4.8 \text{ lb DD}$$

3. Determine the quantity of BOD, SS, and FC washed off using the total curb length and the ratio of contaminant to dust and dirt.

(a) BOD

$$\text{BOD} = \frac{4.8 \text{ lb DD}}{100 \text{ ft curb}} \times \frac{3.4 \text{ mg BOD}}{\text{g DD}} \times \frac{454 \text{ g}}{\text{lb}} \times \frac{1 \text{ lb}}{454,000 \text{ mg}} \times 22,300 \text{ ft curb} = 3.6 \text{ lb}$$

(b) SS

$$\text{SS} = \frac{4.8 \text{ lb DD}}{100 \text{ ft curb}} \times \frac{950 \text{ mg SS}}{\text{g DD}} \times \frac{454 \text{ g}}{\text{lb}} \times \frac{1 \text{ lb}}{454,000 \text{ mg}} \times 22,300 \text{ ft curb} = 1017 \text{ lb}$$

(c) FC

$$\text{FC} = \frac{4.8 \text{ lb DD}}{100 \text{ ft curb}} \times \frac{38,800}{\text{g DD}} \times \frac{454 \text{ g}}{\text{lb}} \times 22,300 \text{ ft curb} = 1.9 \times 10^{10} \text{ organisms}$$

15-4 METHODS FOR CONTROLLING OVERFLOWS

A variety of CSO control technologies are currently in use or are being tested (see Table 15-4). Some of these methods, such as the separation of sewers, have been used for a long time. Other technologies such as swirl concentrators and microscreens have been used only recently. In the following discussion, each of the major categories of these CSO control methods and technologies is briefly described. Further information on CSO control methods can be found in Refs. 6, 7, and 13.

Source Controls

Source control measures, sometimes called best management practices, are measures that can be implemented within a drainage basin to reduce stormwater flows. Source controls do not usually require large capital expenditures. However, they are generally labor intensive; therefore, the associated maintenance costs can be high. Porous pavement and use of pervious areas for groundwater recharge prevent runoff from entering the collection system by directing it into the underlying soil. Flow detention and rooftop storage delay the entry of runoff into the collection system by storing it temporarily and releasing it at a controlled rate. Flows from area drains and roof leaders can be diverted and rerouted to either separate storm drains or pervious areas.

The implementation of these controls on a scale necessary to control CSOs is impractical in most areas because of the presently existing level of development. Such controls, however, can be a requirement for future development or reconstruction to avoid increasing stormwater flows in the combined system. Control implementation can be done through incorporation of the appropriate requirements into sewer use regulations and through strict review of proposed development plans.

Other source control measures are designed to minimize accumulation of contaminants on streets, on other tributary land areas, and in catchbasins. Implementation of these measures will decrease contaminant loadings of stormwater runoff, thereby decreasing contaminant loadings of CSOs, although usually a substantial reduction cannot be obtained. For example, in a number of projects, increased street sweeping could not be correlated with any reduction in contaminant loads [12].

In summary, source controls can affect CSO quantity and contaminant concentrations. However, they often cannot be relied upon to provide a consistent reduction in CSO loading or improvement in aesthetic characteristics.

Collection System Controls

Collection system controls include those techniques and methods that can be used to control the stormwater discharges. Improved management of existing collection system facilities, infiltration/inflow control, sewer separation, flow regulation and diversion, modification of facilities to reduce CSOs, and improved operation and maintenance are the collection system controls used most commonly. The use of regulators and gates has been discussed earlier in this chapter.

TABLE 15-4
List of CSO Control Methods

I. Source Controls (Best Management Practices)
  1. Porous pavements
  2. Flow detention
  3. Rooftop storage
  4. Area drain and roof leader disconnection
  5. Utilization of pervious areas for recharge
  6. Air pollution reduction
  7. Solid waste management
  8. Street sweeping
  9. Fertilizer and pesticide control
  10. Snow removal and deicing control
  11. Soil erosion control
  12. Commercial/industrial runoff control
  13. Animal waste removal
  14. Sewer line flushing
  15. Catchbasin cleaning
  16. Identifying and/or eliminating sewer system cross-connections
  17. Public education programs

II. Collection System Controls
  1. Existing system management and in-system modifications
  2. Complete or partial sewer separation
  3. Infiltration/inflow control
  4. Polymer injection
  5. Regulating devices and backwater gates
  6. Remote monitoring and real-time control
  7. Flow diversion

III. Storage
  1. In-system storage
    a. Inflatable dams
    b. Manual and automatic valves and gates
  2. Surface storage
  3. Off-line storage
    a. Storage tanks
    b. Lagoons
    c. Deep tunnels
    d. Abandoned pipelines
    e. In-receiving water flow balance method
    f. Street storage

IV. Physical Treatment
  1. Sedimentation
  2. Dissolved air flotation
  3. Screens
    a. Bar screens and coarse screens
    b. Fine screens and microstrainers
  4. Filtration
  5. Flow concentrators

V. Biological Treatment
  1. Activated sludge
  2. Trickling filtration
  3. Rotating biological contactors
  4. Treatment lagoons
    a. Oxidation ponds
    b. Aerated lagoons
    c. Facultative lagoons
  5. Land treatment

VI. Physical-Chemical Treatment
  1. Chemical clarification
  2. Filtration
  3. Carbon adsorption
  4. High gradient magnetic separation

VII. Chemical Treatment (disinfection)
  1. Chemical
  2. Radiation

Improved Management. Improved management and in-system modifications can be implemented to use the existing collection system more effectively and to treat the maximum quantity of flow possible, thereby minimizing overflows. To obtain the benefits of improved management, a continual program of maintenance and inspection of the collection system, particularly of the flow regulators, is required. In addition, there are often minor modifications or repairs to the system that will significantly increase the volume of storm flow retained in the system. Real-time control of combined sewer systems has also been used to control the capacity of combined sewers or interceptors [4].

The use of polymers to increase the hydraulic capacity of pipelines is an innovative method used to correct specific capacity deficiencies in a transport system. The injection of polymer slurries into sewers can reduce pipe friction and thereby increase pipe capacity. In certain cases, this capacity increase can be significant and may reduce system surcharging and backups during wet weather. However, the increase in flow capacity attributable to polymer injection will often be insignificant when compared to the magnitude of wet-weather flows.

Control of Infiltration/Inflow. Excessive infiltration/inflow (I/I) in a wastewater collection system, whether separate or combined, can cause operating and mainte- nance problems in both the collection and treatment systems, and it uses up hydraulic capacity that was intended for other purposes in the system. (I/I is defined in Chap. 2.) Combined sewer systems are designed to collect inflow from surface drainage. Occasionally, combined sewers have been designed to carry brooks or streams. Infiltration is normally a constant influx of water of substantially lower volume than inflow. Control of infiltration problems is difficult and often expensive. Control of infiltration often has a limited impact on CSO reduction because infiltration is sub- stantially smaller than inflow.

Sewer Separation. Sewer separation is the conversion of a combined sewer system into separate stormwater and sanitary wastewater collection systems. Historically considered the best answer to combined sewer overflow pollution, sewer separation has been reconsidered in recent years because separation still results in contam- ination from stormwater runoff (see Table 15-3) being discharged to receiving waters. Sewer separation is relatively expensive and can also cause major disruptions to traffic and other daily community activities during construction. However, sewer separation is a positive means of eliminating combined sewer overflows by (1) preventing dry-weather sanitary flow from entering receiving waters dur- ing dry and wet weather periods, (2) reducing the volume of flow to be treated at the wastewater treatment plant, thus reducing operation and maintenance costs, and (3) reducing infiltration if new sanitary sewers are constructed to replace old combined sewers. In some large collection systems, combined sewers in the up- stream portions may exist. For these cases, "partial separation" of the combined portions of the system may be a cost-effective method of removing wastewater flow.

Storage

Methods of CSO control by storage include in-system, surface, and off-line storage. Storage in the collection system provides for flow equalization and reduces the peak flow rate. Storage provides CSO treatment by settling or skimming of stored flows and primarily by diverting stored flows (after flows recede) to dry-weather treatment facilities. Advantages of these storage methods include (1) simplicity of design and operation, (2) rapid response to changes in flow, and (3) full treatment of the stored flow at dry-weather treatment facilities, thus eliminating the discharge of contaminants associated with stormwater. Some of the disadvantages are (1) the large area requirements for off-line or surface storage facilities, (2) high operation and maintenance costs associated with the cleanup of the aeration equipment if long holding times and dewatering pumping operations are required, and (3) the increased flow to be treated at dry-weather treatment facilities, thus increasing operation and maintenance costs.

In-System Storage. For in-system storage, advantage is taken of existing capacity in the combined sewer system piping network and the interceptor to store flows during wet weather. In-system storage can often be used to capture smaller storm flows completely and to regulate and partially capture runoff from large storms. In-system storage can be accomplished by using such devices as regulators, inflatable dams, or automatic gates and valves, and it can best be used in large-diameter conduits with flat slopes.

Inflatable dams, operated by low-pressure air, can be installed in combined sewers to store wet-weather flows. They are normally kept fully inflated just downstream of a dry-weather connection so that dry-weather flows are diverted to an interceptor and so they will be ready at all times to store wet-weather flow. The dams have the capability of changing overflow height during a storm. When flows are large enough to cause upstream flooding, they can be deflated to release stored flows.

Automatic sluice gates and bascule-type gates can also be used for in-system storage. These have some advantages over inflatable dams: They can control water surface elevations more accurately and have been utilized more extensively. In-system storage is suitable in areas where unused capacity exists.

Surface Storage. Surface storage of stormwater refers to the construction of open basins to collect stormwater before it enters the collection system. These open basins are often constructed as part of new subdivisions to retard peak storm runoff rates to predevelopment levels. Surface storage of combined wastewater in open basins is usually not done because of the potential threats to public health and safety.

Off-Line Storage. Off-line storage facilities include storage tanks, deep tunnels, and abandoned pipelines as well as more exotic methods such as underwater bags and the in–receiving water flow balance method. The stored flow is often fed back into the collection system following the storm for subsequent treatment. Off-line storage tanks may require flushing facilities for the removal of settleable solids, ventilation for personnel safety and odor control, and pumping facilities to return the stored flow.

The in–receiving water flow balance method involves using floating pontoons and flexible curtains to create an in-water storage facility. The CSO flowing into the facility displaces the clean water that remains in the facility when not in use. Following the storm, the CSO is pumped to the collection system for subsequent treatment. The in–receiving water flow balance technology has been used successfully for separate stormwater in several lakes in Europe, and is currently being tested in the United States.

15-5 TREATMENT OF COMBINED SEWER OVERFLOWS

Treatment plants that are served by combined sewer systems are designed to provide treatment of the dry-weather wastewater flow, plus some portion of the wet-weather combined wastewater flow. Downstream treatment normally consists of the standard facilities for wastewater treatment described in earlier chapters of this book. In some cases, increased grit removal and handling facilities are employed to handle the high grit loads associated with stormwater runoff. In addition, facilities may be sited near the combined sewer outlets to either (1) treat the combined wastewater before it is discharged to the receiving water or (2) store the combined wastewater so that it may flow (or be pumped) back into the sewer system during dry weather, when sufficient capacity for downstream treatment becomes available. Treatment methods can be classified as physical, biological, physical-chemical, and chemical.

Physical Treatment

Physical treatment alternatives include sedimentation, dissolved air flotation, screening, and filtration. Most of these physical unit operations have been in use for many years and are considered reliable. Physical treatment operations are usually flexible enough to be readily automated and can operate over a wide range of flows. Also, they can stand idle for long periods of time without affecting treatment efficiencies. The principal physical methods used for the treatment of CSOs are considered in greater detail in Chaps. 6 and 9 of this textbook.

Solids separation devices such as swirl concentrators and vortex separators have been used in Europe and, to a lesser extent, in the United States. These devices are small, compact solids separation units with no moving parts [13]. A typical vortex-type CSO solids separation unit is illustrated in Fig. 15-9. During wet weather, the outflow from the unit is throttled, causing the unit to fill and to self-induce a swirling vortex-like flow regime. Secondary flow currents rapidly separate first flush settleable grit and floatable matter. Concentrated foul matter is intercepted for treatment while the cleaner, treated flow discharges to receiving waters. The device is intended to operate under extremely high flow regimes.

Biological and Physical-Chemical Treatment

The use of biological and physical-chemical treatment processes for the treatment of combined wastewater has some serious limitations:

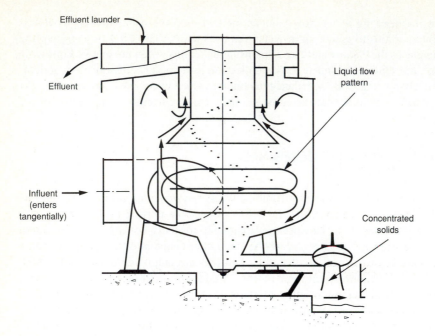

FIGURE 15-9
Cross section through a typical vortex-type solids separation device [13].

1. The biomass used to assimilate the nutrients in the combined wastewater must be kept alive during dry weather, which can be difficult except at an existing treatment plant.
2. Biological processes are subject to upset when subjected to erratic loading conditions.
3. The land requirements for this type of plant can be excessive in an urban area.
4. Operation and maintenance can be costly, and facilities require highly skilled operators.

It is feasible and frequent in practice, however, to treat a portion of the wet-weather flow at the treatment plant. In some treatment facilities the wet-weather flow receives full secondary treatment, whereas in others the flow is split, with some receiving primary treatment and disinfection only and the remainder receiving full secondary treatment.

Chemical Treatment (Disinfection)

Chemical disinfection with various forms of chlorine is the most common method of disinfection in use today for CSOs. Although ozone and UV irradiation have been used successfully for the disinfection of wastewater, the use of these means for the disinfection of CSOs is limited. Ozone, due to its rapid decay characteristics, must be generated at the point of application. The need for on-site generation at the point of application, coupled with the intermittent nature of CSO flows, makes ozone difficult to utilize for CSO disinfection. Ultraviolet (UV) irradiation has been used (although not frequently) for disinfection of treatment municipal wastewater (see Chaps. 7 and 9). In CSO treatment, UV performance would be poor unless the CSO flows were treated prior to irradiation to reduce suspended solids and turbidity, which greatly diminish the effectiveness of UV disinfection.

The most common approach to chlorination at wastewater treatment plants, as described in Chaps. 7 and 9, involves the use of liquid chlorine converted to gas and fed as a liquid. The intermittent nature of CSO flows, combined with the need to have chlorine available nearly instantaneously when storms occur, makes effective use of traditional liquid chlorine systems difficult. Also, handling and storage of pressurized liquid chlorine present a safety hazard, particularly if used at a CSO facility located in an urban area. To overcome these problems, either calcium hypochlorite or sodium hypochlorite may be used for chlorination (see Chap. 9). Both are available in liquid form and may be stored in tanks and fed with proportioning pumps. Sodium hypochlorite is used more frequently because it is easier to handle and more readily available. Both hypochlorites deteriorate over time, with sodium hypochlorite decomposing more rapidly. Hypochlorite in general is more expensive than liquid chlorine, but it can be instantly available for use and does not pose the safety hazards that liquid chlorine does. A typical disinfection facility employing sodium hypochlorite is shown in Fig. 15-10.

As described in Section 7-6, chlorination of waters contaminated with humic and fulvic acids can produce trihalomethanes, which are known carcinogens. Further, it is known that low concentrations of chlorine residuals can affect shellfish reproduction. Hence, dechlorination facilities may be required when chlorination of CSOs is employed. For intermittent operation, the use of sodium metabisulfite or sodium bisulfite is preferred over sulfur dioxide as a dechlorination agent. Sulfur dioxide is similar to chlorine in that it is a highly corrosive gas requiring the use of evaporators, whereas sodium metabisulfite is available in a powder form and sodium bisulfite is available in a liquid form, which can be fed with a chemical pump.

15-6 FUTURE DIRECTIONS IN THE MANAGEMENT OF COMBINED SEWER OVERFLOWS

As "point source" discharges from wastewater treatment facilities have received increasing amounts of treatment, it has become apparent that the discharge of combined sewer overflows is preventing a large number of receiving waters from attaining receiving water standards and desired uses. It is clear that continued regulatory attention will be placed on control of CSOs as well as stormwaters. Greater application of existing technologies and the development of new control devices and technologies are anticipated in the 1990s. The increasing speed and power of microcomputers will allow easier and more flexible use of combined sewer system models by engineers. Calculations for long time periods and numerous control options can be performed rapidly and cost-effectively.

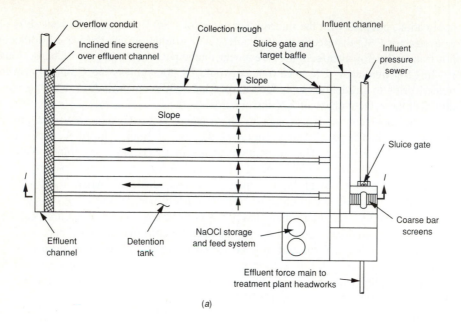

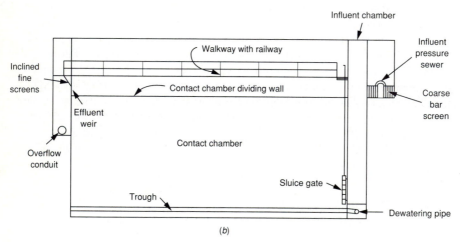

FIGURE 15-10
Typical screening and disinfection facility: (a) plan view and (b) Section 1-1.

DISCUSSION TOPICS AND PROBLEMS

15-1. Dry-weather flow monitoring was conducted in Manhole 1 shown in the accompanying sketch. An average flow rate of 3.65 ft³/s and an average peak flow rate of 7.30 ft³/s were observed over a five-day dry-weather period. Determine the wet to dry weather capacity ratio for the 24 in interceptor.

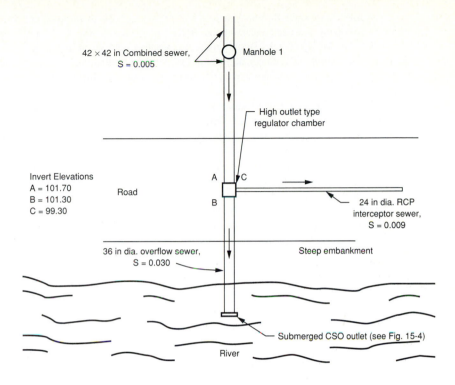

42 × 42 in Combined sewer,
S = 0.005

Manhole 1

High outlet type
regulator chamber

Invert Elevations
A = 101.70
B = 101.30
C = 99.30

Road

A   C
B

24 in dia. RCP
interceptor sewer,
S = 0.009

36 in dia. overflow sewer,
S = 0.030

Steep embankment

Submerged CSO outlet (see Fig. 15-4)

River

15-2. Wet-weather flow and quality were monitored in Manhole 1 in the above sketch during a rainstorm on July 5, 1989. The storm started at approximately 1210 hours and lasted 50 minutes. Based on the information given in the definition sketch provided in Prob. 15-1 and on the following wet-weather flow and quality data, estimate (*a*) how long the overflow event lasted, and (*b*) the total overflow volume and quantity of each contaminant discharged to the river during the storm.

Flow and quality data collected at Manhole 1

Time, hrs	Flow, ft³/s	Biochemical oxygen demand, BOD_5, mg/L	Suspended solids, SS, mg/L	Fecal coliform, MPN/100 mL
1205	7.1	181	140	12.1×10^6
1215	12.7	147	167	4.0×10^6
1225	18.6	110	200	1.1×10^6
1235	31.9	90	241	0.7×10^6
1245	52.4	65	311	0.2×10^6
1255	34.3	71	212	0.08×10^6
1305	20.1	128	110	0.6×10^6
1315	8.2	173	135	6.6×10^6
1325	7.4	194	151	7.8×10^6

15-3. The dry-weather wastewater flow in Manhole 1 of Problem 15-2 was sampled and found to have the following average concentrations: BOD_5 = 180 mg/L, SS = 145 mg/L, and FC = 1.5×10^6 MPN/100 mL. Based on the wet weather data given in Problem 15-2, estimate (a) the quantity of each contaminant in the overflow originating from stormwater runoff and the base flow, and (b) average concentrations of the stormwater component of the overflow.

15-4. For the mechanical regulator in Example 15-1, both the 48 in combined sewer and the 24 in interceptor sewer are flowing full. Determine (a) the flow passing to the interceptor through the regulator, and (b) the overflow rate to the receiving water. Assume all pipes shown are concrete.

15-5. Dust and dirt (DD) was found to accumulate on a 10-acre drainage area at the following rate: DD = $10.7(1 - e^{-0.6t})$, where t = time in days and DD = dust and dirt load, lb/100 ft curb. The drainage area is 54 percent impervious with a total curb length of 1800 ft. Using the ratio of contaminant to DD and the wash-off expressions used in Example 15-2, and knowing that it has been five days since a complete wash-off occurred, determine (a) the quantity of BOD_5, SS, and FC washed off during a rainstorm with the following runoff rates:

Time, min	Average runoff rate, in/hr
0–10	0.3
10–15	0.5
15–25	0.1

(b) During which runoff period would the surface runoff have the highest average concentrations? Estimate the average concentrations.

15-6. A city plans to construct a screening and disinfection facility to treat overflow from three CSOs located in close proximity to each other near a river. Two city-owned parcels of land of about equal size are being considered for the facility. Site A is located directly adjacent to the CSOs, and site B is located approximately 500 ft downriver. Soil borings were performed at both sites. At site A, bedrock was encountered at an average depth of 5 ft below the ground surface. At site B, no bedrock was found to a depth of 25 ft. Assume that a facility at either site would have similar requirements except that for site B, an additional 500 ft of 5.0 ft diameter conduit would be required. Using the following information, determine (a) the outside dimensions of the tank and (b) which site should the city choose for the facility.

Peak overflow rate = 75 ft³/s

Minimum contact time required = 15 min

Available surface area dimensions for tank = 52 ft × 77 ft

Required clearance of outlet weir = 2 ft

Tank walls, floor, and roof are 1 ft thick.

Excavation would extend 1 ft below the tank bottom and 4 ft beyond the walls.

Top of tank may be no higher than the existing ground surface.

Excavation of bedrock is expected to cost $60 per cubic yard higher than normal excavation costs.

Estimated construction cost of 5.0 ft diameter pipe = \$260/ft (assumes no sheeting, pylons, or in-line manholes).

15-7. A city would like to control the pollution from several CSOs that discharge to a scenic receiving water and result in state water quality standards violations. The options have been narrowed to three control methods: (1) source control measures, (2) sewer separation, and (3) off-line storage. Describe what each control method is and the major advantages and disadvantages of each method.

15-8. Describe the "first flush" effect and how it typically occurs in combined sewer systems. What measures could be taken to reduce the effects of the "first flush" phenomenon on receiving waters?

REFERENCES

1. American Public Works Association: *Combined Sewer Regulator Overflow Facilities,* Report to the Federal Water Quality Administration, U.S. Department of the Interior, Report No. 11022 DMU, July 1970.

2. American Public Works Association: *Problems of Combined Sewer Facilities and Overflows,* Report to the Federal Water Pollution Control Administration, U.S. Department of the Interior, Report No. WP-20-11, December 1967.

3. Huber, W. C.: *Storm Water Management Model, User's Manual, Version III,* Report to the U.S. Environmental Protection Agency, Project No. CR-805664, September 1984.

4. Leisure, C. P.: *Computer Management of Combined Sewer Systems,* U.S. Environmental Protection Agency, Report No. EPA-670/2-74-022, 1974.

5. Metcalf & Eddy, Inc.: *Wastewater Engineering, Collection and Pumping of Wastewater,* McGraw-Hill, New York, 1981.

6. Metcalf & Eddy, Inc.: *Urban Stormwater Management and Technology: Update and Users Guide,* Report to the U.S. Environmental Protection Agency, Report No. EPA-600/8-77-014, September 1977.

7. Metcalf & Eddy, Inc.: *Urban Stormwater Management and Technology,* Report to the U.S. Environmental Protection Agency, Report No. EPA 670/2-74-040, December 1974.

8. Metcalf & Eddy, Inc., University of Florida, and Water Resources Engineers, Inc.: *Storm Water Management Model,* Report to the U.S. Environmental Protection Agency, Contract Nos. 14-12-501, 502, and 503, July 1971.

9. Metcalf & Eddy, Inc.: *American Sewerage Practice,* McGraw-Hill, New York, 1928.

10. Roesner, L. A., R. P. Shubinski, and J. A. Aldrich: *Storm Water Management Model User's Manual Version III: Addendum I, EXTRAN,* Report to the U.S. Environmental Protection Agency, 1981.

11. U.S. Environmental Protection Agency: *1988 Needs Survey Report to Congress, Assessment of Needed Publicly Owned Wastewater Treatment Facilities in the United States,* Report No. EPA 430/09-89-001, February 1989.

12. U.S. Environmental Protection Agency: *Results of the Nationwide Urban Runoff Program, Volume 1. Final Report,* NTIS PB84-185552, December 1983.

13. Water Pollution Control Federation: *Combined Sewer Overflow Pollution Abatement,* Manual of Practice FD-17, 1989.

14. Watt, T. R., R. G. Skrentnen, and A. C. Davanzo: *Sewer System Monitoring and Remote Control,* U.S. Environmental Protection Agency, Report No. EPA-670/2-75-020, 1975.

CHAPTER 16

WASTEWATER RECLAMATION AND REUSE

Continued population growth, contamination of both surface and groundwaters, uneven distribution of water resources, and periodic droughts have forced water agencies to search for innovative sources of water supply. Use of highly treated wastewater effluent, now discharged to the environment from municipal wastewater treatment plants, is receiving more attention as a reliable source of water. In many parts of the country, wastewater reuse is already an important element in water resources planning. Wastewater reuse is a viable option, but water conservation, efficient use of existing water supplies, and development of new water resources are other alternatives that must be evaluated.

The purpose of this chapter is to introduce the basic concepts and issues involved in wastewater reclamation and reuse. The chapter is organized in four sections: (1) a general introduction to the subject, including the definition of terms commonly used, (2) a discussion of wastewater reuse applications, (3) a brief review of the principal treatment technologies used for wastewater reclamation, and (4) a short section in which the important planning considerations for wastewater reclamation and reuse are summarized.

16-1 WASTEWATER RECLAMATION AND REUSE: AN INTRODUCTION

This section provides a brief historical perspective on the subject of wastewater reclamation and reuse. Technical terms unique to wastewater reuse practices are defined and included.

Historical Perspective

Early developments in the field of wastewater reuse are synonymous with the historical practice of land treatment and disposal of wastewater. The development of land treatment systems is considered in Chap. 13. With the advent of sewerage systems in the nineteenth century, domestic wastewater was used at "sewage farms," and by 1900 there were numerous sewage farms in Europe and in the United States [35,45]. Although these sewage farms were used primarily for waste disposal, incidental use was made of the water for crop production or for other beneficial uses.

More recently, a number of wastewater reclamation and reuse projects have been developed as a matter of necessity to meet growing water needs. In 1926, at the Grand Canyon National Park in Arizona, treated wastewater was first used in a dual water supply system for toilets, lawn sprinklers, cooling water, and boiler feed water. In 1929, the city of Pomona, California, initiated a project utilizing reclaimed wastewater for irrigation of lawns and gardens [31]. As early as 1912, wastewater (first untreated, then treated in septic tanks) was used in the Golden Gate Park in San Francisco for watering lawns and supplying ornamental lakes. In 1932, a conventional wastewater treatment plant was built near the park and the reuse of treated wastewater continued until 1985.

In 1942, the use of chlorinated wastewater effluent from Baltimore, Maryland, was initiated at Bethlehem Steel Company, which now uses over 100 Mgal/d of secondary effluent for primary metals cooling and steel processing. The impetus for most industrial users to implement wastewater reuse programs has been a lack of alternative water supplies. In 1960, a dual water supply system was implemented at Colorado Springs, Colorado that now supplies reclaimed municipal wastewater principally for landscape irrigation at golf courses, parks, cemeteries, and freeways [49]. A similar urban wastewater reuse system was initiated in St. Petersburg, Florida in 1977 [49] as an essential part of that city's water pollution abatement program. Reclaimed wastewater is now distributed through a 200-mile dual water system for irrigation of public parks, golf courses, school yards, residential lawns, and cooling tower make-up water. Beginning in 1962, the first major groundwater recharge project with reclaimed municipal wastewater was undertaken at Whittier Narrows in Los Angeles County, California. After the extensive health effects evaluation of more than 20 years of records, researchers concluded that there was no measurable adverse impact on the groundwater in the area or on the population ingesting the groundwater affected by the recharge operation [1,29].

According to the only available national survey on wastewater reclamation and reuse projects [11], 536 wastewater reuse projects were in existence in the United States in 1975, as reported in Table 16-1. The estimated total wastewater reuse was 679 Mgal/d. Most of the wastewater reuse sites are located in the arid and semiarid western and southwestern states, including Arizona, California, Colorado, and Texas. However, an increasing number of wastewater reuse projects are being implemented in the humid regions of the United States, which include Florida and South Carolina, for water pollution abatement as well as for water supply purposes. Because of health and safety concerns, nonpotable water reuse for irrigation of crops, parks,

TABLE 16-1
Municipal wastewater reuse projects in the United States[a]

Category	Number of projects	Reclaimed water, Mgal/d
Irrigation-total	470	420
Agriculture	150	199
Landscape	60	33
Not defined	260	188
Industrial-total	29	215
Process		66
Cooling		142
Boiler feed		7
Groundwater recharge	11	34
Other (Recreation, etc.)	26	10
Total	536	679

[a] Based on the only available national survey of wastewater reclamation and reuse projects for the year 1975 [11].

Note: Mgal/d $\times$ 3.7854 $\times$ 10^3 = m^3/d

and golf courses has become a standard practice for planned reuse of municipal wastewater [10]. Some communities, however, are developing plans for potable water reuse where no other possibilities exist for expanding freshwater supplies [21,22,30]. The quantities of wastewater involved in these potable water reuse projects are small, but the technological and public health issues are of great significance.

In the preceding explanation of the historical perspective, the broad scope of wastewater reclamation and reuse was presented, and the various categories of wastewater reuse were described. In recent years, the desirability and benefits of wastewater reuse have been well recognized by several states; for example, the California State Water Code clearly notes that "it is the intention of the Legislature that the State undertake all possible steps to encourage development of water reclamation facilities so that reclaimed water may be made available to help meet the growing water requirements of the State [43]." Today, technically proven wastewater treatment processes exist to prepare water of almost any quality desired. Thus, wastewater reuse has a rightful place and an important role in the optimal planning for efficient use of water resources.

Definition of Terms

The following terms, used frequently in the field of wastewater reclamation and reuse, are important in understanding the concepts discussed in this chapter:

Beneficial uses are the many ways water can be used, either directly by people or for their overall benefit. Examples include municipal water supply, agricultural and industrial applications, navigation, and water contact recreation.

Direct potable reuse is a form of reuse that involves the incorporation of reclaimed wastewater directly into a potable water supply system, often implying the blending of reclaimed wastewater.

Direct reuse is the use of reclaimed wastewater that has been transported from a wastewater reclamation plant to the water reuse site without intervening discharge to a natural body of water. It includes such uses as agricultural and landscape irrigation.

Indirect potable reuse is the potable reuse by incorporation of reclaimed wastewater into a raw water supply. It allows mixing and assimilation by discharge into an impoundment or natural body of water, such as in domestic water supply reservoir or groundwater.

Indirect reuse is the use of wastewater reclaimed indirectly by passing it through a natural body of water or use of groundwater that has been recharged with reclaimed wastewater.

Planned reuse is the deliberate direct or indirect use of reclaimed wastewater without relinquishing control over the water during its delivery.

Potable water reuse is a direct or indirect augmentation of drinking water with reclaimed wastewater that is normally highly treated to protect public health.

Reclaimed wastewater is wastewater that, as a result of wastewater reclamation, is suitable for a direct beneficial use or a controlled use that would not otherwise occur.

Unplanned reuse is the incidental use of wastewater after surrendering control of the water after discharge, such as in the diversion of water from a river downstream of a discharge of treated wastewater.

Wastewater reclamation is the treatment or processing of wastewater to make it reusable. This term is also often used to include delivery of reclaimed wastewater to its place of use and its actual use.

Wastewater recycling is the use of wastewater that is captured and redirected back into the same water-use scheme. Recycling is practiced predominantly in industries such as manufacturing, and it normally involves only one industrial plant or one user.

Wastewater reuse is the use of treated wastewater, for a beneficial use such as agricultural irrigation or industrial cooling.

Potential and Status of Wastewater Reuse

To understand the significance of wastewater reuse, it is helpful to compare the wastewater reuse potential with total water use on a national scale. Water withdrawals in the United States during 1985 were estimated to be an average of 399 Bgal/d (billion gallons per day) of fresh and saline water for off-stream uses — 10 percent less than the 1980 estimate. However, public water supply withdrawals were 7 percent more than they were during 1980. Average per-capita use for all off-stream uses during 1985 was 1650 gal/d of freshwater and saline water combined, and 1400 gal/d of freshwater alone. Off-stream uses include (1) public supply (domestic, public, commercial, and industrial uses), (2) rural (domestic and livestock uses), (3) irrigation, and (4) self-supplied industrial uses including thermoelectric power. The estimates of freshwater withdrawals and wastewater recycling and reuse are reported in Table 16-2. The relatively small municipal wastewater reclamation and reuse compared with water recycling and total freshwater withdrawals in 1985 is expected to remain about the same in the future. However, the actual quantity of wastewater reuse will increase significantly and will become more important geographically for water-short regions of the United States.

A comparison of water withdrawals of both surface water and groundwater by states in 1980 is shown in Fig. 16-1. California accounted for the most water withdrawn for off-stream use with 49.7 Bgal/d, more than double the amount of water withdrawn in either Texas or Idaho, the next largest users. It is estimated that agricultural activities and steam electric plants will continue to use over 75 percent of all freshwater withdrawn in the United States [40,52].

The seven principal categories of municipal wastewater reuse are listed in Table 16-3 in descending order of projected volume of use. Potentially large quantities of reclaimed municipal wastewater can be used in the first four categories. Agricultural and landscape irrigation, the largest current and projected use of water, offers significant opportunities for wastewater reuse. Reviewing the base data for agricultural and landscape irrigation shows that California is by far the largest user of irrigation water, withdrawing about 31 Bgal/d, 22 percent of the national total. This is more than the next largest users of irrigation water, Idaho and Colorado, combined [40].

The second major use of reclaimed municipal wastewater is in industrial activities, primarily for cooling and process needs. Industrial uses vary greatly, and to provide adequate water quality, additional treatment is often required beyond conventional secondary wastewater treatment.

The third reuse application for reclaimed wastewater is groundwater recharge, by way of either spreading basins or direct injection to groundwater aquifers.

TABLE 16-2
Estimates of freshwater withdrawals and wastewater recycling and reuse in the United States in years 1975, 1985, and 2000[a]

Category	Quantity, Bgal/d		
	1975	1985	2000
Total freshwater withdrawals	362.7	356.3	330.9
Wastewater recycling (Industrial)	139.1	386.7	865.5
Steam electric	57	—	517.3
Manufacturing	61	—	316.2
Minerals	21	—	32.0
Wastewater reuse (Municipal)	0.7	2.1	4.8

[a] The Second National Water Assessment published in 1978 by the U.S. Water Resources Council [52] provided the 1975 data base. Water quantity values for both the years 1985 and 2000 are the projected values from Ref. 11.

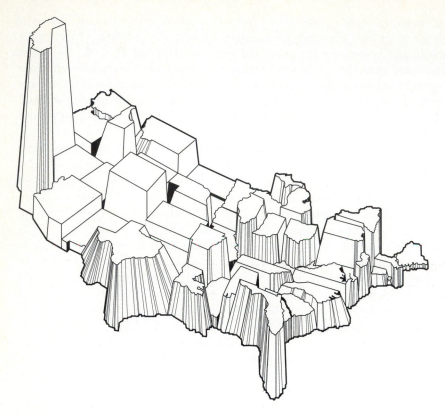

FIGURE 16-1
Comparison of water withdrawals of both surface water and groundwater, by states in 1980. Adapted from Ref. 40.

Groundwater recharge involves assimilation of reclaimed wastewater for replenishment and storage in groundwater aquifers, or establishment of hydraulic barriers to saltwater intrusion. The factors of time-in-storage and separation-in-space between points of application and withdrawal are important public health considerations. Perhaps the greatest single advantage of including groundwater recharge in any program of wastewater reuse is the loss of identity through assimilation that groundwater recharge provides for reclaimed wastewater.

A fourth use of reclaimed wastewater is characterized as miscellaneous subpotable uses for recreational lakes, aquaculture, toilet flushing, and so forth. These subpotable uses are minor reclaimed wastewater applications that presently account for less than 5 percent of total wastewater reuse.

16-2 WASTEWATER REUSE APPLICATIONS

In the planning and implementation of wastewater reclamation and reuse, the reuse application (see Table 16-3) will usually govern the wastewater treatment needed and

TABLE 16-3
Categories of municipal wastewater reuse and potential constraints[a]

Wastewater reuse categories	Potential constraints

Agricultural irrigation
  Crop irrigation
  Commercial nurseries
	Surface- and groundwater pollution if not properly managed
	
	Marketability of crops and public acceptance

Landscape irrigation
  Park
  School yard
  Freeway median
  Golf course
  Cemetery
  Greenbelt
  Residential
	Effect of water quality, particularly salts, on soils and crops
	
	Public health concerns related to pathogens (bacteria, viruses, and parasites)
	
	Use area control including buffer zone. May result in high user costs.

Industrial recycling and reuse
  Cooling
  Boiler feed
  Process water
  Heavy construction
	Constituents in reclaimed wastewater related to scaling, corrosion, biological growth, and fouling
	
	Public health concerns, particularly aerosol transmission of pathogens in cooling water

Groundwater recharge
  Groundwater replenishment
  Salt water intrusion control
  Subsidence control
	Organic chemicals in reclaimed wastewater and their toxicological effects. Total dissolved solids, nitrates, and pathogens in reclaimed wastewater

Recreational/environmental uses
  Lakes and ponds
  Marsh enhancement
  Streamflow augmentation
  Fisheries
  Snowmaking
	Health concerns of bacteria and viruses
	
	Eutrophication due to N and P in receiving water
	
	Toxicity to aquatic life

Nonpotable urban uses
  Fire protection
  Air conditioning
  Toilet flushing
	Public health concerns on pathogens transmitted by aerosols
	
	Effects of water quality on scaling, corrosion, biological growth, and fouling
	
	Cross-connection

Potable reuse
  Blending in water supply reservoir
  Pipe to pipe water supply
	Constituents in reclaimed wastewater, especially trace organic chemicals and their toxicological effects
	
	Aesthetics and public acceptance
	
	Health concerns about pathogen transmission, particularly viruses

[a] Arranged in descending order of projected volume of use.

the degree of reliability required for the treatment processes and operations. Because wastewater reclamation entails the provision of a continuous supply of water with consistent water quality, the reliability of the existing or proposed treatment processes and operations must be evaluated in the planning stage.

The purpose of this section is to present and discuss wastewater reuse applications and to emphasize the water quality requirements that protect the environment and mitigate health risks. The principal reuse categories considered are (1) agricultural and landscape irrigation, (2) industrial applications, (3) groundwater recharge, and (4) potable reuse. Although the quantities of water involved in potable reuse are small, some of the technological advancements are discussed in this section because of their public health interest.

Agricultural and Landscape Irrigation

The quantity of freshwater used for irrigation in the United States in 1985 was estimated at 137 Bgal/d or about 154 million ac-ft/yr. The water was used on approximately 57 million acres of farmland. Irrigation represents 34 percent of total nationwide use for all off-stream categories. Irrigation is by far the largest water use in the west. The nine western water resources regions, led by the California region, accounted for 91 percent of the total water withdrawn for irrigation during 1980 and 1985 [40].

Irrigation of crops developed along with the settlement of the arid West because irrigation was needed to raise crops. In the humid eastern United States, irrigation is used to supplement natural rainfall, to increase the number of plantings per year and the yield of crops, and to reduce the risk of crop failures during drought periods. Irrigation is now also used to maintain recreational lands such as parks and golf courses. The irrigation of landscaped areas and golf courses in the urban environment has become an important use of reclaimed wastewater in recent years.

Evaluation of Irrigation Water Quality. Although irrigation has been practiced throughout the world for several millennia, it is only in this century that the importance of the quality of irrigation water has been recognized. The design approach to irrigation with reclaimed municipal wastewater depends upon whether emphasis is placed on providing a water supply or wastewater treatment (see Chap. 13).

Physical and chemical water quality. The quality of irrigation water is of particular importance in arid zones where extremes of temperature and low humidity result in high rates of evapotranspiration (ET). Evapotranspiration refers to water lost through evaporation from the soil and surface water bodies and transpiration from plants. Water used for irrigation can vary greatly in quality depending upon the type and quantity of dissolved salts. The consequence of evapotranspiration is salt deposition from the applied water, which tends to accumulate in the soil profile. The physical and mechanical properties of the soil, such as degree of dispersion of the soil particles, stability of aggregates, soil structure, and permeability, are sensitive to the types of exchangeable ions present in irrigation water. Thus, when irrigation with reclaimed wastewater is being planned, crop yield and soil properties must both be taken into consideration. The problems, however, are no different from those caused by salinity or trace elements in any water supply and are of concern only if they restrict the use of the water or require special management to maintain acceptable crop yields.

A number of different irrigation water quality guidelines have been proposed. The guidelines presented in Table 16-4 were developed by the University of California Committee of Consultants and were subsequently expanded by Ayers and Westcot [4, 33]. The long-term influence of water quality on crop production, soil conditions, and farm management is emphasized, and the guidelines are applicable to both freshwater and reclaimed wastewater. Four categories of potential management problems associated with water quality are: (1) salinity, (2) specific ion toxicity, (3) water infiltration rate, and (4) miscellaneous problems.

Salinity. Salinity of an irrigation water is determined by measuring its electrical conductivity and is the most important parameter in determining the suitability of a water for irrigation. The electrical conductivity (EC) of a water is used as surrogate measure of total dissolved solids (TDS) concentration. The electrical conductivity is expressed as mmho/cm or decisiemens per meter (dS/m). It should be noted that one mmho/cm is equivalent to one dS/m. Values for salinity are also reported as TDS in mg/L. For most agricultural irrigation purposes, the values for EC and TDS are directly related and convertable within an accuracy of about 10 percent. Equation 16-1 can be used to convert EC values to corresponding TDS values [33].

$$\text{TDS (mg/L)} \approx \text{EC (mmho/cm or dS/m)} \times 640 \tag{16-1}$$

The presence of salts affects plant growth in three ways: (1) osmotic effects, caused by the total dissolved salt concentration in the soil water; (2) specific ion toxicity, caused by the concentration of an individual ion; and (3) soil particle dispersion, caused by high sodium and low salinity. With increasing soil salinity in the root zone, plants expend more of their available energy on adjusting the salt concentration within the tissue (osmotic adjustment) to obtain needed water from the soil. Consequently, less energy is available for plant growth.

In irrigated areas, salts originate from the local groundwater or from salts in the applied irrigation water. Salts tend to concentrate in the root zone owing to evapotranspiration, and plant damage is tied closely to an increase in soil salinity. Establishing a net downward flux of water and salt through the root zone is the only practical way to manage a salinity problem. Under such conditions, good drainage is essential to allow a continuous movement of water and salt below the root zone. Long-term use of reclaimed wastewater for irrigation is not possible without adequate drainage.

If more water is applied than the plant uses, the excesss water will percolate below the root zone, carrying with it a portion of the accumulated salts. Consequently, the soil salinity will reach some constant value dependent on the leaching fraction. The fraction of applied water that passes through the entire rooting depth and percolates below is called the leaching fraction (LF).

TABLE 16-4
Guidelines for interpretations of water quality for irrigation[a]

Potential irrigation problem	Units	Degree of restriction on use		
		None	Slight to moderate	Severe
Salinity (affects crop water availability)				
EC_w	dS/m or mmho/cm	<0.7	0.7–3.0	>3.0
TDS	mg/L	<450	450–2,000	>2,000
Permeability (affects infiltration rate of water into the soil. Evaluate using EC_w and SAR or adj R_{Na} together)[b]				
SAR = 0–3		and $EC_w \geq$ 0.7	0.7–0.2	<0.2
3–6		≥ 1.2	1.2–0.3	<0.3
6–12		≥ 1.9	1.9–0.5	<0.5
12–20		≥ 2.9	2.9–1.3	<1.3
20–40		≥ 5.0	5.0–2.9	<2.9
Specific ion toxicity (affects sensitive crops)[c]				
Sodium (Na)				
Surface irrigation	SAR	<3	3–9	>9
Sprinkler irrigation	mg/L	<70	>70	
Chloride (Cl)				
Surface irrigation	mg/L	<140	140–350	>350
Sprinkler irrigation	mg/L	<100	>100	
Boron (B)	mg/L	<0.7	0.7–3.0	>3.0
Trace elements (see Table 16-5)				
Miscellaneous effects (affects susceptible crops)				
Nitrogen (Total-N)	mg/L	<5	5–30	>30
Bicarbonate (HCO_3) (overhead sprinkling only)	mg/L	<90	90–500	>500
pH	unit	Normal range 6.5–8.4		
Residual chlorine (overhead sprinkling only)	mg/L	<1.0	1.0–5.0	>5.0

[a] Adapted from Refs. 4 and 33.

[b] For wastewater irrigation, it is recommended that SAR be adjusted to include a more correct estimate of calcium in the soil water. A procedure is given in Eq. 16-5 and Table 16-6. The adjusted sodium adsorption ratio (adj R_{Na}) calculated by this procedure is to be substituted for the SAR value in this table.

[c] See also Table 16-5.

$$LF = \frac{D_d}{D_i} = \frac{(D_i - ET_c)}{D_i} \qquad (16\text{-}2)$$

where LF = leaching fraction
D_d = depth of water leached below the root zone, inches
D_i = depth of water applied at the surface, inches
ET_c = crop evapotranspiration, inches

A high leaching fraction results in less salt accumulation in the root zone. If the salinity of irrigation water (EC_w) and the leaching fraction are known, salinity of the drainage water that percolates below the rooting depth can be estimated by using Eq. 16-3:

$$EC_{dw} = \frac{EC_w}{LF} \qquad (16\text{-}3)$$

where EC_{dw} = salinity of the drainage water percolating below the root zone
EC_w = salinity of irrigation water

The EC_{dw} value can be used to assess the potential effects on crop yield and on groundwater. For salinity management, it is often assumed that EC_{dw} is equal to the salinity of the saturation extract of the soil sample, EC_e. This assumption is conservative, however, in that EC_{dw} occurs at the soil-water potential of field capacity and EC_e occurs at a potential of zero, by definition, at laboratory condition. For a quick check, the value of EC_{dw} can be estimated as twice the value of EC_e for most soils.

Example 16-1   Calculation of drainage water quality. A crop is irrigated with reclaimed wastewater whose salinity (EC_w), measured by electrical conductivity, is 1.0 dS/m. If the crop is irrigated to achieve a leaching fraction of 0.15 (i.e., 85 percent of the applied water is used by the crop or lost through evapotranspiration) determine the following: (1) the salinity of the deep percolate water and (2) the appropriate leaching fraction to maintain crop yield. The crop is known to suffer significant loss in yield when TDS of the soil water exceeds 5000 mg/L.

Solution

1. After many successive irrigations, the salt accumulation in the soil will approach an equilibrium concentration based on the salinity of the applied irrigation water and the leaching fraction. Thus, the salinity of the water that percolates below the root zone (drainage water) can be estimated using Eq. 16-3.

$$EC_{dw} = \frac{EC_w}{LF} = \frac{1.0}{0.15} = 6.7 \text{ dS/m}$$

2. Estimate the value of the TDS using Eq. 16-1

$$TDS \text{ (mg/L)} \approx EC \text{ (mmho/cm or dS/m)} \times 640$$

$$TDS \text{ (mg/L)} \approx 6.7 \times 640 = 4290 \text{ mg/L}$$

3. Determine the leaching fraction using Eq. 16-3.

$$LF = \frac{EC_w}{EC_{dw}} = \frac{1.0 \times 640}{5000} = 0.13$$

Thus, to prevent loss in yield, 13 percent of the applied water will be needed to carry salts below root zone and 87 percent will be consumed by evapotranspiration.

Specific ion toxicity. If the decline of crop growth is due to excessive concentrations of specific ions rather than to osmotic effects alone, it is referred to as "specific ion toxicity." As shown in Table 16-4, the ions of most concern in wastewater are sodium, chloride, and boron. The most prevalent toxicity from the use of reclaimed municipal wastewater is from boron. The source of boron is usually household detergents or discharges from industrial plants. The quantities of chloride and sodium also increase as a result of domestic usage, especially where water softeners are used (see Table 3-19, Chap. 3).

For sensitive crops, specific ion toxicity is difficult to correct without changing the crop or the water supply. The problem is accentuated by hot and dry climatic conditions caused by high evapotranspiration rates. The suggested maximum trace element concentrations for irrigation waters are presented in Table 16-5. In severe cases where water is used that contains elemental concentrations above these levels, these elements may accumulate in plants and soils and can result in human and animal health hazards or cause phytotoxicity in plants [4,33].

Water infiltration rate. Another indirect effect of high sodium content is the deterioration of the physical condition of a soil (formation of crusts, water-logging, and reduced soil permeability). If the infiltration rate is greatly reduced, it may be impossible to supply the crop or landscape plant with enough water for vigorous growth. In addition, reclaimed wastewater irrigation systems are often located on less desirable soils or soils already having permeability and management problems. It may be necessary in these cases to modify soil profiles by excavating and rearranging the affected land.

The water infiltration problem occurs within the top few inches of the soil and is mainly related to the structural stability of the surface soil. To predict a potential infiltration problem, the sodium adsorption ratio (SAR) is often used [4,33,35].

$$SAR = \frac{Na}{\sqrt{(Ca + Mg)/2}} \tag{16-4}$$

where the cation concentrations are expressed in meq/L.

The adjusted sodium adsorption ratio (adj R_{Na}) is a recent modification of Eq. 16-4 that takes into account changes in calcium solubility in the soil water [4,33,46].

$$adj\ R_{Na} = \frac{Na}{\sqrt{(Ca_x + Mg)/2}} \tag{16-5}$$

Here Na and Mg concentrations are expressed in meq/L, and the value of Ca_x, also expressed in meq/L, is obtained from Table 16-6.

TABLE 16-5
Recommended maximum concentrations of trace elements in irrigation waters[a]

Element	Recommended maximum concentration[b], mg/L	Remarks
Al (aluminum)	5.0	Can cause nonproductivity in acid soils (pH < 5.5), but more alkaline soils at pH > 5.5 will precipitate the ion and eliminate any toxicity.
As (arsenic)	0.10	Toxicity to plants varies widely, ranging from 12 mg/L for Sudan grass to less than 0.05 mg/L for rice.
Be (beryllium)	0.10	Toxicity to plants varies widely, ranging from 5 mg/L for kale to 0.5 mg/L for bush beans.
Cd (cadmium)	0.010	Toxic to beans, beets, and turnips at concentrations as low as 0.1 mg/L in nutrient solutions. Conservative limits recommended because of its potential for accumulation in plants and soils to concentrations that may be harmful to humans.
Co (cobalt)	0.050	Toxic to tomato plants at 0.1 mg/L in nutrient solution. Tends to be inactivated by neutral and alkaline soils.
Cr (chromium)	0.10	Not generally recognized as an essential growth element. Conservative limits recommended because of lack of knowledge on toxicity to plants.
Cu (copper)	0.20	Toxic to a number of plants at 0.1 to 1.0 mg/L in nutrient solutions.
F (fluoride)	1.0	Inactivated by neutral and alkaline soils.
Fe (iron)	5.0	Not toxic to plants in aerated soils but can contribute to soil acidification and loss of reduced availability of essential phosphorus and molybdenum. Overhead sprinkling may result in unsightly deposits on plants, equipment, and buildings.
Li (lithium)	2.5	Tolerated by most crops up to 5 mg/L; mobile in soil. Toxic to citrus at low levels (> 0.075 mg/L). Acts similar to boron.
Mn (manganese)	0.20	Toxic to a number of crops at a few tenths mg to a few mg/L, but usually only in acid soils.

(continued on next page)

TABLE 16-5
(continued)

Element	Recommended maximum concentration,[b] mg/L	Remarks
Mo (molybdenum)	0.010	Not toxic to plants at normal concentrations in soil and water. Can be toxic to livestock if forage is grown in soils with high levels of available molybdenum.
Ni (nickel)	0.20	Toxic to a number of plants at 0.5 to 1.0 mg/L; reduced toxicity at neutral or alkaline pH.
Pb (lead)	5.00	Can inhibit plant cell growth at very high concentrations.
Se (selenium)	0.020	Toxic to plants at concentrations as low as 0.025 mg/L and toxic to livestock if forage is grown in soils with relatively high levels of added selenium. An essential element for animals but in very low concentrations.
Sn (tin)	—	Effectively excluded by plants; specific tolerance unknown.
Ti (titanium)	—	(See remark for tin)
W (tungsten)	—	(See remark for tin)
V (vanadium)	0.10	Toxic to many plants at relatively low concentrations.
Zn (zinc)	2.0	Toxic to many plants at widely varying concentrations; reduced toxicity at pH > 6.0 and in fine-textured or organic soils.

[a] Adapted from Refs. 4 and 25.

[b] The maximum concentration is based on a water application rate that is consistent with good agricultural practices (4 ft/yr).

The adj R_{Na} value is preferred in irrigation applications with reclaimed municipal wastewater because it reflects the changes in calcium in the soil water more accurately. At a given sodium adsorption ratio, the infiltration rate increases as salinity increases, or it decreases as salinity decreases. Therefore, SAR or adj R_{Na} should be used in combination with the electrical conductivity (EC_w) of irrigation water to evaluate the potential permeability problem as shown in Table 16-4.

Reclaimed municipal wastewater is normally high in calcium, and there is little concern that the water will dissolve and leach too much calcium from the surface soil (See Example 16-2). However, reclaimed wastewater is sometimes high in sodium; the resulting high SAR is a major concern in planning irrigation projects with reclaimed municipal wastewater.

TABLE 16-6
Values of Ca$_x$ used in Eq. 16-5 as a function of the HCO$_3$/Ca ratio and solubility[a]

Ratio of HCO$_3$/Ca in meq/L	Values of Ca$_x$, meq/L											
	Salinity of applied water (EC$_w$), dS/m or mmhos/cm											
	0.1	0.2	0.3	0.5	0.7	1.0	1.5	2.0	3.0	4.0	6.0	8.0
0.05	13.20	13.61	13.92	14.40	14.79	15.26	15.91	16.43	17.28	17.97	19.07	19.94
0.10	8.31	8.57	8.77	9.07	9.31	9.62	10.02	10.35	10.89	11.32	12.01	12.56
0.15	6.34	6.54	6.69	6.92	7.11	7.34	7.65	7.90	8.31	8.64	9.17	9.58
0.20	5.24	5.40	5.52	5.71	5.87	6.06	6.31	6.52	6.86	7.13	7.57	7.91
0.25	4.51	4.65	4.76	4.92	5.06	5.22	5.44	5.62	5.91	6.15	6.52	6.82
0.30	4.00	4.12	4.21	4.36	4.48	4.62	4.82	4.98	5.24	5.44	5.77	6.04
0.35	3.61	3.72	3.80	3.94	4.04	4.17	4.35	4.49	4.72	4.91	5.21	5.45
0.40	3.30	3.40	3.48	3.60	3.70	3.82	3.98	4.11	4.32	4.49	4.77	4.98
0.45	3.05	3.14	3.22	3.33	3.42	3.53	3.68	3.80	4.00	4.15	4.41	4.61
0.50	2.84	2.93	3.00	3.10	3.19	3.29	3.43	3.54	3.72	3.87	4.11	4.30
0.75	2.17	2.24	2.29	2.37	2.43	2.51	2.62	2.70	2.84	2.95	3.14	3.28
1.00	1.79	1.85	1.89	1.96	2.01	2.09	2.16	2.23	2.35	2.44	2.59	2.71
1.25	1.54	1.59	1.63	1.68	1.73	1.78	1.86	1.92	2.02	2.10	2.23	2.33
1.50	1.37	1.41	1.44	1.49	1.53	1.58	1.65	1.70	1.79	1.86	1.97	2.07
1.75	1.23	1.27	1.30	1.35	1.38	1.43	1.49	1.54	1.62	1.68	1.78	1.86
2.00	1.13	1.16	1.19	1.23	1.26	1.31	1.36	1.40	1.48	1.54	1.63	1.70
2.25	1.04	1.08	1.10	1.14	1.17	1.21	1.26	1.30	1.37	1.42	1.51	1.58
2.50	0.97	1.00	1.02	1.06	1.09	1.12	1.17	1.21	1.27	1.32	1.40	1.47
3.00	0.85	0.89	0.91	0.94	0.96	1.00	1.04	1.07	1.13	1.17	1.24	1.30
3.50	0.78	0.80	0.82	0.85	0.87	0.90	0.94	0.97	1.02	1.06	1.12	1.17
4.00	0.71	0.73	0.75	0.78	0.80	0.82	0.86	0.88	0.93	0.97	1.03	1.07
4.50	0.66	0.68	0.69	0.72	0.74	0.76	0.79	0.82	0.86	0.90	0.95	0.99
5.00	0.61	0.63	0.65	0.67	0.69	0.71	0.74	0.76	0.80	0.83	0.88	0.93
7.00	0.49	0.50	0.52	0.53	0.55	0.57	0.59	0.61	0.64	0.67	0.71	0.74
10.00	0.39	0.40	0.41	0.42	0.43	0.45	0.47	0.48	0.51	0.53	0.56	0.58
20.00	0.24	0.25	0.26	0.26	0.27	0.28	0.29	0.30	0.32	0.33	0.35	0.37

[a] Adapted from Refs. 33 and 46.

Example 16-2   Calculation of adjusted sodium adsorption ratio and evaluation of potential water infiltration problems. The following water quality analysis was reported for an aerated lagoon effluent that will be used for irrigating agricultural land. Using the reported water quality data, (1) calculate adj R_{Na}, and (2) determine whether an infiltration problem may develop by using this effluent for irrigation.

Water quality parameter	Concentration, mg/L
BOD	39
SS	160
Total N	4.4
Total P	5.5
pH[a]	7.7
Cations:	
Ca	37
Mg	46
Na	410
K	27
Anions:	
HCO_3	295
SO_4	66
Cl	526
Electrical conductivity (dS/m)	2.4
TDS	1,536
Boron	1.2
Alkalinity (total, as $CaCO_3$)	242
Hardness (total, as $CaCO_3$)	281

[a] Unitless

Solution

1. Calculate the adj R_{Na} using Eq. 16-5 and the values given in Table 16-6.

   (a) Convert the concentrations of the related water quality parameters to meq/L

$$Ca = \frac{37}{20.04} = 1.9$$

$$Mg = \frac{46}{12.15} = 3.8$$

$$Na = \frac{410}{23} = 17.8$$

$$HCO_3 = \frac{295}{61} = 4.8$$

   (b) Determine the value of Ca_x in Eq. 16-5 using the given water quality data.

      i. Salinity of applied water $(EC_w) = 2.4$ dS/m

      ii. Ratio of $HCO_3/Ca = 4.8/1.9 = 2.5$

      iii. $Ca_x = 1.2$ meq/L (from Table 16-6)

   (c) The adj R_{Na} is

$$\text{adj } R_{Na} = \frac{Na}{\sqrt{(Ca_x + Mg)/2}} = \frac{17.8}{\sqrt{(1.2 + 3.8)/2}} = 11.3$$

2. Determine whether infiltration problem will develop. Entering Table 16-4 with adj R_{Na} = 11.3 and EC_w = 2.4 dS/m shows that no infiltration problems are expected for this reclaimed wastewater.

Nutrients. The nutrients in reclaimed municipal wastewater provide fertilizer value for crop or landscape production. However, nutrients can cause problems in certain instances, when they are in excess of plant needs. Nutrients that are important to agriculture and landscape management include N, P, and occasionally K, Zn, B, and S. The most beneficial and the most frequently excessive nutrient in reclaimed municipal wastewater is nitrogen.

The nitrogen in reclaimed wastewater can replace an equal amount of commercial fertilizer during the early to mid-season crop growing period. Excessive nitrogen in the latter part of the growing period may be detrimental to many crops, causing excessive vegetative growth, delayed or uneven maturity, or reduced crop quality. If an alternate low-nitrogen water is available, a switch in water supplies or a blend of reclaimed wastewater and other water supplies has been used to keep nitrogen under control. The fate of nitrogen and phosphorus in soil and groundwater is discussed in Chap. 13 and in Refs. 33 and 35.

Miscellaneous problems. Clogging problems with sprinkler and drip irrigation systems have been reported, particularly with primary and oxidation ponds effluents. Biological growth (slimes) in the sprinkler head, emitter orifice, or supply line causes plugging, as do heavy concentrations of algae and suspended solids. The most frequent clogging problems occur with drip irrigation systems. From the standpoint of public health, such systems are often considered ideal, because they are totally enclosed and so minimize the problems of worker exposure to reclaimed wastewater or spray drift.

In treated wastewater that is chlorinated, chlorine residual of less than 1 mg/L does not affect plant foliage, but chlorine residuals in excess of 5 mg/L can cause severe plant damage when reclaimed wastewater is sprayed directly on foliage [33].

Example 16-3 Suitability and effects of various irrigation waters. Analyses of four representative waters in California are presented in the following data table. The waters are (1) the relatively unpolluted Sacramento River, (2) a moderately saline groundwater in San Joaquin County, and (3) two reclaimed municipal wastewaters from the cities of Fresno and Bakersfield. Assuming that the following conditions are applicable, determine the suitability of these waters for irrigation.

1. Daily crop water demand varies during the growing season and among crop types. Water demand may range from a low of 0.08 in/d to a high of 0.3 to 0.4 in/d.

2. On-farm management of reclaimed wastewater must take crop water demand into account, and the irrigation objective should be to use the reclaimed wastewater efficiently to produce a crop.

Constituents[a,b]	Sacramento River	San Joaquin County groundwater	Fresno wastewater effluent	Bakersfield wastewater effluent
EC (dS/m)	0.11	1.25	0.69	0.77
pH	7.1	7.7	8.6	7.0
Ca	10	100	24.0	47
Mg	5	33	12.8	5
Na	6	92	80	109
K	1.5	3.9	13.8	26
SAR	0.4	2.0	3.3	4.1
HCO_3	42	190	236	218
SO_4	7.3	110	—	62
Cl	2.2	200	70	107
NO_3-N + NH_3-N	0.08	5.9	14[c]	0.5[d]
B	—	1.4	0.43	0.38
TDS[e]	72	800	442	477
As		< 0.002		
Cd		< 0.002		< 0.01
Cr		< 0.02		
Pb		< 0.05		

[a] Adapted from Ref. 3.

[b] All concentrations are expressed in mg/L except electrical conductivity (EC) and pH, which is unitless.

[c] Total Kjeldahl-N.

[d] NO_3-N. The total nitrogen reported was 20-25 mg/L.

[e] TDS values estimated using Eq. 16-1

Solution

1. Evaluation for the suitability of various irrigation waters is made based on the information given in Table 16-4. The results of the analysis are presented in the following table.

Problem area	Degree of problem[a]			
	Sacramento River	San Joaquin County groundwater	Fresno wastewater effluent	Bakersfield wastewater effluent
Salinity	N	S–M	N	S–M
Infiltration	SV	N	S–M	S–M
Toxicity (sensitive crops only)				
Na				
Surface irrigation	N	N	S–M	S–M
Sprinkler irrigation	N	S–M	S–M	S–M
Cl				
Surface irrigation	N	S–M	N	N
Sprinkler irrigation	N	S–M	N	S–M
B	—	S–M	N	N
Miscellaneous (susceptible crops only)				
N	N	S–M	S–M	N
HCO_3	N	S–M	S–M	S–M

[a] N = no problem, S–M = slight to moderate problem, and SV = severe problems are expected when using respective water for a long period of time.

Comment. Although the water quality of the two reclaimed wastewaters is such that minor water infiltration, toxicity, and miscellaneous problems may be anticipated from use, normal agronomic practice used in the area has proven to be adequate to allow full production of adapted crops [3]. The cities of Fresno and Bakersfield have a long history of using municipal wastewater for irrigation.

Health and Regulatory Requirements. The contaminants in reclaimed wastewater that are of health significance may be classified as biological and chemical agents. Where reclaimed wastewater is used for irrigation, biological agents including bacterial pathogens, helminths, protozoa, and viruses pose the greatest health risks.

To protect public health, considerable efforts have been made to establish conditions and regulations that would allow for safe use of reclaimed wastewater for irrigation. Although there is no uniform set of federal standards in the United States for wastewater reclamation and reuse, several states have developed wastewater reclamation regulations, often in conjunction with the development of regulations for land treatment and disposal of wastewater. Reclaimed wastewater regulations for specific irrigation uses are based on the expected degree of human contact with the reclaimed wastewater and the intended use of the irrigated crops. For example, the state of California requires that reclaimed wastewater used for landscape irrigation of areas with unlimited public access must be "adequately oxidized, filtered, and disinfected prior to use," with median total coliform count of no more than 2.2/100 mL [41].

A summary of the California requirements for reclaimed wastewater used for irrigation and recreational impoundments is contained in Table 16-7. In reviewing the criteria in this table, it should be noted that these are basically health-related requirements. The potential effects of reclaimed wastewater on crops or soils are not considered. For detailed information on water quality for soils and crops, reference should be made to Table 16-4. The median number of total coliform count and turbidity have been used for the assessment of the treatment reliability of wastewater reclamation facilities. Wastewater reuse regulations adopted by the state of Arizona contain enteric virus limits; for example, wastewater is not to exceed 1 plaque forming unit (PFU) per 40 L for applications such as spray irrigation of food crops [7]. The state of Florida requires no detectable fecal coliform per 100 mL. This level of disinfection is achieved by requiring tertiary filtration and by maintaining a 1.0 mg/L total chlorine residual after a contact time of 30 minutes at average daily flow. It should also be noted that there are many reuse applications that do not require a high degree of wastewater treatment.

Although these wastewater reuse criteria reported in Table 16-7 lack explicit epidemiological evidence to assess the associated health risks fully, they have been adopted as the attainable and enforceable regulations in the planning and implementation of wastewater reclamation and reuse in California. Additional safety measures that have been used for nonpotable water reuse applications include (1) installation of separate storage and distribution systems for potable water, (2) use of color-coded tapes to distinguish potable and nonpotable distribution piping, (3) cross-

TABLE 16-7
State of California wastewater reclamation criteria for irrigation and recreational impoundments[a]

Use of reclaimed wastewater	Description of minimum treatment requirements			
	Primary [b]	Secondary and disinfected	Secondary coagulated filtered [c] and disinfected	Coliform, MPN/100 ml median (daily sampling)
Irrigation				
Fodder crops	X			No requirement
Fiber	X			No requirement
Seed crops	X			No requirement
Produce eaten raw, surface irrigated		X		2.2
Produce eaten raw, spray irrigated			X	2.2
Processed produce, surface irrigated	X			No requirement
Processed produce, spray irrigated		X		23
Landscapes: golf course, cemeteries, freeways		X		23
Landscapes: parks, play-grounds, schoolyards			X	2.2
Recreational impoundments				
No public contact		X		23
Boating & fishing only		X		2.2
Body-contact (bathing)			X	2.2

[a] Adapted from Ref. 41.

[b] Effluent not containing more than 0.5 mL/L · h settleable solids.

[c] Effluent not containing more than 2 turbidity units.

connection and backflow prevention devices, (4) periodic use of tracer dyes to detect the occurrence of cross contamination in potable supply lines, and (5) irrigation during off-hours to further minimize the potential for human contact.

Wastewater Reclamation Criteria in Other Countries. Reclaimed water quality criteria for protecting health in developing countries are often established in relation to the limited resources available for public works, and other health delivery systems may yield greater health benefits for the funds spent. Confined wastewater collection systems and wastewater treatment are often nonexistent, and reclaimed wastewater often provides an essential water and fertilizer source. For most developing countries, the greatest concern with the use of wastewater for irrigation is that untreated or inadequately treated wastewater contains numerous enteric helminths such as hookworm, ascaris, trichuris, and, under certain circumstances, the beef tapeworm. These infectious agents as well as other microbiological pathogens, can damage the health of the general public consuming the contaminated crops and can also harm farm workers and their families [14,32,39,51].

The World Health Organization (WHO) has recommended that crops that will be eaten raw should be irrigated with treated wastewater only after it has undergone biological treatment and disinfection to achieve a coliform level of not more than 100/100 mL in 80 percent of the samples [50,51]. The criteria recommended by WHO for irrigation with reclaimed wastewater have been accepted as reasonable goals for the design of such facilities in many Mediterranean countries. In some Middle East countries that have recently developed facilities for wastewater reuse, the tendency has been to adopt more stringent wastewater reuse criteria, similar to the California regulations. Adoption of more stringent regulations is done to protect an already high standard of public health by preventing, at any expense, the introduction of pathogens into the human food chain [32].

Industrial Water Reuse

In the United States, approximately 20 million people (almost 25 percent of the work force) are employed in the 300,000 manufacturing facilities. These facilities provide 27 percent of the total U. S. earnings, followed by energy production and the minerals industry [52].

Industrial Water Withdrawals. Total water requirements for in-plant manufacturing are projected to increase to 312 Bgal/d by the year 2000. However, because of pollution control limitations on waste discharge, the technology of water management within plants is expected to change, and the result will be a sixfold increase in in-plant recycling from the 1975 value [52] (see Table 16-2).

As a result of advances in cooling tower technology, freshwater withdrawals for steam electric generation are projected to decrease by 11 percent (79.5 Bgal/d) by the year 2000, but they will still constitute 94 percent of the freshwater withdrawals for energy production. Other types of energy production (mining and processing of coal and oil shale, and extracting and refining of oil and natural gas) account for the remaining 6 percent of the total freshwater withdrawals.

In the minerals industry, water is used for mining metals, nonmetals, and fuels. Water withdrawals for the minerals industry are projected to increase 61 percent by the year 2000 to an average of 11.3 Bgal/d, or 3.7 percent of the nation's freshwater withdrawals.

Cooling Tower Make-Up Water. Cooling tower make-up water represents a significant water use for many industries. For industries such as electric power generating stations, oil refining, and many other types of manufacturing plants, one-quarter to more than one-half of the total water use may be cooling tower make-up. Because a cooling tower normally operates as a closed-loop system, it can be viewed as a separate water system with its own specific set of water quality requirements, largely independent of the particular industry involved. Thus, using reclaimed municipal wastewater for cooling tower make-up water is relatively easy and is practiced in many locations in the United States.

The on-site wastewater reclamation plant for the cooling tower operations at the Palo Verde Nuclear Generating Station in Arizona is shown in Fig. 16-2. Secondary effluent from the cities of Tolleson and Phoenix is pumped 38 miles to this site. Before the effluent is used, it is subjected to advanced treatment consisting of (1) biological nitrification, (2) lime and soda ash addition for softening and phosphorus removal, (3) filtration, (4) pH adjustment, and (5) chlorination. The purpose of the advanced treatment is to reduce corrosion and scaling in the cooling tower systems.

Water and Salt Balances in Cooling Tower. The basic principle of cooling tower operation is that of evaporative condensation and exchange of sensible heat. The air and water mixture releases latent heat of vaporization. Water exposed to the

FIGURE 16-2
On-site wastewater reclamation plant for the cooling tower operations at the Palo Verde Nuclear Generating Station. The secondary treated effluents from cities of Tolleson and Phoenix, Arizona, are pumped 38 miles to this site to undergo advanced wastewater treatment. (Photo from Arizona Public Service Co.).

atmosphere evaporates, and as the water changes to vapor, heat is consumed that amounts to approximately 1000 BTU per pound of water evaporated [9].

Under the normal operating conditions, the loss of water, discharged from the cooling tower to the atmosphere as hot moist vapor, amounts to approximately 1.2 percent for each 10°F of cooling range. Drift, or water lost from the top of the tower to the wind, is the second mechanism by which water is lost from the cooling system. About 0.005 percent of the recirculating water is lost in this way. Although evaporation results in a loss of water from the system, the salt concentration is increased because salts are not removed by evaporation. To prevent the formation of precipitates in the resulting higher-concentration tower water, a portion of the concentrated cooling water is bled off and replaced with low salt make-up water to maintain a proper salt balance. This highly saline water that is bled off from the cooling tower system is called blowdown [9,42].

The total make-up water flow for the cooling tower system includes all three of these water losses. The definition sketch for a recirculating evaporative cooling tower is shown in Fig. 16-3.

The water balance around the cooling tower is

$$Q_m = Q_b + Q_d + Q_e \qquad (16\text{-}6)$$

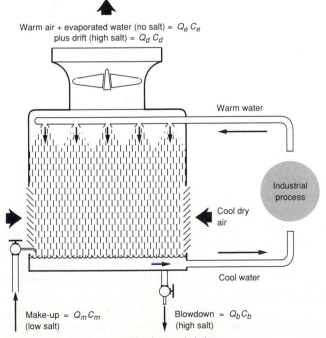

FIGURE 16-3
Definition sketch for salt balance in the recirculating, evaporative cooling tower. Adapted from Ref. 42.

where Q_m = make-up water flow, gal/min

$\quad Q_b$ = blowdown flow, gal/min

$\quad Q_d$ = drift flow, gal/min

$\quad Q_e$ = evaporation loss, gal/min

Drift flow, Q_d, is normally small enough to be ignored, as noted previously (< 0.005 percent).

In a similar way, the salt balance in the cooling tower is

$$Q_m C_m = Q_b C_b + Q_d C_d + Q_e C_e \qquad (16\text{-}7)$$

where C_m = salt concentration in make-up water flow, mg/L

$\quad C_b$ = salt concentration in blowdown flow, mg/L

$\quad C_d$ = salt concentration in drift flow, mg/L

$\quad C_e$ = salt concentration in evaporation loss, mg/L

Because Q_d is negligible, the term $Q_d C_d$ can be omitted without serious error. Further, because the concentration of salt in the evaporation water is also negligible under normal operating conditions, Eq. 16-7 can be reduced to Eq. 16-8.

$$Q_m C_m = Q_b C_b \qquad (16\text{-}8)$$

The magnitude of the blowdown flow (and thus the make-up flow) is dependent upon the concentration of potential precipitants in the make-up water. The ratio of the concentration of the salt C_b in the blowdown to its concentration C_m in the make-up water is known as the cycles of concentration.

$$\text{Cycles of concentration} = \frac{C_b}{C_m} \qquad (16\text{-}9)$$

Combining Eqs. 16-8 and 16-9 yields

$$\text{Cycles of concentration} = \frac{Q_m}{Q_b} \qquad (16\text{-}10)$$

It can be seen in Eq. 16-10 that the cycles of concentration also equal the ratio of the make-up flow to the blowdown flow.

When the cycles of concentration are in the order of 3 to 7, some of the dissolved solids in the circulating water can exceed their solubility limits and precipitate, causing scale formation in pipes and coolers. To avoid scale formation, sulfuric acid is often used to convert calcium and magnesium carbonates into more soluble sulfate compounds. The amount of acid used must be limited to maintain some residual alkalinity in the system. If the pH of the system is reduced to far below 7, accelerated corrosion can occur (see Example 16-4).

Common Water Quality Problems in Cooling Tower Systems. Four general water quality problems are encountered in industrial cooling tower operations: (1) scaling, (2) metallic corrosion, (3) biological growths, and (4) fouling in heat exchanger and condensers [34,42]. Both freshwater and reclaimed municipal wastewater contain contaminants that can cause these problems, but their concentrations in reclaimed wastewater are generally higher.

Scaling. Scaling refers to the formation of hard deposits, usually on hot surfaces, which reduce the efficiency of heat exchange. Calcium scales (calcium carbonate, calcium sulfate, and calcium phosphate) are the principal causes of cooling tower scaling problems. Magnesium scales (magnesium carbonate and phosphate) can also be a problem. Silica deposits are particularly difficult to remove from heat exchanger surfaces; however, most waters contain relatively small quantities of silica.

Reducing the potential for scaling in wastewater is achieved by controlling the formation of calcium phosphate, which is the first calcium salt to precipitate if phosphate is present. Treatment is usually accomplished by removing phosphates by precipitation (see Chap. 11). Other treatment methods such as ion exchange reduce scale formation by removing calcium and magnesium; however, these techniques are comparatively expensive, and their use is limited.

Metallic corrosion. In cooling systems, corrosion can occur when an electrical potential between dissimilar metal surfaces is created. The corrosion cell consists of an anode, where oxidation of one metal occurs, and a cathode, where reduction of another metal takes place. Water quality greatly affects metallic corrosion. Contaminants such as TDS increase the electrical conductivity of the solution and thereby accelerate the corrosion reaction. Dissolved oxygen and certain metals (manganese, iron, and aluminum) promote corrosion because of their relatively high oxidation potential.

The corrosion potential of cooling water can be controlled by the addition of chemical corrosion inhibitors. The chemical requirements to control corrosion are usually much higher for reclaimed wastewater than for freshwater because the concentration of TDS is often two to five times higher in wastewater.

Biological growth. The warm, moist environment inside the cooling tower makes an ideal environment for promoting biological growth. Nutrients, particularly N and P, and available organics further encourage the growth of microorganisms that can attach and deposit on heat exchanger surfaces, inhibiting heat transfer and water flow. Biological growths may also settle and bind other debris present in the cooling water, which may further inhibit effective heat transfer. Certain microorganisms also create corrosive by-products during their growth. Biological growths are usually controlled by the addition of biocides as part of the internal chemical treatment process, which may include the addition of acid for pH control, the use of biocides, and scale and biofoul inhibitors. Because reclaimed wastewater contains a higher concentration of organic matter, it may require larger dosages of biocides. It is possible, however, that most of the nutrients and available organic matter are removed from the reclaimed wastewater during biological and chemical treatment.

When reclaimed water is used for cooling, the assurance of adequate disinfection is a primary concern to protect the health of workers. The disinfection requirements for use of reclaimed water in industrial processes are made on a case-by-case basis. As for unrestricted reclaimed wastewater use in food crop irrigation, the most stringent requirements would be appropriate if there exists a potential for exposure to spray. Protection of the neighboring public as well as the plant operators is of prime importance.

Fouling. Fouling refers to the process of attachment and growth of deposits of various kinds in cooling tower recirculation systems. These deposits consist of biological growths, suspended solids, silt, corrosion products, and inorganic scales. The resulting operational problem is inhibition of heat transfer in the heat exchangers. Control of fouling is achieved by the addition of chemical dispersants that prevent particles from aggregating and subsequently settling. Dispersants are also added at the point of use, as is the usual case for freshwater cooling systems. Also, the chemical coagulation and filtration processes required for phosphorus removal are effective in reducing the concentration of contaminants that contribute to fouling.

TABLE 16-8
Processes used in treating water for cooling or boiler make-up[a]

	Cooling		
Processes	Once through	Recirculated	Boiler make-up
Suspended solids and colloids removal:			
Straining	X	X	X
Sedimentation	X	X	X
Coagulation		X	X
Filtration		X	X
Aeration		X	X
Dissolved-solids modification softening:			
Cold lime		X	X
Hot lime soda			X
Hot lime zeolite			X
Cation exchange sodium		X	X
Alkalinity reduction cation exchange:			
Hydrogen		X	X
Cation exchange hydrogen and sodium		X	X
Anion exchange			X
Dissolved-solids removal:			
Evaporation			X
Demineralization		X	X
Dissolved-gases removal:			
Degasification			
Mechanical		X	X
Vacuum	X		X
Heat			X
Internal conditioning:			
pH adjustment	X	X	X
Hardness sequestering	X	X	X
Hardness precipitation			X
Corrosion inhibition general		X	X
Embrittlement			X
Oxygen reduction			X
Sludge dispersal	X	X	X
Biological control	X	X	

[a] Adapted from Ref. 25

TABLE 16-9
Water quality requirements at point of use for steam generation and cooling in heat exchangers [25]

| Characteristic | Boiler feedwater — Quality of water prior to the addition of chemicals used for internal conditioning | | | | Cooling water | | | |
	Low Pressure	Industrial Intermediate pressure	High pressure	Electrical utilities	Once through Fresh	Once through Brackish[a]	Make-up for recirculation Fresh	Make-up for recirculation Brackish[a]
Silica (SiO$_2$)	30	10	0.7	0.01	50	25	50	25
Aluminum (Al)	5	0.1	0.01	0.01	b	b	0.1	0.1
Iron (Fe)	1	0.3	0.05	0.01	b	b	0.5	0.5
Manganese (Mn)	0.3	0.1	0.01	0.01	b	b	0.5	0.02
Calcium (Ca)	b	0.4	0.01	0.01	200	420	50	420
Magnesium (Mg)	b	0.25	0.01	0.01	b	b	b	b
Ammonia (NH$_4$)	0.1	0.1	0.1	0.07	b	b	b	b
Bicarbonate (HCO$_3$)	170	120	48	0.5	600	140	24	140
Sulfate (SO$_4$)	b	b	b	d	680	2,700	200	2,700
Chloride (Cl)	b	b	b	b,d	600	19,000	500	19,000
Dissolved solids	700	500	200	0.5	1,000	35,000	500	35,000
Copper (Cu)	0.5	0.05	0.05	0.01	b	b	b	b
Zinc (Zn)	b	0.01	0.01	0.01	b	b	b	b
Hardness (CaCO$_3$)	350	1.0	0.07	0.07	850	6,250	650	6,250
Alkalinity (CaCO$_3$)	350	100	40	1	500	115	350	115
pH, units	7.0–10.0	8.2–10.0	8.2–9.0	8.8–9.4	5.0–8.3	6.0–8.3	b	b
Organics:								
Methylene blue active substances	1	1	0.5	0.1	b	b	1	1
Carbon tetrachloride extract	1	1	0.5	b,c	e	e	1	2
Chemical oxygen demand (COD)	5	5	1.0	1.0	75	75	75	75
Hydrogen sulfide (H$_2$S)	b	b	b	b	—	—	b	b
Dissolved oxygen (O$_2$)	2.5	0.007	0.007	0.007	present	present	b	b
Temperature	b	b	b	b	b	b	b	b
Suspended solids	10	5	0.5	0.05	5,000	2,500	100	100

Note: Unless otherwise indicated, units are mg/l and values that normally should not be exceeded. No one water will have all the maximum values shown.
[a] Brackish water—dissolved solids more than 1000 mg/L.
[b] Accepted as received (if meeting other limiting values); has never been a problem at concentrations encountered.
[c] Zero, not detectable by test.
[d] Controlled by treatment for other constituents.
[e] No floating oil.

In most cases, disinfected secondary effluent is supplied to noncritical, once-through cooling. For recirculating cooling tower operation, most wastewaters contain constituents which, if not removed, would limit industries to very low cycles of concentration in their cooling towers. Additional treatment includes lime clarification, alum precipitation, or ion exchange [9,19,34]. Treatment processes used for both external and internal treatment of cooling or boiler make-up water are summarized in Table 16-8. In many cases, the water quality requirements for the use of reclaimed municipal wastewater are the same as those for freshwater. Water quality requirements at the point of use for cooling waters for both once-through and make-up for recirculation are reported in Table 16-9.

Example 16-4 Estimation of blowdown water composition. Reclaimed wastewater with the chemical characteristics given below is being considered for use as make-up water for a cooling tower. Calculate the composition of the blowdown flow if 5 cycles of concentration are to be used. Assume that the temperature of the hot water entering the cooling tower is 120°F and the solubility of $CaSO_4$ is 2200 mg/L as $CaCO_3$ at this temperature.

Parameter	Concentration, mg/L
Total hardness (as $CaCO_3$)	118
Ca (as $CaCO_3$)	85
Mg (as $CaCO_3$)	33
Total alkalinity (as $CaCO_3$)	90
SO_4	20
Cl	19
SiO_2	2

Where the molecular weight of $CaCO_3$ = 100, $CaSO_4$ = 136, H_2SO_4 = 98, and SO_4 = 96.

Solution

1. Determine the total hardness in the circulating water

   (*a*) When the total alkalinity is less than total hardness, Ca and Mg are also present in forms other than carbonate hardness.

   (*b*) Setting the cycles of concentration to 5 and using Eq. 16-9, the total hardness in circulating water is

   $$C_b = (\text{cycles of concentration}) \, (C_m)$$
   $$= 5 \times 118 = 590 \text{ mg/L } CaCO_3$$

2. Determine the total amount of H_2SO_4 that must be added to convert the $CaCO_3$ to $CaSO_4$.

   (*a*) To convert from $CaCO_3$ to $CaSO_4$, sulfuric acid is injected into the circulating water and the following reaction occurs.

   $$CaCO_3 + H_2SO_4 \rightarrow CaSO_4 + H_2O + CO_2$$
   $$100 \qquad 98 \qquad 136$$

(*b*) The alkalinity in the circulating water, if not converted into sulfates, is

$$5 \times 90 = 450 \text{ mg/L as CaCO}_3$$

(*c*) If 10 percent of the alkalinity is left unconverted to avoid corrosion, the amount of alkalinity remaining is

$$0.1 \times 450 = 45 \text{ mg/L as CaCO}_3$$

(*d*) The amount of alkalinity that must be converted is

$$450 - (0.1 \times 450) = 405 \text{ mg/L as CaCO}_3$$

(*e*) The amount of sulfate that must be added for the conversion is

$$SO_4 = 405\frac{96}{100} = 389 \text{ mg/L}$$

(*f*) Converting to mg/L CaSO$_4$ yields:

$$CaSO_4 = (389 \text{ mg/L SO}_4)\frac{136}{96} = 551 \text{ mg/L}$$

3. Determine the required sulfuric acid concentration in the circulating water

$$H_2SO_4 = (389 \text{ mg/L SO}_4)\frac{98}{96} = 397 \text{ mg/L}$$

4. Determine the sulfate concentration in the circulating water contributed by the make-up water.

   (*a*) Sulfate from make-up water is

$$5 \times 20 = 100 \text{ mg/L as SO}_4$$

   (*b*) If combined with Ca, the concentration is

$$CaSO_4 = (100 \text{ mg/L SO}_4)\frac{136}{96} = 142 \text{ mg/L}$$

5. The solubility of CaSO$_4$ at 120°F is 2200 mg/L. In the circulating water, 142 mg/L CaSO$_4$ was originally present after 5 cycles of concentration, and 551 mg/L were formed by the addition of sulfuric acid. Therefore, 1507 mg/L [2200 − (142 + 551)] of additional CaSO$_4$ formation are theoretically permissible before the solubility limit is exceeded. More cycles of concentration could have been used before CaSO$_4$ would precipitate in the system.

6. Determine the concentrations of Cl and SiO$_2$ in the circulating water

   (*a*) Chloride

$$Cl = 5 \times 19 = 95 \text{ mg/L}$$

   (*b*) Silica

$$SiO_2 = 5 \times 2 = 10 \text{ mg/L}$$

7. Summarize the composition of the blowdown flow after 5 cycles of concentration.

Parameter	Concentration, mg/L	
	Initial	Final
Total hardness (as $CaCO_3$)	118	590
Total alkalinity (as $CaCO_3$)	90	45
SO_4	20	489
Cl	19	95
SiO_2	2	10

Groundwater Recharge with Reclaimed Wastewater

Groundwater recharge has been used (1) to reduce, stop, or even reverse declines of groundwater levels, (2) to protect underground freshwater in coastal aquifers against saltwater intrusion from the ocean, and (3) to store reclaimed wastewater and surface water, including flood or other surplus water for future use [47]. Groundwater recharge is also achieved incidentally in land treatment and disposal systems where municipal and industrial wastewater is disposed of via percolation and infiltration (see Chap. 13).

Groundwater recharge with reclaimed wastewater is an approach to wastewater reuse that results in the planned augmentation of groundwater supplies. There are several advantages to storing water underground [1,6,47]: (1) the cost of artificial recharge may be less than the cost of equivalent surface reservoirs; (2) the aquifer serves as an eventual distribution system and may eliminate the need for surface pipelines or canals; (3) water stored in surface reservoirs is subject to evaporation, to potential taste and odor problems caused by algae and other aquatic growth, and to pollution; (4) suitable sites for surface reservoirs may not be available or environmentally acceptable; and (5) the inclusion of groundwater recharge in a wastewater reuse project may also provide psychological and aesthetic secondary benefits as a result of the transition between reclaimed wastewater and groundwater.

Groundwater Recharge Methods. Two groundwater recharge methods are commonly used with reclaimed municipal wastewater: (1) surface spreading in basins, and (2) direct injection into groundwater aquifers.

Groundwater recharge by surface spreading. Surface spreading is the simplest, oldest, and most widely used method of groundwater recharge. In surface spreading, recharge waters percolate from the spreading basins through an unsaturated groundwater (vadose) zone. Infiltration basins are the most favored methods of recharge because they allow efficient use of space and require relatively little maintenance.

If hydrologeological conditions are favorable for groundwater recharge with spreading basins, wastewater reclamation can be implemented relatively simply by rapid infiltration (also known as soil-aquifer treatment (SAT) system; see also Chap. 13). A typical SAT system is shown in Fig. 16-4. Here the necessary treatment can

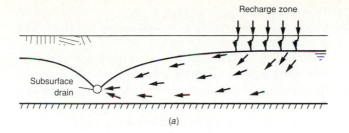

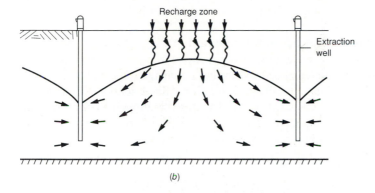

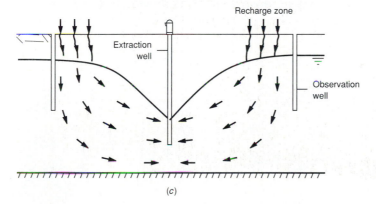

FIGURE 16-4
Schematic of soil-aquifer treatment (SAT) system with: (a) recovery of renovated water by drains, (b) wells surrounding the basins, and (c) wells midway between two parallel strips of basins. Adapted from Ref. 8.

be obtained by filtration as the wastewater percolates through the soil and the vadose zone down to the groundwater and then some distance through the aquifer [6,8]. The extracted groundwater can be used for irrigation for a variety of food crops.

Because recharged groundwater is an eventual source of potable water supply, groundwater recharge with reclaimed municipal wastewater often involves treatment beyond the conventional secondary treatment. For surface spreading operations practiced in California, common wastewater reclamation processes prior to recharge include primary and secondary wastewater treatment and tertiary granular-medium filtration followed by disinfection with chlorine. Groundwater recharge basins located in the Santa Ana River in southern California are shown in Fig. 16-5.

Groundwater recharge by direct injection. Direct subsurface recharge is achieved when water is conveyed and injected directly into a groundwater aquifer. In direct injection, highly treated reclaimed wastewater is generally injected directly into the saturated groundwater zone, usually into a well-confined aquifer. Groundwater recharge by direct injection is practiced, in most cases, where groundwater is deep or where the topography or existing land use makes surface spreading impractical or too expensive. This method of groundwater recharge is particularly effective in creating freshwater barriers against intrusion of saltwater from the sea in coastal aquifers [6,47].

Preapplication Treatment Requirements for Groundwater Recharge. Preapplication treatment requirements for groundwater recharge vary considerably,

FIGURE 16-5
Groundwater recharge basins located in the Santa Ana River in southern California.

depending on the purpose of groundwater recharge, sources of reclaimed wastewater, recharge methods, and location. For example, preapplication treatment for municipal wastewater for the soil-aquifer treatment system may include only primary treatment or treatment in a stabilization pond. However, preapplication treatment processes that leave high algal concentrations in the recharge water should be avoided. Algae can severely clog the soil of infiltration basins. Although renovated water from the SAT system is much better water quality than the influent wastewater, it could be lower quality than the native groundwater. Thus, the SAT system should be designed and managed to avoid encroachment into the native groundwater and to use only a portion of the aquifer (as shown in Fig. 16-4). The distance between infiltration basins and wells or drains should be as large as possible, usually at least 150 to 350 feet to allow for adequate soil-aquifer treatment [8].

More detailed discussions regarding the capability of various advanced wastewater treatment process combinations are found in Section 16-3. To minimize potential health risks, careful attention must be paid to groundwater recharge operations when a possibility exists to augment substantial portions of potable groundwater supplies. Both in surface spreading and in direct injection, locating the extraction wells as great a distance as possible from the spreading basins or the injection wells increases the flow path length and residence time of the recharged water. These separations in space and in time contribute to the assimilation of the recharged reclaimed wastewater with the other aquifer contents and to the loss of identity of the recharged water originated from wastewater.

Fate of Contaminants in Groundwater. An understanding of the behavior of stable organic contaminants and bacterial and viral pathogens is crucial in evaluating the feasibility of groundwater recharge using reclaimed wastewater. Treated effluents contain trace quantities of organic contaminants even when the most advanced treatment technology is used. The transport and fate of these substances in the subsurface environment are governed by various mechanisms that include biodegradation by microorganisms, chemical oxidation and reduction, sorption and ion exchange, filtration, chemical precipitation, dilution, volatilization, and photochemical reactions (in spreading basins) [23,36,48].

Particulate contaminants. Particulate contaminants including microorganisms in reclaimed wastewater are removed by filtration and retained effectively by the soil matrix. Factors affecting the movement, removal, and inactivation of viruses are summarized in Refs. 5, 39, and 45.

Dissolved inorganic and organic contaminants. In addition to the common dissolved mineral constituents, reclaimed wastewater contains many dissolved trace elements. The physical action of filtration does not, however, accomplish the removal of these dissolved inorganic contaminants. For trace metals to be retained in the soil matrix, physical, chemical, or microbiological reactions are required to immobilize the dissolved contaminants. In a groundwater recharge system, the impact of microbial activities on the attenuation of inorganic contaminants is small. Physical and chemical reactions in the soil that are important with respect to trace metal elements include cation exchange, precipitation, surface adsorption, and chelation and complexation [36,48]. Although soils do not possess unlimited capability in attenuating inorganic contaminants, experimental studies have shown that soils do have capacities for retaining large amounts of trace metal elements. Therefore, it is conceivable that a site used for groundwater recharge may be effective in retaining trace metals for extended periods of time [1].

Removal of dissolved organic contaminants is affected primarily by biodegradation and adsorption during groundwater recharge operations. Biodegradation offers the potential of permanent conversion of toxic organic substances into harmless products. The rate and extent of biodegradation are strongly influenced by the nature of the organic substances as well as by the presence of electron acceptors such as dissolved oxygen, nitrate, and sulfate. Biodegradation of easily degradable substances takes place almost exclusively in the first few feet of travel. The fate of some of the more resistant organic compounds found in the water that is recharged is still poorly understood. The effects of dispersion, sorption, and biodecomposition on the time change in concentration of an organic compound at an aquifer observation well are illustrated in Fig. 16-6. The observed concentration is C, and C_0 represents the concentration in the injection water.

Among the end products of complete biodegradation of dissolved organic contaminants are carbon dioxide and water under aerobic conditions, or carbon dioxide, nitrogen, hydrogen sulfide, and methane under anaerobic conditions. However, the degradation process does not necessarily proceed to completion. Degradation may terminate at an intermediate stage and leave a residual organic product that, under the particular conditions, cannot be degraded further at an appreciable rate.

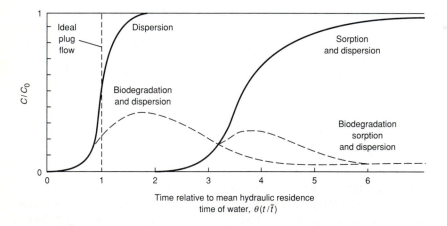

FIGURE 16-6

Effect of dispersion, sorption, and biodecomposition on the time change in concentration of an or organic compound. Expected responses to a step change in concentration are shown. Adapted from Ref. 23.

Pathogen removal. Groundwater contamination by pathogenic microorganisms has not received as much attention as surface water contamination. It has been generally assumed that groundwater is free of pathogenic microorganisms. However, a number of well-documented disease outbreaks have been traced to contaminated groundwater [1,5,12]. The fate of bacterial pathogens and viruses in the subsurface environment is determined by their survival characteristics and their retention in the soil matrix. Both survival and retention are largely determined by (1) the climate, (2) the nature of the soil, and (3) the nature of microorganisms [5].

Temperature and rainfall are two important climatic factors that will affect viral and bacterial survival and movement. At higher temperatures, inactivation and natural die-off are fairly rapid. In the case of bacteria and probably viruses, the die-off rate is approximately doubled with each 10°C rise in temperature between 5 and 30°C. Rainwater, because of its lower pH value, can elute adsorbed virus particles that may then move with the groundwater. The physical and chemical characteristics of the soil will also play a major role in determining survival and retention of microorganisms. Soil properties influence moisture-holding capacity, pH, and organic matter. All of these factors control the survival of bacteria and viruses in the soil.

Resistance of microorganisms to environmental factors varies among different species and strains. Bacteria are believed to be removed largely by filtration processes in the soil, but adsorption is the major factor controlling virus retention [1,5].

Groundwater Recharge Guidelines. The major concern with groundwater recharge with reclaimed municipal wastewater is that potentially adverse health effects may be caused by the introduction of pathogens or trace amounts of toxic contaminants. Because of the increasing concern for long-term health effects, every effort should be made to reduce the number of chemical species and the concentration of specific organic constituents in the recharge water [1,29,44].

A source control program to limit the quantities of potentially harmful constituents entering the sewer system must be an integral part of any groundwater recharge project. Extreme caution is necessary in controlling the quality of the wastewater to be recharged, because restoring a groundwater basin once it is contaminated is difficult and expensive. Additional cost would be incurred if groundwater quality changes resulting from recharge necessitated the treatment of extracted groundwater or the development of additional water sources. The level of municipal wastewater treatment necessary to produce a suitable reclaimed wastewater for groundwater recharge depends upon the groundwater quality objectives, hydrogeologic characteristics of the groundwater basin, and the amount of reclaimed wastewater recharged in relation to other waters recharged. Factors to be considered in the formulation of the groundwater recharge guidelines are summarized in Table 16-10.

In the United States, federal requirements for groundwater recharge in the context of wastewater reclamation and reuse have not been established. As a consequence, wastewater reclamation and reuse requirements for groundwater recharge are presently regulated by the state agencies on a case-by-case determination. Many states require considerably higher wastewater treatment prior to groundwater recharge than preapplication treatment for the rapid infiltration process.

TABLE 16-10
Factors to be considered in the formulation of groundwater recharge guidelines in the United States[a]

<table>
<tr><td colspan="2" align="center">**Surface spreading**</td></tr>
<tr><td>Treatment</td><td>Source control of toxic chemicals</td></tr>
<tr><td></td><td>Primary sedimentation and secondary biological treatment</td></tr>
<tr><td></td><td>Tertiary granular-medium filtration (possibly, activated-carbon adsorption for organics removal)</td></tr>
<tr><td></td><td>Disinfection</td></tr>
<tr><td>Depth to groundwater</td><td>Percolation through an unsaturated zone of undisturbed soil</td></tr>
<tr><td></td><td>Depth to groundwater in the range of 10 to 50 ft depending on percolation rate of the soils</td></tr>
<tr><td>Retention time in ground</td><td>6 to 12 months depending on the type of pretreatment</td></tr>
<tr><td>Maximum percent reclaimed wastewater</td><td>20 to 50% on the annual basis at extraction wells, depending on organics removal</td></tr>
<tr><td>Horizontal distance</td><td>500 to 1000 ft depending on pretreatment</td></tr>
<tr><td>Monitoring</td><td>Extensive including the contaminants in the drinking water regulations</td></tr>
<tr><td colspan="2" align="center">**Direct injection**</td></tr>
<tr><td>Treatment</td><td>Source control of toxic chemicals</td></tr>
<tr><td></td><td>Primary sedimentation and secondary biological treatment</td></tr>
<tr><td></td><td>Chemical coagulation, clarification, and granular-medium filtration</td></tr>
<tr><td></td><td>Activated-carbon adsorption</td></tr>
<tr><td></td><td>Volatile organics removal</td></tr>
<tr><td></td><td>Reverse osmosis or other membrane process</td></tr>
<tr><td></td><td>Disinfection</td></tr>
<tr><td>Depth to groundwater</td><td>Not applicable (direct injection to groundwater aquifers)</td></tr>
<tr><td>Retention time in ground</td><td>12 months</td></tr>
<tr><td>Maximum percent of reclaimed wastewater</td><td>20% on the annual basis at extraction wells</td></tr>
<tr><td>Horizontal distance</td><td>1000 to 2000 ft</td></tr>
<tr><td>Monitoring</td><td>Quite extensive including the contaminants in the drinking water regulations</td></tr>
</table>

[a] Compiled from Refs. 1, 11, 29, and 44.

Potable Water Reuse

The attitude towards using reclaimed wastewater for potable water has been cautious because of health and safety and aesthetic concerns. Nevertheless, some communities are developing or implementing plans for direct or indirect potable reuse where little possibility exists for supplemental freshwater supplies [21,22,30]. Although the quantities involved in potable reuse are small, the technological and public health interests are great, hence, considerable research has been directed toward potable water reuse.

Indirect Potable Water Reuse. The first well-documented episode of potable reuse occurred at Chanute, Kansas, in 1956–57 during the severe drought period of 1952–57 [24]. During a 5-month period, chlorinated secondary effluent was collected behind the dam on the Neosho River and used as intake water for the city's water treatment plant. The tap water met the drinking water standards of the time for bacteriological quality during the entire recirculation period. It was thought, however, that the margin of safety was uncomfortably narrow. The most serious problem was that of public acceptance or, more accurately, public rejection of the water owing to a pale yellow color, an unpleasant musty taste and odor, and foaming [24].

In almost all the cases in which potable water reuse has been considered, alternative sources of water have been developed in the ensuing years, and the need to adopt direct (e.g., pipe-to-pipe) potable water reuse has been avoided. Planned indirect potable reuse systems in operation today include such groundwater recharge operations as the Whittier Narrows Groundwater Recharge Project in Los Angeles County, California [1,29], and the project in El Paso, Texas [1,20]. Indirect potable reuse also occurs in the Occoquan Reservoir in northern Virginia [26,49]. Highly treated effluent from the 15 Mgal/d Manassas, Virginia, advanced wastewater treatment plant operated by the Upper Occoquan Wastewater Authority is discharged directly into the Occoquan Reservoir, a principal drinking water reservoir for more than 660,000 people.

In proposing direct potable water reuse, serious consideration must be given to whether the water is needed for a short-term emergency, as seen in the Chanute episode, or for normal use over a prolonged period. The major emphasis placed upon the potable water reuse today concerns the chronic health effects that might result from ingesting the mixture of inorganic and organic contaminants that remain in water even after it has been subjected to the most advanced treatment methods [15,26].

Potable Water Reuse Criteria. It has been argued that there should be a single water quality standard for potable water and that if reclaimed wastewater can meet this standard, it should be acceptable. It must be recognized, however, that current drinking water standards have evolved with the presumption that water supplies are derived from relatively unpolluted freshwater sources. Although great advances have been made in analytical methods for identifying chemical contaminants in water, only a small fraction of the contaminants present in the surface and groundwater can be identified. This analytical limitation has frustrated attempts to develop comprehensive potable water reuse criteria for various sources of water. In assessing water that is being considered for potable reuse, comparison should be made with the highest quality water locally available [16,18,26–28].

Although the implementation of direct potable use of reclaimed municipal wastewater is obviously limited to extreme situations, research relating to potable reuse has continued in several locations. As the proportional quantities of treated wastewater discharged into the nation's waters increase, much of the research that addressed only potable reuse is becoming equally relevant to the treatment of municipal water supplies derived from receiving waters that have been used for the disposal of treated wastewater [22,37,38].

16-3   WASTEWATER RECLAMATION TECHNOLOGIES

The required water quality for reclaimed wastewater varies with each reuse application (see Table 16-3). Most of the current wastewater reclamation technologies are essentially the same as those used for water and wastewater treatment. In certain cases, however, additional treatment processes may be required for the removal of selected physical and chemical contaminants and for inactivation and removal of microbiological pathogens. In evaluating wastewater reclamation technologies, the overriding considerations are the operational reliability of each unit process and the overall performance of the complete treatment system in providing a reclaimed wastewater that meets the established wastewater reclamation criteria. A summary of the unit operations and processes commonly used in wastewater reclamation and the principal contaminants removed are presented in Table 16-11 (see also Table 11-2).

The focus of this section is to consider a few concepts and treatment technologies that are of particular importance to wastewater reuse. The topics considered are (1) treatment process reliability, (2) suspended solids and turbidity removal, and (3) special treatment considerations and examples of advanced wastewater reclamation process combinations.

Treatment Process Reliability

The reliability of a wastewater reclamation plant can be assessed in terms of its ability to produce consistently acceptable reclaimed wastewater (see also the discussion of reliability presented in Chap. 5). There are two categories of problems that can affect the performance and reliability of a wastewater reclamation plant: (1) problems caused by mechanical breakdown, design deficiencies, and operational failures, and (2) problems caused by the influent wastewater variability, even though the wastewater reclamation plant is properly designed, operated, and maintained. With respect to the first category of problems, operation and maintenance failures are cited most frequently as the leading cause of poor plant performance. For the second category of problems, evaluation of the influent water quality variability and the corresponding operational reliability is of particular importance in the design of wastewater reclamation and reuse systems.

TABLE 16-11
Unit processes and operations used in wastewater reclamation and potential for contaminant removal[a]

Constituent	Primary treatment	Activated sludge	Nitrification	Denitrification	Trickling filter	RBC	Coag.-Floc.-Sed.	Filtration after A/S	Carbon adsorption	Ammonia stripping	Selective ion exchange	Breakpoint chlorination	Reverse osmosis	Overland flow	Irrigation	Infiltration-percolation	Chlorination	Ozone
BOD	X	+	+	O	+	+	+	X	+		X		+	+	+	+		O
COD	X	+	+	O	O	+	+	X	X	O	X		+	+	+	+		+
TSS	+	+	+	O	+	+	+	+	+		+		+	+	+	+		
NH3-N	O	+	+	X		+	O	X	X	+	+	+	+	+	+	+		
NO3-N				+					X	O				X				
Phosphorus	O	X	+	+			+	+	+				+	+	+	+		
Alkalinity		X					X	+								X		
Oil & grease	+	+	+				X		X					+	+	+		
Total coliform		+	+	O			+		+			+		+	+	+	+	+
TDS													+			+		
Arsenic	X	X	X				X	+	O									
Barium		X	O				X	O										
Cadmium	X	+	+	O	X		+	X	O							O		
Chromium	X	+	+	O	+	+	+	X	X							+		
Copper	X	+	+		+	+	+	O	X							+		
Fluoride							X		O							X		
Iron	X	+	+		X	+	+	+	O	+								
Lead	+	+	+		X	+	+	+	O	X						X		
Manganese	O	X	X		O		X	+	X					+				
Mercury	O	O	O		O	+	O	X	O									
Selenium	O	O	O				O	+	O									
Silver	+	+	+		X		+		X									
Zinc	X	X	+		+	+	+		+							+		
Color	O	X	X		O		+	X	+					+	+	+		+
Foaming agents	X	+	+		+		X		+					+	+	+		O
Turbidity	X	+	+	O	X		+	+	+					+	+	+		
TOC	X	+	+	O	X		+	X	+	O	O			+	+	+		+

Symbols: O = 25% removal of influent concentration
X = 25–50%
+ = > 50%
Blank denotes no data, inconclusive results, or an increase

[a] Adapted from Ref. 11.

Reclaimed wastewater quality variability may be taken as an indication of an inherent in-plant treatment problem or a problem caused by diurnal or seasonal variations in influent wastewater flow and characteristics as well as process control practices. A statistical analysis is often used to evaluate the variability of treatment plant performance. Plant performance data are typically plotted on log-normal probability paper (see Chap. 5). From such an analysis, long-term trends and the reliability of the wastewater reclamation plant can be assessed.

An example of a log-normal probability plot for suspended solids for both secondary effluent and the effluent following various forms of advanced wastewater treatment is presented in Fig. 16-7. As shown, the frequency distribution plot is a convenient method of evaluating reliability of treatment systems. System A, for

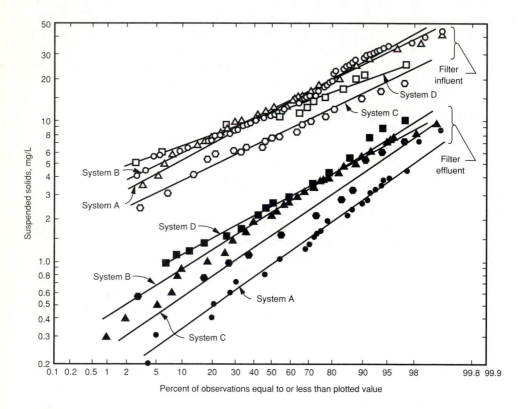

FIGURE 16-7
Frequency distribution of tertiary filter influent and effluent suspended solids. Adapted from Ref. 13. Major operating parameters for System A through D are as follows: System A—150 mg/L alum, 0.2 mg/L anionic polymer, 1 h flocculation, sedimentation at 800 gal/d.ft^2, dual media filtration at 5 gal/min · ft^2, 2 h chlorine contact. System B—5 mg/L alum, 0.06 mg/L anionic polymer, dual media filtration at 5 gal/ min · ft^2, 2 h chlorine contact. System C—10 min carbon contact in downflow column at 3.5 gal/min · ft^2, 2 h chlorine contact. System D—Same as System B except 2 h free residual chlorination.

example, produced an effluent suspended solids concentration of less than or equal to 2 mg/L approximately 85 percent of the time and of less than 1 mg/L about 50 percent of the time [13].

If a low reliability is observed or anticipated for the removal of a constituent critical to a given wastewater reuse application, a series of remedial steps can be incorporated in the design of wastewater reclamation system. The system could include stand-by treatment units, redundancy in contaminant removal processes, flow equalization, emergency storage, and alternative disposal.

Suspended Solids and Turbidity Removal

The recent trend toward the use of reclaimed municipal wastewater in the urban environment rather than in agricultural areas has resulted in a greater exposure of the public to reclaimed wastewater. The urban uses of reclaimed wastewater include golf course, landscape irrigation, groundwater recharge, recreational impoundment, and industrial cooling. Thus, the major concerns for such wastewater reuse applications are health risks caused by pathogens and organics; and by aesthetics related to public acceptance. To achieve efficient inactivation and removal of bacterial and viral pathogens, two key operating criteria must be met: (1) the effluent must be low in suspended solids and turbidity prior to disinfection to reduce shielding of pathogens and also to reduce chlorine demand, and (2) a sufficient disinfectant dose and contact time must be provided for reclaimed wastewater.

To satisfy the first criterion, tertiary granular-medium filtration is frequently installed (1) to remove residual suspended solids found in secondary effluents that may interfere with subsequent disinfection, (2) to reduce the concentration of organic matter that can react with disinfectant, and (3) to improve the aesthetic quality of the reclaimed wastewater by reducing its turbidity. In wastewater reclamation, filtration has been used both as a final step preceding disinfection and as one of the intermediate steps of an advanced wastewater treatment system. Typical flow diagrams employing filtration to produce a high quality effluent for a variety of reuse applications are shown in Fig. 16-8. Performance of the chemical coagulation, flocculation, sedimentation, and filtration system (see Fig. 16-8a) was presented in Fig. 16-7 (System A) in conjunction with the treatment process reliability data.

To produce an essentially virus-free effluent using direct or contact filtration (see Figs. 16-8b and 16-8c), the secondary effluent must be of high quality. For example, to meet the stringent turbidity requirement of less than 2 NTU imposed by the California Wastewater Reclamation Criteria [41], the quality of secondary effluent must be in the range of: turbidity 7 to 9 NTU, suspended solids 14 to 22 mg/L, and total chemical oxygen demand 40 to 80 mg/L. In full-scale wastewater reclamation plants, it has been found that the secondary effluent turbidity should be in the range of 7 to 9 NTU to meet an average filtered effluent turbidity of 2 NTU using direct filtration without chemical addition (see Fig. 11-2). Direct filtration with chemical addition has been used in cases in which the effluent turbidity occasionally exceeds 10 NTU.

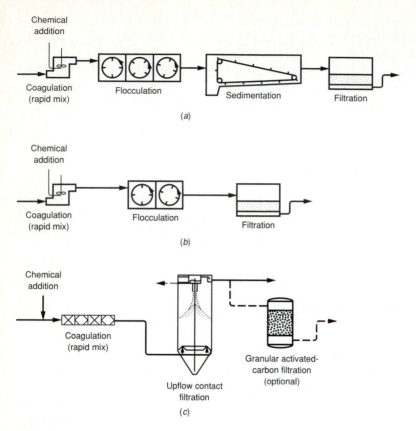

FIGURE 16-8
Comparison of tertiary treatment systems for wastewater reclamation: (a) complete treatment, (b) direct filtration, and (c) contact filtration with optional granular activated carbon adsorption.

A secondary effluent turbidity value of 10 NTU is often taken as the economic dividing line for the application of direct or contact filtration. When secondary effluent turbidity levels are consistently above 10 NTU, improvements to the secondary treatment system are often considered more cost-effective. If a secondary effluent does not meet the water quality requirements discussed in the preceding section, more costly complete treatment must be employed. The complete treatment normally includes chemical coagulation, flocculation, sedimentation, and filtration followed by disinfection with chlorine (see Fig. 16-7, System A footnote, and Fig.16-8a). If complete treatment (i.e., coagulation, flocculation, sedimentation and filtration) is required following conventional wastewater treatment for wastewater reuse, wastewater reuse may be ruled out because of cost considerations.

The overall effectiveness of various tertiary treatment systems on the inactivation and removal of the seeded virus (vaccine-grade, poliovirus) is illustrated in Fig. 16-9. System A produced an average virus removal of 5.2 logs compared to 4.7,

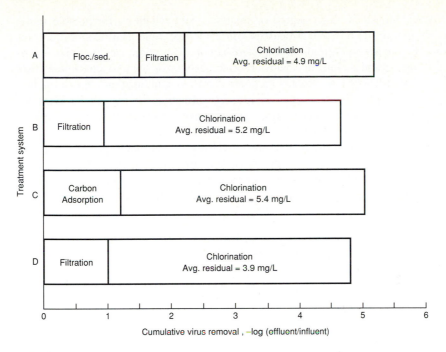

FIGURE 16-9
Virus inactivation and removal efficiency of advanced wastewater treatment system. Polio virus seeding experiment. Major operating parameters for System A through D are found in Fig. 16-7 footnotes. Adapted from Ref. 13.

5.1, and 4.9 logs for Systems B, C, and D, respectively, when an average combined chlorine residual of 5.0 mg/L and a two-hour contact time were used. When a chlorine residual of 10.0 mg/L was maintained, no difference was observed: 5.2 logs of virus removal were observed for all of the systems [13]. It is difficult, however, to assess the actual health significance of these minute differences in virus inactivation and removal, because of the lack of applicable epidemiological data.

Advanced Wastewater Reclamation Process Combinations

With the increased attention paid to drinking water quality, much of the research on drinking water is relevant to indirect potable reuse. Three classes of contaminants are of special concern in wastewater reclamation and reuse in cases in which the potable water supply may be affected: (1) viruses, (2) organic contaminants including pesticides, and (3) heavy metals. The ramifications of these contaminants with respect to health effects are still not understood entirely, and as a result, regulatory agencies are proceeding with caution in permitting wastewater reuse applications that affect potable water supply.

To accomplish the high degree of treatment and reliability for potable reuse, advanced unit operations and processes often include lime clarification, nutrient removal, recarbonation, filtration, activated carbon adsorption, demineralization by reverse osmosis; and disinfection with chlorine, ozone, or both. A conceptual flow diagram of an advanced wastewater treatment process combination capable of producing potable water from municipal wastewater is illustrated in Fig. 16-10. Expected reliability data and average process train performance for the flow diagram depicted in Fig. 16-10 are presented in Table 16-12.

Several examples of existing advanced wastewater treatment process flow diagrams for wastewater reclamation and reuse are illustrated in Figs. 16-11 through 16-13. A 10 Mgal/d advanced wastewater treatment system in operation at El Paso, Texas is shown in Fig. 16-11, in which reclaimed wastewater is injected directly

TABLE 16-12
Reliability data and average process train performance for carbon adsorption of lime-treated activated sludge effluent shown in Fig. 16-10[a]

| Constituent | Average removal, % | Average reliability | | | Average effluent concentration, mg/L |
		10%	50%	90%	
BOD	100	100	100	89	0
COD	100	100	100	97	0
TSS	100	100	99	87	0
NH_3-N	100	97	81	48	0
Phosphorus	100	100	100	99	0
Oil & grease	97	100	98	73	2
Arsenic	61	93	63	0	0.003
Barium	79	95	79	52	0.092
Cadmium	98	100	98	87	0.0002
Chromium	100	100	98	84	0
Copper	98	100	99	98	0.002
Fluoride	x	x	x	x	x
Iron	99	100	100	94	0.023
Lead	99	100	98	78	0.001
Manganese	98	100	98	86	0.002
Mercury	23	31	18	0	0.028
Selenium	7	26	12	0	0.006
Silver	82	100	99	80	0.004
Zinc	98	100	95	58	0.008
TOC	100	100	98	83	0
Turbidity	100	100	100	95	0[b]
Color	93	100	94	56	5[c]
Foaming agents	92	—	84	—	0.17
TDS	95	—	—	—	129

[a] Adapted from Ref. 11

[b] Turbidity units.

[c] Color units.

Symbols: x = data inconclusive
          — = insufficient data

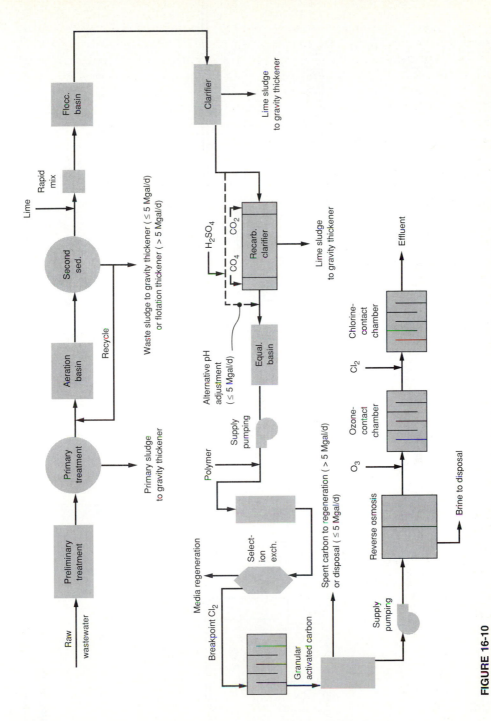

FIGURE 16-10
Conceptual flow diagram of advanced wastewater treatment system capable of producing potable quality water supply. Adapted from Ref. 11.

1181

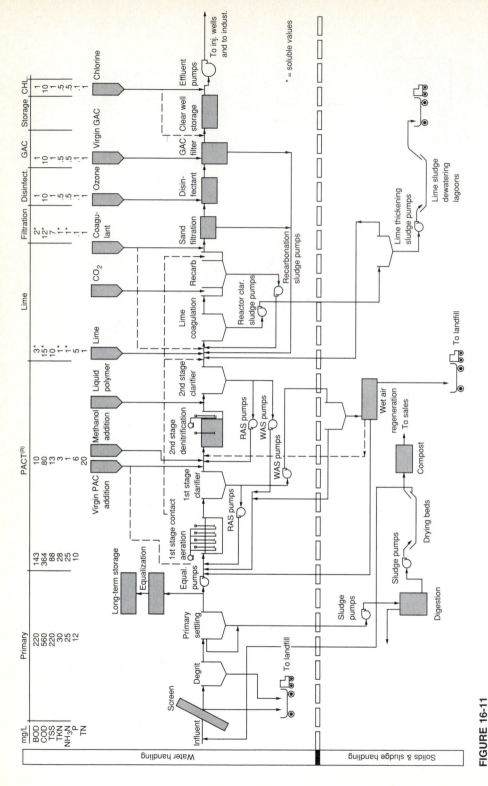

FIGURE 16-11

Multistage wastewater reclamation processes used in El Paso, Texas, for direct injection of reclaimed municipal wastewater. Adapted from Refs. 1 and 20.

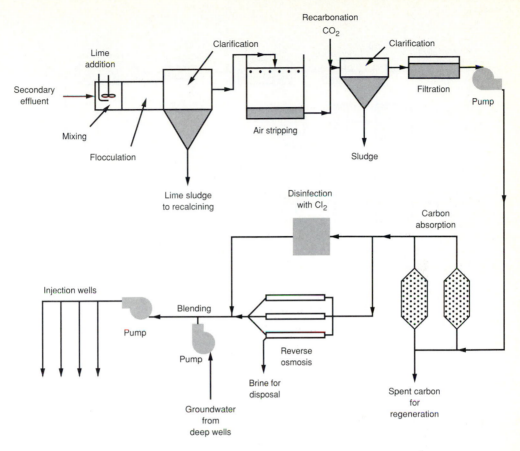

FIGURE 16-12
Schematic flow diagram of treatment processes used at Water Factory 21, Orange County Water District, California. Adapted from Refs. 1 and 49.

into a potable water aquifer. In Orange County, California, groundwater recharge with direct injection of reclaimed wastewater has been in operation since 1976. A schematic flow diagram of the 15 Mgal/d Water Factory 21 reclamation facility is shown in Fig. 16-12. As shown, the advanced wastewater treatment process includes lime clarification with nutrient removal, recarbonation, mixed media filtration, activated carbon adsorption, demineralization by reverse osmosis, and chlorination [1,49]. The 1 Mgal/d Denver Potable Water Reuse Demonstration Plant is illustrated in Fig. 16-13. Health effects studies are currently being conducted on the reclaimed water [37,38]. In another health effects study under way in San Diego, California, effluent from a wastewater reclamation facility comprised of an aquatic floating plant treatment system followed by an advanced treatment system is being compared to the drinking water supply used for the city (see Ref. 19 in Chap. 13).

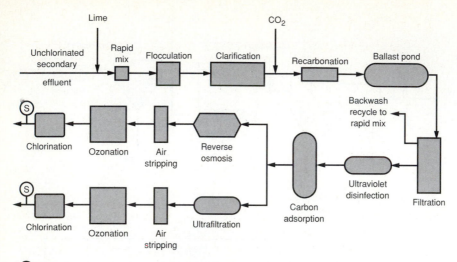

Ⓢ Sampling location for toxicological health effects study

FIGURE 16-13
Flow diagram for the Denver Potable Water Reuse Demonstration Plant for the Health Effects Study.
Adapted from Refs. 37 and 38.

Combining the various unit processes and operations described in Table 16-11 and using the experience obtained with the advanced wastewater reclamation plants shown in Figs. 16-11 through 16-13, it is now possible to produce high-quality water from municipal wastewater for any reuse application. However, the feasibility of such wastewater reuse programs will depend on public acceptance and cost.

16-4 PLANNING CONSIDERATIONS IN WASTEWATER RECLAMATION AND REUSE

In effective planning for wastewater reclamation and reuse, the objectives and the basis for conducting the planning study should be well-defined. The optimum wastewater reclamation and reuse project is best achieved by integrating both wastewater treatment and water supply needs into one plan. This integrated approach is somewhat different from the planning for conventional wastewater treatment facilities, in which cases planning is done only for conveyance, treatment and disposal of municipal wastewater.

The desirable wastewater reclamation and reuse facilities plan should include the following analyses: (1) assessment of wastewater treatment and disposal needs, (2) assessment of water supply and demand, (3) assessment of water supply benefits based on wastewater reuse potential, (4) analysis of reclaimed wastewater market, (5) engineering and economic analyses of alternatives, and (6) implementation plan with financial analysis. Important factors to consider in planning are briefly discussed in this section.

Planning Basis [2]

Two critical components that are the basis of the planning task are the project objectives and study area.

Project Objectives. Wastewater reclamation and reuse can serve the functions of both water pollution control and water supply. Only in the last decade has increased attention been given to the water supply benefits in the facilities planning process. Ignoring the water supply potential of municipal wastewater has often resulted in facilities that now hinder the development of this alternative water resource. For example, in some locations optimum wastewater reuse would have been better achieved if smaller satellite wastewater treatment plants had been constructed with reuse in mind, instead of a large regional wastewater treatment facility.

Because most water and wastewater agencies are established for a single-purpose function, planning by these agencies tends to be single-purpose as well. Optimum wastewater reclamation and reuse is best achieved in a framework of multiple purpose planning and with cooperative efforts of both wastewater management and water supply agencies. Once these multiple benefits and beneficiaries of wastewater reuse are recognized, additional options may be available for sharing project responsibilities and costs among project sponsors.

Project Study Area. The project study area is another critical planning issue. There are two study area horizons to consider in project planning. The first planning horizon is established based on the direct service area of the project facilities plan. The second horizon extends to the area that accrues less direct costs or benefits from a project, which should be accounted for in evaluating the project. Thus, the study area for facilities design includes (1) the collection system area to be served by the wastewater treatment facilities, and (2) the area that can potentially be served by reclaimed wastewater. To evaluate project benefits and costs, the project study area must include (1) the area affected by the environmental effects of the wastewater, and (2) the area benefitting from the supplemental or alternative water supply of the reclaimed wastewater.

The traditional approach to planning is to equate the study area with the project sponsor's jurisdictional boundaries. Such boundaries, however, may not suit the optimum design of a wastewater reclamation and reuse project nor include the areas of benefit. Because water supply is typically dependent on regional water resources outside of the facilities study area, it is essential to look beyond this study area to obtain an understanding of the water resources situation. For instance, overdrafted groundwater basins may be having their most serious impacts on communities great distances beyond the project area. Thus, implementing a wastewater reuse project in one community that reduces groundwater overdraft could result in savings in a water supply in another community.

Market Assessment

In planning a wastewater reclamation project, it is essential to find potential customers that are capable and willing to use reclaimed wastewater. The success of wastewater reuse projects is largely dependent on securing markets for the reclaimed wastewater. A market assessment consists of two parts: (1) determination of background information, including potential uses of reclaimed wastewater, and (2) a survey of potential reclaimed wastewater users and their needs. Background and survey information necessary for a market assessment for reclaimed wastewater is listed in Table 16-13. The results of this assessment form the basis for developing alternatives and determining financial feasibility of a project.

Monetary Analyses

At present, monetary factors tend to be the overriding concern in determining whether to implement a wastewater reuse project and how to go about it, even though technical, environmental, and social factors are important in project planning. In the future, however, environmental considerations and public policy issues may be of greater

TABLE 16-13
Reclaimed wastewater market assessment: Background information and survey[a]

1. Inventory potential users and uses of reclaimed wastewater.
2. Determine health-related requirements regarding water quality and application requirements (e.g., treatment reliability, back-flow prevention, use area controls, irrigation methods) for each type of application of reclaimed wastewater.
3. Determine regulatory requirements to prevent nuisance and water quality problems such as restrictions to protect groundwater.
4. Develop assumptions regarding probable water quality that would be available in the future with various levels of treatment and compare those to regulatory and user requirements.
5. Develop an estimate of future freshwater supply costs to potential users of reclaimed wastewater.
6. Survey potential reclaimed wastewater users, obtaining the following information:
   (a) Specific potential uses of reclaimed wastewater.
   (b) Present and future quantity needs.
   (c) Timing and reliability of needs.
   (d) Water quality needs.
   (e) On-site facilities modifications to convert to reclaimed wastewater and meet regulatory requirements for protection of public health and prevention of pollution problems from reclaimed wastewater.
   (f) Capital investment of the user for on-site facilities modifications, changes in operational costs, desired pay-back period or rate of return, and desired water cost savings.
   (g) Plans for changing use of site in future.
7. Inform potential users of applicable regulatory restrictions, probable water quality available with different levels of treatment, reliability of the reclaimed water, future costs, and quality of freshwater compared to reclaimed water.
8. Determine the willingness of potential users to use reclaimed wastewater now or in the future.

[a] Developed from Ref. 2.

importance than mere cost-effectiveness as a measure of the feasibility of a wastewater reuse project.

Economic and Financial Analyses. Monetary analyses, based on established water resources economics, fall into two categories: economic analysis and financial analysis [17]. The economic analysis is focused on the value of the resources invested in a project to construct and operate it, measured in monetary terms. On the other hand, the financial analysis is focused on the perceived costs and benefits of a project from the viewpoints of project sponsor and participants and others affected by a project. These perceived costs and benefits may not reflect the actual value of resources invested because of subsidies or monetary transfers.

Whereas economic analysis evaluates wastewater reclamation and reuse projects in the context of impacts on society, financial analysis focuses on the local ability to raise money from project revenues, government grants, loans, and bonds to pay for the project. The basic result of the economic analysis should answer the question, "*Should* a reuse project be constructed?" Equally important, however, is the question, "*Can* a reuse project be constructed?" Both orientations, therefore, are necessary, but, only wastewater reclamation and reuse projects that are viable in the economic context are given further consideration for a financial analysis.

Cost and Price of Water. An important factor in monetary analysis of wastewater reclamation and reuse projects is the difference between the cost and the price of water. In an economic analysis, only the future flow of resources invested in or derived from a project is considered. Past resources investments are considered sunk costs that are irrelevant to future investment decisions. Thus, debt service on past investments is not included in an economic analysis. The price of water is the purchase price paid to a water wholesaler or retailer to purchase water. The water price usually reflects a melding of current and past expenditures for a combination of project costs and water system administration costs, which are generally fixed. Only the costs for future construction, operations, and maintenance are relevant to an economic analysis.

To determine the water supply benefit of a wastewater reclamation and reuse project in an economic analysis, the project is usually compared to the development of a new freshwater supply. In performing such an analysis, the relevant costs for comparison are the future stream of costs (1) to construct new freshwater facilities, and (2) to operate and maintain all of the facilities needed to treat and deliver the new increment of water supply developed. Therefore, the present and future price charged for freshwater would not provide a valid basis of comparison to judge the water supply benefit of a wastewater reuse project.

On the other hand, consideration of prices charged for freshwater and reclaimed wastewater is important to determine financial feasibility. The price charged to customers is the perceived cost of water, and, thus, prices will be evaluated by potential reclaimed wastewater users in determining willingness to participate in a wastewater reuse project.

Other Planning Factors

In addition to the monetary analyses, a number of factors have to be evaluated during the planning for a wastewater reclamation and reuse project. Factors of particular significance in project development are related to (1) water demand characteristics, (2) supplemental water supply and emergency backup systems, (3) water quality requirements, and (4) determination of optimum project size. A wastewater reuse project is a relatively small-scale water supply project with considerations of matching supply and demand, appropriate levels of wastewater treatment, reclaimed water storage, and supplemental or backup freshwater supply.

TABLE 16-14
Outline for wastewater reclamation and reuse facilities plan[a]

1. Study area characteristics: Geography, geology, climate, groundwater basins, surface waters, land use, population growth.

2. Water supply characteristics and facilities: Agency jurisdictions, sources and qualities of supply, description of major facilities, water-use trends, future facilities needs, groundwater management and problems, present and future freshwater costs, subsidies, and customer prices.

3. Wastewater characteristics and facilities: Agency jurisdictions, description of major facilities, quantity and quality of treated effluent, seasonal and hourly flow and quality variations, future facilities needs, need for source control of constituents affecting reuse, and description of existing reuse (users, quantities, contractual and pricing agreements).

4. Treatment requirements for discharge and reuse and other restrictions: health and water quality related requirements, user-specific water quality requirements, and use area controls.

5. Potential water reuse customers: Description of market analysis procedures, inventory of potential reclaimed water users, and results of user survey.

6. Project alternative analysis: Capital and operation and maintenance costs, engineering feasibility, economic analysis, financial analysis, energy analysis, water quality impacts, public and market acceptance, water rights impacts, environmental and social impacts, and comparison of alternatives and selection.
   (a) Treatment alternatives
   (b) Alternative markets: based on different levels of treatment and service areas
   (c) Pipeline route alternatives
   (d) Alternative reclaimed water storage locations and options
   (e) Freshwater alternatives
   (f) Water pollution control alternatives
   (g) No project alternative

7. Recommended plan: Description of proposed facilities, preliminary design criteria, projected cost, list of potential users and commitments, quantity and variation of reclaimed water demand in relation to suply, reliability of supply and need for supplemental or back-up water supply, implementation plan, and operational plan.

8. Construction financing plan and revenue program: Sources and timing of funds for design and construction; pricing policy of reclaimed water; cost allocation between water supply benefits and pollution control purposes; projection of future reclaimed water use, freshwater prices, reclamation project costs, unit costs, unit prices, total revenue, subsidies; sunk costs and indebtedness; and analysis of sensitivity to changed conditions.

[a] Developed from Ref. 2.

Planning Report

The results of the completed planning effort are documented in a facilities planning report on wastewater reclamation and reuse. An outline is shown in Table 16-14, which also serves as a checklist for planning considerations. All of the items listed in Table 16-14 have been found at one time or another to affect the evaluation of water reclamation and reuse projects. Thus, all of the factors shown do not deserve an in-depth analysis, but, they should at least be considered. Although the emphasis on the wastewater or water supply aspects will vary depending on whether a project is single or multiple purpose, the nature of wastewater reclamation and reuse is such that both aspects must be considered.

DISCUSSION TOPICS AND PROBLEMS

16-1. The following water quality data from an activated sludge plant are reported.

Water quality parameter[a]	Concentration, mg/L
BOD	11
SS	13
NH_3-N	1.4
NO_3-N	5
Total-P	6
pH	7.6
Cations:	
Ca	82
Mg	33
Na	220
Anions:	
HCO_3	136
SO_4	192
Cl	245
EC (mmhos/cm)	1.4
TDS	910
B	0.7

[a] All concentrations are expressed in mg/L except electrical conductivity and pH, which is unitless.

Determine the SAR and adj R_{Na} for this water and discuss possible reasons for difference in calculated values. Determine if infiltration problems may develop in using this effluent for irrigation. Estimate the salinity (expressed as EC) of drainage water if the crop is irrigated to achieve a leaching fraction of 0.1. Estimate also the TDS concentration of the drainage water.

16-2. Which federal, state, or local agencies would be involved in setting wastewater reclamation regulations? Discuss the pros and cons of a federal government role in establishing wastewater reclamation criteria.

16-3. Conduct literature search and list health and regulatory factors affecting implementation of wastewater reuse projects. What is the basis for setting less stringent microbiological standards in developing countries where enteric diseases are rampant among population?

16-4. A common chemical precipitation reaction in water softening is shown in the following equation.

$$Ca(HCO_3)_2 + Ca(OH)_2 \rightarrow 2CaCO_3 + 2H_2O$$

Given a flow of 5 Mgal/d, determine how much $Ca(OH)_2$ is required each day to reduce the Ca^{+2} concentration from 150 mg/L as $CaCO_3$ to 100 mg/L as $CaCO_3$ using the above equation. Determine also the pounds of $CaCO_3$ and the sludge volume produced each day if the solids are settled to 1 percent by weight and the specific gravity of $CaCO_3$ is 2.8.

16-5. Review Figs. 16-10 through 16-13 in the text and group common unit processes and operations for the removal of suspended solids, organics, bacteria and viruses, and heavy metals. Which wastewater reclamation system is most reliable and why?

16-6. Using Table 16-10 as a guide, develop wastewater reclamation criteria for groundwater recharge where groundwater is used for potable purposes.

16-7. Outline a model wastewater reuse plan for your community. Discuss proposed implementation plan for your community including economic and financial analyses. Why are wastewater reclamation and reuse options adopted or rejected?

16-8. The overall virus inactivation and removal can be expressed in terms of log virus removal, as shown in Fig. 16-9. In conducting a virus seeding experiment to assess virus removal efficiency of a treatment process, a concentration of 10^6 PFU/L vaccine-grade poliovirus is introduced at the head of a wastewater treatment system. Calculate the overall percent inactivation and removal of the seeded virus if 5.2 log virus removal is achieved by the treatment system. Also, calculate the concentration of virus remaining after treatment.

16-9. The question of adequate public health protection can be evaluated by a risk analysis in which the risk and the extent of exposure to pathogens is assessed. Outline a test protocol for a virus risk assessment for the use of reclaimed municipal wastewater in food crop irrigation, golf courses, and groundwater recharge. Refer to the current literature and cite at least three recent (since 1980) references.

16-10. A water reclamation and reuse project is proposed with a capital cost of $1.7 million, including design and construction costs. This cost includes $1.0 million allocated to advanced waste treatment and pump facilities and $0.7 million to distribution pipelines to reclaimed water users. Deliveries of reclaimed water will be 200 ac-ft during the first year of operation, 270 ac-ft during the second year, and 450 ac-ft each year thereafter. Operation and maintenance (O & M) costs will be $40,000 during the first year, $60,000 the second year, and $85,000 each year thereafter. The fertilizer value of the nutrients in the reclaimed water is $40/ac-ft. The useful life of pipelines is 50 years, and for other facilities in this project it is 20 years. Assume that the costs of design, construction management, and legal and other administrative services amount to 20 percent of construction costs, and there is no salvage value associated with these costs at the end of the 20-year planning period.

This project would replace fresh water development which costs $800/ac-ft. (a) Assuming a planning period of 20 years and a discount rate of 6 percent, determine whether this project is economically justified as a water supply project. (b) Calculate the net unit cost per acre-foot of reclaimed water produced by the project.

16-11. For the project described in Prob. 16-10, determine the financial feasability of the proposed project. Capital costs will be financed at 8 percent annual interest. Operation and maintenance costs are predicted to escalate at a rate of 4 percent per year. The price of the reclaimed water will be set at 85 percent of potable water prices. The potable water price will be $400/ac-ft during the first year of operation and will escalate at a rate of 2 percent per year. Also determine the amount of subsidies from other sources, such as a government grant, that would be needed to reduce the local share of capital costs to ensure that there would be a positive cash flow each year and, thus, be financially solvent.

REFERENCES

1. Asano, T. (ed.): *Artificial Recharge of Groundwater,* Butterworth Publishers, Boston, MA, 1985.
2. Asano, T., and R. A. Mills: "Planning and Analysis for Water Reuse Projects," *Journal AWWA,* vol. 82, no. 1, 1990.
3. Ayers, R. S., and K. K. Tanji: "Agronomic Aspects of Crop Irrigation with Wastewater," *Water Forum '81,* vol. 1, p. 578, Am. Soc. of Civil Engrs., 1981.
4. Ayers, R. S., and D. W. Westcot: "Water Quality for Agriculture," FAO Irrigation and Drainage Paper 29, Rev. 1, Food and Agriculture Organization of the United Nations, Rome, 1985.
5. Bitton, G., and C. P. Gerba: *Groundwater Pollution Microbiology,* John Wiley & Sons, New York, 1984.
6. Bouwer, H.: *Groundwater Hydrology,* McGraw-Hill Book Co., New York, 1978.
7. Bouwer, H., and W. L. Chase: "Water Reuse in Phoenix, Arizona," *Future of Water Reuse,* Water Reuse Symposium III, vol. 1, 337, AWWA Research Foundation, Denver, CO, 1985.
8. Bouwer, H.: "Groundwater Recharge as a Treatment of Sewage Effluent for Unrestricted Irrigation," in *Treatment and Use of Sewage Effluent for Irrigation,* M. B. Pescod and A. Arar (eds.), Butterworths, London, 1988.
9. Burger, R.: *Cooling Tower Technology: Maintenance, Upgrading and Rebuilding,* Cooling Tower Institute, Houston, TX, 1979.
10. Crook, J., and D. A. Okun: "The Place of Nonpotable Reuse in Water Management," *Journal WPCF,* vol. 59, no. 5, 1987.
11. Culp/Wesner/Culp: *Water Reuse and Recycling,* Office of Water Research and Technology, U.S. Department of the Interior, Washington DC, July 1979.
12. Dean, R. B., and E. Lund: *Water Reuse: Problems and Solutions,* Academic Press, London, 1981.
13. Dryden, F. D., C. L. Chen, and M. W. Selna: "Virus Removal in Advanced Wastewater Treatment Systems," *Journal WPCF,* vol. 51, no. 8, 1979.
14. Feachem, R. G., D. J. Bradley, H. Garelick, and D. D. Mara: *Sanitation and Disease—Health Aspects of Excreta and Wastewater Management, World Bank Studies in Water Supply and Sanitation 3,* Published for the World Bank by John Wiley & Sons, Chichester, UK, 1983.
15. *Guidelines for Drinking-Water Quality,* Vol. 1. Recommendations, World Health Organization, Geneva, Switzerland, 1984.
16. Isaacson, M., and A. R. Sayed: "Human Consumption of Reclaimed Water—The Namibian Experience," Proceedings of Water Reuse Symposium IV, *Implementing Water Reuse,* 1047, AWWA Research Foundation, Denver, CO, 1988.
17. James, L. D., and R. R. Lee: *Economics of Water Resources Planning,* McGraw-Hill Book Co., New York, 1971.
18. James M. Montgomery, Consulting Engineers, Inc.: *Water Treatment Principles and Design,* John Wiley & Sons, New York, 1985.
19. Kemmer, F. N. (ed.): *The Nalco Water Handbook,* 2nd ed., McGraw-Hill Book Co., New York, 1988.
20. Knorr, D. B., J. Hernandez, and W. M. Copa: "Wastewater Treatment and Groundwater Recharge: A Learning Experience at El Paso, TX" in Proceedings of the Water Reuse Symposium IV, *Implementing Water Reuse,* 211, AWWA Research Foundation, Denver, CO, 1988.

21. Lauer, W. C., S. E. Rogers, and J. M. Ray: "Denver's Potable Water Reuse Project - Current Status," *Proceedings of the Water Reuse Symposium III, Future of Water Reuse,* 1, 316, AWWA Research Foundation, Denver, CO, 1985.

22. Linsted, K. D., and M. R. Rothberg: "Potable Water Reuse," E. J. Middlebrooks (ed.), *Water Reuse,* Ann Arbor Science, Ann Arbor, MI, 1982.

23. McCarty, P. L., B. E. Rittmann, and M. Reinhard: "Processes Affecting the Movement and Fate of Trace Organics in the Subsurface Environment," *Environmental Science and Technology,* vol. 15, no. 1, 1981.

24. Metzler, D. F., et al.: "Emergency Use of Reclaimed Water for Potable Supply at Chanute, KS," *Journal AWWA,* vol. 50, no. 8, 1958.

25. National Academy of Science, National Academy of Engineering: *Water Quality Criteria,* A Report of the Committee on Water Quality Criteria, Superintendent of Documents, U.S. Government Printing Office, Washington, DC, 1972.

26. National Research Council: *Quality Criteria for Water Reuse,* National Academy Press, Washington DC, 1982.

27. National Research Council: *The Potomac Estuary Experimental Water Treatment Plant,* A Review of the U.S. Army Corps of Engineers, Evaluation of the Operation, Maintenance and Performance of the Experimental Estuary Water Treatment Plant, National Academy Press, Washington DC, 1984.

28. National Research Council: *Water for the Future of the Nation's Capital Area,* A Review of the U.S. Army Corps of Engineers Metropolitan Washington Area Water Supply Study, National Academy Press, Washington DC, 1984.

29. Nellor, M. H., R. B. Baired, and J. R. Smyth: "Health Effects of Indirect Potable Water Reuse," *Journal AWWA,* vol. 77, no. 7, 1985.

30. Odendaal, P. E., and W. H. Hattingh: "The Status of Potable Reuse Research in South Africa," The Proceedings of Water Reuse Symposium IV, *Implementing Water Reuse,* 1339, AWWA Research Foundation, Denver, CO, 1988.

31. Ongerth, H. J., and J. E. Ongerth: "Health Consequences of Wastewater Reuse," *Ann. Rev. of Public Health,* vol. 3, 1982.

32. Pescod, M. B., and A. Arar (eds.): *Treatment and Use of Sewage Effluent for Irrigation,* Butterworths, London, 1988.

33. Pettygrove, G. S., and T. Asano (eds.): *Irrigation with Reclaimed Municipal Wastewater—A Guidance Manual,* Lewis Publishers, Inc., Chelsea, MI, 1985.

34. *Principles of Industrial Water Treatment,* 3rd ed., Drew Chemical Corp., Boonton, NJ, 1979.

35. Reed, S. C., and R. W. Crites: *Handbook of Land Treatment Systems for Industrial and Municipal Wastes,* Noyes Publications, Park Ridge, NJ, 1984.

36. Roberts, P. V.: "Water Reuse for Groundwater Recharge: An Overview," *Journal AWWA,* vol. 72, no. 7, 1980.

37. Rogers, S. E., and W. C. Lauer: "Disinfection for Potable Reuse," *Journal WPCF,* vol. 58, no. 3, 1986.

38. Rogers, S. E., et al.: "Organic Contaminants Removal for Potable Reuse," *Journal WPCF,* vol. 59, no. 7, p. 722, 1987.

39. Shuval, H. I., A. Adin, B. Fattal, E. Rawitz, and P. Tekutiel: "Wastewater Irrigation in Developing Countries—Health Effects and Technical Solutions," World Bank Technical Paper 51, The World Bank, Washington DC, 1986.

40. Solley, W. B., C. F. Merk, and R. R. Pierce: "Estimated Use of Water in the United States in 1985," U.S. Geological Survey Circular 1004, U.S. Geological Survey, Federal Center, Denver, CO, 1988.

41. State of California: "Wastewater Reclamation Criteria," California Administrative Code, Title 22, Division 4, Environmental Health, Department of Health Services, Berkeley, CA, 1978.

42. State of California: "Evaluation of Industrial Cooling Systems Using Reclaimed Municipal Wastewater: Applications for Potential Users," State Water Resources Control Board, Sacramento, CA, November 1980.

43. State of California: "The Porter-Cologne Water Quality Control Act and Related Code Sections (including 1988 Amendments)," California State Water Resources Control Board, Sacramento, CA, January 1989.

44. State of California: "Report of the Scientific Advisory Panel on Groundwater Recharge with Reclaimed Wastewater," Prepared for State Water Resources Control Board, Department of Water Resources, and Department of Health Services., Sacramento, CA, November 1987.

45. Sterritt, R. M., and J. N. Lester: *Microbiology for Environmental and Public Health Engineers,* E. & F. N. Spon Ltd., New York, 1988.

46. Suarez, D. L.: "Relation between pH_c and Sodium Adsorption Ratio (SAR) and Alternative Method of Estimating SAR of Soil or Drainage Waters," *Soil Science Society of America Journal,* vol. 45, p. 469, 1981.

47. Todd, D. K.: *Groundwater Hydrology,* 2nd ed., John Wiley & Sons, New York, 1980.

48. Ward, C. H., W. Orger, and P. L. McCarty, (eds.): *Ground Water Quality,* John Wiley & Sons, New York, 1985.

49. *Water Reuse,* Manual of Practice SM-3, 2nd ed., Water Pollution Control Federation, Washington DC, 1989.

50. World Health Organization: *Reuse of Effluent: Methods of Wastewater Treatment and Health Safeguards,* Technical Report Series No. 517, Geneva, Switzerland, 1973.

51. World Health Organization, *Health Aspects of Treated Sewage Re-Use,* Report on a WHO Seminar, EURO Reports and Studies 42, Regional Office for Europe, Copenhagen, Denmark, 1980.

52. U.S. Water Resources Council: "The Nation's Water Resources, The Second National Assessment," Washington DC, 1978.

CHAPTER 17

EFFLUENT DISPOSAL

After treatment, wastewater is either reused, as discussed in the previous chapter, or disposed of in the environment, where it re-enters the hydrologic cycle. Disposal can thus be viewed as the first step in a very indirect and long-term reuse. The most common means of treated wastewater disposal is by discharge and dilution into ambient waters, the subject of this chapter. Another means of disposal is land application, where the wastewater seeps into the ground and recharges underlying groundwater aquifers. Part of the wastewater destined for infiltration also evaporates, and in desert areas this evaporated fraction can be substantial. Land application is covered in Chap. 13.

A fundamental element of wastewater disposal is the associated environmental impact. Numerous environmental regulations, criteria, policies, and reviews now ensure that the environmental impacts of treated wastewater discharges to ambient waters are acceptable. This regulatory framework affects not only the selection of discharge locations and outfall structures but also the level of treatment required. Treatment and disposal are thus linked and cannot be considered independently. For example, to achieve environmental acceptability, a choice may be available between enhanced treatment for one or several wastewater constituents or increased effluent dilution by, for example, moving the discharge further offshore or using a multiport diffuser outfall. Another means is source reduction, whereby individual dischargers are required to decrease their contribution of specific contaminants to the sewers by process changes or pretreatment.

The emphasis of environmental impact evaluations of wastewater discharges used to be on dissolved oxygen. The assimilative capacity of receiving waters, representing the amount of BOD_5 that can be assimilated without excessively taxing ambient dissolved-oxygen levels, was of major concern. This emphasis on dissolved

1195

oxygen lead to requirements for secondary treatment of wastewaters. Recently, the attention has broadened to a wider range of wastewater constituents including nutrients, toxic compounds, and a variety of organic compounds. The environmental impacts of these constituents are diverse and often complex. The first element in evaluating these impacts, however, is the determination of the distribution and fate of these constituents in the water column and bottom sediments. The emphasis of this chapter is on the determination of constituent concentrations as a function of effluent flows and composition, ambient water characteristics, and discharge structure design. Frequently, environmental criteria or standards exist regulating concentrations directly. In some cases, particularly for large discharges, additional environmental analyses are required; but the starting point is the distribution of constituent concentrations.

The purpose of this chapter is to introduce the reader to the general subject of effluent disposal in the aquatic environment. This chapter is not meant to be an exhaustive treatment of the subject; instead, it provides an exposure to the main issues and approaches used to evaluate environmental impacts and to design facilities used for the effective discharge of treated wastewater. The topics discussed are (1) water quality parameters and criteria, (2) fate processes, (3) lake and reservoir disposal, (4) river and estuary disposal, and (5) ocean disposal.

17-1 WATER QUALITY PARAMETERS AND CRITERIA

Water quality parameters that need to be considered relative to the discharge of wastewater to the environment are reviewed in this section. The general nature of the standards and criteria used to measure and regulate water quality are also considered.

Water Quality Parameters

Important water quality parameters relating to wastewater discharges are dissolved oxygen (DO), suspended solids, bacteria, nutrients, pH, and toxic chemicals including volatile organics, acid/base neutrals, metals, pesticides, and PCBs.

Dissolved oxygen is important to aquatic life because detrimental effects can occur when DO levels drop below 4 to 5 mg/L, depending on the aquatic species. Suspended solids affect water column turbidity and ultimately settle to the bottom, leading to possible benthic enrichment, toxicity, and sediment oxygen demand. Coliform bacteria are used as an indicator of other pathogenic organisms of fecal origin and as such provide a measure of the safety of the water for recreational and other uses. Nutrients can lead to eutrophication and DO depletion. The acidity of water, measured by its pH, affects the chemical and ecological balance of ambient waters. Toxic chemicals include a range of compounds that, at different concentrations, have detrimental effects on aquatic life or on humans, upon ingestion of water and/or fish and shellfish. Toxic effects on aquatic life are, as noted in Chap. 3, characterized as acute if they occur after a short exposure (on the order of a few hours) to the toxic constituent or as chronic if effects require a longer term exposure.

Benthic : at the bottom of a depth of water body

Standards and Criteria

Regulations and procedures affecting watewater discharges are diverse and subject to change. Basic approaches, however, are less variable and are briefly reviewed here because engineering aspects of effluent disposal cannot be divorced from the regulatory environment. The emphasis is on current U.S. practices, and thus adaptations may be needed for different times and places.

Wastewater discharges are most commonly controlled through effluent standards and discharge permits. In the United States, the National Pollution Discharge Elimination System (NPDES), administered by the individual states, with Federal EPA oversight, is used for the control of wastewater discharges. Under this system, discharge permits are issued with limits on the quantity and quality of effluents. These limits are based on a case-by-case evaluation of potential environmental impacts and, in the case of multiple dischargers, on wasteload allocation studies aimed at distributing discharge allowances fairly. Discharge permits are designed as an enforcement tool with the ultimate goal of meeting ambient water quality standards (see Chap. 4).

Water quality standards are sets of qualitative and quantitative criteria designed to maintain or enhance the quality of receiving waters. In the United States, these standards are promulgated by the individual states. Receiving waters are divided into several classes depending on their uses, current or intended, with different sets of criteria designed to protect these uses. An example is provided in Table 17-1.

For toxic compounds, chemical-specific or whole-effluent approaches can be taken. In the chemical-specific approach, individual criteria are used for each of the toxic chemicals detected in the wastewater. Criteria can be developed in laboratory experiments to protect aquatic life against acute and chronic effects and to safeguard humans against deleterious health effects including cancer [31]. Toxic effects are a function of exposure concentrations as well as their duration. For example, acute toxicity levels for aquatic life should not be exceeded even for a short time, currently estimated to be on the order of one hour [30]. Chronic toxicity levels can be tolerated for longer times, and a limit currently proposed is based on a four-day average concentration [30]. Human health criteria are typically based on long-term exposures of up to a lifetime so that the corresponding concentration limits are applicable to average values. Allowable frequencies for exceeding the limits may also be specified for the aquatic life criteria, recognizing that it is statistically impossible to ensure that a criterion will never be exceeded and also that ecological communities are able to recover from stress. The chemical-specific approach, however, does not consider the possible additive, antagonistic, or synergistic effects of multiple chemicals. Nor does it consider the biological availability of the compound, which depends on its form in the wastewater.

The whole-effluent approach (see Chap. 3) can be used to overcome the shortcomings of the chemical-specific approach involving the use of toxicity or bioassay tests to determine the concentration at which the wastewater induces acute or chronic toxicity effects. In the whole-effluent approach, selected organisms are exposed to effluent diluted in various ratios with samples of receiving water. At various points during the test, the organisms showing various effects, such as lower reproduction rates, reduced growth, or death, are quantified. Toxicity can then be measured in

TABLE 17-1
Example water body classification

Class	Uses
	Inland waters
AA	Existing or proposed drinking water supply impoundment and tributary surface water.
A	May be suitable for drinking water supply and/or bathing; suitable for all other water uses; character uniformly excellent; may be subject to absolute restrictions on discharges of pollutants.
B	Suitable for bathing, other recreational purposes, agricultural uses, certain industrial processes and cooling; excellent fish and wildlife habitat; good aesthetic value.
C	Suitable for fish and wildlife habitat, recreational boating, and certain industrial processes and cooling; good aesthetic value.
D	May be suitable for bathing or other recreational purposes, certain fish and wildlife habitat, certain industrial processes and cooling water; may have good aesthetic value. Present condition, however, severely inhibits or precludes one or more of the above uses.
	Coastal and marine waters
SA	Suitable for all seawater uses including shellfish harvesting for direct human consumption, bathing, and other contact sports; may be subject to absolute restrictions on the discharge of pollutants.
SB	Suitable for bathing, other recreational purposes, industrial cooling and shellfish harvesting for human consumption after depuration; excellent fish and wildlife habitat; good aesthetic value.
SC	Suitable for fish, shellfish, and wildlife habitat: suitable for recreational boating and industrial cooling; good aesthetic value.
SD	May be suitable for bathing or other recreational purposes, fish and wildlife habitat, and industrial cooling; may have good aesthetic value. Present conditions, however, severely inhibit or preclude one or more of the above uses.

Source: Connecticut Water Quality Standards and Classifications.

several ways such as the effluent concentration at which 50 percent of the organisms are killed (LC_{50}) or the No Observed Effect Level (NOEL), defined as the highest effluent concentration at which no unacceptable effect will occur even at continuous exposure [30].

17-2   FATE PROCESSES

The physical, chemical, and biological processes that control the fate of the water quality parameters previously mentioned are discussed in this section. These processes are numerous and varied. It is convenient to divide them into transport processes, which affect all water quality parameters similarly, and transformation processes, which are constituent-specific. Many of these transformation processes, however, have comparable kinetics so that a different formulation is not required for each constituent.

It is important at this point to reintroduce the conservation of mass equation (see Chap. 5), which is the basis of practically all further analyses. This equation is based on a bookkeeping of the mass of any water quality constituent in a stationary volume of fixed dimensions, called "control volume." The general form of the constituent mass balance can be expressed as follows:

$$\begin{matrix} \text{Rate of mass} & & \text{Rate of mass} & & \text{Rate of mass} & & \text{Rate of mass} & & \text{Rate of mass} \\ \text{increase in} & = & \text{entering} & - & \text{leaving} & + & \text{generated within} & - & \text{lost within} \\ \text{control volume} & & \text{control volume} & & \text{control volume} & & \text{control volume} & & \text{control volume} \end{matrix} \quad (17\text{-}1)$$

Each of the terms in Eq. 17-1 has units of mass per unit time: MT^{-1}. The mass conservation equation is applicable whether the discharge is in a lake, stream, or coastal area. However, the different physical characteristics of these settings require different approaches and approximations to solve for constituent concentrations.

Transport Processes

There are two basic transport processes: (1) advection, or transport of a constituent resulting from the flow of the water in which the constituent is dissolved or suspended, and (2) diffusion, or transport due to turbulence in the water.

Advection. Using the infinitesimally small, box-like control volume shown in Fig. 17-1, the terms of the mass conservation equation relevant to advection in the x direction are

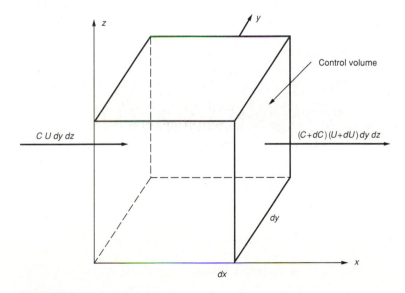

FIGURE 17-1
Masses of constituent entering and leaving control volume per unit time due to advection in x-direction.

$$\text{Rate of mass increase in control volume} = \frac{\partial C}{\partial t} dx\, dy\, dz$$

$$\text{Rate of mass entering control volume} = C U\, dy\, dz$$

$$\text{Rate of mass leaving control volume} = \left(C + \frac{\partial C}{\partial x} dx\right)\left(U + \frac{\partial U}{\partial x} dx\right) dy\, dz$$

where
C = mass concentration of constituent, M/L^3
U = water velocity in x direction, L/T
dx, dy, dz = dimensions of control volume in x, y, and z directions, L
t = time, T

General dimensions rather than specific units are indicated above for the various parameters, as the equations are dimensionally homogeneous. Typically, concentrations are expressed in mg/L (which is equivalent to ppm in water). Substituting the terms given in Eq. 17-1 and simplifying yields

$$\frac{\partial C}{\partial t} = -U\frac{\partial C}{\partial x} - C\frac{\partial U}{\partial x} - \frac{\partial C}{\partial x}\frac{\partial U}{\partial x}dx \tag{17-2}$$

The last term on the right-hand side (after the last minus sign) is negligible compared to the others and can be omitted (because it contains a second-order term). The contributions of flow components in the y and z directions, which are similar to that of flow in the x direction, must be added in:

$$\frac{\partial C}{\partial t} = -U\frac{\partial C}{\partial x} - C\frac{\partial U}{\partial x} - V\frac{\partial C}{\partial y} - C\frac{\partial V}{\partial y} - W\frac{\partial C}{\partial z} - C\frac{\partial W}{\partial z} \tag{17-3}$$

where V and W are the velocity components in the y and z directions. A final simplification results when the equation of continuity from fluid mechanics ($\partial U/\partial x + \partial V/\partial y + \partial W/\partial z = 0$) is introduced in Eq. 17-3. In the resulting equation, the effect of advection on concentration changes with time is defined:

$$\frac{\partial C}{\partial t} = -U\frac{\partial C}{\partial x} - V\frac{\partial C}{\partial y} - W\frac{\partial C}{\partial z} \tag{17-4}$$

Diffusion. Turbulent velocity fluctuations in conjunction with concentration gradients lead to a mass transport phenomenon called "diffusion," which can be described as local mixing by turbulent eddies. The rate of mass transport is proportional to the concentration gradient (or longitudinal rate of concentration variation). Thus, returning to Fig. 17-1, the terms of the mass conservation equation relevant to diffusion in the x direction are

$$\text{Rate of mass increase in control volume} = \frac{\partial C}{\partial t} dx\, dy\, dz$$

$$\text{Rate of mass entering control volume} = -E_x\frac{\partial C}{\partial x} dy\, dz$$

$$\text{Rate of mass leaving control volume} = -\left[E_x\frac{\partial C}{\partial x} + \frac{\partial}{\partial x}\left(E_x\frac{\partial C}{\partial x}\right)dx\right]dy\,dz$$

where E_x = diffusion coefficient (also called diffusivity) in the x direction, L^2/T

Substituting into Eq. 17-1 gives the time rate of change of concentration in the control volume due to diffusion in the x direction:

$$\frac{\partial C}{\partial t} = \frac{\partial}{\partial x}\left[E_x\frac{\partial C}{\partial x}\right] \tag{17-5}$$

Adding the equivalent terms for diffusion in the y and z directions yields

$$\frac{\partial C}{\partial t} = \frac{\partial}{\partial x}\left[E_x\frac{\partial C}{\partial x}\right] + \frac{\partial}{\partial y}\left[E_x\frac{\partial C}{\partial y}\right] + \frac{\partial}{\partial z}\left[E_x\frac{\partial C}{\partial z}\right] \tag{17-6}$$

A difficulty with turbulent diffusion is that the corresponding diffusion coefficients are dependent on the flow and thus are non-uniform (variable in space) and anisotropic (direction dependent) [7].

Transformation Processes

The processes discussed in this section are dependent on the constituent under consideration. For input into the conservation of mass equation, rates of mass gain or loss within the control volume are required. Rates expressions for the major transformation processes relevant to wastewater discharges are presented and reviewed in this section.

BOD Oxidation. As discussed in Sec. 3-3, the oxidation of BOD consumes oxygen and thus represents an oxygen sink for the ambient water. Carbonaceous BOD is oxidized first, followed by nitrogenous BOD after about 8 to 12 days (see Fig. 3-14). Both carbonaceous and nitrogenous BOD oxidation are first-order processes, with the rate of oxidation (equal to the rate of BOD exertion) proportional to the amount of BOD present:

$$r_C = -K_C L_C \text{ and } r_N = -K_N L_N \tag{17-7}$$

where r_C = rate of carbonaceous BOD loss per unit time per unit volume of water, M/TL^3

r_N = rate of nitrogenous BOD loss per unit time per unit volume of water, M/TL^3

L_C = carbonaceous BOD concentration, M/L^3

L_N = nitrogenous BOD concentration, M/L^3

K_C = rate constant for carbonaceous BOD oxidation, T^{-1}

K_N = rate constant for nitrogenous BOD oxidation, T^{-1}

t = time, T

Determination of BOD oxidation rate constants is discussed in Sec. 3-3. The constants are often expressed in 1/days. For ambient water quality predictions, the same time unit must be used for these rates as other quantities, such as velocities, and conversion may be required.

For dissolved-oxygen analyses, BOD oxidation represents an oxygen loss, or sink, which occurs at the same rate as the BOD decay. Accounting for both carbonaceous and nitrogenous components gives

$$r_O = r_C + r_N \tag{17-8}$$

where r_O = rate of oxygen loss per unit time per unit volume of water due to BOD oxidation, M/TL^3

There are other sources and sinks for dissolved oxygen that need to be considered. These include surface reaeration, sediments oxygen demand, photosynthesis, and respiration, which are discussed below.

Surface Reaeration. When the dissolved-oxygen concentration in a body of water with a free surface is below the saturation concentration (Appendix D), a net flux of oxygen occurs from the atmosphere to the water. This flux (mass per unit time per unit surface area) is proportional to the amount by which the dissolved oxygen is below saturation. For a control volume with a free surface area, the rate of dissolved-oxygen increase due to surface reaeration is therefore

$$r_R = k_R \frac{A}{V}(C_s - C) = \frac{k_R}{H}(C_s - C) = K_2(C_s - C) \tag{17-9}$$

where r_R = rate of oxygen gain due to reaeration per unit time per unit volume of water, M/TL^3
k_R = reaeration flux rate, L/T
A = free surface area of control volume, L^2
V = volume of control volume, L^3
C_s = saturation dissolved-oxygen concentration, M/L^3
C = dissolved-oxygen concentration, M/L^3
H = control volume depth, L
K_2 = surface reaeration rate, 1/T

Note that the control volume used above does not necessarily extend down to the bottom of the water body. The control volume must be small enough so that the dissolved-oxygen concentration is approximately uniform. Different control volume depths are thus appropriate for different situations, as will be seen later. Therefore, the reaeration rate of greater physical significance is K_R, whereas K_2 depends on the control volume depth. However, for historical reasons, K_2 is used more frequently. A number of empirical and semi-empirical formulae have been proposed to calculate the reaeration rate [12]. Most of these relationships were devised for streams but are frequently applied in lakes and coastal areas. A commonly used formula is that of O'Connor and Dobbins [19]:

$$K_2 = \frac{(D_o U)^{1/2}}{H^{3/2}} \qquad (17\text{-}10)$$

where D_o = molecular diffusion coefficient for oxygen in water, L^2/T;
   = 18.95×10^{-4} ft^2/d (1.76×10^{-4} m^2/d) at 20°C, to be multiplied by $1.037^{T-20°C}$ for other temperatures, T
   U = current speed, L/T

Estimates of K_2, based on the surface renewal model of reaeration (see Eq. 17-10), are often low by a factor of up to 3, particularly for swift streams [35]. Another approach that can be used to determine K_2 is based on energy dissipation [27]:

$$K_2 = C_e \frac{\Delta h}{t_f} \qquad (17\text{-}11)$$

where Δh = change in surface elevation, L
   t_f = travel time, T
   C_e = escape coefficient = 0.054 ft^{-1} (0.177 m^{-1}) at 20°C, to be adjusted downward for relatively large streams with flows greater than about 250 ft^3/s (7 m^3/s) toward a limiting value of 0.027 ft^{-1} (0.09 m^{-1}); for temperatures other than 20°C multiply by $1.022^{(T-20°C)}$

Sediment Oxygen Demand (SOD). The solids discharged with treated wastewater are partly organic. Upon settling to the bottom, they decompose anaerobically as well as aerobically, depending on conditions. The oxygen consumed in aerobic decomposition represents another dissolved-oxygen sink for the water body. For a control volume in contact with the bottom, the rate of dissolved-oxygen depletion due to sediment oxygen demand is given by

$$r_S = \frac{k_S}{H} \qquad (17\text{-}12)$$

where r_S = rate of oxygen consumption due to SOD per unit time per unit volume of water, M/TL^3
   k_S = sediment oxygen uptake rate, M/L^2T
   H = depth of control volume, L

The major factors affecting k_S are the organic content of the sediments, temperature, dissolved oxygen at the sediment-water interface, makeup of the biological community, and current speed [4].
   Measurement of k_S can be accomplished using a flux chamber to isolate the sediments from the overlying water. Dissolved oxygen in the chamber is measured versus time, from which k_S can be determined. In-situ measurements are preferable, but their reliability is often questionable because of spacial variability and because the bottom shear exerted by the flow is difficult to reproduce in the flux chamber. For preliminary analyses the following order of magnitude values can be used for k_S [25]: 0.2–1.0 g/ft² · d (2–10 g/m² · d) in the vicinity of municipal wastewater outfalls;

0.1–0.2 g/ft^2 · d (1–2 g/m^2 · d) for areas downstream of municipal outfalls and natural estuarine mud; 0.02–0.1 g/ft^2 · d (0.2–1 g/m^2 · d) for sandy bottom; and 0.005–0.01 g/ft^2 · d (0.05–0.1 g/m^2 · d) for mineral soils. The uptake rate k_S can also be estimated based on solids deposition rates, by assuming that the rate of decomposition equals that of deposition [29].

$$k_S = r_o a R_d \qquad (17\text{-}13)$$

where r_o = oxidizable organic content of discharged solids, typically 0.5 to 0.6 for secondary effluent and 0.8 for primary effluent

a = oxygen/sediment stoichiometric ratio = 1.07

R_d = solids deposition rate, M/L^{2}T

Photosynthesis and Respiration. Ambient DO levels can be affected by the growth of algae (phytoplankton, primary productivity) and weeds (macrophytes) feeding on ammonia and nitrate. In this context, these nitrogenous compounds are nutrients. Algae and weeds constitute an oxygen source during daylight hours due to photosynthesis and a continuous oxygen sink due to respiration. For moderate nutrient enrichment levels, photosynthesis and respiration tend to compensate for each other with small overall impact. Higher enrichment levels, however, lead to high productivity (a situation called "eutrophication") with potentially strong effects on DO. Diurnal fluctuations can develop with supersaturated DO levels during daylight hours due to photosynthesis and very low DO levels at night due to respiration. Longer term fluctuations result from photosynthesis/respiration imbalances during high biomass growth and decay periods.

In addition to nitrogen, other nutrients are needed for biomass growth, notably phosphorus and silica. The average molar ratios of nitrogen to phosphorus to carbon in algal protoplasm (Redfield ratios) are approximately N:P:C = 15:1:105 [16]. If one of these nutrients is available in a smaller proportion to the others, it tends to limit growth and any addition of this nutrient will result in a direct increase of biomass. For example, in lakes, phosphorus is typically the limiting nutrient so that addition of phosphorus will spur growth but addition of nitrogen will have minimal effects.

Determining the impact of photosynthesis and respiration on dissolved oxygen requires detailed analyses. One approach is to simulate the fate of the limiting nutrient through an element cycle in which the different forms under which the element can be present are evaluated. For example, an element cycle for nitrogen in the aquatic environment is shown in Fig. 17-2. The transfer of nitrogen between the different boxes and attendant oxygen consumption are simulated [18]. The nitrogen cycle shown in Fig. 17-2 involves fourteen transformations with rates dependent on concentrations of nitrogen within the various boxes. This nitrogen cycle would be applicable to estuaries where nitrogen is the limiting nutrient; for lakes, a phosphorus cycle would likely be required. Clearly, this approach requires the use of computer methods to solve the equations involved. If eutrophication is a problem, existing or suspected, this type of approach will be required; otherwise, it can often be assumed that photosynthesis and respiration are in balance, and their contribution to the oxygen balance may be neglected.

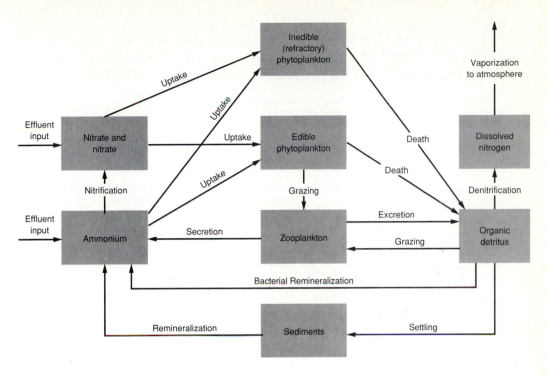

FIGURE 17-2
Nitrogen cycle in the aquatic environment.

Solids Deposition. The suspended solids discharged with treated wastewater ultimately settle to the bottom of the receiving water body. This settling is enhanced by flocculation and hindered by ambient turbulence. In rivers and coastal areas, turbulence is often sufficient to distribute the suspended solids over the entire water depth.

The rate of decay of the suspended-solids concentration due to settling is given by

$$r_D = \frac{w}{H}S \tag{17-14}$$

where r_D = rate of deposition per unit time per unit volume of water, $M/T \cdot L^3$
S = suspended-solids concentration, M/L^3
w = settling velocity of the solids, L/T
H = water depth (if vertically mixed) or depth of control volume, L

Because of the very low solids concentrations obtained after initial effluent mixing and because of ambient turbulence, settling velocities are very low and difficult to measure. Typical settling velocities obtained from settling column tests are given in Table 17-2 [32]. Holographic measurements indicate lower settling velocities with a median of less than 3.33×10^{-6} ft/s (for samples of blended primary/secondary effluent) [34].

TABLE 17-2
Settling velocities of suspended solids in primary and secondary effluents

Settling velocity		Percent of total	
ft/s	cm/s	Primary	Secondary
3.3×10^{-4}	10^{-2}	20	16
3.3×10^{-5}	10^{-3}	30	34
$\leq 3.3 \times 10^{-6}$	$\leq 10^{-4}$	50	50

Bacterial Die-Off. The rate of disappearance of pathogenic bacteria and viruses due to die-off approximately follows first-order kinetics:

$$r_B = -K_B C_B \tag{17-15}$$

where r_B = rate of bacteria die-off per unit time per unit volume of water, count/L^3

C_B = bacteria concentration, count/L^3

K_B = die-off constant, T^{-1}

The die-off constant, K_B, depends on the bacteria or viruses and on the salinity, temperature, and light intensity. For fresh water, decay rates of 0.12 to 26 d^{-1} with a median of 1.0 d^{-1} were measured for coliform in 30 separate in-situ studies [4]. In seawater, bacterial decay is more rapid. The following average expression was derived from about 100 in-situ and laboratory measurements: $K_B = [0.8 + 0.006(\%$ seawater)] $\times 1.07^{(T-20^\circ C)} d^{-1}$ [15]. The dependence on light intensity is more marked in seawater, with an up to 20-fold variation in K_B between daylight and night time [3]. The decay rate is sometimes expressed in terms of T_{90}, the time required for 90 percent of the initial bacteria to die. The relationship between T_{90} and K_B is

$$T_{90} = \frac{-\ln(0.1)}{K_B} = \frac{2.30}{K_B} \tag{17-16}$$

Adsorption. Many chemical constituents tend to attach or sorb to solids. Adsorption was discussed in Sec. 7-2 relative to carbon adsorption as a treatment process. The implication for wastewater discharges is that a substantial fraction of some toxic chemicals is associated with the suspended solids in the effluent. As discussed in Sec. 7-2, the relationship between the equilibrium concentrations of a constituent in the liquid and solid phases is governed by its adsorption isotherm. For environmental applications, equilibrium conditions can often be assumed, and concentrations are often small enough that the linear isotherm, a special case of the Freundlich isotherm, can be used:

$$F = K_d C \tag{17-17}$$

where F = mass of constituent per unit mass of solid, M/M

C = concentration of constituent in liquid, M/L^3

K_d = distribution coefficient, L^3/M

A consequence of Eq. 17-17 is that the ratio of the mass of a constituent in the solid and liquid phases is equal to the distribution coefficient multiplied by the solids concentration ($M_s/M_l = K_dS$). Distribution coefficients, K_d, can range over six orders of magnitude, depending on the chemical constituent. Much information on distribution coefficients is available in the groundwater contamination literature [33], where adsorption plays a paramount role (because of the high solids concentrations in groundwater). Empirical relationships are available to calculate distribution coefficients as a function of the organic content of the soil and other properties of the chemical [14]. Distribution coefficients are high for heavy metals and synthetic organics and particularly high for pesticides, PCBs (polychlorinated biphenyls), and PAHs (polycyclic aromatic hydrocarbons).

Adsorption combined with solids settling results in a removal from the water column of constituents, such as metals, that might not otherwise decay. Environmental criteria usually apply to the total concentration of chemical constituents, defined as the total mass (dissolved plus adsorbed) divided by the water volume. Settling therefore leads to a rate of concentration decrease per unit volume of water, r_A (M/TL^3), given by

$$r_A = -\frac{K_dwS}{H(K_dS + 1)}C \tag{17-18}$$

Volatilization. Some constituents, such as VOCs (volatile organic compounds), are subject to volatilization. The physics of this phenomenon are very similar to surface reaeration, except that the net flux is out of the water surface. Also, for most applications, the equivalent to C_s in Eq. 17-9 is practically zero because the partial pressure of the chemical in the atmosphere is practically zero. The equation giving the rate of decrease of the constituent concentration due to volatilization is

$$r_V = -K_VC \tag{17-19}$$

where r_V = rate of volatilization per unit time per unit volume of water, M/TL^3
K_V = volatilization constant, 1/T
C = concentration of constituent in liquid, M/L^3

Different methods have been proposed to evaluate K_V [26]. For high-volatility compounds with a Henry's law constant larger than 10^{-3} atm-m^3/mol (see Table 6-16), the vaporization constant can be related to the reaeration constant given by Eqs. 17-10 or 17-11:

$$K_V = K_2\frac{d_o}{d_c} \approx K_2\frac{D_o}{D_c} \tag{17-20}$$

where d_o = molecular diameter of oxygen, L
d_c = molecular diameter of compound, L
D_o = diffusion coefficient of oxygen in water (see Eq. 17-10), L^2/T
D_c = diffusion coefficient of compound in water, L^2/T

Conservation of Mass Equation

In general, several of the transformation processes discussed above may affect the concentration of a constituent in addition to transport. When these transformation processes are independent, their effects are additive, and the corresponding transformation rates can be simply summed in the conservation of mass equation. An additional factor not discussed previously is external inputs, which include wastewater discharges and nonpoint sources. These external inputs can be accounted for by an appropriate source term. Combining transport, transformation, and sources into the framework provided by Eq. 17-1 yields the general form of the conservation of mass equation:

$$\frac{\partial C}{\partial t} = -U\frac{\partial C}{\partial x} - V\frac{\partial C}{\partial y} - W\frac{\partial C}{\partial x} + \frac{\partial}{\partial x}\left[E_x\frac{\partial C}{\partial x}\right]$$

$$+ \frac{\partial}{\partial y}\left[E_y\frac{\partial C}{\partial y}\right] + \frac{\partial}{\partial z}\left[E_z\frac{\partial C}{\partial z}\right] + \Sigma r_i + \Sigma I_j \quad (17\text{-}21)$$

where I = external input rate: mass injected per unit time per unit volume of water, M/TL^3

i = transformation process index

j = input identification index

As an example, when the constituent of concern is dissolved oxygen, the transformation term is

$$\Sigma r_i = -r_O + r_R - r_S + r_P - r_{Rp} \quad (17\text{-}22)$$

where r_P = rate of oxygen production by photosynthesis per unit time per unit volume of water, M/TL^3

r_{Rp} = rate of oxygen consumption due to respiration per unit time per unit volume of water, M/TL^3

Many constituents such as BOD, suspended solids, and bacteria are subject to a single transformation process, which follows a first-order decay process. In this case, $\Sigma r = -KC$. And thus, a single form of the conservation of mass equation can be used to analyze several different constituents by appropriately adjusting the decay constant K.

The conservation of mass equation forms the basis of practically all water quality modeling. It is a second-order, partial differential equation, which is difficult to solve in the general case. Numerous simplifications are, however, possible, for which exact solutions exist; a number of these are discussed in the following sections. For more complex situations, numerical methods and computers must be used.

Two important comments should be made. First, the conservation of mass equation is linear so that sums and differences of solutions (C as a function of x, y, z, and t) are also solutions. For example, constituent concentrations corresponding to two separate discharges operating simultaneously are the sum of the concentrations corresponding to each discharge operating individually. Also, if the concentration of a constituent in the effluent is reduced or increased by a certain factor, so are the concentrations at every point in the receiving water. Note that, if there is a background concentration, these propositions are true for the excess concentration over the background. The second comment on the conservation of mass equation is that its solution requires knowledge of the velocity field throughout the affected area as a function of time. Determination of the velocity field can be accomplished through measurements or modeling.

17-3 DISPOSAL INTO LAKES AND RESERVOIRS

In many locations where streams are not available, it may be necessary to discharge treated wastewater into lakes or reservoirs. Other frequent inputs to lakes and reservoirs are septic system leachates and stormwater runoff, which may contain BOD, nutrients, and other pollutants.

Fully Mixed Analysis

Small and shallow lakes and reservoirs tend to remain well-mixed due to wind-induced turbulence. Deeper lakes generally stratify during the summer, but, for many of these, overturning occurs twice a year, mixing upper and lower strata. From a long-term point of view, it may thus be justifiable to use fully mixed analyses for stratifying lakes also. In the fully mixed approach, constituent concentrations are assumed to be approximately uniform in the lake or reservoir. In this case, Eq. 17-21 can be integrated over the lake volume to yield the following expression for constituents subject to a single first-order decay process:

$$\frac{dC}{dt} = -KC + \Sigma I_j + \frac{1}{V}\frac{dV}{dt}C \tag{17-23}$$

The last term in the above equation accounts for changes of reservoir volume due to imbalances of inflows, outflows, rainfall input, and evaporation. The external inputs to be considered are precipitation, tributary streams, surface runoff, groundwater discharge, and waste flows with respective flowrates Q_p, Q_s, Q_r, Q_g, and Q_w and concentrations C_p, C_s, C_r, C_g, and C_w for any constituents of interest (see Fig. 17-3a). Outflows from the lake result in a negative external input of flowrate Q_o and concentration C, equal to the lake concentration. Evaporation from the water surface generally does not result in any constituent loss. The net external input term is thus

$$\Sigma I_j = \frac{Q_p C_p + Q_r C_r + Q_g C_g + Q_w C_w - Q_o C}{V}$$

$$= \frac{M' - Q_o C}{V} \tag{17-24}$$

where $\quad V$ = lake volume, L^3

$\quad M'$ = constituent loading to the lake, M/T

$\quad = Q_p C_p + Q_s C_s + Q_r C_r + Q_g C_g + Q_w C_w$

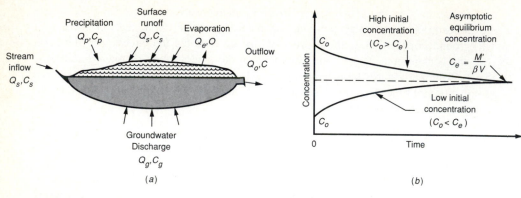

FIGURE 17-3
Fully mixed model for lakes and reservoirs.

Substituting into Eq. 17-23 and integrating gives

$$C = \frac{M'}{V\beta}\left[1 - \exp\left(-\int_0^t \beta\,d\tau\right)\right] + C_o \exp\left(-\int_0^t \beta\,d\tau\right) \qquad (17\text{-}25)$$

where $\beta = K + (Q_o + dV/dt)/V, \mathrm{T}^{-1}$
$\quad C_o$ = initial concentration (at time $t = 0$), $\mathrm{M/L}^3$

When the lake volume remains constant, then $dV/dt = 0$ and β is constant so that the integrals in the exponents simply become

$$\int_0^t \beta\,d\tau = \beta t \text{ (see also Appendix G)}$$

Equation 17-25 is plotted in Fig. 17-3b for two different values of the initial concentration. The concentration converges asymptotically towards an equilibrium value, $C_e = M'/\beta V$, obtained by letting $t \rightarrow \infty$ in Eq. 17-25.

Example 17-1   Lake phosphorus-loading analysis. A phosphorus balance analysis is to be prepared for a lake showing signs of eutrophication. The outflow from the lake is controlled so that the lake level is approximately constant on an average basis. The annual rainfall is 20 in (0.5 m) and the evaporation from the lake is 28 in (0.7 m). Runoff to the lake is equivalent to 5 in (0.125 m) of water per year over the watershed area of 890 mi² (2300 km²). The phosphorus content of rainwater is 0.01 mg/L. A total of 135 ft³/s (3.8 m³/s) of water is withdrawn from the lake for water supply, and 70 percent of it is returned to the lake with an added amount of phosphorus of 2.2 mg/L. Phosphorus loss to sediments is equivalent to a first-order process with a constant of 0.003/d. The phosphorus concentration in the lake was measured at 0.09 mg/L.

If the surface area and average depth of the lake are 50 mi² (130 km²) and 50 ft (15 m) respectively, estimate the phosphorus loading due to runoff. By how much should the wastewater loading be reduced to yield a phosphorus concentration of 0.03 mg/L in the lake?

Solution

1. Determine the average phosphorus concentration in the runoff.

   (a) The average runoff to the lake is

   $$Q_r = (890 \text{ mi}^2) \times (5280 \text{ ft/mi})^2 \times \left[\frac{(5 \text{ in/yr})/(12 \text{ in/ft})}{(365 \text{ d/yr})(86,400 \text{ s/d})} \right] = 328 \text{ ft}^3/\text{s}$$

   (b) The direct precipitation inflow to the lake is

   $$Q_p = 50 \times 5280^2 \times \frac{20/12}{365 \times 86,400} = 74 \text{ ft}^3/\text{s}$$

   (c) The rate of evaporation from the lake is

   $$Q_e = 50 \times 5280^2 \times \frac{28/12}{365 \times 86,400} = 103 \text{ ft}^3/\text{s}$$

   (d) The net inflow to the lake is

   $$Q_o = Q_r + Q_p - Q_e - 0.30 \ Q_{ws} = 328 + 74 - 103 - 41 = 258 \text{ ft}^3/\text{s}$$

   (e) The lake time constant for phosphorus is

   $$\beta = K + \frac{Q_o}{V} = \frac{0.003}{86,400} + \frac{258}{50 \times 5280^2 \times 50} = 3.84 \times 10^{-8} \text{ s}^{-1}$$

   (f) The present phosphorus loading to the lake is thus

   $$M' = C_c \beta V = (0.09 \text{ mg/L})(3.84 \times 10^{-8} \text{s}^{-1})(50 \text{ mi}^2)(5280 \text{ ft/mi})^2(50 \text{ ft})(28.32 \text{ L/ft}^3)$$

   $$= 6821 \text{ mg/s}$$

   (g) The phosphorus loading from runoff is then

   $$Q_r C_r = M' - Q_p C_p - Q_w C_w$$

   $$= 6821 - 74 \times 28.32 \times 0.01 - 0.7 \times 135 \times 28.32 \times 2.2 = 912 \text{ mg/s}$$

   (h) The average phosphorus concentration in runoff water is

   $$C_r = \frac{(912 \text{ mg/s})}{(328 \text{ ft}^3/\text{s})(28.32 \text{ L/ft}^3)} = 0.098 \text{ mg/L}$$

2. Determine by how much the phosphorus loading in the wastewater discharge must be reduced to decrease the phosphorus concentration in the lake to 0.03 mg/L.

   (a) To reduce the lake concentration to 0.03 mg/L, the phosphorus loading must be reduced to

   $$M' = 0.03 \times 3.84 \times 10^{-8} \times 50 \times 5280^2 \times 50 \times 28.32 = 2274 \text{ mg/s}$$

   (b) The concentration of phosphorus in the wastewater returned to the lake must be reduced to

   $$C = 2.2 - \frac{6821 - 2274}{0.7 \times 135 \times 28.32} = 0.50 \text{ mg/L}$$

   Comment. The reduction required (2.2 to 0.5 mg/L) is significant, indicating that lakes are very sensitive to wastewater discharges.

Stratification

Almost all lakes and reservoirs with a depth of 15 ft (5 m) or more stratify during a substantial part of the year. The exception is run-of-the-river reservoirs with a residence time of a month or less [10]. Stratification develops during the spring due to surface heating by solar and atmospheric radiation. Because the density of water decreases with temperature, a hydrodynamically stable situation develops with lighter fluid overlying heavier fluid. If, for example, a parcel of fluid is moved downward into denser fluid, a buoyant upward restoring force results with the net effect of resisting vertical mixing.

Thus, a vertical thermal structure develops with a well-mixed upper layer, the epilimnion, above a region of rapid temperature decrease, the thermocline, which is above a layer of cooler, denser water, the hypolimnion. The typical development of stratification in a lake is shown in Fig. 17-4. In late spring, the epilimnion is thin and not much warmer than the rest of the lake. As summer progresses, the epilimnion thickens and increases in temperature from the combination of surface heating and wind mixing. During the fall, surface cooling and wind mixing result in a decrease of the epilimnion temperature but a continued increase of its thickness. When the epilimnion reaches 4°C, the point of maximum water density, the fall overturn occurs, mixing the lake over its full depth. In colder climates, a weak reverse stratification can develop during the winter, with surface temperatures below 4°C and again lighter than deeper waters. In this case, a spring overturn also occurs.

Other phenomena associated with stratification are density currents and selective withdrawal. Density currents are associated with stream inflows, which tend to flow along the bottom of the lake down to the level of matching density, where they spread horizontally. A consequence is the rapid distribution of inflows over the entire lake area at a particular depth interval. Selective withdrawal occurs at outlets that tend to withdraw from the layer of fluid at the same level as the outlet. Selective withdrawal is of importance for downstream releases from reservoirs. For example, water withdrawn from deep outlets will tend to be cold and may be depleted of oxygen, with possible downstream environmental impacts.

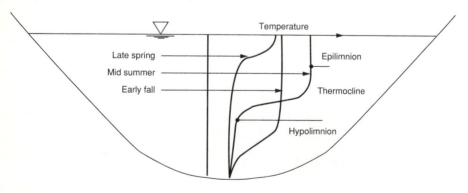

FIGURE 17-4
Stratification cycle in a deep lake.

While rapid changes of temperature occur over the depth of stratified lakes and reservoirs, surprisingly high horizontal uniformity often exists even though horizontal distances are usually many times the depth. This horizontal uniformity is in part due to density currents, which are driven by and eventually eliminate any horizontal non-uniformity. A consequence is that one-dimensional modeling is often appropriate. In this approach, only vertical variations of temperature are resolved. The conservation of heat equation, which is similar to the conservation of mass equation, is applied to horizontal layers of the lake. Bookkeeping of heat and volume fluxes is carried out to determine changes of temperature for each layer. The calculations are driven by surface heat fluxes calculated from meteorological conditions, usually with a daily time step. These calculations can only be performed efficiently using a computer [10].

Eutrophication

A water quality concern with lakes and reservoirs is eutrophication, a natural aging process in which the water becomes organically enriched, leading to increasing domination by aquatic weeds, transformation to marsh land, and eventually to dry land. Eutrophication can be accelerated by human input of nutrients. Die-off and settling of plant growth results in sediment oxygen demand, which tend to decrease dissolved-oxygen levels. The effects of eutrophication, which may be detrimental to aquatic life, are compounded by large day-night excursions in dissolved oxygen due to photosynthesis and respiration.

The process of eutrophication and its relationship to nutrient inputs is complex. In lakes and reservoirs, phosphorus is typically the limiting nutrient, although the presence of nitrogen is also important. A simple criterion, which can be used in conjunction with the fully mixed analysis described previously, is that algal blooms will tend to occur if the concentration of inorganic nitrogen and phosphorus exceed respective values of 0.3 mg/L and 0.01 mg/L [21]. In reservoirs that stratify, early signs of eutrophication are low dissolved-oxygen levels in the hypolimnion, which does not receive any direct reaeration. Dissolved-oxygen predictions for stratifying lakes and reservoirs can be accomplished using one- or two-dimensional computer models simulating nitrogen and phosphorus cycles as well as temperature [28].

17-4 DISPOSAL INTO RIVERS AND ESTUARIES

The disposal of wastewater in rivers and estuaries is considered in this section.

One-Dimensional Modeling Approach

Rivers and estuaries are generally many times longer than they are wide or deep. As a result, inputs from wastewater treatment plants or other sources rapidly mix over the cross section, and a one-dimensional approach is often justified. In the one-dimensional approach, only longitudinal variations of constituent concentrations are resolved in the form of cross-section-averaged values. The general mass conservation equation is averaged over the cross section of the stream giving, for constituents subject to a single first-order decay process,

$$\frac{\partial C}{\partial t} = -U\frac{\partial C}{\partial x} + \frac{\partial}{\partial x}\left[(E_x + E_L)\frac{\partial C}{\partial x}\right] - KC + \Sigma I \tag{17-26}$$

where x = longitudinal distance along river or estuary, L
E_L = longitudinal dispersion coefficient, L^2/T

Equation 17-26 is almost identical to Eq. 17-21 without the terms containing y and z derivatives, except for the appearance of dispersion, which is distinct and separate from turbulent diffusion. The dispersion term arises during the averaging process (which is somewhat involved mathematically) due to the correlation of cross-sectional velocity and concentration variations [8,24]. Dispersion in natural streams is predominantly due to lateral velocity variations, and the following formula can be used to estimate dispersion coefficient [8]:

$$E_L = 0.011\frac{U^2 B^2}{H u_*} \tag{17-27}$$

where E_L = longitudinal dispersion coefficient, L^2/T
U = cross-section-averaged velocity, L/T
B = stream width, L
H = stream depth, L
u_* = shear velocity, L/T = $\sqrt{gHs}$
g = acceleration due to gravity, L/T^2
s = stream slope, L/L

Equation 17-27 remains approximate because it does not account for dead zones in which matter can get trapped, thereby increasing the effective dispersion coefficient. Bends can increase or decrease dispersion depending on their configuration; in particular, successive bends can increase dispersion if their separation is small. In estuaries, tidal flow reversals as well as secondary currents driven by salinity gradients tend to increase dispersion [11]. Dispersion is typically much larger than turbulent diffusion so that E_x can be neglected, compared to E_L in Eq. 17-26.

Instantaneous Source. Instantaneous release of a constituent at a point in the stream may occur as a result of an accident. It can also be used as a means of determining dispersion coefficients, with a deliberate release of a tracer constituent such as dye. The solution of Eq. 17-26 for an instantaneous release at $x = 0$ is

$$C = \frac{Me^{-Kt}}{A\sqrt{4\pi E_L t}}e^{-[(x-Ut)^2/4E_L t]} \tag{17-28}$$

where M = mass released, M
A = stream cross-sectional area, L^2

Corresponding longitudinal concentration distributions at different times as a function of distance after the release are shown in Fig. 17-5. Each instantaneous

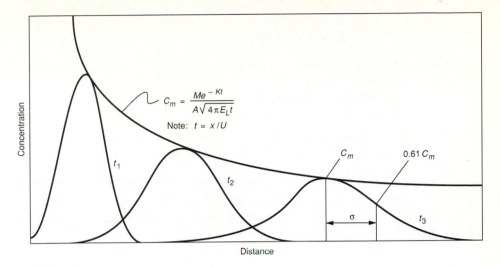

FIGURE 17-5
Instantaneous source in a river.

concentration distribution has the shape of a Gaussian curve, the general form of which is

$$C = C_m e^{-x^2/2\sigma^2} \qquad (17\text{-}29)$$

where C_m = maximum concentration value, obtained at $x = 0$.
   σ = standard deviation = half width of curve at point where $C = 0.61C_m$

The similarity between Eqs. 17-28 and 17-29 is evident. The maximum concentration is

$$C_m = \frac{Me^{-Kt}}{A\sqrt{4\pi E_L t}} \qquad (17\text{-}30)$$

The maximum concentration decreases with time due to decay (exponential term in the numerator) and due to dispersion (square root term in the denominator). The center of the patch is located at $x = Ut$, thus moving downstream at the speed of the flow. The width of the patch, measured by its standard deviation, $\sigma = \sqrt{2E_L t}$, increases with time.

In a field dye study, the dispersion coefficient can be determined from measured maximum concentrations and Eq. 17-30, with $K = 0$. Alternatively, E_L can be determined by matching the standard deviation of measured concentration profiles with the expression given above. In tidal estuaries, the measurements should preferably correspond to the same time during the tide cycle, although Eq. 17-28 is valid for time varying currents also.

Continuous Discharge. The solution of Eq. 17-26 for a continuous discharge at $x = 0$ is

$$C = \frac{M'}{A \sqrt{U^2 + 4KE_L}} e^{(xU/2E_L)(1\pm\sqrt{1+4KE_L/U^2})} \tag{17-31}$$

where M' = discharge rate, M/T = $Q_D C_D$
Q_D = discharge flowrate, L^3/T
C_D = discharge concentration, M/L^3
$\pm$ = + for $x < 0$ and − for $x > 0$

Equation 17-31 is plotted in Fig. 17-6. Note that for a conservative substance $(K = 0)$ the concentration is uniform and equal to M'/AU downstream of the discharge point. The upstream intrusion is not greatly affected by the decay coefficient. In many cases, the value of the term $4KE_L/U^2$ is small compared to 1. For example, the term $4KE_L/U^2$ equals 0.0028 for the following typical values: $U = 1$ ft/s, $K = 0.30$ d^{-1} = 3.5×10^{-6} s^{-1}, and $E_L = 200$ ft^2/s. In this case, concentrations downstream of the source are very closely given by

$$C = \frac{M'}{AU} e^{-Kx/U}; \quad (x > 0) \tag{17-32}$$

which is independent of the dispersion coefficient. Thus, it is generally true that dispersion can be neglected for continuous discharges in rivers.

Dissolved-Oxygen Sag Analysis. For a continuous discharge of wastewater into a river, Eq. 17-32 is applicable to the BOD, which undergoes a first-order decay process:

$$L = L_o e^{-Kx/U} \tag{17-33}$$

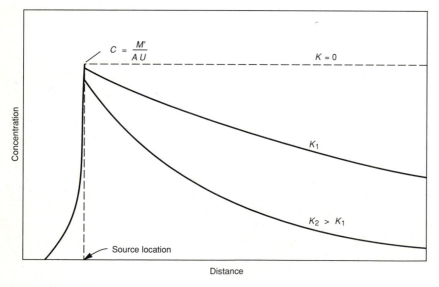

FIGURE 17-6
Continuous source in a river.

where $L_o = M'/AU$

This approach is valid for both carbonaceous BOD and for nitrogenous BOD, provided that x is referenced to the point where nitrogenous BOD begins to be exerted. Thus, nitrogenous BOD can be considered as a fictitious source of BOD at a point downstream of the real source, separated by the travel time, $t = x/U$, equal to the nitrogenous delay time.

For dissolved oxygen, the one-dimensional mass conservation equation is easily deduced from Eq. 17-26. The time derivative term is removed because steady-state concentrations are sought, and the dispersion term is also removed, as it was shown above to have negligible effects for a continuous discharge. The appropriate source and sink terms are added from Eq. 17-22 giving

$$O = -U\frac{dC}{dx} - KL + K_2(C_s - C) + \left(r_p - r_{Rp} - \frac{k_s}{H} + \Sigma I\right) \tag{17-34}$$

The solution of this equation for a continuous discharge at $x = 0$ is

$$D = \underbrace{\frac{KL_0}{K_2 - K}(e^{-Kt} - e^{-K_2 t})}_{\text{BOD exertion}} + \underbrace{D_0 e^{-K_2 t}}_{\text{Initial deficit}} - \underbrace{\frac{r_p - r_{Rp} - k_s/H}{K_2}(1 - e^{-K_2 t})}_{\substack{\text{Photosynthesis, respiration, and} \\ \text{sediment oxygen demand}}} \tag{17-35}$$

where D = dissolved-oxygen deficit = $C_s - C$
  D_o = dissolved-oxygen deficit at $x = 0$
  t = travel time = x/U

Because Eq. 17-35 is general, it can be adapted to different situations. For example, if photosynthesis, respiration, and sediment oxygen demand are not significant, the last term in the equation is equal to zero and can be removed. In this case, Eq. 17-35 gives the classic sag curve [23], shown in Fig. 17-7. Downstream of the discharge point, BOD exertion results in a decrease of dissolved oxygen. Concurrently, dissolved oxygen is replenished through surface reaeration at a rate proportional to the DO deficit (see Eq. 17-9). At a certain distance from the discharge point, the input from reaeration equals the BOD consumption and the DO deficit reaches a maximum. Downstream of this point, input exceeds consumption and the deficit decreases. The point of maximum dissolved-oxygen deficit is obtained by differentiating Eq. 17-35 with respect to travel time and setting the derivative to zero which yields

$$t_{max} = \frac{1}{K_2 - K}\ln\left[\frac{K_2}{K}\left(1 - \frac{D_o(K_2 - K)}{KL_o}\right)\right] \tag{17-36}$$

and

$$D_{max} = \frac{K}{K_2}L_o e^{-Kt_{max}} \tag{17-37}$$

Equation 17-35 is valid for a stretch of river without sources or tributaries. It can, however, be applied sequentially for stretches between sources and tributaries by adjusting L_o and D_o for each stretch. For example, conditions downstream of a

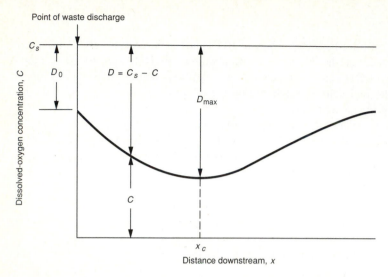

Point of waste discharge

FIGURE 17-7
Characteristic oxygen sag curve obtained using the Streeter-Phelps equation.

discharge are given below as a function of the conditions upstream and the discharge's characteristics:

	River upstream of discharge	Discharge	River downstream of discharge
Flowrate	Q_x	Q_D	$Q = Q_x + Q_D$
BOD	L_x	L_D	$L_0 = (Q_x L_x + Q_D L_D)/Q$
DO deficit	D_x	D_D	$D_0 = (Q_x D_x + Q_D D_D)/Q$

Equation 17-35 should be applied with care, preferably after calibration of the various coefficients using measured data. When significant photosynthesis-respiration effects are expected, a model accounting for those in a more thorough manner may be justified. Several computer models that simulate nutrients cycling have been developed and continue to be improved. Examples of models for rivers are QUAL2E [6] and RIV1 [2].

Example 17-2 Oxygen sag analysis in a river. A city discharges 30 Mgal/d of sewage into a stream whose minimum rate of flow is 300 ft³/s. The velocity of the stream is about 2 mi/h. The temperature of the wastewater is 20°C, while that of the stream is 15°C. The 20°C BOD₅ of the wastewater is 200 mg/L, while that of the stream is 1.0 mg/L. The wastewater contains no dissolved oxygen, but the stream is 90 percent saturated upstream of the discharge. At 20°C, K is estimated to be 0.3 d⁻¹ while K_2 is 0.7 d⁻¹. Determine the critical oxygen deficit and its location. Also estimate the 20°C BOD₅ of a sample taken at the critical point. Use temperature coefficients of 1.135 for K and 1.024 for K_2. Also plot the dissolved-oxygen sag curve.

K = deox coeff
K_2 = reox coeff

Solution

1. Determine the dissolved oxygen in the stream before discharge.

   Saturation concentration 15°C (see Appendix E) = 10.07 mg/L

   Dissolved oxygen in stream = 0.9(10.07) = 9.06 mg/L

2. Determine the temperature, dissolved oxygen, and BOD of the mixture.

$$\text{Temperature of mixture} = \frac{(30 \text{ Mgal/d})[1.55 \text{ ft}^3/\text{s (Mgal/d)}](20°C) + (300 \text{ ft}^3/\text{s})(15°C)}{(30 \text{ Mgal})[1.55 \text{ ft}^3/\text{s} \cdot (\text{Mgal/d})] + 300 \text{ ft}^3/\text{s}}$$

$$= 15.7°C$$

$$\text{Dissolved oxygen of mixture} = \frac{30(1.55)(0) + 300(9.06)}{30(1.55) + 300} = 7.8 \text{ mg/L}$$

$$\text{BOD}_5 \text{ of mixture} = \frac{30(1.55)(200) + 300(1)}{30(1.55) + 300} = 27.7 \text{ mg/L}$$

$$\text{BOD}_L \text{ of mixture} = \frac{27.7}{1 - e^{-0.3(5)}} = 35.6 \text{ mg/L}$$

3. Correct the rate constants to 15.7°C.

$$K = 0.3(1.135)^{15.7-20} = 0.174 \text{ d}^{-1}$$

$$K_2 = 0.7(1.024)^{15.7-20} = 0.63 \text{ d}^{-1}$$

4. Determine t_c and x_c. Saturation concentration, 15.7°C = 9.9 mg/L, D_0 = 9.9 − 7.8 = 2.1 mg/L.

$$t_c = \frac{1}{K_2 - K} \ln \frac{K_2}{K}\left[1 - \frac{D_o(K_2 - K)}{K\text{BOD}_L}\right]$$

$$= \frac{1}{0.63 - 0.174} \ln \frac{0.63}{0.174}\left[1 - \frac{2.1(0.63 - 0.174)}{0.174(35.6)}\right]$$

$$= 2.45 \text{ d}$$

$$x_c = vt_c = (2 \text{ mi/h})(24 \text{ h/d})(2.45 \text{ d}) = 117 \text{ mi}$$

5. Determine D_c.

$$D_c = \frac{0.174}{0.63}(35.6)[e^{-0.174(2.45)}] = 6.4 \text{ mg/L}$$

$$\text{DO}_c = 10.1 - 6.4 = 3.7 \text{ mg/L}$$

6. Determine the BOD$_5$ of a sample taken at x_c.

$$L_t = 35.6e^{-0.174(2.45)} = 23.3 \text{ mg/L}$$

$$20°C \text{ BOD}_5 = 23.3[1 - e^{-0.3(5)}] = 18.1 \text{ mg/L}$$

7. The dissolved-oxygen sag curve is plotted in the figure shown on the following page.

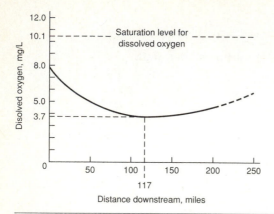

Distance downstream, miles

Estuaries

The term "estuary" generally refers to the portion of rivers near their discharge into the sea, where the effects of tides or salt water are observed. Estuaries can be classified geomorphologically as coastal plain estuaries, such as the Potomac downstream of Washington D.C., or the fjords, or large embayments such as San Francisco Bay. From a hydrodynamic point of view, estuaries can be of the salt wedge type, partially mixed, or well-mixed. These terms refer to the salinity structure of the estuary. In salt wedge estuaries, a well defined wedge of sea water extends upstream into the estuary under the flowing fresh water. In partially mixed estuaries, a gradual salinity gradient occurs between the less saline surface waters and more saline bottom waters. In well-mixed estuaries, salinity is practically uniform vertically but increases toward the sea. Whether an estuary falls in one or the other of these categories depends on its geometry, the fresh water flow, and tidal amplitude. The salt wedge structure tends to occur for larger freshwater flows and narrow estuaries.

The one-dimensional approach can be applied to narrow, well-mixed estuaries. As noted above, the longitudinal dispersion is usually larger than in rivers so that dispersion cannot be neglected, even for steady discharges. There is no exact analytical solution of the mass conservation equation for continuous discharges into estuaries. Thus, in the general case, numerical solution of the conservation of mass equation using a computer is needed. As a first approximation, Eqs. 17-31 and 17-35 can be used with U equal to the net advective velocity (i.e., the fresh water flowrate divided by the cross-sectional area). This tidally averaged approach does not resolve the time variations during the tide cycle or the upstream transport during the flood part of the tide cycle.

In partially mixed estuaries, the vertical salinity gradient results in a secondary circulation superimposed upon the cross-section-averaged longitudinal flow. In this estuarine circulation, saltier water moves slowly upstream along the bottom and is gradually entrained into the seaward-flowing fresher water above. To account for this circulation, a two-dimensional (longitudinal-vertical) or two-layer modeling approach must be used. These types of simulations can only be conducted practically on a computer.

Mixing Zones

In the one-dimensional analysis, it was assumed that the wastewater is mixed fully with the stream flow. Near the discharge point, the fully mixed assumption may not be valid, and a mixing zone may exist where constituent concentrations are between the effluent and fully mixed values. This mixing zone is not significant for dissolved-oxygen analyses because BOD exertion is slow compared to the travel time within the mixing zone. Relative to other constituents, such as toxic chemicals, the characteristics of the mixing zone are important. Clearly, rapid mixing of the effluent with the ambient flow is desirable to decrease toxicity.

For small river depths, vertically mixed conditions are reached within a short distance of the discharge point. Mixing zone analyses can be conducted by simulating the discharge as a plane source extending over the full river depth. For pipe discharges, as a first approximation, the width of the source can be assumed to be equal to the river depth, whereas for multiport diffusers a source width equal to the diffuser length is appropriate. The corresponding distribution of constituent concentrations is given by

$$C = \frac{C_1}{2} e^{-Kx/U} \left[\text{erf} \left(\frac{y + b/2}{2} \sqrt{\frac{U}{E_y x}} \right) - \text{erf} \left(\frac{y - b/2}{2} \sqrt{\frac{U}{E_y x}} \right) \right] \qquad (17\text{-}38)$$

where C_1 = effective source concentration $= Q_D C_D / UbH$
H = river depth, L
y = lateral coordinate from center of source, L
b = source width, L
E_y = lateral diffusion coefficient, $L^2/T \approx 0.6\, Hu_*$ [9]; see Eq. 17-27 for definition of u_*
erf = error function

The error function is available in mathematical tables. A convenient approximation is

$$\text{erf}(x) \approx 1 - \frac{1}{(1 + a_1 x + a_2 x^2 + a_3 x^3 + a_4 x^4)^4} \qquad (17\text{-}39)$$

where a_1 = 0.278393
a_2 = 0.230389
a_3 = 0.000972
a_4 = 0.078108

In the concentration distribution given by Eq. 17-38, the limited river width, which constrains the lateral spreading of the plume, is not considered. This limited river width can be accounted for by introducing image sources symmetrically from the real source with respect to the river banks, as illustrated in Fig. 17-8. Image sources of the image sources are in turn needed to account for the other river bank. Thus, in theory, an infinite number of image sources is required. In practice a few images are usually sufficient. The concentration distribution is then equal to the sum of the

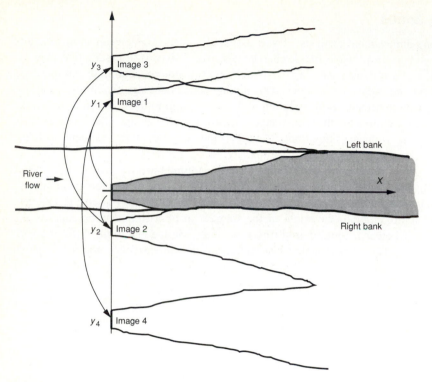

FIGURE 17-8
Mixing zone analysis with method of images in a shallow stream.

concentrations corresponding to the real source, given by Eq. 17-38, and its images. The concentrations due to the image sources are given by Eq. 17-38 with y replaced by $y - y_i$, where y_i is the lateral distance of the image source.

For discharges into deeper rivers, the effluent may not become mixed over the water depth for long distances. In this case, the approaches described below for ocean discharges can be used. For existing discharges, effluent dilution can be determined by field dye studies in which a stable dye is discharged at a known rate into the effluent. Measurements of dye concentration at various points in the receiving water permit calculations of dilution.

The size of mixing zones depends on the effluent characteristics but also on the river flow. Typically, lower effluent dilutions and therefore larger mixing zones are achieved for low river flows. Analysis of flow records are therefore needed to establish the magnitude, duration, and frequency of low-flow periods. Regulations in force in the United States often prescribe the duration and frequency of the low flow that must be used for mixing zone analyses. For example, the flow corresponding to the 7-day average low flow with a recurrence interval of 10 years (noted 7Q10) is often used.

River Outfalls

Many wastewater discharges into rivers and estuaries are through open-ended pipes that achieve minimal initial mixing. In shallow streams, open-ended discharges on the bank sometimes free-fall into the water surface, with the potential for foaming problems. Those can be eliminated by a submerged discharge farther into the stream. For navigable rivers and estuaries, outfall design requires special attention and government permits.

Rapid mixing of wastewater effluent with a river can be achieved by using a multiport diffuser. A diffuser is a structure that is used to discharge the effluent through a series of holes or ports along a pipe extending into the river, preferably perpendicular to the bank. For shallow rivers, vertical mixing of the effluent over the full river depth is achieved rapidly. In this case, the momentum of the discharge may attract into the effluent plume ambient river flow which would otherwise not have

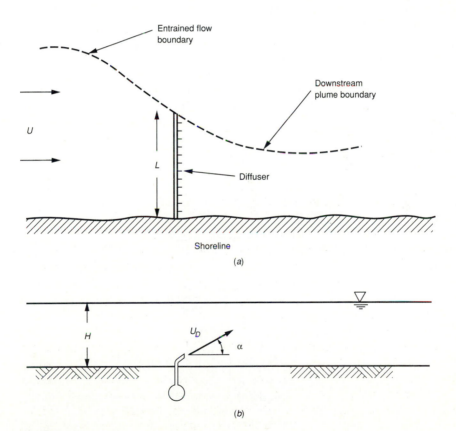

FIGURE 17-9
Typical diffuser in a river: (a) plan view and (b) elevation.

passed over the diffuser, as shown schematically in Fig. 17-9 [1]. The initial dilution, S, achieved within approximately one diffuser length is given by

$$S = \frac{UHL}{2Q_D}\left(1 + \sqrt{1 + \frac{2Q_D U_D \cos\alpha}{U^2 LH}}\right) \qquad (17\text{-}40)$$

where U = river velocity, L/T
    H = river depth, L
    L = diffuser length, L
    U_D = discharge velocity through each port, L/T
    α = orientation of the ports above the horizontal

Equation 17-40 can be applied to shore-attached as well as midstream diffusers. It can be used to determine the length of the diffuser needed to achieve a required dilution. The diffuser length is often the most important parameter relative to cost. High port discharge velocities increase dilution but may cause erosion or navigation problems. In practice, port velocities in excess of 10 ft/s (3 m/s) are rarely used. This guideline can be exceeded in certain cases and during infrequent high-flow events. The port spacing is typically selected to be on the same order as the water depth. A typical river diffuser arrangement is shown in Fig. 17-10. The conduit diameter is decreased along the diffuser to ensure equal flow out of all the ports. The specifics of the diameter reduction should be determined based on manifold hydraulics calculations [9]. A larger port, normally closed, is often provided at the end of the diffuser to allow cleaning.

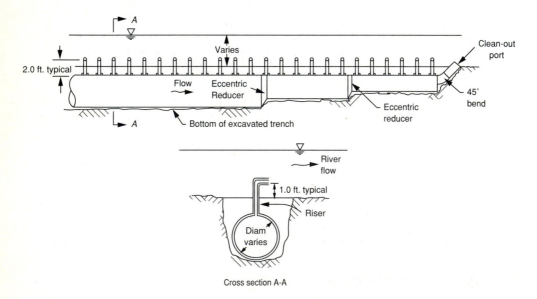

FIGURE 17-10
Typical river diffuser outfall.

Example 17-3 River diffuser design. Determine the length and number of discharge ports for a multiport diffuser that will provide a nearfield dilution of 10 when discharging a maximum flow of 53 ft³/s (1.5 m³/s) in a river. Under low-flow conditions, the river water depth is 4 ft (1.2 m) and the current speed is 2.0 ft/s (0.6 m/s).

Solution

1. Determine the required length of the diffuser.

   (a) For the shallow water conditions prevalent under low river flow conditions, the maximum discharge velocity, U_D, should be lower than the value of 10 ft/s, recommended in the text to reduce the risk of bottom erosion and hazards to boaters. A value of 7 ft/s is selected. Because of the shallow depth, the ports will discharge horizontally ($\alpha = 0$) in the same direction as the river flow.

   (b) The diffuser length is obtained by solving Eq. 17-40 by trial. Substituting the given values and solving by trial, a value of 58 ft is obtained for L:

   $$S = \frac{(2.0 \text{ ft/s})(4 \text{ ft})(58 \text{ ft})}{2(53 \text{ ft}^3/\text{s})} \left[1 + \sqrt{\frac{2(53 \text{ ft}^3/\text{s})(7 \text{ ft/s}) \cos(0.0)}{(2.0 \text{ ft/s})^2(58 \text{ ft})(4 \text{ ft})}} \right] = 10.3$$

2. Determine the required number of ports.

   (a) The port spacing should be on the same order as the water depth. Therefore, the number of ports should be approximately

   $$N = \frac{L}{H} + 1 = \frac{58 \text{ ft}}{4 \text{ ft}} + 1 = 15.5$$

   Selecting 15 ports for N determines their diameter:

   $$D_o = \sqrt{\frac{4Q_D}{\pi N U_D}} = \sqrt{\frac{4 \times 53 \text{ ft}^3/s}{\pi \times 15 \times 7 \text{ ft/s}}} = 0.80 \text{ ft} = 9.6 \text{ in}$$

   (b) Most probably, the port diameter obtained above is nonstandard and needs to be adjusted. For example, if the closest standard pipe diameter is 10 in, the revised number of ports should be

   $$N = \frac{4Q_D}{\pi D_D^2 U_D} = \frac{4 \times 53 \text{ ft}^3/s}{\pi \times (0.833 \text{ ft})^2 \times 7 \text{ ft/s}} = 13.8$$

   The number of ports is therefore revised to 13, which is a little less than the value obtained above so that the actual discharge velocity will be slightly more than 7 ft/s and the dilution slightly above 10.

17-5 OCEAN DISPOSAL

Oceans and large lakes, such as the Great Lakes, provide extensive assimilation capacity and are used for wastewater disposal by many communities. The wastewater is typically carried to an offshore discharge point by a pipe laid on or buried in the ocean floor, or by a tunnel. The discharge can be through a single-port or multiport outfall structure. For discharges in the ocean, the wastewater is buoyant relative to the ambient water. The density of seawater is frequently expressed in σ_t units, which is equal to the density of the water in g/L minus 1000. For example, a seawater density of 1002.4 g/L is equivalent to 2.4 σ_t. Seawater density depends on salinity and temperature and can be obtained from Fig. 17-11 [9]. The density

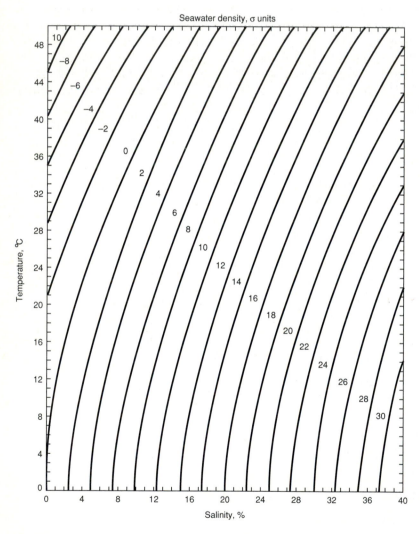

FIGURE 17-11
Density of seawater as a function of temperature and salinity (from Ref. 9).

of wastewater depends on its temperature and to a smaller extent on the suspended-solids concentration.

   The configuration of a wastewater plume in the ocean is shown schematically in Fig. 17-12. In a first region, called the "initial mixing region," or "discharge nearfield," the effluent forms a buoyant plume, rapidly rising in the water column. This plume entrains large amounts of ambient water, thereby diluting the effluent. When the water column is stratified, the ambient water that is first entrained is deep, denser water, which reduces the plume buoyancy as it rises into less dense ambient water. At some point in this ascent, the plume density may become equal to that of the ambient water, and further rise is impeded. The plume reaches an intermediate equilibrium height of rise. When the water column is weakly stratified or not stratified, as in the winter, the plume rises up to the water surface. Beyond the initial mixing region, in a region called the "farfield," the wastewater field is carried away by ambient currents and further diluted by diffusion. The dilution mechanisms acting in the nearfield and farfield are extremely different, and for that reason these two regions are treated separately.

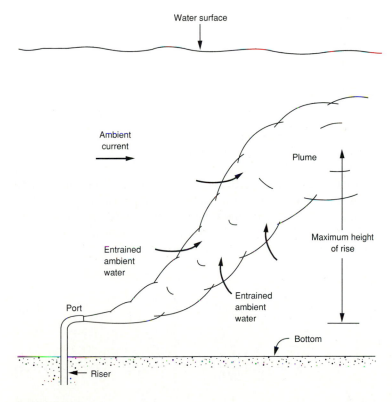

FIGURE 17-12
Wastewater discharge plume in the ocean.

Initial Mixing

Effluent dilution in the initial mixing region has been the subject of innumerable studies in the laboratory, in the field, and with mathematical models. Because of the combined effects of the effluent buoyancy, ambient stratification, and current, the prediction of initial dilution can be involved. Results for simple configurations are presented below. For more complex cases, use of computer models is required [17].

Vertical Single-Port Discharges. For vertical single-port discharges, dilutions at the end of the initial mixing region can be calculated with the following equation based on dimensional analysis and laboratory measurements [17]:

1. Stagnant ambient

$$S_a = 0.13 \ g_D'^{1/3} Q_D^{-2/3} H^{5/3} \tag{17-41}$$

2. Flowing ambient

$$S_a = 0.29 (U/Q_D) H^2 \tag{17-42}$$

where S_a = average plume dilution
g_D' = discharge buoyancy = $g(\Delta \rho_D / \rho)$, L/T^2
$\Delta \rho_D$ = discharge density difference, M/L^3
ρ = ambient water density, M/L^3
Q_D = discharge flowrate, L^3/T
U = ambient current speed, L/T^3
H = water depth, L

These results are for the average plume dilution, which is approximately 1.8 times the minimum dilution, S_m, at the plume centerline. When the plume hits the water surface, it spreads horizontally and additional dilution, by a factor of up to five, occurs [36]. For horizontally discharged effluent, dilutions on the order of 20 to 50 percent greater are achieved at the water surface than for vertical discharges [13]. When the receiving water is stratified linearly (constant vertical density gradient, $d\rho/dz$), the above formula is still applicable, with the water depth replaced by the equilibrium height of rise of the plume, z_e, given by the following expressions:

1. Stagnant ambient

$$z_e = 2.91 \ g_D'^{1/4} Q_D^{1/4} N^{-3/4} \tag{17-43}$$

2. Flowing ambient

$$z_e = 1.85 \left[\frac{Q_D g_D'}{N^2 U} \right]^{1/3} \tag{17-44}$$

where N = $\left(-\frac{g}{\rho} \frac{d\rho}{dz} \right)^{1/2}$ = buoyancy frequency, T^{-1}

Multiport Discharges. Multiport diffuser outfalls for wastewater often have ports that discharge perpendicular to the diffuser axis in both directions, as shown in Fig. 17-13 [20]. Minimum (centerline) dilutions at the end of the initial mixing region as well as the height of rise are given in Fig. 17-14 for the case of a linearly stratified ambient [20]. For this type of plume, the average dilution, S_a, is approximately equal to the minimum dilution, S_m, multiplied by 2.0. As shown in Fig. 17-14, for low current speeds ($F < 0.1$), the dilution is independent of the current speed and direction. For higher current speeds, the dilution increases and is larger when the diffuser is oriented perpendicular to the current.

The characteristics of a diffuser providing a required amount of dilution can be determined using Fig. 17-14. It may also be desirable to obtain a submerged wastefield, and this can also be ascertained using Fig. 17-14 (see Example 17-4). When the ambient stratification cannot be schematized as linear, computer models accounting for actual density profiles need to be used [17].

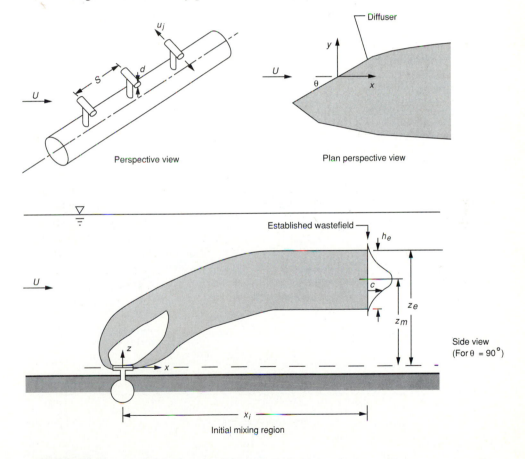

FIGURE 17-13
Ocean diffuser configuration (from Ref. 20).

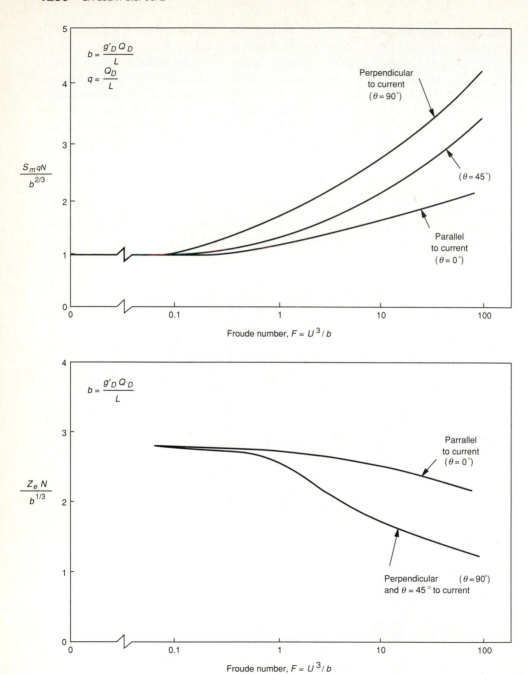

FIGURE 17-14
Minimum initial dilution and equilibrium height of plume rise from ocean diffuser (from Ref. 20).

Example 17-4 Preliminary sizing of an ocean diffuser. Determine the basic dimensions of an ocean multiport diffuser that will create a nearfield dilution of at least 50. The discharge flowrate will vary between 230 and 685 Mgal/d (10 and 30 m³/s), and the current speed varies between zero and 0.66 ft/s. In the winter, the ocean is not stratified, but in the summer stratification develops with a maximum vertical density gradient of approximately 0.028 g/L·ft. The relative discharge density deficiency of the effluent is $\Delta\rho_D/\rho = 0.027$. It is desired that the wastefield remain submerged during the stratified summer time. Also determine the highest nearfield dilution that will be obtained during the stratified summer period.

Solution

1. Determine the minimum nearfield dilution.

   (a) The lowest nearfield dilution is obtained for low current speeds and maximum discharge flowrate. Referring to Fig. 17-14, it can be seen that for low current speeds, specifically for $F < 0.1$, the minimum nearfield dilution, S_m, is such that

   $$\frac{S_m q N}{b^{2/3}} = 1$$

   Substituting the expressions for q and b given in Fig. 17-14 gives

   $$\frac{S_m Q_D^{1/3} N}{g_D'^{1/3} L^{1/3}} = 1 \qquad L = \frac{S_m^3 Q_D N^3}{g_D'^2}$$

   The buoyancy frequency is

   $$N = \sqrt{\frac{-g}{\rho}\frac{d\rho}{dz}} = \sqrt{\frac{32 \text{ ft/s}^2}{1000 \text{ g/L}} 0.0028 \text{ g/L} \cdot \text{ft}} = 0.030 \text{ s}^{-1}$$

   and the discharge buoyancy is

   $$g_D' = g\frac{\Delta\rho_D}{\rho} = 32.2 \text{ ft/s}^2 \times 0.027 = 0.869 \text{ ft/s}^2$$

   (b) The required length of the diffuser is

   $$L = \frac{(50)^3(685 \text{ Mgal/d})[1.55 \text{ ft}^3/\text{s} \cdot (\text{Mgal/d})](0.030 \text{ /s})^3}{(0.869 \text{ ft/s}^2)^2}$$

   $$= 4.736 \text{ ft}$$

2. Determine whether the disposal field will remain submerged.

   (a) The equilibrium height of rise is obtained from Fig. 17-14 (bottom graph). For low current speeds ($F < 0.1$),

   $$Z_e = 2.8\frac{b^{1/3}}{N} = \frac{(g_D'Q_D/L)^{1/3}}{N} = \frac{[(0.869 \times 685 \times 1.55)/4736]^{1/3}}{0.030} = 54.1 \text{ ft}$$

   Note that units are given above.

   (b) To ensure that the wastefield will remain submerged, the diffuser depth should exceed 54.1 ft by a sufficient margin of safety.

3. Determine the maximum nearfield dilution.

   (a) The highest dilution will be obtained for the lowest discharge flowrate and highest current speed. In this case, the Froude number is

$$F = \frac{U^3 L}{g_b' Q_D} = \frac{(0.66 \text{ ft/s})^3 (4736 \text{ ft})}{(0.869 \text{ ft/s}^2)(230 \text{ Mgal/d})[1.55 \text{ft}^3/\text{s} \cdot (\text{Mgal/d})]} = 4.40$$

(b) To achieve the highest nearfield dilution, the diffuser should be oriented perpendicular to the dominant current direction. Assuming that such an orientation is possible and for the above value of the Froude number, Fig. 17-14 gives a dimensionless dilution $S_m qN/b^{2/3} = 2.3$. This is equivalent to

$$S_m = 2.3 \frac{b^{2/3}}{qN} = 2.3 \frac{g_b'^{2/3} L^{1/3}}{Q_D^{1/3} N} = 165$$

Farfield Modeling

Beyond the initial mixing region, the plume, carried by the ambient current, undergoes additional mixing by turbulent diffusion in the transition and farfield regions, which are considered in the following discussion.

Transition Region. In the transition region between the nearfield and the farfield, the flowrate in the plume is given by

$$Q_1 = S_a Q_D \tag{17-45}$$

and the maximum (centerline) concentration of a constituent is

$$C_1 = \frac{C_D}{S_m} \tag{17-46}$$

where C_D = discharge concentration

The plume width, b_1, and thickness, h_1, are related by the following relationship, which simply states that the plume is now moving at the same speed as the ambient current:

$$Q_1 = U b_1 h_1 \tag{17-47}$$

For the case of a diffuser discharge, the plume width is very close to the length of the diffuser ($b_1 = L$). Eq. 17-47 can be used to determine the plume thickness. For the case of a single-port discharge, the plume thickness depends on the residual plume buoyancy. As a first approximation, a thickness of one-tenth the discharge depth can be assumed, and Eq. 17-47 can be used to determine the plume width.

Farfield with Spacially Uniform Current. An estimate of the further reduction of concentrations in the farfield due to diffusion and decay can be obtained by simulating the discharge as a continuous vertical source of width, b_1, and height, h_1. If vertical diffusion is neglected, an exact solution of the mass conservation equation (Eq. 17-21) can be obtained [5]. Neglecting vertical diffusion can be justified when the ambient water is stratified or when the plume occupies the full water depth. From field measurements, it has been found that turbulent diffusion coefficients increase with the size of the plume because larger and larger turbulent eddies participate in the diffusion. A commonly observed variation of the diffusion coefficient is with the plume width raised to the $\frac{4}{3}$ power. For this case, the centerline concentration and plume width are given by the following expressions [5]:

$$C_m = C_1 e^{-Kx/U} \operatorname{erf} \sqrt{\frac{3/2}{\left[1 + (8E_{y1}x/Ub_1^2)\right]^3 - 1}} \qquad (17\text{-}48)$$

$$b_x = b_1\left(1 + \frac{8E_{y1}x}{Ub_1^2}\right)^{3/2} \qquad (17\text{-}49)$$

where C_m = plume centerline concentration
  E_{y1} = initial transverse diffusion coefficient = ft^2/s
    = $0.001 \, b_1^{4/3}$ (when b_1 is ft)
  erf = error function (see Eq. 17-39)
  b_x = plume width at distance x

For ease of application, the above equations are plotted functionally in Fig. 17-15. Several diffuser lengths away from the outfall, the lateral concentration profile in the plume becomes approximately Gaussian (Eq. 17-29) with a standard deviation $\sigma = b_x/2\sqrt{3}$.

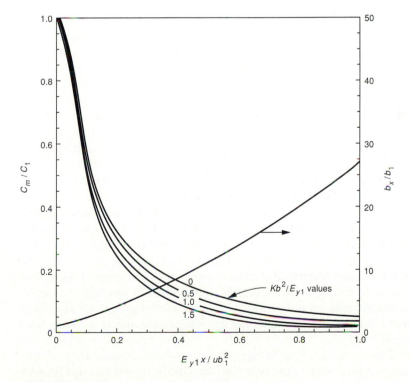

FIGURE 17-15
Farfield plume characteristics.

Example 17-5   Farfield mixing analysis. Consider the outfall designed in Example 17-4 and determine the minimum farfield dilution obtained 10 miles from the discharge point.

Solution

1. Determine the value of the dimensionless parameter governing farfield diffusion (see Eq. 17-49).

   (a) The initial diffusion coefficient is

   $$E_{y1} = 0.001b_1^{4/3} = 0.001 \times b_1/(4736 \text{ ft})^{4/3} = 79.5 \text{ ft}^2/\text{s}$$

   (b) The dimensionless parameter governing farfield diffusion is

   $$\frac{E_{y1}x}{Ub_1^2} = \frac{(79.5 \text{ ft}^2/\text{s})(10 \text{ mi})(5280 \text{ ft/mi})}{(0.66 \text{ ft/s})(4736 \text{ ft})^2} = 0.284$$

2. Determine the dilution factor using Eq. 17-48 or Fig. 17-15.

   (a) Using Fig. 17-15, the centerline concentration 10 miles from the outfall is

   $$C_m/C_1 = 0.23$$

   (b) The corresponding dilution factor is $1/0.23 = 4.35$.

Farfield with Complex Current Patterns. In the expressions presented above, it was assumed that the ambient current, U, is uniform. The uniform current assumption is a legitimate approximation in many cases. However, complex current patterns may need to be accounted for, particularly in tidal conditions where reversals of current direction can occur. In this case, previously discharged effluent may return to the discharge area, which may result in elevated background concentrations. In this case, computer simulations may be required to provide a numerical solution of the conservation of mass (Eq. 17-21), with boundary conditions representative of the actual discharge configuration. The flow equations also need to be solved to provide the distribution of velocities throughout the study domain. The numerical methods typically used to solve these partial differential equations are the finite difference and finite element methods. Although a detailed discussion of these methods is beyond the scope of this book, an example application will be presented.

For the siting of the wastewater outfall for Boston, a two-dimensional finite element model was used to solve for depth-averaged tidal flow velocities and constituent concentrations at the nodes of a grid of triangular elements [32]. The grid used for the Boston simulations is shown in Fig. 17-16. The grid covered Boston Harbor and the entire Massachusetts Bay. The outer boundary of the grid, on a line between Cape Ann and the tip of Cape Cod, Massachusetts, was selected because of its physical significance and because it was sufficiently far from the discharge point to minimize errors that could result from approximate representation of fluxes into and out of the grid.

Simulations were conducted for BOD_5, DO, SS, and toxic chemicals by adjusting the decay terms appropriately. An example of the type of results obtained for dissolved-oxygen deficit at high tide is shown in Fig. 17-17. Simulations were conducted with diffuser discharges at several alternate locations, and for each location compli-

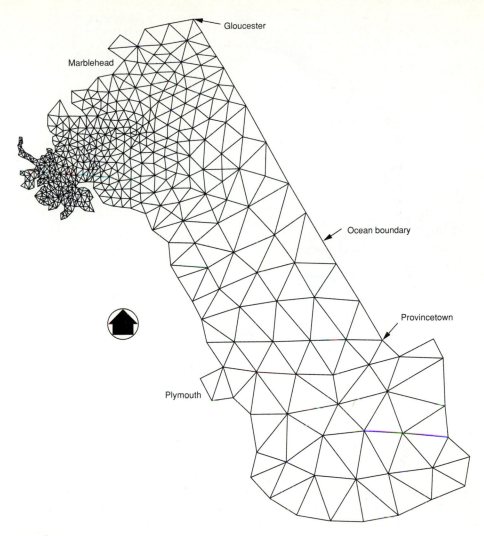

FIGURE 17-16
Finite element grid used for analysis of Boston wastewater discharge in Massachusetts Bay.

ance with water quality criteria was evaluated. Prior to the modeling, an extensive field data collection program was conducted to characterize the currents, stratification cycle, and ambient water quality over several seasons. Current velocity data were used to calibrate the hydrodynamic model. Measured distributions of volatile halogenated organic compounds were used to calibrate the water quality model relative to dispersion.

Computer models allow realistic simulations of candidate options for wastewater disposal. The accuracy of the predictions, however, depends on thorough calibration and verification with measured data. The importance of calibration and verification

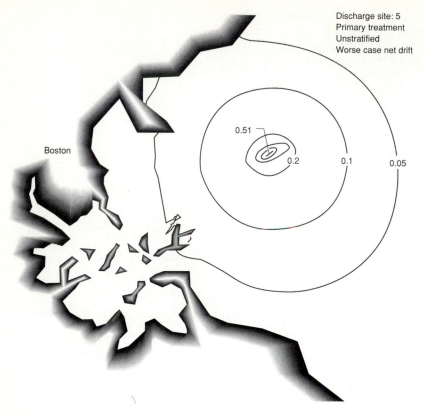

Discharge site: 5
Primary treatment
Unstratified
Worse case net drift

0.51

Boston

0.2   0.1   0.05

FIGURE 17-17
Computed oxygen deficits in mg/L for discharge of primary treated wastewater in Massachusetts Bay.

is also true for simpler approaches based on the equations presented earlier. Even in complex situations, simple analytical solutions are often useful for initial assessments and to provide a check on more complex computer simulations.

PROBLEMS

17-1. Treated wastewater containing 25 mg/L of BOD_5 is discharged continuously at a rate of 1.5 ft³/s into a small lake. The lake has a surface area of 50 acres, a drainage area of 10 mi², and an average depth of 10 ft. Its contents can be considered to be completely mixed. Aerobic conditions prevail throughout its depth. The runoff from the drainage area, containing 1 mg/L of BOD_5, varies from 14 in/yr in the spring to 1.4 in/yr in the summer. The temperature of the lake contents is 5°C in the spring and 25°C in the summer. Determine the BOD_5 of the outlet stream in the spring and the summer, where $K = 0.3 \, d^{-1}$.

17-2. Determine the reaeration coefficient K_2 using the O'Connor-Dobbins formula. The mean velocity of flow for a river is 0.2 ft/s, the depth is 10 ft, and the temperature of the river is 15°C.

17-3. A wastewater containing 130 mg/L of BOD_5 after preliminary treatment is discharged to a river at a rate of 20 Mgal/d. The river has a minimum flowrate of 210 ft³/s, a BOD_5 of 2 mg/L, and a velocity of 0.7 ft/s. After the wastewater is mixed with the river contents, the temperature is 20°C and the dissolved oxygen is 75 percent of saturation. Determine the oxygen sag at the critical point, X_c, and at distances of $X_c/2$ above and below the critical point, and plot the curve. Use $K = 0.25$ d⁻¹ and $K_2 = 0.40$ d⁻¹.

17-4. Wastewater from a small industry is discharged continuously into a nearby river. Using the following data, find (a) the DO deficit at a point 35 miles downstream, (b) the location of the critical point on the oxygen sag curve and the minimum DO in the river at that point, and (c) BOD_5 at a point 12 miles downstream. Assume that BOD_5 in the river upstream of the point of waste discharge is equal to zero. The river characteristics just downstream of the point of waste discharge are

$$
\begin{aligned}
\text{DO} &= 6.0 \text{ mg/L} \\
\text{Velocity} &= 1 \text{ ft/s} \\
\text{Depth} &= 6.5 \text{ ft} \\
\text{Width} &= 33 \text{ ft} \\
\text{Chloride concentration} &= 50,000 \text{ mg/L} \\
\text{Temperature} &= 25°C
\end{aligned}
$$

The wastewater characteristics are $BOD_5 = 12,000$ lb/d and $K = 0.25$ d⁻¹ at 20°C.

17-5. A stream with $K_2 = 0.58$ d⁻¹, temperature = 15°C, and minimum flow = 350 ft³/s receives 9 Mgal/d of wastewater from a city. The river water upstream of the point of waste discharge is 95 percent saturated with oxygen. What is the maximum permissible BOD_5 of the wastewater if the dissolved-oxygen content of the stream is never to go below 4 mg/L? Assume that K of the river-waste mixture equals 0.35 d⁻¹ at 20°C. What would the minimum DO in the stream be if the wastewater received secondary treatment as specified by EPA (BOD_5 in effluent = 30 mg/L)?

17-6. The oxygen resources for a small stream have been investigated and the following coefficients for oxygen production and consumption have been determined:

$$
\begin{aligned}
\text{Organic degradation} &= K = 0.025 \text{ d}^{-1} \\
\text{Reaeration} &= K = 0.45 \text{ d}^{-1} \\
\text{Nitrification} &= N = -3.0 \text{ mg/L} \cdot \text{d} \\
\text{Photosynthesis} &= P_{max} = 5 \text{ mg/L} \cdot \text{d} \\
\text{Respiration} &= R = 1 \text{ mg/L} \cdot \text{d}
\end{aligned}
$$

At some point, X, along the stream, the concentration of ultimate BOD_5 present in 10 mg/L and the dissolved-oxygen concentration is 5 mg/L. If the saturation value for dissolved oxygen is 10 mg/L, determine the following:

(a) The rate of dissolved-oxygen change in mg/L · d at point X at midday when maximum photosynthesis occurs.

(b) The rate of dissolved-oxygen change at the same point but during the night when $P = 0$. Assuming that the rate of dissolved-oxygen change remains constant between point X and another point, Y, situated 1 h of stream flow time downstream from point X, determine the following:

    i. The dissolved-oxygen concentration at point Y near midday.

    ii. The dissolved-oxygen concentration at point Y during the night.

17-7. The freshwater outflow in an estuary is 100 ft³/s and the average cross-sectional area is 1000 ft². Assuming that seawater has a chloride concentration of 18,000 mg/L, determine E from the following data:

X, mi	1.25	2.5	3.75	5.0	6.25
C, mg/L	16,000	11,500	8,350	6,000	4,350

17-8. A wastewater discharge is planned in a river with a width of 480 ft, a depth of 10 ft, a current speed of 1.5 ft/s. To estimate the turbulent diffusion characteristics of the river, an experiment is conducted in which dye is released continuously at the proposed discharge location at the rate of 13.4 lb/h. Dye concentration is measured 12,000 ft downstream of the discharge on the same shore as the discharge. Once steady state has been reached, the concentration measured is 27 ppb. Assuming fully mixed conditions over the depth and using the method of images to account for the shore opposite the discharge, determine the lateral diffusion coefficient, E_y. Assume a source width of 5 ft.

17-9. A study of horizontal diffusion in a body of water consisted of determining the distribution of particles on the second and third days after their release from an initial point ($t = 0$). The particles assumed a Gaussian, or normal, distribution, centered, as they diffused outward, about the initial point. Using the particle distribution data given in the following figure, determine the diffusion coefficient using the following formula:

$$D = \frac{1}{2}\frac{d\sigma^2}{dt}$$

where $D =$ coefficient of diffusion
   $\sigma =$ standard deviation of distribution curve

Express your answer in units of centimeters and seconds. (Courtesy of G. T. Orlob.)

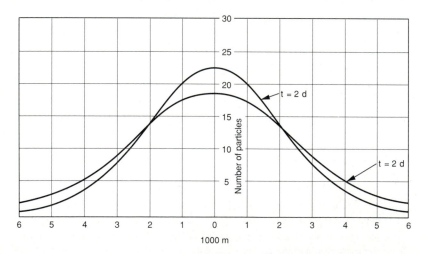

17-10. The freshwater runoff to an estuary has a chloride concentration of 3 mg/L and amounts to 1000 ft³/s. Assume that 70 Mgal/d of wastewater with an average chloride concentration of 50 mg/L is discharged to the estuary and that the chloride concentration at that point is 9000 mg/L. Determine the dilution available and the chloride concentration after mixing.

17-11. Design an ocean outfall for an average wastewater flowrate of 30 Mgal/d and a peak flowrate of 67 Mgal/d. The bottom slope is 1/50 along the route of the outfall. The diffusers are to be located in 80 ft of water. The prevailing current is 0.2 ft/s parallel to the shore. Determine the dilution and coliform content at distances of 2500 and 5000 ft from the diffuser, assuming that (*a*) the wastewater has had primary treatment and (*b*) the wastewater has had primary treatment plus chlorination. Use 10 ft of diffuser length per Mgal/d and assume that $T_{90} = 3$ h.

REFERENCES

1. Adams, E. E.: "Dilution Analysis for Unidirectional Diffusers," *Journal of the Hydraulic Division, ASCE,* vol. 108, no. HY3, 1982.
2. Bedford, K. W., R. M. Sykes, and C. Libicki: "Dynamic Advective Water Quality Model for Rivers," *Journal of Environmental Engineering, ASCE,* vol. 109, no. 3, 1983.
3. Bellair, J. T., G. A. Parr-Smith, and I. G. Wallis: "Significance of Diurnal Variations in Fecal Coliform Die-off Rates in the Design of Ocean Outfalls," *Journal WPCF,* vol. 49, no. 9, 1977.
4. Bowie, G. L., et al.: *Rates, Constants and Kinetics Formulations in Surface Quality Modeling,* (2nd ed.), Report EPA/600/3-85/040, Office of Research and Development, U.S. Environmental Protection Agency, 1985.
5. Brooks, N. H.: "Diffusion of Sewage Effluent in an Ocean Current," *Proceedings, First International Conference on Waste Disposal in the Marine Environment,* University of California, Berkeley, Pergamon Press, NY, 1960.
6. Brown, L. C., and T. O. Barnwell: *The Enhanced Stream Water Quality Models QUAL2E and QUAL2E-UNCAS: Documentation and User's Manual,* Report EPA/600/3-87/007, Environmental Protection Agency, Office of Research and Development, 1987.
7. Csanady, G. T.: *Turbulent Diffusion in the Environment,* Reidel Publishing Company, Boston, MA, 1973.
8. Fischer, H. B.: "The Mechanics of Dispersion in Natural Streams," *Journal of the Hydraulics Division, ASCE,* vol. 93, no. HY6, 1967.
9. Fischer, H. B., E. J. List, R. C. Y. Koh, J. Imberger, and N. H. Brooks: *Mixing in Inland and Coastal Waters,* Academic Press, NY, 1979.
10. Harleman, D. R. F.: "Hydrothermal Analysis of Lakes and Reservoirs," *Journal of the Hydraulics Division, ASCE,* vol. 108, no. HY 3, 1982.
11. Holley, E. R., D. R. F. Harleman, and H. B. Fischer: "Dispersion in Homogeneous Estuary Flow," *Journal of the Hydraulic Division, ASCE,* vol. 98, no. HY8, 1970.
12. Krenkel, P. A., and V. Novotny: *Water Quality Management,* Academic Press, NY, 1984.
13. Lee, J. H. W., and P. Neville-Jones: "Initial Dilution of Horizontal Jet in Crossflow," *Journal of Hydraulic Engineering, ASCE,* vol. 113, no. 5, 1987.
14. Lyman, W. J.: "Adsorption Coefficient for Soil and Sediments," in W. J. Lyman, W. F. Reehl, and D. H. Rosenblatt (eds.), *Handbook of Chemical Property Estimation Methods,* McGraw-Hill, New York, 1982.
15. Mancini, J. L.: "Numerical Estimates of Coliform Mortality Rates under Various Conditions," *Journal WPCF,* vol. 50, no. 11, 1978.
16. Mitchell, R. (ed): *Water Pollution Microbiology,* Wiley Interscience, New York, 1972.
17. Muellenhoff, W. P., A. M. Soldate, Jr., D. J. Baumgartner, M. D. Schuldt, L. R. Davis, and W. E. Frick: *Initial Mixing Characteristics of Municipal Ocean Discharges,* Report EPA/600/3-85/073, U.S. Environmental Protection Agency, Washington, DC, 1985.
18. Najarian, T. O., P. J. Kaneta, J. L. Taft, and M. L. Thatcher: "Application of Nitrogen-Cycle Model to Manasquan Estuary." *Journal of Environmental Engineering, ASCE,* vol. 110, no. 1, 1984.
19. O'Connor, D. J., and W. E. Dobins: "Mechanism of Reaeration in Natural Streams," *Transactions of the American Society of Civil Engineers,* vol. 123, pp. 641–666, 1958.
20. Roberts, P. J. W., W. H. Snyder, and D. J. Baumgartner: "Ocean Outfalls," *Journal of Hydraulic Engineering, ASCE,* vol. 115, no. 1, 1989.

21. Sawyer, C. N.: "Fertilization of Lakes by Agricultural and Urban Drainages," *Journal of the New England Waterworks Association,* vol. 51, pp. 109–127, 1947.

22. Smith, J. H., D. C. Bomberger, and D. L. Haynes: "Prediction of the Volatilization Rate of High Volatility Chemicals from Natural Water Bodies," *Environmental Science and Technology,* vol. 14, pp. 1332–37, 1980.

23. Streeter, H. W., and E. B. Phelps: "A Study of the Pollution and Natural Purification of the Ohio River," *Public Health Bulletin,* vol. 146, U. S. Public Health Service, Washington DC, 1925.

24. Taylor, G. I.: "The Dispersion of Matter in Turbulent Flow through a Pipe," *Proceedings of the Royal Society, A,* vol. CCXXIII, pp. 446–68, 1954.

25. Thomann, R. V.: *System Analysis and Water Quality Management,* McGraw-Hill, New York, 1972.

26. Thomas, R. G.: "Volatilization from Water," in W. J. Lyman, W. F. Reehl, and D. H. Rosenblatt (eds.), *Handbook of Chemical Property Estimation Methods,* McGraw-Hill, New York, 1982.

27. Tsivoglou, E. C., and L. A. Neal: "Tracer Measurement of Reaeration, Predicting the Reaeration Capacity of Inland Streams," *Journal WPCF,* vol. 48, no. 12, 1976.

29. U.S. Environmental Protection Agency: *Revised Section 301(h) Technical Support Document,* Office of Water, Washington, DC, 1982.

30. U.S. Environmental Protection Agency: *Technical Support Document for Water Quality-based Toxics Control,* Office of Water, Washington, DC, 1985.

31. U.S. Environmental Protection Agency: *Criteria for Water 1986,* EPA 440/5-86-001, Office of Water Regulations and Standards, Washington, DC, 1986.

32. U.S. Environmental Protection Agency: *Boston Harbor Wastewater Conveyance System,* Supplemental Environmental Impact Statement, Washington, DC, 1988.

33. Walton, W. C.: *Practical Aspects of Groundwater Modeling,* National Water Well Association, Worthington, OH, 1984.

34. Wang, R-F. T.: *Laboratory Analysis of Settling Velocities of Wastewater Particles in Seawater using Holography,* Report no. 27. Environmental Quality Laboratory, California Institute of Technology, Pasadena, CA, 1988.

35. Wilcock, R. J.: "Study of River Reaeration at Different Flow Rates," *Journal of Environmental Engineering, ASCE,* vol. 114, no. 1, 1988.

36. Wright, S. J., P. J. W. Roberts, Y. Zhongmin, and N. E. Bradley: "Surface Dilution of Submerged Buoyant Jets," *Journal of Hydraulic Research,* IAHR (in press), 1990.

TABLE A-1
Metric conversion factors (U.S. customary units to SI units)

Multiply the U.S. customary unit			To obtain the SI unit	
Name	Symbol	By	Symbol	Name
Acceleration				
feet per second squared	ft/s^2	0.3048^a	m/s^2	meters per second squared
inches per second squared	in/s^2	0.0254^a	m/s^2	meters per second squared
Area				
acre	acre	0.4047	ha	hectare
acre	acre	4.0469×10^{-3}	km^2	square kilometer
square foot	ft^2	9.2903×10^{-2}	m^2	square meter
square inch	in^2	6.4516^a	cm^2	square centimeter
square mile	mi^2	2.5900	km^2	square kilometer
square yard	yd^2	0.8361	m^2	square meter
Energy				
British thermal unit	Btu	1.0551	kJ	kilojoule
foot-pound (force)	$ft \cdot lb_f$	1.3558	J	joule
horsepower-hour	$hp \cdot h$	2.6845	MJ	megajoule
kilowatt-hour	$kW \cdot h$	3600^a	kJ	kilojoule
kilowatt-hour	$kW \cdot h$	3.600×10^{6a}	J	joule
watt-hour	$W \cdot h$	3.600^a	kJ	kilojoule
watt-second	$W \cdot s$	1.000^a	J	joule

(continued)

TABLE A-1
(continued)

Multiply the U.S. customary unit			To obtain the SI unit	
Name	Symbol	By	Name	Symbol
Force				
pound force	lb_f	4.4482	newton	N
Flowrate				
cubic feet per second	ft³/s	2.8317×10^{-2}	cubic meters per second	m³/s
gallons per day	gal/d	4.3813×10^{-5}	liters per second	L/s
gallons per day	gal/d	3.7854×10^{-3}	cubic meters per day	m³/d
gallons per minute	gal/min	6.3090×10^{-5}	cubic meters per second	m³/s
gallons per minute	gal/min	6.3090×10^{-2}	liters per second	L/s
million gallons per day	Mgal/d	43.8126	liters per second	L/s
million gallons per day	Mgal/d	3.7854×10^{3}	cubic meters per day	m³/d
million gallons per day	Mgal/d	4.3813×10^{-2}	cubic meters per second	m³/s
Length				
foot	ft	0.3048[a]	meter	m
inch	in	2.54[a]	centimeter	cm
inch	in	0.0254[a]	meter	m
inch	in	25.4[a]	millimeter	mm
mile	mi	1.6093	kilometer	km
yard	yd	0.9144[a]	meter	m
Mass				
ounce	oz	28.3495	gram	g
pound	lb	4.5359×10^{2}	gram	g
pound	lb	0.4536	kilogram	kg
ton (short: 2000 lb)	ton	0.9072	megagram (10³ kilogram)	Mg (metric ton)
ton (long: 2240 lb)	ton	1.0160	megagram (10³ kilogram)	Mg (metric ton)

Power

British thermal units per second	Btu/s	1.0551	kW	kilowatt
foot-pounds (force) per second	ft · lb$_f$/s	1.3558	W	watt
horsepower	hp	0.7457	kW	kilowatt

Pressure (force/area)

atmosphere (standard)	atm	1.0133×10^2	kPa (kN/m^2)	kilopascal (kilonewtons per square meter)
inches of mercury (60°F)	in Hg (60°F)	3.3768×10^3	Pa (N/m^2)	pascal (newtons per square meter)
inches of water (60°F)	in H$_2$O (60°F)	2.4884×10^2	Pa (N/m^2)	pascal (newtons per square meter)
pounds (force) per square foot	lb$_f$/ft^2	47.8803	Pa (N/m^2)	pascal (newtons per square meter)
pounds (force) per square inch	lb$_f$/in^2	6.8948×10^3	Pa (N/m^2)	pascal (newtons per square meter)
pounds (force) per square inch	lb$_f$/in^2	6.8948	kPa (kN/m^2)	kilopascal (kilonewtons per square meter)

Temperature

degrees Fahrenheit	°F	0.555(°F − 32)	°C	degrees Celsius (centigrade)
degrees Fahrenheit	°F	0.555(°F + 459.67)	°K	degrees Kelvin

Velocity

feet per second	ft/s	0.3048^a	m/s	meters per second
miles per hour	mi/h	4.4704×10^{-1a}	m/s	kilometers per second

Volume

acre-foot	acre-ft	1.2335×10^3	m^3	cubic meter
cubic foot	ft^3	28.3168	L	liter
cubic foot	ft^3	2.8317×10^{-2}	m^3	cubic meter
cubic inch	in^3	16.3871	cm^3	cubic centimeter
cubic yard	yd^3	0.7646	m^3	cubic meter
gallon	gal	3.7854×10^{-3}	m^3	cubic meter
gallon	gal	3.7854	L	liter
ounce (U.S. fluid)	oz (U.S. fluid)	2.9573×10^{-2}	L	liter

^{a} Indicates exact conversion.

TABLE A-2
Metric conversion factors (SI units to U.S. customary units)

| Multiply the SI unit | | By | To obtain the U.S. customary unit | |
Name	Symbol		Symbol	Name
Acceleration				
meters per second squared	m/s^2	3.2808	ft/s^2	feet per second squared
meters per second squared	m/s^2	39.3701	in/s^2	inches per second squared
Area				
hectare ($10,000\ m^2$)	ha	2.4711	acre	acre
square centimeter	cm^2	0.1550	in^2	square inch
square kilometer	km^2	0.3861	mi^2	square mile
square kilometer	km^2	247.1054	acre	acre
square meter	m^2	10.7639	ft^2	square foot
square meter	m^2	1.1960	yd^2	square yard
Energy				
kilojoule	kJ	0.9478	Btu	British thermal unit
joule	J	2.7778×10^{-7}	$kW \cdot h$	kilowatt-hour
joule	J	0.7376	$ft \cdot lb_f$	foot-pound (force)
joule	J	1.0000	$W \cdot s$	watt-second
joule	J	0.2388	cal	calorie
kilojoule	kJ	2.7778×10^{-4}	$kW \cdot h$	kilowatt-hour
kilojoule	kJ	0.2778	$W \cdot h$	watt-hour
megajoule	MJ	0.3725	$hp \cdot h$	horsepower-hour
Force				
newton	N	0.2248	lb_f	pound force
Flowrate				
cubic meters per day	m^3/d	264.1720	gal/d	gallons per day
cubic meters per day	m^3/d	2.6417×10^{-4}	Mgal/d	million gallons per day
cubic meters per second	m^3/s	35.3147	ft^3/s	cubic feet per second
cubic meters per second	m^3/s	22.8245	Mgal/d	million gallons per day
cubic meters per second	m^3/s	15,850.3	gal/min	gallons per minute

Unit	Symbol	Multiply by	To symbol	To unit
liters per second	L/s	22,824.5	gal/d	gallons per day
liters per second	L/s	0.0228	Mgal/d	million gallons per day
liters per second	L/s	15.8508	gal/min	gallons per minute
Length				
centimeter	cm	0.3937	in	inch
kilometer	km	0.6214	mi	mile
meter	m	39.3701	in	inch
meter	m	3.2808	ft	foot
meter	m	1.0936	yd	yard
millimeter	mm	0.03937	in	inch
Mass				
gram	g	0.0353	oz	ounce
gram	g	0.0022	lb	pound
kilogram	kg	2.2046	lb	pound
megagram (10^3 kg)	Mg	1.1023	ton	ton (short: 2000 lb)
megagram (10^3 kg)	Mg	0.9842	ton	ton (long: 2240 lb)
Power				
kilowatt	kW	0.9478	Btu/s	British thermal units per second
kilowatt	kW	1.3410	hp	horsepower
watt	W	0.7376	ft · lb$_f$/s	foot-pounds (force) per second
Pressure (force/area)				
pascal (newtons per square meter)	Pa (N/m^2)	1.4504×10^{-4}	lb$_f$/in^2	pounds (force) per square inch
pascal (newtons per square meter)	Pa (N/m^2)	2.0885×10^{-2}	lb$_f$/ft^2	pounds (force) per square foot
pascal (newtons per square meter)	Pa (N/m^2)	2.9613×10^{-4}	in Hg	inches of mercury (60°F)
pascal (newtons per square meter)	Pa (N/m^2)	4.0187×10^{-3}	in H$_2$O	inches of water (60°F)
kilopascal (kilonewtons per square meter)	kPa (kN/m^2)	0.1450	lb$_f$/in^2	pounds (force) per square inch
kilopascal (kilonewtons per square meter)	kPa (kN/m^2)	0.0099	atm	atmosphere (standard)
Temperature				
degrees Celsius (centigrade)	°C	$1.8(°C)+32$	°F	degrees Fahrenheit
degrees Kelvin	°K	$1.8(°K)-459.67$	°F	degrees Fahrenheit

(continued)

TABLE A-2
(continued)

Multiply the SI unit			To obtain the U.S. customary unit	
Name	**Symbol**	**By**	**Symbol**	**Name**
Velocity				
kilometers per second	km/s	2.2369	mi/h	miles per hour
meters per second	m/s	3.2808	ft/s	feet per second
Volume				
cubic centimeter	cm^3	0.0610	in^3	cubic inch
cubic meter	m^3	35.3147	ft^3	cubic foot
cubic meter	m^3	1.3079	yd^3	cubic yard
cubic meter	m^3	264.1720	gal	gallon
cubic meter	m^3	8.1071×10^{-4}	acre · ft	acre · foot
liter	L	0.2642	gal	gallon
liter	L	0.0353	ft^3	cubic foot
liter	L	33.8150	oz	ounce (U.S. fluid)

TABLE A-3
Conversion factors for commonly used wastewater treatment plant design parameters

U.S. units	To convert, multiply in direction shown by arrows →	←	SI units
acre/(Mgal/d)	0.1069	9.3536	ha/(10^3m^3/d)
Btu	1.0551	0.9478	kJ
Btu/lb	2.3241	0.4303	kJ/kg
Btu/ft^2 · °F · h	5.6735	0.1763	W/m^2 · °C
bu/acre · yr	2.4711	0.4047	bu/ha · yr
ft/h	0.3048	3.2808	m/h
ft/min	18.2880	0.0547	m/h
ft^2/capita	0.0929	10.7639	m^2/capita
ft^3/capita	0.0283	35.3147	m^3/capita
ft^3/gal	7.4805	0.1337	m^3/m^3
ft^3/ft · min	0.0929	10.7639	m^3/m · min
ft^3/lb	0.0624	16.0185	m^3/kg
ft^3/Mgal	7.04805 × 10^{-3}	133.6805	m^3/10^{3}m^3
ft^2/Mgal · d	407.4611	0.0025	m^2/10^{3}m^3 · d
ft^3/ft^2 · h	0.3048	3.2808	m^3/m^2 · h
ft^3/10^3 gal · min	7.04805 × 10^{-3}	133.6805	m^3/m^3 · min
ft^3/min	1.6990	0.5886	m^3/h
ft^3/s	2.8317 × 10^{-2}	35.3145	m^3/s
ft^3/10^3 ft^3 · min	0.001	1,000.0	m^3/m^3 · min
gal	3.7854	0.2642	L
gal/acre · d	0.0094	106.9064	m^3/ha · d
gal/ft · d	0.0124	80.5196	m^3/m · d
gal/ft^2 · d	0.0407	24.5424	m^3/m^2 · d
gal/ft^2 · d	0.0017	589.0173	m^3/m^2 · h
gal/ft^2 · d	0.0283	35.3420	L/m^2 · min
gal/ft^2 · d	40.7458	2.4542 × 10^{-2}	L/m^2 · d
gal/ft^2 · min	2.4448	0.4090	m/h
gal/ft^2 · min	40.7458	0.0245	L/m^2 · min
gal/ft^2 · min	58.6740	0.0170	m^3/m^2 · d
gal/min · ft	12.4193	8.052 × 10^{-2}	L/min · m
hp/10^3 gal	0.1970	5.0763	kW/m^3
hp/10^3 ft^3	26.3342	0.0380	kW/10^3 m^3
in	25.4	3.9370 × 10^{-2}	mm
in Hg (60°F)	3.3768	0.2961	kPa Hg (60°F)
lb	0.4536	2.2046	kg
lb/acre	1.1209	0.8922	kg/ha
lb/10^3 gal	0.1198	8.3452	kg/m^3
lb/hp · h	0.6083	1.6440	kg/kW · h
lb/Mgal	0.1198	8.3454	g/m^3
lb/Mgal	1.1983 × 10^{-4}	8345.4	kg/m^3
lb/ft^2	4.8824	0.2048	kg/m^2
lb$_f$/in^2 (gage)	6.8948	0.1450	kPa (gage)
lb/ft^3 · h	16.0185	0.0624	kg/m^3 · h
lb/10^3 ft^3 · d	0.0160	62.4280	kg/m^3 · d
lb/ton	0.5000	2.0000	kg/tonne
Mgal/acre · d	0.9354	1.0691	m^3/m^2 · d
Mgal/d	3.7854 × 10^3	0.264 × 10^{-3}	m^3/d
Mgal/d	4.3813 × 10^{-2}	22.8245	m^3/s
min/in	3.9370	0.2540	min/10^2 mm
tons/acre	2.2417	0.4461	Mg/ha
yd^3	0.7646	1.3079	m^3

PHYSICAL PROPERTIES OF AIR

TABLE B-1
Atmospheric pressure (U.S. customary units)[a]

| Elevation above sea level, ft | Atmospheric pressure[b] | | | Specific weight (γ) of air at 68°F, lb/ft^{3c} |
| | | Expressed as a column of | | |
	lb/in^2	Water, ft	Mercury, mm	
0	14.7	33.9	760	0.0752
1,000	14.2	32.7	734	0.0726
2,000	13.7	31.6	708	0.0700
3,000	13.2	30.4	681	0.0673
4,000	12.7	29.4	658	0.0651
5,000	12.1	28.2	633	0.0626
6,000	11.8	27.2	610	0.0604
7,000	11.4	26.3	589	0.0583
8,000	10.7	24.8	556	0.0550
9,000	10.5	24.2	543	0.0537
10,000	10.1	23.4	524	0.0518

[a] From Sanks, R. L., G. Tchobanoglous, D. Newton, B. E. Bossermann II, and G. M. Jones (eds.): *Pumping Station Design*, Butterworths, Stoneham, MA, 1989.

[b] Storms commonly reduce atmospheric pressure by about 1.7%.

[c] At other temperatures and pressures, use $p_1 v_1/T_1 = p_2 v_2/T_2$; or use the general formula for atmospheric pressure:

$$\frac{p_b}{p_a} = \exp\left[-\frac{gM(z_b - z_a)}{g_c RT}\right]$$

where g_c = 32.2 ft · lb/(lb$_m$ · s)^2
g = 32.2 ft/s^2
M = 29 lb$_m$/lb$_{mol}$

$R = 1545\dfrac{\text{ft} \cdot \text{lb}}{\text{lb}_{mol} \cdot T}$

T = 460 + °F

TABLE B-2
Atmospheric pressure (SI units)[a]

Elevation above sea level, m	Atmospheric pressure[b]			Specific weight (γ) of air at 20°C, kN/m^{3c}
		Expressed as a column of		
	kPa	Water, m	Mercury, mm	
0	101.3	10.33	760	0.0118
500	95.6	9.74	717	0.0111
1000	90.1	9.19	676	0.0105
1500	84.8	8.64	636	0.0099
2000	79.8	8.13	598	0.0093
2500	73.3	7.47	550	0.0085
3000	70.3	7.17	527	0.0082
3500	66.1	6.74	496	0.0077

[a] From Sanks, R. L., G. Tchobanoglous, D. Newton, B. E. Bosserman II, and G. M. Jones (eds.): *Pumping Station Design,* Butterworths, Stoneham, MA, 1989.

[b] Storms commonly reduce atmospheric pressure by about 1.7%.

[c] At other temperatures and pressures use $p_1 v_1 / K_1 = p_2 v_2 / K_2$ where p is pressure, v is volume, and K is degrees kelvin (°C + 273).

The principal physical properties of water are summarized in Table C-1 in U.S. customary units and in Table C-2 in SI units. They are briefly described below [1].

SPECIFIC WEIGHT

The specific weight of a fluid is its weight per unit volume. In U.S. customary units, it is expressed in pounds per cubic ft. The relationship between γ, ρ, and the acceleration due to gravity g is $\gamma = \rho g$. At normal temperatures γ is 62.4 lb_f/ft^3 (9.81 kN/m^3).

DENSITY

The density of ρ of a fluid is its mass per unit volume. In U.S. customary units, it is expressed in slugs per cubic feet. For water, ρ is 1.940 $slugs/ft^3$ at 32°F. There is a slight decrease in density with increasing temperature.

MODULUS OF ELASTICITY

For most practical purposes, liquids may be regarded as incompressible. The bulk modulus of elasticity E is given by

$$E = \frac{\Delta p}{\Delta V/V}$$

where Δp is the increase in pressure, which when applied to a volume V, results in a decrease in volume ΔV.

TABLE C-1
Physical properties of water (U.S. customary units)[a]

Temperature, °F	Specific weight, γ, lb/ft³	Density,[b] ρ, slug/ft³	Modulus of elasticity,[b] $E/10^3$, lb$_f$/in²	Dynamic viscosity, $\mu \times 10^5$, lb·s/ft²	Kinematic viscosity, $\nu \times 10^5$, ft²/s	Surface tension,[c] σ, lb/ft	Vapor pressure, p_v, lb$_f$/in²
32	62.42	1.940	287	3.746	1.931	0.00518	0.09
40	62.43	1.940	296	3.229	1.664	0.00614	0.12
50	62.41	1.940	305	2.735	1.410	0.00509	0.18
60	62.37	1.938	313	2.359	1.217	0.00504	0.26
70	62.30	1.936	319	2.050	1.059	0.00498	0.36
80	62.22	1.934	324	1.799	0.930	0.00492	0.51
90	62.11	1.931	328	1.595	0.826	0.00486	0.70
100	62.00	1.927	331	1.424	0.739	0.00480	0.95
110	61.86	1.923	332	1.284	0.667	0.00473	1.27
120	61.71	1.918	332	1.168	0.609	0.00467	1.69
130	61.55	1.913	331	1.069	0.558	0.00460	2.22
140	61.38	1.908	330	0.981	0.514	0.00454	2.89
150	61.20	1.902	328	0.905	0.476	0.00447	3.72
160	61.00	1.896	326	0.838	0.442	0.00441	4.74
170	60.80	1.890	322	0.780	0.413	0.00434	5.99
180	60.58	1.883	318	0.726	0.385	0.00427	7.51
190	60.36	1.876	313	0.678	0.362	0.00420	9.34
200	60.12	1.868	308	0.637	0.341	0.00413	11.52
212	59.83	1.860	300	0.593	0.319	0.00404	14.70

[a] Adapted from Ref. 2.
[b] At atmospheric pressure.
[c] In contact with the air.

TABLE C-2
Physical properties of water (SI units)[a]

Temperature, °C	Specific weight, γ, kN/m³	Density,[b] ρ, kg/m³	Modulus of elasticity,[b] $E/10^6$, kN/m²	Dynamic viscosity, $\mu \times 10^3$, N·s/m²	Kinematic viscosity, $\nu \times 10^6$, m²/s	Surface tension,[c] σ, N/m	Vapor pressure, p_v, kN/m²
0	9.805	999.8	1.98	1.781	1.785	0.0765	0.61
5	9.807	1000.0	2.05	1.518	1.519	0.0749	0.87
10	9.804	999.7	2.10	1.307	1.306	0.0742	1.23
15	9.798	999.1	2.15	1.139	1.139	0.0735	1.70
20	9.789	998.2	2.17	1.002	1.003	0.0728	2.34
25	9.777	997.0	2.22	0.890	0.893	0.0720	3.17
30	9.764	995.7	2.25	0.798	0.800	0.0712	4.24
40	9.730	992.2	2.28	0.653	0.658	0.0696	7.38
50	9.689	988.0	2.29	0.547	0.553	0.0679	12.33
60	9.642	983.2	2.28	0.466	0.474	0.0662	19.92
70	9.589	977.8	2.25	0.404	0.413	0.0644	31.16
80	9.530	971.8	2.20	0.354	0.364	0.0626	47.34
90	9.466	965.3	2.14	0.315	0.326	0.0608	70.10
100	9.399	958.4	2.07	0.282	0.294	0.0589	101.33

[a] Adapted from Ref. 2.
[b] At atmospheric pressure.
[c] In contact with air.

DYNAMIC VISCOSITY

The viscosity of a fluid μ is a measure of its resistance to tangential or shear stress. Viscosity in U.S. customary units is expressed in pound seconds per square foot.

KINEMATIC VISCOSITY

In many problems concerning fluid motion, the viscosity appears with the density in the form μ/ρ, and it is convenient to use a single term ν, known as the *kinematic viscosity* and expressed in square feet per second or stokes in U.S. customary units. The kinematic viscosity of a liquid diminishes with increasing temperature.

SURFACE TENSION

Surface tension is the physical property that enables a drop of water to be held in suspension at a tap, a glass to be filled with liquid slightly above the brim and yet not spill, or a needle to float on the surface of a liquid. The surface-tension force across any imaginary line at a free surface is proportional to the length of the line and acts in a direction perpendicular to it. The surface tension per unit length σ is expressed in pounds per foot. There is a slight decrease in surface tension with increasing temperature.

VAPOR PRESSURE

Liquid molecules that possess sufficient kinetic energy are projected out of the main body of a liquid at its free surface and pass into the vapor. The pressure exerted by this vapor is known as the vapor pressure p_v. The vapor pressure of water at 32°F is 0.09 $\mathrm{lb}_f/\mathrm{in}^2$.

REFERENCES

1. Webber, N. B.: *Fluid Mechanics for Civil Engineers,* SI ed., Chapman and Hall, London, 1971.
2. Vennard, J. K., and R. L. Street: *Elementary Fluid Mechanics,* 5th ed., Wiley, New York, 1975.

SOLUBILITY OF GASES DISSOLVED IN WATER

The equilibrium or saturation concentration of gas dissolved in a liquid is a function of the type of gas and the partial pressure of the gas adjacent to the liquid. The relationship between the partial pressure of the gas in the atmosphere above the liquid and the concentration of the gas in the liquid is given by Henry's law:

$$P_g = Hx_g \qquad \text{(D-1)}$$

where P_g = partial pressure of gas, atm
H = Henry's law constant
x_g = equilibrium mole fraction of dissolved gas
$$= \frac{\text{mol gas}(n_g)}{\text{mol gas}(n_n) + \text{mol water}(n_w)}$$

Henry's law constant is a function of the type, temperature, and constituents of the liquid. Values of H for various gases are listed in Table D-1. Use of the data in Table D-1 is illustrated in the following example.

TABLE D-1
Henry's law constants for several gases that are slightly soluble in water

	$H \times 10^{-4}$, atm/mol fraction							
T, °C	Air	CO_2	CO	H_2	H_2S	CH_4	N_2	O_2
0	4.32	0.0728	3.52	5.79	0.0268	2.24	5.29	2.55
10	5.49	0.104	4.42	6.36	0.0367	2.97	6.68	3.27
20	6.64	0.142	5.36	6.83	0.0483	3.76	8.04	4.01
30	7.71	0.186	6.20	7.29	0.0609	4.49	9.24	4.75
40	8.70	0.233	6.96	7.51	0.0745	5.20	10.4	5.35
50	9.46	0.283	7.61	7.65	0.0884	5.77	11.3	5.88
60	10.1	0.341	8.21	7.65	0.1030	6.26	12.0	6.29

Example D-1 Saturation concentration of nitrogen in water. What is the saturation of nitrogen in water in contact with dry air at 1 atm and 20°C?

Solution

1. Dry air contains about 79 percent nitrogen. Therefore $p_g = 0.79$.
2. From Table D-1, at 20°C, $H = 8.04 \times 10^4$, and

$$X_g = \frac{p_g}{H} = \frac{0.79}{8.04 \times 10^4}$$
$$= 9.84 \times 10^{-6}$$

3. One liter of water contains $1000/18 = 55.6$ g mol; thus,

$$\frac{n_g}{n_g + n_w} = 9.84 \times 10^{-6}$$

$$n_g = (n_g + 55.6)9.84 \times 10^{-6}$$

Because the quantity $(n_g)9.84 \times 10^{-6}$ is very much less than n_g,

$$n_g \approx (55.6)9.84 \times 10^{-6}$$
$$\approx 5.47 \times 10^{-4} \text{ mol/L nitrogen}$$

4. Determine the saturation concentration of nitrogen.

$$C_s \approx (5.47 \times 10^{-4} \text{ mol/L})(28 \text{ g/mol})(10^3 \text{ mg/g})$$
$$\approx 15.3 \text{ mg/L}$$

DISSOLVED-OXYGEN CONCENTRATION IN WATER AS A FUNCTION OF TEMPERATURE, SALINITY, AND BAROMETRIC PRESSURE

TABLE E-1
Dissolved-oxygen concentration in water as a function of temperature and salinity (barometric perssure = 760 mm Hg)[a]

Temp, °C	Dissolved-oxygen concentration, mg/L									
	Salinity, parts per thousand									
	0	5	10	15	20	25	30	35	40	45
0	14.60	14.11	13.64	13.18	12.74	12.31	11.90	11.50	11.11	10.74
1	14.20	13.73	13.27	12.83	12.40	11.98	11.58	11.20	10.83	10.46
2	13.81	13.36	12.91	12.49	12.07	11.67	11.29	10.91	10.55	10.20
3	13.45	13.00	12.58	12.16	11.76	11.38	11.00	10.64	10.29	9.95
4	13.09	12.67	12.25	11.85	11.47	11.09	10.73	10.38	10.04	9.71
5	12.76	12.34	11.94	11.56	11.18	10.82	10.47	10.13	9.80	9.48
6	12.44	12.04	11.65	11.27	10.91	10.56	10.22	9.89	9.57	9.27
7	12.13	11.74	11.37	11.00	10.65	10.31	9.98	9.66	9.35	9.06
8	11.83	11.46	11.09	10.74	10.40	10.07	9.75	9.44	9.14	8.85
9	11.55	11.19	10.83	10.49	10.16	9.84	9.53	9.23	8.94	8.66
10	11.28	10.92	10.58	10.25	9.93	9.62	9.32	9.03	8.75	8.47
11	11.02	10.67	10.34	10.02	9.71	9.41	9.12	8.83	8.56	8.30
12	10.77	10.43	10.11	9.80	9.50	9.21	8.92	8.65	8.38	8.12
13	10.53	10.20	9.89	9.59	9.30	9.01	8.74	8.47	8.21	7.96
14	10.29	9.98	9.68	9.38	9.10	8.82	8.55	8.30	8.04	7.80
15	10.07	9.77	9.47	9.19	8.91	8.64	8.38	8.13	7.88	7.65
16	9.86	9.56	9.28	9.00	8.73	8.47	8.21	7.97	7.73	7.50
17	9.65	9.36	9.09	8.82	8.55	8.30	8.05	7.81	7.58	7.36
18	9.45	9.17	8.90	8.64	8.39	8.14	7.90	7.66	7.44	7.22
19	9.26	8.99	8.73	8.47	8.22	7.98	7.75	7.52	7.30	7.09
20	9.08	8.81	8.56	8.31	8.07	7.83	7.60	7.38	7.17	6.96
21	8.90	8.64	8.39	8.15	7.91	7.69	7.46	7.25	7.04	6.84
22	8.73	8.48	8.23	8.00	7.77	7.54	7.33	7.12	6.91	6.72
23	8.56	8.32	8.08	7.85	7.63	7.41	7.20	6.99	6.79	6.60
24	8.40	8.16	7.93	7.71	7.49	7.28	7.07	6.87	6.68	6.49
25	8.24	8.01	7.79	7.57	7.36	7.15	6.95	6.75	6.56	6.38
26	8.09	7.87	7.65	7.44	7.23	7.03	6.83	6.64	6.46	6.28
27	7.95	7.73	7.51	7.31	7.10	6.91	6.72	6.53	6.35	6.17
28	7.81	7.59	7.38	7.18	6.98	6.79	6.61	6.42	6.25	6.08
29	7.67	7.46	7.26	7.06	6.87	6.68	6.50	6.32	6.15	5.98
30	7.54	7.33	7.14	6.94	6.75	6.57	6.39	6.22	6.05	5.89
31	7.41	7.21	7.02	6.83	6.65	6.47	6.29	6.12	5.96	5.80
32	7.29	7.09	6.90	6.72	6.54	6.36	6.19	6.03	5.87	5.71
33	7.17	6.98	6.79	6.61	6.44	6.26	6.10	5.94	5.78	5.63
34	7.05	6.86	6.68	6.51	6.33	6.17	6.01	5.85	5.69	5.54
35	6.93	6.75	6.58	6.40	6.24	6.07	5.92	5.76	5.61	5.46
36	6.82	6.65	6.47	6.31	6.14	5.98	5.83	5.68	5.53	5.39
37	6.72	6.54	6.37	6.21	6.05	5.89	5.74	5.59	5.45	5.31
38	6.61	6.44	6.28	6.12	5.96	5.81	5.66	5.51	5.37	5.24
39	6.51	6.34	6.18	6.03	5.87	5.72	5.58	5.44	5.30	5.16
40	6.41	6.25	6.09	5.94	5.79	5.64	5.50	5.36	5.22	5.09

[a] From Colt, J.: "Computation of Dissolved Gas Concentrations in Water as Functions of Temperature, Salinity, and Pressure," *American Fisheries Society Special Publication 14,* Bethesda, MD, 1984.

TABLE E-2
Dissolved-oxygen concentration in water as a function of temperature and barometric pressure (salinity = 0 ppt)[a]

Temp, °C	Dissolved-oxygen concentration, mg/L									
	Barometric pressure, millimeters of mercury									
	735	740	745	750	755	760	765	770	775	780
0	14.12	14.22	14.31	14.41	14.51	14.60	14.70	14.80	14.89	14.99
1	13.73	13.82	13.92	14.01	14.10	14.20	14.29	14.39	14.48	14.57
2	13.36	13.45	13.54	13.63	13.72	13.81	13.90	14.00	14.09	14.18
3	13.00	13.09	13.18	13.27	13.36	13.45	13.53	13.62	13.71	13.80
4	12.66	12.75	12.83	12.92	13.01	13.09	13.18	13.27	13.35	13.44
5	12.33	12.42	12.50	12.59	12.67	12.76	12.84	12.93	13.01	13.10
6	12.02	12.11	12.19	12.27	12.35	12.44	12.52	12.60	12.68	12.77
7	11.72	11.80	11.89	11.97	12.05	12.13	12.21	12.29	12.37	12.45
8	11.44	11.52	11.60	11.67	11.75	11.83	11.91	11.99	12.07	12.15
9	11.16	11.24	11.32	11.40	11.47	11.55	11.63	11.70	11.78	11.86
10	10.90	10.98	11.05	11.13	11.20	11.28	11.35	11.43	11.50	11.58
11	10.65	10.72	10.80	10.87	10.94	11.02	11.09	11.16	11.24	11.31
12	10.41	10.48	10.55	10.62	10.69	10.77	10.84	10.91	10.98	11.05
13	10.17	10.24	10.31	10.38	10.46	10.53	10.60	10.67	10.74	10.81
14	9.95	10.02	10.09	10.16	10.23	10.29	10.36	10.43	10.50	10.57
15	9.73	9.80	9.87	9.94	10.00	10.07	10.14	10.21	10.27	10.34
16	9.53	9.59	9.66	9.73	9.79	9.86	9.92	9.99	10.06	10.12
17	9.33	9.39	9.46	9.52	9.59	9.65	9.72	9.78	9.85	9.91
18	9.14	9.20	9.26	9.33	9.39	9.45	9.52	9.58	9.64	9.71
19	8.95	9.01	9.07	9.14	9.20	9.26	9.32	9.39	9.45	9.51
20	8.77	8.83	8.89	8.95	9.02	9.08	9.14	9.20	9.26	9.32
21	8.60	8.66	8.72	8.78	8.84	8.90	8.96	9.02	9.08	9.14
22	8.43	8.49	8.55	8.61	8.67	8.73	8.79	8.84	8.90	8.96
23	8.27	8.33	8.39	8.44	8.50	8.56	8.62	8.68	8.73	8.79
24	8.11	8.17	8.23	8.29	8.34	8.40	8.46	8.51	8.57	8.63
25	7.96	8.02	8.08	8.13	8.19	8.24	8.30	8.36	8.41	8.47
26	7.82	7.87	7.93	7.98	8.04	8.09	8.15	8.20	8.26	8.31
27	7.68	7.73	7.79	7.84	7.89	7.95	8.00	8.06	8.11	8.17
28	7.54	7.59	7.65	7.70	7.75	7.81	7.86	7.91	7.97	8.02
29	7.41	7.46	7.51	7.57	7.62	7.67	7.72	7.78	7.83	7.88
30	7.28	7.33	7.38	7.44	7.49	7.54	7.59	7.64	7.69	7.75
31	7.16	7.21	7.26	7.31	7.36	7.41	7.46	7.51	7.46	7.62
32	7.04	7.09	7.14	7.19	7.24	7.29	7.34	7.39	7.44	7.49
33	6.92	6.97	7.02	7.07	7.12	7.17	7.22	7.27	7.31	7.36
34	6.80	6.85	6.90	6.95	7.00	7.05	7.10	7.15	7.20	7.24
35	6.69	6.74	6.79	6.84	6.89	6.93	6.98	7.03	7.08	7.13
36	6.59	6.63	6.68	6.73	6.78	6.82	6.87	6.92	6.97	7.01
37	6.48	6.53	6.57	6.62	6.67	6.72	6.76	6.81	6.86	6.90
38	6.38	6.43	6.47	6.52	6.56	6.61	6.66	6.70	6.75	6.80
39	6.28	6.33	6.37	6.42	6.46	6.51	6.56	6.60	6.65	6.69
40	6.18	6.23	6.27	6.32	6.36	6.41	6.46	6.50	6.55	6.59

[a] From Colt, J.: "Computation of Dissolved Gas Concentrations in Water as Functions of Temperature, Salinity, and Pressure," *American Fisheries Society Special Publication 14*, Bethesda, MD, 1984.

Note: ppt = parts per thousand

When three serial sample volumes (e.g., dilutions) are used in the bacteriological testing of water, the resulting MPN (most probable number) values per 100 mL can be determined using Table F-1. The MPN values given there are based on serial sample volumes of 10, 1, and 0.1 mL. If lower or higher serial sample volumes are used, the MPN values given in Table F-1 must be adjusted accordingly. For example, if sample volumes used are 100, 10, and 1 ml, the MPN values from the table are multiplied by 0.1. Similarly, if the sample volumes are 1, 0.1, and 0.01 ml, the MPN values from the table are multiplied by 10.

In situations where more than three test dilutions have been run, the following rule is applied to select the three dilutions to be used in determining the MPN value [1]: choose the highest dilution that gives positive results in all five portions tested (no lower dilution giving any negative results) and the two next higher dilutions. Use the results at these three volumes in computing the MPN value. In the examples given in the accompanying table, the significant dilution results are shown in boldface. The number in the numerator represents positive tubes; that in the denominator represents the total tubes planted.

TABLE F-1
Most probable number (MPN) of coliforms per 100 mL of sample

Number of positive tubes				Number of positive tubes				Number of positive tubes				Number of positive tubes				Number of positive tubes				Number of positive tubes			
10 mL	1 mL	0.1 mL	MPN	10 mL	1 mL	0.1 mL	MPN	10 mL	1 mL	0.1 mL	MPN	10 mL	1 mL	0.1 mL	MPN	10 mL	1 mL	0.1 mL	MPN	10 mL	1 mL	0.1 mL	MPN
0	0	0		1	0	0	2.0	2	0	0	4.5	3	0	0	7.8	4	0	0	13	5	0	0	23
0	0	1	1.8	1	0	1	4.0	2	0	1	6.8	3	0	1	11	4	0	1	17	5	0	1	31
0	0	2	3.6	1	0	2	6.0	2	0	2	9.1	3	0	2	13	4	0	2	21	5	0	2	43
0	0	3	5.4	1	0	3	8.0	2	0	3	12	3	0	3	16	4	0	3	25	5	0	3	58
0	0	4	7.2	1	0	4	10	2	0	4	14	3	0	4	20	4	0	4	30	5	0	4	76
0	0	5	9.0	1	0	5	12	2	0	5	16	3	0	5	23	4	0	5	36	5	0	5	95
0	1	0	1.8	1	1	0	4.0	2	1	0	6.8	3	1	0	11	4	1	0	17	5	1	0	33
0	1	1	3.6	1	1	1	6.1	2	1	1	9.2	3	1	1	14	4	1	1	21	5	1	1	46
0	1	2	5.5	1	1	2	8.1	2	1	2	12	3	1	2	17	4	1	2	26	5	1	2	64
0	1	3	7.3	1	1	3	10	2	1	3	14	3	1	3	20	4	1	3	31	5	1	3	84
0	1	4	9.1	1	1	4	12	2	1	4	17	3	1	4	23	4	1	4	36	5	1	4	110
0	1	5	11	1	1	5	14	2	1	5	19	3	1	5	27	4	1	5	42	5	1	5	130
0	2	0	3.7	1	2	0	6.1	2	2	0	9.3	3	2	0	14	4	2	0	22	5	2	0	49
0	2	1	5.5	1	2	1	8.2	2	2	1	12	3	2	1	17	4	2	1	26	5	2	1	70
0	2	2	7.4	1	2	2	10	2	2	2	14	3	2	2	20	4	2	2	32	5	2	2	95
0	2	3	9.2	1	2	3	12	2	2	3	17	3	2	3	24	4	2	3	38	5	2	3	120
0	2	4	11	1	2	4	15	2	2	4	19	3	2	4	27	4	2	4	44	5	2	4	150
0	2	5	13	1	2	5	17	2	2	5	22	3	2	5	31	4	2	5	50	5	2	5	180

79	0	3	5	27	0	3	4	17	0	3	3	12	0	3	2	8.3	0	3	1	5.6	0	3	0
110	1	3	5	33	1	3	4	21	1	3	3	14	1	3	2	10	1	3	1	7.4	1	3	0
140	2	3	5	39	2	3	4	24	2	3	3	17	2	3	2	13	2	3	1	9.3	2	3	0
180	3	3	5	45	3	3	4	28	3	3	3	20	3	3	2	15	3	3	1	11	3	3	0
210	4	3	5	52	4	3	4	31	4	3	3	22	4	3	2	17	4	3	1	13	4	3	0
250	5	3	5	59	5	3	4	35	5	3	3	25	5	3	2	19	5	3	1	15	5	3	0
130	0	4	5	34	0	4	4	21	0	4	3	15	0	4	2	11	0	4	1	7.5	0	4	0
170	1	4	5	40	1	4	4	24	1	4	3	17	1	4	2	13	1	4	1	9.4	1	4	0
220	2	4	5	47	2	4	4	28	2	4	3	20	2	4	2	15	2	4	1	11	2	4	0
280	3	4	5	54	3	4	4	32	3	4	3	23	3	4	2	17	3	4	1	13	3	4	0
350	4	4	5	62	4	4	4	36	4	4	3	25	4	4	2	19	4	4	1	15	4	4	0
430	5	4	5	69	5	4	4	40	5	4	3	28	5	4	2	22	5	4	1	17	5	4	0
240	0	5	5	41	0	5	4	25	0	5	3	17	0	5	2	13	0	5	1	9.4	0	5	0
350	1	5	5	48	1	5	4	29	1	5	3	20	1	5	2	15	1	5	1	11	1	5	0
540	2	5	5	56	2	5	4	32	2	5	3	23	2	5	2	17	2	5	1	13	2	5	0
920	3	5	5	64	3	5	4	37	3	5	3	26	3	5	2	19	3	5	1	15	3	5	0
1600	4	5	5	72	4	5	4	41	4	5	3	29	4	5	2	22	4	5	1	17	4	5	0
				81	5	5	4	45	5	5	3	32	5	5	2	24	5	5	1	19	5	5	0

Example	1.0 mL	0.1 mL	0.01 mL	0.001 mL	0.0001 mL	Combination of positives	MPN/ 100 mL
a	5/5	5/5	2/5	0/5		5 – 2 – 0	4,900
b	5/5	5/5	4/5	2/5	0/5	5 – 4 – 2	22,000
c	5/5	0/5	1/5	0/5	0/5	0 – 1 – 0	180
d	5/5	5/5	3/5	1/5	1/5		
d[a]	5/5	5/5	3/5	2/5	0/5	5 – 3 – 2	14,000
e	5/5	4/5	1/5	1/5	0/5		
e[a]	5/5	4/5	2/5	0/5	0/5	5 – 4 – 2	2,200

[a] Adjusted values used to determine the MPN using Table F-1.

In example c, the first three dilutions are used so as to throw the positive result in the middle dilution. Where positive results occur in dilutions higher than the three chosen according to the above rule, they are incorporated into the result of the highest chosen dilution up to a total of five. The results of applying this procedure to the data are illustrated in examples d and e.

REFERENCE

1. *Standard Methods for the Examination of Water and Wastewater,* 17th ed., American Public Health Association, New York, 1989.

GENERAL SOLUTION
PROCEDURE FOR
MATERIALS-BALANCE
EQUATIONS
FOR A BATCH,
COMPLETE-MIX,
AND PLUG-FLOW
REACTOR

Derivation of the time-variant solution for the materials-balance equation for a batch, complete-mix, and plug-flow reactor is illustrated in this appendix [1]. The hydraulic analysis of complete-mix reactors in series is also considered.

BATCH REACTOR

The derivation of the time-variant materials-balance equation for a batch reactor can be illustrated by considering the reactor shown in Fig. G-1a. A materials balance on a reactive constituent C is written as follows:

$$\frac{dC}{dt}V = QC_0 - QC + r_cV \tag{G-1}$$

Accumulation = Inflow − Outflow + Generation

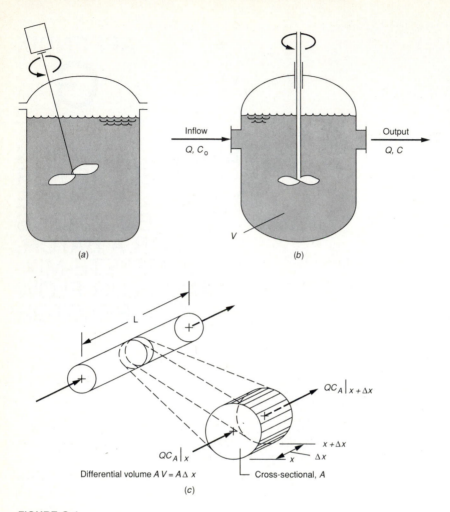

FIGURE G-1
Definition sketches used for the analysis of various reactor types: (*a*) Batch reactor, (*b*) complete-mix reactor, and (*c*) plug-flow reactor [1].

Because $Q = 0$, the resulting equation for a batch reactor is

$$\frac{dC}{dt}V = r_C V \tag{G-2}$$

If the rate of reaction is defined as $r_C = -kC^n$, integrating between the limits $C = C_0$ and $C = C$ and $t = 0$ and $t = t$ yields

$$\int_{C=C_0}^{C=C} \frac{dC}{kC^n} = \int_{t=0}^{t=t} dt = t \tag{G-3}$$

If first-order kinetics are assumed (i.e., $r_C = kC$), the resulting expression is

$$\frac{C}{C_0} = e^{-kt} \tag{G-4}$$

Equation G-4 is the same as the BOD equation (Eq. 3-6) considered in Chap. 3.

COMPLETE-MIX REACTOR

The derivation of the time-variant materials-balance equation for a complete-mix reactor can be illustrated by considering the reactor shown in Fig. G-1b. A materials balance on a reactive constituent C is written as follows:

$$\frac{dC}{dt}V = QC_0 - QC + r_C V \tag{G-5}$$

$$\text{Accumulation} = \text{Inflow} - \text{Outflow} + \text{Generation}$$

Assuming that $r_C = -kC$, Eq. G-5 can be written as

$$C' + \beta C = \frac{Q}{V}C_0 \tag{G-6}$$

where $C' = dC/dt$
$\beta = k + Q/V$

Solution Procedures

The analytical procedures adopted for the solution of mass-balance equations usually are governed by the mathematical form of the final expression. For example, the general nonsteady-state solution for Eq. G-5 is obtained by first noting that Eq. G-5 has the form of the standard first-order ordinary linear differential equation given below:

$$\frac{dy}{dt} + P(t)y = Q(t) \tag{G-7}$$

Although a variety of methods can be used to solve Eq. G-7, the method involving the use of an integrating factor of the form $\exp(\int P\,dt)$ is used most commonly.

Because Eq. G-6 is of the same form as Eq. G-7, the appropriate integrating factor for the solution of Eq. G-5 is

$$e^{\int P\,dt} = e^{\int \beta\,dt} = e^{\beta t} \tag{G-8}$$

Multiplying both sides of Eq. G-5 by the above integrating factor yields

$$e^{\beta t}(C' + \beta C) = \frac{Q}{V}C_0 e^{\beta t} \tag{G-9}$$

The left-hand side of the above equation can be written as $(e^{\beta t}C)'$ by noting that

$$(e^{\beta t}C)' = e^{\beta t}C' + \beta Ce^{\beta t} \tag{G-10}$$

Thus, Eq. G-9 can be written as

$$(e^{\beta t}C)' = \frac{Q}{V}C_0 e^{\beta t} \tag{G-11}$$

The differential sign can be removed by integrating Eq. G-11 as follows:

$$e^{\beta t}C = \frac{Q}{V}C_0 \int e^{\beta t}\,dt \tag{G-12}$$

Integration of Eq. G-12 yields

$$e^{\beta t}C = \frac{QC_0}{V\beta}e^{\beta t} + K \tag{G-13}$$

where K is the constant of integration. Dividing by $e^{\beta t}$ results in

$$C = \frac{QC_0}{V\beta} + Ke^{-\beta t} \tag{G-14}$$

But when $t = 0$ and $C = C_0$, the following equation is the result:

$$K = C_0 - \frac{QC_0}{V\beta} \tag{G-15}$$

Substituting for K in Eq. G-14 results in the following equation, which is the time-variant (nonsteady-state) solution of Eq. G-14.

$$C = \frac{QC_0}{V\beta}(1 - e^{-\beta t}) + C_0 e^{-\beta t} \tag{G-16}$$

When $t \to \infty$, it will be noted that Eq. G-16 becomes

$$C = \frac{QC_0}{V\beta} = \frac{C_0}{1 + k(V/Q)} \tag{G-17}$$

Steady-State Simplification

Fortunately, in most applications in the field of wastewater treatment, the solution of mass-balance equations, such as the one given by Eq. G-16, can be simplified by noting that the long-term (so-called steady-state) concentration is of principal concern. If it is assumed that only the steady-state effluent concentration is desired, then Eq. G-1 can be simplified by noting that, under steady-state conditions, the rate accumulation term is equal to zero ($dC/dt = 0$). Using this fact, Eq. G-1 can be written as

$$O = QC_0 - QC - kCV \tag{G-18}$$

When solved for C, Eq. G-18 yields the following expression, which is the same as Eq. G-17 given above.

$$C = \frac{C_o}{1 + k(V/Q)} \tag{G-19}$$

PLUG-FLOW REACTOR

The derivation of the time-variant materials-balance equation for a plug-flow reactor can be illustrated by considering the reactor shown in Fig. G-1c. For the differential volume element ΔV, the materials balance on a reactive constituent C is written as follows:

$$\frac{\partial C}{\partial t}\Delta V = QC|_x - QC|_{x+\Delta x} + r_C\Delta V \tag{G-20}$$

$$\text{Accumulation} = \text{Inflow} - \text{Outflow} + \text{Generation}$$

where
C = concentration of constituent C, g/m^3
ΔV = differential volume element, m^3
Q = volumetric flowrate, m^3/s
r_C = reaction rate for constituent C, g/m$^3 \cdot$ s

Substituting the differential form for the term $QC|_{x+\Delta x}$ in Eq. G-20 results in

$$\frac{\partial C}{\partial t}\Delta V = QC - Q\left(C + \frac{\Delta C}{\Delta x}\Delta x\right) + r_C\Delta V \tag{G-21}$$

Substituting $A\Delta x$ for ΔV yields

$$\frac{\partial C}{\partial t}A\Delta x = -Q\frac{\Delta C}{\Delta x}\Delta x + r_C\Delta V \tag{G-22}$$

Dividing by A and Δx yields

$$\frac{\partial C}{\partial t} = -\frac{Q}{A}\frac{\Delta C}{\Delta x} + r_C \tag{G-23}$$

Taking the limit as Δx approaches zero yields

$$\frac{\partial C}{\partial t} = -\frac{Q}{A}\frac{\partial C}{\partial x} + r_C \tag{G-24}$$

If steady-state conditions are assumed ($\partial C/\partial t = 0$) and the rate of reaction is defined as $r_C = -kC^n$, integrating between the limits $C = C_o$ and $C = C$ and $x = 0$ and $x = L$ yields

$$\int_{C=C_o}^{C=C} \frac{dC}{kC^n} = -\frac{A}{Q}\int_0^L dx = -\frac{AL}{Q} = \frac{V}{Q} = -\theta_H \tag{G-25}$$

where θ_H is the hydraulic detention time.

Equation G-25 is the basic steady-state solution to the materials-balance equation for a plug-flow reactor without dispersion. The same approach is used for a plug-flow reactor with dispersion [2].

COMPLETE-MIX REACTORS IN SERIES

In some situations, the use of a series of complete-mix reactors may have certain treatment advantages. It is therefore important to understand the hydraulic characteristics of reactors in series such as those shown in Fig. G-2.

Assume that a slug of dye is placed into the first reactor of a series of equally sized reactors so that the resulting concentration of dye in the first reactor is C_1. The total volume of all the reactors is V and the volume of an individual reactor is V/n. Writing a materials balance for the second reactor results in the following:

$$\frac{V}{n}\frac{dC_2}{dt} = QC_1 - QC_2 \text{ or } \frac{dC_2}{dt} + \frac{nQ}{V}C_2 = \frac{nQ}{V}C_1 \tag{G-26}$$

$$\text{Accumulation} = \text{Inflow} - \text{Outflow}$$

Using Eq. G-4, the effluent concentration from the first reactor is given by

$$C_1 = C_0 e^{-n(Q/V)t} = C_0 e^{-nt/t_0} = C_0 e^{-n\theta} \tag{G-27}$$

Substituting this expression for C_1 in Eq. G-26 results in

$$\frac{dC_2}{dt} + \frac{nQ}{V}C_2 = \frac{nQ}{V}C_0 e^{-n(Q/V)t} \tag{G-28}$$

Equation G-28 can be solved using exactly the same procedure as outlined for the solution of Eq. G-5. Carrying through the necessary steps, the result expressed in terms of θ is

$$C_2 = C_0 n\,\theta e^{-n\theta} \tag{G-29}$$

The generalized expression for the effluent concentration for the ith reactor in a series of n reactors is

$$C_i = \frac{C_0}{(i-1)!}(n\,\theta)^{i-1}e^{-n\theta} \tag{G-30}$$

The effluent-concentration curves obtained using Eq. G-30 for one, two, three, or four reactors in series are shown in Fig. G-3.

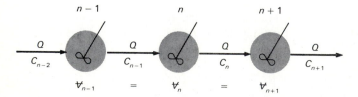

FIGURE G-2
Schematic of identical complete-mix reactors in series.

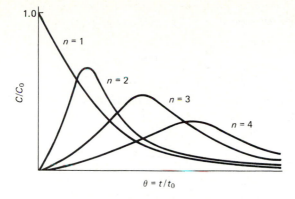

FIGURE G-3
Effluent concentration curves for each of four complete-mix reactors in series.

NONIDEAL PLUG-FLOW REACTOR

In most full-scale plug-flow reactors, the flow usually is nonideal because of entrance and exit flow disturbances and axial dispersion. Depending on the magnitude of these effects, the ideal effluent-tracer curves may look like the curves shown in Fig. G-4.

Because it is difficult to model these effects, the combined nonideal effects are often analytically simulated by replacing the plug-flow reactor with a series of complete-mix reactors, as shown in Fig. G-5. In this situation, the hydraulic characteristics of the simulated plug-flow reactor are modeled by plotting the fraction of material remaining in the series of complete-mix reactors versus the dimensionless detention-time parameter θ. The fraction of tracer remaining in the system F, at any time t, is equal to

$$F = \frac{(V/n)C_1 + (V/n)C_2 + \cdots + (V/n)C_n}{(V/n)C_0}$$

$$F = \frac{C_1 + C_2 + \cdots + C_n}{C_0} \tag{G-31}$$

Using Eq. G-31 to obtain the effluent concentration for a series of three complete-mix reactors, the corresponding expression is

$$F_{3\theta} = \frac{C_0 e^{-3\theta} + C_0(3\theta)e^{-3\theta} + C_0(3\theta)^2 e^{-3\theta}/2}{C_0}$$

$$F_{3\theta} = \left[1 + 3\theta + \frac{(3\theta)^2}{2}\right] e^{-3\theta} \tag{G-32}$$

Curves of the fraction of tracer remaining in the series of complete-mix reactors made up of one, three, and six complete-mix reactors in series are shown in Fig. G-6. For example, using six reactors, about 91 percent of the flow remains in the system for a time period equal to at least $\theta = 0.5$.

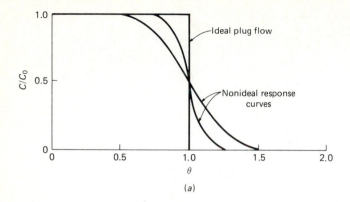

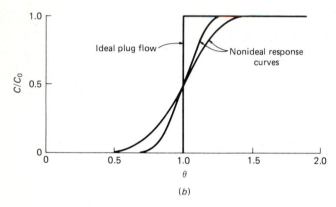

FIGURE G-4
Theoretical and generalized nonideal response curves for plug-flow reactor: (a) continuous purging of tracer and (b) continuous input of tracer.

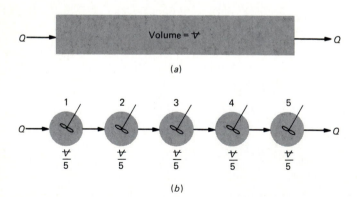

FIGURE G-5
Definition sketch for the hydraulic analysis of a plug-flow reactor with dispersion using complete-mix reactors in series: (a) original plug-flow reactor and (b) substituted reactor composed of a number of complete-mix reactors in series.

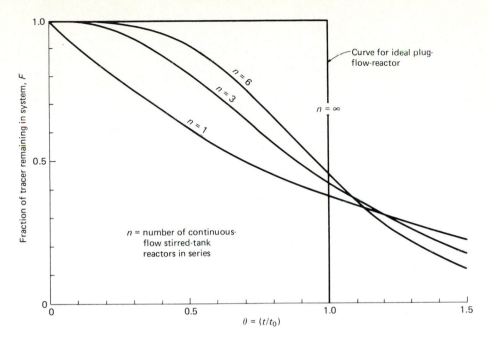

FIGURE G-6
Response curves for complete-mix reactors in series.

REFERENCES

1. Tchobanoglous, G., and E. D. Schroeder: *Water Quality: Characteristics, Modeling, Modification,* Addison-Wesley Publishing Company, Reading, MA, 1985.
2. Thomann, R. V., and J. A. Mueller: *Principles Of Surface Water Quality Modeling And Control,* Harper & Row, New York, 1987.

DETERMINATION OF KINETIC COEFFICIENTS

Values for the parameters Y, k, K_s, and k_d must be available to use biological kinetic models. To determine these coefficients, bench-scale reactors, such as those shown in Figs. H-1 and H-2, or pilot-scale systems are used.

In determining these coefficients, the usual procedure is to operate the units over a range of effluent substrate concentrations; therefore, several different θ_c (at least five) should be selected for operation ranging from 1 to 10 days. Using the data collected at steady-state conditions, mean values should be determined for Q, S_o, S, X, and r_{su}. Equating the value of r_{su} given by Eq. 8-8 to the value of r_{su} given by Eq. 8-41 results in the following expression:

$$r_{su} = -\frac{kXS}{K_s + S} = -\frac{S_o - S}{\theta} \tag{H-1}$$

Dividing by X yields

$$\frac{kS}{K_s + S} = \frac{S_o - S}{\theta X} \tag{H-2}$$

The linearized form of Eq. H-2, obtained by taking its inverse, is

$$\frac{X\theta}{S_o - S} = \frac{K_s}{k}\frac{1}{S} + \frac{1}{k} \tag{H-3}$$

1275

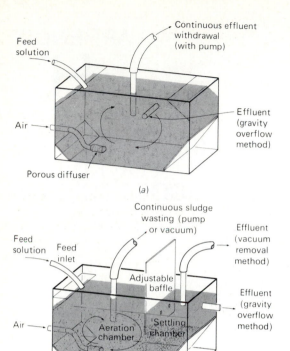

Feed solution

Continuous effluent withdrawal (with pump)

Air

Effluent (gravity overflow method)

Porous diffuser

(a)

Continuous sludge wasting (pump or vacuum)

Feed solution Feed inlet

Effluent (vacuum removal method)

Adjustable baffle

Effluent (gravity overflow method)

Air

Aeration chamber

Settling chamber

Porous diffuser

(b)

FIGURE H-1
Bench-scale continuous-flow stirred-tank reactors used for the determination of kinetic coefficients (a) without solids recycle and (b) with solids recycle.

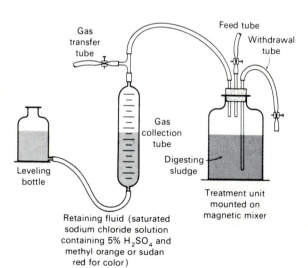

Gas transfer tube

Feed tube

Withdrawal tube

Gas collection tube

Digesting sludge

Leveling bottle

Treatment unit mounted on magnetic mixer

Retaining fluid (saturated sodium chloride solution containing 5% H_2SO_4 and methyl orange or sudan red for color)

FIGURE H-2
Laboratory reactor used for the conduct of anaerobic treatment studies.

The values of K_s and k can be determined by plotting the term $[X\theta/(S_o - S)]$ versus $(1/S)$. The values of Y and k_d may be determined using Eq. 8-40, by plotting $(1/\theta_c)$ versus $(-r_{su}/X)$.

$$\frac{1}{\theta_c} = -Y\frac{r_{su}}{X} - k_d \qquad (8\text{-}40)$$

The slope of the straight line passing through the plotted experimental data points is equal to Y, and the intercept is equal to k_d. The procedure is illustrated in the following example.

Example H-1 Determination of kinetic coefficients from laboratory data. Determine the values of the coefficients k, K_s, μ_m, Y, and k_d using the following data derived from a bench-scale activated-sludge complete-mix reactor without recycle (see Fig. 8-13a).

Unit no.	S_o, mg/L BOD$_5$	S, mg/L BOD$_5$	$\theta = \theta_c$, d	X, mg VSS/L
1	300	7	3.2	128
2	300	13	2.0	125
3	300	18	1.6	133
4	300	30	1.1	129
5	300	41	1.1	121

Solution

1. Determine the coefficients K_s and k.

   (*a*) Set up a computation table to determine the coefficients K_s and k using Eq. H-3.

$$\frac{X\theta}{S_o - S} = \frac{K_s}{k}\frac{1}{S} + \frac{1}{k}$$

$S_o - S$, mg/L	$X\theta$, mg VSS/d/L	$X\theta/(S_o - S)$, d	$1/S$, (mg/L)$^{-1}$
293	409.6	1.398	0.143
287	250.0	0.865	0.077
282	212.8	0.755	0.056
270	141.9	0.526	0.033
259	133.1	0.514	0.024

   (*b*) Plot the term $(X\theta/S_o - S)$ versus $(1/S)$, as shown in the figure at the top of the following page.

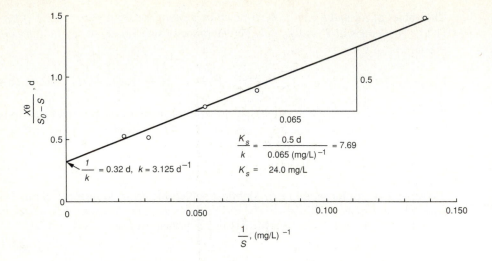

i. From Eq. H-3, the y intercept equals $(1/k)$.

$$\frac{1}{k} = 0.32 \text{ d}, \; k = 3.125 \text{ d}^{-1}$$

ii. From Eq. H-3, the slope of the curve in Fig. H-1 equals K_s/k.

$$\frac{K_s}{k} = \frac{0.5 \text{ d}}{0.065(\text{mg/L})^{-1}} = 7.692 \text{ mg/L} \cdot \text{d}$$

$$K_s = 7.692 \text{ mg/L} \cdot \text{d} \times 3.125 \text{ d}^{-1}$$
$$= 24.0 \text{ mg/L}$$

2. Determine the coefficients Y and k_d.
   (a) Set up a computation to determine the coefficients using Eq. 8-40.

$$\frac{1}{\theta_c} = -Y\frac{r_{su}}{X} - k_d$$

$$\frac{1}{\theta_c} = Y\frac{S_0 - S}{X\theta} - k_d$$

Unit no.	$1/\theta_c$, d^{-1}	$(S_0 - S)/\theta X$, d^{-1}
1	0.313	0.715
2	0.500	1.156
3	0.625	1.325
4	0.909	1.901
5	0.909	1.946

(*b*) Plot the term $(1/\theta_c)$ versus $(S_0 - S/X\theta)$, as shown in the accompanying figure.

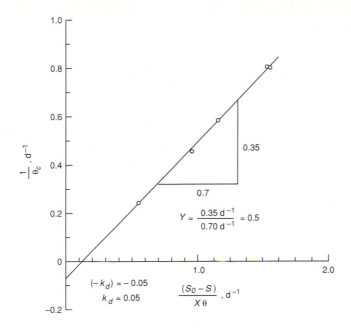

i. The *y* intercept equals $(-k_d)$.

$$-k_d = -0.05 \text{ d}^{-1}$$

$$k_d = 0.05 \text{ d}^{-1}$$

ii. The value of the slope of the curve equals Y.

$$Y = \frac{0.35 \text{ d}^{-1}}{0.70 \text{ d}^{-1}} = 0.5$$

3. Determine the value of the coefficient μ_m using Eq. 8-7.

$$\mu_m = kY$$
$$= 3.125 \text{ d}^{-1} \times 0.5$$
$$= 1.563 \text{ d}^{-1}$$

Comment. In this example, the kinetic coefficients were derived from data obtained using bench-scale, complete-mix reactors without recycle. Similar data can be obtained using complete-mix reactors with recycle. An advantage of using reactors with recycle is that the mean cell-residence time can be varied independently of the hydraulic detention time. A disadvantage is that small bench-scale reactors operated with solids recycle are difficult to control.

MOODY DIAGRAMS FOR THE ANALYSIS OF FLOW IN PIPES

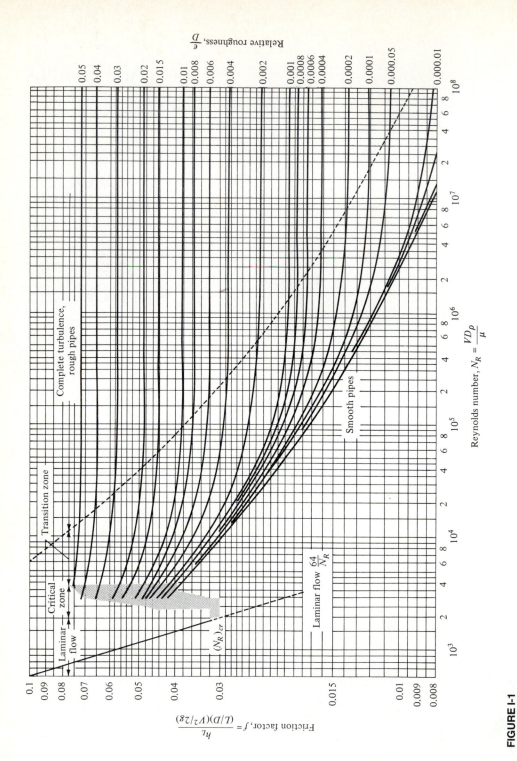

FIGURE I-1
Moody diagram for friction factor in pipes versus Reynolds number and relative roughness [12].

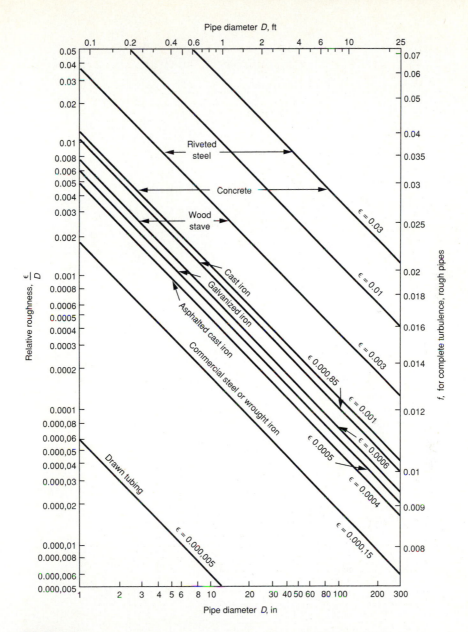

FIGURE I-2
Moody diagram for relative roughness as a function of diameter for pipes constructed of various materials [adapted from 12].

Atomic numbers and atomic masses[*]

Actinium	Ac	89	227.0278	Mercury	Hg	80	200.59
Aluminum	Al	13	26.98154	Molybdenum	Mo	42	95.94
Americium	Am	95	(243)	Neodymium	Nd	60	144.24
Antimony	Sb	51	121.75	Neon	Ne	10	20.179
Argon	Ar	18	39.948	Neptunium	Np	93	237.0482
Arsenic	As	33	74.9216	Nickel	Ni	28	58.70
Astatine	At	85	(210)	Niobium	Nb	41	92.9064
Barium	Ba	56	137.33	Nitrogen	N	7	14.0067
Berkelium	Bk	97	(247)	Nobelium	No	102	(259)
Beryllium	Be	4	9.01218	Osmium	Os	76	190.2
Bismuth	Bi	83	208.9804	Oxygen	O	8	15.9994
Boron	B	5	10.81	Palladium	Pd	46	106.4
Bromine	Br	35	79.904	Phosphorous	P	15	30.97376
Cadmium	Cd	48	112.41	Platinum	Pt	78	195.09
Calcium	Ca	20	40.08	Plutonium	Pu	94	(244)
Californium	Cf	98	(251)	Polonium	Pu	84	(209)
Carbon	C	6	12.011	Potassium	K	19	39.0983
Cerium	Ce	58	140.12	Praseodymium	Pr	59	140.9077
Cesium	Cs	55	132.9054	Promethium	Pm	61	(145)
Chlorine	Cl	17	35.453	Protactinium	Pa	91	231.0389
Chromium	Cr	24	51.996	Radium	Ra	88	226.0254
Cobalt	Co	27	58.9332	Radon	Rn	86	(222)
Copper	Cu	29	63.546	Rhenium	Re	75	186.207
Curium	Cm	96	(247)	Rhodium	Rh	45	102.9055
Dysprosium	Dy	66	162.50	Rubidium	Rb	37	85.4678
Einsteinium	Es	99	(254)	Ruthenium	Ru	44	101.07
Erbium	Er	68	167.26	Samarium	Sm	62	150.4
Europium	Eu	63	151.96	Scandium	Sc	21	44.9559
Fermium	Fm	100	(257)	Selenium	Se	34	78.96
Fluorine	F	9	18.99840	Silicon	Si	14	28.0855
Francium	Fr	87	(223)	Silver	Ag	47	107.868
Gadolinium	Gd	64	157.25	Sodium	Na	11	22.98977
Gallium	Ga	31	69.72	Strontium	Sr	38	87.62
Germanium	Ge	32	72.59	Sulfur	S	16	32.06
Gold	Au	79	196.9665	Tantalum	Ta	73	180.9479
Hafnium	Hf	72	178.49	Technetium	Tc	43	(97)
Helium	He	2	4.00260	Tellurium	Te	52	127.60
Holmium	Ho	67	164.9304	Terbium	Tb	65	158.9254
Hydrogen	H	1	1.0079	Thallium	Tl	81	204.37
Indium	In	49	114.82	Thorium	Th	90	232.0381
Iodine	I	53	126.9045	Thulium	Tm	69	168.9342
Iridium	Ir	77	192.22	Tin	Sn	50	118.69
Iron	Fe	26	55.847	Titanium	Ti	22	47.90
Krypton	Kr	36	83.80	Tungsten	W	74	183.85
Lanthanum	La	57	138.9055	Uranium	U	92	238.029
Lawrencium	Lr	103	(260)	Vanadium	V	23	50.9414
Lead	Pb	82	207.2	Xenon	Xe	54	131.30
Lithium	Li	3	6.941	Ytterbium	Yb	70	173.04
Lutetium	Lu	71	174.97	Yttrium	Y	39	88.9059
Magnesium	Mg	12	24.305	Zinc	Zn	30	65.38
Manganese	Mn	25	54.9380	Zirconium	Zr	40	91.22
Mendelevium	Md	101	(258)				

[*]From *Pure Appl. Chem.*, vol. 47, p. 75 (1976). A value in parentheses is the mass number of the longest lived isotope of the element.